D0605880

FIFTH EDITION

SYSTEMS ANALYSIS AND DESIGN

IN A CHANGING WORLD

John W. Satzinger
Missouri State University

Robert B. Jackson
RBJ and Associates

Stephen D. Burd
University of New Mexico

COURSE TECHNOLOGY
CENGAGE Learning™

Australia • Brazil • Japan • Korea • Mexico • Singapore • Spain • United Kingdom • United States

Systems Analysis and Design in a Changing World, Fifth Edition
John W. Satzinger, Robert B. Jackson, Stephen D. Burd

Editor-in-Chief: Alex von Rosenberg

Acquisitions Editor: Charles McCormick

Product Manager: Kate Hennessy

Development Editor: Dan Seiter

Editorial Assistant: Bryn Lathrop

Marketing Director: Brian Joyner

Marketing Manager: Bryant Chrzan

Content Project Manager: Matt Hutchinson, GEX Publishing Services

Art Director: Marissa Falco

Manufacturing Coordinator: Justin Palmeiro

Cover Photo: © Radius Images/RF/PhotoLibrary

ISBN-13: 978-1-4239-0228-7

ISBN-10: 1-4239-0228-9

Course Technology
25 Thomson Place
Boston, MA 02210
USA

Cengage Learning is a leading provider of customized learning solutions with office locations around the globe, including Singapore, the United Kingdom, Australia, Mexico, Brazil, and Japan. Locate your local office at: **international.cengage.com/region**

Cengage Learning products are represented in Canada by Nelson Education, Ltd.

For your lifelong learning solutions, visit **course.cengage.com**

Visit our corporate website at **cengage.com**

Printed in Canada
2 3 4 5 6 7 12 11 10 09

DEDICATION

To JoAnn, Brian, Kevin, LaVone, and Arnie—JWS

To my immediate and extended family—RBJ

To Dee, Amelia, and Alex—SDB

BRIEF CONTENTS

PART 1: The Systems Analyst

Chapter 1 The World of the Information Systems Analyst 2

Chapter 2 Approaches to System Development 36

Chapter 3 The Analyst as a Project Manager 72

PART 2: Systems Analysis Activities

Chapter 4 Investigating System Requirements 116

Chapter 5 Modeling System Requirements 158

Chapter 6 The Traditional Approach to Requirements 202

Chapter 7 The Object-Oriented Approach to Requirements 238

Chapter 8 Evaluating Alternatives for Requirements, Environment, and Implementation 280

PART 3: Systems Design Tasks

Chapter 9 Elements of Systems Design 314

Chapter 10 The Traditional Approach to Design 352

Chapter 11 Object-Oriented Design: Principles 386

Chapter 12 Object-Oriented Design: Use Case Realizations 428

Chapter 13 Designing Databases 486

Chapter 14 Designing the User Interface 528

Chapter 15 Designing System Interfaces, Controls, and Security 568

PART 4: Implementation and Support

Chapter 16 Making the System Operational 616

Chapter 17 Current Trends in System Development 660

Index **701**

Online Supplemental Web Resources

Online Supplemental Chapter 1 Packages and Enterprise Resource Planning

Online Appendices A, B, C, D, and E

Glossary

TABLE OF CONTENTS

Note that more material is available at the book's Web site, including an online chapter and appendices. For information, see the "Student Companion Web Site" section in this preface.

PART 1 The Systems Analyst

Chapter 1	**The World of the Information Systems Analyst**	**2**
	A Systems Analyst at Consolidated Refineries	3
	Overview	4
	The Analyst as a Business Problem Solver	4
	Systems That Solve Business Problems	6
	Required Skills of the Systems Analyst	10
	Analysis-Related Careers	14
	The Analyst's Role in Strategic Planning	16
	Rocky Mountain Outfitters and Its Strategic Information Systems Plan	18
	The Analyst as a System Developer (the Heart of the Course)	27
	Summary	31
	Key Terms	31
	Review Questions	32
	Thinking Critically	32
	Experiential Exercises	32
	Case Studies	33
	Further Resources	35
Chapter 2	**Approaches to System Development**	**36**
	Development Approaches at Ajax Corporation, Consolidated Concepts, and Pinnacle Manufacturing	37
	Overview	37
	The Systems Development Life Cycle	38
	Activities of Each SDLC "Phase"	45
	Methodologies, Models, Tools, and Techniques	49
	Two Approaches to System Development	53
	Current Trends in Development	61
	Tools to Support System Development	63
	Summary	67
	Key Terms	67
	Review Questions	68
	Thinking Critically	68

	Experiential Exercises	69
	Case Studies	69
	Further Resources	71
Chapter 3	**The Analyst as a Project Manager**	**72**
	Bestway Fuel Systems: Moving to an Adaptive SDLC	73
	Overview	73
	Project Management	74
	Project Initiation and Project Planning	83
	Defining the Problem	87
	Producing the Project Schedule	90
	Identifying Project Risks and Confirming Project Feasibility	99
	Staffing and Launching the Project	107
	Recap of Project Planning for RMO	109
	Summary	111
	Key Terms	111
	Review Questions	112
	Thinking Critically	112
	Experiential Exercises	113
	Case Studies	113
	Further Resources	114

PART 2 Systems Analysis Activities

Chapter 4	**Investigating System Requirements**	**116**
	Mountain States Motor Sports	117
	Overview	118
	Analysis Activities in More Detail	119
	System Requirements	122
	Models and Modeling	124
	Stakeholders—The Source of System Requirements	128
	Techniques for Information Gathering	133
	Validating the Requirements	150
	Summary	153
	Key Terms	154
	Review Questions	154

	Thinking Critically	154
	Experiential Exercises	155
	Case Studies	156
	Further Resources	157
Chapter 5	**Modeling System Requirements**	**158**
	Waiters on Call Meal-Delivery System	159
	Overview	160
	User Goals, Events, and Use Cases	160
	Use Case Descriptions	171
	"Things" in the Problem Domain	176
	The Entity-Relationship Diagram	182
	The Domain Model Class Diagram	187
	Where You Are Headed	194
	Summary	195
	Key Terms	195
	Review Questions	196
	Thinking Critically	196
	Experiential Exercises	197
	Case Studies	198
	Further Resources	201
Chapter 6	**The Traditional Approach to Requirements**	**202**
	San Diego Periodicals: Following the Data Flow	203
	Overview	204
	Traditional and Object-Oriented Views of Activities/Use Cases	205
	Data Flow Diagrams	205
	Documentation of DFD Components	221
	Locations and Communication through Networks	230
	Summary	234
	Key Terms	234
	Review Questions	234
	Thinking Critically	235
	Experiential Exercises	235
	Case Studies	235
	Further Resources	237

TABLE OF CONTENTS

Chapter 7	**The Object-Oriented Approach to Requirements**	**238**
	Electronics Unlimited, Inc.: Integrating the Supply Chain	239
	Overview	239
	Object-Oriented Requirements	240
	The System Activities—A Use Case/Scenario View	242
	Identifying Inputs and Outputs—The System Sequence Diagram	252
	Identifying Object Behavior—The State Machine Diagram	260
	Integrating Object-Oriented Models	269
	Summary	271
	Key Terms	271
	Review Questions	271
	Thinking Critically	272
	Experiential Exercises	275
	Case Studies	276
	Further Resources	279
Chapter 8	**Evaluating Alternatives for Requirements, Environment, and Implementation**	**280**
	Tropic Fish Tales: Netting the Right System	281
	Overview	281
	Project Management Perspective	283
	Deciding on Scope and Level of Automation	284
	Defining the Application Deployment Environment	291
	Choosing Implementation Alternatives	297
	Contracting with Vendors	305
	Presenting the Results and Making the Decisions	307
	Summary	309
	Key Terms	309
	Review Questions	309
	Thinking Critically	310
	Experiential Exercises	310
	Case Studies	311
	Further Resources	312

TABLE OF CONTENTS

PART 3 Systems Design Tasks

Chapter 9 Elements of Systems Design 314

Fairchild Pharmaceuticals: Finalizing Architectural Design for a Production System 315
Overview 316
Project Management Revisited: Execution and Control of Projects 317
Understanding the Elements of Design 324
Design Activities 330
Network Design 334
The Deployment Environment and Application Architecture 339
Summary 349
Key Terms 349
Review Questions 350
Thinking Critically 350
Experiential Exercises 350
Case Studies 351
Further Resources 351

Chapter 10 The Traditional Approach to Design 352

Theatre Systems, Inc.: Something Old, Something New 353
Overview 354
The Structured Approach to Designing the Application Architecture 354
The Automation System Boundary 355
The System Flowchart 357
The Structure Chart 360
Module Algorithm Design: Pseudocode 371
Integrating Structured Application Design with Other Design Tasks 373
Three-Layer Design 374
Summary 379
Key Terms 379
Review Questions 379
Thinking Critically 380
Experiential Exercises 384
Case Studies 384
Further Resources 385

TABLE OF CONTENTS

Chapter 11	**Object-Oriented Design: Principles**	**386**
	New Capital Bank: Part 1	387
	Overview	388
	Object-Oriented Design: Bridging from Analysis to Implementation	388
	Object-Oriented Architectural Design	392
	Fundamental Principles of Object-Oriented Detailed Design	404
	Design Classes and the Design Class Diagram	409
	Detailed Design with CRC Cards	416
	Fundamental Detailed Design Principles	419
	Summary	423
	Key Terms	423
	Review Questions	424
	Thinking Critically	424
	Experiential Exercises	425
	Case Studies	425
	Further Resources	427
Chapter 12	**Object-Oriented Design: Use Case Realizations**	**428**
	New Capital Bank: Part 2	429
	Overview	429
	Detailed Design of Multilayer Systems	430
	Use Case Realization with Sequence Diagrams	433
	Designing with Communication Diagrams	454
	Updating and Packaging the Design Classes	457
	Design Patterns	463
	Summary	473
	Key Terms	473
	Review Questions	474
	Thinking Critically	475
	Experiential Exercises	483
	Case Studies	484
	Further Resources	485

Chapter 13 Designing Databases **486**

Nationwide Books: Designing a New Database 487
Overview 488
Databases and Database Management Systems 488
Relational Databases 490
Object-Oriented Databases 503
Hybrid Object-Relational Database Design 510
Data Types 514
Distributed Databases 516
Summary 524
Key Terms 524
Review Questions 524
Thinking Critically 525
Experiential Exercises 526
Case Studies 526
Further Resources 527

Chapter 14 Designing the User Interface **528**

Interface Design at Aviation Electronics 529
Overview 529
Identifying and Classifying Inputs and Outputs 530
Understanding the User Interface 532
Guidelines for Designing User Interfaces 540
Documenting Dialog Designs 544
Guidelines for Designing Windows and Browser Forms 549
Guidelines for Designing Web Sites 552
Designing Dialogs for Rocky Mountain Outfitters 554
Summary 562
Key Terms 562
Review Questions 563
Thinking Critically 563
Experiential Exercises 564
Case Studies 564
Further Resources 567

TABLE OF CONTENTS

Chapter 15 Designing System Interfaces, Controls, and Security 568

Downslope Ski Company: Designing a Secure Supplier System Interface 569
Overview 570
Identifying System Interfaces 570
Designing System Inputs 574
Designing System Outputs 582
Designing Integrity Controls 592
Designing Security Controls 599
Summary 607
Key Terms 607
Review Questions 608
Thinking Critically 609
Experiential Exercises 611
Case Studies 611
Further Resources 613

PART 4 Implementation and Support

Chapter 16 Making the System Operational 616

Tri-State Heating Oil: Juggling Priorities to Begin Operation 617
Overview 618
Program Development 619
Quality Assurance 631
Data Conversion 639
Installation 641
Documentation 646
Training and User Support 650
Maintenance and System Enhancement 652
Summary 656
Key Terms 656
Review Questions 656
Thinking Critically 657
Experiential Exercises 658
Case Studies 658
Further Resources 659

Chapter 17 Current Trends in System Development **660**
Valley Regional Hospital: Measuring a Project's Progress 661
Overview 661
Software Principles and Practices 662
Adaptive Methodologies to Development 666
Model-Driven Architecture—Generalizing Solutions 684
Frameworks, Components, and Services 687
Summary 695
Key Terms 695
Review Questions 696
Thinking Critically 696
Experiential Exercises 697
Case Studies 698
Further Resources 699

Index **701**

Supplemental Web Resources

Online Supplemental Chapter 1 Packages and Enterprise Resource Planning
Premier Candy Corp.: The Possible Pitfalls of ERP
Overview
Packaged Software
Enterprise Resource Planning
A Closer Look at One ERP Package: SAP R/3
Summary
Key Terms
Review Questions
Thinking Critically
Experiential Exercises
Case Studies
Further Resources

Online Appendix A Principles of Project Management
Online Appendix B Project Schedules with PERT/CPM Charts
Online Appendix C Calculating Net Present Value, Payback Period, and Return on Investment
Online Appendix D Presenting the Results to Management
Online Appendix E Guide to Using Microsoft Project
Glossary

FEATURES

Systems Analysis and Design in a Changing World, Fifth Edition, was written and developed with both instructor and student needs in mind. Here is just a sample of the unique and exciting features that help bring the field of systems analysis and design to life.

Figure 1-9
Current RMO catalog cover (Fall 2010)

Rocky Mountain Outfitters
ROCKY MOUNTAIN OUTFITTERS
2010 CATALOG

John and Liz had considered making a major commitment to business-to-consumer (B2C) e-commerce in the early 2000s. They worried about the risk of sudden and potentially explosive growth, but felt that they had to develop an online ordering system to remain competitive. At the time, in-house staff was not trained in Web technologies, so John and Liz decided to outsource development and operation of the Web site.

By 2007, they realized that the Web-based ordering system was substantially underperforming against the competition for many reasons, including the following:

- Slow and cumbersome updates to online content
- Poor coordination with in-house customer service functions
- Poor coordination between Web-based ordering and supply chain management functions
- Poor technical support and other support by the site operator
- Deteriorating relations with RMO management

In late 2006, RMO performed a detailed market analysis that showed alarming trends, including the following:

- RMO sales growth was slower than the industry average, resulting in decreasing market share.
- The average age of customers ordering by phone and mail was increasing, and was much higher than the industry average age of all customers.
- Compared to competitors, RMO's Web-based sales were a much smaller percentage of total sales, and the average order amount was lower than the industry average.

The analysis painted a disturbing picture of declining performance. Continued strong sales to older customers via traditional channels were offset by weak sales to younger customers via the Web. RMO was failing to attract and retain the customers who represented the bulk of present and future business.

20 ◆ PART 1 THE SYSTEMS ANALYST

The text uses an **integrated case study** of moderate complexity—Rocky Mountain Outfitters (RMO)—to illustrate key concepts and techniques.

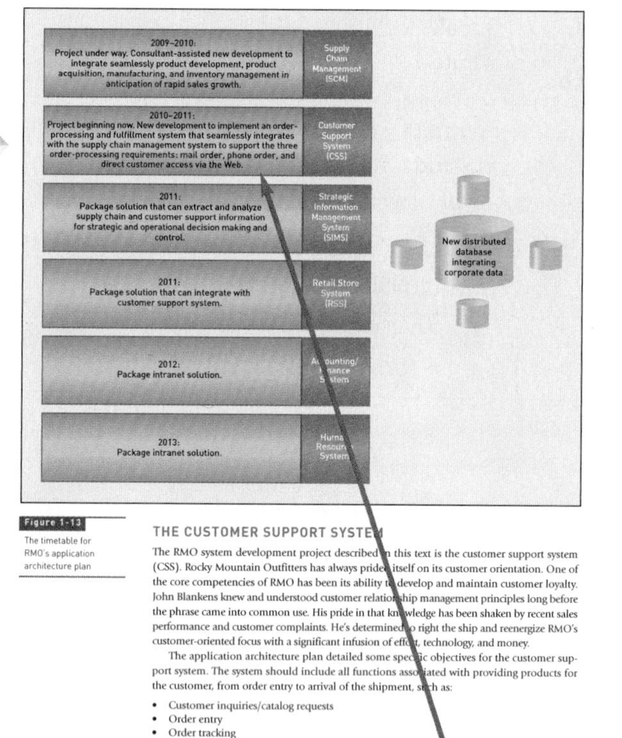

2009–2010:
Project under way. Consultant-assisted new development to integrate seamlessly product development, product acquisition, manufacturing, and inventory management in anticipation of rapid sales growth.
Supply Chain Management (SCM)
2010–2011:
Project beginning now. New development to implement an order-processing and fulfillment system that seamlessly integrates with the supply chain management system to support the three order-processing requirements: mail order, phone order, and direct customer access via the Web.
Customer Support System (CSS)
2011:
Package solution that can extract and analyze supply chain and customer support information for strategic and operational decision making and control.
Strategic Information Management System (SIMS)
2011:
Package solution that can integrate with customer support system.
Retail Store System (RSS)
2012:
Package intranet solution.
2013:
Package intranet solution.
New distributed database integrating corporate data

Figure 1-13
The timetable for RMO's application architecture plan

THE CUSTOMER SUPPORT SYSTEM

The RMO system development project described in this text is the customer support system (CSS). Rocky Mountain Outfitters has always prided itself on its customer orientation. One of the core competencies of RMO has been its ability to develop and maintain customer loyalty. John Blankens knew and understood customer relationship management principles long before the phrase came into common use. His pride in that knowledge has been shaken by recent sales performance and customer complaints. He's determined to right the ship and reenergize RMO's customer-oriented focus with a significant infusion of effort, technology, and money.

The application architecture plan detailed some specific objectives for the customer support system. The system should include all functions associated with providing products for the customer, from order entry to arrival of the shipment, such as:

- Customer inquiries/catalog requests
- Order entry
- Order tracking
- Shipping
- Back ordering
- Returns
- Sales analysis

26 ◆ PART 1 THE SYSTEMS ANALYST

An overview of the **strategic systems plan** for RMO is presented in Chapter 1 to place the project in context. The planned system architecture provides for rich examples—a client/server Windows-based component, as well as a Web-based, e-commerce component with direct customer interaction via the Internet.

The **new customer support system (CSS)** is the system development project used throughout the text for examples and explanations. It is strategically important to RMO, and the company must integrate the new system with legacy systems and other planned systems.

FEATURES

The text describes both **predictive** and **adaptive approaches to the SDLC,** and recommends iterative development for many projects.

Each chapter includes several **Best Practice** features that highlight the latest thinking on techniques and tools.

most of this textbook—we will focus on the initial development project and not on the support projects. In other words, our primary concern is with getting the system developed and deployed the very first time.

In today's diverse development environment, many different approaches to developing systems are used, and they are based on different SDLCs. As you might suppose, some approaches have been used for a long time and have varying rates of success. In the ever-changing world of information technology, new and unique approaches to building systems have emerged, which also have varying success rates. Although it is difficult to find a single, comprehensive classification system that encompasses all of the approaches, one useful technique is to categorize SDLC approaches according to whether they are more predictive or adaptive. These two classifications represent the end points of a continuum from completely predictive to completely adaptive (see Figure 2-1).

Figure 2-1
Predictive versus adaptive approaches to the SDLC

The choice of SDLC varies depending on the project

Predictive SDLC | Adaptive SDLC

Requirements well understood and well defined. Low technical risk.

Requirements and needs uncertain. High technical risk.

predictive approach
an SDLC approach that assumes the development project can be planned and organized in advance and that the new information system can be developed according to the plan

adaptive approach
an SDLC approach that is more flexible, assuming that the project cannot be planned out completely in advance but must be modified as it progresses

A predictive approach to the SDLC is an approach that assumes that the development project can be planned and organized in advance and that the new information system can be developed according to the plan. Predictive SDLCs are useful for building systems that are well understood and defined. For example, a company may want to convert its old, mainframe inventory system to a newer networked client/server system. In this type of project, the staff already understands the requirements very well, and no new processes need to be added. So, the project can typically be planned carefully, and the system can be built according to the specifications.

At the other end of the scale, an adaptive approach to the SDLC is used when the exact requirements of a system or the users' needs are not well understood. In this situation, the project cannot be planned completely in advance. Some requirements of the system may yet need to be determined, after some preliminary development work. Developers should still be able to build the solution, but they must be flexible and adapt the project as it progresses.

In practice, any project could have—and most do have—both predictive and adaptive elements. That is why Figure 2-1 shows the characteristics as end points on a sliding scale—not as two mutually exclusive categories. The predictive approaches are more traditional and were invented from the 1970s to the 1990s. Many of the newer, adaptive approaches have evolved along with the object-oriented approach and were created during the 1990s and into the twenty-first century. Let's first look at some of the more predictive approaches and then examine some of the newer adaptive approaches.

BEST PRACTICE
Recognize that any specific project you work on will have some predictive and some adaptive elements.

THE TRADITIONAL PREDICTIVE APPROACHES TO THE SDLC

The development of a new information system requires several different, but related, activities. In predictive approaches, we first have a group of activities that plan, organize, and schedule the project, usually called *project planning activities*. These activities map out the overall

CHAPTER 2 Approaches to System Development ◆ 39

Details about the RMO case are integrated directly into each chapter to make a point or to illustrate a concept—**just-in-time examples**—rather than isolating the case study in separate sections of the chapters.

Project management aspects of the case are reinforced throughout by use of **RMO memos** describing the status of the project in every chapter. The same system project is used to illustrate traditional and object-oriented models and solutions, so both approaches can be understood and directly compared.

RECAP OF PROJECT PLANNING FOR RMO

Barbara and Steve spent the entire month of February putting together the schedule and plans for the CSS. Even though Barbara was the project manager, she and Steve worked together as peers. As a team, they could brainstorm and double-check each other's work. They had worked together before and had an excellent relationship—one based on mutual respect and trust. They could be candid and knew how to work through disagreements as well as how to come to consensus on important issues. Barbara also knew that the work Steve produced was always well thought out and very professionally done. He was a skilled systems analyst and would help make sure that the work done in the planning phase was solid.

The success of the overall project depended heavily on the planning Barbara and Steve did during this phase. The foundation for all other project activities is established during project planning. As Barbara planned for the kickoff meeting to launch the project officially, she reviewed the areas of project management to make sure that she had addressed all of the critical issues.

For project scope management, she developed a list of business benefits, a list of system capabilities, and a context diagram. At this point in the project, the scope definition was still very general. She would make sure the project's scope was precisely defined during the information-gathering activities of the analysis phase.

She and Steve had developed a detailed work breakdown structure and entered the information into Microsoft Project. The schedule was very detailed for the analysis phase, but less so for the design and implementation phases. She would add those details as decisions were made about the implementation approach. She thought that her approach to project time management had been established, and she would have the tools necessary to track the schedule as the project progressed.

The costs and potential benefits had been estimated and used to develop an NPV estimate. She would redo the NPV when she redid the schedule at the end of the analysis phase to ensure that the costs and schedule were within the allowed budget. The other part of cost management was to monitor the costs during the life of the project. Microsoft Project would help her track the costs of each task.

Steve had done a lot of the work to identify and assess risks during the feasibility analysis. Barbara knew that they would both continue to look for risks and assess potential problems during the project. She asked Steve to take time each week to assess the risks and update the list of the highest risks for the project. She felt confident that she would not be blindsided by some unexpected problem.

For project communication and project quality, Barbara established procedures for the project. She set up a central database to post the project's status, decisions, and working documents to make sure that all the team members were kept well informed. She established a routine and format for weekly status reports from the team leaders and a status report to the oversight committee. An example of one of her status report memos to the oversight committee is shown. These status reports all follow a standard format. In addition to the formal status memos, she would also write more informal memos to John MacMurty. For project quality, internal procedures required that team members and RMO users review all work products. Other quality procedures, such as the test plan, would be established as the project progressed.

CHAPTER 3 The Analyst as a Project Manager ◆ 109

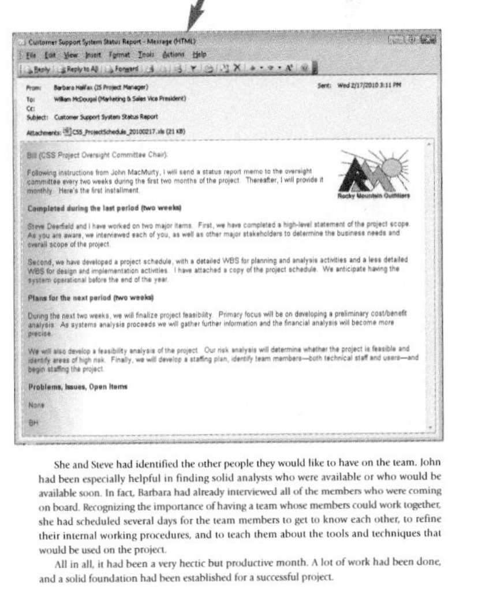
Customer Support System Status Report - Message (HTML)

File Edit View Insert Format Tools Actions Help

Reply | Reply to All | Forward

From: Barbara Halifax (IS Project Manager) Sent: Wed 2/17/2010 3:11 PM
To: William McDougal (Marketing & Sales Vice President)
Cc:
Subject: Customer Support System Status Report
Attachments: CSS_ProjectSchedule_20100217.xls (21 KB)

Bill (CSS Project Oversight Committee Chair):

Following instructions from John MacMurty, I will send a status report memo to the oversight committee every two weeks during the first two months of the project. Thereafter, I will provide it monthly. Here's the first installment.

Rocky Mountain Outfitters

Completed during the last period (two weeks)

Steve Deerfield and I have worked on two major items. First, we have completed a high-level statement of the project scope. As you are aware, we interviewed each of you, as well as other major stakeholders to determine the business needs and overall scope of the project.

Second, we have developed a project schedule, with a detailed WBS for planning and analysis activities and a less detailed WBS for design and implementation activities. I have attached a copy of the project schedule. We anticipate having the system operational before the end of the year.

Plans for the next period (two weeks)

During the next two weeks, we will finalize project feasibility. Primary focus will be on developing a preliminary cost/benefit analysis. As systems analysis proceeds we will gather further information and the financial analysis will become more precise.

We will also develop a feasibility analysis of the project. Our risk analysis will determine whether the project is feasible and identify areas of high risk. Finally, we will develop a staffing plan, identify team members—both technical staff and users—and begin staffing the project.

Problems, Issues, Open Items

None

BH

She and Steve had identified the other people they would like to have on the team. John had been especially helpful in finding solid analysts who were available or who would be available soon. In fact, Barbara had already interviewed all of the members who were coming on board. Recognizing the importance of having a team whose members could work together, she had scheduled several days for the team members to get to know each other, to refine their internal working procedures, and to teach them about the tools and techniques that would be used on the project.

All in all, it had been a very hectic but productive month. A lot of work had been done, and a solid foundation had been established for a successful project.

110 ◆ PART 1 THE SYSTEMS ANALYST

FEATURES

Every chapter follows up on the RMO case details by adding an end-of-chapter case study named **Rethinking Rocky Mountain Outfitters**. Each case extends an example in the chapter or poses additional questions to consider about the RMO system project.

This edition also includes the **Focusing on Reliable Pharmaceutical Service** case study, which is included at the end of every chapter to provide additional experience with problem-solving techniques and issues addressed in the chapter. Reliable is a smaller company than RMO, and the strategic information system plan and specific system development project provide a different perspective to analysis and design.

As Monica reviewed Stewart's record, she found that he had done an excellent job as a team leader on his last project. His last assignment was as a combination team leader/systems analyst on a four-person team. He had been involved in systems analysis, design, and programming, and he also managed the work of the other three team members. He had assisted in the development of the project schedule and had been able to keep his team right on schedule. It also appeared that the quality of his team's work was as good as, if not better than, other teams on the project. She wondered what advice she should give him to help him advance his career. She was also wondering if now was the time to give him his own project.

1. Do you think the decision by CLT to build its own project managers from the existing employee base is a good one? What advice would you give to CLT to make sure that it has strong project management skills in the company?
2. What kind of criteria would you develop for Monica to use to measure whether Stewart (or any other potential project manager) is ready for project management responsibility?
3. How would you structure the job for new project managers to ensure, or at least increase the possibility of, a high level of success?
4. If you were Monica, what kind of advice would you give to Stewart about managing his career and attaining his immediate goal to become a project manager?

RETHINKING ROCKY MOUNTAIN OUTFITTERS

ROCKY MOUNTAIN OUTFITTERS

The chapter identified six areas of project feasibility that need to be evaluated for any new project. However, as indicated, each of these areas of feasibility can also be considered an evaluation of the potential risks of the project. Based on your understanding of Rocky Mountain Outfitters, both from this chapter and the information provided in Chapter 1, build a table that summarizes the risks faced by RMO for this new project. Include four columns titled (1) Project risk, (2) Type of risk, (3) Probability of risk, and (4) Steps to alleviate risk.

Identify as many risks to the project as you can. Type of risk means the category or area of the project feasibility that is at risk. It might help you think about risks in the different categories, for example (1) risk management, (2) economic, (3) organizational and cultural, (4) technological, (5) schedule, and (6) resources. The chapter provided a few examples of risk in each of these areas. However, many other risks can cause project failures. Think as broadly as possible and expand the list of potential risks in each area.

Obviously, other kinds of risks are associated with a project of the magnitude of the customer support system. You might want to consider some risks external to the company, such as economic, marketplace, legal, environment, and so forth. Other types of internal risks might also be associated with components that are purchased or outsourced, such as development tools, learning curves, poor quality of purchased components, and failure of vendors.

A common risk management technique is to build a table and identify the top 10 risks to the project. Contingency plans can then be built for the top 10 risks. Periodically, the project management team reevaluates the risk list to determine the current top 10 risks. After you build the table, identify which risks you would classify as the top 10 risks.

FOCUSING ON RELIABLE PHARMACEUTICAL SERVICE

Reliable PHARMACEUTICALS

Chapter 2 discussed Reliable Pharmaceutical Service's Web-based application to connect its client nursing homes directly with a new prescription and billing system. You considered both the risks of a sequential, waterfall approach to the SDLC and the risks of an iterative and incremental approach to the SDLC for its development.

1. Now consider the way the project was probably initiated. To what extent is the project the result of (a) an opportunity, (b) a problem, or (c) a directive?
2. Many of the system users (such as employees at health-care facilities) are not Reliable employees. What risks of project failure are associated with the mixed user community? What would you, as a project manager, do to minimize those risks?
3. What are some of the tangible benefits to the project? What are some of the intangible benefits? What are some of the tangible and intangible costs? How would you handle the project's benefits and costs that will accrue to the health-care facilities—would you include tangible benefits and costs to the nursing homes in the cost/benefit analyses? Why or why not?
4. Overall, do you think the approach taken to the project (sequential waterfall versus iterative and incremental) would make a difference in the tangible and intangible costs and benefits? Discuss.
5. Overall, do you think the approach taken to the project would make a difference in minimizing the risks of project failure? Discuss.

FURTHER RESOURCES

Scott W. Ambler, *Agile Modeling: Effective Practices for XP and the RUP*. John Wiley and Sons, 2004.

Jim Highsmith, *Agile Project Management: Creating Innovative Products*. John Wiley and Sons, 2004.

Gopal K. Kapur, *Project Management for Information, Technology, Business, and Certification*. Prentice-Hall, 2005.

Jack R. Meredith and Samuel J. Mantel Jr., *Project Management: A Managerial Approach* (6th ed.). John Wiley and Sons, Inc., 2004.

Joseph Phillips, *IT Project Management: On Track from Start to Finish*. McGraw-Hill, 2002.

Project Management Institute, *A Guide to the Project Management Body of Knowledge*, 3rd edition. Project Management Institute, 2004.

Walker Royce, *Software Project Management: A Unified Framework*. Addison-Wesley, 1998.

Kathy Schwalbe, *Information Technology Project Management, Fifth Edition*. Course Technology, 2008.

Each chapter includes an opening case study, states clear **learning objectives**, and introduces the **chapter outline**.

CHAPTER 2

APPROACHES TO SYSTEM DEVELOPMENT

LEARNING OBJECTIVES

After reading this chapter, you should be able to:

- Explain the purpose and various phases of the traditional systems development life cycle (SDLC)
- Explain when to use an adaptive approach to the SDLC in place of the more predictive traditional SDLC
- Explain the differences between a model, a tool, a technique, and a methodology
- Describe the two overall approaches used to develop information systems: the traditional approach and the object-oriented approach
- Describe the key features of current trends in system development: the Unified Process (UP), Extreme Programming (XP), and Scrum
- Explain how automated tools are used in system development

CHAPTER OUTLINE

The Systems Development Life Cycle
Activities of Each SDLC Phase
Methodologies, Models, Tools, and Techniques
Two Approaches to System Development
Current Trends in Development
Tools to Support System Development

DEVELOPMENT APPROACHES AT AJAX CORPORATION, CONSOLIDATED CONCEPTS, AND PINNACLE MANUFACTURING

Kim, Mary, and Bob, graduating seniors, were discussing their recent interview visits to different companies that recruited computer information system (CIS) majors on their campus. All agreed that they had learned a lot by visiting the companies, but they also all felt somewhat overwhelmed at first.

"At first I wasn't sure I knew what they were talking about," Kim cautiously volunteered. During her on-campus interview, Kim had impressed Ajax Corporation with her knowledge of data modeling. When she visited the Ajax home office data center for the second interview, the interviewers spent quite a lot of time describing the company's system development methodology.

"A few people said to forget everything I learned in school," continued Kim. Ajax Corporation had purchased a complete development methodology called *IM One* from a small consulting firm. Most employees agreed it works fairly well. The people who had worked for Ajax for quite a while thought *IM One* was unique, and they were very proud of it. They had invested a lot of time and money learning and adapting to it.

"Well, that got my attention when they said forget what I learned in school," noted Kim, "but then they started telling me about their SDLC, about iterations, about business events, about data flow diagrams, and about entity-relationship diagrams, and things like that." Kim had recognized that many of the key concepts in the *IM One* methodology were fairly standard models and techniques from the structured approach to system development.

"I know what you mean," said Mary, a very talented programmer who knew just about every new programming language available. "Consolidated Concepts went on and on about things like OMG and UML and UP and some people named Booch, Rumbaugh, and Jacobson. But then it turned out that they were using the object-oriented approach to develop systems, and they liked the fact that I knew Java and VB .NET. No problem once I got past all of the terminology they used. They said they'd send me out for training on Rational Software Architect, a visual modeling tool for the object-oriented approach."

Bob had a different story. "A few people said analysis and design were no longer a big deal. I'm thinking, 'Knowing that would have saved me some time in school.'" Bob had visited Pinnacle Manufacturing, which had a small system development group supporting manufacturing and inventory control. "They said they try to just jump in and get to the code as soon as possible. Little documentation. Not much of a project plan. Then they showed me some books on their desks, and it looked like they had been doing a lot of reading about analysis and design. I could see they were using Extreme Programming and agile modeling techniques and focusing only on best practices needed for their small projects. It turns out they just organize their work differently by looking at risk and writing user stories while building prototypes. I recognized some sketches of class diagrams and sequence diagrams on the boss's whiteboard, so I felt fairly comfortable."

Kim, Mary, and Bob all agreed that there was much to learn in these work environments but also that many different terms and points of view are used to describe the same key concepts and techniques they learned in school. They were all glad they focused on the fundamentals in their CIS classes and that they had been exposed to a variety of approaches to system development.

OVERVIEW

As the experiences of Kim, Mary, and Bob demonstrate, there are many ways to develop an information system, and doing so is very complex. Project managers rely on a variety of aids to help them with every step of the process. The systems development life cycle (SDLC) introduced in this chapter provides an overall framework for managing the process of system

FEATURES

Margin definitions of key terms are placed in the text when the term is first used.

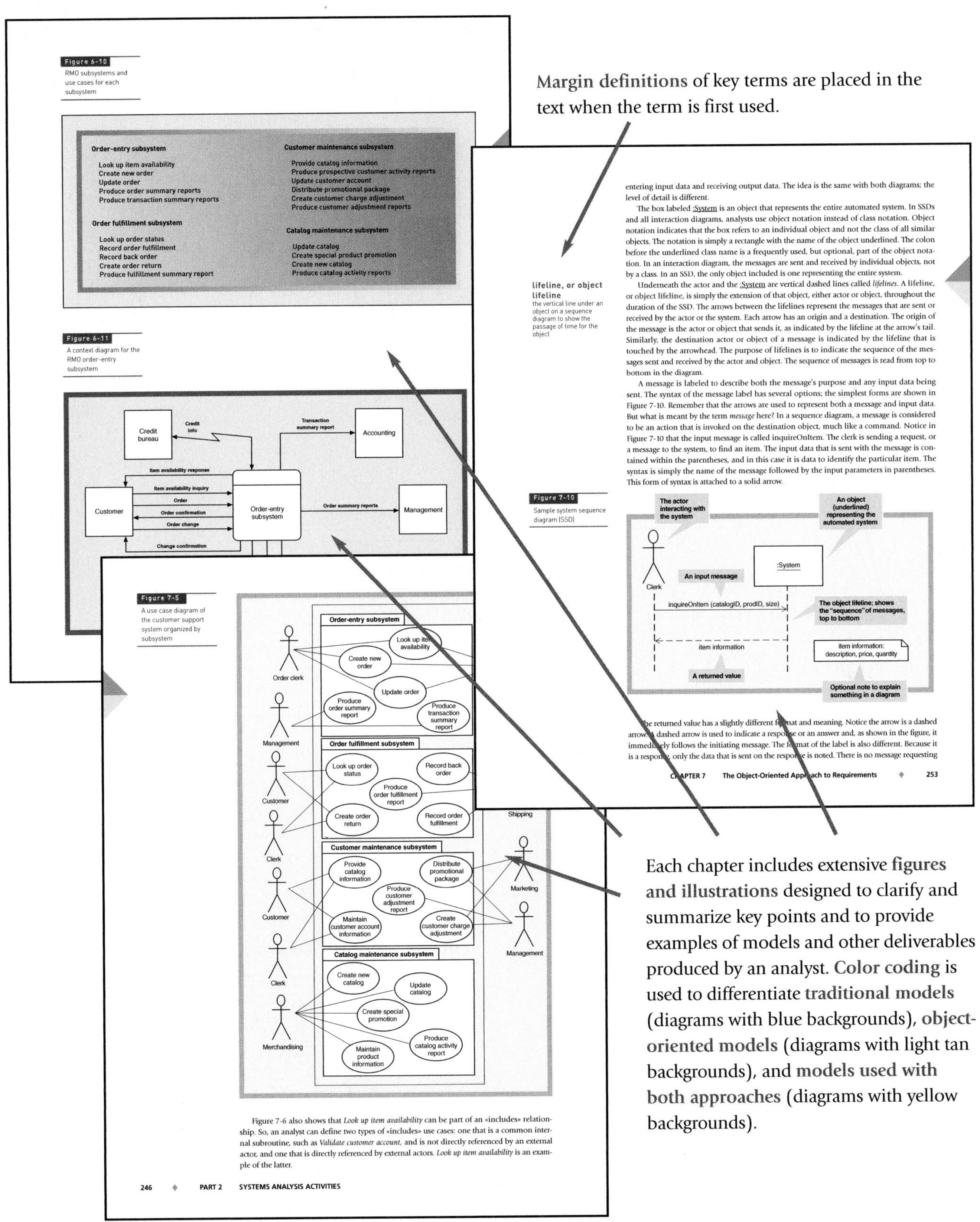

Each chapter includes extensive figures and illustrations designed to clarify and summarize key points and to provide examples of models and other deliverables produced by an analyst. Color coding is used to differentiate traditional models (diagrams with blue backgrounds), object-oriented models (diagrams with light tan backgrounds), and models used with both approaches (diagrams with yellow backgrounds).

FEATURES

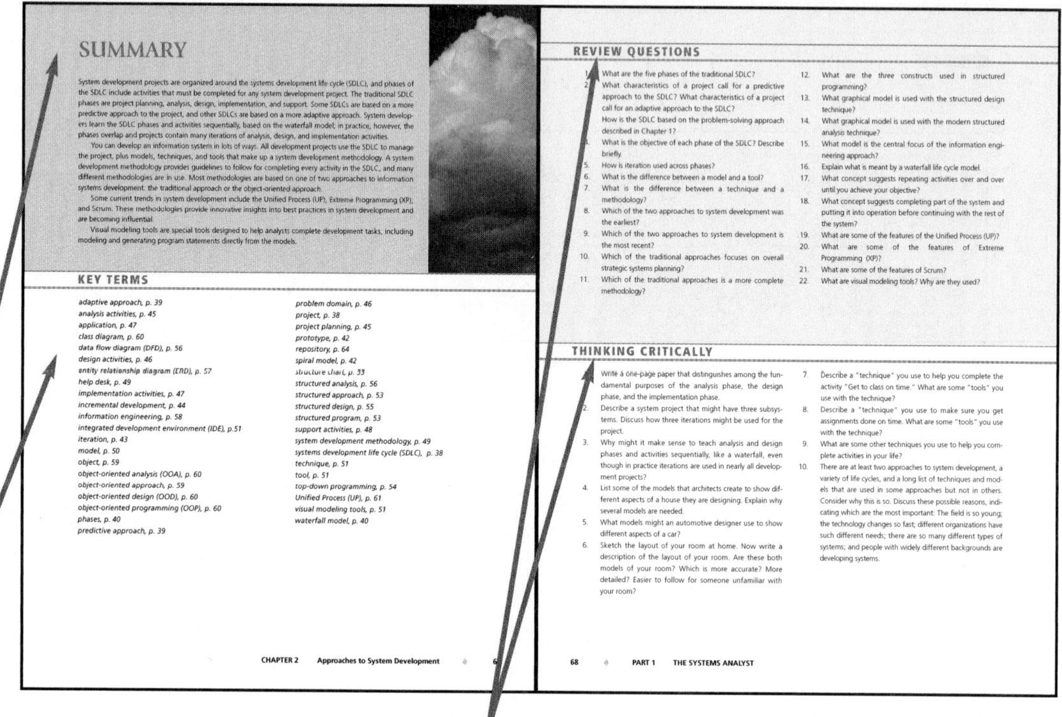

Each chapter also includes ample review questions, problems and exercises to get the student thinking critically, a collection of experiential exercises involving additional research or problem solving, end-of-chapter case studies that invite students to practice completing analysis and design tasks appropriate to the chapter, and a list of further resources and references.

End-of-chapter material includes a detailed summary, and an indexed list of key terms.

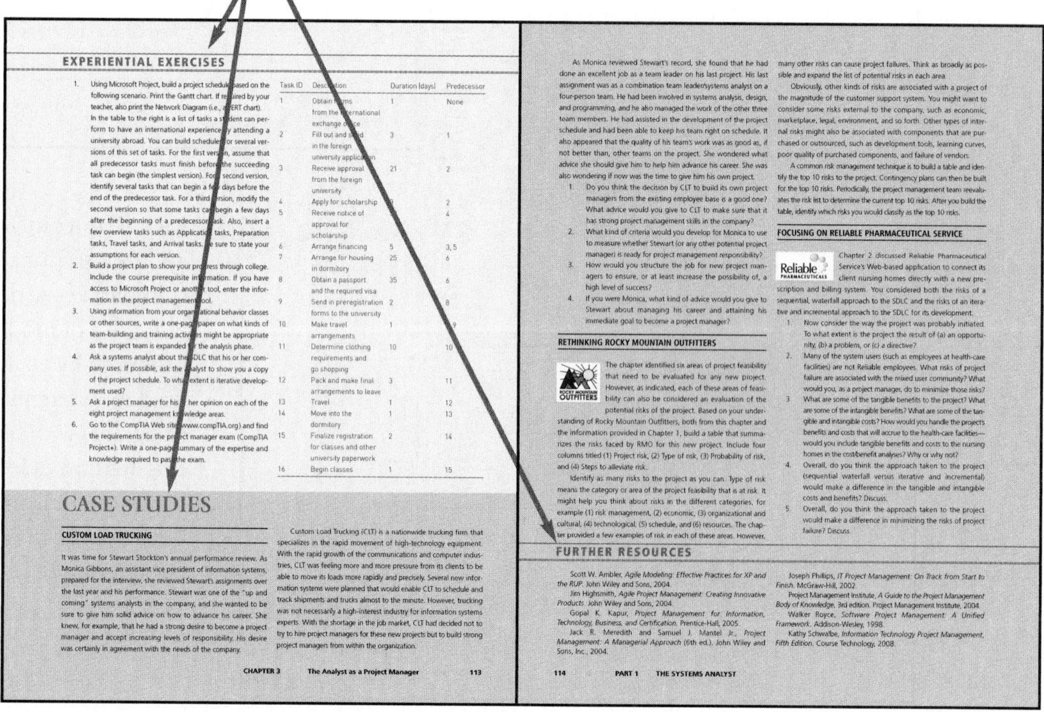

PREFACE

We have been very gratified as authors to receive so many supportive and enthusiastic comments about *Systems Analysis and Design in a Changing World*. Students and instructors in the United States and Canada have found our text to be the most up to date and flexible book available. The book has also been translated into many languages and is now used productively in Europe, Australia, New Zealand, India, China, and elsewhere. Our innovative and truly balanced coverage of traditional structured approaches and newer object-oriented approaches has continued to keep pace with changes in the field. The IS 2002 model curriculum suggests including a balanced coverage of both traditional and objected-oriented analysis and design, something this text has supported from the very beginning. The proposed IS 2008 model curriculum continues to place systems analysis and design in the core of IS/IT. This content is essential for system development majors as well as careers in business intelligence, business process management, ERP/package selection and support, and information technology service management. In this fifth edition, we continue to lead the way by making it feasible to cover object-oriented analysis and design in much greater depth using the latest OO models and design patterns. We also provide up-to-date coverage of adaptive and agile techniques and processes, and emphasize layered system architectures and Web development. Finally, we include substantial coverage of project management tools and techniques, including coverage of iterative and agile project management.

OBJECTIVES AND VISION

This text is designed for use in undergraduate and graduate courses that teach systems analysis and design. Systems analysis and design is a practical field that relies on a core set of concepts and principles, as well as what sometimes seems an eclectic collection of rapidly evolving tools and techniques. Therefore, learning analysis and design today requires an appreciation of the tried-and-true techniques widely embraced by experienced analysts, plus mastery of new and emerging tools and techniques that recent graduates are increasingly expected to apply on the job. It is not easy to develop and support information systems in today's rapidly changing environment, but the satisfaction and rewards for a job well done are substantial.

This text was developed by a team who was committed to producing an analysis and design text that was different—a text that is flexible and innovative, yet comprehensive and deep. We were guided by the belief that the text must be flexible enough to appeal both to instructors emphasizing more traditional approaches to systems analysis and design and to those emphasizing the latest object-oriented techniques. At the same time, we did not want to oversimplify the problem of system development. Many new developments affect systems analysis and design, and we wanted to include key trends—use cases, predictive and adaptive life cycles, agile development, UML, Web development, packaged solutions, enterprise resource planning (ERP), components, and so on.

We also wanted the text to teach the key concepts and techniques, not just describe them. Therefore, we focus on fundamentals of lasting value and then show how these fundamentals apply to all development approaches. We explore both traditional structured analysis and design and object-oriented analysis and design in depth. Flexible and innovative? Comprehensive and deep? We think you will agree that our text achieves these objectives.

INNOVATIONS

This text is unique in its integration of key systems modeling concepts that apply to both the traditional structured approach and the newer object-oriented approach—events that trigger system use cases and objects/entities that are part of the system's problem domain. We devote one chapter to identifying use cases and modeling key objects/entities. After completing that chapter, instructors can emphasize structured analysis and design or object-oriented analysis and design, or both. The object-oriented approach is not added as an afterthought—it is assumed from the beginning that everyone should understand the key object-oriented concepts. The traditional approach is not discarded—it is assumed from the beginning that everyone should understand the key structured concepts.

FULL COVERAGE OF OO APPROACH

The object-oriented approach presented in this text is based on the Unified Modeling Language (UML 2.0) from the Object Management Group, as originated by Grady Booch, James Rumbaugh, and Ivar Jacobson. A model-driven approach to analysis starts with use cases and scenarios and then defines problem domain classes involved in the users' work. We include requirements modeling with use case diagrams, use case descriptions, activity diagrams, and system sequence diagrams. Design models are also discussed in detail, with particular attention to use case realization with sequence diagrams, design class diagrams, and package diagrams. Design principles and design patterns are discussed throughout. Our database design chapter covers two approaches to object persistence—a hybrid approach using relational database management and a pure approach using object database management systems (ODBMS). An iterative, adaptive, and agile approach to OO development is emphasized throughout. Instructors who emphasize the object-oriented approach will not be disappointed by the presentation and depth of coverage in this text.

FULL COVERAGE OF TRADITIONAL APPROACH

The traditional approach presented in this text is based on modern structured analysis and design as refined by Stephen McMenamin and John Palmer, Ed Yourdon, and Meilir Page-Jones. Modern structured analysis is an integrated, model-driven approach that includes event partitioning, data modeling with entity-relationship diagrams (ERDs), and process modeling with data flow diagrams (DFDs). Modern structured design is also based on event partitioning and uses the structure chart for software design. Database design using relational database management techniques is featured. In this edition, we encourage students to try use cases and use case descriptions as an alternative approach to defining business processing requirements. Instructors who emphasize the structured approach to development will be pleased by the presentation and depth of coverage in this text.

EMERGING TOOLS AND TRENDS

Additional concepts and techniques are included in response to the realities of system development today. First, system development and the system development life cycle (SDLC) are explicitly defined as highly iterative. Although the text is organized as a sequential series of phases, the actual development project and the project plan are iterative. Second, emerging techniques and methodologies that use an iterative approach are introduced, including the Unified Process (UP), Extreme Programming (XP), Agile Development, and Scrum. Finally, packaged solutions and enterprise resource planning (ERP) are described as alternatives to custom development throughout the book. They are also described in detail in a separate ERP chapter on the book's Web site.

EMPHASIS ON ITERATION AND ARCHITECTURE

We did not reduce the amount of attention paid to the traditional approach to development. Many instructors choose to emphasize the traditional approach, but they also now cover the object-oriented approach to varying degrees. For both the traditional and the OO approach, however, we emphasize iterative development and three-layer architecture throughout. Web architectures and patterns are also included and modeled with UML. Predictive and adaptive approaches to the SDLC are discussed in relation to both approaches.

PROJECT MANAGEMENT COVERAGE AND SOFTWARE TOOLS

Many undergraduate programs depend on their systems analysis and design course to teach project management principles. To satisfy this need, we cover project management by taking a three-pronged approach. First, specific project management techniques, skills, and tasks are included and highlighted throughout chapters of the book. This integration teaches students how to apply specific project management tasks to the various phases and activities of the systems development life cycle, including iterative development. Second, we include a 120-day trial version of Microsoft Project 2007 Professional in the back of the book so that students can obtain hands-on experience with this important tool. Third, a fairly extensive treatment of project management concepts and principles is provided in an appendix on the book's Web site. This information is based on the Project Management Body of Knowledge (PMBOK), as developed by the Project Management Institute—the primary professional organization for project managers in the United States.

CHANGES FOR THE FIFTH EDITION

As we began considering updates to include in the fifth edition, we focused on refining some of the presentation and pedagogy, tightening some of the examples, and updating the material to reflect ongoing changes in analysis and design theory and practice. We also made some major changes based on our current research and feedback from instructors using the book.

The balanced coverage of the structured approaches and newer object-oriented approaches remains intact. This text can be used to emphasize the traditional structured approach with data flow diagrams or use case modeling, entity-relationship diagrams, structure charts, and relational databases; to focus on the object-oriented approach with use case modeling, domain and design class diagrams, interaction diagrams, package diagrams, and state machine diagrams; or to cover and compare both approaches in depth. We expanded the coverage of use cases to include them as a requirements model for the traditional approach as well as for the object-oriented approach. More and more development teams that work with traditional approaches and architectures are finding use cases and use case descriptions helpful. We did not remove the discussion of data flow diagrams, but we suggest that some instructors might cover use cases instead.

IMPROVED ORGANIZATION

We changed the organization and order of some material within some chapters. We left some material on the book's Web site as an online supplemental chapter and appendices. This gives instructors more flexibility in designing their courses, and it also makes the book more manageable. In this edition we made substantial changes to the OO design coverage by including two OO design chapters.

PREDICTIVE VERSUS ADAPTIVE APPROACHES TO THE SDLC

Another key change is the emphasis on both predictive and adaptive approaches to the SDLC as a way to define a continuum between sequential and highly iterative life cycles. Project managers should be able to tailor the SDLC to meet specific project needs.

ENHANCED OO DESIGN COVERAGE

Probably the most noticeable change in the last edition was the extensive enhancement and expanded coverage of the object-oriented approach. In this edition, we continue to refine the discussion and examples to make them as accessible as possible without sacrificing depth. Chapter 11 is all new, emphasizing the OO design process, design architectures, and design principles. Chapter 12 is based on the fourth edition's Chapter 11, but it received extensive updates to the examples. As a result, the OO coverage is improved and greatly expanded.

ENHANCED COVERAGE OF IMPLEMENTATION AND SUPPORT

In this edition, we extensively updated our chapter on implementation and support (Chapter 16). Although analysis and design courses have traditionally surveyed implementation, iterative approaches call for more emphasis on programmers, implementation and integration techniques, and testing in early iterations. Therefore, it becomes impossible to consider analysis and design without considering implementation and testing throughout the project.

EXPANDED COVERAGE OF EMERGING APPROACHES

Our text has always presented emerging concepts and approaches to analysis and design and system development. In this edition, we more fully integrate some specific agile methodologies within the discussion of adaptive approaches to the SDLC in Chapter 2. Then in Chapter 17, we discuss agile development, the Unified Process (UP), Extreme Programming (XP), and Scrum.

STUDENT COMPANION WEB SITE

We have created an exciting online companion for students as they work through the fifth edition of *Systems Analysis and Design in a Changing World*. In the back of this text, you will find a key code that provides full access to a robust Web site, *www.course.com/mis/sad5*. This Web resource includes the following features:

- **Practice Quizzes.** New quizzes, created specifically for this text, allow users to test themselves on the content of each chapter and immediately see what questions they answered correctly and incorrectly. For each question answered incorrectly, users are given the correct answer and the page in the text where that information is covered. Special testing software randomly compiles a selection of questions from a large database, so quizzes can be taken multiple times on a given chapter, with some new questions included each time.

- **Case Project**. The Web site offers an additional case project that is similar in scope and complexity to the Reliable Pharmaceuticals case in the book. This case project gives students the opportunity to sharpen their skills. It has installments for each chapter as well as corresponding solutions.
- **PowerPoint Slides.** Students can view the book's PowerPoint presentations, which cover the key points from each chapter. These presentations are a useful study tool.
- **Online Chapters and Appendices.** Students can access the following features on the site:
  - Online Supplemental Chapter 1, Packages and Enterprise Resource Planning
  - Appendix A, Principles of Project Management
  - Appendix B, Project Schedules with PERT/CPM Charts
  - Appendix C, Calculating Net Present Value, Payback Period, and Return on Investment
  - Appendix D, Presenting the Results to Management
  - Appendix E, Guide to Using Microsoft Project
- **Useful Web Links.** The site offers a repository of links to various Web sites where students can find more information about systems analysis and design in industry, possible careers, and other interesting resources for further learning.

ORGANIZATION AND USE

As in the fourth edition, the fifth edition is organized into four parts. Because of the increased separation of traditional and OO materials for system design and the expanded coverage of OO concepts, this print edition includes 17 chapters, supported by an additional chapter and five appendices on the book's Web site. Depending on the approach taken by the instructor, many chapters or sections of chapters can be skipped without loss of continuity. Some chapters are entirely optional. We begin with an overview of the entire text. Later, we discuss different approaches to using the text in analysis and design courses, and include suggested course outlines for instructors that emphasize either the traditional structured approach or the object-oriented approach. These outlines are also useful for instructors who teach graduate courses on analysis and design.

PART 1: THE SYSTEMS ANALYST

Chapter 1 discusses the work of an information systems analyst, including a streamlined discussion of systems and the role of the systems analyst as a problem solver in a modern business organization. The strategic information systems plan for Rocky Mountain Outfitters is discussed, and the customer support system is identified as the planned project ready to start development. Chapter 2 then asks, Now that we have a project, what do we have to do to get this system built? That is, what are the methodologies, models, tools, and techniques that can be used to develop systems? Predictive and adaptive approaches to the system development life cycle (SDLC) and iterative variations are introduced. We make it clear that a variety of approaches exist for system development and that today's analysts need to be familiar with all of them. Even if students specialize in one approach in their course or later in their job, they should be able to distinguish among structured, object-oriented, and several agile methodologies in a meaningful way. Chapter 3 moves right to the heart of the course—the system development project—introduced while describing the project planning phase of the SDLC in detail. Project planning, feasibility assessment, and project management techniques are covered. Students are drawn quickly into the RMO project so that the material has a meaningful context.

PART 2: SYSTEMS ANALYSIS ACTIVITIES

Part 2 moves ahead with systems analysis techniques. Chapter 4 describes the activities of the analysis phase of the SDLC in more detail. Then it focuses on investigating system requirements, including gathering information and interviewing system owners and users. Chapter 5 covers modeling system requirements, which includes event partitioning, use cases, and modeling objects/entities, as described earlier. Chapter 6 continues requirements modeling using the traditional approach, including data flow diagrams (DFDs), data flow definitions, and process descriptions. Chapter 7 continues the discussion begun in Chapter 5 using the object-oriented approach to requirements. Instructors can choose to emphasize Chapter 6 or Chapter 7 to focus the course on either the traditional or the object-oriented approach, or both. Chapter 8 presents an overview of technical environments that affect the generation of alternative system solutions. Then, a comprehensive guide to generating and evaluating alternatives is presented, including the reality that a packaged solution is always a viable option.

PART 3: SYSTEMS DESIGN TASKS

Chapter 9 introduces systems design and discusses the activities of the systems design phase of the SDLC in more detail. Details of the technological environment that affect design are reviewed, including networks, client/server architecture, and three-layer design. Chapter 10 discusses the traditional approach to design, including the latest thinking on three-layer designs. Chapter 11 and Chapter 12 address object-oriented design. Chapter 11 teaches design concepts, UML design models, and architectural design in depth. Chapter 12 teaches students how to design the interaction details for each use case—use case realization using sequence diagrams, communication diagrams, design class diagrams, and package diagrams. Instructors can choose to emphasize Chapter 10 or Chapter 11 to focus the course on either the traditional or the object-oriented approach, or both. More depth in OO design can be provided by covering Chapter 12 in addition to Chapter 11. Chapter 13 covers database design—relational, hybrid, and object-oriented databases. Chapter 14 covers user interfaces and human-computer interaction; we include general principles and concepts of dialog design in addition to using UML diagrams to model the dialog. Chapter 15 discusses system interfaces, with particular attention to system controls and system security.

PART 4: IMPLEMENTATION AND SUPPORT

Systems implementation is increasingly technology specific, and because of the diverse development environments in the real world, we decided to streamline the discussion of implementation. Chapter 16 provides an overview of implementation and support that addresses traditional technology and object technology. We also include a comprehensive discussion of some emerging approaches to system development in Chapter 17, including agile development, the Unified Process (UP), Extreme Programming (XP), and others. Similarly, although packaged solutions are discussed as viable alternatives throughout, we include a detailed discussion of packages and enterprise resource planning (ERP) in Online Supplemental Chapter 1, including specific examples from SAP.

DESIGNING YOUR ANALYSIS AND DESIGN COURSE

As discussed earlier, there are many approaches to teaching analysis and design courses, and the objectives of the course differ considerably from college to college. In some IS departments, the analysis and design course is a capstone course in which students apply the material learned in prior database, telecommunications, and programming courses to a real analysis and design project. In other IS departments, analysis and design is used as an introduction to the field of system development, taken prior to more specialized courses. Some IS departments offer a two-course sequence emphasizing analysis in the first semester and design and implementation in the second semester. Some IS departments have only one course that covers both analysis and design.

The design of the analysis and design course, always difficult, is complicated even more by the choice of emphasizing either the traditional structured approach or the newer object-oriented approach, again depending on local curriculum priorities. Additionally, the more iterative approach to development, in general, has made choices about sequencing the analysis and design topics more difficult. For example, with iterative development, a two-course sequence cannot be divided into analysis and then design as easily.

Given these issues, it is not practical to offer sample syllabi that will work for all of these options. The objectives, course content, assignments, and projects have too many variations. What we can offer are some suggestions for using the text in various approaches to the course.

TRADITIONAL ANALYSIS AND DESIGN COURSE

A traditional systems analysis and design course provides coverage of activities and tasks using structured analysis and structured design, with database design, input/output/controls design, and dialog (interface) design. It is usually assumed that the project will use custom development, including Web development. The course emphasizes the SDLC, project management, information gathering, and management reporting. One-semester courses are usually limited to completing some prototypes of the user interface to give students closure. Sometimes this course is spread over two semesters, with some implementation of an actual system in the second semester for a more complete development experience.

For this approach to the analysis and design course, a reasonable outline would omit chapters and sections detailing OO, current trends, and packages (these concepts are introduced throughout the text, however, so students will still be familiar with them). Additionally, because of the amount of material to cover, the appendices detailing project management, financial feasibility, scheduling, and presentations might be omitted.

A suggested outline for a course emphasizing the traditional approach follows:

Chapter 1: The World of the Information Systems Analyst
Chapter 2: Approaches to System Development
Chapter 3: The Analyst as a Project Manager
Chapter 4: Investigating System Requirements
Chapter 5: Modeling System Requirements
Chapter 6: The Traditional Approach to Requirements
Chapter 8: Evaluating Alternatives for Requirements, Environment, and Implementation
Chapter 9: Elements of System Design
Chapter 10: The Traditional Approach to Design
Chapter 13: Designing Databases (skip OO design sections)

Chapter 14: Designing the User Interface (skip UML examples)
Chapter 15: Designing System Interfaces, Controls, and Security (skip OO sections)
Chapter 16: Making the System Operational (skip OO sections)

OBJECT-ORIENTED ANALYSIS AND DESIGN COURSE

This course is similar to the coverage of both analysis and design in the traditional course, except that object-oriented models and techniques are emphasized exclusively. The course covers object-oriented analysis and design, with database design, input/output/controls design, and dialog (interface) design. It is usually assumed that the projects will use custom development, including Web development. The course emphasizes iterative development with three-layer architecture, project management, information gathering, and management reporting. One-semester courses are usually limited to completing some prototypes of the user interface to give students closure. Sometimes this course is spread over two semesters, with some implementation of an actual system in the second semester for a more complete development experience. Iterative development is usually emphasized.

For this approach to the analysis and design course, a reasonable outline would omit chapters and sections detailing structured analysis and structured design. Chapter 17 might be included to cover components and iteration, but packages probably would not be covered. Additionally, because of the amount of material to cover, the appendices detailing project management, financial feasibility, scheduling, and presentations might be omitted.

A suggested outline for a course emphasizing object-oriented development follows:

Chapter 1: The World of the Information Systems Analyst
Chapter 2: Approaches to System Development
Chapter 3: The Analyst as a Project Manager
Chapter 4: Investigating System Requirements
Chapter 5: Modeling System Requirements
Chapter 7: The Object-Oriented Approach to Requirements
Chapter 8: Evaluating Alternatives for Requirements, Environment, and Implementation
Chapter 9: Elements of System Design
Chapter 11: Object-Oriented Design: Principles
Chapter 12: Object-Oriented Design: Use Case Realizations
Chapter 13: Designing Databases
Chapter 14: Designing the User Interface
Chapter 15: Designing System Interfaces, Controls, and Security
Chapter 16: Making the System Operational
Chapter 17: Current Trends in System Development

TRADITIONAL COURSE WITH IN-DEPTH ANALYSIS AND PROJECT MANAGEMENT

Some courses delve more deeply into systems analysis methods and emphasize project management. Sometimes these courses are graduate courses, and sometimes they assume design and implementation are covered in more technical courses. In some cases, it might be assumed that packages are likely solutions rather than custom development, so defining requirements and managing the process are more important than design activities.

The appendices covering project management, financial feasibility, scheduling, and presentations should be included. Chapters on detailed design might be omitted. The packages/ERP chapter (Online Supplemental Chapter 1) might be included, if appropriate.

A suggested outline for courses emphasizing the traditional approach, with in-depth coverage of analysis and project management, follows:

Chapter 1: The World of the Information Systems Analyst
Chapter 2: Approaches to System Development
Chapter 3: The Analyst as a Project Manager
Online Appendix A: Principles of Project Management
Online Appendix B: Project Schedules with PERT/CPM Charts
Online Appendix C: Calculating Net Present Value, Payback Period, and Return on Investment
Chapter 4: Investigating System Requirements
Online Appendix D: Presenting the Results to Management
Chapter 5: Modeling System Requirements
Chapter 6: The Traditional Approach to Requirements
Chapter 8: Evaluating Alternatives for Requirements, Environment, and Implementation
Chapter 9: Elements of System Design
Online Supplemental Chapter 1: Packages and Enterprise Resource Planning

OBJECT-ORIENTED COURSE WITH IN-DEPTH ANALYSIS AND PROJECT MANAGEMENT

Some courses cover object-oriented systems analysis methods in more depth—but not OO design—and emphasize project management. Sometimes these courses are graduate courses, and sometimes they assume design and implementation are covered in more technical courses. In some cases, it might be assumed that packages are likely solutions rather than custom development, so defining requirements and managing the process are more important than design activities.

The appendices covering project management, financial feasibility, scheduling, and presentations should be included. Chapters on detailed design might be omitted. The packages/ERP chapter (Online Supplemental Chapter 1) might be included, if appropriate.

A suggested outline for a course covering object-oriented analysis, with in-depth coverage of project management, follows:

Chapter 1: The World of the Information Systems Analyst
Chapter 2: Approaches to System Development
Chapter 3: The Analyst as a Project Manager
Online Appendix A: Principles of Project Management
Online Appendix B: Project Schedules with PERT/CPM Charts
Online Appendix C: Calculating Net Present Value, Payback Period, and Return on Investment
Chapter 4: Investigating System Requirements
Online Appendix D: Presenting the Results to Management
Chapter 5: Modeling System Requirements
Chapter 7: The Object-Oriented Approach to Requirements
Chapter 8: Evaluating Alternatives for Requirements, Environment, and Implementation
Chapter 9: Elements of System Design
Online Supplemental Chapter 1: Packages and Enterprise Resource Planning

COMPARATIVE ANALYSIS AND DESIGN COURSE

Some courses survey the field of analysis and design to provide a comprehensive exposure to major approaches. Sometimes these courses are graduate courses for experienced developers, and sometimes they emphasize concepts over detailed hands-on experience with techniques. A reading knowledge of the key models might be the objective. However, the instructor often will require hands-on projects using both traditional and object-oriented techniques for the same system in the same course.

The entire book can be covered for the most complete treatment. Alternatively, sections of material that cover details about some of the techniques can be omitted. A fast-paced survey course can cover the chapters quickly for recognition and reading knowledge of models. Chapter 17 and Online Supplemental Chapter 1 might directly follow Chapter 8, as shown in the following outline, and then the course can continue surveying design. If the comparative course emphasizes systems analysis and project management, it might end after Online Supplemental Chapter 1 without covering design. There are many possibilities to consider.

A suggested outline for a comparative course follows:

Chapter 1: The World of the Information Systems Analyst
Chapter 2: Approaches to System Development
Chapter 3: The Analyst as a Project Manager
Chapter 4: Investigating System Requirements
Chapter 5: Modeling System Requirements
Chapter 6: The Traditional Approach to Requirements
Chapter 7: The Object-Oriented Approach to Requirements
Chapter 8: Evaluating Alternatives for Requirements, Environment, and Implementation
Chapter 17: Current Trends in System Development
Online Supplemental Chapter 1: Packages and Enterprise Resource Planning
Chapter 9: Elements of System Design
Chapter 10: The Traditional Approach to Design
Chapter 11: Object-Oriented Design: Principles
Chapter 13: Designing Databases
Chapter 14: Designing the User Interface
Chapter 15: Designing System Interfaces, Controls, and Security
Chapter 16: Making the System Operational

AN ITERATIVE APPROACH TO THE ANALYSIS AND DESIGN COURSE

One of the biggest challenges facing analysis and design instructors is how to handle iterative development. This is an issue for both the traditional approach and the object-oriented approach. Textbooks can teach analysis techniques and then design techniques sequentially, but that is not the way the techniques are used in practice. Students do not always appreciate this point. One way to make the course resemble real-world practice is to teach iteratively. The idea that no one gets it right the first time certainly applies to learning analysis and design.

As with iterative development, the course could survey analysis and design techniques rapidly, perhaps so students could obtain a reading knowledge of the models and then go back over the analysis and design material in more depth. Some sections of chapters might be skipped the first time through. But there is a great difference between understanding and interpreting analysis and design models and actually creating analysis and design models. Therefore, it might make sense to go through the techniques and make a reading knowledge a goal for the first iteration. Then students can be asked to reconsider the models as they create new ones based on a course project.

It would be difficult to ask students to read everything once and then to reread it all again. Therefore, one approach might be to rapidly survey the field without digressing into specifics. Then the second iteration could add new material while teaching prior material in depth. For example, the first iteration might emphasize Chapter 5 but skim through either Chapter 6 or 7 (depending on whether traditional or OO is emphasized). You might cover the overview of design in Chapter 9, but the rest of the design chapters might be limited to Chapter 10 or Chapter 11 (depending on whether traditional or OO is emphasized). The second iteration could go into requirements models and design chapters in depth.

There are many other possibilities to consider. What is important is to consider the iterative approach in some way when designing your course. We would appreciate any feedback you can provide on ideas you have considered or tried with an iterative approach to teaching analysis and design.

AVAILABLE SUPPORT

Systems Analysis and Design in a Changing World, Fifth Edition, includes teaching tools to support instructors in the classroom. The ancillary materials that accompany the textbook include an Instructor's Manual, solutions, Test Banks and Test Engine, Distance Learning content, PowerPoint presentations, and Figure Files. Please contact your Cengage Course Technology sales representative to request the Teaching Tools CD-ROM, if you have not already received it. Or, go to the Web page for this book at *www.course.com* to download many of these items.

THE INSTRUCTOR'S MANUAL

The Instructor's Manual includes suggestions and strategies for using the text, including course outlines for instructors that emphasize the traditional structured approach or the object-oriented approach. The manual is also helpful for those teaching graduate courses on analysis and design.

SOLUTIONS

We provide instructors with answers to review questions and suggested solutions to chapter exercises and cases. Detailed traditional and UML OO models are included for all exercises and cases that ask for modeling solutions.

EXAMVIEW®

This objective-based test generator lets the instructor create paper, LAN, or Web-based tests from test banks designed specifically for this Course Technology text. Instructors can use the QuickTest Wizard to create tests in fewer than five minutes by taking advantage of Course Technology's question banks, or instructors can create customized exams.

DISTANCE LEARNING CONTENT

Course Technology, the premiere innovator in management information systems publishing, is proud to present online courses in WebCT and Blackboard.

- **Blackboard and WebCT Level 1 Online Content.** If you use Blackboard or WebCT, the test bank for this textbook is available at no cost in a simple, ready-to-use format. Go to *www.course.com* and search for this textbook to download the test bank.
- **Blackboard and WebCT Level 2 Online Content.** Blackboard Level 2 and WebCT Level 2 are also available for *Systems Analysis and Design in a Changing World*. Level 2 offers course management and access to a Web site that is fully populated with content for this book.

For more information on how to bring distance learning to their course, instructors should contact their Course Technology marketing representative.

POWERPOINT PRESENTATIONS

Microsoft PowerPoint slides are included for each chapter. Instructors might use the slides in a variety of ways, such as teaching aids during classroom presentations or as printed handouts for classroom distribution. Instructors can add their own slides for additional topics they introduce to the class.

FIGURE FILES

Figure files allow instructors to create their own presentations using figures taken directly from the text.

SOFTWARE BUNDLING OPTIONS

Many instructors like to include software for students to use for exercises and course projects, and this text offers many bundling possibilities. Some instructors like to emphasize visual modeling tools, and Course Technology can bundle several popular tools with the text.

CREDITS AND ACKNOWLEDGMENTS

This book was originally launched following some extensive brainstorming by senior vice president and publisher Kristen Duerr of Course Technology and lead author John Satzinger. We agreed that an analysis and design text required a major commitment from the publisher to be competitive. We also agreed that no one person could complete a text that met the objectives—flexible and innovative, yet comprehensive and deep. Therefore, Course Technology took an active role in assembling a team of authors who shared the vision. The managing editor brought in to direct the initial project was Jennifer Locke, who had a major role in bringing the authors together and shaping the direction and final form of the text. We were also fortunate to have Barrie Tysko placed in charge of managing the second and third editions.

We were very fortunate to have senior product manager Eunice Yeates-Fogle placed in charge of managing the fourth edition. Eunice also managed the development of our other Course Technology text—*Object-Oriented Analysis and Design with the Unified Process*. This fifth edition was managed by Kate Hennessy, who was charged with recruiting a new developmental editor, negotiating with the production department for an accelerated writing and editing schedule, and dealing with numerous author uncertainties and scheduling conflicts that made the project quite daunting from the publisher's point of view.

Another essential member of the team was developmental editor Dan Seiter. Dan jumped in and quickly adapted to the styles and personalities of the author team, and he rapidly digested and mastered the complex content and objectives of the text. Our previous developmental editor, Karen Hill of Elm Street Publishing Services, guided us through the first four editions. She collected and digested the comments and reactions of initial reviewers, provided guidance and design for the features and chapter pedagogy, suggested improvements and refinements to the organization and content, and edited the chapters to provide a consistent style.

We are grateful for the forward thinking and continuing support of Course Technology executive editors Mac Mendelsohn, Bob Woodbury, and David Boelio. Many other people were involved in the production of this text. Amanda Young Shelton of Course Technology provided substantial support for the first edition. Marisa Taylor and her production team at GEX Publishing Services came through with every commitment on schedule and produced a beautiful and functional text.

We also want to thank some other key people for their specific contributions—Richard A. Johnson of Missouri State University for writing Online Supplemental Chapter 1 on packages and ERP, and William Baker for contributing material on presentation techniques. Many other colleagues and friends at Missouri State University, Brigham Young University, the University of New Mexico, and elsewhere contributed to and supported our work in one way or another. Special thanks also go to Lavette Teague, Lorne Olfman, and Paul Gray for guidance and inspiration.

Last, but certainly not least, we want to thank all of the reviewers who worked so hard for us, beginning with an initial proposal and continuing throughout the completion of all five editions of this text. We were lucky enough to have reviewers with broad perspectives, in-depth knowledge, and diverse preferences. We listened very carefully, and the text is much better as a result of their input. Reviewers for the various editions included:

Rob Anson, *Boise State University*
Marsha Baddeley, *Niagara College*
Teri Barnes, *DeVry Institute—Phoenix*
Robert Beatty, *University of Wisconsin—Milwaukee*
Anthony Cameron, *Fayetteville Technical Community College*
Genard Catalano, *Columbia College*
Paul H. Cheney, *University of Central Florida*
Jung Choi, *Wright State University*
Jon D. Clark, *Colorado State University*
Lawrence E. Domine, *Milwaukee Area Technical College*
Jeff Hedrington, *University of Phoenix*
Ellen D. Hoadley, *Loyola College in Maryland*
Norman Jobes, *Conestoga College, Waterloo, Ontario*
Gerald Karush, *Southern New Hampshire University*
Robert Keim, *Arizona State University*
Rajiv Kishore, *The State University of New York, Buffalo*
Rebecca Koop, *Wright State University*
Hsiang-Jui Kung, *Georgia Southern University*
James E. LaBarre, *University of Wisconsin—Eau Claire*
Tsun-Yin Law, *Seneca College*

David Little, *High Point University*
George M. Marakas, *Indiana University*
Roger McHaney, *Kansas State University*
Cindi A. Nadelman, *New England College*
Bruce Neubauer, *Pittsburgh State University*
Michael Nicholas, *Davenport University—Grand Rapids*
George Pennells
Julian-Mark Pettigrew
Mary Prescott, *University of South Florida*
Alex Ramirez, *Carleton University*
Eliot Rich, *The State University of New York, Albany*
Robert Saldarini, *Bergen Community College*
Laurie Schatzberg, *University of New Mexico*
Deborah Stockbridge, *Quincy College*
Jean Smith, *Technical College of the Lowcountry*
Peter Tarasewich, *Northeastern University*
Craig VanLengen, *Northern Arizona University*
Bruce Vanstone, *Bond University*
Terence M. Waterman, *Golden Gate University*

All of us involved in the development of this text wish you all the best as you take on the challenge of analysis and design in a changing world.

—John Satzinger
—Robert Jackson
—Steve Burd

PART 1

THE SYSTEMS ANALYST

CHAPTER 1
The World of the Information Systems Analyst

CHAPTER 2
Approaches to System Development

CHAPTER 3
The Analyst as a Project Manager

CHAPTER 1

THE WORLD OF THE INFORMATION SYSTEMS ANALYST

LEARNING OBJECTIVES

After reading this chapter, you should be able to:

- Explain the key role of a systems analyst in business
- Describe the various types of systems and technology an analyst might use
- Explain the importance of technical skills, people skills, and business skills for an analyst
- Explain why ethical behavior is crucial for a systems analyst's career
- Describe various job titles in the field and places of employment where analysis and design work is done
- Discuss the analyst's role in strategic planning for an organization
- Describe the analyst's role in a system development project

CHAPTER OUTLINE

The Analyst as a Business Problem Solver

Systems That Solve Business Problems

Required Skills of the Systems Analyst

Analysis-Related Careers

The Analyst's Role in Strategic Planning

Rocky Mountain Outfitters and Its Strategic Information Systems Plan

The Analyst as a System Developer (the Heart of the Course)

A SYSTEMS ANALYST AT CONSOLIDATED REFINERIES

Mary Wright thought back about her two-year career as a programmer analyst. She had been asked to talk to visiting computer information system (CIS) students about life on the job. "It seems like yesterday that I finally graduated from college and loaded up a U-Haul to start my new job at Consolidated," she began.

Consolidated Refineries is an independent petroleum refining company in west Texas. Consolidated buys crude oil from freelance petroleum producers and refines it into gasoline and other petroleum products for sale to independent distributors. Demand for refined petroleum products had been increasing rapidly, and Consolidated was producing at maximum capacity. Capacity planning systems and refining operations systems were particularly important computer information systems for Consolidated, because careful planning and process monitoring resulted in increased production at reduced costs. This increasing demand, and other competitive changes in the energy industry, made information systems particularly important to Consolidated.

Mary continued her informal talk to visiting students. "At first I did programming, mainly fixing things that end users wanted done. I completed some training on Java and object-oriented analysis to round out my experience. The job was pretty much what I had expected at first until everything went crazy over the IPCS project."

The Integrated Process Control System (IPCS) project was part of the company's information systems plan drawn up the year before. Edward King, the CEO of Consolidated Refineries, had pushed for more strategic planning at the company from the beginning, including drawing up a five-year strategic plan for information systems. The IPCS development project was scheduled to begin in the third or fourth year of the plan, but suddenly priorities changed. Demand for petroleum products had never been higher, and supplies of crude oil were becoming scarce. At the same time, political pressure was making price increases an unpopular option.

Something had to be done to increase production and reduce costs. It would be years before an additional refinery could be built, and additional crude oil supplies from new oil fields were years away. The only option for Consolidated's growth and increased profits was to do a better job with the plants and supplies it had. So, top executives decided to make a major commitment to implementing the IPCS project, with the goal of radically improving capacity planning and process monitoring. Everyone at Consolidated also wanted access to this information anywhere and anytime.

"It seemed like the IPCS project was the only thing the company cared about," continued Mary. "I was assigned to the project as the junior analyst assisting the project manager, so I got in on everything. Suddenly I was in meeting after meeting, and I had to digest all kinds of information about refining and distribution, as if I were a petroleum engineer. I met with production supervisors, suppliers, and marketing managers to learn about the oil business, just as if I were taking business school courses. I traveled all over to visit oil fields and pipelines—including a four-day trip to Alaska on about two days' notice! I interviewed technology vendors' representatives and consultants who specialized in capacity planning and process control systems. I've been spending a lot of time at my computer, too, writing reports, letters, and memos—not programming!

"We've been working on the project for seven months now, and every time I turn around, Mr. King, our CEO, is saying something about how important the IPCS project is to the future of the company. He repeats the story to employees and to the stockholders. Mr. King attends many of our status meetings, and he even sat next to me the day I presented a list of key requirements for the system to the top management team.

"This is not at all the way I thought it would be."

OVERVIEW

As Mary Wright's story about Consolidated Refineries illustrates, information systems with strategic value are critical to the success of business organizations and their top executives. Most of the activities and tasks completed by a system developer, even a new graduate like Mary, involve much more than programming. Systems analysis is really more about understanding the business and its goals and strategies, defining requirements for information systems that support those goals and strategies, and supporting the business. It's not at all what most college students imagine it to be.

People today are attracted to information systems careers because information technology (IT) can have a dramatic impact on productivity and profits. Most of you regularly use the latest technologies for online purchases and reservations, online auctions and customer support, and e-mail and wireless messaging. But it is not the technology itself that increases productivity and profits; it is the people who develop information system solutions that harness the power of the technology that makes these benefits possible. The challenges are great because more and more people expect to have information systems that provide access to information anywhere and anytime.

The key to successful system development is thorough systems analysis and design to understand what the business requires from the information system. **Systems analysis** means understanding and specifying in detail what the information system should accomplish. **Systems design** means specifying in detail how the many components of the information system should be physically implemented. This text is about systems analysis and design techniques used by a **systems analyst**, a business professional who develops information systems.

systems analysis
the process of understanding and specifying in detail what the information system should accomplish

systems design
the process of specifying in detail how the many components of the information system should be physically implemented

systems analyst
a business professional who uses analysis and design techniques to solve business problems using information technology

This chapter describes the world of the systems analyst—the nature of the work, the knowledge and skills that are important, and the types of systems and special projects an analyst works on. First, we define the analyst's work as problem solving for an organization, so the problem-solving process the analyst follows is described. Next, because most problems an analyst works on are solved in part by an information system, the chapter reviews the types of information systems that businesses use. A systems analyst is a business professional who requires extensive technical, business, and people knowledge and skills, so these skills are reviewed next. Then we survey the variety of workplaces and positions in which analysis work is done. Sometimes an analyst works on special projects such as strategic planning, business process reengineering, and enterprise resource planning. An analyst's work is really not at all the way most CIS students think it will be.

Finally, the chapter introduces Rocky Mountain Outfitters (RMO), a regional sports clothing distributor headquartered in Park City, Utah. RMO is following a strategic information systems plan that calls for a series of information system development and integration projects over the next several years. The project that RMO is about to launch is a system development project for a new customer support system that will integrate phone, mail, and Web-based orders. The Rocky Mountain Outfitters case is used throughout the text to illustrate analysis and design techniques.

THE ANALYST AS A BUSINESS PROBLEM SOLVER

Systems analysis and design is, first and foremost, a practical field grounded in time-tested and rapidly evolving knowledge and techniques. Analysts must certainly know about computers and computer programs. They possess special skills and develop expertise in programming. But they must also bring to the job a fundamental curiosity to explore how things are done and the determination to make them work better.

Developing information systems is not just about writing programs. Information systems are developed to solve problems for organizations, as the opening case study demonstrated, and a systems analyst is often thought of as a problem solver rather than a programmer. So, what kinds of problems does an analyst typically solve?

- Customers want to order products any time of the day or night. So, the problem is how to process those orders around the clock without adding to the selling cost.
- Production needs to plan very carefully the amount of each type of product to produce each week. So, the problem is how to estimate the dozens of parameters that affect production and then allow planners to explore different scenarios before committing to a specific plan.
- Suppliers want to minimize their inventory holding costs by shipping parts used in the manufacturing process in smaller daily batches. So, the problem is how to order in smaller lots and accept daily shipments to take advantage of supplier discounts.
- Marketing wants to anticipate customer needs better by tracking purchasing patterns and buyer trends. So, the problem is how to collect and analyze information on customer behavior that marketing can put to use.
- Management continually wants to know the current financial picture of the company, including profit and loss, cash flow, and stock market forecasts. So, the problem is how to collect, analyze, and present all of the financial information management wants.
- Employees demand more flexibility in their benefits programs, and management wants to build loyalty and morale. So, the problem is how to process transactions for flexible health plans, wellness programs, employee investment options, retirement accounts, and other benefit programs offered to employees.

Figure 1-1

The analyst's approach to problem solving

Research and understand the problem

Verify that the benefits of solving the problem outweigh the costs

Define the requirements for solving the problem

Develop a set of possible solutions (alternatives)

Decide which solution is best, and make a recommendation

Define the details of the chosen solution

Implement the solution

Monitor to make sure that you obtain the desired results

Information system developers work on problems such as these—and many more. Some of these problems are large and strategically important. Some are much smaller, affecting fewer people, but important in their own way. All programming for the information system that solves the business problem is important, but solving each of these problems involves more than programming.

How does an analyst solve problems? Systems analysis and design focuses on understanding the business problem and outlining the approach to be taken to solve it. Figure 1-1 shows a general approach to problem solving that can be adapted to solving business problems using information technology. Obviously, part of the solution is a new information system, but that is just part of the story.

The analyst must first understand the problem and learn everything possible about it—who is involved, what business processes come into play, and what other systems would be affected by solving the problem. Then the analyst needs to confirm for management that the benefits of solving the problem outweigh the costs. Sometimes it would cost a fortune to solve the problem, so it might not be worth solving.

If solving the problem is feasible, the analyst defines in detail what is required to solve it—what specific objectives must be satisfied, what data needs to be stored and used, what processing must be done to the data, and what outputs must be produced. *What* needs to be done must be defined first. *How* it will be done is not important yet.

After detailed requirements are defined, the analyst develops a set of possible solutions. Each possible solution (an alternative) needs to be thought through carefully. Usually, an information system alternative is defined as a set of choices about physical components that make up an information system—*how* it will be done. Many choices must be made, involving questions such as these:

- What are the needed components?
- What technology should be used to build the different components?
- Where are the components located?
- How will components communicate over networks?
- How are components configured into a system?
- How will people interact with the system?

- Which components are custom-made, and which are purchased from vendors?
- Who should build the custom-made components?
- Who should assemble and support the components?

Many different alternatives must be considered, and the challenge is to select the best—that is, the solution with the fewest risks and most benefits. Alternatives for solving the problem must be cost-effective, but they also must be consistent with the corporate strategic plan. Does the alternative contribute to the basic goals and objectives of the organization? Will it integrate seamlessly with other planned systems? Does it use technology that fits the strategic direction that management has defined? Will end users be receptive to it? Analysts must consider many factors and make tough decisions.

After the systems analyst has determined, in consultation with management, which alternative to recommend and management has approved the recommendation, the design details must be worked out. Here the analyst is concerned with creating a blueprint (design specifications) for how the new system will work. Systems design specifications describe the construction details of all parts of the system, including databases, user interfaces, networks, operating procedures, conversion plans, and, of course, program modules.

Thus far we haven't mentioned programming, even though we're near the end of the steps outlined in Figure 1-1. Inexperienced developers have a tendency to rush into programming without completing the earlier steps. Sometimes early programming may be needed to evaluate technical feasibility or to help users understand how a completed system might look and behave. But much of the time, early programming results in wasted time and money because key system requirements or design constraints are not well understood. Building a system based on incomplete or misunderstood requirements ensures that the project will be over budget, late, and will deliver a system that doesn't fully solve the problems it was intended to address. An information system can cost a lot of money to build and install—perhaps millions of dollars. It is not unusual for dozens of programmers to work on programs to get a system up and running, and those programmers need to know exactly what the system is to accomplish—thus, detailed specifications are required. This text presents the tools and techniques that an analyst uses during system development to create the detailed specifications. Some of these specifications are the result of systems analysis, and some are the result of systems design.

Although this text is oriented toward potential systems analysts, it also provides a good foundation for others who will deal with business problems that could be solved with the help of an information system. Managers throughout business must become more and more knowledgeable about using information technology to solve business problems. Many general business students take a systems analysis and design course to round out their background in two-year and four-year degree programs. Many graduate programs, such as master of business administration (MBA) and master of accountancy (M.Acc) programs, have technology tracks with courses that use this book. Remember that systems analysis and design work is not just about developing systems; it is really about solving business problems using information technology. So even though they never build information systems, managers need to gain expertise in these concepts to be effective in their jobs.

system
a collection of interrelated components that function together to achieve some outcome

SYSTEMS THAT SOLVE BUSINESS PROBLEMS

information system
a collection of interrelated components that collect, process, store, and provide as output the information needed to complete business tasks

We described the systems analyst as a business problem solver. We said that the solution to the problem is usually an information system. Before we talk about how you learn to be a systems analyst, let's quickly review some information systems concepts.

INFORMATION SYSTEMS

A **system** is a collection of interrelated components that function together to achieve some outcome. An **information system** is a collection of interrelated components that collect, process, store, and provide as output the information needed to complete a business task. Completing a business task is usually the "problem" we talked about earlier.

A payroll system, for example, collects information on employees and their work, processes and stores that information, and then produces paychecks and payroll reports (among other things) for the organization. A sales management system collects information about customers, sales, products, and inventory levels. It enables customers and sales personnel to create and modify sales orders, select payment methods, and output sales information for tasks such as generating financial statements, computing bonuses, and scheduling production.

subsystem
a system that is part of a larger system

What are the interrelated components of an information system? You can think about components in several ways. Any system can have subsystems. A subsystem is a system that is part of another system, so subsystems might be one way to think about the components of a system. For example, a sales management system might be one subsystem of a customer relationship management (CRM) system. Another CRM subsystem might enable customers to view past and current orders, track order fulfillment and shipping, and modify their account information. A third CRM subsystem might maintain the product catalog database and provide Web-based access to product specifications and manuals. A fourth CRM subsystem might provide technical support via telephone and a Web site with detailed tracking of customer support requests and related reporting to improve call center management and product quality. When looking at the business as a single system, the CRM system is only one subsystem among others, including the accounting and financial management system, the manufacturing management system, and the human resources management system.

functional decomposition
dividing a system into components based on subsystems that are further divided into smaller subsystems

The view of a system as a collection of subsystems is very useful to the analyst. It enables the analyst to focus attention on a single area of a business or organization, a group of related areas, or the interfaces among areas. Figure 1-2 shows how one system can be divided, or decomposed, into subsystems, which in turn can be further decomposed into subsystems. This approach to dividing a system into components is referred to as functional decomposition.

Figure 1-2
information systems and subsystems

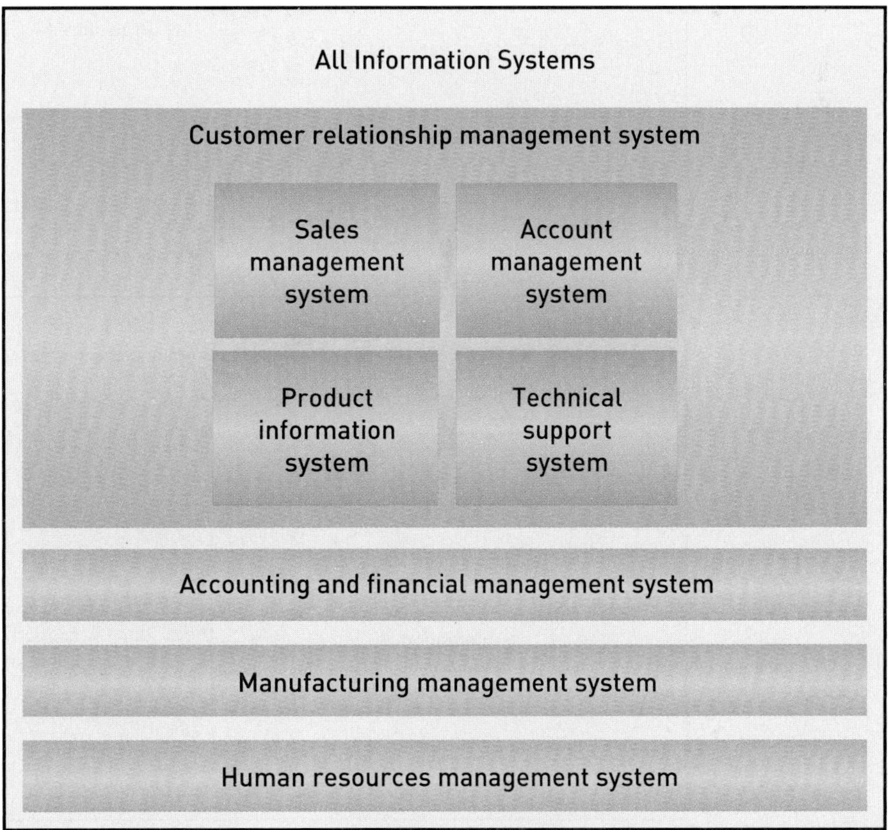

system boundary

the separation between a system and its environment that inputs and outputs must cross

automation boundary

the separation between the automated part of a system and the manual part of a system

Another way to think about the components of a system is to list the parts that interact. For example, an information system includes hardware, software, inputs, outputs, data, people, and procedures. This view is also very useful to the analyst. These interrelated components function together in a system, as shown in Figure 1-3.

Every system has a boundary between it and its environment. Any inputs or outputs must cross the **system boundary**. Defining these inputs and outputs is an important part of systems analysis and design. In an information system, people are also key components, and these people do some of the system's work. So there is another boundary that is important to a systems analyst—the **automation boundary**. On one side of the automation boundary is the automated part of the system, where work is done by computers. On the other side is the manual part of the system, where work is done by people (see Figure 1-4).

Figure 1-3

Information systems and component parts

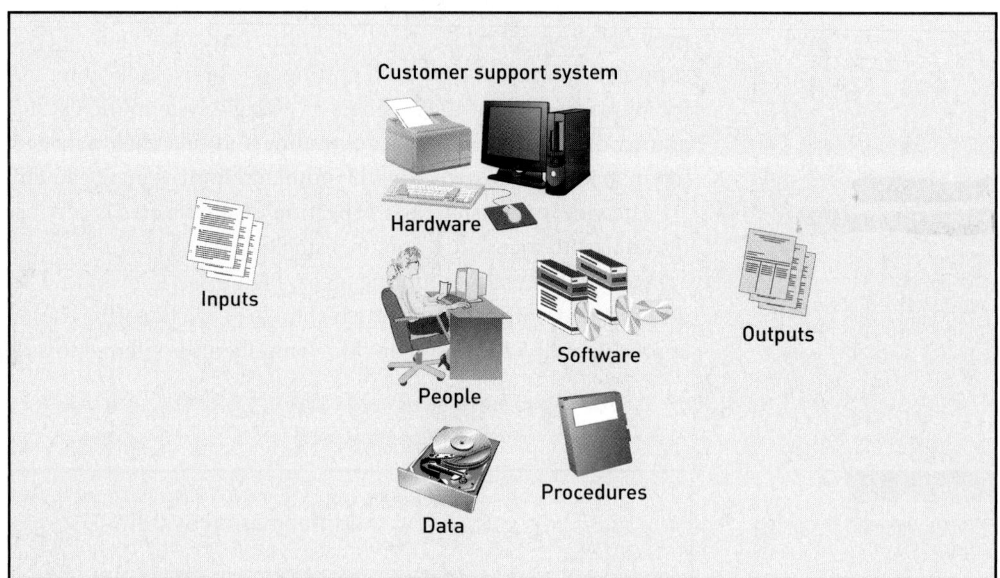

Figure 1-4

The system boundary versus the automation boundary

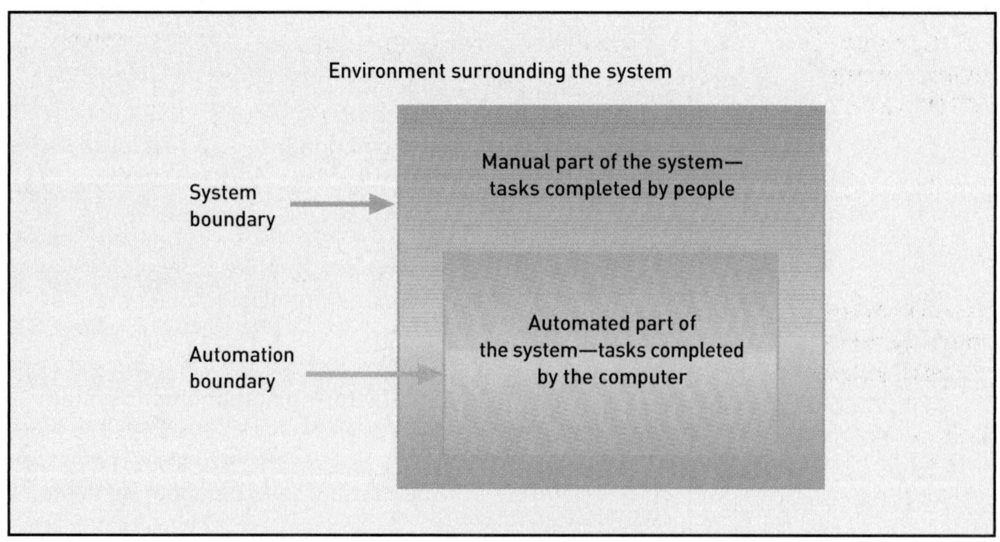

TYPES OF INFORMATION SYSTEMS

Because organizations perform many different types of activities, many types of information systems exist—all of which can be innovative and use the latest technologies. The types of information systems found in most businesses are shown in Figure 1-5. You learned about these types of systems in your introductory information systems course, so we briefly review only the most common ones here.

Figure 1-5

Types of information systems

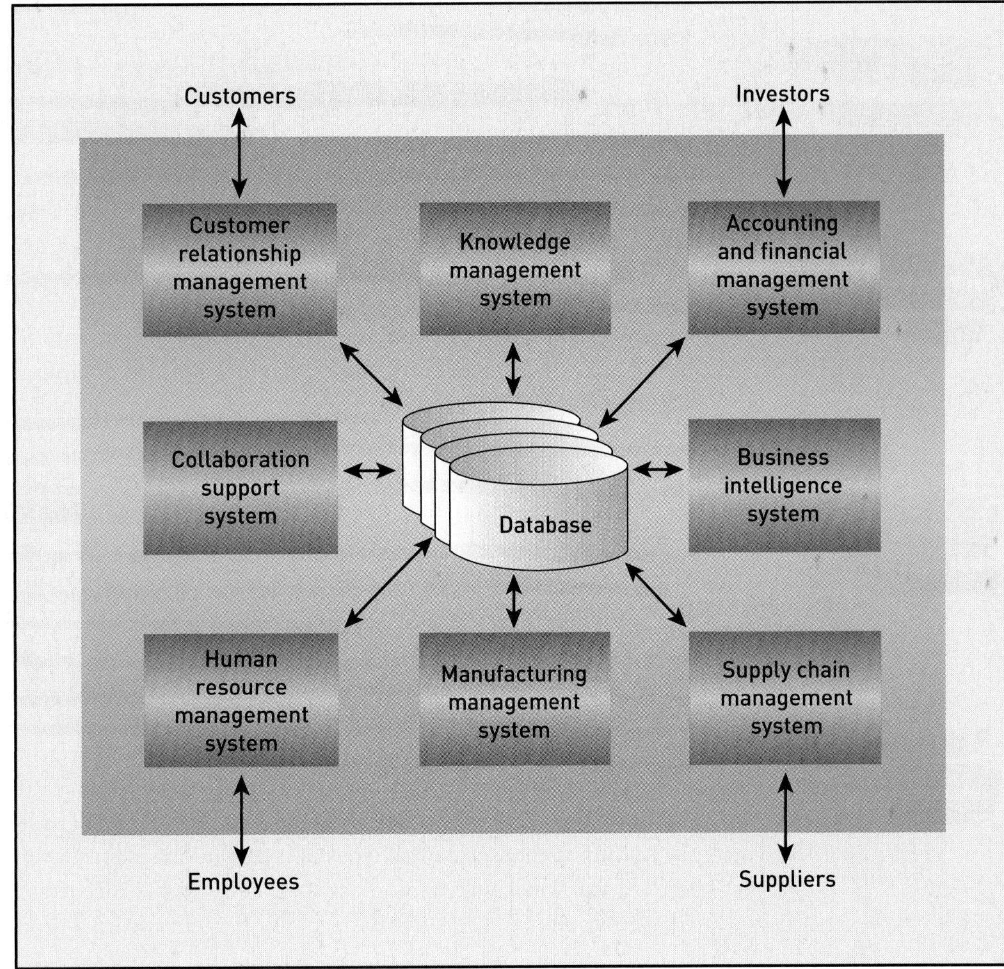

customer relationship management (CRM) system

a system that supports marketing, sales, and service operations involving direct and indirect customer interaction

supply chain management (SCM) system

a system that seamlessly integrates product development, product acquisition, manufacturing, and inventory management

A customer relationship management (CRM) system incorporates processes that support marketing, sales, and service operations involving direct and indirect customer interaction. A supply chain management (SCM) system incorporates processes that seamlessly integrate product development, product acquisition, manufacturing, and inventory management. Both systems are important because they are part of the interface between the organization and key external entities. Both types of systems have had rapid changes over the last two decades, including expanded scope and functionality, significant application of Web-based technologies, and increased integration across organizational boundaries. For example, most modern organizations now manage sales and service via a single system, enable Web-based ordering and account management via consumer-oriented Web sites, and employ automated interfaces for business customers that directly connect one organization's SCM to other organizations' CRMs. Integration across organizational boundaries has increased the speed and efficiency of business transactions and enabled modern business practices such as just-in-time delivery of raw materials in manufacturing organizations and direct shipment from manufacturers to end users by third-party resellers.

Other systems that interface with external entities include accounting and financial management (AFM) systems and human resource management (HRM) systems. AFM systems record accounting information needed to produce financial statements and other reports used by investors and creditors. AFM systems also include financial functions such as cash management, cash flow forecasting, and securities management. HRM systems include processes concerned with employees, such as payroll, health insurance, pensions, hiring, and training. AFM and HRM systems are partly governed by external regulations and must frequently interact with regulatory authorities in areas such as taxes, public financial markets, and occupational health and safety.

accounting and financial management (AFM) system
a system that records accounting information needed to produce financial statements and other reports used by investors and creditors

human resource management (HRM) system
a system that supports employee-related tasks such as payroll, benefits, hiring, and training

Organizations also have information systems with few or no interactions with external entities. A manufacturing management system controls internal production processes that turn raw materials into finished goods. A knowledge management system (KMS) supports the storage of and access to documents from all parts of the organization. It enables rapid communication of policies, procedures, and data and helps ensure continuity of knowledge despite changes in personnel assignments.

manufacturing management system
a system that controls internal production processes that turn raw materials into finished goods

knowledge management system (KMS)
a system that supports the storage of and access to documents from all parts of the organization

A collaboration support system (CSS) enables geographically distributed personnel to collaborate on projects and tasks. CSSs encompass a variety of technologies, including voice communications, video-conferencing, project management and scheduling tools, and Wiki technology that enables Web-based management of documents by project participants. A business intelligence system supports strategic planning and executive decision making. It enables users to organize internal and external data about customers, suppliers, competitors, and economic conditions for use in statistical analysis, simulations, and other forms of planning.

collaboration support system (CSS)
a system that enables geographically distributed personnel to collaborate on projects and tasks

business intelligence system
a system that supports strategic planning and executive decision making

Today, many companies use enterprise resource planning (ERP) systems that incorporate most or all of the system types described previously in this section. Software vendors such as SAP, Oracle, and IBM offer comprehensive packages for companies in specific industries. To adopt an ERP solution, the company must carefully study its existing processes and information needs and then determine which ERP vendor provides the best match. ERP systems are so complex that an organization must often commit nearly everyone in the information systems department and throughout the organization to research options. They are also very expensive, both in initial costs and support costs. Extensive change is involved for management and for staff. After the decision is made to adopt an ERP system, it is very difficult to return to the old ways of doing business, or to the old systems.

An important aspect of all types of information systems is their data integration. For example, order data originally captured by the CRM system is needed by the SCM system to drive purchasing, the manufacturing management system to drive production scheduling, the AFM system for accounting and to help determine near-term financing requirements, and the business intelligence system to drive estimates of future sales and profitability. Data sharing among all these systems is made possible by databases—centrally managed collections of data that can store large amounts of information and make it accessible to many users and systems at the same time. Databases and database technology are further discussed in Chapter 13.

REQUIRED SKILLS OF THE SYSTEMS ANALYST

Systems analysts (or any professionals doing systems analysis and design work) need a great variety of special skills. First, they need to be able to understand how to build information systems, which requires quite a bit of technical knowledge. Then, as discussed previously, they have to understand the business they are working for and how the business uses each of the types of systems. Finally, the analysts need to understand quite a bit about people and the way they work. People are the source of information about requirements, the labor that builds systems, and the ultimate users of the information system. Figure 1-6 summarizes the analyst's knowledge and skill requirements.

Figure 1-6

Required skills of the systems analyst

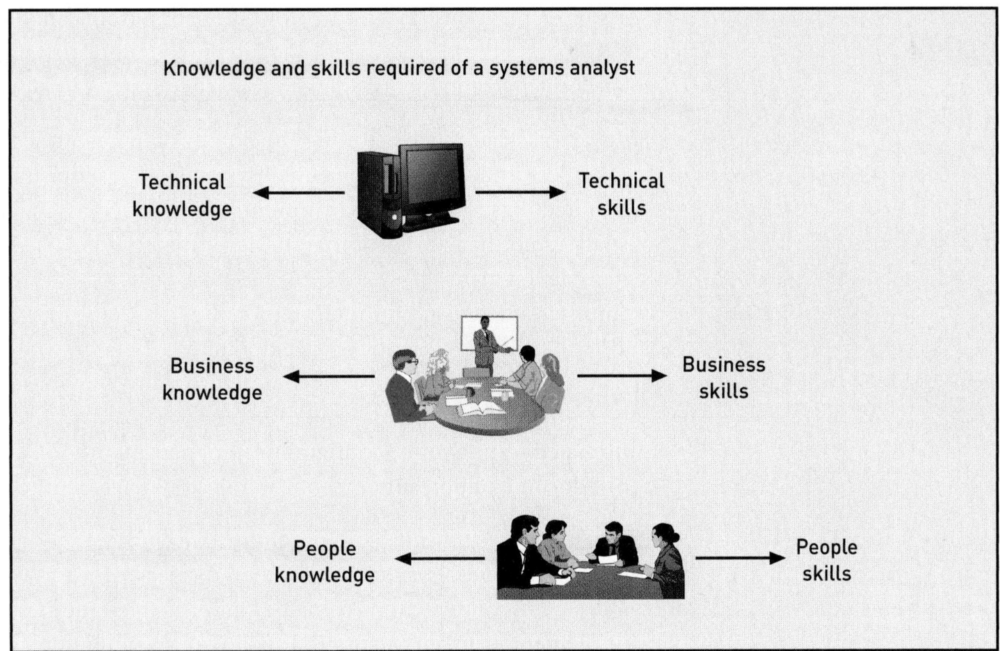

enterprise resource planning (ERP)
a process in which an organization commits to using an integrated set of software packages for key information systems

database
a centrally managed collection of data that is accessible to many users and systems at the same time

TECHNICAL KNOWLEDGE AND SKILLS

It should not be surprising that a systems analyst needs technical expertise. The scope, breadth, and depth of technology employed in medium- and large-scale organizations are vast. A company's "simple" online order-processing application might involve a system with thousands of users spread over hundreds of locations. The database might contain hundreds of tables with millions of records in each table. The system might have taken years to construct, cost millions of dollars, and be supported by global networks, hundreds of servers, and dozens of support staff. If the system fails for even an hour, the company could lose millions of dollars in sales and disrupt its entire supply chain. Such a system is a critical business resource, so the staff that support and maintain it work in round-the-clock shifts and are on call day and night in case of a problem. The importance of technology to modern organizations cannot be overstated.

Even if an analyst is not involved in activities such as programming, network design, or hardware configuration, it is still crucial to have an understanding of different types of technology—what they are used for, how they work, and how they are evolving. No one person can be an expert at all types of technology; there are technical specialists to consult for the details. But a systems analyst should understand the fundamentals about the following:

- Computers and how they work
- File, database, and storage hardware and software
- Input and output hardware and software
- Computer networks and protocols
- Programming languages, operating systems, and utilities
- Communication and collaboration technology such as digital telephones, videoconferencing, and Web-based document management systems

Just as an organization's business environment continually changes, so does the technology used for its information systems. The rapid change in technology often drives other needed changes. Thus, all participants in information system development should upgrade their knowledge and skills continually. Those who don't will be left behind.

tools software products used to help develop analysis and design specifications and completed system components

A systems analyst also needs to know a lot about tools and techniques for developing systems. Tools are software products that are used to develop analysis and design specifications and completed system components. Some tools used in system development include the following:

- Software packages such as Intuit QuickBooks, Microsoft Access, and Adobe Dreamweaver that can be used to implement small systems or develop subsystems
- Integrated development environments (IDEs) such as Oracle JDeveloper and Microsoft Visual Studio that support program development, database design, software testing, and system deployment
- Computer-aided visual modeling tools, such as Rational XDE Modeler, Visible Analyst, and Embarcadero Describe, that help analysts create, store, modify, and manage system specifications and sometimes generate programs, databases, Web-based interfaces, and other software components
- Automated testing tools, configuration management tools, software library management tools, documentation support tools, project management tools, and so on

techniques strategies for completing specific system development activities

Techniques are strategies for completing specific system development activities. How do you plan and manage a system development project? How do you define requirements? How do you design user interactions using design principles and best practices? How do you complete implementation and testing? How do you install and support a new information system? Much of this text explains how to use specific techniques for project planning, defining requirements, and designing system components. But it also covers some aspects of implementation and support. Some examples of techniques include the following:

- Project planning techniques
- Cost/benefit analysis techniques
- Interviewing techniques
- Requirements modeling techniques
- Architectural design techniques
- Network configuration techniques
- Database design techniques

BUSINESS KNOWLEDGE AND SKILLS

Other knowledge and skills that are crucial for an analyst include those that apply to understanding business organizations in general. After all, the problem to be solved is a business problem. What does the analyst need to know? The following are examples:

- What business functions do organizations perform?
- How are organizations structured?
- How are organizations managed?
- What type of work goes on in organizations (finance, manufacturing, marketing, customer service, and so on)?

Systems analysts benefit from a fairly broad understanding of businesses in general, so they typically study business administration in college. In fact, computer information systems (CIS) or management information systems (MIS) majors are often included in the college of business for that reason. The accounting, marketing, management, and operations courses taken in a CIS or MIS degree program serve a very important purpose of preparing the graduate for the workplace. Project management techniques such as planning, scheduling, budgeting, feasibility analysis, and management reporting are particularly important.

Systems analysts also need to understand the type of organization for which they work. Some analysts specialize in a specific industry for their entire career—perhaps in manufacturing, retailing, financial services, or aerospace. The reason for this business focus is simple: It takes a long time to understand the problems of a specific industry. An analyst with deep understanding of a specific industry can solve complex problems for companies in the industry.

Familiarity with a specific company also provides important guidance on system needs and changes. Often, just knowing the people who work for a company and understanding subtleties of the company culture can make a big difference in the effectiveness of an analyst. It takes years of experience working for a company to really understand what is going on. The more an analyst knows about how an organization works, the more effective he can be. Some specifics the analyst needs to know about the company include the following:

- What the specific organization does
- What makes it successful
- What its strategies and plans are
- What its traditions and values are

BEST PRACTICE

Be sure you understand the organization, its culture, its mission, and its objectives before jumping to conclusions about system solutions.

PEOPLE KNOWLEDGE AND SKILLS

Interpersonal skills are perhaps the analyst's most important skills, because analysts rely on others, including managers, users, programmers, technical specialists, customers, and vendors, to take a system from initial idea to final implementation. The analyst is a translator for all project participants, translating business objectives into functional requirements, user needs into system specifications, and technical jargon and details into terms that nontechnical personnel can easily understand. The analyst must be an effective communicator in many contexts, including conversations, interviews, technical reviews, and formal presentations.

Required interpersonal skills go well beyond oral and written communication. For example, the analyst must develop rapport with users who may be resistant to change, negotiate with management for resources such as budget, time, and personnel, and manage development personnel with many different skills, capabilities, and attitudes. The analyst must be an effective teacher, mentor, confidant, collaborator, manager, and leader, shifting easily among those roles many times over the course of a typical work day. In an increasing multinational environment, the analyst must effectively interact with people of diverse backgrounds, customs, and beliefs.

All of these interpersonal skills are critical to project success. The wrong system is acquired or constructed when business and user requirements are misunderstood or ignored. Projects fail without support from managers, users, and development staff. Critical subsystems don't interact correctly when technical specifications are incorrectly communicated or documented. The development team can't adapt to new information and change without effective feedback among all project participants.

BEST PRACTICE

Analysts typically devote several weeks per year to training and continuing education. An analyst should devote time to developing so-called "**soft skills**" such as interviewing, team management, and leadership, and should develop **hard skills** such as database design, programming, and telecommunications.

soft skills
skills in nontechnical areas such as interviewing, team management, and leadership

hard skills
skills in technical areas such as database design, programming, and telecommunications

A FEW WORDS ABOUT INTEGRITY AND ETHICS

One aspect of a career in information systems that students often underestimate is the importance of personal integrity and ethics. A systems analyst is asked to look into problems that involve information from many different parts of an organization. Especially if it involves individual employees, the information might be private, such as salary, health, and job performance. The analyst must have the integrity to keep this information private.

The problems the analyst works on can also involve confidential corporate information, including proprietary information about products or planned products, strategic plans or tactics, and even top-secret information involving government military contracts. Sometimes a company's security processes or specific security systems can be involved in the analyst's work. Analysts are expected to uphold the highest ethical standards when it comes to private proprietary information, whether the analysts are employees or outside consultants.

Ethics and integrity also include follow-through on commitments, dealing directly with mistakes and gaps in relevant knowledge and skills, and practicing open and honest communication. As a pivotal member of the development team, an analyst's lack of follow-through or task completion can cause problems that reverberate throughout the project. No one can be highly skilled in every aspect of system development across all application areas and organizational contexts. An analyst must take honest stock of his or her strengths, weaknesses, and performance, ask for needed help and resources, and be ready to provide the same to others. The analyst must also balance organizational privacy needs and the reluctance of some project participants to provide complete information with the improved outcomes that arise from free exchange of information and ideas. It is a difficult balance to strike, but one that is critical to project success.

ANALYSIS-RELATED CAREERS

Employment in the fields of information systems and computer technology spans a wide variety of skills, organizations, and roles. Rapid changes in technology, business practices, and the structure of the global economy have changed related jobs. Typical information system graduates of the late twentieth century were employed as programmer analysts. Job tasks consisted primarily of programming with some analysis and design. As employees moved "up the ladder" the mix of activities changed, the breadth and importance of analysis and design activities increased, and supervisory responsibilities for maintenance and development project teams were gradually added. Employees typically worked within a dedicated information systems department of a business or government organization or for a company that developed and maintained information systems under contract to other organizations. The "career ladder" was usually well defined and skills were easily transferred among jobs.

The employment picture is much more complex in the twenty-first century. The number of programmer analysts employed by "brick and mortar" companies has decreased due to increased productivity and outsourcing. Many software development jobs have shifted to companies that produce and sell ERP software, and many of those companies have moved some or all operations out of North America and Western Europe to India, China, and countries of the old Soviet bloc. Given the significant changes that have occurred, is there really a need for analysis and design skills and are there any related jobs in North America and Western Europe?

The answer is yes, but the number and nature of the jobs, their titles, and the organizations that fill those positions are much more complex than in the past. Despite the widespread use of ERP software, many businesses still have smaller in-house development staffs that concentrate on areas of strategic importance, competitive advantage, and unique firm requirements. In-house development, including analysis and design, is especially common in security-sensitive industries, national defense, and research and development in national laboratories. Thus, employment of analysts and software developers within traditional industries continues, but at a slower pace than in the past.

Changes in software development, technology, and business practices have created many new career opportunities for analysts, including:

- Sales and support of ERP software
- Business analysts for user organizations
- Auditing, compliance, and security
- Web development

Companies that produce and sell ERP software have become a significant part of the information systems employment picture. Large companies such as SAP, Oracle, and IBM have significant ERP market share, though there are many smaller and more specialized competitors. Selling and supporting ERP software requires many analysis and design skills. ERP systems are complex combinations of hardware and software components. Determining the component mix that best matches a particular customer and deploying and supporting that solution requires considerable analysis and design skills. Thus, the job of account representative for many ERP firms requires considerable skill in analysis and sometimes in design. In addition, ERP firms employ many analysts and designers to support account representatives and to continually improve their products to match changing technology and customer needs.

User organizations in "line areas" such as finance, customer service, and logistics often employ personnel with significant analysis and design responsibilities. These employees evaluate changing business needs, redesign business processes to better satisfy those needs, and research, evaluate, purchase, deploy, and support new technology to support the redesigned processes. They often work closely with ERP firms and act as user representatives and contract managers for their employers. Although such a position entails many different skills, analysis and design skills are essential. Unlike traditional programmer analyst jobs, these positions are difficult to outsource and less likely to be moved offshore, though they are often globally distributed in large multinational organizations.

Accounting is an area of rapid job growth for information systems professionals, especially within large accounting and auditing firms and within the accounting and internal audit staffs of their clients. The Sarbanes-Oxley Act in the United States, and similar legislation and regulation in other countries, requires publicly traded companies to continually evaluate the adequacy of their financial reporting and internal control systems. Auditors must also certify the adequacy of business processes and evaluate whether the firm is at risk of near-term failure due to financial, legal, market, or other problems. Because businesses rely heavily on automated systems to support business processes and financial reporting, accountants and auditors work closely with technical personnel who understand those systems. The core skill set required for those jobs is analysis and design. Employees with experience and skills both in accounting and information systems are in high demand.

As Web technology has permeated modern organizations, the demand for employees with related skills has skyrocketed. Most medium- and large-scale organizations have in-house staff that develop and maintain Web sites, build Web-based application software, and serve as internal Web consultants to other parts of the organization. Many consulting firms specialize in developing and maintaining a Web presence for other organizations. Analysis and design skills are an important part of developing and maintaining Web-based applications and Web presence. To employ such systems to maximal advantage, developers must analyze business needs and design appropriate systems deployed with appropriate technology.

As you've probably surmised by now, career opportunities for analysts and people with significant analysis and design skills are as varied as the related job titles and descriptions. Here are some job titles you might encounter:

- Programmer analyst
- Business systems analyst
- System liaison
- End-user analyst

- Business consultant
- Systems consultant
- Systems support analyst
- Systems designer
- Software engineer
- System architect
- Web architect
- Webmaster
- Web developer

Sometimes systems analysts might also be called project leaders or project managers. Be prepared to hear all kinds of titles for people who are involved in analysis and design work.

In sum, the career prospects for analysts are bright, but the nature of related jobs, their location, and the typical career development path for analysts and other information system professionals has changed significantly over the last two decades. As in many other areas of the economy, large numbers of employees doing similar tasks for a single company is no longer the norm. Similar tasks are now more automated and more dispersed, resulting in jobs in a greater variety of organizations with broader responsibilities and rapidly changing requirements. Analysis and design skills are at the core of many of these new jobs. Employees who can understand business processes, user needs, and the technology that supports those processes and needs are in high demand. Continuing penetration of information technology into every aspect of modern organizations ensures that demand will be strong far into the future.

THE ANALYST'S ROLE IN STRATEGIC PLANNING

We have described a systems analyst as someone who solves specific business problems by developing or maintaining information systems. The analyst might also be involved with senior managers on strategic management problems—that is, problems involving the future of the organization and plans and processes to ensure its survival and growth. Sometimes an analyst who is only a few years out of college can be summoned to meet with top-level executives and even be asked to present recommendations to achieve corporate goals. How might this happen?

SPECIAL PROJECTS

First, the analyst might be working to solve a problem that affects executives, such as designing an MIS to provide information to executives. The analyst might interview the executives to find out what information they need to do their work. An analyst might be asked to spend a day with an executive or even travel with an executive to get a feel for the nature of the executive's work. Then the analyst might develop and demonstrate prototypes of the system to get more insight into the needs of the executives.

Another situation that could involve an analyst in strategic management problems is a business process reengineering study. Business process reengineering seeks to alter radically the nature of the work done in a business function. The objective is radical improvement in performance, not just incremental improvement. Therefore, the analyst might be asked to participate in a study that carefully examines existing business processes and procedures and then to propose information system solutions that can have a radical impact. Many tools and techniques of analysis and design are used to analyze business processes, redesign them, and then provide computer support to make them work.

business process reengineering
a technique that seeks to alter the nature of the work done in a business function, with the objective of radically improving performance

STRATEGIC PLANNING

strategic planning

a process during which executives try to answer questions about the company, such as where the business is now, where they want the business to be, and what they have to do to get there

Most business organizations invest considerable time and energy completing strategic plans that typically cover five or more years. During the **strategic planning** process, executives ask themselves such fundamental questions about the company as where the business is now, where they want the business to be, and what they have to do to get there. A typical strategic planning process can take months or even years, and often plans are continually updated. Many people from throughout the organization are involved, all completing forecasts and analyses, which are combined into the overall strategic plan. After a strategic plan is set, it drives all of the organization's processes, so all areas of the organization must participate and coordinate their activities. Therefore, a marketing strategic plan and a production strategic plan must fit within the overall strategic plan.

INFORMATION SYSTEMS STRATEGIC PLANNING

information systems strategic plan

the plan defining the technology and applications that the information systems function needs to support the organization's strategic plan

One major component of the strategic plan is the **information systems strategic plan**. Today, information systems are so tightly integrated into an organization that nearly any planned change calls for new or improved information systems. Beyond that, the information systems themselves often drive the strategic plan. For example, after some chaotic early years involving the Internet, many new Internet-based companies have survived (such as Amazon.com and eBay), and many other companies have altered their business processes and developed new markets in which to compete. In other cases, the opportunities presented by new information systems technology have led to new products and markets with more subtle impacts. Information systems and the possibilities that they present play a large role in the strategic plans of most organizations.

Information systems strategic planning sometimes involves the whole organization. Usually at the recommendation of the chief information systems executive, top management will authorize a major project to plan the information systems for the entire organization.

In developing the information systems strategic plan, members of the staff look at the overall organization to anticipate problems rather than react to systems problems as they come up. Several techniques help the organization complete an information systems strategic planning project. A consulting firm is often hired to help with the project. Consultants can offer experience with strategic planning techniques and can train managers and analysts to complete the plan.

Usually managers and staff from all areas of the organization are involved, but the project team is generally led by information systems managers with the assistance of consultants. Systems analysts often become involved in collecting information and interviewing people.

application architecture plan

a description of the integrated information systems that the organization needs to carry out its business functions

technology architecture plan

a description of the hardware, software, and communications networks required to implement planned information systems

Many documents and existing systems are reviewed. Then the team tries to create a model of the entire organization—to map the business functions it performs. Another model, one that shows the types of data the entire organization creates and uses, is also developed. The team examines all of the locations where business functions are performed and data is created and used. From these models, the team puts together a list of integrated information systems for the organization, called the **application architecture plan**. Then, given the existing systems and other factors, the team outlines the sequence needed to implement the required systems.

Using the list of information systems needed, the team defines the **technology architecture plan**—that is, the types of hardware, software, and communications networks required to implement all of the planned systems. The team must look at trends in technology and make commitments to specific technologies and possibly even technology vendors. The components of the information systems strategic plan are shown in Figure 1-7.

In an ideal world, a comprehensive information systems planning project would solve all of the problems that information systems managers face. Unfortunately, the world continues to change at such a rate that plans must be continually updated. Unplanned information system projects come up all the time, and priorities must be continually evaluated.

Figure 1-7

Components of an information systems strategic plan

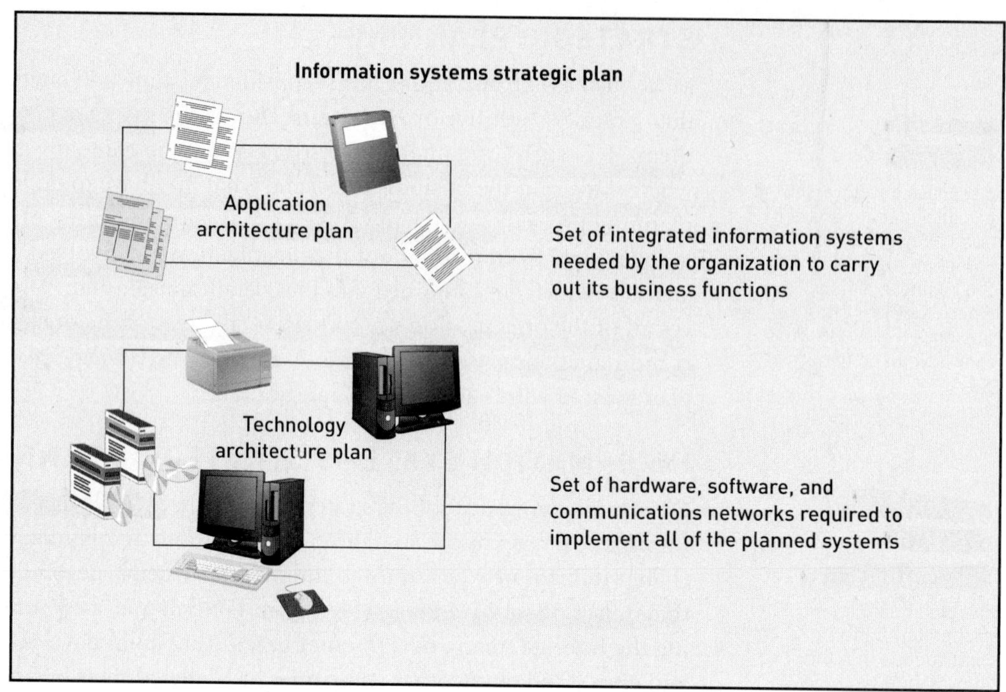

ROCKY MOUNTAIN OUTFITTERS AND ITS STRATEGIC INFORMATION SYSTEMS PLAN

To demonstrate the important systems analysis and design techniques in this text, we follow a system development project for a company named Rocky Mountain Outfitters (RMO). RMO is a sports clothing manufacturer and distributor that is about to begin development of a new customer support system. You will encounter RMO customer support system examples in all chapters of this book. For now, try to get a feel for the nature of the business, the approach the company took to define the information systems strategic plan, and the basic objectives of the customer support system that is part of the plan.

INTRODUCING ROCKY MOUNTAIN OUTFITTERS (RMO)

RMO started in 1978 as the dream of John and Liz Blankens of Park City, Utah. Liz had always been interested in fashion and clothing and had worked her way through college by designing, sewing, and selling winter sports clothes to the local ski shops in Park City. She continued with this side business even after she graduated, and soon it was taking all of her time.

Liz had been dating John Blankens since they met at a fashion merchandising convention. John had worked for several years for a retail department store chain after college and had just completed his MBA. Together they decided to try to expand Liz's business into retailing to reach a larger customer base.

The first step in their expansion involved direct mail-order sales to customers using a small catalog (see Figure 1-8). Liz immediately had to expand the manufacturing operations by adding a designer and production supervisor. As interest in the catalog increased, Liz and John sought out additional lines of clothing and accessories to sell along with their own product lines. They also opened a retail store in Park City.

Figure 1-8

Early RMO catalog cover (Fall 1978)

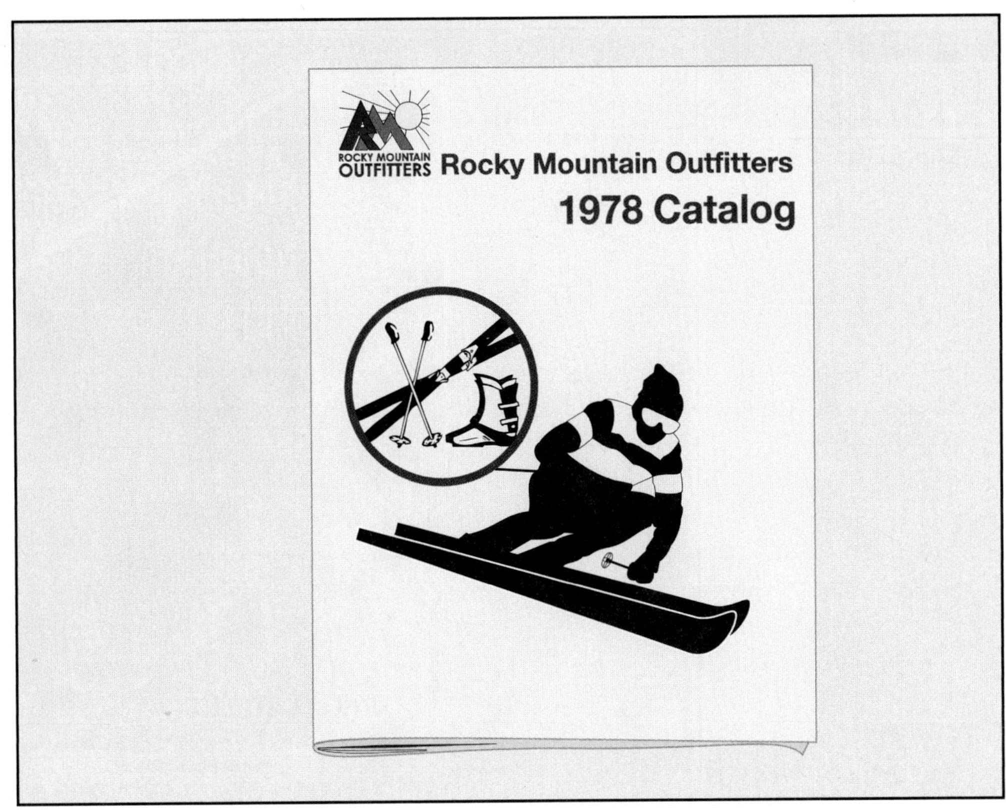

By the early 2000s, RMO had grown to become a significant regional sports clothing distributor in the Rocky Mountain and Western states. The states of Arizona, New Mexico, Colorado, Utah, Wyoming, Idaho, Oregon, Washington, and Nevada, and the eastern edge of California had seen tremendous growth in recreation activities. Along with the increased interest in outdoor sports, the market for both winter and summer sports clothes had exploded. Skiing, snowboarding, mountain biking, water skiing, jet skiing, river running, jogging, hiking, ATV biking, camping, mountain climbing, and rappelling had all seen a tremendous increase in interest in these states. Of course, people needed appropriate sports clothes for their activities, so RMO expanded its line of sportswear to respond to this market. It also added a line of high-fashion active wear and accessories to round out its offerings to the expanding market of active people. The current RMO catalog offers an extensive selection (see Figure 1-9).

Rocky Mountain Outfitters now employs more than 600 people and generates almost $180 million annually in sales. The mail-order operation is still the major source of revenue, at $90 million. Phone-order sales are $50 million. In-store retail sales have remained a modest part of the business, with sales of $5 million at the Park City retail store and $5 million at the recently opened Denver store. In 2004, John and Liz contracted with an outside firm to develop and host a Web-based ordering system. Though the system has been working since early 2005, it accounts for a disappointing $30 million in sales.

RMO STRATEGIC ISSUES

Rocky Mountain Outfitters was one of the first sports clothing distributors to provide a Web site featuring its products. The site originally gave RMO a simple Web presence to enhance its image and to allow potential customers to request a copy of the catalog. It also served as a portal for links to all sorts of outdoor sports Web sites. The first RMO Web site enhancement added more specific product information, including weekly specials that could be ordered by phone. Eventually, nearly all product offerings were included in an online catalog posted at the Web site. But orders could only be placed by mail or by phone.

Figure 1-9

Current RMO catalog cover (Fall 2010)

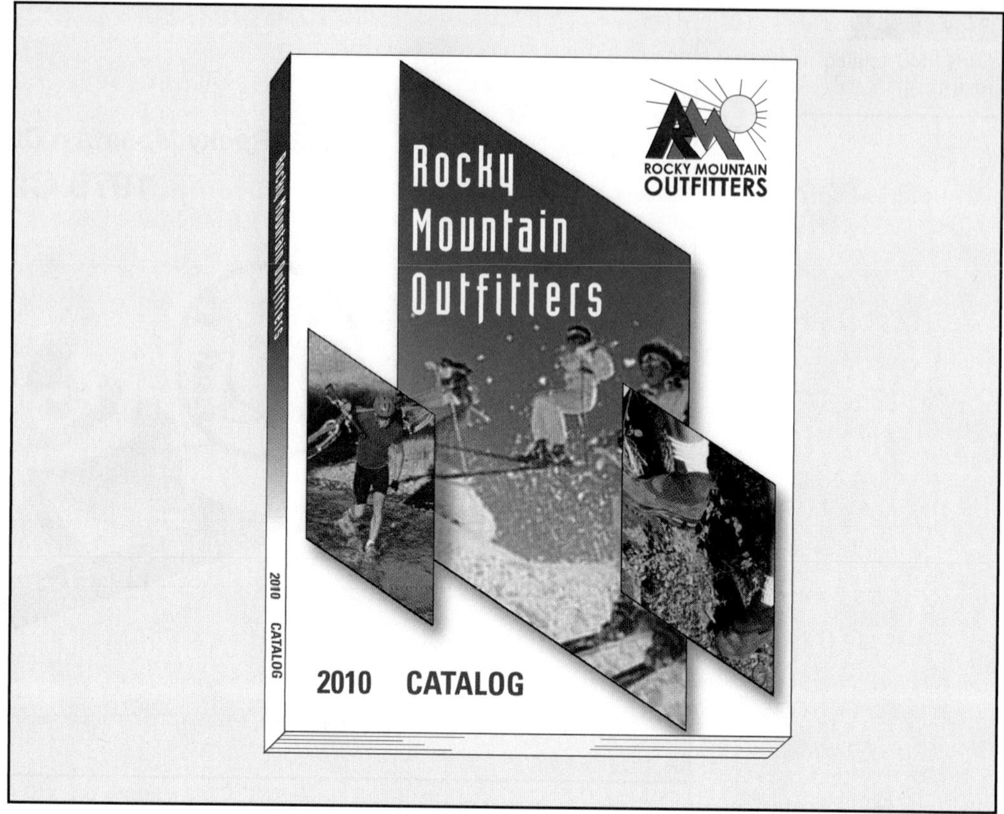

John and Liz had considered making a major commitment to business-to-consumer (B2C) e-commerce in the early 2000s. They worried about the risk of sudden and potentially explosive growth, but felt that they had to develop an online ordering system to remain competitive. At the time, in-house staff was not trained in Web technologies, so John and Liz decided to outsource development and operation of the Web site.

By 2007, they realized that the Web-based ordering system was substantially underperforming against the competition for many reasons, including the following:

- Slow and cumbersome updates to online content
- Poor coordination with in-house customer service functions
- Poor coordination between Web-based ordering and supply chain management functions
- Poor technical support and other support by the site operator
- Deteriorating relations with RMO management

In late 2006, RMO performed a detailed market analysis that showed alarming trends, including the following:

- RMO sales growth was slower than the industry average, resulting in decreasing market share.
- The average age of customers ordering by phone and mail was increasing, and was much higher than the industry average age of all customers.
- Compared to competitors, RMO's Web-based sales were a much smaller percentage of total sales, and the average order amount was lower than the industry average.

The analysis painted a disturbing picture of declining performance. Continued strong sales to older customers via traditional channels were offset by weak sales to younger customers via the Web. RMO was failing to attract and retain the customers who represented the bulk of present and future business.

In early 2007, John and Liz hired a consulting firm to evaluate their current application and IT architecture and to help them with strategic IS planning. The consultants recommended modernizing much of the RMO technology infrastructure and implementing completely new supply chain management and Web-based ordering systems. Upgrading the existing Web-ordering system was ruled out due to outdated technology, numerous flaws, and poor vendor performance.

BEST PRACTICE

Most business executives understand that information systems are strategically important, and they usually have excellent ideas and insights. Be sure to ask for their input.

The next section provides some additional background on RMO and summarizes the overall information systems plan it is currently following. Subsequent chapters of this book focus on one of the crucial information systems that is part of the plan—the customer support system.

RMO'S ORGANIZATIONAL STRUCTURE AND LOCATIONS

Rocky Mountain Outfitters is still managed on a daily basis by John and Liz Blankens. John is president, and Liz is vice president of merchandising and distribution (see Figure 1-10). Other top managers include William McDougal, vice president of marketing and sales, and JoAnn White, vice president of finance and systems. The systems department reports to JoAnn White.

One hundred thirteen employees work in human resources, merchandising, accounting and finance, marketing, and information systems in the corporate offices in Park City, Utah. There are two retail stores: the original Park City store and the newer Denver store. Manufacturing facilities for high-fashion clothing and accessories are located in Salt Lake City and more recently in Portland, Oregon. Most other products are manufactured under contract in Central America and Asia. There are three distribution/warehouse facilities: Salt Lake City, Albuquerque, and Portland. All mail-order processing is done in a facility in Provo, Utah, employing 58 people. The phone-sales center, employing 20, is located in Salt Lake City. Figure 1-11 shows the locations of these facilities.

THE RMO INFORMATION SYSTEMS DEPARTMENT

The information systems department is headed by Mac Preston, an assistant vice president with the title chief information officer (CIO), and there are nearly 50 employees in the department (see Figure 1-12). Since 2007, RMO has aggressively modernized development and other IT skills through extensive training and new hires. Mac's title of CIO reflects a promotion following the successful completion of the information systems strategic planning project. He is not quite equal to a full vice president, but his position is considered increasingly important to the future of the company. Mac reports to the finance and systems vice president, whose background is in finance and accounting. The information systems department will eventually report directly to the CEO if Mac has success implementing the new strategic information systems plan.

Mac organized information systems into two areas—system support and system development. Ann Hamilton is director of system support. System support involves such functions as telecommunications, database administration, operations, and user support. John MacMurty is director of system development. System development includes four project managers, six systems analysts, 10 programmer analysts, and a couple of clerical support employees.

Figure 1-10

Rocky Mountain Outfitters' organizational structure

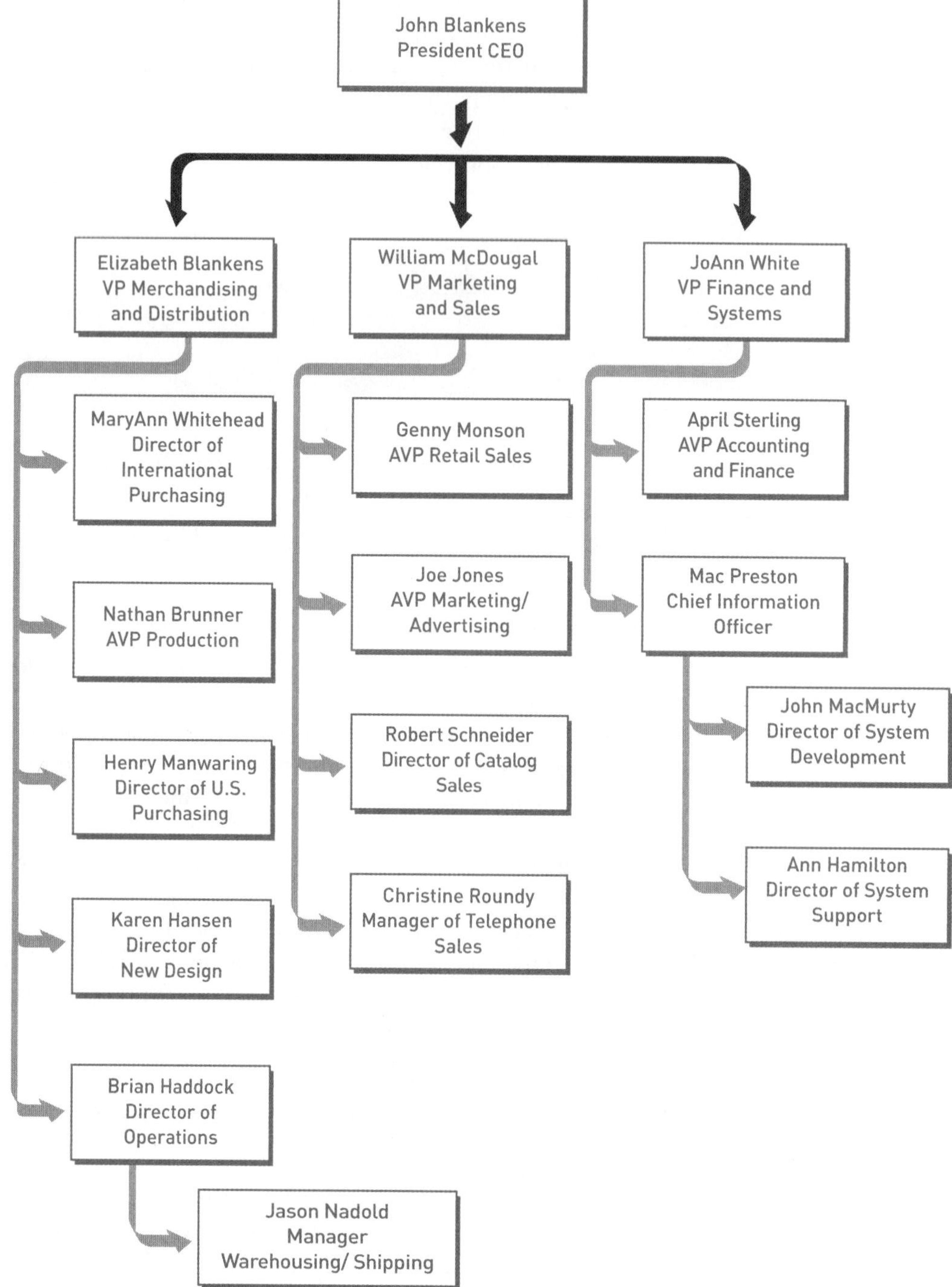

John Blankens
President CEO
Elizabeth Blankens
VP Merchandising
and Distribution
William McDougal
VP Marketing
and Sales
JoAnn White
VP Finance and
Systems
MaryAnn Whitehead
Director of
International
Purchasing
Nathan Brunner
AVP Production
Henry Manwaring
Director of U.S.
Purchasing
Karen Hansen
Director of
New Design
Brian Haddock
Director of
Operations
Jason Nadold
Manager
Warehousing/ Shipping
Genny Monson
AVP Retail Sales
Joe Jones
AVP Marketing/
Advertising
Robert Schneider
Director of Catalog
Sales
Christine Roundy
Manager of Telephone
Sales
April Sterling
AVP Accounting
and Finance
Mac Preston
Chief Information
Officer
John MacMurty
Director of System
Development
Ann Hamilton
Director of System
Support

Figure 1-11

Rocky Mountain Outfitters' locations

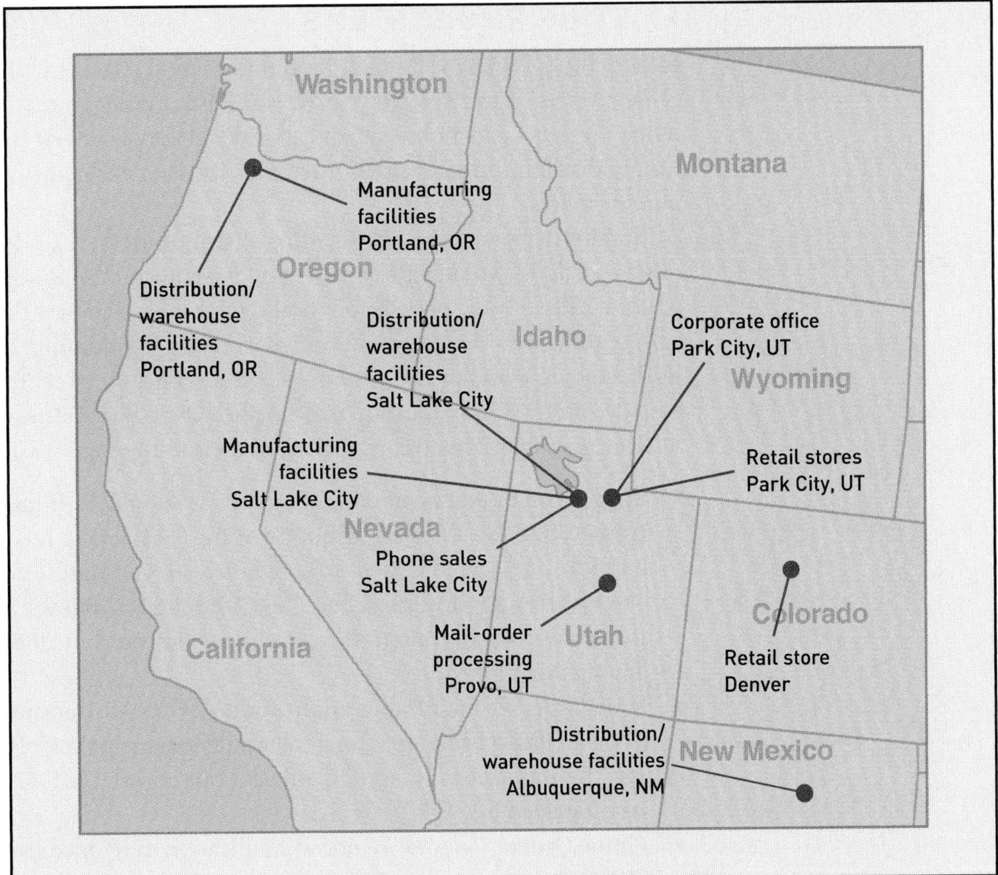

Figure 1-12

RMO information systems department staffing

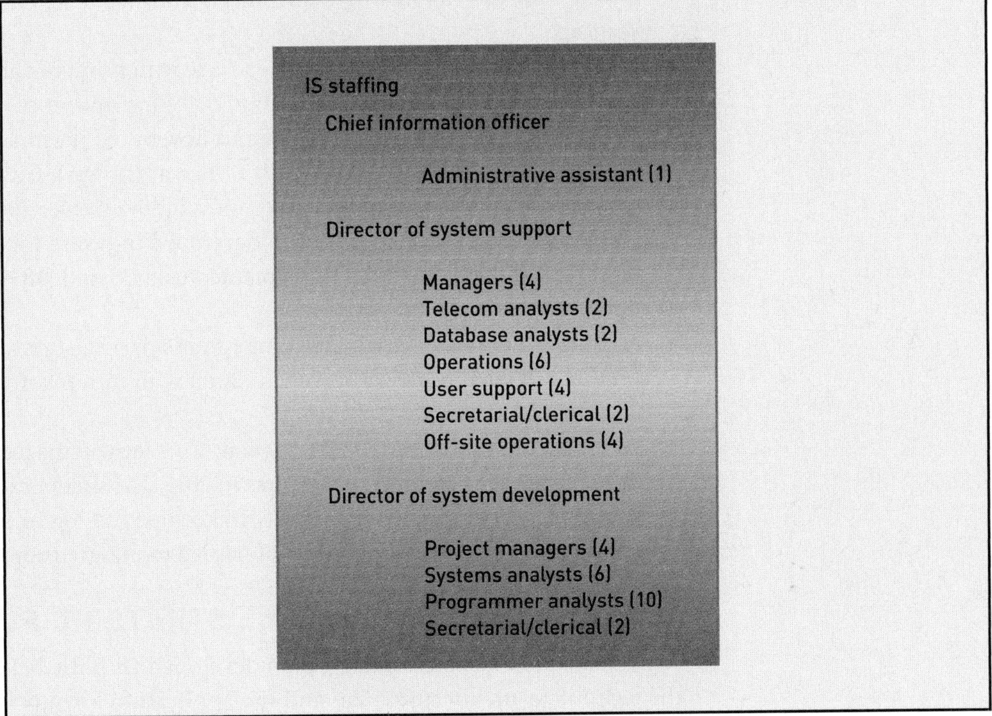

EXISTING RMO SYSTEMS

Most of the computer technology and information systems staff at RMO is located at the data center in Park City. A server cluster supports the inventory, mail-order, accounting, and human resource functions. A high-capacity network connects the data center to the manufacturing, distribution, and mail-order sites for data exchange, video-conferencing, and telephone services.

Smaller servers at the home office, distribution sites, and manufacturing sites support office functions and provide core networking and communication services. Retail stores run a point-of-sale software package on local servers that exchanges data directly with the inventory system at the data center. The phone-sales center has multiple servers supporting office functions, core networking and communication services, call center management software, and an order-processing application that interacts directly with the data center servers.

The existing information systems and their technology are organized as follows:

- **Supply Chain Management.** Originally developed in-house as a mainframe application using COBOL/CICS with some VSAM files and a DB2 relational database. In 2002, the application was migrated to upgraded server hardware and reimplemented as a client/server application using C++, DB2, and Windows terminal services. It supports inventory control, purchasing, and distribution, but does not have good integration among those functions.
- **Mail Order.** A mainframe application developed in-house using COBOL. Mail-order clerks in Provo use simple microcomputers that emulate older IBM terminals. Despite its age, the application is fast and efficient but unsuitable for handling phone orders. It was last upgraded in 1999.
- **Phone Order.** A modest Windows application developed using Visual Basic and Microsoft SQL Server as a quick solution to customer demand for phone orders. It is poorly integrated with merchandising/distribution and has reached capacity. Implemented nine years ago.
- **Retail Store Systems.** A retail store package with point-of-sale processing. It was upgraded eight years ago from overnight batch to real-time inventory updates to the data center.
- **Office Systems.** Small local servers and networked personal and laptop computers support applications such as Microsoft Office at the Park City offices and other sites. The servers were all recently upgraded to run Windows Server 2008.
- **Human Resources.** An application developed in-house for payroll and benefits running on servers at the data center. Implemented using C and DB2 17 years ago with continuing minor updates.
- **Accounting/Finance.** Originally a mainframe package from a leading accounting package vendor. Migrated to newer servers in 2003 with minor software upgrades, but otherwise unchanged.
- **Web-based Catalog and Order System.** Currently managed and operated by an outside company under a contract that expires in 2011. The system exchanges data in real time with data center servers, but has difficulty in keeping catalog content current, irregular performance, and a poor user interface that has led to many customer complaints and lost sales.

THE INFORMATION SYSTEMS STRATEGIC PLAN

The information systems strategic plan developed with the help of the consultants includes the technology architecture plan and the application architecture plan. The planning team looked closely at existing systems and at the business objectives of RMO. As initially proposed, supply chain management and customer relationship management provided a vision for the plan. These ideas support the strategic objectives of RMO to build more direct customer relationships and to expand the marketing presence beyond the Western states.

The main features of the plans include the following:

Technology architecture plan

1. Further distribute business applications across multiple locations and computer systems, reserving the data center for Web server, database, and telecommunications functions to allow incremental and rapid growth in capacity.
2. Migrate strategic business processes to the Internet and Web, first supporting supply chain management, next supporting direct customer ordering on a new, dynamic Web site, and finally supporting additional customer relationship management (CRM) functions that link internal systems and databases.
3. Anticipate the eventual move toward Web-based intranet solutions for business functions such as human resources, accounting, finance, and information management, using purchased software to the greatest extent possible.

Application architecture plan

1. Supply chain management (SCM): Implement systems that seamlessly integrate product development, product acquisition, manufacturing, and inventory management in anticipation of rapid sales growth. Custom development with support of consultants.
2. Customer support system (CSS): Implement an order-processing and fulfillment system that seamlessly integrates with the supply chain management systems to support mail, phone, and Web-based ordering. Custom in-house development.
3. Strategic information management system (SIMS): Implement an information system that can extract and analyze supply chain and customer support information for strategic and operational decision making and control. Package solution.
4. Retail store system (RSS): Replace the existing retail store system with a system that can integrate with the customer support system. Package solution.
5. Accounting/finance: Purchase a package solution, definitely an intranet application, to maximize employee access to financial data for planning and control.
6. Human resources: Purchase a package solution, definitely an intranet application, to maximize employee access to human resource (HR) forms, procedures, and benefits information.

The timetable for implementing the application architecture plan is shown in Figure 1-13. Key components of the supply chain management system, particularly inventory management components, must be defined before the customer support system project can be started. The customer support system project must be started as soon as possible, though, as it is the core system supporting customer relationship management.

John and Liz consider the SCM system and the CSS to be their core business processes. They've decided to develop and maintain those systems in-house to ensure the fulfillment of specific RMO requirements. Consultants have been called in to help define requirements and develop the integration plan for supply chain management. Several leading consulting firms specialize in supply chain management.

The customer support system will also be developed in-house, although limited use of purchased components is anticipated.

The other systems in the plan will probably be software package solutions selected from among the best-rated software currently available. Using packaged solutions for these business functions will free in-house IS staff to concentrate on the core supply chain and customer relationship management systems. The key requirement for packaged solutions is to integrate seamlessly with other RMO systems and use modern intranet and Web-based technology.

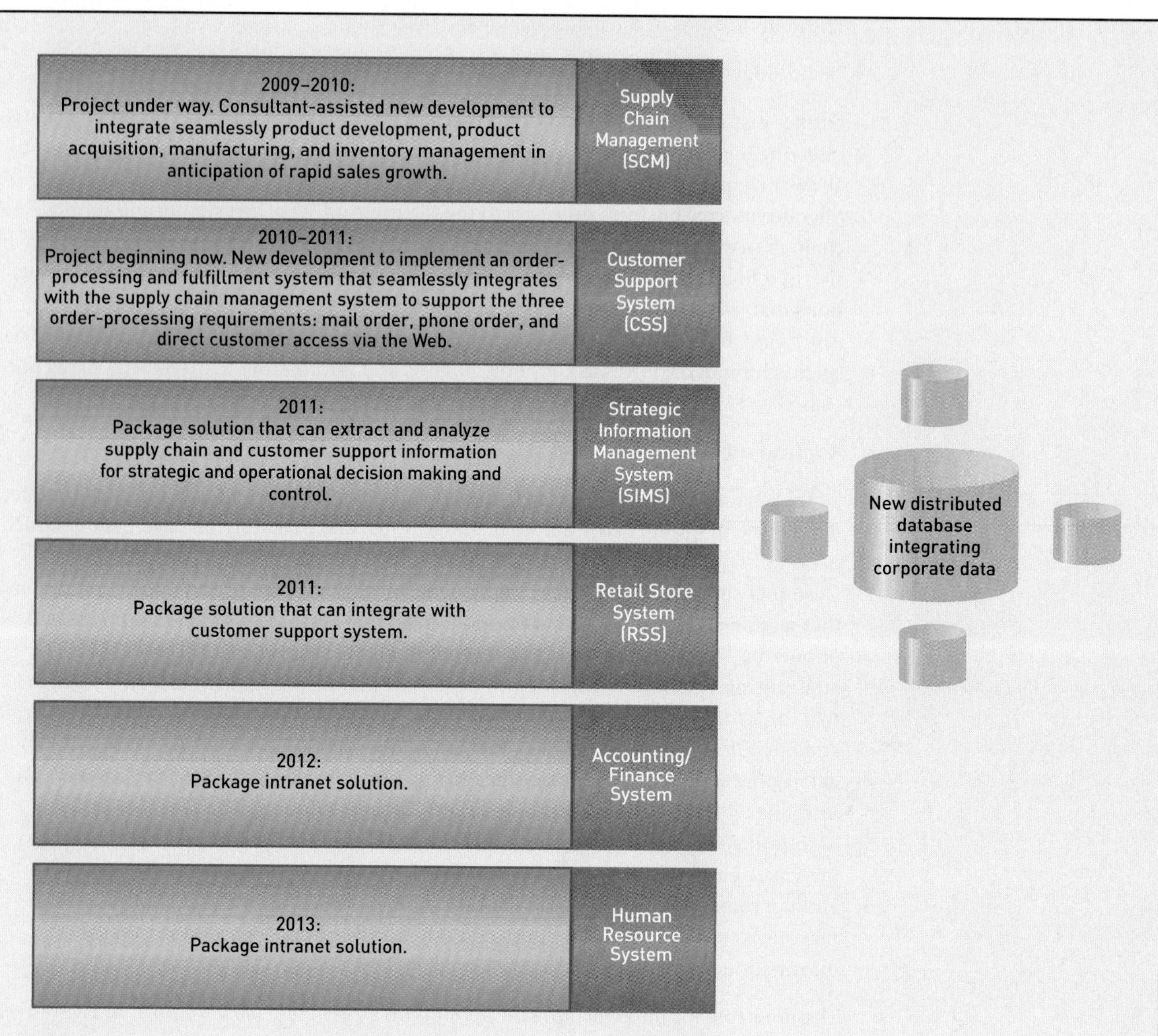

2009–2010:
Project under way. Consultant-assisted new development to integrate seamlessly product development, product acquisition, manufacturing, and inventory management in anticipation of rapid sales growth.
Supply Chain Management (SCM)
2010–2011:
Project beginning now. New development to implement an order-processing and fulfillment system that seamlessly integrates with the supply chain management system to support the three order-processing requirements: mail order, phone order, and direct customer access via the Web.
Customer Support System (CSS)
2011:
Package solution that can extract and analyze supply chain and customer support information for strategic and operational decision making and control.
Strategic Information Management System (SIMS)
2011:
Package solution that can integrate with customer support system.
Retail Store System (RSS)
2012:
Package intranet solution.
Accounting/ Finance System
2013:
Package intranet solution.
Human Resource System
New distributed database integrating corporate data

Figure 1-13

The timetable for RMO's application architecture plan

THE CUSTOMER SUPPORT SYSTEM

The RMO system development project described in this text is the customer support system (CSS). Rocky Mountain Outfitters has always prided itself on its customer orientation. One of the core competencies of RMO has been its ability to develop and maintain customer loyalty. John Blankens knew and understood customer relationship management principles long before the phrase came into common use. His pride in that knowledge has been shaken by recent sales performance and customer complaints. He's determined to right the ship and reenergize RMO's customer-oriented focus with a significant infusion of effort, technology, and money.

The application architecture plan detailed some specific objectives for the customer support system. The system should include all functions associated with providing products for the customer, from order entry to arrival of the shipment, such as:

- Customer inquiries/catalog requests
- Order entry
- Order tracking
- Shipping
- Back ordering
- Returns
- Sales analysis

Customers should be able to order by telephone, mail, or the Web with equal ease. All catalog items would also be available through a sophisticated RMO Web catalog, and the Web catalog must be consistent with printed catalogs so that customers can browse the printed catalog and then order at the RMO Web site, if they choose. In addition, customers might find an item in the printed catalog and search for more information at the Web site.

Order-entry processing needs to support the graphical, self-help style of a customer-oriented Web interface, as well as a streamlined, rapid-response interface required by trained sales representatives. Every sales employee must have rapid access to all information about products, inventory, orders, and customers and be able to apply the information in a way that provides customers with the best possible service.

Although some objectives are defined for the system, a complete systems analysis will define the requirements for the system in detail. These objectives only form some guidelines to keep in mind as the project gets under way.

THE ANALYST AS A SYSTEM DEVELOPER (THE HEART OF THE COURSE)

We have discussed many roles that a systems analyst can play in an organization, including strategic planning and helping identify the major information systems projects that the business will pursue. However, the main job of an analyst is working on a specific information systems development project. This text is about planning and executing an information systems project—in other words, working as a system developer. The text is organized around this theme. In this section, we provide an overview of the text—a preview of what system development involves—as exemplified by the development process ahead for Barbara Halifax, who is in charge of the RMO customer support system project that is about to start (see memo).

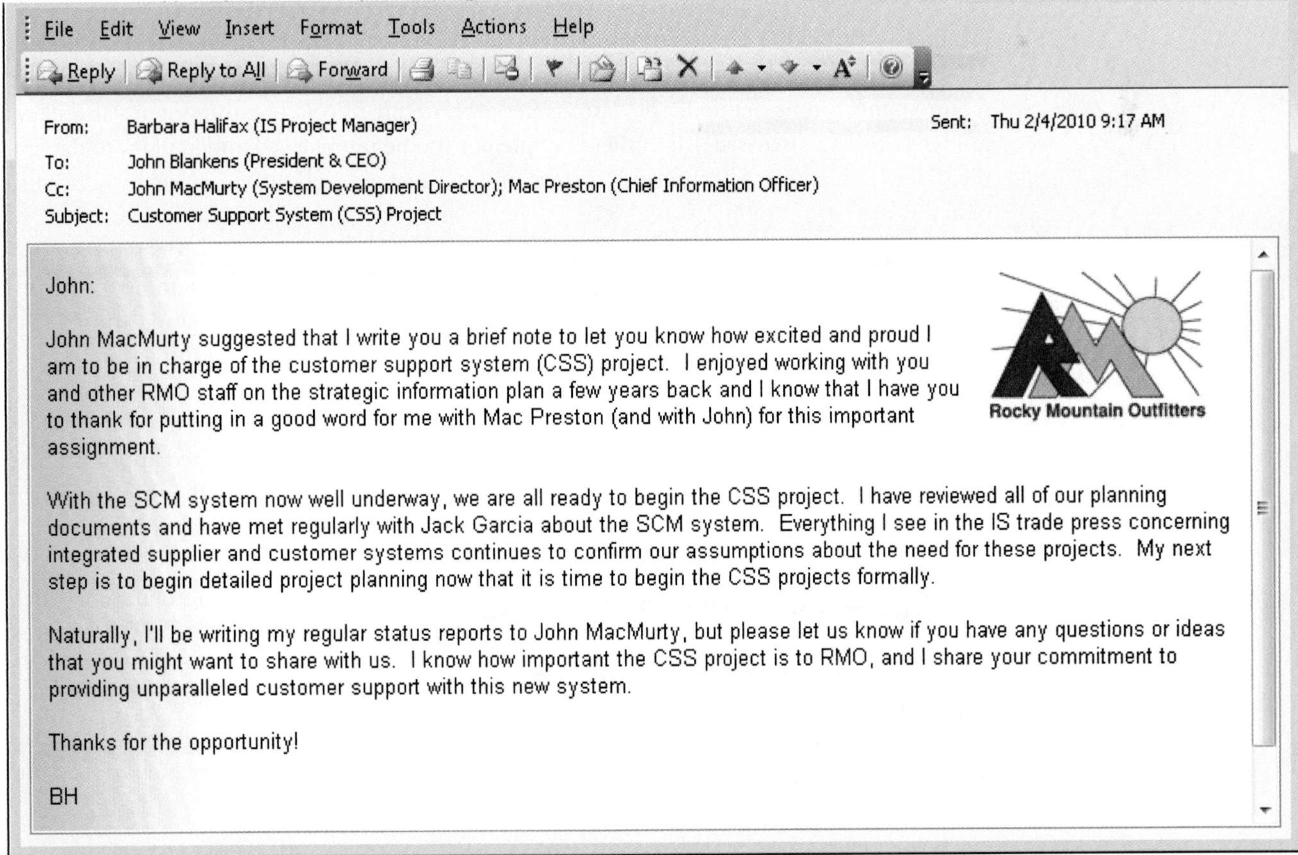

From: Barbara Halifax (IS Project Manager)
Sent: Thu 2/4/2010 9:17 AM
To: John Blankens (President & CEO)
Cc: John MacMurty (System Development Director); Mac Preston (Chief Information Officer)
Subject: Customer Support System (CSS) Project

John:

John MacMurty suggested that I write you a brief note to let you know how excited and proud I am to be in charge of the customer support system (CSS) project. I enjoyed working with you and other RMO staff on the strategic information plan a few years back and I know that I have you to thank for putting in a good word for me with Mac Preston (and with John) for this important assignment.

With the SCM system now well underway, we are all ready to begin the CSS project. I have reviewed all of our planning documents and have met regularly with Jack Garcia about the SCM system. Everything I see in the IS trade press concerning integrated supplier and customer systems continues to confirm our assumptions about the need for these projects. My next step is to begin detailed project planning now that it is time to begin the CSS projects formally.

Naturally, I'll be writing my regular status reports to John MacMurty, but please let us know if you have any questions or ideas that you might want to share with us. I know how important the CSS project is to RMO, and I share your commitment to providing unparalleled customer support with this new system.

Thanks for the opportunity!

BH

PART 1: THE SYSTEMS ANALYST

The first part of the text describes the work of the systems analyst. This chapter (Chapter 1) describes the nature of the analyst's work in terms of types of problems solved, the required skills, and the job titles and places where an analyst might work. We hope it is clear so far that the analyst does much more than think about and write programs. The rest of the text is organized around the problem-solving approach we described at the beginning of the chapter.

Not only does the analyst get involved with business problems, but he or she can work on very high-level strategic issues and with people at all levels of the organization relatively early in his or her career. This chapter also described Rocky Mountain Outfitters and its strategic information systems plan. The rest of the text focuses on one of the planned new systems—the customer support system—and its development.

Chapter 2 focuses on the variety of approaches available for developing an information system. The system development life cycle (SDLC) is introduced as a technique for managing and controlling a project. A variety of tools, techniques, and methodologies are discussed, including the traditional structured approach and the newer object-oriented approach. System developers should be familiar with the fundamental concepts of both approaches. This text covers both approaches throughout, pointing out where they are similar and where they are different.

Chapter 3 gets to the heart of the system development project by describing how a project is planned and managed. The SDLC provides the structure used for project management. Other project management tools and techniques are also introduced, including feasibility studies, project scheduling, and project staffing. An information systems project is just like any other project in these respects. It is important for an analyst to understand the role of project management. Specific issues also arise when planning an information systems project, and an analyst needs to be familiar with them and the way they relate to the larger context of the activities of project planning for an information systems project.

PART 2: SYSTEMS ANALYSIS TASKS

Chapters 4 through 8 cover systems analysis in detail. Chapter 4 discusses techniques for gathering information about the problem that the new system is to solve so that the system requirements can be defined. The various people who are affected by the system (the stakeholders) are also discussed. All of these people need to be interviewed and kept up to date on the status of the project. Techniques such as prototyping and walkthroughs are introduced to help the analyst communicate with everyone involved.

Chapter 5 introduces the concept of models and modeling to record the detailed requirements for the system in a useful form. When discussing an information system, two key concepts are particularly useful: "events" that cause the system to respond and "things" the system needs to store information about. These two concepts, events and things, are important no matter which approach to system development you are using—either the traditional structured approach or the object-oriented approach. Business events are used to identify system activities in the traditional approach, and use cases in the object-oriented approach. The entity-relationship diagram (ERD) is introduced as a model for showing the things affected in the traditional approach, and the class diagram is introduced as a model of things in the object-oriented approach.

Chapters 6 and 7 continue the discussion of modeling system requirements, at which point the traditional structured approach begins to look different from the object-oriented approach. Chapter 6 covers the traditional approach to requirements, which focuses on processes. Data flow diagrams (DFDs), structured English, and data flow definitions are emphasized. Chapter 7 covers the object-oriented approach to requirements, which focuses on objects and user interactions. Use cases, use case diagrams, and system sequence diagrams are emphasized. System developers should be familiar with both approaches to defining systems requirements, but it is important to recognize that in a given system development project, one approach or the other will typically be used. However, many developers are now finding use cases and the use case diagram helpful for defining functional requirements even though they plan to design the system using the traditional approach. So, all readers will benefit from studying the use case section of Chapter 7.

Chapter 8 demonstrates techniques for generating alternatives for actually implementing the system. Each alternative is described and evaluated carefully for feasibility. Then, the best alternative is recommended to management. The final approval of the recommended alternative is a key decision point for the project.

PART 3: SYSTEMS DESIGN TASKS

After one of the alternatives is approved, work on the actual design details begins. Chapters 9 through 15 cover system design issues. Chapter 9 provides an overview of systems design, including the activities completed during the design phase and the general technical environments that are used to implement the system. The three-layer design approach used with both the traditional and object-oriented approaches is introduced. Chapter 10 discusses the traditional approach to system design, showing the types of models used (system flowcharts, structure charts, and pseudocode). Chapters 11 and 12 discuss the object-oriented approach to design, showing the types of models used (sequence diagrams, communication diagrams, design class diagrams, and package diagrams). Important design patterns and approaches to evaluating the quality of object-oriented designs are also discussed.

Chapter 13 describes the issues involved in designing the database for the system, using either a relational database, an object-oriented database, or a hybrid approach that combines relational databases with object technology.

Chapter 14 discusses the user interface to the system, providing an overview of the field of human-computer interaction (HCI) and guidelines for developing user-friendly systems. The chapter covers Windows graphical user interfaces and browser-based interfaces used in Web development. These design concepts apply to both the traditional approach and the object-oriented approach.

Chapter 15 covers the design of system interfaces, system controls, and security. System interfaces include output design of various types of reports that are typically produced online and on paper. Information systems controls are discussed, including the importance of ensuring that inputs are accurate and complete and that processing is done correctly. Techniques for protecting the system from unauthorized access are also discussed. These concepts also apply to both the traditional approach and the object-oriented approach.

PART 4: IMPLEMENTATION AND SUPPORT

Chapter 16 describes the fourth and fifth phases of the SDLC: system implementation and system support. No matter how the system is obtained, a major part of the project is making the system operational and keeping it that way. The analyst's role in implementing the system includes quality control, testing, training users, and conversion to operating the new system. Maintenance and support of the system can continue for many years, involving fixing problems and enhancing the system over time.

Often, a new programmer analyst is involved in maintenance and support of an existing system. Maintenance and support of the system are also the most expensive parts of the project, and decisions made during analysis and design can have a big impact on the ease of maintenance and the overall cost of the system over its lifetime.

This text emphasizes systems analysis and design using a view of the system development process that makes extensive use of iteration and modeling. But you should also be familiar with current trends that focus more explicitly on iteration, risk, and other techniques. The

Unified Process (UP), Agile Modeling, Extreme Programming (XP), Scrum, and object frameworks and component-based development are all discussed in Chapter 17.

ADDITIONAL MATERIALS ON WEB SITE

This edition includes some important additional materials on the book's Web site, *www.course.com/mis/sad5*. Implementing a software package instead of custom development of a system is almost always a viable alternative, as discussed in Chapter 1 and in more detail in Chapter 8. Packages and enterprise resource planning are discussed in Online Supplemental Chapter 1. Also available are online appendices that cover additional material on project management, project planning, financial feasibility, interviewing, and using Microsoft Project.

SUMMARY

A systems analyst is someone who solves business problems using information systems technology. Problem solving means looking into the problem in great detail, understanding everything about the problem, generating several alternatives for solving the problem, and then picking the best solution. Information systems are usually part of the solution, and information systems development is much more than writing programs.

A system is a collection of interrelated components that function together to achieve some outcome. Information systems, like other systems, contain components, and an information systems outcome is the solution to a business problem. Information systems components can be thought of as subsystems that interact or as hardware, software, inputs, outputs, data, people, and procedures. Many different types of systems solve organizational problems, including customer relationship management systems, supply chain management systems, human resource management systems, manufacturing management systems, accounting and financial management systems, and purchased software that integrates these systems, often referred to as enterprise resource planning systems. In addition, organizations use collaboration support systems and business intelligence systems.

A systems analyst needs broad knowledge and a variety of skills, including technical, business, and people knowledge and skills. Integrity and ethical behavior are crucial to the success of the analyst. Analysts encounter a variety of technologies that often change rapidly. Systems analysis and design work is done by people with a variety of job titles: not only systems analyst but programmer analyst, systems consultant, systems engineer, and Web developer, among others. Analysts also work for consulting firms, as independent contractors, and for companies that produce software packages.

A systems analyst can become involved in strategic planning by working with executives on special projects, by helping with business process reengineering projects, and by working on company strategic plans. Analysts also assist businesses in their efforts to select and implement enterprise resource planning systems. Sometimes an information systems strategic planning project is conducted for the entire organization, and analysts often are involved. The Rocky Mountain Outfitters planning project described in this chapter is an example.

Usually the systems analyst works on a system development project, one that solves a business problem identified by strategic planning. That is the emphasis in the rest of this text: how the analyst works on a system development project, completing project planning, systems analysis, systems design, systems implementation, and system support activities. The Rocky Mountain Outfitters customer support system project is used to illustrate the system development process.

KEY TERMS

accounting and financial management (AFM) systems, p. 10
application architecture plan, p. 17
automation boundary, p. 8
business intelligence system, p. 10
business process reengineering, p. 16
collaboration support system (CSS), p. 10
customer relationship management (CRM) system, p. 9
database, p. 11
enterprise resource planning (ERP), p. 11
functional decomposition, p. 7
hard skills, p. 13
human resource management (HRM) system, p. 10
information system, p. 6
information systems strategic plan, p. 17
knowledge management system, p. 10
manufacturing management system, p. 10
soft skills, p. 13
strategic planning, p. 17
subsystem, p. 7
supply chain management (SCM) system, p. 9
system, p. 6
system boundary, p. 8
systems analysis, p. 4
systems analyst, p. 4
systems design, p. 4
techniques, p. 12
technology architecture plan, p. 17
tools, p. 12

REVIEW QUESTIONS

1. Give an example of a business problem.
2. What are the main steps followed when solving a problem?
3. Define *system*.
4. Define *information system*.
5. What types of information systems are found in most organizations?
6. List the six fundamental technologies an analyst needs to understand.
7. List four types of tools the analyst needs to use to develop systems.
8. List five types of techniques used during system development.
9. What are some of the things an analyst needs to understand about businesses and organizations in general?
10. What are some of the things an analyst needs to understand about people?
11. List 10 job titles that involve analysis and design work.
12. How might an analyst become involved with executives and strategic planning relatively early in his career?

THINKING CRITICALLY

1. Describe a business problem your university has that you would like to see solved. How can information technology help solve it?
2. Describe how you would go about solving a problem you face. Is the approach taken by a systems analyst, as described in the text, any different?
3. Many different types of information systems were described in this chapter. Give an example of each type of system that might be used by a university.
4. What is the difference between technical skills and business skills? Explain how a computer science graduate might be strong in one area and weak in another. Discuss how the preparation for a CIS or MIS graduate is different from that for a computer science graduate.
5. Explain why an analyst needs to understand how people think, how they learn, how they react to change, how they communicate, and how they work.
6. Who needs greater integrity to be successful, a salesperson or a systems analyst? Or does every working professional need integrity and ethical behavior to be successful? Discuss.
7. Explain why developing a small information system for use by a single department requires different skills than developing a large information system with many internal and external users.
8. How might working for a consulting firm for a variety of companies make it difficult for the consultant to understand the business problem a particular company faces? What might be easier for the consultant to understand about a business problem?
9. Explain why a strategic information systems planning project must involve people outside the information systems department. Why would a consulting firm be called in to help organize the project?
10. Explain why a commitment to enterprise resource planning (ERP) would be very difficult to undo after it has been made.

EXPERIENTIAL EXERCISES

1. It is important to understand the nature of the business you work for as an analyst. Contact some information systems developers and ask them about their employers. Do they seem to know a lot about the nature of the business? If so, how did the developers gain that knowledge—for example, was it through self-study, formal training or course work, or on-the-job training via participation in system development projects? What are the developers' plans for the future—for example, do the plans involve more training, more courses, or working on projects in specific business areas?
2. Think about the type of position you want (for example, working for a specific company, working for a consulting firm, or working for a software vendor). Do some research on each job by looking at companies' recruiting brochures or Web sites. What do they indicate are the key skills they look for in a new hire? Are there any noticeable differences between consulting firms and the other organizations?
3. You have read an overview of the Rocky Mountain Outfitters' strategic information systems plan, including the technology architecture plan and the application architecture plan. Research system planning at your university. Is there a plan for how information technology will be used over the next few years? If so, describe some of the key provisions of the technology architecture plan and the application architecture plan.

CASE STUDIES

ASSOCIATION FOR INFORMATION TECHNOLOGY PROFESSIONALS MEETING

"I'll tell you exactly what I look for when I interview a new college grad," Alice Adams volunteered. Alice, a system development manager at a local bank, was talking with several professional acquaintances at a monthly dinner meeting of the Association for Information Technology Professionals (AITP). AITP provides opportunities for information systems professionals to get together occasionally and share experiences. Usually a few dozen professionals from information systems departments at a variety of companies attend the monthly meetings.

"When I interview students, I look for problem-solving skills," continued Alice. "Every student I interview claims to know all about Java and .NET and Dreamweaver and XYZ, or whatever the latest development package is. But I always ask interviewees one thing: 'How do you generally approach solving problems?' And then I want to know if they have even thought much about banks like mine and financial services generally, so I ask, 'What would you say are the greatest problems facing the banking industry these days?'"

Jim Parsons, a database administrator for the local hospital, laughed. "Yes, I know what you mean. It really impresses me if they seem to appreciate how a hospital functions, what the problems are for us—how information technology can help solve some of our problems. It is the ability to see the big picture that really gets my attention."

"Yeah, I'm with you," added Sam Young, the manager of marketing systems for a retail store chain. "I am not that impressed with the specific technical skills an applicant has. I assume they have the aptitude and some skills. I do want to know how well they can communicate. I do want to know how much they know about the nature of our business. I do want to know how interested they are in retail stores and the problems we face."

"Exactly," confirmed Alice.

1. Do you agree with Alice and the others about the importance of problem-solving skills? Industry-specific insight? Communication skills? Discuss.
2. Should you research how a hospital is managed before interviewing for a position with an information systems manager at a hospital? Discuss.
3. In terms of your career, do you think it really makes a difference whether you work for a bank, a hospital, or a retail chain? Or is an information systems job going to be the same no matter where you work? Discuss.

RETHINKING ROCKY MOUNTAIN OUTFITTERS

RMO's strategic information systems plan calls for building a new supply chain management (SCM) system prior to building the customer support system (CSS). John Blankens has stated often that customer orientation is the key to success. If that is so, why not build the CSS first, so customers can immediately benefit from improved customer ordering and fulfillment? Wouldn't that increase sales and profits faster? RMO already has factories that produce many items RMO sells, and RMO has long-standing relationships with suppliers around the globe. The product catalog is well established, and the business has existing customers who appear eager and willing to shop online. Why wait? Perhaps John Blankens has made a mistake in planning.

1. What are some of the reasons that RMO decided to build the supply chain management system prior to the customer support system?
2. What are some of the consequences to RMO if it is wrong to wait to build the customer support system?
3. What are some of the consequences to RMO if the owners change their minds and start with the customer support system before building the supply chain management system?
4. What are some other changes that you might make to the RMO strategic information systems plan (both the application architecture plan and the technology architecture plan)? Discuss.

FOCUSING ON RELIABLE PHARMACEUTICAL SERVICE

The Reliable Pharmaceutical Service is a privately held company incorporated in 1975 in Albuquerque, New Mexico. It provides pharmacy services to health-care delivery organizations that are too small to have their own in-house pharmacy. Reliable grew rapidly in its first decade, and by the late 1980s its clients included two dozen nursing homes, three residential rehabilitation facilities, two small psychiatric hospitals, and four small specialty medical hospitals. In 1990, Reliable expanded its Albuquerque service area to include Santa Fe and started two new service areas in Las Cruces and Gallup.

Reliable accepts pharmacy orders for patients in client facilities and delivers the orders in locked cases every 12 hours. In the Albuquerque and Santa Fe service area, Reliable employs approximately 12 delivery personnel, 20 pharmacist's assistants (PAs), 6 licensed pharmacists, and 10 office and clerical staff. Another 15 employees work in the Las Cruces and Gallup service areas. The management team includes another six people, mainly company owners.

Personnel at each health-care facility submit patient prescription orders by telephone. Many prescriptions are standing orders, which are filled during every delivery cycle until specifically canceled. Orders are logged into a computer as they are received. At the start of each 12-hour shift, the computer generates case manifests for each floor or wing of each client facility. A case manifest identifies each patient and the drugs he or she has been prescribed, including when and how often the drugs should be administered. The shift supervisor assigns the case manifests to pharmacists, who in turn assign tasks to PAs. Pharmacists supervise and coordinate the PAs' work.

All drugs for a single patient are collected in one plastic drawer of a locking case. Each case is marked with the institution's name, floor number, and wing number (if applicable). Each drawer is marked with the patient's name and room number. Dividers are inserted within a drawer to separate multiple prescriptions for the same patient. When all of the individual components of an order have been assembled, a pharmacist makes a final check of the contents, signs each page of the manifest, and places two copies of the manifest in the bottom of the case, one copy in a file cabinet in the assembly area, and the final copy in a mail basket for billing. When all of the cases have been assembled, they are loaded onto a truck and delivered to the health-care facilities.

Order entry, billing, and inventory management procedures are a hodgepodge of manual and computer-assisted methods. Reliable uses a combination of Excel spreadsheets, an Access database, and antiquated custom-developed billing software running on personal computers. Pharmacy assistants use the custom-developed billing software to enter orders received by telephone and to produce case manifests. The system has become increasingly unwieldy as facility contracts and Medicare and Medicaid reimbursement procedures have become more complex. Some costs are billed to the health-care facilities, some to insurance companies, some to Medicare and Medicaid, and some directly to patients. The company that developed and maintained the billing software has gone out of business, and the office staff has had to work around software shortcomings and limitations with cumbersome procedures. Inventory management is done manually.

In 2004, Reliable's revenues leveled off at $40 million and profits plateaued at $5.5 million. By 2008, revenue was declining approximately 4 percent per year, and profit was declining at over 8 percent per year. Several reasons for the decline included the following:

- Price controls in both Medicare and Medicaid reimbursements and contracts with facilities managed by health maintenance organizations (HMOs) and large national health-care companies
- Increasing competition from national retail pharmacy chains such as Walgreens and in-house pharmacies at large local hospitals
- Inefficient operating procedures, which haven't received a comprehensive review or overhaul in almost two decades

Reliable's management team spent most of the last year developing a strategic plan, the key element of which is a major effort to streamline operations to improve service and reduce costs. Management sees this effort as the only hope of surviving in a future dominated by large health-care companies that can dictate price and outsource pharmaceutical services to whomever they choose. Management plans a significant expansion into neighboring states after the system is up and running to recoup its costs and increase economies of scale.

Reliable is much smaller than Rocky Mountain Outfitters, the company discussed in this chapter. But the organization still requires a comprehensive set of information systems to support its operations and management. We will include a case study at the end of each chapter that applies chapter concepts to Reliable Pharmaceutical Service.

1. How many information systems staff members do you think Reliable can reasonably afford to employ? What mix of skills would they require? How flexible would they have to be in terms of the work they do each day?
2. What impact should Web and wireless technology have on the way Reliable deploys its systems? Should the Web and wireless technology change the way Reliable does business?
3. Create an application architecture plan and a technology architecture plan for Reliable Pharmaceutical Service to follow for the next five years. What system projects come first in your plan? What system projects come later?

FURTHER RESOURCES

For a comprehensive review of information systems concepts, see:

Effy Oz, *Management Information Systems, Fifth Edition*. Course Technology, 2006.

Kathy Schwalbe, *Information Technology Project Management, Fifth Edition*. Course Technology, 2007.

Ralph M. Stair and George W. Reynolds, *Principles of Information Systems, Eighth Edition*. Course Technology, 2007.

CHAPTER

2

APPROACHES TO SYSTEM DEVELOPMENT

LEARNING OBJECTIVES

After reading this chapter, you should be able to:

- Explain the purpose and various phases of the traditional systems development life cycle (SDLC)
- Explain when to use an adaptive approach to the SDLC in place of the more predictive traditional SDLC
- Explain the differences between a model, a tool, a technique, and a methodology
- Describe the two overall approaches used to develop information systems: the traditional approach and the object-oriented approach
- Describe the key features of current trends in system development: the Unified Process (UP), Extreme Programming (XP), and Scrum
- Explain how automated tools are used in system development

CHAPTER OUTLINE

The Systems Development Life Cycle

Activities of Each SDLC Phase

Methodologies, Models, Tools, and Techniques

Two Approaches to System Development

Current Trends in Development

Tools to Support System Development

DEVELOPMENT APPROACHES AT AJAX CORPORATION, CONSOLIDATED CONCEPTS, AND PINNACLE MANUFACTURING

Kim, Mary, and Bob, graduating seniors, were discussing their recent interview visits to different companies that recruited computer information system (CIS) majors on their campus. All agreed that they had learned a lot by visiting the companies, but they also all felt somewhat overwhelmed at first.

"At first I wasn't sure I knew what they were talking about," Kim cautiously volunteered. During her on-campus interview, Kim had impressed Ajax Corporation with her knowledge of data modeling. When she visited the Ajax home office data center for the second interview, the interviewers spent quite a lot of time describing the company's system development methodology.

"A few people said to forget everything I learned in school," continued Kim. Ajax Corporation had purchased a complete development methodology called *IM One* from a small consulting firm. Most employees agreed it works fairly well. The people who had worked for Ajax for quite a while thought *IM One* was unique, and they were very proud of it. They had invested a lot of time and money learning and adapting to it.

"Well, that got my attention when they said forget what I learned in school," noted Kim, "but then they started telling me about their SDLC, about iterations, about business events, about data flow diagrams, and about entity-relationship diagrams, and things like that." Kim had recognized that many of the key concepts in the *IM One* methodology were fairly standard models and techniques from the structured approach to system development.

"I know what you mean," said Mary, a very talented programmer who knew just about every new programming language available. "Consolidated Concepts went on and on about things like OMG and UML and UP and some people named Booch, Rumbaugh, and Jacobson. But then it turned out that they were using the object-oriented approach to develop systems, and they liked the fact that I knew Java and VB .NET. No problem once I got past all of the terminology they used. They said they'd send me out for training on Rational Software Architect, a visual modeling tool for the object-oriented approach."

Bob had a different story. "A few people said analysis and design were no longer a big deal. I'm thinking, 'Knowing that would have saved me some time in school.'" Bob had visited Pinnacle Manufacturing, which had a small system development group supporting manufacturing and inventory control. "They said they try to just jump in and get to the code as soon as possible. Little documentation. Not much of a project plan. Then they showed me some books on their desks, and it looked like they had been doing a lot of reading about analysis and design. I could see they were using Extreme Programming and agile modeling techniques and focusing only on best practices needed for their small projects. It turns out they just organize their work differently by looking at risk and writing user stories while building prototypes. I recognized some sketches of class diagrams and sequence diagrams on the boss's whiteboard, so I felt fairly comfortable."

Kim, Mary, and Bob all agreed that there was much to learn in these work environments but also that many different terms and points of view are used to describe the same key concepts and techniques they learned in school. They were all glad they focused on the fundamentals in their CIS classes and that they had been exposed to a variety of approaches to system development.

OVERVIEW

As the experiences of Kim, Mary, and Bob demonstrate, there are many ways to develop an information system, and doing so is very complex. Project managers rely on a variety of aids to help them with every step of the process. The systems development life cycle (SDLC) introduced in this chapter provides an overall framework for managing the process of system development. But the developer relies on many more concepts for help, including methodologies, models, tools, and techniques. It is very important for you to understand what these concepts are before exploring system development in any detail.

This chapter reviews two main approaches to system development that are currently used to develop business systems: the traditional approach and the object-oriented approach. The traditional approach refers to both structured system development (structured analysis, structured design, and structured programming) and information engineering (IE). The object-oriented approach refers to system development using newer object technologies that require a different approach to analysis, design, and programming.

Traditional and object-oriented approaches use the SDLC as a project management framework, and this chapter describes some variations of the SDLC. In addition, an analyst needs to be familiar with some current trends in system development that may continue to influence analysis and design. Finally, system developers need computer support tools to complete work tasks, including programming tools and specially designed drawing tools. This chapter presents some examples of these software tools. Most of the models, tools, and techniques discussed in this chapter are used during the analysis and design phases of the SDLC.

At Rocky Mountain Outfitters, one of Barbara Halifax's initial jobs as the project manager for the customer support system project is to make decisions about the approach used to develop the system. All of the options described in this chapter are open to her. We will not describe her final decisions, though, because we use the customer support system example throughout this text as we present more details about all approaches.

THE SYSTEMS DEVELOPMENT LIFE CYCLE

project
a planned undertaking that has a beginning and an end and that produces a desired result or product

Chapter 1 explained that systems analysts solve business problems. For problem-solving work to be productive, it needs to be organized and goal oriented. Analysts achieve these results by organizing the work into projects. A **project** is a planned undertaking that has a beginning and an end and that produces a desired result or product. The term *system development project* describes a planned undertaking that produces a new information system. Some system development projects are very large, requiring thousands of hours of work by many people and spanning several calendar years. In the RMO case study introduced in Chapter 1, the system being developed will be a moderately sized computer-based information system, requiring a moderately sized project lasting less than a year. Many system development projects are smaller, lasting a month or two. For a system development project to be successful, the people developing the system must have a detailed plan to follow. Success depends heavily on having a plan that includes an organized, methodical sequence of tasks and activities that culminate with an information system that is reliable, robust, and efficient.

One of the key, fundamental concepts in information system development is the systems development life cycle. Businesses and organizations use information systems to support all the many, varied processes that a business needs to carry out its functions. As explained in Chapter 1, there are many different kinds of information systems, and each has its own focus and purpose in supporting business processes. Each one of these information systems has a life of its own, and we, as system developers, refer to this idea as the *life cycle of a system*. During the life of an information system, it is first conceived as an idea; then it is designed, built, and deployed during a development project; and finally it is put into production and used to support the business. However, even during its productive use, a system is still a dynamic, living entity that is updated, modified, and repaired through smaller projects. This entire process of building, deploying, using, and updating an information system is called the **systems development life cycle**, or SDLC.

systems development life cycle (SDLC)
the entire process of building, deploying, using, and updating an information system

As noted previously, several different projects may be required during the life of a system, first to develop the original system and then to upgrade it later. In this chapter—and in fact in most of this textbook—we will focus on the initial development project and not on the support projects. In other words, our primary concern is with getting the system developed and deployed the very first time.

In today's diverse development environment, many different approaches to developing systems are used, and they are based on different SDLCs. As you might suppose, some approaches have been used for a long time and have varying rates of success. In the ever-changing world of information technology, new and unique approaches to building systems have emerged, which also have varying success rates. Although it is difficult to find a single, comprehensive classification system that encompasses all of the approaches, one useful technique is to categorize SDLC approaches according to whether they are more predictive or adaptive. These two classifications represent the end points of a continuum from completely predictive to completely adaptive (see Figure 2-1).

Figure 2-1

Predictive versus adaptive approaches to the SDLC

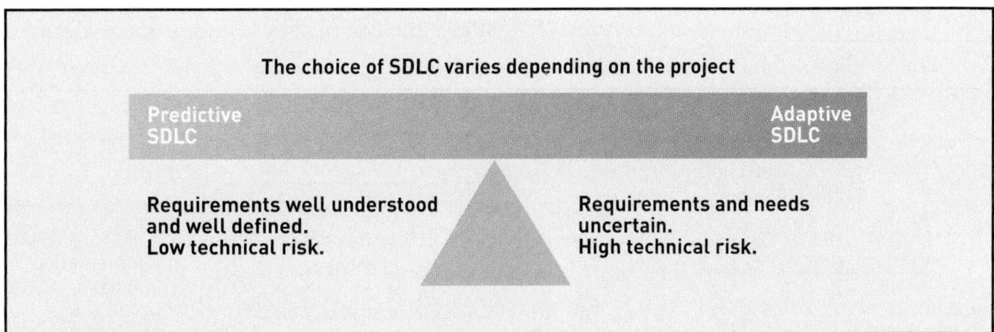

predictive approach

an SDLC approach that assumes the development project can be planned and organized in advance and that the new information system can be developed according to the plan

adaptive approach

an SDLC approach that is more flexible, assuming that the project cannot be planned out completely in advance but must be modified as it progresses

A **predictive approach** to the SDLC is an approach that assumes that the development project can be planned and organized in advance and that the new information system can be developed according to the plan. Predictive SDLCs are useful for building systems that are well understood and defined. For example, a company may want to convert its old, mainframe inventory system to a newer networked client/server system. In this type of project, the staff already understands the requirements very well, and no new processes need to be added. So, the project can typically be planned carefully, and the system can be built according to the specifications.

At the other end of the scale, an **adaptive approach** to the SDLC is used when the exact requirements of a system or the users' needs are not well understood. In this situation, the project cannot be planned completely in advance. Some requirements of the system may yet need to be determined, after some preliminary development work. Developers should still be able to build the solution, but they must be flexible and adapt the project as it progresses.

In practice, any project could have—and most do have—both predictive and adaptive elements. That is why Figure 2-1 shows the characteristics as end points on a sliding scale—not as two mutually exclusive categories. The predictive approaches are more traditional and were invented from the 1970s to the 1990s. Many of the newer, adaptive approaches have evolved along with the object-oriented approach and were created during the 1990s and into the twenty-first century. Let's first look at some of the more predictive approaches and then examine some of the newer adaptive approaches.

BEST PRACTICE

Recognize that any specific project you work on will have some predictive and some adaptive elements.

THE TRADITIONAL PREDICTIVE APPROACHES TO THE SDLC

The development of a new information system requires several different, but related, activities. In predictive approaches, we first have a group of activities that plan, organize, and schedule the project, usually called *project planning activities*. These activities map out the overall

phases
related system development activities, which are grouped into categories of project planning, analysis, design, implementation, and support

structure of the project. Next, a group of activities must focus on understanding the business problem that needs to be solved and on defining the business requirements. We refer to this set of activities as *analysis activities*. The intent is to understand exactly what the system must do to support the business processes. A third group of activities is focused on designing the new system. Those activities, called *design activities*, use the requirements that were defined earlier to develop the program structure and algorithms for the new system. Yet another group of activities is necessary to build the system. We call those activities *implementation activities*, and they include programming, testing, and installing the system for the business users.

These four groups of activities—planning, analysis, design, and implementation—are sometimes referred to as **phases**, and they are the elements that provide the framework for managing the project. Another phase, called the *support phase*, includes the activities needed to upgrade and maintain the system after it has been deployed. The support phase is part of the overall SDLC, but it is not normally considered to be part of the initial development project. Figure 2-2 illustrates the five phases of a traditional SDLC.

Figure 2-2
Information system development phases

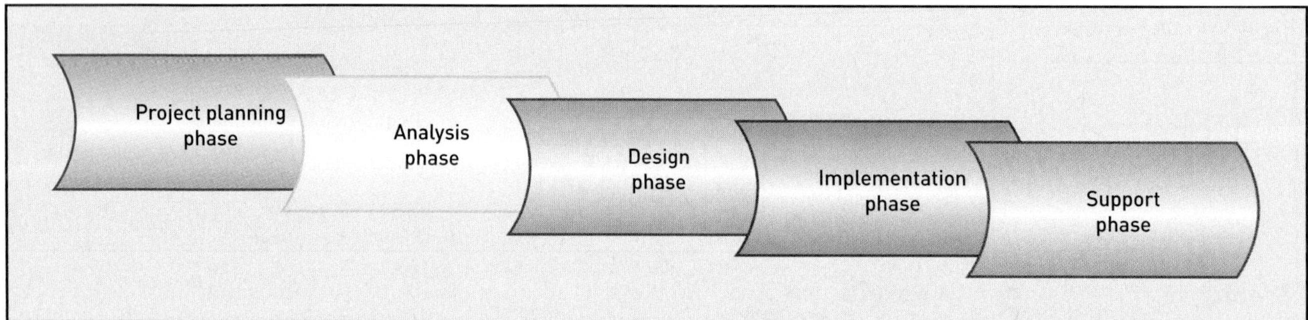

The five phases are quite similar to the steps in the general problem-solving approach outlined in Chapter 1. First, the organization recognizes it has a problem to solve (project planning). Next, the project team investigates and thoroughly understands the problem and the requirements for a solution (analysis). After the problem is understood, a solution is specified in detail (design). The system that solves the problem is then built and installed (implementation). As long as the system is being used by the organization, it is maintained and enhanced to make sure it continues to provide the intended benefits (support). See Figure 2-3.

Figure 2-3
SDLC phases and objectives

SDLC phase	Objective
Project planning	To identify the scope of the new system, ensure that the project is feasible, and develop a schedule, resource plan, and budget for the remainder of the project
Analysis	To understand and document in detail the business needs and the processing requirements of the new system
Design	To design the solution system based on the requirements defined and decisions made during analysis
Implementation	To build, test, and install a reliable information system with trained users ready to benefit as expected from use of the system
Support	To keep the system running productively, both initially and during the many years of the system's lifetime

waterfall model
an SDLC approach that assumes the various phases of a project can be completed sequentially—one phase leads (falls) into the next phase

The SDLC approach that is farthest to the left on the predictive/adaptive scale—that is, most predictive—is called a *waterfall model*. As shown in Figure 2-4, the **waterfall model** assumes that the various phases of a project can be carried out and completed entirely sequentially. A detailed plan is first developed, then the requirements are thoroughly specified, then the system is designed down to the last algorithm, then it is programmed, tested, and installed. After a project drops over the waterfall into the next phase, there is no going back. In practice, the waterfall model requires rigid planning and final decision making at each step of the development project. You can probably guess that a pure waterfall model does not work very well. Developers, being human, have never been able to complete a phase without making mistakes or leaving out important components that had to be added later. Even though we do not use the waterfall model in its purest form anymore, it still provides a valuable foundation to understand development. No matter what system is being developed, we still need to include planning activities, analysis activities, design activities, and implementation activities. We just cannot do them in rigid waterfall steps.

Figure 2-4

The waterfall model of the SDLC

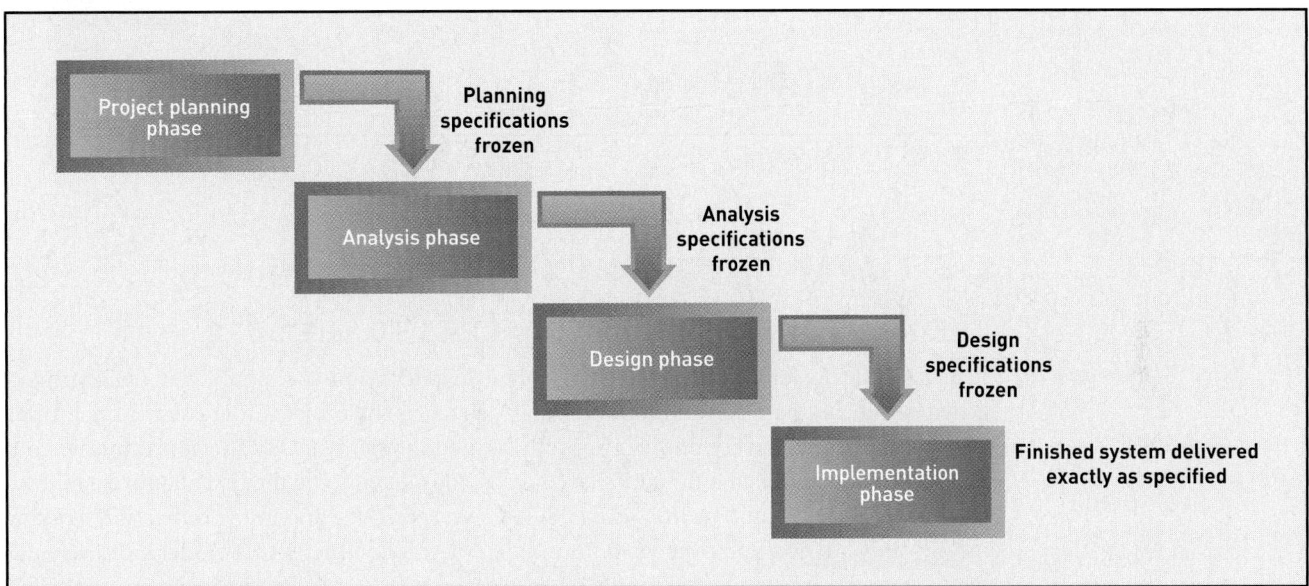

Moving to the right on the predictive/adaptive scale, we find modified waterfall models. We still want to be predictive—that is, still develop a fairly thorough plan—but we recognize that the phases of projects must overlap, influencing and depending on each other. Some analysis must be done before the design can start, but during the design, we discover that we need more detail in the requirements, or even that some of the requirements cannot be met in the manner originally requested. Figure 2-5 illustrates how these activities can overlap.

Another reason phases overlap is efficiency. While the team members are analyzing needs, they may be thinking about and designing various forms or reports. To help them understand the needs of the users, they may want to design some of the final system. But when they do early design, they will frequently throw some components away and save others for later inclusion in the final system. In addition, many components of a computer system are interdependent, which requires analysts to do both analysis and some design at the same time.

Then why not overlap all activities completely? The answer is dependency. Some activities naturally depend on the results of prior work. Analysts cannot get very far into design without a basic understanding of the nature of the problem. Thus, some analysis must happen before design. It would also be inefficient for programmers to write program code before having an overall design structure, because they would have to throw too much away.

Many companies' information systems and many projects today are based on a modified waterfall model. For projects that build well-understood applications, a modified waterfall model is appropriate. Even systems based on an object-oriented approach can be built with modified waterfall models.

Figure 2-5

The overlap of system development phases

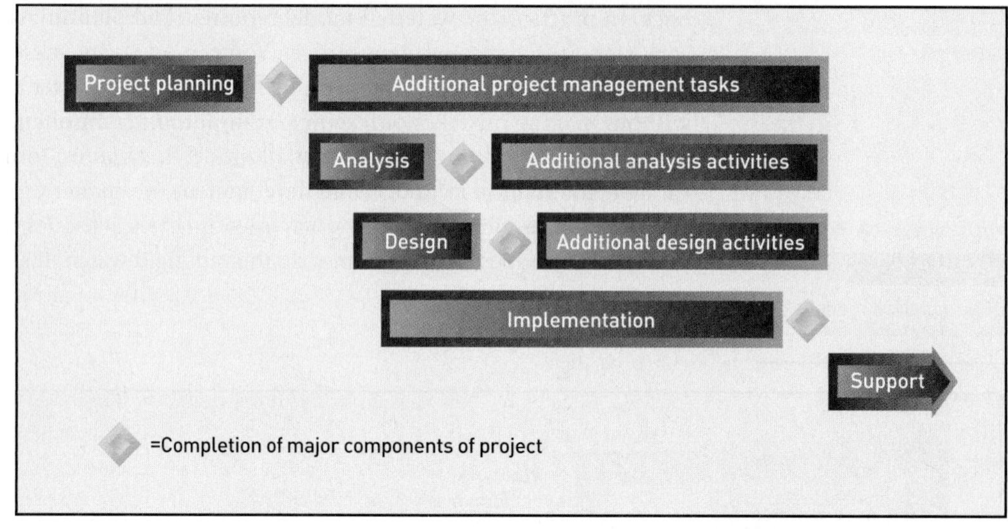

THE NEWER ADAPTIVE APPROACHES TO THE SDLC

spiral model
an adaptive SDLC approach that cycles over and over again through development activities until a project is complete

Remember that by an adaptive approach, we mean a development approach in which project activities—including plans and models—are adjusted as the project progresses. Farther to the right on the scale is a very popular approach called the *spiral model*. The **spiral model** contains many adaptive elements, and it is generally considered to be the first adaptive approach to system development. The life cycle is shown as a spiral, starting in the center and working its way outward, over and over again, until the project is complete. This model looks very different from the static waterfall model and sets the tone for the project to be managed differently. Figure 2-6 shows the spiral model graphically.

You can implement a spiral approach in many different ways. The example in Figure 2-6 begins with an initial planning phase, as shown in the center of the figure. The purpose of this phase is to gather just enough information to begin developing an initial prototype (discussed next). Planning phase activities include a feasibility study, a high-level user requirements survey, generation of implementation alternatives, and choice of an overall design and implementation strategy.

prototype
a preliminary working model showing some aspect of a larger system

After the initial planning is completed, work begins in earnest on the first prototype (the blue ring in the figure). A **prototype** is a preliminary working model of a larger system. For each prototype, the development process follows a sequential path through analysis, design, construction, testing, integration with previous prototype components, and planning for the next prototype. When planning for the next prototype is completed, the cycle of activities begins again. Although the figure shows four prototypes, the spiral model approach can be adapted for any number of prototypes.

A key concept of the spiral approach is the focus on *risk*. Although there are many choices about what to focus on in each iteration, the spiral model recommends identifying risk factors that must be studied and mitigated. The part of the system that appears to have the greatest risk should be addressed in the first iteration. Sometimes the greatest risk is not one subsystem or one set of system functions; rather, the greatest risk might be the technological feasibility of new technology. If so, the first iteration might focus on a prototype that proves the technology will work as planned. Then the second iteration might begin work on a prototype that addresses risk associated with the system requirements or other issues. Another time, the greatest risk might be user acceptance of change. So the first iteration might focus on producing a prototype to show the users that their working lives will be enriched by the new system.

Figure 2-6
The spiral life cycle model

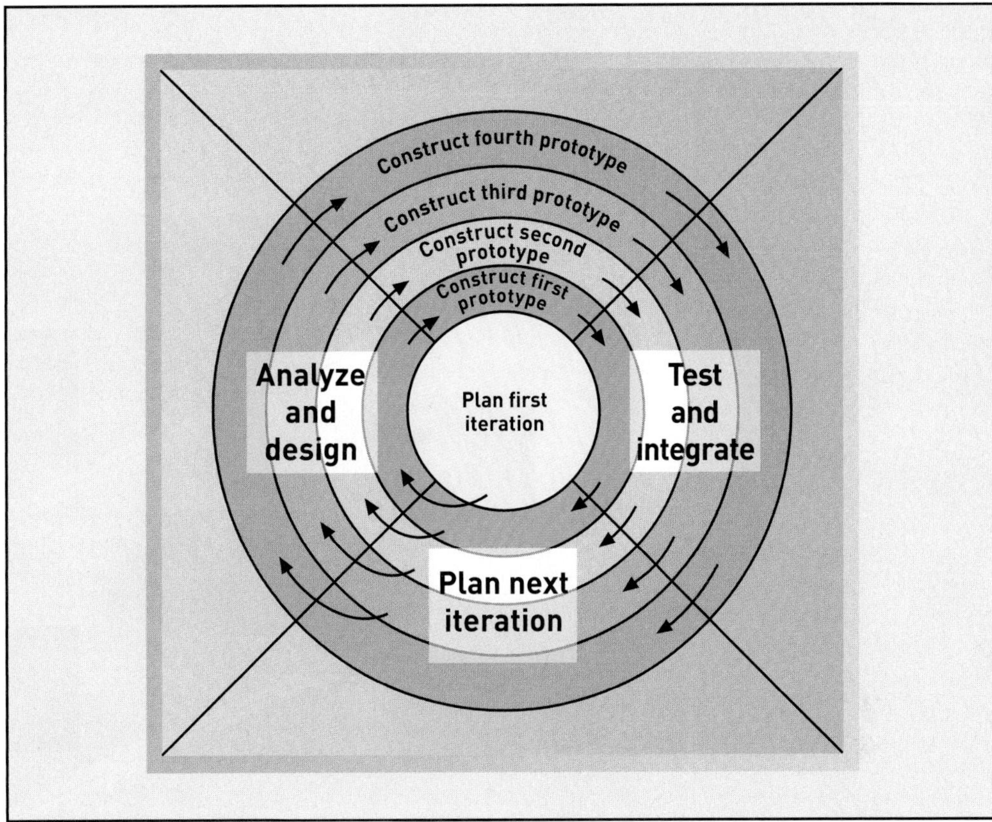

iteration
system development process in which work activities—analysis, design, implementation—are done once, then again, and yet again on different system components; they are repeated until the system is closer to what is ultimately needed

Figure 2-6, which shows the spiral model, uses the term **iteration**. In problem solving, iterations are used to divide a very large, complex problem into smaller, more easily managed problems. Each small problem is solved in turn until the large problem is solved. System development uses iteration for the same purpose. We take a large system and figure out some way to partition it, or divide it into smaller components. Then we plan, analyze, design, and implement each smaller component. Of course, we also add an integration step to combine the smaller components into a comprehensive solution. This approach is frequently called an *iterative approach* to the SDLC. Many of the more popular adaptive approaches today use iteration as a fundamental element of the approach. Figure 2-7 illustrates how an iterative approach works.

Iteration means that work activities—analysis, design, implementation—are done once, then again, and yet again; they are repeated. With each iteration, the developers refine the result so that it is closer to what is ultimately needed. Iteration assumes that no one gets the right result the first time. With an information system, you need to do some analysis and then some design before you really know whether the system will work and accomplish its goals. Then you do more analysis and design to make improvements. In this view, it is not realistic to complete analysis (define all of the requirements) before starting work on the design. Similarly, completing the design is very difficult unless you know how the implementation will work (particularly with constantly changing technology). So you complete some design, then some implementation, and the iteration process continues—more analysis, more design, and more implementation. Naturally, the approach to or the amount of iteration depends on the complexity of the project.

You can organize iterations in several ways. One approach is to define the key functions that the system must include and then implement those key functions in the first iteration. After they are completed, the next set of required, but less crucial, system functions are implemented. Finally, optional system functions, those that would be "nice to have," are implemented in the last iteration. Another approach is to focus on one subsystem at a time. The first subsystem implemented contains the core functions and data on which the other subsystems depend. Then the next iteration includes an additional subsystem, and so on.

Figure 2-7
Iteration of system development activities

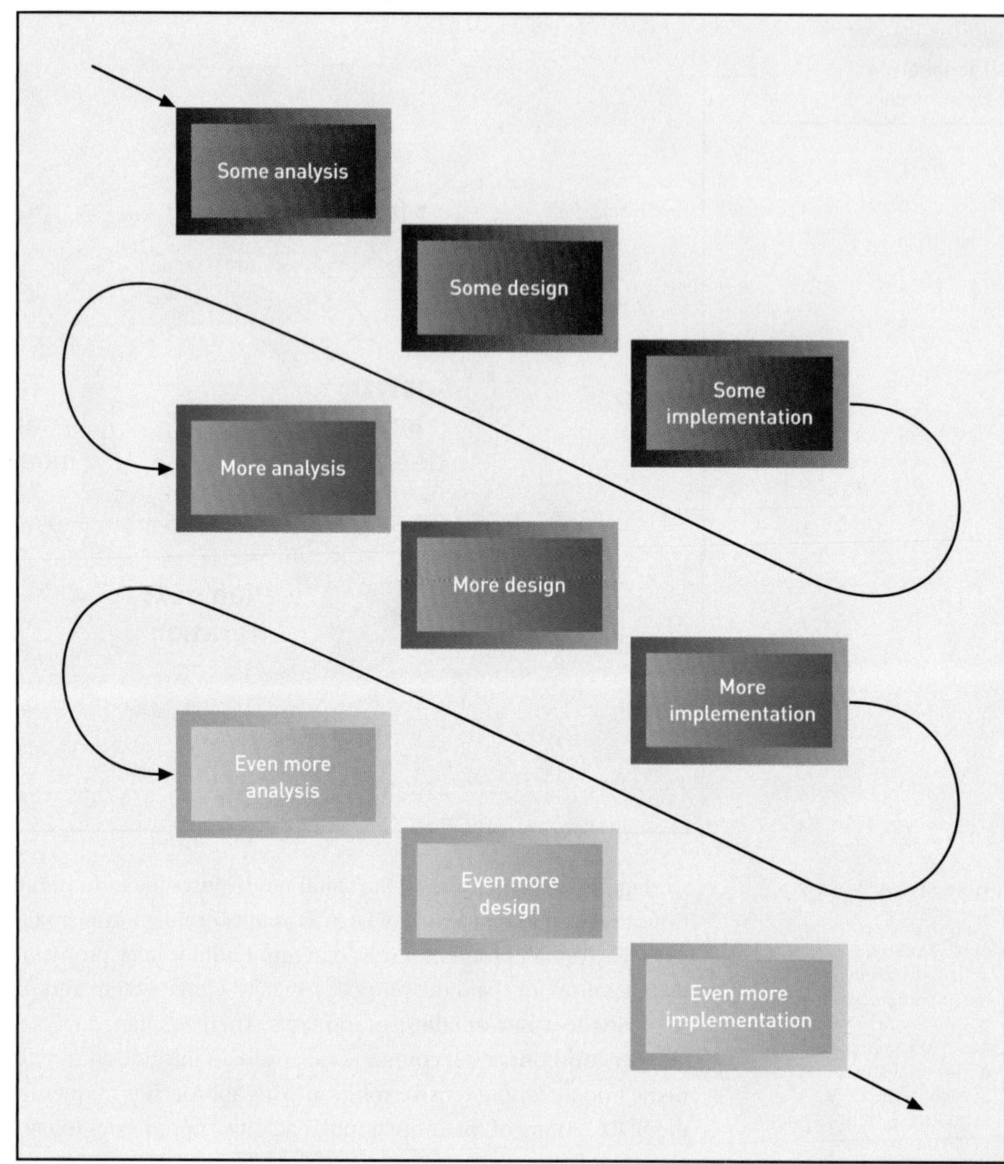

Sometimes iterations are defined according to the complexity or risk of certain components. Often, the most complex or risky parts of the system are addressed first because plans can be changed earlier in the project without huge consequences. Other times, some of the simplest parts are handled first to get as much of the system finished as quickly as possible. How the iterations are defined depends on many factors and might be different with every project you encounter. Most adaptive approaches suggest tackling the toughest problems with the highest risk first.

BEST PRACTICE

Address the aspects of the project that pose the greatest risk in early project iterations.

incremental development
a development approach that completes parts of a system in several iterations and then puts them into operation for users

A related approach, which is a type of iterative approach, is called **incremental development**. With this approach, you complete parts of the system in a few iterations and then put the system into operation for users. This approach gets part of the system into users' hands as early as possible so they can benefit from it. Then you complete a few more iterations to develop another part of the system, integrate it with the first part, and again put it into operation. Finally, you complete the last part and integrate it with the rest. Today, much of system development uses varying degrees of iteration. The object-oriented approach is always described as highly iterative.

ACTIVITIES OF EACH SDLC "PHASE"

We described each SDLC phase generally and demonstrated how the activities of each phase are often carried out iteratively. Next, we discuss the activities of the SDLC phases—project planning, analysis, design, implementation, and support—in more detail.

PROJECT PLANNING

project planning

the initial activities of the SDLC, whose objective is to identify the scope of the new system and plan the project

The primary objectives of **project planning** are to identify the scope of the new system, ensure that the project is feasible, and develop a schedule, resource plan, and budget for the remainder of the project. We identify five activities in project planning:

- Define the problem.
- Produce the project schedule.
- Confirm project feasibility.
- Staff the project.
- Launch the project.

The most important activity of project planning is to define precisely the business problem and the scope of the required solution. At this stage in the project, you will not know all of the functions or processes that will be included within the system. However, it is important to identify the major uses of the new system and the business problems that the new system must address.

The two activities of producing the project schedule and staffing the project are clearly closely related. A detailed project schedule listing tasks, activities, and required staff is developed. Fortunately, some excellent methods and tools are available to provide support for this activity, which are explained in the next chapter. Large projects require elaborate schedules with specific, identifiable milestones and control procedures, and a critical part of this phase is identifying the necessary human resources and planning to acquire them at the required times during the project.

The next major element is to confirm that the project is feasible. Many projects are initiated as part of an enterprise-wide strategic plan. Within the overall plan, each project must also stand on its own merit. Feasibility analysis investigates economic, organizational, technical, resource, and schedule feasibility. Each of these types of feasibility analysis is explained in more detail in the next chapter.

Finally, the total plan for the project is reviewed with upper management, and the project is initiated. Initiation of the project entails allocating funds, assigning project members, and obtaining other necessary resources such as office and development tools. An official announcement often communicates the project launch.

ANALYSIS ACTIVITIES

analysis activities

the activities of the SDLC whose objective is to understand the user needs and develop requirements

The primary objective of the **analysis activities** is to understand and document the business needs and the processing requirements of the new system. Analysis is essentially a discovery process. The key words that drive the activities during analysis are *discovery* and *understanding*. Six primary activities are considered part of this phase:

- Gather information.
- Define system requirements.
- Build prototypes for discovery of requirements.

- Prioritize requirements.
- Generate and evaluate alternatives.
- Review recommendations with management.

problem domain
the area of the user's business for which a system is being developed

Gathering information is a fundamental part of analysis. During this activity, the systems analysts meet with users to learn as much as possible about the **problem domain**—the area of the user's business that needs an information system solution and that is being researched. The analysts obtain information about the problem domain by observing the users as they do their work; by interviewing and asking questions of the users; by reading existing documents about procedures, business rules, and job responsibilities; and by reviewing existing automated systems. In addition to gathering information from the users of the system, the analysts should consult other interested parties. They may include middle management, senior executives, and at times even external customers. Gathering information is the core activity for discovery and understanding.

But it is not sufficient simply to gather information. Analysts must review, analyze, and structure the information obtained so that they can develop an overall understanding of the new system's requirements. This activity is called defining the system requirements, and the primary technique that is used is drawing diagrams to express and model the new system's processing requirements.

As we discussed earlier, one important activity that can help an analyst gather and understand the requirements is to build a prototype of pieces of the new system. Then users can review them. Users often find it easier to express their needs by reviewing working prototypes of alternatives. "A picture is worth a thousand words" is as true in defining system requirements as it is in general, and a prototype is the "picture" that can elicit valuable insights from end users.

As the processing requirements are uncovered, each must be prioritized. There are always more requests for automation support than there is budget or resources to provide it. Thus, the most important needs must be identified and given priority for development. As the analysts prioritize the requirements, they also research various alternatives for implementing the system. Implementation alternatives include building the system in-house, buying a software package, or contracting to a third party to develop and install a new system.

Finally, the team selects and recommends an alternative to upper management. The recommendation recaps the results of the analysis phase activities, and together the team makes firm decisions about an alternative.

DESIGN ACTIVITIES

design activities
the activities of the SDLC during which the system and programs are designed

The objective of the **design activities** is to design the solution system based on the requirements defined and decisions made during analysis. High-level design consists of developing an architectural structure for the software components, databases, user interface, and operating environment. Low-level design entails developing the detailed algorithms and data structures that are required for software development. Seven major activities must be completed during the design phase:

- Design and integrate the network.
- Design the application architecture.
- Design the user interfaces.
- Design the system interfaces.
- Design and integrate the database.
- Prototype for design details.
- Design and integrate the system controls.

Design activities are closely interrelated and generally are all done with substantial overlap.

The network consists of the computer equipment, network, and operating system platforms that will house the new information system. Many of today's new systems are being installed in network and client/server environments. Design includes configuring these network environments. Sometimes the design is already complete based on an existing operating environment and strategic IT plans. At other times, substantial work must be done to develop an operating environment to provide the level of service the new system requires.

application

the portion of the new information system that satisfies the user's needs in the problem domain

The **application** is the portion of the new information system that satisfies user needs with regard to the problem domain. In other words, the application provides the processing functions for the business requirements. Designing the appropriate computer programs for the application consists of using the diagrams showing the system's requirements that were developed during analysis.

The user interface is a critical component of any new system. During the analysis activities, prototyping may have defined some elements of the user interface. During design, these elements are all combined to yield an integrated user interface consisting of forms, reports, screens, and sequences of interactions.

Most new information systems must also communicate with other, existing systems, so the design of the method and details of these communication links must also be precisely defined. These are called *system interfaces*.

Databases and information files are an integral part of information systems for business. The diagrams of the new system's data storage requirements, developed during analysis, are used to design the database that will support the application portion of the new system. At times, the database for the specific system must also be integrated with information databases of other systems already in use.

During design, it is often necessary to verify the correctness or workability of the proposed design. Again, one important verification method is to build working prototypes of parts of the system to ensure that it will function correctly in the operating environment. In addition, analysts can test and verify alternative design strategies by building prototypes of the new system. Sometimes, if the prototypes are built correctly, they can be saved and used as part of the final system.

Finally, every system must have sufficient controls to protect the integrity of the database and the application program. Because of the highly competitive nature of the global economy and the risks associated with technology and security, every new system must include adequate mechanisms to protect the information and assets of the organization. These controls should be integrated into the new system while it is being designed, not after it has been constructed.

IMPLEMENTATION ACTIVITIES

implementation activities

the activities of the SDLC during which the new system is programmed and installed

Implementation activities result in the final system being built, tested, and installed. The objective is not only to produce a reliable, fully functional information system, but also to ensure that the users are all trained and that the organization is ready to benefit as expected from use of the system. All the prior activities must come together to culminate in an operational system. Five major activities make up the implementation phase:

- Construct software components.
- Verify and test.
- Convert data.
- Train users and document the system.
- Install the system.

The software can be constructed through various techniques. The conventional approach is to write computer programs using a language such as Visual Basic, C#, or Java. Other techniques, based on development tools and existing components, are becoming popular today. The software must also be tested, and the first kind of testing verifies that the system actually works. Additional testing is also required to make sure that the new system meets the needs of the system's users.

During implementation, the analysts may also build additional prototypes. These prototypes are used to verify different implementation strategies and to ensure that the system can handle the volumes of transactions that will exist after it is placed in production.

Almost every new system replaces an existing system, either a completely manual system or an earlier automated system. Normally, the existing information is important and needs to be converted to the format required in the new system. The activity to convert the data often becomes a small project of its own, with analysis, design, and implementation of procedures to clean and convert the data to the new system.

No system is successful unless the users understand it and can use it appropriately. A critical activity during implementation is to train the users on the new system so that they will be productive as soon as possible.

Finally, the actual changeover is the culminating activity. The new equipment must be in place and functioning, the new computer programs must be installed and working, and the database must be populated and available. The individual pieces of the new system must be up and running before the system can be used for its intended purpose. In today's widely dispersed organizations, the system must frequently be installed in many locations and integrated throughout the organization.

SUPPORT ACTIVITIES

support activities

the activities of the SDLC whose objective is to keep the system running productively after it is installed

The objective of the **support activities** is to keep the system running productively during the years following its initial installation. The support activities begin only after the new system has been installed and put into production, and it lasts throughout the productive life of the system. The expectation for most business systems is that the system will last for years. During support, upgrades or enhancements may be carried out to expand the system's capabilities, and they will require their own development projects. Three major activities occur during support:

- Maintain the system.
- Enhance the system.
- Support the users.

Every system, especially a new one, contains components that do not function correctly. Software development is complex and difficult, so it is never error-free. Of course, the objective of a well-organized and carefully executed project is to deliver a system that is robust and complete and that gives correct results. However, because of the complexity of software and the impossibility of testing every possible combination of processing requirements, there will always be conditions that have not been fully tested and thus are subject to errors. In addition, business needs and user requirements change over time. Key tasks in maintaining the system include both fixing the errors (also known as *fixing bugs*) and making minor adjustments to processing requirements. Usually a system support team is assigned responsibility for maintaining the system.

Most newly hired programmer analysts begin their careers working on system maintenance projects. Tasks typically completed include changing the information provided in a report, adding an attribute to a table in a database, or changing the design of Windows or browser forms. These changes are requested and approved before the work is assigned, so a change request approval process is always part of the system support phase.

During the productive life of a system, it is also common to make major modifications. At times, government regulations require new data to be maintained or information to be provided. Also, changes in the business environment—new market opportunities, new competition, or new system infrastructure—necessitate major changes to the system. To implement these major system enhancements, the company must approve and initiate an upgrade development project. An upgrade project often results in a new version of the system. During your career, you may have the opportunity to participate in several upgrade projects.

help desk
the availability of support staff to assist users with any technical or processing problem associated with an information system

The other major activity during support is to provide assistance to the system users. A **help desk**, consisting of knowledgeable technicians, is a popular method to answer users' questions quickly and help increase their productivity. Training new users and maintaining current documentation are important elements of this activity. As a new systems analyst, you may have the opportunity to conduct training or to staff the help desk to gain experience with user problems and needs. Many newly hired programmer analysts start their careers working at a help desk for part of their workweek.

METHODOLOGIES, MODELS, TOOLS, AND TECHNIQUES

Systems analysts have a variety of aids at their disposal to help them complete activities and tasks in the SDLC. Among them are methodologies, models, tools, and techniques. The following sections discuss each of these aids.

system development methodology
comprehensive guidelines to follow for completing every activity in the systems development life cycle, including specific models, tools, and techniques

METHODOLOGIES

A **system development methodology** provides guidelines to follow for completing every activity in the systems development life cycle, including specific models, tools, and techniques (see Figure 2-8). Some methodologies are homegrown, developed by systems professionals in the company based on their experience. Some methodologies are purchased from consulting firms or other vendors.

Figure 2-8
Relationships among components of a methodology

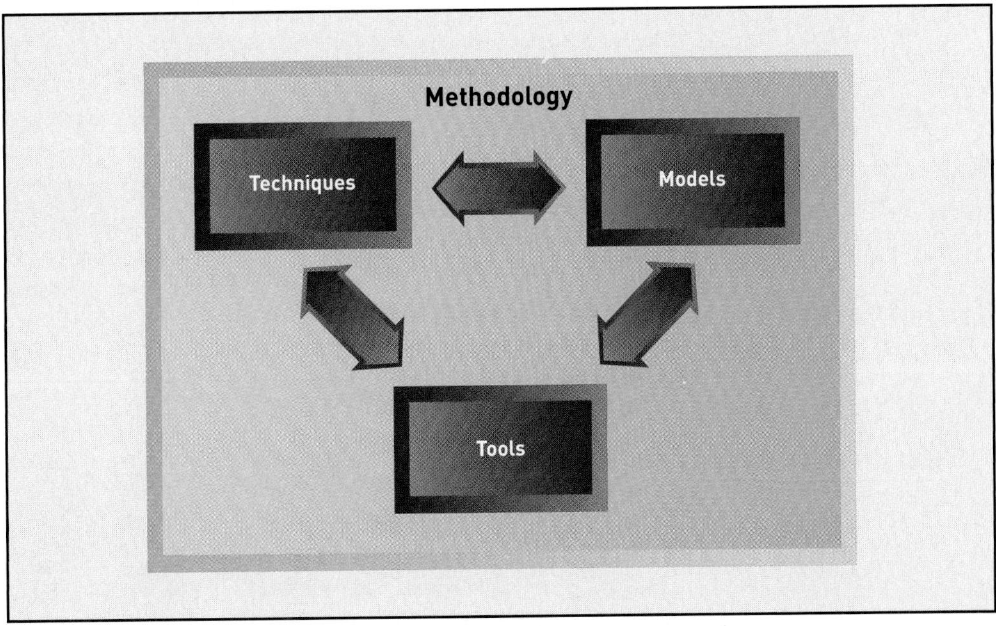

Some methodologies (whether homegrown or purchased) contain written documentation that can fill a bookcase. The documentation defines everything the developers might need to produce at any point in the project, including how documentation should look and what reports to management should contain. Other methodologies are much more informal—one document will contain general descriptions of what should be done. Sometimes the methodology that a company adopts is "just follow some sort of methodology," but such freedom of choice is becoming rare. Most people want the methodology to be flexible, though, so that it can be adapted to many different types of projects and systems. The methodology used by the organization determines how prescriptive or adaptive the approach to a system development project should be.

Because a methodology contains instructions about how to use models, tools, and techniques, you must understand what models, tools, and techniques are.

MODELS

model

a representation of an important aspect of the real world

Anytime people need to record or communicate information, in any context, it is very useful to create a model—and a model in information systems development has the same purpose as any other model. A **model** is a representation of an important aspect of the real world. Sometimes the term *abstraction* is used because we abstract (separate out) an aspect of particular importance to us. Consider a model of an airplane. To talk about the aerodynamics of the airplane, it is useful to have a small model that shows the plane's overall shape in three dimensions. Sometimes a drawing showing the cross-sectional details of the wing of the plane is what is needed. In other cases, a list of mathematical characteristics of the plane might be necessary to understand how the plane will behave. All of these are models of the same plane.

Some models are physically similar to the real product. Some models are graphical representations of important details. Some models are abstract mathematical notations. Each emphasizes a different type of information. In airplane design, aerospace engineers use lots of different models. Learning to be an aerospace engineer involves learning how to create and use all of the models. It is the same for an information system developer, although models for information systems are not yet as standardized or precise as aerospace models. But system developers are making progress. First, it is important to recognize that the field is very young, and many senior analysts were self-taught. More importantly, though, an information system is much less tangible than an airplane—you can't really see, hold, or touch it. Therefore, the models of the information system can seem much less tangible, too.

What sort of models do developers make of aspects of an information system? Models used in system development include representations of inputs, outputs, processes, data, objects, object interactions, locations, networks, and devices, among other things. Most of the models are graphical models, which are drawn representations that employ agreed-upon symbols and conventions. These are often called *diagrams* and *charts*. You have probably drawn models showing program logic using flowcharts. Much of this text describes how to read and create a variety of models that represent an information system.

Another kind of model important to develop and use is a project-planning model, such as Gantt charts, which are shown in Chapter 3. These models represent the system development project itself, highlighting its tasks and task completion dates. Another model related to project management is a chart showing all of the people assigned to the project. Figure 2-9 lists some models used in system development.

Figure 2-9

Some models used in system development

Some models of system components

- Flowchart
- Data flow diagram (DFD)
- Entity-relationship diagram (ERD)
- Structure chart
- Use case diagram
- Class diagram
- Sequence diagram

Some models used to manage the development process

- Gantt chart
- Organizational hierarchy chart
- Financial analysis models – NPV, ROI

TOOLS

tool
software support that helps create models or other components required in the project

integrated development environments (IDE)
tools that help programmers with a variety of programming tasks

visual modeling tools
tools that help the analyst create and verify important system models, often generating program code

A **tool** in the context of system development is software support that helps create models or other components required in the project. Tools might be simple drawing programs for creating diagrams. They might include a database application that stores information about the project, such as data flow definitions or written descriptions of processes. A project management software tool, such as Microsoft Project (described in Chapter 3), is another example of a tool used to create models. The project management tool creates a model of the project tasks and task dependencies.

Tools have been specifically designed to help system developers. Programmers should be familiar with **integrated development environments (IDEs)** that include many tools to help with programming tasks—smart editors, context-sensitive help, and debugging tools. Some tools can generate program code for the developer. Some tools reverse-engineer old programs—generating a model from the code so that the developer can determine what the program does, in case the documentation is missing (or was never done). **Visual modeling tools** are available to systems analysts to help them create and verify important system models, often generating program code. Visual modeling tools are described in more detail later in this chapter. Figure 2-10 lists types of tools used in system development.

Figure 2-10
Some tools used in system development

Project management application
Drawing/graphics application
Word processor/text editor
Visual modeling tool
Integrated development environment (IDE)
Database management application
Reverse-engineering tool
Code generator tool

TECHNIQUES

technique
a collection of guidelines that help an analyst complete a system development activity or task

A **technique** in system development is a collection of guidelines that help an analyst complete a system development activity or task. A technique often includes step-by-step instructions for creating a model, or it might include more general advice for collecting information from system users. Some examples include data-modeling techniques, software-testing techniques, user-interviewing techniques, and relational database design techniques.

Sometimes a technique applies to an entire life cycle phase and helps you create several models and other documents. The modern structured analysis technique (discussed later) is an example. Even the strategic system planning techniques discussed in Chapter 1 and project management techniques discussed in Chapter 3 fit this definition. Figure 2-11 lists some techniques commonly used in system development.

Figure 2-11

Some techniques used in system development

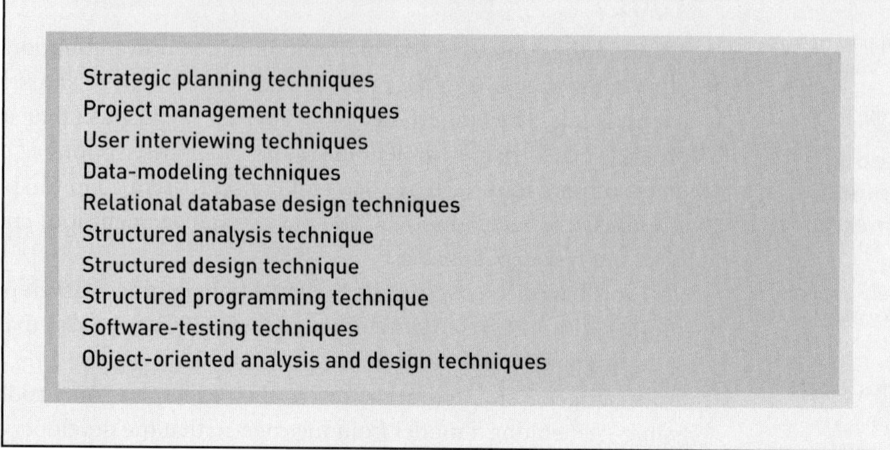

Strategic planning techniques
Project management techniques
User interviewing techniques
Data-modeling techniques
Relational database design techniques
Structured analysis technique
Structured design technique
Structured programming technique
Software-testing techniques
Object-oriented analysis and design techniques

How do all these components fit together? A *methodology* includes a collection of *techniques* that are used to complete activities within each phase of the systems development life cycle. The activities include completion of a variety of *models* as well as other documents and deliverables. Like any other professionals, system developers use software *tools* to help them complete their activities.

As part of her responsibility as project manager for the new customer support system for Rocky Mountain Outfitters, Barbara Halifax has to make decisions about the methodology to use to develop the system (see Barbara's memo).

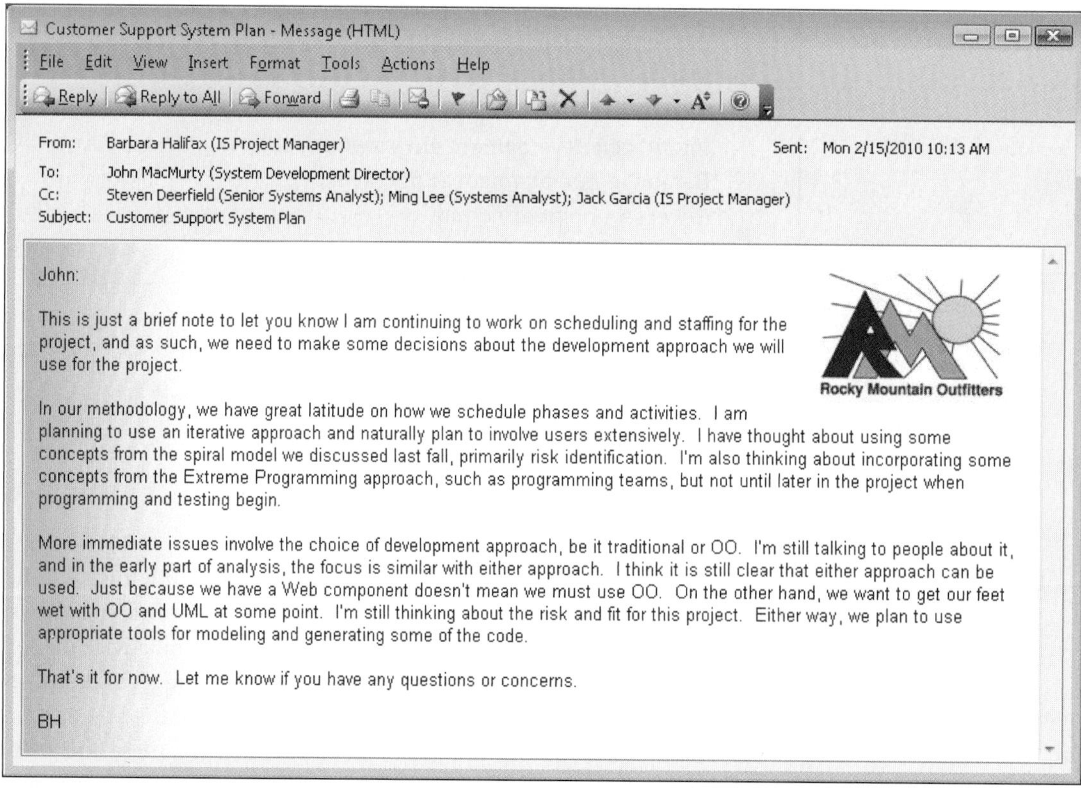

Customer Support System Plan - Message (HTML)

From: Barbara Halifax (IS Project Manager) Sent: Mon 2/15/2010 10:13 AM
To: John MacMurty (System Development Director)
Cc: Steven Deerfield (Senior Systems Analyst); Ming Lee (Systems Analyst); Jack Garcia (IS Project Manager)
Subject: Customer Support System Plan

John:

This is just a brief note to let you know I am continuing to work on scheduling and staffing for the project, and as such, we need to make some decisions about the development approach we will use for the project.

In our methodology, we have great latitude on how we schedule phases and activities. I am planning to use an iterative approach and naturally plan to involve users extensively. I have thought about using some concepts from the spiral model we discussed last fall, primarily risk identification. I'm also thinking about incorporating some concepts from the Extreme Programming approach, such as programming teams, but not until later in the project when programming and testing begin.

More immediate issues involve the choice of development approach, be it traditional or OO. I'm still talking to people about it, and in the early part of analysis, the focus is similar with either approach. I think it is still clear that either approach can be used. Just because we have a Web component doesn't mean we must use OO. On the other hand, we want to get our feet wet with OO and UML at some point. I'm still thinking about the risk and fit for this project. Either way, we plan to use appropriate tools for modeling and generating some of the code.

That's it for now. Let me know if you have any questions or concerns.

BH

System development is done in many different ways. This diversity can confuse new employees when they go to work as system developers. Sometimes it seems every company that develops information systems has its own methodology. Sometimes different development groups within the same company use different methodologies, and each person in the company might have his own way of developing systems.

Yet, as you have already seen in the opening case, there are many common concepts. In virtually all development groups, some variation of the systems development life cycle is used, with phases for project planning, analysis, design, implementation, and support. In addition, virtually every development group uses models, tools, and techniques that make up an overall system development methodology.

All system developers should be familiar with two very general approaches to system development, because they form the basis of virtually all methodologies: the *traditional approach* and the *object-oriented approach*. This section reviews the major characteristics of both approaches and provides a bit of history.

THE TRADITIONAL APPROACH

The traditional approach includes many variations based on techniques used to develop information systems with structured and modular programming. This approach is often referred to as *structured system development*. A refinement to structured development, called *information engineering* (IE), is a popular variation.

Structured System Development

structured approach
system development using structured analysis, structured design, and structured programming techniques

Structured analysis, structured design, and structured programming are the three techniques that make up the **structured approach**. Sometimes these techniques are collectively referred to as the *structured analysis and design technique* (*SADT*). The structured programming technique, developed in the 1960s, was the first attempt to provide guidelines to improve the quality of computer programs. You certainly learned the basic principles of structured programming in your first programming course. The structured design technique was developed in the 1970s to make it possible to combine separate programs into more complex information systems. The structured analysis technique evolved in the early 1980s to help clarify requirements for a computer system before developers designed the programs.

structured program
a program or program module that has one beginning and one ending, and for which each step in the program execution consists of sequence, decision, or repetition constructs

Structured Programming High-quality programs not only produce the correct outputs each time the program runs, they make it easy for other programmers to read and modify the program later. And programs need to be modified all the time. A **structured program** is one that has one beginning and one ending, and each step in the program execution consists of one of three programming constructs:

- A sequence of program statements
- A decision where one set of statements or another set of statements executes
- A repetition of a set of statements

Figure 2-12 shows these three structured programming constructs.

Figure 2-12

Three structured programming constructs

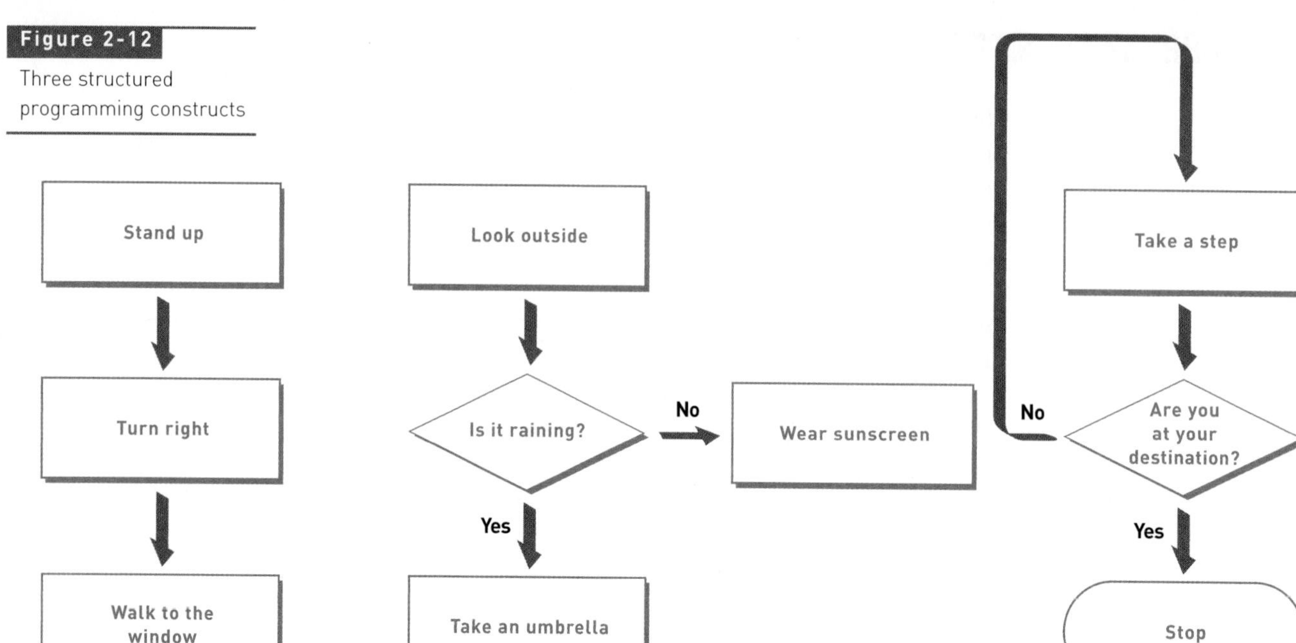

Before these rules were developed, programmers made up programming techniques as they went along, which resulted in some very convoluted programs. Most programmers were happy if the programs ran at all, and they were even happier if the programs produced the right outputs. But following these simple rules made it much easier to read and interpret what a program does.

top-down programming

dividing more complex programs into a hierarchy of program modules

Another concept related to structured programming is top-down programming. **Top-down programming** divides more complex programs into a hierarchy of program modules (see Figure 2-13). One module at the top of the hierarchy controls program execution by "calling" lower-level modules as required. Sometimes the modules are part of the same program. For example, in COBOL, one main paragraph calls another paragraph using the Perform keyword. In Visual Basic, a statement in an event procedure can call a general procedure. The programmer writes each program module (paragraph or procedure) using the rules of structured programming (one beginning, one end, and sequence, decision, and repetition constructs).

Figure 2-13

Top-down, or modular, programming

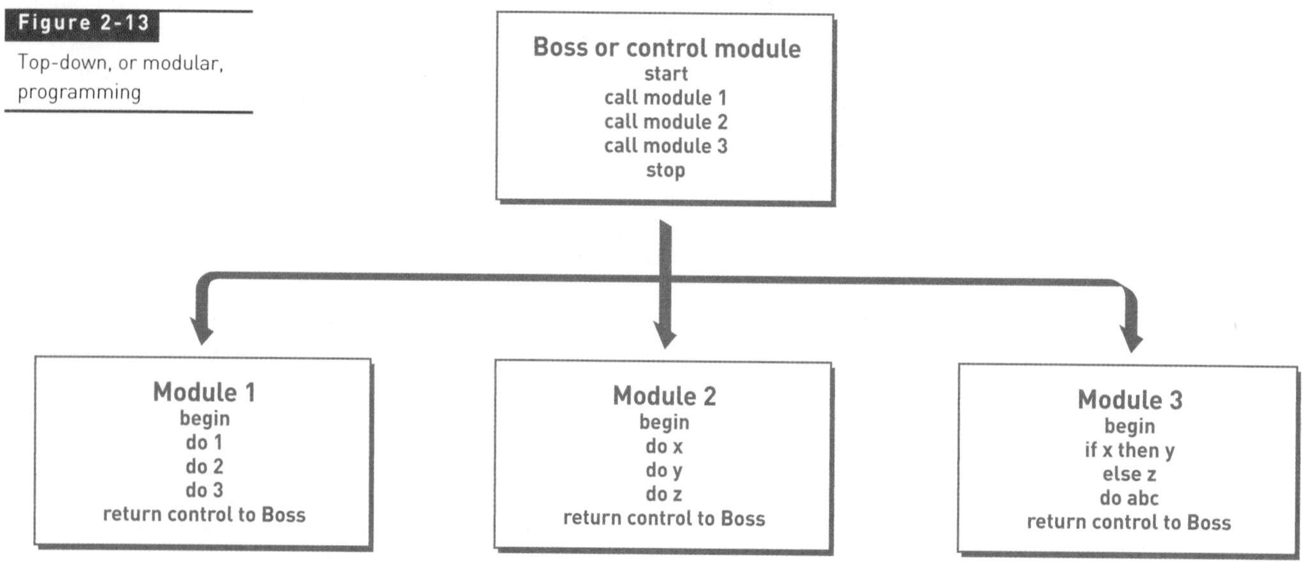

Sometimes separate programs are produced that work together as one "system." Each of these programs follows top-down programming and structured programming rules, but the programs themselves are organized into a hierarchy, as with top-down programming. One program calls other programs. When the hierarchy involves multiple programs, such an arrangement is sometimes called *modular programming*.

Structured Design As information systems continued to become increasingly complex through the 1970s, each system involved many different functions. Each function performed by the system might be made up of dozens of separate programs. The **structured design** technique was developed to provide some guidelines for deciding what the set of programs should be, what each program should accomplish, and how the programs should be organized into a hierarchy. The modules and the arrangement of modules are shown graphically using a model called a **structure chart** (see Figure 2-14).

structured design

a technique providing guidelines for deciding what the set of programs should be, what each program should accomplish, and how the programs should be organized into a hierarchy

structure chart

a graphical model showing the hierarchy of program modules produced by the structured design technique

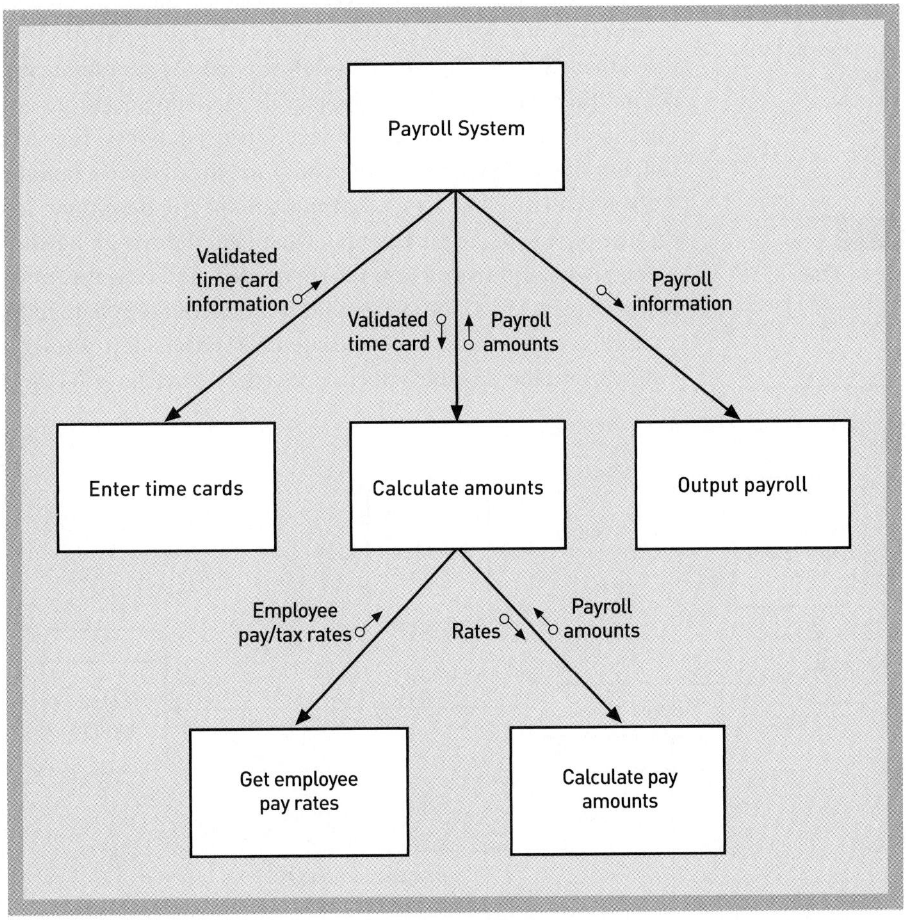

Figure 2-14

A structure chart created using the structured design technique

Two main principles of structured design are that program modules should be designed so they are (1) loosely coupled and (2) highly cohesive. *Loosely coupled* means each module is as independent of the other modules as possible, which allows each module to be designed and later modified without interfering with the performance of the other modules. *Highly cohesive* means that each module accomplishes one clear task. That way, it is easier to understand what each module does and to ensure that if changes to the module are required, none will accidentally affect other modules.

The structured design technique defines different degrees of coupling and cohesion and provides a way of evaluating the quality of the design before the programs are actually written. As with structured programming, quality is defined in terms of how easily the design can be understood and modified later when the need arises.

Structured design assumes the designer knows what the system needs to do—what the main system functions are, what the required data are, and what the needed outputs are. Designing the system is obviously much more than designing the organization of the program modules. Therefore, it is important to realize that the structured design technique helps the designer complete part of but not the entire design life cycle phase.

structured analysis
a technique used to define what processing the system needs to do, what data it needs to store and use, and what inputs and outputs are needed

data flow diagram (DFD)
a structured analysis model showing the inputs, processes, storage, and outputs of a system

By the 1980s, file and database design techniques were developed to be used along with structured design. Newer versions of structured design assume database management systems are used in the system, and program modules are designed to interact with the database. In addition, because more nontechnical people were becoming involved with information systems, user-interface design techniques were developed. For example, menus in an interactive system determine which program in the hierarchy gets called. Therefore, a key aspect of user-interface design is done in conjunction with structured design.

Modern Structured Analysis Because the structured design technique requires the designer to know what the system should do, techniques for defining system requirements were developed. System requirements define what the system must do in great detail, but without committing to one specific technology. By deferring decisions about technology, the developers can sharply focus their efforts on what is needed, not on how to do it. If these requirements are not fully and clearly worked out in advance, the designers cannot possibly know what to design.

The **structured analysis** technique helps the developer define what the system needs to do (the processing requirements), what data the system needs to store and use (data requirements), what inputs and outputs are needed, and how the functions work together as a whole to accomplish tasks. The key graphical model of the system requirements used with structured analysis is called the **data flow diagram (DFD)**, and it shows inputs, processes, storage, and outputs, and the way they function together (see Figure 2-15).

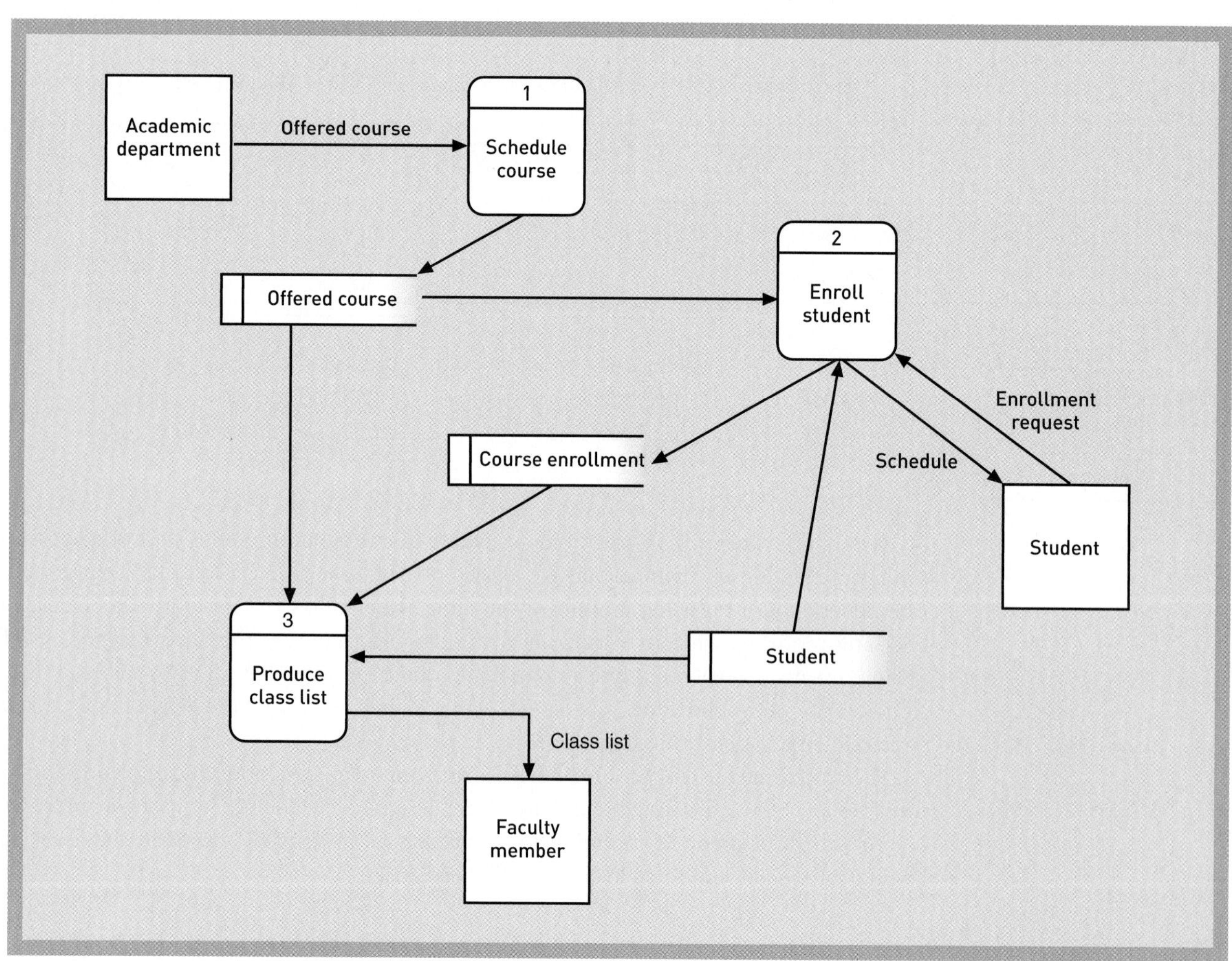

Figure 2-15
A data flow diagram (DFD) created using the structured analysis technique

The most recent variation of structured analysis defines systems processing requirements by identifying all of the events that will cause the system to react in some way. For example, in an order-entry system, if a customer orders an item, the system must process a new order (a major system activity). Each event leads to a different system activity. The analyst takes each of these activities and creates a data flow diagram showing the processing details, including inputs and outputs.

entity-relationship diagram (ERD)
a structured analysis and information engineering model of the data needed by a system

A model of the needed data is also created based on the types of things about which the system needs to store information (data entities). For example, to process a new order, the system needs to know about the customer, the items wanted, and the details about the order. This model is called an **entity-relationship diagram (ERD)**. The data entities from the entity-relationship diagram correspond to the data storage shown on data flow diagrams. Figure 2-16 shows an example of an entity-relationship diagram. Figure 2-17 illustrates the sequence followed when developing a system using structured analysis, structured design, and structured programming.

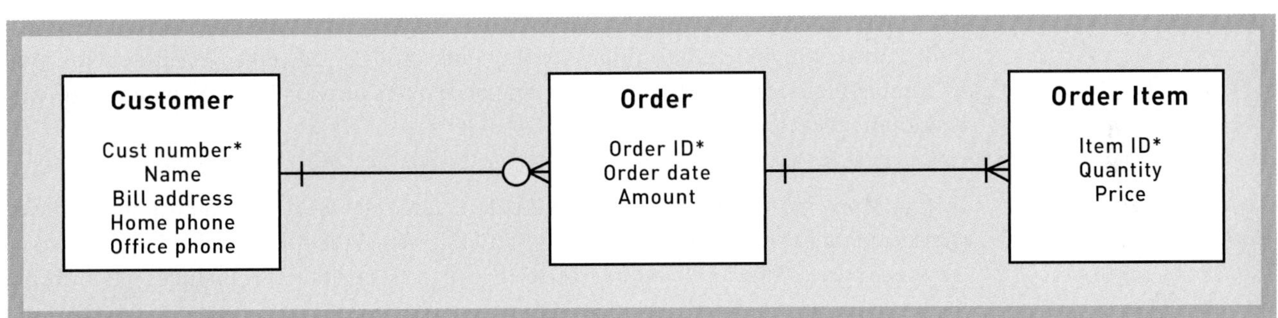

Figure 2-16
An entity-relationship diagram (ERD) created using the structured analysis technique

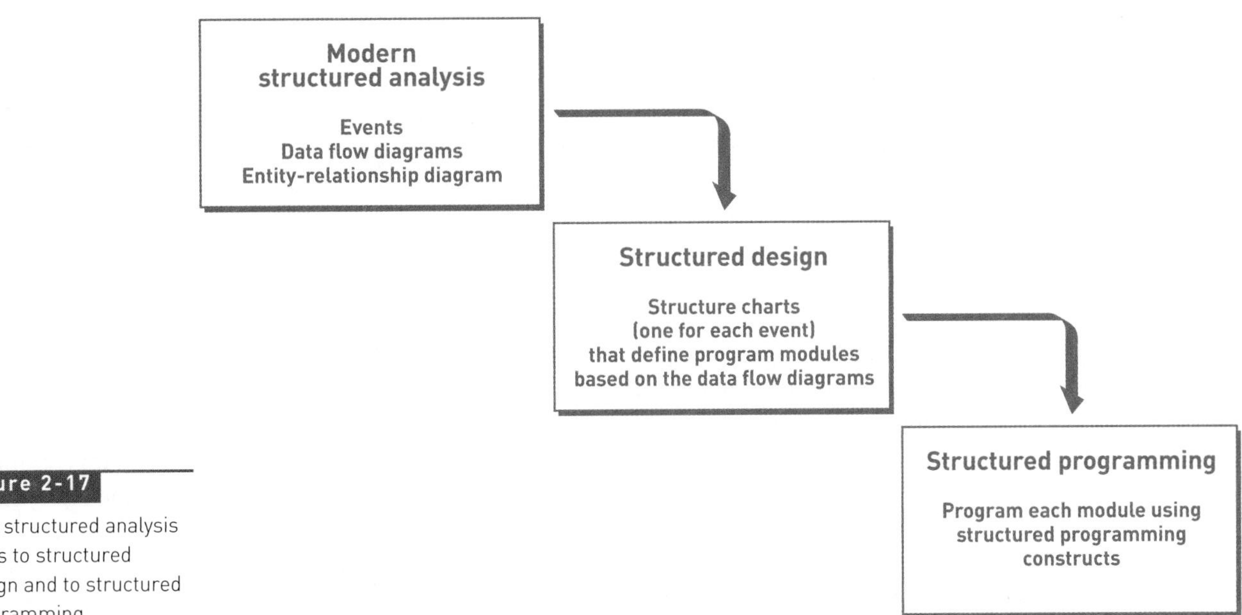

Figure 2-17
How structured analysis leads to structured design and to structured programming

Weaknesses of the Structured Approach Because the structured approach to system development evolved over time, many variations can be found in practice. Some people are still following the original versions of structured analysis and structured design they learned years ago, ignoring many improvements. Others picked up bits and pieces of the techniques on the job and never formally studied the details.

Many people have considered the structured approach to be weak because the techniques address only some, but not all, of the activities of analysis and design. Critics desired a more comprehensive and rigorous set of techniques to make system development more like an engineering discipline and less like an art. In addition, many people thought the transition from the data flow diagram (in structured analysis) to the structure chart (in structured design) did not work well in practice. Others thought that data modeling and the entity-relationship diagram were much more important than modeling processes with the data flow diagram. The structured approach, despite its inclusion of data modeling and database design, still made processes rather than data the central focus of the system.

Finally, many people thought that to ensure that systems are comprehensive and coordinated, the development of a system should begin only after the organization completed an overall strategic system planning effort. Therefore, they wanted the approach to development to include a strategic system planning technique, both to determine which systems should be built and to provide some initial requirements models that ensured all systems would be compatible. Because of these goals, some developers turned to a refinement of structured development: information engineering.

Information Engineering

information engineering

a traditional system development methodology thought to be more rigorous and complete than the structured approach, because of its focus on strategic planning, data modeling, and automated tools

Information engineering is a refinement to structured development that begins with overall strategic planning to define all of the information systems that the organization needs to conduct its business (the application architecture plan). The plan also includes a definition of the business functions and activities that the systems need to support, the data entities about which the systems need to store information, and the technological infrastructure that the organization plans to use to support the information systems.

Each new system project begins by using the defined activities and data entities created during strategic systems planning. Then the activities and data are refined as the project progresses. At each step, the project team creates models of the processes, the data, and the ways they are integrated.

The type of data needed to conduct the business changes very little over time, but the processes followed to collect data change frequently. Therefore, the information engineering approach focuses much more on data than the structured approach. Just as the structured approach includes data requirements, information engineering includes processes, too. The processing model of information engineering—the process dependency diagram—is similar to a data flow diagram, but it focuses more on which processes are dependent on other processes and less on data inputs and outputs. Events trigger the processes, as with modern structured analysis.

A final major difference with information engineering is the more complete life cycle support it provides through the use of an integrated tool. The tool helps automate as much of the work as possible. It also forces the analyst to follow the information engineering approach faithfully, sometimes at the expense of flexibility.

Information engineering is mainly credited to James Martin, who wrote several books on information engineering and developed tools to support it. By the late 1980s, information engineering was very popular for large, mainframe systems. But because they lacked flexibility, the tools that supported information engineering were less useful with smaller desktop applications and client/server applications. By the 1990s, fewer companies were using information engineering exclusively, although many of the concepts and techniques continue to be used, particularly the approach to planning and the emphasis on data modeling.

The information engineering approach refines many of the concepts of the structured approach into a rigorous and comprehensive methodology. Both approaches define information systems requirements, design information systems, and construct information systems by looking at processes, data, and the interaction of the two. This text merges key concepts from these two approaches into one, which we will refer to hereafter as *the traditional approach*. The traditional approach, in one version or another, is still widely used for information system development, although many information systems projects are now using object-oriented technology—which requires a completely different approach.

THE OBJECT-ORIENTED APPROACH

object-oriented approach

an approach to system development that views an information system as a collection of interacting objects that work together to accomplish tasks

object

a thing in the computer system that can respond to messages

An entirely different approach to information systems, the **object-oriented approach**, views an information system as a collection of interacting objects that work together to accomplish tasks (see Figure 2-18). Conceptually, there are no processes or programs; there are no data entities or files. The system consists of objects. An **object** is a thing in the computer system that is capable of responding to messages. This radically different view of a computer system requires a different approach to systems analysis, systems design, and programming.

The object-oriented approach began with the development of the Simula programming language in Norway in the 1960s. Simula was used to create computer simulations involving "objects" such as ships, buoys, and tides in fjords. It is very difficult to write procedural programs that simulate ship movement, but a new way of programming simplified the problem. In the 1970s, the Smalltalk language was developed to solve the problem of creating graphical user interfaces (GUIs) that involved "objects" such as pull-down menus, buttons, check boxes, and dialog boxes. More recent object-oriented languages include C++, Java, and C#. These languages focus on writing definitions of the types of objects needed in a system, and as a result, all parts of a system can be thought of as objects, not just the graphical user interface.

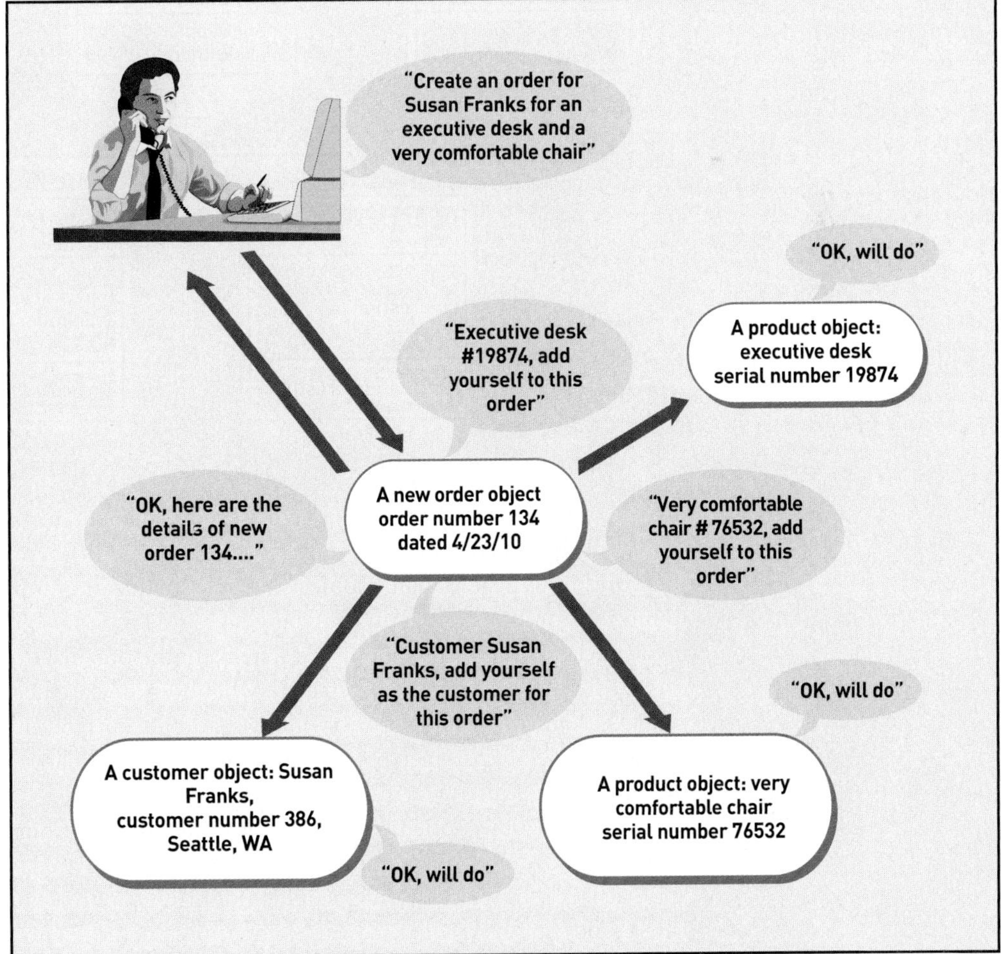

Figure 2-18

The object-oriented approach to systems (read clockwise starting with user)

object-oriented analysis (OOA)

defining all of the types of objects that do the work in the system and showing what use cases are required to complete tasks

object-oriented design (OOD)

defining all of the types of objects necessary to communicate with people and devices in the system, showing how objects interact to complete tasks, and refining the definition of each type of object so it can be implemented with a specific language or environment

object-oriented programming (OOP)

writing statements in a programming language to define what each type of object does, including the messages that the objects send to each other

class diagram

a graphical model used in the object-oriented approach to show classes of objects in the system

Because the object-oriented approach views information systems as collections of interacting objects, **object-oriented analysis (OOA)** defines all of the types of objects that do the work in the system and shows what user interactions, called *use cases*, are required to complete tasks. **Object-oriented design (OOD)** defines all of the additional types of objects necessary to communicate with people and devices in the system, shows how the objects interact to complete tasks, and refines the definition of each type of object so it can be implemented with a specific language or environment. **Object-oriented programming (OOP)** consists of writing statements in a programming language to define what each type of object does.

An object is a type of thing—a customer or an employee, as well as a button or a menu. Identifying types of objects means classifying things. Some things, such as customers, exist both outside the system (the real customer) and separately inside the system (a computer representation of a customer). A classification or "class" represents a collection of similar objects; therefore, object-oriented development uses a **class diagram** to show all of the classes of objects in the system (see Figure 2-19). For every class, there may be more specialized subclasses. For example, a savings account and a checking account are two special types of accounts (two subclasses of the class Account). Similarly, a pull-down menu and a pop-up menu are two special types of menus. Subclasses exhibit or "inherit" characteristics of the class above them.

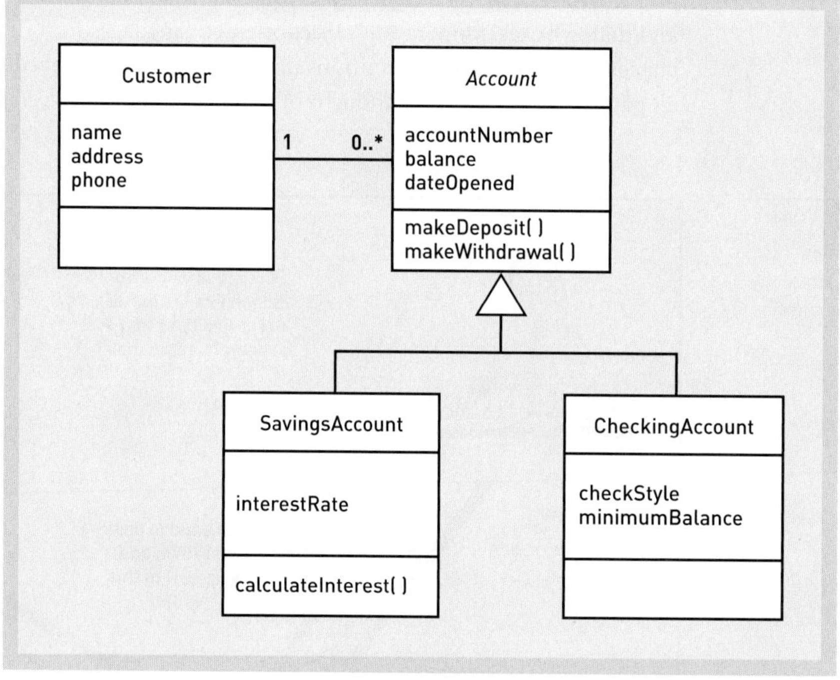

Figure 2-19

A class diagram created during object-oriented development

The object-oriented approach yields several key benefits, among them naturalness and reuse. The approach is natural—or intuitive—for people, because they tend to think about the world in terms of tangible objects. It is less natural to think about complex procedures found in procedural programming languages. Also, because the object-oriented approach involves classes of objects, and many systems in the organization use the same objects, these classes can be used over and over again whenever they are needed. For example, almost all systems use menus, dialog boxes, windows, and buttons, but many systems within the same company also use customer, product, and invoice classes that can be reused. There is less need to "reinvent the wheel" to create an object.

Clearly, the object-oriented approach is quite different from the traditional approach. But in other ways, quite a few traditional concepts are simply repackaged in the object-oriented approach. For this reason, some people find the OO approach difficult to understand at first. Parts 2 and 3 of this book discuss the similarities and differences in detail to help clarify each approach's strengths.

Many systems being developed today combine both traditional and object-oriented technology. Some integrated development environments (IDEs) also combine traditional and object-oriented technology in the same tool—for example, object-oriented programming is used for the user interface, and procedural programming for the rest. Many system projects are also exclusively traditional in analysis and design, and others are exclusively object-oriented, even within the same information systems department. These are some of the reasons that it is important to cover both traditional approaches and newer object-oriented approaches in this text. Everyone should know the basic concepts of each, but your coursework might emphasize one approach over the other.

CURRENT TRENDS IN DEVELOPMENT

One thing that never changes in the information systems field is that things are always changing. New tools and techniques are always appearing—sometimes with much publicity and anticipation—and system developers are always looking for new and better ways to work. The techniques and life cycles discussed previously are examples of ongoing changes to system development methodologies. A few important current trends in system development are discussed in this section. Any one of these trends could become common and even dominate system development in the future. Or, as discussed previously, system developers might adapt key concepts or techniques from each of these trends and use them when appropriate.

THE UNIFIED PROCESS (UP)

Unified Process (UP)
an object-oriented system development methodology offered by IBM's Rational Software

You learned that some companies obtain complete system development methodologies from consulting firms, either by purchasing rights to the methodology or by contracting for extensive training services from the consulting firm to learn the methodology. The **Unified Process (UP)** is an object-oriented system development methodology offered by IBM's Rational Software, originated by the three proponents of the Unified Modeling Language (UML): Grady Booch, James Rumbaugh, and Ivar Jacobson. The UP is their attempt to define a complete methodology that, in addition to providing several unique features, uses UML for system models. In the UP, the term development *process* is synonymous with development *methodology*. The UP is an example of an SDLC that is in the middle of the predictive versus adaptive scale.

Although you will learn much about UML because it is a standard modeling notation for the object-oriented (OO) approach, the UP is *not* a standard OO development methodology. UML models described in this text can be used in a variety of ways with any OO development methodology, but because of the stature of Booch, Rumbaugh, and Jacobson, the UP is gaining a lot of attention. Certainly the UP includes many useful and innovative techniques. Booch, Rumbaugh, and Jacobson have written several books about the UP and have endorsed other books about it written by colleagues, so it is possible to learn and to use the UP without purchasing services from Rational.

The UP is designed to reinforce six "best practices" of system development that are common to many system development methodologies:

- Develop iteratively.
- Define and manage system requirements.
- Use component architectures.
- Create visual models.
- Verify quality.
- Control changes.

The UP defines four life cycle phases: inception, elaboration, construction, and transition. The UP life cycle is shown in Figure 2-20. The inception phase defines the scope of the project by specifying *use cases* as with any development approach. You will learn how to identify use cases and create use case diagrams in Chapters 5 and 7 of this text. The project team also completes a feasibility study to determine whether resources should be invested in the project.

Figure 2-20

UP life cycle with phases, iterations, and disciplines

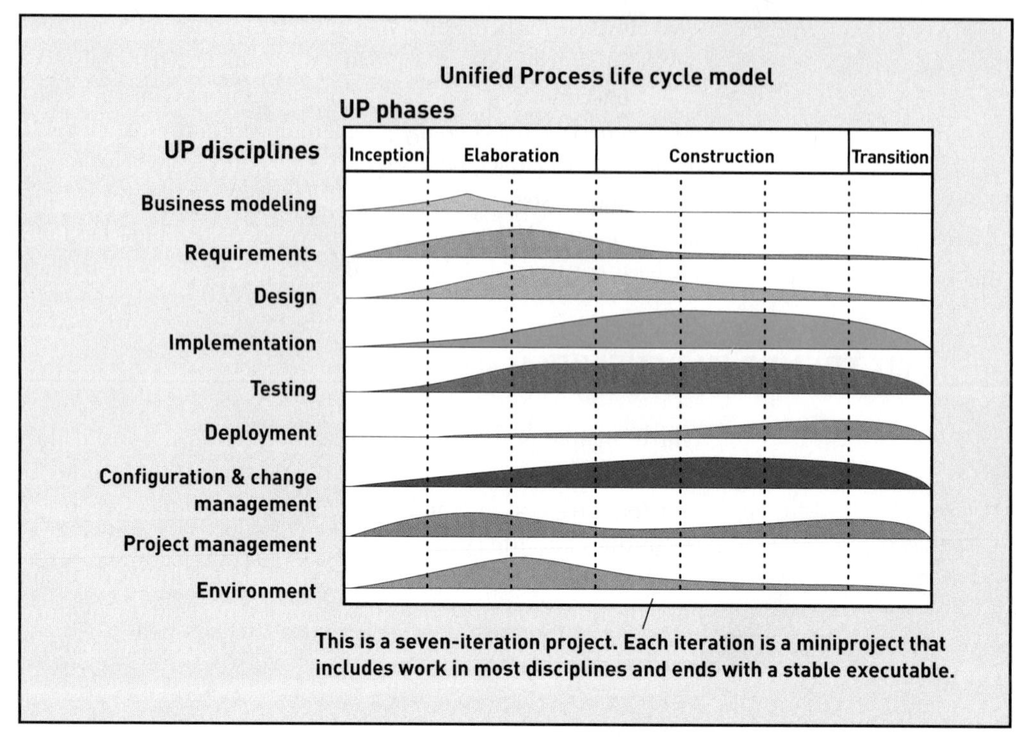

The elaboration phase focuses on several iterations that take part of the system and define the requirements, design the solution, and implement the solution. The team defines the requirements and the design by creating use case diagrams, class diagrams, sequence diagrams, and other UML diagrams. Final cost and benefit estimates are also completed by the end of the elaboration phase.

During the construction phase, you continue to build the system using additional iterations that also include design, implementation, and testing, possibly creating multiple releases of the system. During the transition phase, you turn the system over to the end users and focus on end-user training, installation, and initial support.

The four UP phases are different from the traditional SDLC because they do not define generic analysis, design, and implementation phases. Instead, they define the project sequentially by indicating the emphasis of the project team at any point in time. To make iterative development manageable, the UP defines disciplines within each phase. They include business modeling, requirements modeling, design, implementation, testing, deployment, configuration and change management, and project management. Each iteration involves activities from all disciplines. The UP also defines many roles played by developers and many models created during the project. Typical roles include designer, use case specifier, systems analyst, implementer, and architect.

As with any methodology, the UP includes very detailed information about what to do and when to do it for every activity involved in system development. The techniques and models presented in this book are consistent with many of the techniques and models included in the UP, but this book does not focus on the UP exclusively. The UP is described in more detail in Chapter 17.

EXTREME PROGRAMMING (XP)

Extreme Programming (XP) is a system development approach recently popularized by Kent Beck. XP adapts techniques from many sources and adds some new ideas. It is sometimes referred to as a "lightweight" system development methodology, meaning it is kept simple and focused on making the development process more efficient for the developer. It is an example of a highly adaptive approach to the SDLC.

The developers begin planning the system project by having the users describe *user stories*, which are similar to use cases. User stories are descriptions of the support the users need from the system—in other words, the required system functionality. The developers document these stories quickly with informal descriptive models. Along with providing the user stories, users describe a set of acceptance tests that will demonstrate that the system provides the required functionality once it is completed.

The developers then plan a series of *releases* for the project, with each release including a working part of the final system, as with incremental development. The project proceeds with work on the first release, which usually takes several iterations to complete. When the first release is completed, the second release is started.

In many ways, XP is much like other iterative and incremental approaches. But XP contains some additional features that make it popular. It requires continuous testing, continuous integration, and heavy user involvement, for example. It also requires that all programming be done by teams, with two programmers working together at one workstation when writing and testing code. This and other features emphasize open and effective communication among team members. A final feature is the firm belief that developers should work no more than 40 hours per week, to prevent burnout but also to demonstrate that system projects can be completed on schedule without overworking the staff if the XP techniques and tools are used for the project. XP is described in more detail in Chapter 17.

SCRUM

Scrum is another new adaptive development methodology. The term *Scrum* refers to rugby's system for getting an out-of-play ball back into play. The name stuck due to many similarities between the sport and the system development approach: both are quick, adaptive, and self-organizing. The basic idea behind Scrum is to respond to a current situation as rapidly and positively as possible.

The Scrum philosophy is responsive to a highly changing, dynamic environment in which users may not know exactly what is needed and may also change priorities frequently. In this type of environment, changes are so numerous that projects can bog down and never reach completion. Scrum excels in these situations because it focuses primarily on the development team and their work. It emphasizes individuals more than processes and describes how teams of developers can work together to build software in a series of short miniprojects. Key to this philosophy is the complete control a team exerts over its own organization and its work processes. Software is developed incrementally, and controls are imposed by focusing on the things that can be accomplished. Scrum is described in more detail in Chapter 17.

TOOLS TO SUPPORT SYSTEM DEVELOPMENT

No matter which methodology you use, it is important to use automated tools to improve the speed and quality of system development work whenever possible. One type of tool discussed earlier is a visual modeling tool. These tools are specifically designed to help systems analysts complete system development tasks. Analysts use a visual modeling tool to create models of the system, many of them graphical models. But a visual modeling tool is much more than a drawing tool.

BEST PRACTICE

Use automated tools whenever possible, but remember that sketches on napkins or envelopes are often enough for a small team on a small project. Don't let the tool create more problems than it solves.

repository

a database that stores information about the system in a visual modeling tool, including models, descriptions, and references that link the various models together

Visual modeling tools contain a database of information about the project, called a **repository**. The repository stores information about the system, including models, descriptions, and references that link the various models together. The tool can check the models to make sure they are complete and follow the correct diagramming rules. The tool also can check one model against another to make sure they are consistent. If you consider how much time an analyst spends creating models, checking them, revising them, and then making sure they all fit together, it is apparent how much help a visual modeling tool can provide. Figure 2-21 shows tool capabilities surrounding the repository. If system information is stored in a repository, the development team can use the information in a variety of ways. Every time a team member adds information about the system, it is immediately available for everyone else.

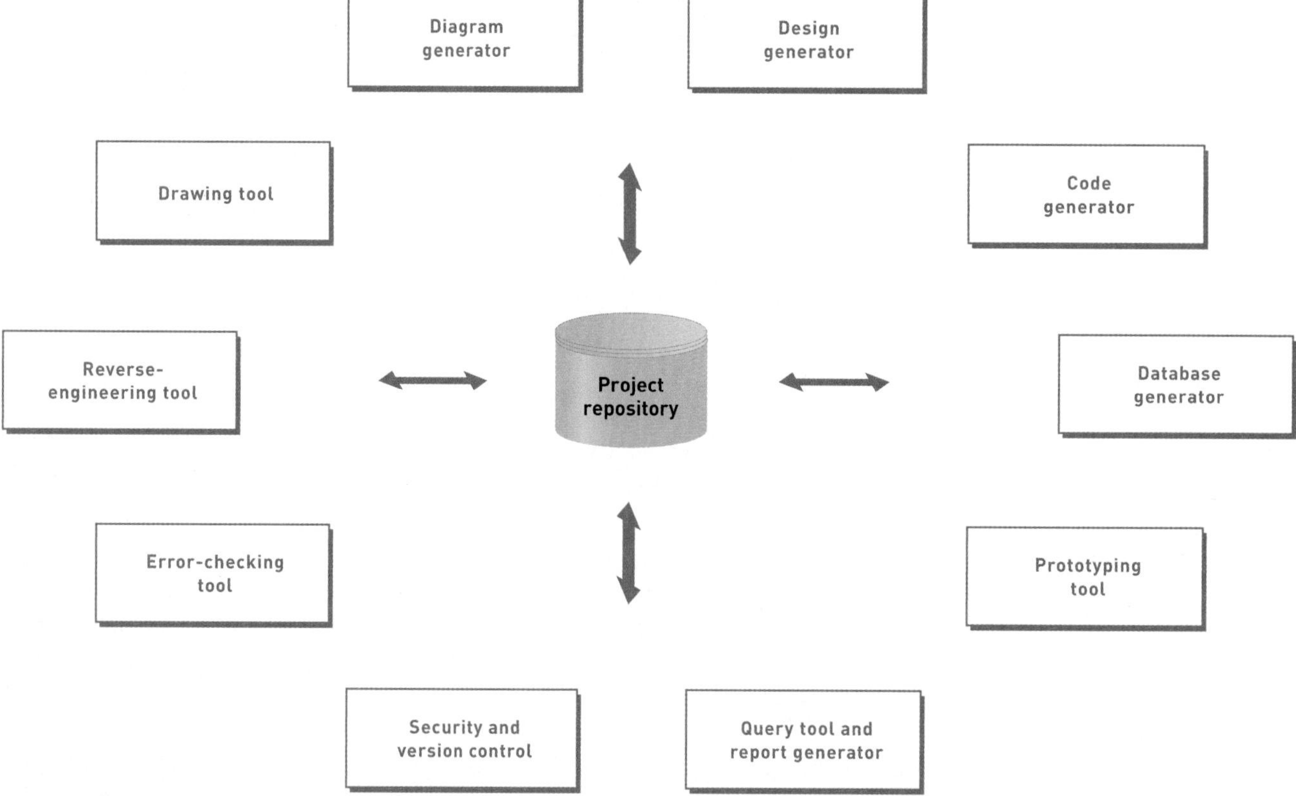

Figure 2-21

A visual modeling tool repository contains all information about the system

Some visual modeling tools are designed to be as flexible as possible, allowing analysts to use any development approach they desire. Other tools are designed for very specific methodologies.

Microsoft Visio is a drawing tool that analysts use to create just about any system model they might need. Visio comes with a collection of drawing templates that include symbols used in a variety of business and engineering applications. Software and system development templates provide symbols for flowcharts, data flow diagrams, entity-relationship diagrams, all of the UML diagrams, and others found throughout this text. The templates provide a limited repository for storing definitions and descriptions of diagram elements, but Visio does not provide a complete repository for a system development project. Many system developers prefer the flexibility that Visio offers for drawing any diagram needed, however.

Figure 2-22 shows Visio displaying several UML diagrams used with the OO approach—a class diagram, a use case diagram, a sequence diagram, and a package diagram. Symbols for these diagrams are selected from the templates listed at the left. Note also that the items shown in the diagrams are listed and defined at the left.

Figure 2-22

Visio for drawing a variety of diagrams and charts

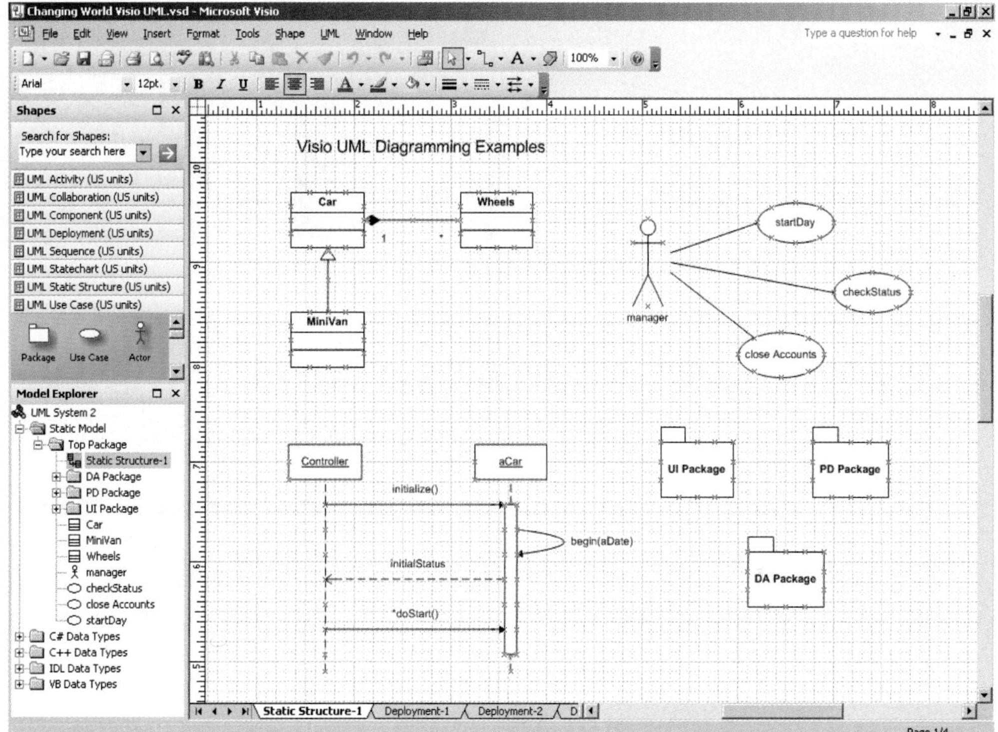

Figure 2-23 shows a flexible tool called Visible Analyst from Visible Systems Corporation (*www.visible.com*). This tool makes it easy to draw typical traditional models, such as data flow diagrams and entity-relationship diagrams, and it also supports object-oriented UML models. The wide variety of diagrams available is shown on the screen by the boxes forming a view of the project. Visible Analyst includes a repository for defining system components and provides error-checking and consistency-checking support. Figure 2-24 shows IBM's Rational Software Development platform. This tool is designed to support the Unified Process methodology with UML diagrams and code generation.

Figure 2-23

Visible Analyst showing the variety of diagrams available to system developers

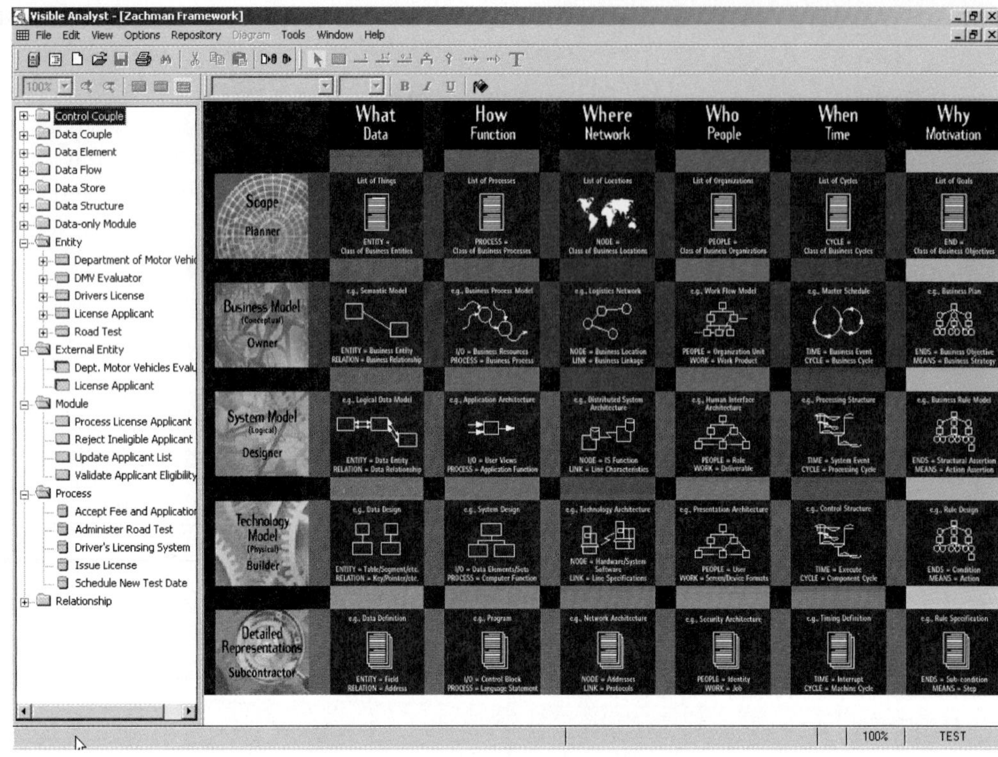

Figure 2-24

IBM's Rational Software Development Platform showing the design of a use case using a sequence diagram

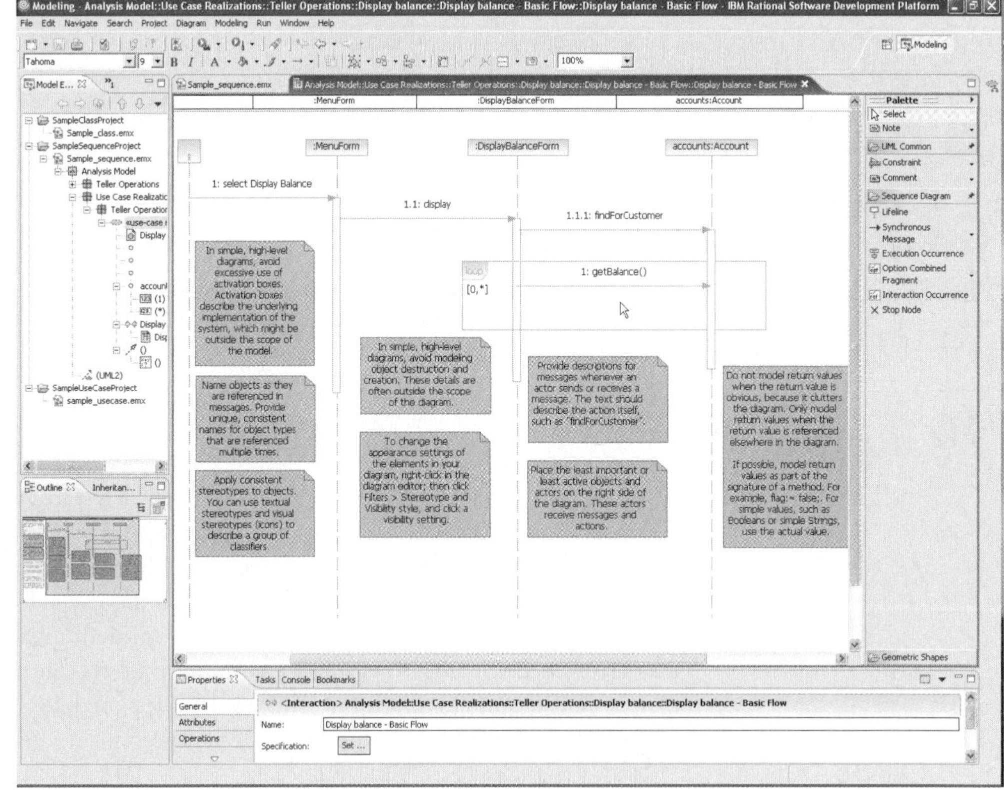

SUMMARY

System development projects are organized around the systems development life cycle (SDLC), and phases of the SDLC include activities that must be completed for any system development project. The traditional SDLC phases are project planning, analysis, design, implementation, and support. Some SDLCs are based on a more predictive approach to the project, and other SDLCs are based on a more adaptive approach. System developers learn the SDLC phases and activities sequentially, based on the waterfall model; in practice, however, the phases overlap and projects contain many iterations of analysis, design, and implementation activities.

You can develop an information system in lots of ways. All development projects use the SDLC to manage the project, plus models, techniques, and tools that make up a system development methodology. A system development methodology provides guidelines to follow for completing every activity in the SDLC, and many different methodologies are in use. Most methodologies are based on one of two approaches to information systems development: the traditional approach or the object-oriented approach.

Some current trends in system development include the Unified Process (UP), Extreme Programming (XP), and Scrum. These methodologies provide innovative insights into best practices in system development and are becoming influential.

Visual modeling tools are special tools designed to help analysts complete development tasks, including modeling and generating program statements directly from the models.

KEY TERMS

adaptive approach, p. 39
analysis activities, p. 45
application, p. 47
class diagram, p. 60
data flow diagram (DFD), p. 56
design activities, p. 46
entity-relationship diagram (ERD), p. 57
help desk, p. 49
implementation activities, p. 47
incremental development, p. 44
information engineering, p. 58
integrated development environment (IDE), p.51
iteration, p. 43
model, p. 50
object, p. 59
object-oriented analysis (OOA), p. 60
object-oriented approach, p. 59
object-oriented design (OOD), p. 60
object-oriented programming (OOP), p. 60
phases, p. 40
predictive approach, p. 39
problem domain, p. 46
project, p. 38
project planning, p. 45
prototype, p. 42
repository, p. 64
spiral model, p. 42
structure chart, p. 55
structured analysis, p. 56
structured approach, p. 53
structured design, p. 55
structured program, p. 53
support activities, p. 48
system development methodology, p. 49
systems development life cycle (SDLC), p. 38
technique, p. 51
tool, p. 51
top-down programming, p. 54
Unified Process (UP), p. 61
visual modeling tools, p. 51
waterfall model, p. 40

REVIEW QUESTIONS

1. What are the five phases of the traditional SDLC?
2. What characteristics of a project call for a predictive approach to the SDLC? What characteristics of a project call for an adaptive approach to the SDLC?
3. How is the SDLC based on the problem-solving approach described in Chapter 1?
4. What is the objective of each phase of the SDLC? Describe briefly.
5. How is iteration used across phases?
6. What is the difference between a model and a tool?
7. What is the difference between a technique and a methodology?
8. Which of the two approaches to system development was the earliest?
9. Which of the two approaches to system development is the most recent?
10. Which of the traditional approaches focuses on overall strategic systems planning?
11. Which of the traditional approaches is a more complete methodology?
12. What are the three constructs used in structured programming?
13. What graphical model is used with the structured design technique?
14. What graphical model is used with the modern structured analysis technique?
15. What model is the central focus of the information engineering approach?
16. Explain what is meant by a waterfall life cycle model.
17. What concept suggests repeating activities over and over until you achieve your objective?
18. What concept suggests completing part of the system and putting it into operation before continuing with the rest of the system?
19. What are some of the features of the Unified Process (UP)?
20. What are some of the features of Extreme Programming (XP)?
21. What are some of the features of Scrum?
22. What are visual modeling tools? Why are they used?

THINKING CRITICALLY

1. Write a one-page paper that distinguishes among the fundamental purposes of the analysis phase, the design phase, and the implementation phase.
2. Describe a system project that might have three subsystems. Discuss how three iterations might be used for the project.
3. Why might it make sense to teach analysis and design phases and activities sequentially, like a waterfall, even though in practice iterations are used in nearly all development projects?
4. List some of the models that architects create to show different aspects of a house they are designing. Explain why several models are needed.
5. What models might an automotive designer use to show different aspects of a car?
6. Sketch the layout of your room at home. Now write a description of the layout of your room. Are these both models of your room? Which is more accurate? More detailed? Easier to follow for someone unfamiliar with your room?
7. Describe a "technique" you use to help you complete the activity "Get to class on time." What are some "tools" you use with the technique?
8. Describe a "technique" you use to make sure you get assignments done on time. What are some "tools" you use with the technique?
9. What are some other techniques you use to help you complete activities in your life?
10. There are at least two approaches to system development, a variety of life cycles, and a long list of techniques and models that are used in some approaches but not in others. Consider why this is so. Discuss these possible reasons, indicating which are the most important: The field is so young; the technology changes so fast; different organizations have such different needs; there are so many different types of systems; and people with widely different backgrounds are developing systems.

EXPERIENTIAL EXERCISES

1. Go to the campus placement office and gather some information on companies that recruit information systems graduates on your campus. Try to find any information about the approach they use to develop systems. Is their SDLC described? Do any mention an IDE or a visual modeling tool? Visit the company Web sites and see whether you can find any more information.
2. Visit the Web sites for a few leading information systems consulting firms. Try to find information about the approach they use to develop systems. Are their SDLCs described? Do their sites mention any tools?

CASE STUDIES

A "COLLEGE EDUCATION COMPLETION" METHODOLOGY

Like many readers of this book, you are probably a college student working on a degree. Think of completing college as a project—a big project, lasting many years and costing more than you might want to admit. Some students do a better job managing the college completion project than others. Many fail entirely (certainly not you), and most students probably complete college late and way over budget (again, certainly not you).

As with any other project, to be successful, you should follow some sort of "college education completion" methodology. That is, you should follow a comprehensive set of guidelines for completing activities and tasks from the beginning of planning for college through to the successful completion.

1. What might be the phases of your personal college education completion life cycle?
2. What are some of the activities of each phase?
3. What are some techniques you use to help complete the activities? What models might you create during the process of completing college? Differentiate models you create that get you through college from those that help you plan and control the process of completing college.
4. What are some of the tools you use to help you complete the models?

FACTORY SYSTEM DEVELOPMENT PROJECT

Sally Jones is assigned to manage a new system development project that will automate some of the work being done in her company's factory. It is fairly clear what is needed: to automate the tracking of the work in progress and the finished goods inventory. What is less clear is the impact of any automated system on the factory workers. Sally has several concerns: How might a new system affect the workers? Will they need a lot of training? Will working with a new system slow down their work or interfere with the way they now work? How receptive will the workers be to the changes the new system will surely bring to the shop floor?

At the same time, Sally recognizes that the factory workers themselves might have some good ideas about what will work and what won't, especially concerning (1) which technology is more likely to survive in the factory environment and (2) what sort of user interface will work best for the workers. Sally doesn't know much about factory operations, although she does understand inventory accounting.

1. Is the proposed system an accounting system? A factory operations system? Or both?
2. Which life cycle variations might be appropriate for Sally to consider using?
3. Which activities of analysis and of design discussed in this chapter should involve factory workers as well as factory management?

RETHINKING ROCKY MOUNTAIN OUTFITTERS

Barbara Halifax wrote her boss that she was still considering many potential approaches to the customer support system development project. She is still completing the project planning phase, so not much time has passed at this point. Consider the training required for the development staff if RMO decides to use an object-oriented approach for the project. How extensive would the training needs be for the RMO staff? What type of training would be required? Is it just about new programming languages, or is it broader than that? How far can the project progress before the decision is made?

Barbara mentions that either approach can be used and that, even though some Web development is involved, the team does not have to use an OO approach. Do you think she is correct? Why or why not? Do some types of projects *require* an OO approach?

Barbara also mentions that she plans to use some iteration and to involve users extensively throughout the project. What life cycle variations are under consideration? What else might she do to speed up the development process? What else might she consider adapting from the United Process, from Extreme Programming, or from Scrum?

FOCUSING ON RELIABLE PHARMACEUTICAL SERVICE

In Chapter 1, you generated some ideas related to Reliable Pharmaceutical Service's five-year information systems plan. Management has placed a high priority on developing a Web-based application to connect client facilities with Reliable. Before the Web component can be implemented, though, Reliable must automate more of the basic information it handles about patients, health-care facilities, and prescriptions.

Next, Reliable must develop an initial informational Web site, which will ultimately evolve into an extranet through which Reliable will share information and link its processes closely with its clients and suppliers. One significant requirement of the extranet is compliance with the Health Insurance Portability and Accountability Act of 1996, better known as HIPAA. HIPAA requires health-care providers and their contractors to protect patient data from unauthorized disclosure. Ensuring compliance with HIPAA will require careful attention to extranet security.

After basic processes are automated and the extranet Web site is in place, the system will enable clients to add patient information and place orders through the Web. The system should streamline processes for both Reliable and its clients. It should also provide useful query and patient management capabilities to distinguish Reliable's services from those of its competitors, possibly including drug interaction and overdose warnings, automated validation of prescriptions with insurance reimbursement policies, and drug and patient cost data and summaries.

1. One approach to system development that Reliable might take is to start one large project that uses a waterfall model to the SDLC to thoroughly plan the project, analyze all requirements in detail, design every component, and then implement the entire system, with all phases completed sequentially. What are some of the risks of taking this approach? What planning and management difficulties would this approach entail?
2. Another approach to system development might be to start with the first required component and get it working. Later, other projects could be undertaken to work on the other identified capabilities. What are some of the risks of taking this approach? What planning and management difficulties would this approach entail?
3. A third approach to system development might be to define one large project that will use an iterative approach to the SDLC. Briefly describe what you would include in each iteration. Describe how incremental development might apply to this project. How would an iterative approach decrease project risks compared with the first approach? How might it decrease risks compared with the second approach? What are some risks the iterative approach might add to the project?

FURTHER RESOURCES

Some classic and more recent texts include the following:

Scott W. Ambler, *Agile Modeling: Effective Practices for Extreme Programming and the Unified Process*. Wiley Computer Publishing, 2002.

D. E. Avison and G. Fitzgerald, *Information Systems Development: Methodologies, Techniques and Tools* (3rd ed.). Maidenhead, McGraw-Hill, 2003.

Kent Beck, *Extreme Programming Explained: Embrace Change*. Addison-Wesley Publishing Company, 2000.

Tom DeMarco, *Structured Analysis and System Specification*. Prentice Hall, 1978.

C. Gane and T. Sarson, *Structured Systems Analysis: Tools and Techniques*. Prentice Hall, 1979.

Ivar Jacobson et al., *Object-Oriented Software Engineering: A Use Case Driven Approach*. Addison-Wesley, 1992.

Ivar Jacobson, Grady Booch, and James Rumbaugh, *The Rational Unified Process*. Addison-Wesley, 1999.

James Martin, *Information Engineering: A Trilogy* (books 1, 2, and 3). Prentice Hall, 1990.

Steve McConnell, *Rapid Development*. Microsoft Press, 1996.

Meilir Page-Jones, *The Practical Guide to Structured System Design* (2nd ed.). Prentice Hall, 1988.

John Satzinger, Robert Jackson, and Stephen Burd, *Object-Oriented Analysis and Design with the Unified Process*. Course Technology, 2005.

John Satzinger and Tore Orvik, *The Object-Oriented Approach: Concepts, System Development, and Modeling with UML* (2nd ed.). Course Technology, 2001.

Ed Yourdon, *Modern Structured Analysis*. Prentice Hall, 1989.

CHAPTER 3

THE ANALYST AS A PROJECT MANAGER

LEARNING OBJECTIVES

After reading this chapter, you should be able to:

- Explain the elements of project management and the responsibilities of a project manager
- Explain project initiation and the project planning activities of the SDLC
- Describe how the scope of the new system is determined
- Develop a project schedule using Gantt charts
- Develop a cost/benefit analysis and assess the feasibility of a proposed project
- Discuss how to staff and launch a project

CHAPTER OUTLINE

Project Management

Project Initiation and Project Planning

Defining the Problem

Producing the Project Schedule

Identifying Project Risks and Confirming Project Feasibility

Staffing and Launching the Project

Recap of Project Planning for RMO

BESTWAY FUEL SYSTEMS: MOVING TO AN ADAPTIVE SDLC

"I feel pretty good about the final decision for the development methodology in this new project. At first I was concerned that we were biting off more than we could chew. But I think the approach we have chosen will work okay, and I think I will be able to adjust to the new techniques. It will be exciting to try out some of these newer "Agile" techniques I've been reading about." Gary Johnson, project manager at Bestway Fuel Systems Inc., was talking to his boss, Sam Brown, Director of Systems Development.

"Yes, I agree. Although our current development methodology has worked pretty well for us over the years, it's time to move into the twenty-first century. I'm glad you're willing to give it a try. This new project, Tracking Employee Profit Sharing (TEPS), is just about the right size to use as a pilot project."

"Although this project does not have all the typical characteristics of an adaptive approach, it is still going to be a good fit. It will be especially good to use as a learning project for us. Here is what I have in mind. I would like to use an iterative, adaptive approach for this project as the basic development method, while at the same time begin to use and to learn these newer Agile techniques. My only concern is that maybe it's too much to try to learn in one project." Although a little apprehensive, Gary's excitement was evident.

"Well, I will assign a couple of our senior developers to work with you on this. Then why don't you see if you can find a good training course for the three of you to attend. I think we can squeeze money out of the budget for you to travel offsite if necessary. I really want this to be successful, and so I don't want you to jump into this project without the necessary support. If this project is successful, then I will depend on you three to help train the rest of our group. I really think it's important that we upgrade our development methodology, so this is an important first step for us." Sam's support was evident. It also appeared that he had already talked to senior executives in the company, and they also were behind this new direction. Strategic directions for Bestway required several new mission-critical systems to be developed over the next five years. So this small (six-month) project was an important pilot for the systems group.

Gary spent a couple of hours that afternoon researching what kind of training he thought would be best for himself and his two teammates. None of the three had much experience with Agile methods. The other two developers were senior employees. Even though he was the manager for the project, the other two developers also functioned as project leaders within the group. He decided it would be good for all of them to have the same training, including training both in Agile concepts and project management techniques for iterative projects.

By the end of the day, he had formulated a rough training plan. First, he found what appeared to be some excellent books. They would be good to have in the library for references and for self-study. He found a couple of online courses to help review project management concepts. He decided that he and his two teammates should attend an Agile development course. He felt good about what he had found out. His next task, first thing in the morning, would be to put together a schedule for the next month, one that included training mixed with planning for the new project. He went home pleased with the progress of the day.

OVERVIEW

Chapter 1 described the business environment, with its insatiable need for information systems in today's competitive and rapidly paced global economy. That chapter also discussed the job duties of systems analysts, including their role in information technology (IT) and IT strategic planning. You also learned about the various types of information systems the analyst might develop and support. Chapter 2 introduced the systems development life cycle (SDLC); the methodologies, models, tools, and techniques used to develop systems; and several approaches to system development that are used generally.

This chapter begins to narrow the focus to teach the specifics of how an information system is developed within a company. The Rocky Mountain Outfitters (RMO) customer support system project is used as the specific example, and this chapter discusses the project planning activities of the SDLC for RMO. Because of their importance in information systems development and project planning, the principles of project management are introduced in this chapter. Project management encompasses the skills and techniques that are necessary to succeed in planning and managing the development of a new system. As a knowledgeable worker and problem solver, you will need both technical and management skills to be a contributing member of a system development team. This chapter provides you with the fundamentals of project management, and later chapters elaborate on key management principles associated with the various phases of the project.

The second section of the chapter discusses how information systems projects are initiated. Projects are started for two primary reasons. First, a project to develop a new information system may be started because the new system is part of an overall strategic plan, as discussed for RMO in Chapter 1. The second reason that new information system projects are started is to respond to an immediate business need. Such a need usually arises from some unforeseen information or processing problem within the company.

An important objective of the chapter is also to describe the major project planning activities of the SDLC, which were listed in Chapter 2. The techniques that are taught in the chapter can be used either for predictive or adaptive SDLC approaches. Examples of both are provided within the chapter. The planning process for a new project entails several important steps, such as defining the scope of the project, comparing the estimated costs and anticipated benefits of the new system, and developing a project schedule. The final sections explain these specific steps and the skills associated with the steps. Because project management, analyzing costs and benefits, and project scheduling are all very large topics, additional information about each of these topics is included in appendices on the book's Web site. You are encouraged to review the appendices for more information on these topics.

PROJECT MANAGEMENT

Many of you may have experience building a Web page with HTML or writing a computer program for yourself or a friend. In those cases, where it was just you working, you were not too concerned about how to organize your work or how to manage the project. However, as soon as two or more developers are working together, the work must be partitioned and organized with specific assignments for each developer. This is true whether the project uses a predictive approach or an adaptive approach. Failing to organize usually causes wasted time and effort, confusion, and it even may cause the project to end in failure.

Even though every project team designates one person as the project manager who has primary responsibility for the functioning of the team, all experienced members contribute to the management of the team. The project manager for the RMO customer support system project is Barbara Halifax, but she has one senior systems analyst helping her every step of the way. As the project proceeds, all team members are involved in aspects of managing the project.

The development of a new software system, the enhancement or upgrade of an existing system, and even the integration and deployment of a software package into an existing system are all accomplished during a development project. As we discussed in Chapter 2, a project is a planned undertaking with a beginning and an end that produces a predetermined result and is usually constrained by a schedule and resources. Information systems projects fit this definition. In addition, they are usually quite complex, with many people and tasks that must be organized and coordinated. Whatever its objective, each project is unique; no two are exactly alike. Different products are produced, different activities are required with varying schedules, and different resources are used. Their uniqueness makes information systems projects difficult to control—each involves new activities that have never been done exactly the same way before.

PROJECT SUCCESS FACTORS

How important is project management for the success of a system development project? In 1994, the Standish Group began studying the success and failure rates of system development projects. The surprising initial results indicated that almost 32 percent of all development projects were canceled before they were completed. In addition, more than half of computer system projects cost almost double the original budget. Less than half (about 42 percent) had the same scope and functionality as originally proposed. In fact, many systems were implemented with only a portion of the original requirements satisfied. Depending on company size, completely successful projects (on time, on budget, with full functionality) ranged from only 9 percent to about 16 percent. As of 2000, the percentage of successful system development projects was still a dismal 28 percent, with 72 percent canceled or completed late, over budget, or with limited functionality. Clearly, system development is a difficult activity requiring very careful planning, control, and execution.

It is interesting to look at the reasons that projects do not fulfill the desired objectives. Some primary reasons that projects fail, or are only partially successful, include the following:

- Incomplete or changing system requirements
- Limited user involvement
- Lack of executive support
- Lack of technical support
- Poor project planning (including inadequate risk assessment)
- Unclear objectives (including unreasonable expectations)
- Lack of required resources

Additional studies of successful projects help to highlight some reasons that projects succeed:

- Clear system requirement definitions
- Substantial user involvement
- Support from upper management
- Thorough and detailed project plans
- Realistic work schedules and milestones

The success factors are, in most cases, just the reverse of those for failures. Note that reasons such as "the technology is too complex" do not appear in the lists. This omission indicates that projects fail most frequently because project management has failed. Successful projects result from strong project management that ensures the preceding success characteristics are an integral part of the project.

The obvious question, then, is "How can we improve the project success rate?" Companies that have achieved greater success have attacked the problem from three different perspectives. First, they incorporate good principles of project management into their projects. They identify best practices in project management, and they train their project managers to use those practices. In this chapter and throughout the book, you will learn many principles of good project management. Second, they adopt a system development methodology. Current trends indicate that iterative, adaptive approaches often help to improve a project's success. Chapter 2 introduced you to the development methodologies that are commonly adopted. All of those methodologies are based on the concepts and techniques covered in this book. Third, successful companies pay particular attention to the factors that influence project success. The organization becomes focused on instituting characteristics of successful projects, and all team members and stakeholders work to incorporate best practices.

project management
organizing and directing other people to achieve a planned result within a predetermined schedule and budget

THE ROLE OF THE PROJECT MANAGER

Project management is organizing and directing other people to achieve a planned result within a predetermined schedule and budget. At the beginning of a project, a plan is developed that specifies the activities that must take place, the deliverables that must be produced, and the resources that are needed. So, project management can also be defined as the processes used to plan the project and then to monitor and control it.

One of the most exciting careers for IT-oriented people is in project management. As projects become more complex because of shorter time frames, distributed project teams (including offshore and cross-cultural teams), rapidly changing technology, and more sophisticated requirements, highly qualified project managers are sought after and well paid. Many universities are adding project management courses to their curricula to respond to the needs of industry.

Many career paths lead to project management. In some companies, the project coordination role is performed by recent college graduates. Other companies recognize the value of a person with strong organizational and people skills, who understands the technology but does not want a highly technical career. Those companies provide opportunities for employees to gain experience in management and business skills and to advance to project management through experience as a coordinator of smaller projects. Other companies take a "lead engineer" approach to project management, in which a person must thoroughly understand the technology to manage a project. Management at these companies believes that project management requires someone with strong development skills to understand the technical issues and to manage other developers.

The project manager defines and executes project management tasks. The success or failure of a given project is directly related to the skills and abilities of the project manager. The project success factors listed earlier—clear requirement definitions, substantial user involvement, upper management support, thorough planning, and realistic schedules and milestones—are the responsibility of the project manager, and he or she must ensure that sufficient attention is given to those details. In fact, a project manager must be an expert in two areas. First, she must be a good manager of people and resources, which is referred to as internal responsibilities. Second, she must have strong communication and public relations skills, or what may be called externally oriented talents.

From the internal team perspective, the project manager serves as the director and locus of control for the project team and all of their activities. The project manager establishes the team's structure so that work can be accomplished. The following list identifies a few of these internal responsibilities:

- Identify project tasks and build a work breakdown structure.
- Develop the project schedule.
- Recruit and train team members.
- Assign team members to tasks.
- Coordinate activities of team members and subteams.
- Assess project risks.
- Monitor and control project deliverables and milestones.
- Verify the quality of project deliverables.

From an external organizational perspective, the project manager is the focal point or main contact for the project. He or she must represent the team to the outside world and communicate team member needs. Some of the major external responsibilities include the following:

- Report the project's status and progress.
- Establish good working relationships with those who identify the needed system requirements (that is, the people who will use the system).
- Work directly with the client (the project's sponsor) and other stakeholders.
- Identify resource needs and obtain resources.

client

the person or group that funds the project

A project manager works with several groups of people. From the external perspective, the **client** will be paying for the development of the new system—in other words, the customer. So, when we speak of project approval and release of funds, we mean they come from the

oversight committee

clients and key managers who review and direct the project

user

the person or group of people who will use the new system

client. For in-house development, the client can be an executive committee or a vice president who is funding the project. For large, mission-critical projects, an **oversight committee** (sometimes called the steering committee) may be formed. This committee consists of clients and other key executives who have a vision of the organization's strategic direction and a strong interest in the project's success. Experience in directing large development projects is helpful but not a prerequisite for a committee member. The **users**, on the other hand, are the people who will actually use the new system. In some cases, the client and user are the same person. Often, however, they are not. The user typically provides information about the detailed functions and operations needed in the new system. The client also provides input on the business framework and strategy, which are important factors that influence the scope and design of a system. In addition, the client approves and oversees the project, along with its funding.

Communication with the client and oversight committee is an important part of the project manager's external responsibilities. Similarly, working with the team leader, team members, and any subcontractors is a normal part of a project manager's internal responsibilities. Some users are very active in the project and can be considered part of the project team. Other users have only part-time involvement. In any event, the project manager must ensure that all internal and external communication is flowing properly. Figure 3-1 depicts the various groups of people involved in a development project.

Obviously, the project manager does not always perform all the tasks involved with these responsibilities; other team members assist the manager. However, the primary responsibility for the project rests with the project manager.

Figure 3-1

Participants in a system development project

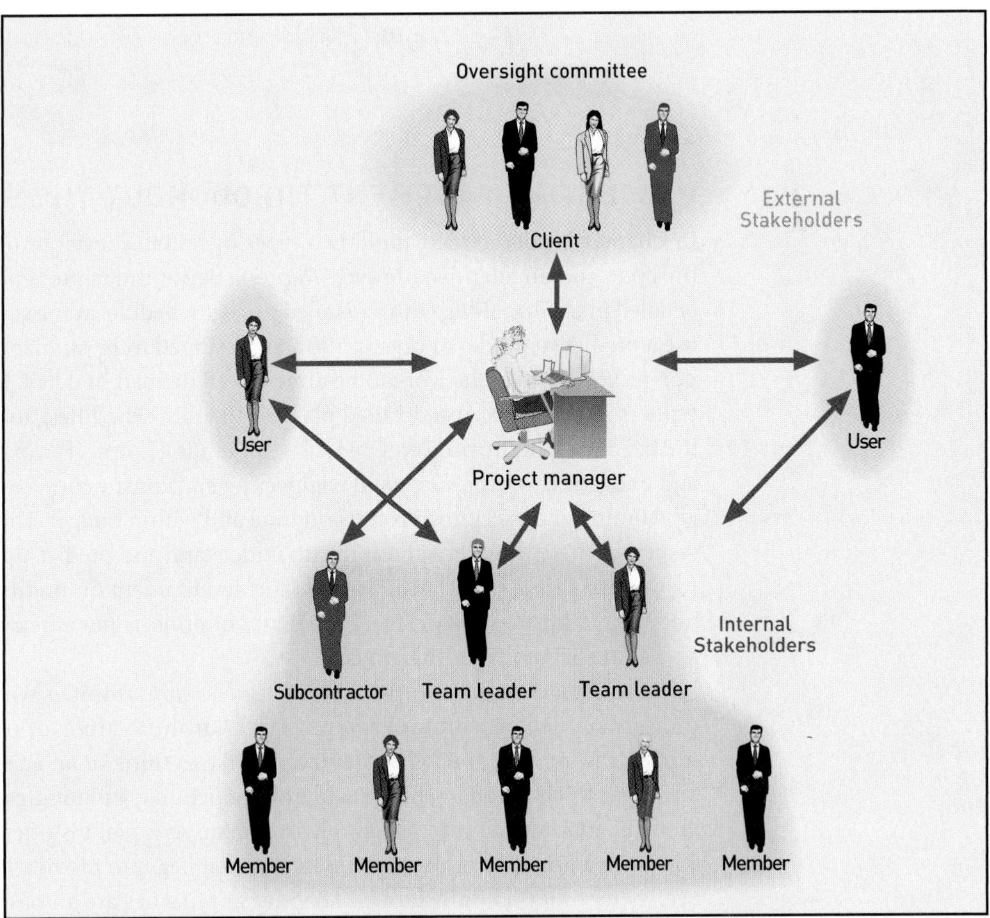

In looking at organizations around the world and the way that development projects are handled, we can see that the role of the project manager and careers in project management vary tremendously. Figure 3-2 lists some of the different positions project managers hold. In some companies, the project manager functions as a coordinator and does not have direct "line" (reporting) authority. At the other end of the spectrum, for big development projects, the project manager may be a very experienced developer with both management skills and a solid understanding of a full range of technical issues. In those situations, the role of the project manager is very much a "line" position with responsibility and authority for other staff members.

Figure 3-2

Various roles of project managers

Title	Power/authority	Organization structure	Description of duties
Project coordinator or project leader	Limited	Projects may be run within the departments, or projects may have a strong "lead developer" who controls the development of the end product.	Develops the plans. Coordinates activities. Keeps people informed of status and progress. Does not have "line" authority on the project deliverables.
Project manager, project officer, or team leader	Moderate	Projects are run within an IT department, but other business functions are independent.	May have both project management duties and some technical duties. Manages projects that are generally medium sized. May share project responsibility with clients.
Project manager or program manager	High to almost total	Project organization is a prime, high-profile part of the company. Company is organized around projects, or there is a large and powerful IT department.	Usually has extensive experience in technical issues as well as project management. Involved in both management decisions and technical issues. Frequently has support staff to do paperwork. Manages projects that can be big.

PROJECT MANAGEMENT THROUGHOUT THE SDLC

In Chapter 2, you learned about two types of systems development life cycles—a predictive life cycle and an adaptive life cycle. A predictive approach to a project requires much more detailed planning, along with a detailed project schedule, at the start of a project. Projects that use a predictive model of organization are assumed to be similar to engineering or construction projects, where the work to be done is well defined in detail. Executives are used to these types of projects because detailed plans, estimates, schedules, and budgets are developed at the beginning of the project. Obviously, all kinds of projects can have overruns in schedules and budgets. However, successful engineering and construction companies become proficient at planning and executing projects on time and within budget. The most important factor for success in these projects is the ability to understand and predict almost every contingency that may occur. This approach can work for the development or modification of software for well-understood business processes. These types of projects put a heavy load on the project manager at the beginning of the project.

In today's fast-paced world of new business opportunities with new technologies, many projects are delving into unknown waters. For those kinds of projects, an adaptive SDLC model is more appropriate. Sometimes we like to think of an adaptive, iterative approach as an "organic" approach—the system grows much like a plant grows. It starts small, and as it grows, it adapts to fit into its new environment. As a plant is often beautiful and useful even while it is immature, homegrown systems often begin to provide benefits to the organization even before they are fully grown. Adaptive approaches are all based on iterations within the project, as was explained in Chapter 2.

However, the negative aspects of growing plants can also occur in adaptive projects. At the beginning, companies do not know which way the project will grow, how long it will take to be fully grown, or even how to know when it is fully grown. Many adaptive projects cause problems for upper management because the total budget is hard to predict. At times the scope of the project is also undefined at the start. Other adaptive projects never seem to end. Our point is that even though adaptive approaches have some strong advantages to software development, they also must be carefully monitored and managed to be successful. Thus, a project manager for an adaptive project must be just as skilled and proficient as in a predictive project. The tasks are slightly different, but the need to plan, monitor, and control is just as important.

It is important to understand the difference between project management tasks and project development tasks. The definition of project management includes the concept that the manager directs other people to achieve a planned result, while project development tasks are "hands-on" tasks directly related to the new system. To better understand the distinction, we can compare project management in a software development project with supervisory tasks on a construction project. The construction manager of a building works with the architect, reads the plans, checks the schedule, and assigns work to the team members according to the schedule. Those tasks are supervisory tasks and are different from the hands-on activities of pouring concrete or laying bricks. Usually, a construction manager does not actually pour concrete or lay bricks. He or she is too busy coordinating the project by assigning work, checking progress, and resolving problems. Software projects, unless they are very small, also require a full-time commitment to project management tasks.

Figure 3-3 is an adaptation of Figure 2-5, which distinguishes between management activities and development activities and which shows the overlap of the various phases in a predictive development project. Notice that in Figure 3-3, project planning involves both project management and SDLC tasks. The overlap occurs because planning for a project requires participation by both the key team members and the project manager. The complexity of software development projects requires team members to be actively involved in identifying activities, estimating work requirements, and building the project schedule. After the detailed plans have been developed, the team members focus their energies on the SDLC tasks and the project manager focuses on management tasks.

As indicated by the figure, three major project management processes overlap the SDLC processes: executing, controlling, and closing the project out. Execution includes tasks that are concerned with following the project schedule, assigning and coordinating the work of project teams, and communicating with all project stakeholders. Control tasks involve determining progress and taking corrective action when necessary, assessing whether requests for scope changes are necessary, maintaining an outstanding issues list, and resolving problems. Project closeout includes tasks targeted to a smooth shutdown of the project, such as releasing team members for other assignments, finalizing the budget and its expenditures, and reviewing or auditing the results of the project. An important point illustrated in the figure is that these project management tasks last throughout a project and happen concurrently with the SDLC activities associated with analysis, design, and implementation.

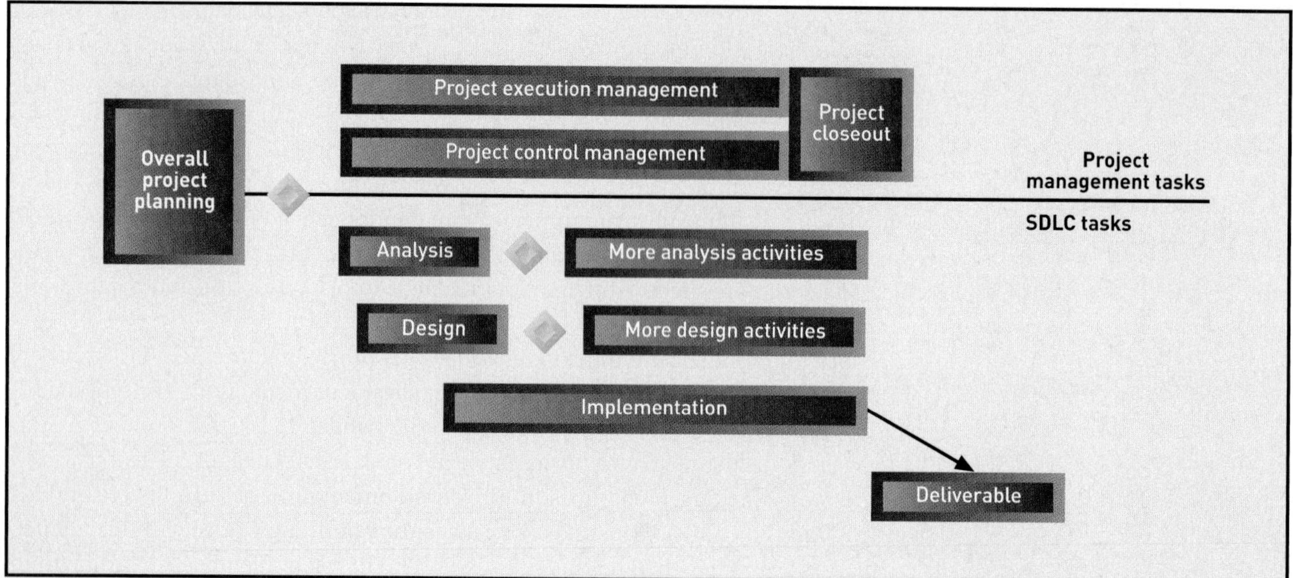

Figure 3-3

Project management and SDLC tasks for a predictive project

Figure 3-4 illustrates how project management tasks are applied in an adaptive project. In the figure, we have taken the spiral life cycle concepts (shown earlier in Figure 2-6) and depicted two iterations in a linear fashion (similar to the iterations shown in Figure 2-7). Even though there are many variations of adaptive approaches, Figure 3-4 represents the key ideas that are common to all of them. A project always starts with a major planning effort. However, in adaptive approaches, there is a dual focus to planning: The first objective is to define the scope of the project in broad terms, and the second is to identify the iterations, or cycles. Planning the project at this point is less detailed; it determines the major steps but leaves the details for later.

After the major planning phase is finished, the detailed iterations—the cycles—proceed. In some adaptive approaches, the major planning phase is also considered to be an iteration. Each cycle requires detailed planning, execution management, control management, and cycle closeout. Each cycle also requires SDLC activities for analysis, design, and implementation. Of course, the exact nature of the SDLC tasks depends on what type of deliverable the client wants from the iteration.

Comparing the predictive and adaptive approaches, we can see that in an adaptive project, planning tasks are more distributed across the entire lifetime of the project. However, the same set of project management skills is required. The only difference is in how and when the planning, executing, and controlling tasks are carried out.

In the next section, we discuss the major areas of knowledge entailed in being a good project manager. As you saw in Chapter 2, project planning is called a "phase" in the waterfall model. In an adaptive, iterative project, however, project management, which includes planning, is often done as part of the first iteration. Hence, project management or project planning activities apply both to predictive projects and adaptive projects. The remainder of the chapter details the specific project planning skills that you will need—whether you are a project manager or a senior systems analyst on a project team.

PROJECT MANAGEMENT AND THE LEVEL OF FORMALITY

Another dimension that has a heavy impact on project management is the level of formality required for a given project. Some projects, particularly small ones, are conducted with a very low level of formality. Status meetings occur in the hallway or around the water cooler. Written documentation, formal specifications, and detailed models are kept to a minimum. Developers and users usually work closely together on a daily basis to define requirements and develop the system. Other projects, usually larger, more complex ones, are executed with

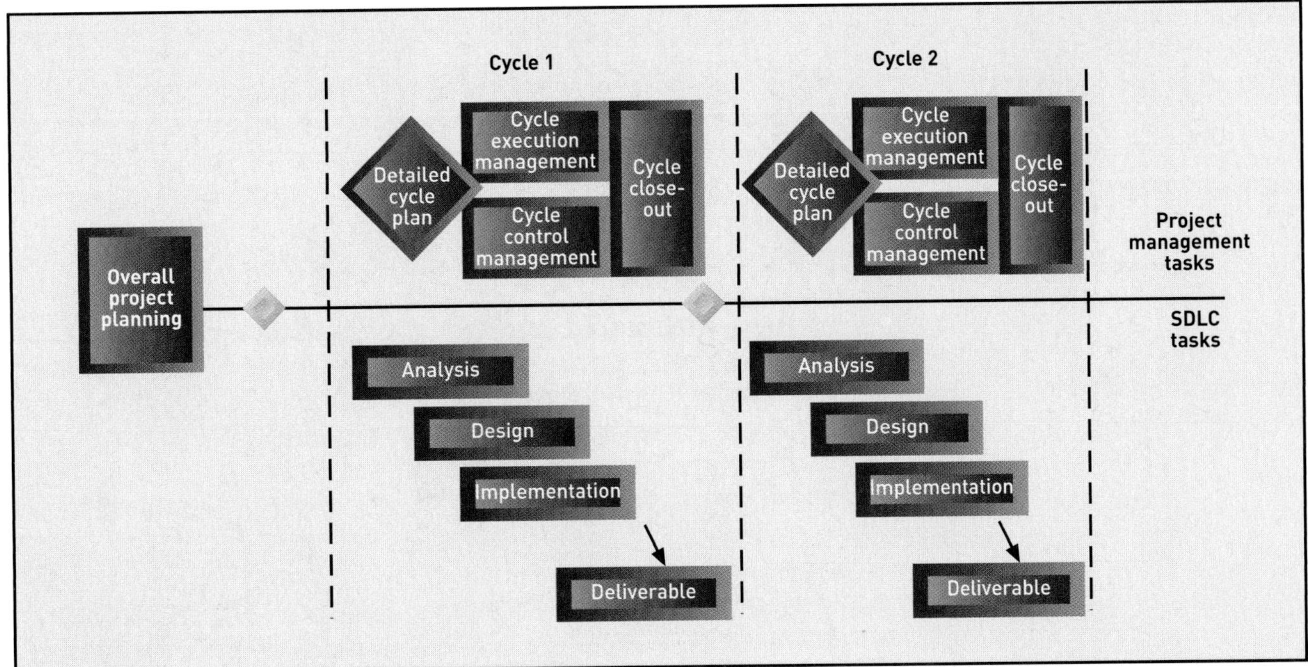

Figure 3-4

Project management and SDLC tasks for an adaptive project

a high level of formality. Status meetings are held on a predefined schedule with specific participants, agendas, minutes, and follow-through. Specifications are formally documented with an abundance of diagrams and documentation. Specifications are frequently verified through formal review meetings between developers and users (see Chapter 4 on Structured Walkthrough).

A project's level of formality is a dimension that is superimposed above whether the project approach is predictive or adaptive. Historically, large predictive projects were also quite formal, with lots of meetings and documentation. Unfortunately, the extensive documentation tended to increase the length of the project and sometimes contributed to cost overruns. Techniques were developed to help manage large predictive projects with less formality. One technique, called Rapid Application Development (RAD), was based on holding intensive meetings with all critical stakeholders (for example, both developers and users), where specifications were hammered out. Running prototypes were built during the meetings to capture the requirements and even begin the design and implementation of the new system. This approach required less documentation and fewer status and review meetings. Even though there was less formality, the skills required of the project manager were just as important. Of course, small predictive projects also could often be managed in a less formal manner.

Adaptive projects can also be either more or less formal in the way they are managed. The Unified Process presented in the previous chapter is an adaptive approach, but in its pristine form it is also quite formal. Each phase is precisely defined with specific outcomes, including specifications, diagrams, prototypes, and deliverables defined for each phase and iteration. However, adaptive, iterative approaches also lend themselves easily to being managed with much less formality. The inherent characteristics of an iterative approach with its "just in time" project plans easily adjust to less documentation, fewer diagrams for specifications, and less formal status reporting.

Agile Software Development

a philosophy of software development that embraces flexibility and agility

In the mid-1990s and early 2000s, various groups of professionals in the field of software development began creating a set of techniques and methods with the objective of managing projects successfully, but with much less overhead. Originally these techniques were called "lightweight methods," but more recently they have been called "agile methods." Today, **Agile Software Development** is a major movement in software development. It shares many of the same objectives as an adaptive approach, and thus is most often used with adaptive projects. In fact, the terms are often used interchangeably. However, Agile Development does have additional characteristics than those identified for adaptiveness. Some of the elements included in the Agile Manifesto include the following:

- Individuals and interactions over processes and tools; for example, don't focus on formal processes or sophisticated tools, but on the team and team interactions.
- Working software over comprehensive documentation; for example, don't spend all your time drawing diagrams and writing specifications. Use modeling to help solve the problem and then write software.
- Customer collaboration over contract negotiation; for example, work closely with the users and clients so that the solution is the "right" solution.
- Responding to change over following a plan; for example, be flexible and accept changes in the business and in the users' requirements, even late in the project. Don't be so tied to the plan that you lose the flexibility to solve users' problems.

Sometimes developers mistakenly interpret the manifesto to mean you don't need to create plans or specification models. But that is incorrect. Agile Development means you create a plan and build models as effective means to the end, but not as the end results themselves. Use specifications and plans as tools.

The opening case of this chapter presented Bestway Fuel Systems and their initial foray into both an adaptive and agile methodology. Many newer methods, such as Agile Unified Process (AUP), SCRUM, XP, Adaptive Software Development (ASD), Crystal Clear, and Dynamic Systems Development Method (DSDM), are inherently both adaptive and agile. The important point to learn in this discussion, however, is that all of these approaches require project management skills to ensure a successful project. In the next section, we identify some of the project management competencies required of all good project managers.

PROJECT MANAGEMENT KNOWLEDGE AREAS

The Project Management Institute (PMI) is a professional organization that promotes project management, primarily within the United States but also throughout the world. In addition, professional organizations in other countries promote project management. If you are interested in strengthening your project management skills, you should consider joining one of these organizations, obtaining their materials, and participating in training. The PMI has a well-respected and rigorous certification program. In fact, many corporations encourage project managers to become certified, and industry articles frequently indicate that project management is one of the most important skills today.

As part of its mission, the PMI has defined a body of knowledge (BOK) for project management. This body of knowledge, referred to as the PMBOK, is a widely accepted foundation of information that every project manager should know. The PMBOK has been organized into nine different knowledge areas. Although these nine areas do not represent all there is to know about project management, they provide an excellent foundation.

- **Project Scope Management.** Defining and controlling the functions that are to be included in the system, as well as the scope of the work to be done by the project team
- **Project Time Management.** Building a detailed schedule of all project tasks and then monitoring the progress of the project against defined milestones
- **Project Cost Management.** Calculating the initial cost/benefit analysis and its later updates and monitoring expenditures as the project progresses
- **Project Quality Management.** Establishing a comprehensive plan for ensuring quality, which includes quality-control activities for every phase of the project
- **Project Human Resource Management.** Recruiting and hiring project team members; training, motivating, and team building; and implementing related activities to ensure a happy, productive team
- **Project Communications Management.** Identifying all stakeholders and the key communications to each; also establishing all communications mechanisms and schedules

- **Project Risk Management.** Identifying and reviewing throughout the project all potential risks for failure and developing plans to reduce these risks
- **Project Procurement Management.** Developing requests for proposals, evaluating bids, writing contracts, and then monitoring vendor performance
- **Project Integration Management.** Integrating all the other knowledge areas into one seamless whole

Additional details of each of these knowledge areas are provided in Appendix A on the book's Web site. Other textbooks that focus exclusively on project management, several of which are listed in the "Further Resources" section at the end of the chapter, also provide in-depth discussion and specific techniques that apply to each knowledge area.

As you progress in your career, you would be wise to keep a record of project management skills you observe in others, as well as those you learn by your own experience. One place to start is with the set of skills for a systems analyst described in Chapter 1. A good project manager knows how to develop a plan, execute it, anticipate problems, and make adjustments. Project management skills *can* be learned. Those who aspire to managing projects are proactive in self-improvement and learn the necessary skills. Build on what you learn in this textbook and continue to practice and hone your project management skills.

PROJECT INITIATION AND PROJECT PLANNING

Information system development projects are initiated for various reasons. Three general driving forces are as follows: (1) to respond to an opportunity, (2) to resolve a problem, and (3) to conform to a directive.

Most companies are continually looking for ways to increase their market share or to open up new markets. One way they create opportunities is through strategic plans, both short term and long term. In many ways, planning is an optimal way to identify new projects. The benefit of this approach is that it provides a more stable and consistent environment in which to develop new systems. As the strategic plans are developed, projects are identified, prioritized, and scheduled. Projects initiated through strategic planning are sometimes described as *top-down* projects.

weighted scoring
a method to prioritize projects based on criteria with unequal weights

To prioritize these projects, companies use a technique called weighted scoring. First, the IT strategic planning committee identifies a set of criteria to judge the importance of new projects. Examples of criteria are "opens a new market" or "provides a high net present value." These criteria are weighted for their importance, and each potential project is rated according to the set of criteria. Projects with the highest scores are given priority for initiation.

Projects are also initiated to solve an immediate business problem. These projects attempt to close the gap between what information processing is required to run the business correctly and what is currently in operation. They can be initiated as part of a strategic plan but more commonly are requested by middle managers to resolve some difficulty in company operations. Obviously, senior executives are also aware of internal problems and can initiate projects to solve them. Sometimes these needs are so critical that they are brought to the attention of the strategic planning committee and integrated into the overall business strategy. At other times, an immediate need cannot wait, such as a new sales commission schedule or a new report needed to assess productivity. In these cases, managers of business functions will request the initiation of individual development projects.

Finally, projects are initiated to respond to outside directives. One common outside pressure is legislative changes that require new information-gathering and external reporting requirements, such as changes in tax laws and labor laws. For example, regulations in the Health Insurance Portability and Accountability Act (HIPAA) are intended to safeguard patients' medical information. This act affected Reliable Pharmaceutical Service, as discussed in the case at the end of Chapter 2. Legislative changes can also expand or contract the range of services and products that an organization can offer in a market. New regulations and laws can affect the strategic plan, resulting in an expedited need for new systems. We have seen many regulatory changes in the telecommunications industry, with cable TV and telephone companies vying for opportunities to provide cellular services, Internet access, and personalized entertainment.

Several steps normally occur with the initiation of a new project. A project charter is typically identified, which describes the purpose of the new system, the potential start and completion dates, and, very importantly, the key stakeholders and sponsors of the new system.

Whatever the reason for project initiation, it usually requires an initial review to ensure that the benefits outweigh the costs and risks of development. Thus, the first activities of almost every project after it is approved are those that precisely define the business problem, determine the scope of the project, and perform a feasibility analysis, including a cost/benefit analysis. We group all of these initial planning activities together as part of the overall project planning component of the SDLC.

BEST PRACTICE

Whatever the source of the new project, be sure to carefully evaluate its feasibility before proceeding.

INITIATING THE CUSTOMER SUPPORT SYSTEM FOR ROCKY MOUNTAIN OUTFITTERS

As described in Chapter 1, RMO senior executives, with help from an outside consulting firm, have developed a well-considered information systems strategic plan. The plan includes both a technology architecture component and an application architecture component. Implementation of the plan had begun with the initiation of the supply chain management (SCM) project. The company founders, John and Liz Blankens, understood clearly that to maintain good customer relations, they need to have systems in place to support the fulfillment of sales as they move to broader geographical and Internet-based sales. The company will not realize the full benefit of the SCM system until the customer support system (CSS) also comes online. The SCM system will provide several cost-reduction efficiencies, but RMO expects the real business benefit to come from a dramatic increase in sales from the expanded sales and marketing capabilities of the new CSS.

The supply chain management system was well under way. The project was to be implemented in several increments because several of RMO's suppliers would also have to upgrade their systems. The first increment was on schedule, the requirements had been finalized, the overall architectural design was firm, and the pieces of the new system were expected to be ready early the next year. John Blankens was really excited about the progress and was anxious to get started with the new customer support system. He called a special meeting of the company's executive committee to assess the progress of the current projects and to evaluate the possibility of moving ahead with the new CSS. Prior to the meeting, he asked VP of Finance and Systems JoAnn White to bring a detailed financial analysis of current system budgets and projections of the financial impacts that RMO could expect from beginning the CSS project in the near future. He also invited Chief Information Officer Mac Preston to evaluate the workload of the system development staff and the availability of staff to begin. Several other assignments were given to committee members to consider his proposal carefully.

After a long discussion, the executive committee decided that it was not only feasible to begin the project now, but critical to do so. Other retailers had proven that Internet sales and marketing, if planned and executed correctly, could provide tremendous benefits to a company. Even though there had been several jerks and sputters, e-commerce was here to stay. It was imperative for future viability that RMO, like other brick-and-mortar retailers, also have a strong presence on the Internet.

As a result of the meeting, the committee directed Mac to start the project. First, he met with Director of System Development John MacMurty and asked him to finalize his plans for a project manager and another experienced systems analyst to get the project started. He also asked John to produce a project charter to confirm the decisions made by the executive committee. John began contacting executives to elicit their participation as members of the oversight committee. He understood well that if he could get strong commitment from senior executives in the company, he would ensure good user involvement in the project. One of the key elements of successful projects is to get broad involvement from the users. After a couple of days of discussion with executives throughout the company, the oversight committee was complete. Vice president of marketing and sales William McDougal, who had requested to be the project sponsor because it supported his area directly, was the committee chair. Other members were Robert Schneider, director of catalog sales, Brian Haddock from operations, and, of course, John Blankens and Mac Preston.

The project began with the assignment of Barbara Halifax as the full-time project manager, as indicated in the RMO memo in Chapter 1. Barbara has been with RMO for several years. Prior to joining RMO, she worked for the information systems consulting division of one of the large accounting firms. Her experience in consulting gave her broad exposure to many different companies and systems. Senior management in RMO had complete confidence in her abilities to manage the CSS project. Steven Deerfield, a senior systems analyst, was also assigned to the project. Deerfield and Halifax had worked together before and had very compatible work styles. Because this project was a critical component of RMO's long-range strategic plan, two of the very best systems analysts in the company were assigned. Figure 3-5 illustrates the project charter, which documents the preliminary activities to get the project initiated.

Figure 3-5

RMO project charter

Project Name: ***Customer Support System***

Project Purpose: To provide increased level of customer support. Should include all customer-related functions from order entry to arrival of the shipment, including customer inquiries/catalog, order entry, order tracking, shipping, back order, returns, and sales analysis.

Anticipated Completion: Within 10 months of project initiation

Approved Budget: Up to $1,500,000

Key Participants:

Participant	***Position***	***Primary responsibilities***
Barbara Halifax	Project manager	Manage the entire project
John MacMurty	Director	Supervise project manager Check status weekly Serve on oversight committee
Mac Preston	Chief information officer (CIO)	Serve on oversight committee
William McDougal	Senior VP marketing/sales	Direct project sponsor Approve budget, schedule Serve on oversight committee
Robert Schneider	Director of catalog sales	Serve on oversight committee Provide user support/resources
Brian Haddock	Director of operations	Serve on oversight committee Provide user support/resources
Jason Nadold	Manager of shipping	Provide user support/resources

As described in Chapter 1, the primary objective of the system is to support RMO's goal of building customer loyalty and of providing all the necessary tools for customer relationship management. The system is to further this objective by supporting all types of customer services—including ordering, returns, and online catalogs—for the ongoing telephone sales and a new capability with Internet sales. Customers not only must have access to the online catalog of RMO products either via a telephone sales representative or the Internet, but must also be able to see their past purchasing history. Managers at RMO would like the system to include several "bells and whistles" to support their vision of RMO customer service.

The following section describes the project planning activities, using examples from the RMO project. As explained previously, these activities are exclusively project management activities, which are used to plan, organize, schedule, and finally obtain approval for the project. Note that even though this project seems to have tacit approval from senior management, it must meet the rigorous evaluation criteria of all RMO projects. Even though only two members of the team have been assigned at this point, Barbara and Steve have extensive experience and excellent project management skills.

PROJECT PLANNING ACTIVITIES

From Figures 3-3 and 3-4, we see that both predictive and adaptive projects begin with overall project planning. The major difference in planning between the two types of projects is the level of detail provided. Predictive project members attempt to plan the entire project, including the schedule, at a fairly detailed level. Adaptive project members plan the overall project, but leave much of the detail to be developed during iteration planning. The remainder of the chapter explains project management techniques that are used for all project management tasks, whether at the beginning of the project or during the iterations. The project planning activities of the SDLC, as depicted in Figure 3-6, consist of the activities required to get the project organized and started. As discussed in Chapter 2, project management activities are as follows:

- Define the problem.
- Produce the project schedule.
- Confirm project feasibility.
- Staff the project.
- Launch the project.

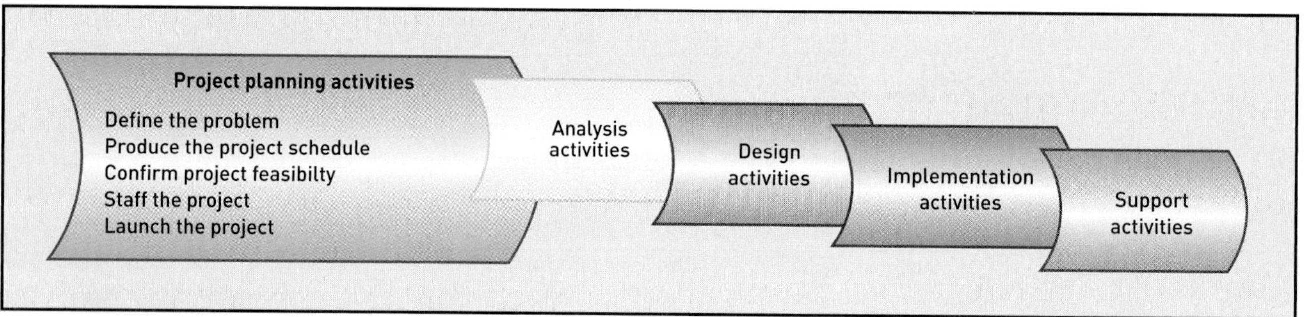

Figure 3-6
Activities required for project planning

These activities are all project management activities. Initial project planning activities are usually staffed with only two or three highly experienced systems analysts, one of whom serves as the project manager. The other systems analysts assigned to the team are experienced developers with strong analytical skills, as well as experience in managing and controlling projects. The first team members that are assigned frequently become the core team leaders around which the rest of the team is built. At the successful conclusion of this phase, the project will have begun with resources, schedules, and a budget.

To help you learn about the activities of project planning, the following sections each describe a project planning activity and then show how it applies to RMO. Figure 3-7 lists each activity with the key question the project team tries to answer when completing the activity. For example, at the end of project planning, the key question to answer for the *Launch the project* activity becomes: *Are we ready to start the project?*

Figure 3-7

Project planning activities and their key questions

Project planning activities	Key questions
Define the problem	Do we understand what we are supposed to be working on?
Produce the project schedule	Can the project be completed on time given the available resources?
Confirm project feasibility	Is it still feasible to begin working on this project?
Staff the project	Are the resources available, trained, and ready to start the project?
Launch the project	Are we ready to start the project?

DEFINING THE PROBLEM

Carefully defining the problem is one of the most important activities of the project. The objective is to define precisely the business problem to be solved and thereby determine the scope of the new system. This activity defines the target that you want to hit. If the target is ill defined, then all subsequent activities will lack focus. As pointed out earlier, one of the primary causes of project failure is an unclear objective.

The first task within this activity is to review the business needs that originally initiated the project. As with RMO, if the project was initiated as part of the strategic plan, then the planning documents are reviewed. If the project originated from departmental needs, then key users are consulted to help the project team understand the business need. As the needs are identified, the team also develops a detailed list of the expected benefits. We define those as the **business benefits**. The list of business benefits contains the results that the organization anticipates it will accrue from a new system. Business benefits are normally described in terms of the influences that can change the financial statements, either by decreasing costs or increasing revenues.

business benefits

the benefits that accrue to the organization; often measured in monetary terms

The second task in this activity is to identify, at a high level, the expected capabilities of the new system. The objective is to define the scope of the problem in terms of the requirements of the information system that can solve the problem. Although at first defining the expected capabilities may not appear to be defining the problem, it is necessary to understand the scope of the new system and hence the project's scope.

Members of the development team combine these three components—the problem description, the business benefits, and the system capabilities—to get a **system scope document**. These members (for example, the systems analysts) work with the users and the client to develop this document. Sometimes this document is combined with the project charter; in other cases, it is independent. Figure 3-8 is an example of the system scope document for RMO. Note the differences between the business benefits and the system capabilities. The business benefits focus on the financial benefit to the company. The system capabilities focus on the system itself. The benefits are achieved through the capabilities provided by the system.

system scope document

a document—containing problem description, business benefits, and system capabilities—to help define the scope of a new system

At times, especially when the new system is an attempt to push the state of the art, it may be necessary to build a preliminary prototype as a proof of the concept. New solutions, particularly those based on new technology, may not be well accepted or well understood. In that situation, the project team can build a **proof of concept prototype** to illustrate that a solution is possible and feasible. When a proof of concept is necessary, the project scope document will refer to the results of the initial prototype's construction, test, and fitness for purpose. For example, RMO senior management may want the system to automatically suggest complementary accessories for Web customers who purchase items over the Internet. The project team may need to build some prototypes to verify that the request is technologically feasible.

proof of concept prototype

a very preliminary prototype built to illustrate that a solution to a business need is feasible

Frequently the project team also develops a diagram to describe the scope of the system in terms of information flowing into and out of the system. This diagram, which is called the **context diagram**, shows the primary users of the system and the information that is

context diagram

a data flow diagram (DFD) showing the scope of a system

System Scope Document
Customer Support System

Problem Description

Catalog sales began in Rocky Mountain Outfitters as a small experiment that soon developed into a rapidly growing division of the company. Support was initially provided by manual procedures with some simple off-the-shelf programs to assist in order taking and fulfillment. By 2006, the growth of catalog sales, including Internet sales, was stretching the capabilities of the current system. As a result of a long-term strategic plan, RMO decided to initiate two major system development projects. The first, the supply chain management (SCM) system, was started in 2006 and is progressing on schedule. The second identified system is a customer support system (CSS) to provide sales, marketing, and a full range of customer support functionality. This project is an integral part of the total long-term strategic plan of RMO to continue to grow and maintain its leadership position in the sportswear industry.

Anticipated Business Benefits

The primary business benefit to be obtained from the new system is for RMO to maintain its leadership position in the sportswear industry. More immediate benefits include the following:

- Reduce errors caused by manual processing of orders.
- Expedite order fulfillment due to more rapid order processing.
- Maintain or reduce staffing levels in mail-order and phone-order processing.
- Dramatically increase Internet sales through a highly interactive Web site.
- Increase turnover by tracking sales of popular items and slow movers.
- Increase level of customer loyalty through extensive customer support and information.

System Capabilities

To obtain the business benefits listed previously, the customer support subsystem shall include the following capabilities:

- Be a high-support system with online customer, order, back-order, and returns information.
- Support traditional telephone and mail catalog sales with rapid-entry screens.
- Include Internet customer and catalog sale capability, including purchase and order tracking.
- Maintain adequate database and history information to support market analysis.
- Provide a history of customer transactions for customer query.
- Be able to handle substantial increases in volume (300 percent or more) without degradation.
- Support 24-hour shipment of new orders.
- Coordinate order shipment from multiple warehouses.
- Maintain history to support analysis of sales and forecasting of market demand.

Figure 3-8

System scope document for the RMO customer support system

exchanged between them and the system. Figure 3-9 is an example of a simplified data flow diagram (DFD) called the context diagram for the RMO customer support system. Because, at this point, the project team is only documenting the scope of the new system, the diagram includes only major information requirements. In fact, the diagram focuses primarily on output information from the system. Chapter 6 provides a more detailed explanation and example of the RMO context diagram. Chapter 7 explains an object-oriented system overview diagram, called a *use case diagram*.

The box with rounded corners in the middle of the diagram represents the customer support system itself. The boxes around the oval are the entities that provide information to the system or that receive information from the system. The lines with arrowheads are the major inputs to and outputs from the system. This diagram identifies only the major information flows into and out of the system. The objective is to get an overview of a proposed solution and not get involved in the details.

Note that the context diagram is also used during the analysis phase. During project planning, the diagram helps define the scope of the problem. This diagram becomes a starting point for the more detailed investigation done during analysis.

Defining the scope carefully is important for establishing an estimate of the amount of effort required to complete the project. The size or scope of the system determines the amount of effort, which then determines the time and cost of the project. Several techniques can be used to measure the size or scope of a proposed system, although accurate estimates

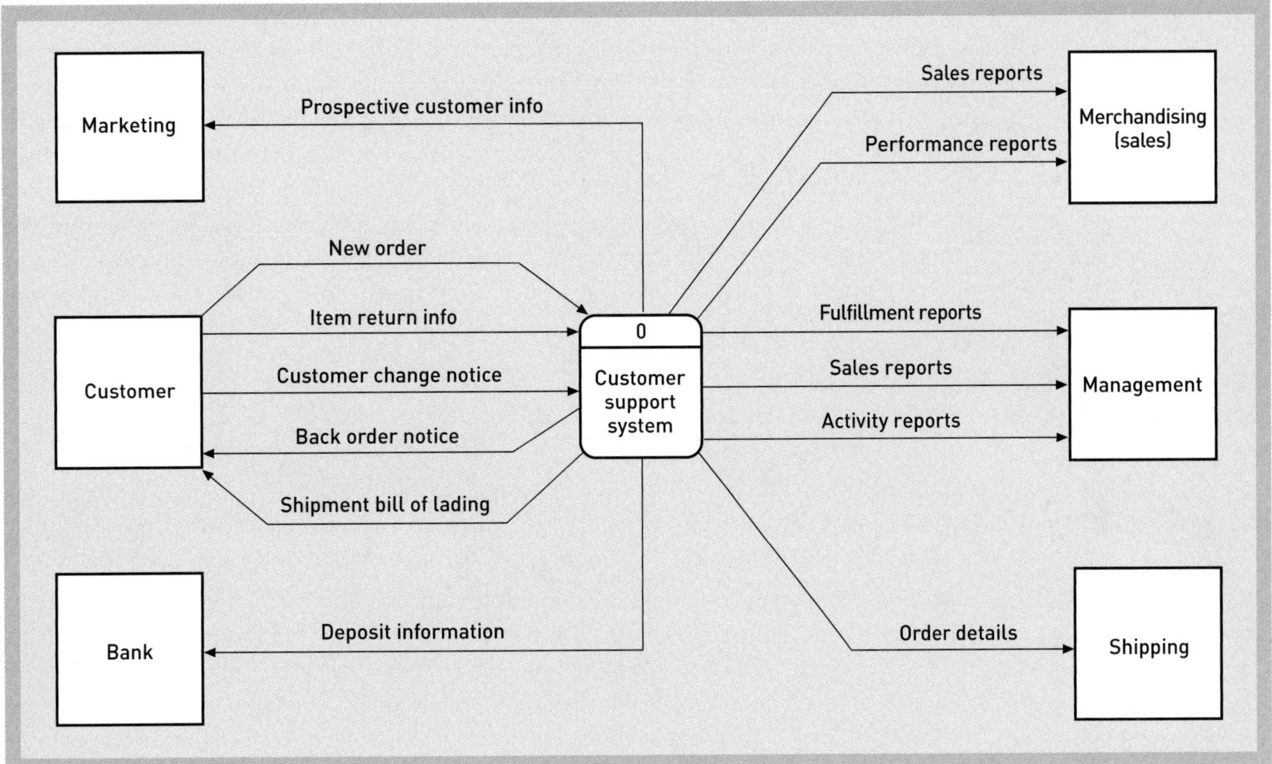

Figure 3-9

Context diagram for the customer support system

are difficult to achieve. Most approaches count the number of activities or use cases the system is required to support. Chapter 5 discusses activities and use cases and presents techniques for identifying them. In addition, data entities and classes that are in the problem domain of the system can be counted to estimate the scope. Chapter 5 also presents these identification techniques. Some approaches count the number of *function points* that can be identified. One technique, called the COnstructive COst MOdel (COCOMO), attempts to count function points as the number of inputs and outputs, the number of files maintained, the number of updates required, and so on.

The key question to be answered when completing the problem definition activity is: *Do we understand what we are supposed to be working on?*

DEFINING THE PROBLEM AT RMO

Barbara and Steve, the CSS project team, developed the lists for the system scope document after talking to William McDougal, vice president of marketing and sales, and his assistants. Chapter 4 explains more about interviewing users and eliciting important information. It is essential to obtain information from the people who will use the system and to involve the people who will benefit most from it. They provide valuable insights to ensure that the system satisfies the business needs. As noted previously, the most critical element in the success of a system development project is user involvement.

One additional task is required to complete the problem definition activity. The project team conducts a preliminary investigation of alternative solutions to reassess the assumptions the team made when the project was initiated. Because the schedule and budget for the remainder of the project inherently assume a particular approach to developing the system, it is critical to make those implicit assumptions explicit so that all participants understand the constraints on the project schedule and the team can perform an accurate feasibility analysis.

For example, if an "off-the-shelf" program is identified as a possible solution, part of the schedule during the analysis phase must include tasks to evaluate the program against the needs being researched. If the most viable solution appears to be a new system developed completely in-house, detailed analysis tasks are planned and scheduled.

While Barbara was finishing the problem definition statement, Steve did some preliminary investigation of possible solutions. He researched the trade magazines, the Internet, and other resources to determine whether sales and customer support systems could be bought and installed rapidly. Although he found several, none seemed to have the exact match of capabilities that RMO needed. He and Barbara, along with William McDougal, had several discussions about how best to proceed. They decided that the best approach was to proceed with the analysis phase of the project before making any final decision about solutions. They would revisit this decision, in much more detail, after the analysis phase activities were completed. For now, Barbara and Steve began developing a schedule, budget, and feasibility statement for the new system.

PRODUCING THE PROJECT SCHEDULE

Before discussing the details of a project schedule, let's clarify two terms: *task* and *activity*. Fundamentally, an *activity* is made up of a group of related tasks or other smaller activities. A *task*, then, is the smallest piece of work that is identified and scheduled. Activities are also identified, named, and scheduled. For example, suppose that you are scheduling the design phase for a waterfall-type project. Within the design phase, you identify activities such as *Design the user interface*, *Design and integrate the database*, and *Complete the application design*. Within the *Design the user interface* activity, you might identify individual tasks such as *Design the customer entry form* and *Design the order-entry form*. The waterfall methodologies group activities together into phases, such as analysis phase or design phase. Iterative, adaptive methodologies group activities together into iterations. You will be able to see the differences in the example schedules provided later in this chapter.

During project planning, it may not be possible to schedule every task in the entire project because it is too early to know all of the tasks that will be necessary. However, one of the requirements of project planning is to provide estimates of the time to complete the project and the total cost of the project. Because one of the major factors in project cost is payment of salaries to the project team, the estimate of the time and labor to complete the project becomes critical. The activity of developing the project schedule is one of the most difficult endeavors of project planning, yet it is one of the most important. The development of a project schedule is divided into three main steps:

- Develop a work breakdown structure.
- Build a schedule using a Gantt chart.
- Develop resource requirements and the staffing plan.

DEVELOPING A WORK BREAKDOWN STRUCTURE

work breakdown structure (WBS)

the hierarchy of phases, activities, and tasks of a project; one method to estimate and schedule the tasks of a project

The first step in building a project schedule is to identify all of the activities and tasks that need to be scheduled. A **work breakdown structure (WBS)** is simply a list of all the required individual activities and tasks for the project. Figure 3-10 is one example of a work breakdown structure that shows the activities and tasks for the overall project planning phase of the RMO project.

The primary activities in this WBS are precisely the activities that were identified earlier as the key activities for project planning. Each activity is further divided into individual tasks to be completed. The WBS identifies a hierarchy, much like an outline for a paper. The project requires a WBS for each phase of the SDLC, and the project planning WBS is shown here because we discuss these activities in detail in this chapter. Obviously, the analysis, design, and implementation phase WBSs would be even more important to define because project planning attempts to schedule the entire project.

There are two general approaches for building a WBS: (1) by deliverable and (2) by a sequential timeline. The first approach identifies every deliverable, both intermediate and

Figure 3-10

Work breakdown structure for planning activities of the RMO project

Task Name
1 Project Planning
1.1 Define the Problem
1.1.1 Meet with users
1.1.2 Determine scope
1.1.3 Write problem description
1.1.4 Identify business benefits
1.1.5 Identify system capabilities
1.1.6 Develop context diagram
1.2 Produce the project schedule
1.2.1 Develop WBS
1.2.2 Estimate durations
1.2.3 Determine sequences
1.2.4 Develop Gantt Chart
1.3 Confirm project feasibility
1.3.1 Identify intangible cost/benefits
1.3.2 Estimate tangible benefits
1.3.3 Calculate cost/benefit
1.3.4 Organizational feasibility
1.3.5 Technical feasibility
1.3.6 Evaluate resource availability
1.3.7 Risk analysis
1.4 Staff the project
1.4.1 Develop resource plan
1.4.2 Procure project team
1.4.3 Procure user liaisons
1.4.4 Conduct training
1.5 Launch the project
1.5.1 Make executive presentation
1.5.2 Procure facilities
1.5.3 Procure support resources
1.5.4 Conduct kickoff meeting

final, that must be developed. Then the WBS identifies every task that is necessary to create that deliverable. For example, if the project is to build a house, one of the intermediate deliverables would be to install the electrical wiring. The tasks for that deliverable relate to hiring an electrical contractor, drilling holes, running wires, connecting junction boxes, connecting fixtures, and so forth. The second approach—the sequential timeline approach—works through the normal sequence of activities that are required for the final deliverable. For our example of building a house, the sequential timeline approach relates to tasks such as surveying the property, digging the foundation, pouring the foundation, framing the walls, and so forth.

The four most effective techniques for identifying the tasks of the WBS are:

- **Top-down:** Identifying major activities first and then listing internal tasks
- **Bottom-up:** Listing all the tasks you can think of and organizing them later
- **Template:** Using a standard template of tasks for projects that are fairly standard
- **Analogy:** Finding a similar, or analogous, project that is finished and copying its tasks

When it is possible, teams try to use the template or analogy approach. Otherwise, a combination of the top-down and bottom-up approaches can be used to brainstorm a good list of tasks. After the team members identify the major activities, they might brainstorm in a bottom-up fashion to try to identify any other tasks that might have to be done. Often, teams will brainstorm using a blank wall and Post-it notes so they can move tasks around and reorganize them. Figure 3-11 is an example of a work breakdown structure in a more graphical, hierarchical format. The figure also shows a project that is using an adaptive, iterative approach, so it encompasses the first iteration, which is planning.

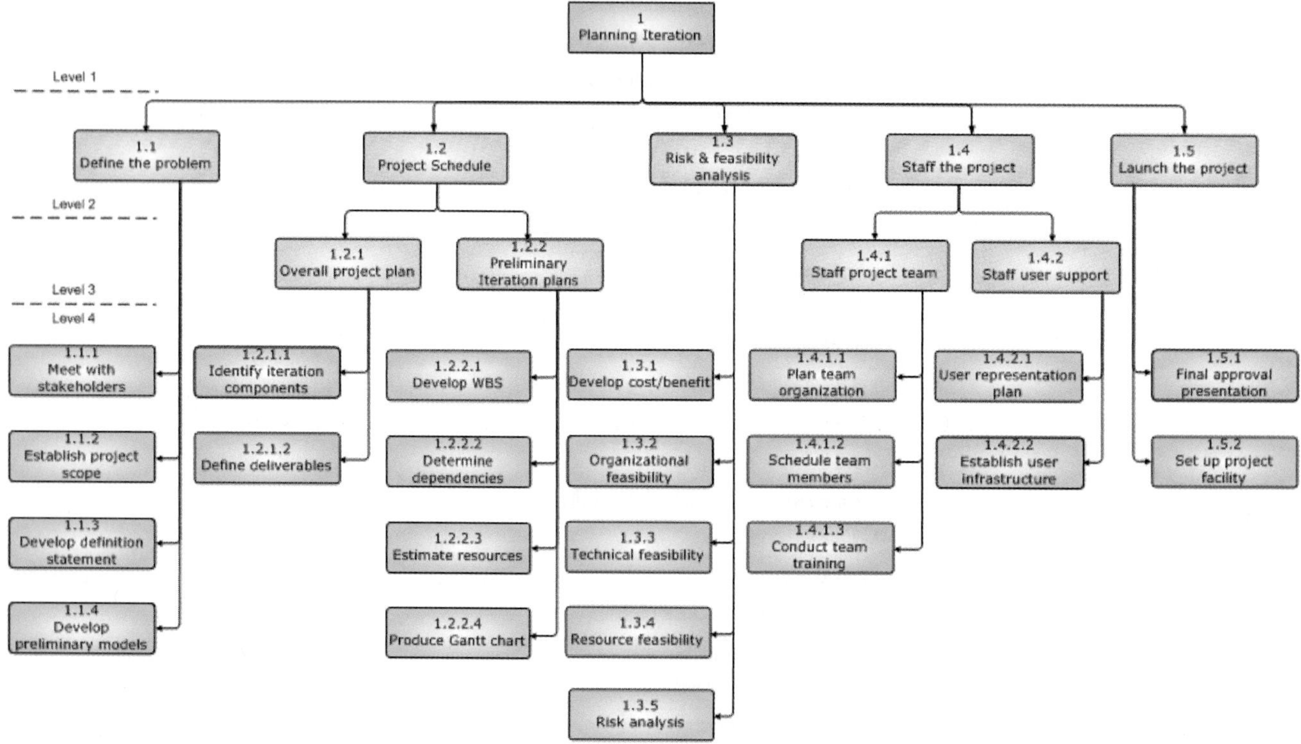

Figure 3-11

Work breakdown structure for the first iteration for an adaptive project for RMO

The left side of the figure shows several levels of the WBS. The top level is the overall iteration. Because this WBS is based on the deliverables for the iteration, the second level lists the major deliverables. The third level denotes major activities and the fourth level contains the individual tasks. It is not necessary to take all the branches to the same level of detail. Some deliverables may only require a single activity, while others may need to be expanded to several more layers.

When developing a WBS, new analysts frequently ask, "How detailed should the individual tasks be?" A few guidelines can help answer that question:

1. There should be a way to recognize when the task is complete.
2. The definition of the task should be clear enough so that someone can estimate the amount of effort required to complete it.
3. As a general rule for software projects, the effort should take 2 to 10 working days.

DEVELOPING THE SCHEDULE

A project schedule lists all project activities and tasks and the order in which they must be completed. To build the schedule, the project team must identify dependencies between the tasks on the WBS and estimate the effort that each task will require. The first step is to identify the dependencies between the tasks—the lowest-level items on each vertical branch. Dependencies identify which tasks must be completed first or must precede other tasks. The terms used for this relationship are *predecessor* and *successor tasks*. For example, before the team can test a system component, obviously it must be programmed or at least partially programmed.

The most common way to relate tasks is to consider the order in which they are completed—that is, as one task finishes, the next one starts. This is called a *finish-start relationship*. Other ways to relate tasks include *start-start relationships*, which means that tasks start at the same time, and *finish-finish relationships*, which means they must finish at the same time. Any of these relationships can be adjusted to include a time span (number of days) of lead or lag times. For example, an analyst may decide that task B starts two days after task A starts. This is a start-start relationship with a two-day lag.

The second step is to estimate the effort required for each task. At this point in the process of building the schedule, we want to estimate the actual employee effort required. The effort can be in hours or days or weeks, but it should be the actual amount of work required to complete the task.

Both the dependencies and the effort are estimated on the lowest-level tasks. The higher-level activities are merely summations of the low-level task times. Trying to estimate at both levels causes inconsistencies. Some developers prefer to estimate the effort before defining the dependencies; however, we recommend that the dependencies be developed first. Estimates of effort are usually more accurate if they are considered in the context of the task dependencies.

Once the tasks have been identified, the relationships determined, and the effort estimated, you can begin building the schedule. Today, we use a project scheduling tool such as Microsoft Project to build schedules. You can use MS Project to help document your work as you define the relationships and effort. We distinguish between the first three tasks of creating the WBS, determining dependencies and estimating effort, and building the schedule because the first three are "human brain" tasks that must be done by the project planner. The tool can be used to build the schedule itself.

Entering the WBS into MS Project

This section provides a brief introduction to MS Project. For more detailed instructions, refer to Appendix E on the book's Web site, which contains a learning guide to MS Project (*www.course.com/mis/sad5*). Two types of charts are used to develop a project schedule: a **PERT/CPM chart** (MS Project calls it a Network Diagram) and a **Gantt chart**. Both charts show essentially the same information, but in different formats. Each chart also has its own strengths and weaknesses. We will use the Gantt chart format, but we recommend that you view the Network diagram in MS Project.

PERT/CPM chart

a technique for scheduling a project based on individual tasks or activities

Gantt chart

a bar chart that represents the tasks and activities of the project schedule

To begin entering your project into MS Project, you create a new project. Click File|New and then enter the date you want the project to start. The first view is the Gantt Chart data-entry view. Normally, the page has three panes open: an icon menu pane on the left, the task data-entry pane in the middle, and a calendar bar chart pane on the right. In the middle panel, we begin by entering the name of each task in the Task Name column. The easiest way to enter tasks is to go from top to bottom down each leg from left to right across the WBS diagram. Enter all the activities and tasks, including the top-level activities. If your WBS reflects a left-to-right sequence of tasks, the Gantt chart will also group the tasks by the order in which they are executed. The only critical issue at this point is to make sure the lower-level tasks are positioned directly below their higher-level activity.

After all of the tasks are entered, you should begin at the lowest level, in this case level 3, and select all of the tasks listed below a given activity. Once they are selected, click the right arrow, which is the second icon from the left on the formatting taskbar. This "demotes" all of the selected tasks to subtasks of the activity. Continue this process for all lowest-level tasks. Next, select all second- and third-level tasks that are associated with a single first-level activity. Demote the entire group. Notice that the relationship between the lower levels remains.

Figure 3-12 illustrates this process. In the figure, you can see that the indentation scheme has been completed for all sublevels beneath the activities of "Define the problem" and "Produce the schedule."

Figure 3-12

Entering the WBS into MS Project

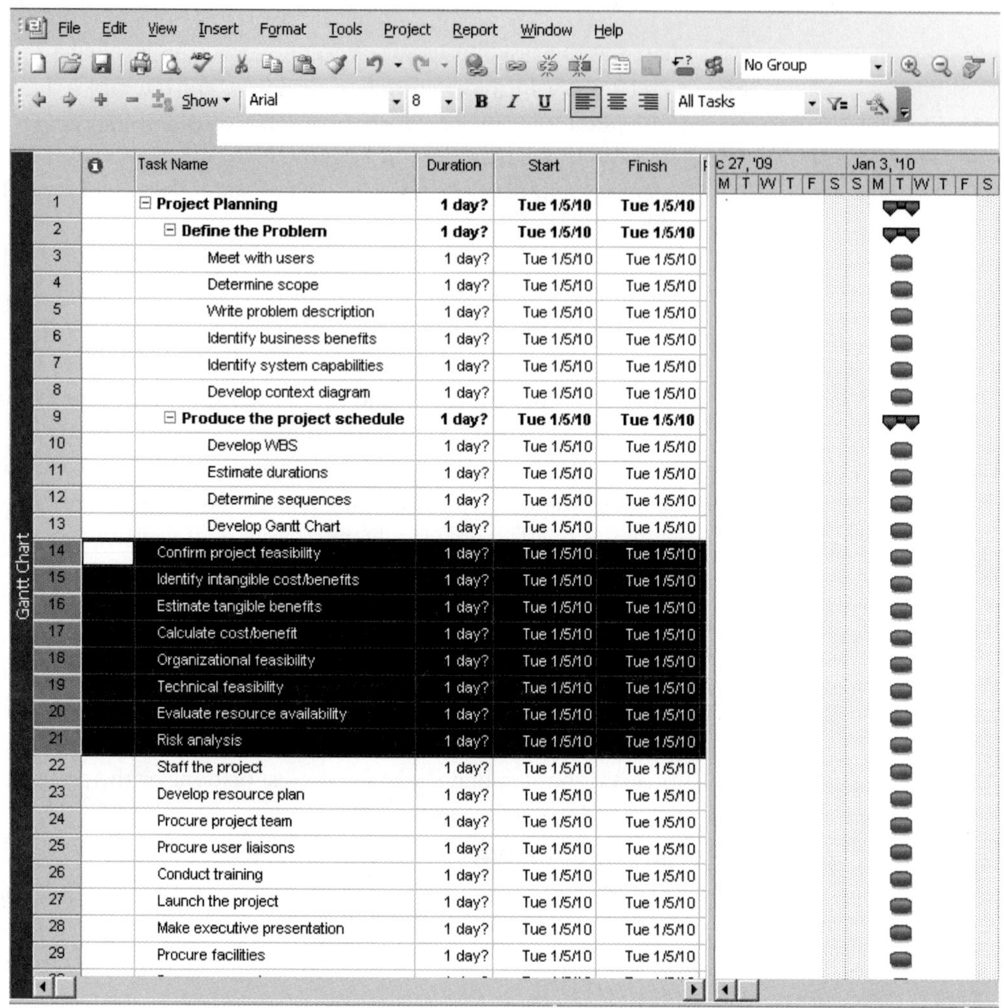

	Task Name	Duration	Start	Finish
1	**Project Planning**	**1 day?**	**Tue 1/5/10**	**Tue 1/5/10**
2	**Define the Problem**	**1 day?**	**Tue 1/5/10**	**Tue 1/5/10**
3	Meet with users	1 day?	Tue 1/5/10	Tue 1/5/10
4	Determine scope	1 day?	Tue 1/5/10	Tue 1/5/10
5	Write problem description	1 day?	Tue 1/5/10	Tue 1/5/10
6	Identify business benefits	1 day?	Tue 1/5/10	Tue 1/5/10
7	Identify system capabilities	1 day?	Tue 1/5/10	Tue 1/5/10
8	Develop context diagram	1 day?	Tue 1/5/10	Tue 1/5/10
9	**Produce the project schedule**	**1 day?**	**Tue 1/5/10**	**Tue 1/5/10**
10	Develop WBS	1 day?	Tue 1/5/10	Tue 1/5/10
11	Estimate durations	1 day?	Tue 1/5/10	Tue 1/5/10
12	Determine sequences	1 day?	Tue 1/5/10	Tue 1/5/10
13	Develop Gantt Chart	1 day?	Tue 1/5/10	Tue 1/5/10
14	Confirm project feasibility	1 day?	Tue 1/5/10	Tue 1/5/10
15	Identify intangible cost/benefits	1 day?	Tue 1/5/10	Tue 1/5/10
16	Estimate tangible benefits	1 day?	Tue 1/5/10	Tue 1/5/10
17	Calculate cost/benefit	1 day?	Tue 1/5/10	Tue 1/5/10
18	Organizational feasibility	1 day?	Tue 1/5/10	Tue 1/5/10
19	Technical feasibility	1 day?	Tue 1/5/10	Tue 1/5/10
20	Evaluate resource availability	1 day?	Tue 1/5/10	Tue 1/5/10
21	Risk analysis	1 day?	Tue 1/5/10	Tue 1/5/10
22	Staff the project	1 day?	Tue 1/5/10	Tue 1/5/10
23	Develop resource plan	1 day?	Tue 1/5/10	Tue 1/5/10
24	Procure project team	1 day?	Tue 1/5/10	Tue 1/5/10
25	Procure user liaisons	1 day?	Tue 1/5/10	Tue 1/5/10
26	Conduct training	1 day?	Tue 1/5/10	Tue 1/5/10
27	Launch the project	1 day?	Tue 1/5/10	Tue 1/5/10
28	Make executive presentation	1 day?	Tue 1/5/10	Tue 1/5/10
29	Procure facilities	1 day?	Tue 1/5/10	Tue 1/5/10

Notice that each line in the schedule is assigned a number. This numbering is fixed by MS Project, and you cannot change it. You can add WBS codes, as shown in Figure 3-11, using "Tools>Options>Show outline numbers." Also, notice that MS Project uses the word *duration* instead of *effort* to measure the size of the task. Duration is the length of time the task takes, and it is related to effort by the following equation:

Duration × Persons = Effort

The default duration is one day.

You should first enter your effort estimates in the Duration column. You may need to modify them later, when you enter resources. Also notice the column that denotes predecessor tasks. You enter the line number (task ID) for the predecessor task(s) in that column. As you enter predecessors and durations, MS Project automatically builds the Gantt chart in the right pane of the window. Notice that the defaults show all the tasks beginning on the first day of the project, with a duration of one day.

After you have entered all the tasks and identified the correct hierarchy relationships, you can enter the durations and the predecessors, but only for the lowest-level tasks. MS Project will calculate the total duration, with beginning and ending dates, for all of the summary-level activities. You can enter this information directly in the data-entry view, as shown in Figure 3-12. You also can enter the information using the "Window>Split" Option. The lower half of the screen splits and presents a form that you can use to enter effort, predecessor, resource, and other information about each task. The task highlighted on the data-entry panel is the one shown in the split window.

After entering the desired information, click OK to apply the changes. You can also navigate the task list using the Previous and Next buttons on the split window. Figure 3-13 shows the Gantt chart with a split window. In this case, the split window shows a drop-down box of all the tasks, which allows you to choose one or more as the predecessor task(s). You can enter the task duration in a box at the top of the split window. After entering the information for predecessors and duration, click OK to apply the update. After the update has been applied, the OK and Cancel buttons change to Previous and Next to allow you to navigate to other tasks.

As you enter the predecessor and duration information, MS Project updates the bars on the Gantt chart to reflect the actual project schedule (see Figure 3-14). Of course, this is an estimated schedule—in other words, it is the plan. The actual project will probably not unfold in exactly this way. This figure shows the Tracking Gantt chart view. The Tracking Gantt chart is normally used after the project has begun and the project team begins tracking progress. We chose this view to illustrate more capabilities of MS Project, as explained below.

Figure 3-13

Using a split window to enter duration and predecessor information

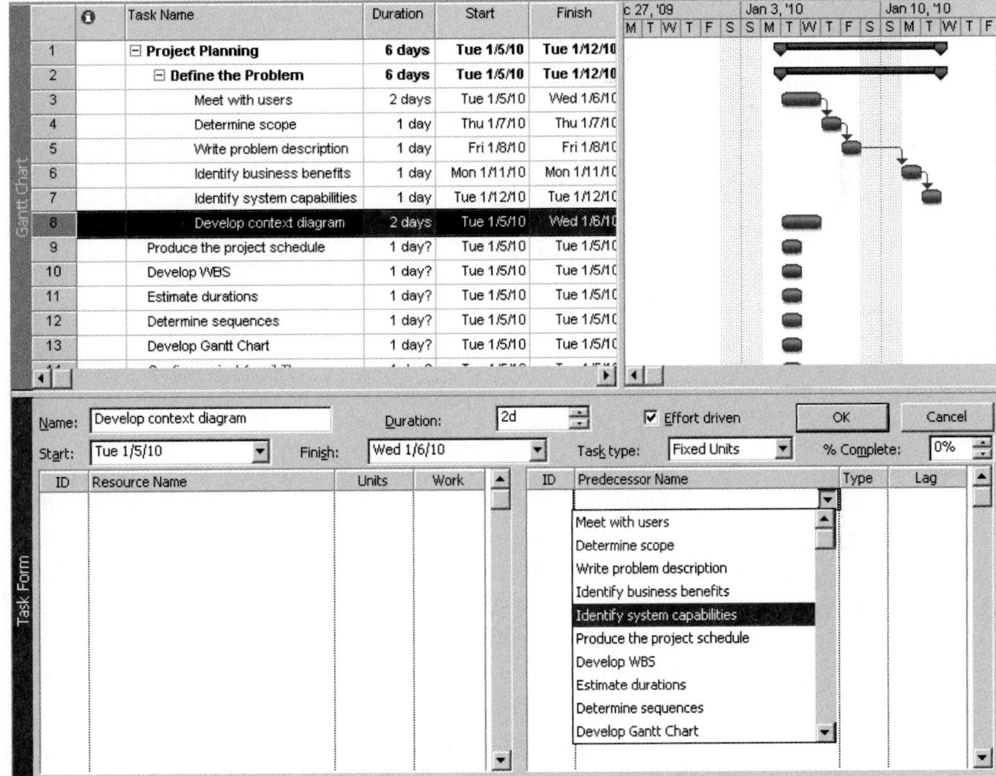

This figure shows the data-entry pane (on the left) and the bar chart pane (on the right). In the data-entry pane, the indentation distinguishes the summary activities from the detailed tasks. On the right, the summary bars also have a different bar representation, with a black roof on the bar. The duration of the summary tasks is automatically derived from the sum of the detailed tasks. The dates are also calculated automatically by MS Project.

critical path

a sequence of tasks that cannot be delayed without causing the project to be completed late

Notice that some taskbars are shown in red and others are shown in blue. The red tasks are on the critical path of the project. The **critical path** is the sequence of tasks that determine the length of the project. The critical path indicates the earliest date that the project can be completed. Another important characteristic of the critical path is that if any tasks on it are delayed, or take longer than expected to finish, then the entire project is delayed. This fact is important for a project manager; he or she should watch the tasks on the critical path carefully and take special steps to see that they are not delayed.

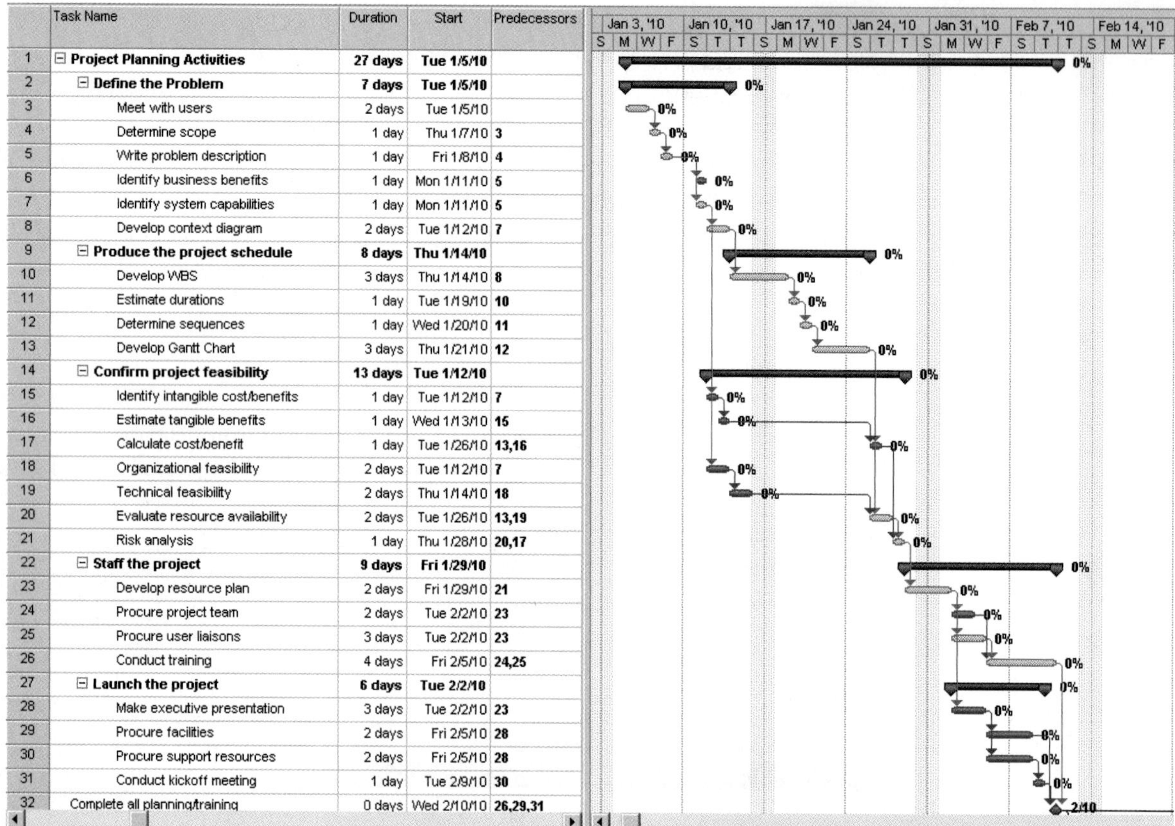

	Task Name	Duration	Start	Predecessors
1	**Project Planning Activities**	**27 days**	**Tue 1/5/10**	
2	**Define the Problem**	**7 days**	**Tue 1/5/10**	
3	Meet with users	2 days	Tue 1/5/10	
4	Determine scope	1 day	Thu 1/7/10	3
5	Write problem description	1 day	Fri 1/8/10	4
6	Identify business benefits	1 day	Mon 1/11/10	5
7	Identify system capabilities	1 day	Mon 1/11/10	5
8	Develop context diagram	2 days	Tue 1/12/10	7
9	**Produce the project schedule**	**8 days**	**Thu 1/14/10**	
10	Develop WBS	3 days	Thu 1/14/10	8
11	Estimate durations	1 day	Tue 1/19/10	10
12	Determine sequences	1 day	Wed 1/20/10	11
13	Develop Gantt Chart	3 days	Thu 1/21/10	12
14	**Confirm project feasibility**	**13 days**	**Tue 1/12/10**	
15	Identify intangible cost/benefits	1 day	Tue 1/12/10	7
16	Estimate tangible benefits	1 day	Wed 1/13/10	15
17	Calculate cost/benefit	1 day	Tue 1/26/10	13,16
18	Organizational feasibility	2 days	Tue 1/12/10	7
19	Technical feasibility	2 days	Thu 1/14/10	18
20	Evaluate resource availability	2 days	Tue 1/26/10	13,19
21	Risk analysis	1 day	Thu 1/28/10	20,17
22	**Staff the project**	**9 days**	**Fri 1/29/10**	
23	Develop resource plan	2 days	Fri 1/29/10	21
24	Procure project team	2 days	Tue 2/2/10	23
25	Procure user liaisons	3 days	Tue 2/2/10	23
26	Conduct training	4 days	Fri 2/5/10	24,25
27	**Launch the project**	**6 days**	**Tue 2/2/10**	
28	Make executive presentation	3 days	Tue 2/2/10	23
29	Procure facilities	2 days	Fri 2/5/10	28
30	Procure support resources	2 days	Fri 2/5/10	28
31	Conduct kickoff meeting	1 day	Tue 2/9/10	30
32	Complete all planning/training	0 days	Wed 2/10/10	26,29,31

Figure 3-14

Tracking Gantt chart of RMO's planning activities

slack time

the amount of time a task can be delayed without affecting the project schedule (also called float)

float

the amount of time a task can be delayed without affecting the project schedule (also called slack time)

milestone

a definite completion point in a schedule that is marked by a specific deliverable or event

The tasks shown with the blue bars are not on the critical path. Each of those tasks has some slack time. **Slack time**, or **float**, is the amount of time any task can be delayed without having a negative impact on the project completion date. For example, task number 19, "Technical feasibility," completes on January 15, 2010, but task number 20, "Evaluate resource availability," does not begin until January 26, 2010. Hence, task 19 has 11 days of slack time. In other words, if the task slipped by 11 days, it would still have no negative impact on the project.

Another important concept from this diagram is a milestone. A **milestone** is a precise point on the project schedule that indicates a specific completion point. Often, a milestone is accompanied by a deliverable or end product. Milestones provide checkpoints for project managers to verify the progress of the project. Figure 3-14 contains one milestone, task 32, indicating the completion of planning activities. In MS Project, milestones are created by entering a duration value of zero days.

Developing the Resource Requirements and the Staffing Plan

Another important activity that planners must complete when developing the project's schedule is a resource and staffing plan. As the project manager and one or two other experienced developers work to create the WBS and estimate the effort required for each task, they normally also try to identify the specific resources needed to complete the task. The core team members usually carry out much of the activities during the planning, because most of the tasks are project management activities.

There are several ways to enter resource information into MS Project. The first step is to identify the specific resources for the project by using the Resource Sheet view, as shown in Figure 3-15. In this figure, we have indicated a project manager for 100 percent availability and have even assigned him a rate for the project. We have also indicated that senior analysts will be part of the project, and have noted that two are needed—for a total of 200 percent availability and a rate of $50 per hour. (These rates may seem high, but we are accounting for benefits and perhaps even "consulting rate" charges.) Resources can be identified by type, as we have done here, or even by people's names. Usually, though, it is better to identify types of resources needed.

		Resource Name	Type	Material Label	Initials	Group	Max. Units	Std. Rate	Ovt. Rate	Cost/Use	Accrue At	Base Calendar
1		Project Manager	Work		P		100%	$75.00/hr	$100.00/hr	$0.00	Prorated	Standard
2		Senior Analyst	Work		S		200%	$50.00/hr	$75.00/hr	$0.00	Prorated	Standard

Figure 3-15

Resource sheet showing two resources

The second step is to assign these resources to the tasks. We prefer to use the split window approach, as explained earlier. You select a task, then use the drop-down box under the Resource Name column. Select the resource and indicate how long it will be available for the selected task. Figure 3-16 illustrates the entry of resources for the Develop Context Diagram task. Notice that we have assigned one and one-half senior analysts to this task.

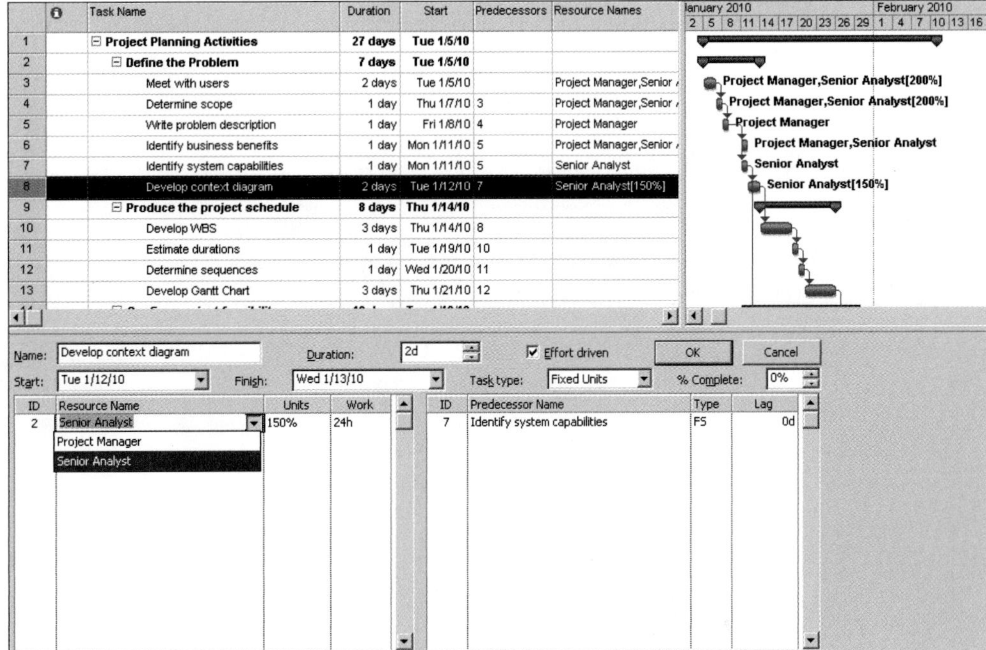

Task Name
Duration
Start
Predecessors
Resource Names
January 2010
February 2010
Project Planning Activities 27 days Tue 1/5/10
Define the Problem 7 days Tue 1/5/10
Meet with users 2 days Tue 1/5/10 Project Manager,Senior
Determine scope 1 day Thu 1/7/10 3 Project Manager,Senior
Write problem description 1 day Fri 1/8/10 4 Project Manager
Identify business benefits 1 day Mon 1/11/10 5 Project Manager,Senior
Identify system capabilities 1 day Mon 1/11/10 5 Senior Analyst
Develop context diagram 2 days Tue 1/12/10 7 Senior Analyst[150%]
Produce the project schedule 8 days Thu 1/14/10
Develop WBS 3 days Thu 1/14/10 8
Estimate durations 1 day Tue 1/19/10 10
Determine sequences 1 day Wed 1/20/10 11
Develop Gantt Chart 3 days Thu 1/21/10 12
Project Manager,Senior Analyst[200%]
Project Manager,Senior Analyst[200%]
Project Manager
Project Manager,Senior Analyst
Senior Analyst
Senior Analyst[150%]
Name: Develop context diagram
Duration: 2d
Effort driven
OK
Cancel
Start: Tue 1/12/10
Finish: Wed 1/13/10
Task type: Fixed Units
% Complete: 0%
ID Resource Name Units Work
2 Senior Analyst 150% 24h
Project Manager
Senior Analyst
ID Predecessor Name Type Lag
7 Identify system capabilities FS 0d

Figure 3-16

Entering resources for the scheduled tasks

Keep one caveat in mind when entering resources. Remember the equation "Duration × Persons = Effort." The first time you enter resources using the split window, MS Project ignores the equation and leaves the duration as you originally estimated it. However, after the first time, MS Project applies the equation every time you modify the resources. The Effort driven check box in the split window indicates to MS Project that the effort should remain constant. So, referring to the equation, if you change the number of resources or the availability of the resources, MS Project will change the task duration so that the effort remains a constant. You can turn this feature off by unchecking the Effort driven check box.

This brief introduction to MS Project illustrates how to use a tool in the development of the WBS and the project schedule. MS Project has many more features that you need to learn if you want to use its full capabilities. Appendix E on the book's Web site (*www.course.com/mis/sad5*) provides a more detailed tutorial on using MS Project.

SCHEDULING THE ENTIRE SDLC

The examples shown in Figures 3-14 through 3-16 only detail the WBS for the project planning activities. Obviously, the other SDLC phases would also need to be scheduled. Each activity within each phase is made up of a list of tasks. A template-based or analogy-based

WBS can be used to provide the detailed list of tasks for each analysis, design, and implementation phase activity. As you learn more about each phase and its activities in this book, you will understand more about the required tasks that need to be scheduled.

If we assume that the project includes overlapping SDLC phases, a Gantt chart showing the entire project at the phase and activity level of detail might look like Figure 3-17. Note that the length of each activity does not imply that the team is working full-time on that activity from start to finish. Rather, the activity starts and continues with varying degrees of effort for the duration. All team members get used to multitasking; that is, working on more than one activity or task at the same time. Therefore, an overlapping view of the project is not useful for calculating total labor cost, but it can show the completion of each phase and the entire project. The elapsed time for the CSS development project is about nine months. After that, the support phase begins.

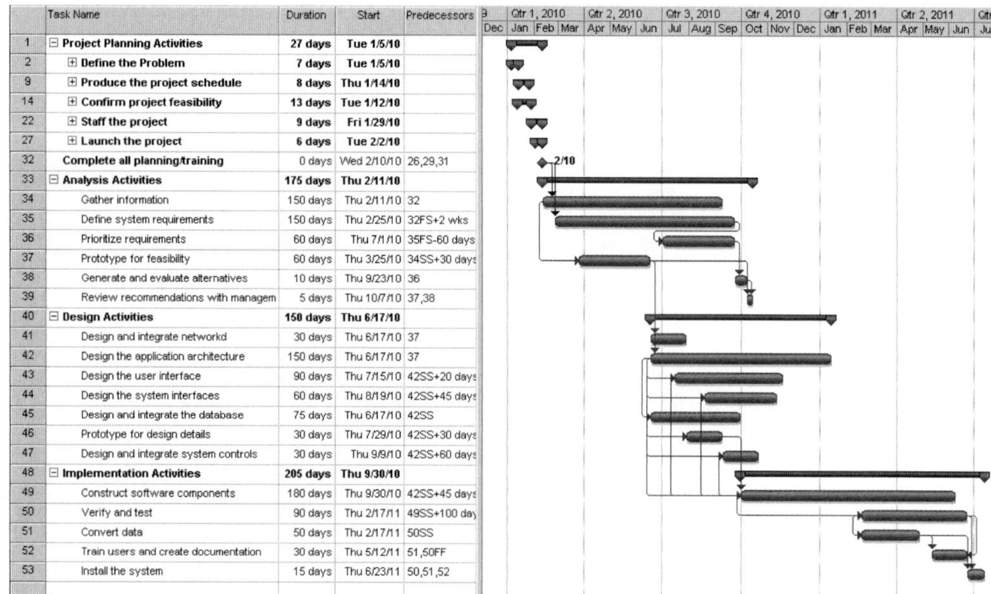

	Task Name	Duration	Start	Predecessors
1	**Project Planning Activities**	**27 days**	**Tue 1/5/10**	
2	**Define the Problem**	**7 days**	**Tue 1/5/10**	
9	**Produce the project schedule**	**8 days**	**Thu 1/14/10**	
14	**Confirm project feasibility**	**13 days**	**Tue 1/12/10**	
22	**Staff the project**	**9 days**	**Fri 1/29/10**	
27	**Launch the project**	**6 days**	**Tue 2/2/10**	
32	**Complete all planning/training**	0 days	Wed 2/10/10	26,29,31
33	**Analysis Activities**	**175 days**	**Thu 2/11/10**	
34	Gather information	150 days	Thu 2/11/10	32
35	Define system requirements	150 days	Thu 2/25/10	32FS+2 wks
36	Prioritize requirements	60 days	Thu 7/1/10	35FS-60 days
37	Prototype for feasibility	60 days	Thu 3/25/10	34SS+30 days
38	Generate and evaluate alternatives	10 days	Thu 9/23/10	36
39	Review recommendations with managem	5 days	Thu 10/7/10	37,38
40	**Design Activities**	**150 days**	**Thu 6/17/10**	
41	Design and integrate networkd	30 days	Thu 6/17/10	37
42	Design the application architecture	150 days	Thu 6/17/10	37
43	Design the user interface	90 days	Thu 7/15/10	42SS+20 days
44	Design the system interfaces	60 days	Thu 8/19/10	42SS+45 days
45	Design and integrate the database	75 days	Thu 6/17/10	42SS
46	Prototype for design details	30 days	Thu 7/29/10	42SS+30 days
47	Design and integrate system controls	30 days	Thu 9/9/10	42SS+60 days
48	**Implementation Activities**	**205 days**	**Thu 9/30/10**	
49	Construct software components	180 days	Thu 9/30/10	42SS+45 days
50	Verify and test	90 days	Thu 2/17/11	49SS+100 day
51	Convert data	50 days	Thu 2/17/11	50SS
52	Train users and create documentation	30 days	Thu 5/12/11	51,50FF
53	Install the system	15 days	Thu 6/23/11	50,51,52

Figure 3-17

Gantt chart for the complete customer support system project

Recall that for an adaptive approach, the project schedule will be based on iterations. For planning and scheduling purposes, many project managers use project management software and Gantt charts to plan and track the activities and tasks within each iteration of the project. Each iteration includes analysis, design, and implementation activities that focus on a portion of the system's functionality. Some analysis activities will be included in every iteration; other activities might only be included in a few. For example, each iteration might include analysis activities *Gather information* and *Define system requirement* and design activities *Design the application architecture, Design the user interfaces*, and *Design and integrate the database*. Similarly, each iteration might include implementation activities *Construct software components* and *Verify and test*. Other activities from each of these phases might be included in some but not all iterations, depending on the project plan. A Gantt chart in Figure 3-18 shows how the RMO project might be scheduled with three iterations.

More detailed information on these scheduling techniques—especially on how to build schedules, including PERT/CPM and Gantt charts using Microsoft Project—is provided in Appendix B on the book's Web site. The key question to be answered when completing this activity is: *Can the project be completed on time given the available resources?*

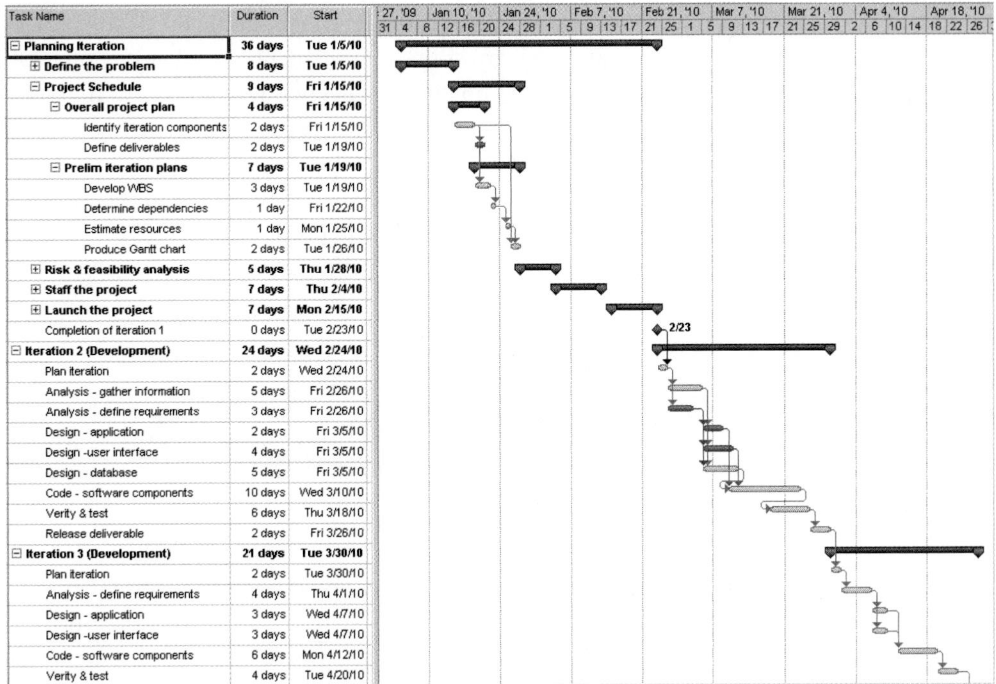

Task Name	Duration	Start
Planning Iteration	36 days	Tue 1/5/10
Define the problem	8 days	Tue 1/5/10
Project Schedule	9 days	Fri 1/15/10
Overall project plan	4 days	Fri 1/15/10
Identify iteration components	2 days	Fri 1/15/10
Define deliverables	2 days	Tue 1/19/10
Prelim iteration plans	7 days	Tue 1/19/10
Develop WBS	3 days	Tue 1/19/10
Determine dependencies	1 day	Fri 1/22/10
Estimate resources	1 day	Mon 1/25/10
Produce Gantt chart	2 days	Tue 1/26/10
Risk & feasibility analysis	5 days	Thu 1/28/10
Staff the project	7 days	Thu 2/4/10
Launch the project	7 days	Mon 2/15/10
Completion of iteration 1	0 days	Tue 2/23/10
Iteration 2 (Development)	24 days	Wed 2/24/10
Plan iteration	2 days	Wed 2/24/10
Analysis - gather information	5 days	Fri 2/26/10
Analysis - define requirements	3 days	Fri 2/26/10
Design - application	2 days	Fri 3/5/10
Design -user interface	4 days	Fri 3/5/10
Design - database	5 days	Fri 3/5/10
Code - software components	10 days	Wed 3/10/10
Verify & test	6 days	Thu 3/18/10
Release deliverable	2 days	Fri 3/26/10
Iteration 3 (Development)	21 days	Tue 3/30/10
Plan iteration	2 days	Tue 3/30/10
Analysis - define requirements	4 days	Thu 4/1/10
Design - application	3 days	Wed 4/7/10
Design -user interface	3 days	Wed 4/7/10
Code - software components	6 days	Mon 4/12/10
Verify & test	4 days	Tue 4/20/10

Figure 3-18

Gantt chart for an iterative approach to the customer support system

IDENTIFYING PROJECT RISKS AND CONFIRMING PROJECT FEASIBILITY

Project feasibility analysis is an activity that verifies whether a project can be started and successfully completed. Because, by definition, a project is a unique endeavor, every project has unique challenges that affect its feasibility. As we learned early in the chapter, information system projects do not have a very good track record. Even well-planned projects sometimes go awry and get into trouble.

The objective in assessing feasibility is to determine whether a development project has a reasonable chance of success. Feasibility analysis essentially identifies all the risks of failure. First, the project team assesses the original assumptions and identifies other risks that could jeopardize the project's success. Then, if necessary, the team establishes plans and procedures to ensure that those risks do not interfere with the success of the project. However, if the team suspects that serious risks could jeopardize the project, members must discover and evaluate them as soon as possible. Generally, the team performs the following activities when confirming a project's feasibility:

- Assess the risk to the project (risk management).
- Determine the organizational and cultural feasibility.
- Evaluate the technological feasibility.
- Determine the schedule feasibility.
- Assess the resource feasibility.
- Determine the economic feasibility.

ASSESSING THE RISKS TO THE PROJECT (RISK MANAGEMENT)

risk management

the project management area in which the team tries to identify potential trouble spots that could jeopardize the success of the project

Feasibility analysis also includes risk management. **Risk management** is the project management area that is forward-looking, during which the team tries to identify potential trouble spots that could jeopardize the success of the project. Sometimes project managers look at the feasibility from various points of view but forget to identify specific risks. We believe that good project management requires both: a careful look at the overall feasibility of the project and at the individual risks. We first present a simple technique for identifying and assessing risks. Then we explain various areas that should be considered for project feasibility and risk management.

Risk management is done throughout the life of the project. During a project's initiation, the primary activity of risk management is to identify potential risks and assess their negative impact. There are no quick and easy ways to identify all of the risks that a project might face. We have found that the best way to identify risks is simply to have a brainstorming session. The core team members should be the primary participants in this session, but selected stakeholders might also participate. The best people to include in the sessions are those who are experienced and have worked on previous projects. As with any brainstorming session, participants should let the ideas flow freely before judging and eliminating the bad ones.

BEST PRACTICE

Brainstorming sessions that include key project members and stakeholders are a good way to identify risks.

After the potential risks have been identified, the team can use a simple matrix to analyze the potential for harm to the project. Figure 3-19 is an example of a simple table that illustrates the technique. The left column is the list of identified risks. The second column, titled "Potential impact on project," provides an assessment of how badly the project will be affected if the risk materializes. The team makes a subjective judgment of three possible values: high, medium, or low. The next column indicates how likely it is that the negative event will really happen. Instead of calculating complex probabilities, again the team just estimates whether the likelihood is high, medium, or low.

Figure 3-19

Simplified risk analysis

Risk description	Potential impact on project (high, medium, low)	Likelihood of occurrence (high, medium, low)	Difficulty of timely anticipation (hard, medium, easy)	Overall threat (high, medium, low)
Critical team member (expert) not available	High	Medium	Medium	High
Changing legal requirements	High	Low	Hard	Low
Organization employees not computer savvy	Medium	Medium	Easy	Medium

The next column records the evaluation of how hard or easy it is to predict that the negative event will happen and whether that prediction can be made in time to take corrective action. For example, the first risk is that a critical expert will not be available to the team. If the project manager finds out on the day the team member is supposed to start work that he or she is not available, that risk is obviously hard to predict and can have a very negative impact on the project. If the project manager expects to have a month's warning that the resource will not be available, the negative event is easy to predict, and some other contingency can be arranged. The team evaluates the risks in this column as hard, medium, or easy.

Finally, given the values in the middle three columns, the team assigns an overall evaluation of each risk. The project manager uses this information to watch and track the potential risks and is often able either to prevent the negative event from happening or to have a backup plan ready when it does occur.

ORGANIZATIONAL AND CULTURAL FEASIBILITY

As discussed in Chapter 1, each company has its own culture, and any new system must be accommodated within that culture. There is always the risk that a new system departs so dramatically from existing norms that it cannot be successfully deployed. The analysts involved with feasibility analysis should evaluate organizational and cultural issues to identify potential risks for the new system. Such issues might include the following:

- A current low level of computer competency
- Substantial computer phobia
- A perceived loss of control by staff or management
- Potential shifting of political and organizational power due to the new system
- Fear of change of job responsibilities
- Fear of loss of employment due to increased automation
- Reversal of long-standing work procedures

It is not possible to enumerate all the potential organizational and cultural risks that exist. The project management team needs to be very sensitive to reluctance within the organization to identify and resolve these risks. The question to ask for operational feasibility is: *What items might prevent the effective use of the new system and the resulting loss of business benefits?*

After identifying the risks, the project management team can take positive steps to counter them. For example, the team can hold additional training sessions to teach new procedures and provide increased computer skills. Higher levels of user involvement in developing the new system will tend to increase user enthusiasm and commitment.

TECHNOLOGICAL FEASIBILITY

Generally, a new system brings new technology into the company. At times the new system stretches the state of the art of the technology. Other projects use existing technology but combine it into new, untested configurations. Also, even existing technology can pose the same challenges as new technology if there is a lack of expertise within the company. If an outside vendor is providing a capability in a certain area, the client organization usually assumes the vendor is expert in that area. However, even an outside vendor is subject to the risk that the requested level of technology is too complicated.

The project management team needs to assess carefully the proposed technological requirements and available expertise. When these risks are identified, the solutions are usually straightforward. The solutions to technological risks include providing additional training, hiring consultants, or hiring more experienced employees. In some cases, the scope and approach of the project may need to be changed to ameliorate technological risk. The important point is that a realistic assessment will identify technological risks early, making it possible to implement corrective measures.

SCHEDULE FEASIBILITY

The development of a project schedule always involves high risk. Every schedule requires many assumptions and estimates without adequate information. For example, the needs, and hence the scope, of the new system are not well known, the time needed to research and finalize requirements must be estimated, and the availability and capability of team members are not completely known.

Another frequent risk in developing the schedule occurs when upper management decides that the new system must be deployed within a certain time. Sometimes there is an important business requirement for defining a fixed deadline, such as RMO's need to complete the CSS in time for online ordering over the holidays. Similarly, universities require the completion of new systems before key dates in the university schedule. For example, if a new admissions system is not completed before the admissions season, then it might as well wait another full year. In cases like these, schedule feasibility can be the most important feasibility factor to consider.

If the deadline appears arbitrary, the tendency is to build the schedule to show that it can be done. Unfortunately, this practice usually spells disaster. The project team should build the schedule without any preconceived notion of required completion dates. After the schedule is completed, comparisons can be done to see whether timetables coincide. If not, the team can take corrective measures, such as reducing the scope of the project, to increase the probability of the project's on-time completion.

One objective of defining milestones during the project schedule is to permit the project manager to assess the ongoing risk of schedule slippage. If the team begins to miss milestones, the manager can possibly implement corrective measures early. Contingency plans can be developed and carried out to reduce the risk of further slippage.

Allocating adequate personnel with the right experience and expertise is always a problem in a project. Any complex project may incur overruns and schedule extensions. It may be difficult to identify the sources of these risks, but a conscious effort to identify them will at least highlight areas of weakness. Long projects are especially subject to difficulties with resource allocation and schedule slippage. Solutions can involve contingency plans in case in-house resources are not available.

RESOURCE FEASIBILITY

The project management team must also assess the availability of resources for the project. The primary resource consists of team members. Development projects require the involvement of systems analysts, system technicians, and users. Required people may not be available to the team at the necessary times. An additional risk is that people assigned to the team may not have the necessary skills for the project. After the team is functioning, members may have to leave the team. This threat can come either from staff who are transferred within the organization if other special projects arise, or from qualified team members who are hired away by other organizations. Although the project manager usually does not like to think about these possibilities, skilled people are in short supply and sometimes do leave projects.

The other resources required for a successful project include adequate computer resources, physical facilities, and support staff. Generally, these resources can be made available, but the schedule can be affected by delays in the availability of these resources.

ECONOMIC FEASIBILITY

Economic feasibility consists of two tests: (1) Is the anticipated value of the benefits greater than projected costs of development? and (2) Does the organization have adequate cash flow to fund the project during the development period? Even though the project may have received initial approval based on the need or strategic plan, final approval usually requires a thorough analysis of the development costs and the anticipated financial benefits. Obviously, the justification for developing a new system is that it will increase income, either through cost savings or by increased revenues. A determination of the economic feasibility of the project always requires a thorough **cost/benefit analysis**.

cost/benefit analysis

the analysis to compare costs and benefits to see whether investing in the development of a new system will be beneficial

Developing a cost/benefit analysis is a three-step process. The first step is to estimate the anticipated development and operational costs. Development costs are those that are incurred during the development of the new system. Operational costs are those that will be incurred after the system is put into production. The second step is to estimate the anticipated financial benefits. Financial benefits are the expected annual savings or increases in revenue derived from the installation of the new system. The third step, the cost/benefit analysis step, is calculated from the detailed estimates of costs and benefits. The most frequent error that inexperienced analysts make during cost/benefit analysis is to try to do the calculations before thoroughly defining costs and benefits. A cost/benefit analysis that does not have thorough and complete supporting detail is valueless.

Development Costs

Although the project manager has final responsibility for estimating the costs of development, senior-level analysts always assist with the calculations. Generally, project costs come in the following categories:

- Salaries and wages
- Equipment and installation

- Software and licenses
- Consulting fees and payments to third parties
- Training
- Facilities
- Utilities and tools
- Support staff
- Travel and miscellaneous

Salaries and wages are calculated based on the staffing requirements for the project. As the project schedule and staffing plans are developed, the estimated cost for salaries and wages can be determined. Figure 3-20 is an example of the estimated cost for salaries for the RMO customer support system. The numbers in this table were calculated by identifying personnel who are needed for the project and the length of time they are to be assigned to the project. As discussed previously, it is not easy to estimate the amount of time a person will be assigned to the project. Some team members work full-time on the project; others work on several projects concurrently. If the WBS is detailed and accurate, the salary and wage costs can be more accurately specified.

Each of the other categories of costs requires detailed calculations to determine the estimated costs. The project manager can make detailed cost estimates of equipment, software licenses, training, and so forth. These details are then combined to provide an estimate of the total costs of development. Figure 3-21 is a summary table of all of the costs. Again, each line in the summary table must be supported with details such as those shown in Figure 3-20.

Figure 3-20

Supporting detail for salaries and wages for RMO customer support system project

Supporting detail for salaries and wages for RMO customer support system project

Team member	Salary/wage for project
Project leader	$101,340.00
Senior systems analyst	$90,080.00
Systems analyst	$84,980.00
Programmer analysts	$112,240.00
Programmers	$58,075.00
Systems programmers	$49,285.00
Total salaries and wages	$496,000.00

Figure 3-21

Summary of development costs for RMO customer support system project

Summary of development costs for RMO customer support system project

Expense category	Amount
Salaries/wages	$496,000.00
Equipment/installation	$385,000.00
Training	$78,000.00
Facilities	$57,000.00
Utilities	$152,000.00
Support staff	$38,000.00
Travel/miscellaneous	$112,000.00
Licenses	$18,000.00
Total	$1,336,000.00

Sources of Ongoing Costs of Operations

After the new system is up and running, normal operating costs are incurred every year. The calculation of the cost and benefit of the new system must also account for these annual operating costs. Generally, analysts do not include the normal costs of running the business in this cost. Only the costs that are directly related to the new system and its maintenance are included. The following list identifies the major categories of costs that might be allocated to the operation of the new system:

- Connectivity
- Equipment maintenance
- Costs to upgrade software licenses
- Computer operations
- Programming support
- Amortization of equipment
- Training and ongoing assistance (the help desk)
- Supplies

Figure 3-22

Summary of estimated annual operating costs for RMO customer support system

Summary of estimated annual operating costs for RMO customer support system	
Recurring expense	**Amount**
Connectivity	$60,000.00
Equipment maintenance	$40,000.00
Programming	$65,000.00
Help desk	$28,000.00
Amortization	$48,000.00
Total recurring costs	$241,000.00

Figure 3-22 is a summary of the estimated annual operating costs for the RMO customer support system. As with the development costs, each entry in the table should be supported with detailed calculations. This figure represents only those costs that are anticipated for the RMO system. Other organizations may have a different set of operating costs.

Sources of Benefits

The project manager and members of the project team can determine most of the development and operational costs. However, the user and the client receive the benefits of the system. Consequently, the client and the user must determine the value of the anticipated benefits. Members of the project team can and do assist, but they should never attempt to determine the value of benefits by themselves.

Benefits usually come from two major sources: decreased costs or increased revenues. Cost savings or decreases in expenses come from increased efficiency in company operations. Areas in which to look for reduced costs include the following:

- Reducing staff by automating manual functions or increasing efficiency
- Maintaining constant staff with increasing volumes of work
- Decreasing operating expenses such as shipping charges for "emergency shipments"
- Reducing error rates through automated editing or validation
- Achieving quicker processing and turnaround of documents or transactions
- Capturing lost discounts on money management
- Reducing bad accounts or bad credit losses
- Reducing inventory or merchandise losses through tighter controls
- Collecting receivables (accounts receivable) more rapidly

- Capturing lost income due to "stock-outs" by implementing better inventory management
- Reducing cost of goods through volume discounts and purchases
- Reducing paperwork costs by implementing electronic data interchange and other automation

This list is just a sampling of the myriad benefits that can accrue. Unlike development costs, there are no "standard" benefits. Each project is different, and the anticipated benefits are different. Figure 3-23 is an example of the benefits that RMO expects from implementing the new customer support system.

Figure 3-23

Sample benefits for RMO

Sample benefits for RMO		
Benefit/cost saving	**Amount**	**Comments**
Increased efficiency in mail-order department	$125,000.00	5 people @ $25,000
Increased efficiency in phone-order department	$25,000.00	1 person @ $25,000
Increased efficiency in warehouse/shipping	$87,000.00	
Increased earnings due to Web presence	$500,000.00	Increasing at 50%/year
Other savings (inventory, supplies, and so on)	$152,000.00	
Total annual benefits	$889,000.00	

Financial Calculations

net present value (NPV)
the present value of dollar benefits and costs for an investment such as a new system

Companies use a combination of methods to measure the overall benefit of the new system. One popular approach is to determine the **net present value (NPV)** of the new system. The two concepts behind net present value are (1) that all benefits and costs are calculated in terms of today's dollars (present value) and (2) that benefits and costs are combined to give a net value. The future stream of benefits and costs are netted together and then discounted by a factor for each year in the future. The discount factor is like an interest rate, except it is used to bring future values back to current values. Appendix C on the book's Web site provides detailed instructions on how to calculate economic feasibility. You should read Appendix C to ensure that you understand the details.

Figure 3-24 shows a copy of the RMO net present value calculation done in Appendix C on the book's Web site (Figure C-1). In this case, the new system gives an NPV of $3,873,334 over a five-year period using a discount rate of 10 percent.

payback period
the time period in which the dollar benefits have offset the dollar costs

breakeven point
the point in time at which the dollar benefits have offset the dollar costs

return on investment (ROI)
a measure of the percentage gain from an investment such as a new system

Another method that organizations use to determine whether an investment will be beneficial is the **payback period**. The payback period, sometimes called the **breakeven point**, is the point in time at which the increased cash flow (benefits) exactly pays off the costs of development and operation. Appendix C on the book's Web site provides the detailed equations necessary for this calculation. Figure 3-24 illustrates the calculations for the payback period. A running accumulated net value is calculated year by year. The year when this value becomes positive is the year in which payback occurs. In the RMO example, this payback happens within the third year.

The **return on investment (ROI)** is another evaluation method used by organizations. The objective of the NPV is to determine a specific value based on a predetermined discount rate. The objective of the ROI is to calculate a percentage return (like an interest rate) so that the costs and the benefits are exactly equal over the specified time period. Figure 3-24 shows an ROI calculation for RMO, as developed in Appendix C. The time period can be the expected life of the investment (such as the productive life of the system), or it can be an arbitrary time period.

For RMO, assuming a five-year benefit period, the ROI is 172.18 percent. In other words, the investment in the development costs returned 172.18 percent on the investment for a

	RMO cost/benefit analysis	Year 0	Year 1	Year 2	Year 3	Year 4	Year 5	Total
1	Value of benefits	$ -	$ 889,000	$ 1,139,000	$ 1,514,000	$ 2,077,000	$ 2,927,000	
2	Discount factor (10%)	1	0.9091	0.8264	0.7513	0.6830	0.6209	
3	Present value of benefits	$ -	$ 808,190	$ 941,270	$ 1,137,468	$ 1,418,591	$ 1,817,374	$6,122,893
4	Development costs	$(1,336,000)						$(1,336,000)
5	Ongoing costs		$ (241,000)	$ (241,000)	$ (241,000)	$ (241,000)	$ (241,000)	
6	Discount factor (10%)	1	0.9091	0.8264	0.7513	0.6830	0.6209	
7	Present value of ongoing costs	$ -	$ (219,093)	$ (199,162)	$ (181,063)	$ (164,603)	$ (149,637)	$(913,559)
8	PV of net of benefits and costs	$(1,336,000)	$ 589,097	$ 742,107	$ 956,405	$ 1,253,988	$ 1,667,737	
9	Cumulative NPV	$(1,336,000)	$(746,903)	$ (4,769)	$951,609	$2,205,597	$ 3,873,334	
10	Payback period		2 years + 4796 / (4796 + 951,609) = 2 + .005 or 2 years and 2 days					
11	5-year return on investment		(6,122,893 - (1,336,000 + 913,559)) / (1,336,000 + 913,559) = 172.18%					

Figure 3-24

Net present value, payback period, and return on investment for RMO

period of five years. Because the system is generating benefits at that point in time, if you assumed that the lifetime was longer, such as 10 years, you would get a much higher ROI.

Intangibles

As indicated in the best practice, many projects are initiated solely for their intangible benefits. Never discount the importance of ascertaining the "behind the scenes" reasons for a project. There may be political reasons for or against the project that override all other feasibility analyses. The previous cost/benefit calculation is dependent on an organization's ability to quantify the costs and the benefits. However, in many instances, an organization cannot measure some costs and benefits and determine a value. If it can estimate a dollar value for a benefit or a cost, the organization treats the value as a **tangible benefit** or cost. If there is no reliable method of estimating or measuring the value, it is considered an **intangible benefit**. In some instances, the importance of the intangible benefits far exceed the tangible costs, at least in the opinion of the client, and the client proceeds to develop the system even though the dollar numbers do not indicate a good investment.

tangible benefits

benefits that can be measured or estimated in terms of dollars and that accrue to the organization

intangible benefits

benefits that accrue to the organization but that cannot be measured quantitatively or estimated accurately

Examples of intangible benefits include the following:

- Increased levels of service (in ways that cannot be measured in dollars)
- Increased customer satisfaction (not measurable in dollars)
- Survival (a standard capability common in the industry, or common to many competitors)
- Need to develop in-house expertise (such as with a pilot program with new technology)

Examples of intangible costs include the following:

- Reduced employee morale
- Lost productivity (the organization may not be able to estimate it)
- Lost customers or sales (during some unknown period of time)

Only tangible benefits and costs are used when calculating NPV, payback, and ROI. Even though the intangibles do not enter into the calculations, they should be considered. In fact, they may be the deciding factor in whether the project proceeds or not.

BEST PRACTICE

Intangible benefits are not included in the cost/benefit analysis, but often they are the most important reason for initiating a project. Be clear about the objectives of management before making a recommendation.

Sources of Funds

As we explained earlier, the project team performs the cost/benefit analysis in conjunction with the development of the project budget. The two components of economic feasibility are concerned with a positive result from the cost/benefit analysis and the source of funds for system development. Organizations can finance development projects in various ways. Frequently, new information systems are financed using a combination of current cash flows and long-term capital. The project team may not be involved in obtaining the financing for the project. However, the results of the cost/benefit analysis will greatly influence the financing decisions.

COMPLETING THE FEASIBILITY ANALYSIS

Each of the preceding feasibility analyses has assumed that the RMO project is feasible. But, not every project is feasible. For a project to be viable, it must pass all of the feasibility tests. In other words, the team must examine each area of the project carefully and make a determination based on relevant data. If the project is not feasible in any one of the categories, they must make adjustments. If adjustments cannot improve the situation, the project should not be initiated. One viable alternative to starting a project that has high risk of failure is simply to do nothing—for now. A project that is not feasible today—for example, due to technical difficulties, high costs, or inadequate expertise—may become feasible in the future. Project managers generally dislike concluding that the project is not feasible and should not be done. The alternative, however, is to begin a project that is destined to fail, harming the company and all involved.

An assessment of each of these six areas of feasibility is an important part of project planning. The key question to be answered when completing this activity is: *Is it still feasible to begin working on this project?*

STAFFING AND LAUNCHING THE PROJECT

The responsibility for staffing the project team falls primarily on the project manager. Human resource management, as explained in Appendix A on the book's Web site, includes finding the right people with the correct skills and then organizing and managing them throughout the project. The staffing activity consists of five tasks:

- Develop a resource plan for the project.
- Identify and request specific technical staff.
- Identify and request specific user staff.
- Organize the project team into workgroups.
- Conduct preliminary training and team-building exercises.

Based on the tasks identified in the project schedule, the project manager can develop a detailed resource plan. In fact, the schedule and the resource requirements are usually developed concurrently. If the project manager is using a tool such as Microsoft Project to build the schedule, then the resources required for each task are part of the total schedule. In developing the plan, the project manager recognizes (1) that resources are usually not available as soon as requested and (2) that a period of time is needed for a person to become acquainted with the project.

After developing the plan, the project manager can then identify specific people and request that they become part of the team. Generally, two sources exist for members of the team: (1) technical staff and (2) user staff. Technical staff means the systems analysts, the programmer analysts, the network specialists, and other technicians. Technical staff expect to move from project to project and find change normal. The project manager will meet with the director or vice president of information systems to identify and schedule the necessary resources. In some instances, it might be necessary to hire additional technical staff, so the human resource department might need to become involved. Even though finding technical people for the team is standard procedure, finding and assigning all of the required team members can take some time.

The user staff are people from the user community who are assigned to the team. Sometimes it is difficult to get users assigned to the team full-time. Being assigned to a project team is not part of the normal job progression of someone in a user department or group. However, projects do progress more smoothly if a few full-time team members can represent the user community and act as liaisons. Referring back to causes of project failure, remember that having users closely associated with the project team or assigned to it will enhance the chances of success.

On small projects, members of the project team may all work together. However, a project team that is larger than four or five members usually is divided into smaller working groups. Each group will have a group leader who coordinates the tasks assigned to the group. The responsibility for dividing the team into groups and assigning group leaders falls on the project manager.

Finally, training and team-building exercises are conducted. Training may be done for the project team as a whole when new technology such as a new database or a new programming language is used. In other cases, new team members who are unfamiliar with the tools and techniques being used may require individual training. The team should conduct appropriate training for both technical people and users. Team-building exercises are especially important when members have not worked together before. The integration of user members of the team with technical people is an important consideration in developing effective teams and workgroups.

The key question to be answered when completing the staffing activity is: *Are the resources available, trained, and ready to start the project?*

After the previous project planning activities are complete, it is time to launch the project. The scope of the new system is defined, the risks have been identified, the project has been found feasible both economically and otherwise, a detailed schedule has been developed, team members have been identified and are ready, and it is now time to start. Two final tasks usually occur at this point. First, the membership of the oversight committee is finalized, and it meets to give final go-ahead for the project, including releasing the necessary funds. Second, the organization makes a formal announcement through its standard communication channels that gives credence to the project and solicits cooperation from all involved parties in the organization. In other words, the project gets the blessing and visible support of the organization's senior executives. No project should begin without these two events.

The key question to be answered when launching the project is: *Are we ready to start the project?*

RECAP OF PROJECT PLANNING FOR RMO

Barbara and Steve spent the entire month of February putting together the schedule and plans for the CSS. Even though Barbara was the project manager, she and Steve worked together as peers. As a team, they could brainstorm and double-check each other's work. They had worked together before and had an excellent relationship—one based on mutual respect and trust. They could be candid and knew how to work through disagreements as well as how to come to consensus on important issues. Barbara also knew that the work Steve produced was always well thought out and very professionally done. He was a skilled systems analyst and would help make sure that the work done in the planning phase was solid.

The success of the overall project depended heavily on the planning Barbara and Steve did during this phase. The foundation for all other project activities is established during project planning. As Barbara planned for the kickoff meeting to launch the project officially, she reviewed the areas of project management to make sure that she had addressed all of the critical issues.

For project scope management, she developed a list of business benefits, a list of system capabilities, and a context diagram. At this point in the project, the scope definition was still very general. She would make sure the project's scope was precisely defined during the information-gathering activities of the analysis phase.

She and Steve had developed a detailed work breakdown structure and entered the information into Microsoft Project. The schedule was very detailed for the analysis phase, but less so for the design and implementation phases. She would add those details as decisions were made about the implementation approach. She thought that her approach to project time management had been established, and she would have the tools necessary to track the schedule as the project progressed.

The costs and potential benefits had been estimated and used to develop an NPV estimate. She would redo the NPV when she redid the schedule at the end of the analysis phase to ensure that the costs and schedule were within the allowed budget. The other part of cost management was to monitor the costs during the life of the project. Microsoft Project would help her track the costs of each task.

Steve had done a lot of the work to identify and assess risks during the feasibility analysis. Barbara knew that they would both continue to look for risks and assess potential problems during the project. She asked Steve to take time each week to assess the risks and update the list of the highest risks for the project. She felt confident that she would not be blindsided by some unexpected problem.

For project communication and project quality, Barbara established procedures for the project. She set up a central database to post the project's status, decisions, and working documents to make sure that all the team members were kept well informed. She established a routine and format for weekly status reports from the team leaders and a status report to the oversight committee. An example of one of her status report memos to the oversight committee is shown. These status reports all follow a standard format. In addition to the formal status memos, she would also write more informal memos to John MacMurty. For project quality, internal procedures required that team members and RMO users review all work products. Other quality procedures, such as the test plan, would be established as the project progressed.

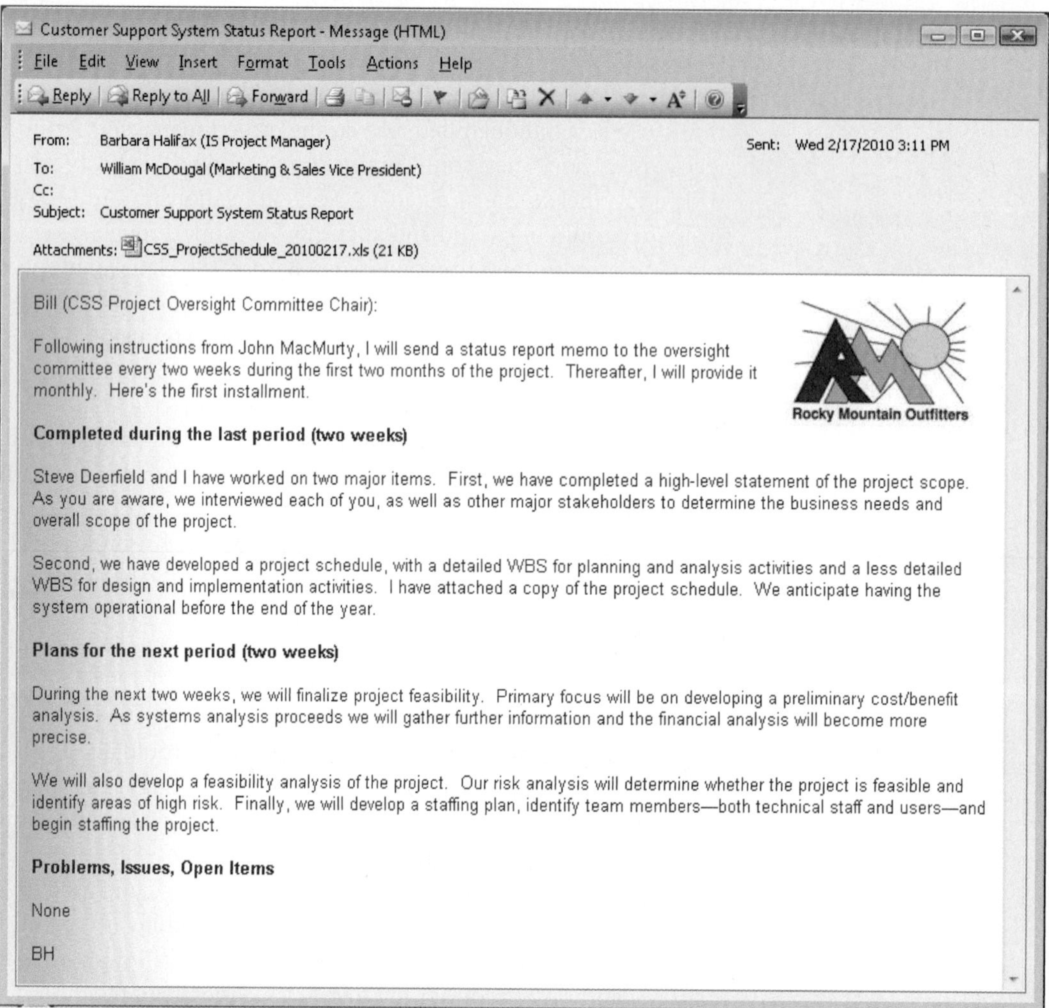

Customer Support System Status Report - Message (HTML)

File Edit View Insert Format Tools Actions Help

Reply | Reply to All | Forward

From: Barbara Halifax (IS Project Manager) Sent: Wed 2/17/2010 3:11 PM
To: William McDougal (Marketing & Sales Vice President)
Cc:
Subject: Customer Support System Status Report
Attachments: CSS_ProjectSchedule_20100217.xls (21 KB)

Bill (CSS Project Oversight Committee Chair):

Following instructions from John MacMurty, I will send a status report memo to the oversight committee every two weeks during the first two months of the project. Thereafter, I will provide it monthly. Here's the first installment.

Completed during the last period (two weeks)

Steve Deerfield and I have worked on two major items. First, we have completed a high-level statement of the project scope. As you are aware, we interviewed each of you, as well as other major stakeholders to determine the business needs and overall scope of the project.

Second, we have developed a project schedule, with a detailed WBS for planning and analysis activities and a less detailed WBS for design and implementation activities. I have attached a copy of the project schedule. We anticipate having the system operational before the end of the year.

Plans for the next period (two weeks)

During the next two weeks, we will finalize project feasibility. Primary focus will be on developing a preliminary cost/benefit analysis. As systems analysis proceeds we will gather further information and the financial analysis will become more precise.

We will also develop a feasibility analysis of the project. Our risk analysis will determine whether the project is feasible and identify areas of high risk. Finally, we will develop a staffing plan, identify team members—both technical staff and users—and begin staffing the project.

Problems, Issues, Open Items

None

BH

She and Steve had identified the other people they would like to have on the team. John had been especially helpful in finding solid analysts who were available or who would be available soon. In fact, Barbara had already interviewed all of the members who were coming on board. Recognizing the importance of having a team whose members could work together, she had scheduled several days for the team members to get to know each other, to refine their internal working procedures, and to teach them about the tools and techniques that would be used on the project.

All in all, it had been a very hectic but productive month. A lot of work had been done, and a solid foundation had been established for a successful project.

SUMMARY

The focus of this chapter is on project management activities that form the basis of project planning activities of the SDLC. The chapter covered three major themes: (1) project management, (2) information system project initiation and project planning, and (3) techniques used by the project manager and analysts for completing the project planning activities of the SDLC.

The development of a new system requires an organized, step-by-step approach. We call this approach the systems development life cycle (SDLC), as discussed in Chapter 2. The SDLC defines the phases, activities, and tasks that require attention during the system development project. Project management tasks are involved in project planning at the beginning of the project, but these tasks continue throughout the project as well.

Project management is the organizing and directing of other people to achieve a planned result within a predetermined schedule and budget. Project management can be divided into eight knowledge areas: scope, time, cost, quality, human resources, communications, risk, and procurement.

Projects are initiated based on information system needs that are identified and prioritized in strategic plans of the organization. They are also initiated on an ad hoc basis as problems or directives arise. After a project is initiated, project planning activities are carried out primarily by the project manager and one or two other senior analysts. Many of the responsibilities of the project manager are carried out via the activities of project planning. Project planning consists of five activities: (1) defining the problem, (2) producing the project schedule, (3) confirming project feasibility, (4) staffing the project, and (5) launching the project.

To define the problem, the project manager investigates the problem and the ideas originally defined for a system solution. The scope of the project is established, and an initial system context diagram is used to graphically model the major inputs and outputs. The project schedule is produced by creating a work breakdown structure of phases, activities, and tasks required to complete the project, based on the SDLC. Scheduling is difficult because phases and activities often overlap and several iterations might be used for the project. Scheduling techniques are used to investigate scheduling bottlenecks and risks. Ultimately, the project schedule is used as the basis for calculating project labor costs, as labor is based on the amount of time spent by project members on project tasks.

Confirming project feasibility requires evaluating risks related to the types of feasibility: risk, organizational and cultural, technological, schedule, resource, and economic. Risk management addresses all sources of project risk. Economic feasibility is confirmed using cost/benefit analyses to compare the costs of the project with the expected benefits. Net present value (NPV), payback period, and return on investment (ROI) calculations are used to determine whether the cost/benefit analysis is favorable for the project, although intangible benefits are often important reasons for moving forward with a project.

Project planning activities are completed by a small team, often just the project manager and one or two key analysts. When the project moves on to the analysis phase, additional team members must be identified and assigned to the project. The staffing plan must address team member needs months into the future as well. When the project is ready to be launched, key management personnel and executive sponsors must be notified and involved to ensure project success.

KEY TERMS

Agile Software Development, p. 81
breakeven point, p. 105
business benefits, p. 87
client, p. 76
context diagram, p. 87
cost/benefit analysis, p. 102
critical path, p. 95
float, p. 96
Gantt chart, p. 93
intangible benefits, p. 106
milestone, p. 96
net present value (NPV), p. 105
oversight committee, p. 77
payback period, p. 105
PERT/CPM chart, p. 93
project management, p. 75
proof of concept prototype, p. 87
return on investment (ROI), p. 105
risk management, p. 99
slack time, p. 96
system scope document, p. 87
tangible benefits, p. 106
user, p. 77
weighted scoring, p. 83
work breakdown structure (WBS), p. 90

...ONS

...ivities of project planning.
...projects fail.
...ojects are successful.
...s projects are initiated?
...gement.
...t by "Agile Software Development."
...ation system project management is similar ... anagement in general.
8. Explain how iterative development makes project scheduling more complex.
9. Describe the six types of feasibility used to evaluate a project.
10. What is the purpose of the cost/benefit analysis used to assess economic feasibility?
11. Explain the difference between tangible and intangible costs and benefits. Which are ignored in cost/benefit analyses?
12. Explain how "just in time" project management is used for adaptive projects.
13. List at least five possible sources of tangible benefits from the installation of a new system.
14. List at least four sources of development costs.
15. What is meant by the critical path?
16. What is the purpose of a system context diagram?
17. Describe the eight knowledge areas of project management.
18. What activities in the planning phase are specifically focused on project management?

THINKING CRITICALLY

1. Write a short paper that discusses how project management techniques can overcome the reasons for project failure listed at the beginning of the chapter.
2. Given the following narrative, make a list of expected business benefits:

   Especially for You Jewelers is a small jewelry company in a college town. Over the last couple of years, Especially for You has experienced a tremendous increase in its business. However, its financial performance has not kept pace with its growth. The current system, which is partially manual and partially automated, does not track accounts receivables sufficiently, and Especially for You is having difficulty determining why the receivables are so high. In addition, Especially for You runs frequent specials to attract customers. It has no idea whether these specials are profitable or whether the benefit, if there is one, comes from associated sales. Especially for You also wants to increase repeat sales to existing customers, and thus needs to develop a customer database. The jewelry company wants to install a new direct sales and accounting system to help solve these problems.
3. Given the following narrative, make a list of system capabilities:

   The new direct sales and accounting system for Especially for You Jewelers is an important element in the future growth and success of the jewelry company. The direct sales portion of the system needs to track every sale and be able to link to the inventory system for cost data to provide a daily profit and loss report. The customer database needs to be able to produce purchase histories to assist management in preparing special mailings and special sales to existing customers. Detailed credit balances and aged accounts for each customer would help solve the problem with the high balance of accounts receivables. Special notice letters and credit history reports would help management reduce accounts receivable.
4. Develop a project charter for Especially for You Jewelers based on your work from problems 3 and 4.
5. Build a Gantt chart based on the following list of tasks and dependencies to build and test a screen form for a new system. Identify the critical path.

Task ID	Description	Duration (days)	Predecessor
0	Start	0	—
1	Meet with user	2	0
2	Review existing forms	1	0
3	Identify and specify fields	3	1, 2
4	Build initial prototype	2	3
5	Develop test data (valid data)	4	3
6	Develop error test data	2	5
7	Test prototype	3	4, 6
8	Make final refinements	3	7

6. Suppose that you work in a dentist's office and are asked to develop a system to track patient appointments. How would you start? What would you do first? What kinds of things would you try to find out first? How does your approach compare with what this chapter has described?

EXPERIENTIAL EXERCISES

1. Using Microsoft Project, build a project schedule based on the following scenario. Print the Gantt chart. If required by your teacher, also print the Network Diagram (i.e., a PERT chart).
   In the table to the right is a list of tasks a student can perform to have an international experience by attending a university abroad. You can build schedules for several versions of this set of tasks. For the first version, assume that all predecessor tasks must finish before the succeeding task can begin (the simplest version). For a second version, identify several tasks that can begin a few days before the end of the predecessor task. For a third version, modify the second version so that some tasks can begin a few days after the beginning of a predecessor task. Also, insert a few overview tasks such as Application tasks, Preparation tasks, Travel tasks, and Arrival tasks. Be sure to state your assumptions for each version.
2. Build a project plan to show your progress through college. Include the course prerequisite information. If you have access to Microsoft Project or another tool, enter the information in the project management tool.
3. Using information from your organizational behavior classes or other sources, write a one-page paper on what kinds of team-building and training activities might be appropriate as the project team is expanded for the analysis phase.
4. Ask a systems analyst about the SDLC that his or her company uses. If possible, ask the analyst to show you a copy of the project schedule. To what extent is iterative development used?
5. Ask a project manager for his or her opinion on each of the eight project management knowledge areas.
6. Go to the CompTIA Web site (www.compTIA.org) and find the requirements for the project manager exam (CompTIA Project+). Write a one-page summary of the expertise and knowledge required to pass the exam.

Task ID	Description	Duration (days)	Predecessor
1	Obtain forms from the international exchange office	1	None
2	Fill out and send in the foreign university application	3	1
3	Receive approval from the foreign university	21	2
4	Apply for scholarship	3	2
5	Receive notice of approval for scholarship	30	4
6	Arrange financing	5	3, 5
7	Arrange for housing in dormitory	25	6
8	Obtain a passport and the required visa	35	6
9	Send in preregistration forms to the university	2	8
10	Make travel arrangements	1	7, 9
11	Determine clothing requirements and go shopping	10	10
12	Pack and make final arrangements to leave	3	11
13	Travel	1	12
14	Move into the dormitory	1	13
15	Finalize registration for classes and other university paperwork	2	14
16	Begin classes	1	15

CASE STUDIES

CUSTOM LOAD TRUCKING

It was time for Stewart Stockton's annual performance review. As Monica Gibbons, an assistant vice president of information systems, prepared for the interview, she reviewed Stewart's assignments over the last year and his performance. Stewart was one of the "up and coming" systems analysts in the company, and she wanted to be sure to give him solid advice on how to advance his career. She knew, for example, that he had a strong desire to become a project manager and accept increasing levels of responsibility. His desire was certainly in agreement with the needs of the company.

Custom Load Trucking (CLT) is a nationwide trucking firm that specializes in the rapid movement of high-technology equipment. With the rapid growth of the communications and computer industries, CLT was feeling more and more pressure from its clients to be able to move its loads more rapidly and precisely. Several new information systems were planned that would enable CLT to schedule and track shipments and trucks almost to the minute. However, trucking was not necessarily a high-interest industry for information systems experts. With the shortage in the job market, CLT had decided not to try to hire project managers for these new projects but to build strong project managers from within the organization.

As Monica reviewed Stewart's record, she found that he had done an excellent job as a team leader on his last project. His last assignment was as a combination team leader/systems analyst on a four-person team. He had been involved in systems analysis, design, and programming, and he also managed the work of the other three team members. He had assisted in the development of the project schedule and had been able to keep his team right on schedule. It also appeared that the quality of his team's work was as good as, if not better than, other teams on the project. She wondered what advice she should give him to help him advance his career. She was also wondering if now was the time to give him his own project.

1. Do you think the decision by CLT to build its own project managers from the existing employee base is a good one? What advice would you give to CLT to make sure that it has strong project management skills in the company?
2. What kind of criteria would you develop for Monica to use to measure whether Stewart (or any other potential project manager) is ready for project management responsibility?
3. How would you structure the job for new project managers to ensure, or at least increase the possibility of, a high level of success?
4. If you were Monica, what kind of advice would you give to Stewart about managing his career and attaining his immediate goal to become a project manager?

RETHINKING ROCKY MOUNTAIN OUTFITTERS

The chapter identified six areas of project feasibility that need to be evaluated for any new project. However, as indicated, each of these areas of feasibility can also be considered an evaluation of the potential risks of the project. Based on your understanding of Rocky Mountain Outfitters, both from this chapter and the information provided in Chapter 1, build a table that summarizes the risks faced by RMO for this new project. Include four columns titled (1) Project risk, (2) Type of risk, (3) Probability of risk, and (4) Steps to alleviate risk.

Identify as many risks to the project as you can. Type of risk means the category or area of the project feasibility that is at risk. It might help you think about risks in the different categories, for example (1) risk management, (2) economic, (3) organizational and cultural, (4) technological, (5) schedule, and (6) resources. The chapter provided a few examples of risk in each of these areas. However, many other risks can cause project failures. Think as broadly as possible and expand the list of potential risks in each area.

Obviously, other kinds of risks are associated with a project of the magnitude of the customer support system. You might want to consider some risks external to the company, such as economic, marketplace, legal, environment, and so forth. Other types of internal risks might also be associated with components that are purchased or outsourced, such as development tools, learning curves, poor quality of purchased components, and failure of vendors.

A common risk management technique is to build a table and identify the top 10 risks to the project. Contingency plans can then be built for the top 10 risks. Periodically, the project management team reevaluates the risk list to determine the current top 10 risks. After you build the table, identify which risks you would classify as the top 10 risks.

FOCUSING ON RELIABLE PHARMACEUTICAL SERVICE

Chapter 2 discussed Reliable Pharmaceutical Service's Web-based application to connect its client nursing homes directly with a new prescription and billing system. You considered both the risks of a sequential, waterfall approach to the SDLC and the risks of an iterative and incremental approach to the SDLC for its development.

1. Now consider the way the project was probably initiated. To what extent is the project the result of (a) an opportunity, (b) a problem, or (c) a directive?
2. Many of the system users (such as employees at health-care facilities) are not Reliable employees. What risks of project failure are associated with the mixed user community? What would you, as a project manager, do to minimize those risks?
3. What are some of the tangible benefits to the project? What are some of the intangible benefits? What are some of the tangible and intangible costs? How would you handle the project's benefits and costs that will accrue to the health-care facilities—would you include tangible benefits and costs to the nursing homes in the cost/benefit analyses? Why or why not?
4. Overall, do you think the approach taken to the project (sequential waterfall versus iterative and incremental) would make a difference in the tangible and intangible costs and benefits? Discuss.
5. Overall, do you think the approach taken to the project would make a difference in minimizing the risks of project failure? Discuss.

FURTHER RESOURCES

Scott W. Ambler, *Agile Modeling: Effective Practices for XP and the RUP*. John Wiley and Sons, 2004.

Jim Highsmith, *Agile Project Management: Creating Innovative Products*. John Wiley and Sons, 2004.

Gopal K. Kapur, *Project Management for Information, Technology, Business, and Certification*. Prentice-Hall, 2005.

Jack R. Meredith and Samuel J. Mantel Jr., *Project Management: A Managerial Approach* (6th ed.). John Wiley and Sons, Inc., 2004.

Joseph Phillips, *IT Project Management: On Track from Start to Finish*. McGraw-Hill, 2002.

Project Management Institute, *A Guide to the Project Management Body of Knowledge*, 3rd edition. Project Management Institute, 2004.

Walker Royce, *Software Project Management: A Unified Framework*. Addison-Wesley, 1998.

Kathy Schwalbe, *Information Technology Project Management, Fifth Edition*. Course Technology, 2008.

PART

2 SYSTEMS ANALYSIS ACTIVITIES

CHAPTER 4
Investigating System Requirements

CHAPTER 5
Modeling System Requirements

CHAPTER 6
The Traditional Approach to Requirements

CHAPTER 7
The Object-Oriented Approach to Requirements

CHAPTER 8
Evaluating Alternatives for Requirements, Environment, and Implementation

CHAPTER 4

INVESTIGATING SYSTEM REQUIREMENTS

LEARNING OBJECTIVES

After reading this chapter, you should be able to:

- Describe the activities of systems analysis
- Explain the difference between functional and nonfunctional system requirements
- Describe three types of models and reasons for creating models
- Identify and understand the different types of users who will be involved in investigating system requirements
- Describe the kind of information that is required to model system requirements
- Determine system requirements through review of documentation, interviews, observation, prototypes, questionnaires, joint application design sessions, and vendor research
- Discuss the need for validation of system requirements to ensure accuracy and completeness and the use of a structured walkthrough

CHAPTER OUTLINE

Analysis Activities in More Detail

Functional and Nonfunctional System Requirements

Models and Modeling

Stakeholders—The Source of System Requirements

Techniques for Information Gathering

Validating the Requirements

MOUNTAIN STATES MOTOR SPORTS

Amanda Lamy, president and majority stockholder of Mountain States Motor Sports (MSMS), is an avid motorcycle enthusiast and businesswoman. Headquartered in Denver, MSMS is a retailer of motorized sporting vehicles, including boats, jet skis, all-terrain vehicles, and motorcycles. The company has 47 stores located in nearly every state west of the Mississippi and in two Canadian provinces.

Over the last decade, the market for motorcycles in general—and custom-built bikes in particular—has boomed. Amanda owned three custom bikes herself and decided a year ago that the time was ripe for MSMS to expand into that market. She began seeking business acquisitions or partnerships with custom motorcycle manufacturers throughout the West.

Amanda finalized a partnership agreement with Abeyta's Custom Choppers (ACC) in Tucson just over a month ago. Other acquisitions and partnerships are planned in the near future. The partnership gave MSMS exclusive rights to distribute ACC's custom bikes and gave ACC funds to enlarge and modernize its production facility and a percentage of all retail sales. As part of the modernization, MSMS would build ACC a new information system and would also use that system in other custom bike shops.

MSMS and ACC faced a significant dilemma in developing the new information system. MSMS had no experience in manufacturing, and ACC's current accounting and production-control systems were a hodgepodge of manual procedures and automated support, primarily through Microsoft Excel spreadsheets. The business experts had little computer knowledge or experience, and the in-house computer experts at MSMS had no understanding of the business for which they would be building a new system. Buying a system off the shelf was not an option. The market was too small; no vendors served it.

After conferring with an experienced software development consulting firm, MSMS decided to conduct a joint application development (JAD) session over a three-day period. Participants in the session included the owner of ACC, an accountant and a salesperson from ACC, Amanda, her chief information officer, her vice president for operations, and a handful of MSMS programmers and developers. Participants from the consulting firm included the session leader, a developer with experience in custom and small lot production-control systems, a technical support staff member, and an administrative assistant. The session was conducted at the consulting company's offices in a computerized meeting room with dedicated servers for prototype development and deployment, and appropriate diagramming and software development tools. All participants stayed in a nearby hotel to maximize available working hours.

The session got off to a rocky start. ACC's representatives lacked computer experience, and they were uncomfortable in the heavily automated meeting room and uncertain they could accomplish the task at hand. Most of the first morning was spent acclimating ACC representatives and other participants to the process. Through the skills of the session leader, who assigned an MSMS staff member to perform all "hands on" computer tasks for ACC personnel for the entire session, camaraderie developed among participants during the first morning and over an extended lunch.

After everyone was comfortable, work proceeded in earnest. On the first day, the participants specified the overall scope and major functions of the system and described the interaction between ACC's and MSMS's accounting functions in detail. Graphical models of the interactions were generated as simple block diagrams. On the second day, attention turned to marketing and production. In the morning, the team created storyboards to describe how a salesperson in an MSMS store would display options for custom bikes as a prelude for creating a detailed order. While most of the participants ate lunch, two MSMS developers scanned pictures of customized bikes to mock up a user interface for an online design program.

In the afternoon, the mock-up was expanded into a more complete order-entry system, which captured most of the details that ACC would need to schedule and complete production. During the second evening, MSMS developers worked late into the night to flesh out the prototype. On the third morning, participants evaluated the prototype and made suggestions for further improvements. They devoted the remainder of the morning and most of the afternoon to developing requirements and design criteria for a production management system that encompassed scheduling, parts management, and accounting for time and materials. The session concluded with a one-hour review of all the requirements that had been specified and the design decisions that had been made, followed by development of an open items list and a rough schedule and budget for the project.

The JAD session really helped provide a "running start" for the entire project. The participation of key stakeholders made information gathering, prototyping, and identifying key requirements possible. Although some of the requirements defined in the JAD session were modified later, many of the requirements were incorporated into the system with little change.

OVERVIEW

In the previous chapters, you learned that system development consists of four major sets of activities: planning, analysis, design, and implementation. This chapter focuses on the activities of systems analysis and the skills and detailed tasks required. As discussed in Chapter 1, an analyst uses many skills in system development. Two key skills that are needed to perform systems analysis are (1) fact-finding for the investigation of system requirements and (2) modeling of business processes based on the system requirements. Even though systems analysis includes many other activities, these two skills are fundamental. In this chapter, you will develop fact-finding and investigation skills. Later chapters cover modeling in greater detail.

During the fact-finding and investigation activities, you learn details of business processes and daily operations. In fact, the objective during these activities is to try to become as knowledgeable about how the business operates as the users you interview. Why become an expert? Because only then can you ensure that the system meets the needs of the business. You bring a fresh perspective to the problem and possess a unique set of skills that you can employ to identify new and better ways to accomplish business objectives with information technology. Many current users are so accustomed to the way they have been performing their tasks that they cannot envision better, more advanced ways to achieve results. Your technical knowledge, combined with your newly acquired problem domain knowledge, can bring unique solutions to business processes—and make a difference in the organization.

An additional benefit to becoming an expert in the problem domain is that you build credibility with the users. Your suggestions will carry more weight because they will meet users' specific needs. During the development of a new system, you will have many suggestions and recommendations about daily business procedures, which usually require major changes in user activities. If you can "walk the walk and talk the talk" of the users' business operations, they are much more likely to accept your recommendations. Otherwise, you may be viewed as an outsider who really does not understand their problems.

The sections that follow first give an overview of the activities of systems analysis. They define system requirements and explore the different types of requirements that analysts encounter. They explain the importance of creating models. Then they explain several techniques analysts use to learn about the business processes and to gather information using both traditional and newer accelerated methods, such as the JAD sessions discussed in the Mountain States Motor Sports case. The chapter ends with a discussion of validation techniques for analysis models.

ANALYSIS ACTIVITIES IN MORE DETAIL

All of the system development approaches described and virtually all of the specific system development methodologies you will encounter in organizations include similar activities in analysis and design (see Figure 4-1). Naturally, different system development methodologies recommend different techniques for completing these activities. In many cases, they just have different names—the underlying tasks are essentially the same. In some cases, different models might be created to complete an activity. But the activities always involve answering the same key questions.

Figure 4-1

Analysis activities

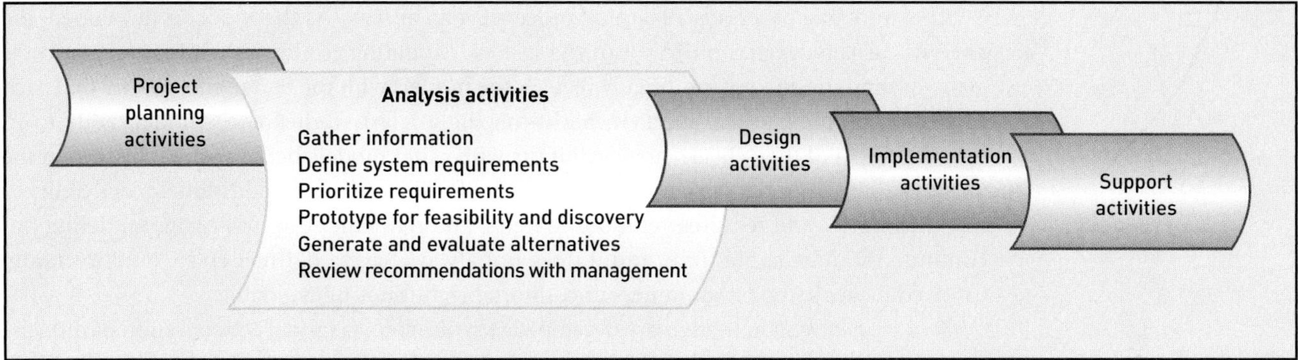

Analysis involves defining in great detail what the information system needs to accomplish to provide the organization with the desired benefits. Many alternative ideas should be proposed and the best design solution should be selected from among them. Later, during systems design, the selected alternative is designed in detail. Six activities must be completed during analysis. These activities are complementary and are usually completed simultaneously. For example, the analyst gathers information continuously and defines requirements based on that information.

GATHER INFORMATION

Analysis involves gathering a considerable amount of information. Systems analysts obtain some information from people who will be using the system, either by interviewing them or by watching them work. They obtain other information by reviewing planning documents and policy statements. Documentation from the existing system should also be studied carefully. Analysts can obtain additional information by looking at what other companies (particularly vendors) have done when faced with a similar business need. In short, analysts need to talk to nearly everyone who will use the new system or has used similar systems, and they must read nearly everything available about the existing system.

Beginning analysts often underestimate how much there is to learn about the work the user performs. The analyst must become an expert in the business area the system will support. For example, if you are implementing an order-entry system, you need to become an expert on the way orders are processed (including accounting). If you are implementing a loan-processing system, you need to become an expert on the rules used for approving credit. If you work for a bank, you need to think of yourself as a banker. The most successful analysts become very involved with their organization's main business.

Analysts also need to collect technical information. They try to understand the existing system by identifying and understanding activities of all current and future users, by identifying all present and future locations where work occurs, and by identifying all system interfaces with other systems both inside and outside the organization. Beyond that, analysts need to identify software packages that might be used to satisfy the system requirements. These specifics are discussed later in the chapter.

The key question to be answered when completing this activity is: *Do we have all of the information (and insight) we need to define what the system must do?*

DEFINE SYSTEM REQUIREMENTS

As all of the necessary information is gathered, it is very important to record it. Some of this information describes technical requirements (for example, facts about needed system performance or expected number of transactions). Other information involves functional requirements—what the system is required to do. Defining functional requirements is not just a matter of writing down facts and figures. Instead, many different types of models are created to help record and communicate what is required.

The modeling process is a learning process for an analyst. As the model is developed, the analyst learns more and more about the system. Modeling continues while information is gathered, and the analyst continually reviews the models with the end users to verify that each model is complete and correct. In addition, the analyst studies each model, adds to it, rearranges it, and then checks how well it fits with other models being created. Just when the analyst is fairly sure the system requirements are fully specified, an additional piece of information surfaces and requires yet more changes, and refinement begins again. Modeling can continue for quite some time, and it does not always have a defined end. The uncertainty involved makes some programmers uncomfortable, but it is unavoidable.

logical model

any model that shows what the system is required to do without committing to any one technology

physical model

any model that shows how the system will actually be implemented

Two types of system models are developed. A requirements model (or collection of models) is a logical model. A **logical model** shows what the system is required to do in great detail, without committing to any one technology. By being neutral about technology, the development team can focus its efforts first on what is needed, not what form it will take. For example, a model might specify an output of the system as a list of data elements without committing to either paper or on-screen formats. The focus of the model is what information the users need. A **physical model**, on the other hand, shows how the system will actually be implemented. A physical model of the output would include details about format.

The difference between logical and physical models is a key concept distinguishing systems analysis and systems design. Generally, systems analysis involves creating detailed logical models, and systems design involves detailed physical models.

The specific models created depend on the technique being used for systems analysis. The modern structured analysis technique uses data flow diagrams (DFDs) and entity-relationship diagrams (ERDs). Information engineering uses process dependency diagrams and entity-relationship diagrams. Object-oriented techniques produce class diagrams and use case diagrams. Specific examples of these models are described in detail in Chapters 5, 6, and 7.

The key question to be answered when completing this activity is: *What (in detail) do we need the system to do?*

PRIORITIZE REQUIREMENTS

After the system requirements are well understood and detailed models of the requirements are completed, it is important to establish which of the functional and technical requirements are most crucial for the system. Sometimes users suggest additional system functions that are desirable but not essential. However, users and analysts need to ask themselves which functions are truly important and which are fairly important but not absolutely required. Again, an analyst who understands the organization and the work done by the users will have more insight for answering these questions.

Why prioritize the functions requested by the users? Resources are always limited, and the analyst must always be prepared to justify the scope of the system. Therefore, it is important to know what is absolutely required. Unless the analyst carefully evaluates priorities, system requirements tend to expand as users make more suggestions (a phenomenon called *scope creep*).

The key question to be answered when completing this activity is: *What are the most important things the system must do?*

PROTOTYPE FOR FEASIBILITY AND DISCOVERY

Creating prototypes of parts of the new system can be very valuable during systems analysis. The primary purpose of building prototypes during analysis—often called *discovery prototypes*—is to better understand the users' needs. Discovery prototypes are not built with the intent of being fully functional but to check the feasibility of certain approaches to the business need. In many cases, users are trying to improve their business processes or streamline procedures. So, to facilitate the investigation of new business processes, analysts can build prototypes. Using sample screens or reports, analysts discuss with users how the new system can support new processes, and they can demonstrate new business procedures for the new system. Prototypes such as these help users discover requirements they might not have thought about otherwise and get them (and the analysts) thinking creatively "outside of the box."

If the system involves new technology, it is also important early in the project to assess whether the new technology will provide the capabilities to address the business need. Then, the team can be sure that the technology is feasible. Prototypes can prove that the technology will do what it is supposed to do. Also, if the system will include new or innovative technology, the users may need help visualizing the possibilities available from the new technology when defining what they require; prototypes can fill that need.

Prototyping helps answer two key questions: *Have we proven that the technology proposed can do what we think we need it to do?* and equally important, *Have we built some prototypes to ensure the users fully understand the potential of what the new system can do?*

GENERATE AND EVALUATE ALTERNATIVES

Many alternatives exist for the final design and implementation of a system. So, it is very important to define carefully and then evaluate all of the possibilities. When requirements are prioritized, the analyst can generate several alternatives by eliminating some of the less important requirements. In addition, technology also raises several alternatives for the system. Beyond those considerations, decisions such as whether to build the system using in-house development staff or a consulting firm affect the outcome. Furthermore, one or more off-the-shelf software packages could possibly satisfy all of the requirements.

Clearly, lots of alternatives are open to the project team, and each needs to be described or modeled at a high (summary) level. Each alternative also has its own costs, benefits, and other characteristics that must be carefully measured and evaluated (as in the feasibility study described in Chapter 3). The best alternative is then chosen. Choosing an alternative is not as easy as it sounds, because costs and benefits are very difficult to measure. And many design details are still uncertain. The analyst evaluates project feasibility once as an early project planning activity, and again later as an analysis activity.

The key question to be answered when completing this activity is: *What is the best way to create the system?*

REVIEW RECOMMENDATIONS WITH MANAGEMENT

All of the preceding activities are done in parallel—gather information, define requirements, prioritize requirements, prototype for feasibility and discovery, and generate and evaluate alternatives. Reviewing recommendations with management is usually done when all of the other analysis activities are complete or nearly complete. Management should be kept informed of progress through regular project reporting. And the project manager must eventually recommend a solution and obtain a decision from management. Questions the analyst must consider are the following: Should the project continue at all? If the project continues, which alternative is the best choice? Given the recommended alternative, what are the revised budget and schedule for completing the project?

Making a recommendation to senior executives is a major management checkpoint in the project. Every alternative—including cancellation—should be explored. Even though quite a bit of work might already have been invested in the project, it is still possible that the best choice is to cancel the project. Perhaps the benefits are not as great as originally thought. Perhaps the costs are much greater than originally thought. Or, because of the rapidly changing business environment, perhaps the organization's objectives have changed since the project was originally proposed, making it less important to the organization. For any of these reasons, it might be best to recommend that the project be canceled.

If the project is worthwhile, the project team has detailed documentation of the system requirements and a proposed design alternative, so the project manager should be able to produce a more accurate estimate of the budget and schedule for the project. If top managers understand the rationale for continuing the project, then they will probably provide the requested resources. The key point to remember is that continuing on to design activities is never automatic. Good project management techniques always require continual reassessment of the feasibility of the project and formal management reviews.

The key question to be answered when completing this activity is: *Should we continue with the design and implement the system we propose?*

Each of the six activities of analysis involves many stakeholders and tasks and involves answering one or more key questions. The activities and key questions are summarized in Figure 4-2.

Figure 4-2

Analysis activities and their key questions

Analysis activities	Key questions
Gather information	Do we have all of the information (and insight) we need to define what the system must do?
Define system requirements	What (in detail) do we need the system to do?
Prioritize requirements	What are the most important things the system must do?
Prototype for feasibility and discovery	Have we proven that the technology proposed can do what we think we need it to do? Have we built some prototypes to ensure the users fully understand the potential of what the new system can do?
Generate and evaluate alternatives	What is the best way to create the system?
Review recommendations with management	Should we continue with the design and implement the system we propose?

SYSTEM REQUIREMENTS

system requirements

specifications that define the functions to be provided by a system

System requirements are all of the capabilities and constraints that the new system must meet. Generally analysts divide system requirements into two categories: functional and nonfunctional requirements. Recall that identifying the system scope is a project planning activity. During that activity, the analyst identifies a set of system capabilities. During analysis, the analyst then defines and describes those capabilities in greater detail. In other words, the analyst expands those high-level capabilities into detailed system requirements.

functional requirement

a system requirement that describes an activity or process that the system must perform

Functional requirements are the activities that the system must perform—that is, the business uses to which the system will be applied. They derive directly from the capabilities identified during project planning. For example, if you are developing a payroll system, the required business uses might include functions such as "write paychecks," "calculate commission amounts," "calculate payroll taxes," "maintain employee-dependent information," and "report year-end tax deductions to the IRS." The new system must handle all of these functions.

Identifying and describing all of these business uses requires a substantial amount of time and effort because the list of functions and their relationships can be very complex.

Functional requirements are based on the procedures and rules that the organization uses to run its business. Sometimes they are well documented and easy to identify and describe. An example might be, "All new employees must fill out a W-4 form to enter information about their dependents in the payroll system." Other business rules might be more obtuse or difficult to find. An example from Rocky Mountain Outfitters might be that "an additional 2 percent commission rate is paid to order takers on telephone sales for 'special promotions' that are added to the order." These special promotions are unadvertised specials that are sold by the telephone order clerk—thus the special commission rate. Discovering rules such as this is critical to the final design of the system. If this rule weren't discovered, you might design a system that allows only fixed commission rates and discover much later in the development process that your design could not accommodate this rule.

nonfunctional requirement
characteristics of the system other than activities it must perform or support, such as technology, performance, usability, reliability, and security

technical requirement
a system requirement that describes an operational characteristic related to an organization's environment, hardware, and software

performance requirement
a system requirement that describes an operational characteristic related to workload measures, such as throughput and response time

usability requirement
a system requirement that describes an operational characteristic related to users, such as the user interface, work procedures, online help, and documentation

reliability requirement
a system requirement that describes the dependability of a system, such as how it handles service outages, incorrect processing, and error detection and recovery

security requirement
a system requirement that describes user access to certain functions and the conditions under which access is granted

Nonfunctional requirements are characteristics of the system other than activities it must perform or support. There are many different types of nonfunctional requirements, including the following:

- **Technical requirements** describe operational characteristics related to the environment, hardware, and software of the organization. For example, the client components of a new system might be required to operate on portable and desktop PCs running the Windows operating system and using Internet Explorer. The server components might have to be written in Java and communicate with one another using a component interaction standard such as CORBA (Common Object Request Broker Architecture) or SOAP (Simple Object Access Protocol).
- **Performance requirements** describe operational characteristics related to measures of workload, such as throughput and response time. For example, the client portion of a system might be required to have one-half-second response time on all screens, and the server components might need to support 100 simultaneous client sessions (with the same response time).
- **Usability requirements** describe operational characteristics related to users, such as the user interface, related work procedures, online help, and documentation. For example, a Web-based interface might be required to follow organization-wide graphic design guidelines, such as menu placement and format, color schemes, use of the organization's logo, and required legal disclaimers.
- **Reliability requirements** describe the dependability of a system—how often a system exhibits behaviors such as service outages and incorrect processing and how it detects and recovers from those problems. Reliability requirements are sometimes considered a subset of performance requirements.
- **Security requirements** describe which users can perform what system functions under what conditions. For example, access to certain system outputs might be limited to managers at a certain level or employees of a specific department. Some access might be authorized from home and others only from within the organization's local network. Security requirements can also apply to areas such as network communications and data storage. For example, an organization might require encryption of all data transmitted over the Internet and control of all database server access through use of a username and password.

Both functional and nonfunctional system requirements are needed for a complete definition of a new system, and both are investigated and documented during systems analysis. Functional requirements are most often documented in graphical and textual models, as described in Chapters 5 through 7. Nonfunctional requirements are usually documented in narrative descriptions that accompany the models.

MODELS AND MODELING

An analyst can best describe the requirements for an information system using a collection of models, as we discussed in Chapter 2. Recall that a model is a representation of some aspect of the system being built. Because a system is so complex, analysts create a variety of models to encompass the detailed information they have collected and digested. Also, the analyst uses many types of models to show the system at different levels of detail (or levels of abstraction), including a high-level overview as well as detailed views of certain aspects of the system. Some models show different parts of the problem and solution; for example, one model might show inputs, while another shows the data stored. Some models show the same problem and solution from different perspectives; one model might show how objects interact from the perspective of outside actors, and another might show how objects interact in terms of sequencing.

THE PURPOSE OF MODELS

Some developers think of a model as documentation produced after the analysis and design work is done. But actually, the process of creating a model helps an analyst clarify and refine the design. The analyst learns as he or she completes and then studies parts of the model. Analysts also raise questions while creating a model and answer them as the modeling process continues. New pieces are added; the consequences of changes are evaluated and again questioned. In this respect, the modeling process itself provides direct benefits to the analyst. The technique used to create the model is valuable in itself even if the analyst never shows a particular model to anyone else. But usually models are shared with others as analysis and design progresses.

Another key reason that modeling is important in system development is the complexity of describing information systems. Information systems are very complex, and parts of the systems are intangible. Models of the various parts help simplify the analyst's efforts and focus them on a few aspects of the system at a time. The reason that an analyst uses so many different models is that each relates to different aspects of the system. In fact, some of the models created by the analyst may serve only to integrate these aspects—showing how the other models fit together.

Because of the amount of information gathered and digested and the length of time each analyst spends on a project, analysts need to review the models frequently to help recall details of work previously completed. People can retain only a limited amount of information, so we all need memory aids. Models provide a way of storing information for later use in a form that can be readily digested.

The support for communication is one of the most often cited reasons for creating the models. Given that the analyst learns while working through the modeling process and that the collection of models reduces the complexity of the information system, the models also serve a critical role in supporting communication among project team members and with system users. If one team member is working on models of inputs and outputs, and another team member is working on models of the processes that convert the inputs to outputs, then they need to communicate and coordinate to make sure these models fit together. The second team member needs to see what outputs are desired before modeling the process that creates them. At the same time, both team members need to know what data is stored (the data model) so they know what inputs are needed and what processes are needed to access the required data. Models support essential communication and teamwork among the project team members.

Models also assist in communication with the system users and foster understanding. Typically, an analyst reviews the models with a variety of users to get feedback on the analyst's understanding of the system requirements. Users need to see clear and complete models to comprehend what the analyst is proposing. In addition, the analyst sometimes works with users to develop the models, so the modeling process helps users better understand the possibilities that the new system can offer. Users also need to communicate among themselves using the models. And the analyst and the users together can use models to relate system capabilities to managers who are responsible for approving the system.

Finally, the requirements models produced by the analyst are used as documentation for future development teams when they maintain or enhance the system. Considering the amount of resources invested in a new system, it is critical for the development team to leave behind a clear record of what was created. An important activity during implementation is to package the documentation accurately, completely, and in a form that future developers can use. Much of the documentation consists of the models created throughout the project. Figure 4-3 summarizes the reasons modeling is important to system development.

Figure 4-3

Reasons for modeling

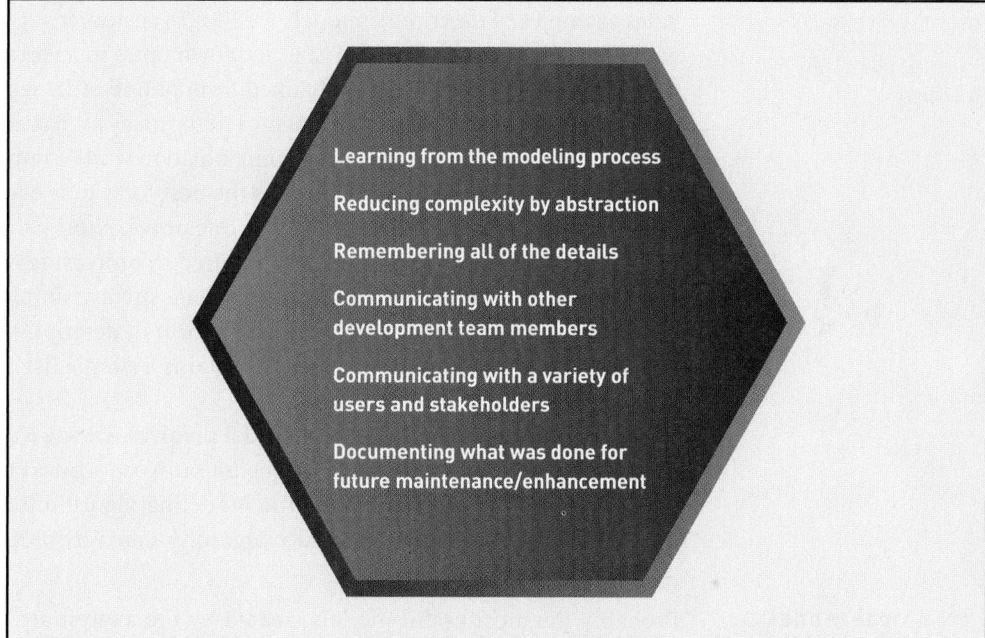

Learning from the modeling process
Reducing complexity by abstraction
Remembering all of the details
Communicating with other development team members
Communicating with a variety of users and stakeholders
Documenting what was done for future maintenance/enhancement

Although this book emphasizes models and techniques for creating models, remember that system projects vary in the number of models required and in their formality. Smaller, simpler system projects do not need models that show every system detail, particularly when the project team has experience with the type of system being built. Sometimes the key models are created informally in a few hours. Although models are often created using powerful visual modeling tools, as discussed in Chapter 2, useful and important models can be drawn quickly over lunch on a paper napkin or in an airport waiting room on the back of an envelope. As with any SDLC activity, an iterative approach is used for creating requirements and design models. The first draft of a model has some but not all details worked out. The next iteration might fill in more details or correct previous misconceptions.

TYPES OF MODELS

Analysts use many types of models when developing information systems. The type of model used depends on the nature of the information being represented. Models can be categorized into three general types: mathematical models, descriptive models, and graphical models.

Mathematical Models

mathematical model

a series of formulas that describe technical aspects of a system

A **mathematical model** is a series of formulas that describe technical aspects of a system. Mathematical models are used to represent precise aspects of the system that can be best represented by using formulas or mathematical notation, such as equations that represent network throughput requirements or a function expressing the response time required for a query. These models are examples of *technical requirements*. In addition, scientific and engineering applications tend to compute results using elaborate mathematical algorithms. The mathematical notation is the most appropriate way to represent these functional requirements, and it is also the most natural way for scientific and engineering users to express those requirements. An analyst working on scientific and engineering applications had better be comfortable with math.

But mathematical notation is also sometimes efficient for simpler requirements for business systems. For example, in a payroll application, it is reasonable to model gross pay as regular pay plus overtime pay. A reorder point for inventory, a discount price for a product, or a salary adjustment for a promotion might be modeled with a simple formula.

Descriptive Models

descriptive model
narrative memos, reports, or lists that describe some aspect of a system

Not all requirements can be precisely defined with mathematics. For these requirements, analysts use **descriptive models**, which can be narrative memos, reports, or lists. Figure 4-4 provides examples of descriptive models for RMO's customer support system. Initial interviews with users might require the analyst to jot down notes in a narrative form, such as the description of the phone-order process obtained from phone-order representatives. Sometimes users describe what they do in reports or memos to the analysts. The analyst might convert these narrative descriptions to a graphical modeling notation while compiling all of the information.

Sometimes a narrative description is the best form to use for recording information. Use case descriptions are often written out as one or two short paragraphs of text. More detailed use case descriptions are lists of steps required in processing between the actor and the system. Many useful models of information systems involve simple lists, such as lists of features, inputs, outputs, events, or users. Lists are a form of descriptive or narrative models that are concise, specific, and useful. Figure 4-4 contains a simple list of inputs to the customer support system.

A final example of a descriptive model involves writing a process or procedure in a very precise way, referred to as *structured English* or pseudocode. Programmers are familiar with structured English or pseudocode for modeling algorithms that, when followed, always obtain the same result. Therefore, such algorithms are very precise models of processing.

Graphical Models

graphical model
diagrams and schematic representations of some aspect of a system

Probably the most useful models created by the analyst are graphical models. **Graphical models** include diagrams and schematic representations of some aspect of a system. Graphical models make it easy to understand complex relationships that are too difficult to follow when described verbally. Recall the old saying that a picture is worth a thousand words. In system development, a carefully constructed graphical model might be worth a million words!

Some graphical models actually look similar to a real-world part of the system, such as a screen design or a report layout design. But for most of the analyst's work, the graphical models use symbols to represent more abstract things, such as external agents, processes, data, objects, messages, and connections. The key graphical models used during systems analysis tend to represent the more abstract aspects of a system, because the analysis focuses on fairly abstract questions about system requirements without indicating the details of how they will be implemented. The more concrete models of screen designs and report layouts are completed during systems design.

A variety of graphical models are used. Each model highlights (or abstracts) important details of some aspect of the information system. Each type of model should ideally use unique and standardized symbols to represent pieces of information. That way, whoever looks at a model can understand it. However, the number of available symbols is limited—circle,

Figure 4-4

Some descriptive models

A narrative description of processing requirements as verbalized by an RMO phone-order representative:

"When customers call in, I first ask if they have ordered by phone with us before, and I try to get them to tell me their customer ID number that they can find on the mailing label on the catalog. Or, if they seem puzzled about the customer number, I need to look them up by name and go through a process of elimination, looking at all of the Smiths in Dayton, for example, until I get the right one. Next, I ask what catalog they are looking at, which sometimes is out of date. If that is the case, then I explain that many items are still offered, but that the prices might be different. Naturally, they point to a page number, which doesn't help me because of the different catalogs, but I get them to tell me the product ID somehow..."

List of inputs for the RMO customer support system:

Item inquiry
New order
Order change request
Order status inquiry
Order fulfillment notice
Back-order notice
Order return notice
Catalog request
Customer account update notice
Promotion package details
Customer charge adjustment
Catalog update details
Special promotion details
New catalog details

square, rectangle, line, and so on—so be careful when you are first learning the symbols of each model. You will also find variations in the notation used for each type of model in practice. The Unified Modeling Language (UML) now provides diagramming standards for models used in the object-oriented approach. However, diagrams used in the traditional approach are less standardized.

OVERVIEW OF MODELS USED IN ANALYSIS AND DESIGN

The analysis activity named *Define system requirements* involves creating a variety of models. They are referred to as logical models because they define in great detail what is required without committing to one specific technology. Analysts create many types of logical models to define system requirements. Figure 4-5 lists some of the more commonly used models. Barbara Halifax currently has her project team working to create requirements models for the customer support system.

Many models are also created during systems design. Design models are physical models because they show how some aspect of the system will be implemented with specific technology. Some of these models are extensions of requirements models created during systems analysis or derive directly from the requirements models. Some models (for example, a class diagram) are used during analysis and during design. Chapters 5, 6, and 7 describe some of the requirements models in detail.

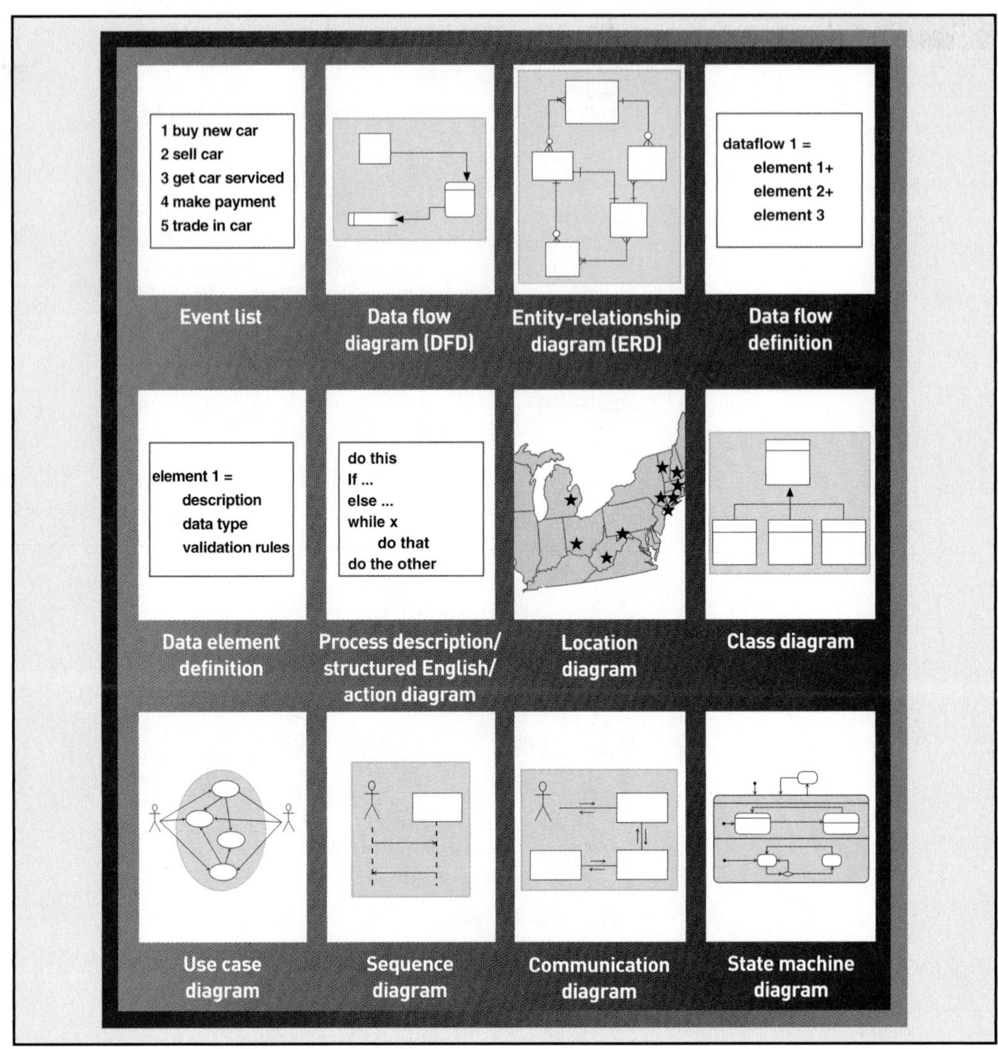

Figure 4-5

Models created during analysis

STAKEHOLDERS—THE SOURCE OF SYSTEM REQUIREMENTS

stakeholders

all the people who have an interest in the success of a new system

Your primary source of information for system requirements is the various stakeholders of the new system. **Stakeholders** are all people who have an interest in the successful implementation of the system. Generally, we categorize stakeholders into one of three groups: (1) the users, those who actually use the system on a daily basis; (2) the clients, those who pay for and own the system; and (3) the technical staff, the people who must ensure that the system operates in the computing environment of the organization. Figure 4-6 illustrates the various kinds of stakeholders who have an interest in a new system. We have discussed earlier the difference between users and clients. During analysis, the analyst also needs to consider the technical staff as well. One of the most important first steps in determining system requirements is to identify these various system stakeholders. In the past, problems have arisen with new systems because only some of the stakeholders were included in the project and the system was built exclusively for them. One of an analyst's first tasks is to identify every type of stakeholder who

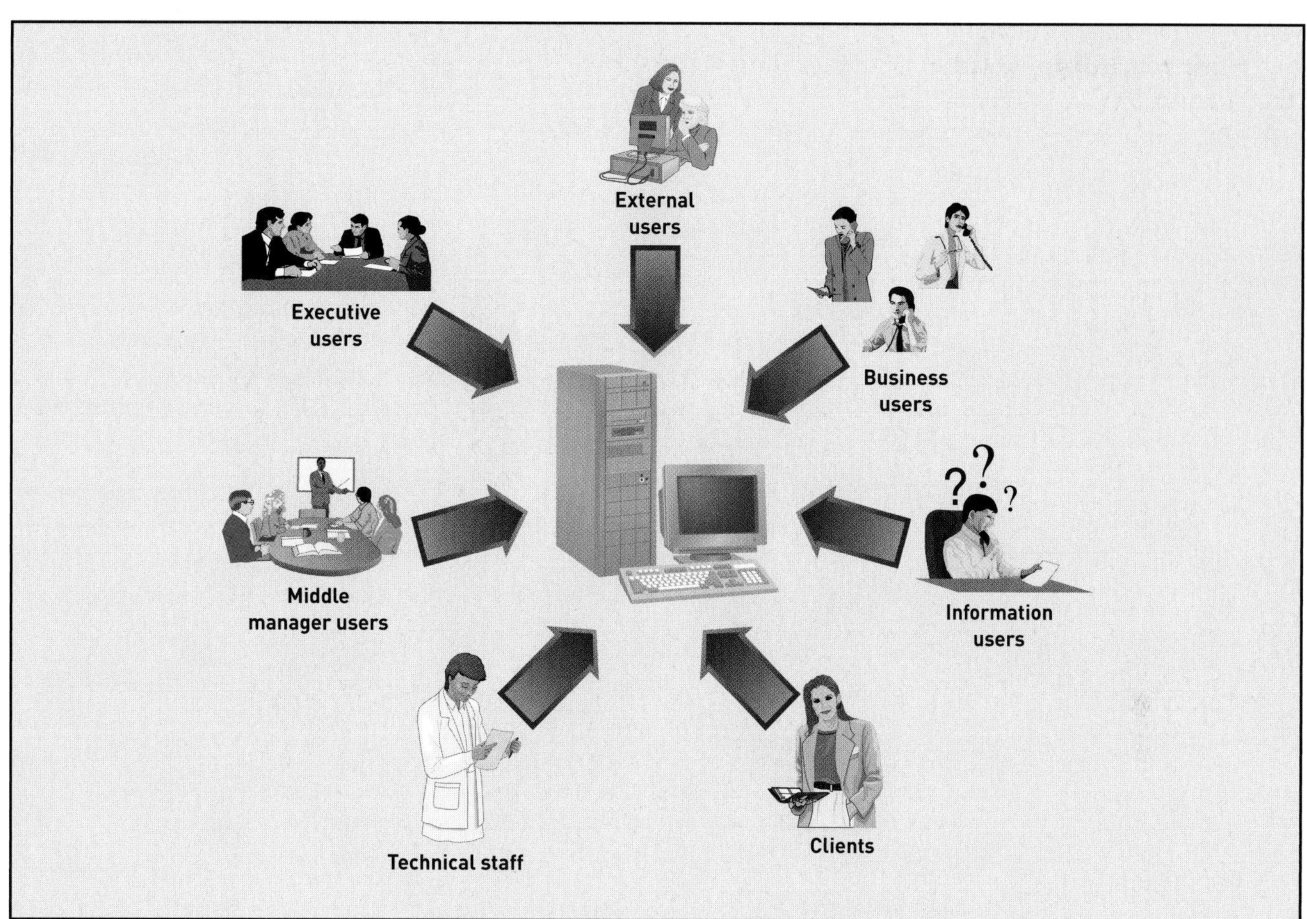

Figure 4-6

Stakeholders with an interest in new system development

has an interest in the new system. The second task is to ensure that critical people from each stakeholder category are available to the project as the business experts.

USERS AS STAKEHOLDERS

User roles—that is, types of system users—should be identified in two dimensions: horizontally and vertically. By horizontally, we mean that the analyst must look for information flow across business departments or functions. For example, a new inventory system may affect receiving, warehousing, sales, and manufacturing. So, individual employees from each of these departments must describe their requirements. The sales department may need to determine when and how to update inventory quantities or to commit inventory at the time of the sale but before it is shipped. Manufacturing may need certain information from the inventory system to assist in scheduling production. So, remembering to include the horizontal dimension in the definition of requirements will ensure that the many different departments, even those that may appear unrelated to the new system, are included.

By vertical dimension, we mean the information needs of clerical staff, of middle management, and of senior executives. Each of these stakeholders has different information requests for the system that must be included in the design. The following sections describe the characteristics and information needs of the various users on the vertical dimension. These same characteristics also apply to each department across the horizontal dimension.

Business Users

Business users are the people who use the system to perform the day-to-day operations of an organization. We often call these operations *transactions*. A **transaction** is a piece of work done in an organization, such as "enter an order." In Chapter 1, you learned that a transaction processing system handles these types of business operations. Business users provide information about the daily operations of the business and ways the system must support them.

transaction

a single occurrence of a piece of work or an activity done in an organization

Information Users

An information user is a person who needs current information from the system. This person might be an operational user or someone else. In some cases, a business might want to make information directly available to customers. However, an information user may not be allowed to enter information on business transactions, just to view specific information. An information user, then, provides an analyst with insight about what kinds of information should be available daily, weekly, monthly, and annually, and about what format is most convenient.

Management Users

Managers are responsible for seeing that the company is performing its daily procedures efficiently and effectively. Consequently, they need statistics and summary information from a system. Management will help an analyst answer the following types of questions:

- What kinds of reports must the system produce?
- What kind of performance statistics must the system maintain?
- What kind of volume information must the system keep, and what volumes of transactions must the new system support?
- Are the controls in the system adequate to prevent errors and fraud?
- How many requests for information will be made and how often?

Executive Users

The top executives of an organization are interested in strategic issues, as well as the daily issues just described. They typically want information from a system so that they can compare overall improvements in resource utilization. They might want the system to interface with other systems to provide strategic information on trends and directions of the industry and the business.

External Users

More and more systems today allow external entities to have direct access to the system. Customers may access the system directly through the Internet. Suppliers may have access to a system to check inventory levels and to initiate billing transactions. These users are more difficult to identify and access because they are not regular members of the organization. However, today they belong to an important group that must be considered in system development.

CLIENT STAKEHOLDERS

Although the project team must meet the information processing needs of the users, it also must satisfy the client. Chapter 3 defined the client as the person or group that is providing the funding for the project. In many cases, the client is the same group as the executive users. However, clients may also be a separate group, such as a board of trustees or executives in a parent company. The project team includes the client in its list of important stakeholders because the team must provide periodic status reviews to the client throughout development. The client or a direct representative on a steering or oversight committee also usually approves stages of the project and releases funds.

TECHNICAL STAKEHOLDERS

Although the technical staff is not a true user group, this group affects many system requirements. The technical staff includes people who establish and maintain the computing environment of the organization. They provide guidance in such areas as programming language, computer platforms, and other equipment. For some projects, the project team includes a member of the technical staff. For other projects, technical personnel are available as needed.

THE STAKEHOLDERS FOR ROCKY MOUNTAIN OUTFITTERS

To demonstrate the different perspectives of stakeholders, let's look at the proposed customer support system for Rocky Mountain Outfitters.

An important part of investigating system requirements is to identify all of the stakeholders. The set of requirements is incomplete if users, clients, external entities, or important technical staff are not consulted as information is being gathered. At RMO, operational users of the new order-processing system include inside sales representatives who take orders over the phone, as well as clerks who process mail orders. They all have different views about what the system should do for them. Sales representatives talk about looking up product information for customers and confirming availability and shipping dates. Mail-order clerks talk about scanning order information into the system to eliminate typing. The warehouse workers who put the shipments together need information about orders that have been shipped, orders to be shipped, and back orders, as well as their normal operational screens that allow them to put orders together into shipments with printed bills of lading.

John and Liz Blankens, as owners, have special interests in reports of the products that have been ordered and shipped. They are interested in watching seasonal trends within and across products. In the sports equipment business, it is critically important to push the trendy items quickly and move on when the trend is past.

The development of the customer support system has been funded in part from internal cash flows. Funds have also been obtained, however, through a special line of credit at the bank. RMO normally has a short-term line of credit for seasonal needs. Because the CSS project is a longer-term investment for a capital good, John and Liz obtained a different line of financing for it. Their banker is extremely interested in the success of the project, so in this case, the project team even met with the bank's staff to see what special formats of financial information the bank would like the system to maintain.

Finally, because this system will involve new technology—the Internet and distributed systems—very heavy involvement is required by the technical staff. Consequently, many stakeholders will have input into the types of information that can be extracted from the system. Figure 4-7 illustrates, from the upper-level RMO organization chart, people who will be involved. The orange positions indicate the executives and middle managers who will be involved as stakeholders. The project manager will build a list of all users who need to be involved in requirements definition. This organization chart is just the beginning. Other department managers and key employees will also be added.

How did the project team identify which stakeholders to include in the interview schedule? This is always a difficult question. The process begins, however, with an analysis of the scope of the new system. After defining the scope, the team must carefully analyze all the people who may require information from the system in any way. At this point, it is better to err on the side of including too many stakeholders rather than missing important sources of requirements. Barbara Halifax sent John MacMurty a memo updating him on her progress in identifying the CSS stakeholders and her upcoming plans for gathering information (see Barbara's memo).

Figure 4-7

RMO stakeholders involved in the CSS requirements definition

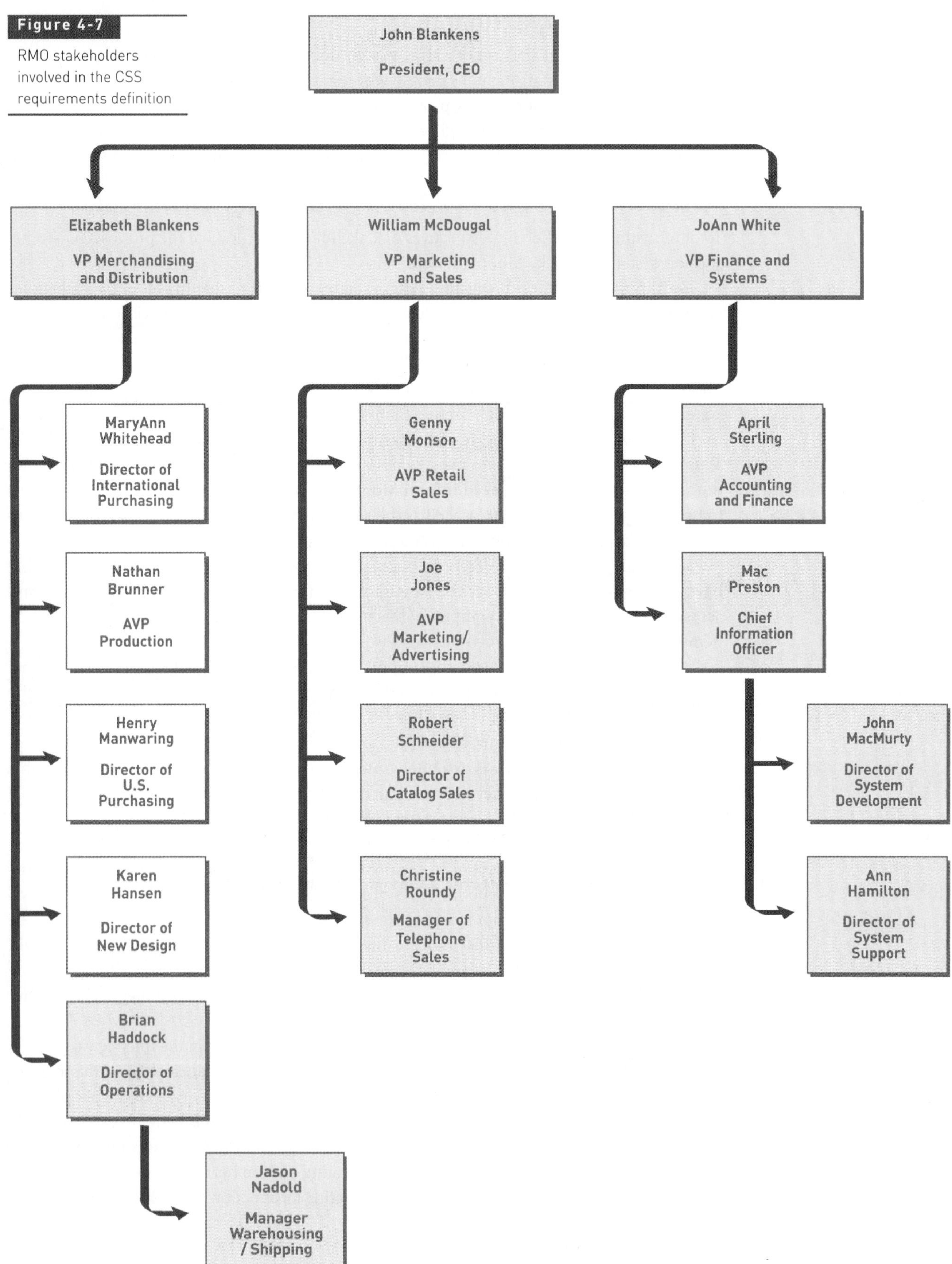

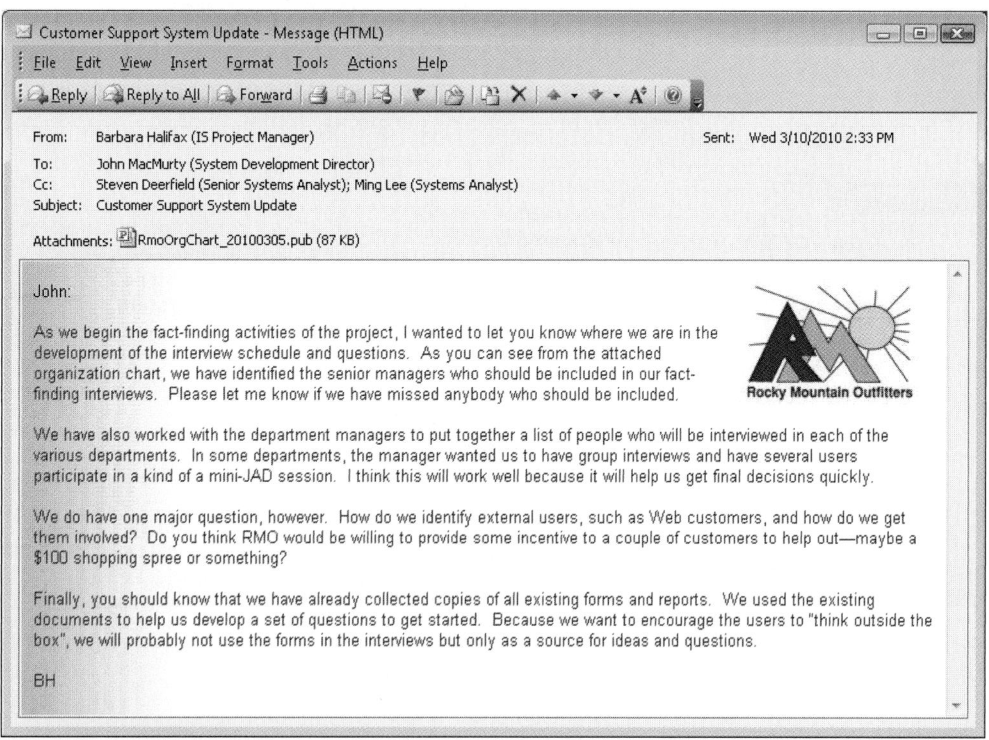
Customer Support System Update - Message (HTML)

File Edit View Insert Format Tools Actions Help

Reply | Reply to All | Forward

From: Barbara Halifax (IS Project Manager) Sent: Wed 3/10/2010 2:33 PM
To: John MacMurty (System Development Director)
Cc: Steven Deerfield (Senior Systems Analyst); Ming Lee (Systems Analyst)
Subject: Customer Support System Update

Attachments: RmoOrgChart_20100305.pub (87 KB)

John:

As we begin the fact-finding activities of the project, I wanted to let you know where we are in the development of the interview schedule and questions. As you can see from the attached organization chart, we have identified the senior managers who should be included in our fact-finding interviews. Please let me know if we have missed anybody who should be included.

We have also worked with the department managers to put together a list of people who will be interviewed in each of the various departments. In some departments, the manager wanted us to have group interviews and have several users participate in a kind of a mini-JAD session. I think this will work well because it will help us get final decisions quickly.

We do have one major question, however. How do we identify external users, such as Web customers, and how do we get them involved? Do you think RMO would be willing to provide some incentive to a couple of customers to help out—maybe a $100 shopping spree or something?

Finally, you should know that we have already collected copies of all existing forms and reports. We used the existing documents to help us develop a set of questions to get started. Because we want to encourage the users to "think outside the box", we will probably not use the forms in the interviews but only as a source for ideas and questions.

BH

TECHNIQUES FOR INFORMATION GATHERING

The objective of the analysis activities is to understand the business functions and develop the system requirements. The question that always arises is whether to study and document the existing system or whether to document only the requirements of the new system. When the structured approach, as well as the other approaches explained in Chapter 2, were first developed, systems analysts would first document the existing system and then extrapolate the requirements of the new system from that documentation. In those days, the development of system requirements was a four-step process: (1) identify the physical processes and activities of the existing system, (2) extract the logical business function that was inherent in each existing physical process, (3) develop the logical business functions for the approach to be used in the new system, and (4) define the physical processing requirements of the new system. One disadvantage of this approach was the inordinate amount of time it took. Another problem, frequently with long-term consequences, was that system developers would often simply automate the existing system—in other words, "pave the cow paths." As a result, no matter how inefficient the current system was, system developers would simply automate the procedures that were already in place.

BEST PRACTICE

Avoid "analysis paralysis" by focusing on the new system requirements from the beginning.

Today, analysts use an accelerated approach by balancing the review of current business functions with the new system requirements. It is still critical to have a complete, correct set of system requirements, but in today's fast-paced world, there is no time or money to review all the old systems and document all the inefficient procedures. As shown in Figure 4-8, the focus of analysis activities today is to develop a set of logical system requirements for the new system immediately. Analysts review the current system only when they need to understand the business needs, not to define the specific processes of the old system. This focus on the new while sometimes referring to the old is a balancing act for system professionals. They need to understand the business needs in extreme detail (remember, "walk the walk and talk the talk"), but they do not want to get caught up in old, inefficient methods. In fact, in today's development environment, one of the most valuable capabilities that a good system developer can bring is a new perspective to the problem.

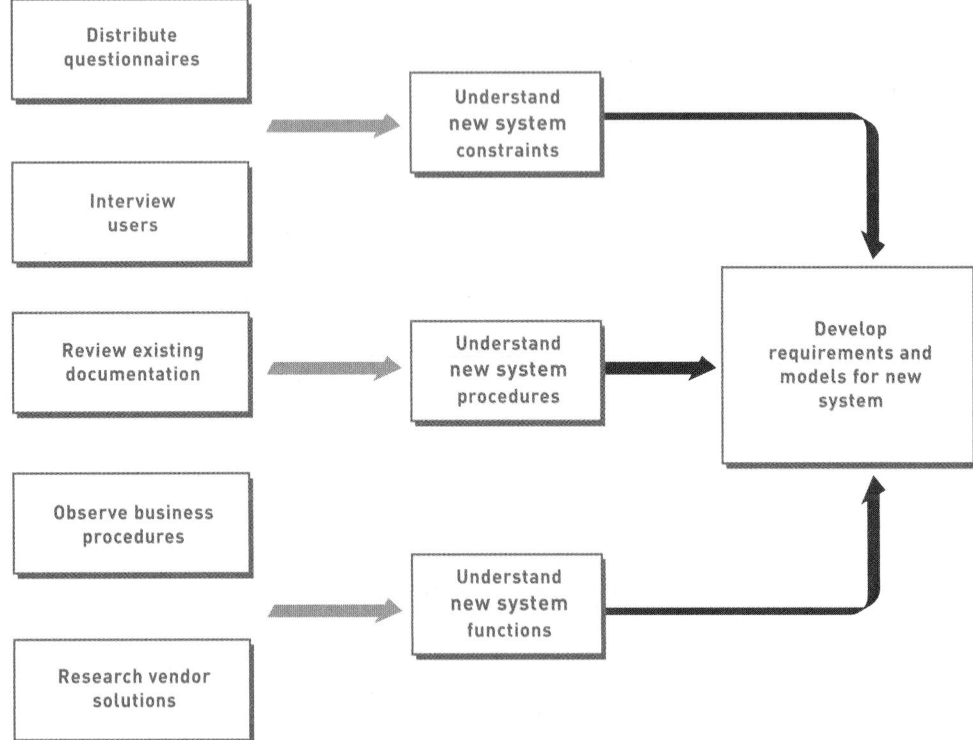

Figure 4-8

The relationship between information gathering and model building

The analysts develop the logical model of the new system as they gather information. The project team creates the physical model (that is, how the system will be built) later as part of systems design. Analysts focus on certain themes and use various techniques to develop the logical model of the system.

QUESTION THEMES

The first questions that new systems analysts ask are, "What kind of information do I need to collect? What is a requirement?" Basically, you want to obtain information that will enable you to build the logical model of the new business system. As shown in Figure 4-9, three major themes should guide you as you pursue your investigation.

Figure 4-9

Themes for information-gathering questions

Theme	Questions to users
What are the business operations and processes?	What do you do?
How should those operations be performed?	How do you do it? What steps do you follow?
What information is needed to perform those operations?	What information do you use? What forms or reports do you use?

What Are the Business Processes?

In the first question—What do you do?—the focus is on understanding the business functions. This question is the first step in being able to walk the walk. The analyst must obtain a comprehensive list of all the business processes. In most cases, the users provide answers in terms of the current system, so the analyst must discern carefully which of those functions are fundamental—which will remain and which may possibly be eliminated with an improved system. For example, sales clerks might indicate that the first thing they do when a customer places an order is to check the customer's credit history. In the new system, sales clerks might never need to perform that function; the system might perform the check automatically. The function remains a system requirement, but the method of carrying out the function moves from the clerks to the computer system.

How Is the Business Process Performed?

The second question—How can it be done?—moves the discussion from the current system to the new system. The focus is on how the new system *should* support the function rather than on how it does now. Thus, the first two questions go hand in hand to discover the need and begin to define the system requirement in terms of the new system. The users most frequently talk about the current system, but it is critical for the systems analyst to go beyond the current process. He or she must be able to help the user visualize new and more efficient approaches to performing the business processes made possible by the new technology.

What Information Is Required?

The final question—What information is needed?—elaborates on the second question by defining specific information that the new system must provide. The answers to the second and third questions form the basis for the definition of the system requirements. One of the shortcomings of many new systems analysts is that they do not identify all of the required pieces of information. In both this question and the previous one, detail is the watchword. An analyst must understand the nitty-gritty detail to develop a correct solution.

Focusing on these three themes helps an analyst ask intelligent, meaningful questions in an investigation. Later, as you learn about models, you will be able to formulate additional meaningful and detailed questions to ask.

As you develop skill in asking questions and building models, your problem-solving and analytical skills will increase. Remember, your value as a systems analyst is not that you know how to build a specific model or how to program in a specific language. Your value is in your ability to analyze and solve business information problems—to gather the correct information. Fundamental to that skill is how effectively and efficiently you can identify and capture these business rules. Effective requirements are complete, comprehensive, and correct. An efficient analyst is one who moves the project ahead rapidly with minimal intrusion on users' time and use of other resources, yet ensures that the information gathered will produce complete, comprehensive, and correct requirements specifications.

The next sections present the various methods of information gathering. All of these methods have been proven to be effective, although some are more efficient than others. In most cases, analysts combine methods to increase both their effectiveness and efficiency and provide a comprehensive fact-finding approach. The most widely used methods are the following:

- Review existing reports, forms, and procedure descriptions
- Conduct interviews and discussions with users
- Observe and document business processes
- Build prototypes
- Distribute and collect questionnaires
- Conduct joint application design (JAD) sessions
- Research vendor solutions

REVIEW EXISTING REPORTS, FORMS, AND PROCEDURE DESCRIPTIONS

This step should probably be the first in fact-finding activities. There are two sources of information for existing procedures and forms. One source is external to the organization—at industry-wide professional organizations and at other companies. It may not be easy to obtain information from other companies, but they are a potential source of important information.

Sometimes, industry journals and magazines report the findings of "best practices" studies. The project team would be negligent in its duties if its members were not familiar with best practice information. Also, with systems crossing organization boundaries more and more, external sources are an important source of system requirements.

The second source of reports, forms, and procedures is the existing business documents and procedure descriptions within the organization. This internal review serves two purposes. First, it is a good way to get a preliminary understanding of the processes. Often new systems analysts need to learn about the industry or the specific application that they are studying. A preliminary review of existing documentation will bring them up to speed fairly rapidly.

To begin the process, the analysts ask users to provide copies of the forms and reports that they currently use. They also request copies of procedural manuals and work descriptions. The review of these materials provides an understanding of the business functions. They also form the basis for the development of detailed interview questions.

The second way to use documents and reports is in the interviews themselves. Forms and reports can serve as visual aids for the interview, and as the working documents for discussion (see Figure 4-10). Discussion can center on the use of each form, its objective, its distribution, and its information content. The discussion should also include specific business events that initiate the use of the form. Several different business events might require the same form, and specific information about the event and the business process is critical. It is also always helpful to have forms that have been filled out with real information to ensure that the analyst obtains a correct understanding of the fields and data content.

Reviewing the documentation of existing procedures helps identify business rules that may not come up in the interviews. Written procedures also help reveal discrepancies and redundancies in the business processes. However, procedure manuals frequently are not kept up to date, and they commonly include errors. To ensure that the assumptions and business rules that derive from the existing documentation are correct, analysts should review them with the users.

Figure 4-10

A sample order form for Rocky Mountain Outfitters

Rocky Mountain Outfitters—Customer Order Form

Rocky Mountain Outfitters

Name and address of person placing order.
(Please verify your mailing address and make correction below.)
Order Date / /

Name

Address Apt. No

City State Zip

Phone: Day ( ) Evening ( )

Gift Order or Ship To: (Use only if different from address at left.)

Name

Address Apt. No

City State Zip

Gift ☐ Address for this Shipment Only ☐ Permanent Change of Address ☐

Gift Card Message

Delivery Phone ( )

Item No.		Description	Style	Color	Size	Sleeve Length	Qty	Monogram	Style	Price Each	Total

MERCHANDISE TOTAL

Regular FedEx shipping $4.50 per U.S. delivery address
(Items are sent within 24 hours for delivery in 2 to 4 days)

Please add $4.50 per each additional U.S. delivery address

FedEx Standard Overnight Service

Any additional freight charges

International Shipping (see shipping information on back)

Method of Payment

Check/Money Order ☐ Gift Certificate(s) ☐ AMOUNT ENCLOSED $

American Express ☐ MasterCard ☐ VISA ☐ Other ☐

Account Number

MO YR
/
Expiration Date

Signature

CONDUCT INTERVIEWS AND DISCUSSIONS WITH USERS

Interviewing stakeholders is by far the most effective way to understand business functions and business rules. It is also the most time-consuming and resource-expensive option. In this method, systems analysts meet with individuals or groups of users. A list of detailed questions is prepared, and discussion continues until all the processing requirements are understood and documented by the project team. Obviously, this process may take some time, so it usually requires multiple sessions with each of the users or user groups.

To conduct effective interviews, analysts need to organize in three areas: (1) preparing for the interview, (2) conducting the interview, and (3) following up the interview. Figure 4-11 is a sample checklist that summarizes the major points to be covered; it is useful in preparing for and conducting an interview.

Checklist for Conducting an Interview

Before

- Establish the objective for the interview
- Determine correct user(s) to be involved
- Determine project team members to participate
- Build a list of questions and issues to be discussed
- Review related documents and materials
- Set the time and location
- Inform all participants of objective, time, and locations

During

- Dress appropriately
- Arrive on time
- Look for exception and error conditions
- Probe for details
- Take thorough notes
- Identify and document unanswered items or open questions

After

- Review notes for accuracy, completeness, and understanding
- Transfer information to appropriate models and documents
- Identify areas needing further clarification
- Send thank-you notes if appropriate

Figure 4-11

A sample checklist to prepare for user interviews

Preparing for the Interview

Every successful interview requires preparation. The first and most important step in preparing for an interview is to establish its objective. In other words, what do you want to accomplish with this interview? Write down the objective so that it is firmly established in your mind. The second step is to determine which users should be involved in the interview. Frequently, the first two steps are so intertwined that both are done together. Even if you don't do anything else to prepare for your interviews, you must at least complete these two steps. The objective and the participants drive everything else in the interview.

The interview participants include both users and project members. Generally, at least two project members are involved in every interview. The two project members help each other during the interview and compare notes afterward to ensure accuracy. The number of users varies depending on the objective of the interview. A small number of users is generally best when the interview objective is narrow or of a fact-finding nature. In such cases, interviewing more than three users at a time tends to cause unnecessarily long discussions. Larger groups are better if the objective is more open-ended, such as when exploring new process alternatives in a BPR project. Larger groups are often better for generating and evaluating new ideas.

However, it can be difficult to manage a large group meeting to ensure high-quality input from all participants. Professional facilitators and formal discovery techniques such as joint application design (discussed later in this chapter) may be employed if the objective is complex or critical and the group is large.

BEST PRACTICE

Make sure at least two project members participate in user interviews.

The next step is to prepare detailed questions to be used in the interview. Write down a list of specific questions and prepare notes based on the forms or reports received earlier. Usually you should prepare a list of questions that are consistent with the objective of the interview. Both open-ended questions and closed-ended questions are appropriate. Open-ended questions such as, "How do you do this function?" encourage discussion and explanation. Closed-ended questions such as, "How many forms a day do you process?" are used to get specific facts. Generally, open-ended questions help get the discussion started and encourage the user to explain all the details of the business process and the rules.

The last step is to make the final interview arrangements and to communicate those arrangements to all participants. A specific time and location should be established. If possible, a quiet location should be chosen to avoid interruptions. Each participant should know the objective of the meeting and, when appropriate, should have a chance to preview the questions or materials to be used. Interviews consume a substantial amount of time, and they can be made more efficient if each participant knows beforehand what is to be accomplished.

Conducting the Interview

New systems analysts are usually quite nervous about conducting interviews. However, in most cases, the users are excited about getting a better system to help them do their jobs. Practicing good manners usually ensures that the interview will go well. Here are a few guidelines.

Dress appropriately. Dress at least as well as the best-dressed user. In many corporate settings such as banks or insurance companies with managers present, business suits are appropriate. In factory or manufacturing settings, work dress may be appropriate. The objective in dressing is to project competence and professionalism without intimidating the user.

Arrive on time. If anything, be a little early. If the session is in a conference room, ensure that it is set up appropriately. For a large group or a long session, plan for refreshment breaks.

Limit the time of the interview. Both the preparation and the interview itself affect the time required. As you set the objective and develop questions, plan for about an hour and a half. If the interview will require more time to cover the questions, it is usually better to break off the discussion and schedule another session. (Other techniques that we discuss later have all-day sessions.) The users have other responsibilities, and the systems analysts can absorb only so much information at one time. It is better to have several shorter interviews than one long marathon. A series of interviews provides an opportunity to absorb the material and to go back to get clarification later. Both the analysts and the users will have better attitudes with several shorter interviews.

Look for exception and error conditions. Look for opportunities to ask "what if" questions. "What if it doesn't arrive? What if the signature is missing? What if the balance is incorrect? What if two order forms are exactly the same?" The essence of good systems analysis is understanding all of the "what ifs." Make a conscious effort to identify all of the exception conditions and ask about them. More than any other skill, the ability to think of the exceptions will strengthen the skill of discovering the detailed business rules. It is a hard skill to teach from a textbook; experience will hone this skill. You will teach yourself this skill by conscientiously practicing it.

Probe for details. In addition to looking for exception conditions, the analyst must probe to ensure a complete understanding of all procedures and rules. One of the most difficult skills to learn as a new systems analyst is to get enough details. Frequently, it is easy to get a general overview of how a process works. But do not be afraid to ask detailed questions until you thoroughly understand how it works and what information is used. You cannot do effective systems analysis by glossing over the details.

Take careful notes. It is a good idea to take handwritten notes. Usually tape recorders make users nervous. Note taking, however, signals that you think the information you are obtaining is important, and the user is complimented. If two analysts conduct each interview, they can compare notes later. Identify and document in your notes any unanswered questions or outstanding issues that were not resolved. A good set of notes provides the basis for building the analysis models as well as establishing a basis for the next interview session.

Figure 4-12 is a sample agenda for an interview session. Obviously, you do not need to conform exactly to a particular agenda. However, as with the interview checklist shown in Figure 4-11, this figure will help prod your memory on issues and items that should be discussed in an interview. Make a copy and use it. As you develop your own style, you can modify the checklist for the way you like to work.

Figure 4-12

Sample interview session agenda

Discussion and Interview Agenda

Setting

Objective of Interview
Determine processing rules for sales commission rates

Date, Time, and Location
April 21, 2010, at 9:00 a.m. in William McDougal's office

User Participants (names and titles/positions)
William McDougal, vice president of marketing and sales, and several of his staff

Project Team Participants
Mary Ellen Green and Jim Williams

Interview/Discussion

1. *Who is eligible for sales commissions?*
2. *What is the basis for commissions? What rates are paid?*
3. *How is commission for returns handled?*
4. *Are there special incentives? Contests? Programs based on time?*
5. *Is there a variable scale for commissions? Are there quotas?*
6. *What are the exceptions?*

Follow-Up

Important decisions or answers to questions
See attached write-up on commission policies

Open items not resolved with assignments for solution
See Item numbers 2 and 3 on open items list

Date and time of next meeting or follow-up session
April 28, 2010, at 9:00 a.m.

Following Up the Interview

Follow-up is an important part of each interview. The first task is to absorb, understand, and document the information that was obtained. Generally, analysts document the details of the interview by constructing models of the business processes and writing textual descriptions of nonfunctional requirements. These tasks should be completed as soon as possible after the interview and the results distributed to the interview participants for validation. If the modeling methods are complex or unfamiliar to the users, the analyst should schedule follow-up meetings to explain and verify the models, as described in the last section of this chapter.

During the interview, you probably asked some "what if" questions that the users could not answer. They are usually policy questions raised by the new system that management has not considered before. It is extremely important that these questions not get lost or forgotten. For example, Figure 4-13 is a sample table for tracking outstanding or unresolved issues for Rocky Mountain Outfitters. The table includes questions posed by users or analysts and responsibilities assigned for resolving the issues. If several teams are working, a combined list can be maintained. Other columns that might be added to the list are an explanation of the problem's resolution and the date resolved.

Figure 4-13

A sample open-items list

Outstanding issues control table

ID	Issue title	Date identified	Target end date	Responsible project person	User contact	Comments
1	Partial shipments	6-12-2010	7-15-2010	Jim Williams	Jason Nadold	Ship partials or wait for full shipment?
2	Returns and commissions	7-01-2010	9-01-2010	Jim Williams	William McDougal	Are commissions recouped on returns?
3	Extra commissions	7-01-2010	8-01-2010	Mary Ellen Green	William McDougal	How to handle commissions on special promotions?

Finally, make a list of new questions based on areas that need further elaboration or that are missing information. This list will prepare you for the next interview.

BEST PRACTICE

Maintain an open-items list for unresolved problems and questions.

OBSERVE AND DOCUMENT BUSINESS PROCESSES

Along with interviews, another extremely useful method of gathering information is to observe users directly at their job sites and to document the processes you observe. This first-hand experience is invaluable to understanding exactly what occurs in business processes.

Observing

The old adage that a picture is worth a thousand words is also true with systems analysis. More than any other activity, observing the business processes in action will help you understand the business functions. However, while observing existing processes, you must also be able to visualize the new system's associated business processes. That is, as you observe the current business processes to understand the fundamental business needs, you should never forget that the processes could, and often should, change to be more efficient. Don't get locked into believing there is only one way of performing the process.

You can observe the work in several ways, from a quick walkthrough of the office or plant to doing the work yourself. A quick walkthrough gives a general understanding of the layout of the office, the need for and use of computer equipment, and the general workflow. Spending several hours observing users at their jobs helps you understand the details of using the computer system and carrying out business functions. By being trained as a user and actually doing the job, you can discover the difficulties of learning new procedures, the importance of a system that is easy to use, and the stumbling blocks and bottlenecks of existing procedures and information sources.

It is not necessary to observe all processes at the same level of detail. A quick walkthrough may be sufficient for one process, whereas another process that is critical or more difficult to understand might require an extended observation period. If you remember that the objective is a complete understanding of the business processes and rules, you can assess where to spend your time to gain that thorough understanding. As with interviewing, it is usually better if two analysts combine their efforts in observing procedures.

Observation often makes the users nervous, so you need to be as unobtrusive as possible. You can put users at ease in several ways, such as working with a user or observing several users at once. Common sense and sensitivity to the needs and feelings of the users will usually result in a positive experience.

Documenting Workflows with Activity Diagrams

As you gather information about business processes, primarily by interviewing the users and by observing the processes, you will need to document your results. One effective way to capture this information is through the use of diagrams. Eventually, you may want to use diagrams to describe the workflows of the new system, but for now, let's just focus on how we would document the current business workflows.

workflow
a sequence of steps to process a business transaction

A **workflow** is the sequence of processing steps that completely handles one business transaction or customer request. Workflows may be simple or complex. Complex workflows can be composed of dozens or hundreds of processing steps and may include participants from different parts of an organization. As an analyst, you may try to depend only on your memory to remember and understand the workflow (a bad idea), you may write it down in a long description, or you can document it with a diagram. The advantages of a simple diagram are that you can visualize it better, and you can review it with the users to make sure it is correct. One of the major benefits of using diagrams and models is that they become a powerful communication mechanism between the project team and the users.

activity diagram
a type of workflow diagram that describes the user activities and their sequential flow

No single diagram is commonly used to model workflows. Diagrams commonly employed include flowcharts, data flow diagrams, and activity diagrams. Data flow diagrams do a good job of capturing the flow of data within a workflow, but they aren't designed to represent control flows. Flowcharts are specifically designed to represent control flow among processing steps, but they don't represent data flow. So, many analysts use a type of workflow diagram called an *activity diagram*. An **activity diagram** is simply a workflow diagram that describes the various user (or system) activities, the person who does each activity, and the sequential flow of these activities. The activity diagram is one of the Unified Modeling Language (UML) diagrams associated with the object-oriented approach, but it can be used with any development approach.

synchronization bar
a symbol in an activity diagram to control the splitting or uniting of sequential paths

swimlane
a rectangular area on an activity diagram representing the activities of a single agent

Figure 4-14 shows the basic symbols used in an activity diagram. The ovals represent the individual activities in a workflow. The connecting arrows represent the sequence between the activities. The black circles are used to denote the beginning and ending of the workflow. The diamond is a decision point at which the flow of the process will either follow one path or the other path. The heavy solid line is a **synchronization bar**, which either splits the path into multiple concurrent paths or recombines concurrent paths. The **swimlane** represents an agent who performs the activities. Because in a workflow it is common to have different agents (that is, people) performing different steps of the workflow process, the swimlane symbol divides the workflow activities into groups showing which agent performs which activity.

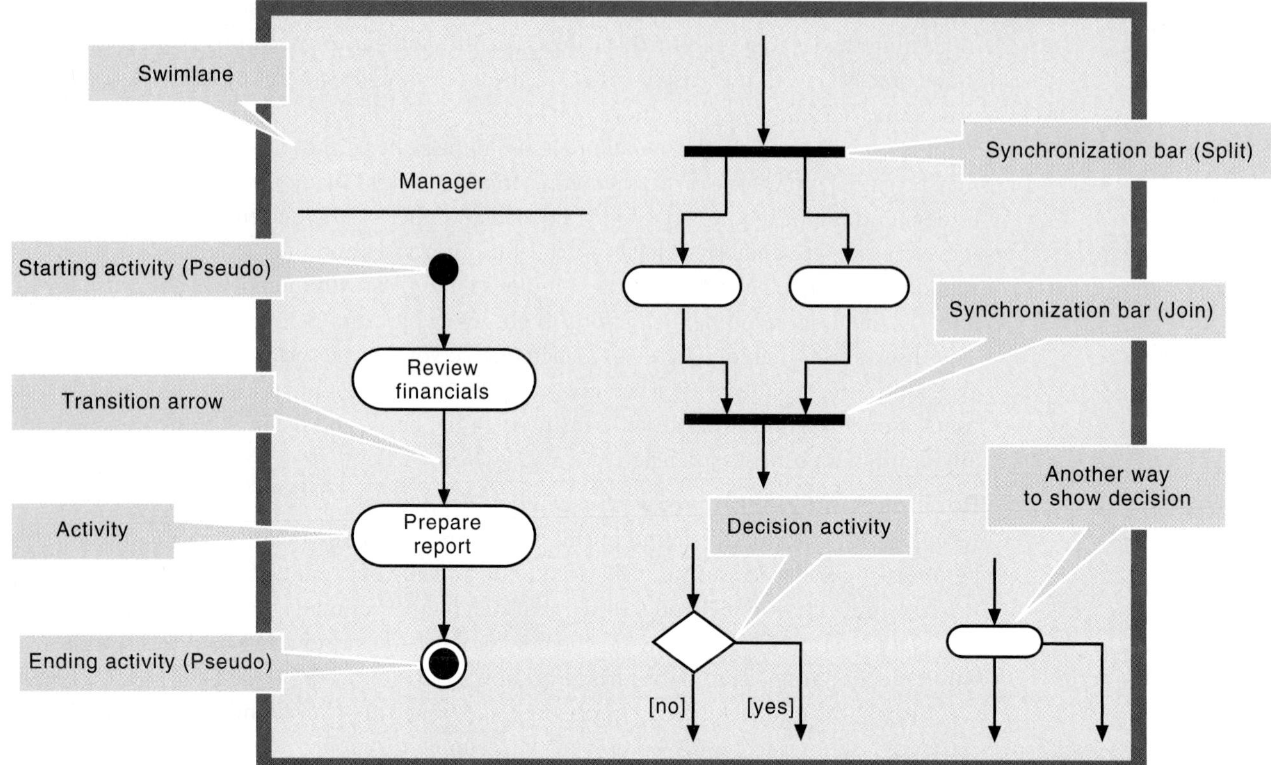

Figure 4-14
Activity diagram symbols

Figure 4-15 is an actual activity diagram for a workflow. This workflow represents a customer requesting a quote from a salesperson. If it is a simple request, the salesperson can enter the data and create the quote. If it is complex, the salesperson requests assistance from a technical expert to generate the quote. In both cases, the computer system calculates the details of the quote.

Suppose in this case that you have interviewed the salesperson and observed the generation of a quote. Looking at Figure 4-15, you can see how the workflow progresses. The customer initiates the first step by requesting a quote. The salesperson performs the next step in the workflow. She writes down the details of the quote request and then decides whether she can do it herself or whether she needs help. If she does not need help, the salesperson enters the information into the computer system. If the salesperson needs help, the technical expert performs the next step. The expert reviews the quote request to make sure that the requested components can be integrated into a functioning computer system. The activity of checking the request is fairly complex, and you could break it down into more detailed steps if desired. For now, let's leave the diagram at this level of detail. The expert then enters the information into the system. At this point, the computer system generates the detailed quote. Notice that no matter which path was taken, they both result in this common activity. Finally, the customer reviews the quote and decides whether it needs changes or is acceptable. In this simple case, the customer always buys something, so this workflow is obviously not completely accurate.

Notice that an activity diagram focuses on the sequence of activities. This diagram is straightforward and quite easy to understand. In fact, one of the strengths of using activity diagrams to document workflows is that users also find them very easy to understand. You can use graphical representations such as this diagram to review your understanding of the particular workflow procedure with the user.

Figure 4-15

A simple activity diagram to demonstrate a workflow

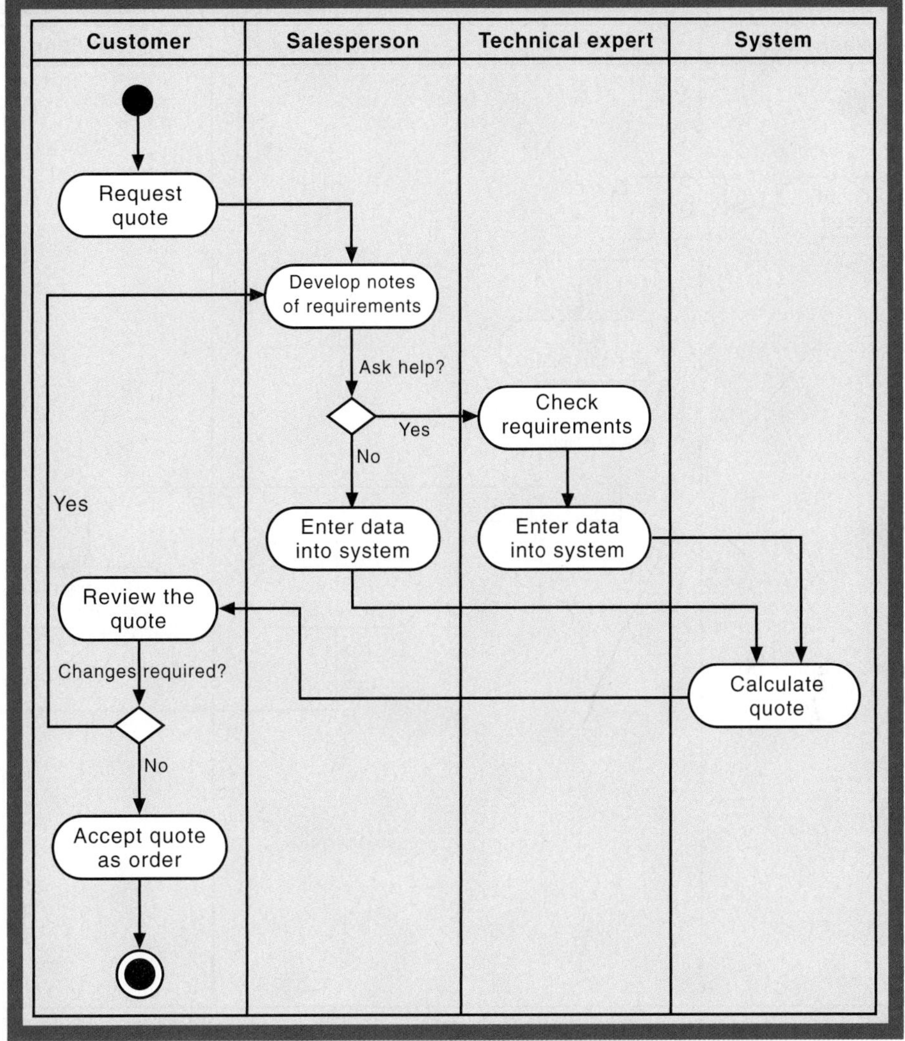

Figure 4-16 illustrates another workflow. This diagram demonstrates some new concepts. Let's assume that the customer from the previous example did want to proceed with an order. Figure 4-16 shows the workflow that is required to get the order scheduled for production. The salesperson sends to engineering the printed quote, which has now become an order. This example emphasizes the fact that a document is being transmitted. To indicate that a document is being passed, you place the document symbol at the end of the connecting arrow, and the arrow now becomes a conduit for transmitting a document, not just a flow of activities. After engineering develops the specifications, two concurrent activities happen: purchasing orders the materials, and production writes the program for the automated milling machines. These two activities are completely independent and can occur at the same time. Notice that one synchronization bar splits the path into two concurrent paths, and another synchronization bar reconnects them. Finally, scheduling puts the order on the production schedule.

Creating activity diagrams to document workflows is straightforward. The first step is to identify the agents to create the appropriate swimlanes. Next, just follow the various steps of

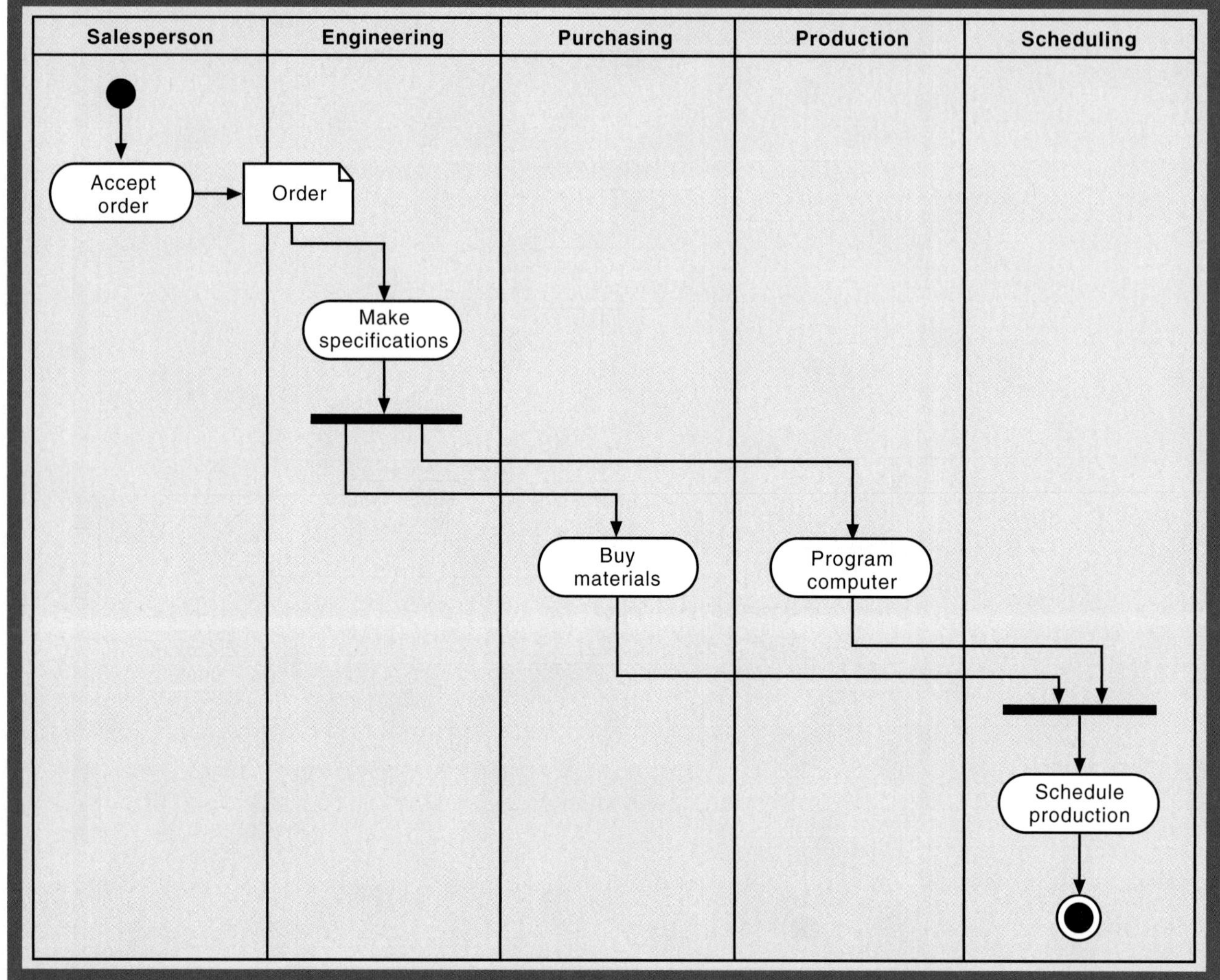

Figure 4-16

An activity diagram showing concurrent paths

the workflow, and make appropriate ovals for the activities. Connect the activity ovals with arrows to show the workflow. Here are a couple of simple guidelines:

- Use a decision symbol to represent an either/or situation—one path *or* the other path, but not both. As a shorthand notation, you can merge an activity (using an oval) and a decision (using a diamond) into a single oval with two exit arrows, as indicated on the right in Figure 4-14. This notation represents a decision (either/or) activity. Wherever you have an activity that reads "verify" or "check," you will probably require a decision—one for the "accept" path and one for the "reject" path. You can merge either/or paths into a common activity (as in *Calculate quote* shown in Figure 4-15) or into other connecting arrows.
- Use synchronization bars for parallel paths—situations in which *both* paths are taken. Include both a beginning and ending synchronization bar. You can also use synchronization bars to represent a loop such as a "do while" programming loop. Put the bar at the beginning of the loop and describe it as "for every." Put another synchronization bar at the end of the loop with the description "end for every."

BUILD PROTOTYPES

prototype
a preliminary working model of a larger system

As we discussed briefly in Chapter 2, a **prototype** is an initial, working model of a larger, more complex entity. Prototypes are used to test and validate ideas, and there are many names to differentiate these uses: throwaway prototypes, discovery prototypes, design prototypes, and evolving prototypes. As already explained, prototypes are used during analysis to test feasibility and to help identify processing requirements. These prototypes may be in the form of simple screens or report programs. During design, prototypes may be built to test various design and interface alternatives. Even during implementation, prototypes may be built to test the effectiveness and efficiency of different programming techniques.

Prototyping is such a strong tool that you will find it used in almost every development project in some way. As mentioned earlier in the chapter, a discovery prototype is used for a single discovery objective and then discarded after the concept has been tested. For example, if you use a prototype to determine screen formats and processing sequences, you would throw away the prototype after finishing that definition. Evolving prototypes, on the other hand, are prototypes that grow and change and may eventually even be used as the final, live system. As you will see later in Chapter 16, one approach to prototyping is to keep modifying the prototype and adding to it until it actually becomes the system that is installed.

The following are characteristics of effective prototypes:

mock-up

an example of a final product that is for viewing only and is not executable

- **Operative**. Generally, a prototype should be a working model, with the emphasis on *working*. A simple start to a prototype, called a **mock-up**, is an electronic form (such as a screen) that shows what an interface or system looks like but cannot execute an activity. Later, a working prototype will actually execute and provide both "look and feel" characteristics, but it may lack some functionality.
- **Focused**. To test a specific concept or verify an approach, a prototype should be focused on a single objective. Extraneous execution capability that is not part of the specific objective should be excluded. Although it might be possible to combine several simple prototypes into a larger prototype, the focused objective still applies. Later, the project team can combine prototypes to test the integration of several components.
- **Quick**. Rapid prototype development requires appropriate tools for creating interfaces and software. A complete application development environment may not be necessary and the environment may not need to support "industrial strength" features. What's important is an efficient developer interface that quickly produces a testable prototype.

Integrating prototyping activities in the project is fairly simple. The important point to keep in mind is to have an overall philosophy and purpose for building prototypes and to maintain a consistent focus across all the prototypes that are built.

DISTRIBUTE AND COLLECT QUESTIONNAIRES

Questionnaires have a limited and specific use in information gathering. The benefit of a questionnaire is that it enables the project team to collect information from a large number of stakeholders. Even if the stakeholders are widely distributed geographically, they can still help define requirements through questionnaires.

Frequently, the project team can use a questionnaire to obtain preliminary insight on the information needs of the various stakeholders. This preliminary information can then be used to help determine the areas that need further research with document reviews, interviews, and observation. Questionnaires are also helpful to answer quantitative questions such as, "What forms are used to enter new customer information?" and, "On the average, how long does it take to enter one standard order?" Finally, questionnaires can be used to determine the users' opinions about various aspects of the system. Such questions as "On a scale of 1 to 7, how important is it to be able to access a customer's past purchase history?" are often called **closed-ended questions**, because they direct the person answering the question to provide a direct response to only that question. They do not invite discussion or elaboration. The strength of closed-ended questions, however, is that the answers are always limited to the set of choices. The project team can tabulate the answers to determine averages or trends.

closed-ended questions

questions that have a simple, definitive answer

Figure 4-17 is a sample questionnaire showing three types of questions. The first part has closed-ended questions to determine quantitative information. The second part consists of opinion questions in which respondents are asked whether they agree or disagree with the statement. Both types of questions are useful for tabulating and determining quantitative averages. The final part requests an explanation of a procedure or problem. Questions such as these are good as a preliminary investigation to help direct further fact-finding activities.

Figure 4-17

A sample questionnaire

RMO Questionnaire

This questionnaire is being sent to all telephone-order sales personnel. As you know, RMO is developing a new customer support system for order taking and customer service.

The purpose of this questionnaire is to obtain preliminary information to assist in defining the requirements for the new system. Follow-up discussions will be held to permit everybody to elaborate on the system requirements.

Part I. Answer these questions based on a typical four-hour shift.

1. How many phone calls do you receive?_______________
2. How many phone calls are necessary to place an order for a product?_______________
3. How many phone calls are for information about RMO products, that is, questions only?_______________
4. Estimate how many times during a shift customers request items that are out of stock._______________
5. Of those out-of-stock requests, what percentage of the time does the customer desire to put the item on back order?_______________%
6. How many times does a customer try to order from an expired catalog?_______________
7. How many times does a customer cancel an order in the middle of the conversation?_______________
8. How many times does an order get denied due to bad credit?_______________

Part II. Circle the appropriate number on the scale from 1 to 7 based on how strongly you agree or disagree with the statement.

Question	Strongly Agree						Strongly Disagree
It would help me do my job better to have longer descriptions of products available while talking to a customer.	1	2	3	4	5	6	7
It would help me do my job better if I had the past purchase history of the customer available.	1	2	3	4	5	6	7
I could provide better service to the customer if I had information about accessories that were appropriate for the items ordered.	1	2	3	4	5	6	7
The computer response time is slow and causes difficulties in responding to customer requests.	1	2	3	4	5	6	7

Part III. Please enter your opinions and comments.

Please briefly identify the problems with the current system that you would like to see resolved in a new system.

____
____
____

BEST PRACTICE

Limit the number of open-ended questions on a questionnaire.

open-ended questions

questions that require discussion and do not necessarily have a simple, short answer

Questionnaires are not well suited to helping you learn about processes, workflows, or techniques. The questions identified earlier, such as "How do you do this process?", are best answered using interviews or observation. Questions that encourage discussion and elaboration are called **open-ended questions**. Although a questionnaire can contain a very limited number of open-ended questions, stakeholders frequently do not return questionnaires that contain many open-ended questions.

CONDUCT JOINT APPLICATION DESIGN SESSIONS

joint application design (JAD)

a technique to define requirements or design a system in a single session by having all necessary people participate

Joint application design (JAD) is a technique used to expedite the investigation of system requirements. The normal interview and discussion approach, as explained earlier, requires a substantial amount of time. The analysts first meet with the users, then document the discussion by writing notes and building models, and then review and revise the models. Unresolved issues are placed on an open-items list and may require several additional meetings and reviews to be finalized. This process can extend from several weeks to months, depending on the size of the system and the availability of user and project team resources.

BEST PRACTICE

A JAD session speeds up the process of defining requirements.

The objective of JAD is to compress all of these activities into a shorter series of JAD sessions with users and project team members. An individual JAD session might last from a single day to a week. During the session, all of the fact-finding, model-building, policy decisions, and verification activities are completed for a particular aspect of the system. If the system is small, the entire analysis might be completed during the JAD session. The critical factor in a successful JAD session is to have all of the important stakeholders present and available to contribute and make decisions. The actual participants vary depending on the objective of the specific JAD session. The following people and groups may be involved:

- **The JAD Session Leader**. One of the more important members of the group, the session leader is experienced or trained in understanding group dynamics and in facilitating group discussion. Normally, a JAD session involves quite a few people. Each session has a detailed agenda with specific objectives that must be met, and the discussion must progress toward meeting those objectives. Maintaining focus requires someone with skills and experience to keep people on task tactfully. Often it is tempting to appoint a systems analyst as the session leader. However, experience indicates that successful JAD sessions are conducted by someone who is trained to lead group decision making.
- **Users**. Earlier, this chapter identified various classes of users. It is important to have all of the appropriate users in the JAD sessions. Frequently, as requirements are discovered, managers must make policy decisions. If managers are not available in the sessions to make those decisions, progress is halted. Because of business pressures, it might be difficult for top executives to be present during the entire session. In that case, arrangements should be made for executives to visit the session once or twice a day to become involved in policy discussions.
- **Technical Staff**. A representative from the technical support staff should also be present in the JAD session. There are always questions and decisions about technical issues that need to be answered. For example, participants might need details of computer and network configurations, operating environments, and security issues.
- **Project Team Members**. Both systems analysts and user experts from the project team should be involved in JAD sessions. These members assist in the discussion, clarify points, control the level of detail needed, build models, document the results, and generally see that the system requirements are defined to the necessary level of detail. The session leader is a facilitator, but often the leader is not the expert on how much detail and definition is required. Members of the project team are the experts on ensuring that the objectives are completely satisfied.

JAD sessions are usually conducted in special rooms with supporting facilities. First, because the process is so intense, it is important to be away from the normal day-to-day interruptions. Sometimes an off-site location may be necessary, or notification that interruptions are not welcome may be posted. On the other hand, it is usually helpful to have telephone access to executives and technical staff who are not involved in the meetings but who may be invited from time to time to finalize policy or technical decisions.

Resources in the JAD session room should include an overhead projector, a black or white board, flip charts, and adequate workspace for the participants. JAD sessions are work sessions, and all the necessary work paraphernalia should be provided.

Recently, JAD sessions have been taking advantage of electronic support to increase their efficiency. Analysis and documentation can be enhanced if participants have personal or laptop computers connected in a network. Then as requirements are documented with narrative descriptions or models, or even as some simple discovery prototypes are built, they can be made available to everybody. Often, an easy-to-use suite of modeling and application development tools are provided to assist in visualization of screen and report layouts and file design. The suite may also provide a central repository for all the requirements developed during the session.

group support system (GSS)

a computer system that enables multiple people to participate with comments at the same time, each from their own computer

Group support systems (GSSs), which also run on the network of computers, allow all participants to post comments (anonymously, if desired) in a common working chat room. This approach helps participants who may be shy in group discussions to become more active and contribute to the group decisions. GSSs also enable the team to store final requirements as decisions are made. Normally JAD sessions are conducted with everyone in the same room. However, GSSs on wider networks provide the opportunity for virtual meetings with participants at geographically dispersed locations.

Figure 4-18 shows an example of a conference room with electronic support. Such a room might be available in larger companies that have development projects in progress more frequently. The room shown in Figure 4-18 has workstations available to develop model diagrams and prototypes during the JAD sessions. This room could even be quite sophisticated by having computer support for collaborative work (CSCW) software on the computers to facilitate comment and discussion. With CSCW software, certain executives could even participate from remote locations, if necessary.

As stated earlier, one of the dangers of JAD is the risk involved in expediting decisions. Because the objective of JAD is to come to a conclusion quickly on policy decisions and requirements details, sometimes decisions are not optimal. At times, details are inappropriately defined or missed altogether. However, JAD sessions have been largely successful in reducing project development efforts and shortening the schedule.

Figure 4-18

A JAD facility

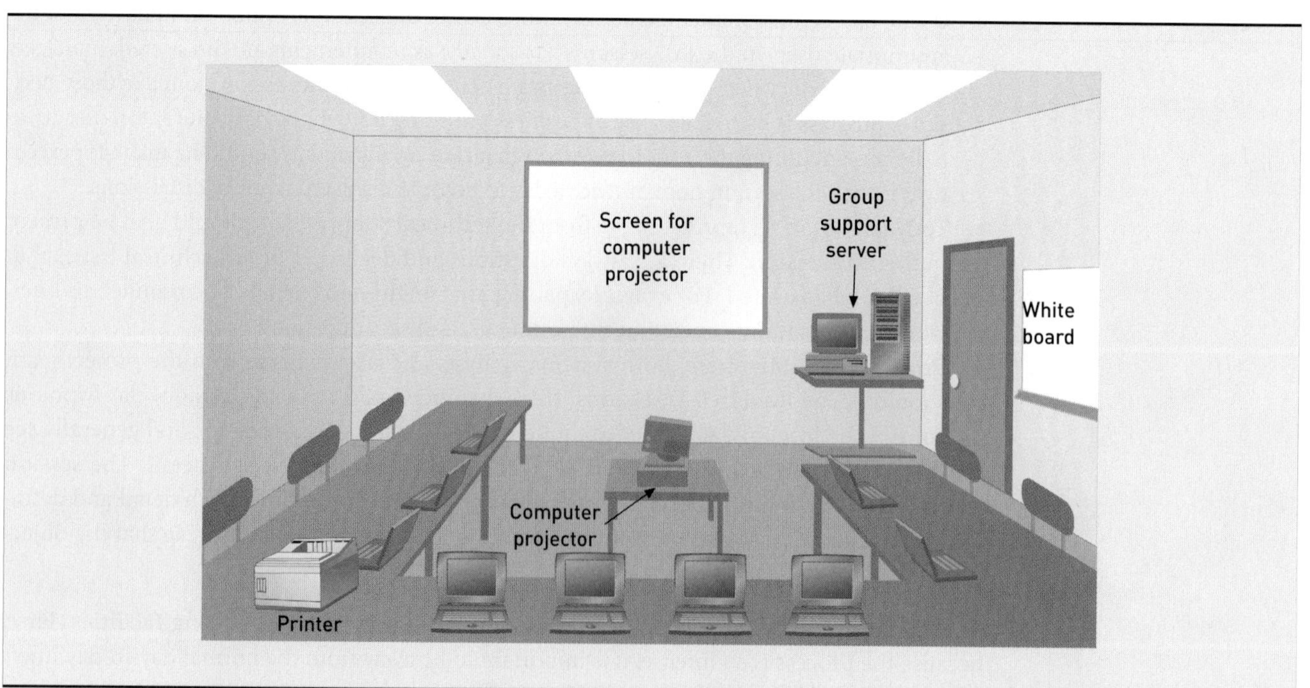

RESEARCH VENDOR SOLUTIONS

Many of the problems and opportunities that companies want to address with new information systems have already been solved by other companies. In many instances, consulting firms have experience with the same problems, and sometimes software firms already have packaged solutions for a particular business need. Directories such as Data Sources also list thousands of hardware, software, consulting, and solution developers—it makes sense to learn about and capitalize on this existing knowledge.

There are three positive contributions and one danger in exploring existing solutions. First, researching alternative solutions will frequently help users generate new ideas for how to better perform their business functions. Seeing how someone else solved a problem, and applying that idea to the culture and structure of the existing organization, will often provide viable alternative solutions for business needs.

Second, some of these solutions are excellent and state of the art. Without this research, the development team may create a system that is obsolete even before it is designed. Companies need solutions that not only solve basic business problems but that are up to date with current competitive practices.

Third, it is often cheaper, and less risky, to buy a solution rather than to build it. If the solution meets the needs of the company and can be purchased, then that is usually a safer, quicker, and less expensive route. There are many ways to buy solutions. Chapter 8 discusses alternative schemes to build and buy. Early in the development project, you want to research other alternatives but not make a final decision until you have investigated all the alternatives.

The danger, or caveat, in this process is that sometimes the users, and even the systems analysts, want to buy one of the alternatives immediately. But if a solution such as a packaged software system is purchased too early in the process, the company's needs may not be thoroughly investigated. Too many companies have bought a system only to find out later that it only supports half the functions that were needed. Don't fall into this dangerous pit.

The first difficulty in researching vendor alternatives is simply to find out who has solutions that fit the business need. Many of the large software and hardware companies, such as Oracle, IBM, Microsoft, and Computer Associates, have specific solution systems. There are also directories of system solutions—of software, hardware, and developer companies. Data Sources is one of the better ones. You can also search the Internet to find more directories. Sometimes these directories can be found in the library—a company technical library, the city library, or a nearby university library.

Other places to look are in trade journals for the industry. For example, the retail industry has several trade journals. System providers frequently advertise in these journals and at trade shows. Another method is word of mouth from other companies in the industry. Generally, users will have friends who work in competing companies. These people are sometimes aware of specific vendors that have helped solve their own business needs. Although companies compete fiercely on the sales and marketing end, it is not unusual for them to belong to a common trade organization that helps to share knowledge about the industry, including knowledge about system solutions.

After a list of possible providers has been developed, the next step is to research the details of each solution. It is easy to get the sales and marketing literature, but it is more difficult to get specifics of the system. Useful resources include: (1) technical specifications, (2) a demo or trial system, (3) references of existing clients who would let you observe their system, (4) an on-site visit, and (5) a printout of the screens and reports.

The final step is to review the details of the information received. Depending on the information obtained, it can be reviewed solely by the project team or with the users. In many cases, a review with key users is the most beneficial in understanding and identifying various approaches to addressing the business need. In any event, researching the solutions that have already been developed is an effective early step in understanding the business and identifying possible courses of action.

VALIDATING THE REQUIREMENTS

Now that you have learned about information-gathering techniques and ways to elicit requirements from the users, you need to make sure that the information you gathered is correct. All too frequently, systems analysts think they understand what the users need but have failed to capture some very important subtleties about the business processes. Obviously, correcting such a mistake after the system has been programmed is very expensive. In fact, various studies have indicated that fixing a requirements error later in the development cycle can cost hundreds of times more than it would have cost to fix during the requirements definition.

If we compare the development of a new information system to the construction of a house, the requirements determined during analysis are like a house's blueprints and construction designs. The construction of the house is dependent on those blueprints. What if there are errors in the blueprints, such as load-bearing walls that are not strong enough or missing structural supports? If these errors are not discovered until the second story is built, it will be extremely expensive to remove and rebuild walls. So, the blueprints have to be correct. How does an architect ensure that they are correct? By not waiting until the house is being built to "test" the correctness of the blueprints.

System requirements have a similar problem. The design and construction of the system depend on correct requirements. It is too late, and very expensive, if the requirements are "tested" only while the programming is being done. Testing and validation of the system requirements must be done as early as possible.

At this point in your project activities, you have collected information about the user requirements. You may have developed some workflow diagrams. In the next chapter, you will learn about building models to describe the system requirements. All of these elements should be thoroughly tested before the actual design and programming begin. When writing a computer program, a programmer must verify the accuracy of the code by conducting various tests. Executing the program on a computer—by entering appropriate input data and observing the output—tests a computer program. Analysts cannot test the requirements that way, so they have to use a different approach.

Various techniques can be used to validate the information from the users and the requirements that are developed from that information. To check internal consistency, analysts build models and verify that they are mathematically consistent. You will learn more about models in the next chapter. One powerful technique, called *structured walkthroughs*, is useful both for validating the requirements against the users' needs as well as verifying internal consistency.

structured walkthrough
a review of the findings from your investigation and of the models built based on those findings

A **structured walkthrough**, sometimes just called a *walkthrough*, is a review of the findings from your investigation and of the models built based on those findings. A walkthrough is considered structured because analysts have formalized the review process into a set procedure. The objective of a structured walkthrough is to find errors and problems. Its purpose is to ensure that the model is correct. The fundamental concept is one of documenting the requirements as you understand them and then reviewing them for any errors, omissions, inconsistencies, or problems. A review of the findings can be done informally with colleagues on the project team, but a structured walkthrough must be more formal.

It is important to note one critical point: A structured walkthrough is not a performance review. Managers should be involved only if they were involved in the original fact-finding and thus are required for verification or validation. The review is of an analyst's work and not the person. To help you understand the more structured approach, this section reviews the what, when, who, and how of a structured walkthrough.

One of the major responsibilities of the project manager, as described in Appendix A on the book's Web site, is to ensure the quality of the final system. Often during the rush and pressure of a project, systems analysts will think, "My work is good. It does not need to be reviewed." But it is very unwise for a project manager to skip the review. Because of the costly consequences, it just does not make sense to exclude from the project plan specific tasks and procedures to ensure that the requirements are complete and accurate. Omissions such as this will always cause problems later in the project. Structured walkthroughs can be performed to validate gathered information; however, they should definitely be performed to validate the specification models (discussed in the next few chapters). This section focuses on the process of a structured walkthrough.

WHAT AND WHEN

The key input to a structured walkthrough is one or more analysis or design models. They can be narratives describing processes, flowcharts showing workflows, or diagrams documenting entire procedures. Normally, it is better to conduct several smaller walkthroughs that review three to six pages of documentation than to cover 30 pages of details. Any written work that is a fairly independent package can be reviewed in a walkthrough. It is not uncommon to hold smaller walkthroughs every week or two with members of the project team. The frequency of the walkthroughs is not as critical as the timing—a walkthrough should be scheduled as soon as possible after the documents have been created.

WHO

The two main parties involved in walkthroughs are the person or people who need their work reviewed and the group that reviews it. For verification—that is, internal consistency and correctness—it is best to have other experienced analysts involved in the walkthrough. They look for inconsistencies and problems. For validation—that is, ensuring that the system satisfies all the needs of the various stakeholders—the appropriate stakeholders should be involved. The nature of the work to be reviewed dictates who the reviewers should be. If it is a diagram showing a business process, the users who supplied the original definition should be involved. If it is a technical specification of design details, the technical staff should be involved in the review. At times, the reviewers may be members of the project team. In other instances, they are external users or technical staff. Those who can validate the correctness of the work are the people who should be invited.

HOW

As with an interview, a structured walkthrough requires preparation, execution, and follow-up.

Preparation

The analyst whose work is being reviewed prepares material for review. Next, he or she identifies the appropriate participants and provides them copies of the material. Finally, the analyst schedules a time and place for the walkthrough and notifies all participants.

Execution

During the walkthrough, the analyst presents the material point by point. If it is a diagram or flowchart, he or she walks through the flow, explaining each component. One effective technique is to define a sample test case and process it through the defined flow. The reviewers look for inconsistencies or problems and point them out. A librarian, a helper for the presenter, documents the comments made by the reviewers. Presenters should never be their own librarians because they should not be distracted from explaining the documentation. To ensure accuracy, someone else should record the errors, comments, and suggestions.

BEST PRACTICE

In a structured walkthrough, have a nonparticipant act as librarian to record all errors, comments, and suggestions.

Corrections and solutions to problems are not made during the walkthrough. At most, some suggested solutions may be provided, but the documentation should not be corrected during the walkthrough. Because presenters are commonly a little nervous, it is unfair to ask them to make wise decisions on the spur of the moment. If a misunderstanding of the user requirements is uncovered, a brief review might be in order. However, if an error is fairly complex, it is better to schedule an additional interview to clarify the misunderstanding. The walkthrough should not get bogged down into a fact-finding session. The reviewer should only provide feedback, and the presenter can integrate it into the material later, when he or she has the entire set of comments and can make corrections without interruptions or further criticism.

Follow-Up

Follow-up consists of making the required corrections. If the reviewed material has major errors and problems, an additional walkthrough may be necessary. Otherwise, the corrections are made, and the project continues to the next activities.

Figure 4-19 is a sample review form that was used in one of the review sessions at Rocky Mountain Outfitters for the sales commission rates and rules. Not shown are several attached sheets, including a couple of flowcharts of procedures. The material reviewed in this case is simply a list of business rules for commission rates. The reviewers are senior managers from the user community. Because sales commission business rules are critical, and these managers make the policy decisions about commissions, they are the obvious choices to review the rules as uncovered in discussions and interview sessions.

Figure 4-19

A structured walkthrough evaluation form

Walkthrough Control Sheet

Project Control Information

Project: *Online Catalog System, Customer Support Subsystem*

Segment of project being reviewed: *Review of business rules for sales commission rates*

Team leader: *Mary Ellen Green*

Author of work: *Jim Williams*

Walkthrough Details

Date, time, and location
April 10, 2010. 10:00 a.m. MIS conference room.

Description of materials being reviewed:
This is a review of the business rules before they are integrated into the diagrams and models. There is a short flowchart attached showing the flow of the commission process. There is another flowchart showing the process to set commission rates. We will also review outstanding issues to ensure that all understand the policy decisions that must be made.

Participating reviewers:
William McDougal, Genny Monson, Robert Schneider

Results of Walkthrough

_____XX__Accept. sign-offs:_______________________________
__________Minor revisions. Description of revisions:

__________Rework and schedule new walkthrough. Description of required rework:
Excellent and thorough. No rework required.

SUMMARY

There are six primary activities of systems analysis:

- Gather information.
- Define system requirements.
- Prioritize requirements.
- Prototype for feasibility and discovery.
- Generate and evaluate alternatives.
- Review recommendations with management.

Generally we divide system requirements into two categories: functional and nonfunctional requirements. The functional requirements are those that explain the basic business functions that the new system must support. Nonfunctional requirements involve the objectives of the system for technology, performance, usability, reliability, and security.

Models are useful for defining requirements and designs. There are three types of models: mathematical, descriptive, and graphical.

To ferret out the requirements, analysts must work with various stakeholders in the new system. We categorize stakeholders into three groups: (1) the users, those who will actually use the system day to day; (2) the clients, those who pay for and own the system; and (3) the technical staff, the people who must ensure that the system operates within the computing environment of the organization. One of the most important first steps in determining systems requirements is to identify these various system stakeholders.

A fundamental question to investigate system requirements is, "What kind of information do I need?" This chapter provides you with some general guidelines. As you learn more about modeling, you should also understand better what information you need. Three major themes of information should be pursued:

- What are the business processes and operations?
- How are the business processes performed?
- What are the information requirements?

Analysts use seven primary techniques to gather this information, and one technique ensures its correctness. The seven fact-finding techniques are the following:

- Review existing reports, forms, and procedure descriptions.
- Conduct interviews and discussions with users.
- Observe and document business processes.
- Build prototypes.
- Distribute and collect questionnaires.
- Conduct JAD sessions.
- Research vendor solutions.

The fundamental idea of a prototype is an initial, working model of a larger, more complex entity. The primary purpose of a prototype is to have a working model that will test a concept or verify an approach. Discovery prototypes are built to define requirements but are then usually discarded or at least not used for the final programming.

Joint application design is a technique used to expedite the investigation of system requirements by holding several marathon sessions with all the critical participants. Discussion results in requirements definition and policy decisions immediately, without the delays of interviewing separate groups and trying to reconcile differences. When done correctly, JAD is a powerful and effective technique.

The review technique to ensure that analysis is accurate and complete is called a *structured walkthrough*. Remember that a structured walkthrough has the objective of reviewing and improving the work. It is not a performance review.

KEY TERMS

activity diagram, p. 141
closed-ended questions, p. 145
descriptive model, p. 126
functional requirement, p. 122
graphical model, p. 126
group support system (GSS), p. 148
joint application design (JAD), p. 147
logical model, p. 120
mathematical model, p. 125
mock-up, p. 145
nonfunctional requirements, p. 123
open-ended questions, p. 146
performance requirement, p. 123
physical model, p. 120
prototype, p. 144
reliability requirement, p. 123
security requirement, p. 123
stakeholders, p. 128
structured walkthrough, p. 150
swimlane, p. 141
synchronization bar, p. 141
system requirements, p. 122
technical requirement, p. 123
transaction, p. 130
usability requirement, p. 123
workflow, p. 141

REVIEW QUESTIONS

1. List the six activities of systems analysis.
2. What are three types of models?
3. What is the difference between functional requirements and nonfunctional requirements?
4. Explain the use of a discovery prototype and an evolutionary prototype.
5. List and describe the three fact-finding themes.
6. What is the objective of a structured walkthrough?
7. Explain the steps in preparing for an interview session.
8. What are the benefits of doing vendor research during information-gathering activities?
9. What categories of stakeholders should you include in fact-finding?
10. What is meant by vertical and horizontal dimensions when determining users to involve?
11. What is JAD? When is it used?
12. What is BPR? What does it have to do with systems analysis?
13. What technique is used to validate user requirements?
14. Describe the open-items list and explain why it is important.
15. What do *correct*, *complete*, and *comprehensive* mean with regard to systems analysis?
16. List and describe the seven information-gathering techniques.
17. What is the purpose of an activity diagram?
18. Draw and explain the symbols used on an activity diagram.

THINKING CRITICALLY

1. Provide an example of each of the three types of models that might apply to designing a car, a house, and an office building. Explain why requirements models are logical models rather than physical models.
2. One of the toughest problems in investigating system requirements is to make sure that they are complete and comprehensive. What things would you do to ensure that you get all of the right information during an interview session?
3. What can you do to ensure that you have included all of the right stakeholders on your list of people to interview? How can you double-check your list?
4. One of the problems you will encounter during your investigation is "scope creep"—that is, user requests for additional features and functions. Scope creep happens because sometimes users have many unsolved problems and the system investigation may be the first time anybody has listened to their needs. How do you keep the system from growing and including new functions that should not be part of the system?
5. It is always difficult to observe users in their jobs. It frequently makes both you and them uncomfortable. What things could you do to ensure that user behavior is not changing because of your visit? How could you make observation more natural?
6. What would you do if you got conflicting answers for the same procedure from two different people you interviewed? What would you do if one was a clerical person and the other was the department manager?
7. You are a team leader of four systems analysts. You have one analyst who has never done a structured walkthrough of her work. How would you help the analyst to get started? How would you ensure that the walkthrough was effective?

8. You have been assigned to resolve several issues on the open-items list, and you are having a hard time getting policy decisions from the user contact. How can you encourage the user to finalize these policies?

9. You are going on your first consulting assignment to do systems analysis. Your client does not like to pay to train new, inexperienced analysts. What should you do to appear competent and well prepared? How should you approach the client?

10. In the running case of Rocky Mountain Outfitters, you have set up an interview with Jason Nadold in the shipping department. Your objective is to determine how shipping works and what the information requirements for the new system will be. Make a list of questions, open-ended and closed-ended, that you would use. Include any questions or techniques you would use to ensure you find out about the exceptions.

11. Develop an activity diagram based on the following narrative. Note any ambiguities or questions that you have as you develop the model. If you need to make assumptions, also note them.

    The purpose of the Open Access Insurance System is to provide automotive insurance to car owners. Initially, prospective customers fill out an insurance application, which provides information about the customer and his or her vehicles. This information is sent to an agent, who sends it to various insurance companies to get quotes for insurance. When the responses return, the agent then determines the best policy for the type and level of coverage desired and gives the customer a copy of the insurance policy proposal and quote.

12. Develop an activity diagram based on the following narrative. Note any ambiguities or questions that you have as you develop the model. If you need to make assumptions, also note them.

    The purchasing department handles purchase requests from other departments in the company. People in the company who initiate the original purchase request are the "customers" of the purchasing department. A case worker within the purchasing department receives the request and monitors it until it is ordered and received.

    Case workers process requests for the purchase of products under $1,500, write a purchase order, and then send it to the approved vendor. Purchase requests over $1,500 must first be sent out for bid from the vendor that supplies the product. When the bids return, the case worker selects one bid. Then, he or she writes a purchase order and sends it to the vendor.

13. Develop an activity diagram based on the following narrative. Note any ambiguities or questions that you have as you develop the model. If you need to make assumptions, also note them.

    The shipping department receives all shipments on outstanding purchase orders. When the clerk in the shipping department receives a shipment, he or she finds the outstanding purchase order for those items. The clerk then sends multiple copies of the shipment packing slip. One copy goes to purchasing, and the department updates its records to indicate that the purchase order has been fulfilled. Another copy goes to accounting so that a payment can be made. A third copy goes to the requesting in-house customer so that he or she can receive the shipment.

    After payment is made, the accounting department sends a notification to purchasing. After the customer receives and accepts the goods, he or she sends notification to purchasing. When purchasing receives these other verifications, it closes the purchase order as fulfilled and paid.

EXPERIENTIAL EXERCISES

1. Conduct a fact-finding interview with someone involved in a procedure that is used in a business or organization. This person could be someone at the university, in a small business in your neighborhood, in the student volunteer office at the university, in a doctor's or dentist's office, in a volunteer organization, or at your local church. Identify a process that is done, such as keeping student records, customer records, or member records. Make a list of questions and conduct the interview. Remember, your objective is to understand that procedure thoroughly—that is, to become an expert on that single procedure.

2. Follow the same instructions as for exercise 1, except make this exercise an observation experience. Either observe the other person do the work or ask to carry out the procedure yourself. Write down the details of the process you observe.

3. Get a group of your fellow students together and conduct a structured walkthrough of your results from exercise 1 or 2. Using the results of your interview or observation, document the procedure in a flowchart with some narrative. Then, conduct a walkthrough with several colleagues. Or take another assignment, such as Thinking Critically question 9, and walk through your preparation for that assignment. Follow the steps outlined in the text.

4. Research and write a one- to two-page research paper using at least three separate library sources on one of the following topics:
    a. Joint application design
    b. Prototyping as a discovery mechanism
    c. Computer support for collaborative work (CSCW)
    d. Workflow systems
    e. Structured walkthrough

5. Using Rocky Mountain Outfitters and the customer support subsystem as your guide, develop a list of all the procedures that may need to be researched. You may want to think about the exercise in the context of your experience with retailers such as L. L. Bean, Lands' End, or Amazon.com. Get some catalogs, check out the Internet marketing done on the retailers' Web sites, and then think about the underlying business procedures that are required to support those sales activities. List the procedures and describe your understanding of each.

CASE STUDIES

JOHN AND JACOB, INC., ONLINE TRADING SYSTEM

John and Jacob, Inc., is a regional brokerage firm that has been successful over the last several years. Competition for customers is intense in this industry. The large national firms have very deep pockets, with many services to offer to clients. Severe competition also comes from discount and Internet trading companies. However, John and Jacob has been able to cultivate a substantial customer base from upper-middle income clients in the northeastern United States. To maintain a competitive edge with its customers, John and Jacob is in the process of developing a new online trading system. The plan for the system identifies many new capabilities that would provide new services to its clients.

Edward Finnigan, the project manager, is in the process of identifying all the groups of people who should be included in the development of the system requirements. He isn't quite sure exactly who should be included. Here are the issues he's considering:

- **Users.** The trading system is to be online to each of the company's 30 trading offices. Obviously, the brokers who are going to use the system need to have input, but how should this be done? Edward also isn't sure what approach would be best to ensure that the requirements are complete, yet not require tremendous amounts of time. Including all of the offices would increase enthusiasm and support for the system, but it would take a lot of time. Involving more brokers would bring divergent opinions that would have to be reconciled.
- **Customers.** The trading system will also include confirmations, reports of trades, and customer statements. Web access is also planned, which will enable customers to effect trades and to check their accounts. Consequently, Edward wonders how to involve John and Jacob customers in the development of system requirements. Normally, customers are not asked to participate in the development of systems. However, it would be nice to know how best to serve John and Jacob's customers. Edward is sensitive to this issue because some brokers have told him that many customers do not like the format of their statements from the current system. He would like to involve customers, but he does not know how.
- **Other Stakeholders**. Edward knows he should involve other stakeholders to help define system requirements. He isn't quite sure whom he should contact. Should he go to senior executives? Should he contact middle management? Should he include back-office functions such as accounting and investing? He isn't quite sure how to get organized or how to decide who should be involved.

1. What is the best method for Edward to involve the brokers (users) in development of the new online trading system? Should he use a questionnaire? Should he interview the brokers in each of the company's 30 offices, or would one or two brokers representing the entire group be better? How can Edward ensure that the information about requirements is complete, yet not lose too much time doing so?
2. Concerning customer input for the new system, how can Edward involve customers in the process? How can he interest them in participating? What methods can Edward use to ensure that the customers he involves are representative of John and Jacob's entire customer group?
3. As Edward considers what other stakeholders he should include, what are some criteria he should use? Develop some guidelines to help him build a list of people to include.

RETHINKING ROCKY MOUNTAIN OUTFITTERS

Barbara Halifax, the project manager for the CSS project, had finished identifying the list of stakeholders in the project. As shown earlier in the chapter, quite a few senior executives would be involved. Most of them would not have major input. Those in Bill McDougal's area would, of course. Not only was he the project sponsor, but all his assistants were excited about this new system and its potential to help the business grow. Barbara had a good working relationship with all of these executives.

Barbara had also identified numerous department managers and senior customer service representatives who would be able to provide detailed processing requirements. She had divided her list of stakeholders into two groups. The first group consisted of all those with primary responsibility to help define user requirements. The second group included those who would not have direct use of the system but would need reports and information from the system. She wanted to make sure the needs of these people were also satisfied.

As an experienced project manager, Barbara had her checklists of things to do. She used a project manager checklist to help her remember all important tasks. Being a project manager was much too critical, and potentially stressful, to do it "by the seat of your pants."

As she reviewed her list, she noticed several activities that she had not yet considered on the CSS project. She was thinking that before she let her project team start to meet with the users, she ought to consider these items and review them with her team. The items that most caught her attention were the following:

- Develop a communications plan with the user.
- Manage user expectations.
- Control the scope and avoid scope creep.

Based on the concepts you learned in this chapter, what would you do if you were Barbara? (You also might want to review Appendix A on the book's Web site). Obviously, you want to provide the best possible solution for the company, but you also need to control the project, the scope, and the users so that the system will be successful and be installed on time.

1. Identify the major points you would include in a communications plan at this point in the project.
2. What advice would you give your project team to help it manage the user expectations?
3. What early planning can you do now to ensure that the scope is realistic—to meet the need but within the time and budget allotted?

FOCUSING ON RELIABLE PHARMACEUTICAL SERVICE

Reliable Pharmaceutical Service plans to develop an extranet that enables its client health-care facilities to order drugs and supplies as if they were ordering from an internal pharmacy. The extranet should enable Reliable's suppliers to function as if they were part of Reliable's internal organization. These views of the final system have significant implications for defining system requirements and for gathering information about those requirements.

1. What information-gathering methods are most appropriate to learn about requirements from Reliable's own management staff and other employees? From client health-care organizations? From suppliers?
2. Should patients in client health-care facilities participate in the information-gathering process? If so, why, and in what ways should they participate?
3. With respect to gathering information from suppliers and clients, how deeply within those organizations should systems analysts look when defining requirements? How might Reliable deal with supplier and client reluctance to provide detailed information about their internal operations?
4. For which user community or communities (internal, supplier, or client) are prototypes likely to be most beneficial? Why?

FURTHER RESOURCES

Vangalur S. Alagar, *Specification of Software Systems*. Springer-Verlag, 1998.

Soren Lauesen, *Software Requirements: Styles and Techniques*. Addison-Wesley, 2002.

Stan Magee, *Guide to Software Engineering Standards and Specifications*. Artech House, 1997.

Suzanne Robertson and James Robertson, *Mastering the Requirements Process*. Addison-Wesley, 2000.

Karl Wiegers, *Software Requirements*. Microsoft Press, 1999.

Jane Wood, *Joint Application Development*. John Wiley & Sons, 1995.

Ralph Young, *Effective Requirements Practices*. Addison-Wesley, 2001.

CHAPTER

5

MODELING SYSTEM REQUIREMENTS

LEARNING OBJECTIVES

After reading this chapter, you should be able to:

- Understand why identifying use cases is the key to defining functional requirements
- Use three techniques for identifying use cases
- Write brief, intermediate, and fully developed use case descriptions
- Explain how the concept of things in the problem domain also defines requirements
- Identify and analyze data entities and domain classes needed in the system
- Read, interpret, and create an entity-relationship diagram
- Read, interpret, and create a domain model class diagram

CHAPTER OUTLINE

User Goals, Events, and Use Cases

Use Case Descriptions

"Things" in the Problem Domain

The Entity-Relationship Diagram

The Domain Model Class Diagram

Where You Are Headed

WAITERS ON CALL MEAL-DELIVERY SYSTEM

Waiters on Call is a restaurant meal-delivery service started in 2003 by Sue and Tom Bickford. The Bickfords both worked for restaurants while in college and always dreamed of opening their own restaurant. But unfortunately, the initial investment was always out of reach. The Bickfords noticed that many restaurants offer takeout food, and some restaurants—primarily pizzerias—offer home delivery service. Many people they met, however, seemed to want home delivery with a wider food selection.

Sue and Tom conceived Waiters on Call as the best of both worlds: a restaurant service without the high initial investment. The Bickfords contracted with a variety of well-known restaurants in town to accept orders from customers and to deliver the complete meals. After preparing the meal to order, the restaurant charges Waiters on Call a wholesale price, and the customer pays retail plus a service charge and tip. Waiters on Call started modestly, with only two restaurants and one delivery driver working the dinner shift. Business rapidly expanded, and the Bickfords realized they needed a custom computer system to support their operations. They hired a consultant, Sam Wells, to help them define what sort of system they needed.

"What sort of events happen when you are running your business that make you want to reach for a computer?" asked Sam. "Tell me about what usually goes on."

"Well," answered Sue, "when a customer calls in wanting to order, I need to record it and get the information to the right restaurant. I need to know which driver to ask to pick up the order, so I need drivers to call in and tell me when they are free. Sometimes customers call back wanting to change their orders, so I need to get my hands on the original order and notify the restaurant to make the change."

"Okay, how do you handle the money?" queried Sam.

Tom jumped in. "The drivers get a copy of the bill directly from the restaurant when they pick up the meal. The bill should agree with our calculations. The drivers collect that amount plus a service charge. When drivers report in at closing, we add up the money they have and compare it with the records we have. After all drivers report in, we need to create a deposit slip for the bank for the day's total receipts. At the end of each week, we calculate what we owe each restaurant at the agreed-to wholesale price and send each a statement and check."

"What other information do you need to get from the system?" continued Sam.

"It would be great to have some information at the end of each week about orders by restaurant and orders by area of town—things like that," Sue said. "That would help us decide about advertising and contracts with restaurants. Then we need monthly statements for our accountant."

Sam made some notes and sketched some diagrams as Sue and Tom talked. Then after spending some time thinking about it, he summarized the situation for Waiters on Call. "It sounds to me like you need a system to use whenever these events occur:

- A customer calls in to place an order, so you need to *record an order*.
- A driver is finished with a delivery, so you need to *record delivery completion*.
- A customer calls back to change an order, so you need to *update an order*.
- A driver reports for work, so you need to *sign in the driver*.
- A driver submits the day's receipts, so you need to *reconcile driver receipts*.

"Then you need the system to produce information at specific points in time—for example, when it is time to:

- Produce an end-of-day deposit slip
- Produce end-of-week restaurant payments
- Produce weekly sales reports
- Produce monthly financial reports

"Based on the way you have described your business operations, I am assuming you will need the system to store information about these types of things, which we call data entities or domain classes:

- Restaurants
- Menu items
- Customers
- Orders
- Order payments
- Drivers

"Then I suppose you will need to maintain information about restaurants and drivers. You'll need to use the system when you add a new restaurant, a restaurant changes the menu, you hire a new driver, or a driver leaves. Am I on the right track?"

Sue and Tom quickly agreed that Sam was talking about the system in a way they could understand. They were confident that they had found the right consultant for the job.

OVERVIEW

The preceding chapter described the systems analysis activities of the SDLC and then introduced the many tasks and techniques involved when completing the first analysis activity—gathering information about the system, its stakeholders, and its requirements. An extensive amount of information is required to properly define the system's functional and nonfunctional requirements. This chapter, along with Chapters 6 and 7, presents techniques for documenting the functional requirements by creating a variety of models. These models are created as part of the analysis activity *Define system requirements*, although remember that the analysis activities actually are done in parallel and in iterations.

In this chapter we focus on two key concepts that help define system requirements in both the traditional and the object-oriented approach: use cases and things in the problem domain of users. This chapter covers specific models for both the traditional approach and the object-oriented approach, including those based on the Rocky Mountain Outfitters (RMO) customer support system. Chapter 6 continues the discussion of requirements models for the traditional approach, and Chapter 7 continues the object-oriented approach.

USER GOALS, EVENTS, AND USE CASES

use case
an activity the system performs

Virtually all newer approaches to system development begin the requirements modeling process with the concept of a use case. A **use case** is an activity the system performs, usually in response to a request by a user. The term *use case* originated with the object-oriented approach, but today it is also used when modeling functional requirements in the traditional approach. If you are focusing on the traditional approach in your studies, remember that a use case is basically the same as an activity or process.

user goal technique
an approach for identifying use cases in which an analyst talks to all users to get them to describe their goals in using the system

Several techniques are recommended for identifying use cases. One approach is to talk to all users to get them to describe their goals in using the system. This approach is called the **user goal technique**. First, list all users and think through what each type of user needs the system to do for their jobs. Then interview each type of user and focus on their goals. By focusing on one type of user at a time, an analyst can systematically address the problem of identifying use cases. In the user goal technique, the analyst might start with the existing system and list all system functions that are currently included, adding any new functionality requested by users. Then the existing system functions and requested functions can be used to establish user goals.

Figure 5-1 lists a few Rocky Mountain Outfitters users (also called *actors*) and some of their work goals for the customer support system. By asking users about their goals, an analyst can focus on new or promising processes rather than just automating existing procedures. Many analysts find this approach useful for getting an initial list of use cases.

Figure 5-1
Identifying use cases with the user goal technique

User/actor	User goal and resulting use case
Order clerk	Look up item availability Create new order Update order
Shipping clerk	Record order fulfillment Record back order
Merchandising manager	Create special promotion Produce catalog activity report

CRUD technique
an approach in which an analyst looks at each type of data and includes use cases that create the data, read or report on the data, update the data, and delete the data

Another important technique for identifying use cases is the **CRUD technique**. CRUD is an acronym for *Create*, *Read* or *Report*, *Update*, and *Delete*. The analyst starts by looking at the types of data stored by the system, which are modeled as data entities or domain classes, as described later in this chapter. Examples of types of data include Customer, Order, Inventory Item, and Shipment in the RMO customer support system example. To identify use cases, the analyst looks at each type of data and includes use cases that create the data, read or report on the data, update the data, and delete the data. Figure 5-2 shows an example of potential use cases for a Customer.

Figure 5-2
Identifying use cases with the CRUD technique

Data entity/class	CRUD	Resulting use case
Customer	Create	Add new customer
	Read/Report	Look up customer Produce customer list List customer orders
	Update	Update customer information
	Delete	Delete inactive customer

The most comprehensive technique for identifying use cases is the event decomposition technique, which is described in detail in the next section. No matter what approach they use to identify use cases, analysts must identify them at the right level of detail. For example, one analyst might identify a use case as typing in a customer name on a form. A second analyst might identify a use case as the entire process of adding a new customer. A third analyst might even define a use case as working with customers all day, which could include adding new customers, updating customer records, deleting customers, following up on late-paying customers, or contacting former customers. The first example is too narrow to be useful. The second example defines a complete user goal, which is the right level of analysis for a use case. Working with customers all day—the third example—is too broad to be useful.

elementary business process (EBP)
a task that is performed by one person, in one place, in response to a business event; it adds measurable business value and leaves the system and its data in a consistent state

The appropriate level of detail for identifying use cases is one that focuses on **elementary business processes (EBPs)**. An EBP is a task that is performed by one person, in one place, in response to a business event; that adds measurable business value; and that leaves the system and its data in a consistent state. (See the 2005 Larman text, as referenced in the "Further Resources" section at the end of the chapter, for additional discussion of EBPs.) In Figure 5-1, some of the RMO customer support system goals that will become use cases are *Create new order*, *Record order fulfillment*, *Create special promotion*, and so forth. These use cases are good examples of elementary business processes. In Figure 5-2, some of the use cases are *Add new customer*, *Produce customer list*, and *Update customer information*; all are good examples of elementary business processes. Each is performed by one user in a set time and place, and after it is completed, the system and its data are in a consistent state.

BEST PRACTICE

Try to use several approaches for identifying use cases, and then cross-check to be sure no use cases have been overlooked.

event

an occurrence at a specific time and place that can be described and is worth remembering

event decomposition

an analysis technique that focuses on identifying the events to which a system must respond and then determining how the system must respond

Note that each EBP (and so each use case) occurs in response to a business event. An event occurs at a specific time and place, can be described, and should be remembered by the system. Events drive or trigger all processing that a system does, so listing events and analyzing them makes sense when you need to define system requirements by identifying use cases.

EVENT DECOMPOSITION TECHNIQUE

In determining the use cases for a system, many analysts use event decomposition, the third technique mentioned earlier in this chapter. This technique focuses on identifying the events to which a system must respond and then determining how a system must respond—the system's use cases. When defining the requirements for a system, it is useful to begin by asking, "What events occur that will require the system to respond?" By asking about the events that affect the system, you direct your attention to the external environment and look at the system as a black box. This initial perspective helps keep your focus on a high-level view of the system (looking at the scope) rather than on the inner workings of the system. It also focuses your attention on the system's interfaces to outside people and other systems. End users—those who will actually use the system—can readily describe system needs in terms of events that affect their work. So, the external focus on events is appropriate when working with users. Finally, focusing on events gives you a way to divide (or decompose) the system requirements into use cases so that you can study each separately.

Some events that are important to a store's charge account processing system are shown in Figure 5-3. The system requirements are decomposed into use cases based on six events. A customer triggers three events: pays a bill, makes a charge, or changes address. The system responds with three use cases: *Record a payment*, *Process a charge*, or *Maintain customer data*. Three other events are triggered inside the system by time: time to send out monthly statements, time to send late notices, and time to produce end-of-week summary reports. The system responds with use cases that carry out what it is time to do: *Produce monthly statements*, *Send late notices*, and *Produce summary reports*. Describing this system in terms of events keeps the focus of the charge account system on the business requirements and the elementary business processes. Then the next step is to divide the work among developers; one analyst might focus on the events triggered by people, and another analyst might focus on events triggered internally. The system is decomposed in a way that allows it to be understood in detail. The result is a list of use cases triggered by business events at the right level of analysis.

The importance of the concept of events for defining functional requirements was first emphasized for modern structured analysis when this concept was adapted to real-time systems in the early 1980s. Real-time systems must react immediately to events in the environment. Early examples of real-time systems include control systems such as manufacturing process control or avionics guidance systems. For example, in process control, if a vat of chemicals is full, then the system needs to turn off the fill valve. The relevant event is "vat is full," and the system needs to respond to that event immediately. In an airplane guidance system, if the plane's altitude drops below 5,000 feet, then the system needs to turn on the low-altitude alarm.

Most information systems now being developed are so interactive that they can be thought of as real-time systems. In fact, people expect a real-time response to almost everything. So, use cases for business systems are now identified by using the event decomposition approach.

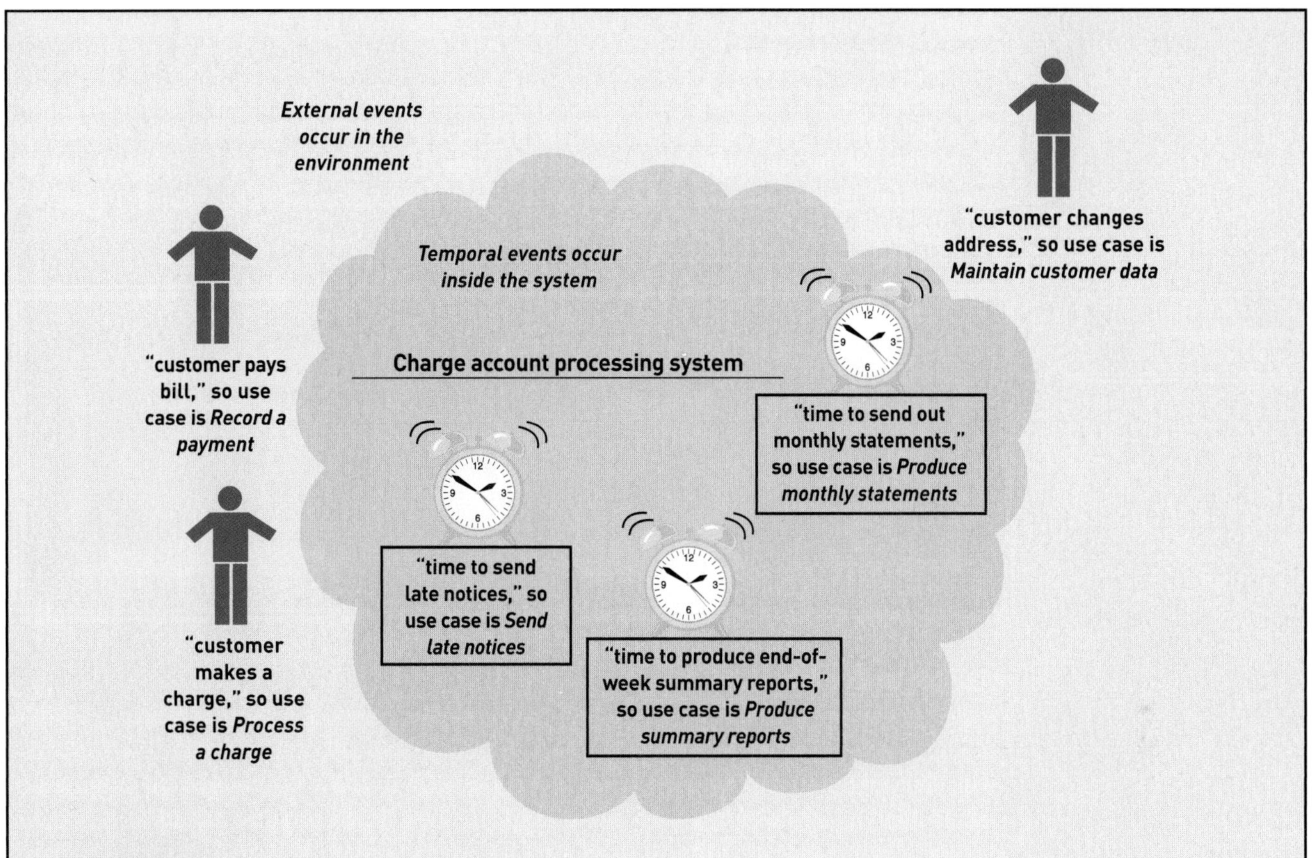
External events occur in the environment
Temporal events occur inside the system
"customer changes address," so use case is Maintain customer data
"customer pays bill," so use case is Record a payment
Charge account processing system
"time to send out monthly statements," so use case is Produce monthly statements
"time to send late notices," so use case is Send late notices
"time to produce end-of-week summary reports," so use case is Produce summary reports
"customer makes a charge," so use case is Process a charge

Figure 5-3

Events affecting a charge account processing system that lead to use cases

TYPES OF EVENTS

There are three types of events to consider when using the event decomposition technique to identify use cases: external events, temporal events, and state events (also called internal events). The analyst begins by trying to identify and list as many of these events as possible, refining the list while talking with system users.

External Events

external event
an event that occurs outside the system, usually initiated by an external agent or actor

An **external event** is an event that occurs outside the system, usually initiated by an external agent or actor. An external agent (or actor) is a person or organizational unit that supplies or receives data from the system. To identify the key external events, the analyst first tries to identify all of the external agents that might want something from the system. A classic example of an external agent is a customer. The customer may want to place an order for one or more products. This event is of fundamental importance to an order-processing system like the one needed by Rocky Mountain Outfitters. But other events are associated with a customer. Sometimes a customer wants to return an ordered product, or a customer needs to pay the invoice for an order. External events such as these are the types that the analyst looks for because they begin to define what the system needs to be able to do. They are events that lead to important transactions that the system must process.

When describing external events, it is important to name the event so that the external agent is clearly defined. The description should also include the action that the external agent wants to pursue. So, the event *Customer places an order* describes the external agent (a customer) and the action that the customer wants to take (to place an order for some products) that directly affects the system. Again, if the system is an order-processing system, the system needs to process the order for the customer.

Important external events can also result from the wants and needs of people or organizational units inside the company—for example, management requests for information. A typical event in an order-processing system might be *Management checks order status*. Perhaps managers want to follow up on an order for a key customer, and the system must routinely provide that information.

Another type of external event occurs when external entities provide new information that the system simply needs to store for later use. For example, a regular customer reports a change in address, phone, or employer. Usually one event for each type of external agent can be described to handle updates to data, such as *Customer updates account information*. Figure 5-4 provides a checklist to help in identifying external events.

Figure 5-4

External event checklist

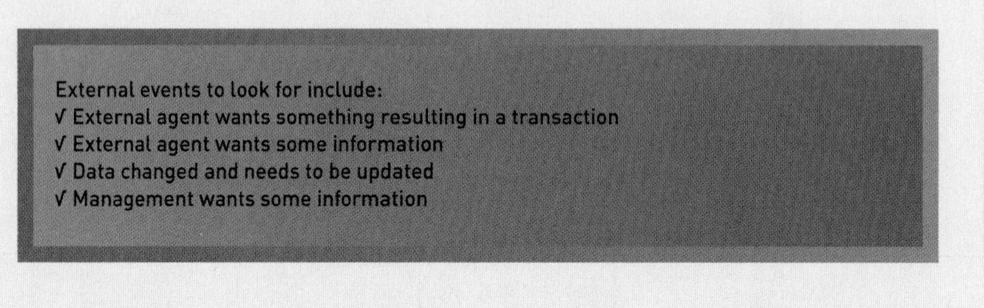
External events to look for include:
√ External agent wants something resulting in a transaction
√ External agent wants some information
√ Data changed and needs to be updated
√ Management wants some information

Temporal Events

temporal event
an event that occurs as a result of reaching a point in time

A second type of event is a **temporal event**, an event that occurs as a result of reaching a point in time. Many information systems produce outputs at defined intervals, such as payroll systems that produce a paycheck every two weeks (or each month). Sometimes the outputs are reports that management wants to receive regularly, such as performance reports or exception reports. These events are different from external events in that the system should automatically produce the required output without being told to do so. In other words, no external agent or actor is making demands, but the system is supposed to generate information or other outputs when they are needed.

The analyst begins identifying temporal events by asking about the specific deadlines that the system must accommodate. What outputs are produced at that deadline? What other processing might be required at that deadline? The analyst usually identifies these events by defining what the system needs to produce at that time. The payroll example discussed previously might be named *Time to produce biweekly payroll.* The event defining the need for a monthly summary report might be named *Time to produce monthly sales summary report.* Figure 5-5 provides a checklist to use in identifying temporal events.

Figure 5-5

Temporal event checklist

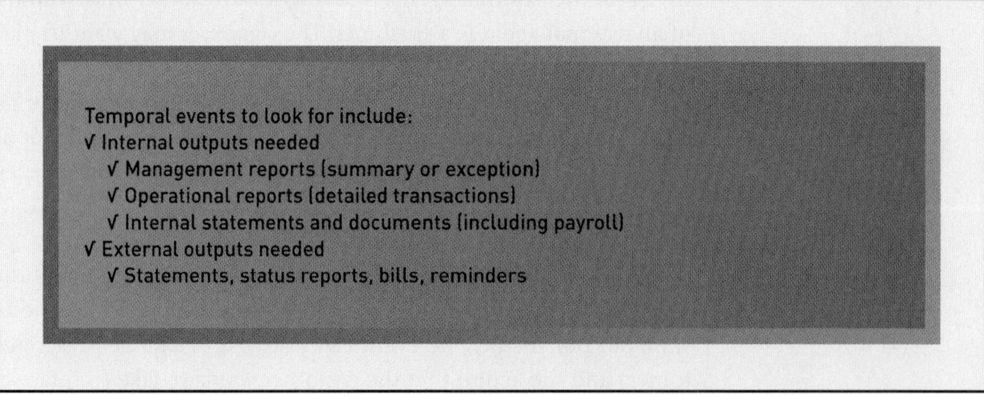
Temporal events to look for include:
√ Internal outputs needed
√ Management reports (summary or exception)
√ Operational reports (detailed transactions)
√ Internal statements and documents (including payroll)
√ External outputs needed
√ Statements, status reports, bills, reminders

Temporal events do not have to occur on a fixed date. They can occur after a defined period of time has elapsed. For example, a bill might be given to a customer when a sale has occurred. If the bill has not been paid within 15 days, the system might send a late notice. The temporal event *Time to send late notice* might be defined as a point 15 days after the billing date.

State Events

state event

an event that occurs when something happens inside the system that triggers the need for processing

A third type of event is a state event, an event that occurs when something happens inside the system that triggers the need for processing. State events are also called *internal events*. For example, if the sale of a product results in an adjustment to an inventory record and the inventory in stock drops below a reorder point, it is necessary to reorder. The state event might be named *Reorder point reached*. Often state events occur as a consequence of external events. Sometimes they are similar to temporal events, except the point in time cannot be defined. The reorder event might be named *Time to reorder inventory*, which sounds like a temporal event.

IDENTIFYING EVENTS

It is not always easy to define the events that affect a system. But some guidelines can help an analyst think through the process.

Events Versus Prior Conditions and Responses

It is sometimes difficult to distinguish between an event and part of a sequence of prior conditions that leads up to the event. Consider an example of a customer buying a shirt from a retail store (see Figure 5-6). From the customer's perspective, this purchase involves a long sequence of events. The first event might be that a customer wants to get dressed. Then the customer wants to wear a striped shirt. Next, his striped shirt appears to be worn out. Then the customer decides to drive to the mall. Then he decides to go into Sears. Then he tries on a striped shirt. Then the customer decides to leave Sears and go to Wal-Mart to try on a shirt. Finally, the customer wants to purchase the shirt. The analyst has to think through such a sequence to arrive at the point at which an event directly affects the system. In this case, the system is not affected until the customer is in the store, has a shirt in hand ready to purchase, and says, "I want to buy this shirt."

Figure 5-6

Sequence of actions that lead up to only one event affecting the system

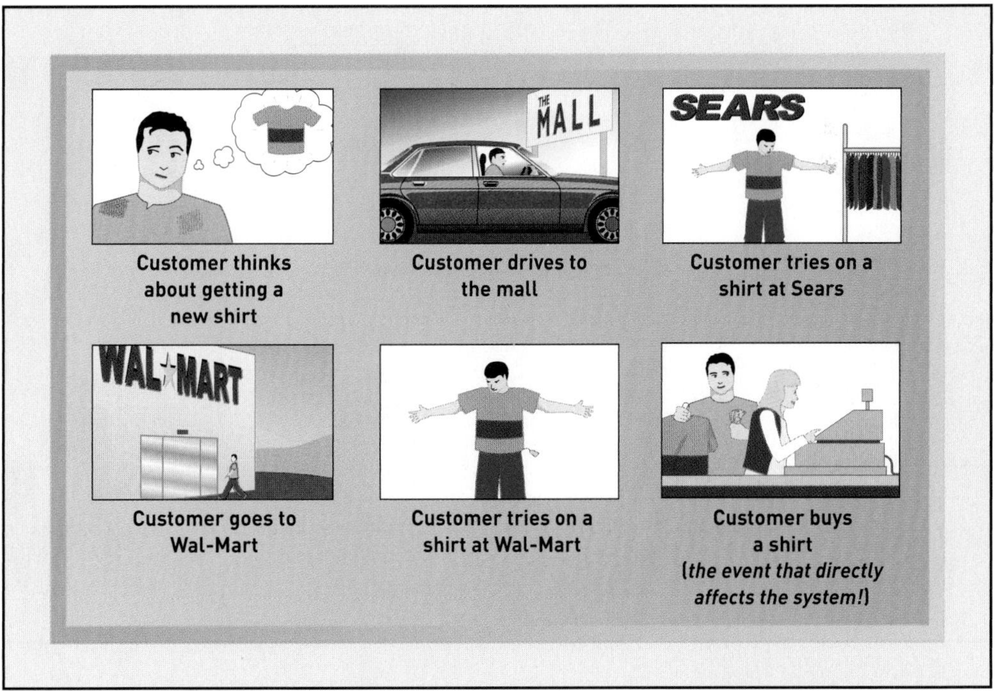

In other situations, it is not easy to distinguish between an external event and the system's response. For example, when the customer buys the shirt, the system requests a credit card number, and the customer supplies the credit card. Is the act of supplying the credit card an event? In this case, no. It is part of the interaction that occurs while completing the original transaction.

The way to determine whether an occurrence is an event or part of the interaction following the event is by asking whether any long pauses or intervals occur—that is, can the system transaction be completed without interruption? Or is the system at rest again waiting for the next transaction? After the customer wants to buy the shirt, the process continues until the transaction is complete. There are no significant stops after the transaction begins. After the transaction is complete, the system is at rest, waiting for the next transaction to begin. The elementary business process (EBP) concept defined earlier describes this as leaving the system and its data in a consistent state.

On the other hand, separate events occur when the customer buys the shirt using his store credit card account. When the customer later pays the bill at the end of the month, is the processing part of the interaction involving the purchase? In this case, no. The system records the transaction and then does other things. It does not halt all processes to wait for the payment. A separate event occurs later that results in sending the customer a bill (this is a temporal event: *Time to send monthly bills*). Eventually, another external event occurs (*Customer pays the bill*).

The Sequence of Events: Tracing a Transaction's Life Cycle

It is often useful in identifying events to trace the sequence of events that might occur for a specific external agent or actor. In the case of Rocky Mountain Outfitters' new customer support system, the analyst might think through all of the possible transactions that might result from one new customer (see Figure 5-7). First, the customer wants a catalog or asks for some information about item availability, resulting in a name and address being added to the database. Next, the customer might want to place an order. Perhaps he or she will want to change the order, correcting the size of the shirt, for example, or buy another shirt. Next, the customer might want to check the status of an order to find out the shipping date. Perhaps the customer has moved and wants an address change recorded for future catalog mailings. Finally, the customer might want to return an item. Thinking through this type of sequence can help identify events.

Figure 5-7

The sequence of "transactions" for one specific customer resulting in many events

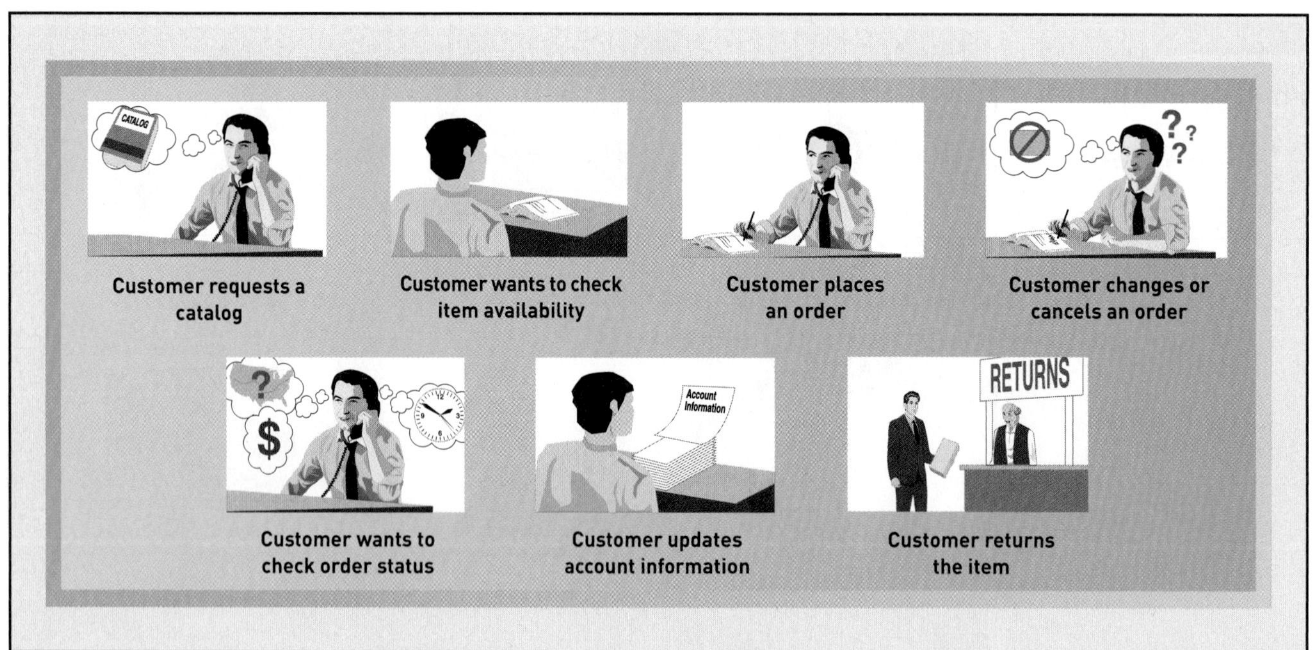

Technology-Dependent Events and System Controls

Sometimes the analyst is concerned about events that are important to the system but do not directly concern users or transactions. Such events typically involve design choices or system controls. During analysis, the analyst should temporarily ignore these events. They are important for design, however.

Some examples of events that affect design issues include external events that involve actually using the physical system, such as logging on. Although important to the final operation of the system, such a detail of implementation should be deferred. At this stage, the analyst should focus only on the functional requirements—the work that the system needs to complete. A logical model does not need to indicate how the system is actually implemented, so the model should omit the implementation details.

system controls
checks or safety procedures put in place to protect the integrity of the system

perfect technology assumption
the assumption that events should be included during analysis only if the system would be required to respond under perfect conditions

Most of these events involve **system controls**, which are checks or safety procedures put in place to protect the integrity of the system. Logging on to a system is required because of system security controls, for example. Other controls protect the integrity of the database, such as backing up the data every day. Both of these controls are important to the system, and they will certainly be added to the system during design. But spending time on these controls during analysis only adds to the requirements model details that the users are not typically very concerned about (they trust information services to take care of such details).

One technique used to help decide which events apply to controls is to assume that technology is perfect. The **perfect technology assumption** states that events should be included during analysis only if the system would be required to respond under perfect conditions—that is, with equipment never breaking down, capacity for processing and storage being unlimited, and people operating the system being completely honest and never making mistakes. By pretending that technology is perfect, analysts can eliminate events like *Time to back up the database* because they can assume that the disk will never crash. Again, during design, the project team adds these controls because technology is obviously not perfect. Figure 5-8 lists some examples of events that can be deferred until the design phase.

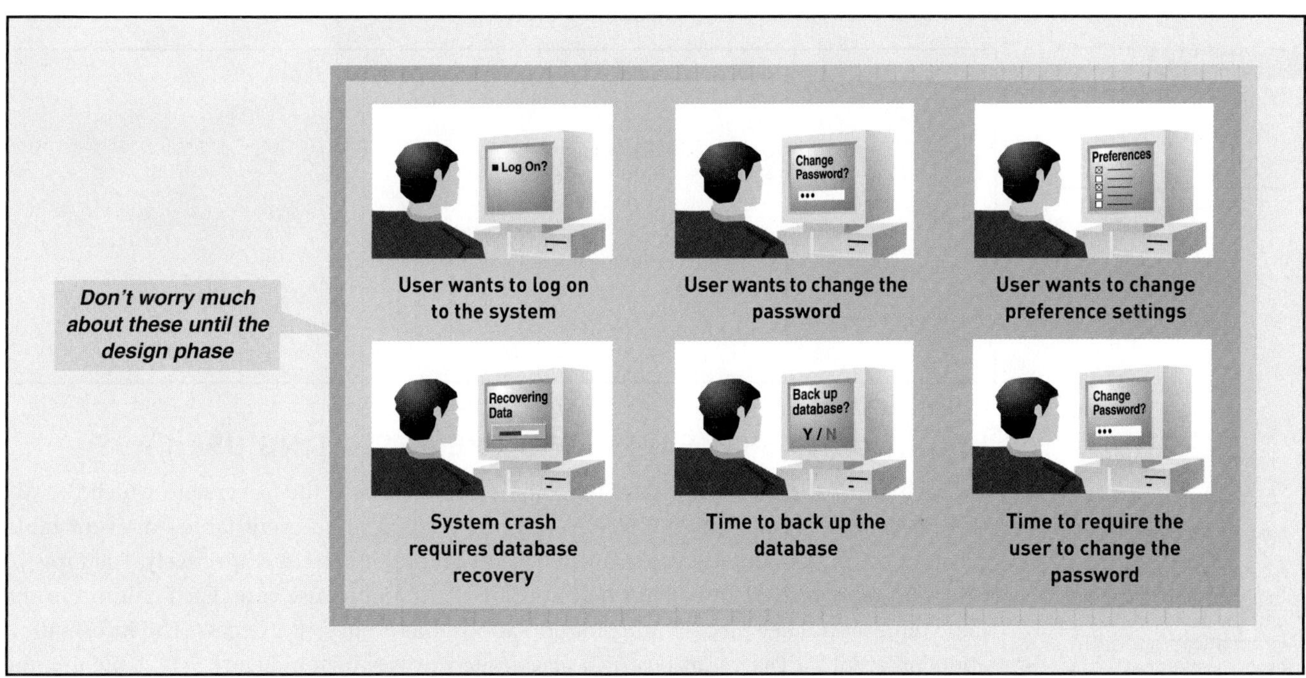
Don't worry much about these until the design phase
Log On?
User wants to log on to the system
Change Password?
User wants to change the password
Preferences
User wants to change preference settings
Recovering Data
System crash requires database recovery
Back up database?
Y / N
Time to back up the database
Change Password?
Time to require the user to change the password

Figure 5-8
Events deferred until design

EVENTS IN THE ROCKY MOUNTAIN OUTFITTERS CASE

The Rocky Mountain Outfitters customer support system involves a variety of events, many of them similar to those just discussed. A list of the external events is shown in Figure 5-9. Some of the most important external events involve customers: *Customer wants to check item availability, Customer places an order, Customer changes or cancels order*. Other external events involve RMO departments: *Shipping fulfills order, Marketing wants to send promotional materials to customers, Merchandising updates catalog*. The analyst can develop this list of external events by looking at all of the people and organizational units that want the system to do something for them.

Figure 5-9

External events for the RMO customer support system

Customer wants to check item availability
Customer places an order
Customer changes or cancels order
Customer or management wants to check order status
Shipping fulfills order
Shipping identifies back order
Customer returns item (defective, changed mind, full or partial returns)
Prospective customer requests catalog
Customer updates account information
Marketing wants to send promotional materials to customers
Management adjusts customer charges (correct errors, make concessions)
Merchandising updates catalog (add, change, delete, change prices)
Merchandising creates special product promotion
Merchandising creates new catalog

The customer support system also includes quite a few temporal events, shown in Figure 5-10. Many of these produce periodic reports for organizational units: *Time to produce order summary reports, Time to produce fulfillment summary reports, Time to produce catalog activity reports.* The analyst can develop the list of temporal events by looking for all of the regular reports and statements that the system must produce at certain times.

Figure 5-10

Temporal events for the RMO customer support system

Time to produce order summary reports
Time to produce transaction summary reports
Time to produce fulfillment summary reports
Time to produce prospective customer activity reports
Time to produce customer adjustment/concession reports
Time to produce catalog activity reports

LOOKING AT EACH EVENT AND THE RESULTING USE CASE

event table

a catalog of use cases that lists events in rows and key pieces of information about each event in columns

For each event, the most important information to identify is the use case to which the system needs to respond. This information can be entered in an event table. An **event table** includes rows and columns, representing events and their details, respectively. Each row in the event table records information about one event and its use case. Each column in the table represents a key piece of information about that event and use case. The information about an event *Customer wants to check item availability* is shown in Figure 5-11. Note that the resulting use case is named *Look up item availability*.

BEST PRACTICE

Use an event table as a catalog of information about the use cases that make up the functional requirements of the system.

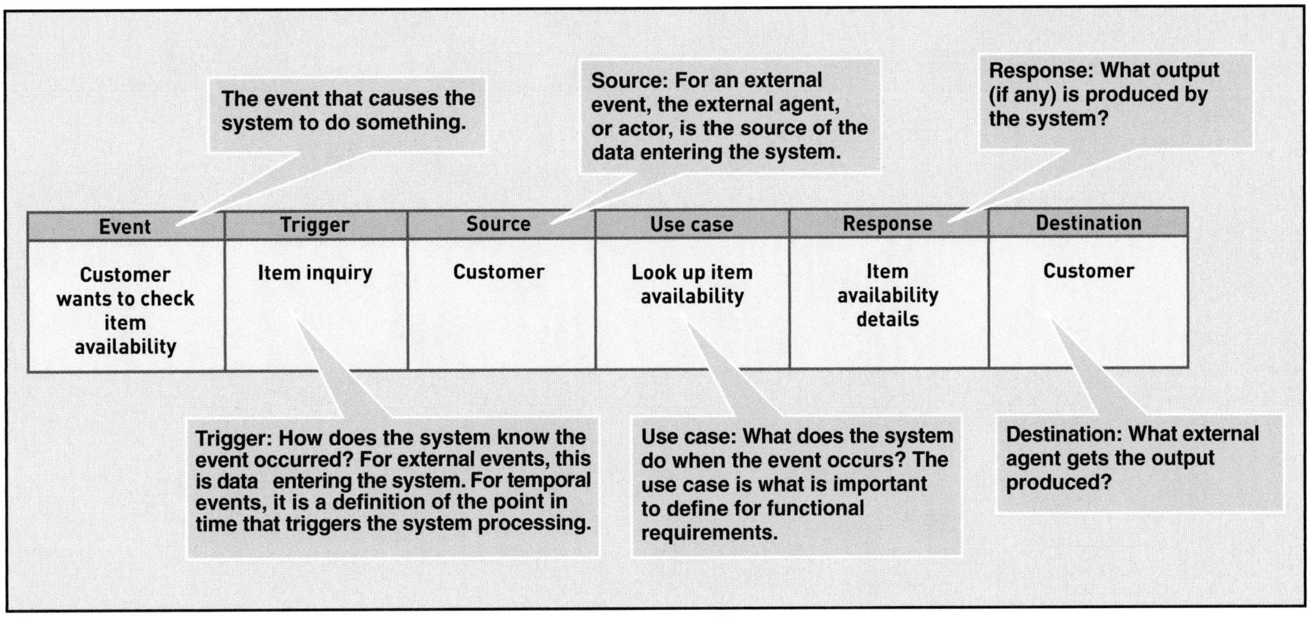

Event	Trigger	Source	Use case	Response	Destination
Customer wants to check item availability	Item inquiry	Customer	Look up item availability	Item availability details	Customer

The event that causes the system to do something.
Source: For an external event, the external agent, or actor, is the source of the data entering the system.
Response: What output (if any) is produced by the system?
Trigger: How does the system know the event occurred? For external events, this is data entering the system. For temporal events, it is a definition of the point in time that triggers the system processing.
Use case: What does the system do when the event occurs? The use case is what is important to define for functional requirements.
Destination: What external agent gets the output produced?

Figure 5-11

Information about each event in an event table

trigger

a signal that tells the system that an event has occurred, either the arrival of data needing processing or a point in time

source

an external agent or actor that supplies data to the system

response

an output, produced by the system, that goes to a destination

destination

an external agent or actor that receives data from the system

Information in the event table documents important aspects of the event and the resulting use case. First, for each event, how does the system know the event has occurred? A signal that tells the system an event has occurred is called the **trigger**. For an external event, the trigger is the arrival of data that the system must process. For example, when a customer places an order, the new order details are provided as input. The **source** of the data is also important to know. In this case, the source of the new order details is the customer—an external agent. For a temporal event, the trigger is a point in time. For example, at the end of each business day, the system knows it is time to produce transaction summary reports.

Next, what does the system do when the event occurs? What the system does (the reaction to the event) is the *use case*. When a customer places an order, the system is used to carry out the use case *Create a new order*. When it is time to produce transaction summary reports, the system is used to carry out the use case *Produce transaction summary reports*.

Finally, what response does the use case produce? A **response** is an output from the system. When the system produces transaction summary reports, those reports are the outputs. One use case can generate several responses. For example, when the system creates a new order, an order confirmation goes to the customer, the order details go to shipping, and a record of the transaction goes to the bank. The **destination** is the place where any response (output) is sent, again an external agent. Sometimes a use case generates no response at all. For example, if the customer wants to update account information, the information is recorded in the database, but no output needs to be produced. Recording information in the database is part of the use case.

The list of events—together with the trigger, source, use case, response(s), and destination(s) for each event—can be placed in an event table so that the analyst can keep track of them for later use. An event table is a convenient way to record key information about the requirements for the information system. The event table for the RMO customer support system is shown in Figure 5-12. Each use case in the event table is further described with use case descriptions, as shown in the next section. Then this event table will later be used in Chapter 6 to draw data flow diagrams to define functional requirements using the traditional approach. In Chapter 7, this event table will be used to draw use case diagrams and system sequence diagrams using the object-oriented approach.

Customer support system event table					
Event	**Trigger**	**Source**	**Use case**	**Response**	**Destination**
1. Customer wants to check item availability	Item inquiry	Customer	Look up item availability	Item availability details	Customer
2. Customer places an order	New order	Customer	Create new order	Real-time link Order confirmation Order details Transaction	Credit bureau Customer Shipping Bank
3. Customer changes or cancels order	Order change request	Customer	Update order	Change confirmation Order change details Transaction	Customer Shipping Bank
4. Time to produce order summary reports	"End of week, month, quarter, and year"		Produce order summary reports	Order summary reports	Management
5. Time to produce transaction summary reports	"End of day"		Produce transaction summary reports	Transaction summary reports	Accounting
6. Customer or management wants to check order status	Order status inquiry	Customer or management	Look up order status	Order status details	Customer or management
7. Shipping fulfills order	Order fulfillment notice	Shipping	Record order fulfillment		
8. Shipping identifies back order	Back-order notice	Shipping	Record back order	Back-order notification	Customer
9. Customer returns item	Order return notice	Customer	Create order return	Return confirmation Transaction	Customer Bank
10. Time to produce fulfillment summary reports	"End of week, month, quarter, and year"		Produce fulfillment summary reports	Fulfillment summary reports	Management
11. Prospective customer requests catalog	Catalog request	Prospective customer	Provide catalog info	Catalog	Prospective customer
12. Time to produce prospective customer activity reports	"End of month"		Produce prospective customer activity reports	Prospective customer activity reports	Marketing
13. Customer updates account information	Customer account update notice	Customer	Update customer account		
14. Marketing wants to send promotional materials to customers	Promotion package details	Marketing	Distribute promotional package	Promotional package	Customer and prospective customer
15. Management adjusts customer charges	Customer charge adjustment	Management	Create customer charge adjustment	Charge adjustment notification Transaction	Customer Bank

Figure 5-12

The complete event table for the RMO customer support system: a catalog of use cases

Event	Trigger	Source	Use case	Response	Destination
Customer support system event table, continued					
16. Time to produce customer adjustment/ concession reports	"End of month"		Produce customer adjustment reports	Customer adjustment reports	Management
17. Merchandising updates catalog	Catalog update details	Merchandising	Update catalog		
18. Merchandising creates special product promotion	Special promotion details	Merchandising	Create special promotion		
19. Merchandising creates new catalog	New catalog details	Merchandising	Create new catalog	Catalog	Customer and prospective customer
20. Time to produce catalog activity reports	"End of month"		Produce catalog activity reports	Catalog activity reports	Merchandising

Figure 5-12 cont.

The complete event table for the RMO customer support system: a catalog of use cases

USE CASE DESCRIPTIONS

use case description
a description that lists the processing details for a use case

actor
in UML diagrams, a person who uses the system

A list of use cases and an event table provide an overview of all the use cases for a system. Detailed information about each use case is described with a **use case description**. A use case description lists and describes the processing details for a use case. Implied in all use cases is a person who uses the system. In UML, that person is called an **actor**. An actor is always outside the automation boundary of the system but may be part of the manual portion of the system. In this respect, an actor is not always the same as the source of the event in the event table. A source of an event is the initiating person or entity that supplies data. In contrast, an actor in a use case is the person who is actually interacting with the computer system itself. By defining actors that way—as those who interact with the system—we can more precisely define the exact interactions to which the automated system must respond. This tighter focus helps define the specific requirements of the automated system itself—to refine them as we move from the event table to the use case details.

BEST PRACTICE

Be sure to remember that actors have direct contact with the automated system.

scenario or use case instance
a unique set of internal activities within a use case that represents a unique path through the use case

Another way to think of an actor is as a role. For example, in the RMO case, the use case *Create new order* might involve an order clerk talking to the customer on the phone. Or, the customer might be the actor if the customer places the order directly, through the Internet.

To create a comprehensive, robust system that truly meets users' needs, we must understand all of the detailed steps of each use case. Internally, a use case includes a whole sequence of steps to complete a business process. For example, frequently several variations of the business steps exist within a single use case. The use case *Create new order* will have a separate flow of activities depending on which actor invokes the use case. The processes for an order clerk creating a new order over the telephone might be quite different from the processes for a customer creating an order over the Internet. Each flow of activities is a valid sequence for the *Create new order* use case. These different flows of activities are called **scenarios**, or sometimes **use case instances**. Thus, a scenario is a unique set of internal activities within a use case and represents a unique path through the use case.

Typically, use case descriptions are written at three separate levels of detail: brief description, intermediate description, and fully developed description, depending on an analyst's needs.

BRIEF DESCRIPTION

A brief description can be used for very simple use cases, especially when the system to be developed is also a small, well-understood application. A simple use case would normally have a single scenario and very few, if any, exception conditions. An example would be *Update customer data*. Generally, a use case such as *Create new order* is complex enough that either an intermediate or fully developed description is developed, although a brief description might be written initially (see Figure 5-13).

Figure 5-13

Brief description of *Create new order* use case

Create new order **description**

When the customer calls to order, the order clerk and system verify customer information, create a new order, add items to the order, verify payment, create the order transaction, and finalize the order.

INTERMEDIATE DESCRIPTION

The intermediate-level use case description expands the brief description to include the internal flow of activities for the use case. If there are multiple scenarios, each flow of activities is described individually. Exception conditions can be documented if they are needed. Figures 5-14 and 5-15 show intermediate descriptions that document the two scenarios of *Order clerk creates telephone order* and *Customer creates Web order*. These two scenarios were identified as separate work flows for the *Create new order* use case. Notice that each describes what the user and the system require to carry out the processing for the scenario. Exception conditions are also listed. Each step is identified with a number to make it easier to read. In many ways, this description is a version of structured English, which can include sequence, decision, and repetition blocks.

Figure 5-14

Intermediate description of the telephone order scenario for *Create new order*

Flow of activities for scenario of *Order Clerk creates telephone order*

Main Flow:

1. Customer calls RMO and gets order clerk.
2 Order clerk verifies customer information. If a new customer, invoke *Maintain customer account information* use case to add a new customer.
3. Clerk initiates the creation of a new order.
4. Customer requests an item be added to the order.
5. Clerk verifies the item and adds it to the order.
6. Repeat steps 4 and 5 until all items are added to the order.
7. Customer indicates end of order; clerk enters end of order; system computes totals.
8. Customer submits payment; clerk enters amount; system verifies payment.
9. System finalizes order.

Exception Conditions:

5. If an item is not in stock, then customer can
   a. choose not to purchase item, or
   b. request item be added as a back-ordered item.
8. If customer payment is rejected due to bad-credit verification, then
   a. order is canceled, or
   b. order is put on hold until check is received.

Flow of activities for scenario of *Customer creates Web order*
Main Flow: 1. Customer connects to the RMO home page and then links to the order page. 2. If this is a new customer, customer links to the customer account page and adds the appropriate information to establish a customer account. 2a If existing customer, customer logs on. 3. The system starts a new order and displays the catalog frame. 4. Customer searches the catalog. 5. When customer finds the correct item, he/she requests it be added to the order; the system adds it to the shopping cart. 6. Customer repeats steps 4 and 5. 7. Customer requests end of order; system displays a summary of the ordered items. 8. Customer makes any changes. 9. Customer requests payment screen; system displays payment screen. 9a Customer enters payment information; system displays summary information and sends confirmation e-mail. 10. System finalizes order.
Exception Conditions: 2a If existing customer forgets password, then a. customer can invoke forgotten password processing, or b. customer can create a new customer account. 9a If customer payment is rejected due to bad-credit verification, then a. customer can cancel the order, or b. order is put on hold until check is received.

Figure 5-15

Intermediate description of the Web order scenario for *Create new order*

FULLY DEVELOPED DESCRIPTION

The fully developed description is the most formal method for documenting a use case. Even though it takes a little more work to define all the components at this level, it is the preferred method of describing the internal flow of activities for a use case. One of the major difficulties for software developers is that they often struggle to obtain a deep understanding of the users' needs. But if you create a fully developed use case description, you increase the probability that you thoroughly understand the business processes and the ways the system must support them. Figure 5-16 is an example of a fully developed use case description of the telephone order scenario for the *Create new order* use case, and Figure 5-17 shows the Web order scenario for the same use case.

Figures 5-16 and 5-17 can also serve as a standard template for documenting a fully developed description for other scenarios and use cases. The first and second compartments are used to identify the use cases and scenarios within use cases, if needed, that are being documented. In larger or more formal projects, a unique identifier can also be added for the use case, with an extension identifying the particular scenario. Sometimes the name of the system developer who produced the form is also added.

The third compartment identifies the triggering event that initiates the use case from the event table. The fourth compartment is a brief description of the use case or scenario. Analysts may just duplicate the brief description they constructed earlier here. The fifth compartment identifies the actor or actors. The sixth compartment identifies other use cases and the way they are related to this use case. These cross references to other use cases help document all aspects of the users' requirements.

The stakeholders compartment identifies interested parties other than specific actors. They might be users who do not actually invoke the use case but who have an interest in results produced from the use case. For example, in Figures 5-16 and 5-17, no one in the marketing department actually creates new orders, but they do perform statistical analysis of the orders that were entered. So, marketers have an interest in the data that is captured and stored from the *Create new order* use case. Considering all stakeholders is an important step for system developers so that they ensure they have understood all requirements.

Use Case Name:	*Create new order*	
Scenario:	Create new telephone order	
Triggering Event:	Customer telephones RMO to purchase items from the catalog.	
Brief Description:	When customer calls to order, the order clerk and system verify customer information, create a new order, add items to the order, verify payment, create the order transaction, and finalize the order.	
Actors:	Telephone sales clerk.	
Related Use Cases:	Includes: *Check item availability.*	
Stakeholders:	Sales department: to provide primary definition. Shipping department: to verify information content is adequate for fulfillment. Marketing department: to collect customer statistics for studies of buying patterns.	
Preconditions:	Customer must exist. Catalog, Products, and Inventory items must exist for requested items.	
Postconditions:	Order and order line items must be created. Order transaction must be created for the order payment. Inventory items must have the quantity on hand updated. The order must be related (associated) to a customer.	
Flow of Activities:	**Actor**	**System**
	1. Sales clerk answers telephone and connects to a customer.	
	2. Clerk verifies customer information.	2.1 Display customer information.
	3. Clerk initiates the creation of a new order.	3.1 Create a new order.
	4. Customer requests an item be added to the order.	
	5. Clerk verifies the item (*Check item availability* use case).	5.1 Display item information.
	6. Clerk adds item to the order.	6.1 Add an order item.
	7. Repeat steps 4, 5, and 6 until all items are added to the order.	
	8. Customer indicates end of order; clerk enters end of order.	8.1 Complete order. 8.2 Compute totals.
	9. Customer submits payment; clerk enters amount.	9.1 Verify payment. 9.2 Create order transaction. 9.3 Finalize order.
Exception Conditions:	2.1 If customer does not exist, then the clerk pauses this use case and invokes *Maintain customer information* use case. 2.2 If customer has a credit hold, then clerk transfers the customer to a customer service representative. 4.1 If an item is not in stock, then customer can a. choose not to purchase item, or b. request item be added as a back-ordered item. 9.1 If customer payment is rejected due to bad-credit verification, then a. order is canceled, or b. order is put on hold until check is received.	

Figure 5-16

Fully developed description of the telephone order scenario for *Create new order*

preconditions

conditions that must be true before a use case begins

The next two compartments, called *preconditions* and *postconditions*, provide critical information about the state of the system before and after the use case executes. **Preconditions** state what conditions must be true before a use case begins. In other words, they identify what the state of the system must be for the use case to begin, including what objects must already exist, what information must be available, and even the condition of the actor prior to beginning the use case.

Use Case Name:	*Create new order*	
Scenario:	Create new Web order	
Triggering Event:	Customer logs on to the RMO Web site and requests to purchase an item.	
Brief Description:	Customer logs on and requests the new order form. The customer searches the catalog online and purchases items from the catalog. The system adds the purchased items to the order. At the end the customer enters credit-card information.	
Actors:	Customer.	
Related Use Cases:	Includes: *Register new customer, Check item availability.*	
Stakeholders:	Sales department: to provide primary definition. Shipping department: to verify information content is adequate for fulfillment. Marketing department: to collect customer statistics for studies of buying patterns.	
Preconditions:	Catalog, Products, and Inventory items must exist for requested items.	
Postconditions:	Order and order line items must be created. Order transaction must be created for the order payment. Inventory items must have the quantity on hand updated. The order must be related (associated) to a customer.	
Flow of Activities:	**Actor**	**System**
	1. Customer connects to the RMO home page and then links to the order page.	
	2. If this is a new customer, then customer links to the customer account page and adds the appropriate information to establish a customer account.	2.1 Create new customer record.
	2a. If existing customer, customer logs on.	2a.1 Validate customer account. 2.2 Create a new shopping cart order; display order form with catalog frame.
	3. Customer searches catalog.	3.1 Display products from catalog based on searches and selections.
	4. When customer finds the correct item, he/she requests it be added to the order.	4.1 Add item to shopping cart order.
	5. Repeat steps 3 and 4.	
	6. Customer requests end of order.	6.1 Display shopping cart items, with totals and amounts due; edit and submit buttons.
	7. Customer makes any changes.	
	8. Customer requests payment screen.	8.1 Display payment details screen.
	9. Customer enters payment information.	9.1 Accept payment, finalize order, send confirmation e-mail.
Exception Conditions:	4.1 If an item is not in stock, then customer can a. choose not to purchase item, or b. request item be added as a back-ordered item. 8.1 If customer payment is rejected due to bad-credit verification, then a. order is canceled, or b. order is put on hold until check is received.	

Figure 5-17

Fully developed description of the Web order scenario for *Create new order*

postconditions

conditions that must be true upon completion of the use case

A **postcondition** identifies what must be true upon completion of the use case. The same items that are used to describe the precondition should be included in the statement of the postcondition. For example, during the processing of a use case that updates various financial accounts, some accounts will be out of balance. So, a postcondition for that use case would be that the updates should be complete for all accounts and that they should all be in balance.

BEST PRACTICE

Preconditions and postconditions are critical to understanding the processing done for a use case.

The final two compartments in the template describe the detailed flow of activities of the use case. In this instance, we have shown a two-column version, identifying the steps performed by the actor and the responses required by the system. The item numbering helps identify the sequence of the steps. Some developers prefer the one-column version, as shown at the intermediate level. Alternative activities and exception conditions are described in the final compartment. The numbering of exception conditions also helps tie the exceptions with specific steps in the use case description.

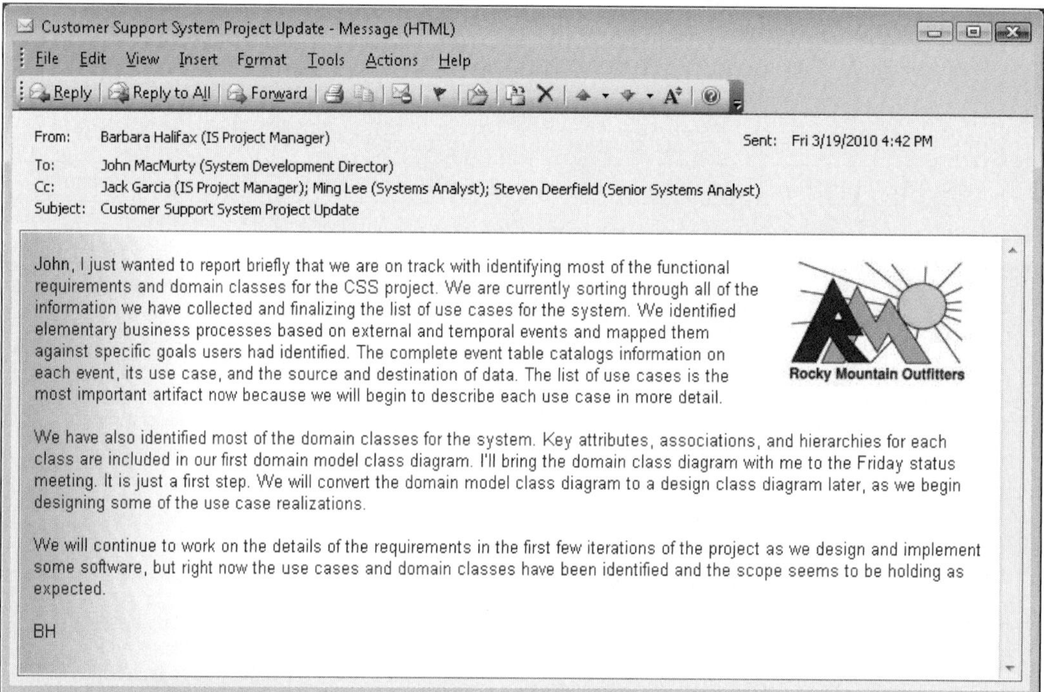

Customer Support System Project Update - Message (HTML)
File Edit View Insert Format Tools Actions Help
Reply Reply to All Forward
From: Barbara Halifax (IS Project Manager)
Sent: Fri 3/19/2010 4:42 PM
To: John MacMurty (System Development Director)
Cc: Jack Garcia (IS Project Manager); Ming Lee (Systems Analyst); Steven Deerfield (Senior Systems Analyst)
Subject: Customer Support System Project Update
John, I just wanted to report briefly that we are on track with identifying most of the functional requirements and domain classes for the CSS project. We are currently sorting through all of the information we have collected and finalizing the list of use cases for the system. We identified elementary business processes based on external and temporal events and mapped them against specific goals users had identified. The complete event table catalogs information on each event, its use case, and the source and destination of data. The list of use cases is the most important artifact now because we will begin to describe each use case in more detail.
Rocky Mountain Outfitters
We have also identified most of the domain classes for the system. Key attributes, associations, and hierarchies for each class are included in our first domain model class diagram. I'll bring the domain class diagram with me to the Friday status meeting. It is just a first step. We will convert the domain model class diagram to a design class diagram later, as we begin designing some of the use case realizations.
We will continue to work on the details of the requirements in the first few iterations of the project as we design and implement some software, but right now the use cases and domain classes have been identified and the scope seems to be holding as expected.
BH

"THINGS" IN THE PROBLEM DOMAIN

The RMO memo from Barbara Halifax shows the importance of events, use cases, and functional requirements, as discussed earlier in the chapter. Another key concept discussed in the memo involves understanding and modeling things about which the system needs to store information. To the users, these items are the things they deal with when they do their work—products, orders, invoices, and customers—that need to be part of the system. They are often referred to as things in the problem domain of the system. For example, an information system needs to store information about customers and products, so it is important for the analyst to identify lots of information about them. Often these things are similar to the external agents or actors that interact with the system. For example, a customer external agent places an order, but the system also needs to store information about the customer. In other cases, these things are distinct from external agents. For example, there is no external agent named *product*, but the system needs to store information about products.

In the traditional approach to development, these things make up the data about which the system stores information. The type of data that needs to be stored is definitely a key aspect of the requirements for any information system. In the object-oriented approach, these things become the objects that interact in the system. No matter which approach you use to develop an information system, identifying and understanding these things are both key initial steps.

TYPES OF THINGS

As with use cases, an analyst should ask the users to discuss the types of things that they work with routinely. The analyst can ask about several types of things to help identify them. Many things are tangible and therefore more easily identified, but others are intangible. Different types of things are important to different users, so it is important to include information from all types of users.

Figure 5-18 shows some types of things to consider. Tangible things are often the most obvious, such as an airplane, book, or vehicle. In the Rocky Mountain Outfitters case, a catalog and an item in the catalog are tangible things of importance. Another common type of thing in an information system is a role played by a person, such as employee, customer, doctor, or patient. A customer is obviously a very important role a person plays in the Rocky Mountain Outfitters case.

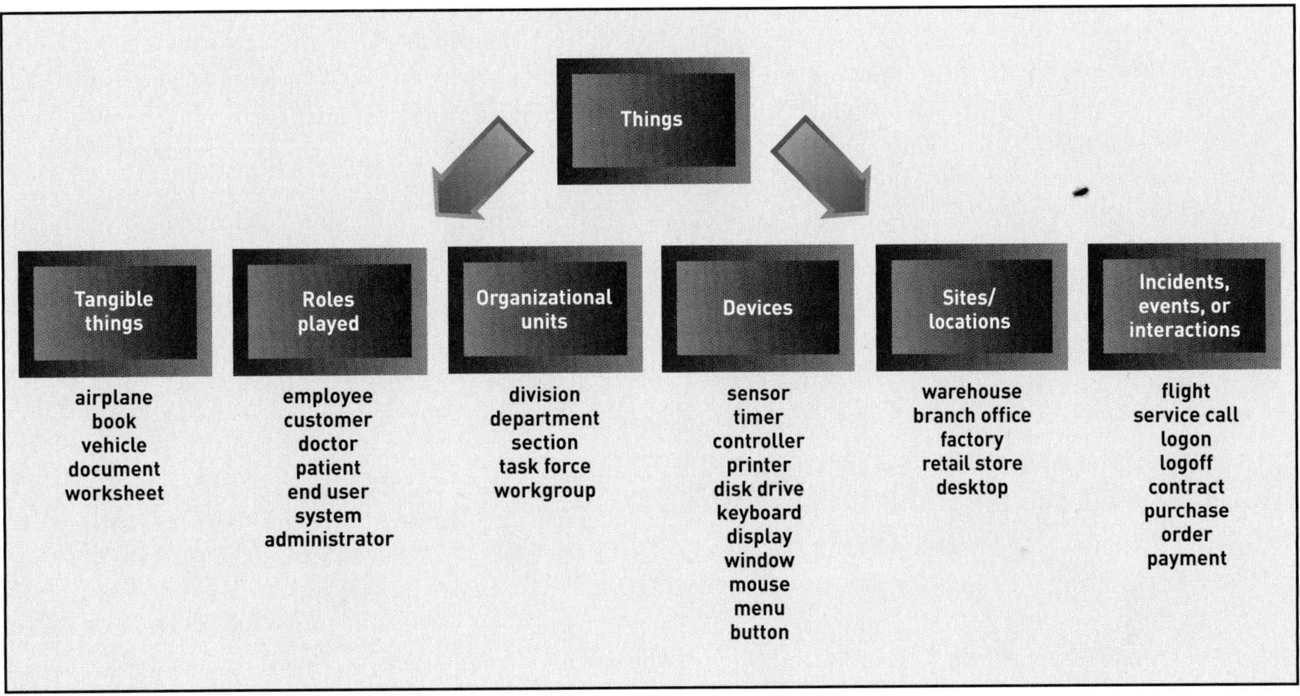

Figure 5-18
Types of things

Other types of things can include organizational units, such as a division, department, or workgroup. Similarly, sites or locations might be important in a particular system, such as a warehouse, a store, or a branch office. Finally, information about an incident or interaction of importance can be considered a thing—information about an order, a service call, a contract, or an airplane flight. An order, a shipment, and a return are important incidents in the Rocky Mountain Outfitters case. Sometimes these incidents are thought of as relationships between things. For example, an order is a relationship between a customer and an item of inventory. Initially, the analyst might simply list all of these as things and then make adjustments that might be required by different approaches to analysis and design.

The analyst identifies these types of things by thinking about each event in the event table and asking questions. For example, for each event, what types of things are affected that the system needs to know about and store information about? When a customer places an order, the system needs to store information about the customer, the items ordered, and the details about the order itself, such as the date and payment terms. Note that for the fully developed use case descriptions written for each use case, the preconditions and postconditions usually list specific things of importance. Figure 5-16 shows the *Create new order* scenario, which includes Customer, Catalog, Product, Inventory Item, Order, Line Item, and Transaction. All are important things in the user's work.

PROCEDURE FOR DEVELOPING AN INITIAL LIST OF THINGS

The general guidelines just discussed reveal that analysts can use many sources of information to develop an initial list of things about which the system needs to store information. Another useful procedure is to begin by listing all of the *nouns* that users mention when talking about the system. Consider the events, the activities or use cases, the external agents or actors, and the triggers and responses from the event table as potential things, for example. Then add to the list any additional nouns that appear in information about the existing system or that come up in discussions with stakeholders.

Step One: Using the event table and information about each event, identify all of the nouns.

For the RMO customer support system, the nouns include RMO, customer, product item, order, confirmation, transaction, shipping, bank, change request, summary report, management, transaction report, accounting, back order, back-order notification, return, return confirmation, fulfillment reports, prospective customer, catalog, marketing, customer account, promotional materials, charge adjustment, catalog details, merchandising, and catalog activity reports.

Step Two: Using other information from existing systems, current procedures, and current reports or forms, add items or categories of information needed.

For the RMO customer support system, these items might include more detailed information, such as price, size, color, style, season, inventory quantity, payment method, shipping address, and so forth. Some of these items might be additional categories, and some might be more specific pieces of information about things you have already identified (called *attributes*).

Step Three: Refine the list and record assumptions or issues to explore.

As this list of nouns builds, it will be necessary to refine it. Ask these questions about each noun to try to decide whether you should *include it:*

- Is it a unique thing the system needs to know about?
- Is it inside the scope of the system I am working on?
- Does the system need to remember more than one of these items?

Ask these questions about each noun to decide whether you should *exclude it:*

- Is it really a synonym for some other thing I have identified?
- Is it really just an output of the system produced from other information I have identified?
- Is it really just an input that results in recording some other information I have identified?

Ask these questions about each noun to decide whether you should *research it:*

- Is it likely to be a specific piece of information (attribute) about some other thing I have identified?
- Is it something that I might need if assumptions change?

Figure 5-19 lists some of the nouns from the RMO customer support system event table and other sources, with some notes about each one.

RELATIONSHIPS AMONG THINGS

relationship

a naturally occurring association among specific things, such as *an order is placed by a customer* and *an employee works in a department*

After recording and refining the list of things, the analyst researches and records additional information. Many important relationships among things are important to the system. A **relationship** is a naturally occurring association among specific things, such as *an order is placed by a customer* and *an employee works in a department* (see Figure 5-20). *Is placed by* and *works in* are two relationships that naturally occur between specific things. Information systems need to store information about employees and about departments, but equally important is storing information about the specific relationships—John works in the accounting department, and Mary works in the marketing department, for example. Similarly, it is quite important to store the fact that Order 1043 for a shirt was placed by John Smith.

Figure 5-19

Partial list of "things" based on nouns for RMO

Identified noun	Notes on including noun as a thing to store
Accounting	We know who they are. No need to store it.
Back order	A special type of order? Or a value of order status? Research.
Back-order information	An output that can be produced from other information.
Bank	Only one of them. No need to store.
Catalog	Yes, need to recall them, for different seasons and years. Include.
Catalog activity reports	An output that can be produced from other information. Not stored.
Catalog details	Same as catalog? Or the same as product items in the catalog? Research.
Change request	An input resulting in remembering changes to an order.
Charge adjustment	An input resulting in a transaction.
Color	One piece of information about a product item.
Confirmation	An output produced from other information. Not stored.
Credit card information	Part of an order? Or part of customer information? Research.
Customer	Yes, a key thing with lots of details required. Include.
Customer account	Possibly required if an RMO payment plan is included. Research.
Fulfillment reports	An output produced from information about shipments. Not stored.
Inventory quantity	One piece of information about a product item. Research.
Management	We know who they are. No need to store.
Marketing	We know who they are. No need to store.
Merchandising	We know who they are. No need to store.
Order	Yes, a key system responsibility. Include.
Payment method	Part of an order. Research.
Price	Part of a product item. Research.
Product item	Yes, what RMO includes in a catalog and sells. Include.
Promotional materials	An output? Or documents stored outside the scope? Research.
Prospective customer	Possibly same as customer. Research.
Return	Yes, the opposite of an order. Include.
Return confirmation	An output produced from information about a return. Not stored.
RMO	There is only one of these! No need to store.
Season	Part of a catalog? Or is there more to it? Research.
Shipment	Yes, a key thing to track. Include.
Shipper	Yes, they vary and we need to track the order. Include.
Shipping	Our department. No need to store.
Shipping address	Part of customer? Or order? Or shipment? Research.
Size	Part of a product item. Research.
Style	Part of a product item. Research.
Summary report	An output produced from other information. Not stored.
Transaction	Yes, each one is important and must be remembered. Include.
Transaction report	An output produced from transaction information. Not stored.

Figure 5-20

Relationships naturally occur among things

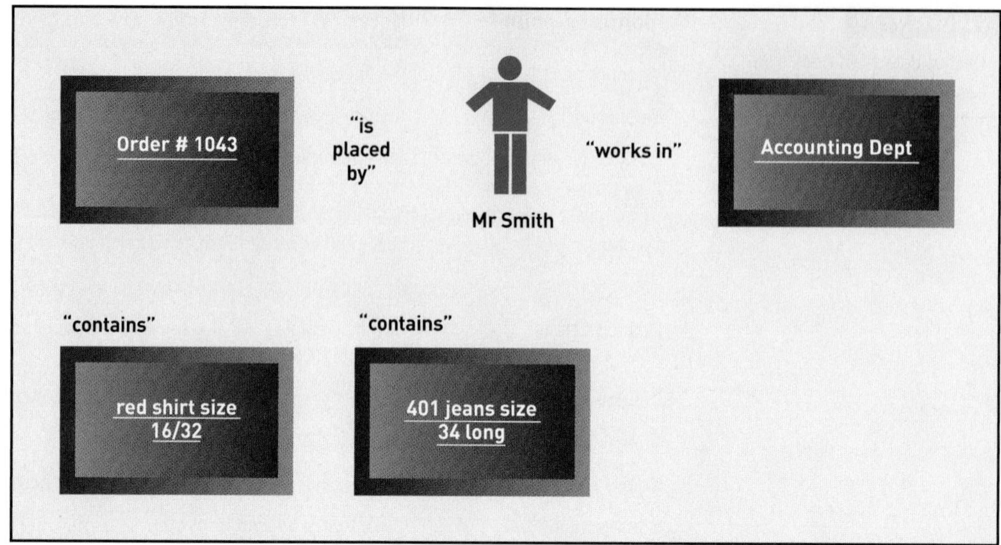

Relationships between things apply in two directions. For example, *a customer places an order* describes the relationship in one direction. Similarly, *an order is placed by a customer* describes the relationship in the other direction. It is important to understand the relationship in both directions because sometimes it might seem more important for the system to record the relationship in one direction than in the other. For example, Rocky Mountain Outfitters definitely needs to know what items a customer ordered so the shipment can be prepared. However, it might not be apparent initially that the company needs to know all of the customers who have ordered a particular item. What if the company needs to notify all customers who ordered a defective or recalled product? Knowing this information would be very important, but the operational users might not immediately recognize that fact.

cardinality

the number of associations that occur among specific things, such as *a customer places many orders* and *an employee works in one department*

multiplicity

a synonym for cardinality (used with the object-oriented approach as defined by UML)

It is also important to understand the nature of each relationship in terms of the number of associations for each thing. For example, a customer might place many different orders, but an order is placed by only one customer. The number of associations that occur is referred to as the **cardinality** of the relationship. Cardinality can be one to one or one to many. Again, cardinality is established for each direction of the relationship. The term **multiplicity** is used to refer to the number of associations in the object-oriented approach, as defined by UML. Figure 5-21 lists examples of cardinality/multiplicity associated with an order.

Figure 5-21

Cardinality/multiplicity of relationships

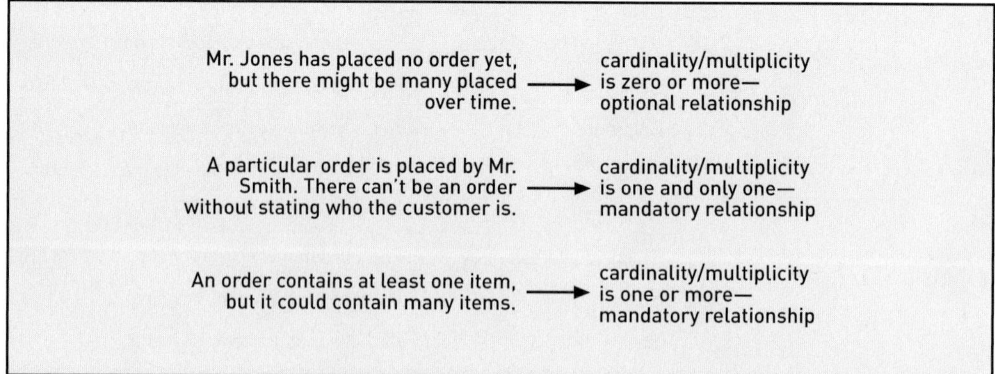

Mr. Jones has placed no order yet, but there might be many placed over time.
cardinality/multiplicity is zero or more—optional relationship
A particular order is placed by Mr. Smith. There can't be an order without stating who the customer is.
cardinality/multiplicity is one and only one—mandatory relationship
An order contains at least one item, but it could contain many items.
cardinality/multiplicity is one or more—mandatory relationship

Sometimes it is important to describe not just the cardinality but also the range of possible values of the cardinality (the minimum and maximum cardinality). For example, a particular customer might not ever place an order. In this case, there are zero associations. Alternatively, the customer may place one order, meaning one association exists. Finally, the customer might place two, three, or even more orders. The relationship for a customer placing an order can have a range of zero, one, or more, usually indicated as zero or more. The zero is the minimum cardinality, and "more" is the maximum cardinality. These terms are referred to as "cardinality constraints."

In some cases, at least one association is required (a mandatory as opposed to optional relationship). For example, the system might not record any information about a customer until the customer places an order. Therefore, the cardinality would read "customer places one or more orders."

A one-to-one relationship can also be refined to include minimum and maximum cardinality. For example, the order is placed by one customer—it is impossible to have an order if there is no customer. Therefore, one is the minimum cardinality, making the relationship mandatory. Because there cannot be more than one customer for each order, one is also the maximum cardinality. Sometimes such a relationship is read as "an order must be placed by one and only one customer."

binary relationships
relationships between two different types of things, such as a customer and an order

unary (recursive) relationship
a relationship among two things of the same type, such as one person being married to another person

ternary relationship
a relationship among three different types of things

The relationships described here are between two different types of things—for example, a customer and an order. These are called **binary relationships**. Sometimes a relationship is between two things of the same type. For example, the relationship *is married to* is between two different people. This type of relationship is called a **unary relationship** (sometimes called a *recursive relationship*). Another example of a unary relationship is an organizational hierarchy in which one organizational unit reports to another organizational unit—the packing department reports to shipping, which reports to distribution, which reports to marketing.

A relationship can also be among three different types of things, called a **ternary relationship**, or any number of different types of things, called an ***n*-ary relationship**. One particular order, for example, might be associated with a specific customer plus a specific sales representative, requiring a ternary relationship.

Storing information about the relationships is just as important as storing information about the specific things. It is important to have information on the name and address of each customer, but it is equally important (or perhaps more so) to know what items each customer has ordered.

BEST PRACTICE

Initially, focus on identifying each "thing" in the problem domain, but also be sure to focus on associations/relationships among them, which are often just as important to the system users.

n-ary relationship
a relationship among *n* (any number of) different types of things

attribute
one piece of specific information about a thing

identifier (key)
an attribute that uniquely identifies a thing

compound attribute
an attribute that contains a collection of related attributes

ATTRIBUTES OF THINGS

Most information systems store and use specific pieces of information about each thing, as discussed in Figure 5-19 earlier. The specific pieces of information are called **attributes**. For example, a customer has a name, a phone number, a credit limit, and so on. Each of these details is an attribute. The analyst needs to identify the attributes of each thing that the system needs to store. One attribute may be used to identify a specific thing, such as a Social Security number for an employee or an order number for a purchase. The attribute that uniquely identifies the thing is called an **identifier**, or **key**. Sometimes the identifier is already established (a Social Security number, vehicle ID number, or product ID number). Sometimes the system needs to assign a specific identifier (an invoice number or transaction number).

A system may need to remember many similar attributes. For example, a customer has several names—a first name, a middle name, a last name, and possibly a nickname. A **compound attribute** is an attribute that contains a collection of related attributes, so an analyst may choose one compound attribute to represent all of these names, perhaps naming it *Customer full name*. A customer might also have several phone numbers—a home phone number, office phone number, fax phone number, and cellular phone number. The analyst might start out by describing the most important attributes but later add to the list. Attribute lists can get quite long. Some examples of attributes of a customer and the values of attributes for specific customers are shown in Figure 5-22.

Figure 5-22

Attributes and values

All customers have these attributes:	Each customer has a value for each attribute:		
Customer ID	101	102	103
First name	John	Mary	Bill
Last name	Smith	Jones	Casper
Home phone	555-9182	423-1298	874-1297
Work phone	555-3425	423-3419	874-8546

THE ENTITY-RELATIONSHIP DIAGRAM

data entities

the things about which the system needs to store information in the traditional approach to information systems

The traditional approach to system development (the structured techniques and information engineering approaches, as described in Chapter 2) places a great deal of emphasis on data storage requirements for a new system. **Data entities** are the things about which the system needs to store information. Data storage requirements include the data entities, their attributes, and the relationships among the data entities. The model used to define the data storage requirements with the traditional approach is called the *entity-relationship diagram* (*ERD*).

EXAMPLES OF ERD NOTATION

On the entity-relationship diagram, rectangles represent data entities, and the lines connecting the rectangles show the relationships among data entities. Figure 5-23 shows an example of a simplified entity-relationship diagram with two data entities, Customer and Order. Each Customer can place many Orders, and each Order is placed by one Customer. The cardinality is one to many in one direction and one to one in the other direction. The "crow's feet" symbol on the line next to the Order data entity indicates "many" orders. But other symbols on the relationship line also represent the minimum and maximum cardinality constraints. See Figure 5-24 for an explanation of relationship symbols. The model in Figure 5-23 actually says that a Customer places a minimum of zero and a maximum of many Orders. Reading in the other direction, the model says an Order is placed by at least one and only one Customer. This notation can express precise details about the system. The constraints reflect the business policies that management has defined, and the analyst must discover what these policies are. The analyst does not determine that two customers cannot share one order; management does.

Figure 5-23

A simple entity-relationship diagram

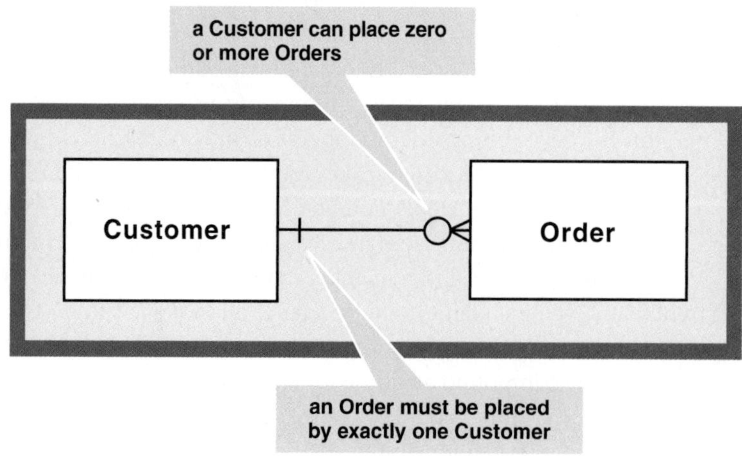
a Customer can place zero or more Orders
Customer
Order
an Order must be placed by exactly one Customer

Figure 5-24

Cardinality symbols of relationships

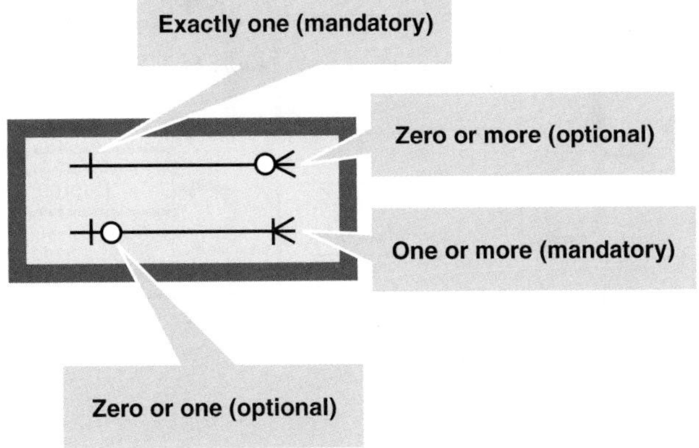

Figure 5-25 shows the model expanded to include the order items (one or more specific items included on the order). Each order contains a minimum of one and a maximum of many items (there could not be an order if it did not contain at least one item). For example, an order might include a shirt, a pair of shoes, and a belt, and each of these items is associated with the order. This example also shows some of the attributes of each data entity: A customer has a customer number, a name, a billing address, and several phone numbers. Each order has an order ID, order date, and so on. Each order item has an item ID, quantity, and price. The attributes of the data entity are listed below the name, with the key identifier listed first.

Figure 5-25

An expanded ERD with attributes shown

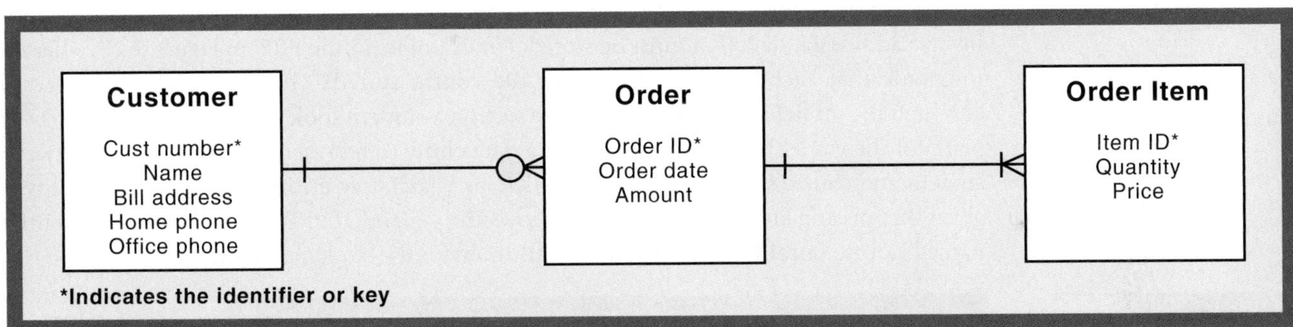

Figure 5-26 shows how the actual data in some transactions might look. John is a customer who has placed two orders. The first order, placed on February 4, was for two shirts and one belt. The second order, placed on March 29, was for one pair of boots and two pairs of sandals. Mary is a customer who has not yet placed an order. Recall that a customer might place zero or more orders. Therefore, Mary is not associated with any orders. Finally, Sara placed an order on March 30 for three pairs of sandals.

While working on the model, the analyst often refines the ERD. One example of refinement is analyzing many-to-many relationships. Figure 5-27 shows an example of a many-to-many relationship. At a university, courses are offered as course sections, and a student enrolls in many course sections. Each course section contains many students. Therefore, the relationship between course section and student is many to many. There are situations in which many-to-many relationships occur naturally, and they can be modeled as shown, with "crow's feet" on both ends of the relationship. If a relational database is designed from an ERD with a many-to-many relationship, a separate table containing keys from both sides of the relationship is created because relational databases cannot directly implement many-to-many relationships. Chapter 13 discusses relational databases in detail.

Figure 5-26

Customers, orders, and order items consistent with the expanded ERD

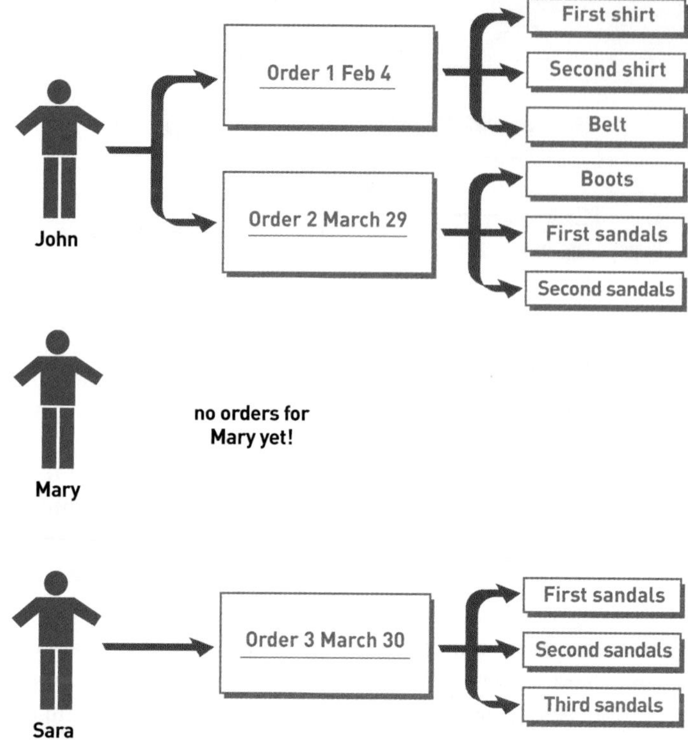

On closer analysis, however, analysts often discover that many-to-many relationships involve additional data that must be stored. For example, in the ERD in Figure 5-27, where is the grade that each student receives for the course stored? This is important data, and although the model indicates which course section a student took, the model does not have a place for the grade. The solution is to add a data entity to represent the relationship between student and course section, sometimes called an **associative entity**. The associative entity is given the missing attribute. Figure 5-28 shows the expanded ERD with an associative entity named Course Enrollment, which has an attribute for the student's grade.

associative entity

a data entity that represents a many-to-many relationship between two other data entities

Figure 5-27

A university course enrollment ERD with a many-to-many relationship

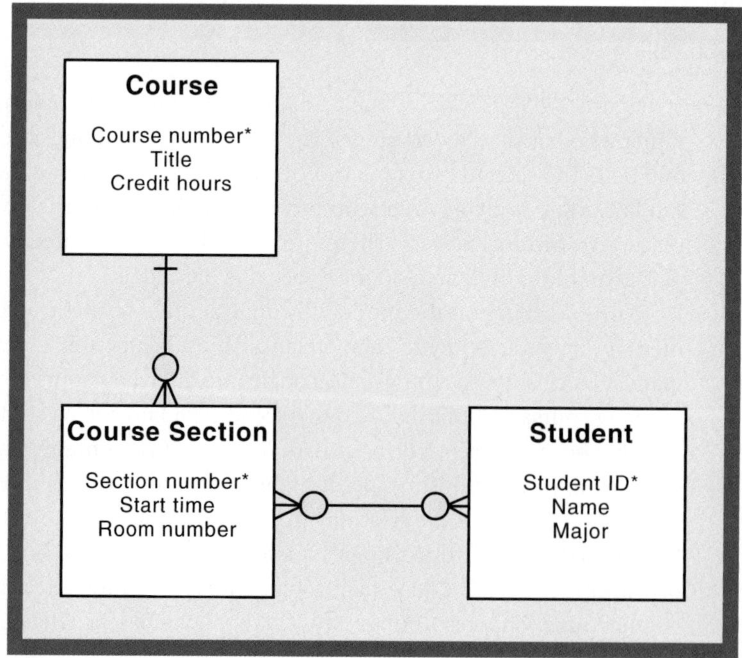

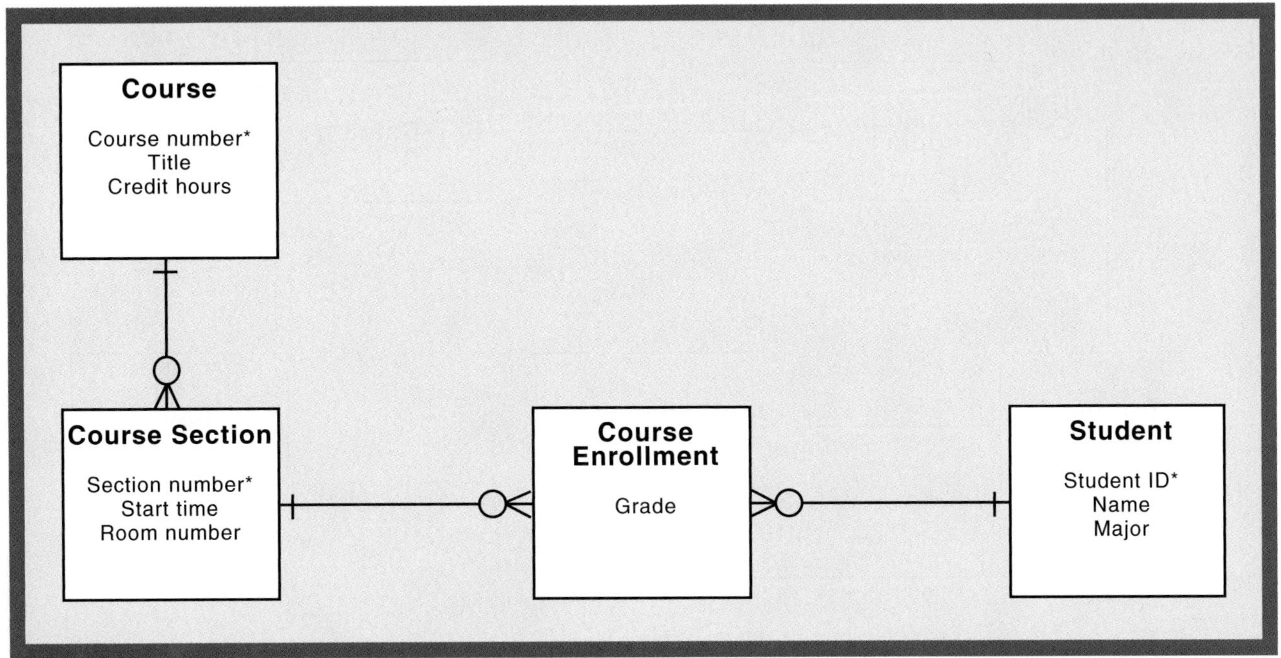

Figure 5-28

A refined university course enrollment ERD with an associative entity

Reading the relationships in Figure 5-28 from left to right, the ERD says that one course section has many course enrollments, each with its own grade, and each course enrollment applies to one specific student. Reading from right to left, it says one student has many course enrollments, each with its own grade, and each course enrollment applies to one specific course section. A database implemented using this model will be able to produce grade lists showing all students and their grades in each course section, as well as grade transcripts showing all grades earned by each student.

Other refinements are made to the ERD during the modeling process. One major refinement process that applies to designing relational databases, called *normalization*, is discussed in Chapter 13.

THE ROCKY MOUNTAIN OUTFITTERS ERD

The Rocky Mountain Outfitters entity-relationship diagram is a variation of the customer and order example already described. Most of the data entities are from the list of things developed in Figure 5-19. Figure 5-29 shows a fairly complete version of the model but without the attributes to make it easier to focus on the data entities and relationships.

Each customer can place zero or more orders. Each order can have one or more order items, meaning the order might be for one shirt and two sweaters. Each order item is for a specific inventory item, meaning a specific size and color of shirt. Although the diagram does not show such an attribute, an inventory item should have an attribute for quantity on hand of that size and color. Because there are many colors and sizes (each with its own quantity), each inventory item is associated with a product item that describes the item generically (vendor, gender, description).

An earlier version of the model showed that each product item is contained in one or more catalogs, and each catalog contains one or more product items, a many-to-many relationship. Therefore, this model adds an associative entity named Catalog Product between Catalog and Product Item because the relationship has some attributes that need to be remembered, specifically the regular and special prices. Each catalog can list a different price for the same product item (ski pants might be cheaper in the spring catalog).

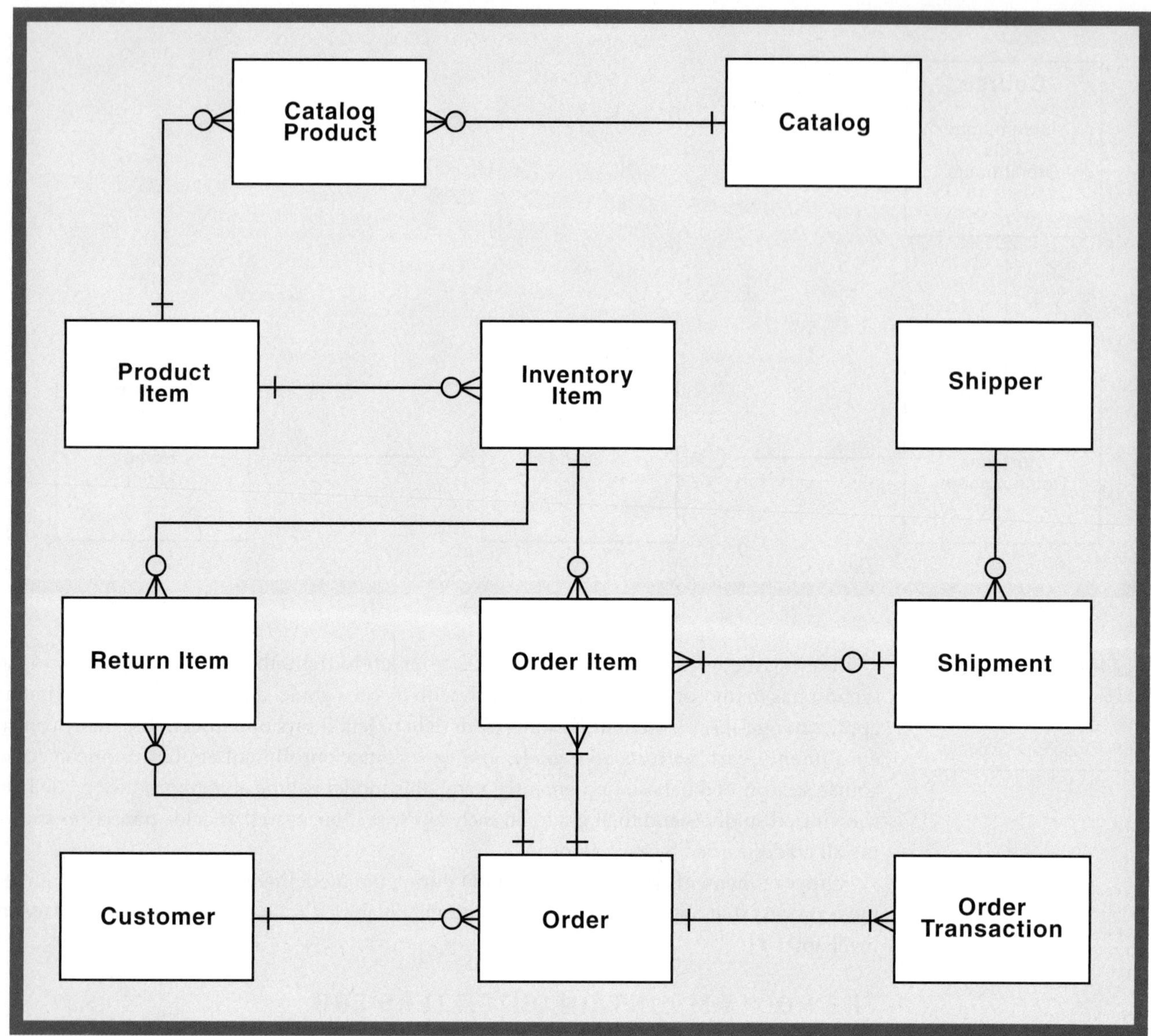

Figure 5-29

Rocky Mountain Outfitters customer support system entity-relationship diagram (ERD) without attributes

The entity-relationship diagram for Rocky Mountain Outfitters also has information about shipments. Because this diagram includes requirements for orders, not retail sales, each order item is part of a shipment. A shipment may contain many order items. Each shipment is shipped by one shipper.

The ERD shown in Figure 5-29 contains a lot of very specific information about the data storage requirements for the system. Be sure that you can trace through all of the relationships shown and try to describe a specific example of each data entity involved in one specific order. Try listing the attributes for each data entity to check your understanding. Draw a sketch similar to that shown in Figure 5-26 to show some actual data this ERD describes. You can check your understanding of the attributes by looking ahead to the domain model class diagram in the next section and to the relational database design in Chapter 13.

After it is developed, a model like this entity-relationship diagram needs to be walked through carefully, as you would walk through the logic of a program. Being able to walk through and "debug" any model is a very important skill in system development, as discussed in Chapter 4.

To test your understanding of the diagram, consider whether one order might have items shipped by different shippers. Is it possible, given the requirements shown in this diagram? The answer is yes. Operationally, some order items might be back ordered, so when they are finally shipped, they are part of a different shipment. A different shipper could handle this shipment.

Other requirements shown in the model include one or more order transactions for each order. An order transaction is a record of a payment or a refund for the order. One order transaction is created when the customer initially pays for the order. Later, though, the customer might add another item to the order, generating an additional charge. This involves a second order transaction. Finally, the customer might return an item, requiring a refund and a third order transaction.

THE DOMAIN MODEL CLASS DIAGRAM

The object-oriented approach also emphasizes understanding the things involved in the user's work. This approach models classes of objects instead of data entities. The classes of objects have attributes and associations, just like the data entities. Multiplicity (called cardinality in the traditional approach) also applies among classes. The sets of requirements models for the traditional and object-oriented approaches eventually diverge, looking quite different because of the object behavior. The design models are definitely very different. But initially, when defining requirements, the approach to modeling is similar with the object-oriented approach.

The class diagram is used to show classes of objects for a system. The notation is from the Unified Modeling Language (UML), which has become the standard for models used with object-oriented system development. One type of UML class diagram shows the things in the users' work domain, referred to as the *domain model class diagram*. Another type of UML class diagram notation is used to create *design class diagrams* when designing software classes (see Chapter 11). On a class diagram, rectangles represent classes, and the lines connecting the rectangles show the associations among classes. Figure 5-30 shows a symbol for one domain class: Customer. The domain class symbol is a rectangle with two sections. The top section contains the name of the class, and the bottom section lists the attributes of the class. Class names always begin with a capital letter, and attribute names always begin with a lowercase letter. Class diagrams are drawn by showing classes and associations among classes. You will first learn about the UML notation for creating the domain model class diagram. Many of the examples used previously for the entity-relationship diagram are redrawn using UML domain class diagram notation so that you can compare them. In fact, many developers now use the UML class diagram in place of the ERD even when using the traditional approach. Later, you will learn about additional hierarchies used in domain class diagrams.

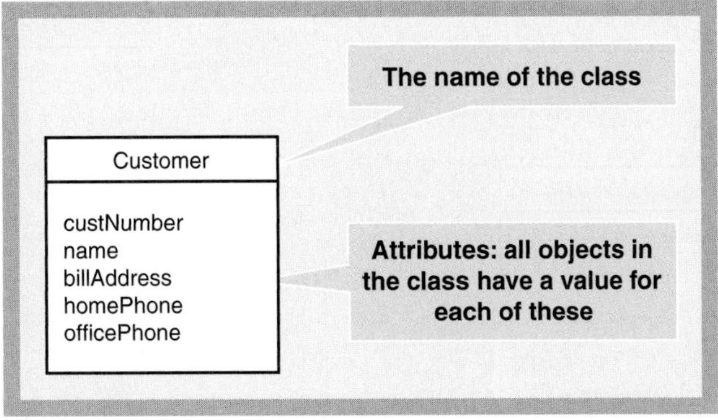

Figure 5-30

The UML domain class symbol with name and attributes

DOMAIN MODEL CLASS DIAGRAM NOTATION

Figure 5-31 shows a simplified domain model class diagram with three classes, Customer, Order, and OrderItem. Here each class symbol includes only two sections. In the diagram notation, we see that each Customer can place many Orders, and each Order is placed by one Customer; the associations "places" and "consists of" can be included on the diagram as shown for clarity, but this detail is optional. The multiplicity is one to many in one direction and one to one in the other direction. The multiplicity notation, shown as an asterisk on the line next to the Order class, indicates "many" orders. See Figure 5-32 for a summary of multiplicity notation. The other association shows that an Order consists of one or more OrderItems, and each OrderItem is associated with one Order.

Figure 5-31

A simple domain model class diagram

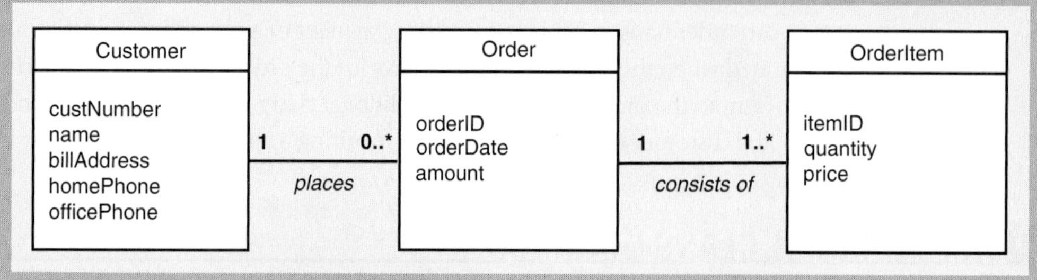

BEST PRACTICE

Focus first on problem domain classes that are "things" in the users' work environment, not on the software classes that you will eventually need to design.

Figure 5-33 shows the initial course enrollment example as a domain model class diagram. Recall that a Course has zero or more CourseSections. Each CourseSection enrolls zero or more Students, and each Student is enrolled in zero or more CourseSections—a many-to-many association. But because each student's grade for the section must be stored, the model must be modified, as it was in the ERD example. The class diagram notation adds an *association class* named CourseEnrollment to hold the grade attribute, as shown in Figure 5-34. A dashed line connects the association class to the association line between CourseSection and Student.

Figure 5-32

Multiplicity of associations

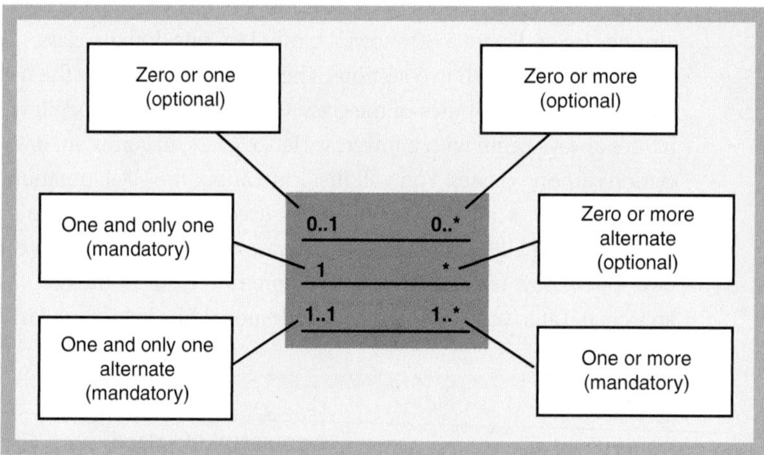

Figure 5-33

A university course enrollment domain model class diagram with a many-to-many association

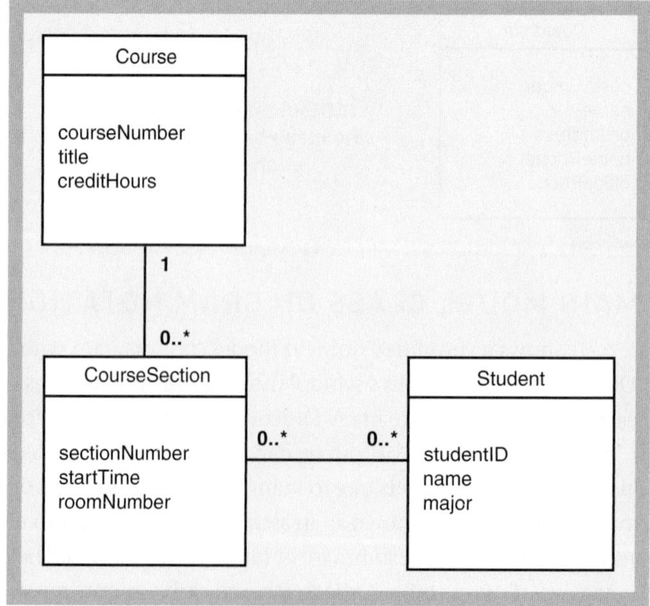

Figure 5-34

A refined university course enrollment domain model class diagram with an association class

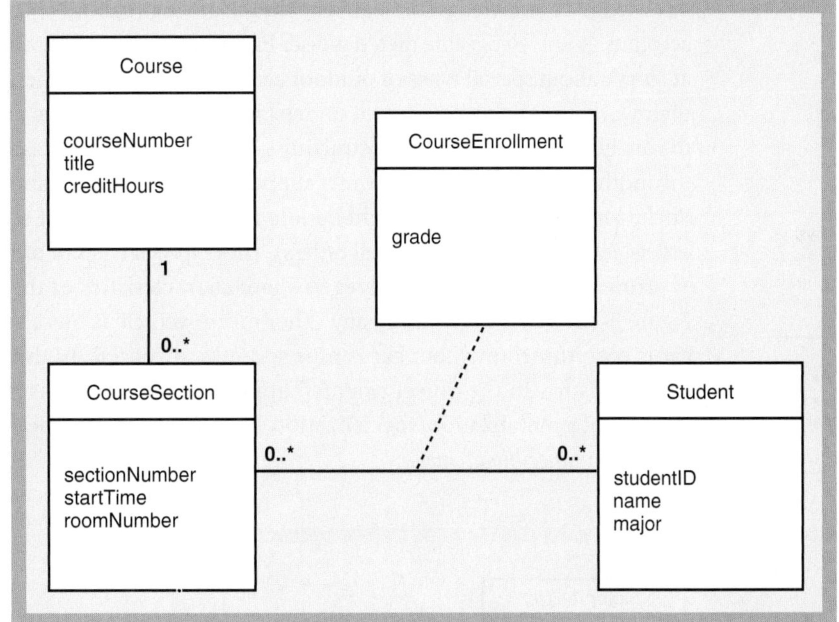

Reading the associations in Figure 5-34 from left to right, the class diagram says that one course section has many course enrollments, each with its own grade, and each course enrollment applies to one specific student. Reading from right to left, it says one student has many course enrollments, each with its own grade, and each course enrollment applies to one specific course section. A system implemented on the basis of this domain model will be able to produce grade lists showing all students and their grades in each course section, as well as grade transcripts showing all grades earned by each student. It is equivalent to the ERD shown in Figure 5-28.

MORE COMPLEX ISSUES ABOUT CLASSES OF OBJECTS

Some issues about the problem domain come up more frequently with the object-oriented approach than with the traditional approach, although the issues are not exclusively object-oriented. These issues are two additional ways that people structure their understanding of things in the real world: generalization/specialization hierarchies and whole-part hierarchies. This section discusses these concepts and shows how the class diagram is used to represent them.

Generalization/Specialization

generalization/specialization hierarchies

hierarchies that structure or rank classes from the more general superclass to the more specialized subclasses; sometimes called *inheritance hierarchies*

Generalization/specialization hierarchies are based on the idea that people classify things in terms of similarities and differences. Generalizations are judgments that group similar types of things; for example, there are many types of motor vehicles—cars, trucks, and tractors. All motor vehicles share certain general characteristics, so a motor vehicle is a more general class. Specializations are judgments that categorize different types of things—for example, special types of cars include sports cars, sedans, and sport utility vehicles. These types of cars are similar in some ways, yet different in other ways. Therefore, a sports car is a special type of car.

A generalization/specialization hierarchy is used to structure or rank these things from the more general to the more special. As discussed previously, classification refers to defining classes of things. Each class of thing in the hierarchy might have a more general class above it, called a *superclass*. At the same time, a class might have a more specialized class below it, called a *subclass*. In Figure 5-35, a car has three subclasses and one superclass (MotorVehicle). UML class diagram notation uses a triangle that points to the superclass to show a generalization/ specialization hierarchy.

We mentioned that people structure their understanding by using generalization/specialization hierarchies. That is, people learn by refining the classifications they make about some field of knowledge. A knowledgeable banker can talk at length about special types of loans and deposit accounts. A knowledgeable merchandiser like John Blankens at Rocky Mountain Outfitters can talk at length about special types of outdoor activities and clothes. Therefore, when asking users about their work, the analyst is trying to understand the knowledge the user has about the work, which the analyst can represent by constructing generalization/specialization hierarchies. At some level, the motivation for the new customer support system at RMO started with John's recognition that Rocky Mountain Outfitters might handle many special types of orders with a new system (Web orders, telephone orders, and mail orders). These special types of orders are shown in Figure 5-36.

inheritance
a concept that allows subclasses to share characteristics of their superclasses

Inheritance allows subclasses to share characteristics of their superclasses. Returning to Figure 5-35, a car is everything any other motor vehicle is but also something special. A sports car is everything any other car is plus something special. In this way, the subclass "inherits" characteristics. In the object-oriented approach, inheritance is a key concept that is possible because of generalization/specialization hierarchies. Sometimes these hierarchies are referred to as *inheritance hierarchies.*

Figure 5-35
A generalization/specialization hierarchy for motor vehicles

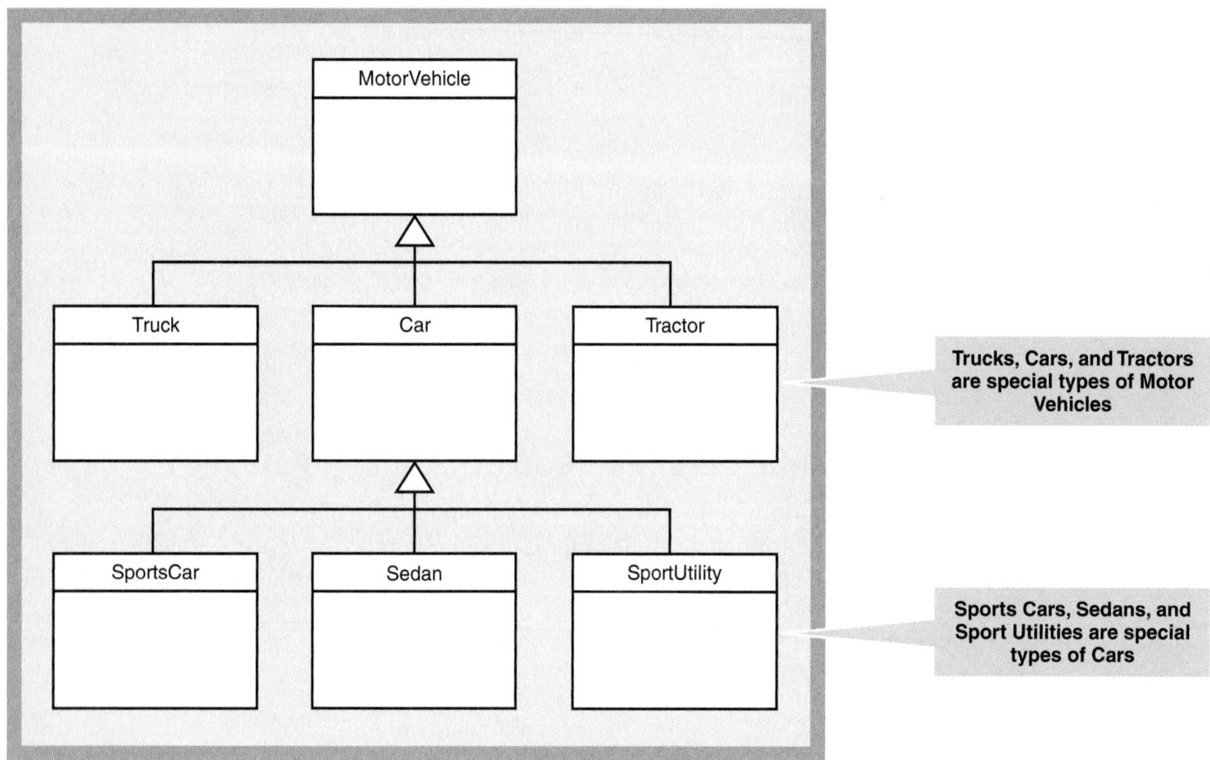

Whole-Part Hierarchies

whole-part hierarchies
hierarchies that structure classes according to their associated components

aggregation
whole-part relationship between an object and its parts

composition
whole-part relationship in which the parts cannot be dissociated from the object

Another way that people structure information about things is by defining them in terms of their parts. For example, learning about a computer system might involve recognizing that the computer is actually a collection of parts—processor, main memory, keyboard, disk storage, and monitor. A keyboard is not a special type of computer; it is part of a computer. Yet, it is also something entirely separate in its own right. **Whole-part hierarchies** capture the relationships that people make when they learn to make associations between an object and its components.

There are two types of whole-part hierarchies: aggregation and composition. The term **aggregation** is used to describe a form of association that specifies a whole-part relationship between the aggregate (whole) and its components (parts) where the parts can exist separately. Figure 5-37 demonstrates the concept of aggregation in a computer system, showing the UML diamond symbol to represent aggregation. The term **composition** is used to describe whole-part relationships that are even stronger, where the parts, once associated, can no longer exist separately. The UML diamond symbol is filled in to represent composition.

Order
WebOrder
TelephoneOrder
MailOrder
Web Orders, Telephone Orders, and Mail Orders are special types of Orders

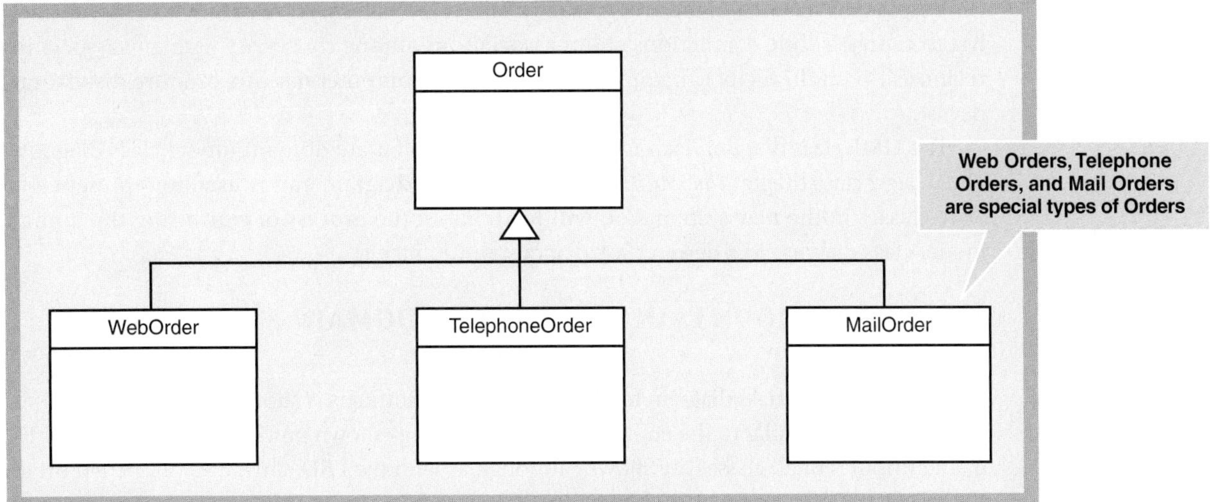

Figure 5-36

A generalization/specialization hierarchy for orders

Figure 5-37

Whole-part (aggregation) relationships between a computer and its parts

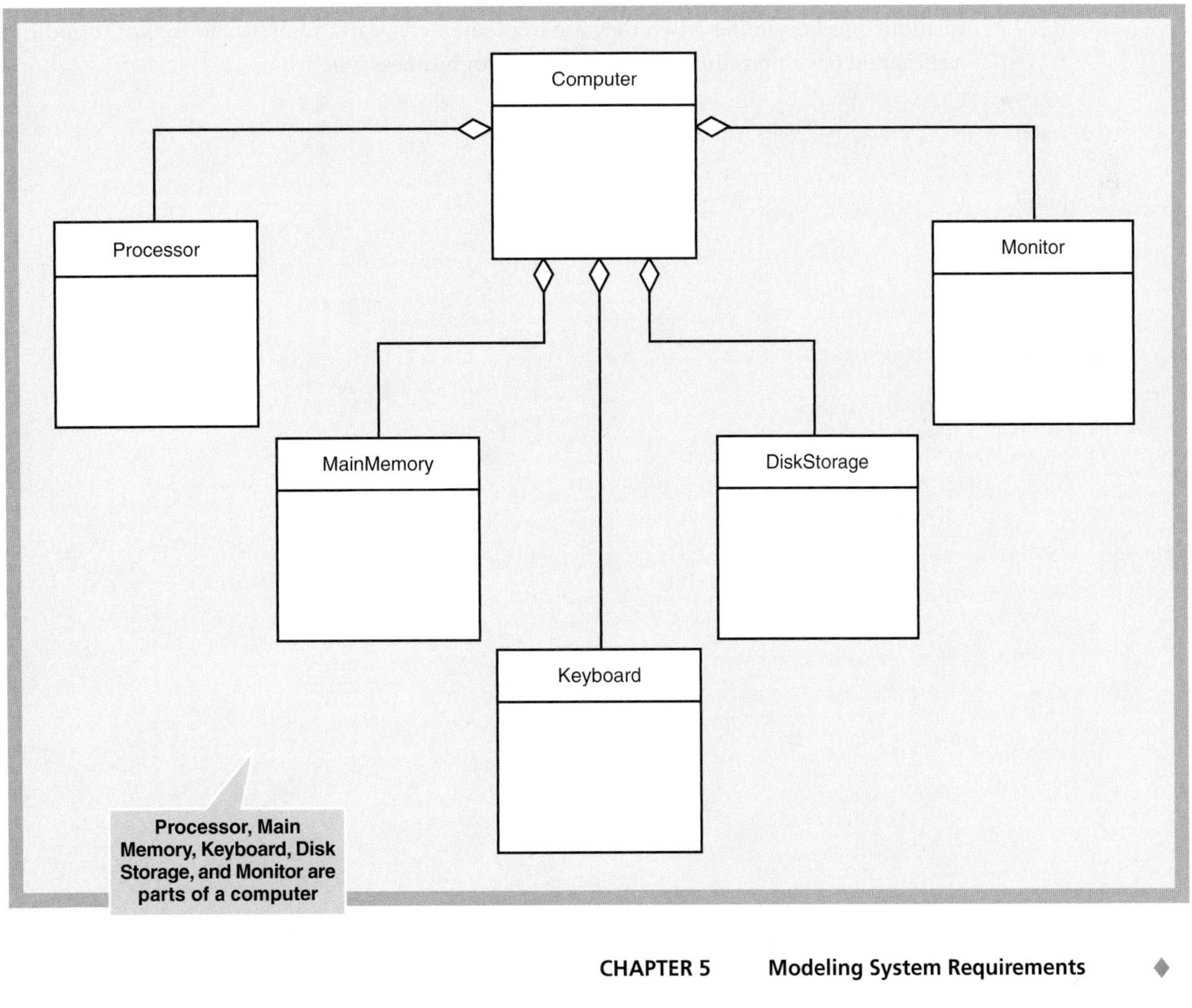

Computer
Processor
Monitor
MainMemory
DiskStorage
Keyboard
Processor, Main Memory, Keyboard, Disk Storage, and Monitor are parts of a computer

Whole-part hierarchies, both aggregation and composition, serve mainly to allow the analyst to express subtle distinctions about associations among classes. As with any association relationship, multiplicity can apply, such as when a computer has one or more disk storage devices.

The UML class diagram examples we have seen so far are domain model class diagrams. The design class diagram is a refinement of the class diagram and is used to represent software classes in the new system. You will learn about the process of converting the domain model class diagram to a design class diagram in Chapter 11.

THE ROCKY MOUNTAIN OUTFITTERS DOMAIN MODEL CLASS DIAGRAM

The domain model class diagram for Rocky Mountain Outfitters is shown in Figure 5-38. As you can see, it is very similar to the entity-relationship diagram shown previously in Figure 5-29. The main attributes of all classes are shown, although as with the ERD, attributes can be left off the class diagram when presenting an overview of the model.

A generalization/specialization hierarchy is included to show that an order can be any one of three types—Web order, telephone order, and mail order—as discussed previously. Note that all types of orders share the attributes listed for Order, but each special type of order has some additional attributes. Order is an abstract class (the name is in italic) because any order must be one of the three special types.

The other classes and associations among classes are similar to the RMO entity-relationship diagram. CatalogProduct is an association class attached to the association between Catalog and ProductItem. Multiplicity for association relationships is indicated with both minimums and maximums. No whole-part associations (aggregation or composition) are shown, although it might be argued that an OrderTransaction is part of an Order or that a ProductItem is part of a Catalog. It does not make much difference in this example because whole-part and association relationships are similar when they are implemented. Many analysts choose not to indicate aggregation or composition on class diagrams for business systems.

Figure 5-38

Rocky Mountain Outfitters domain model class diagram

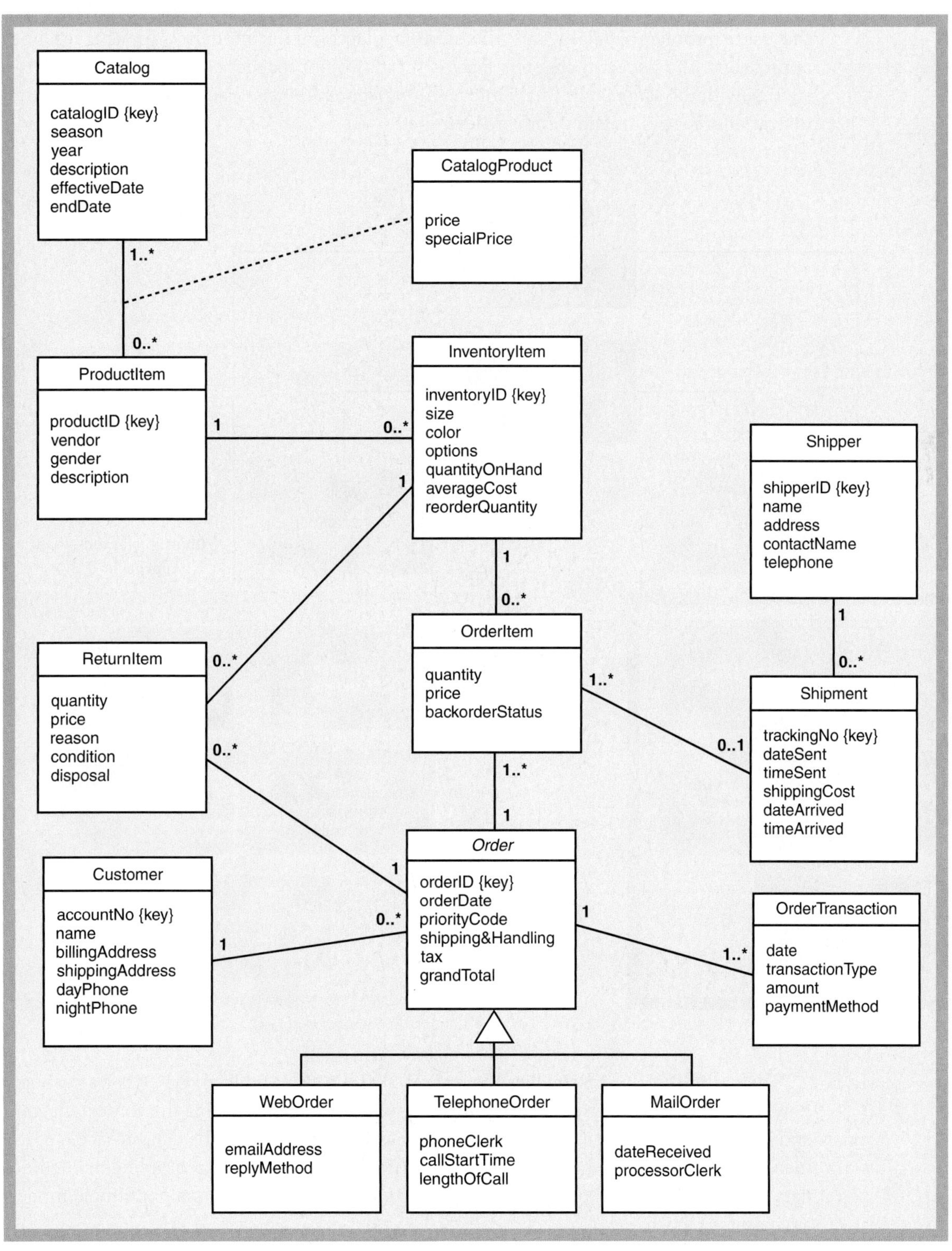

WHERE YOU ARE HEADED

The requirements models for a new system created using analysis activities become quite different depending on whether the project team uses the traditional approach or the object-oriented approach. The two key concepts discussed in this chapter—use cases and things in the user's problem domain—are the starting places in the modeling process for both approaches. The next two chapters discuss these two approaches separately, in both cases starting with the same preliminary information. Figure 5-39 shows how the two approaches diverge after the events and things are identified.

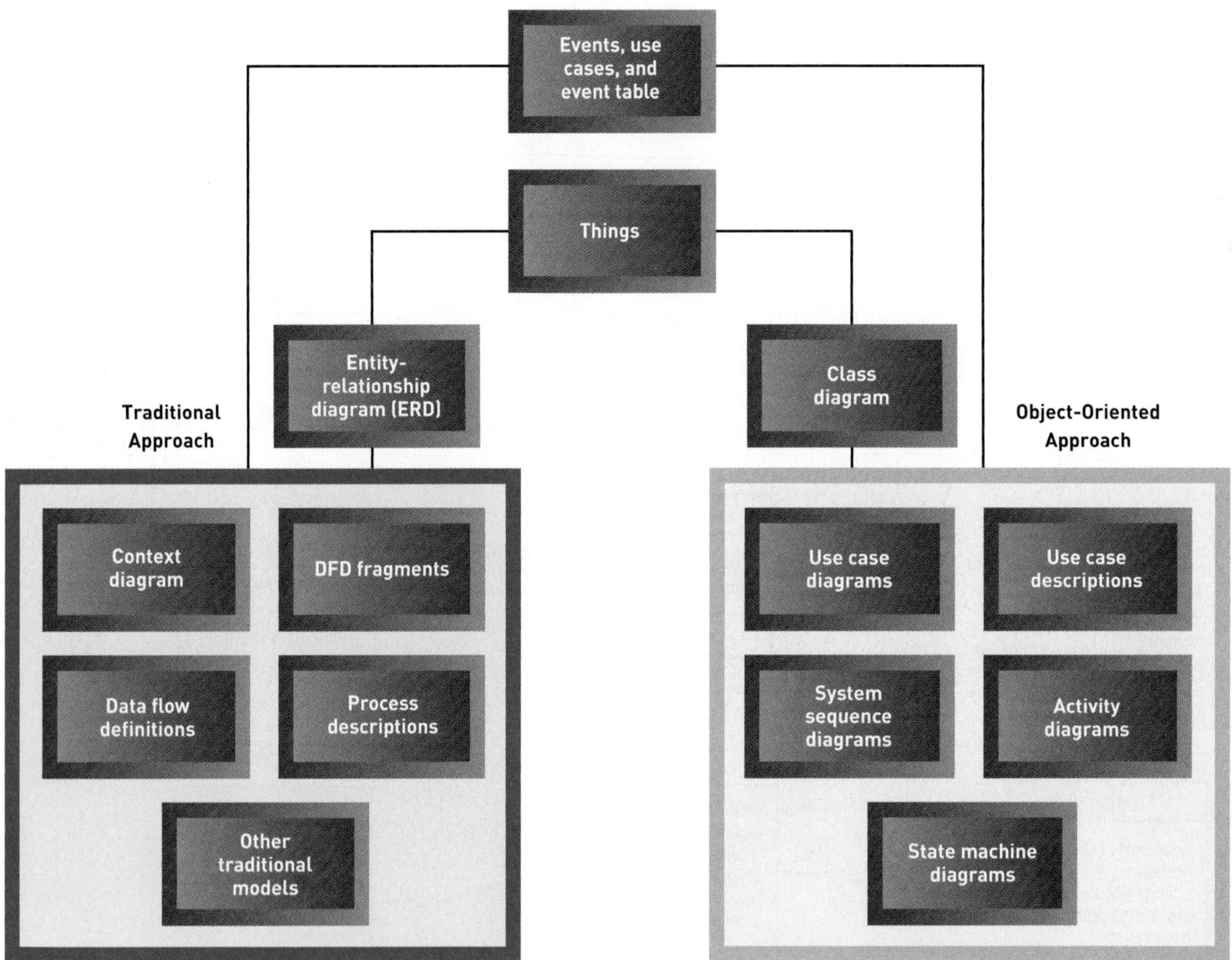

Figure 5-39
Requirements models for the traditional approach and the object-oriented approach

The traditional approach takes the use cases in the event table and creates a set of data flow diagrams (DFDs) based on the information in the table, including the context diagram and DFD fragments. The entity-relationship diagram (ERD) defines the data storage requirements that are included in the DFDs. Other information about the requirements includes data flow definitions and process descriptions. These and some additional traditional models are discussed in Chapter 6.

The object-oriented approach takes the event table and use case descriptions and creates use case diagrams, activity diagrams, system sequence diagrams, and state machine diagrams. These models are discussed in Chapter 7.

SUMMARY

This chapter is the first of three chapters that present techniques for modeling a system's functional requirements, highlighting the tasks that are completed during the analysis activity named *Define system requirements*. Use cases and things in the user's work environment are key concepts common to all approaches to system development. The traditional approach uses entity-relationship diagrams (ERD), and the object-oriented approach uses class diagrams as key models of the problem domain.

A key early step in the modeling process is to identify and list the use cases that define the functional requirements for the system. Use cases can be identified using the user goal technique, the CRUD technique, and the event decomposition technique. The event decomposition technique begins by identifying the events that require a response from the system. An event is something that can be described, something that occurs at a specific time and place, and something worth remembering. External events occur outside the system, usually triggered by someone who interacts with the system. Temporal events occur at a defined point in time, such as the end of a work day or the end of every month. State or internal events occur based on an internal system change. Information about each event is recorded in an event table, which lists the event, the trigger for the event, the source of the trigger, the use case that the system must carry out, the response produced as system output, and the destination for the response.

Each use case identified by the analyst is further documented by a use case description. A use case description can be brief, intermediate, or fully developed. Use case actors, scenarios, stakeholders, preconditions, postconditions, flows of activities, and exception conditions are identified and documented.

The other key concept involves the things users deal with in their work that the system needs to remember, such as products, orders, invoices, and customers. There are many naturally occurring relationships among things the user works with: A customer places an order, and an order requires an invoice. Cardinality (or multiplicity) of a relationship refers to the number of associations involved in a relationship: A customer might place many orders, and each order is placed by one customer. Attributes are specific pieces of information about a thing, such as a name and an address for a customer. The traditional approach models these things as data entities that represent data that is stored. The object-oriented approach models these things as objects belonging to a domain class. The traditional approach uses the entity-relationship diagram to show data entities, attributes of data entities, and relationships. The object-oriented approach uses the UML class diagram to show the same information, calling it the *domain model class diagram*. Two additional concepts are used in class diagrams (although they are sometimes used in entity-relationship diagrams, too): generalization/specialization hierarchies, which allow inheritance from a superclass to a subclass, and whole-part hierarchies, which allow a collection of objects to be associated as a whole and its parts. The UML class diagram is also used to model design classes, which will be explored in Chapter 11.

The next two chapters discuss requirements models produced by the traditional approach and the object-oriented approach, respectively.

KEY TERMS

actor, p. 171
aggregation, p. 190
associative entity, p. 184
attribute, p. 181
binary relationships, p. 181
cardinality, p. 180
composition, p. 190
compound attribute, p. 181
CRUD technique, p. 161
data entities, p. 182
destination, p. 169
elementary business process (EBP), p. 161
event, p. 162
event decomposition, p. 162
event table, p. 168
external event, p. 163
generalization/specialization hierarchies, p. 189
identifier (key), p. 181
inheritance, p. 190
multiplicity, p. 180
n-ary relationship, p. 181
perfect technology assumption, p. 167
postconditions, p. 175
preconditions, p. 174
relationship, p. 178
response, p. 169
scenario, p. 171
source, p. 169
state event, p. 165
system controls, p. 167
temporal event, p. 164
ternary relationship, p. 181
trigger, p. 169
unary (recursive) relationship, p. 181
use case, p. 160
use case description, p. 171
use case instances, p. 171
user goal technique, p. 160
whole-part hierarchies, p. 190

REVIEW QUESTIONS

1. What are the two key concepts used to begin defining system requirements?
2. What is a use case?
3. What are three techniques used to identify use cases?
4. What is an event and what is an elementary business process (EBP)?
5. What are the three types of events?
6. Which type of event results in data entering the system?
7. Which type of event occurs at a defined point in time?
8. Which type of event does not result in data entering the system but always results in an output?
9. What type of event would be named *Employee quits job?*
10. What type of event would be named *Time to produce paychecks*?
11. What are some examples of system controls?
12. What does the perfect technology assumption state?
13. What are the columns in an event table?
14. What is a trigger? A source? A use case? A response? A destination?
15. What is the difference between a use case and a scenario? Give an example of each.
16. What are the three types of use case descriptions? Which one is usually sufficient for simple use cases?
17. What are preconditions and postconditions? Give an example of each.
18. What are exception conditions? Give an example.
19. What communicates back and forth in a two-column flow of activities?
20. What is a "thing" called in models used in the traditional approach?
21. What is a "thing" called in the object-oriented approach?
22. What is a relationship?
23. What is cardinality or multiplicity of a relationship?
24. Describe how an entity-relationship diagram shows the minimum and maximum cardinality.
25. What are unary, binary, and *n*-ary relationships?
26. What are attributes and compound attributes?
27. What is an associative entity?
28. What symbols are shown in an entity-relationship diagram?
29. What symbols are shown in a domain model class diagram?
30. How is multiplicity shown on a domain model class diagram?
31. What is a generalization/specialization hierarchy?
32. From what type of class do subclasses inherit?
33. What are two types of whole-part hierarchies?
34. What does the triangle symbol indicate on a line connecting classes on the class diagram?
35. How is an association class shown on a class diagram?
36. What type of classes are shown in a domain model class diagram?

THINKING CRITICALLY

1. Explain how a user goal can be used as a technique to identify use cases.
2. Explain how the CRUD technique can be used to identify use cases.
3. Explain the importance of elementary business processes (EBPs) in identifying use cases.
4. Review the external event checklist in Figure 5-4, and think about a university course registration system. What is an example of an event of each type in the checklist? Name each event using the guidelines for naming an external event.
5. Review the temporal event checklist in Figure 5-5. Would a student grade report be an internal or external output? Would a class list for the instructor be an internal or external output? What are some other internal and external outputs for a course registration system? Using the guidelines for naming temporal events, what would you name the events that trigger these outputs?
6. In a course registration system, for the event *Student registers for classes*, create an event table entry listing the event, trigger, source, use case, response(s), and destination(s). For the event *Time to produce grade reports*, create another event table entry.
7. Consider the following sequence of actions taken by a customer at a bank. Which action is the event the analyst should define for a bank account transaction-processing system? (1) Kevin gets a check from Grandma for his birthday. (2) Kevin wants a car. (3) Kevin decides to save his money. (4) Kevin goes to the bank. (5) Kevin waits in line. (6) Kevin makes a deposit in his savings account. (7) Kevin grabs the deposit receipt. (8) Kevin asks for a brochure on auto loans.
8. Consider the perfect technology assumption, which states that events should be included during analysis only if the system would be required to respond under perfect conditions. Could any of the events in the event table for Rocky Mountain Outfitters be eliminated based on this assumption? Explain. Why are events such as *User logs on to system* and *Time to back up the data* required only under imperfect conditions?
9. Draw an entity-relationship diagram, including minimum and maximum cardinality for the following: The system stores information about two things: cars and owners. A car has attributes for make, model, and year. The owner has attributes for name and address. Assume that a car must be owned by one owner, and an owner can own many cars, but an owner might not own any cars (perhaps she just sold them all, but you still want a record of her in the system).
10. Draw a class diagram for the cars and owners described in exercise 9 but include subclasses for sports car, sedan, and minivan with appropriate attributes.

11. Consider the entity-relationship diagram shown in Figure 5-28, the refined ERD showing course enrollment with an associative entity. Does this model allow a student to enroll in more than one course section at a time? Does the model allow a course section to contain more than one student? Does the model allow a student to enroll in several sections of the same course and get a grade for each enrollment? Does the model store information about all grades earned by all students in all sections?

12. Again consider the entity-relationship diagram shown in Figure 5-28. Add the following to the diagram and list any assumptions you had to make. A faculty member usually teaches many course sections, but some semesters a faculty member may not teach any. Each course section must have at least one faculty member teaching it, but sometimes teams teach course sections. Furthermore, to make sure that all course sections are similar, one faculty member is assigned as course coordinator to oversee the course, and each faculty member can be coordinator of many courses.

13. If the entity-relationship diagram you drew in exercise 12 showed a many-to-many relationship between faculty member and course section, a further look at the relationship might reveal the need to store some additional information. What might this information include? (Hint: Does the instructor have specific office hours for each course section? Do you give an instructor some sort of evaluation for each course section?) Expand the ERD to allow the system to store this additional information.

14. Draw a class diagram for the course enrollment system completed in exercise 13. Be sure to use the correct notation for association classes.

15. Consider a system that needs to store information about computers in a computer lab at a university, such as the features and location of each computer. What are the things that might be included in a model? What are some of the relationships among these things? What are some of the attributes of these things? Draw an entity-relationship model for this system.

16. Draw a domain model class diagram for the computer lab system described in exercise 15.

17. Consider the domain model class diagram for Rocky Mountain Outfitters shown in Figure 5-38. If a Web order is created, how many attributes does it have? If a telephone order is created, how many attributes does it have? If an existing customer places a phone order for one item, how many new objects are created overall for this transaction?

18. A product item for RMO is not the same as an inventory item. A product item is something like a men's leather hunting jacket supplied by Leather 'R' Us. An inventory item is a specific size and color of the jacket—like a size medium brown leather hunting jacket. If RMO adds a new jacket to its catalog, and six sizes and three colors are available in inventory, how many objects need to be added overall?

19. Consider the following domain model class diagram showing college, department, and faculty members.
    a. What kind of relationships are shown in the model?
    b. How many attributes does a "faculty member" have? Which (if any) have been inherited from another class?
    c. If you add information about one college, one department, and four faculty members, how many objects do you add to the system?
    d. Can a faculty member work in more than one department at the same time? Explain.
    e. Can a faculty member work in two departments at the same time, where one department is in the college of business and the other department is in the college of arts and sciences? Explain.

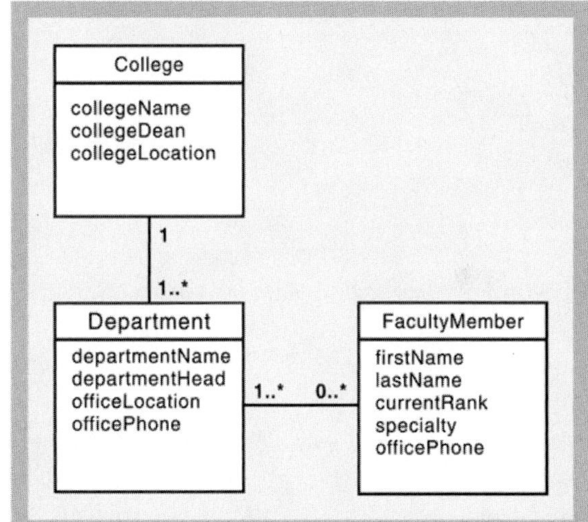

EXPERIENTIAL EXERCISES

1. Visit some Web sites of car manufacturers such as Honda, BMW, Toyota, and Acura. Many of these sites have a use case that is typically named *Build and price a car*. As a potential customer, you can select a car model, select features and options, and get the car's suggested price and list of specifications. Try one of these use cases and write a fully developed use case description based on what you see. Include the use case name, triggering event, stakeholders, actors, preconditions, postconditions, a two-column flow of activities, and exception conditions.

2. Set up a meeting with a librarian. During your meeting, ask the librarian to describe the situations that come up in the library to which the book checkout system needs to respond. List these external events. Now ask about points in time, or deadlines, that require the system to produce a statement, notice, report, or other output. List these temporal events. Does it seem natural for the librarian to describe the system in this way? Similarly, ask the librarian to describe the things about which the system needs to store information. See whether you can get the librarian to list the important attributes and describe relationships among things. Does it seem natural for

the librarian to describe these things? Create either an ERD or a class diagram based on what you learn.

3. Visit a restaurant or the college food service and talk to a server (or talk with a friend who is a food server). Ask about the external events, temporal events, and data entities or objects, as you did in exercise 1. What are the events for order processing at a restaurant? Complete an event table and either an ERD or class diagram.
4. Review the procedures for course registration at your university and talk with the staff in advising, in registration, and in your major department. Think about the sequence that goes on over an entire semester. What are the events that students trigger? What are the events that your major department triggers? What are the temporal events that result in information going to students? What are the temporal events that result in information going to instructors or departments?
5. Again review information about your own university. Create generalization/specialization hierarchies using the domain model class diagram notation for (1) types of faculty, (2) types of students, (3) types of courses, (4) types of financial aid, and (5) types of housing. Include attributes for the superclass and the subclasses in each case.

CASE STUDIES

THE SPRING BREAKS 'R' US TRAVEL SERVICE BOOKING SYSTEM

Spring Breaks 'R' Us Travel Service (SBRU) books spring-break trips at resorts for college students. During the fall, resorts submit availability information to SBRU indicating rooms, room capacity, and room rates for each week of the spring-break season. Each resort offers bookings for a different number of weeks each season, and rooms have different rates depending on the week. Usually, the resorts make a variety of rooms with different capacities available so students can book the right room size. Couples can book a two-person room, for example, and four people can book a room for four.

In December, SBRU generates a list of resorts, available weeks, and room rates that is distributed to college campus representatives all over the country. When a group of students submits a reservation request for a week at a particular resort, SBRU assigns the students to a room with sufficient capacity and sends each student a confirmation notice. When the cutoff date for a week arrives, SBRU sends each resort a list of students booked in each room for the following week. When the students arrive at the resort, they pay the resort directly for the room. Resorts send commission checks directly to the SBRU accounting system, which is separate from the booking system. When spring break is over, students return to their schools and hit the books.

1. To what events must the SBRU booking system respond? Create a complete event table listing the event, trigger, source, use case, response, and destination for each event. Be sure to consider only the events that trigger processing in the booking system, not the SBRU accounting system or the systems operated by the resorts.
2. List the data entities (or classes) that are mentioned. List the attributes of each data entity (or class). List the relationships among data entities (or classes).
3. Which classes might be refined into a generalization/specialization hierarchy? List the superclass and any subclasses for each of them.

THE REAL ESTATE MULTIPLE LISTING SERVICE SYSTEM

The Real Estate Multiple Listing Service system supplies information that local real estate agents use to help them sell houses to their customers. During the month, agents list houses for sale (listings) by contracting with homeowners. The agent works for a real estate office, which sends information on the listing to the multiple listing service. Therefore, any agent in the community can get information on the listing.

Information on a listing includes the address, year built, square feet, number of bedrooms, number of bathrooms, owner name, owner phone number, asking price, and status code. At any time during the month, an agent might directly request information on listings that match customer requirements, so the agent contacts the multiple listing service with the request. Information is provided on the house, on the agent who listed the house, and on the real estate office for which the agent works. For example, an agent might want to call the listing agent to ask additional questions or call the homeowner directly to make an appointment to show the house. Twice each month (on the 15th and 30th), the multiple listing service produces a listing book that contains information on all listings. These books are sent to all of the real estate agents. Many real estate agents want the books (which are easier to flip through), so they are provided even though the information is often out of date. Sometimes agents and owners decide to change information about a listing, such as reducing the price, correcting previous information on the house, or indicating that the house is sold. The real estate office sends in these change requests to the multiple listing service when the agent asks the office to do so.

1. To what events must the multiple listing service system respond? Create a complete event table listing the event, trigger, source, use case, response, and destination for each event.
2. Draw an entity-relationship diagram to represent the data storage requirements for the multiple listing service system, including the attributes mentioned. Does your model include data entities for offer, buyer, and closing? If so,

reconsider. Include information that the multiple listing service needs to store, which might be different from information the real estate office needs to store.

3. Draw a domain model class diagram that corresponds to the ERD but shows that different types of listings have different attributes. The description in the case assumes all listings are for single-family houses. What about multifamily listings or commercial property listings?

THE STATE PATROL TICKET PROCESSING SYSTEM

The purpose of the State Patrol ticket processing system is to record driver violations, to keep records of the fines paid by drivers when they plead guilty or are found guilty of moving violations by the courts, and to notify the court that a warrant for arrest should be issued when such fines are not paid in a timely manner. A separate State Patrol system records accidents and verification of financial responsibility (insurance). Yet a third system produces driving record reports from the ticket and accident records for insurance companies. Finally, a fourth system issues, renews, or suspends driver's licenses. These four systems are obviously integrated in that they share access to the same database, but otherwise, they are operated separately by different departments of the State Patrol. State Patrol operations (what the officers do) are entirely separate.

The portion of the database used with the ticket processing system involves driver data, ticket data, officer data, and court data. Driver data, officer data, and court data are used by the system. The system creates and maintains ticket data. Driver attributes include license number, name, address, date of birth, date licensed, and so on. Ticket attributes include ticket number (each is unique and preprinted on each sheet of the officer's ticket book), location, ticket type, ticket date, ticket time, plea, trial date, verdict, fine amount, and date paid. Court and officer data include the name and address of each, respectively. Each driver may have zero or more tickets, and each ticket applies to only one driver. Officers write quite a few tickets.

When an officer gives a ticket to a driver, a copy of the ticket is turned in and entered into the system. A new ticket record is created, and relationships to the correct driver, officer, and court are established in the database. If the driver pleads guilty, he or she mails in the fine in a preprinted envelope with the ticket number on it. In some cases, the driver claims innocence and wants a court date. When the envelope is returned without a check and the trial request box has an "X" in it, the system notes the plea on the ticket record; looks up driver, ticket, and officer information; and sends a ticket details report to the appropriate court. A trial date questionnaire form is also produced at the same time and is mailed to the driver. The instructions on the questionnaire tell the driver to fill in convenient dates and mail the questionnaire directly to the court. Upon receiving this information, the court schedules a trial date and notifies the driver of the date and time.

When the trial is completed, the court sends the verdict to the ticketing system. The verdict and trial date are recorded for the ticket. If the verdict is innocent, the system that produces driving record reports for insurance companies will ignore the ticket. If the verdict is guilty, the court gives the driver another envelope with the ticket number on it for mailing in the fine.

If the driver fails to pay the fine within the required period, the ticket processing system produces a warrant request notice and sends it to the court. This happens if the driver does not return the original envelope within two weeks or does not return the court-supplied envelope within two weeks of the trial date. What happens then is in the hands of the court. Sometimes the court requests that the driver's license be suspended, and the system that processes drivers' licenses handles the suspension.

1. To what events must the ticket processing system respond? Create a complete event table listing the event, trigger, source, use case, response, and destination for each event.
2. For the use case *Record new ticket*, complete a fully developed use case description based on the information in the case study.
3. Draw an entity-relationship diagram to represent the data storage requirements for the ticket processing system, including the attributes mentioned. Explain why it is important to understand how the system is integrated with other State Patrol systems.
4. Draw a domain model class diagram that corresponds to the ERD but assumes there are different types of drivers. Classifications of types of drivers vary by state. Some states have restricted licenses for minors, for example, and special licenses for commercial vehicle operators. Research your state's requirements, and create a generalization/specialization hierarchy for the class Driver, showing the different attributes each special type of driver might have. Consider the same issues for types of tickets. Include some special types of tickets in a generalization/specialization hierarchy in the class diagram.
5. Use the CRUD technique to verify that all domain classes are provided for in the use cases identified in the event table. In an integrated system like the ticket processing system, some domain classes are created by and updated by another system. Create a table with domain classes down the rows and each state patrol system in the case across the columns. Indicate C, R, U, or D for each class and each system.

RETHINKING ROCKY MOUNTAIN OUTFITTERS

When listing nouns and making some decisions about the initial list of things (see Figure 5-19), the RMO team decided to research Customer Account as a possible data entity or class if the system included an RMO payment plan (similar to a company charge account plan). Many retail store chains have their own charge accounts for the convenience of the customer—to increase sales to the customer and to better track customer purchase behavior.

Consider the implications to the system if management decided to incorporate an RMO charge account and payment plan as part of the customer support system.

1. Discuss the implications that such a change would have on the scope of the project. How might this new capability

change the list of stakeholders the team would involve when collecting information and defining the requirements? Would the change have any effect on other RMO systems or system projects planned or under way? Would the change have any effect on the project plan originally developed by Barbara Halifax? In other words, is this a minor change or a major change?

2. What events need to be added to the event table? Complete the event table entries for these additional events. What activities or use cases for existing events might be changed because of a charge account and payment plan? Explain.
3. What are some additional things and relationships among things that the system would be required to store because of the charge account and payment plan? Modify the entity-relationship diagram and the class diagram to reflect these charges.

FOCUSING ON RELIABLE PHARMACEUTICAL SERVICE

In Chapter 1, you learned about the background and prescription-processing operations for Reliable Pharmaceutical Service. As discussed in this chapter, defining the requirements for the new system starts by taking the information gathered about the needed system and then focusing on the events that require system processing and on the things about which the system needs to store information. The full system would involve many events and things. In this chapter's case exercise, we focus on only a subset of events for the system and a subset of data entities or classes. Exercises in later chapters will add to the scope and complexity of the requirements for Reliable.

1. Create an event table that lists information about system requirements based on the following specific system processing: When a nursing home needs to fill prescriptions for its patients, it provides order details to Reliable. Reliable immediately records information about the order and prescriptions. Prescription orders come in from all of Reliable's nursing-home clients throughout the day. At the start of each 12-hour shift, Reliable prepares a case manifest, detailing all recent orders, which is given to one of the pharmacists. When the pharmacist has assembled the orders for each client, the pharmacist records the order fulfillment. (Review the Reliable case description at the end of Chapter 1 for more details.) In addition, the system needs to add or update patient information, add or update drug inventory information, produce purchase orders to replenish the drug inventory, record inventory adjustments, and generate various management reports. For now, ignore any billing, payments, or insurance processing.
2. Create an entity-relationship diagram that shows the data storage requirements for the following portion of the system: Add a few attributes to each data entity and show minimum and maximum cardinality. To process the prescription order, Reliable needs to know about the patients, the nursing home, and the nursing-home unit where each patient resides. Each nursing home has at least one, but possibly many, units. A patient is assigned to a specific unit. An order consists of one or more prescriptions, each for one specific drug and for one specific patient. An order, therefore, consists of prescriptions for more than one patient. Careful tracking and record keeping is obviously crucial. In addition, each patient has many prescriptions. One pharmacist fills each order.
3. Create a domain model class diagram for the object-oriented approach that shows the same requirements as described in step 2. Include a few attributes for each class and show minimum and maximum multiplicity. Be sure to identify any association classes and use the correct notation.
4. How important is it to understand that each order includes prescriptions for more than one patient? Is this the type of information that is difficult to sort out at first? Did you see the implications initially, or did you have to work through the model until it made sense to you? Discuss.

FURTHER RESOURCES

Some classic and more recent texts include the following:

Peter Rob and Carlos Coronel, *Database Systems: Design, Implementation, and Management, Seventh Edition*. Course Technology, 2007.

Craig Larman, *Applying UML and Patterns* (3rd ed.). Prentice-Hall, 2005.

Grady Booch, Ivar Jacobson, and James Rumbaugh, *The Unified Modeling Language User Guide*. Addison-Wesley, 1999.

Ed Yourdon, *Modern Structured Analysis*. Prentice Hall, 1989.

Stephen McMenamin and John Palmer, *Essential Systems Analysis*. Prentice Hall, 1984.

CHAPTER 6

THE TRADITIONAL APPROACH TO REQUIREMENTS

LEARNING OBJECTIVES

After reading this chapter, you should be able to:

- Explain how the traditional approach and the object-oriented approach differ when modeling the details of a use case
- List the components of a traditional system and the symbols representing them on a data flow diagram
- Describe how data flow diagrams can show the system at various levels of abstraction
- Develop data flow diagrams, data element definitions, data store definitions, and process descriptions
- Develop tables to show the distribution of processing and data access across system locations

CHAPTER OUTLINE

Traditional and Object-Oriented Views of Activities/Use Cases

Data Flow Diagrams

Documentation of DFD Components

Locations and Communication through Networks

SAN DIEGO PERIODICALS: FOLLOWING THE DATA FLOW

Arturo Romero and Lei Xu were meeting to review a first draft of several data flow diagrams for San Diego Periodicals' new advertising billing system. Arturo was the analyst assigned to define the new system requirements. Lei was the manager in charge of advertising accounts—she knew virtually all there was to know about how the current system operated. The two had met several times before. Their most recent meeting (just last week) reviewed details of the current system events, ad request processing, and process participants. Arturo left that meeting with pages of notes and sample forms and reports from the current system. Arturo had telephoned Lei several times since that meeting to ask additional questions.

Arturo began the review by saying, "The materials and information that you gave me last week took me quite awhile to absorb, but I think I was able to understand and document all of the system activities or use cases that we discussed. The purpose of this meeting is to ensure that the processing requirements that I've written down are complete and accurate. Let's start with a few of the diagrams that I've created."

Arturo laid out three diagrams on the table. Lei looked at them briefly and said, "I've never seen diagrams like this before—they look like blueprints for playing a game of marbles. And I thought the entity-relationship diagrams were strange!"

Arturo replied, "I expect this review will go slowly because this is your first look at this style of documentation. I'll explain how to interpret the diagrams as we go along. Ask as many questions as you like. The quality of our work depends on your understanding the diagrams, so don't be shy."

Arturo continued, "The pictures are called data flow diagrams, or DFDs for short. They divide your system into processing functions represented by the rectangles with rounded-off corners. The arrows show data movement among processes and between processes and files." Lei pointed to a square on one of the diagrams and said, "I assume that this is a company purchasing ad space?"

Arturo replied, "Yes, the squares represent people or organizations that supply inputs or expect output data from the system."

Lei said, "I think I can get the hang of this. I recognize most of the names that you've used for the processes and data. I'm not sure what these other symbols are—they're named for things that we store in our manual files and database, but they don't seem to correspond exactly to our system."

"They don't," Arturo replied. "They're entities from the entity-relationship diagram that we developed a couple of weeks ago. But let's skip over those for the moment. Why don't we walk through the processing sequence for booking an ad, and we'll discuss the entities as we get to them?"

Arturo and Lei continued reviewing the DFDs, and the next thing they knew, over an hour had passed. Several pages of Arturo's notepad had been filled, and 25 corrections and comments were noted in red on the data flow diagrams. Lei said, "My brain feels completely drained. I don't think that I can do any more of this today."

Arturo replied, "You've given me plenty of things to work on, so let's call it quits for now. Can we meet for two hours at nine o'clock on Thursday?"

Lei replied, "Yes, I'm free then. So, will you be bringing more data flow diagrams, or do you have something even weirder up your sleeve?"

Arturo smiled and said, "The toughest stuff is behind us, but you should expect a few more surprises."

OVERVIEW

Chapter 5 described two key concepts associated with modeling system requirements in both the traditional and the object-oriented (OO) approaches to information systems development: events that trigger use cases and things in the users' work domain. In this chapter, the focus turns to the details of what the system does when an event occurs: activities and interactions between computer processes and data.

This chapter describes the traditional structured approach to representing activities and interactions. We describe and present the diagrams and other models of the traditional approach, and we provide examples from the Rocky Mountain Outfitters customer support system to show how each model is related. Chapter 7 describes details of the OO approach to representing activities and interactions.

Modeling activities and interactions is a difficult process with either the traditional or OO approach. Building models is a challenging and time-consuming task. Activities and interactions must be specified in exacting detail. Analysts and users must jointly evaluate model completeness, correctness, and quality. As illustrated in the accompanying RMO progress memo, coordinating the efforts of project participants and building a consensus about detailed system requirements are complex project management activities (see memo).

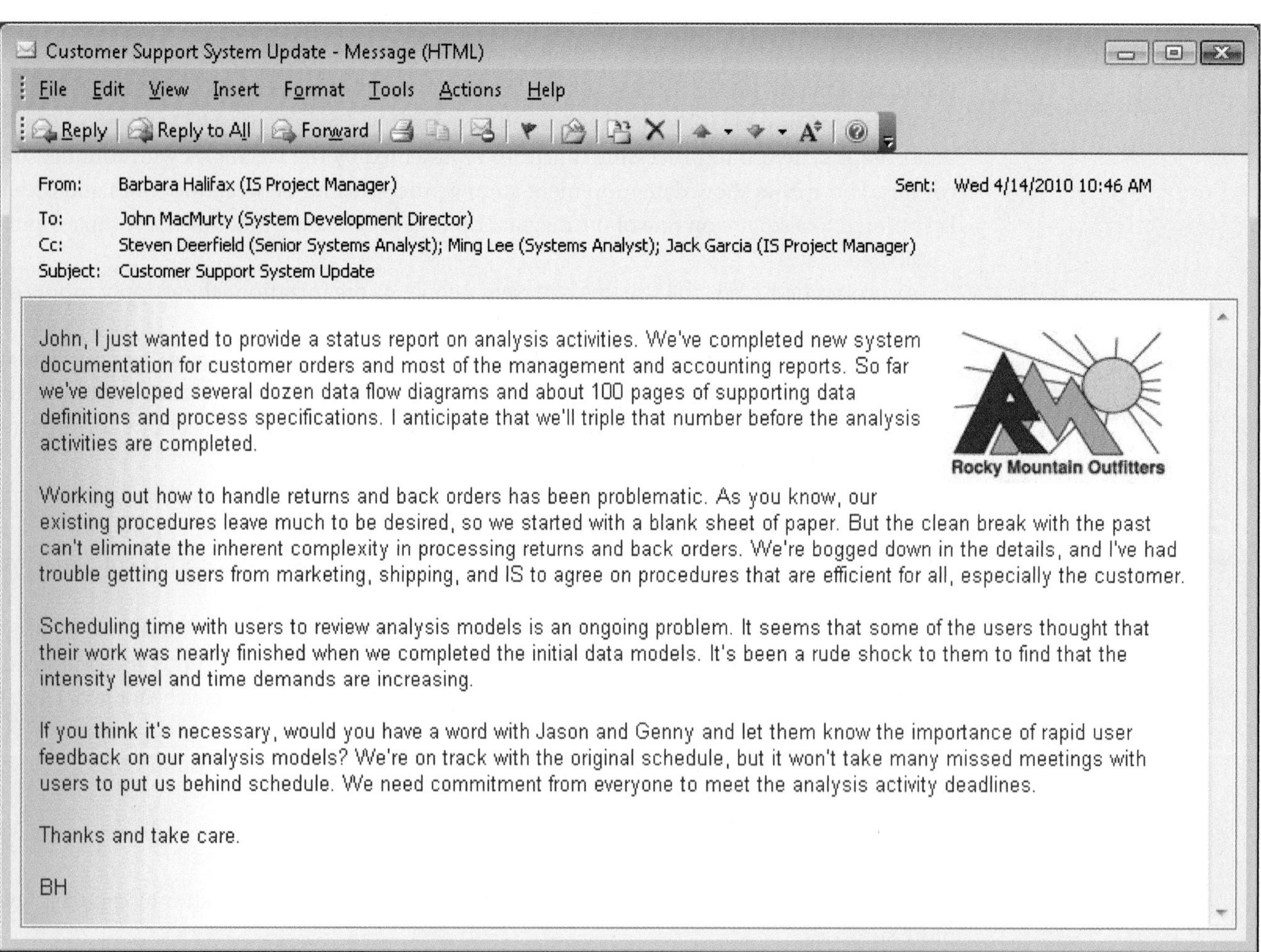
Customer Support System Update - Message (HTML)

File Edit View Insert Format Tools Actions Help

Reply | Reply to All | Forward

From: Barbara Halifax (IS Project Manager) Sent: Wed 4/14/2010 10:46 AM

To: John MacMurty (System Development Director)

Cc: Steven Deerfield (Senior Systems Analyst); Ming Lee (Systems Analyst); Jack Garcia (IS Project Manager)

Subject: Customer Support System Update

John, I just wanted to provide a status report on analysis activities. We've completed new system documentation for customer orders and most of the management and accounting reports. So far we've developed several dozen data flow diagrams and about 100 pages of supporting data definitions and process specifications. I anticipate that we'll triple that number before the analysis activities are completed.

Working out how to handle returns and back orders has been problematic. As you know, our existing procedures leave much to be desired, so we started with a blank sheet of paper. But the clean break with the past can't eliminate the inherent complexity in processing returns and back orders. We're bogged down in the details, and I've had trouble getting users from marketing, shipping, and IS to agree on procedures that are efficient for all, especially the customer.

Scheduling time with users to review analysis models is an ongoing problem. It seems that some of the users thought that their work was nearly finished when we completed the initial data models. It's been a rude shock to them to find that the intensity level and time demands are increasing.

If you think it's necessary, would you have a word with Jason and Genny and let them know the importance of rapid user feedback on our analysis models? We're on track with the original schedule, but it won't take many missed meetings with users to put us behind schedule. We need commitment from everyone to meet the analysis activity deadlines.

Thanks and take care.

BH

TRADITIONAL AND OBJECT-ORIENTED VIEWS OF ACTIVITIES/USE CASES

The traditional and OO approaches to system development differ in how a system's response to an event is modeled and implemented. The traditional approach views a system as a collection of processes, some performed by people and some performed by computers. Traditional computer processes are much like procedural computer programs—they contain instructions that execute in a sequence. When the process executes, it interacts with stored data, reading data values and then writing other data values back to the data file. The process might also interact with people, such as when an instruction asks the user to input a value or it displays information to the user on the computer screen. The traditional approach to systems, then, involves processes, stored data, inputs, and outputs. When modeling what the system does in response to an event, the traditional approach includes processing models that emphasize these system features.

In contrast, the OO approach views a system as a collection of interacting objects. The objects are based on the things in the problem domain discussed in Chapter 5. Objects are capable of behaviors (called *methods*) that allow them to interact with each other and with people using the system. One object asks another object to do something by sending it a message. There are no conventional computer processes or data files per se. Objects carry out the activities and remember the data values. When modeling what the system does in response to an event, the OO approach includes models that show objects, their behavior, and their interactions with other objects.

Figure 6-1 summarizes the differences between traditional and OO approaches to systems. Because of these differences, the traditional and OO approaches to requirements employ different models, as summarized in Figure 6-2. The remainder of this chapter explores the traditional models on the left side of Figure 6-2.

Figure 6-1
Traditional versus OO approaches

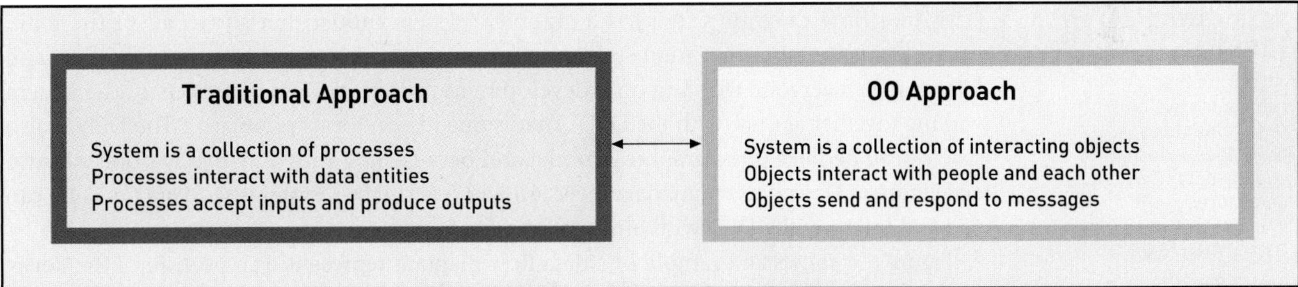

DATA FLOW DIAGRAMS

The traditional approach to information system development describes activities as processes carried out by people or computers. A graphical model that has proven to be quite valuable for modeling processes is the data flow diagram. There are other process models, such as the activity diagrams used with business process reengineering, but the data flow diagram is the most commonly used process model.

Figure 6-2

Requirements models for the traditional and OO approaches

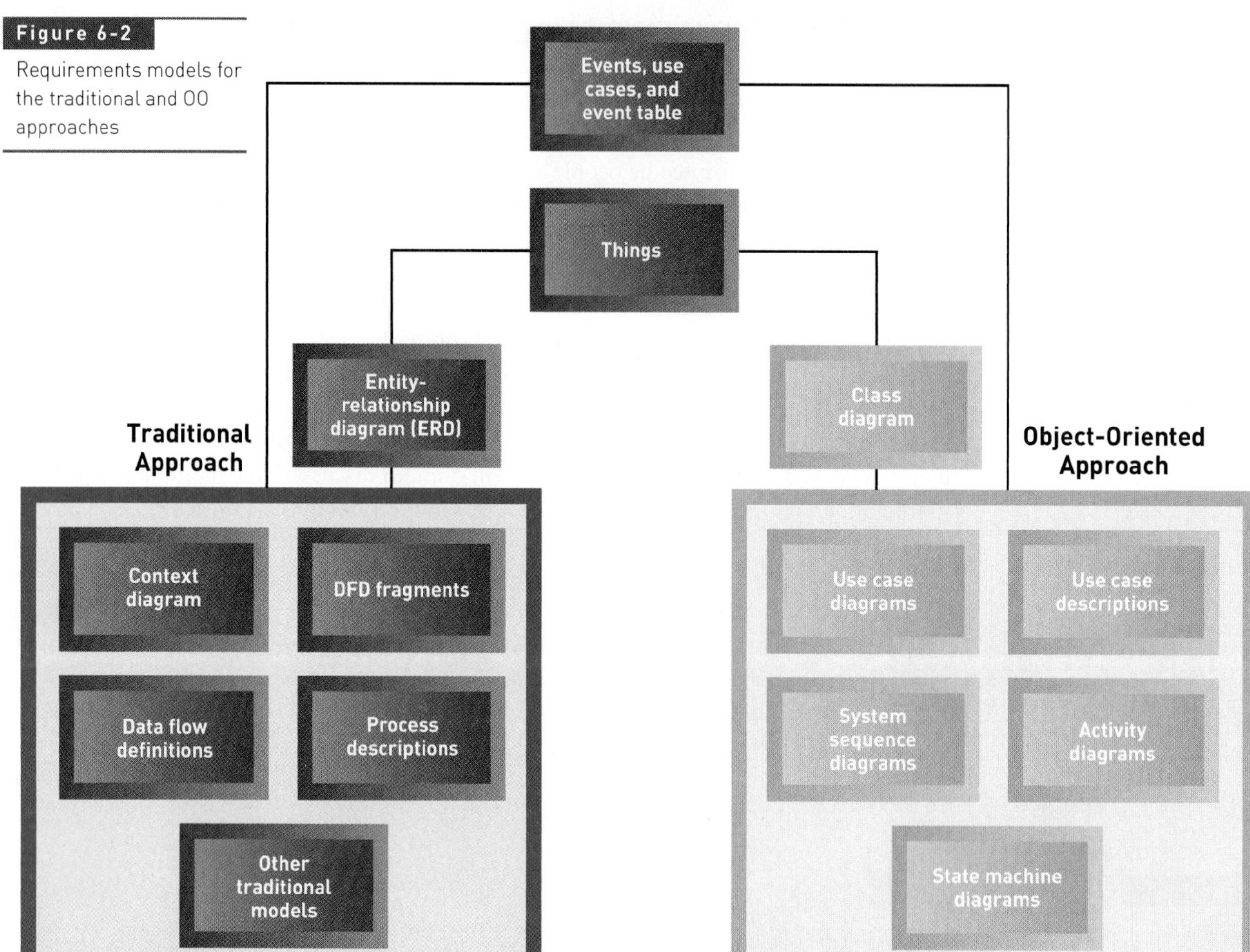

data flow diagram (DFD)
a diagram that represents system requirements as processes, external agents, data flows, and data stores

external agent
a person or organization, outside the system boundary, that supplies data inputs or accepts data outputs

process
a symbol on a DFD that represents an algorithm or procedure by which data inputs are transformed into data outputs

data flow
an arrow on a DFD that represents data movement among processes, data stores, and external agents

A **data flow diagram (DFD)** is a graphical system model that shows all of the main requirements for an information system in one diagram: inputs and outputs, processes, and data storage. Everyone working on a development project can see all aspects of the system working together at once with the DFD. That is one reason for its popularity. The DFD is also easy to read because it is a graphical model and because there are only five symbols to learn (see Figure 6-3). End users, management, and all information systems workers typically can read and interpret the DFD with minimal training.

Figure 6-4 shows an example of a data flow diagram representing a portion of the Rocky Mountain Outfitters (RMO) customer support system. The square is an **external agent**, Customer, the source and destination for some data outside the system. The rectangle with rounded corners is a **process** named *Look up item availability* that can also be referred to by its number, 1. A process defines rules for transforming inputs to outputs. The lines with arrows are **data flows**. Figure 6-4 shows two data flows between Customer and process 1: a process input named Item inquiry and a process output named Item availability details. The final symbol—the flat, open-ended rectangle—is a **data store**. Each data store represents a file or part of a database that stores information about a data entity. In this example, data flows (lines with arrows) point from the data stores to the process, meaning that the process looks up information in the data stores named Catalog, Product item, and Inventory item.

You might recognize that the process in Figure 6-4 corresponds to a use case in the event table for RMO shown in Chapter 5 (see Figure 5-12). The event was *Customer wants to check item availability*, the trigger was Item inquiry, the source was Customer, the response was Item availability details, and the destination for the response was Customer. Therefore, the data flow diagram shows the system use case in response to this one event in graphical form.

Figure 6-3
Data flow diagram symbols

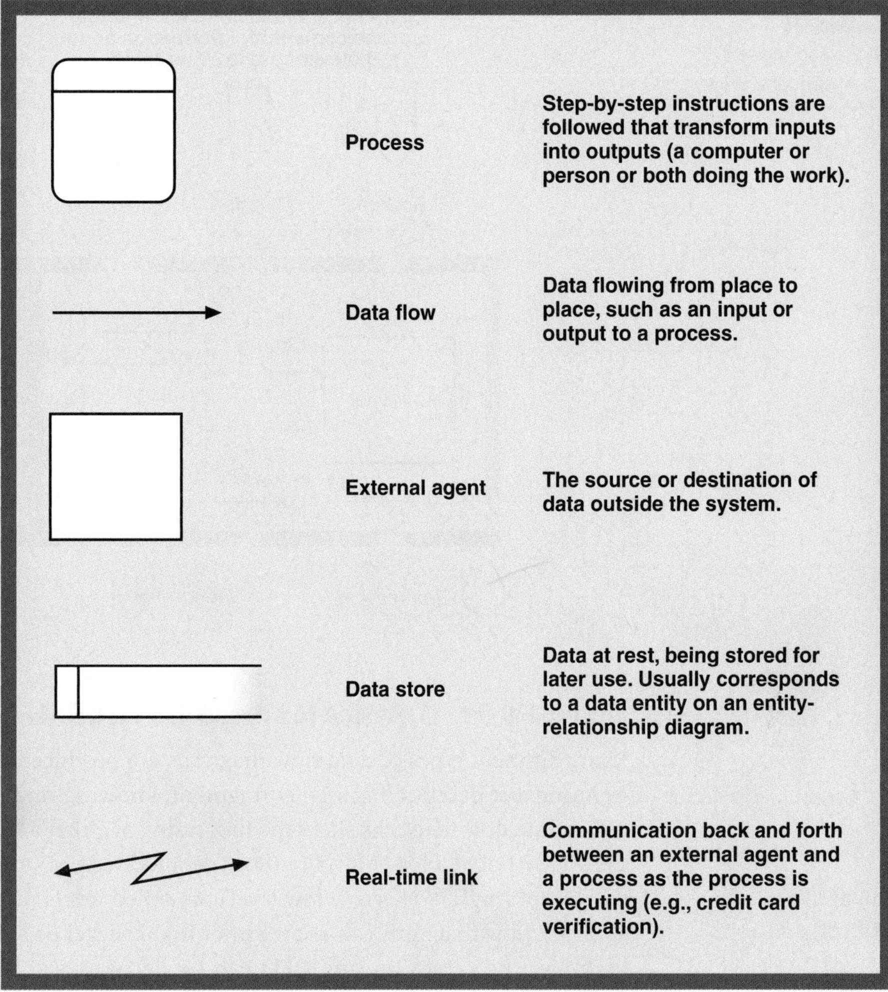

Process
Step-by-step instructions are followed that transform inputs into outputs (a computer or person or both doing the work).
Data flow
Data flowing from place to place, such as an input or output to a process.
External agent
The source or destination of data outside the system.
Data store
Data at rest, being stored for later use. Usually corresponds to a data entity on an entity-relationship diagram.
Real-time link
Communication back and forth between an external agent and a process as the process is executing (e.g., credit card verification).

Figure 6-4
A DFD showing the process *Look up item availability* (a DFD fragment from the RMO case)

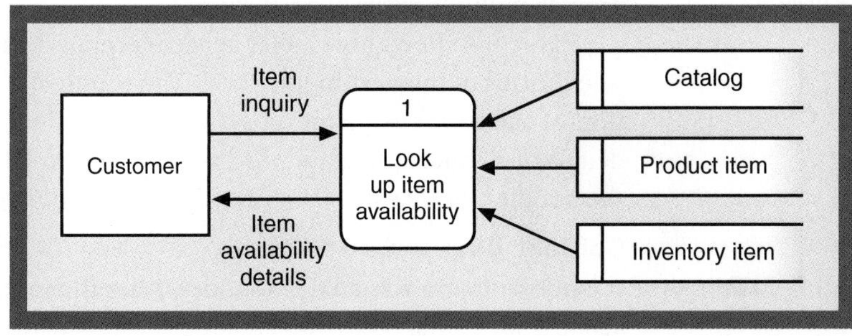

But another piece of information on the DFD is not in the event table: the data stores containing information about an item's availability. Each data store in Figure 6-4 represents a data entity from the entity-relationship diagram (ERD) shown in Chapter 5 (see Figure 5-29). The process on the DFD uses information that we provided by including these data entities and their attributes in the ERD for the system. Therefore, the data flow diagram integrates processing triggered by events with the data entities modeled using the ERD. Figure 6-5 summarizes the correspondences among components of the DFD, events described in the event table, and entities defined in the ERD.

data store
a place where data is held pending future access by one or more processes

BEST PRACTICE

When employing the traditional approach, identify use cases and then model the details of each use case with a data flow diagram fragment.

Figure 6-5
The DFD integrates the event table and the ERD

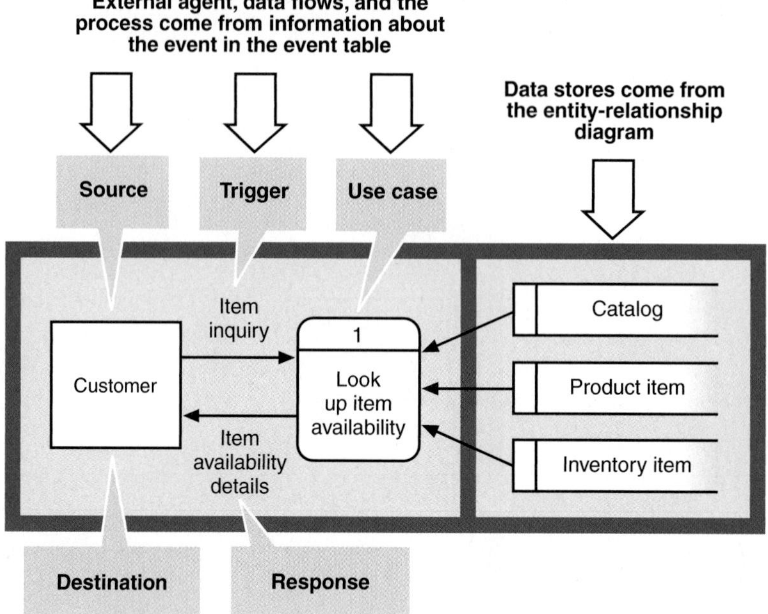

DATA FLOW DIAGRAMS AND LEVELS OF ABSTRACTION

level of abstraction
any modeling technique that breaks the system into a hierarchical set of increasingly more detailed models

Many different types of data flow diagrams are produced to show system requirements. The example just described is a DFD fragment, showing one process in response to one event. Other data flow diagrams show the processing at either a higher level (a more general view of the system) or at a lower level (a more detailed view of one process). These differing views of the system (high level versus low level) are called **levels of abstraction**.

Data flow diagrams can show either higher-level or lower-level views of the system. The high-level processes on one DFD can be decomposed into separate lower-level, detailed DFDs. Processes on the detailed DFDs can also be decomposed into additional diagrams to provide multiple levels of abstraction.

Figure 6-6 shows how DFDs at each level of detail provide additional information about one process at the next higher level. The topmost DFD shows the most abstract representation of the course registration system as a single process. The middle DFD shows internal details of a context diagram process. The bottom DFD shows internal details of process 1 in the middle DFD. Each DFD abstraction level is described further in the following sections.

Context Diagram

context diagram
a DFD that summarizes all processing activity within the system in a single process symbol

A **context diagram** is a DFD that describes the most abstract view of a system. All external agents and all data flows into and out of the system are shown in one diagram, with the entire system represented as one process. The topmost DFD in Figure 6-6 is a context diagram for a simple university course registration system that interacts with three external agents: Academic department, Student, and Faculty member. Academic departments supply information on offered courses, students request enrollment in offered courses, and faculty members receive class lists when the registration period is complete.

A context diagram clearly shows the system boundary. The system scope is defined by what is represented within the single process and what is represented as external agents. External agents that supply or receive data from the system are outside the system scope, and everything else is inside the system scope. The context diagram does not usually show data stores because all of the system's data stores are considered to be within the system scope (that is, part of the internal implementation of the process that represents the system). However, data stores may be shown when they are shared by the system being modeled and another system.

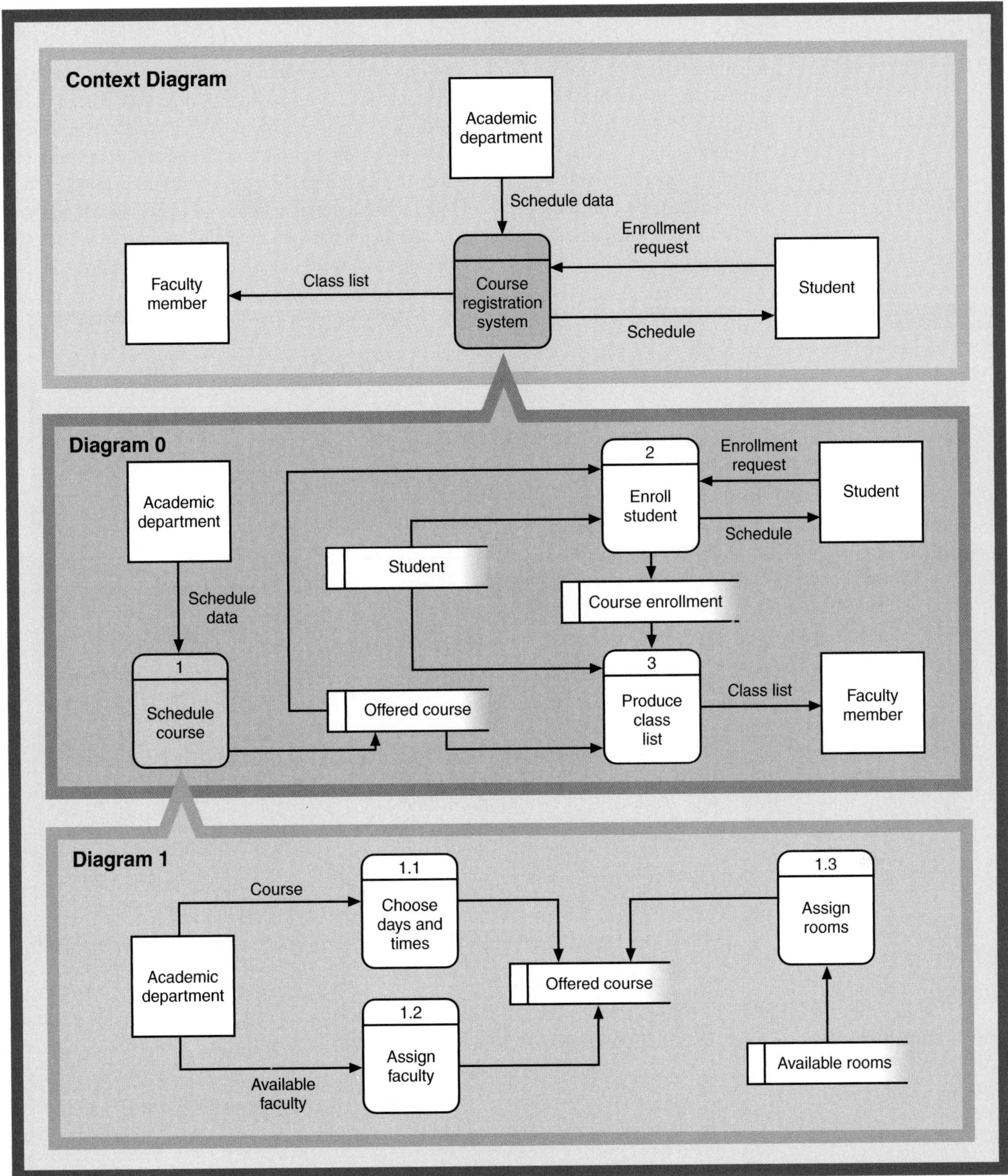

Figure 6-6

Layers of DFD abstraction for a course registration system

The context diagram is usually created in parallel with the event table described in Chapter 5. Each trigger for an external event becomes an input data flow, and the source becomes an external agent. Each response becomes an output data flow, and the destination becomes an external agent. Triggers for temporal events are not data flows, so there are no input data flows for temporal events. Note that the context diagram DFD can be created directly from the event table. The two models provide alternative views of the same system requirements information.

DFD fragment
a DFD that represents the system response to one event within a single process symbol

DFD Fragments

A **DFD fragment** is created for each use case triggered by an event in the event table. Each DFD fragment is a self-contained model showing how the system responds to a single event. The analyst usually creates DFD fragments one at a time, focusing attention on each part of the system. The DFD fragments are drawn after the event table and context diagram are complete.

Figure 6-7 shows the three DFD fragments for the simple course registration system. Each DFD fragment represents all processing for a use case triggered by an event within a single process symbol. The fragments show details of interactions among the process, external agents, and internal data stores. The data stores used on a DFD fragment represent entities on the ERD. Each DFD fragment shows only those data stores that are actually needed to respond to the event.

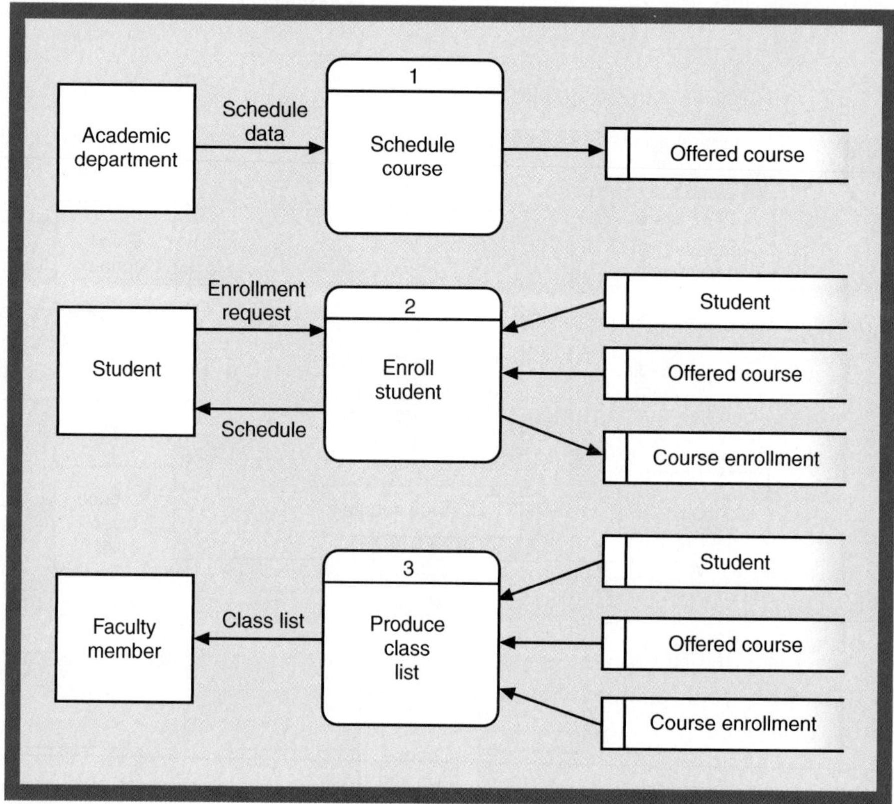

Figure 6-7
Three DFD fragments for the course registration system

event-partitioned system model, or diagram 0
a DFD that models system requirements using a single process for each event in a system or subsystem

The Event-Partitioned System Model

All of the DFD fragments for a system or subsystem can be combined on a single DFD called the **event-partitioned system model**, or **diagram 0**. Figure 6-8 shows how the three course registration system DFD fragments shown in Figure 6-7 are combined to create diagram 0.

Diagram 0 is used primarily as a presentation tool. It summarizes an entire system or subsystem in greater detail than does a context diagram. However, analysts often avoid developing diagram 0 because:

- The information content duplicates the set of DFD fragments.
- The diagram is often complex and unwieldy, particularly for large systems that respond to many events.

As we'll discuss later in the chapter, redundancy and complexity are two DFD characteristics that analysts should avoid whenever possible.

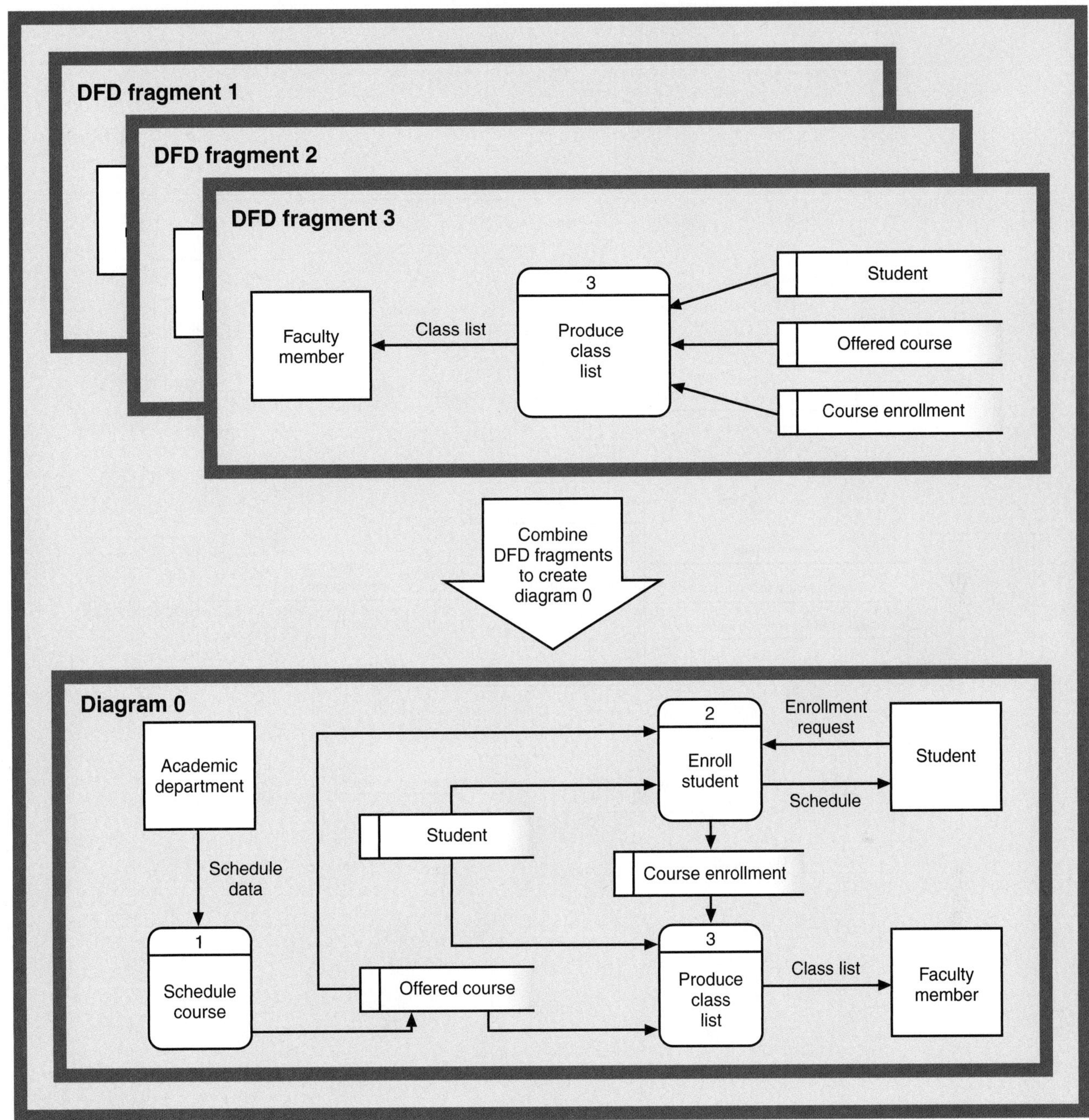

Figure 6-8
Combining DFD fragments to create the event-partitioned system model for the course registration system

RMO DATA FLOW DIAGRAMS

Figure 6-9 shows a context diagram for the Rocky Mountain Outfitters customer support system. Normally, data flows and external agents on the context diagram are taken directly from the event table as discussed previously, but because the RMO customer support system responds to 20 events, this figure combines data flows for some events for simplicity. In a smaller system example with 10 to 15 events, you should include all data flows on the context diagram.

When a system responds to many events, it is commonly divided into subsystems, and a context diagram is created for each subsystem. Figure 6-10 divides the RMO customer support

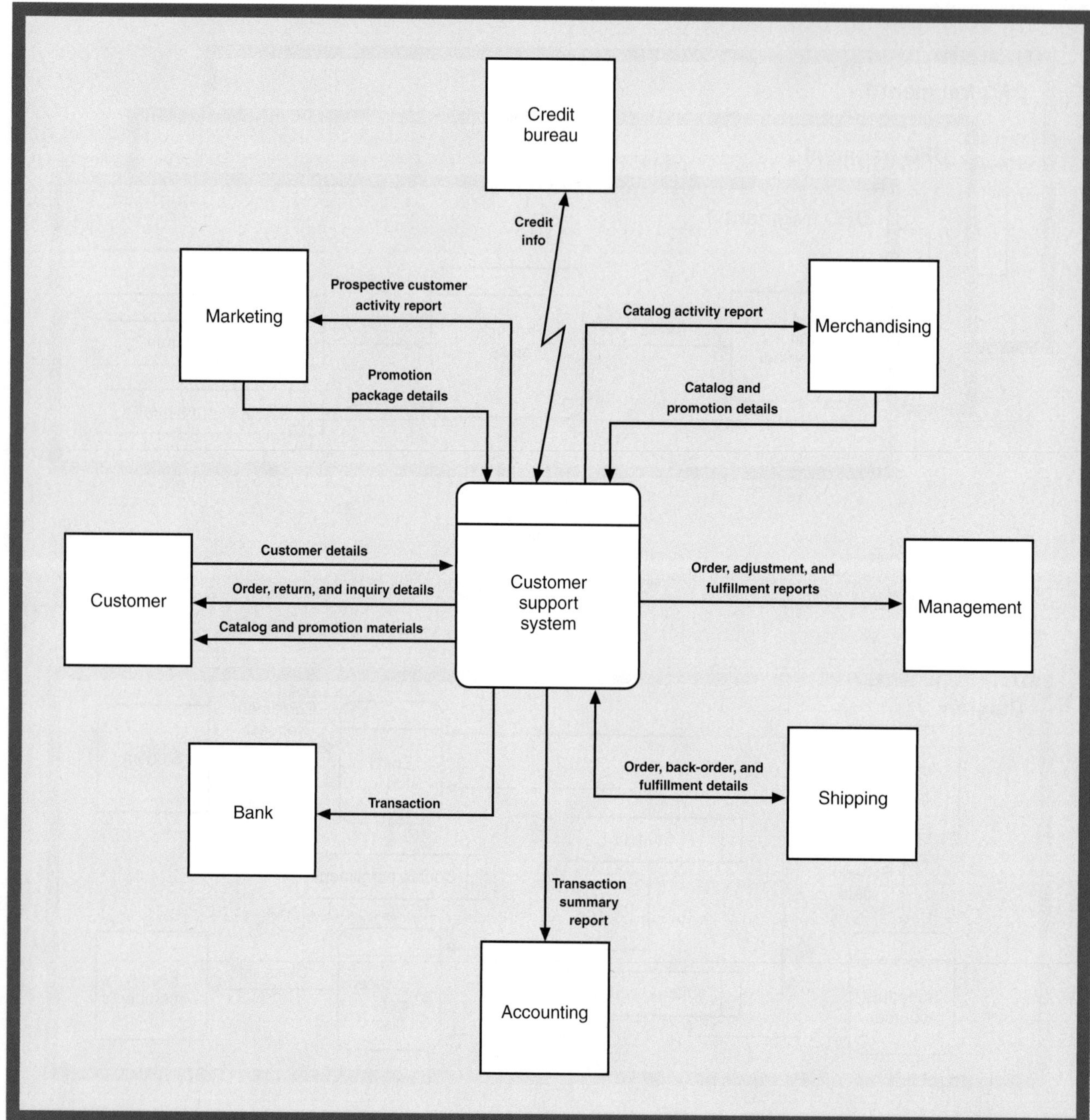

Figure 6-9

A context diagram for the RMO customer support system

system into subsystems based on use case similarities, including interactions with external agents, interactions with data stores, and similarities in required processing. Figure 6-11 shows the context diagram for the order-entry subsystem. Note that all data flows from the event table for this subsystem are shown on the DFD.

Figure 6-10

RMO subsystems and use cases for each subsystem

Order-entry subsystem

- Look up item availability
- Create new order
- Update order
- Produce order summary reports
- Produce transaction summary reports

Order fulfillment subsystem

- Look up order status
- Record order fulfillment
- Record back order
- Create order return
- Produce fulfillment summary report

Customer maintenance subsystem

- Provide catalog information
- Produce prospective customer activity reports
- Update customer account
- Distribute promotional package
- Create customer charge adjustment
- Produce customer adjustment reports

Catalog maintenance subsystem

- Update catalog
- Create special product promotion
- Create new catalog
- Produce catalog activity reports

Figure 6-11

A context diagram for the RMO order-entry subsystem

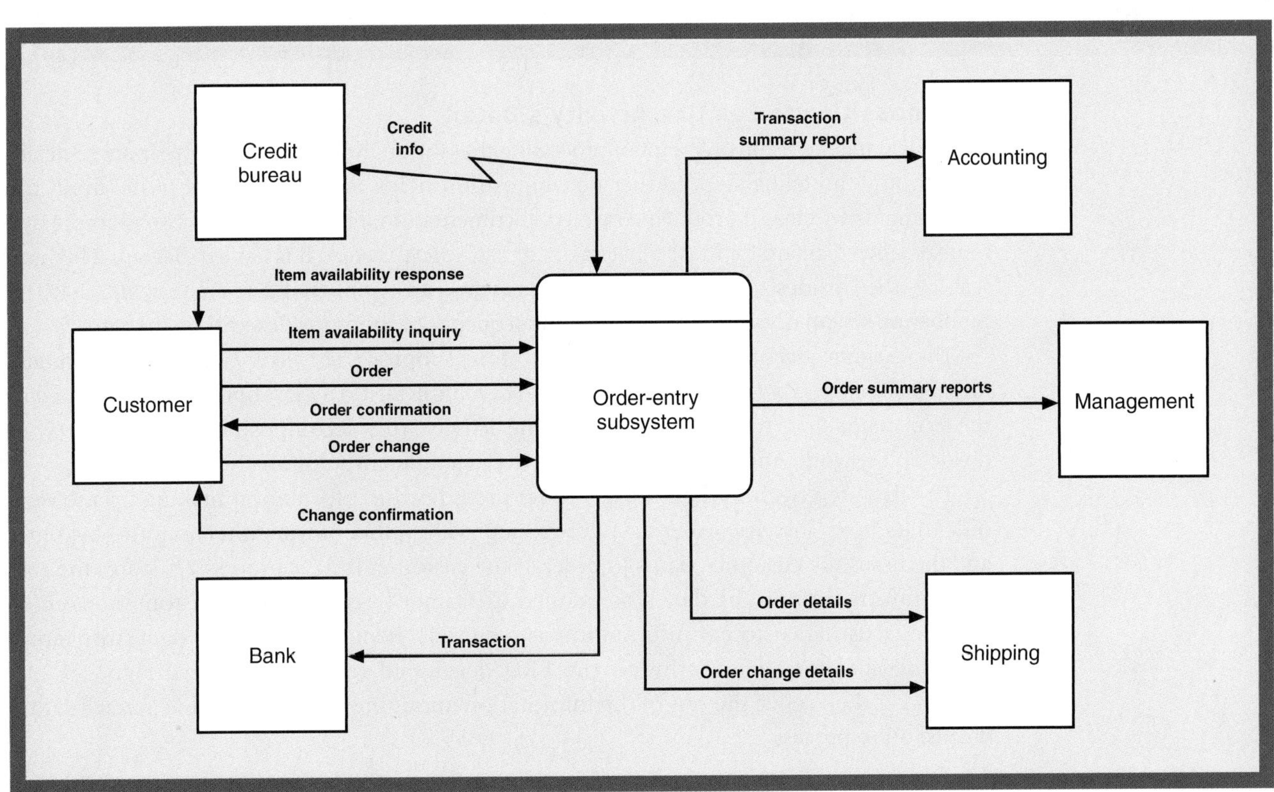

Figure 6-12 shows the DFD fragments for the RMO order-entry subsystem. Note that there are five DFD fragments, one for each order-entry subsystem use case listed in Figure 6-10.

Similarly, Figure 6-13 shows the RMO order-entry subsystem diagram 0, the result of combining the DFD fragments from Figure 6-12. To simplify the diagram and make it more readable, the seven data stores in Figure 6-12 are collapsed into a single data store in Figure 6-13. Recall that diagram 0 is just used as a presentation aid. The DFD fragments show which processes interact with which individual data stores.

Figure 6-12
DFD fragments for the RMO order-entry subsystem

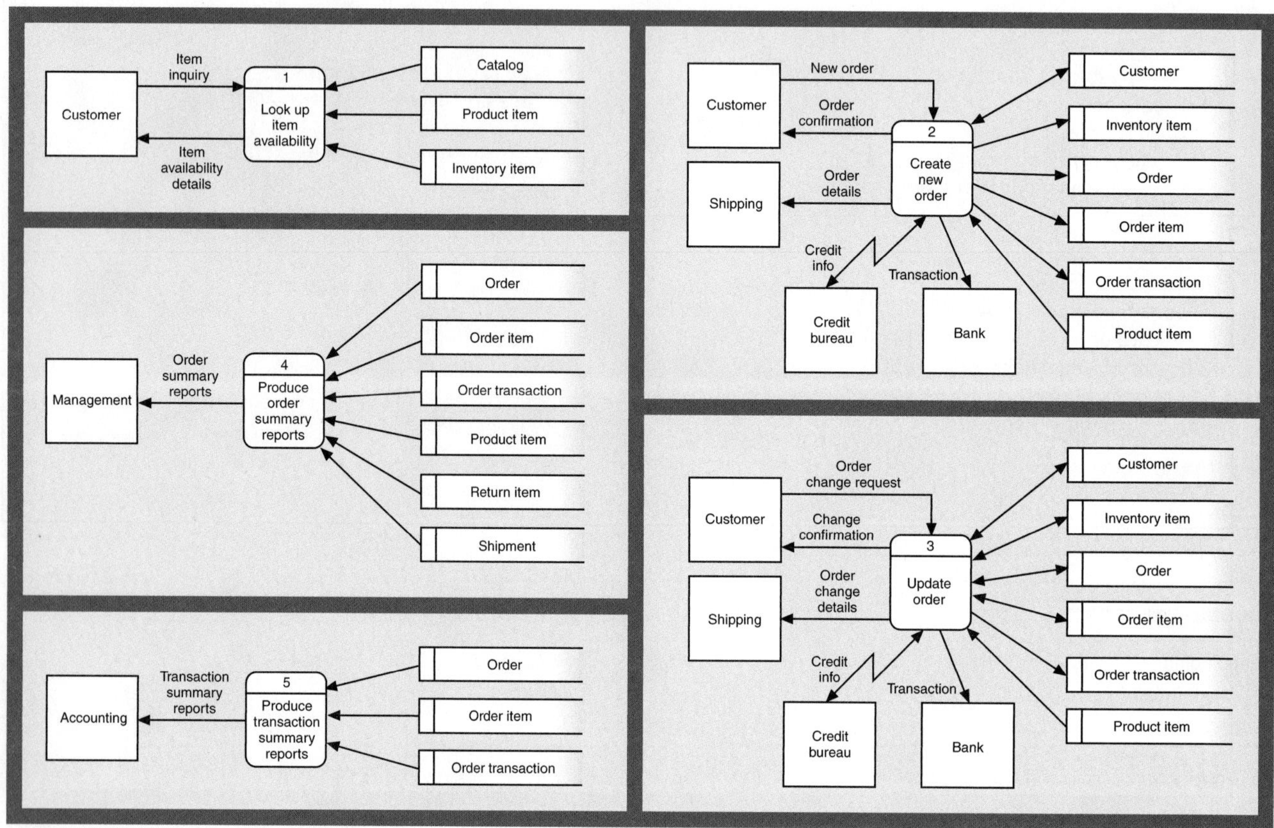

Decomposition to See One Activity's Detail

Some DFD fragments involve a lot of processing that the analyst needs to explore in more detail. As with any modeling step, further decomposition helps the analyst learn more about the requirements while also producing needed documentation. Figure 6-14 shows an example of a more detailed diagram for RMO DFD fragment 2, *Create new order*. It is named diagram 2 because it shows the "insides" of process 2. The subprocesses are numbered 2.1, 2.2, 2.3, and 2.4. The numbering system does not necessarily imply sequence of subprocess execution, though.

The diagram decomposes process 2 into four subprocesses: *Record customer information, Record order, Process order transaction,* and *Produce confirmation*. These subprocesses are viewed as the four major steps required to complete the activity. This decomposition is just one way to divide up the work. Another analyst might arrive at a different solution.

The first step begins when the customer provides the information making up the New order data flow. The New order data flow contains all of the information about the customer and the items the customer wants to order. If the customer is new, process 2.1 stores the customer information in the data store named Customer (creating a new customer record or updating existing customer information as required). Remember that the data store represents the customer data entity on the ERD developed in Chapter 5 (see Figure 5-29). Process 2.1 then sends the rest of the information about the order, a data flow named Order details, on to process 2.2.

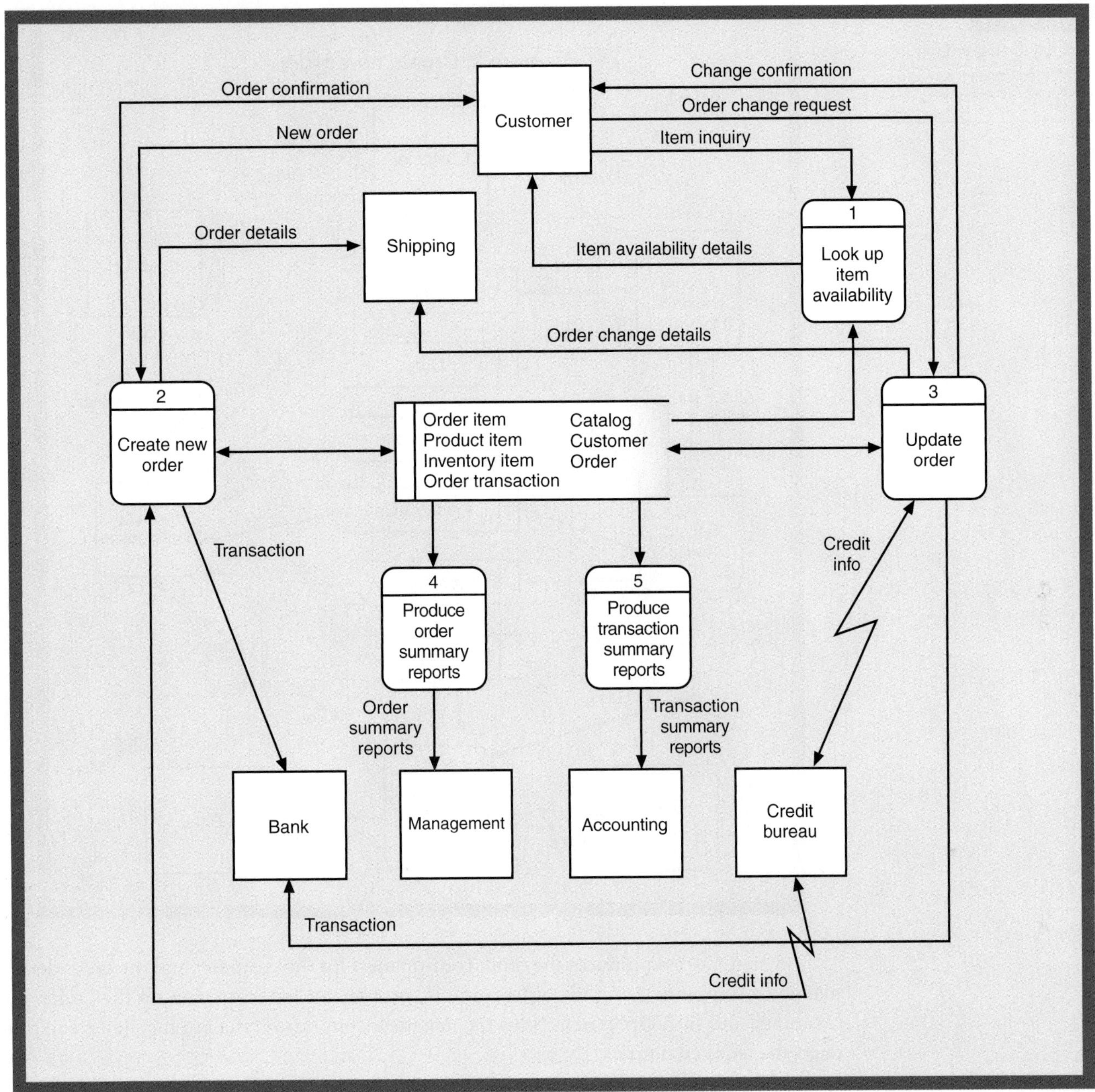

Figure 6-13

An event-partitioned model of the order-entry subsystem (diagram 0)

Process 2.2 takes the Order details data flow and creates a new order record by adding data to the Order data store. Then for each item ordered, the stock on hand and the current price are looked up in the Product item and Inventory item data stores. If adequate stock is on hand, an order item record is created for that item, and the stock on hand for the inventory item record is changed. If three items are ordered, one order record is created and three order item records are created.

Process 2.2 adds up the total amount due for the order (price times quantity for each item) and sends the data flow named Transaction details to process 2.3 to record the transaction. Transaction details include the order number, amount, and credit card information. Process 2.3 needs a real-time link to a credit bureau to get a credit authorization for the customer's credit card. This needs to be a real-time link rather than a data flow because data needs to flow back and forth rapidly while the process is executing. If the credit card is approved, a record of the transaction is created in the Order transaction data store, and a data flow for the transaction goes directly to the bank.

Figure 6-14

A detailed diagram for *Create new order* (diagram 2)

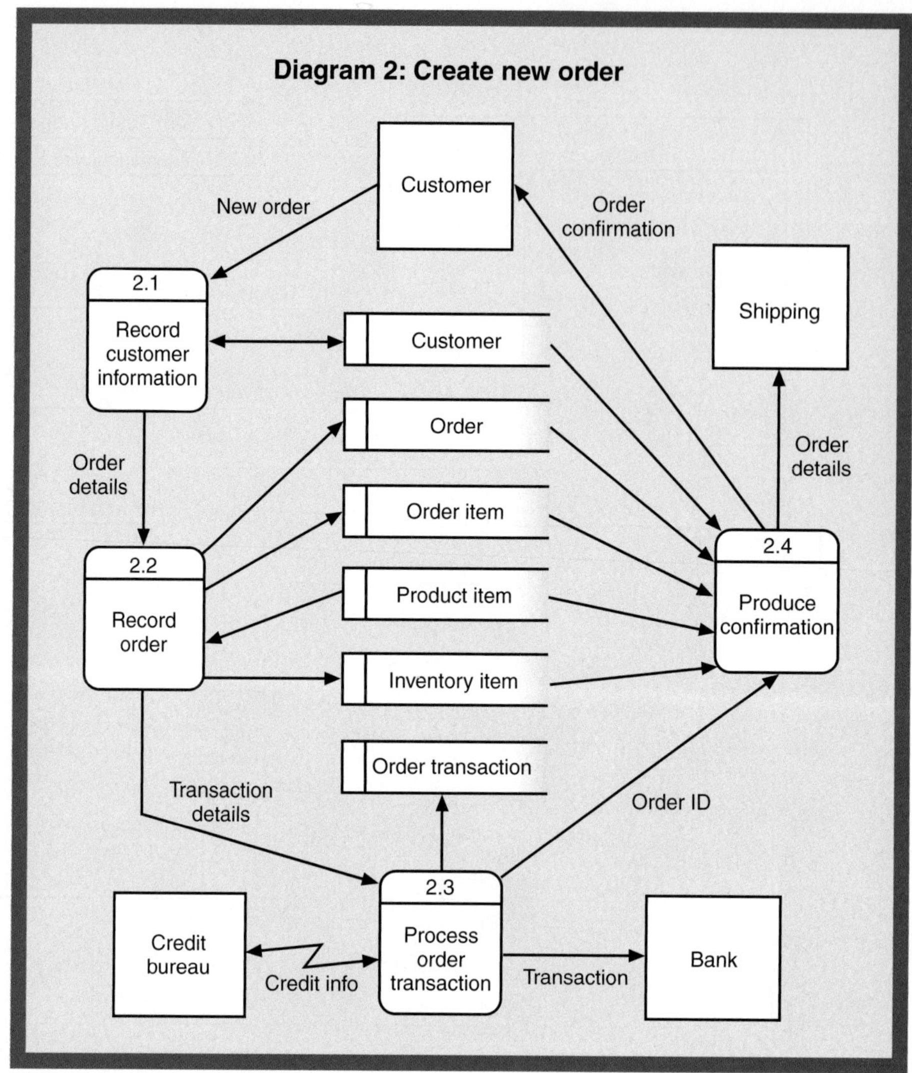

The final process produces the order confirmation for the customer and the order details that go to shipping. Using the order number, process 2.4 looks up data on the Order, the Customer, and each Order item (plus the item description from the Product item) and produces the required outputs.

PHYSICAL AND LOGICAL DFDS

A DFD can be a physical system model, a logical system model, or a blend of the two. If the DFD is a logical model, it assumes that the system might be implemented with any technology, as described in Chapter 4. If the DFD is a physical model, one or more assumptions about implementation technology will be embedded in the DFD. These assumptions can take many forms and might be very difficult to spot.

Consider whether diagram 2 in Figure 6-14 is a logical model. First, is it clear what type of computer is doing the processing? Could it be a desktop system? A centralized mainframe system? A networked client/server system? Or, could the entire process as just described be carried out by people without any computer at all? Similarly, are the data stores sequential computer files? Are they tables in a relational database? Or are they files of paper in a file cabinet? How does the system get the data flow New order from the customer so it can be processed? By clicking check boxes and list boxes in a Windows application? Or on a Web page? Or by manually filling out a form that a clerk types into the system? Or by talking to the clerk over the phone? Or by talking to a speech recognition program over the phone?

All of the alternatives described are possible, and if the model is a logical model, you should not be able to tell how the system is implemented. At the same time, the processing requirements (what must go on) should be fairly detailed, down to indicating what attribute values are needed. The model could be even more detailed and still be a logical model.

Now consider whether the DFD in Figure 6-15 is a physical system model by comparing it with diagram 1 in Figure 6-6. A number of elements indicate assumptions about implementation technology, including:

- Technology-specific processes
- Actor-specific process names
- Technology- or actor-specific process orders
- Redundant processes, data flows, and files

The most obvious technology assumption is embedded in the name of process 1.1. Making copies is an inherently manual task, which implies that the data store Old schedules and the data flows into and out of process 1.1 are paper. It is possible that the data store and flows are electronic, but if so, the question arises why a process would be needed to make electronic copies.

Figure 6-15

A physical DFD for scheduling courses (resist the temptation to create physical DFDs during analysis)

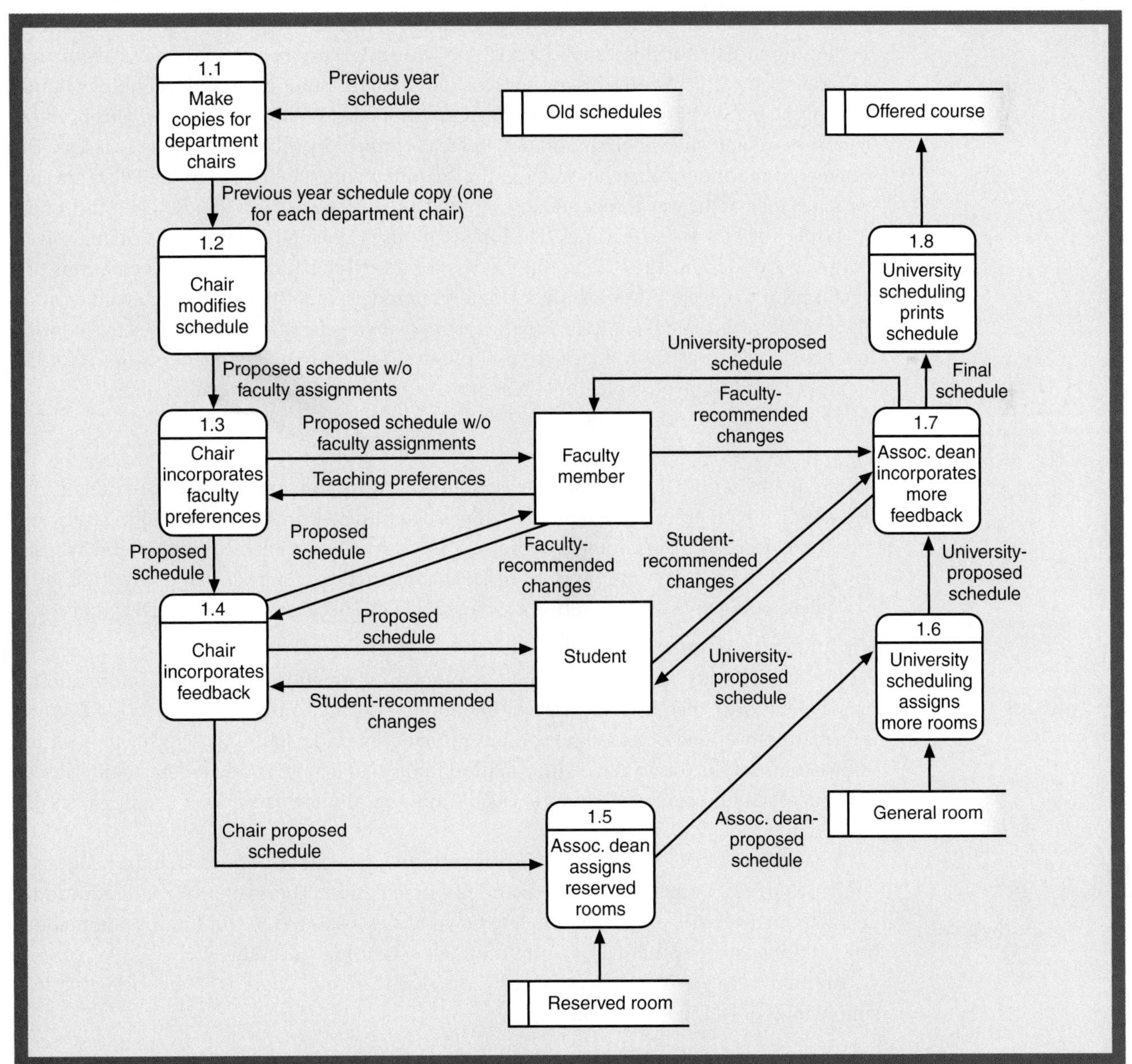

Many of the process names include actors in the system. References to Chair, Assoc. dean, and University scheduling all indicate that a particular individual or department performs a process. The sequential flow of data among the processes is a by-product of the person or department that carries out each process. One can imagine alternate implementations with fewer processes, different process orders, or different assignment of processes to individuals and departments. The DFD clearly models one very specific set of decisions about process ordering and responsibility.

The DFD also includes processes with similar or redundant processing logic. For example, faculty input is accepted early, but faculty members later perform error checking twice (the data flows from processes 1.4 and 1.7). Also, rooms are assigned at two different times from two different data stores (Reserved room for process 1.5, and General room for process 1.6). As before, these features indicate very specific assumptions about the technology and division of responsibility. The redundant error checking indicates that it is possible for previous processes to make mistakes. A system implemented with perfect technology needs no internal error checking. The partitioning of room assignment between two files and processes may be related to technology (for example, no one process could successfully assign all rooms at once), or it could indicate a historic division of responsibility for room assignment.

Inexperienced analysts often develop DFDs such as the one in Figure 6-15. The path to developing such a model is simple: Model everything the current system does exactly the way it does it. The problem with this approach is that design assumptions and technology limitations of the old system can become inadvertently embedded in the new system. This problem is most prevalent when analysis and design are performed by different persons or teams. The designer(s) may not realize that some of the "requirements" embedded in the DFDs are simply reflections of the way things are now, not the way they necessarily should be in the future.

Physical DFDs are sometimes developed and used during the last stages of analysis or early stages of design. They are useful models for describing alternate implementations of a system prior to developing more detailed design models. But analysts should avoid creating physical DFDs during all analysis activities, except when generating alternatives. Even during that activity, analysts should clearly label physical DFDs as such so readers know that the model represents one possible implementation of the logical system requirements.

EVALUATING DFD QUALITY

A high-quality set of DFDs is readable, is internally consistent, and accurately represents system requirements. Accuracy of representation is determined primarily by consulting users and other knowledgeable stakeholders. A project team can ensure readability and internal consistency by applying a few simple rules to DFD construction. Analysts can apply these rules while developing the DFDs or during a separate quality check after preparing DFD drafts.

Minimizing Complexity

information overload
difficulty in understanding that occurs when a reader receives too much information at one time

People have a limited ability to manipulate complex information. If too much information is presented at once, people experience a phenomenon called **information overload**. When information overload occurs, a person has difficulty in understanding. The key to avoiding information overload is to divide information into small and relatively independent subsets. Each subset should contain a comprehensible amount of information that people can examine and understand in isolation.

A layered set of DFDs is an example of dividing a large set of information into small, independent subsets. Each DFD can be examined in isolation. The reader can find additional detail about a specific process by moving down to the next level, or find information about how a DFD relates to other DFDs by examining the next-higher-level DFD.

An analyst can avoid information overload within any single DFD by following two simple rules of DFD construction:

- 7 ± 2
- Interface minimization

rule of 7 ± 2
the rule of model design that limits the number of model components or connections among components to no more than nine

The rule of 7 ± 2 (also known as Miller's Number) derives from psychology research, which shows that the number of information "chunks" that a person can remember and manipulate at one time varies between five and nine. A larger number of chunks causes information overload. Information chunks can be many things, including names, words in a list, digits, or components of a picture.

Some applications of the rule of 7 ± 2 to DFDs include the following:

- A single DFD should have no more than 7 ± 2 processes.
- No more than 7 ± 2 data flows should enter or leave a process, data store, or data element on a single DFD.

These rules are general guidelines, not unbreakable laws. DFDs that violate these rules may still be readable, but violations should be considered a warning of potential problems.

minimization of interfaces
a principle of model design that seeks simplicity by limiting the number of connections among model components

Minimization of interfaces is directly related to the rule of 7 ± 2. An interface is a connection to some other part of a problem or description. As with information chunks, the number of connections that a person can remember and manipulate is limited, so the number of connections should be kept to a minimum. Processes on a DFD represent chunks of business or processing logic. They are related to other processes, entities, and data stores by data flows. A single process with a large number of interfaces (data flows) may be too complex to understand. This complexity may show up directly on a process decomposition as a violation of the rule of 7 ± 2. An analyst can usually correct the problem by dividing the process into two or more subprocesses, each of which should have fewer interfaces.

Pairs or groups of processes with a large number of data flows between them are another violation of the interface minimization rule. Such a condition usually indicates a poor partitioning of processing tasks among the processes. The way to fix the problem is to reallocate the processing tasks so that fewer interfaces are required. The best division of work among processes is the simplest, and the simplest division is one that requires the fewest interfaces among processes.

Ensuring Data Flow Consistency

An analyst can often detect errors and omissions in a set of DFDs by looking for specific types of inconsistency. Three common and easily identifiable consistency errors are as follows:

- Differences in data flow content between a process and its process decomposition
- Data outflows without corresponding data inflows
- Data inflows without corresponding outflows

balancing
equivalence of data content between data flows entering and leaving a process and data flows entering and leaving a process decomposition DFD

A process decomposition shows the internal details of a higher-level process in a more detailed form. In most cases, the data content of flows to and from a process at one DFD level should be equivalent to the content of data flows to and from all processes in a decomposition. This equivalency is called balancing, and the higher-level DFD and the process decomposition DFD are said to be "in balance."

Note the use of the term *data content* in the previous paragraph. Data flow names can vary among DFD levels for a number of reasons, including decomposition of one combined data flow into several smaller flows. Thus, the analyst must be careful to look at the *components* of data flows, not just data flow names. For this reason, detailed analysis of balancing should not be undertaken until data flows have been fully defined.

Unbalanced DFDs may be acceptable when the imbalance is due to data flows that were ignored at the higher levels. For example, diagram 0 for a large system usually ignores details of error handling, such as when an item is ordered but is later determined to be out of stock and discontinued by its manufacturer. A process called *Fulfill order* on diagram 0 would not have any data flows associated with this condition. In the process decomposition of *Fulfill order*, the analyst might add a process and data flows to handle discontinued items.

Another type of DFD inconsistency can occur between the data inflows and outflows of a single process or data store. By definition, a process transforms data inflows into data outflows.

In a logical DFD, data should not be needlessly passed into a process. The following consistency rules can be derived from these facts:

- All data that flows into a process must flow out of the process or be used to generate data that flows out of the process.
- All data that flows out of a process must have flowed into the process or have been generated from data that flowed into the process.

black hole
a process or data store with a data input that is never used to produce a data output

Figure 6-16 shows an example that violates the first rule. Compare Figure 6-16 with the first DFD fragment in Figure 6-12, and note the difference in the data inflows to the process. Looking up item availability requires only information to identify the item and access to corresponding data stores. In Figure 6-16, excess data input (an entire order) flows into the process, and the process accesses more data stores than needed to generate the data outflow Item availability details. A process such as the one shown in Figure 6-16 is sometimes called a **black hole** because some or all of the data that enters never leaves.

Figure 6-16
A process with unnecessary data input—a black hole

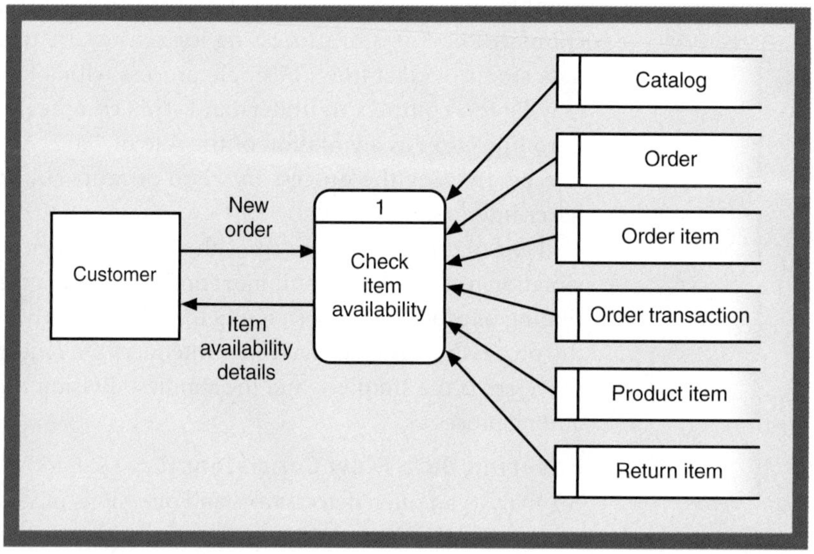

miracle
a process or data store with a data element that is created out of nothing

Figure 6-17 shows an example that violates the second rule. Compare Figure 6-17 with the bottom DFD fragment in Figure 6-7, and note the difference in the data inflows to the process. In Figure 6-17, insufficient data enters the process to produce the data output. Required data inputs from the Offered course and Course enrollment are missing. A process such as the one shown in Figure 6-17 is sometimes called a **miracle** because data emerges from the process without any apparent source.

Figure 6-17
A process with an impossible data output—a miracle

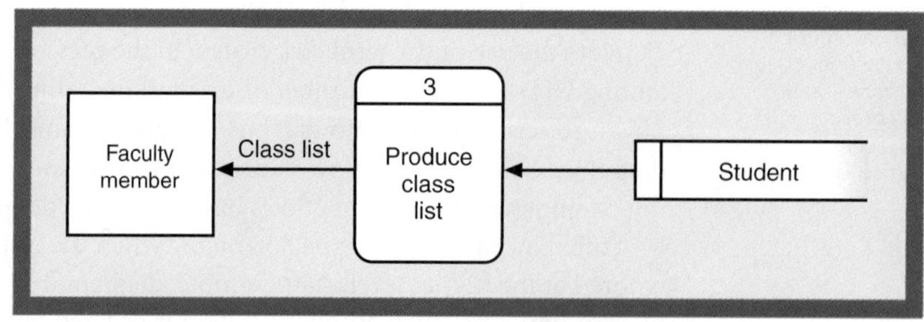

Analysts sometimes can spot black holes and miracles simply by examining the DFD. In other cases, close examination of the data dictionary or process descriptions is required. In Figure 6-18, data elements A, B, and C flow into the process but do not flow out. Data element A is used to determine what formula to apply to recompute the value of X, so that data

Figure 6-18

A process with unnecessary data input

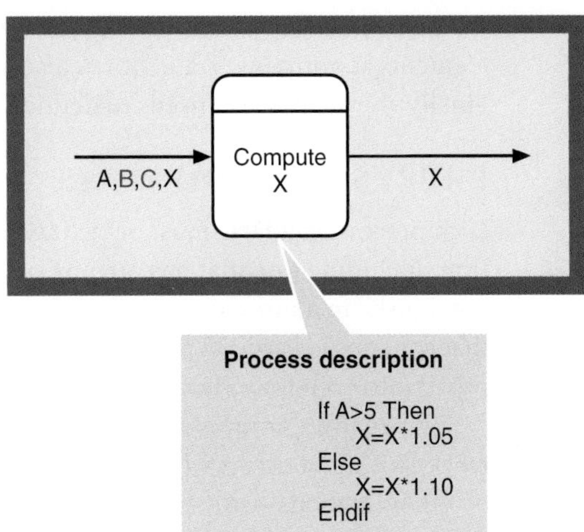

element is a necessary input. However, data elements B and C play no role in generating process output and thus should be eliminated as unnecessary inflows.

In Figure 6-19, data elements A, B, and Y flow out of the process. Data element A flows into the process. Data element Y is computed by an algorithm based on data element A. However, data element B does not flow into the process and is not computed by internal processing logic. Thus, data element B indicates either an error in the data outflow (B should be eliminated) or an omission in the internal processing logic (the rule that determines B is missing).

Note that both consistency rules apply to data stores as well as processes. Any data element that is read from a data store must have been previously written to that data store. Similarly, any data element that is written to a data store eventually must be read from the data store. Examining the consistency of data flows to and from a data store is complicated by the fact that a data element may flow into and out of a data store on completely different DFDs.

Figure 6-19

A process with an impossible data output

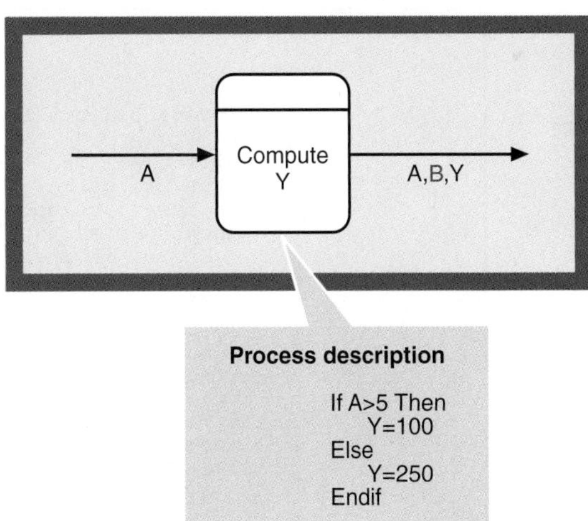

Evaluating data flow consistency is a straightforward but tedious process. Fortunately, most analysis modeling tools automatically perform data flow consistency checking. But those tools place rigorous requirements on the analyst to specify the internal logic of processes precisely. Without precise process descriptions, it is impossible for the tool (or a human being) to know what data elements are used as input or generated as output by internal processing logic.

DOCUMENTATION OF DFD COMPONENTS

In the traditional approach, data flow diagrams show all three types of internal system components—processes, data flows, and data stores—on one diagram, but additional details about each component need to be described. First, each lowest-level process needs to be described in detail. In addition, the analyst needs to define each data flow in terms of the data elements it contains. Data stores also need to be defined in terms of the data elements. Finally, the analyst also needs to define each data element.

PROCESS DESCRIPTIONS

Each process on a DFD must be defined formally. There are several options for process definition, including one that has already been discussed—process decomposition. As discussed previously, in a process decomposition, a higher-level process is formally defined by a DFD that contains lower-level processes. These lower-level processes may in turn be further decomposed into even lower-level DFDs.

Eventually a point is reached at which a process doesn't need to be defined further by a DFD. This point occurs when a process becomes so simple that it can be described adequately by other methods—structured English, decision tables, or decision trees. With each method, the process is described as an algorithm, and an analyst chooses the most appropriate presentation format by determining which is most compact, readable, and unambiguous. In most cases, structured English is the preferred method.

structured English
a method of writing process specifications that combines structured programming techniques with narrative English

Structured English uses brief statements to describe a process very carefully. Structured English looks a bit like programming statements, but without references to computer concepts. Rules of structured programming are followed, and indentation is used for clarity. For example, a simple set of instructions for processing ballots after a vote is shown in Figure 6-20. Some statements are simply instructions. Other statements repeat instructions. Still other statements direct the program to execute one set of instructions or the other. The procedure always starts at the top and ends at the bottom. Therefore, the rules of structured programming apply. Note, though, that a process described by structured English is not necessarily a computer program—it might be done by a person—so it is a logical model. It is unambiguous, so anyone following the instructions will arrive at the same result.

Figure 6-20
A structured English example

```
Process Ballots Procedure

Collect all ballots
Place all ballots in a stack
Set Yes count and No count to zero
Repeat for each ballot in the stack
   If Yes is checked then
      Add one to Yes count
   Else
      Add one to No count
   Endif
   Place ballot on counted ballot stack
Endrepeat
If Yes count is greater than No count then
   Declare Yes the winner
Else
   Declare No the winner
Endif
Store the counted ballot stack in a safe place
End Process Ballots Procedure
```

An example of a process description for Rocky Mountain Outfitters is shown in Figure 6-21. Note how the process description provides more specific details about what the process does. If one process description method becomes too complex, the analyst should choose another. Excess length (for example, more than 20 lines) or multiple levels of indentation (indicating complex decision logic) indicate that a structured English description may be too complex. An analyst can sometimes address excess indentation by converting the description to an equivalent decision table or decision tree. In other cases, a process decomposition may be required.

Figure 6-21

RMO process 2.1 *(Record customer information)* and its structured English process description

```
Process 2.1 - Record Customer Information

Ask if customer has an account (or has made a previous order)
If customer has an account then
   Ask for identification information
   Query database with identifying information
   Copy query response data to Order details
Else
   Create an empty Customer record in the database
   Ask customer for Customer attributes
   Update empty Customer record with Customer attributes
Endif
Ask customer for order information for first item
While more order items Do
   Update Order details with order information
Endwhile
```

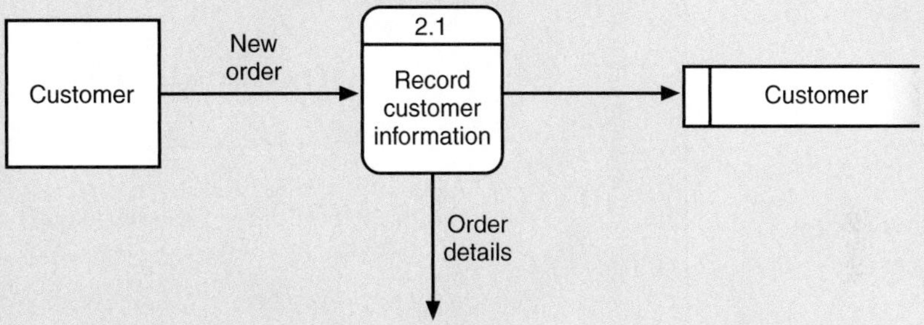

Structured English is well suited to describing processes with many sequential processing steps and relatively simple control logic (such as a single loop or an *if-then-else* statement). Structured English is not well suited for describing processes with the following characteristics:

- Complex decision logic
- Few (or no) sequential processing steps

Decision logic is complex when multiple decision variables and a large number of possible combinations of those variables need to be considered. When a process with complex decision logic is described with structured English, the result is typically a long and difficult-to-read description. For example, consider the structured English description for calculating shipping costs shown in Figure 6-22. Note that the description is relatively long and consists mostly of control structures (*if, else,* and *endif* statements).

decision table
a tabular representation of processing logic containing decision variables, decision variable values, and actions or formulas

decision tree
a graphical description of process logic that uses lines organized like branches of a tree

Decision tables and **decision trees** can summarize complex decision logic more concisely than structured English. Figures 6-23 and 6-24 show a decision table and decision tree that represent the same logic as the structured English example in Figure 6-22. Both incorporate decision logic into the structure of the table or tree to make the descriptions more readable than their structured English equivalent. The decision table is more compact, but the decision tree is easier to read. Sometimes an analyst needs to describe a process all three ways before deciding which approach describes a particular process best.

The following steps are used to construct a decision table:

1. Identify each decision variable and its allowable values (or value ranges).
2. Compute the number of decision variable combinations as the product of the number of values (or value ranges) of each decision variable.
3. Construct a table with one more column than the number of decision variable combinations computed in step 2 (the extra column is for decision variable names and process action or computation descriptions). The table should have a row for each decision variable and a row for each process action or computation.

Figure 6-22

A structured English process description for determining delivery charges

```
If YTD purchases > $250 then
    If number of items ordered < 4 then
        If delivery date is next day then
            delivery charge is $25
        Endif
        If delivery date is second day then
            delivery charge is $10
        Endif
        If delivery date is seventh day then
            delivery charge is $1.50 per item
        Endif
    Else
        If delivery date is next day then
            delivery charge is $6 per item
        Endif
        If delivery date is second day then
            delivery charge is $2.50 per item
        Endif
        If delivery date is seventh day then
            delivery charge is zero (free)
        Endif
    Endif
Else
    If number of items ordered < 4 then
        If delivery date is next day then
            delivery charge is $35
        Endif
        If delivery date is second day then
            delivery charge is $15
        Endif
        If delivery date is seventh day then
            delivery charge is $10
        Endif
    Else
        If delivery date is next day then
            delivery charge is $7.50 per item
        Endif
        If delivery date is second day then
            delivery charge is $3.50 per item
        Endif
        If delivery date is seventh day then
            delivery charge is $2.50 per item
        Endif
    Endif
Endif
```

YTD purchases > $250	YES						NO					
Number of Items (N)	N≤3			N≥4			N≤3			N≥4		
Delivery Day	Next	2nd	7th	Next	2nd	7th	Next	2nd	7th	Next	2nd	7th
Shipping Charge ($)	25	10	N*1.50	N*6.00	N*2.50	Free	35	15	10	N*7.50	N*3.50	N*2.50

Figure 6-23

A decision table for calculating shipping charges

4. Assign the decision variable with the fewest values (or value ranges) to the first row of the table. Put the decision variable name in the first column. Divide the remaining columns into sets of columns for each decision variable value (or value range).
5. Choose the next decision variable with the fewest values (or value ranges) for the second row. Put the variable's name in the first column. Compute the number of column groups as the product of the number of values (or value ranges) of this variable and all

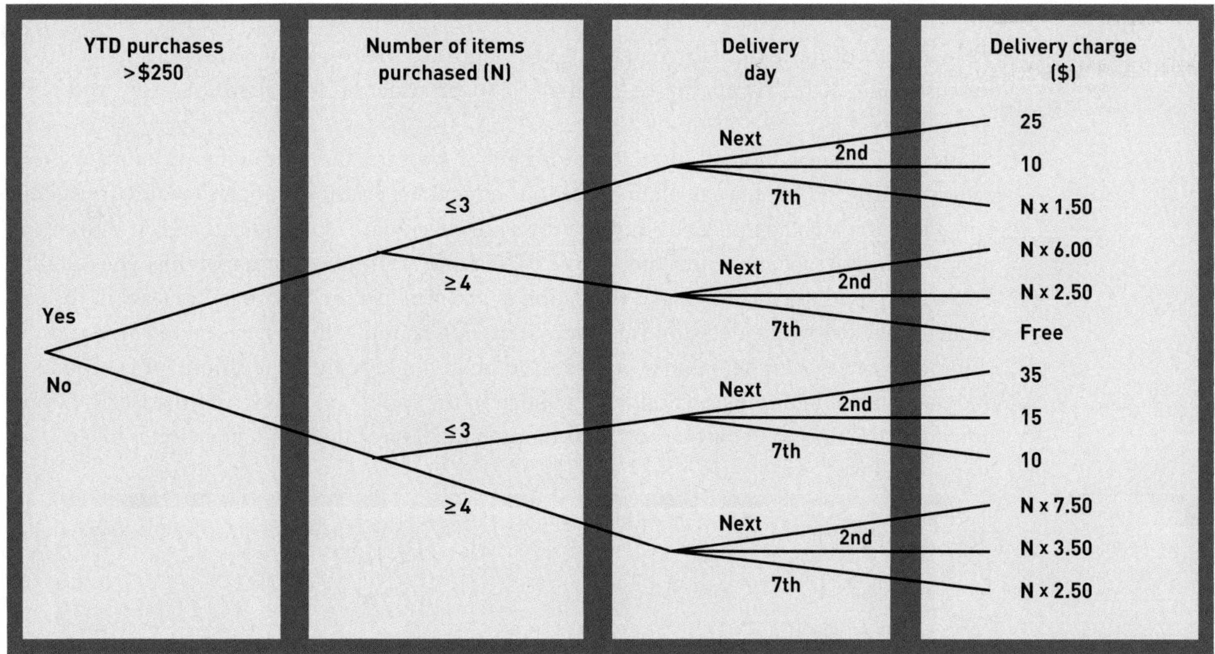

Figure 6-24

A decision tree for calculating shipping charges

the variables above it in the table. Divide the remaining columns into the computed number of groups, and insert values (or value ranges) in a regular pattern.

6. Continue inserting rows as instructed in step 5 until all decision variables have been included in the table.
7. Add a row for each calculation or action. For each calculation cell, insert the appropriate constant value or formula for the combination of decision variable values that appear above the cell in the same column. For each action cell, place a check mark in the cell if that action is performed when the decision variables have the values shown in the column above the cell.

Now let's follow these steps to show how the decision table in Figure 6-23 was constructed. There are three decision variables: year to date (YTD) purchases, number of items ordered, and delivery day. YTD purchases has two relevant ranges: less than \$250, and greater than or equal to \$250. Note that decision variable ranges must be mutually exclusive and collectively exhaustive. Number of items ordered also has two relevant ranges: less than or equal to three, and greater than or equal to four. Delivery day has three possible values: next day, second day, and seventh day. There are $2 \times 2 \times 3 = 12$ combinations of values, so there are 13 columns in the table to allow for a decision variable name, the formula, and the action names.

Both YTD purchases and number of items have two relevant value ranges, so either can occupy the first row. We chose YTD purchases. It has two value ranges, so we created two groups of $12 \div 2 = 6$ columns, and labeled one for each possible value. The next row is for number of items. It has two ranges, so we need four groups of three columns—that is, $12 \div 2$ value ranges for number of items $\div$ 2 value ranges for YTD purchases. We insert the value ranges for number of items into the column groups in a regular pattern, as shown in the sample figure. The delivery day is now inserted into the table. Because it is the last decision variable, we don't need to group any columns beneath it. We simply insert the values of the delivery date into individual columns in a regular pattern, as we did for the other decision variables.

The final step is to insert the row containing formulas and values for the shipping charges. Each cell contains a value or formula for the combination of decision variable values in the columns above. For example, shipping is free for customers with YTD purchases greater than $250, orders of more than three items, and seventh-day delivery. The shipping charge is $35 for customers with YTD purchases less than $250, an order of three items or fewer, and next-day delivery.

If the decision table is used to represent a process that implements one or more actions—instead of value calculations, as in the previous example—then the table must contain a row for each action. Cells in these rows are checkmarked to indicate which actions are performed under which conditions. Figure 6-25 shows a simple example of this type of table. Two action rows are included, and the action is performed if a check mark appears in the cell immediately below the decision variable values. For example, if the customer is new and the shipment contains an item back-ordered more than 25 days, then the shipment is expedited and the detailed return instructions are included in the container. If the customer isn't new and the order contains no items back-ordered more than 25 days, then neither action is taken.

Figure 6-25

A simple decision table with multiple action rows

New customer	Yes		No	
Item back order ≥ 25 days	Yes	No	Yes	No
Include detailed return instructions	✓	✓		
Expedite delivery	✓		✓	

You can construct a decision tree using almost the same steps as listed previously for constructing a decision table. The primary difference is that rows in a decision table are columns in a decision tree, and vice versa. To see this for yourself, draw an imaginary line through the table in Figure 6-23 from the top-left to bottom-right. Then flip the table along the imaginary line and compare the structure of the flipped table to the decision tree in Figure 6-24. The only other significant difference between a table and a tree is that a tree uses labeled branches instead of grouped columns to represent decision variable values.

BEST PRACTICE

Functional requirements documentation for the traditional approach includes (1) an ERD with attributes, (2) the set of DFDs (context DFD, DFD fragments, and any needed detailed DFDs), (3) process descriptions, and (4) data flow definitions, data store definitions, and data element definitions.

DATA FLOW DEFINITIONS

data flow definition
a textual description of a data flow's content and internal structure

A data flow is a collection of data elements, so **data flow definitions** list all the elements. For example, a simplified New order data flow (to process 2.1 in Figure 6-14) consists of a customer name, credit card number, and list of catalog item numbers and quantities. Some of these elements are actually structures of other elements, such as a customer name consisting of first name, middle initial, and last name. The system stores most of these data elements, so they coincide with the attributes of data entities included in the ERD.

Sometimes data flow definitions contain a more complex structure. In the New order example, each data flow consists of many catalog items and quantities (a repeating group). It is important to document this structure. The notations for data flow definitions vary. One approach is simply to list the data elements, as shown in Figure 6-26. The elements that can have many values are indicated. Another approach uses an algebraic notation such as that shown in Figure 6-27. The data flow "equals" or "consists of" one element plus another element, and so on. Groups of elements that can have many values are enclosed in curly braces. This example shows New order "equals" the customer name "plus" customer address "plus"

Figure 6-26

Data flow definitions simply listing elements

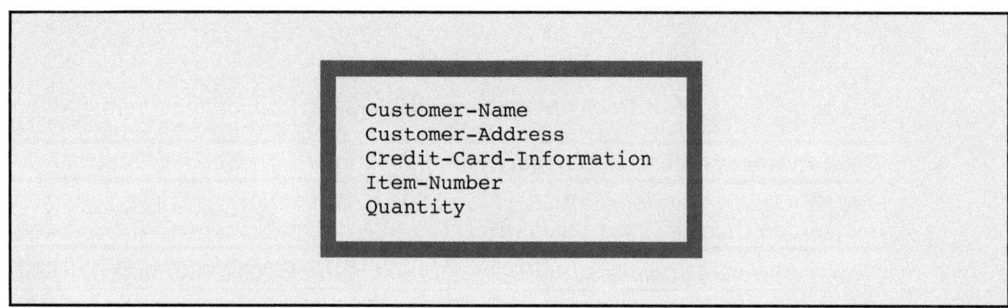

Figure 6-27

Algebraic notation for data flow definition (New-Order)

New-Order = Customer-Name + Customer-Address + Credit-Card-Information + $_{1}^{N}${ Item-Number + Quantity }

credit card information "plus" "one or more" inventory item number and quantity. In this example, the customer name can be defined separately as a structure of elements.

Figures 6-28 and 6-29 show a complex report and its corresponding data flow definition. The structure of the report is a repeating group of products with an embedded repeating group of inventory items. The data flow definition captures this structure by embedding the item repeating group within one set of curly braces and the product repeating group within the outermost set of curly braces.

DATA STORE DEFINITIONS

Because a data store on the DFD represents a data entity on the ERD, no separate definition is typically needed (except perhaps a note referring the reader to the ERD). If data stores are not linked to an ERD, the analyst simply defines the data store as a collection of elements (possibly with a structure) in the same way that data flows are defined.

DATA ELEMENT DEFINITIONS

Data element definitions describe a data type, such as string, integer, floating point, or Boolean. Each element should also be described to indicate specifically what it represents. Sometimes these descriptions are very specific. A date of sale might be defined as the date the payment for the order was received. Alternately, the date of sale might be the date an order is placed. Sometimes different departments in the same company have different definitions for the same element, so it is very important for the analyst to confirm exactly what the element means to users.

Other parts of a data element definition vary depending on the type of data. A length is usually defined for a string. For example, a middle initial might be one character maximum, but how long should a first name be? Numeric values usually have a minimum and maximum value that can be defined as a valid range. Sometimes specific values are allowed for the element, such as valid codes. If the element is a code, it is important to define the valid codes and their meaning. For example, code A might mean ship immediately, code B might mean hold for one day, and code C might mean hold shipment pending confirmation. Some sample data element definitions are shown in Figure 6-30.

Analysts need to maintain a central store of all these definitions as a project reference and to ensure consistency. A **data dictionary** is a repository for definitions of data flows, data stores, and data elements. A data dictionary may be a simple loose-leaf notebook or word-processing file in smaller development projects. In larger projects, a project management or documentation tool usually holds the data dictionary. The data dictionary may also hold process descriptions.

data dictionary

a repository for definitions of data flows, data elements, and data stores

Rocky Mountain Outfitters — Products and Items

ID	Name	Season	Category	Supplier	Unit Price	Special	Special Price	Discontinued
RM0125	*Outdoor Field*	*Spr/Fall*	*Mens C*	*8201*	*$39.00*	▦	*$0.00*	*No*

Description *Outdoor Nylon Jacket with Lining*

Size	Color	Style	Units in Stock	Reorder Level	Units on Order
Large	Blue		1500	150	
Large	Green		1500	150	
Large	Red		1500	150	
Large	Yellow		1500	150	
Medium	Blue		1500	150	
Medium	Green		1500	150	
Medium	Red		1500	150	
Medium	Yellow		1500	150	
Small	Blue		1500	150	
Small	Green		1500	150	
Small	Red		1500	150	
Small	Yellow		1500	150	
Xlarge	Blue		1500	150	
Xlarge	Green		1500	150	
Xlarge	Red		1500	150	
Xlarge	Yellow		1500	150	

ID	Name	Season	Category	Supplier	Unit Price	Special	Special Price	Discontinued
RM0125	*Hiking Walkers*	*All*	*Footwear*	*7993*	*$49.95*	▦	*$0.00*	*No*

Description *Hiking Walkers with Patterned Tread Durable Uppers*

Size	Color	Style	Units in Stock	Reorder Level	Units on Order
10	Brown		1000	100	
10	Tan		1000	100	
11	Brown		1000	100	
11	Tan		1000	100	
12	Brown		1000	100	
12	Tan		1000	100	
13	Brown		1000	100	
13	Tan		1000	100	
7	Brown		1000	100	
7	Tan		1000	100	
8	Brown		1000	100	
8	Tan		1000	100	
9	Brown		1000	100	
9	Tan		1000	100	

Figure 6-28

A sample report produced by the RMO customer support system

DFD SUMMARY

Figure 6-31 shows each of the components of a traditional analysis model—an entity-relationship diagram, data flow diagrams, process definitions, and data definitions. The four components form an interlocking set of specifications for most system requirements. The data flow diagram provides the highest-level view of the system, summarizing processes, external agents, data stores, and the flow of data among them. Each of the other components describes some aspect of the data flow diagram in greater detail.

The models described thus far were developed in the 1970s and 1980s as part of the traditional structured analysis methodology (see Yourdon 1989 in the "Further Resources" section). They were designed to document completely the logical requirements of a system.

Figure 6-29

A data flow definition for the RMO products and items report

```
products-and-items-report =

 N{ product-id + product-name + season + category +
 1  supplier + unit-price + special + special-price +
    discontinued + description +
     N{ size + color + style + units-in-stock +
     1  reorder-level + units-on-order
      }
  }
```

Figure 6-30

Data element definitions

```
units-in-stock =
  a positive integer

supplier =
  a four digit numeric code

unit-price =
  a positive real number accurate to two decimal places,
  always in U.S. dollars

description =
  a text field containing a maximum of 50 printable characters

special =
  a coded field with one of the following values
  0: item is not "on special"
  1: item is "on special"
```

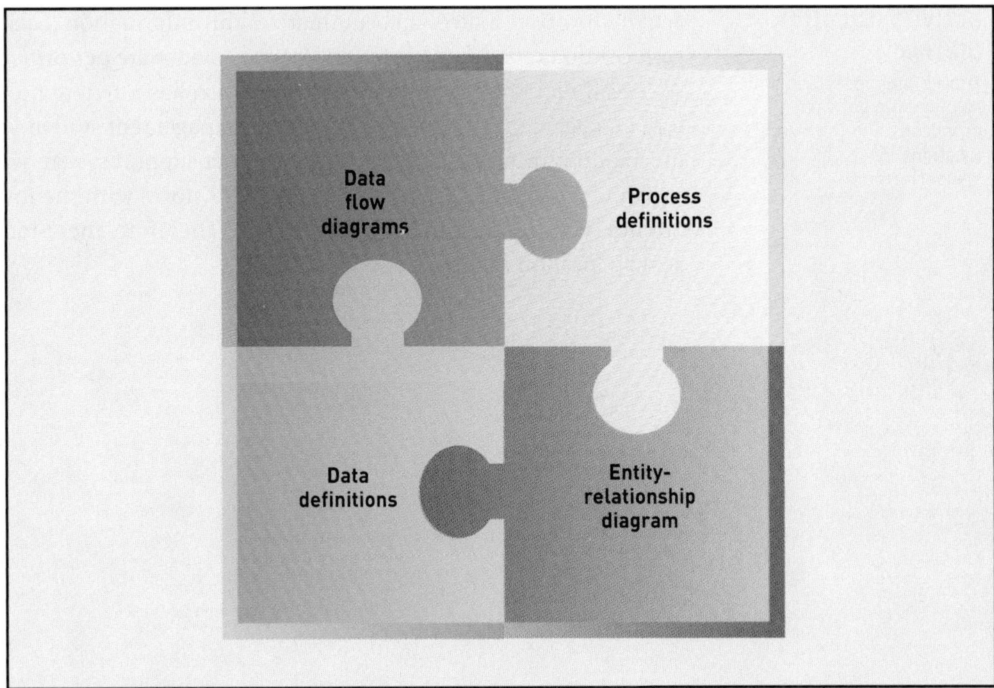

Figure 6-31

The components of a traditional systems analysis model

However, some analysts choose to augment the structured models with models borrowed from other methodologies. Such models may be used to describe information not captured by the structured models. Or they may be used to present similar information in a slightly different form. The remainder of this chapter describes some of these "borrowed" models and ways they can be used to augment the traditional structured analysis models.

LOCATIONS AND COMMUNICATION THROUGH NETWORKS

Because structured systems analysis concentrates on logical modeling, physical issues such as processing locations and networks are sometimes ignored during analysis. However, a great deal of information about process, data, and user distribution is needed during the early stages of design. Examples include the following:

- Number of locations of users
- Processing and data access requirements of users at specific locations
- Volume and timing of processing and data access requests

Gathering location information during analysis enables analysts to make better decisions during the last two analysis activities—*Generate and evaluate alternatives* and *Review recommendations with management.* Location information is also useful in many design activities, including *Design and integrate the network, Design the application architecture,* and *Design and integrate the database.*

The first step in gathering location information is to identify and describe the locations where work is being done or where it will be performed. Possible locations include business offices, warehouses, and manufacturing facilities, and less obvious locations such as customer or supplier offices, employee homes, hotel rooms, and automobiles. All of these locations should be listed, and a **location diagram** should be drawn to summarize the locations graphically. A location diagram for Rocky Mountain Outfitters is shown in Figure 6-32. The location diagram shows the analyst what network connections might be required, but it also has the added benefit of reminding everyone that users at all locations should be consulted about the system.

location diagram
a diagram or map that identifies all of the processing locations of a system

The next step is to list the functions that are performed by users at each location. Using the event table, the analyst can list where each activity is performed. Figure 6-33 shows an **activity-location matrix** that summarizes this information. Each row is a system activity, and each column represents a location. Many activities are performed at multiple locations.

activity-location matrix
a table that describes the relationship between processes and the locations in which they are performed

Recall that Rocky Mountain Outfitters also has a system project under way for the inventory management system. The inventory management system will involve many activities at the manufacturing facilities, but the customer support system will not. In addition, RMO has a plan for integrating the system at the retail stores with the inventory management system, but not with the customer support system. Therefore, these locations are not shown on the activity-location matrix.

Figure 6-32
The Rocky Mountain Outfitters location diagram

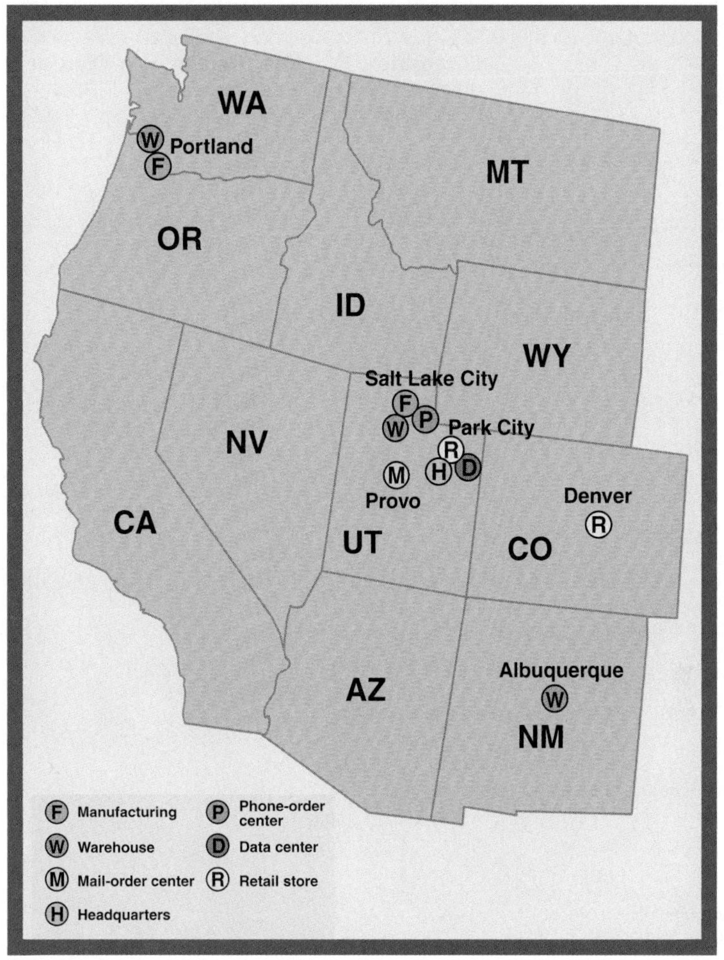

activity-data matrix
a table that describes stored data entities, the locations from which they are accessed, and the nature of the accesses

CRUD
acronym of create, read, update, and delete

Other matrices can be created to highlight access requirements. One approach is to list activities and data entities (or classes of objects) in an **activity-data matrix**. This matrix shows which activities require access to the data or objects. This information can be found on the DFD fragments for the traditional approach and on the sequence diagrams for the OO approach. In either approach, creating a matrix to summarize this information can be useful.

Figure 6-34 shows an activity-data matrix for Rocky Mountain Outfitters. The cells of the matrix show additional information to clarify what the activity does to the data. The letter *C* means the activity creates new data, *R* means the activity reads data, *U* means the activity updates data, and *D* means the activity might delete data. The acronym **CRUD** (create, read, update, and delete) is often used to describe this type of matrix.

ACTIVITY	LOCATION				
	Corporate offices (Park City)	**Distribution warehouses** (Salt Lake City, Albuquerque, Portland)	**Mail-order** (Provo)	**Phone sales** (Salt Lake City)	**Customer direct interaction** (Anticipated)
Look up item availability	X	X	X	X	X
Create new order			X	X	X
Update order				X	X
Look up order status	X	X		X	X
Record order fulfillment		X			
Record back order		X			
Create order return		X			
Provide catalog info			X	X	X
Update customer account	X		X	X	X
Distribute promotional package	X				
Create customer charge adjustment	X				
Update catalog	X				
Create special product promotion	X				
Create new catalog	X				

Figure 6-33

Activity-location matrix for the Rocky Mountain Outfitters customer support system

Figure 6-34

Rocky Mountain Outfitters activity-data matrix

ACTIVITIES	DATA ENTITIES										
	Catalog	Customer	Inventory item	Order	Order item	Order transaction	Package	Product item	Return item	Shipment	Shipper
Look up item availability			R								
Create new order		CRU	RU	C	C	C	R	R		C	R
Update order		RU	RU	RUD	RUD	RUD	R	R		CRUD	R
Look up order status		R		R	R	R				R	R
Record order fulfillment					RU					RU	
Record back order					RU					CRU	
Create order return		CRU		RU		C			C		
Provide catalog info	R		R				R	R			
Update customer account		CRUD									
Distribute promotional package	R	R	R				R	R			
Create customer charge adjustment		RU				CRUD					
Update catalog	RU		R				RU	R			
Create special product promotion	R		R				R	R			
Create new catalog	C		R				CRU	R			

C = Creates new data, R = Reads existing data, U = Updates existing data, D = Deletes existing data

SUMMARY

Data flow diagrams (DFDs) are used in combination with the event table and entity-relationship diagram (ERD) to model system requirements. DFDs model a system as a set of processes, data flows, external agents, and data stores. DFDs are relatively easy to read because they graphically represent key features of the system using a small set of symbols. Because there are many features to be represented, many types of DFDs are developed, including context diagrams, DFD fragments, subsystem DFDs, event-partitioned DFDs, and process decomposition DFDs.

Each process, data flow, and data store requires a detailed definition. Analysts may define processes in a number of ways, including a structured English process specification, a decision table, a decision tree, or a process decomposition DFD. Process decomposition DFDs are used when internal process complexity is too great to allow the creation of a readable, one-page definition by any other means. Data flows are defined in terms of their component data elements and their internal structure. Data elements may be further defined in terms of their type and allowable content. Data stores correspond to entities on the ERD, and thus require no additional definition.

The location diagram, activity-location matrix, and activity-data matrix describe important information about system locations. The location diagram summarizes geographic locations where the system is to be used. The activity-location matrix describes which processes are implemented at which locations. The activity-data matrix summarizes where and how each data store is used.

We've now covered all of the models that are used to document system requirements in the traditional approach to systems analysis. Chapter 7 covers the models used to document system requirements in the OO approach to systems analysis. Chapter 8 covers the transition from systems analysis to systems design.

KEY TERMS

activity-data matrix, p. 231
activity-location matrix, p. 230
balancing, p. 219
black hole, p. 220
context diagram, p. 208
CRUD, p. 231
data dictionary, p. 227
data flow, p. 206
data flow definition, p. 226
data flow diagram (DFD), p. 206
data store, p. 207
decision table, p. 223
decision tree, p. 223
DFD fragment, p. 210
event-partitioned system model, or diagram 0, p. 210
external agent, p. 206
information overload, p. 218
level of abstraction, p. 208
location diagram, p. 230
minimization of interfaces, p. 219
miracle, p. 220
process, p. 206
rule of 7 ± 2, p. 219
structured English, p. 222

REVIEW QUESTIONS

1. List at least three different types of DFDs. What is each diagram type used to represent?
2. List the five component parts (symbols) of a DFD. Briefly describe what each symbol represents.
3. How does an analyst determine whether a person or organization should be represented on a DFD as an external agent or by one or more processes?
4. Processes on an event-partitioned DFD can be described by a detailed DFD or a process specification. How does an analyst determine which is the most appropriate form of description?
5. Describe how each column of an event table is represented on a DFD (that is, what symbols are used?).
6. How are entities from the ERD represented on a DFD? How are relationships from the ERD represented on a DFD?
7. What features may be present on a physical DFD that should never be present on a logical DFD?
8. What DFD characteristics does an analyst examine when evaluating DFD quality?
9. What is a black hole? What is a miracle? How can each be detected?
10. Why might an analyst describe a process with a decision table or tree instead of structured English?
11. What is an activity-location matrix? How is it related to DFDs?
12. What is an activity-data matrix? How is it related to DFDs and the ERD?

THINKING CRITICALLY

1. Assume that you are preparing a DFD to describe the process of creating, approving, and closing a mortgage loan by a mortgage broker. Should the broker be represented as an external agent or by one or more processes? Why? What about the closing agent, the credit bureau, and the bank that issues the mortgage note?
2. Examine the course registration system described in Figure 6-6. Are there any other processes that would be required to implement a fully functioning system? Hint: Black holes and miracles may indicate processing steps that were left out of the DFD.
3. Assume that the transaction summary report for the RMO order-entry subsystem (see process 5 in Figure 6-12) contains a listing of every order that was created during a date range entered by the user. The report title page contains the report name, the date range, and the date and time the report was prepared. For each order, the report lists the order number, order date, order total, and form of payment. Within each order, the report lists all order items and returns, including item number, quantity ordered (or returned), and price. Report totals include the sum of all order totals, average order total, average item price, and average return price. Write a data flow definition entry for the report, and write a process specification for the process that produces the report.
4. Create an activity-data (CRUD) matrix for the course registration system in Figure 6-6.

EXPERIENTIAL EXERCISES

1. Develop a physical DFD that models the process of grocery shopping, from the time you write down a shopping list until the time you store purchased groceries in your home. Construct your DFD as a linear sequence of processes. Now develop a logical DFD to describe the same scenario. Try to develop a diagram that is equally valid as a logical description of the way you currently buy groceries and as a logical description of ways you might buy groceries without ever leaving your home.
2. Consider the admissions requirements for a degree program, major, or concentration at your school. Look up the requirements in the school catalog and rewrite them in structured English. Develop an equivalent decision table and/or decision tree. Which is easier to understand? Why?
3. Get a copy of your school transcript. Write a data definition that describes its contents. Write data element definitions for the fields Grade, Credits, and Degree.
4. Define process 2 in Figure 6-7 as it is implemented at your school. Use whatever combination of process decomposition and process specification is appropriate. If you develop any process decomposition DFDs, be sure to define all data flows.

CASE STUDIES

THE REAL ESTATE MULTIPLE LISTING SERVICE SYSTEM

Refer to the description of the Real Estate Multiple Listing Service system in the Chapter 5 case studies. Use the event list and ERD for that system as a starting point for the following exercises:

1. Draw a context DFD.
2. Draw an event-partitioned DFD.
3. Draw any required process decomposition DFDs.

STATE PATROL TICKET PROCESSING SYSTEM

Refer to the description of the State Patrol ticket processing system in the Chapter 5 case studies. Use the event list and ERD for that system as a starting point for the following exercises:

1. Draw a context DFD.
2. Draw an event-partitioned DFD.
3. Draw any required process decomposition DFDs.
4. Create data flow definitions for any data flows that are fully described in the written system description.

RETHINKING ROCKY MOUNTAIN OUTFITTERS

This chapter contains many DFDs describing the RMO order-entry subsystem but no DFDs describing the RMO order fulfillment subsystem, customer maintenance subsystem, or catalog maintenance subsystem (see the subsystem event lists in Figure 6-10). Review the RMO event table (Figure 5-12) and ERD (Figure 5-29) and perform the following tasks:

1. Develop DFD fragments for all of the events not documented in Figure 6-12.
2. Develop a single DFD that shows processing for all events, using one process for each subsystem and showing all

needed data stores. To simplify the diagram, place all external agents along the outer edge, and duplicate them as necessary to minimize long or crossing data flows. Place all data stores in the middle of the diagram.

3. Develop a data flow definition for the RMO customer order form in Figure 6-35.

FOCUSING ON RELIABLE PHARMACEUTICAL SERVICE

Continue your modeling efforts for the Reliable Pharmaceutical Service case by performing the following tasks:

1. Create a context diagram for the Reliable Pharmaceutical case based on the system description in Chapter 1 and the event table that you developed in Chapter 5.
2. Create DFD fragments for each event from the event table and ERD that you developed in Chapter 5.
3. Create an event-partitioned model (diagram 0) by combining the DFD fragments you created for question 2.
4. Create a logical DFD showing the processing details for the event *Time to generate orders (shipments)* based on the description in Chapter 1. Pay careful attention to modeling data movement and processing, not the movement and processing of physical goods (for example, drugs). Create any process descriptions and data definitions needed to fully specify system requirements.
5. Consider the problem of modeling the billing procedures briefly described in Chapter 1. Should a physical DFD of billing procedures be developed? Why, or why not?

Rocky Mountain Outfitters—Customer Order Form

ROCKY MOUNTAIN OUTFITTERS

Name and address of person placing order.
(Please verify your mailing address and make correction below.)
Order date __/__/__

Name
Address Apt. no.
City State Zip

Phone: Day ( ) ______ Evening ( ) ______

Gift Order or Ship To: (Use only if different from address at left.)

Name
Address Apt. No
City State Zip

Gift ☐ Address for this Shipment Only ☐ Permanent Change of Address ☐

Gift Card Message

Delivery Phone ( ) ______

	Item no.	Description	Style	Color	Size	Sleeve Length	Qty	Monogram	Style	Price Each	Total

MERCHANDISE TOTAL ______

Total Monogramming charges ($5.00 per line per item) ______

Sales tax on merchandise delivered in Colorado and Utah ______

Regular FedEx shipping $4.50 per U.S. delivery address (Items are sent within 24 hours for delivery in 2 to 4 days) $4.50

Please add $4.50 per each additional U.S. delivery address ______

FedEx Standard Overnight Service
Add $6.00 for delivery in 2 business days to confirm U.S. address* ______

Any additional freight charges ______

International Shipping (see shipping information on back) ______

Method of Payment

Check/Money Order ☐ Gift Certificate(s) ☐ AMOUNT ENCLOSED $ ______

American Express ☐ MasterCard ☐ VISA ☐ Other ☐

Account Number

MO YR
__/__
Expiration Date

Signature ______

Figure 6-35
RMO catalog order form

FURTHER RESOURCES

J. Martin, *Information Engineering: Book I Introduction*. Prentice Hall, 1988.

J. Martin, *Information Engineering: Book II Planning and Analysis*. Prentice Hall, 1989.

Stephen M. McMenamin and John F. Palmer, *Essential Systems Analysis*. Yourdon Press, 1984.

G. A. Miller, "The magical number seven, plus or minus two: Some limits on our capacity for processing information." *Psychological Review*, volume 63 (1956), pp. 81–97.

Edward Yourdon, *Modern Structured Analysis*. Yourdon Press, 1989.

CHAPTER

7

THE OBJECT-ORIENTED APPROACH TO REQUIREMENTS

LEARNING OBJECTIVES

After reading this chapter, you should be able to:

- Understand the models and processes of defining object-oriented requirements
- Develop use case diagrams and activity diagrams
- Develop system sequence diagrams
- Develop state machine diagrams to model object behavior
- Explain how use case descriptions and UML diagrams work together to define functional requirements for the object-oriented approach

CHAPTER OUTLINE

Object-Oriented Requirements

The System Activities—A Use Case/Scenario View

Identifying Inputs and Outputs—The System Sequence Diagram

Identifying Object Behavior—The State Machine Diagram

Integrating Object-Oriented Models

ELECTRONICS UNLIMITED, INC.: INTEGRATING THE SUPPLY CHAIN

Electronics Unlimited is a warehousing distributor that buys electronic equipment from various suppliers and sells it to retailers throughout the United States and Canada. It has operations and warehouses in Los Angeles, Houston, Baltimore, Atlanta, New York, Denver, and Minneapolis. Its customers range from large nationwide retailers, such as Target, to medium-sized independent electronics stores.

Many of the larger retailers are moving toward integrated supply chains. Information systems used to be focused on processing internal data; however, today these retail chains want suppliers to become part of a totally integrated supply chain system. In other words, the systems need to communicate between companies to make the supply chain more efficient.

To maintain its position as a leading wholesale distributor, Electronics Unlimited has to convert its system to link both with its suppliers (the manufacturers of the electronic equipment) and its customers (the retailers). It is developing a completely new system that uses object-oriented techniques to provide these links. Object-oriented techniques facilitate system-to-system interfaces by using predefined components and objects to accelerate the development process. Fortunately, many of the system development staff have recently begun learning about object-oriented development and are eager to apply the techniques and models to a system development project.

William Jones is explaining object-oriented development to the group of systems analysts who are being trained in this approach. "We're developing most of our new systems using object-oriented principles. The complexity of the new system, along with its interactivity, makes the object-oriented approach a natural way to develop requirements. It takes a little different thought process than you may be used to, but the object-oriented models track very closely with the new object-oriented programming languages."

William continued, "This way of thinking about a system in terms of objects is very interesting. It also is consistent with the object-oriented programming techniques you learned in your programming classes. You probably first learned to think about objects when you developed screens for the user interface. All of the controls on the screen, such as buttons, text boxes, and drop-down boxes, are objects. Each has its own set of trigger events that activate its program functions.

"Now you just extend that same thought process so that you think of things like purchase orders and employees as objects, too. We can call them *problem domain* or sometimes *business objects* to differentiate them from screen objects such as windows and buttons. During analysis, we have to find out all of the trigger events and methods associated with each business object."

"How do we do that?" one of the analysts asked.

"You continue with your fact-finding activities and build a scenario for each business process. The way the business objects interact with each other in the scenario determines how you identify the initiating activity. We refer to those activities as the *messages* between objects. The tricky part is that you need to think in terms of objects instead of just processes. Sometimes it helps me to pretend I am an object. I will say, 'I am a purchase order object. What functions and services are other objects going to ask me to do?' After you get the hang of it, it works very well, and it is enlightening to see how the system requirements unfold as you develop the diagrams."

OVERVIEW

The basic objective of requirements definition is understanding—understanding users' needs, understanding how the business processes are carried out, and understanding how the system will be used to support those business processes. As we indicated in Chapter 2, system developers use a set of tools and techniques to discover and understand the requirements for a new system. This activity is a key part of the systems analysis activities of the systems development life cycle. In object-oriented development, the set of analysis activities is more specifically referred to as *object-oriented analysis* (OOA). The first step in the process for developing this understanding requires the fact-finding skills you learned in Chapter 4. Fact-finding activities are also called *discovery activities*, and obviously discovery must precede understanding. In this chapter, you learn to take discovery to the next level—to build understanding.

Chapter 4 introduced the concepts of models and modeling activities as a way to define and document system requirements. The models introduced in Chapter 5 focus on two primary aspects of functional requirements: the use cases and the things involved in users' work. As you learned, use cases are triggered by events in the business's environment to which the system must respond. Those events are identified and documented in an event table. Use cases are also identified with the user goal technique and the CRUD technique.

A new system also needs to record and store information about things involved in the business processes. In a manual system, the information would be recorded on paper and stored in a filing cabinet. In an automated system, the information is stored in electronic files or a database. The information storage requirements of a system are documented either with entity-relationship diagrams (ERDs) in the traditional approach or with domain model class diagrams in the object-oriented approach.

In this chapter, you learn how to understand and define the requirements for a new system using object-oriented analysis models and techniques. You should be aware that the line between object-oriented analysis and object-oriented design is somewhat fuzzy because the models that are built to define requirements during analysis are refined and extended to produce a systems design. Recall that we mentioned the object-oriented approach almost always uses an iterative approach to development, which identifies some of the requirements, then does some preliminary design and implementation, then iterates again and again through requirements, design, and implementation. So, even though we do not focus here on the iterative nature of requirements definition, it is a normal part of the object-oriented approach. Chapters 11 and 12 extend the requirements into a complete object-oriented design that can serve as the foundation for programming the new system.

OBJECT-ORIENTED REQUIREMENTS

As discussed in Chapter 4, one of the great benefits of using models to document requirements is that it helps you, as the system developer, to think clearly and carefully about the details of the processing and information needs of the stakeholders. As you read this chapter and work the exercises associated with it, you should pay careful attention to how the models require you to search out and understand user needs. Because of the benefit derived from developing models, object-oriented system requirements are specified and documented through the process of building models.

The object-oriented (OO) modeling notation that we present in this textbook is based on the Unified Modeling Language (UML) version 2.0. UML is the accepted, standard OO modeling language of the industry. The UML standard is maintained by the Object Management Group (OMG), which is a consortium of more than 800 software vendors, developers, and organizations that have combined efforts to develop and foster uniformity in object-oriented systems. Established in 1989, OMG's mission is to promote the theory and practice of object technology in the development of distributed computing systems. The OMG maintains and approves any changes to the standards for OO modeling. As a result, UML standards continue to evolve but will remain standardized for the benefit of system developers—and students. More details about UML and the OMG can be found on the OMG Web site at *www.omg.org*.

As shown in Figure 7-1, the system development process starts with the identification of events that trigger elementary business processes called *use cases,* and things that are problem domain classes involved in the elementary business process. The problem domain classes are important in both the development of the new system itself as well as the design of the database. New developers frequently ask which to define first, the use cases or the classes of objects. In reality, both aspects are closely related and are usually defined together. Experienced developers often move back and forth between identifying classes and use cases, and they make several passes before completing a set of requirements. Do not be discouraged if you find yourself changing your diagrams and models as you work to define requirements.

Figure 7-1

Requirements diagrams for traditional and object-oriented models

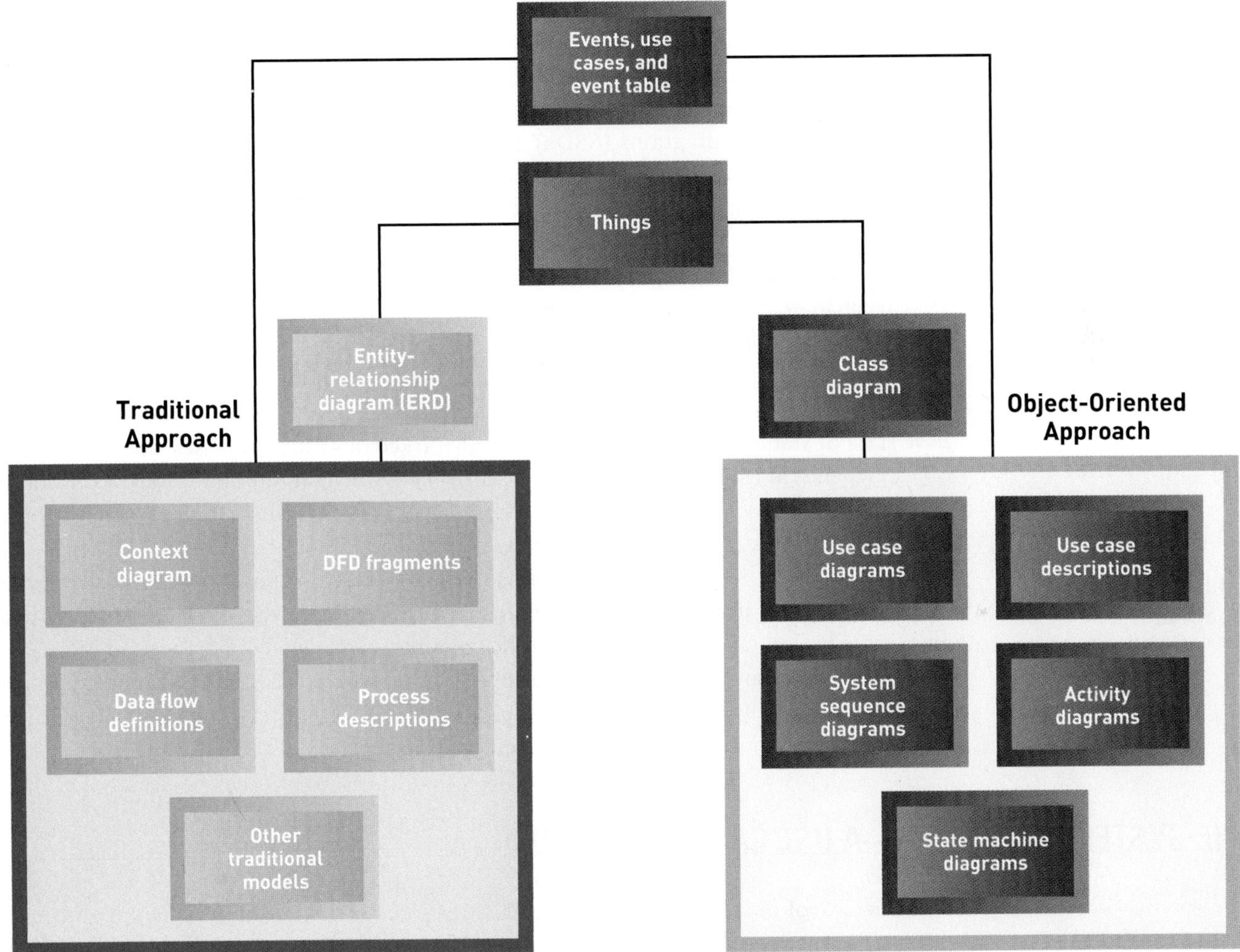

The object-oriented approach requires several interrelated models to create a complete set of specifications. Even though it might seem complex at first to have so many different types of diagrams, as you use them, you will learn to appreciate how they all fit together like a puzzle to produce a complete specification. Essentially, the object-oriented approach "divides and conquers" complex systems. Each model describes a different aspect of the system, so you only focus on one aspect at a time. But you must learn all the different models and the way they fit together. Later, at the end of the chapter, we discuss how all of the diagrams unite to form a complete view of a system's functional requirements. As a beginner with UML, you should concentrate now on learning each new model and understanding its role in specifying the total system.

use case model
a collection of models that can be used to capture system requirements based on use cases with the object-oriented approach

use case diagram
a diagram to show the various user roles and how those roles use the system

This chapter focuses on a collection of models that can be used to capture system requirements based on use cases with the object-oriented approach. Called the **use case model**, it includes use case diagrams, use case descriptions (discussed in Chapter 5), activity diagrams, and system sequence diagrams. The purpose of a **use case diagram** is to identify the "uses," or use cases, of the new system—in other words, to identify how the system will be used. The use case diagram can be derived directly from the column titled "Use case" in the event table. A use case diagram is a convenient way to document the system activities. Sometimes a single, comprehensive diagram is used to identify all use cases for an entire system. At other times, a set of smaller use case diagrams is used.

Each use case must be described either in brief or fully developed detail, as discussed in Chapter 5. Each use case can also be defined using an activity diagram. As you learned in Chapter 4, activity diagrams can be used to describe any business processes done by people in an organization. However, they are also used to describe processes that include both manual and automated system activities, so they can be used to define a use case.

system sequence diagram
a diagram showing the sequence of messages between an external actor and the system during a use case or scenario

message
the communication between objects within a use case

domain model
a model that describes classes of objects and their states

state machine diagram
a diagram showing the life of an object in states and transitions

System sequence diagrams (SSDs) are used to define the inputs and outputs and the sequence of interactions between the user and the system for a use case. They are used in conjunction with detailed descriptions or with activity diagrams. In a sequence diagram, these information flows in and out of a system are called **messages**. The users are identified, and the detailed messages are described.

This chapter also focuses on the **domain model**, which describes classes of objects and their states. The *domain model class diagram* (discussed in Chapter 5) is used to define the classes of objects in the problem domain, and the **state machine diagram** introduced in this chapter details possible object states. Some objects that are identified in the class diagram have state or status conditions that need to be tracked, and the processes allowed for that object depend on its status. A customer order may have several important status conditions that control the processing of that order—for example, an order that is not complete should not be shipped. A state machine diagram identifies these status conditions and specifies the processes allowed. State machine diagrams are also used during design to identify various states of the system itself and allowable events that can be processed. So, as with the class diagram, state machine diagrams can be considered either an analysis tool or a design tool.

In many cases, analysts use all the models included in the use case model and the domain model to completely define the system requirements. However, sometimes only two or three models may be required to specify the requirements accurately.

THE SYSTEM ACTIVITIES—A USE CASE/SCENARIO VIEW

The objective of the use case model is to identify and define all of the elementary business processes that the system must support. Analysts define the use cases at two levels—an overview level and a detailed level. The event table and the use case diagrams provide an overview of all the use cases for a system. Detailed information about each use case is described with a use case description, an activity diagram, and a system sequence diagram, or a combination of these models.

USE CASES AND ACTORS

A use case is an activity the system carries out, usually in response to a request by a user of the system. You can think of a use case as a situation in which the system must accomplish some goal of a user. For example, consider the RMO system. One of the processes that the RMO system must perform is to process new customer orders. So, one use case for this system is *Create new order*. Notice that the focus is on the automated system—on the activities that the *system* must perform to create an order.

Implied in all use cases is a person who uses the system. In UML, that person is called an *actor*. An actor is always outside the automation boundary of the system but may be part of the manual portion of the system. In this respect, an actor is not always the same as the source of the event in the event table. A source of an event is the initiating person who supplied data, such as a customer, and is usually external to the system, including the manual system. In contrast, an actor in use case analysis is the person who is actually interacting with the computer system itself. By defining actors that way—as those who interact with the system—we can more precisely define the exact interactions to which the automated system must respond. This tighter focus helps define the specific requirements of the automated system itself—to refine them as we move from the event table to the use case details. One way to help identify actors at the right level of detail is to assume that actors must have hands. Thinking of actors as having hands encourages us to define actors as those who actually touch the automated system. But remember that some actors are not people. They can also be other systems or other devices that receive services from the system.

BEST PRACTICE

Be sure that actors have direct contact with the automated system.

Another way to think of an actor is as a role. For example, in the RMO case, the use case *Create new order* might involve an order clerk talking to the customer on the phone. Or, the customer might be the actor if the customer places the order directly, through the Internet. One final way to think about an actor and a use case is that a use case is a goal that the actor wants to achieve. One way to state this goal is to say, "The order clerk uses the system to create a new order." Notice that in this sentence both the actor (the *order clerk*) and the use case (*Create a new order*) are identified. In fact, stating the use cases in sentence form is a good technique to understand the relationship between use cases and actors.

THE USE CASE DIAGRAM

Figure 7-2 shows how a use case is documented in a use case diagram. A simple stick figure is used to represent an actor. The stick figure is given a name that characterizes the role the actor is playing. The use case itself is symbolized by an oval with the name of the use case inside. The connecting lines between actors and use cases indicate which actors invoke which use cases. Although hands are not part of the standard UML notation, the actor in this figure is drawn with hands to help you remember that this actor must have direct access to the automated system.

Figure 7-2

A simple use case with an actor

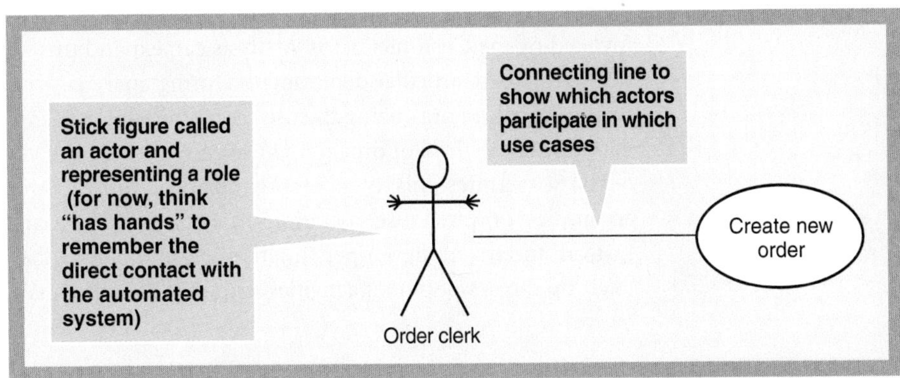

A use case diagram is a graphical model that summarizes the information about the actors and use cases. To do use case analysis, a system developer looks at the system as a whole and tries to identify all of its major uses.

Automation Boundary and Organization

Figure 7-3 expands the use case diagram shown in Figure 7-2 to include additional use cases and additional actors. In this instance, both the order clerk and the customer are allowed to access the system directly. As indicated by the relationship lines, each person actor can use every use case. A rectangle is used to indicate an actor that is not a person. In this instance, the Inventory system actor can invoke the use case *Look up item availability*. A boundary line is also drawn around the entire set of use cases. This boundary is the automation boundary. It denotes the boundary between the environment, where the actors reside, and the internal components of the computer system.

Figure 7-3

A use case diagram of the Order-entry subsystem for RMO, showing a system boundary

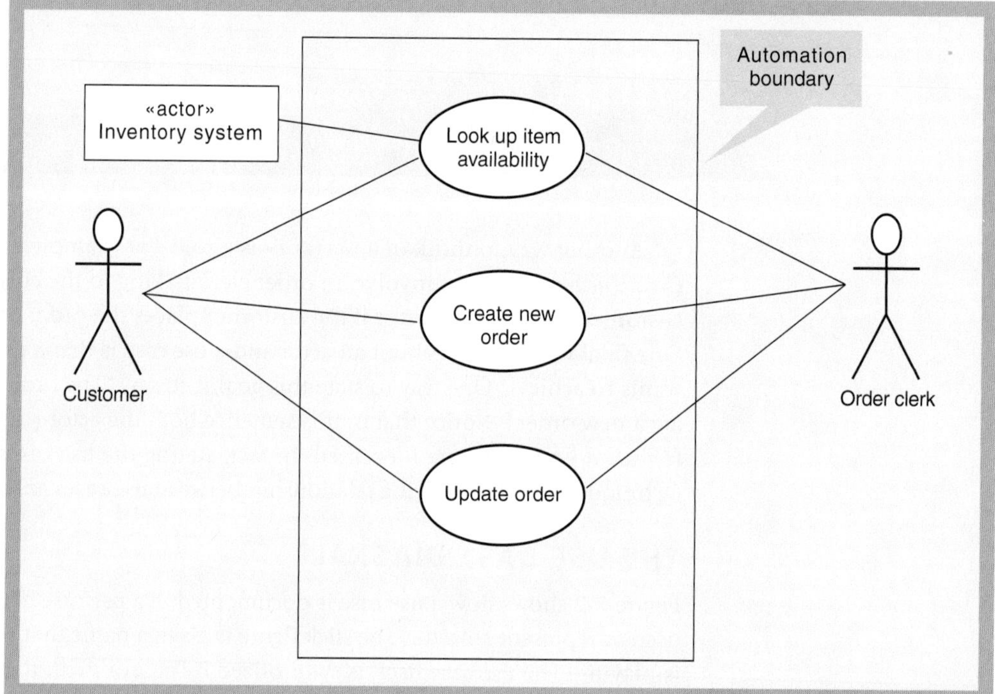

There are many ways to organize the use cases for ease of understanding and development. One way is to show all use cases that are invoked by a particular actor—that is, from the user's viewpoint. This approach is often used during requirements definition because the systems analyst may be working with a particular user and identifying all of the functions that user performs with the system. Figure 7-4 illustrates this point of view, showing all of the use cases invoked by the Customer actor. Analysts can expand this approach to include all the use cases belonging to a particular department. During analysis, analysts focus on determining the user requirements, so organizing the use cases from the user's viewpoint is quite beneficial.

Another method of organizing use cases is from the viewpoint of a system and its subsystems. Sometimes this type of organization mirrors the user departments—focusing on accounting or warehouse operations one at a time, for example—but it does not have to do so. Instead, the system developers might want to organize the use cases by a system's subsystems to group the development activities and team assignments. Figure 7-5 illustrates this approach,

package
a symbol used to denote a group of similar elements

showing many of the RMO use cases organized by subsystem. In this figure, we introduce a new notation, called a **package**. A package groups similar components together. The package notation is a tabbed rectangle with the name of the package in the tab. In Figure 7-5, the packages indicate subsystems. This figure contains four separate subsystems, each shown as a package, and their corresponding use cases. Actors are duplicated to make the diagram easy to read; however, use cases are not duplicated because each use case belongs to only one subsystem. We discuss more details about packages and package diagrams in Chapters 11 and 12.

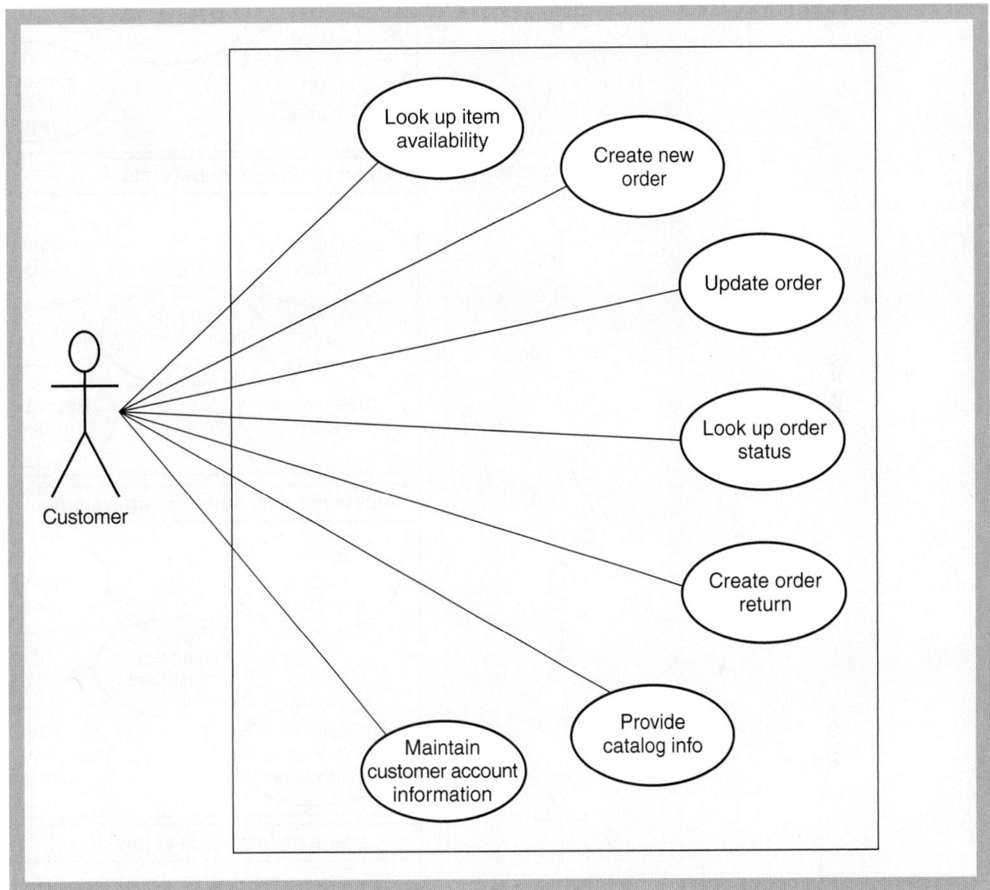

Figure 7-4
All use cases involving the Customer actor

«Includes» Relationships

Frequently during the development of a use case diagram, it is reasonable for one use case to use the services of a common subroutine. For example, two of the Order-entry subsystem use cases are *Create new order* and *Update order*. Each of these use cases may need to validate the customer account. A common subroutine may be defined to carry out this function, and it becomes an additional use case. Figure 7-6 shows the additional use case, named *Validate customer account*, which is used by both the other use cases. The relationship between these use cases is denoted by the dashed connecting line with the arrow. The direction of the arrow indicates which use case is included as a part of the major use case. The relationship is read *Create new order «includes» Validate customer account*. Sometimes this relationship is referred to as the «includes» relationship, or sometimes as the «uses» relationship.

Figure 7-5

A use case diagram of the customer support system organized by subsystem

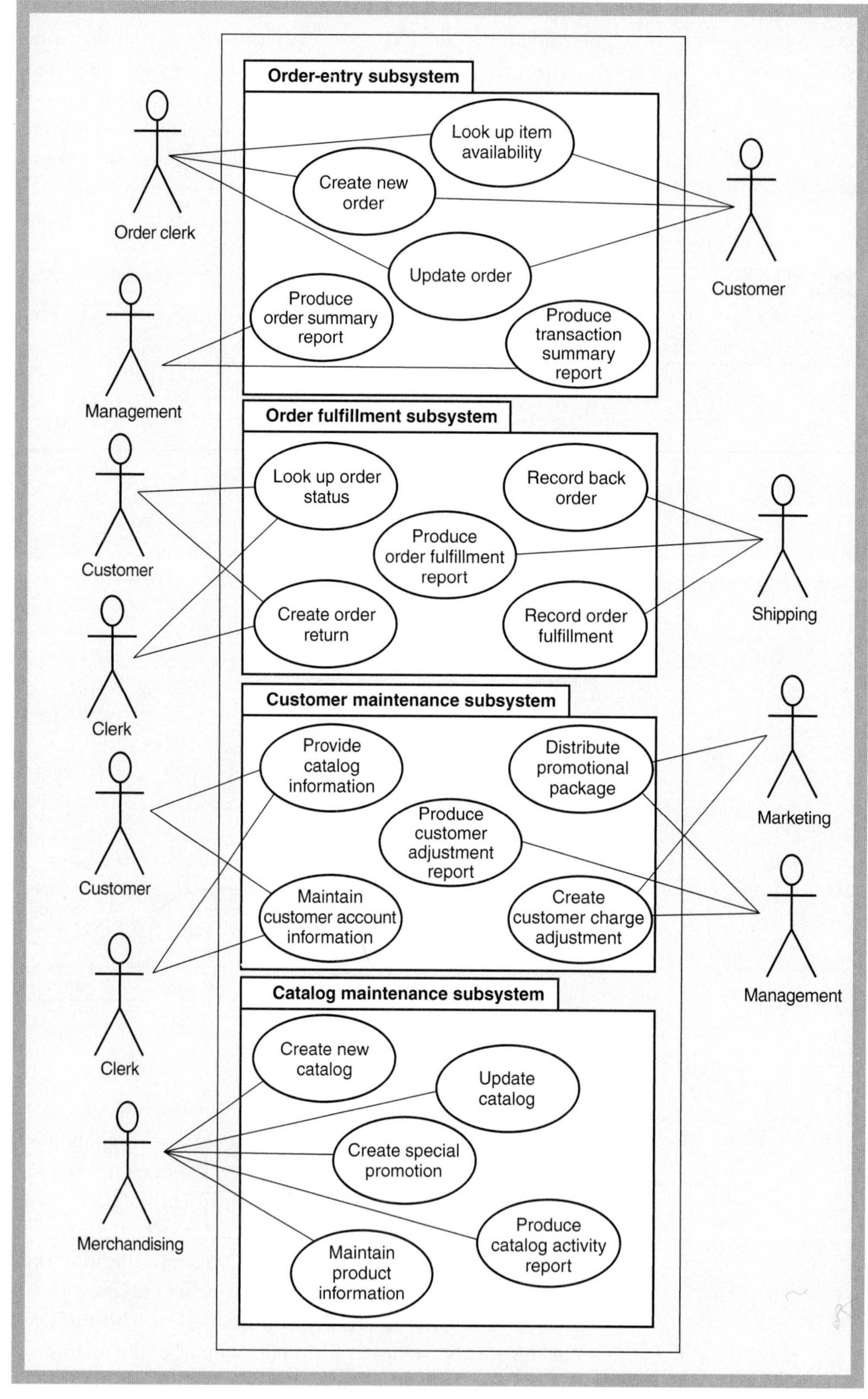

Figure 7-6 also shows that *Look up item availability* can be part of an «includes» relationship. So, an analyst can define two types of «includes» use cases: one that is a common internal subroutine, such as *Validate customer account,* and is not directly referenced by an external actor, and one that is directly referenced by external actors. *Look up item availability* is an example of the latter.

Figure 7-6

An example of the Order-entry subsystem with «includes» use cases

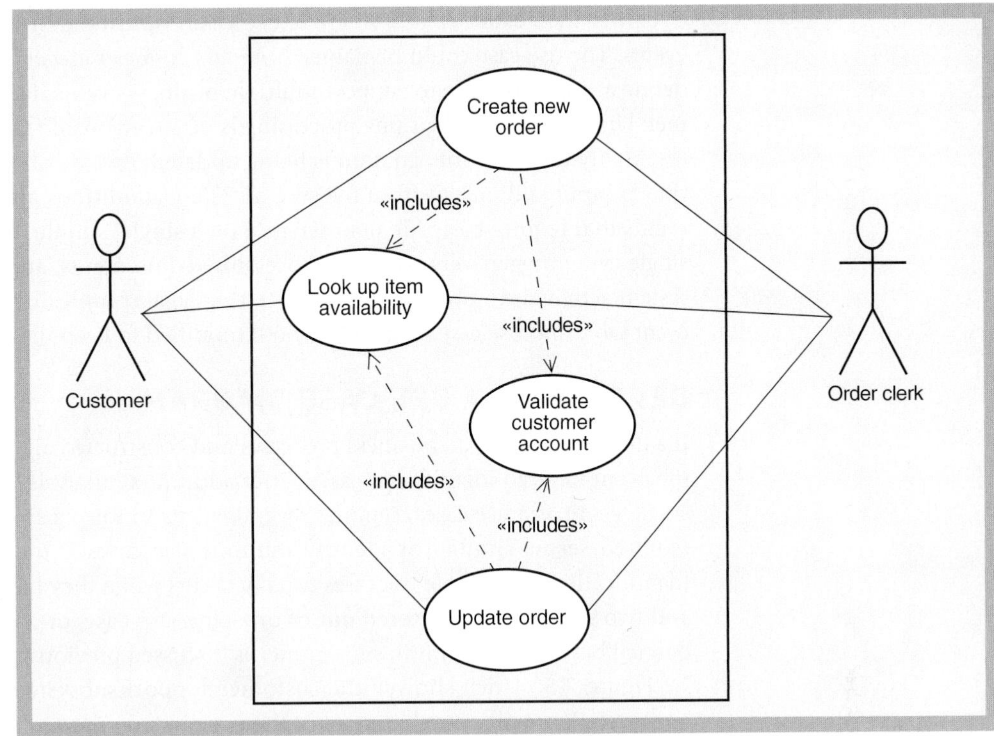

The Use Case Diagram Compared with the Event Table

As indicated earlier, the event table and the use case diagram contain much of the same information, and the event table is really a catalog of information about all the use cases. One of the questions you might be asking yourself is, "If they are so similar, do I need to develop both models?" In fact, for any given project, you might not develop both models. Some analysts prefer to start by listing use cases rather than events, and they move directly to the use case diagram. The user goal technique and the CRUD technique are often used this way. The event table can be used as the foundation for either traditional structured development or object-oriented development, just as use case descriptions can be used for either approach.

However, some differences do exist between the two models. First, the point of view of each is slightly different. An event table always focuses on the business processes. It does so by identifying business events and external, initiating sources for those events. These external sources are the ones that cause the business event to be initiated, and they can be somewhat removed from the automated system. On the other hand, a use case diagram emphasizes the automated system. Because it is concerned only with the automated system, the actors actually have contact with the automated system and might not necessarily be the original initiators of the business event.

Another difference between the two models can be seen when identifying temporal and state events. Because use cases are usually initiated by actors, temporal and state events are often overlooked if the analyst does not carefully identify all events. This is a deficiency of use case modeling if use cases are defined too narrowly. As discussed in Chapter 14, online system menus typically include menu options representing each temporal event from the event table so that such an event can be triggered by a user as well as being a purely temporal event. Therefore, we recommend including a use case for each temporal and state event to ensure these requirements are not overlooked.

It is important to remember that the analyst will be completing the event table and the use case diagrams concurrently. The analyst will also continually refine and update events and use cases. The refinements that occur usually involve adjustments to balance the scope of each use case. For example, during the development of the event table, two events called *Add new customer* and *Update customer information* may have been identified. From the system's point of view, the use case for both business events is almost the same because they both involve updating the customer file. A single use case could be defined to support both business events. The use case could be named *Maintain customer account information*. It is common to define a single use case to support multiple business events if the following three criteria are met: First, essentially the same processing is occurring inside the automated system. Second, essentially the same information is being updated. And third, essentially the same information is input and output from the system. These conditions are frequently met for business events that require basic file maintenance on a single, simple data file or table. Sometimes a single event triggers very complex processing requirements, and it makes sense to divide the system activity into two use cases to better manage complexity. In all of these situations, the event table and use case diagram are both modified to keep the models synchronized.

DEVELOPING A USE CASE DIAGRAM

If a developer analyzed business processes and constructed an event table, he or she will use the event table to identify use cases. After additional analysis, the developer may identify a single event as a use case, combine several events to form a single use case if the processing required seems similar, or identify multiple use cases if the processing seems complex. Identification of multiple use cases usually occurs when they have the «includes» relationship and two use cases are factored out of one large use case, or when an additional use case is defined based on a common subroutine, as discussed previously.

Figure 7-5, which showed the customer support subsystems, was developed using this approach. You will note that most of the use cases defined in the figure come directly from the event table shown in Figure 5-12. In fact, the names of the use cases in Figure 7-5 come from the description provided in the Use case column of the event table. There are a couple of exceptions to this pattern. Because temporal events normally also can be initiated manually, we have used the option of identifying an external actor for each temporal use case. The other exception is with event number 13, *Customer updates account information*. In this instance, the use case definition is expanded to include all scenarios having to do with maintaining customer information. The use case is titled *Maintain customer account information* to denote that it will include additions, updates, and deletions. These examples show when the use case diagram could refine the event table.

If an event table has not been created, the other starting point to develop a use case diagram is to identify the actors and the elementary business processes with the user goal technique. To do so, you must remember two preconditions. First, you must make the system boundary an automated system so that the actors you identify actually contact the system—that is, have hands. Second, you must assume perfect technology. Be sure that the use cases are based on business events and not technical activities like logging on to the system or changing passwords. Given those preconditions, you can develop the use case diagram in two steps, which are done in iteration.

1. Identify the actors of the system. Note that actors are actually roles played by users. Instead of listing the actors as Bob, Mary, or Mr. Hendricks, you should identify the specific roles that these people play. Remember that the same person may play various roles as he or she uses the system. Those roles become such titles as order clerk, department manager, auditor, and so forth. It is important to be comprehensive and to identify every possible role that will use the system. Other systems can also be actors of a system, as indicated in Figure 7-3.
2. After the actor roles have been identified, the next step is to develop the list of goals those roles have in the use of the automated system. A goal is a task performed by an actor to accomplish some business function that adds value to the business. Goals are such tasks as "process a sale," "accept a return," or "ship an order." Goals are units of work that can be identified and described. At the completion of the goal, the data of the system should be stable for some time.

These two steps are performed in brainstorming sessions with project team members and users. There is no magical way to find or identify use cases. Even though the focus is on the automated system, a thorough analysis of the business processes is required to understand all ways that actors will need to use the system.

Another important technique that you should use when developing the use case diagram directly is the CRUD technique, which compares the identified use cases with the domain model class diagram. Analysts use the CRUD technique after making an initial use case diagram to double-check their work. Recall that CRUD stands for create, read (or report), update, and delete. The CRUD technique was first introduced in Chapter 5, and it is a technique originally associated with Information Engineering (IE). The CRUD technique requires that every class in the class diagram have sufficient use cases to support creating new object instances, reading or reporting on those objects, updating those objects, and in many cases deleting object instances. The use case may not be named *create or update*, but the underlying process should add a new instance or update an existing instance. For example, a use case named *Record payment* does not explicitly indicate that a new payment object is created, but a detailed description of the use case will indicate that a new payment is created. The use case *Create new order* might create OrderItem objects and update InventoryItem objects. In other cases, many of the use cases are named beginning with the word *maintain* to cover routine additions, updates, reads, and deletions. Keep in mind, though, that with integrated systems, one system might be responsible for creating objects and another system might only update them. The CRUD technique provides a cross-check, not a final solution, and it also provides an opportunity to confirm important system integration requirements that otherwise might not be obvious.

ACTIVITY DIAGRAMS FOR DESCRIBING USE CASES

In Chapter 5 you learned how to document each use case or scenario with written descriptions. Use case descriptions can be brief, intermediate, or fully developed. Figure 7-7 reproduces a fully developed use case description for the use case *Create new order*, which was first shown in Chapter 5. Recall that the template for fully developed use case descriptions includes use case name, scenario, triggering event, brief description, actors, related use cases, stakeholders, preconditions, postconditions, flow of activities, and exception conditions.

The other way to document a use case scenario is with an activity diagram. In Chapter 4, you learned about activity diagrams as a form of workflow diagram. You learned that an activity diagram is an easily understood diagram to document the workflows of the business processes. Activity diagrams are a standard UML diagram. In this instance, activity diagrams are an effective technique to document the flow of activities for each use case scenario.

Figures 7-8 and 7-9 are the activity diagrams that document the same two scenarios as shown in Chapter 5. In Figure 7-8, the customer interacts with the order clerk, who in turn uses the system. Because the purpose of a use case is to specify the interaction of an actor (with hands) with the system, the figure includes swimlanes for the Order Clerk and the Computer System. However, to aid in understanding the total flow of activities for the scenario, the Customer—the one who initiates the steps—is also included. Note that the Customer swimlane is an optional addition in Figure 7-8 that simply aids in understanding the total workflow. We also see a new use for the synchronization bar. It is used in this figure to define the end points of a repeated section; that is, a loop. In Figure 7-9, the customer is the actor who interacts with the computer system, so only two swimlanes are required to describe the steps in the scenario.

Figure 7-7

Fully developed description of the telephone order scenario for *Create new order*

Use Case Name:	*Create new order*	
Scenario:	Create new telephone order	
Triggering Event:	Customer telephones RMO to purchase items from the catalog.	
Brief Description:	When customer calls to order, the order clerk and system verify customer information, create a new order, add items to the order, verify payment, create the order transaction, and finalize the order.	
Actors:	Telephone sales clerk.	
Related Use Cases:	Includes: *Check item availability.*	
Stakeholders:	Sales department: to provide primary definition. Shipping department: to verify information content is adequate for fulfillment. Marketing department: to collect customer statistics for studies of buying patterns.	
Preconditions:	Customer must exist. Catalog, Products, and Inventory items must exist for requested items.	
Postconditions:	Order and order line items must be created. Order transaction must be created for the order payment. Inventory items must have the quantity on hand updated. The order must be related (associated) to a customer.	
Flow of Activities:	**Actor**	**System**
	1. Sales clerk answers telephone and connects to a customer.	
	2. Clerk verifies customer information.	2.1 Display customer information.
	3. Clerk initiates the creation of a new order.	3.1 Create a new order.
	4. Customer requests an item be added to the order.	
	5. Clerk verifies the item (*Check item availability* use case).	5.1 Display item information.
	6. Clerk adds item to the order.	6.1 Add an order item.
	7. Repeat steps 4, 5, and 6 until all items are added to the order.	
	8. Customer indicates end of order; clerk enters end of order.	8.1 Complete order. 8.2 Compute totals.
	9. Customer submits payment; clerk enters amount.	9.1 Verify payment. 9.2 Create order transaction. 9.3 Finalize order.
Exception Conditions:	2.1 If customer does not exist, then the clerk pauses this use case and invokes *Maintain customer information* use case. 2.2 If customer has a credit hold, then clerk transfers the customer to a customer service representative. 4.1 If an item is not in stock, then customer can a. choose not to purchase item, or b. request item be added as a back-ordered item. 9.1 If customer payment is rejected due to bad-credit verification, then a. order is canceled, or b. order is put on hold until check is received.	

An activity diagram can be used to support any level of use case descriptions. As you can see, activity diagrams are very similar to the two-column description in the fully developed description. The benefit of creating an activity diagram is that it is more visual and makes it easier to understand the overall flow of activity. These two instances show only the main flow of activity—without the exception conditions. The exception conditions can also be shown by adding more activity ovals. Early termination of the workflow can also be indicated by an exit arrow going to an exception end activity. An exception end activity is depicted much the same as a normal end activity, except that the circle encloses a large X instead of a black dot.

As a quick glance at Figures 7-8 and 7-9 demonstrates, the two scenarios of the *Create new order* use case are quite different. Even though the scenarios carry out the same basic function, the set of screens and options on the screens might be quite different for each. Activity diagrams are also helpful in developing system sequence diagrams, as explained in the next section.

Figure 7-8

Activity diagram of the telephone order scenario

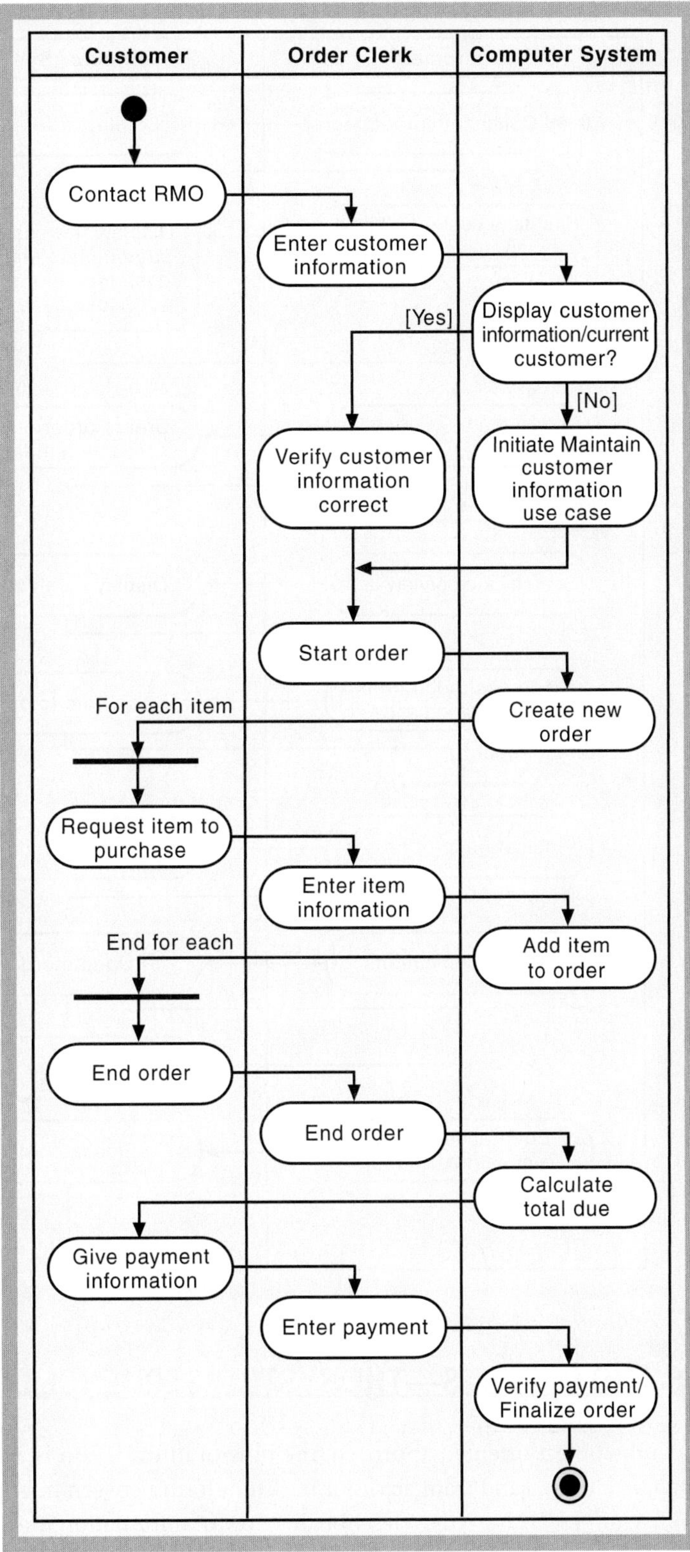

Figure 7-9

Activity diagram of the Web order scenario

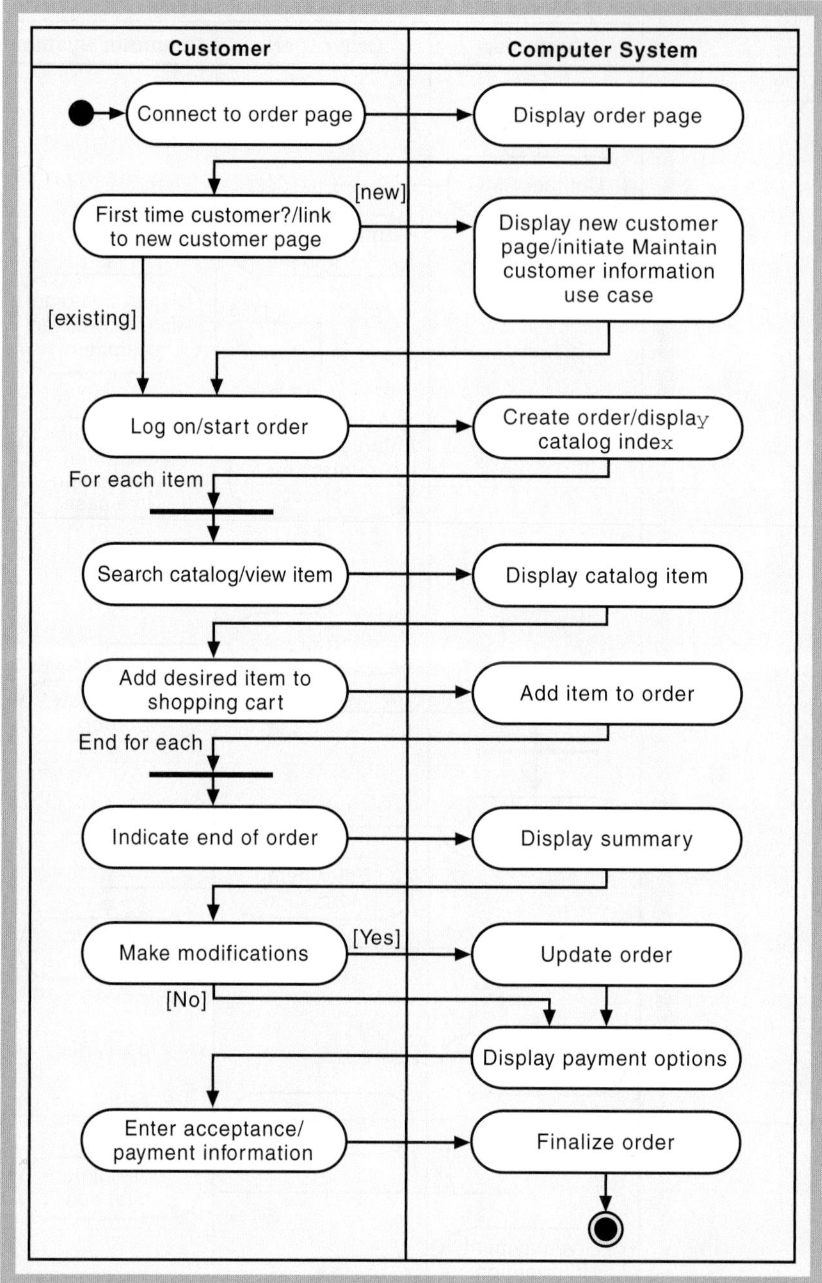

IDENTIFYING INPUTS AND OUTPUTS—THE SYSTEM SEQUENCE DIAGRAM

In the object-oriented approach, the flow of information is achieved through sending messages either to and from actors or back and forth between internal objects. A system sequence diagram (SSD) is used to describe this flow of information into and out of the automated system. So, an SSD documents the inputs and the outputs and identifies the interaction between actors and the system. An SSD is a type of **interaction diagram**. In the following sections, and in industry practice, we often use the terms *interaction* and *message* interchangeably.

interaction diagram either a communication diagram or a sequence diagram that shows the interactions between objects

SSD NOTATION

Figure 7-10 shows a generic SSD. As with a use case diagram, the stick figure represents an actor—a person (or role) that interacts with the system. In a use case diagram, the actor "uses" the system, but the emphasis in an SSD is on how the actor "interacts" with the system by entering input data and receiving output data. The idea is the same with both diagrams; the level of detail is different.

The box labeled :System is an object that represents the entire automated system. In SSDs and all interaction diagrams, analysts use object notation instead of class notation. Object notation indicates that the box refers to an individual object and not the class of all similar objects. The notation is simply a rectangle with the name of the object underlined. The colon before the underlined class name is a frequently used, but optional, part of the object notation. In an interaction diagram, the messages are sent and received by individual objects, not by a class. In an SSD, the only object included is one representing the entire system.

lifeline, or object lifeline
the vertical line under an object on a sequence diagram to show the passage of time for the object

Underneath the actor and the :System are vertical dashed lines called *lifelines*. A **lifeline**, or **object lifeline**, is simply the extension of that object, either actor or object, throughout the duration of the SSD. The arrows between the lifelines represent the messages that are sent or received by the actor or the system. Each arrow has an origin and a destination. The origin of the message is the actor or object that sends it, as indicated by the lifeline at the arrow's tail. Similarly, the destination actor or object of a message is indicated by the lifeline that is touched by the arrowhead. The purpose of lifelines is to indicate the sequence of the messages sent and received by the actor and object. The sequence of messages is read from top to bottom in the diagram.

A message is labeled to describe both the message's purpose and any input data being sent. The syntax of the message label has several options; the simplest forms are shown in Figure 7-10. Remember that the arrows are used to represent both a message and input data. But what is meant by the term *message* here? In a sequence diagram, a message is considered to be an action that is invoked on the destination object, much like a command. Notice in Figure 7-10 that the input message is called inquireOnItem. The clerk is sending a request, or a message to the system, to find an item. The input data that is sent with the message is contained within the parentheses, and in this case it is data to identify the particular item. The syntax is simply the name of the message followed by the input parameters in parentheses. This form of syntax is attached to a solid arrow.

Figure 7-10
Sample system sequence diagram (SSD)

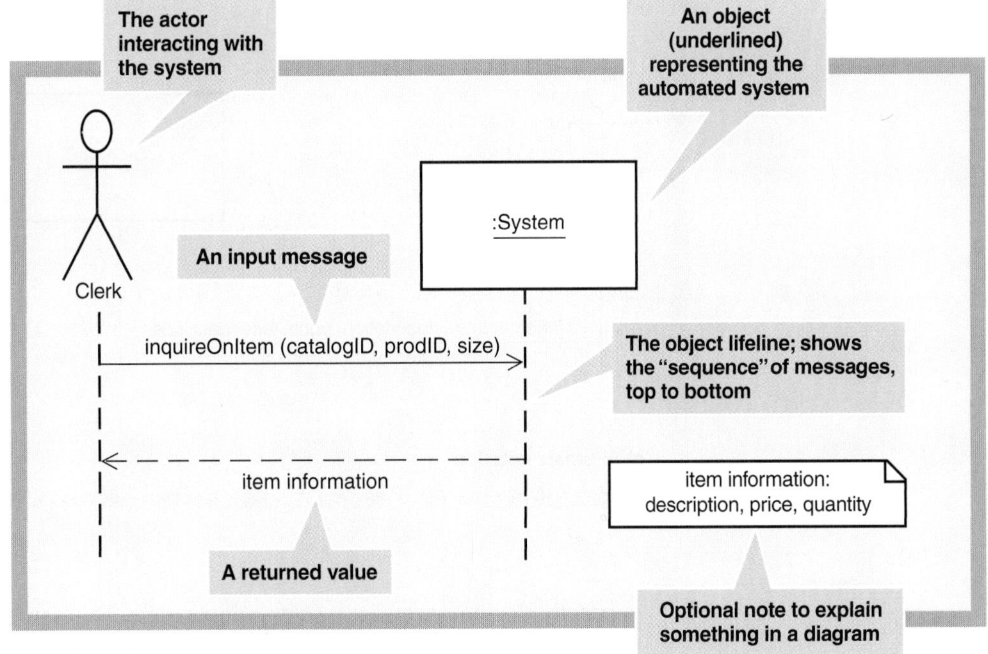

The returned value has a slightly different format and meaning. Notice the arrow is a dashed arrow. A dashed arrow is used to indicate a response or an answer and, as shown in the figure, it immediately follows the initiating message. The format of the label is also different. Because it is a response, only the data that is sent on the response is noted. There is no message requesting a service, only the data being returned. In this case, a valid response might be a list of all the information returned, such as description, price, and quantity of an item. However, an abbreviated version is also satisfactory. In this case, the information returned is named item information. Additional documentation is required to show the details. In Figure 7-10, this additional information is shown as a note. A note can be added to any UML diagram to add explanations. The details of item information could also be documented in supporting narratives or even simply referenced by the attributes in the Customer class.

Figure 7-11

Repeating message
(a) Detailed notation
(b) Alternate notation

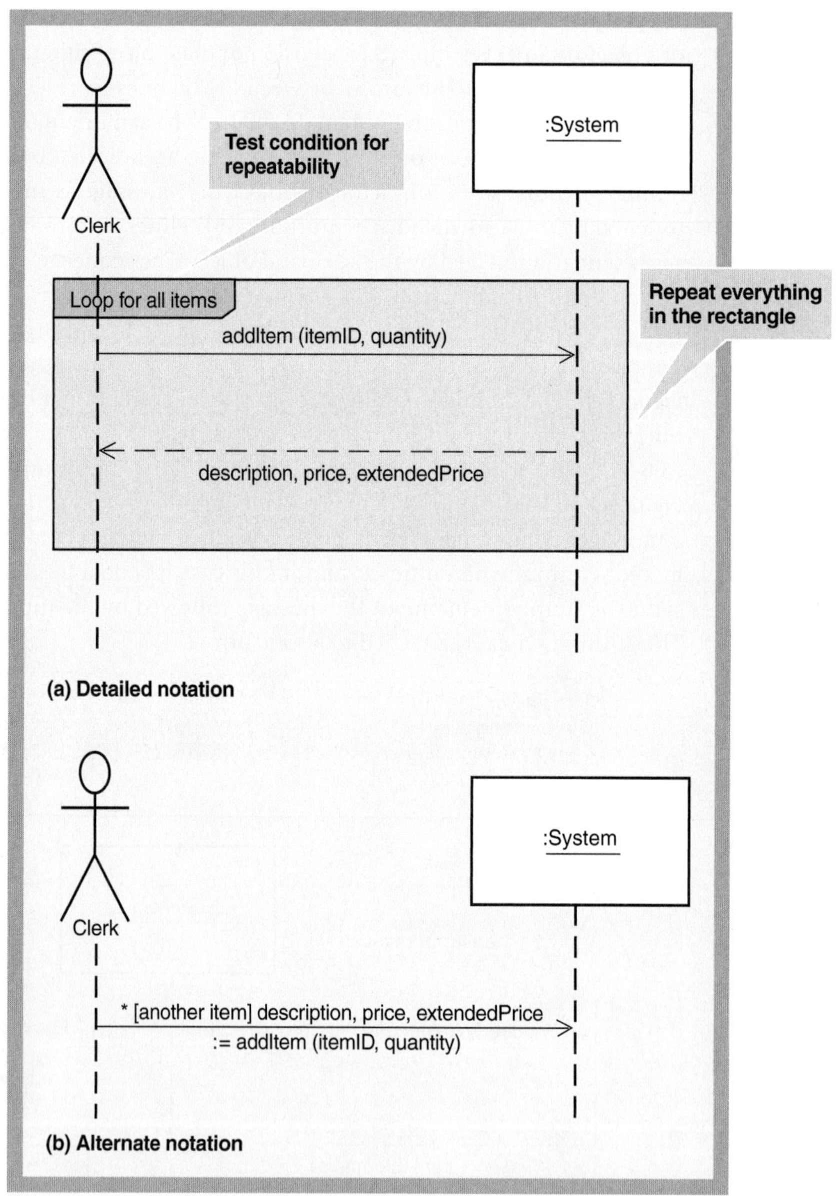

Frequently, the same message is sent multiple times. For example, when an actor enters items on an order, the message to add an item to an order may be sent multiple times. Figure 7-11(a) illustrates the notation to show this repeating operation. The message and its return are located inside a larger rectangle. In a smaller rectangle at the top of the large rectangle is the descriptive text to control the behavior of the messages within the larger rectangle. The condition loop for all items indicates that the messages in the box repeat many times or are associated with many instances.

true/false condition part of a message between objects that is evaluated prior to transmission to determine whether the message can be sent

Figure 7-11(b) shows an alternate notation. The square brackets and text inside them are called a true/false condition for the messages. The asterisk (*) preceding the true/false condition indicates that the message repeats as long as the true/false condition evaluates to true. Analysts use this abbreviated notation for several reasons. First, a message and the returned data can be shown in one step. Note that the return data is identified as a return value on the left side of an assignment operator—the := sign. This alternative simply shows a value that is returned. Second, the true/false condition is placed on the message itself. Note that in this example, the true/false condition is used for the control of the loop. True/false conditions are also used to evaluate any type of test that determines whether a message is sent. For example, [credit card payment] might be used to control whether a message is sent to the system to verify a credit-card number. Finally, an asterisk is also placed on the message itself. So, for simple repeating messages, the alternate notation is shorter. However, if several messages are included within the repeat or there are multiple messages, each with its own true/false condition, the more detailed notation is more explicit and precise.

The complete notation for a message is the following:

* *[true/false condition] return-value := message-name (parameter-list)*

Any part of the message can be omitted. In brief, the notation components are the following:

- An asterisk (*) indicates repeating or looping of the message.
- Brackets [ ] indicate a true/false condition. It is a test for that message only. If it evaluates to true, the message is sent. If it evaluates to false, the message is not sent.
- Message-name is the description of the requested service. It is omitted on dashed-line return messages, which only show the return data parameters.
- Parameter-list (with parentheses on initiating messages and without parentheses on return messages) shows the data that is passed with the message.
- Return-value on the same line as the message (requires :=) is used to describe data being returned from the destination object to the source object in response to the message.

BEST PRACTICE

Develop SSDs carefully and correctly. They become critical components for detailed design and user interface design.

DEVELOPING A SYSTEM SEQUENCE DIAGRAM

An SSD is normally used in conjunction with the use case descriptions to help document the details of a single use case or scenario within a use case. To develop an SSD, you will need to have a detailed description of the use case, either in the fully developed form, as shown in Figure 7-7, or as activity diagrams, as shown in Figures 7-8 and 7-9. These two models identify the series of activities within a use case, but they do not explicitly identify the inputs and outputs. An SSD will provide this explicit identification of inputs and outputs. One advantage of using activity diagrams is that it is easy to identify when an input or output occurs. Inputs and outputs occur whenever an arrow in an activity diagram goes from an external actor to the computer system. Figure 7-12 is a simplified version of Figure 7-8 for the telephone order scenario of the RMO *Create new order* use case. Obviously, the simplified version has many things missing, but it allows us to focus on the process without having to consider all of the complexity of the real world, and to focus on the basics of SSD development.

In this simplified activity diagram, there are three swimlanes: the Customer, the Order Clerk, and the Computer System. Before beginning the SSD, you must first determine the system boundary. In this instance, the system boundary coincides with the vertical line between the Order Clerk swimlane and the Computer System swimlane. Because the purpose of the SSD is to describe the inputs to and outputs from the automated computer system, only the Order Clerk and the Computer System will be included in the SSD. It is not wrong to include both actors in the SSD, but it is more focused to show only the system and the actor who sends the inputs and receives the outputs.

The development of an SSD based on an activity diagram can be divided into four steps:

1. **Identify the input messages.** In Figure 7-12, there are three locations with a workflow arrow crossing the boundary line between the clerk and the system. At each location that the workflow crosses the automation boundary, input data is required; therefore, a message is needed.
2. **Describe the message from the external actor to the system using the message notation described earlier.** In most cases, you will need a message name that describes the service requested from the system and the input parameters being passed. Figure 7-13, the SSD for the *Create new order* use case, illustrates the three messages. Notice that the names of the messages reflect the services that the actor is requesting of the system: startOrder, addItem, and completeOrder. Other names could also have been used. For example, instead of addItem, the name could be enterItemInformation.
   The other information required is the parameter list for each message. Determining exactly which data items must be passed in is more difficult. In fact, developers frequently find that determining the data parameters requires several iterations before a correct, complete list is obtained. The important principle for identifying data parameters is to base it on the class diagram. In other words, the appropriate attributes from the classes are listed as parameters. Looking at the attributes, along with an understanding of what the system needs to do, will help you find the right attributes.
   In the example of the first message, startOrder, the precondition for this use case states that a customer should exist. A postcondition is that the order must be connected to the customer. So, for this simplified version of the use case, the first message passes in the accountNo, which is the identifier in the customer class. Other than the accountNo, no other parameters are needed for the system to locate the existing customer details.

Figure 7-12

A simplified activity diagram of the telephone order scenario

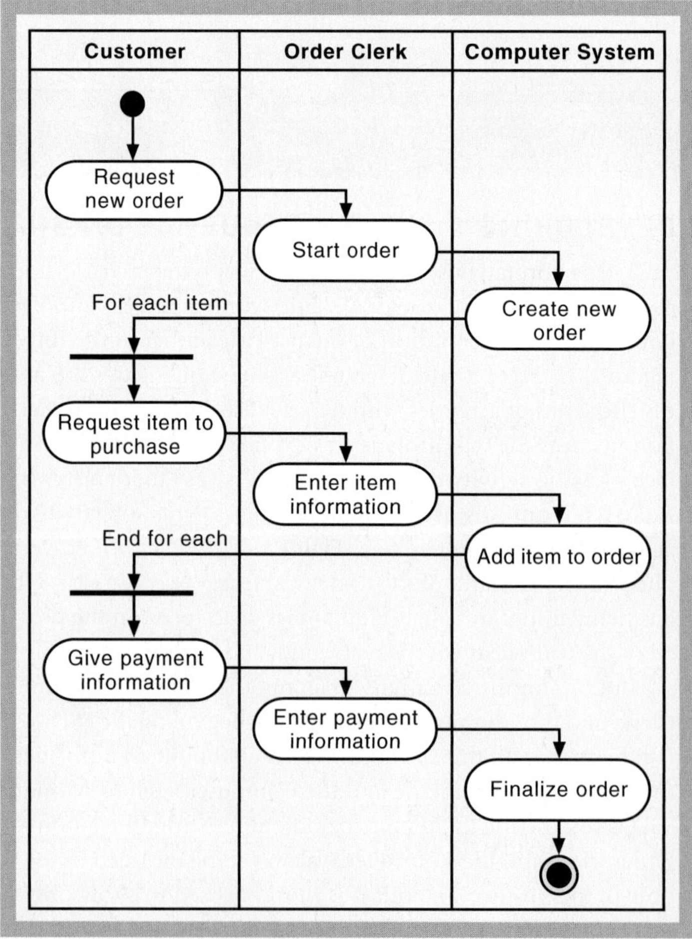

In the second message, addItem, parameters are needed to identify the item from the catalog and the quantity to be purchased. The parameters catalogID, prodID, and size are used to describe the inventory item that will be added to the order. The quantity field, of course, simply identifies how many.

The third message, based on the activity diagram, enters the payment amount. This parameter corresponds to the amount attribute in the OrderTransaction class.

3. **Identify and add any special conditions on the input messages, including iteration and true/false conditions.** In this instance, the iteration box and the true/false condition associated with it are shown in square brackets.
4. **Identify and add the output return messages.** Remember, there are two options to show return information: either as a return value on the message itself or as a separate return message with a dashed-line arrow. The activity diagram can provide some clues about return messages, but there is no standard rule that when a transition arrow in the workflow goes from the system to an external actor, an output always occurs. In Figure 7-12, there are two arrows from the Computer System swimlane to the Customer swimlane. However, in Figure 7-13, only one output message is required. The arrow from the *Create new order* activity in Figure 7-12 does not require output data. In this instance, the only output identified is on the middle message showing the details of the item added to the order—the description, the price, and the extended price (the price times quantity). The other messages could possibly have shown output information such as customer name and address for the first input message, and order confirmation for the third one.

Figure 7-13

An SSD of the simplified telephone order scenario for the *Create new order* use case

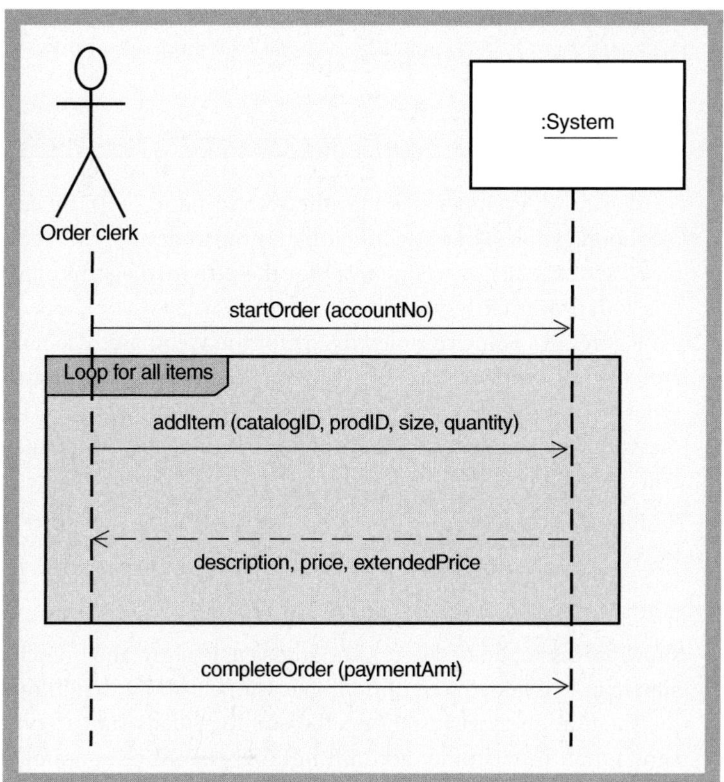

Remember that the objective is discovery and understanding, so you should be working closely with users to define exactly how the workflow proceeds and exactly what information needs to be passed in and provided as output. This is an iterative process, and you will probably need to refine these diagrams several times before they accurately reflect the needs of the users. During Rocky Mountain Outfitters' development project, Barbara Halifax, the project manager, has reviewed many diagrams with the users (see Barbara's status memo).

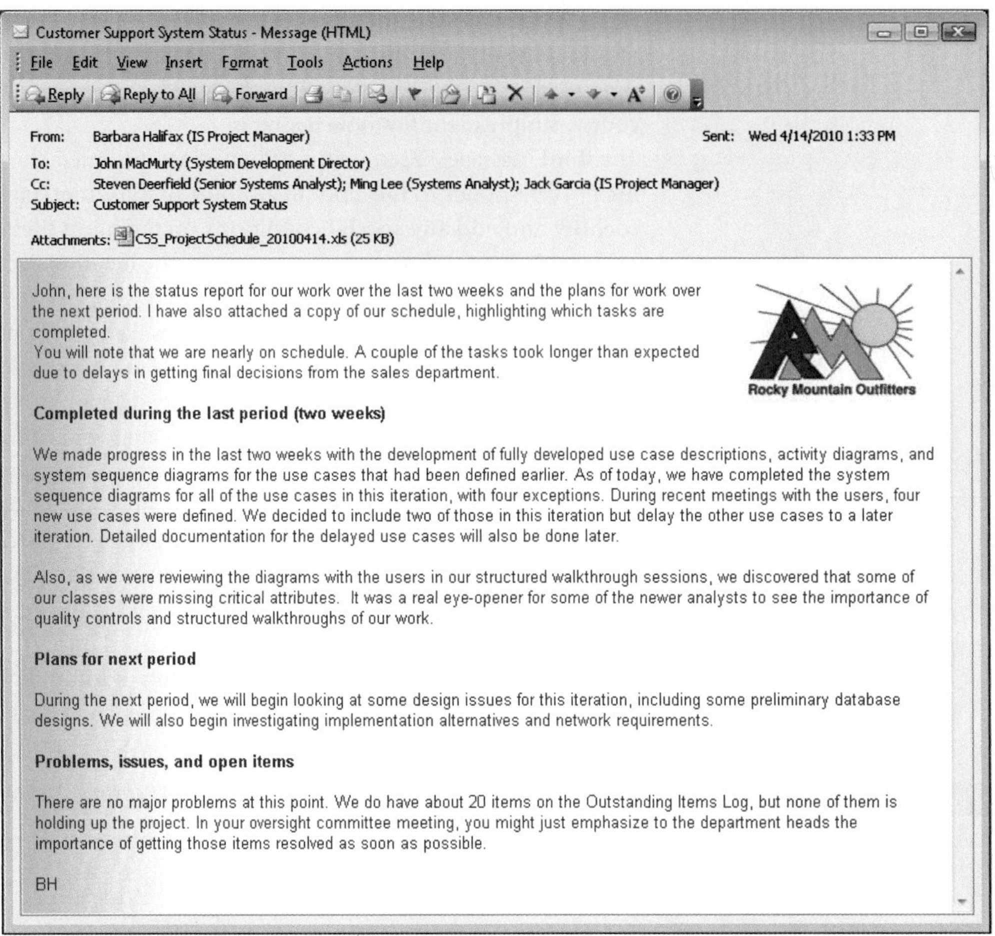
Customer Support System Status - Message (HTML)

File Edit View Insert Format Tools Actions Help

Reply | Reply to All | Forward

From: Barbara Halifax (IS Project Manager) Sent: Wed 4/14/2010 1:33 PM
To: John MacMurty (System Development Director)
Cc: Steven Deerfield (Senior Systems Analyst); Ming Lee (Systems Analyst); Jack Garcia (IS Project Manager)
Subject: Customer Support System Status
Attachments: CSS_ProjectSchedule_20100414.xls (25 KB)

John, here is the status report for our work over the last two weeks and the plans for work over the next period. I have also attached a copy of our schedule, highlighting which tasks are completed.
You will note that we are nearly on schedule. A couple of the tasks took longer than expected due to delays in getting final decisions from the sales department.

Rocky Mountain Outfitters

Completed during the last period (two weeks)

We made progress in the last two weeks with the development of fully developed use case descriptions, activity diagrams, and system sequence diagrams for the use cases that had been defined earlier. As of today, we have completed the system sequence diagrams for all of the use cases in this iteration, with four exceptions. During recent meetings with the users, four new use cases were defined. We decided to include two of those in this iteration but delay the other use cases to a later iteration. Detailed documentation for the delayed use cases will also be done later.

Also, as we were reviewing the diagrams with the users in our structured walkthrough sessions, we discovered that some of our classes were missing critical attributes. It was a real eye-opener for some of the newer analysts to see the importance of quality controls and structured walkthroughs of our work.

Plans for next period

During the next period, we will begin looking at some design issues for this iteration, including some preliminary database designs. We will also begin investigating implementation alternatives and network requirements.

Problems, issues, and open items

There are no major problems at this point. We do have about 20 items on the Outstanding Items Log, but none of them is holding up the project. In your oversight committee meeting, you might just emphasize to the department heads the importance of getting those items resolved as soon as possible.

BH

Let's now develop an SSD for the Web scenario of *Create new order*. Not only is this example more complex, but it will highlight how to develop the requirements for deploying Web-based systems. Refer to Figure 7-9 for the activity diagram of a Web-based order. Notice that this workflow is fairly complex.

Figure 7-14 is the completed SSD for the Web-based scenario. In Figure 7-9, the workflow crosses the automated system boundary from the Customer to the Computer System eight times, some of which are optional flows. In Figure 7-14, the first message, with its response message, begins the use case by requesting the new order page (requestNewOrder). The system does not need input data to perform the processes requested by these two messages, so no input parameters are required. The next input message is a request for the new customer page (newCustomerPage). On this message, there is a true/false condition to test whether this is a new customer. Thus, the message only fires if the new customer condition evaluates to true. Because the objective of a sequence diagram is only to show the messages and not to show processing logic, there is no message to show the branching out to another use case; a simple note is added to remind the developers about that jump.

The third message just allows the user to actually start an order (beginOrder). The message shows that the customer account number is an input parameter. When the user interface is actually developed, this information may already be in the system because it may be on the screen from adding a new customer. However, by showing it as an input parameter, the developers will know that it has to be available, either from the user or captured from another page.

The next process is one of adding items to the order. The activity diagram in Figure 7-9 shows a loop to add items, which is captured by the iteration box. However, one of the activities in the workflow is *Search catalog/view item*. Even though a loop is not explicitly shown, a search normally implies a loop of some type. So, on the input message to view a product in Figure 7-14, an asterisk has been added for iteration. The iteration box and the asterisk on the input message create a nested loop condition. Note that on these two messages, the return-value method is used to return data. The remaining messages and responses follow the activity diagram.

These first sections of the chapter have explained the set of models that are used in object-oriented development to specify the processing aspects of the new system. The use case diagram provides an overview of all of the events that must be supported. The scenario descriptions, as provided by written narratives or activity diagrams, give the details of the internal steps within each use case. Precondition and postcondition statements help define the context for the use case—that is, what must exist before and after processing. Finally, the system sequence diagram describes the inputs and outputs that occur within a use case. Together, these models provide a comprehensive description of the system processing requirements and give the foundation for system design.

Now that the use cases have been explained, let's find out how to capture important object status information.

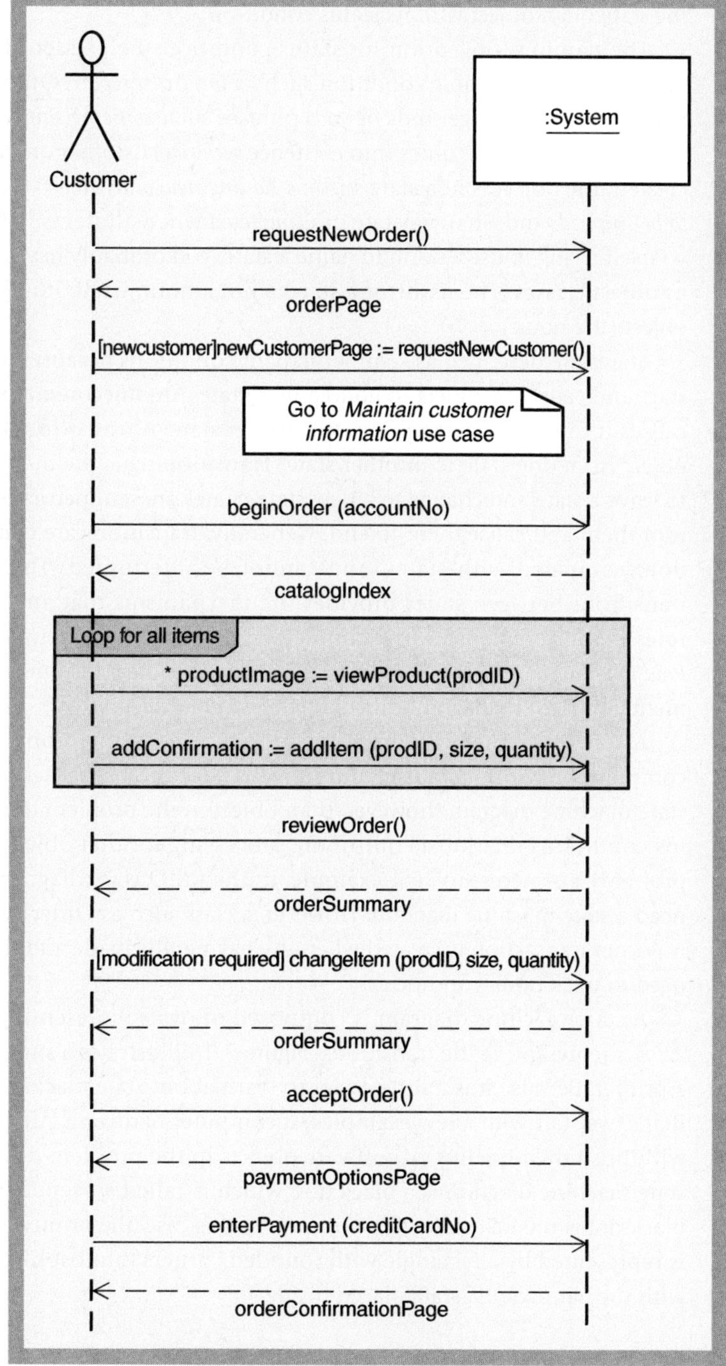

Figure 7-14

An SSD of the Web order scenario for the *Create new order* use case

IDENTIFYING OBJECT BEHAVIOR—THE STATE MACHINE DIAGRAM

Sometimes it is important for a computer system to maintain information about the status of problem domain objects. For example, a customer might want to know whether a particular order has been shipped. A manager might also ask about a customer order and might want to know if it has been paid for. So, the system needs to be able to track the status of customer orders. During requirements definition, analysts need to identify and document which domain objects require status checking and what business rules determine valid status conditions. Referring back to RMO, an example of a business rule is that a customer order should not be shipped until it has been paid for.

state
a condition during an object's life when it satisfies some criterion, performs some action, or waits for an event

The status condition for a real-world object is often referred to as the state of the object. Defined precisely, a **state** of an object is a condition that occurs during its life when it satisfies some criterion, performs some action, or waits for an event. For real-world objects, we equate the state of an object with its status condition.

The naming convention for status conditions helps identify valid states. A state might have a name of a simple condition such as *On* or *In repair*. Other states are more active, with names consisting of gerunds or verb phrases such as *Being shipped* or *Working*. For example, a specific Order object comes into existence when a customer orders something. Right after it is created, the object is in a state such as *Adding new order items*, then a state of *Waiting for items to be shipped*, and finally a state of *Completed* when all items have been shipped. If you find yourself trying to use a noun to name a state, you probably have an incorrect idea about states or object classes. The name of a state is not a noun itself; it is something that describes the object (the noun).

transition
the movement of an object from one state to another state

States are described as semipermanent conditions because external events can interrupt a state and cause the object to go to a new state. An object remains in a state until some event causes it to move, or transition, to another state. A **transition**, then, is the movement of an object from one state to another state. Transitioning is the mechanism that causes an object to leave a state and change to a new state. States are semipermanent because transitions interrupt them and cause them to end. Generally, transitions are considered to be short in duration, compared with states, and cannot be interrupted. The combination of states and transitions between states provides the mechanisms that analysts use to capture business rules. In our previous RMO example, we would say that a customer order must first be in a *Paid for* state before it can transition to a *Shipped* state. This information is captured and documented in a UML diagram called a *state machine diagram*.

A state machine diagram can be developed for any problem domain classes that have complex behavior or status conditions that need to be tracked. Not all classes will require a state machine diagram, however. If an object in the problem domain class does not have status conditions that must control the processing for that object, a state machine diagram is probably not necessary. For example, in the RMO class diagram, a class such as Order may need a state machine diagram. However, a class such as OrderTransaction probably does not. An order transaction is created when the payment is made and then just sits there; it does not need to track other conditions.

pseudostate
the starting point of a state machine diagram, indicated by a black dot

A state machine diagram is composed of ovals representing the states of an object and arrows representing the transitions. Figure 7-15 illustrates a simple state machine diagram for a printer. Because it is a little easier to learn about state machine diagrams by using tangible items, we start with a few examples of computer hardware. After the basics are explained, we will illustrate modeling of software objects in the problem domain. The starting point of a state machine diagram is a black dot, which is called a **pseudostate**. The first shape after the black dot is the first state of the printer. In this case, the printer begins in the *Off* state. A state is represented by a rectangle with rounded corners (almost like an oval, but more squared), with the name of the state placed inside.

destination state
for a particular transition, the state to which an object moves after the completion of a transition

As shown in Figure 7-15, the arrow leaving the *Off* state is called a *transition*. The firing of the transition causes the object to leave the *Off* state and make a transition to the *On* state. After a transition begins, it runs to completion by taking the object to the new state, called the **destination state**. A transition begins with an arrow from an **origin state**—the state prior to the transition—to a destination state, and is labeled with a string to describe the components of the transition.

Figure 7-15

Simple state machine diagram for a printer

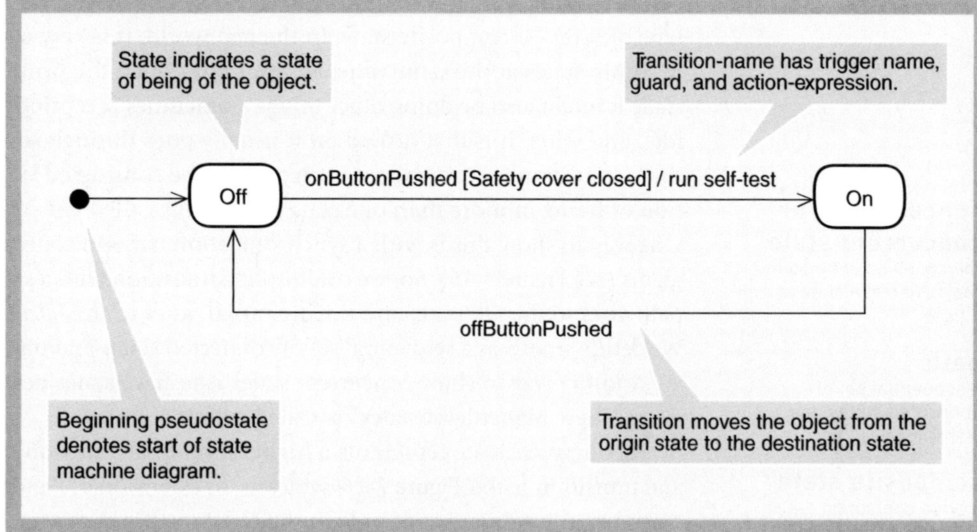

origin state
for a particular transition, the original state of an object from which the transition occurs

The transition label consists of three components:

transition-name (parameters, ...) [guard-condition] / action-expression

In Figure 7-15, the transition-name is onButtonPushed. The transition is like a trigger that fires or an event that occurs. The name should reflect the action of a triggering event. In Figure 7-15, no parameters are being sent to the printer. The guard-condition is Safety cover closed. For the transition to fire, the guard must be true. The forward slash divides the firing mechanism from the actions or processes. Action-expressions indicate some process that must occur before the transition is completed and the object arrives in the destination state. In this case, the printer will run a self-test before it goes into the On state.

message event
the trigger for a transition, which causes the object to leave the origin state

The transition-name is the name of a **message event** that triggers the transition and causes the object to leave the origin state. Notice that the format is very similar to a message in a system sequence diagram. In fact, you will find that the message event names and transition-names use almost the same syntax. One other relationship exists between the messages and the transitions; transitions are caused by messages coming to the object. The parameter portion of the message name comes directly from the message parameters.

guard-condition
a true/false test to see whether a transition can fire

The **guard-condition** is a qualifier or test on the transition, and it is simply a true/false condition that must be satisfied before the transition can fire. For a transition to fire, first the trigger must occur, and then the guard must evaluate to true. Sometimes a transition has only a guard-condition and no triggering event. In that case, the trigger is constantly firing, and whenever the guard becomes true, the transition occurs.

Recall from the discussion of sequence diagrams that messages have a similar test, which is called a *true/false condition*. This true/false condition is a test on the sending side of the message, and before a message can be sent, the true/false condition must be true. In contrast, the guard-condition is on the receiving side of the message. The message may be received, but the transition fires only if the guard-condition is also true. This combination of tests, messages, and transitions provides tremendous flexibility in defining complex behavior.

action-expression
a description of the activities performed as part of a transition

The action-expression is a procedural expression that executes when the transition fires. In other words, it describes the action to be performed. Any of the three components—transition-name, guard-condition, or action-expression—may be empty. If either the transition-name or the guard-condition is empty, it automatically evaluates to true. Either of them may also be complex, with AND and OR connectives.

COMPOSITE STATES AND CONCURRENCY

Before teaching you how to develop a state machine diagram, we need to introduce one other type of state—a composite state. In the real world, it is very common for an object to be in multiple states at the same time. For example, when the printer in Figure 7-15 is in the on state, it might also be doing other things. Sometimes it is printing, sometimes it is just sitting idle, and when it is first turned on it usually goes through some self-checking steps. All of these conditions occur while it is on and can be considered simultaneous states. The condition of being in more than one state at a time is called concurrency, or concurrent states. One way to show this is with a synchronization bar and concurrent paths, as in activity diagrams (see Figure 4-16). So, we could split a transition with a synchronization bar so that one path goes to the *On* state, and the other path goes to the *Idle*, *Printing*, and *Self-check* states. We define a path as a sequential set of connected states and transitions.

concurrency, or concurrent state
the condition of being in more than one state at a time

path
a sequential set of connected states and transitions

composite state
a state containing other states and transitions (that is, a path)

Another way to show concurrent states is to have states nested inside other, higher-level, states. These higher-level states are called composite states.

A composite state represents a higher level of abstraction and can contain nested states and transition paths. Figure 7-16, which is an extension of Figure 7-15, illustrates this idea for a printer. The printer is not only in the *On* state, but is concurrently also in either an *Idle* or *Working* state. The rounded rectangle for the *On* state is divided into two compartments. The top compartment contains the name, and the lower compartment contains the nested states and transition paths.

When the printer enters the *On* state, it automatically begins at the nested black dot and moves to the *Idle* state. So, the printer is in both the *On* state and the *Idle* state. When the print message is received, the printer makes the transition to the *Working* state but also remains in the *On* state. Some new notation is also introduced for the *Working* state. In this instance, the lower compartment contains the action-expressions; that is, the activities that occur while the printer is in the *Working* state.

Figure 7-16
Sample composite states for the printer object

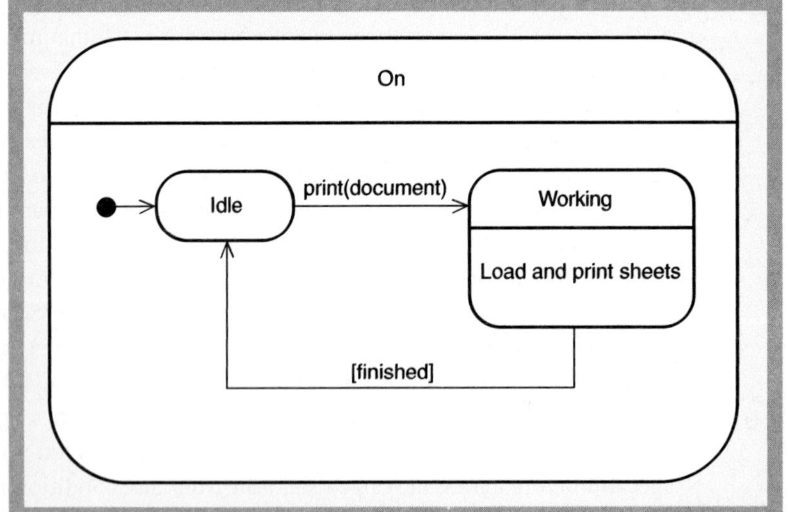

We can extend this idea of composite states and concurrency one step further by allowing multiple paths within a composite state. Perhaps an object has entire sets of states and transitions—multiple paths—that are active concurrently. To document concurrent multiple paths for a single object, we draw a composite state with the lower portion divided into multiple compartments, one for each concurrent path of behavior. For example, imagine a printer that has an input bin to hold the paper. This printer also cycles between two states in its work cycle of *Idle* and *Working*. We may want to describe two separate paths, one representing the states of the input paper tray and the other the states of the printing mechanism. The first path will have states of *Empty*, *Full*, and *Low*. The second path will contain the two states *Idle* and *Working*. These two paths are independent—the movement between states in one compartment is completely independent of movement between states in the other compartment.

As before, there are two ways to document this concurrent behavior. First, we could use a synchronization bar with one path becoming three paths. Second, we can use a composite state. Figure 7-17 extends the printer example from Figure 7-16. In this example, there are two concurrent paths within the composite state. The upper concurrent path represents the paper tray part of the printer. The two paths are completely independent, and the printer moves through the states and transitions in each path independently. When the Off button is pushed, the printer leaves the *On* state. Obviously, when the printer leaves the *On* state, it also leaves all of the paths in the nested states. It does not matter whether the printer is in a state or in the middle of a transition. When the Off button is pushed, all activity is stopped, and the printer exits the *On* state. Now that you know the basic notation of state machine diagrams, we turn next to how to develop a state machine diagram.

Figure 7-17

Concurrent paths for a printer in the *On* state

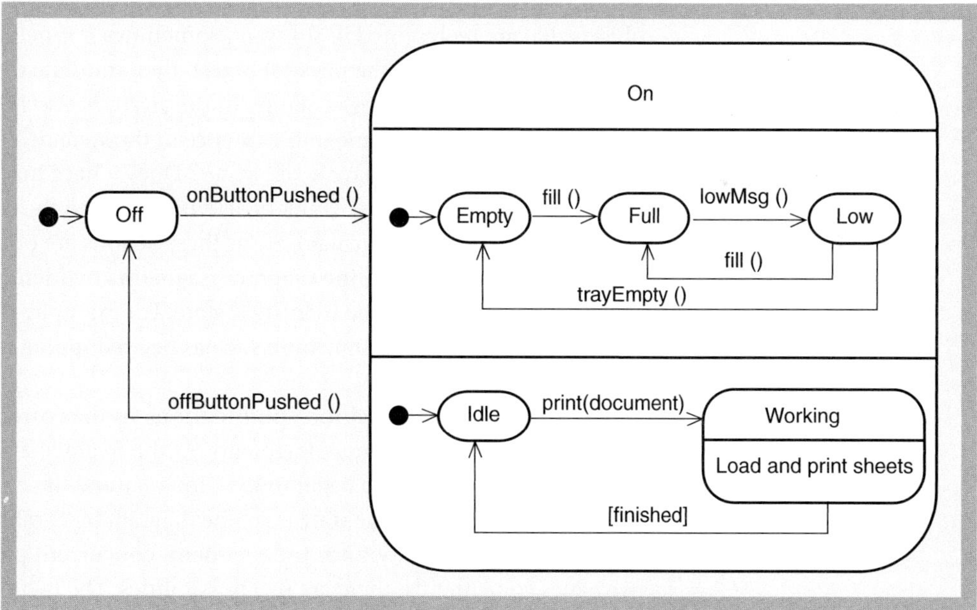

RULES FOR DEVELOPING STATE MACHINE DIAGRAMS

State machine diagram development follows a set of rules. The rules help you to develop state machine diagrams for classes in the problem domain. Usually the primary challenge in building a state machine diagram is to identify the right states for the object. It might be helpful to pretend that you are the object itself. It is easy to pretend to be a customer but a little more difficult to say, "I am an order," or, "I am a shipment. How do I come into existence? What states am I in?" However, if you can begin to think this way, it will help you develop state machine diagrams.

The other major area of difficulty for new analysts is to identify and handle composite states with nested threads. Usually the primary cause of this difficulty is a lack of experience in thinking about concurrent behavior. The best solution is to remember that developing state machine diagrams is an iterative behavior, more so than developing any other type of diagram. Analysts seldom get a state machine diagram right the first time. They always draw it and then refine it again and again. Also, remember that when you are defining requirements, you are only getting a general idea of the behavior of an object. During design, as you build detailed sequence diagrams, you will have an opportunity to refine and correct important state machine diagrams.

Finally, don't forget to ask about an exception condition—especially when you see the words *verify* or *check*. Normally, there will be two transitions out of states that verify something—one for acceptance and one for rejection.

Here is a list of steps that will help you get started in developing state machine diagrams:

1. **Review the class diagram and select the classes that will require state machine diagrams.** Remember, we normally include only those that have multiple status conditions that are important for the system to track. Then begin with the classes that appear to have the simplest state machine diagrams, such as the OrderItem class for RMO, discussed later.
2. **For each selected class in the group, make a list of all the status conditions you can identify.** At this point, simply brainstorm. If you are working on a team, have a brainstorming session with the whole team. Remember that you are defining states of being of the software classes. However, these states must also reflect the states for the real-world objects that are represented in software. Sometimes it is helpful to think of the physical object, identify states of the physical object, then translate those that are appropriate into corresponding system states or status conditions. It is also helpful to think of the life of the object. How does it come into existence in the system? When and how is it deleted from the system? Does it have active states? Does it have inactive states? Does it have states in which it is waiting? Think of activities done to the object or by the object. Often, the object will be in a particular state as these actions are occurring.
3. **Begin building state machine diagram fragments by identifying the transitions that cause an object to leave the identified state.** For example, if an Order is in a state of *Ready to be shipped*, then a transition such as beginShipping will cause the Order to leave that state.
4. **Sequence these state-transition combinations in the correct order.** Then aggregate these combinations into larger fragments. As the fragments are being aggregated into larger paths, it is natural to begin to look for a natural life cycle for the object. Continue to build longer paths in the state machine diagram by combining the fragments.
5. **Review the paths and look for independent, concurrent paths.** When an item can be in two states concurrently, there are two possibilities. The two states may be on independent paths, as in the printer example of *Working* and *Full*. This occurs when the states and paths are independent, and one can change without affecting the other. Alternately, one state may be a composite state, so the two states should be nested, one inside the other. One way to identify a candidate for a composite state is to determine whether it is concurrent with several other states and whether these other states depend on the original state. For example, the *On* state has several other states and paths that can occur while the printer is in the *On* state, and those states depend on the printer being in the *On* state.
6. **Look for additional transitions.** Often, during a first iteration, several of the possible combinations of state-transition-state are missed. One method to identify them is to take every paired combination of states and ask whether there is a valid transition between the states. Test for transitions in both directions.

7. **Expand each transition with the appropriate message event, guard-condition, and action-expression.** Include with each state appropriate action-expressions. Much of this work may have been done as the state machine diagram fragments were being built.
8. **Review and test each state machine diagram.** We test state machine diagrams by reviewing them carefully. Review each of your state machine diagrams by doing the following:
   a. Make sure your states are really states of the object in the class. Ensure that the names of states truly describe states of being of the object.
   b. Follow the life cycle of an object from its coming into existence to its being deleted from the system. Be sure that all possible combinations are covered and that the paths on the state machine diagram are accurate.
   c. Be sure your diagram covers all exception conditions, as well as the normal expected flow of behavior.
   d. Look again for concurrent behavior (multiple paths) and the possibility of nested paths (composite states).

DEVELOPING RMO STATE MACHINE DIAGRAMS

Let's practice these steps by developing two state machine diagrams for RMO. Step 1 is to review the domain class diagram and select the classes that may have status conditions that need to be tracked. In this case, we select the Order and OrderItem classes. We assume that customers will want to know the status of their orders and the status of individual items on the order. Other classes that are candidates for state machine diagrams are InventoryItem, to track in-stock or out-of-stock items; Shipment, to track arrivals; and possibly Customer, to track active and inactive customers. For our purposes here, we focus on the Order and OrderItem classes. We use the OrderItem class because it is simpler, and it is always best to start with the simplest class. Also, it is a dependent class—it depends on Order. Finally, it is best to use a bottom-up approach, starting with the lower items on a hierarchy, which usually have less ripple effect.

Developing the OrderItem State Machine Diagram

Start by identifying the possible status conditions that might be of interest. Some necessary status conditions are *Ready to be shipped*, *On back order*, and *Shipped*. An interesting question comes to mind at this point: Can an order item be partially shipped? In other words, if the customer ordered 10 of a single item, but there are only five in inventory, should RMO ship those five and put the other five on back order? You should see the ramifications of this decision. The system and the database would need to be designed to track and monitor detailed information to support this capability. The domain class diagram for RMO (see Figure 5-38) indicates that an OrderItem can be associated with either zero (not yet shipped) shipments or one (totally shipped) shipment. Based on the current specification, the definition does not allow partial shipments of OrderItems.

This is just another example of the benefit of building models. Had we not been developing the state machine diagram model, this question might never have been asked. The development of detailed models and diagrams is one of the most important activities that a system developer can perform. It forces analysts to ask fundamental questions. Sometimes new system developers think that model development is a waste of time, especially for small systems. However, truly understanding the users' needs before writing the program always saves time in the long run.

The next step is to identify exit transitions for each of the status conditions. Figure 7-18 is a table showing the states that have been defined and the exit transitions for each of those states. One additional state has been added to the list, *Newly added*, which covers the condition that occurs when an item has been added to the order, but the order is not complete or paid for, so the item is not ready for shipping.

Figure 7-18

States and exit transitions for OrderItem

State	Transition causing exit from state
Newly added	finishedAdding
Ready to ship	shipItem
On back order	itemArrived
Shipped	No exit transition defined

The fourth step is to combine the state-transition pairs into fragments and to build a state machine diagram with the states in the correct sequence. Figure 7-19 illustrates the partially completed state machine diagram. The flow from beginning to end for OrderItem is quite obvious. However, at least one transition seems to be missing. There should be some path to allow entry into the *On back order* state, so we recognize that this first-cut state machine diagram needs some refinement. We will fix that in a moment.

Figure 7-19

Partial state machine diagram for OrderItem

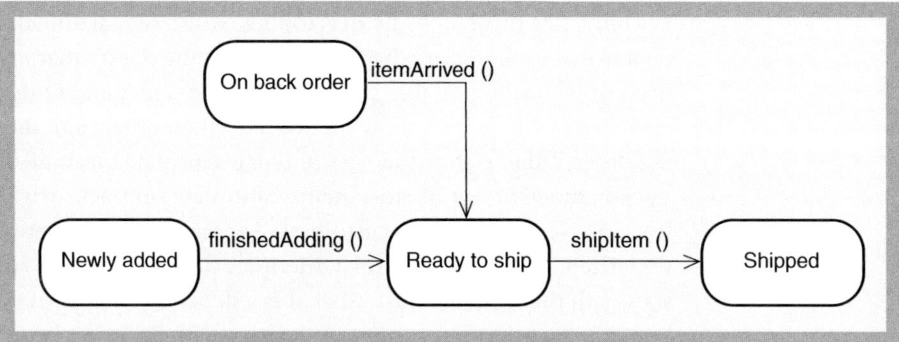

The fifth step is to look for concurrent paths. In this case, it does not appear that an OrderItem can be in any two of the identified states at the same time. Of course, because we chose to begin with a simple state machine diagram, that was expected.

The sixth step is to look for additional transitions. This step is where we flesh out other necessary transitions. The first addition is to have a transition from *Newly added* to *On back order*. To continue, examine every pair of states to see whether there are other possible combinations. In particular, look for backward transitions. For example, can an OrderItem go from *Ready to ship* to *On back order*? This would happen if the shipping clerk found that there were not enough items in the warehouse, even though the system indicated that there should have been. Other backward loops, such as from *Shipped* to *Ready to ship*, or from *On back order* to *Newly added*, do not make sense and are not included.

The seventh step is to complete all the transitions with correct names, guard-conditions, and action-expressions. Two new transition-names are added. The first is the transition from the beginning black dot to the *Newly added* state. That transition causes the creation, or in system terms the *instantiation*, of a new OrderItem object. It is given the same name as the message into the system that adds it—addItem ( ). The final transition is the one that causes the order item to be removed from the system. This transition goes from the *Shipped* state to a final circled black dot, which is a final pseudostate. On the assumption that it is archived to a backup tape when it is deleted from the active system, that transition is named archive ( ).

Action-expressions are added to the transitions to indicate any special action that is initiated by the object or on the object. In this case, only one action is required. When an item that was *Ready to ship* moves to *On back order*, the system should initiate a new purchase order to the supplier to buy more items. So, on the markBackOrdered ( ) transition, an action-expression is noted to place a purchase order. Figure 7-20 illustrates the final state machine diagram for OrderItem.

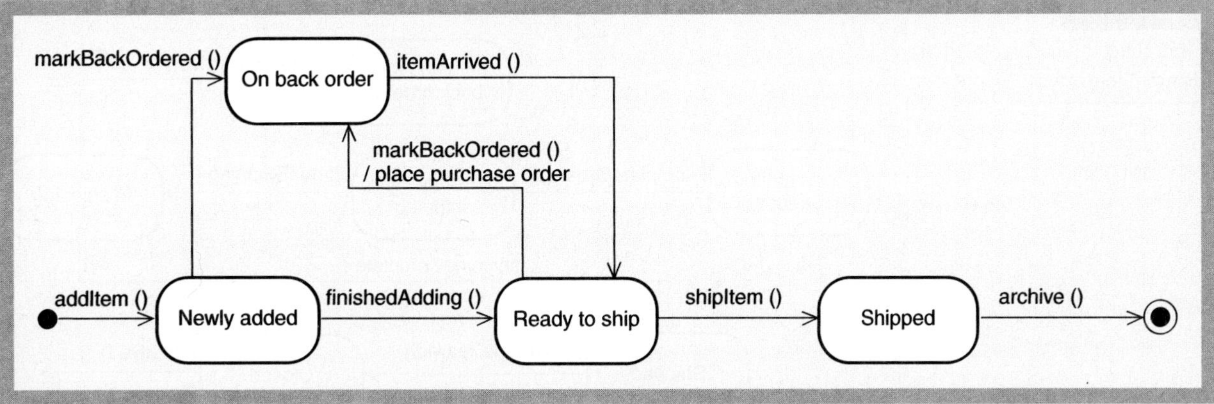

Figure 7-20
Final state machine diagram for OrderItem

The final step, reviewing and testing the state machine diagram, is the quality-review step. It is always tempting to omit this step; however, a good project manager ensures that the systems analysts have time in the schedule to do a quick quality check of their models. A walkthrough at this point in the project is very appropriate.

Developing the Order State Machine Diagram

An Order object is a little more complex than the OrderItem objects. In this example, you will see some additional features of state machine diagrams that support more complex objects.

Figure 7-21 is a table of the defined states and exit transitions that, on first iteration, appear to be required. Reading from top to bottom, the states describe the life cycle of an order (for example, the status conditions). First, an Order comes into existence and is ready to have items added to it—*Open for item adds*. The users in RMO indicated that they wanted an order to remain in this state for 24 hours in case the customer wanted to add more items. After all the items are added, the order is *Ready for shipping*. Next, it goes to shipping and is in the *In shipping* state. At this point, it is not quite clear how *In shipping* and *Waiting for back orders* relate to each other. That relationship will have to be sorted out as the state machine diagram is being built. Finally, the order is *Shipped*, and after the payment clears, it is *Closed*.

Figure 7-21
States and exit transitions for Order

State	Exit transition
Open for item adds	completeOrder
Ready for shipping	beginShipping
In shipping	shippingComplete
Waiting for back orders	backOrdersArrive
Shipped	paymentCleared
Closed	archive

In step 4, fragments are built and combined to yield the first-cut state machine diagram. Figure 7-22 illustrates the first-cut state machine diagram. The state machine diagram built from the fragments appears to be correct for the most part. However, we note some problems with the *Waiting for back orders* state.

After some analysis, we decide that being *In shipping* and *Waiting for back orders* are concurrent states. And another state is needed, called *Being shipped*, for the state in which the shipping clerk is actively shipping items. One way to show the life of an order is to put it in the *In shipping* state when shipping begins. It also enters the *Being shipped* state at that point. The order can cycle between *Being shipped* and *Waiting for back orders*. The exit out of the composite state only occurs from the *Being shipped* state, which is inside the *In shipping* state. Obviously, upon leaving the inside state, the order also leaves the composite *In shipping* state.

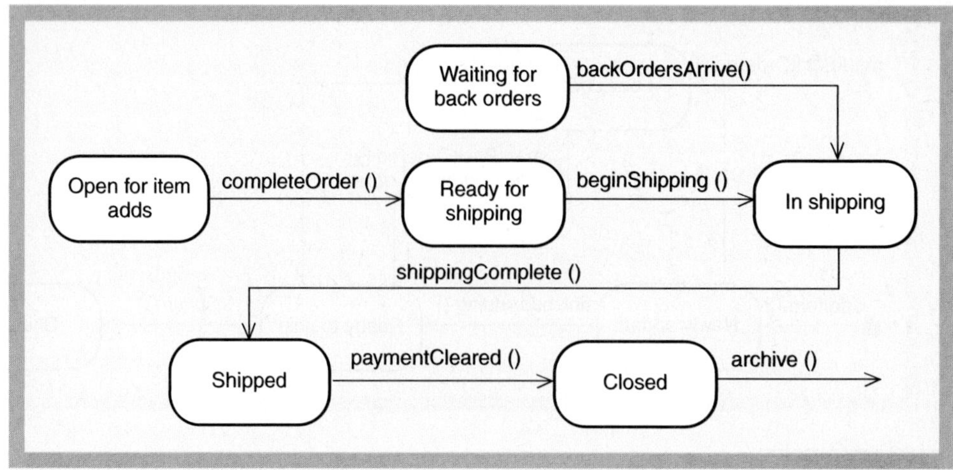

Figure 7-22

First-cut state machine diagram for Order

As we go through steps 5, 6, and 7, we note that new transitions must be added. The creation transition from the initial pseudostate is required. Also, transitions must be included to show when items are being added and when they are being shipped. Usually we put these looping activities on transitions that leave a state and return to the same state. In this case, the transition is called addItem ( ). Note how it leaves the *Open for item adds* state and returns to the same state. Figure 7-23 takes the state machine diagram to this level of completion.

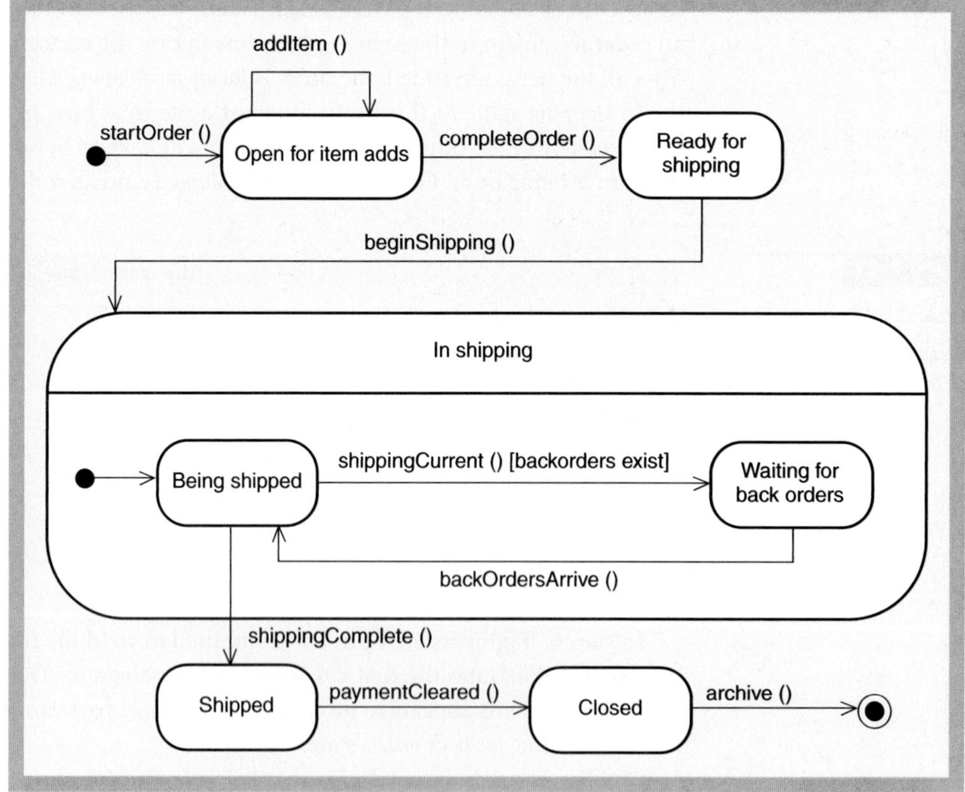

Figure 7-23

Second-cut state machine diagram for Order

The benefit of developing a state machine diagram for an object, even a business object, is that it helps to capture and clarify business rules. From the state machine diagram, we can see that shipping cannot commence while the order is in the *Open for item adds* state. New items cannot be added to the order after it has been placed in the *Ready for shipping* state. The order is not considered shipped until all items are shipped. If the order has the status of *In shipping*, we know that it is either actively being worked on or waiting for back orders.

As always, the benefits of careful model building help us gain a true understanding of the system requirements. Let's now look at the big picture and pull the different models into a whole to see how they fit together.

INTEGRATING OBJECT-ORIENTED MODELS

The diagrams described in this chapter allow analysts to completely specify the system requirements. If you were developing a system using a waterfall systems development life cycle, you would develop the complete set of diagrams to represent all system requirements before continuing with design. However, because you are using an iterative approach, you would only construct the diagrams that are necessary for a given iteration. A complete use case diagram would be important to get an idea of the total scope of the new system. But the supporting details included in use case descriptions, activity diagrams, and system sequence diagrams need only be done for use cases in the specific iteration.

BEST PRACTICE

Developing and integrating models are critical to ensure that you understand the business requirements.

The domain model class diagram is a special case. Much like the entire use case diagram, the domain model class diagram should be as complete as possible for the entire system, as shown for RMO in Chapter 5. The number of problem domain classes for the system provides an additional indicator of the total scope of the system. Refinement and actual implementation of many classes will wait for later iterations, but the domain model should be fairly complete. The domain model is necessary to identify all of the domain classes that are required in the new system. Although we do not focus on database design in this chapter, the domain model is also used to design the database.

Throughout the chapter, you have seen how the construction of a diagram depends on information provided by another diagram. You have also seen that the development of a new diagram often helps refine and correct a previous diagram. You should also have noted that the development of detailed diagrams is critical to gain a thorough understanding of the user requirements. Figure 7-24 illustrates the primary relationships among the requirements models for OO development. The use case diagram and other diagrams on the left are used to capture the processes of the new system. The class diagram and its dependent diagrams capture information about the classes for the new system. The solid arrows represent major dependencies, and the dashed arrows show a minor dependency. The dependencies generally flow from top to bottom, but some arrows have two heads to illustrate that influence goes in both directions.

Figure 7-24

Relationships among OO requirements models

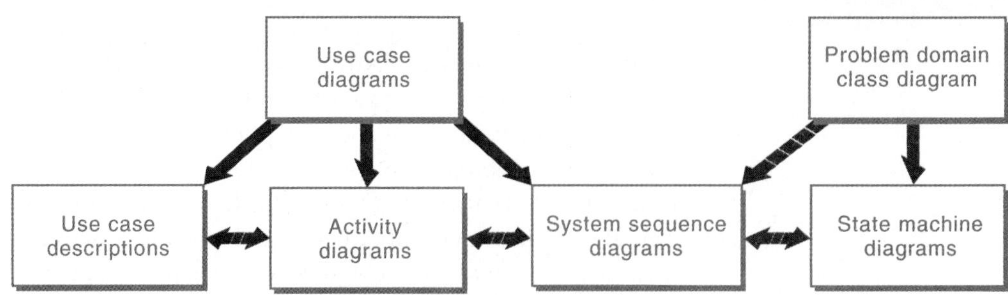

Note that the use case diagram and the problem model domain class diagram are the primary models from which others draw information. You should develop those two diagrams as completely as possible. In this chapter, we noted that a CRUD analysis performed between the class diagram and use case diagram helps ensure that they are as complete as possible. The detailed descriptions, either in narrative format or in activity diagrams, are important internal documentation of the use cases and must completely support the use case diagram. Internal descriptions such as preconditions and postconditions use information from the class diagram. These detailed descriptions are also important for development of system sequence diagrams. So, the detailed descriptions, activity diagrams, and system sequence diagrams must all be consistent with regard to the steps of a particular use case. As you progress in developing the system, and especially as you begin doing detailed system design, you will find that understanding the relationships among these models is an important element in the quality of your models.

SUMMARY

The object-oriented approach has a complete set of diagrams that together document the user's needs and define the system requirements. These requirements are specified using the following models:

- Domain model: class diagrams and state machine diagrams
- Use case model: use case diagrams, detailed models (description format or activity diagram), and system sequence diagrams (SSDs)

A use case diagram documents the various ways that the system can be used. It can be developed independently or in conjunction with the event table, where one event triggers one use case. A use case diagram consists of actors, use cases, and connecting lines. A use case identifies a single function that the system supports. An actor represents a role of someone or something that uses the system. The connecting lines indicate which actors invoke which use cases. Use cases can also invoke other use cases as a common subroutine. This type of connection between use cases is called the «includes» relationship.

The internal activities of a use case are first described by an internal flow of activities. It is possible to have several different internal flows, which represent different scenarios of the same use case. Thus, a use case may have several scenarios. These details are documented either in use case descriptions or with activity diagrams.

Another diagram that provides more details of the use case's processing requirements is a system sequence diagram, or SSD. An SSD documents the inputs and outputs of the system. The scope of each SSD is usually a use case or a scenario within a use case. The components of an SSD are the actor—the same actor identified in the use case—and the system. The system is treated as a black box, in that the internal processing is not addressed. Messages, which represent the inputs, are sent from the actor to the system. Output messages are returned from the system to the actor. The sequence of messages is indicated from top to bottom.

The domain model class diagram continues to be refined when defining requirements. The behavior of business objects represented in the class diagram is an aspect of the requirements that is also studied and modeled. The state machine diagram is used to model object states and state transitions that occur in a use case. All of the models discussed in this chapter are interrelated, and information in one model explains information in others.

KEY TERMS

action-expression, p. 262
composite state, p. 262
concurrency, or concurrent state, p. 262
destination state, p. 261
domain model, p. 242
guard-condition, p. 261
interaction diagram, p. 252
lifeline, or object lifeline, p. 253
message, p. 242
message event, p. 261
origin state, p. 261
package, p. 245
path, p. 262
pseudostate, p. 260
state, p. 260
state machine diagram, p. 242
system sequence diagram, p. 242
transition, p. 260
true/false condition, p. 255
use case diagram, p. 242
use case model, p. 242

REVIEW QUESTIONS

1. What is the OMG?
2. What is UML? What type of modeling is it used for?
3. What are the four basic parts of a use case model? What is its purpose or objective?

4. What are the two basic parts of the domain model? What is its purpose or objective?
5. What is the difference between a use case description and an activity diagram?
6. What is the «includes» relationship used for?
7. What is the difference in the focus on the boundary condition of a use case diagram and an event table?
8. With regard to a use case, what is an activity diagram used for?
9. What is the purpose of a system sequence diagram? What symbols are used in a system sequence diagram?
10. What are the steps required to develop a system sequence diagram?
11. What is the purpose of a state machine diagram?
12. List the primary steps for developing a state machine diagram.
13. List the elements that make up a transition description. Which elements are optional?
14. What is a composite state? What is it used for?
15. What is meant by the term *path*?
16. What is the purpose of a guard-condition?
17. Identify the models explained in this chapter and their relationship to each other.

THINKING CRITICALLY

1. To review your skills in developing a class diagram, develop a domain model class diagram, including associations and multiplicities, based on the following narrative.

   This case is a simplified (initial draft) version of a new system for the University Library. Of course, the library system must keep track of books. Information is maintained both about book titles and the individual book copies. Book titles maintain information about title, author, publisher, and catalog number. Individual copies maintain copy number, edition, publication year, ISBN, book status (whether it is on the shelf or loaned out), and date due back in.

   The library also keeps track of its patrons. Because it is a university library, there are several types of patrons, each with different privileges. There are faculty patrons, graduate student patrons, and undergraduate student patrons. Basic information about all patrons is name, address, and telephone number. For faculty patrons, additional information is office address and telephone number. For graduate students, information such as graduate program and advisor information is maintained. For undergraduate students, program and total credit hours are maintained.

   The library also keeps information about library loans. A library loan is a somewhat abstract object. A loan occurs when a patron approaches the circulation desk with a stack of books to check out. Over time a patron can have many loans. A loan can have many physical books associated with it. (And a physical book can be on many loans over a period of time. Information about past loans is kept in the database.) So, in this case, an association class should probably be created for loaned books.

   If a patron wants a book that is already checked out, the patron can put that title on reserve. This is another class that does not represent a concrete object. Each reservation is for only one title and one patron. Information such as date reserved, priority, and date fulfilled is maintained. When a book is fulfilled, the system associates it with the loan on which it was checked out.

2. Develop a use case diagram for the university library system.
   Part a. Based on the following descriptions, list the use cases and actors.

   Patrons have access to the library information to search for book titles and to see whether a book is available. A patron can also reserve a title if all copies are checked out. When patrons bring books to the circulation desk, a clerk checks out the books on a loan. Clerks also check books in. When books are dropped in the return slot, clerks check in the books. Stocking clerks keep track of the arrival of new books.

   The managers in the library have their own activities. They will print reports of book titles by category. They also like to see (online) all overdue books. When books get damaged or destroyed, managers delete information about book copies. Managers also like to see what books are on reserve.

   Part b. Given your list of use cases and actors, develop a use case diagram.

   Part c. Given the domain model class diagram you developed in question 1, do a CRUD analysis and list any new use cases you discover. Or, if you change the name of any use cases, indicate that as well. In this case, patron information can be accessed and downloaded from another university database.

3. To review your skills in developing a class diagram, develop a domain model class diagram, including associations and multiplicities, based on the following narrative.

   A clinic with three dentists and several dental hygienists needs a system to help administer patient records. This system does not keep any medical records. It only processes patient administration.

Each patient has a record with his or her name, date of birth, gender, date of first visit, and date of last visit. Patient records are grouped together under a household. A household has attributes such as name of head of household, address, and telephone number. Each household is also associated with an insurance carrier record. The insurance carrier record contains name of insurance company, address, billing contact person, and telephone number.

In the clinic, each dental staff person also has a record that tracks who works with a patient (dentist, dental hygienist, x-ray technician). Because the system focuses on patient administration records, only minimal information is kept about each dental staff person, such as name, address, and telephone number. Information is maintained about each office visit, such as date, insurance copay amount (amount paid by the patient), paid code, and amount actually paid. Each visit is for a single patient, but, of course, a patient will have many office visits in the system. During each visit, more than one dental staff person may be involved by doing a procedure. For example, the x-ray technician, dentist, and dental hygienist may all be involved on a single visit. In fact, some dentists are specialists in such things as crown work, and even multiple dentists may be involved with a patient. For each *staff person does procedure in a visit* combination (many-to-many), detailed information is kept about the procedure. This information includes the type of procedure, a description, the tooth involved, the copay amount, the total charge, the amount paid, and the amount the insurance company denied.

Finally, the system also keeps track of invoices. There are two types of invoices: invoices to insurance companies and invoices to heads of household. Both types of invoices are fairly similar, listing each visit, the procedures involved, the patient copay amount, and the total due. Obviously, the totals for the insurance company are different from the patient amounts owed. Even though an invoice is a report (when printed), it also maintains some information such as date sent, total amount, amount already paid, amount due and the total received, date received, and total denied. (Insurance companies do not always pay all they are billed.)

4. Develop a use case diagram for the dental clinic.

Part a. Based on the following descriptions, list the use cases and actors.

The receptionist keeps track of patient and head-of-household information, and will enter this information in the system. The receptionist will also keep track of office visits by the patients. Patient information is also entered and maintained by the office business manager. In addition, the business manager maintains the information about the dental staff.

The business manager also prints the invoices. Patient invoices are printed monthly and sent to the head of household. Insurance invoices are printed weekly. When the invoices are printed, the business manager double-checks a few invoices against information in the system to make sure it is being aggregated correctly. She also enters the payment information when it is received.

Dental staff are responsible for entering information about the dental procedures they perform.

The business manager also prints an overdue invoice report that shows heads of household who are behind on their payments. Sometimes dentists like to see a list of the procedures they performed during a week or month, and they can request that report.

Part b. Given your list of use cases and actors, develop a use case diagram.

Part c. Expand the use case diagram you have developed based on a CRUD analysis of the class diagram you developed in the previous problem.

5. Interpret and explain the use case diagram in Figure 7-25. Explain the various roles of those using the system and the functions that each role requires. Explain the relationships and the ways the use cases are related to each other.

6. Given the following narrative, do the following:

a. Develop an activity diagram for each scenario, and

b. Complete a fully developed use case description for each scenario.

Quality Building Supply has two kinds of customers: contractors and the general public. Sales to each are slightly different.

A contractor buys materials by taking them to the contractor's checkout desk. The clerk enters the contractor's name into the system. The system displays the contractor's information, including current credit standing.

The clerk then opens up a new ticket (sale) for the contractor. Next, the clerk scans in each item to be purchased. The system finds the price of the item and adds the item to the ticket. At the end of the purchase, the clerk indicates the end of the sale. The system compares the total amount against the contractor's current credit limit and, if it is acceptable, finalizes the sale. The system creates an electronic ticket for the items, and the contractor's credit limit is reduced by the amount of the sale. Some contractors like to keep a record of their purchases, so they request that the ticket details be printed. Others aren't interested in a printout.

A sale to the general public is simply entered into the cash register, and a paper ticket is printed as the items are identified. Payment can be by cash, check, or credit card. The clerk must enter the type of payment to ensure that the cash register balances at the end of the shift. For credit-card payments, the system prints a credit-card voucher that the customer must sign.

7. Given the following narrative, develop either an activity diagram or a fully developed description for a use case of *Add a new vehicle to an existing policy* in a car insurance system.

Figure 7-25

A use case diagram for the inventory system

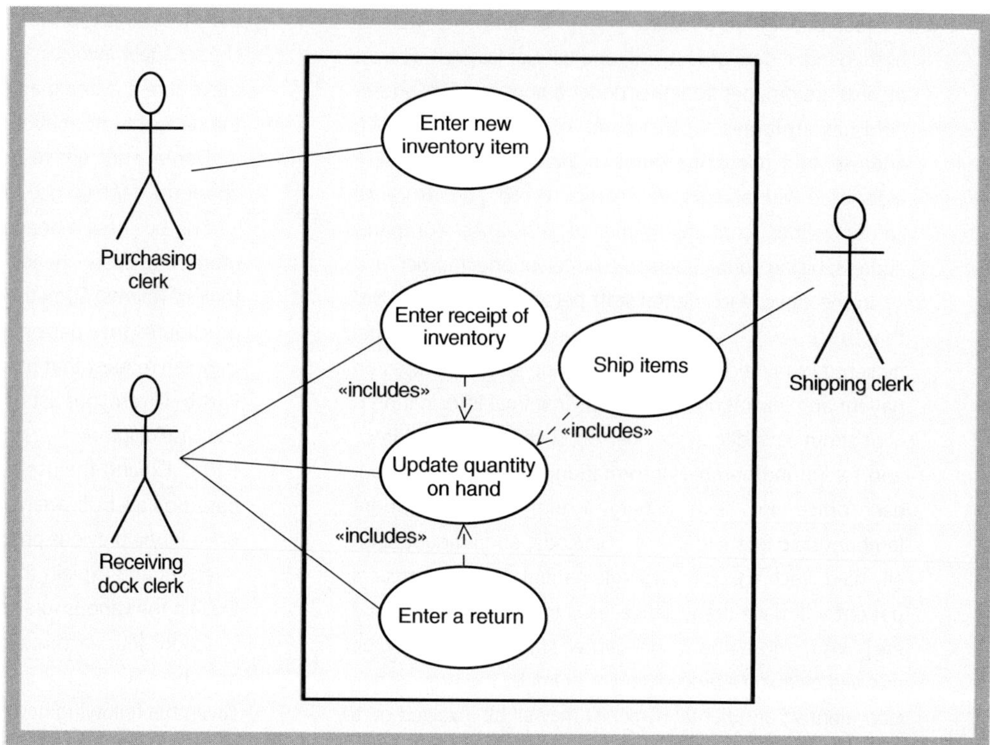

A customer calls a clerk at the insurance company and gives his policy number. The clerk enters this information, and the system displays the basic insurance policy. The clerk then checks the information to make sure the premiums are current and the policy is in force.

The customer gives the make, model, year, and vehicle identification number (VIN) of the car to be added. The clerk enters this information, and the system ensures that the given data is valid. Next, the customer selects the types of coverage desired and the amount of each. The clerk enters the information, and the system records it and validates the requested amount against the policy limits. After all of the coverages have been entered, the system ensures the total coverage against all other ranges, including other cars on the policy.

Finally, the customer must identify all drivers and the percentage of time they drive the car. If a new driver is to be added, then another use case, *Add new driver*, is invoked.

At the end of the process, the system updates the policy, calculates a new premium amount, and prints the updated policy statement to be mailed to the policy owner.

8. Given the following list of classes and relationships for the previous car insurance system, list the preconditions and the postconditions for the *Add a new vehicle to an existing policy* use case.

   Classes in the system:
   - Policy
   - InsuredPerson
   - InsuredVehicle
   - Coverage
   - StandardCoverage (lists standard insurance coverages with prices by rating category)
   - StandardVehicle (lists all types of vehicles ever made)

   Relationships in the system:
   - Policy has InsuredPersons (one to many)
   - Policy has InsuredVehicles (one to many)
   - Vehicle has Coverages (one to many)
   - Coverage is a type of StandardCoverage
   - Vehicle is a StandardVehicle

9. Develop a system sequence diagram based on the narrative and your activity diagram for problem 6 in this section.
10. Develop a system sequence diagram based on the narrative or your activity diagram for problem 7 in this section.
11. Review the cellular telephone state machine diagram in Figure 7-26 and then answer the following questions. (Note that this telephone has unique characteristics that are not found in ordinary telephones. Base your answers only on the state machine diagram.)
    a. What happens to turn on the telephone?
    b. What states does the telephone go into when it is turned on?
    c. What are the three ways that the telephone can be turned off?
    d. Can the telephone turn off in the middle of the *Active* (*Talking*) state?
    e. How can the telephone get to the *Active* (*Talking*) state?
    f. Can the telephone be plugged in while someone is talking?

Figure 7-26

Cellular telephone state machine diagram

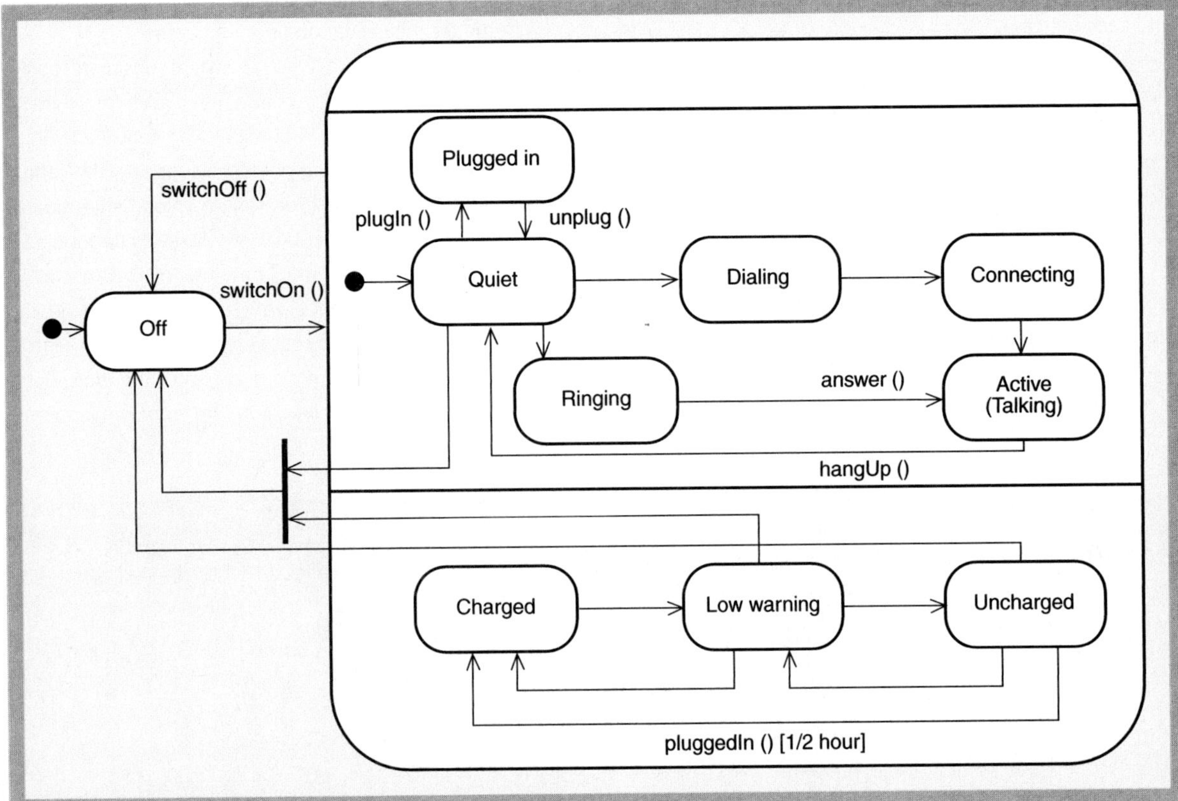

g. Can the telephone change battery states while someone is talking? Explain which movement is allowed, and which is not allowed.

h. What states are concurrent with what other states? Make a two-column table showing the concurrent states.

12. Given the following description of a shipment by Union Parcel Shipments, first identify all of the states and exit transitions, then develop a state machine diagram.

A shipment is first recognized after it has been picked up from a customer. After it is in the system, it is considered to be *active* and *in transit*. Every time it goes through a checkpoint, such as arrival at an intermediate destination, it is scanned and a record is created indicating the time and place of the checkpoint scan. The status changes when it is placed on the delivery truck. It is still active, but now it is also considered to have a status of *delivery pending*. Of course, after it is delivered, the status changes again.

From time to time, a shipment has a destination that is outside the area served by Union. In those cases, Union has working relationships with other courier services. After a package is handed off to another courier, it is noted as being *handed over*. In those instances, a tracking number for the new courier is recorded (if it is provided). Union also asks the new courier to provide a status change notice after the package has been delivered.

Unfortunately, from time to time a package gets lost. In that case, it remains in an *active* state for two weeks but is also marked as *misplaced*. If after two weeks the package has not been found, it is considered *lost*. At that point, the customer can initiate lost-package procedures to recover any damages.

EXPERIENTIAL EXERCISES

1. The functionality required by Rocky Mountain Outfitters' customer support system is also found in several real-world companies. Based on your experience with online shopping and shopping carts, build a use case diagram of functions that a Web customer can perform (similar to Figure 7-4). Web sites that you might refer to include L.L. Bean (*www.llbean.com/*), Lands' End (*www.landsend.com/*), Amazon.com (*www.amazon.com/*), and Barnes and Noble Booksellers (*www.barnesandnoble.com/*).

2. Create fully developed use case descriptions for each of the use cases you defined in exercise 1.
3. Based on the flow of activities you developed in exercise 2, develop system sequence diagrams for those same use cases and scenarios. Add preconditions and postconditions to each use case.
4. Analyze the information requirements of the Web site from exercise 1. Doing a reverse CRUD analysis (going from the use case diagram to the domain model class diagram) will help you identify classes. Develop a domain model class diagram.
5. Locate a company in your area that develops software. Consulting companies or companies with a large staff of information systems professionals tend to be more rigorous in their approach to software development. Set up an interview. Determine the development approaches that the company uses. Many companies still use traditional structured techniques combined with some object-oriented development. In other companies, some projects are structured, while other projects are object oriented. Find out what kinds of modeling the company does for requirements specification. Compare your findings with the techniques taught in this chapter.
6. IBM Rational is a wholly owned subsidiary of IBM. The authors of UML have also been executives in Rational. Consequently, IBM Rational was an early leader in developing visual modeling tools to support UML and object-oriented modeling. You can download an evaluation copy of IBM Rational's UML tool (IBM Rational Software Architect) and use it to draw the RMO diagrams. This will give you experience with a widely used industry tool. Alternatively, your college or university can enroll in the Seed program and provide copies of the tools in its laboratories. The URL is *www-306.ibm.com/software/rational*.

CASE STUDIES

THE REAL ESTATE MULTIPLE LISTING SERVICE SYSTEM

Refer to the description of the Real Estate Multiple Listing Service system in the Chapter 5 case studies. Using the event list and ERD for that system as a starting point, develop the following object-oriented models:

1. Convert your ERD to a domain class diagram.
2. Develop a use case diagram.
3. Create a fully developed use case description or an activity diagram for each use case.
4. Develop a system sequence diagram for each use case.

THE STATE PATROL TICKET PROCESSING SYSTEM

Refer to the description of the State Patrol ticket processing system in the Chapter 5 case studies. Using the event list and ERD for that system as a starting point, develop the following object-oriented models:

1. Convert your ERD to a class diagram.
2. Develop a use case diagram.
3. Create fully developed use case descriptions for two of the primary use cases, such as *Recording a traffic ticket* and *Scheduling a court date*.
4. Develop system sequence diagrams for those same use cases.
5. Develop a state machine diagram for a ticket.

THE DOWNTOWN VIDEOS RENTAL SYSTEM

DownTown Videos is a chain of 11 video stores scattered throughout a major metropolitan area in the Midwest. The chain started with a single store several years ago and has grown to its present size. Paul Lowes, the owner of the chain, knows that competing with the national chains will require a state-of-the-art movie rental system. You have been asked to develop the system requirements for the new system.

Each store has a stock of movies and video games for rent. For this first iteration, just focus on the movies. It is important to keep track of each movie title and to identify its category (classical, drama, comedy, and so on), its rental type (new release, standard), movie rating, and other general information such as producer, release date, and cost. In addition to tracking each title, the business must track individual copies to note their purchase date, their condition, their type (VHS or DVD), and their rental status. User functions must be provided to maintain this inventory information.

Customers, the lifeblood of the business, are also tracked. DownTown considers each household to be a customer, so special mailings and promotions are offered to each household. For any given customer, several people may be authorized to rent videos and games. The primary contact for each customer can also establish rental parameters for other members of the household. For example, if a parent wants to limit a child's rental authorization to only PG and PG-13 movies, the system will track that.

Each time a movie is rented, the system must keep track of which copies of which movies are rented; the rental date and time and the return date and time; and the household and person renting the movie. Each rental is considered to be open until all of the movies and games have been returned. Customers pay for rentals when checking out videos at the store.

For this case, develop the following diagrams:

1. A domain model class diagram.
2. A use case diagram. Analyze user functions. Also do a CRUD analysis based on the class diagram.

3. An activity diagram for each of the use cases that involve renting and checking in movies, and each of the use cases that maintain customer and family member information.
4. A system sequence diagram for each of the use cases from problem 3.
5. A state machine diagram identifying the possible states (status conditions) for a physical copy of a movie, based on the use case descriptions provided earlier in the chapter and your knowledge of how a video store might work.

THEEYESHAVEIT.COM BOOK EXCHANGE

TheEyesHaveIt.com Book Exchange is a type of e-business exchange that does business entirely on the Internet. The company acts as a clearinghouse for both buyers and sellers of used books.

To offer books for sale, a person must register with EyesHaveIt. The person must provide a current physical address and telephone number, as well as a current e-mail address. The system will then maintain an open account for this person. Access to the system as a seller is through a secure, authenticated portal.

A seller can list books on the system through a special Internet form. Information required includes all of the pertinent information about the book, its category, its general condition, and the asking price. A seller may list as many books as desired. The system maintains an index of all books in the system so that buyers can use the search engine to search for books. The search engine allows searches by title, author, category, and keyword.

People who want to buy books come to the site and search for the books they want. When they decide to buy, they must open an account with a credit card to pay for the books. The system maintains all of this information on secure servers.

When a request to purchase is made, along with the payment, TheEyesHaveIt.com sends an e-mail notice to the seller of the book that was chosen. It also marks the book as sold. The system maintains an open order until it receives notice that the books have been shipped. After the seller receives notice that a listed book has been sold, the seller must notify the buyer via e-mail within 48 hours that the purchase is noted. Shipment of the order must be made within 24 hours after the seller sends the notification e-mail. The seller sends a notification to both the buyer and TheEyesHaveIt.com when the shipment is made.

After receiving notice of shipment, TheEyesHaveIt.com maintains the order in a shipped status. At the end of each month, a check is mailed to each seller for the book orders that have been in a shipped status for 30 days. The 30-day waiting period exists to allow the buyer to notify TheEyesHaveIt.com if the shipment does not arrive for some reason, or if the book is not in the same condition as advertised.

If they want, buyers can enter a service code for the seller. The service code is an indication of how well the seller is servicing book purchases. Some sellers are very active and use TheEyesHaveIt.com as a major outlet for selling books. So, a service code is an important indicator to potential buyers.

For this case, develop the following diagrams:

1. A domain model class diagram
2. A use case diagram
3. A fully developed description for two use cases such as *Add a seller* and *Record a book order*
4. A system sequence diagram for each of the two use cases in problem 3

RETHINKING ROCKY MOUNTAIN OUTFITTERS

The event table for RMO is shown in Figure 5-12. Based on this event table, the use case diagram in Figure 7-5 was developed. The chapter illustrates detailed models (activity and system sequence diagrams) for *Create new order*.

Using the information provided in the RMO case descriptions and the figures in the book (Figures 5-12 and 7-5), create a fully developed use case description and system sequence diagram for each of the following Customer actor use cases: (1) *Update order* and (2) *Create order return*. Now do the same for both of the Shipping actor use cases.

FOCUSING ON RELIABLE PHARMACEUTICAL SERVICE

Previous chapters have described the activities and processes of Reliable Pharmaceutical Service. Use the previous descriptions, particularly the basic description in Chapter 1 and the detailed descriptions from Chapter 5, as well as the following additional description of the case, to develop object-oriented requirements models.

Company processes (for use case development):
There are several points in the order-fulfillment process at which information must be recorded in the system. Obviously, new orders must be recorded. Case manifests must be printed at the start of each shift. In fact, because a prescription itself may take a fairly long time to be completely used, as in the case of long-standing prescriptions, information must be entered into the system each time a medication is sent (prescription fulfillment), noting the quantity of medication that was sent and which pharmacist filled the prescription for that shift.

As explained in Chapter 5, basic information about all of the patients, nursing homes, staff, insurance companies, and so forth must also be recorded in the system.

Information requirements (for class diagram requirements):
Reliable needs to know about the patients, the nursing home, and the nursing-home unit where each patient resides. Each nursing home has at least one but possibly many units. A patient is assigned to a specific unit.

Prescriptions are rather complex entities. They contain basic information such as ID number, original date of order, drug, unit of dosage (pill, teaspoon, suppository), size of dosage (milligrams, number of teaspoons), frequency or period of dosage (daily, twice a day, every other day, every 4 hours), and special considerations (take with food, take before meals). In addition, there are several types of prescriptions, each with unique characteristics. Some orders are for a single, one-time-only prescription. Some orders are for a certain number of dosages (pills). Some orders are for a time period (start date, end date). Information about the prescription order must be maintained. An order occurs when the nursing home phones in the needed prescriptions. Because prescriptions may last for an extended period of time, a prescription is a separate entity from the order itself. The system records which employee accepted and entered the original order.

The system also has basic data about all drugs. Each drug has generic information such as name, chemical, and manufacturer. However, more detailed information for each type of dosage, such as the size of each pill, is also kept. A single drug may have many different dosage sizes and types.

In addition, information about the fulfillment of orders must also be maintained. For example, on a prescription for a number of pills, the system must keep a record of each time a pill or a number of pills is dispensed. A record is also maintained of which pharmacist or assistant fulfilled the order. Assume all prescriptions are dispensed only as needed for a 12-hour shift.

Basic data is kept about prescription payers, such as name, address, and contact person. For this first iteration, do not worry about billing or payments. Those capabilities may be added in a later iteration.

Based on your previous work, the cases from prior chapters, and the description here, do the following:

1. Refine and extend the domain model class diagram you developed in Chapter 5 as necessary.
2. Develop a use case diagram. Base it directly on the event table you created for Chapter 5. Be sure to include a CRUD analysis with your class diagram from question 1 and discuss what additional use cases might be needed based on your CRUD analysis.
3. Develop an activity diagram for each use case related to entering new orders, creating case manifests, and fulfilling orders. You should have at least three activity diagrams. Write a fully developed use case description for each of these use cases.
4. Develop a system sequence diagram for each use case you developed in question 3.
5. Develop a state machine diagram for an order.

FURTHER RESOURCES

Grady Booch, James Rumbaugh, and Ivar Jacobson, *The Unified Modeling Language User Guide*. Addison-Wesley, 1999.

E. Reed Doke, J.W. Satzinger, and S.R. Williams, *Object-Oriented Application Development Using Java.* Course Technology, 2002.

Hans-Erik Eriksson, Magnus Penker, Brian Lyons, and David Fado, *UML 2 Toolkit*. John Wiley & Sons, 2004.

Martin Fowler, *UML Distilled Third Edition: A Brief Guide to the Standard Object Modeling Language*. Addison-Wesley, 2004.

Ivar Jacobson, Grady Booch, and James Rumbaugh, *The Unified Software Development Process.* Addison-Wesley, 1999.

Philippe Kruchten, *The Rational Unified Process, An Introduction (3rd Edition)*. Addison-Wesley, 2005.

Craig Larman, *Applying UML and Patterns: An Introduction to Object-Oriented Analysis and Design and the Unified Process, 3rd Edition*. Prentice-Hall, 2005.

Object Management Group, *UML 2.0 Superstructure Specification*, 2004.

James Rumbaugh, Ivar Jacobson, Grady Booch, *The Unified Modeling Language Reference Manual.* Addison-Wesley, 1999.

CHAPTER 8

EVALUATING ALTERNATIVES FOR REQUIREMENTS, ENVIRONMENT, AND IMPLEMENTATION

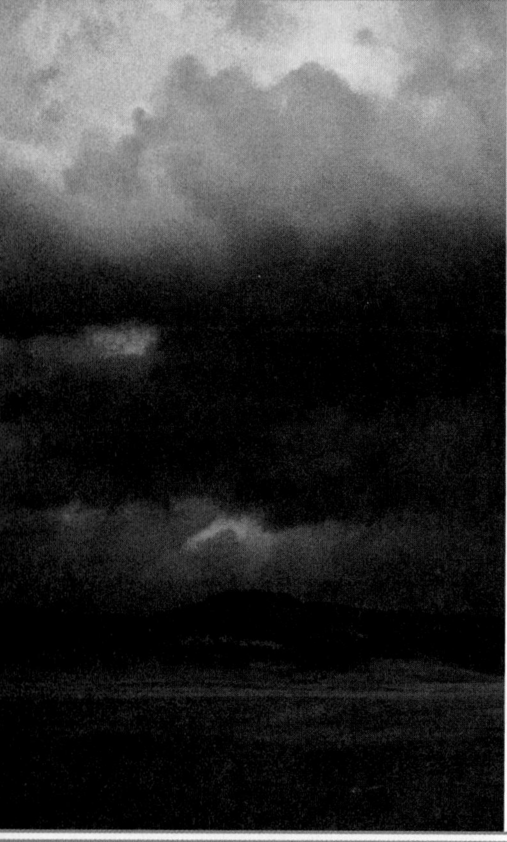

LEARNING OBJECTIVES

After reading this chapter, you should be able to:

- Prioritize the system requirements based on the desired scope and level of automation for the new system
- Describe the strategic decisions that integrate the application deployment environment and the design approach for the new system
- Determine alternative approaches for system implementation
- Evaluate and select an implementation approach based on the needs and resources of the organization
- Describe key elements of a request for proposal (RFP) and evaluate vendors' proposals for outsourced alternatives
- Develop a professional presentation of findings to management

CHAPTER OUTLINE

Project Management Perspective

Deciding on Scope and Level of Automation

Defining the Application Deployment Environment

Choosing Implementation Alternatives

Contracting with Vendors

Presenting the Results and Making the Decisions

TROPIC FISH TALES: NETTING THE RIGHT SYSTEM

Robert Holmes wasn't exactly sure how to proceed with his project. He had six proposals from software vendors to develop an Internet-based ordering system for his company, Tropic Fish Tales. He and his project team had to figure out some way to make a meaningful comparison among the proposals to determine which alternative best fit the needs of the company. Then he had to make a presentation of his analysis and recommendations.

The problem was that none of the six proposals was the same. He and his team had spent a tremendous amount of time developing a request for proposal (RFP) that they had sent to several firms providing custom solutions. They had worked hard on the RFP to make sure that it contained a very precise definition of business requirements. Even with this well-designed RFP, none of the six proposals looked the same. He was going to have to devise a method to do a fair comparison among the proposals. Otherwise, how would he know which solution was the best for Tropic Fish Tales?

His company had made an early decision to develop an RFP and obtain outside assistance with the development. The project appeared to be pretty large, and the information system staff was quite small and inexperienced. The least-expensive solution was from a company that had a standard off-the-shelf ordering system. The advantage was that it would be quick and fairly inexpensive to install and get working. However, the disadvantage was that it did not quite fit all of the requirements. Robert wasn't sure how important the missing functionality was to his company. The system could be made to work with some modifications to work procedures and forms.

At the other end of the spectrum was a proposal for a completely new state-of-the-art system for Internet sales, with electronic interfaces to suppliers and shippers. This system was a complete electronic commerce solution with fully automated support. The proposal also indicated that substantial transaction, customer, and order history information would be retained and available in real time. The system also contained automated inventory management functions. Although the system had more capability than the company really needed, it would certainly bring Tropic Fish Tales to the forefront of high-technology solutions. Robert wondered whether the company could afford the price, however, which was about three times the cost of the low-cost solution.

The other proposals ranged between the two extremes. One company proposed to develop a system from the ground up, working very closely with Robert's firm to ensure that the system fit the requirements perfectly. Another company had a base system that it proposed to modify. The base system was for a different industry and was not currently Internet based, so substantial modifications would be necessary. One solution ran only on UNIX machines. Even though the system appeared to have most of the desired functionality, it would take some work to modify it for a Windows server network, which is the current environment for the company.

Robert was scheduled to meet Bill Williams, the director of information systems, later in the day. He hoped Bill would have some suggestions about how to address this problem.

OVERVIEW

As we discussed in previous chapters, the six major analysis activities of system development are the following:

- Gather information
- Define system requirements
- Prototype for feasibility and discovery
- Prioritize requirements
- Generate and evaluate alternatives
- Review recommendations with management

You have already learned about fact finding, defining system requirements, and prototyping activities. You learned how to define system requirements using either a traditional approach or an object-oriented approach. This chapter explains the last three analysis activities—that is, the transitional activities that refocus the project from discovery and analysis to solutions and design. These final activities are pivotal in the project; they set the direction for the design and implementation of the solution system.

We first discuss the project management orientation that underlies all three activities. Recall from Chapter 3 that one of the major responsibilities of the project manager is to define and control the scope of the new system. The objective of prioritizing the requirements is to define the scope of the system precisely, and the scope directly affects the project cost and schedule, which are also the project manager's responsibility. Evaluating implementation alternatives guides the rest of the project. The outcome of these final activities determines the detailed schedule for the final phases of the project.

Next, we discuss evaluating and prioritizing the system requirements. It is normal during analysis to uncover many more requirements and needs than can reasonably be included within the system, so the development team must categorize and prioritize the requirements to determine what to include. Frequently, two or three alternative combinations of requirements will be developed, along with their required resources, and then an oversight committee of executives, users, and technical managers decides which approach is most viable. This chapter discusses various strategies for prioritizing requirements and selecting a scope and level of automation.

Then we discuss the various alternatives for the production environment, including alternatives for the hardware configuration and operating systems. The existing or planned environment for the new system is a critical consideration—that is, what hardware, system software, networks, and standards will support the new system? The chapter contains a brief overview of choices and constraints for the deployment and development environments. We also demonstrate the important points to consider in the environment by discussing them in the context of Rocky Mountain Outfitters and its new system.

Next, we discuss the alternatives for design and implementation. The focus is on the various options for actually building and installing the system. After the system scope is determined and a decision on the environment has been made, several alternative methods of development are reviewed. These alternatives can range from building the new system completely from scratch to buying a system from someone else to outsourcing the entire development and daily operation. We review the most popular alternatives and discuss steps used to make a selection. Included within this discussion are instructions on how to develop and use a request for proposal (RFP).

Although we present these activities as the last three activities associated with analysis, in most cases they are done in parallel with other analysis activities. In predictive types of projects the requirements can be prioritized as detailed specifications are developed. Ongoing consideration and evaluation of the deployment environment, including hardware and system software, is done early in the project. For adaptive types of projects, decisions about overall capability are made early. Detailed decisions about the exact nature of the various functions can be deferred until later in the project. However, deployment environment decisions must be made early in the project because early iterations will implement portions of the system. Hence, even though the textbook presents them as final analysis activities, they are actually parallel activities for analysis.

Finally, this chapter discusses the concepts associated with presenting initial findings to upper management with recommendations in order to obtain approval and funding for the remainder of the project. In predictive projects, this decision milestone may be a major milestone and may only need to be done once. For adaptive projects, milestones may be encountered on several occasions, which require corresponding evaluation, presentation, and decisions.

PROJECT MANAGEMENT PERSPECTIVE

In Chapter 3 we saw that the activities during project planning required the project manager's heavy involvement. The analysis activities described in this chapter are no different in that regard. But in addition to a management component, they also have a critical technical requirement. Thus, both the project manager and senior technical members of the project team must work together to complete these activities successfully.

System development projects come in all sizes and complexities and vary in the level of their formality. One effective technique for managing large or complex projects is to develop decision metrics—systems to measure the alternatives and to evaluate them based on their relative scores. Other projects are smaller and less complex and can use more informal techniques with fewer metrics. In this chapter, we discuss several techniques to help in evaluating alternatives.

In Chapter 3, and in Appendix A on the textbook's Web site, nine knowledge areas of project management are identified: scope, time, cost, quality, human resources, communications, risk, procurement, and integration. The three activities of prioritizing requirements, evaluating alternatives, and reviewing recommendations with management involve seven of the nine knowledge areas.

A project's scope is directly affected by the priorities established for the system requirements. While prioritizing requirements, the project manager precisely defines the functions that will be included in the project and sets a baseline, which he or she can use to control and direct the rest of the project. A firm list of functions that users and project staff have agreed to can control the scope of the project and keep it manageable. If no firm decisions are ever made about what should be in the new system, it is almost impossible for the project manager to control the size of the project.

The schedule, which is part of project time management, is further developed at this time, as decisions are made about the scope, environment, and implementation. In fact, in many projects, the schedule is not completed until these decisions are made. For example, if the team decides to purchase components for the new system or to hire outside programmers, the project schedule must reflect those decisions.

Project cost management involves both estimating the project costs and controlling them. Costs and schedule profoundly affect decisions regarding a project's scope, environment, and implementation. Frequently, a project manager must recalculate the cost/benefit ratio to confirm a project's financial feasibility. On many projects, a go/no-go decision is made at this stage of analysis—when the project manager recalculates costs and benefits.

Presenting findings to the oversight committee is a key responsibility of a project manager. Project communications management involves collecting and explaining all of the key decisions, feasibility analyses, risks, benefits, schedules, and costs to the stakeholders who are funding the project.

As the team makes decisions, particularly technical decisions about environment and implementation, the project manager must determine and evaluate the various risks associated with each alternative. A complete risk analysis and feasibility assessment is done for each of the alternatives being considered. Because key project decisions are being made, it is important for the project manager to conduct a thorough risk analysis.

As implementation alternatives are evaluated, the project manager begins activities associated with procurement management. Vendors must be identified and evaluated. Requests for proposals are developed, and proposals are evaluated. Contract negotiations may even begin. An effective project manager must have good procurement skills to ensure that reliable and professional vendor relationships are established and good purchase decisions are made.

Finally, even though specific tasks might not be directly associated with project quality management, it should be obvious that quality is the objective of all activities.

Project management runs throughout a project's lifetime, but the two times when project management tasks are most prevalent are during the initial planning activities and during the evaluation of system alternatives. The skills of the project manager are most evident as critical decisions and project directions are established.

The decisions affecting requirements, environment, and implementation approach are made together because they are interdependent. In the following sections, we treat each topic separately, but in reality they are all intertwined. First, we address the requirements and project scope.

DECIDING ON SCOPE AND LEVEL OF AUTOMATION

Prioritizing requirements includes tasks to define both the scope and the level of automation for the new system. Scope and level of automation are two very closely related aspects of the new application system. The scope of the system defines which business functions will be included in the system. For example, in the current Rocky Mountain Outfitters (RMO) point-of-sale system, the scope includes handling mail and telephone sales, but not Internet sales. The level of automation is how much computer support exists for the functions that are included. In the new system, a very low level of automation for telephone sales would be to require telephone clerks to use printed catalogs at their desks to verify customer requests. The system would then support only simple data entry of the order information. A higher level of automation for telephone sales would be to have the catalog and customer information online so that telephone clerks get automated entry and verification of inventory items and customer name and address information.

CONTROLLING A PROJECT'S SCOPE

One common problem with development projects is scope creep. As the name implies, the development team may receive requests to add new system functions after the requirements have been defined and decisions finalized. One way to help control this problem is by formalizing the process to identify, categorize, and prioritize the functions that will be included within the new system so that everyone agrees to and signs off on system functions. In Chapter 5, you learned that the event table describes all of the business events that the system must support. Continuing to use the event table to control which business functions will be supported by the new system is an effective technique to control the project's scope.

During analysis, users usually request many more business functions than the schedule and budget can allow. The team needs to decide which functions are critical and must be included and which can be deferred until later. A common approach to determine the scope is to list each requested function and rate its importance, using such categories as "mandatory," "important," and "desirable." Determining the priority of each function is usually done in conjunction with a description of the level of automation for each function.

Remember that predictive projects are better adapted to applications that are fairly well defined. Thus, scope decisions are usually made and even finalized fairly early in the project. A scope decision milestone, with accompanying presentation, is often completed as requirements are finalized. Adaptive projects, however, often require partial decisions at various points during the life of the project. In many ways, scope decisions, which are dispersed throughout the project, make it much more difficult to control the scope in adaptive projects. It is too easy to spend scarce resources such as time, money, and human effort on capabilities that are less important to the organization's business need and hence to the project.

DETERMINING THE LEVEL OF AUTOMATION

The level of automation describes the support the system will provide for each function. For most functions of an application system, at least three levels of automation can be defined: low, middle, and high. At the lowest level, the computer system only provides simple record keeping. Data input screens allow employees to capture information and insert it into a database. Simple field edits and validation of input data are also included. For example, a low level of automation for an order-entry function has a data-entry screen to enter customer and order information. The system date may be used for the order date. The user manually enters each line item for the order. The system might or might not automatically calculate the price.

Usually, stock on hand and anticipated shipment dates cannot be verified. At the end of order entry, the information is stored in the database and the function is concluded.

Analysts also define a middle-range level of automation for each function, which may be a single midrange point or various midrange alternatives. Usually, the midrange alternative is a combination of features from the high-level and the low-level automation alternatives. Analysts make their best guess of what is necessary and what is justified at the current stage of technology and within the budget.

A high level of automation occurs when the system takes over, as much as possible, the processing of a function. Usually, it is more difficult for an analyst to define high-end automation than low-end automation because low-end automation is basically an automated version of a current manual procedure. However, generating a high-level automation alternative requires brainstorming and thinking "outside the box" to create new processes and procedures. Business Process Management is a discipline whose objective is to evaluate the effectiveness of current business processes with the potential of eliminating or completely revamping workflows into highly efficient processes. Often radical approaches are designed that provide dramatic increases in processing speeds and levels of service. In almost all cases, well-designed computer systems with high levels of automation are necessary to achieve "order of magnitude" improvements.

Figure 8-1 is a table that contains both scoping and level of automation information for each function of the RMO customer support system. The figure contains all of the business events from the original event table (see Figure 5-12) as well as seven new functions that were identified during systems analysis. The objective of this table is to identify all of the potential events and functions that the new system needs to perform. Each business function is prioritized as mandatory, important, or desirable. Users and clients prioritize the functions based on the needs of the business and the objectives of the new system. For example, if one objective of the system is to increase customer support, functions that allow RMO to respond to customer requests will be mandatory functions, at least at some level of automation.

Figure 8-1

RMO's CSS functions with priorities and three levels of automation

Functions (expanded from event list)	Priority (mandatory, important, desirable)	Low-end automation	Medium (most probable) automation	High-end automation (medium level plus . . . when + appears)
Check item availability	Important	Periodic listing of quantity on hand	Real-time; internal and Web	+ Sales prompting
Place order	Mandatory	Clerk data entry	Clerk real-time and customer via Web	+ Promotion prompting and stock-out alternatives
Change or cancel order	Important	Clerk overnight	Clerk real-time and customer via Web for 24 hours	Clerk real-time and customer via Web up to shipment
Check order status	Important	Clerk overnight	Clerk real-time and customer via Web	+ Automatic notification
Fulfill order	Mandatory	Print pull list and shipping label	Pull list, shipping label, real-time update	Automated warehouse Real-time update
Create back order	Important	Clerk data entry	Real-time	+ System automatic and notify supplier
Return item	Important	Clerk data entry	Real-time, clerk update restock, and customer	Automatic inventory and account update
Mail catalog	Mandatory	Print labels	Personalize cover letter	+ Personalize throughout
Correct customer account	Important	Data entry	Real-time	+ Automatic from activity
Send promotional material	Important	Print labels	Personalized cover page	Personalized based on buying history

Functions (expanded from event list)	Priority (mandatory, important, desirable)	Low-end automation	Medium (most probable) automation	High-end automation (medium level plus . . . when + appears)
Adjust customer charges	Mandatory	Data entry	Real-time update	+ Automatic from activity
Update catalog	Mandatory	Data entry	Real-time	+ Automatic suggestions from sales history
Create promotional materials	Important	Data entry	Real-time	Recommendations based on sales history
Create new catalog	Mandatory	Record keeping of products, prices, and so on	Record keeping of products, prices, pictures, and layouts	Digital scan and page layout
System Reports				
Produce order summary reports	Important	Printed on request	Online view and real-time	Data visualization tools
Produce activity reports	Important	Printed on request	Online view and real-time	Data visualization tools
Produce transaction summary reports	Important	Printed on request	Online view and real-time	Data visualization tools
Produce customer adjustment report	Important	Printed on request	Online view and real-time	Data visualization tools
Produce fulfillment reports	Important	Printed on request	Online view and real-time	Data visualization tools
Produce catalog activity reports	Important	Printed on request	Online view and real-time	Data visualization tools
Newly Identified Events				
Maintain customer purchase history	Important	Archive files with summary reports	Archive, printed promotional notices	Automatic, real-time for sales prompting
Provide ongoing feed to manufacturing	Desirable	Printed reports	Daily update	Real-time and trend analysis
Provide EDI feed to suppliers from sales data	Desirable	Printed reports and history	Daily update	Real-time and trend analysis
Tie in to shipper system	Desirable	No link	Daily update and e-mail notification to customer	Automatic feed and shipment tracking via Web link
Perform data warehousing and conduct data analysis	Desirable		Trend analysis	Trend analysis, data visualization tools
Prompt automated sales	Desirable		Based on promotions	Based on sales promotions and history
Conduct expanded sales analysis with DSS	Desirable		Printed reports	Data visualization tools

Figure 8-1 cont.

RMO's CSS functions with priorities and three levels of automation

The table also includes various levels of automation for each function. The analysts do not attempt to describe every characteristic of each level of automation in the table. Supporting descriptions will describe each cell in more detail. The table provides an overview of the functions, the priority of each, and the various methods to implement each function at the different levels of automation.

Let's take the order-entry function within RMO as an example. We start by identifying the best customer service possible. We ask the question, "Why does a customer need to be able to order exactly what she wants at the most convenient time?" Also, "What does a customer want to know about her order at the time she has finished ordering?" (Normally, we would try to reengineer the entire process and would also ask questions such as, "When, or how soon, does the customer want to receive the items she has ordered?" But let's limit our discussion only to the order-entry component of the process.)

To answer the order-entry questions, RMO staff decide that they want to provide their online customers all the benefits of catalog shopping: convenience, the ability to order anytime of day or night, no crowds, the ability to order from home, wide selection, simplicity of ordering, and privacy. RMO also wants to provide, as much as it is feasible, the benefits of store shopping: being able to inspect the items; trying them on and comparing sizes, colors, and patterns; having items and related accessories near each other; examining several products together for match and compatibility; and so forth.

Given the stated desires, a high-end system might have the following characteristics:

- Customers can access the catalog online, with full-color, three-dimensional pictures. For more technical products, the catalog should include detailed descriptions and diagrams showing their construction and other details. This service can be provided for Internet customers through the Web. For telephone customers, the catalog can be provided through the phone line, directly to a television set.
- The catalog is also interactive and allows the customer to combine several items with graphical imaging that displays them together (for example, showing a shirt, jacket, and shorts on a simulated person).
- The user interface to the catalog and order system is either voice activated or keypad activated.
- The system should make suggestions of related items that customers may need or desire to purchase at the same time.
- The system should verify that all items are in stock and establish a firm time when shipment will occur. (The fulfillment portion of the system should support shipment within 24 hours or less or, even better, guarantee same-day delivery.)
- Items not in stock should be immediately ordered from the manufacturer or other supply source (the system will immediately send the transaction to the other systems), enabling RMO to ensure delivery to the customer at a future date.
- Payment is verified online, just as in a store.
- The customer can see a history of all prior orders and can check the status of any individual order either with the telephone or on the Web.

Interestingly, all of these capabilities can be supported with current technology. The question, of course, is whether RMO can justify the cost at this time. In any event, we have defined high-end automation for the order-entry portion of the new system.

SELECTING ALTERNATIVES

After the identified functions have been prioritized and the levels of automation have been analyzed, the project team reviews all the alternatives. Preliminary decisions might have been made based on individual need or importance, but the entire group of alternatives is normally evaluated together. This provides a more global, or "big picture," view of the proposed system. This is true of both predictive and adaptive projects. Adaptive projects may have less detail at this point, but an overall view of system scope is important for any type of new development. In recent years, companies have been building new systems to gain competitive advantage in the marketplace. In addition, more and more companies are entering into e-commerce ventures on both the supply and delivery sides. By establishing more global and strategic criteria, companies can make better long-term decisions for their new systems. The following list identifies some of the key criteria that are used:

- **Strategic Plan.** The initial decision to develop a new information system is frequently an outgrowth of a long-term strategic plan. As discussed previously, strategic planning occurs both for the long-term organizational strategy and for information technology to support organizational plans. As decisions are made concerning individual capabilities of the new system, the strategic plan is frequently used as a global measuring rod. For example, if an organization's long-term goal is to develop a supply chain system with automatic interfaces between itself and its suppliers, the system must be designed to support these interfaces even though they might not be implemented in a first phase.
- **Economic Feasibility.** Obviously, higher levels of automation require substantially more funds to implement. Frequently, development teams generate several groups of capabilities and levels of automation and then project costs to develop those different packages. With more detailed information about requirements and assessments of the difficulty of developing certain capabilities, a more accurate cost/benefit analysis can be generated.
- **Schedule and Resource Feasibility.** Including more advanced features in a system not only costs more but also lengthens the schedule. One effective method to minimize the immediate impact on a project is to plan for future system upgrades. All commercial software developers work this way, and it is a viable alternative for in-house development. A new system often has less capability than the organization ultimately desires. But as users gain experience with the new system and information systems staff learn from past experience, together they can enhance the system until they achieve the desired level of automation.
- **Technological Feasibility.** Not only must project teams review the technical feasibility of desired alternatives, but they must also carefully consider whether the organization has in-house expertise to develop and implement the system. Frequently, organizations hire additional staff or contract with outside resources to obtain technical expertise. Usually it is more prudent to select alternatives that do not require pushing the state of the art. Yet, some companies with substantial funds and broad access to resources do so. For cutting-edge projects, a detailed risk analysis is critical.
- **Operational, Organizational, and Cultural Feasibility.** Changes in business processes also involve risk—risk that must be managed. Broader scope and higher levels of automation usually require organizations to reengineer their business functions and manual processes. Benefits can be substantial and dramatic; however, users need support for the change to maintain morale and commitment to the new system. Typically, information systems staff underestimate the difficulties of changing people's work procedures and job activities. For that reason, it is usually a good idea to involve people who have been trained in organizational behavior to assist in managing changes.

BEST PRACTICE

Feasibility factors—economic feasibility, schedule feasibility, resource feasibility, technological feasibility, organizational feasibility, and, increasingly, *risk*—are used to evaluate the initial feasibility of a project and the feasibility of each alternative.

EVALUATING ALTERNATIVES FOR RMO

Rocky Mountain Outfitters is in the preliminary stages of selecting functions and automation levels for the customer support system. A final decision depends on the alternatives the development team chooses for implementation. We explore those alternatives next.

Based on preliminary budget and resource availability, the project team at RMO determined that it is possible to include all functions that were categorized as either mandatory or important in the table in Figure 8-1. For each of those functions, the team does a detailed analysis for the desired level of automation. Fundamental decisions about level of automation affect several functions at the same time. For example, three levels of automation affect *Check item availability*, *Place order*, *Change or cancel order*, and so forth. The three basic alternatives listed in Figure 8-1 are (1) data entry of information with overnight processing, (2) real-time entry for both employee clerks and for customers via the Web, and (3) the same system as the medium level with added sales prompting based on promotions and even personal customer purchase history. In fact, these three alternatives could even be divided into more alternatives, such as Web versus no Web and sales prompting for promotions but not purchase history. A fundamental decision on the automation level will then need to be consistent across the listed functions.

Figure 8-2 lists the functions and shows by shading which functions are to be included and at what level of automation. The low level of automation was not acceptable to RMO management. Most of the current systems already provided that level of automation. At first, RMO management thought that the medium level of automation was sufficient for the first version of the system. It also did not put undue strain on the budget. However, upon further discussion and consideration, RMO decided to move into high-end automation as rapidly as possible. The technical support group was able to show that current trends in hardware advances, such as processing speeds and storage capabilities, would allow RMO to acquire sufficient computing capabilities to support many advanced functions. In this instance, the excellent working relationship between Barbara Halifax, the project manager, and the technical support group helped to configure a solution that moved RMO forward. As shown in the figure, RMO management chose the high-end alternative for many of the functions. Management felt that company growth and competitive advantage would depend on sophisticated and advanced sales support provided by the high-end automation.

Of the seven newly identified functions, RMO management decided to include three in the project—two of which are at the high-end level of automation. The first addition is to prepare for the high-end support by including the function to maintain customer history and use it to develop special promotions. Feasibility analysis indicates that this alternative does not require substantial increases in cost or length of the project schedule.

The second addition is a more rapid update of inventory levels to the manufacturing facilities. The cost/benefit analysis of this alternative indicates an immediate return by a reduction of back orders and stock-outs.

The third addition is a subsystem to provide more sophisticated analysis of sales trends. This subsystem will utilize the database of sales orders and time series data based on customer histories. Thus, it builds on the data being included for individual customers. It was difficult for the project team to calculate a precise cost/benefit ratio for this new capability, but the sales manager convinced the oversight committee that the capability was critical to the future competitiveness of RMO. So, resources will be dedicated to add this capability. This subsystem is somewhat independent, so to minimize its impact on the schedule, the project team will implement the subsystem several months after the rest of the system.

Functions (expanded from event list)	Priority (mandatory, important, desirable)	Low-end automation	Medium (most probable) automation	High-end automation (medium level plus . . . when + appears)
Check item availability	Important	Periodic listing of quantity on hand	Real-time: internal and Web	+ Sales prompting
Place order	Mandatory	Clerk data entry	Clerk real-time and customer via Web	+ Promotion prompting and stock-out alternatives
Change or cancel order	Important	Clerk overnight	Clerk real-time and customer via Web for 24 hours	Clerk real-time and customer via Web up to shipment
Check order status	Important	Clerk overnignt	Clerk real-time and customer via Web	+ Automatic notification
Fulfill order	Mandatory	Print pull list and shipping label	Pull list, shipping label, real-time update	Automated warehouse Real-time update
Create back order	Important	Clerk data entry	Real-time	+ System automatic and notify supplier
Return item	Important	Clerk data entry	Real-time, clerk update restock, and customer	Automatic inventory and account update
Mail catalog	Mandatory	Print labels	Personalize cover letter	+ Personalize throughout
Correct customer account	Important	Data entry	Real-time	+ Automatic from activity
Send promotional material	Important	Print labels	Personalized cover page	Personalized based on buying history
Adjust customer charges	Mandatory	Data entry	Real-time update	+ Automatic from activity
Update catalog	Mandatory	Data entry	Real-time	+ Automatic suggestions from sales history
Create promotional materials	Important	Data entry	Real-time	Recommendations on sales history
Create new catalog	Mandatory	Record keeping of products, prices, and so on	Record keeping of products, prices, pictures, and layouts	Digital scan and page layout
System Reports				
Produce order summary reports	Important	Printed on request	Online view and real-time	Data visualization tools
Produce activity reports	Important	Printed on request	Online view and real-time	Data visualization tools
Produce transaction summary reports	Important	Printed on request	Online view and real-time	Data visualization tools
Produce customer adjustment report	Important	Printed on request	Online view and real-time	Data visualization tools
Produce fulfillment reports	Important	Printed on request	Online view and real-time	Data visualization tools

Figure 8-2

Preliminary selection of alternative functions and level of automation for RMO (selections are shaded)

Functions (expanded from event list)	Priority (mandatory, important, desirable)	Low-end automation	Medium (most probable) automation	High-end automation (medium level plus . . . when + appears)
Produce catalog activity reports	Important	Printed on request	Online view and real-time	Data visualization tools
Newly Identified Events				
Maintain customer purchase history	Important	Archive files with summary reports	Archive, printed promotional notices	Automatic, real-time for sales prompting
Provide ongoing feed to manufacturing	Desirable	Printed reports	Daily update	Real-time and trend analysis
Provide EDI feed to suppliers from sales data	Desirable	Printed reports and history	Daily update	Real-time and trend analysis
Tie in to shipper system	Desirable	No link	Daily update and e-mail notification to customer	Automatic feed and shipment tracking via Web link
Perform data warehousing and conduct data analysis	Desirable		Trend analysis	Trend analysis, data visualization tools
Prompt automated sales	Desirable		Based on promotions	Based on sales promotions and history
Conduct expanded sales analysis with DSS	Desirable		Printed reports	Data visualization tools

Figure 8-2 cont.

Preliminary selection of alternative functions and level of automation for RMO (selections are shaded)

DEFINING THE APPLICATION DEPLOYMENT ENVIRONMENT

application deployment environment

the configuration of computer equipment, system software, and networks for the new system

One of the primary considerations in developing a new information system is the application deployment environment. The **application deployment environment** is the configuration of computer hardware, system software, and networks in which the new application software will operate. An important part of any project is ensuring that the application deployment environment is defined and well matched to application requirements. At this life cycle stage, the analyst's goal is to define the environment in sufficient detail to be able to choose from among competing alternatives and to provide sufficient information for design to begin. Additional details are added as design proceeds.

HARDWARE, SYSTEM SOFTWARE, AND NETWORKS

In the early years of computer applications, there was only one application type and one deployment environment: a batch-mode application executing on a centralized mainframe using files stored on disk or tape, with offline data-entry devices such as keypunch machines. As computing technology has matured, the range of application types has grown to include the following:

- Stand-alone applications on desktop or laptop computers, small server computers, and PDA devices
- Online interactive applications with wired or wireless connectivity
- Distributed applications spread over various computing platforms and databases
- Internet-based applications

Just as the number of application types has proliferated, so has the variety of hardware, system software, and networks that support them. Computers now range in size from handheld devices to large supercomputers. In addition, analysts are faced with many choices in supporting software such as operating systems (for example, UNIX and Windows), database management systems (for example, Oracle and DB2), component infrastructure software and standards (for example, Java 2 Enterprise Edition [J2EE] and Microsoft .NET), and Web server software (for example, Internet Information Server and Apache). Modern application software relies on a complex infrastructure that includes the client and server hardware, supporting system software, computer networks, and standards that enable them to operate smoothly together.

When choosing or defining the deployment environment, analysts are concerned with several important characteristics, including the following:

- **Compatibility with System Requirements.** Requirements such as user locations, speed of access and update, security, and transaction volume have a significant impact on environmental requirements. For example, high-volume transaction processing systems such as credit-card payment-processing systems require secure high-speed networks, powerful servers, and compatible operating systems and database management systems (DBMSs).
- **Compatibility among Hardware and System Software.** Although hardware and system software compatibility has generally improved over time, it is still a significant consideration. For example, because Oracle and Sun Microsystems are frequent partners in software and standards development, it is no surprise that the Oracle DBMS performs well on Sun servers running Solaris (Sun's version of UNIX). Similarly, Microsoft operating systems and database management systems are well suited to computers using Intel processors. Ensuring good compatibility of hardware and system software simplifies a system's installation and configuration, improves performance, and minimizes long-term operating costs.
- **Required Interfaces to External Systems.** Modern applications often interact with external systems operated by entities such as credit-reporting agencies, customers, suppliers, and the government. Implementing external interfaces may require a certain system software and, less frequently, specific hardware. For example, a credit-reporting agency might provide services via Web-based XML requests or a J2EE component. An application that interacts with the credit-reporting system must support one or both of those interfaces and include whatever system software is compatible with the interfaces.
- **Conformity with the IT Strategic Plan and Architecture Plans.** Because there are so many choices in hardware and system software, organizations find it difficult and expensive to support many different types. Most medium- and large-scale organizations have strategic application and technology architecture plans that focus their efforts on a limited set of hardware and software alternatives. For example, an organization might choose to emphasize a standard platform consisting of UNIX, Oracle, J2EE, and compatible hardware from Sun Microsystems and Hewlett-Packard. Although that environment might not be best for every application type, sticking to it whenever possible will minimize the total cost of infrastructure maintenance and maximize the long-term compatibility among systems for that particular organization.
- **Cost and Schedule.** Deployment environment alternatives may vary in their impact on project cost and schedule. Typically, environment choices that match the IT strategic plan and existing systems are the fastest and least expensive to acquire, configure, and support.

In sum, the analyst must define an application deployment environment that enables the application to meet stated requirements, fits within the organization's IT plans, and can be acquired and configured within acceptable limits of budget and schedule.

DEVELOPMENT TOOLS

development environment
the programming languages, CASE tools, and other software used to develop application software

Analysts must also consider and select development tools. The **development environment** consists of the programming language(s), CASE (computer-assisted software engineering) tool(s), and other software used to develop application software. The specific deployment environment usually limits development environment choices. For example, choosing a deployment environment based on Microsoft .NET limits the set of compatible development tools to those provided by Microsoft (for example, Visual Studio .NET) and a relatively small number of third-party vendors. System software choices will also be limited to those most compatible with the deployment and development environment (for example, Microsoft server operating systems, Internet Information Services, and SQL Server, for a .NET application).

Normally, companies have a preferred language for system development, and their analysts are familiar with its features. However, as technology changes, newer languages frequently provide additional capabilities. Analysts can choose from numerous development languages—from structured languages such as COBOL to object-oriented languages such as Smalltalk, C++, and Java to Web-based languages such as JavaScript and PHP. Using a new language does require additional commitment and funding to provide the development team with necessary training.

The choice of development tools, such as compilers, debuggers, and integrated development environments, is usually limited by the target operating system, database management system, and component or Web service standards. For example, a deployment environment consisting of UNIX, Oracle, and J2EE would usually lead developers to choose the Java programming language and a tool suite such as Oracle JDeveloper, Sun ONE Studio, or IBM WebSphere.

Many corporations have committed to a particular database management system, and it can limit tool selection also. Most DBMS vendors also supply a compatible set of development tools, which can substantially accelerate the development of some application types compared with development tools not optimized to a particular DBMS. Examples include Microsoft Access and Visual Basic, Microsoft SQL Server and Visual Studio .NET, and Oracle Application Server and JDeveloper.

In sum, application deployment environment choices, particularly the operating system, DBMS, and distributed software standard, tend to limit development tool choices. Thus, an analyst should consider the deployment and development environments together when determining their fit to a particular application.

THE ENVIRONMENT AT ROCKY MOUNTAIN OUTFITTERS

The systems environment at RMO had been built piecemeal over the life of the company to support the business functions at the various locations. Currently, there are two major manufacturing plants, which provide products for three warehouses. The warehouses also stock items from other manufacturers. Most applications are hosted at the Park City data center.

The Current Environment

Figure 8-3 illustrates the current hardware and software environment at RMO. RMO uses modern servers, operating systems, and database management systems and connects them with a high-speed network that supports data, voice, and video-conferencing. With the exception of basic office software, most applications are relatively old. Some applications were purchased and others were developed in-house. They are supported by a hodgepodge of system software and are written in a variety of programming languages. All use terminal-based interfaces and lack any Web-based interfaces. The current Web site is contracted to a vendor in Salt Lake City and the RMO network extends to the Web hosting site.

Facility and location	Computer hardware	System software	Application software
Data Center (Park City)	Server cluster	UNIX Windows Server 2008 DB2 database management system Microsoft Exchange	Supply chain management package Human resources application (C) Accounting/finance package
Headquarters (Park City)	Midrange server	Windows Server 2008 Windows terminal services Microsoft	Microsoft Office
Mail Order Center (Salt Lake City)	Mainframe server	UNIX Windows Server 2008 Windows terminal services Mail order application (COBOL)	Mail order application (COBOL) Microsoft Office
Manufacturing (Portland and Salt Lake City)	Midrange server	Windows Server 2008 Windows terminal services Microsoft Windows	Microsoft Office
Phone Order Center (Salt Lake City)	Midrange servers	Windows Server 2008 SQL Server database management system Windows terminal services Microsoft Windows	Phone order application (Visual Basic) Microsoft Office
Retail (Park City and Denver)	Small server Point of sale terminals	UNIX	Point of sale software package
Warehouses (Albuquerque, Portland, and Salt Lake City)	Midrange server	Windows Server 2008 Windows terminal services Microsoft Windows	Microsoft Office

Figure 8-3
The existing processing environment at RMO

The Proposed Environment

Many of the decisions associated with the target environment are made during strategic planning, which establishes long-term directions for an organization. In other situations, the strategic plan is modified as new systems are developed to use the latest technological advancements. In RMO's case, many technical decisions were made during the initial phases of the supply chain management (SCM) project that is well under way. Because the new customer support system (CSS) must integrate seamlessly with the SCM, technical decisions must be consistent with prior decisions as well as the long-term technology plan.

Because environment decisions are corporate-wide strategic decisions, RMO convened a meeting to discuss the technology alternatives and to make decisions. Attendees consisted of Mac Preston, chief information officer; John MacMurty, director of system development; and Barbara Halifax, project manager. Additional technical staff were also included in the meeting to provide details as needed.

To ensure that all participants were aware of potential alternatives, Barbara presented and reviewed the information shown in Figure 8-4. This figure identifies potential implementation alternatives and was similar to the one used to make decisions for the SCM project. The

Figure 8-4
Processing environment alternatives

Alternative	Description
1. Implement browser-based interfaces with Active-X or Java applets for internal applications.	This solution provides a consistent interface, extends applications to some remote locations, and facilitates e-commerce growth. Clients must support the applet environment. Virtual private network (VPN) technology is needed for remote access.
2. Implement browser-based interfaces for internal applications with all processing on internal servers.	This solution also provides a consistent interface and requires VPN technology. Applications are available from more locations due to less dependence on the client environment. More server capacity is required because clients perform no processing functions.
3. Use a mix of alternatives 1 and 2.	Use applets for applications with minimal remote access requirements, such as human resources, and a thin client model for applications with substantial remote access requirements, such as supply chain management.
4. Centralize the database.	Supports high-volume transaction processing for centralized applications and provides high security, control, and consistency.
5. Distribute the database.	Distributed data provides rapid response for distributed applications and improves fault tolerance. It increases total hardware costs and administrative complexity and carries a higher risk, given RMO's lack of experience with the technology.
6. Use complete OO components, such as J2EE objects, supported by an OO database management system.	This solution would make seamless interfaces between applications—SCM, CSS, and other systems. It would position RMO for future OO migration. This solution requires middleware integration software.
7. Use OO for the user interface and business processing layers with a traditional relational database.	Use Visual Basic or Java to develop the applications. Use DB2 or relational Oracle for database processing. This solution would be low-risk and very efficient for high volumes.

alternatives are listed by type of technology and degree of centralization. The first three alternatives considered are whether to:

- Move to browser-based interfaces with some client-side processing.
- Use thin client browser-based interfaces.
- Use a mix of the two options.

The next two alternatives focus on supporting hardware for the database—whether to:

- Use a centralized database on a large server or server cluster.
- Distribute the database across several servers in multiple locations.

Finally, the location and type of database are considered. The decision is whether to use more traditional relational database technology or to move to more advanced object-oriented databases. Any decisions made for the CSS would need to be consistent with prior decisions for the SCM.

RMO wants its system to be state of the art, but it also does not want to have a high-risk project and attempt new technology that is not yet proven or for which it lacks needed skills or experience. Figure 8-5 lists the major components of the strategic direction for RMO.

Current, well-tested technology can provide client/server processing on a rack of multiprocessor servers to support high-volume Internet transactions. Microsoft's Internet Server will provide Internet support. The existing DB2 database on the server cluster is a very viable option to provide efficient back-end processing. The database will require redesign and must be rebuilt for the new system, but the fundamental processing environment is solid.

All of the COBOL applications will be replaced with new systems that will be written using Java, Visual Basic, VBScript, and PHP as appropriate.

In this approach, the server cluster will remain as the central database server. The other two tiers will be application servers. The users will have individual client personal computers that are connected to the application servers. Barbara Halifax attached the table shown in Figure 8-5 to her biweekly status report to John MacMurty (see the accompanying memo). She also identified the other open issues that needed to be decided. Even though the operating environment decisions are critical to the progress of the project, other important issues also still need to be addressed.

Figure 8-5

Strategic directions for the processing environment at RMO

Issue	Direction(s)
Required interfaces to other systems	1. Automatic feed to SCM system 2. Interface to feed the accounting general ledger 3. Interface to provide automatic feed to external systems—credit-card verification and package shipping 4. Potential move to XML for a common interface language
Equipment configuration	1. Servers with multiple CPU configuration for front-end applications 2. Database support provided with the existing server cluster
Operating system	1. Windows Server 2008 front-end servers 2. UNIX for server cluster
Network configuration	1. Windows network 2. IIS for Web servers
Language environment	1. Visual Basic, Java, and PHP for application and Web development
Database environment	1. Maintain DB2 database on the server cluster 2. Reevaluate long-term strategy for the OO database

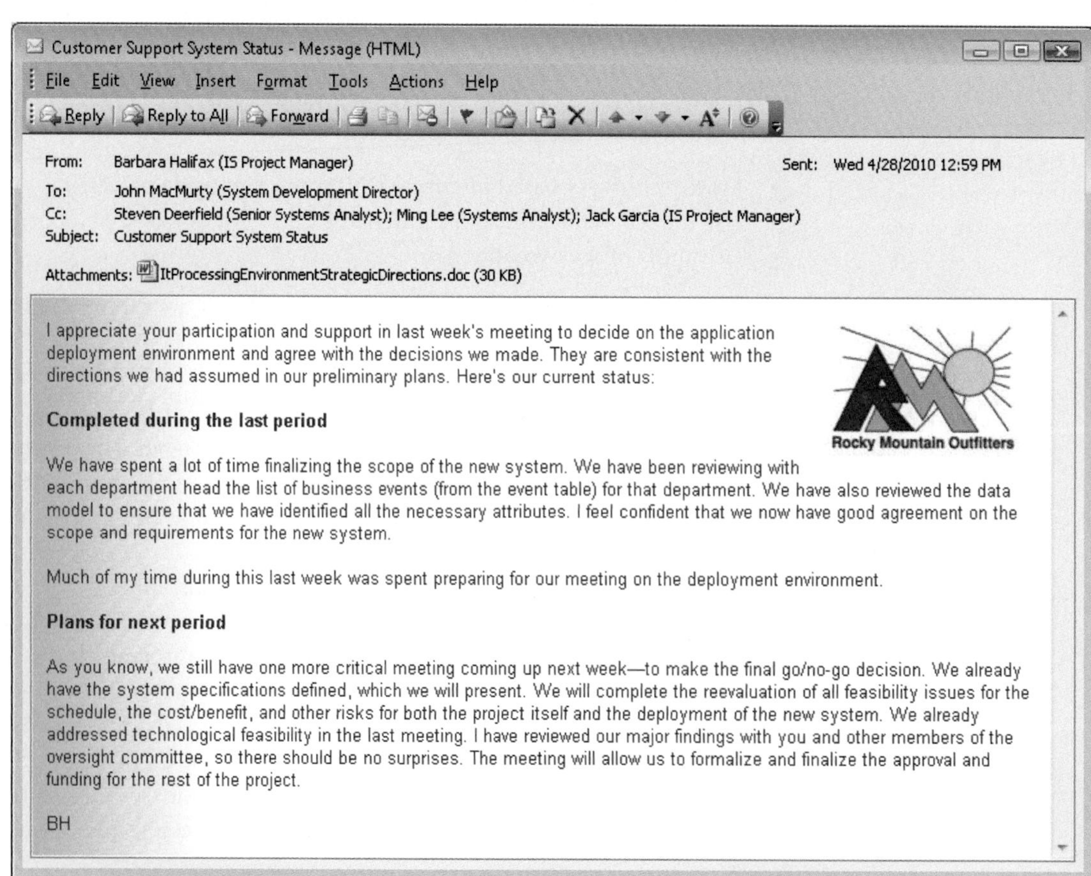

Customer Support System Status - Message (HTML)

File Edit View Insert Format Tools Actions Help

Reply | Reply to All | Forward

From: Barbara Halifax (IS Project Manager) Sent: Wed 4/28/2010 12:59 PM

To: John MacMurty (System Development Director)

Cc: Steven Deerfield (Senior Systems Analyst); Ming Lee (Systems Analyst); Jack Garcia (IS Project Manager)

Subject: Customer Support System Status

Attachments: ItProcessingEnvironmentStrategicDirections.doc (30 KB)

I appreciate your participation and support in last week's meeting to decide on the application deployment environment and agree with the decisions we made. They are consistent with the directions we had assumed in our preliminary plans. Here's our current status:

Completed during the last period

We have spent a lot of time finalizing the scope of the new system. We have been reviewing with each department head the list of business events (from the event table) for that department. We have also reviewed the data model to ensure that we have identified all the necessary attributes. I feel confident that we now have good agreement on the scope and requirements for the new system.

Much of my time during this last week was spent preparing for our meeting on the deployment environment.

Plans for next period

As you know, we still have one more critical meeting coming up next week—to make the final go/no-go decision. We already have the system specifications defined, which we will present. We will complete the reevaluation of all feasibility issues for the schedule, the cost/benefit, and other risks for both the project itself and the deployment of the new system. We already addressed technological feasibility in the last meeting. I have reviewed our major findings with you and other members of the oversight committee, so there should be no surprises. The meeting will allow us to formalize and finalize the approval and funding for the rest of the project.

BH

CHOOSING IMPLEMENTATION ALTERNATIVES

So far, we have described the analysis and fact-finding activities in the development project. As the project team makes decisions about the scope, level of automation, and processing environment, it also makes related decisions about the actual approach to designing, programming, and installing the system. There are numerous ways to implement a solution. For example, if an application is fairly standard, perhaps the organization could just buy a computer program or system to support it. Even for more sophisticated systems, other companies may have already developed standard systems that can be purchased. If purchasing is not an option, the organization might decide to build the system in-house, and even then there are various alternatives. Outside programmers can be contracted for a range of services or specific technical expertise. The point is that the organization must plan how to actually implement the system, and there are a multitude of options.

Figure 8-6 presents some variations on implementing a system. The left axis represents the build-versus-buy options. The bottom axis shows the alternatives of developing the system in-house versus outsourcing the project. Each axis represents a continuum. For example, an entire system can be bought, or the entire solution can be built. But between those extremes are systems in which portions are purchased and portions are built. In other words, a basic solution may be purchased, but it might require modification or programming of some components to interface with existing systems. Similarly, many options exist for all or part of the solution to be developed in-house or outsourced.

Figure 8-6
Implementation alternatives

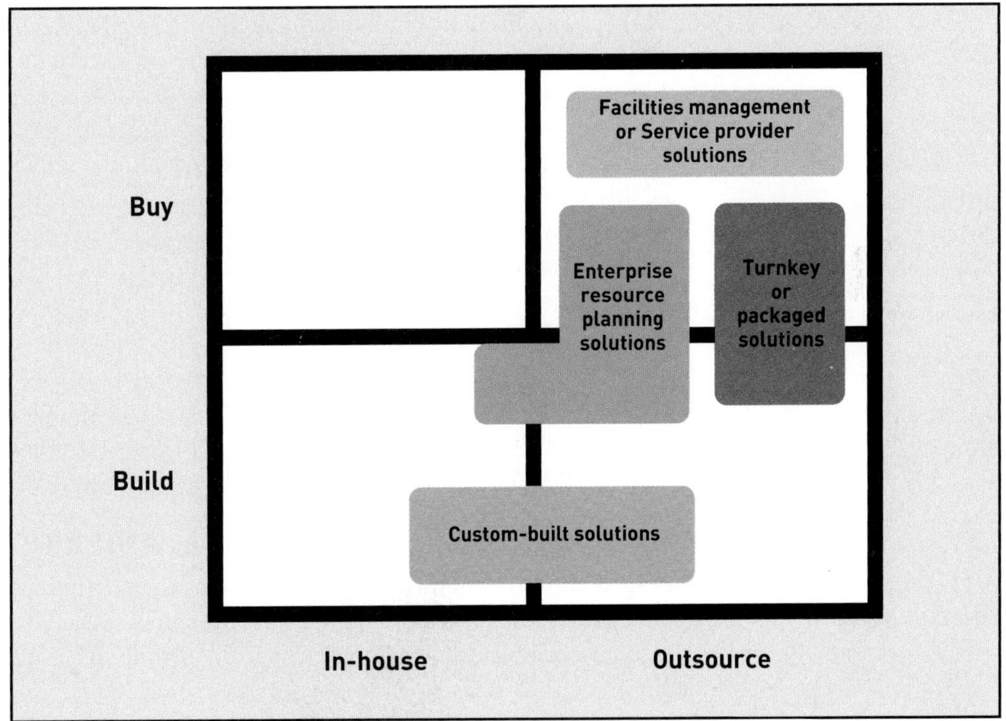

The shapes between the two axes show various general approaches to obtaining a system. Facilities management occurs when the entire system, including development and operation, is contracted to another company. Below that is packaged software or a turnkey system. Although slightly different approaches—packaged software is shrink-wrapped, off-the-shelf, whereas a turnkey system is a customized package—both usually require some modification to fit the existing environment. Thus, these options usually have a "build" component. Enterprise resource planning (ERP) solutions begin with a standard system, but they require substantial integration with a company's business processes. ERP solutions are integrated so tightly with the entire organization and all its systems that the implementation frequently requires a substantial effort. Custom-built systems require substantial programming by either in-house staff or outsourced consultants and programmers. Each alternative is explained in more detail in the following paragraphs.

FACILITIES MANAGEMENT OR SERVICE PROVIDER SOLUTIONS

facilities management
the outsourcing of all data processing and information technology to an outside vendor

Facilities management is the outsourcing of the entire data processing and information support capability for an entire organization. Facilities management is not an actual development technique or implementation alternative. Instead, it is the result of an organization's strategic decision to move all system development, implementation, and operation to an outside provider. For example, a bank may hire a facilities management firm to provide all of its data-processing capability. The computers, software systems, networks—even the technical staff—all belong to the outside firm. The bank in essence has opted to let another firm become its information systems department.

Outsourcing of all IS functions is a long-term, strategic decision. It applies to an entire organization and not just a single development project. So, even though we discuss it as one of the alternatives for implementation, this decision is not typically made by any project team. It is usually a top executive decision. Normally, a facilities management contract between an organization and a provider is a multimillion-dollar contract that covers services for 8 to 10 years. Electronic Data Systems (EDS), a multibillion-dollar company, is one company that obtains the majority of its revenues by providing facilities management services to many industries. EDS supports the banking, health insurance (such as Blue Cross and Blue Shield), grocery, insurance, and retailing industries, as well as governments. EDS can provide high-quality facilities management services in these various industries by employing a staff of highly experienced industry specialists.

packaged software
software that is already built and can be purchased as a package

Service provider solutions also require a long-term strategic decision. In a service provider solution, a company only buys the required technology services. No in-house computing capability is required, at least not for the purchased service. For example, many small companies contract with other companies to host a store Web site. The service might include building and presenting the Web site, as well as the sales functions for the business, such as catalog presentation, shopping cart and ordering support, and credit card and payment processing. The business does not need computer expertise or computer personnel; it is all provided as a service by the hosting company. Even for larger companies, there are service companies that provide complete support for accounting, human resources, and payroll. For both approaches, facilities management or service provider, the company does not build its own in-house technology group, but depends entirely on outside providers.

PACKAGED, TURNKEY SOFTWARE, AND ERP SYSTEMS

turnkey system
a complete system solution, including software and hardware, that can be turned over to the purchasing organization

Packaged software comprises software systems that are purchased to support a particular application. A strict definition implies that the software is used as is, with no modifications. We all have packaged software on our personal computers, such as a word processor or an accounting/general ledger package. We buy the software components without the source code but with documentation, install it, and use it. We don't modify it or try to add new capabilities. We use it exactly as it comes, with only the built-in options. The advantages of this software are that it works well and is inexpensive for the amount of capability provided. It is also usually well documented, relatively error free, and stable.

Packaged software has its place in the overall scheme of an organization's IS strategy. First, many packages can become part of a larger project. For example, a standard reporting system package may provide reporting capabilities to users. Generally, whenever possible, companies try to find packaged software to perform those standard functions.

A **turnkey system** is provided by an outside company as a complete solution, including hardware and software, and the organization only needs to turn it on. In most cases, the outside vendor specializes in a particular industry and its application software. Literally hundreds of firms, many of them small to medium-sized, specialize in systems for particular needs. These turnkey system firms advertise in trade journals for an industry. A few examples are legal systems for law firms, video systems for video stores, patient record systems for dentists and doctors, point-of-sale systems for small retail firms, construction management systems for construction firms, library systems for libraries, and so forth. The list is almost endless.

One critical problem with turnkey systems is that they often do not exactly meet the needs of an organization, and the organization frequently has the onerous task of modifying the way it does business to conform to the computer system. Some turnkey system vendors will modify their systems to suit particular customers. An organization normally purchases the base system, a certain number of customized changes, and a service agreement. The vendor firm analyzes the unique requirements of the organization and makes those changes to the program code. The service agreement can range from simple input form and report modifications to more extensive modifications over a period of months or years. In some cases, only executable code is provided; in others, both executable and source code are provided so that the organization can also make its own modifications. Sometimes the vendor firm makes all modifications; other times the purchasing organization may have programmers work with the vendor's project team to reduce the cost of customization and to gain experience on the new system. Again, numerous combinations are possible, and this method is very popular for obtaining software for small and medium-sized applications that are somewhat, but not completely, standard.

In the past, turnkey systems were used only for specialized systems within an organization. However, recently, several large firms have introduced this approach for enterprisewide systems. These systems, called *enterprise resource planning (ERP)* systems, support all operational functions of an entire organization. Companies such as SAP and Oracle have had good success introducing ERP systems into organizations. Obviously, when the support is enterprisewide, the deployment is a major undertaking. Many of these projects take longer than a year to install and cost millions of dollars.

The advantage of ERP systems is that a new system can usually be obtained at a much lower cost and risk than through in-house development. The cost is lower because 60 to 80 percent of the application already exists in the base system. Risk is lower because the base system is usually well developed and tested. In addition, other organizations are already using it, so it has a track record of success.

The disadvantage is that the ERP system might not do exactly what the organization needs, even after the system has been customized. Frequently, a gap exists between the exact needs of the organization and the functionality the system provides. The company then must modify its internal processes and train its users to conform to the new system. ERP systems are discussed in more detail in Online Supplemental Chapter 1, "Packages and Enterprise Resource Planning," on the book's Web site.

CUSTOM-BUILT SOFTWARE SYSTEMS

Custom-built software systems are those that are developed partly or completely by an outside organization and tailored to the exact needs of an organization. The new system is developed from scratch, based on the systems development life cycle. In some cases, the project team is staffed entirely by a consulting firm; in others, the project team is a combination of in-house staff and outside consultants.

The advantage of custom development is that an organization purchases a tremendous amount of experience and expertise to build a new system. Usually, the consulting firm has developed similar systems in the past and has extensive domain knowledge for a particular industry and application. It also will have a large pool of very experienced staff to solve complex technical problems. In addition, a large, experienced staff can be brought to the project rapidly to meet schedules and deadlines. Outsourcing and contract development are the fastest-growing segments of the IS industry.

The major disadvantage of custom development, of course, is the cost. Not only is the organization paying for the development of a new system, but it is paying for it in hourly wages for consultants. Typically, organizations opt for custom development when they do not have in-house expertise or have very aggressive schedules that must be met. Normally, the anticipated return on investment for the new system must be quite high to justify the cost of this approach. Custom systems are usually large systems with very high transaction volumes. One example is a health-care system with millions of claims to process. When the system can reduce the cost of processing a claim by one or two dollars, the total savings reach millions of dollars quickly due to the high volume.

Most large and medium-sized companies have an in-house information systems development staff. In fact, you might find excellent employment opportunities as a member of the development staff in such companies. One of the main problems with in-house development, particularly in medium-sized firms, is that a portion of a project may require special technical expertise beyond employees' experience. As a result, one alternative is to use company employees to manage and staff the project but to hire special consultants to assist in areas in which extra expertise is required. That way, the organization can maintain control and ensure progress but still obtain assistance when needed.

The advantages of this approach are primarily control of the project and knowledge of the project team. Company staff also have a better understanding of the internal culture of the organization and the specific processing needs of various business groups. One other major benefit is that the organization can build internal expertise by developing the system in-house.

The major disadvantage is that the in-house staff may not recognize when they need assistance. At times, the "not invented here" syndrome—the notion that "if we did not think of it or develop it, it is no good"—complicates development because perfectly good, reasonably priced solutions are not utilized. Sometimes the technical problems are more complex than anticipated, and in-house people do not recognize the need to obtain expert assistance.

SELECTING AN IMPLEMENTATION ALTERNATIVE

At times, selecting an implementation alternative is straightforward. At other times, deciding among alternatives can be difficult, especially when outside providers are included. For example, one solution may have some of the required functions but not all. Another solution may have the requisite functions but may only run on an undesirable platform and operating system. Some solutions may provide a quick, inexpensive solution for existing problems but may be limited for future growth; others offer long-term capabilities but are very expensive and take a long time to develop. One vendor may propose a turnkey system, another custom development, another a turnkey system with a particular database management system and platform, and yet another a joint development project. The problem in selecting is the proverbial comparison of apples and oranges. Frequently, there is very little in common among the solutions proposed by outside vendors because each vendor proposes a system that fits its own strength. The systems analyst must establish a set of common criteria to compare the alternatives with as much consistency as possible.

BEST PRACTICE

Remember that selecting a new system is not just a matter of "make or buy" or "outsourcing." There are many combinations of implementation and support approaches to consider.

Identifying Criteria

To begin selection, you must identify the criteria that you will use to compare the various alternatives. You will use these criteria to compare all viable alternatives, although differences among alternatives may make some criteria more or less applicable to those proposals. In particular, there are usually some differences in criteria or evaluation methods in comparing packaged and turnkey systems with custom-built systems. For example, packaged and turnkey systems typically have an existing base of users who can be queried regarding system functionality, reliability, and other important characteristics. For custom-built systems, it may be more important to ascertain the technical skills of the vendor staff. For alternatives that combine purchase of existing solutions with substantial customization or new development, both criteria would merit careful scrutiny.

Different criteria and evaluation methods can also be applied to alternatives presented by outside vendors and those presented by an internal IS department. However, there is an inherent danger of bias toward internal providers when applying different criteria for internally generated alternatives. In theory, criteria such as "vendor reliability" and long-term costs should be evaluated similarly for both internal and external providers. But as a practical matter, some criteria are often ignored or given less emphasis due to the perception of lower risk and greater control over internal IS staff and departments. The project manager and oversight committee must carefully examine selection criteria and measurement methods to ensure fair and complete comparison of internal and external alternatives.

There are three major areas to consider in selecting an implementation alternative:

- General requirements
- Technical requirements
- Functional requirements

General requirements include considerations that are important but not directly associated with the computer system itself. The first major component of general requirements is the feasibility assessment, which was discussed earlier in the context of selecting the scope and level of automation. Each of the implementation alternatives under consideration must meet the requirements for cost, technology, operations, and schedule defined in the feasibility analysis. The following list identifies several criteria that can be included in this section:

- The performance record of the provider
- Level of technical support from the provider
- Availability of experienced staff
- Development cost
- Expected value of benefits
- Length of time (schedule) until deployment
- Impact on internal resources
- Requirements for internal expertise
- Organizational impacts (retraining, skill levels)
- Expected cost of data conversion
- Warranties and support services (from outside vendors)

Obviously, some criteria are more important to the organization than others. For example, in the preceding list, we might want to purchase only from a very reputable, stable, and experienced provider. So, the performance record of the provider is extremely important. On the other hand, we might have some leeway in the schedule, so a very short deployment schedule might not be critical. The relative importance of each item in the list can be weighted with a numbering scale. Figure 8-7 provides a sample table of general criteria and weighting factors for RMO. That table uses a five-point weighting scale. Criteria that are more important are given a higher number, such as a five or maybe a four. Those that are less important are assigned lower numbers. The extended score is the weight times the raw score for each category.

The four alternatives along the top of the table represent various implementation options. The first alternative is to develop the system in-house. The second and third alternatives are different turnkey systems that start with a basic package and modify it. The last option is to contract with a consulting firm to develop a completely new system from the ground up. The four alternatives are for illustration only—to show the various weighting values possible.

Functional requirements represent the functions that must be included within the system. These requirements are developed during the analysis activities, identified in the event table, and described in the data flow diagrams or use case diagrams. Each project has a unique set of functional requirements based on the needs of the system.

General requirements criteria	Weight (5=high, 1=low)	Alternative 1 In-house		Alternative 2 Package #1 + modify		Alternative 3 Package #2 + modify		Alternative 4 Custom development	
		Raw	Extended	Raw	Extended	Raw	Extended	Raw	Extended
Availability of experienced staff	4	3	12	3	12	3	12	5	20
Developmental cost	3	5	15	5	15	3	9	1	3
Expected value of benefits	5	5	25	3	15	4	20	3	15
Length of time until deployment	4	2	8	5	20	4	16	2	8
Low impact on internal resources	2	2	4	4	8	5	10	4	8
Requirements for internal expertise	2	2	4	4	8	5	10	4	8
Minimal organizational impacts	3	4	12	3	9	4	12	4	12
Performance record of the provider	5	5	25	4	20	4	20	4	20
Level of technical support provided	4	5	20	3	12	3	12	3	12
Warranties and support services provided	4	5	20	4	16	4	16	4	16
Total			**145**		**135**		**137**		**122**

Figure 8-7

A matrix showing a partial list of general requirements

Figure 8-8 illustrates a partial list of functional requirements for the RMO customer support system. The weighting technique is the same as is used for general requirements.

In addition to the functional and general requirements, each new system normally has a set of technical requirements that must be met. Technical requirements are also system constraints—the constraints under which the system must operate. This category includes all other requirements that are placed on the system, its method of operation, its performance, its utility, and so forth. The following list indicates some of the items that should be considered under technical requirements:

- Robustness (the software does not crash)
- Programming errors (the software calculates correctly)
- Quality of code (maintainability)
- Documentation (user and system, online and written)
- Ease of installation
- Flexibility (the software makes it easy to adjust to new functionality and new environments)
- Structure (maintainable, easy to understand)
- User-friendliness (natural and intuitive use)
- Performance (response time)
- Scalability (ability to handle large volumes)
- Compatibility with operating environment (hardware, operating system)

Functional requirements criteria	Weight (5=high, 1=low)	Alternative 1 In-house		Alternative 2 Package #1 + modify		Alternative 3 Package #2 + modify		Alternative 4 Custom development	
		Raw	Extended	Raw	Extended	Raw	Extended	Raw	Extended
Make inquiry on items	4	5	20	4	16	5	20	5	20
Create customer order	5	5	25	5	25	5	25	5	25
Change order	4	5	20	5	20	5	20	5	20
Make inquiry on orders	4	5	20	5	20	4	16	5	20
Package order	5	5	25	5	25	5	25	5	25
Ship order	5	5	25	5	25	5	25	5	25
Create back order	4	5	20	5	20	5	20	5	20
Accept return	4	5	20	5	20	4	16	5	20
Correct customer account	4	5	20	3	12	4	16	5	20
Update catalog	5	5	25	2	10	3	15	5	25
Create special promotions	3	5	15	0	0	2	6	5	15
Initiate a promotion mailing	3	5	15	0	0	2	6	5	15
Create sales summaries	3	5	15	3	9	3	9	5	15
Create order summaries	2	5	10	3	6	3	6	5	10
Create shipment summaries	2	5	10	2	4	5	10	5	10
Total			**285**		**212**		**235**		**285**

Figure 8-8

A matrix showing a partial list of functional requirements

Figure 8-9 shows possible weighting factors and scores for technical requirements. For alternatives that are already built, such as packages or ERP systems, scores can usually be derived. However, for custom-built alternatives, such as in-house projects, these points become objectives for the new system. In other words, because nothing is built yet, these items cannot be measured or evaluated. However, they do become criteria for the construction of the new system. In Figure 8-9, to make balanced comparisons between the alternatives, we have assigned values to "build" alternatives that are the averages of the "buy" alternatives (the values are marked by asterisks).

Probably the most difficult part of this exercise is establishing the weighting factors. The client, system users, and project team should all have a voice in establishing the weighting factors. Consideration must be given not only to the relative importance of each criterion within each area—general, functional, or technical—but also to the balance among all major areas. In other words, the rating team must ensure that the relative weight of general requirements compared with functional requirements truly represents the desires of the client.

Technical requirements criteria	Weight (5=high, 1=low)	Alternative 1 In-house		Alternative 2 Package #1 + modify		Alternative 3 Package #2 + modify		Alternative 4 Custom development	
		Raw	Extended	Raw	Extended	Raw	Extended	Raw	Extended
Robustness	5	?	*18	3	15	4	20	?	*18
Programming errors	4	?	*16	4	16	4	16	?	*16
Quality of code	4	?	*18	4	16	5	20	?	*18
Documentation	3	5	15	3	9	4	12	4	12
Easy installation	3	5	15	5	15	4	12	4	12
Flexibility	3	4	12	3	9	4	12	5	15
Structure	3	4	12	4	12	4	12	4	12
User-friendliness	4	5	20	3	12	4	16	5	20
Total			**126**		**104**		**120**		**123**

Figure 8-9

A matrix showing a partial list of technical requirements

Making the Selection

After requirements have been considered and rated, each alternative can then be evaluated with a raw score based on how well it meets the criteria. Ranges for raw scores generally vary from a simple three-point scale to a more finely ratcheted six-point scale. For example, a three-point scale could contain these ratings: Fully Satisfy (2), Partially Satisfy (1), and Not Satisfy (0). A six-point scale could represent these ratings: Superior (5), Excellent (4), Good (3), Fair (2), Poor (1), and Disqualify (0). To calculate a weighted score in each criterion for each alternative, staff would multiply the raw score by the weighting factor. An overall score, which is the sum of the individual criteria scores, determines a ranking among the various alternatives for this category.

RMO decided to undertake most of the CSS development with in-house staff. As seen in Figure 8-7, the first three alternatives rank very close together on general requirements, with in-house development having a slight advantage. In Figure 8-8, alternatives 1 and 4 are approximately equal and better than alternatives 2 and 3. In Figure 8-9, which shows technical requirements, alternatives 1, 3, and 4 are very close. So, overall, in-house development does provide a slight advantage. RMO's in-house systems analysts and technical staff had proven several times in the past that they could handle development of complex systems. In addition, this approach would enable RMO to continue to build in-house expertise. But although the approach selected was to do development in-house, the information systems group was not averse to hiring specialists when needed.

RMO has sufficient in-house expertise to develop the networks and the database portions of the system. The Web-based development will also be done in-house, but RMO will probably have to hire several specialists. Because RMO wanted to keep the expertise within the company, hiring some new specialists seemed to be a viable method.

Some of the integration issues, such as integrating new hardware and system software with existing systems, could possibly become quite complicated. Some very experienced consultants were available, and RMO anticipated that it would need to retain a couple to oversee this portion of the project.

RMO staff is now ready to proceed with the CSS project. A review of the feasibility constraints identified no serious problems. The project is still on schedule and within budget, and the attitude within the company is very positive. With the availability of these additional resources, the project also appears to be technically feasible.

For RMO's customer support system, in-house development was the chosen alternative for implementation. But to get the information needed to evaluate all the other alternatives, the CSS team sent out a formal request for proposal to each prospective vendor.

GENERATING A REQUEST FOR PROPOSAL

request for proposal (RFP)
a formal document, containing details on the system requirements, sent to vendors to request that they bid on supplying hardware, software, and/or support services

As just mentioned, a **request for proposal (RFP)** is a formal document sent to vendors. Its basic purpose is to state requirements and solicit proposals from vendors to meet those requirements. Use of RFPs is almost universal in government contracts and fairly common in private industry. The project manager has primary responsibility for developing the RFP and evaluating submitted proposals.

Often, particularly in governmental purchasing, an RFP is a legal document. Vendors rely on information and procedures specified in the RFP. That is, they invest resources in responding with the expectation that certain procedures will be followed consistently and completely. Thus, an RFP is often considered to be a contractual offer, and a vendor's response represents an acceptance of that offer.

A good RFP includes a detailed explanation of the information needs of an organization and the processing requirements that must be fulfilled. Chapter 4 defined system requirements as consisting of functional and technical requirements. A good RFP will provide detailed explanations of both these types of system requirements. If the early project assumptions indicated that the most viable option for the new system would be to purchase a turnkey solution or outsource for custom development, the analysis activities are geared toward developing an RFP. When the outside firm is selected, it will then ensure that its staff obtains in-depth knowledge about the problem domain before beginning detailed design or customization.

To develop and distribute a good RFP, the purchasing organization must do an in-depth analysis. This work is not usually a problem for firms that have in-house information systems staff. However, for firms that do not have information systems staff, determining processing requirements can be a problem. They may also have difficulty generating a meaningful RFP or evaluating the various purchase alternatives. Smaller, unsophisticated firms tend to ignore this problem and simply try to make the best decision they can, often in ignorance. A wiser approach is to hire an independent consultant, one who will not be involved in the development, to help establish the selection criteria and decide on a vendor. The same criteria for choosing a final vendor should be used in selecting this independent consultant.

Figure 8-10 shows an outline of a generic RFP. Obviously, each RFP must be tailored to the specific needs of the organization and the requirements of the project. The first part of the RFP, comprising items I and II, provides background information on the company and the need for a new system. Next, items III through V describe in detail all of the requirements that the new system must meet. In this example, we have divided the requirements into technical, functional, and general requirements. Section VI requests information on the provider's background and experience. The final two sections, VII and VIII, indicate how the proposal should be submitted and how it will be evaluated.

The RFP should clearly state the procedural requirements for submitting a valid proposal. When possible, the organization should include an outline of a valid proposal, along with a statement of the contents of each section. In addition, the RFP should clearly state deadlines for questions, proposal delivery, and other important events.

Figure 8-10

A sample RFP table of contents

Request for Proposal
Table of Contents

I. Introduction and background
   A. Background on company
   B. Overview of industry/business

II. Overview of need
   A. Description of business need
   B. Expected business benefits
   C. Overview of system requirements

III. Description of technical requirements
   A. Operating environment
   B. Performance requirements
   C. Integration, interfaces, and compatibility
   D. Hardware specifications
   E. Expansion and growth requirements
   F. Maintainability requirements

IV. Description of functional requirements
   A. Specification of primary functions
   B. Specification of information outputs
   C. Specification of the user interface
   D. Identification of optional functions and enhancements

V. Description of general requirements
   A. Maintenance and support
   B. Documentation and training
   C. Future releases
   D. Other contractual requirements

VI. Requested provider and project information
   A. Request for statement of work and project schedule
   B. Request for reference list of provider
   C. Request for project personnel information

VII. Details for submitting the proposal
   A. Time requirements
   B. Format requirements

VIII. Evaluation criteria and process
   A. Expected timetable of evaluation
   B. Method of evaluation of technical, functional, and general requirements

The requirements statement constitutes the majority of the RFP. The body of the RFP can formalize and state the guidelines previously described. Requirements should be separated into those that are absolute (essential) and those that are optional or subject to negotiation. This categorization is a more formal version of the prioritization discussed earlier. The RFP also should state explicitly the evaluation criteria—for example, the categories that are to be evaluated as well as the weighting factors.

BENCHMARKING AND CHOOSING A VENDOR

One method to evaluate the quality of a vendor's system is either to observe it in use or to install it on a trial basis and test it out. However productive this approach may be, it is also expensive and difficult. The format of the data, the forms, and even the platform may be alien to existing configurations. If the system is complex, people must be trained to use it. Staff must also be available to do testing. Although this approach can be quite expensive, it may be less expensive than making a bad decision.

benchmark
an evaluation of a system against some standard

Some applications are amenable to a more rigorous evaluation called a **benchmark**. A benchmark is a performance evaluation of application software (or test programs) using actual hardware and systems software under realistic processing conditions. In years past, benchmarking was often difficult to perform because of the expense of the hardware and software configurations and the length and cost of installation. Currently, these problems are less

severe because hardware is cheaper, installation procedures have been streamlined, and competition among vendors is fierce.

Another way to observe a system in use is to visit another company. Sometimes the vendor will have demonstration versions already installed in its own facilities. Potential purchasers are permitted to go to the vendor's plant and test the system. Vendors also have previous clients who are using the system. It is almost always a good idea to visit these previous clients. When prospective clients make a site visit, they are not permitted to test the system themselves with their own data, but they will be permitted to observe the other company using the system. They can also talk to previous clients about their experience with the system and the vendor. It is always a good idea to get references and to talk with other companies that have done business with the vendor.

Frequently, companies want to add capabilities to the software as the business environment changes. To do so, they must make the enhancements themselves or have their vendor supply upgrades and fixes. Organizations should ascertain the level of ongoing research and development being done by the vendor. How compatible are these new capabilities to a particular system after it has been customized? Is there an active users group that can suggest new enhancements and modifications to the vendor? The company is making a major investment, and it is important that the investment be considered in the long term. The largest investment in any new system is the long-term cost of maintenance. A good vendor will help its clients leverage their maintenance dollars by providing upgrade support and new capabilities based on feedback from existing clients.

DEVELOPING A CONTRACT

After a final decision is made on which proposal provides the best solution and value, a contract is written. Contract development and negotiation are usually a team effort involving the project manager, legal counsel, and frequently other senior executives. The project manager's involvement is essential to ensure that the contract meets the needs of the project and that important performance and termination clauses are included.

Contracts can be divided into several different types, which shift the risk either to the purchaser or the vendor. Fixed-dollar contracts put most of the risk on the vendor. The advantage to the purchasing company is that the vendor assumes the burden of project delays and overruns. However, the vendor usually sets a high price to compensate for the risk. Cost-plus-percentage contracts put the risk on the purchaser. In fact, cost-plus-percentage contracts encourage the vendor to spend more because its income is directly proportional to the costs of the project. A middle ground, with both sharing the risk, is a cost-plus-fixed-fee or cost-plus-incentive contract. In this case, both the purchaser and the vendor benefit if the project finishes as quickly as possible. With a fixed fee, the profit margin for the vendor is high if the project progresses quickly. If the project drags on, however, the fixed fee results in a lower profit margin.

PRESENTING THE RESULTS AND MAKING THE DECISIONS

The results of the investigation and analysis activities described in this chapter are normally summarized in a written report and presented orally to executives. The intended audience is the executive oversight committee, which has decision-making and funding responsibility for the project. The objective of the documentation and presentation is to provide the necessary background so that informed decisions can be made.

The responsibility of the project team, including both technical and user members, is to do the detailed investigation and calculations to enable an informed analysis of all of the alternatives. However, the final decision of which alternative, or mix of alternatives, is chosen rests with the executive oversight committee. This committee not only controls the budget and provides the funding but also is responsible for the overall strategic direction of the company.

One of the more difficult tasks for the project team is to compile, organize, and present the alternatives and critical issues in a way that is easy to understand yet accurate and complete. The executive oversight committee usually consists of people who are, first and foremost, business executives. They generally are not technical experts, yet they need to make decisions that affect the entire organization. So, presenting findings is one of the most difficult tasks that the project team will have. It requires careful consideration to find the right balance of detail. At one extreme is so much technical detail that the oversight committee cannot understand or follow the logic and becomes lost or bored. At the other end are recommendations without sufficient supporting detail or logic.

The formality of the presentation varies from organization to organization. Some companies require very formal written reports and oral presentations. Other organizations require nothing written and only informal discussion between the client (the person funding the project) and the project team leader. Smaller organizations tend to be less formal, and large corporations typically have standard policies and procedures for approval.

The format of the document and report varies considerably, depending on the desires of the audience. A detailed description of how to develop this presentation is provided in Appendix D. (Appendix D is available for download at *www.course.com*.) Generally, the written documentation will follow the same format as the presentation.

SUMMARY

The activities explained in this chapter are primarily project manager responsibilities. The focus of the project changes at this point from one of discovering the requirements to that of developing a solution system. So, the activities described are pivotal for the project to change emphasis. Obviously, the project manager takes primary responsibility for this change in direction. These activities involve seven of the eight project management knowledge areas that are described in Appendix A on the book's Web site.

One of the important activities during analysis is to prioritize the system requirements based on the scope and level of automation desired. The scope of the new system determines which functions it will support. The level of automation is a measure of how automated the selected functions will be. Highly automated functions have sophisticated computer systems such as expert systems to help carry out the business functions.

The application deployment environment is the configuration of computer hardware, systems software, and networks in which the new system must operate. It determines the constraints that are imposed on the system development alternatives. The analyst must define an environment, or multiple environmental choices, that match application requirements and the organization's strategic application and technology architecture plans.

Another activity that is done in conjunction with prioritizing the requirements is determining what alternatives are possible for developing the solution and then selecting one of those alternatives. Implementation alternatives include such options as building the system in-house, buying a packaged or turnkey solution, or contracting with a developer to build it (outsourcing). When outsourcing is anticipated, a request for proposal (RFP) is developed and sent out. The RFPs are then evaluated for how well they match the requirements. Selecting from the various alternatives should be a careful process. The evaluation includes consideration of such factors as the match of the proposed system to the functional and technical requirements and the reputation and performance record of the submitting vendor.

One of the final analysis activities is to develop recommendations and present them to management. After the analysis is complete, a more knowledgeable decision can be made about the direction, cost, feasibility, and approach of the rest of the project. The systems analyst documents the results of the analysis activities and presents them in a logical fashion that is focused toward the executives who make funding decisions.

KEY TERMS

application deployment environment, p. 291
benchmark, p. 306
development environment, p. 292
facilities management, p. 298
packaged software, p. 298
request for proposal (RFP), p. 305
turnkey system, p. 298

REVIEW QUESTIONS

1. What is meant by the *application deployment environment*? Why is it important in the consideration of a development approach?
2. List and briefly describe the characteristics that an analyst examines when choosing or defining the deployment environment.
3. Describe the relationship between the application deployment and development environments.
4. Explain the fundamentals of facilities management.
5. What is the difference between scope and level of automation?
6. What is meant by the make-versus-buy decision?
7. Define a *packaged solution*. Explain what is entailed in the packaged solution approach.
8. What is meant by *ERP*? How does an ERP approach affect acquiring a new solution?
9. What does *outsourcing* mean? How does it affect a project?
10. Define *benchmark*. Why is it useful in selecting a new system?
11. What is an *RFP*? Why is it developed at the end of the analysis activities instead of at the beginning?
12. What is the difference between general requirements, technical requirements, and functional requirements?

THINKING CRITICALLY

1. What are the advantages of purchasing a packaged solution? What are the disadvantages or dangers?
2. What are the advantages of building a solution from the ground up? What are the disadvantages?
3. What are the advantages to outsourcing a development project? What are the disadvantages?
4. Discuss the importance of developing a formal technique and specific criteria for evaluation alternatives.
5. Given the following narrative, identify the functions to be included within the scope of the system. Also identify several levels of automation for each function. The purpose of this question is to give you an opportunity to think creatively, especially to identify high-level automation alternatives for the various functions.

   Conference Coordinators (CC) assists organizations or corporations in coordinating and organizing conferences and meetings. It provides such services as designing and printing brochures, handling registration of attendees, fielding questions from attendees, securing meeting spaces and hotel rooms, and planning extracurricular activities. CC gets its business in two ways: by following up on leads that a company is going to be holding a conference and by having the company contact CC directly. When a contact is made, the client is asked for basic information about the desired event: city, dates, anticipated number of attendees, price range, and external activities desired. From this information, CC prepares a bid. CC likes to keep its turnaround time on bids to under five working days. Each project is assigned to a project manager, who will gather information from the support staff to prepare the bid. If necessary, he or she may also request information from the visitors' center for the desired city.
6. What are important points that determine weighting factors for the functional requirements listed in the system requirements for a proposed system?
7. List the important points that determine weighting factors in the general and technical requirements for a proposed system.
8. Given the following matrix of various technical requirements, develop your own weighting factors for an inventory management system at a small plumbing supplier. Justify your weights. Extend the raw scores to the Extended column and calculate the totals. Which would you choose? Justify your selection: Did you go strictly by the numbers, or are there other factors you might consider? How do you handle a number that is not given: give it an average of the others, pick the best of the others, guess a value, or assign a zero? (Raw numbers use a six-point scale.)

Category	Weight	Alternative 1 Build in-house		Alternative 2 Buy turnkey		Alternative 3 Buy package	
		Raw	Extended	Raw	Extended	Raw	Extended
Robustness		5		3		3	
Programming errors		?		4		4	
Quality of code		?		4		5	
Documentation		4		4		3	
Easy installation		5		5		4	
Flexibility		5		4		3	
User-friendliness		5		5		5	
Total							

EXPERIENTIAL EXERCISES

1. Assume that the deployment environment for a high-volume payment processing system consists of the following:
   - Oracle DBMS running under the UNIX operating system on a cluster of HP servers
   - WebSphere application server running under the Z/OS operating system on an IBM zSeries 900 mainframe
   - J2EE application software that will be executed by other internal and external systems

   Investigate possible development environments for this deployment environment. Describe their advantages and disadvantages and recommend a specific set of development tools.

2. Set up an interview with an organization that uses information systems. Ask for an example of an RFP for a software system. Identify the parts of the RFP. Compare them with the recommended components discussed in this chapter.
3. From a news article or Internet information, find an example of a company that is installing an ERP package (SAP, Oracle, or another company). If possible, get a copy of the overall project plan and analyze the various activities. Compare them with a standard SDLC. Find out the total budget for the project.
4. Develop an RFP for RMO to be sent out to various vendors.
5. Develop a recommended implementation approach for RMO. Also develop a presentation of your recommendation to upper management.
6. Look through some trade magazines (*Software*, *CIO*, *Datamation*, *Infoweek*, and so on) to find examples of companies that have done an evaluation of vendors. Describe their methods and comment on their strengths and weaknesses.

CASE STUDIES

TROPIC FISH TALES' RFPS

Now that you have read and studied the chapter, review the opening case on Tropic Fish Tales. Your job is to provide specific advice for Robert Holmes or Bill Williams on how to evaluate the various RFPs.

Assuming that you can build some matrices that measure relative strengths among the proposals, comment on the applicability of doing an evaluation based strictly on the numbers. In other words, assume that Robert and Bill were able to create criteria and weights to measure the benefit to the company of the different alternatives.

1. Do you think it would be possible to sum up the resulting values and make a decision based only on the numbers? Support your answer.
2. What factors, other than those in the matrix of weighted criteria, might Robert and Bill need to consider in making a decision? Can these other factors influence the decision as strongly as the quantified criteria?
3. What if the values of several alternatives are very close? What other factors might Robert and Bill need to consider?

THE REAL ESTATE MULTIPLE LISTING SERVICE SYSTEM

Consider the requirements of the multiple listing service system developed in Chapters 5, 6, and 7. Assume that you're the project manager and that you work for a consulting firm hired by the multiple listing service to perform only the survey and analysis activities.

1. Assume that system users and owners have indicated a strong desire for a system that can be accessed "anytime, anywhere." Discuss the implications of their desire for the system scope. Given the preferences of the system users and owners, should you prepare a table similar to Figure 8-2? Why or why not?
2. Discuss the implications of the anytime, anywhere requirement for the application deployment environment. What type(s) of hardware, network, and software architecture will be required to fulfill that requirement?
3. Investigate the availability of packaged and turnkey systems for multiple listing services. Search the Internet and real estate trade magazines and Web sites. Discuss the pros and cons of choosing a packaged or turnkey system.
4. Develop an RFP outline that covers packaged, turnkey, and custom-developed systems. What are the difficulties of writing one RFP that covers all three scenarios? Who should be involved in evaluating RFP responses?

RETHINKING ROCKY MOUNTAIN OUTFITTERS

Various application deployment environments would actually be acceptable for RMO's strategic plan. The staff's current thinking was to move more toward a Microsoft solution, using the latest version of Microsoft Server with Microsoft's IIS as the Web server. However, Linux with Apache servers offers another large installed base of servers. Considering that RMO could also take that approach, do the following:

1. Describe a viable configuration using Apache/Linux.
2. Compare the relative market penetration of Microsoft and Apache/Linux (a good starting place is *http://news.netcraft.com*).

The database issue is another potential controversy for RMO. The current decision is to keep the mainframe and run DB2, a very efficient relational database. However, another alternative would be to implement an Oracle database. Oracle is also very strong in the marketplace. Given these two alternatives, do the following:

3. Compare the relative market penetration of these two solutions.
4. List the strengths and weaknesses of each approach: that of the DB2 mainframe approach and that of Oracle running on some type of multiple processor server computer.

FOCUSING ON RELIABLE PHARMACEUTICAL SERVICE

Assume that Reliable has completed a thorough analysis of system requirements (part of which you worked on as case exercises in Chapters 4 through 7). Management is now confronted with the task of choosing a system scope and implementation approach. To summarize the alternatives, you have prepared the following table, which divides the requirements into functional subsets, estimates the duration of design and implementation for each function if software is custom-built, and categorizes the risk for each function based on software complexity, technology maturity, and certainty about requirements.

Function	Project duration	Risk
Inventory and purchasing	9 months	Moderate
Order fulfillment (manual data entry)	6 months	Low
Web-based order entry	9 months	High
Prescription warning	12 months	High
Billing	18 months	High

Top executives have evaluated the table and determined that all of the functions are high-priority needs. The project is critical to restoring profitability and maintaining market share. Reliable is well behind the technology curve for its industry, and it needs to modernize to reduce costs and to provide expected levels of service. Unfortunately, overlap and dependency among the functions makes it difficult to consider implementing only a subset of the functions. Executives would prefer to implement all functions in a single project, but they consider the combined project duration for all functions to be much too long.

Significant parts of the proposed system, such as inventory, purchasing, and prescription warning, are similar to systems used by retail pharmacies and in-house pharmacies in large hospitals and health maintenance organizations. But some significant differences exist in Reliable's requirements for order entry, product delivery, and billing. There are a handful of large vendors and several dozen smaller vendors that specialize in pharmacy systems.

Assume that management has identified the following options for proceeding with the system development or acquisition:

- Contract with a vendor to modify a packaged prescription software system to suit Reliable's needs.
- Contract with a vendor to purchase the generic parts of a prescription system and extend the system to address Reliable's unique needs with custom-built software.
- Contract with a system development firm to custom-build a system, possibly making use of some off-the-shelf components for inventory management and prescription warning.

Reliable's executives have assigned you the following tasks:

1. Develop an RFP outline that addresses the options identified by the executives. List and briefly describe each general, technical, and functional requirement.
2. Assume that you have already developed a complete set (over 100 printed pages) of analysis documents using either the traditional or object-oriented approach. Should those be included in the RFP? Why or why not?
3. Develop matrices (similar to Figures 8-7, 8-8, and 8-9) for evaluating RFP responses.
4. Develop a list of vendors to whom the RFP should be sent.

FURTHER RESOURCES

Scott E. Donaldson and Stanley G. Siegel, *Cultivating Successful Software Development: A Practitioner's View.* Prentice Hall, 1997.

Ralph L. Kliem and Irwin S. Ludin, *Project Management Practitioner's Handbook.* American Management Association, 1998.

Sanjiv Purba, David Sawh, and Bharat Shah, *How to Manage a Successful Software Project, Methodologies, Techniques, Tools.* John Wiley & Sons, 1995.

John J. Rakos, *Software Project Management for Small to Medium Sized Projects.* Prentice Hall, 1990.

Kathy Schwalbe, *Information Technology Project Management, Fifth Edition.* Course Technology, 2008.

Neal Whitten, *Managing Software Development Projects: Formula for Success.* John Wiley & Sons, 1995.

PART 3 SYSTEMS DESIGN TASKS

CHAPTER 9
Elements of Systems Design

CHAPTER 10
The Traditional Approach to Design

CHAPTER 11
Object-Oriented Design: Principles

CHAPTER 12
Object-Oriented Design: Use Case Realizations

CHAPTER 13
Designing Databases

CHAPTER 14
Designing the User Interface

CHAPTER 15
Designing System Interfaces, Controls, and Security

CHAPTER 9

ELEMENTS OF SYSTEMS DESIGN

LEARNING OBJECTIVES

After reading this chapter, you should be able to:

- Discuss the issues related to managing and coordinating the activities of the SDLC
- Explain the major components and levels of design
- Describe each major design activity
- Develop a simple network diagram
- Describe common deployment environments and matching application architectures

CHAPTER OUTLINE

Project Management Revisited: Execution and Control of Projects

Understanding the Elements of Design

Design Activities

Network Design

The Deployment Environment and Application Architecture

FAIRCHILD PHARMACEUTICALS: FINALIZING ARCHITECTURAL DESIGN FOR A PRODUCTION SYSTEM

James Schultz is a summer intern with Fairchild Pharmaceuticals. For the past two weeks, he has been assigned to an ongoing development project for a production scheduling and control system. The project was nearing a point when critical design and deployment decisions were needed. Sufficient requirements had been identified through analysis activities to know what was required in the solution. Before proceeding with more detailed analysis or detailed design, some basic design decisions were required.

James works for Carla Sanchez, the chief analyst and project manager. For the past two weeks, James has been shadowing Carla as she finished prioritizing the identified functional requirements. He helped Carla prepare presentations to the project oversight committee and system users. He has absorbed a lot of information about the project in a short time, but the details and overall project direction haven't yet formed a coherent picture in his mind. Yesterday, the oversight committee signed off on the work to date, so Carla asked James to stop by her office the next morning to discuss his assignment for the next project phase.

James knocked on Carla's open door and asked, "Is this a good time or should I come back later?"

"I have time now," she said. "Come in and have a seat. Let's start by reviewing the results of yesterday's meetings, and I'll answer any questions you have. Then we'll narrow down your tasks for the next few weeks of the project."

James said, "I have two questions that I think are related. The first is, Which implementation details have been decided and which haven't? The discussions with the users and oversight committee left me with the impression that the decision to go with a full-blown Web-based system had already been made. Yet none of the supporting infrastructure for a Web-based system currently exists, or does it?"

Carla replied, "A few elements are in place, but most of it will need to be designed and acquired. But let's hear your other question before we get into that."

James continued, "Well, you've sort of anticipated my next question, which is, What do we need to do next? There seem to be several important tasks that need to be started now, such as choosing system software to support Web services, determining what changes will be needed to the company network, and designing the database. But I suspect that I've left out a few important pieces. Also, the tasks and decisions are so interdependent that I don't know which should be tackled first."

Carla smiled before she replied, "Well, you really were awake during all of the meetings! You should take some pride in knowing that you're confused about exactly the right things at this point in the project. The transition from focusing on requirements definition to designing solutions is an important but uncertain step in all projects, and this one is no exception. It's hard to move from a detailed knowledge of what the user wants and needs to a precise blueprint of a system that will satisfy those wants and needs. As you've correctly observed, many important decisions need to be made very quickly, and they overlap. They're also heavily constrained by available time, budget, and existing systems, skills, and infrastructure."

Looking a bit relieved, James replied, "So what's up first, and where do I fit in?"

Carla replied, "The generic name for the next step is systems design, with the first part being architectural design. It's where we'll finalize all of the big-picture decisions, such as what hardware will support the new system, what operating systems we'll use, how we'll store and access data, and what languages and tools we'll use. Some of these issues were briefly addressed at the start of the project, and some decisions were implicit in the choice of deployment environment and automation scope approved by the oversight committee yesterday. What we need to do now is to lay all of them on the table, make sure they're compatible with one another and with existing systems and capabilities, and parcel out the detailed tasks associated with each."

Carla continued, "I spent yesterday afternoon dividing the work into major categories, including hardware and operating systems, Web support services, database design, application software design, and user interface design. I summarized the choices made so far and the remaining decisions we need to discuss. Key players will meet as a group for the rest of the week to discuss options in each area and develop the system architecture. For example, we'll decide whether to extend our existing database to support the new system or develop a new database with a new DBMS. By the end of the week, we'll have made all of the critical architectural decisions, ensured that the pieces all fit together, and developed plans to tackle each area with personnel assignments and time lines. From that point forward, work in each area can proceed in parallel. Professor Chen told me that you've done an independent study in Web services support software, right?"

"Yes," James replied. "I did a comparative study of infrastructure requirements and communication protocols for Web services using CORBA, Microsoft .NET, and Java 2 Web Services. I did an in-depth technology review of each and visited two sites using each technology to see how they worked in practice."

"Good," said Carla. "That knowledge will come in handy because we need to decide whether to base the new system on Web services and, if so, what supporting infrastructure and development tools to use. I think that you'll learn a lot by working with me for another week or two as we hammer out the architectural design. After we get the detailed design tasks rolling, we'll choose one for you that suits your interests and abilities. There'll be plenty of interesting tasks from which to choose and more than enough work to keep you busy for the next month or two."

OVERVIEW

We begin this chapter by revisiting the principles of good project management. Chapter 3 and Appendix A provide a substantive discussion of project management principles as well as detailed discussions of project planning. Chapter 8 provided more discussion of project management tools to determine project scope and implementation alternatives. In this chapter we will explain project management concepts for how to monitor and control an ongoing project. These concepts are based on sound principles. However, the actual skills to know how to run a project come only with experience. What you learn from the text will give you a solid foundation to observe and learn before you manage your own project.

Chapter 8 described the activities and decisions associated with finalizing the major elements of the user's requirements. Those activities were focused on finalizing the major functional components of the system to meet the business need. This chapter is an extension of those activities, but the focus changes to the solution system. In other words, during analysis the focus is on *understanding* what the system should do—for example, the requirements. Design is oriented toward the *solution*—in other words, specifying how the system will be built and the structural components of the new system. Such activities as defining the deployment environment and determining levels of automation are direct inputs to the design processes described in this chapter.

A normal question new developers ask is: *"When are these tasks carried out in a real project?"* Unfortunately, there is not an easy answer. We have the issue of whether the project is predictive or adaptive, but even within the first part of predictive projects these tasks are often spread across many weeks. Many projects begin with some of these decisions already made, particularly when companies already have a strong technology infrastructure in place. For other projects, the new system may be the result of a new thrust for the organization, and the decisions are wide open. However, it is normal for the project team to start thinking about these issues very early on and to begin making preliminary decisions as requirements are being defined. The important point to understand, however, is that the topics discussed in

this chapter and the following chapters are design topics. In other words, they are solution oriented. Realizing that these are design issues will help you to know that you should not try to come up with a solution until you understand the problem.

This chapter is the first of seven chapters that discuss design. In this chapter, we briefly describe all design activities and discuss the first activity (network design) in more detail. Later chapters explore other design activities using both traditional and object-oriented models and techniques.

BEST PRACTICE

Don't design the solution before you understand the problem.

PROJECT MANAGEMENT REVISITED: EXECUTION AND CONTROL OF PROJECTS

In Chapter 3 we introduced the basic concepts of project management, and explained many of the fundamental skills required for planning the project. Appendix A teaches you about the Project Management Body of Knowledge (PMBOK) by providing a detailed explanation of each of its nine areas. You should study Appendix A before trying to understand the concepts presented in this chapter. In this section, we build on the concepts taught in the appendix to provide more detailed ideas about actually running a project. Here we address project management issues related to the execution, monitoring, and control of an ongoing project. In many ways, this part of project management is the most difficult. In this activity, the project is actually run and the project is moved forward. Running a project requires an entire set of project management skills and talents, almost all of which are learned on the job. In other words, even though we have a few techniques to teach, much of this knowledge must come from mentoring and experience.

This section provides explanations in four areas of running a project:

- Organizing teams and assigning work
- Communicating status and information
- Monitoring and controlling project progress
- Controlling project issues and risks

Figure 3-3, for predictive projects, and Figure 3-4, for adaptive projects, identified two major project management activities: project execution management and project control management. These two activities work together to allow the project manager to schedule and execute the identified project tasks, and to monitor progress and take corrective action when things get out of control. Certain issues of project execution and control depend on whether the project is a predictive project or an adaptive project.

ORGANIZING PROJECT TEAMS AND ASSIGNING WORK

Some project managers see themselves as "the boss," and believe it is their job to supervise and direct team members. Other project managers see themselves as facilitators. They know that the actual work of the project is done by team members, so these project managers consider that their job is to clear all the obstacles so team members can get the work done. Although the second approach sounds good in theory, it can be quite difficult. How does the project manager ensure that a group of individuals, each with different skills, knowledge, motivations, and desires, learns to work together in a highly collaborative manner to achieve high performance? Extensive research on teams has found that an effective team can sometimes outperform a regular team by more than a factor of ten. One of the unique opportunities of a project

manager is to help build a high-performance team. Appendix A addresses various human resource issues in a software team setting. This section briefly introduces two project execution topics related to teams. First, what are some key considerations in organizing the teams? Second, what are the key issues in assigning tasks on the project schedule to teams or team members?

Team Organization

Every project is different, both in the kinds of tasks that make up the project and in the individual team members assigned to the project. Consequently, it is not possible to specify a particular team organization that works best in all situations. Over time, most project managers develop preferences for organizing successful teams. Several key questions that should be considered in organizing the team are identified in the following list:

- Assign a team leader for each subteam or let the subteams organize themselves?
- Assign members permanently to a team or have floating team assignments?
- Assign team members to subteams to achieve a balance based on (1) skills, (2) experience, or (3) personality traits?
- Balance team membership based on permanent (core) members or transient members?
- Provide formal team training or on-the-job training for team members?
- Place team members in a large, common work area or individual cubicles or offices?

Assigning Tasks to Team Members

As with the organization of the project team, there are many different methods for partitioning the work among team members. Each project manager will tend to develop a certain style of working with team members to ensure that all scheduled tasks are completed in a timely fashion. The organization structure of the team also affects how tasks are assigned. Nevertheless, we have listed a few key questions that must be decided and implemented to move the project forward:

- What is the formality (versus informality) of the project (schedule, assignments, status, and so on)?
- Should tasks be assigned to subteams or to key individuals?
- Should tasks be assigned well in advance or using a just-in-time approach?
- Is the project schedule stable or is it a changing schedule?
- How do the number and duration of critical-path tasks compare to the number of tasks that are not on the critical path?
- Should tasks be assigned based on specific skills or on availability?
- Should tasks be assigned so that team members are fully scheduled or should open times be provided on people's schedules?

MANAGING THE COMMUNICATION PROCESS

Pervasive to all types of working relationships and work activities is the problem of communication. Employees who are at the bottom of the communication hierarchy are very sensitive to a lack of communication. For some reason, when people rise within the corporate hierarchy, they seem to forget the importance and need to keep people informed.

On the other hand, some managers are so caught up with keeping everyone informed that they waste lots of time in long status meetings and progress reviews. Sometimes there are so many meetings that team members become frustrated because they cannot get their regular work done. Most of us have had to sit through a meeting in which everybody made a detailed report of their status and problems. Such meetings can take hours and are usually a highly inefficient use of time.

A good project manager is sensitive to communication needs and establishes processes that ensure team members and outside stakeholders have the information they need, when they need it. The communication process is actually an information management issue. Before thinking about how to communicate, a project manager must consider all information issues in and around the project.

The first step in communications management is to determine, at a detailed level, what kind of information is required for project success. There are two types of information on a project. The first is project-related information, which includes the project schedule, work and team assignments, status and progress reports, presentation materials, and so on. The second kind of information is system information, which concerns the business needs and the new system being built. System information includes requirements definitions, specifications, design models, open issues logs, and finally program code. The two types of information are quite different, yet each is critical for project success.

The second step concerns how to manage the information. For each information item, a mechanism should be put in place to: 1) collect the information, 2) store the information, and 3) distribute the information. Related to the first and last of these tasks are the questions of who, what, and how: Who is the source of which information, how is it collected, who needs which information, and how is it distributed? These questions can be answered early in the project when a stakeholder analysis is conducted. The team must answer the questions of what information should be stored and how it should be stored.

Figure 9-1 is a summary of the communications issues. A good project manager will set up the communication mechanisms early in the project. Decisions about how much information to collect, store, and distribute will be influenced greatly by the formality of the project, as discussed in Chapter 3. In today's world, many electronic tools are available to enable good communications. If the processes are established correctly, with good electronic tools, the project manager no longer becomes the information bottleneck and information is collected, stored, and distributed almost automatically.

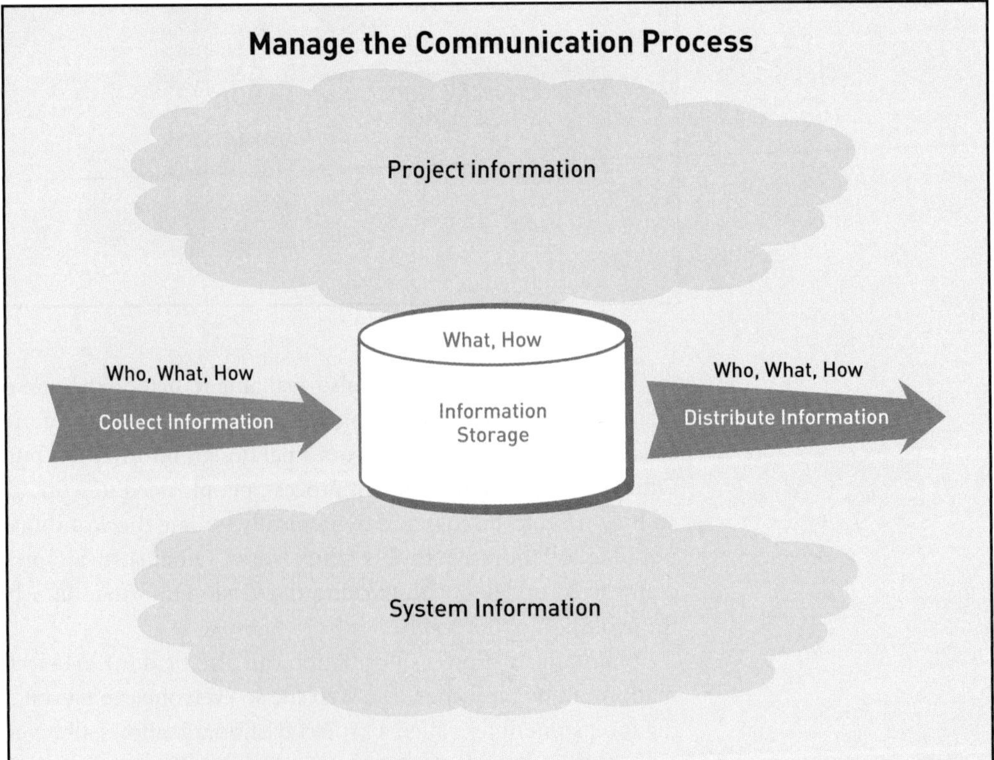

Figure 9-1
Communication processes within a project

A common and widespread technique to record and track system information is to use a central repository of system information. Most developer support tools have a central repository to capture information. The central repository not only records all design information, it is normally configured so that all teams can view project information to facilitate communication among the project teams. Figure 9-2 illustrates various information components that may exist within a data repository. One distinct advantage of using such a repository is that all system information is available to every member of the team. Especially on large projects, it is helpful for members of one team to be able to access design decisions that affect their work, but which are the responsibility of another team.

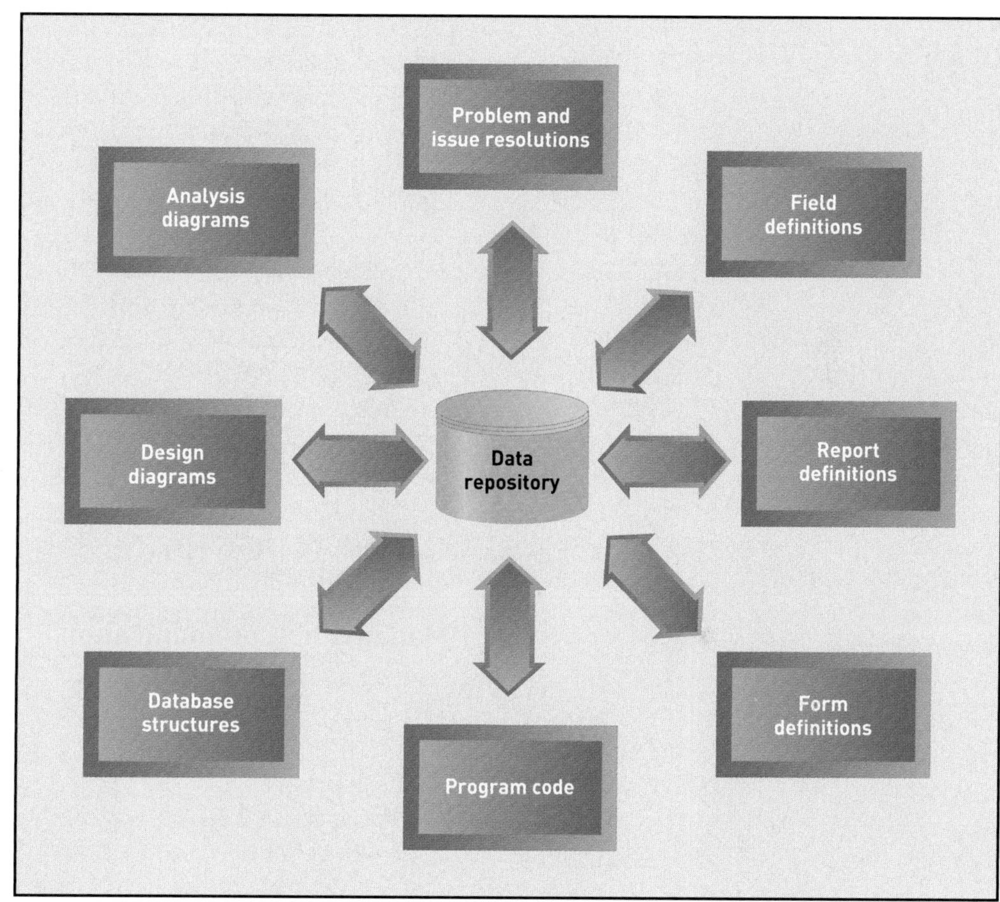

Figure 9-2
System information stored in a data repository

Other electronic tools are also available to help with team communication and information coordination. These tools and techniques, often referred to as *computer support for collaborative work*, not only record final design information but assist in team collaboration. Often during the development process, people need to work together to develop the design, so they need to discuss and dynamically update the working documents or diagrams. One popular collaborative tool is Lotus Notes. Other software programs allow figures and diagrams to be updated with tracking and version information, which helps the team document the evolution of the result.

Maintaining project information can also be done via electronic means. Schedule information can be published to a Web site so everyone can view it. Another type of project tracking tool, sometimes called a project dashboard, allows all types of project information to be posted and viewed by Web browsers. Figure 9-3 is an example of a project dashboard system that allows easy access to project information. Spreadsheets, e-mails, newsletters, and list servers all provide capable means to maintain, collect, and distribute information. Web page

management systems such as content management systems or wiki systems make posting and monitoring project information easy. Wikipedia is an example of a wiki system that can be updated by any interested party and easily viewed. Effective project managers set up the information and communication process at the beginning of the project. Once it is set up correctly, it often will take care of itself. Each team member can update his or her information at the appropriate times. Everyone should have access to all information, with special e-mails or notices of critical information postings to those who require it.

Figure 9-3

Sample dashboard showing project overview and status information

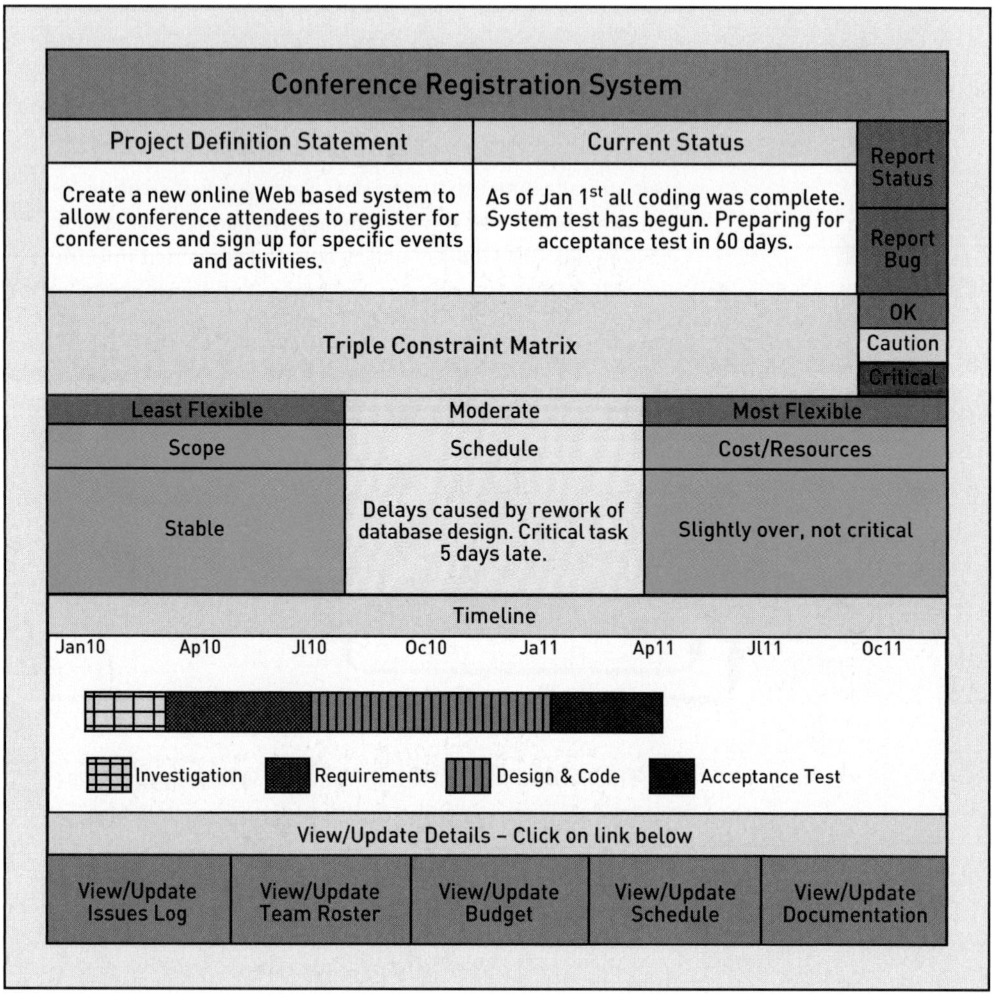

Conference Registration System

Project Definition Statement	Current Status
Create a new online Web based system to allow conference attendees to register for conferences and sign up for specific events and activities.	As of Jan 1st all coding was complete. System test has begun. Preparing for acceptance test in 60 days.

Report Status

Report Bug

Triple Constraint Matrix

OK

Caution

Critical

Least Flexible	Moderate	Most Flexible
Scope	Schedule	Cost/Resources
Stable	Delays caused by rework of database design. Critical task 5 days late.	Slightly over, not critical

Timeline

Jan10 Ap10 Jl10 Oc10 Ja11 Ap11 Jl11 Oc11

Investigation Requirements Design & Code Acceptance Test

View/Update Details – Click on link below

View/Update Issues Log	View/Update Team Roster	View/Update Budget	View/Update Schedule	View/Update Documentation

MONITORING THE PROJECT PLAN

In theory, executing and controlling the project plan sounds easy, but in fact it is quite complicated. The basic premise of executing any project is that you have some type of project plan. In Chapter 3, you learned how to use Microsoft Project to build a project plan. What we did not teach, and what you must learn from experience, is how to build a realistic and workable project plan. As you have opportunities to work with other project managers, you will better learn how to make good project plans. Assuming that you have a good project plan, and have been able to identify and recruit the right team members, executing the plan becomes easier.

How a team builds and executes project plans will vary depending on whether the project structure is based on a predictive approach or an adaptive approach. In the predictive approach, the team tries to lay out all of the project details at the beginning. This approach requires project plans that are usually quite large and complex. The adaptive approach is less daunting because the detailed project plan is done for each iteration. At the beginning of an

iteration, the first step is to plan the work. Because the piece of work is smaller and often is better understood, these schedules tend to be smaller and less complex. In either case, the first step in executing the project is to assign an appropriate project member or team to the task.

Figure 9-4 is a high-level flowchart that illustrates the basic process for monitoring and controlling the project. The first box, *Assign work to person or team,* is a complex task by itself. During the life of a project the makeup of the project personnel often changes. Analysis activities require a lot of user involvement, and team members who can understand the user's needs are most effective. As design and implementation activities begin, programmers and technical staff are often added to the team. Whenever a new member is added, time must be allocated for training and educating the new person. Often the size of the team will increase as design and programming tasks are done in parallel. It is not unusual for various subteams to be formed for specific internal miniprojects. For example, after some requirements have been specified, acceptance test data can be identified and created. A subteam comprised of a user and a testing expert can start the process of creating test data. As another example, a subteam can be formed to begin data conversion after the new database structure is defined. The existing data or database will need to be converted into the new format. The data conversion subproject can often become complex and require substantial effort. The skills required to assign teams to specific tasks is usually learned from experience.

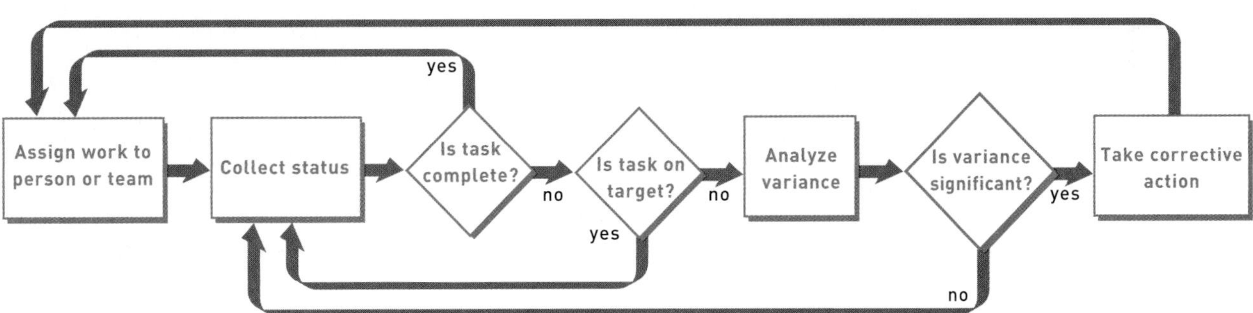

Figure 9-4
Workflow to monitor and control project execution

The second box in the flowchart, *Collect status,* is less complex. We offer the following guidelines. First, providing status information should be a standard process for all team members. Generally it is a waste of time to collect status in a "status meeting." If status information needs to be communicated specifically, then e-mails can be sent. If project teams need to coordinate their tasks or results, then a coordination meeting can be held between the affected team members. But general status meetings are an ineffective use of time. Status information should be collected and posted electronically for all to see. If the project tasks have been identified at the correct level of detail, then status information can be reported at milestones as complete or not complete. The "Percent complete" figure never seems to work as one would anticipate. For example, tasks might reach 90% completion and then remain there for weeks.

The next box, *Analyze variance,* is primarily a judgment decision. If a delayed task affects the overall schedule because it is a critical-path task, then it is a significant variance. Sometimes a variance also raises the risk of a potential delay or defect, which also should be considered significant.

The final box, *Take corrective action,* can also be complex. Experienced project managers have a whole set of tools they can use to try to correct the variance. Sometimes the correction is as simple as reassigning team members to get more people working on the task, or maybe

it just requires some extra hours of overtime. At other times, tasks may have to be rearranged. In more serious instances, the entire schedule may have to be reworked or more team members may need to be recruited for the team. The objective of corrective action is to get the project back to a known and predictable schedule.

CONTROLLING ISSUES AND RISKS

Every development project, whether it is predictive or adaptive, always has lots of questions that need answers and many decisions to be made. From the first fact-finding session, through architectural and program design, and until the final acceptance test is conducted, new issues will arise. In many cases, these issues are quickly resolved and the project moves rapidly forward. However, in other instances, the answer to a question or the resolution of an open issue will require additional research. Some open issues have an impact on various parts of the system or project and need to be resolved in a timely manner. For example, a set of business rules for sales commissions includes when and how commissions are calculated, what happens to commissions on merchandise returns, when commissions are paid, how the commission schedule varies to encourage sales of high-margin items and sale items, and so on. These business rules must be defined to design the database properly and to develop the commission programs. However, what if management is still making decisions about these business rules? Research and executive discussion will be needed before final decisions are made. You would not want to hold up the entire project for a few of these decisions. On the other hand, you want to make sure that they do not fall through the cracks. Plus, these decisions must be made before the database can be finalized and the program structure can be designed.

During the project planning activities the project manager also identified potential risks that could have a negative impact on the project schedule. As the project progresses some of those risks may disappear. Other risks, however, may turn into real problems, and new risks may appear during the project. A good project manager will establish procedures to track the identified risks and document any newly identified risks.

Finally, as the project progresses, new items or requests from users will be identified that cannot be immediately incorporated into the new system. If the project uses a predictive approach, sometimes new requests are generated after the system is almost complete or even into acceptance testing. Those requested changes should be documented and put on the list for the next version of the system. If the project uses an adaptive approach, some requests will need to wait until the next iteration, or possibly a later version. So, for either type of project, a change log should be maintained to itemize detail change requests.

The monitoring and control of open issues and risks for a project is usually nothing more complex than building various tracking logs. These logs can be built in a simple spreadsheet and posted on the project Web site or central repository. It is a good idea to make these logs available to all team members. Figure 9-5 is an example of a tracking log. The column headings will vary depending on the type of log you use. For example, a log for tracking user requests will have slightly different columns than one that tracks open technical design issues. Figure 9-5 is an open-issue tracking log with items that need to be resolved by a certain date and that have a person responsible for resolving the issue.

BEST PRACTICE

Maintain an open-items list for unresolved problems and questions.

	A	B	C	D	E	F	G	H	I
1	Issue Log#	Issue Date	Issue Description	Priority	Issue Impact	Person Responsible	Target Fix Date	Resolution Description	Actual Fix Date
2		1/18/2010	Commission structure for sales promotion is undefined	Urgent	Database structure may need to be modified	William Henry	2/1/2010		
3									
4									
5									

Figure 9-5

Sample issue tracking log

THE PROJECT TEAM AT RMO

As the customer support system project moves forward into design at RMO, two new members have been added to the project team. Consistent with the earlier discussion, RMO has initiated two new subprojects: one for data conversion and one for the system and acceptance test plans. To integrate new people into the team, Barbara Halifax reorganized the structure of the project team. Those who had been on the team throughout the analysis activities are now key players in getting the new team members up to speed. The accompanying RMO memo highlights some of the current changes in the project.

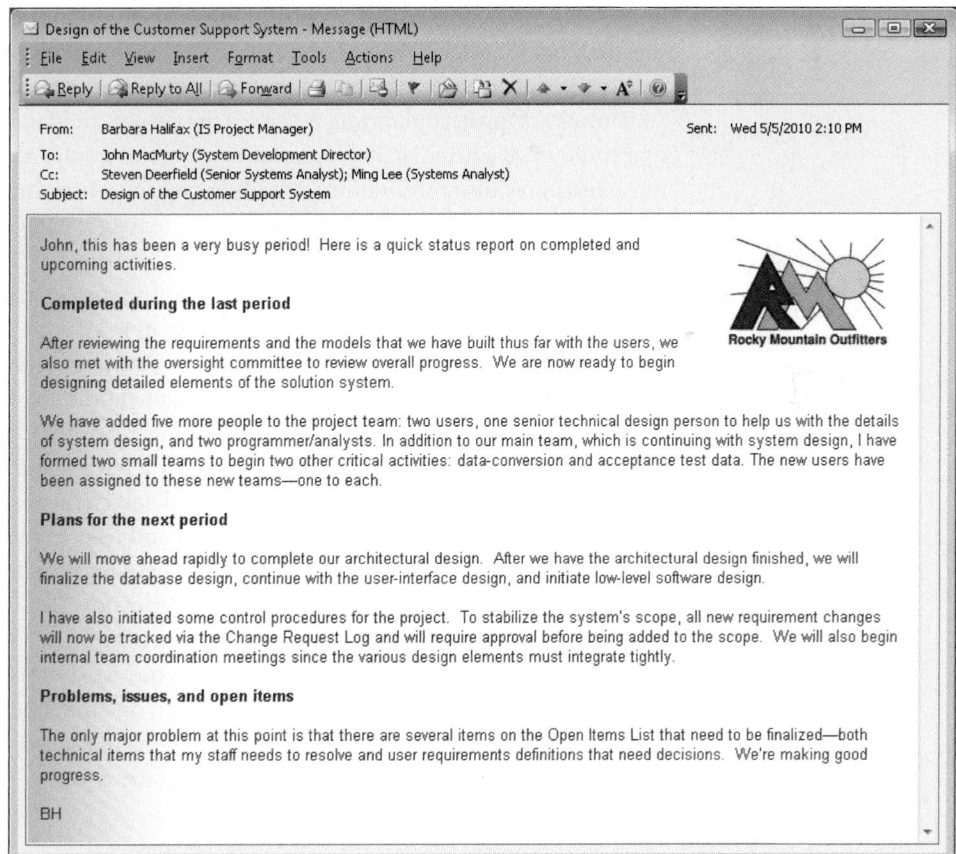

Design of the Customer Support System - Message (HTML)

File Edit View Insert Format Tools Actions Help

Reply | Reply to All | Forward

From: Barbara Halifax (IS Project Manager) Sent: Wed 5/5/2010 2:10 PM

To: John MacMurty (System Development Director)

Cc: Steven Deerfield (Senior Systems Analyst); Ming Lee (Systems Analyst)

Subject: Design of the Customer Support System

John, this has been a very busy period! Here is a quick status report on completed and upcoming activities.

Rocky Mountain Outfitters

Completed during the last period

After reviewing the requirements and the models that we have built thus far with the users, we also met with the oversight committee to review overall progress. We are now ready to begin designing detailed elements of the solution system.

We have added five more people to the project team: two users, one senior technical design person to help us with the details of system design, and two programmer/analysts. In addition to our main team, which is continuing with system design, I have formed two small teams to begin two other critical activities: data-conversion and acceptance test data. The new users have been assigned to these new teams—one to each.

Plans for the next period

We will move ahead rapidly to complete our architectural design. After we have the architectural design finished, we will finalize the database design, continue with the user-interface design, and initiate low-level software design.

I have also initiated some control procedures for the project. To stabilize the system's scope, all new requirement changes will now be tracked via the Change Request Log and will require approval before being added to the scope. We will also begin internal team coordination meetings since the various design elements must integrate tightly.

Problems, issues, and open items

The only major problem at this point is that there are several items on the Open Items List that need to be finalized—both technical items that my staff needs to resolve and user requirements definitions that need decisions. We're making good progress.

BH

UNDERSTANDING THE ELEMENTS OF DESIGN

Systems design is the process of describing, organizing, and structuring the components of a system at both the architectural level and a detailed level, with a view toward constructing the proposed system. Systems design is like a set of blueprints used to build a house. The blueprints are organized by the different components of the house, and describe the rooms,

stories, walls, windows, doors, wiring, plumbing, and all other details. We do the same organizing in systems design, although the components we are describing are those of the new system. We design and specify various components of the solution.

To understand the various elements of systems design, we must consider two questions:

- What are the components that require systems design?
- What are the inputs to and outputs of the design process?

MAJOR COMPONENTS AND LEVELS OF DESIGN

To perform design, analysts first partition the entire system into its major components because an information system is too complex to design all at once. Figure 9-6 depicts how these various components fit together. The icons in the figure are pieces of hardware, and inside the hardware are the software components. The cloud represents the entire system, and the various icons show the parts of the system that must work together to make the system functional. Information systems professionals must ensure that they develop a total solution for the users—they have not done their job if they haven't provided an integrated, complete solution.

As we will see in an upcoming section, the design activities of the SDLC support this partitioning of the final system into design components. Basically, each design activity is focused around designing one of the identified components shown in Figure 9-6.

Figure 9-6

System components requiring systems design

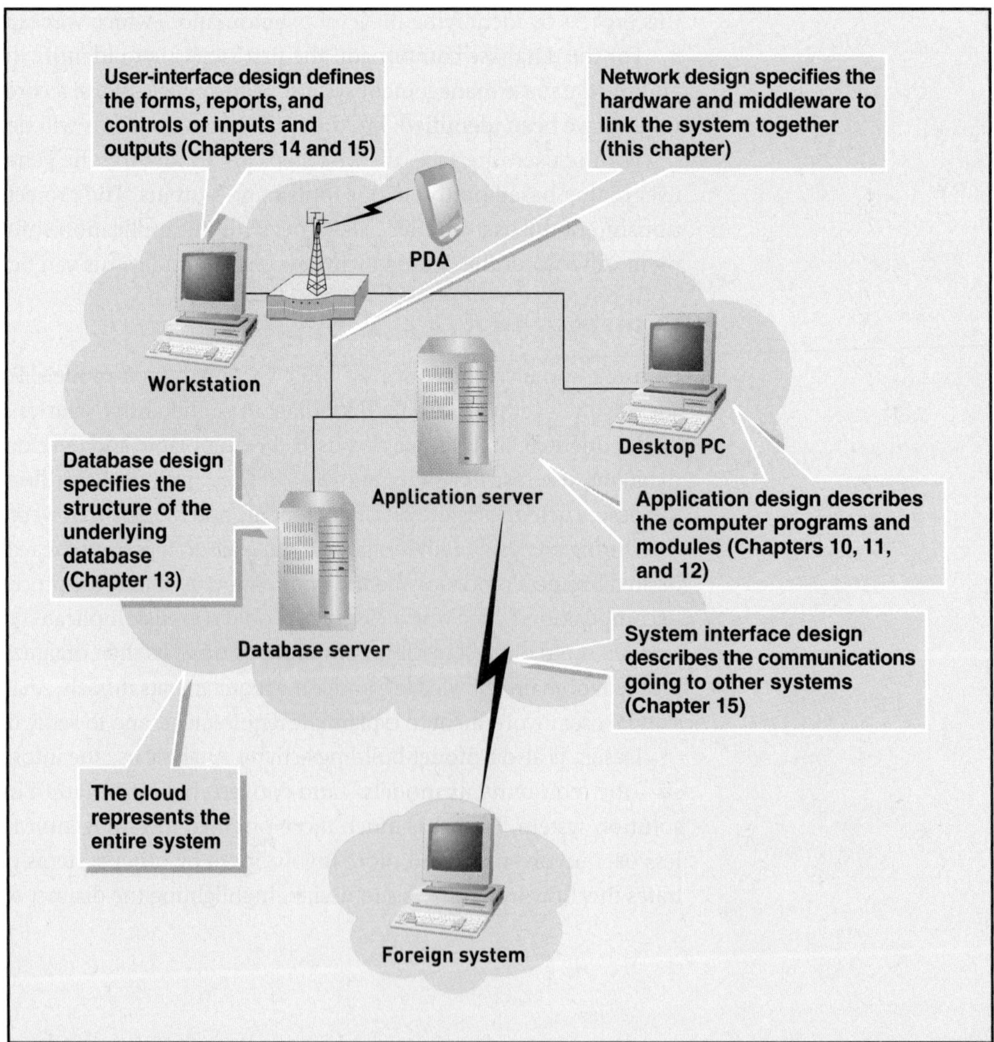
User-interface design defines the forms, reports, and controls of inputs and outputs (Chapters 14 and 15)
Network design specifies the hardware and middleware to link the system together (this chapter)
PDA
Workstation
Desktop PC
Database design specifies the structure of the underlying database (Chapter 13)
Application server
Application design describes the computer programs and modules (Chapters 10, 11, and 12)
Database server
System interface design describes the communications going to other systems (Chapter 15)
The cloud represents the entire system
Foreign system

A second important idea underlying systems design is that of the different levels of design. During analysis, we first identified the scope of the problem before we tried to understand the details. We called this step top-down analysis. Analysis, as it was presented, included both top-down activities (for example, scope first, then details) and bottom-up activities (for example, DFD fragments first, then the middle-level diagram). The same ideas apply during design.

As you begin working in industry, you will find that various names are given to the design at the highest level, including architectural design, general design, and conceptual design. We will use the term **architectural design**. During architectural design, you first determine the overall structure and form of the solution before trying to design the details. Designing the details is usually called **detail design**. It is not so important at this point to distinguish which activities are architectural design and which are detail design. Neither is it important to identify which models or documents belong to architectural design or to detail design. What is important is to recognize that design should proceed in a top-down fashion. Let's review the implications of this approach for each of the design components identified in Figure 9-6.

architectural design

broad design of the overall system structure; also called *general design* or *conceptual design*

detail design

low-level design that includes the design of specific program details

For the entire system, the analysts first identify the overall application deployment environment. They determine the overall architectural requirements and structure of the network before specifying the details of the routers, firewalls, servers, workstations, and other components. This approach was introduced in Chapter 8 and is expanded in this chapter.

For the application software, the first steps are to identify the various subsystems and their relationships to the network, the database, and the user-interface components. Part of that early design is the automation system boundary. The system boundary identifies which functions are included within the automated system and which are manual procedures. Notice that we began this process by identifying the level of automation, which was explained in Chapter 8.

For the database component, the first steps are to identify the type of database to be used and the database management system. Some details of the record structures and the data fields might have been identified, but the final design decisions will depend on the architecture.

For the user interface, the first steps are to identify the general form and structure of the user dialog based on the major inputs and outputs. The project team also describes the relationship of the user-interface elements with the application software and the hardware equipment. Afterward, the detailed window and report layouts can be developed.

INPUTS FOR SYSTEM DESIGN

During the analysis activities, we built documents and models. For traditional analysis, models such as the event table, data flow diagrams, and entity-relationship diagrams were built. For object-oriented analysis, we also used the event table and developed other models such as class diagrams, use case diagrams, and use case descriptions. Regardless of the approach, the input to the design activities is the set of documents and models that were built during earlier activities.

During analysis, analysts also built models to represent the real world and to understand the desired business processes and the information used in those processes. Basically, analysis involves decomposition—breaking a complex problem with complicated information requirements into smaller, more understandable components. Analysts then organize, structure, and document the problem domain knowledge by building requirements models. Analysis and modeling require substantial user involvement to explain the requirements and to verify that the models are accurate.

Design is also a model-building activity. Analysts use the information gathered during analysis—the requirements models—and convert that information into models that represent the solution system. Design is much more oriented toward technical issues and therefore requires less user involvement and more involvement by other systems professionals. Figure 9-7 illustrates this flow from analysis to design, highlighting the distinct objectives of each phase.

The original definition of design indicates that it involves describing, organizing, and structuring the system solution. The output of the design activities is a set of diagrams and documents that achieves this objective. These diagrams model and document various aspects of the solution system. As with the analysis models, some components are similar for structured and OO approaches, but other components are very different.

Figure 9-7

Analysis objectives versus design objectives

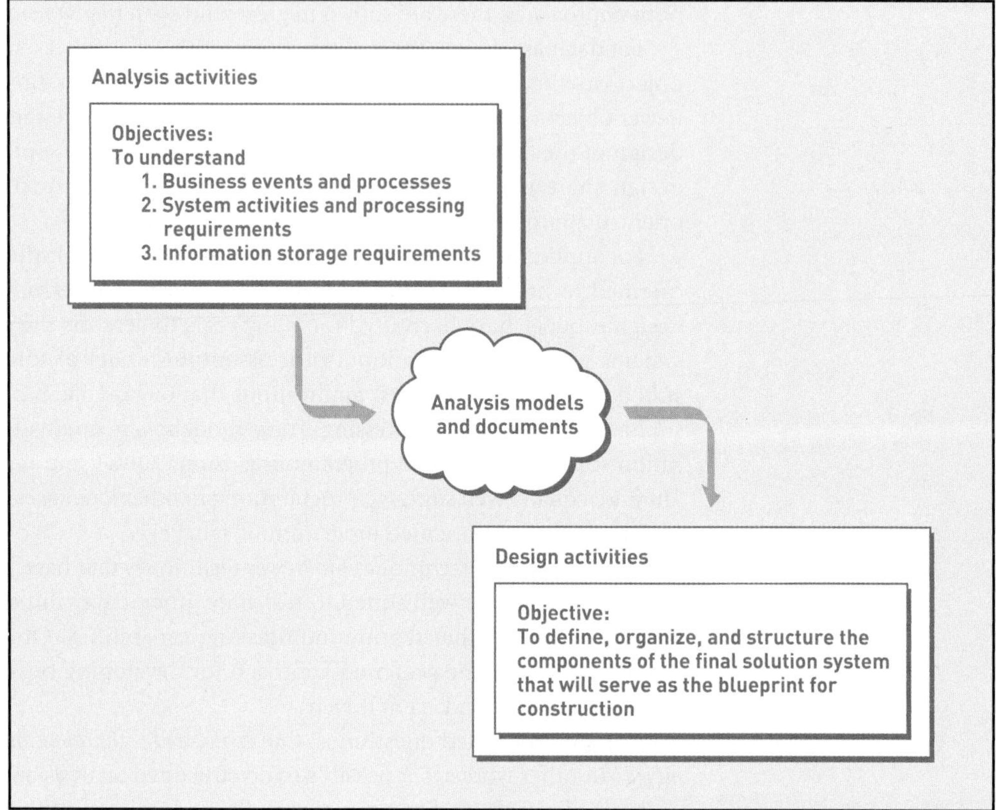

We should note how the structure of the project affects the design activities, and particularly the models and documents that are produced. Predictive projects usually tend to have a pronounced change in focus from analysis to design. Even with the overlap of analysis and design, as we saw in Figure 3-3, generally analysis activities and design activities are very distinct. Adaptive projects often use a "just-in-time" approach to design, with analysis flowing right into design and then into programming. For adaptive projects, it is not always easy to distinguish when a developer transitions from understanding the problem to configuring a solution, but it is important for developers to recognize when they change the focus toward a solution.

The formality of the project also affects design. Formal projects usually require well-developed design documents, which are often reviewed in structured walkthrough meetings. Developers on informal projects often create their designs with notepads and pencils, and then throw away the design once the program is coded. In other words, design in informal projects, such as in many Agile projects, is used as the means to the end, which is actual program code. However, we emphasize that even though outsiders do not see the design documents, the design process must still be followed. A programmer who jumps into code without carefully thinking it through ends up with errors, patches, and poorly structured systems. We often hear this approach referred to as cowboy coding. Our point? Learn how to design and build design models, even if you just draw them on the back of an envelope.

Figure 9-8 duplicates the information about traditional and OO requirements models originally shown in Figure 5-39 and extends it with the design models for both traditional structured design and object-oriented design. As noted in the figure, the models developed during the analysis activities feed directly into the models built for design—the traditional analysis models feed the traditional design models, and the object-oriented analysis models feed the object-oriented design models. Note also that several design models are common to both approaches; these are shown in green and span the two sides of the figure.

For database design, the traditional approach usually uses a relational database model. The object-oriented technique can require the design of either a relational database model or a newer object-oriented database model. For user-interface design, both techniques include the design of the human-computer dialog, forms, and reports. Both database and user-interface design share many of the same techniques, whether a structured approach or an object-oriented approach is used.

For application architecture design, however, traditional structured techniques and object-oriented techniques do differ substantially. Structured techniques, including analysis and design models, have been used for many years to describe the structure and organization of systems written using the input-process-output model of software. These models are well suited to describing business applications that rely on databases or files and do not require sophisticated real-time processing. These models were originally developed to support application software design and programming using COBOL and BASIC programming languages. They are equally well suited to programming in other languages, such as C, FORTRAN, Pascal, and other business-oriented programming languages.

Object-oriented techniques are newer techniques that have become widely used since the late 1980s. They are well suited to real-time, interactive, and event-driven software such as operating systems that require multitasking capabilities. Object-oriented development is rapidly becoming the preferred approach for developing business applications, which are usually interactive and event driven.

A frequently asked question is: *Can structured techniques and object-oriented techniques be mixed?* In other words, is it possible to do structured analysis and then object-oriented design of the application, or vice versa? Generally, such mixing and matching do not work well for application design because the basic philosophies of the two approaches are so fundamentally different. However, in some situations, it might be possible to mix and match, such as when designing and implementing the interface using OO after completing traditional structured analysis. The design of the application software using a traditional approach provides an architectural structure based on the top-down procedural functions of the system. A system designed using object-oriented techniques has an architectural structure based on the set of interacting objects for each use case.

After analysts have addressed the major components of a system, have considered its architectural design, and have in hand the documents and models developed during analysis, they can begin to consider how to design the system. We turn next to design activities.

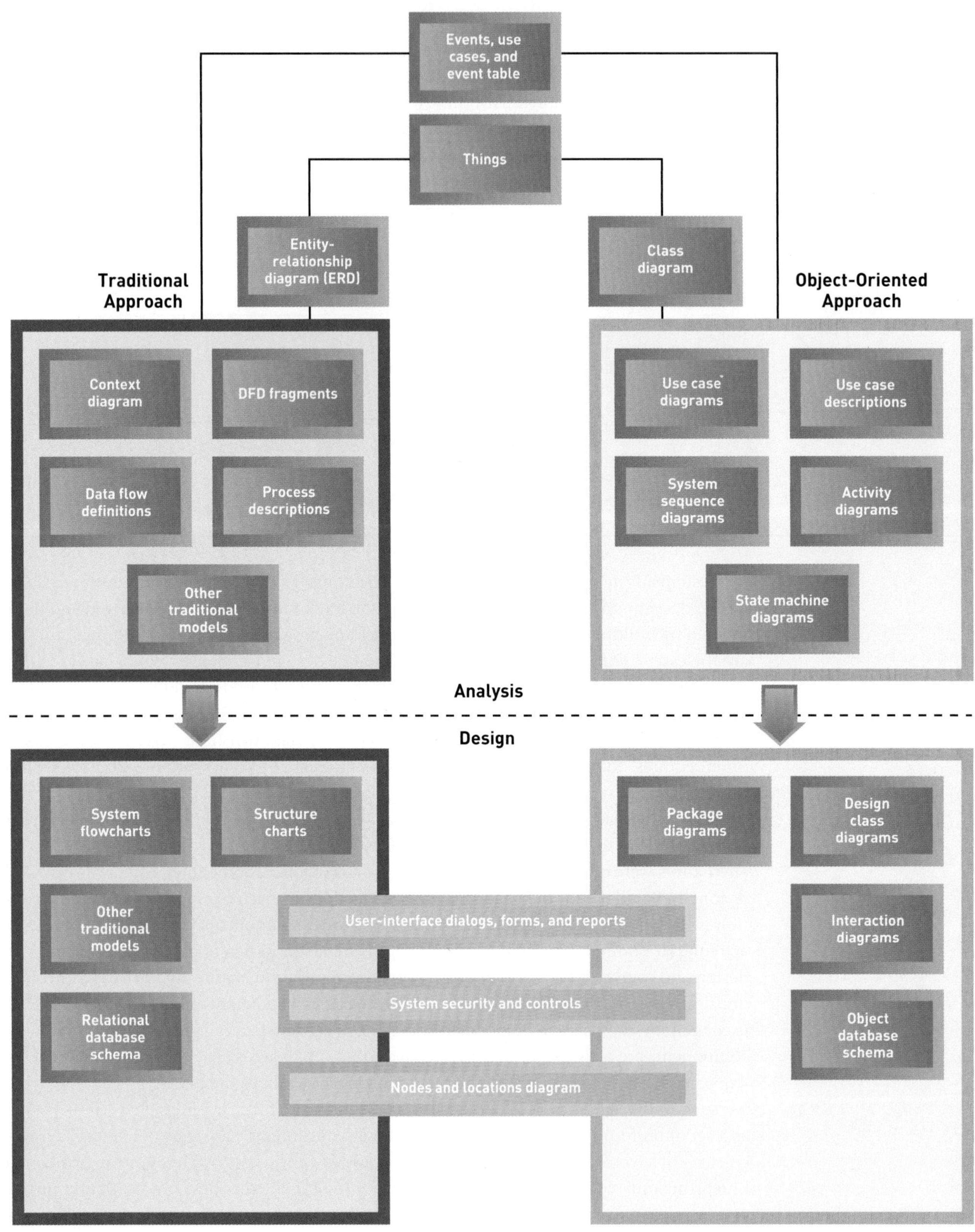
Events, use cases, and event table
Things
Entity-relationship diagram (ERD)
Class diagram
Traditional Approach
Object-Oriented Approach
Context diagram
DFD fragments
Data flow definitions
Process descriptions
Other traditional models
Use case diagrams
Use case descriptions
System sequence diagrams
Activity diagrams
State machine diagrams
Analysis
Design
System flowcharts
Structure charts
Other traditional models
Relational database schema
Package diagrams
Design class diagrams
Interaction diagrams
Object database schema
User-interface dialogs, forms, and reports
System security and controls
Nodes and locations diagram

Figure 9-8

Traditional structured and object-oriented models

DESIGN ACTIVITIES

The activities required to complete design in the SDLC provide an overview of the design process. As indicated previously, these activities provide the design for each of the components illustrated in Figure 9-6. More details about design processes are explained later in this chapter and in subsequent chapters as we discuss each of the design activities. Figure 9-9 identifies the activities that are associated with design.

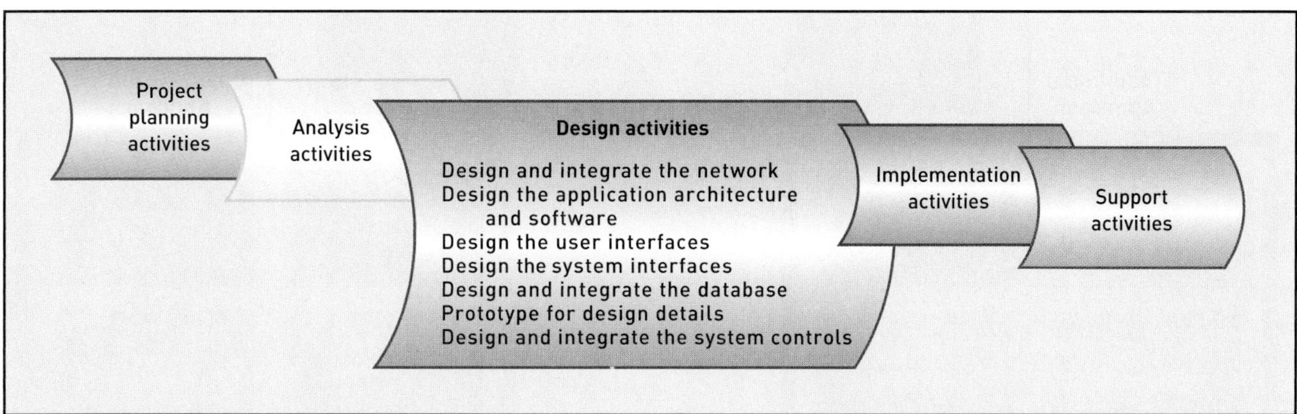

Figure 9-9
SDLC components with design activities

Systems design is a model building endeavor, just as it was during systems analysis. As design decisions are made, especially at the detail level, those decisions are derived and documented by building models. As indicated earlier, the models may be quite informal, but they are the essence of design. For example, in database design, we identify which tables will be required and what fields will be in which table before we begin to build the tables with SQL statements. In software design, we decide which classes are the core classes and which are utility classes and what responsibilities (methods) each class will have. User interface design often requires storyboards or other visual models to make efficient workflow decisions. All of these systems design tasks are model building tasks.

Systems design involves specifying in detail how a system will work using a particular technology. Some of the design details will have been developed during systems analysis, when the alternatives were described. But much more detail is required. Sometimes systems design work is done in parallel with the analysis activities. In addition, each component of the final solution is heavily influenced by the design of all the other components. Thus, all systems design activities are done in parallel. For example, the database design is used heavily in software design and even affects user interface design. The application architecture drives many of the decisions for how the network must be configured. When an iterative approach to the SDLC is used, major design decisions are made in the first or second iteration; however, many designed components are revisited during later iterations. As with analysis activities, each activity of systems design can be summarized with a question, as shown in Figure 9-10.

Each of the activities develops a specific portion of the final set of design documents. Just as a set of building blueprints has several different documents, a systems design package also consists of several sets of documents that specify the entire system. In addition, just as the blueprints must all be consistent and integrated to describe the same physical building, the various systems design documents also must be consistent and integrated to provide a comprehensive set of specifications for the complete system. For example, if one analyst is working on the user interface without consulting the database designer, the analyst could build an interface with the wrong fields or wrong field types and lengths. Internal consistency is a mandatory element of effective system modeling and design.

Figure 9-10

Design activities and key questions

Design activity	Key question
Design and integrate the network	*Have we specified in detail how the various parts of the system will communicate with each other throughout the organization?*
Design the application architecture and software	*Have we specified in detail how each system activity is actually carried out by the people and computers?*
Design the user interface(s)	*Have we specified in detail how all users will interact with the system?*
Design the system interface(s)	*Have we specified in detail how the system will work with all other systems inside and outside our organization?*
Design and integrate the database	*Have we specified in detail how and where the system will store all of the information needed by the organization?*
Prototype for design details	*Have we created prototypes to ensure all detailed design decisions have been fully understood?*
Design and integrate the system controls	*Have we specified in detail how we can be sure that the system operates correctly and the data maintained by the system is safe and secure?*

DESIGN AND INTEGRATE THE NETWORK

Sometimes a new system is implemented along with a new network. If this is the case, the network needs to be designed. More often, though, network specialists have established the network based on an overall strategic plan, and designers choose an alternative that fits the existing network. So rather than designing a network, the project team typically must integrate the system into an existing network.

Important technical issues arise when making the system operate over a network, such as reliability, security, throughput, and synchronization. Again, specialists are often brought in to help with the technical details. The requirements developed during systems analysis specify what work goes on at what locations, so these locations need to be connected. Technical requirements (as opposed to functional requirements) often have to do with communication via networks.

Later in this chapter, we highlight critical issues in network design and planning. The key question to be answered when completing the *Design and integrate the network* activity is: *Have we specified in detail how the various parts of the system will communicate with each other throughout the organization?*

DESIGN THE APPLICATION ARCHITECTURE AND SOFTWARE

In this activity we include decisions about the structure and configuration of the new system and the design of the actual computer software. Although we indicated that all components of system design depend on each other, the desired configuration of the application architecture drives all other design decisions, including network design. Designing the application architecture involves specifying in detail how all system activities will actually be carried out. For example, should users be able to access the new system only at work on their desktops, or should they also be able to work from home via an Internet connection? Is it necessary to allow remote wireless devices to connect to the system? What kind of transactions (use cases) and what volume of transactions must the new system be able to handle? These kinds of application decisions will drive the application architecture, the network, and other hardware requirements.

These user tasks are described during systems analysis in great detail as logical models with use case diagrams or data flow diagrams and descriptions, without indicating what specific technology is to be used. During systems design, these user tasks become system activities or transactions. The objective of systems design is to determine the exact way to support each of these transactions, including the architectural structure of the solution system and the design of the software components. After specific architectural design alternatives are chosen, the detailed computer processing models can be built. Models created include physical data flow diagrams, structure charts, sequence diagrams, design class diagrams, and other physical models.

The approach to application design and the design models created vary depending on the development and deployment environments. If the programming language is Visual Basic, for example, the type and nature of the models developed will be different than if the language were COBOL. If client/server architecture is used, the models used are different than with a centralized architecture. If object-oriented technology is used, the models are quite different than for process-based technology. In addition, some activities are carried out by people rather than computers, so manual procedures need to be designed.

The key question to be answered when completing the *Design the application architecture and software* activity is: *Have we specified in detail how each system activity is actually carried out by the users and computers?*

DESIGN THE USER INTERFACES

A critical aspect of the information system is the quality of the user interface. The design of the user interface defines how the user will interact with the system. To most users, the interface is a graphical user interface with windows, dialog boxes, and mouse interaction. Increasingly, it can include sound, video, and voice commands. Users' abilities and needs differ widely; each user interacts with the system in different ways. In addition, different approaches to the interface might be needed for different parts of the system. Therefore, you have many user interfaces to consider. And as information systems become increasingly interactive and accessible, the user interface is becoming a larger part of the system.

Analysts should remember that to the user of the system, the user interface *is* the system. The user interface is more than just the screens—it is everything the user comes into contact with while using the system, conceptually, perceptually, and physically. So, the user interface is not just an add-on to the system. New technology also has led to many new requirements for the user interface. For example, will users only use computers with large screens, or will they also use PDAs and other remote devices with small graphical areas? Will other devices be used for entering information such as text, verbal commands, pictures, and graphics? The elements and requirements of the user interface need to be considered throughout the development process.

The nature of the user interface begins to emerge very early in the development process, when requirements are being defined. The specification of the tasks the users complete begins to define the user interface. Then when alternatives are being defined, a key aspect of each alternative is its type of user interface. The activity of designing the user interface in detail, however, occurs during systems design.

Sometimes specialists in user-interface design are brought in to help with the project. These specialists might be called **interface designers**, *usability consultants*, or *human factors engineers*. The visual programming environments now available make it easy for developers to create graphical user interfaces for applications. But it is still very difficult to make a graphical user interface friendly or intuitive.

interface designers
specialists in user-interface design; also called *usability consultants* or *human factors engineers*

The processes associated with user-interface design are discussed in Chapter 14. The key question to be answered when completing the *Design the user interfaces* activity is: *Have we specified in detail how all users will interact with the system?*

DESIGN THE SYSTEM INTERFACES

No system exists in a vacuum. A new information system will affect many other information systems. Sometimes one system provides information that is later used by another system, and sometimes systems exchange information continuously as they run. The component that enables systems to share information is the system interface, and each system interface needs to be designed in detail. From the very beginning of systems design, analysts must ensure that all of the systems work together well.

In some cases, the new system needs to interface with a system outside the organization—for example, at a supplier's site or customer's home. Increasingly, organizations are linking systems together across organizational boundaries. At RMO, for example, the new supply chain management system will have information flows from RMO to its key suppliers. The new customer service management system will have real-time links to banks and other credit verification organizations.

Some system interfaces link internal organizational systems, so the analyst may have information available about other systems. Internally at RMO, the sales subsystem must have access to the warehouse database to know which items are in stock and which are not available. In other cases, the new system needs to interface with an application that the organization has purchased and installed.

System interfaces can become quite complex, particularly with so many types of technology available today. Often, an organization needs people with very specialized technical skills to work on these interfaces. System interface design is discussed in more detail in Chapter 15. The key question to be answered when completing the *Design the system interfaces* activity is: *Have we specified in detail how the system will work with all other systems inside and outside our organization?*

DESIGN AND INTEGRATE THE DATABASE

Designing the database for the system is another key design activity. The data model (a logical model) created during systems analysis is used to create the implementation model of the database. Usually the first decision is to determine the database structure. Sometimes the database is a collection of traditional computer files. More often, it is a relational database consisting of dozens or even hundreds of tables. Sometimes files and relational databases are used in the same system. At other times object-oriented databases might be the most appropriate design. Other decisions include whether the database is centralized or distributed. The internal properties of the database must also be designed, including such things as tables, attributes, and links.

Analysts must consider many important technical issues when designing the database. Many of the technical (as opposed to functional) requirements defined during systems analysis concern database performance needs (such as response times). Much of the design work might involve performance tuning to make sure the system actually works fast enough. Security and encryption issues, which are important aspects of information integrity, must be addressed and designed into the solution. Another key aspect of designing the database is making sure that new databases are properly integrated with existing databases.

Chapter 13 describes database design in detail. The key question to be answered when completing the *Design and integrate the database* activity is: *Have we specified in detail how and where the system will store all of the information needed by the organization?*

PROTOTYPE FOR DESIGN DETAILS

The basic idea of a prototype is to test some new or risky aspect of the new system before committing major resources to a particular configuration of the new solution. In fact, prototypes are used not only to verify a design decision, but to confirm that a particular approach will satisfy the

user's business needs. Prototyping can be used to confirm design choices about user interfaces, the database, network architecture, controls, or even programming environments being used. Because many design decisions are made early in a predictive project, prototyping is often a critical tool to ensure that correct decisions are made and to reduce risk. Therefore, when analysts consider all of the design activities, they think about how prototypes might be used to help understand a variety of design decisions. It is also important to recognize that rapid application development (RAD) approaches develop prototypes during design that evolve into the finished system. In those cases, the prototype is the system. In adaptive projects, prototyping may not be characterized in the same way. Frequently an iteration will be used to "try out some new technology." Even though it is not specifically called prototyping, the objective is the same.

The key question to be answered when completing the *Prototype for design details* activity is: *Have we created prototypes to ensure that all detailed design decisions have been fully understood?*

DESIGN AND INTEGRATE THE SYSTEM CONTROLS

A final design activity involves ensuring that the system has adequate safeguards to protect organizational assets. These safeguards are referred to as system controls. This activity is not listed last because it is less important than the others. On the contrary, especially in today's environment, where outsiders can potentially cause severe damage to a system and its data, designing system controls is a crucial activity. It is listed last because controls have to be considered for all other design activities—user interface, system interface, application architecture, database, and network design.

User-interface controls limit access to the system to authorized users. System interface controls ensure that other systems cause no harm to this system. Application controls ensure that transactions are recorded precisely and that other work done by the system is done correctly. Database controls ensure that data is protected from unauthorized access and from accidental loss due to software or hardware failure. Finally, and of increasing importance, network controls ensure that communication through networks is protected. All of these controls need to be designed into the system, based on the existing technology. Specialists are often brought in to work on some controls, and all system controls need to be thoroughly tested.

Control issues are addressed in several chapters but more explicitly in Chapter 15. The key question to be answered when completing the *Design and integrate the system controls* activity is: *Have we specified in detail how we can be sure that the system operates correctly and the data maintained by the system is safe and secure?*

NETWORK DESIGN

The first activity in the list of design activities is to design the network. Here we provide only a brief introduction to network design. In real projects in large companies, network design is also frequently done by full-time system programmers and technical support staff. Of course, in a small company the analyst also may be the technical staff, so he or she may need good network design skills. Network design is taught at the detailed level in a networking class in a CIS or MIS curriculum.

Networks are used throughout organizations today. As a result, many new development projects involve network design. Network planning and design are critical issues that must be dealt with early in the design of any multitiered system. The key design issues are as follows:

- Integrating network needs of the new system with existing network infrastructure
- Describing the processing activity and network connectivity at each system location
- Describing the communication protocols and middleware that connect layers
- Ensuring that sufficient network capacity is available

BEST PRACTICE

Consult with in-house experts to determine whether the network can support the new system without disrupting existing services.

COMPUTER NETWORKS

computer network
a set of transmission lines, equipment, and communication protocols to permit sharing of information and resources

local area network (LAN)
a computer network in which the distances are local, such as within the same building

A **computer network** is a set of transmission lines, specialized hardware, and communication protocols that enable communication among different users and computer systems. Computer networks are divided into two classes depending on the distance they span. A **local area network (LAN)** is typically less than one kilometer and connects computers within a single building or floor. The term **wide area network (WAN)** can describe any network over one kilometer, though the term typically implies much greater distances spanning cities, countries, continents, or the entire globe.

Figure 9-11 shows a possible computer network for RMO. A single LAN serves each geographic location, and all LANs are connected by a WAN. Users and computers in a single location communicate via their LAN. Communication among geographically dispersed sites uses the LANs at both sites and the WAN. A **router** connects each LAN to the WAN. A router scans messages on the LAN and copies them to the WAN if they are addressed to a user or computer on another LAN. The router also scans messages on the WAN and copies them to the LAN if they are addressed to a local user or computer.

Figure 9-11
A possible network configuration for RMO

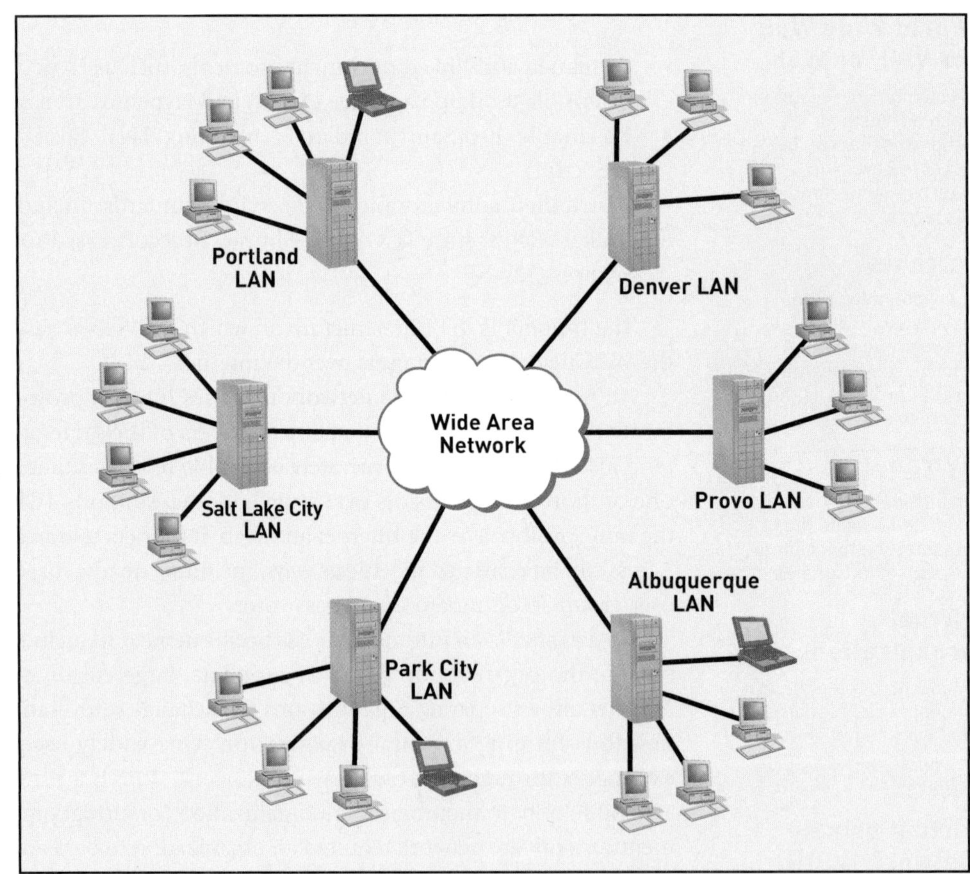

wide area network (WAN)
a computer network spread across large distances, such as a city, state, or nation

router
network equipment that directs information within the network

Technologies such as Ethernet are typically used to implement LANs. They provide low to moderate amounts of message-carrying capacity at relatively low cost. WAN technologies such as asynchronous transmission mode are more complex and expensive, though they typically provide higher message-carrying capacity and greater reliability. WANs may be constructed using purchased equipment and leased long-distance transmission lines. WAN setup and

operation may also be subcontracted from a long-distance telecommunications vendor such as AT&T or Sprint.

Computer networks provide a generic communication capability among computer systems and users. This generic capability can support many services, including direct communications (such as telephone service and video conferencing), message-based communications (such as e-mail), and resource sharing (such as access to electronic documents, application programs, and databases). A single network can simultaneously support multiple services with appropriate hardware and sufficient transmission capacity.

There are many ways to distribute information system resources across a computer network. Users, application programs, and databases can be placed on the same computer system, on different computer systems on the same LAN, or on different computer systems on different LANs. Application programs and databases can also be subdivided and each distributed separately.

THE INTERNET, INTRANETS, AND EXTRANETS

Internet
a global collection of networks that use the same networking protocol—TCP/IP

World Wide Web (WWW), or Web
a collection of resources such as files and programs that can be accessed over the Internet using standard protocols

intranet
a private network that is accessible to a limited number of users, but which uses the same TCP/IP protocol as the Internet

extranet
an intranet that has been extended outside the organization to facilitate the flow of information

virtual organization
a loosely coupled group of people and resources that work together as though they were an organization

virtual private network (VPN)
a network with security and controlled access for a private group but built on top of a public network such as the Internet

The **Internet** is a global collection of networks that are interconnected using a common low-level networking standard called TCP/IP (Transmission Control Protocol/Internet Protocol). The **World Wide Web (WWW)**, also called simply the *Web*, is a collection of resources (programs, files, and services) that can be accessed over the Internet by a number of standard protocols, including the following:

- Formatted and linked document protocols, such as Hypertext Markup Language (HTML), eXtensible Markup Language (XML), and Hypertext Transfer Protocol (HTTP)
- Executable program standards, including Java, JavaScript, and Visual Basic Script (VBScript)
- Distributed software and Web-service standards, including Common Object Request Broker Architecture (CORBA), Simple Object Access Protocol (SOAP), and Java 2 Web Services (J2WS)

The Internet is the infrastructure on which the Web is based. In other words, resources of the Web are delivered to users over the Internet.

An **intranet** is a private network that uses Internet protocols but is accessible only by a limited set of internal users (usually members of the same organization or workgroup). The term also describes a set of privately accessible resources that are organized and delivered via one or more Web protocols over a network that supports TCP/IP. Although an intranet uses the same protocols as the Internet and Web, it restricts resource access to a limited set of users. Access can be restricted in various ways, including unadvertised resource names, firewalls, and user/group account names and passwords.

An **extranet** is an intranet that has been extended to include directly related business users outside the organization (such as suppliers, large customers, and strategic partners). An extranet allows separate organizations to exchange information and coordinate their activities, thus forming a **virtual organization**. One widely used method of implementing an extranet is through a **virtual private network (VPN)**, a private network that is secure and accessible only to members of an organization (or virtual organization). Historically, implementing a private network required an organization to own and operate its own network lines

or leased, dedicated telephone lines. A VPN sends encrypted messages through public Internet service providers.

NETWORK INTEGRATION

Modern organizations rely on networks to support many different applications. Thus, the majority of new systems must be integrated into existing networks without disrupting existing applications. Network design and management are highly technical tasks, and most organizations have permanent in-house staff, contractors, or consultants to handle network administration.

The analyst for a new project begins network design by consulting with the organization's network administrators to determine whether the existing network can accommodate the new system. In some cases, the existing network capacity is sufficient, and only minimal changes are required, such as adding connections for new servers or modifying routing and firewall configuration to enable new application layers to communicate.

Planning for more extensive changes—such as significant capacity expansion, new communication protocols, or modified security protocols—is much more complex. Typically, the network administrator assumes the responsibility of acquiring new capacity and making any configuration changes to support the new system because he or she understands the existing network and the way other network-dependent applications operate. The analyst's role for the new system in these cases is to provide the network administrator with sufficient information and time to enable system development, testing, and deployment.

NETWORK DESCRIPTION

Location-related information gathered during analysis may have been documented using location diagrams (such as Figure 6-32), activity-location matrices (such as Figure 6-33), and activity-data matrices (such as Figure 6-34). This information is important for the design of the application architecture and the deployment environment, including the network. Usually network design is done in conjunction with the application architecture, which is explained in the next section. During network design, the analyst expands the information content of these documents to include processing locations, communication protocols, middleware, and communication capacity.

network diagram

a model that shows how application layers are distributed across locations and computer systems

There are many different ways to describe the network infrastructure for a specific application. Figure 9-12 shows a **network diagram** that describes how application layers are distributed across locations and computer systems for the RMO customer support system. The diagram summarizes key architectural decisions from Figure 8-5 and combines them with specific assumptions about where application software will execute, where servers and workstations will be located, and how network resources will be organized.

The diagram embodies specific assumptions about server locations, which would be decided in consultation with network administrators. The Web/application servers could have been distributed outside the Salt Lake City data center, which might have improved system response time and reduced data communication capacity requirements on the private WAN. However, distributing the servers would also entail duplication of server administration at multiple locations, which would increase operational complexity and cost. Decisions such as server locations, communication routes, and network security options are determined both by application requirements and organization-wide policies.

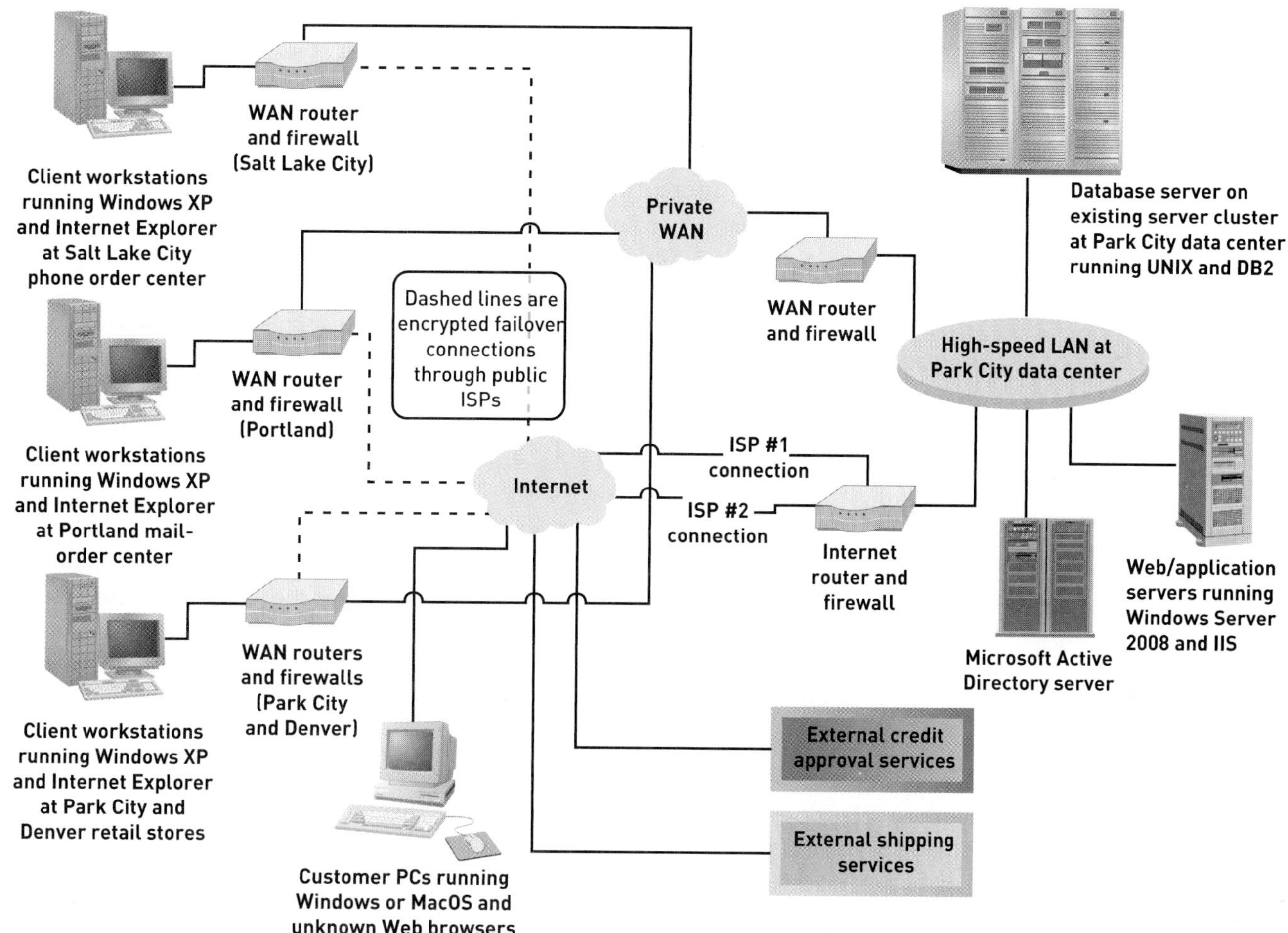

Figure 9-12

A network diagram for the RMO customer support system

COMMUNICATION PROTOCOLS

The network diagram is also a starting point for specifying protocol requirements. For example, the private WAN connections must support protocols required to process Microsoft Active Directory logins and queries. If the WAN fails, messages are routed through encrypted (VPN) connections over the Internet, so those connections must support the same protocols as the private WAN. All clients must be able to send HTTP requests and receive active content such as HTML forms and embedded scripts. Application servers must be able to communicate with credit verification and shipping services via the Internet. Firewalls and routers must be configured to support all interactions among the workstations, customer PCs, Web/application servers, the Active Directory server, and external credit and shipping services. The Park City data center LAN must support at least one protocol for transmitting database queries and responses among the mainframe and Web/application servers.

NETWORK CAPACITY

Information from activity-location and activity-data matrices is the starting point for estimating communication capacity requirements for various LAN, WAN, and Internet connections. Figure 9-13 reproduces data from the RMO activity-data matrix (see Figure 6-34), which covers two activities (*Look up item availability* and *Create new order*) and three data entities (Customer, Inventory Item, and Order). Similar tables would be required for all combinations of activity, data entity, and location. Figure 9-13 includes estimates of data size per access type and the average and peak number of access per minute or hour.

	DATA ENTITIES		
Activities and locations	**Customer**	**Inventory item**	**Order**
Look up item availability (Salt Lake City phone order center)		R (125 bytes, 25/min average, 250/min peak)	
Look up item availability (Park City retail store)		R (125 bytes, 5/hr average, 15/hr peak)	
Look up item availability (Denver retail store)		R (125 bytes, 5/hr average, 15/hr peak)	
Create new order (Salt Lake City phone order center)	C (500 bytes, 2/min average, 10/min peak) R (500 bytes, 8/min average, 80/min peak) U (500 bytes, 2/min average, 10/min peak)	R (60 bytes, 30/min average, 300/min peak) U (60 bytes, 30/min average, 300/min peak)	C (200 bytes, 10/min average, 100/min peak)
Create new order (Portland mail-order center)	C (500 bytes, 1/min average, 10/min peak) R (500 bytes, 4/min average, 40/min peak) U (500 bytes, 1/min average, 10/min peak)	R (60 bytes, 15/min average, 150/min peak) U (60 bytes, 15/min average, 150/min peak)	C (200 bytes, 5/min average, 50/min peak)
Create new order (Park City retail store)	C (500 bytes, 1/hr average, 5/hr peak) R (500 bytes, 4/hr average, 20/hr peak) U (500 bytes, 1/hr average, 5/hr peak)	R (60 bytes, 15/hr average, 75/hr peak) U (60 bytes, 15/hr average, 75/hr peak)	C (200 bytes, 5/hr average, 25/hr peak)
Create new order (Denver retail store)	C (500 bytes, 1/hr average, 5/hr peak) R (500 bytes, 4/hr average, 20/hr peak) U (500 bytes, 1/hr average, 5/hr peak)	R (60 bytes, 15/hr average, 75/hr peak) U (60 bytes, 15/hr average, 75/hr peak)	C (200 bytes, 5/hr average, 25/hr peak)

C = Creates new data, R = Reads existing data, U = Updates existing data, D = Deletes existing data

Figure 9-13

Partial activity-data matrix for RMO customer support system updated with data access size and volume

Data size per access type is an educated guess at this point in the system design because none of the software layers, interlayer communication dialogs, or databases have been designed. After these components have been designed in more detail or implemented, analysts can refine their estimates or actually sample and measure real data transmissions. Actual data transmission capacity includes communication protocols in addition to raw data.

THE DEPLOYMENT ENVIRONMENT AND APPLICATION ARCHITECTURE

The second activity identified in Figure 9-9 is to design the application architecture and software. This section provides an introduction to various options for the application architecture. Then the next three chapters will provide detailed explanations to design the application software.

The application architecture involves the structure and organization of the new software system—for example, is it a desktop system, is it a networked system, or is it an Internet-based system? Obviously the architectural structure of the software system will have to be supported by the computer equipment and its configuration. Hence, this section introduces concepts related to both the application architecture and deployment environment (computer hardware, operating systems, and other system components). Of course, all of the architectural decisions influence, and must be consistent with, the network environment discussed in the previous section.

SINGLE-COMPUTER AND MULTITIER COMPUTER ARCHITECTURE

single-computer architecture

architecture that employs a single computer system executing all application-related software

As its name implies, **single-computer architecture** employs a single computer system and its directly attached peripheral devices, as shown in Figure 9-14a. Even though a single-computer architecture can refer to a stand-alone PC, a single PC has limited capabilities even at today's computer speeds. Hence, in this context we are discussing large mainframe computers with extensive computing capability derived by multiple internal CPUs, multiple virtual machines (each with its own operating system), large online data storage, and tremendous throughput capabilities. The primary advantage of single-computer architecture is its simplicity. Information systems deployed on a single-computer system, even though the software may be complex, usually do not have complex interactions with other systems and thus operate in a less complex and less cluttered environment. The other major advantage of a mainframe architecture is the extremely high-volume processing that is supported. A single high-speed mainframe can often do the same work as a cluster of computer servers. Companies that use mainframe computers usually have business needs that require high volumes of online transactions or heavy workloads throughout the day and night. Historically mainframe computers have been very expensive. Today, however, the price of mainframe computers is competitive with that of large server computers.

At first, servers functioned as less powerful mainframes in a single computer environment. However, as operating systems and communication software became more sophisticated, it soon became possible to connect several servers into a cluster of computers that could work together to share the workload. Nowadays, clusters of servers can handle even greater workloads than a single mainframe computer. Many systems are so large that even the largest mainframe computer cannot perform all the required processing, data storage, and data retrieval tasks in a network environment. Such systems require another architectural approach.

multitier architecture

architecture that distributes application-related software or processing load across multiple computer systems

clustered architecture

a group of computers of the same type that share processing load and act as a single large computer system

multicomputer architecture

a group of dissimilar computers that share processing load through specialization of function

Multitier architecture employs multiple computer systems in a cooperative effort to meet information-processing needs. Multitier architecture can be further subdivided into two types:

- **Clustered architecture**, shown in Figure 9-14b, employs a group (or cluster) of computers, usually from the same manufacturer and model family. Programs are allocated to the least utilized computer when they execute so that the processing load can be balanced across all machines. In effect, a cluster acts as a single large computer system. Clustered computer systems are normally located near one another so that they can be connected with short high-capacity communication links.
- **Multicomputer architecture**, shown in Figure 9-14c, also employs multiple computer systems. But hardware and operating systems are not required to be as similar as in a clustered architecture. A suite of application or system programs and data resources is exclusively assigned to each computer system. Each computer system is optimized to the role that it will play in the combined system, such as database or application server.

Clustered computers are usually configured to be servers in that they provide support (for example, data access or program execution) for a larger network of independent computing devices.

Figure 9-14

Single-computer, clustered, and multicomputer architectures

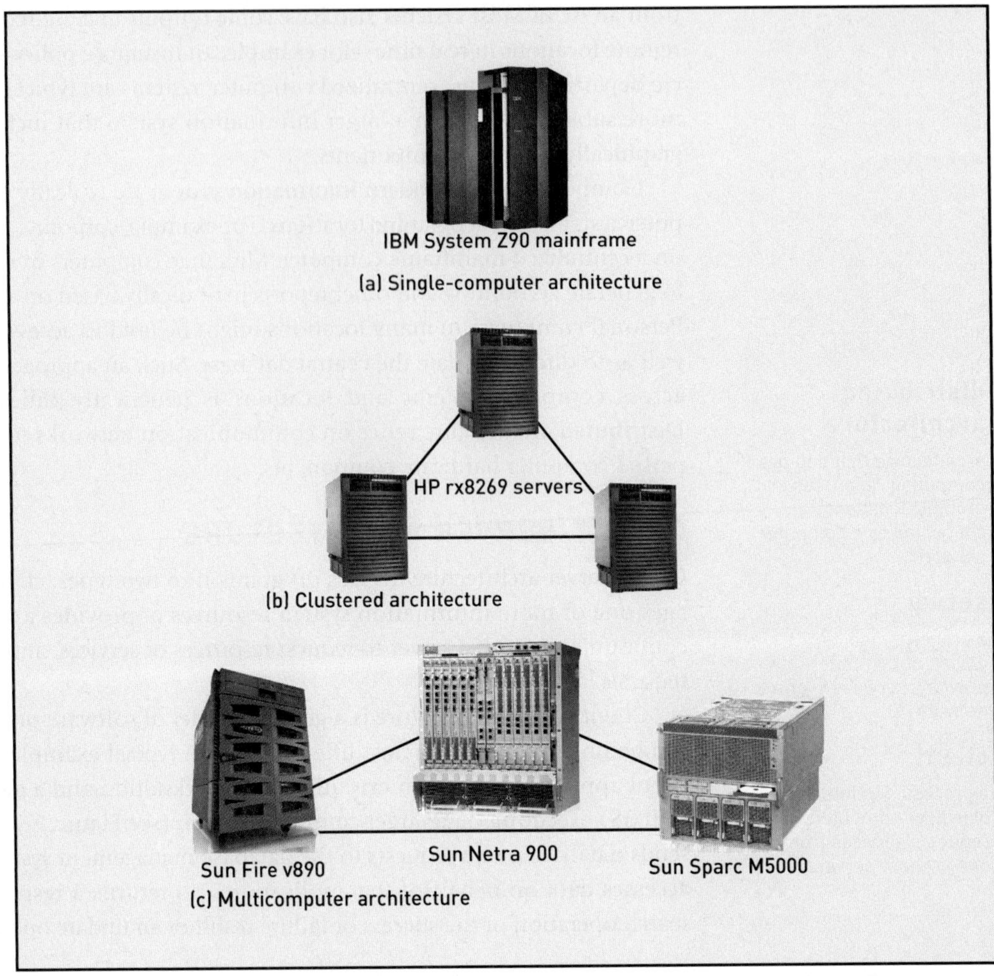

CENTRALIZED AND DISTRIBUTED ARCHITECTURE

centralized architecture

architecture that locates all computing resources in a central location

The term centralized architecture describes deployment of all computer systems in a single location. Centralized architecture is generally used for large-scale processing applications, including both batch and real-time applications. Such applications are common in industries such as banking, insurance, and catalog sales. Information systems in such industries often have the following characteristics:

- Some input transactions do not need to be processed in real time (for example, out-of-state checks delivered in large nightly batches from central bank clearinghouses).
- Online data-entry personnel can be centrally located (for example, a centrally located group of telephone order takers can serve geographically dispersed customers).
- The system produces a large volume of periodic outputs (for example, monthly credit-card statements mailed to customers).
- A high volume of transactions occurs between high-speed computers (for example, business-to-business processing for supply chain management).

Any application that has two or three of these characteristics is a viable candidate for implementation on a centralized configuration of either a mainframe or server cluster. Current trends in conducting e-business have instilled new life into centralized computing because of the transaction volumes of many business-to-business (B2B) processes.

Centralized computer systems are seldom used as the sole hardware platform for an information system. Most systems have some transaction inputs that must be accepted from geographically dispersed locations and processed in real time—for example, a cash withdrawal

from an ATM. Most systems also have some outputs that are requested from and delivered to remote locations in real time—for example, an insurance policy inquiry by a state motor vehicle department. Thus, centralized computer systems are typically used to implement one or more subsystems within a larger information system that includes online, batch, and geographically dispersed components.

Components of a modern information system are typically distributed across many computer systems and geographic locations. For example, corporate financial data might be stored on a centralized mainframe computer. Midrange computers in regional offices might be used to generate accounting and other reports periodically based on data stored on the mainframe. Personal computers in many locations might be used to access and view periodic reports as well as to directly update the central database. Such an approach to distributing components across computer systems and locations is generically called **distributed architecture**. Distributed architecture relies on communication networks to connect geographically dispersed computer hardware components.

distributed architecture
architecture that deploys computing resources in multiple locations connected by a computer network

CLIENT/SERVER ARCHITECTURE

Client/server architecture divides programs into two types: client and server. A **server** manages one or more information system resources or provides a well-defined service. A **client** communicates with a server to request resources or services, and the server responds to those requests.

server
a process, module, object, or computer that provides services over a network

client
a process, module, object, or computer that requests services from one or more servers

Client/server architecture is a general model of software organization and behavior that can be implemented in many different ways. A typical example is the interaction between a client application program executing on a workstation and a database management system (DBMS) executing on a larger computer system (see Figure 9-15). The application program sends database access requests to the database management system via a network. The DBMS accesses data on behalf of the application and returns a response such as the results of a search operation or the success or failure result of an update operation.

Figure 9-15
Client/server architecture with a shared database

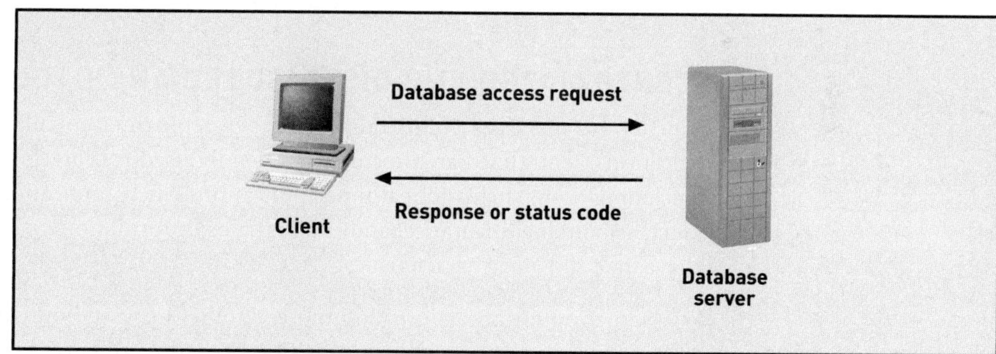

When designing client/server software, the following architectural issues must be addressed:

- Decomposing the application into client and server programs, modules, or objects
- Determining which clients and servers will execute on which computer systems
- Describing the communication protocols and networks that connect clients and servers

BEST PRACTICE

Identify resources or services that can be managed by independent software units.

The key to decomposing the application into clients and servers is identifying resources or services that can be centrally managed by independent software units. Examples of centrally managed services include security authentication and authorization, credit verification, and scheduling. In each case, a service provides a set of well-defined processes such as retrieval, update, and approval based on a data store that is hidden from the client, as shown in Figure 9-16.

Figure 9-16

Interaction among client, server, and a service-related data store

Client and server software can execute on any computer system. But the most typical arrangement is to place server software on separate server computer systems and to distribute client software to computer systems "close" to end users, such as desktop workstations. Figure 9-17 shows a typical arrangement for an order-processing application. Credit verification, delivery scheduling, and database server processes execute on a centrally located midrange or mainframe computer, and users execute multiple copies of the client software on workstations.

Figure 9-17

Interaction among multiple clients and a single server

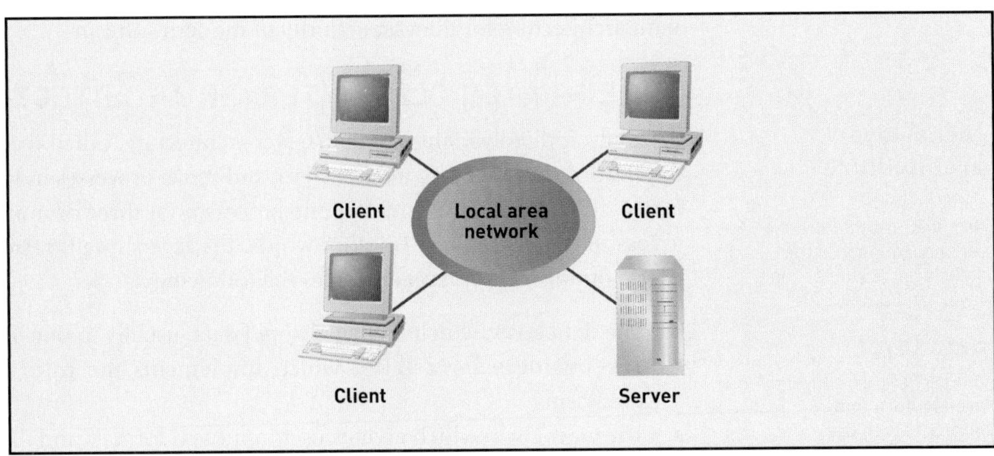

The client and server communicate via well-defined communication protocols over a physical network. In Figure 9-17, the network is a LAN, and an appropriate low-level network protocol such as TCP/IP provides basic communication services. But the system designer must also specify higher-level protocols, or languages, by which client and server exchange service requests, responses, and data. In some cases, such as communication with a DBMS, standard protocols and software may be employed, such as Structured Query Language (SQL) via an Open Database Connectivity (ODBC) database connection. But in other cases, the designer must define the exact format and content of valid messages and responses. If a service is provided by an external organization (for example, credit verification), the external organization will have already designed an appropriate protocol, and the application designer must ensure that clients adhere to the protocol.

The primary advantage of client/server architecture is deployment flexibility. Client/server architecture arose as an approach to distributing software across networked computers. It provides the inherent advantages of a networked environment, including the following:

- **Location flexibility.** The ability to "move" system components without "disturbing" other system components, in response to changing organizational parameters, such as size and physical location
- **Scalability.** The ability to increase system capacity by upgrading or changing the hardware on which key software components execute
- **Maintainability.** The ability to update the internal implementation of one part of a system without needing to change other parts (for example, the credit verification software can be rewritten or replaced as long as the new software uses the existing client/server protocol)

The primary disadvantages of client/server architecture are the additional complexity introduced by the client/server protocols and the potential performance, security, and reliability issues that arise from communication over networks. A centralized application executing as one large program on a single computer needs no client/server protocols, and all communication within the application occurs within the relatively secure, reliable, and efficient confines of a single machine.

For most organizations, the flexibility advantages of client/server far outweigh the disadvantages. As a result, client/server architecture and its newer variants have become the dominant architecture for the vast majority of modern software.

THREE-LAYER CLIENT/SERVER ARCHITECTURE

three-layer architecture
a client/server architecture that divides an application into the view layer, business logic layer, and data layer

data layer
the part of three-layer architecture that interacts with the database

business logic layer
the part of three-layer architecture that contains the programs that implement the business rules of the application

view layer
the part of three-layer architecture that contains the user interface

A widely applied variant of client/server architecture, called **three-layer architecture**, divides application software into a set of client and server processes independent of hardware or locations. All layers might reside on one processor, or three or more layers might be distributed across many processors. In other words, the layers might reside on one or more tiers. The most common set of layers includes the following:

- The **data layer**, which manages stored data, usually in one or more databases
- The **business logic layer**, which implements the rules and procedures of business processing
- The **view layer**, which accepts user input and formats and displays processing results

Figure 9-18 illustrates the interaction of the three layers. The view layer acts as a client of the business logic layer, which, in turn, acts as a client of the data layer.

Like earlier forms of client/server architecture, three-layer architecture is inherently flexible. Interactions among the layers are always requests or responses, which makes the layers relatively independent of one another. It doesn't matter where other layers are implemented or on what type of computer or operating system they execute. The only interlayer dependencies are a common language for requests and responses and a reliable network with sufficient communication capacity.

Multiple layers can execute on the same computer, or each layer can operate on a separate computer. Complex layers can be split across two or more computers. System capacity can be increased by splitting layer functions across computers or by load sharing across redundant computers. In the event of a malfunction, redundancy improves system reliability if the server load can be shifted from one computer to another. In sum, three-layer architecture provides the flexibility needed by modern organizations to deploy and redeploy information-processing resources in response to rapidly changing conditions.

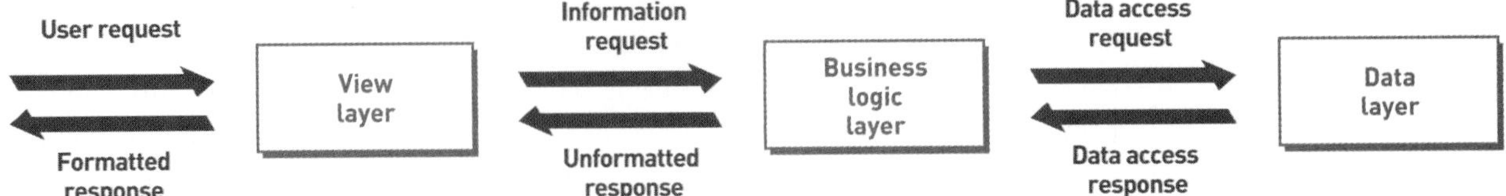

Figure 9-18

Three-layer architecture

Three-layer architecture is currently a widely applied architectural design pattern with both the traditional approach and the object-oriented approach. As with other forms of client/server architecture, the key design tasks are decomposing the application into layers, clients, and servers, distributing the "pieces" across hardware platforms, and defining the physical network and protocols.

The business logic layer is the core of the application software and is constructed according to the requirements models developed during analysis, as described in Chapters 5 through 7. For example, in the traditional approach, all of the business logic defined for system activities within the RMO data flow diagrams would be implemented as functions or procedures in the business logic layer. The window or browser forms making up the view layer would not contain much procedural code. In the object-oriented approach, the classes of objects in the RMO class diagram (see Figure 5-38) would be implemented within the business logic layer as classes of objects that interact to complete user tasks. In either case, the business logic layer is a server for the view layer and is a client of the data layer. However, the business logic layer may itself be decomposed into multiple clients and services. Three-layer architecture is usually implemented with object-oriented techniques and tools, as described in Chapter 11, though it is also implemented with traditional design techniques and programming languages, as described in Chapter 10. In this respect, three-layer architecture is a prominent architectural design pattern that applies to both traditional and OO approaches.

In this text, Chapters 10 and 11 describe how the view and data layers are designed with traditional and OO approaches. Chapter 13 describes the details of the database that is accessed by the data access layer. Chapter 14 describes user-interface design techniques and guidelines that are independent of the software that implements the view layer, such as the arrangement of interface elements on a video display and the dialog between user and computer that supports a specific application task.

INTERNET AND WEB-BASED APPLICATION ARCHITECTURE

The Web is a complex example of client/server architecture. Web resources are managed by server processes that can execute on dedicated server computers or on multipurpose computer systems. Clients are programs that send requests to servers using one or more of the standard Web resource request protocols. Web protocols define valid resource formats and a standard means of requesting resources and services. Any program, not just a Web browser, can use Web protocols. Thus, Web-like capabilities can be embedded in ordinary application programs.

Internet and Web technologies present an attractive alternative for implementing information systems. For example, consider the problem of data entry and access by an RMO buyer when purchasing items from the company's suppliers. Buyers are typically on the road for several months a year, often for weeks at a time. A traveling buyer needs some means of remotely interacting with RMO's supply chain management (SCM) system to record purchasing agreements and query inventory status.

One way of providing these capabilities would be to design custom application software and a private network to connect to the software. The primary portion of the system could be installed on a server at RMO. The client portion of the application—for data entry—would then be installed on the buyers' laptop computers. A buyer would then connect to the system from remote locations to gain access to the application server, make queries to the database, and enter data.

Another alternative for implementing remote access for buyers would be to construct an application that uses a Web browser interface. The application would execute on a Web server, communicate with a Web browser using HTML or XML, and be accessible from any computer with an Internet connection. Buyers could use a Web browser on their laptop computer and connect to the application via a local Internet service provider. Buyers could also access the application from any other computer with Internet access (for example, a computer in a vendor's office, hotel business suite, or copy center such as FedEx Kinko's).

Flexibility is the key to the Internet alternative. Implementing the application via the Internet greatly expands the application's accessibility and eliminates the need to install custom client software on buyers' laptop computers. With Internet technology, client software can be updated simply by updating the version stored on the Web server. The application is relatively cheap to develop and deploy because existing Web standards and networking resources are employed. Custom software and private access via modems require more complex development and maintenance of a greater number of customized resources.

Implementing an application via the Web, an intranet, or an extranet has a number of advantages over traditional client/server applications, including the following:

- **Accessibility.** Web browsers and Internet connections are nearly ubiquitous. Internet, intranet, and extranet applications are accessible to a large number of potential users (including customers, suppliers, and off-site employees).
- **Low-cost communication.** The high-capacity WANs that form the Internet backbone were funded primarily by governments. Traffic on the backbone networks travels free of charge to the user, at least for the present. Connections between private LANs and the Internet can be purchased from a variety of private Internet service providers at relatively low cost. In essence, a company can use the Internet as a low-cost WAN.
- **Widely implemented standards.** Web standards are well known, and many computing professionals are already trained in their use. Server, client, and application development software is widely available and relatively cheap.

Information resource delivery via an intranet or extranet enjoys all of the advantages of Web delivery because they use Web standards. In many ways, intranets, extranets, and the Web represent the logical evolution of client/server computing into an off-the-shelf technology. Organizations that once shied away from client/server computing because of the costs and required learning curve can now enjoy client/server benefits at substantially reduced complexity and cost.

Of course, there are negative aspects of application delivery via the Internet and Web technologies, including the following:

- **Security.** Web servers are a well-defined target for security breaches because Web standards are open and widely known. Wide-scale interconnection of networks and the use of Internet and Web standards make servers accessible to a global pool of hackers.
- **Reliability.** Internet protocols do not guarantee a minimum level of network throughput or even that a message will be received by its intended recipient. Standards have been proposed to address these shortcomings, but they have yet to be widely adopted.
- **Throughput.** The data transfer capacity of many home users is limited by analog modems to under 56 kilobits per second. Internet service providers and backbone WANs can become overloaded during high-traffic periods, resulting in slow response time for all users and long delays when accessing large resources.
- **Volatile standards.** Web standards change rapidly. Client software is updated every few months. Developers of widely used applications are faced with a dilemma: Use the latest standards to increase functionality or use older standards to ensure greater compatibility with older user software.

For RMO, the primary disadvantages of implementing the customer order application via the Internet are security, performance, and reliability. If a buyer can access the system via the Web, then so can anyone else. Access to sensitive parts of the system can be restricted by a number of means, including user accounts and passwords. But the risk of a security breach will always be present. Performance and reliability are limited by the buyer's Internet connection point and the available Internet capacity between that connection and the application server. Unreliable or overloaded local Internet connections can render the application unusable. RMO has no control over these factors.

The key architectural design issues for Web-based applications are similar to those for other client/server architectures: defining client and server processes or objects, distributing them across hardware platforms, and connecting them with appropriate networks, middleware, and protocols. However, for Web-based applications, the choices for middleware and protocols tend to be much more limited than for other forms of client/server architecture.

Now that we've discussed common approaches to application architecture, we turn our attention to designing the networking infrastructure that connects parts of a modern information system.

WEB SERVICES ARCHITECTURE

Web services architecture
a client/server architecture that packages software into server processes that can be accessed via Web protocols

Web services architecture is another modern variant of client/server architecture. Web services architecture packages software into server processes that can be accessed via Web protocols, including XML, SOAP, Web Services Description Language (WSDL), and Universal Description, Discover, and Integration (UDDI). Figure 9-19 shows how a client interacts with a Web service via a Web services directory. Information about a Web service, such as server and service names and port numbers, XML data formats, and security requirements, is described in WSDL and published in a Web services directory. The client interacts with the directory to determine what services are available and how to interact with them. The client then initiates a connection with the Web service using SOAP and XML.

Figure 9-19
Web services architecture

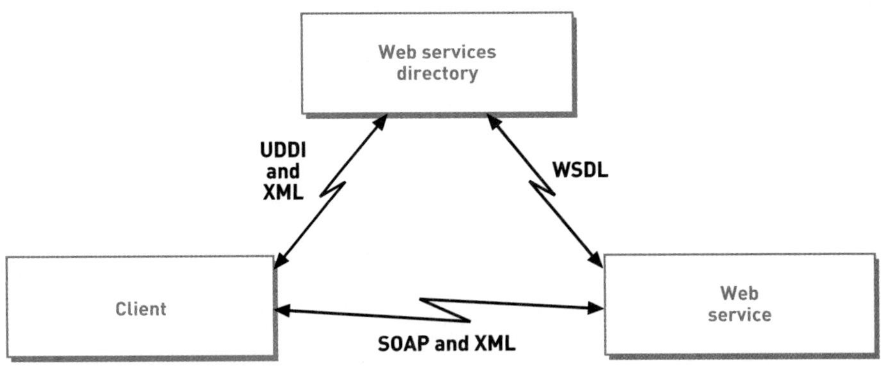

The credit verification service depicted in Figure 9-16 could be implemented as a Web service. The credit bureau would implement one or more services and make them accessible via SOAP on an application or Web server. The credit bureau would publish service information in one or more Web services directories, which would enable clients to discover and interact with those services. The published service description would include the required inputs—such as credit-card name, number, expiration date, and charge amount—and outputs such as approval or denial and an authorization code. The internal implementation of the service would be irrelevant to the client as long as it matched the WSDL description in the directory.

Web services architecture provides a flexible mechanism for making software services available to both internal and external clients. It is widely used to create a unified system by combining software services distributed across multiple organizations and computers. For example, RMO could employ external credit verification, shipment, and inventory replenishment Web services in its new online ordering system. RMO might also structure some internal functions such as querying inventory quantities or posting customer transactions as Web services to make them easily accessible from multiple applications and locations.

MIDDLEWARE

middleware

computer software that implements communication protocols on the network and helps different systems communicate

Client/server and three-layer architecture relies on special programs to enable communication between the various layers. Software that implements this communication interface is usually called middleware. Middleware connects parts of an application and enables requests and data to pass between them. There are various methods to implement the middleware functions. Some common types of middleware include transaction processing monitors, object request brokers (ORBs), and Web service directories. Each type of middleware has its own set of protocols to facilitate communication between various components of an information system.

When specifying the protocols to be used for client/server or interlayer communication, the designer usually relies on standard frameworks and protocols incorporated into middleware. For example, interactions with DBMSs usually employ standard protocols such as ODBC and SQL with supporting software obtained from the DBMS vendor or a third party. Third-party service providers such as credit bureaus and electronic purchasing or bidding services usually employ a standard Web protocol such as HTTP or XML. Industry-specific protocols have been developed in many industries such as health care and banking.

Complex OO software distributed across multiple layers and hardware platforms relies on an ORB based on a distributed object interface standard such as CORBA. Distributed non-OO software relies on different middleware products based on standards such as DCE or Microsoft's COM+. Web-based applications rely on Web-oriented protocols such as Microsoft's .NET and Sun's J2WS and specific middleware products that implement and support those protocols.

SUMMARY

Managing a live project after it gets past the planning stage is a complex and often stressful job. Nevertheless, some people thrive on this kind of work and are very good at managing. Four key areas must be carefully controlled for successful ongoing project management:

- Organizing the team and assigning tasks to team members
- Monitoring the communications for the project
- Monitoring and controlling project progress
- Keeping track of all the open items during the life of the project

Systems design is the process of organizing and structuring the components of a system to allow the construction (that is, programming) of the new system. The design of a new system consists of activities that relate specifically to the design of various new system components. The components that need to be designed include the network, application architecture and software, the user interfaces, the system interfaces, the database, and the system controls. Prototyping may be required to fully specify any part or all of the design.

The inputs to the design activities are the diagrams, or models, that were built during analysis. The outputs of the design are also a set of diagrams, or models, that describe the architecture of the new system and the detailed logic within various programming components. The inputs, design activities, and outputs are different depending on whether a structured approach or an object-oriented approach is used.

Designing the application architecture can be subdivided into architectural and detail design. Detail design often refers to the design of the software programs. Architectural design adapts the application to the deployment environment, including hardware, software, and networks. Modern application software is usually deployed in a distributed multicomputer environment and is organized according to client/server architecture or a variant such as three-layer architecture or Web services architecture. Architectural design decisions include decomposing the application into clients, servers, or layers, distributing software across hardware platforms, and specifying required protocols, middleware, and networks.

Architectural design decisions can be documented in a network diagram. The network diagram describes the organization of computing and network resources and specifies details such as the required protocols and which application software and middleware execute on which computer systems. Required network capacity can be determined by expanding the activity-location and activity-data matrices to include estimates of message size and volume.

KEY TERMS

architectural design, p. 326
business logic layer, p. 344
centralized architecture, p. 341
client, p. 342
clustered architecture, p. 340
computer network, p. 335
data layer, p. 344
detail design, p. 326
distributed architecture, p. 342
extranet, p. 336
interface designers, p. 332
Internet, p. 336
intranet, p. 336
local area network (LAN), p. 335
middleware, p. 348
multicomputer architecture, p. 340
multitier architecture, p. 340
network diagram, p. 337
router, p. 335
server, p. 342
single-computer architecture, p. 340
three-layer architecture, p. 344
view layer, p. 344
virtual organization, p. 336
virtual private network (VPN), p. 336
Web services architecture, p. 347
wide area network (WAN), p. 335
World Wide Web (WWW), or Web, p. 336

REVIEW QUESTIONS

1. What are some key issues to consider when organizing a project team?
2. What is the difference between project information and system information?
3. What is the sequence of steps that will help monitor and control project progress?
4. What is the primary objective of systems design?
5. What is the difference between analysis and design? List the design activities of the SDLC.
6. Why is project management so critical during design? What tools can a project manager use during design?
7. Explain the difference between centralized architecture and distributed architecture.
8. Explain the difference between clustered architecture and multicomputer architecture in a centralized system.
9. How are the Internet, intranets, and extranets similar? How are they different?
10. Describe client/server architecture and list the key architecture design issues that must be addressed when developing a client/server information system.
11. List and briefly describe the function of each layer in three-layer architecture.
12. What role does middleware play?
13 Describe the process of network design.
14. What roles do systems analysts and network administrators play in network design?
15. What is a network diagram? What information does it convey and where does the analyst gather that information?
16. How does the analyst generate estimates of required communication capacity? What analysis activity models are used as input?
17. What is Web services architecture? What are some examples of its potential use for business systems?

THINKING CRITICALLY

1. Discuss the evolution of client/server computing from file server to multilayer applications to Web-based applications. What has been the driving force behind this evolution? Where do you think network computing will be in the next five years? Ten years?
2. Assume that the deployment environment for a high-volume payment processing system consists of the following (these assumptions are from the scenario in Chapter 8's first Experiential Exercise):
   - Oracle DBMS running under the UNIX operating system on a cluster of HP servers
   - WebSphere application server running under the Z/OS operating system on an IBM zSeries 900 mainframe
   - J2EE application software that will be executed by other internal and external systems

   What key architectural design decisions must be made for the system? When should the decisions be made and who should make them? Outline the subsequent design tasks that should occur after the key architectural design decisions are made. To what extent can the subsequent steps be performed in parallel?
3. Develop a network diagram that supports the architectural design decisions in your answer to problem 2.

EXPERIENTIAL EXERCISES

1. Set up a meeting with the chief analysts of a medium- or large-scale development project and discuss the transition from analysis to design for that project. How and when were key architectural decisions made about the automation boundary, network design, and supporting infrastructure? Who made the decisions? Were the early architectural decisions modified later in the project? If so, how and why?
2. Find an example of an application system that is browser-based and uses TCP/IP standards. Explain how it works, showing sample screens and reports. List each middleware component and describe its function. List each protocol employed and identify the standard family or families to which the protocol belongs.
3. Examine the RMO network diagram in Figure 9-12 and note the connections to external service providers for credit verification and shipping services. Identify at least three companies that can provide each service. Investigate their online Web-based service capabilities and describe the protocols used by clients to interact with their services.
4. Find a local company or a systems development team from the information systems department at your college or university and meet with a project manager. Ask him how he manages his projects. Specifically ask him about the four areas discussed in this chapter: assigning tasks to team members, establishing communication protocols and using electronic tools, monitoring and controlling progress, and tracking open issues.

CASE STUDIES

THE REAL ESTATE MULTIPLE LISTING SERVICE SYSTEM

In Chapter 8, you were asked to discuss the implications of the "anytime, anywhere" requirement for the application deployment environment and to describe the type(s) of hardware, network, and software architecture needed to fulfill that requirement. Assume that you addressed that question by specifying a three-layer architecture using ordinary PCs running Web browsers to implement the view layer. Draw a network diagram that represents your chosen solution.

Today's computer-based real estate listings typically include graphical data, such as still and moving pictures, in addition to text descriptions of properties. What is the impact of such data on data communication requirements within your network design, assuming 10 listing accesses per hour? 100 listing accesses per hour? 1,000 listing accesses per hour?

RETHINKING ROCKY MOUNTAIN OUTFITTERS

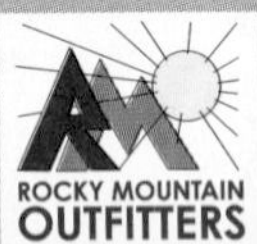

In Chapter 8, you were asked to consider an alternative deployment scenario for RMO based on Apache Web servers running under Linux and an Oracle database server. Modify the network diagram in Figure 9-12 to reflect the alternative deployment scenario. What changes, if any, are required for the client workstations and customer PCs? What changes, if any, are required in middleware and communication protocols? Will there be any change in the estimates of required data-communication capacity among client workstations and servers at the Park City data center? Why or why not?

FOCUSING ON RELIABLE PHARMACEUTICAL SERVICE

Assume the same facts as presented in the Chapter 8 Reliable Pharmaceutical case. Also assume that you are the project manager for the selected vendor's development team. Your company, RxTechSys, develops and markets software to retail and hospital pharmacies and has decided to take on the Reliable project to expand potential market share. RxTechSys and Reliable will jointly develop the new software. RxTechSys will then market the finished product to other companies and pay a royalty to Reliable for each sale.

RxTechSys has been in the pharmacy software business for 20 years. The latest version of the software is a Web-based application built on the Microsoft .NET platform. Major functions such as inventory control, purchasing, billing, and prescription warning are implemented as separate .NET Web services. As part of the team that prepared the response to Reliable's RFP, you determined that RxTechSys's current system can be adapted to Reliable's needs as follows:

- Existing browser-based prescription entry can be modified to handle data input from multiple customer locations over a VPN. This is a significant modification due to expanded data content and greater security requirements.
- Order fulfillment software will have to be written from scratch.
- Billing software will require significant modification because your current system assumes that all patients have their health care managed by a single institution, with possible third-party reimbursements through Medicaid/Medicare.
- Other parts of your existing system can be used with little or no modification.

Reliable has provided you with a complete set of object-oriented analysis models, the quality of which you approved during contract negotiations. Your task is to move the project forward through design and implementation.

Reliable has assigned an operational manager with some computer experience to your team full-time, and she is authorized to assign other Reliable personnel to your project as needed. You have been assigned a full-time staff of four developers, two of whom have substantial design experience and all of whom participated in developing the most recent version of RxPharmSys software.

Develop a design plan and schedule that covers the next 4 to 6 weeks (your expected project duration is 10 months). What design decisions must be made within the next two weeks? Who should make them? How will design and development proceed thereafter—what tasks must be performed and in what order? How will you manage and control the project?

FURTHER RESOURCES

Robert Orfali, Dan Harkey, and Jeri Edwards, *Client/Server Survival Guide, Third Edition*. Wiley, 1999.

CHAPTER

10

THE TRADITIONAL APPROACH TO DESIGN

LEARNING OBJECTIVES

After reading this chapter, you should be able to:

- Describe the steps involved in the traditional approach to designing the application architecture
- Develop a system flowchart
- Develop a structure chart using transaction analysis and transform analysis
- Write pseudocode for structured modules
- Explain how to use three-layer design with the traditional approach

CHAPTER OUTLINE

The Structured Approach to Designing the Application Architecture

The Automation System Boundary

The System Flowchart

The Structure Chart

Module Algorithm Design: Pseudocode

Integrating Structured Application Design with Other Design Tasks

Three-Layer Design

THEATRE SYSTEMS, INC.: SOMETHING OLD, SOMETHING NEW

Bernard closed the office door and spoke to his officemate, Stana, in an exasperated voice. "I don't understand why Jim insists that I update these system flowcharts and structure charts. We should throw this all out and start from scratch with object-oriented (OO) design. I drew a few of these traditional diagrams in school, but we spent most of our time with object-oriented diagrams and techniques. I feel as if I'm being asked to fix a computer with a hammer and saw!"

Bernard, a recent MIS college graduate, was a new employee at Theatre Systems, Inc., which sells and supports financial reporting software for small and medium-sized U.S. theatre chains. He was hired just as an upgrade project was moving from analysis to design. Although the company's software had been updated regularly with improvements and new features, it lacked some modern capabilities, such as a Web-based interface and scalable multilayer architecture.

Stana, who had worked for the company for almost four years, replied, "Well, you need to remember two things. First, many of the IS staff here don't fully understand OO analysis and design techniques. Version 1 of our software and its analysis and design documentation was developed in the early 1980s. All of the upgrades since then have been incremental, so there's been no need to burn the old design models and start from scratch. Significant chunks of the system are unchanged in over a decade.

"Second, there's a question of suitability of the tools to the task. If our goal was to develop and implement an entirely new system that could scale from the smallest mom-and-pop theatre to the largest nationwide theatre chain and be deployed and redeployed at whim, then we'd almost certainly be using the latest distributed software technologies, OO programming languages, and the OO analysis and design tools that best match them. We'd also throw away most of our existing source code and redevelop the entire system from scratch. But our current project calls for grafting some Web browser front-end interfaces onto a system written in C with as little change as possible to the existing code. Structured design models work very well for the existing C programs and functions."

"So how do I represent Web interfaces and client/server interactions with structured techniques?" Bernard asked.

Stana replied, "The trick is to think of the Web server as a container for application software programs that communicate with the Web browser over a real-time link—the Internet or an intranet. In structured design, the primary software units are programs and modules. So in the current system, the modules are C functions, which are packaged into a small number of complex programs that do many things, with all-encompassing menu-based front ends. One of your most important tasks for this upgrade is to decompose those large programs into smaller ones and move functions that implement the existing user interface out of the C code and into Web-page code. The remaining functions are application logic that you can package into small programs that can be called from Web-server scripts. Each of those small programs should be one structure chart and one box on a system flowchart. You should be able to cut and paste from the existing structure charts to create rough drafts as a starting point."

Bernard looked relieved at first, but then confusion and concern crossed his face. "Jim is going to review my work at the end of the week. I'm worried that I'll make some huge mistakes and that he'll think he made a mistake in hiring me. Would you look over some of my work before I meet with him?"

Stana gave Bernard a reassuring smile. "Jim put you in this office with me. And even though I'm assigned to other project tasks, he asked me to help you when you needed it. Software development only succeeds when everyone works as a team. People who *don't* ask for help are the ones who get fired. So why don't you spend the rest of the morning designing the entry and verification modules for the snack bar receipts, and we'll sit down after lunch and go over them?"

OVERVIEW

This chapter describes the traditional approach to designing software. The chapter begins with an overview of the structured models, model development process, and related terminology. We then describe how data flow diagrams are annotated with automation boundary information. Next, we explore how information from analysis models is transformed into design models using system flowcharts, structure charts, and module pseudocode. Then we discuss how traditional software design is integrated with other design activities. The chapter concludes with an examination of how the traditional approach is applied to designing a three-layer architecture.

As described in the opening case, traditional software design and structured design models have been in use for many years. They are commonly used with systems developed using procedural programming languages and are well suited to describing systems with both batch and online components. Most new systems are developed with object-oriented programming languages, so traditional systems design models are decreasing in popularity. However, as illustrated in the case, many older systems in use today were designed and documented using traditional methods and models. Also, traditional design concepts such as coupling, cohesion, and top-down partitioning underlie both traditional and object-oriented design methods, so it is important to understand those concepts. Finally, traditional models are sometimes adapted to newer software development methods and paradigms such as multilayered software. So, analysts should be knowledgeable about the traditional approach to design.

THE STRUCTURED APPROACH TO DESIGNING THE APPLICATION ARCHITECTURE

The application architecture consists of the application software programs that will carry out the functions of the system. Application design must be done in conjunction with the design of both the database and the user interface. However, we focus exclusively on the design of the computer software itself for ease of understanding.

module
an identifiable component of a computer program that performs a defined function

computer program
an executable entity made up of a set of modules

Third-generation programming languages such as Visual Basic, C, COBOL, or Pascal are organized around **modules** that are arranged in a hierarchy like a tree. The top module is often called the *boss module* or the *main module*. The middle-level modules are *control modules*, and the *leaf modules* (those at the ends of the branches) are the detailed modules that contain most of the algorithms and logic for the program. A module, then, is a small section of program code that carries out a single function. A **computer program** is a set of modules that are compiled into a single executable entity. The design of a computer program is specified with a structure chart, which is discussed in detail later in this section.

In large systems, a single program usually cannot perform all of the required functions. Sometimes one program is written to perform online activities, and another program carries out periodic functions that are executed once a day. Other programs may have specialized functions, such as backing up the data or producing year-end financial reports. All these individually executable entities, or programs, comprise the entire system.

system flowchart
a diagram that describes the overall flow of control between computer programs in a system

pseudocode
statements similar to that of a programming language that describe the logic of a module

Both the structure of the overall system and any individual subsystems are documented using a **system flowchart**. The system flowchart identifies each program, along with the data it accesses. The system flowchart also shows the relationships among the various programs, subsystems, and their files or databases. It documents the architectural structure of the overall system. We describe how to design a system flowchart later in this chapter.

Finally, the project team must also design the internal logic of individual modules. The internal algorithms that comprise the logic of the modules are usually documented using **pseudocode**. If you have taken programming classes, you probably had to write algorithms in pseudocode before you actually coded your programs. Pseudocode is very much like the

structured English described in Chapter 6. Pseudocode describes the logic of a module using statements similar to that of a programming language.

In general, analysts use a top-down approach for design. The inputs for the development of the design models and documents are the data flow diagrams and their detailed documentation, in structured English, and the detailed data flow definitions. Figure 10-1 illustrates the process. Analysts use an intermediate form of the DFD called the *DFD with automation system boundary*, which divides the computerized portions of the system from the manual portions. So, this diagram determines which portions of the system need to be included in the design. This enhanced DFD is actually used as the source for the design models, as shown in the figure.

Figure 10-1

Structured design models

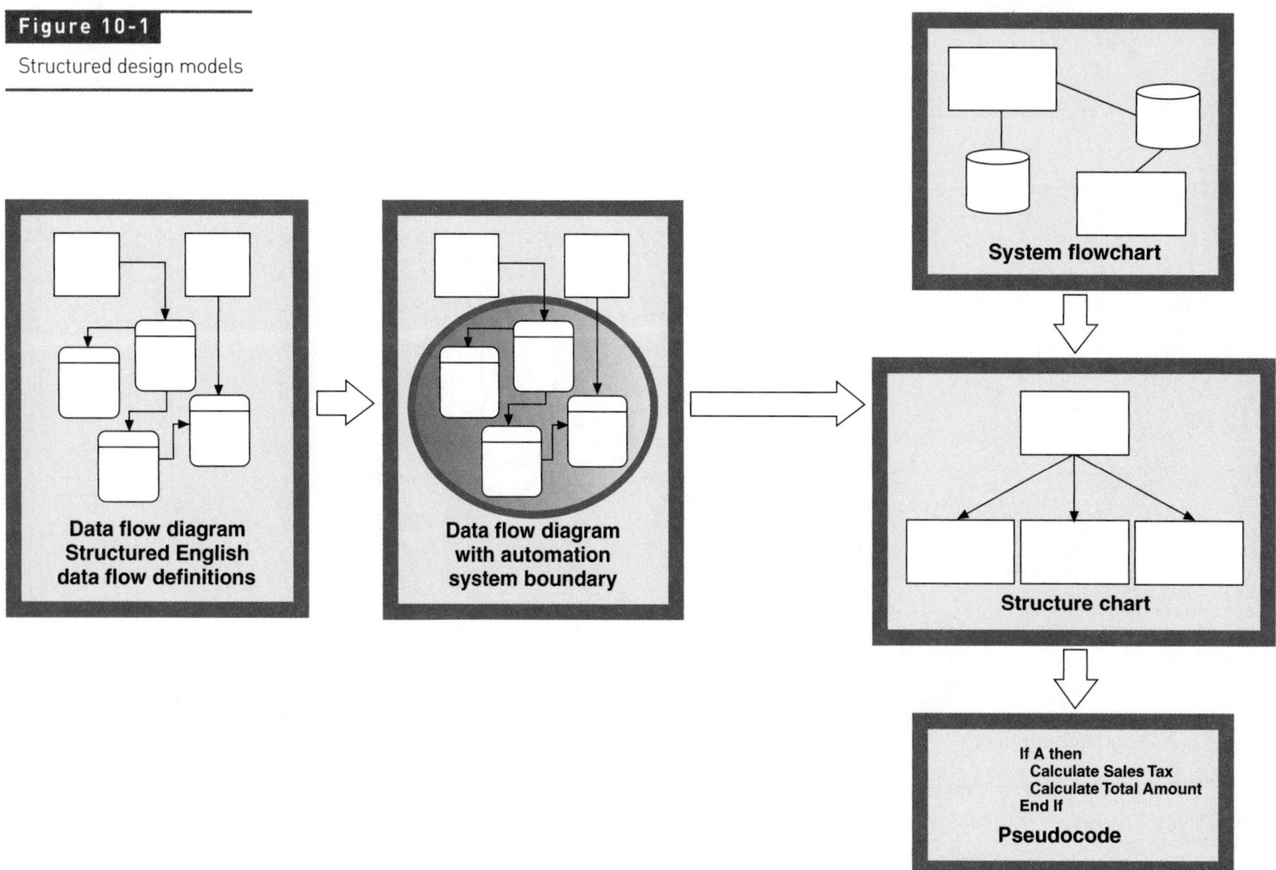

The following sections of the chapter trace the sequence of activities shown in Figure 10-1. First, the automation system boundary is discussed. Then we explain the development of the system flowchart. Next, we discuss the approach to the design of the structure chart. Finally, we describe the form and method of writing pseudocode.

THE AUTOMATION SYSTEM BOUNDARY

The automation system boundary partitions the data flow diagram processes into manual processes and those that are to be included in the computer system. During the analysis activities, we looked at the business events and all of the processes that were triggered by those events. At that time, we did not try to distinguish between manual and automated processes, but to develop the computer system's design, we must identify the processes that will be automated.

Figure 10-2 illustrates a typical data flow diagram with the automation system boundary added. This figure shows both the system boundary, which identifies the entire automated system, and program boundary lines, which partition the DFD into separate programs. This

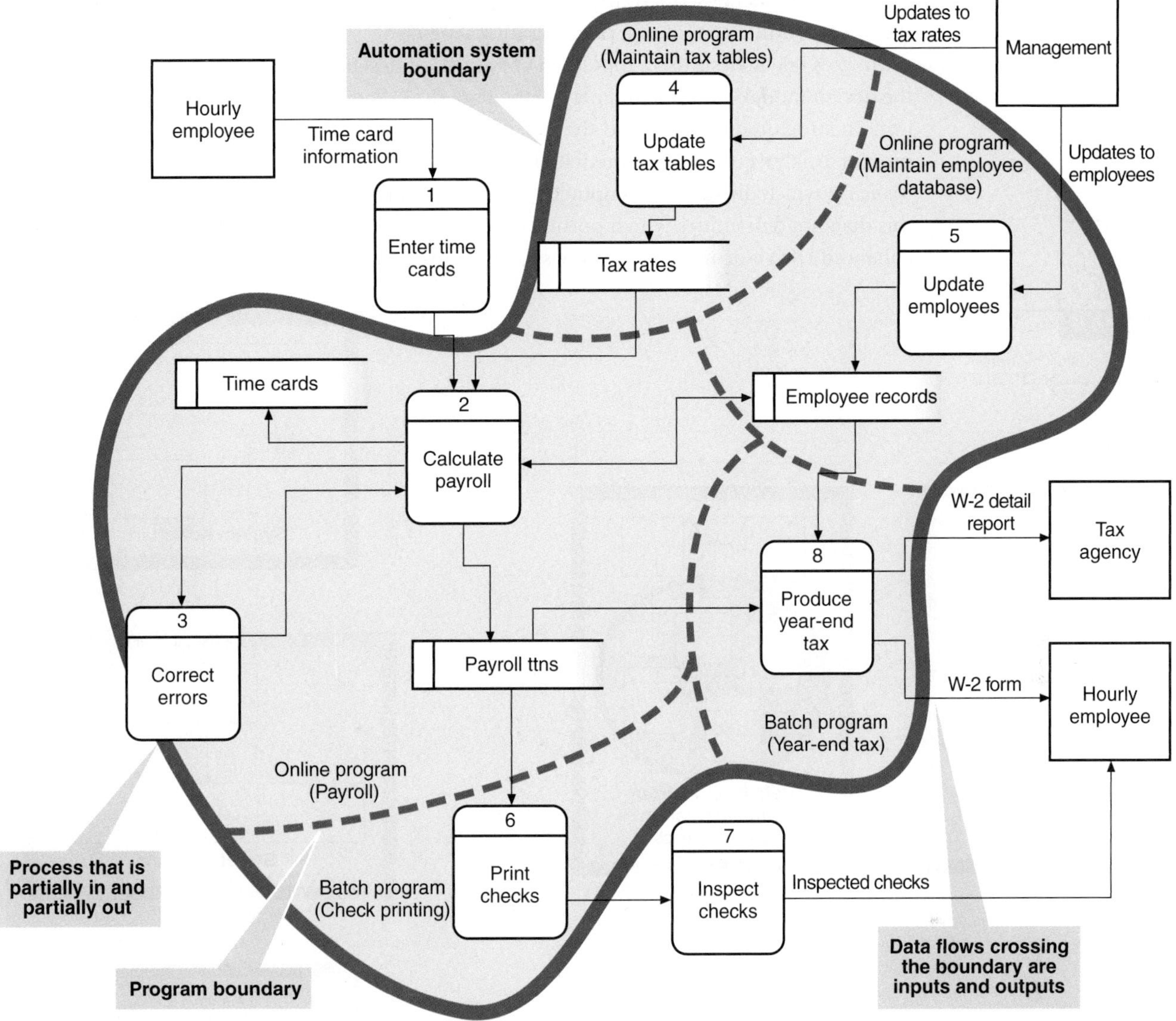

Figure 10-2

The data flow diagram with an automation system boundary

diagram is the first step in design, and it determines what the programs are and what processes are included within those programs. In this context, we define a program as a separate executing entity.

Processes can either be inside or outside the system boundary. Processes that are outside are manual processes, such as sorting and inspecting paper documents, entering customer orders, or visually inspecting incoming shipments. The processes that are inside the boundary may be carried out in online or batch modes. Online processes are usually active every day during working hours. Batch processes may be activated each night or only periodically. In some cases, the system boundary goes right through a process, which indicates that the process is mid- or high-level and is exploded on a more detailed diagram. Some of the processes in the exploded detail will be inside the system boundary, and some will be outside.

Data flows are found inside, outside, or crossing the system boundary and the program boundaries. The data flows that cross the system boundary are particularly important; they represent inputs and outputs of the system. In other words, the design of the program interfaces, including both the user interface and transmittals to other systems, is defined by the boundary-crossing data flows. In the final system, these data flows will be forms or reports in the user interface, or files or telecommunication transmittals between systems. Data flows that cross the boundaries between programs represent program-to-program communication. In the final system, these data flows will also be files or telecommunication transmittals between programs.

Figure 10-2 is the high-level data flow diagram showing all of the major processes for a payroll program. The system boundary can also be drawn on each data flow fragment to show more detail about which processes are internal or external and which low-level data flows cross the boundary.

THE SYSTEM FLOWCHART

The system flowchart is the representation of various computer programs, files, databases, and associated manual procedures that make up a complete system. Processes are grouped into programs and subsystems based on similarities such as shared timing (for example, a process performed monthly), shared access to stored data (for example, all processes that update employee data), and shared users (for example, processes that produce reports for the marketing department). The programs and subsystems thus created have complex interdependencies, including flow of data, flow of control, and interaction with permanent data stores. The system flowchart is frequently constructed during the analysis activities. For example, the subsystems of the RMO customer support system were defined during the analysis activities (see Chapter 6), and the set of activities or use cases allocated to each subsystem makes up the program modules.

BEST PRACTICE

The system flowchart helps to document the application architecture, showing subsystems, inputs, outputs, and data storage.

A system flowchart graphically describes the organization of the subsystems into automated and manual components, showing the flow of data and control among them. System flowcharts are used primarily to describe large information systems consisting of distinct subsystems and dozens or more programs. They are also used to describe systems that perform batch processing (for example, systems used to process bank transactions, payroll checks, and utility bills). A common characteristic of such systems is the division of processing into discrete steps (such as validation of input transactions, update of a master file with transaction data, and production of periodic reports) with a fixed execution sequence. Many batch systems also make extensive use of files in addition to or instead of databases.

System flowcharts first came into widespread use to document the processing and data flow between programs that processed information through batch transaction files. Frequently, in these systems, one program would produce a file of all the daily transactions. Then another program would process the transactions and update a master file. Yet another program would be used to produce the various reports required from the system.

Today's newer systems perform much processing in real time, as each transaction is entered. These systems also usually make updates to a relational database system instead of a master file. Centralized database management systems also include many of the processes that were previously done by individual programs. System flowcharts also may be drawn for these newer systems, although the diagrams tend to be much simpler because the processing is more centralized to a single program or subsystem. But because the systems developed these days are generally much more complex overall, you will still see system flowcharts used to represent how all of the pieces fit together.

Many business systems today also have both real-time and batch components. For example, today your credit-card purchases are at least verified in real time and may even be posted in real time. However, monthly account statements and customer payments typically are processed in batches. A system flowchart is used to describe the overall organization of this type of system and show the relationship between the real-time components and the batch processing.

Figure 10-3 illustrates the most common symbols that are used in system flowcharts. These symbols are fairly common throughout the industry, although from time to time you will see variations.

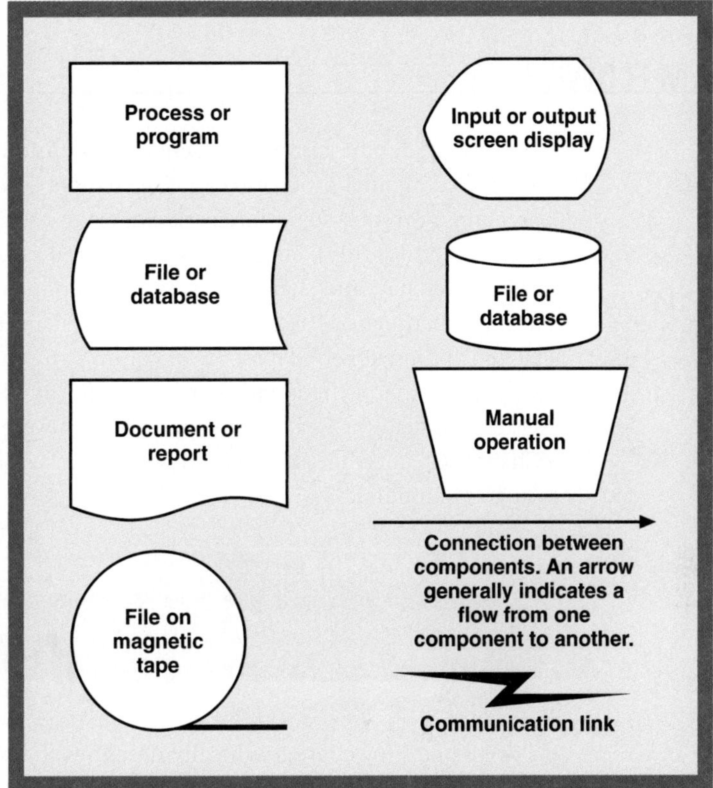

Figure 10-3
Common system flowchart symbols

Figure 10-4 is a system flowchart for the payroll system shown on the DFD in Figure 10-2. Note that the system flowchart identifies the files, programs, and manual processes of the total system. We have added physical implementation descriptions by identifying the file media, disk, or tape. Frequently, we also include additional system functions and files such as backups and history files. Even though the information shown in Figures 10-2 and 10-4 is very similar, the focus of the diagrams is different. The system flowchart focuses on the implementation of physical objects, such as executable programs, files, and documents.

Figure 10-4 shows that the payroll program has four inputs and produces three outputs. The inputs are the time cards, the tax rate table file, the employee database, and corrections. Outputs produced are an error report, a payroll transaction file, and an updated payroll history file. Other programs (that is, independent executables) are the two programs to maintain the tax rate tables and the employee database, and the other two programs to write checks and to produce year-end income tax reports.

Figure 10-5 is an example of a system flowchart for Rocky Mountain Outfitters. The four main programs correspond to the subsystems identified in the list of subsystems in Figure 6-10. As in the example of the order-entry subsystem shown in Figure 6-12, each subsystem will

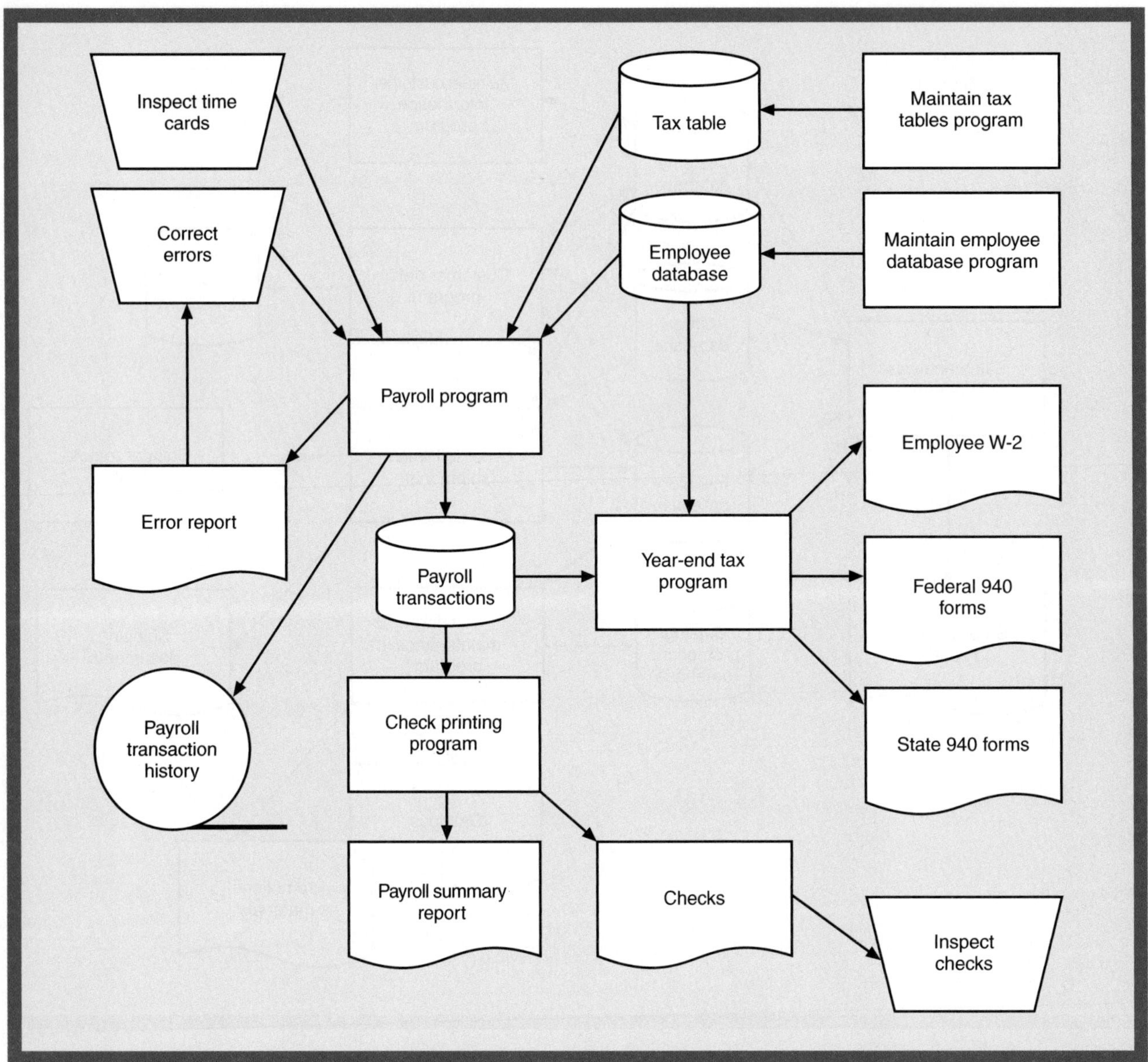

Figure 10-4

A sample system flowchart for a payroll system

include DFD fragments for each of the activities in the subsystem. In the system flowchart, the individual data stores have been converted into database files. As you can see, the creation of the system flowchart requires the architectural design of the major program steps, the architectural design of the databases, the identification of the major interfaces, and the identification of the primary outputs.

Figure 10-5 also shows one additional subsystem that is not identified in Figure 6-10. During the scoping and level of automation review discussed in Chapter 8, RMO decided that it needed a higher level of automation for a couple of sales analysis reports. In this instance, instead of adding the reports to an existing subsystem, the project team defined a new subsystem.

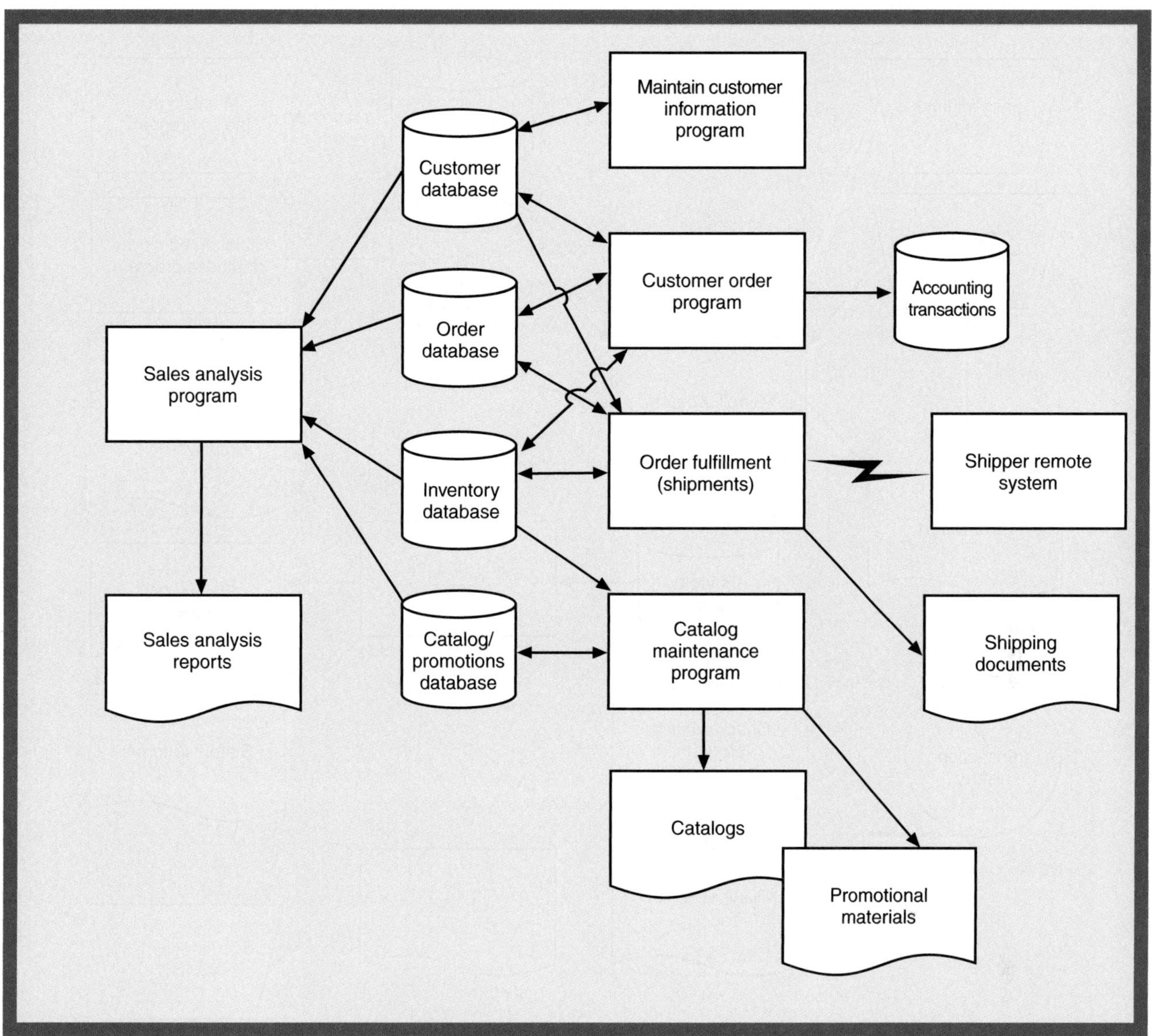

Figure 10-5

The system flowchart for Rocky Mountain Outfitters

THE STRUCTURE CHART

The primary objective of structured design is to create a top-down decomposition of the functions to be performed by a given program in a new system. Each independent program shown in the system flowchart performs a set of functions. Using structure chart techniques always provides a hierarchical organization of program functions. First, this section explains what a structure chart is and how to interpret one. We explain how each structure chart is related to the DFDs created during systems analysis, and then we show how to use a detailed data flow diagram to develop one.

structure chart
a hierarchical diagram showing the relationships between the modules of a computer program

A **structure chart** hierarchy describes the functions and the subfunctions of each part of a system. For example, the program may have a function called *Calculate pay amounts*. Some subfunctions are *Calculate base amount*, *Calculate overtime amount*, and *Calculate taxes*. In a structure chart, these functions are drawn as a rectangle. Each rectangle represents a module.

BEST PRACTICE

Use a structure chart to document the modular design of each program shown in a system flowchart. Each high-level module is usually based on one activity or use case triggered by an event. So, traditional structured design can be described as use case driven.

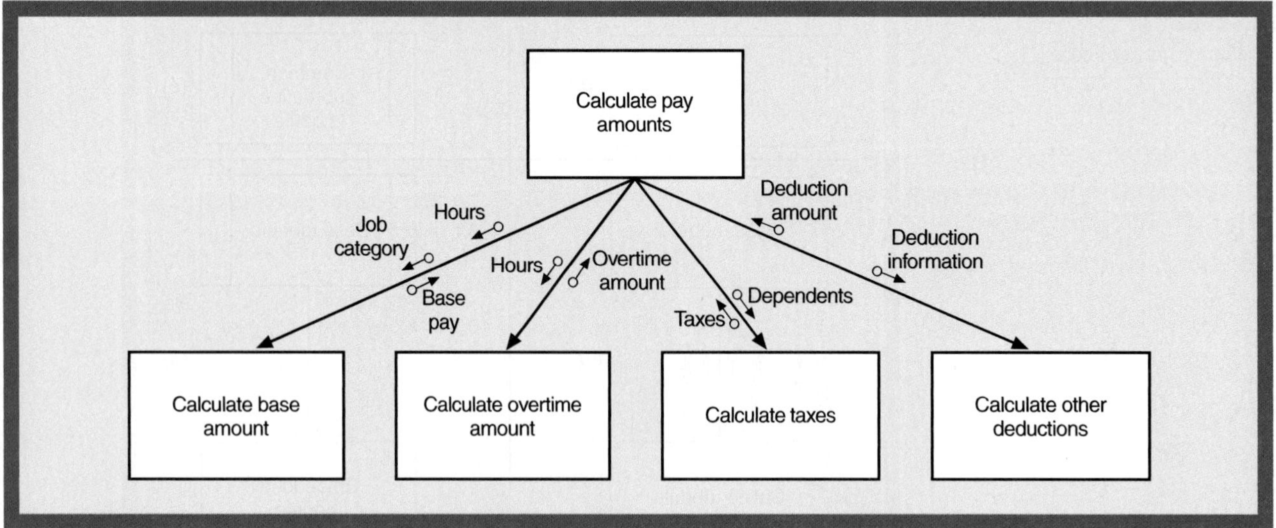

Figure 10-6

A simple structure chart for the *Calculate pay amounts* module

A module is the basic component of a structure chart and is used to identify a function. Figure 10-6 shows a simple structure chart for the payroll example. Modules, as shown by the rectangles, are relatively simple and independent components. Higher-level modules are control modules that control the flow of execution. Lower-level modules are "worker bee" modules and contain the program logic to actually perform the functions. In Figure 10-6, for example, the *Calculate pay amounts* module may do nothing more than call the lower-level modules in the correct sequence to carry out the logic of calculating payroll.

Notice how a structure chart provides a simple and direct organization to a computer program whose purpose is to calculate payroll amounts. Modular programming is a time-tested method to write computer programs that are easy to understand and maintain. Breaking a complex program into small modules makes initial programming and maintenance programming easier. The development of a structure chart is based on rules and guidelines. The key points are that the program is a hierarchy and that the modules are well-formed with high internal cohesiveness and minimal data coupling. Later, we describe in more detail the characteristics of good modules.

The lines connecting the modules indicate the calling structure from the higher-level modules to the lower-level modules. The little arrows next to the lines show the data that is passed between modules and represent the inputs and outputs of each module. At the structure chart level, we are not yet concerned with what is happening inside the module. We only want to know that somehow the module does the function indicated by its name, using the input data and producing the output data.

Figure 10-7 shows the common symbols used to draw structure charts. The rectangle represents a module. In a structure chart, a module can represent something as simple as a block of code, such as a paragraph or section in a COBOL program. In other languages, a module typically represents a function, procedure, or subroutine. Examples of modules as program fragments include subroutines (as in FORTRAN and BASIC), paragraphs or subprograms (as in COBOL), procedures (as in Pascal), and functions (as in FORTRAN, C, and C++). A module also can be a separately compiled entity such as a complete C program. The rectangle with the double bars is simply an existing module or a module that is used in several places. Use of the double bar notation is optional.

program call
the transfer of control from a module to a subordinate module to perform a requested service

Part c of Figure 10-7 shows a call from a higher-level module to the lower-level module. A **program call** occurs when one module invokes a lower-level module to perform a needed service or calculation. The implementation of a program call varies among programming languages. For example, a program call can be implemented as a function call in C or C++, a procedure call in Pascal, or a subroutine call in FORTRAN. In each case, the program passes control to the called module, the called module executes a series of program statements, the

Figure 10-7

Structure chart symbols

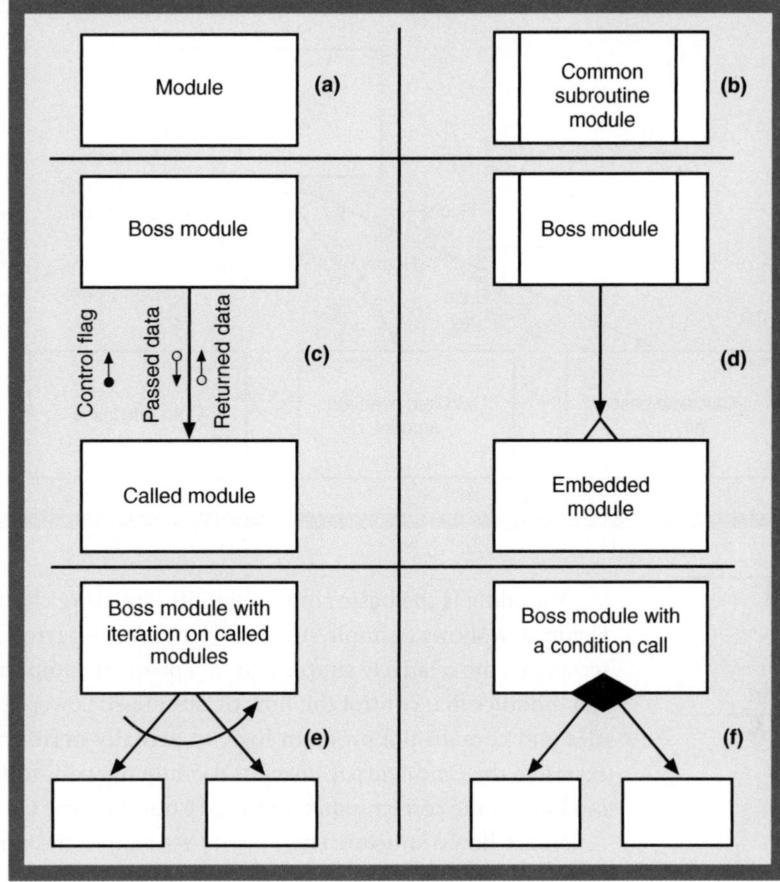

data couples
the individual data items that are passed between modules in a program call

called module passes control back to the calling module, and execution resumes with the statement or instruction immediately following the call.

Figure 10-7c also indicates how data is passed between modules. The arrows with the open circle, called **data couples**, represent data being passed into and out of the module. A data couple can be an individual data item (such as a customer account number) or a higher-level data structure (such as an array, record, or other data structure). The type of coupling used at each level of the structure chart usually follows the principle of layering of detail. That is, coupling between modules near the top of a structure chart typically consists of data structures representing high-level aggregations of data. Coupling between modules at the bottom of the structure chart typically consists of single data items, flags, and relatively small data structures.

The arrow with the darkened circle is a control flag. A flag is purely internal information that is used between modules to indicate some result. Flags originating from lower-level modules often indicate a result, such as a record passing a validation test. Another common use is to indicate that the end of a file was reached.

Figure 10-7d illustrates a lower-level module that is broken out on the structure chart but that in all probability will be subsumed into the calling module for programming. This documentation technique primarily ensures that the function performed by the module is highlighted. Figure 10-7e and f show two alternatives for program calls. In 10-7e we show the notation used to indicate iteration through several modules. In 10-7f we show conditional calling of low-level modules—that is, the program calls modules only when certain conditions exist.

Figure 10-8 is a more complete view of the Payroll program, including the original *Calculate pay amounts* function from Figure 10-6. Notice that the entire structure chart shown in Figure 10-8 is based on the system activities following the temporal event *Time to produce payroll*. During systems analysis for the payroll system, the analyst would have identified this event as one that occurs at the end of every week for hourly employees. Many other events that trigger activities or use cases would have been identified as well.

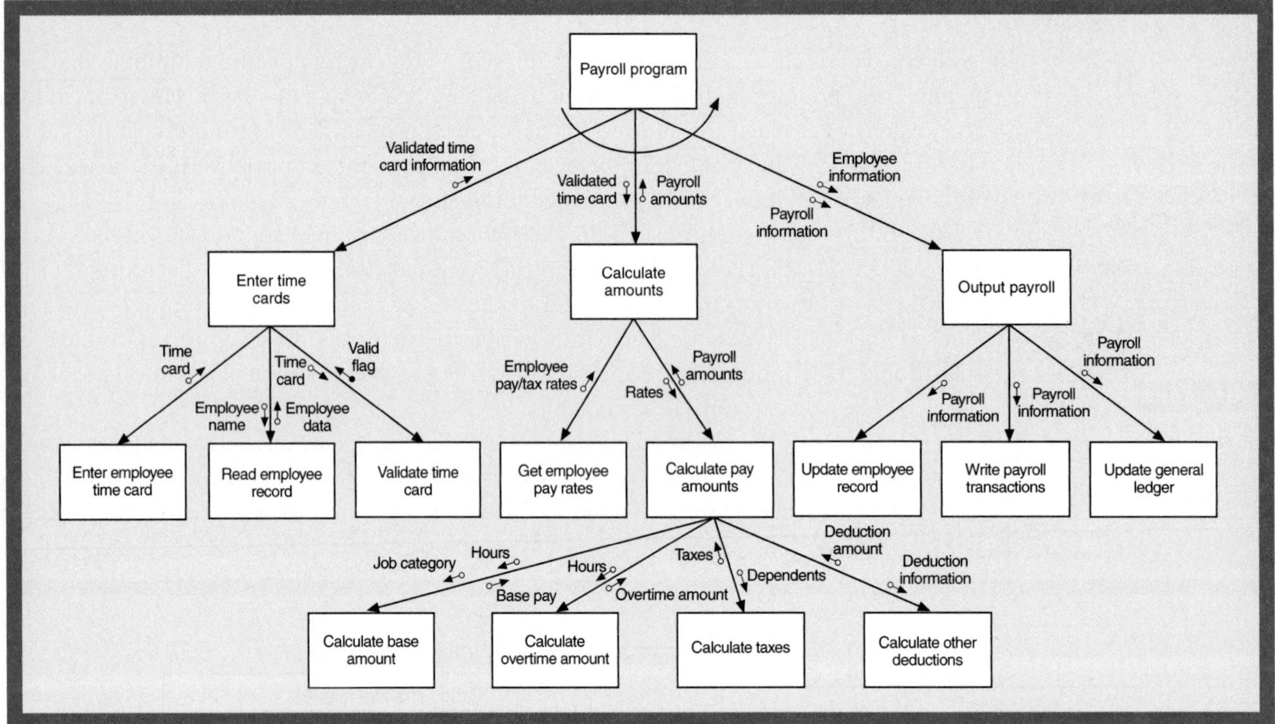

Figure 10-8

A structure chart for the entire Payroll program

A basic idea of structured programming is that each module only has to do a very specific function. The module at the very top of the tree is the boss module. Its functions are to call the modules on the next tier, pass information to them, and receive information back. The function of each middle-level module is to control the processing of the modules below it. Each has control logic and any error-handling logic that is not handled by the lower-level module. The modules at the extremities, or the leaves, contain the actual algorithms to carry out the functions of the program. This approach to programming separates program control logic from business algorithm logic and makes programming much easier.

The arrows from the higher-level modules to the lower-level modules indicate the program call. The direction of the call is always from left to right. Notice that the structure chart maintains a strict hierarchy in the calling structure. A lower module never calls a higher module. The curved arrow immediately below the boss module indicates a loop across all three calls. In other words, the main module will have an internal loop that includes calls to all three lower-level modules within the same loop.

In the example, you can see the flow of information downward and back up. Usually, a higher-level module requests a service from a lower-level module and passes the necessary input information. The lower-level module then returns the requested information or some control information, as a flag, to notify the higher-level module of the successful completion of the task. Looking at the *Enter time cards* subhierarchy, you can observe that the employee time card information is passed to the boss. Then the employee name is passed to the next module, which reads some employee information and passes it back up. Finally, employee data and time card information are passed to the rightmost module, which validates the time card. This module returns a control flag, indicating success or failure of the validation. If the validation fails, the program sends error messages and goes into its error-handling routines. We have not shown all the complexities required in a real program, especially the error-handling modules.

Included within the structure chart will be the modules that access data from the outside world. It is important that the design of these modules be consistent with the design of the user interface, the interface to other systems, and the database design. The structure charts that have been developed should also be consistent with the system flowchart. If changes were made during this design activity, the project team should update the system flowchart accordingly.

DEVELOPING A STRUCTURE CHART

Structure charts create a hierarchy of modules for a program. A structure chart looks like a tree with a root module and branches. A subtree is simply a branch that has been separated from the overall tree. When the subtree is placed back in the larger tree, the root of the subtree becomes just another branch in the overall tree. Why is this important? The structure chart can be developed in pieces and combined for the final diagram.

transaction analysis
the development of a structure chart based on a DFD that describes the processing for several types of transactions

Figure 10-5 showed the system flowchart for RMO's customer support system. Each major program corresponds to a subsystem in the event-partitioned diagram. Each program will have its own structure chart. However, as you can see in Figure 6-10, each program—that is, subsystem—consists of several activities. Each activity corresponds to a process on the event-partitioned DFD, and each process will be detailed in a DFD fragment based on the activity triggered by an event from the event table.

You can develop structure charts using one of two methods: transaction analysis and transform analysis. **Transaction analysis** uses as input the system flowchart and the event table to develop the top level of the tree—that is, the main program boss module and the first level

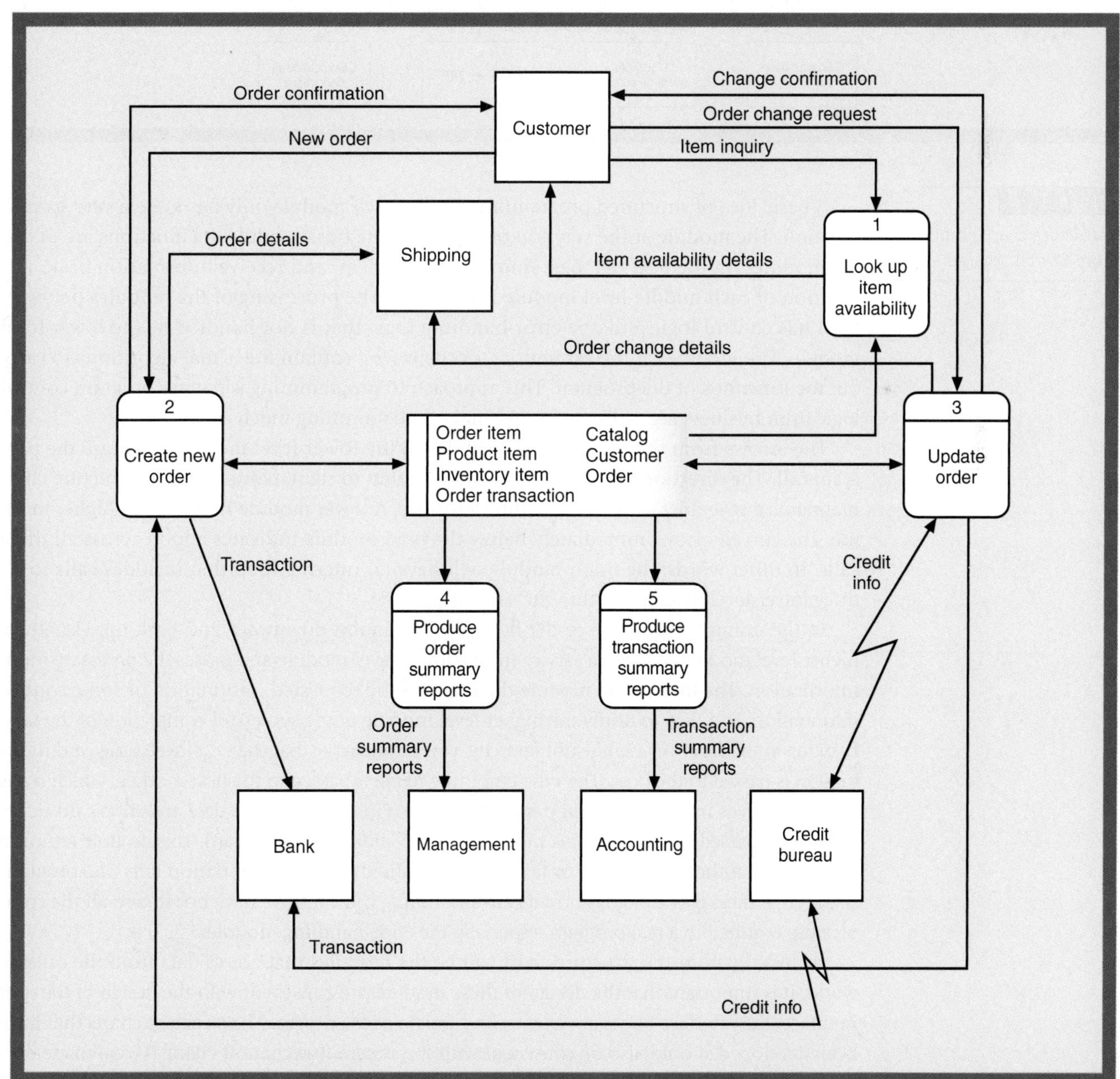

Figure 10-9
Event-partitioned DFD for the order-entry subsystem

transform analysis the development of a structure chart based on a DFD that describes the input-process-output data flow

of called modules. Transform analysis uses as input the DFD fragments to develop the subtrees. Each subtree root module corresponds to the first-level branch of the main program structure chart. We discuss each method in turn.

Transaction Analysis

In transaction analysis, the first step is to examine the system flowchart and identify each major program, such as the Customer order program in Figure 10-5. Figure 10-9 duplicates Figure 6-13, the event-partitioned DFD for the order-entry subsystem. This DFD shows the five processes derived from the five events of this subsystem. In this subsystem, the five primary processes are five different transactions that must be supported. These transactions are *Look up item availability, Create new order, Update order, Produce order summary reports,* and *Produce transaction summary reports.*

Figure 10-10 shows the structure chart based on transaction analysis for this program. As already mentioned, transaction analysis is the process of identifying each separate transaction that must be supported by the program and constructing a branch for each one. In essence, this program, at least at the highest level, is simply a module to display a screen for the user to enter a transaction choice and then to invoke the appropriate module to process the transaction. This diagram does not show the additional detail below each of the transaction modules. Each of the transaction modules, which are named after the transactions, will be the main boss module for a subtree to process the transaction. Each subtree will be developed based on the DFD fragment for that activity and will be developed utilizing transform analysis.

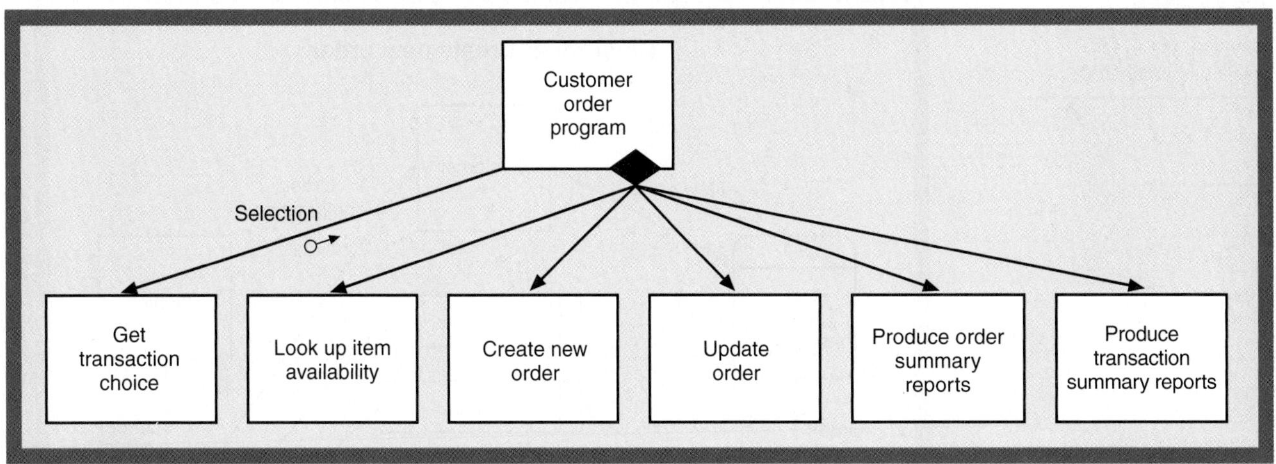

Figure 10-10 High-level structure chart for the Customer order program

This structure chart also has very few data couples. Essentially, the only information passed is the transaction selection from the *Get transaction choice* module. That information is used by the control module to select the correct processing module. The subtree beneath the processing module will display the appropriate screens to accept and pass the detailed information required.

Transform Analysis

Transform analysis is based on the idea that the computer program "transforms" input data into output information. Structure charts developed with transform analysis usually have three major subtrees: an input subtree to get the data, a calculate subtree to perform the logic, and an output subtree to write the results. Figure 10-8 is a good example of a structure chart that was developed using transform analysis, for the process of transforming time card inputs into payroll outputs following one event. Note that a DFD fragment usually follows this pattern of input-process-output, and the structure chart converts the processing on the DFD fragment to a top-down structure of program modules.

Sometimes DFD fragments are decomposed into detailed diagrams. The detailed diagrams provide more detail than can be used for the structure chart. Figures 10-11 through 10-14 provide an example of transform analysis from Rocky Mountain Outfitters. Figure 10-11 shows the DFD fragment created for the *Create new order* activity. Figure 10-12 contains the

Figure 10-11

The *Create new order* DFD fragment

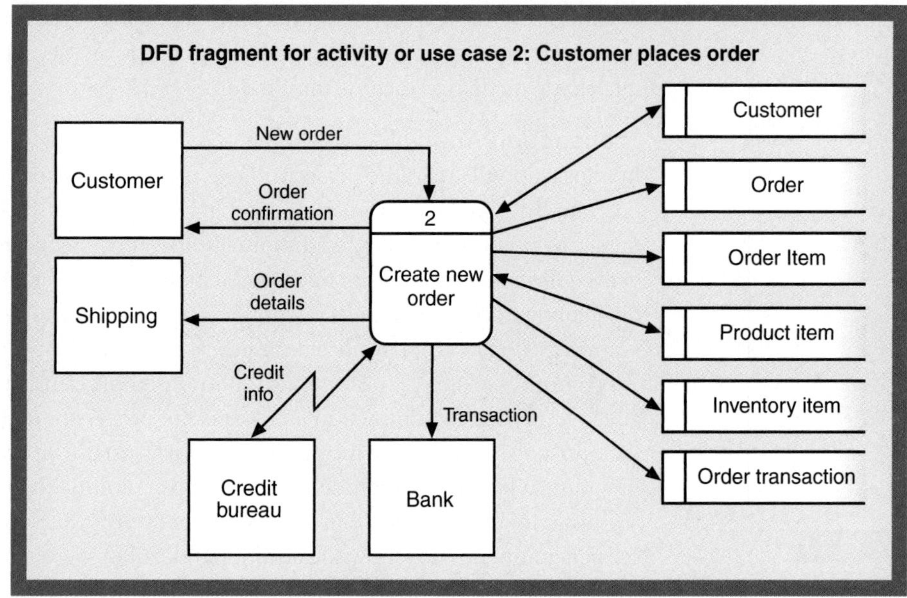

Figure 10-12

Exploded view of the *Create new order* DFD

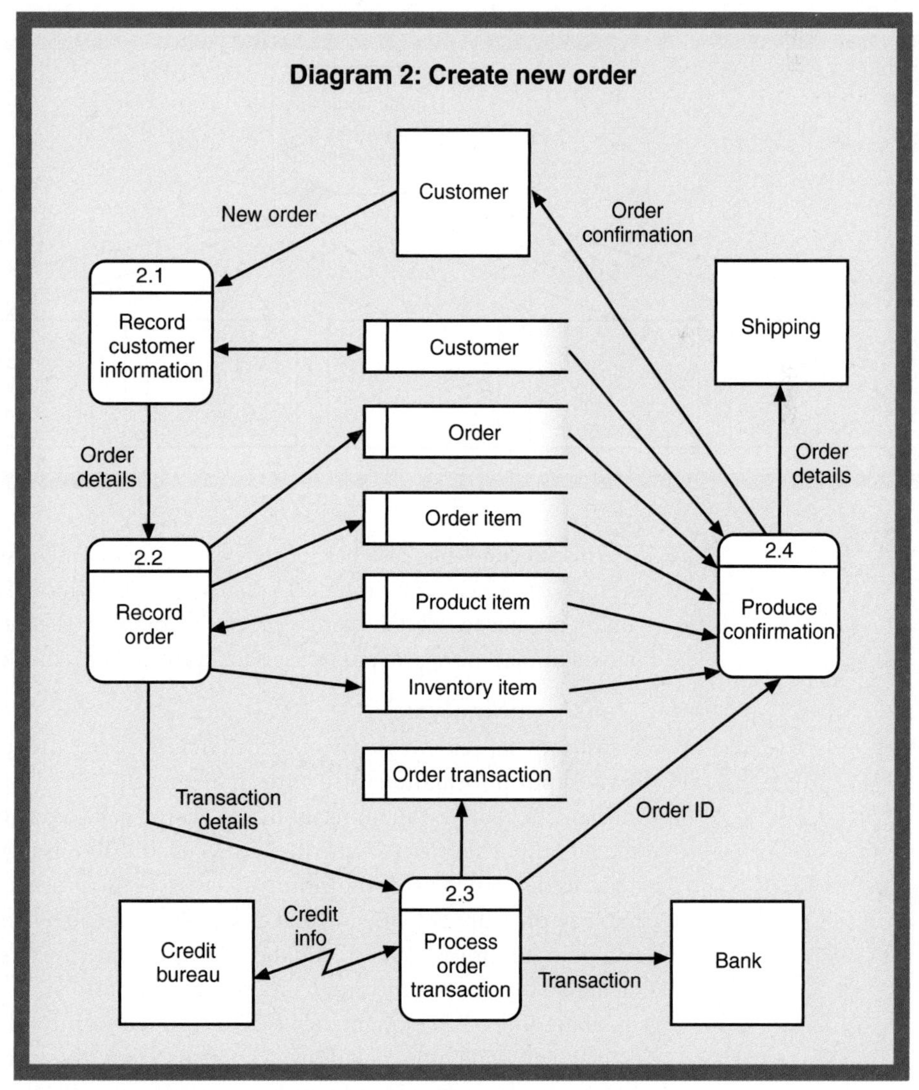

exploded view for that activity, which is used for transform analysis. The structure chart is developed directly from the data flow diagram. The fundamental idea is that the detailed diagram processes from the data flow diagram become the leaf modules in the structure charts. The mid-level processes—that is, the processes that were exploded to derive the low-level processes—become the intermediate-level boss modules on the structure chart. Thus, the hierarchy of the structure chart directly reflects the organization of the set of nested, or leveled, data flow diagrams. Additional boss modules might need to be developed to provide the correct structure to the structure chart.

As stated previously, the general form of a structure chart developed with transform analysis is input-process-output. The method to develop a structure chart from a data flow diagram fragment consists of the following steps:

afferent data flow
the incoming data flow from a sequential set of DFD processes

efferent data flow
the outgoing data flow in a sequential set of DFD processes

central transform
set of DFD processes that are located between the input and output processes

1. Determine the primary information flow. This flow is the main stream of data that is transformed from some input form to the output form.
2. Find the process that represents the most fundamental change from an input stream to an output stream (see Figure 10-13). The input data stream is called the **afferent data flow**. The output data stream is called the **efferent data flow**. The center process is called the **central transform**.
3. Redraw the data flow diagram with the input to the left and the output to the right. The central transform process goes in the middle. If this diagram is an exploded-view data flow diagram, add the parent process to the diagram. You can omit nonprimary data flows to simplify the drawing. An example of this redrawn data flow diagram is shown in Figure 10-13.

Figure 10-13
Rearranged view of the *Create new order* DFD

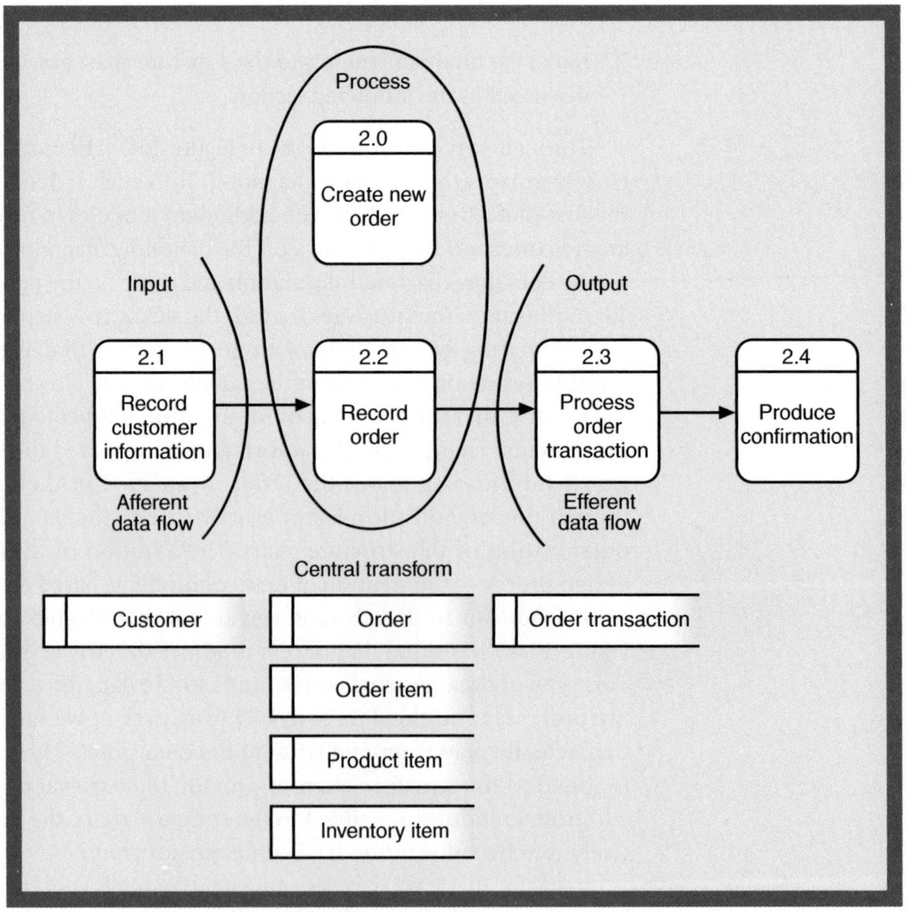

Figure 10-14

First draft of the structure chart

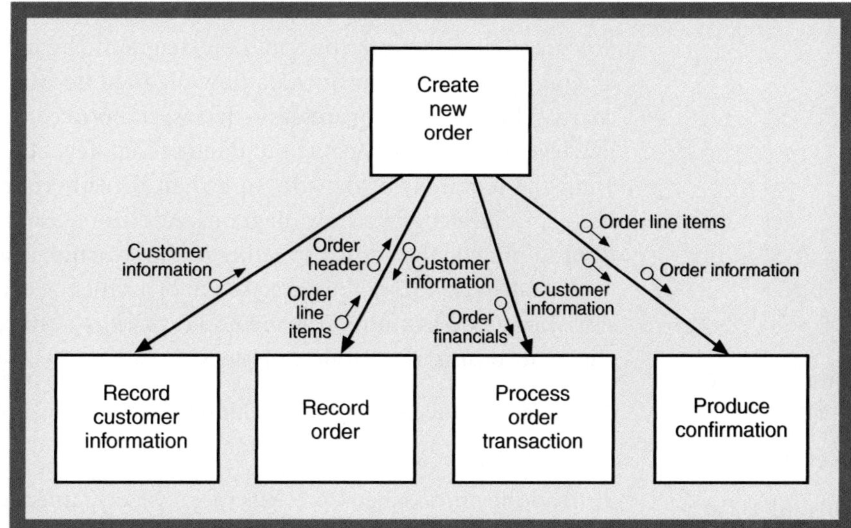

4. Generate the first-draft structure chart, based on the redrawn data flow, with the calling hierarchy and the required data couples. An example of this diagram is shown in Figure 10-14.

5. Add other modules as necessary to get input data via the user-interface screens, read from and write to the data stores, and write output data or reports. Usually, these modules are lower-level, or utility, modules. Add the appropriate data couples based on the data flows to and from these data stores.

6. Using any structured English or decision table documentation as a basis, add other required intermodule relationships such as looping and decision symbols.

7. Make the final refinements to the structure chart based on the quality-control concepts discussed in the following section.

Through step 4, as you can see in Figure 10-14, the organization of the structure chart very closely mirrors the data flow diagram from which it derives. In step 5, we begin to enhance the first-draft structure chart with additional modules to read and write data. Frequently, there are no corresponding processes on the data flow diagram. Thus, at this point we depend less on the data flow diagram information and more on the requirements of a good design. Figure 10-15 illustrates the structure chart for the next step—step 5.

Comparing Figure 10-14 with 10-15, notice that Figure 10-14 indicates that all the input information comes from the far-left module, *Record customer information*. In Figure 10-15, observe that the accessing of information has been distributed across other branches of the structure chart. Customer information is retrieved through the far-left branch, but additional information about the order is retrieved in the second branch of the chart. Even though this organization is not exactly true to the data flow diagram, it is a more logical organization of the structure chart. The addition of these data access modules is truly a design process—the creation of new components based on systems design principles.

In addition to distributing the access of customer input data, the structure chart in Figure 10-15 has other data access modules to retrieve product and inventory information. This type of data retrieval corresponds to the data flows on the data flow diagram between the processes and the data stores. During design, we must explicitly identify the modules that actually read from and write to the data stores. The module *Get product/inventory items* is added to the structure chart to provide the retrieval of the product information. As the additional modules are added to the structure chart, the data couples are defined more precisely to reflect this more detailed design structure.

In Figure 10-15, we have also added the symbols concerning looping and optional calls. The black diamond indicates that the call to create a customer record is optional and, in fact, is required only if the customer is a new customer. The general form of the structure chart has

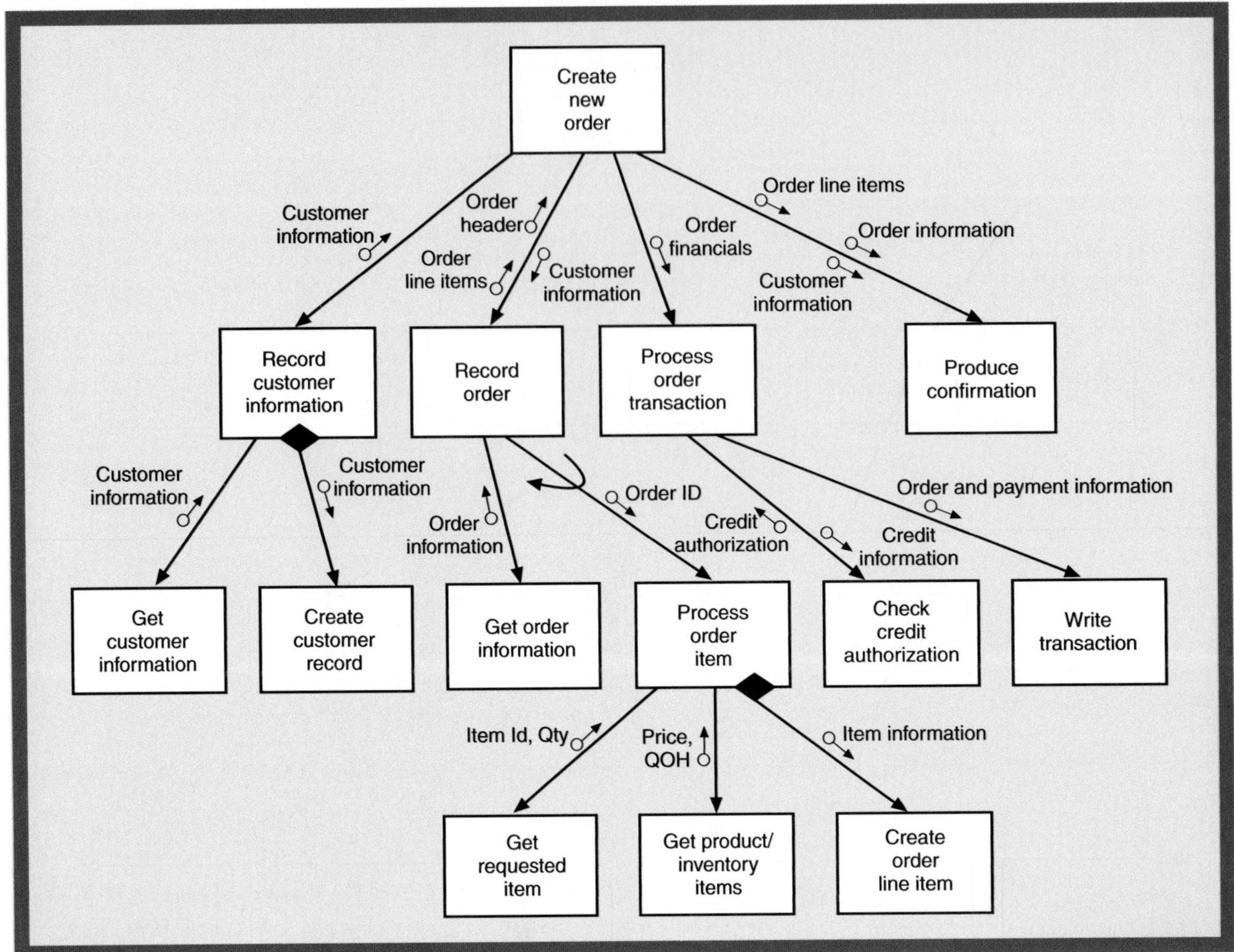

Figure 10-15

The structure chart for the *Create new order* program

the inputs to the left and the outputs to the right. The black diamond on the call to *Create order line item* indicates a situation in which an item is out of stock, so the call to *Create order line item* is conditional.

The high-level boss module, *Create new order*, and its tree of modules can be plugged into the transaction structure chart in Figure 10-10. Figure 10-16 illustrates the process of combining the top-level structure chart, developed using transaction analysis, with the lower-level subtrees, developed with transform analysis.

EVALUATING THE QUALITY OF A STRUCTURE CHART

module coupling
the manner in which modules relate to each other; the preferred method is data coupling

module cohesion
a measure of the internal strength of a module

The process of developing structure charts from DFDs can become rather involved. Rules and guidelines can be used to test the quality of the final structure chart, however. Two measures of quality are **module coupling** and **module cohesion**. Generally, it is desirable to have highly cohesive and loosely coupled modules.

The principle of coupling is a measure of how a module is connected to the other modules in the program. The objective is to make modules as independent as possible because a module that is independent can execute in almost any environment. An independent module has a well-defined interface of several predefined data fields, and it passes back a well-defined result in predefined data fields. The module does not need to know who invoked it and, in fact, can be invoked by any module that conforms to the input and output data structure. The best coupling is through simple data coupling. In other words, the module is called, and a specific set of data items is passed. The module performs its function and returns the output data items. This kind of module can be reused in any structure chart that needs the specific function performed.

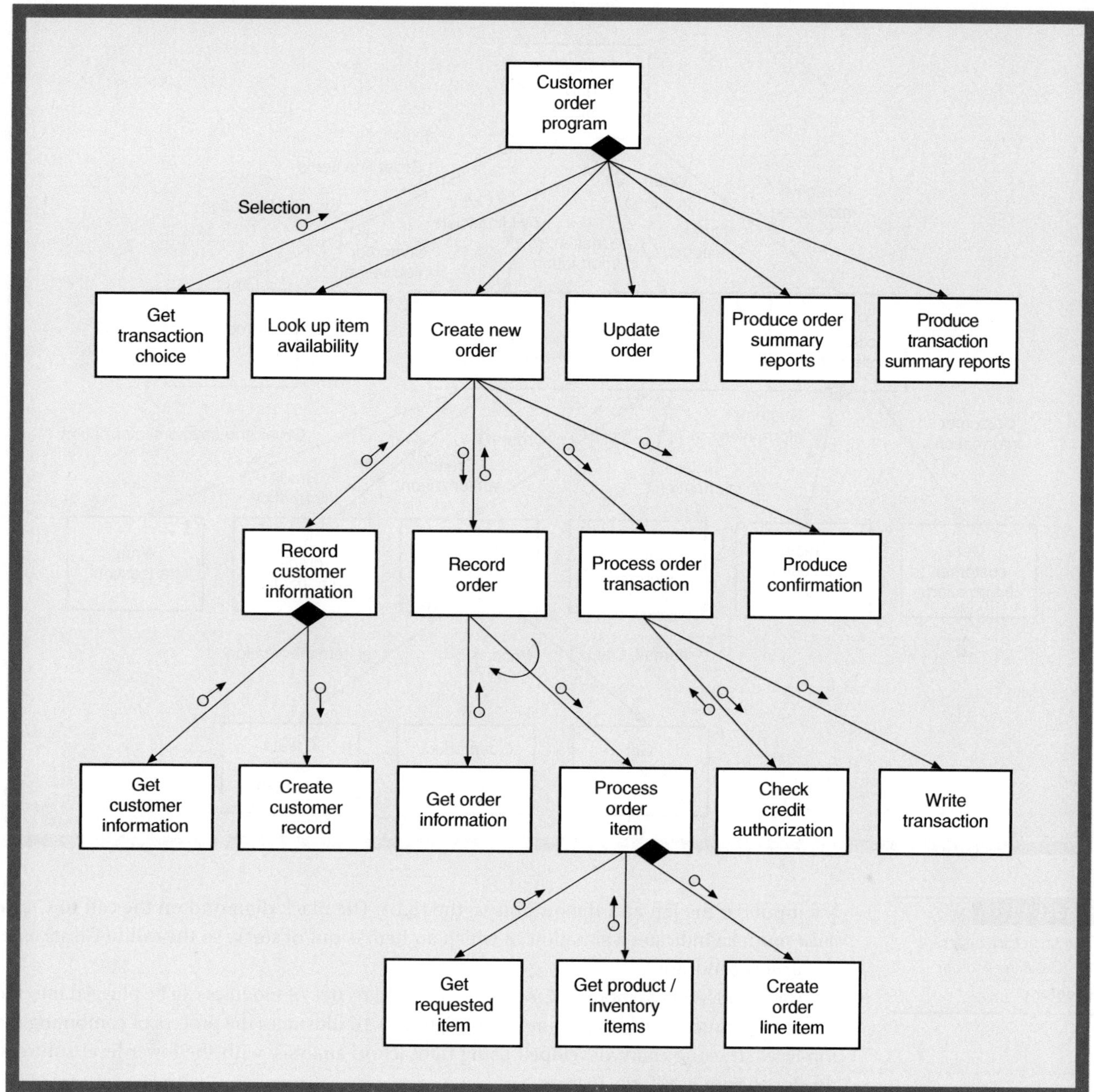

Figure 10-16

Combination of structure charts (data couple labels are not shown)

Cohesion refers to the degree to which all of the code within a module contributes to implementing one well-defined task. Modules with high cohesion implement a single function. All of the instructions within the module are part of that function, and all are required for the function. Modules with low (or poor) cohesion implement multiple or loosely related functions.

Note that the amount of coupling and the specific data items being passed are good indicators of the degree of module cohesion. Modules that implement a single task tend to have relatively low coupling because all of their internal code acts on the same data item(s). Modules with poor cohesion tend to have high coupling because loosely related tasks typically are performed on different data items. Thus, a module with low cohesion generally has several unrelated data items passed by its superior.

BEST PRACTICE

Coupling and cohesion are also the two key design goals for object-oriented design, in which objects are loosely coupled and each object is highly cohesive.

A flag passed down the structure chart is also an indicator of poor cohesion in the lower-level module. Flags passed into a module are used typically to select the part of the recipient module's code that will be executed. Part a of Figure 10-17 shows an example of poor cohesion. A project team can improve cohesion by partitioning the module into separate modules, one for each value of the flag, as shown in part b of Figure 10-17. The code of the superior module is programmed to use the flag to decide which of the partitioned subordinate modules to call.

Figure 10-17

Examples of module cohesion

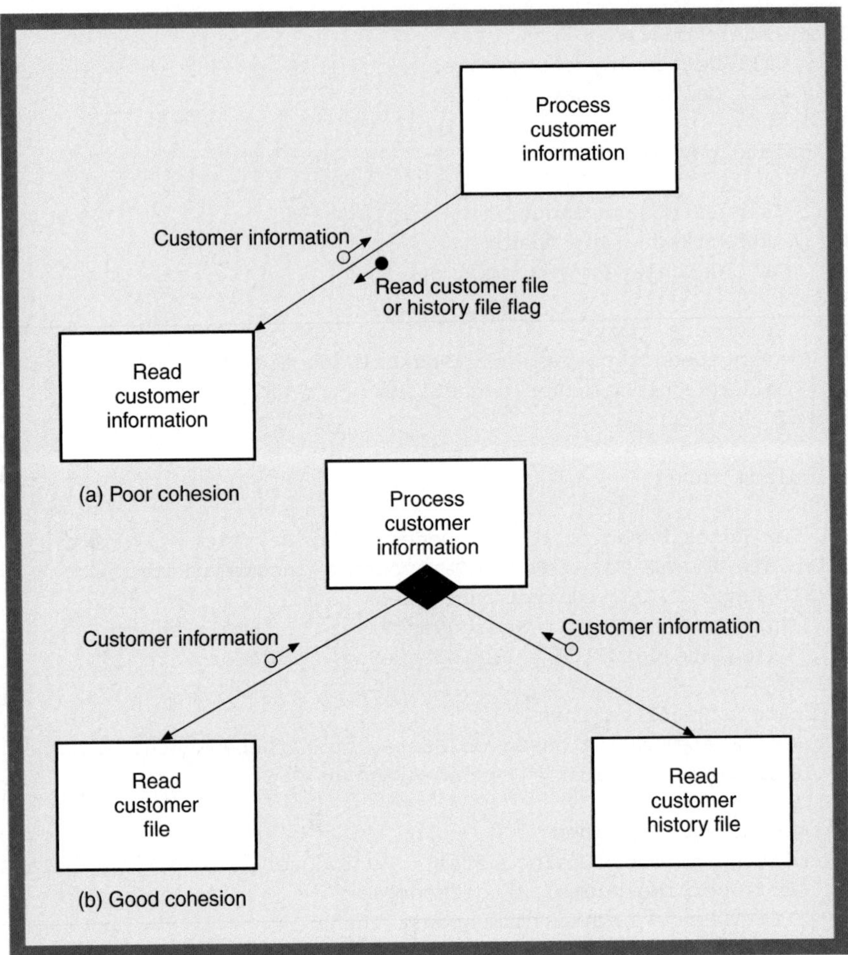

MODULE ALGORITHM DESIGN: PSEUDOCODE

The previous two models, the system flowchart and the structure chart, provide the overall structure of the system and the structure within each program. The next requirement of design is to describe the internal logic within each module. Three common methods are used to describe module logic: flowcharts, structured English, and pseudocode. All three methods are equivalent in their ability to describe logic. Flowcharting is a visual method that uses boxes and lines to describe the flow of logic in a program. In the early days of computing, flowcharting was used almost exclusively. Today, however, versions of pseudocode and structured English have replaced flowcharting. You learned about structured English in Chapter 6. Pseudocode is a variation of structured English that is closer to a programming language. Frequently, analysts write pseudocode using statements that are very similar to the target language. If they are writing to COBOL, they use COBOL-like syntax. If they are writing in Visual Basic or C, they use a syntax that mirrors those languages.

Figure 10-18 shows a simple example of the logic of the payroll system. Pseudocode statements for the *Payroll program*, *Calculate amounts*, *Calculate pay amounts*, and *Calculate taxes* modules are shown. This figure shows examples of each of the three types of control statements used in structured programming: sequence, a sequence of executable statements; decision, if-then-else logic; and iteration, do-until or do-while.

```
Payroll program
DoUntil No more time cards
     Call Enter time cards
     Call Calculate amounts
     Call Output payroll
End Until

Calculate amounts
     Call Get employee pay rates
     Call Calculate pay amounts

Calculate pay amounts

Call Calculate base amount
If (HoursWorked > 40) Then
     Call Calculate overtime amount
End If
Call Calculate taxes
If (SavingsDeduction=yes) or (MedicalDeduction=yes) or (UnitedWay=yes) Then
     Call Calculate other deductions
End if

Calculate taxes

Get Tax Rates based on Number Dependents, Payrate
Calculate Income Tax = PeriodPayAmount * IncomeTaxRate
If YTD Pay < FICA MaximumAmount Then
     Calculate EmpFICA = PeriodBasePay * FICAEmployeeRate
     Calculate CorpFICA = PeriodBasePay * FICACorpRate
End If
If StateTaxRequired Then
     Get StateTaxRate based on State, NumberDependents, Payrate
     Calculate StateTax = PeriodPayAmount * StateTaxRate
End If
If OvertimePay > 0 Then
     Calculate OvertimeIncomeTax = PeriodOvertimePay * IncomeTaxRate
     Add OvertimeIncomeTax to Incometax
     If YTDPay < FICAMaximum Amount Then
          Calculate EmpOvertimeFICA = PeriodOvertimePay * FICAEmployeeRate
          Calculate CorpOvertimeFICA = PeriodOvertimePay * FICACorpRate
     End If
     If StateTaxRequired Then
          Calculate StateOvertimeTax = PeriodOvertimePay * StateTaxRate
     End If
End If
```

Figure 10-18

Pseudocode for the *Calculate pay amounts* hierarchy

INTEGRATING STRUCTURED APPLICATION DESIGN WITH OTHER DESIGN TASKS

So far, you've learned how to develop a structure chart based on the information in a data flow diagram. The primary focus was on capturing the information in the structural relationship between the processes. The structure chart developed from either transaction analysis or transform analysis will be correct, but it may not be complete. Before the structure chart can be considered complete, it usually must be modified or enhanced to integrate the design of the user interface, the database, and the network, as Barbara Halifax discusses in the accompanying RMO memo. Because user-interface and database design are discussed in later chapters, this section and the next section on three-layer design only briefly address the types of changes that need to be made.

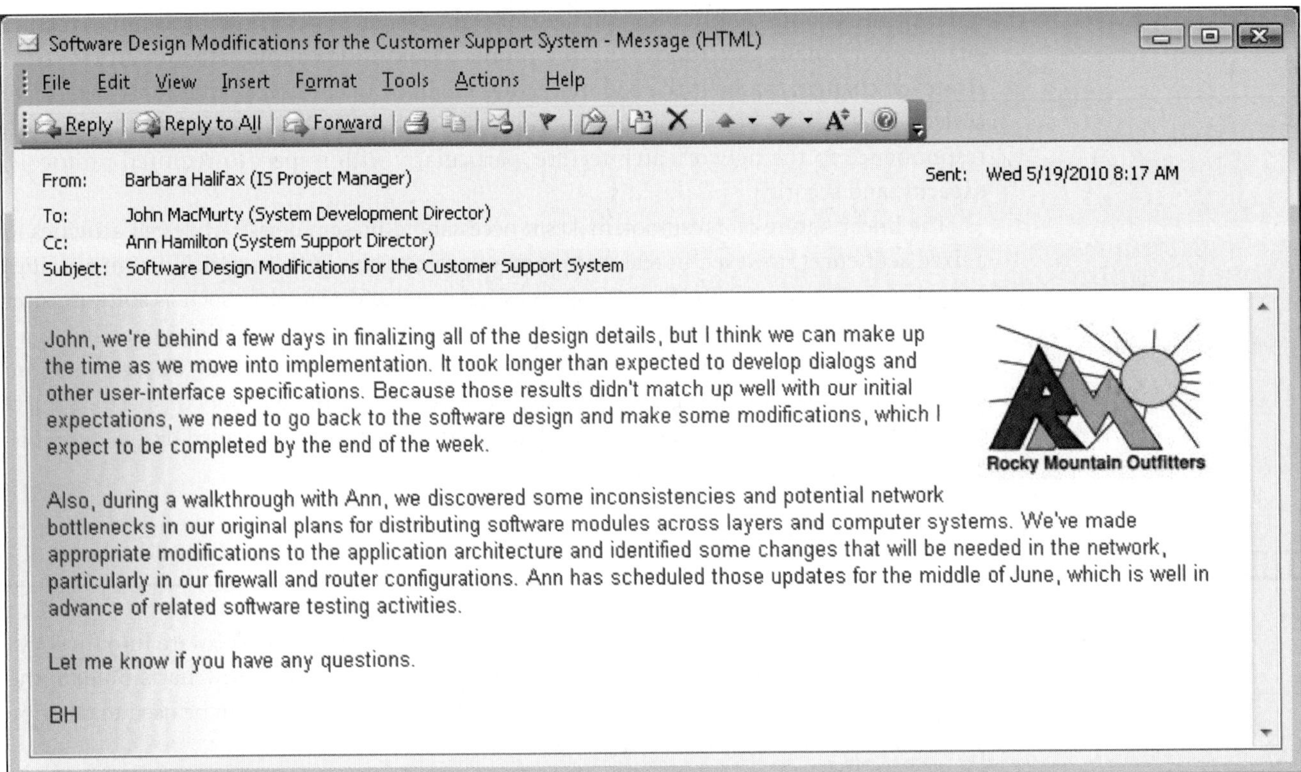
Software Design Modifications for the Customer Support System - Message (HTML)

From: Barbara Halifax (IS Project Manager)
Sent: Wed 5/19/2010 8:17 AM
To: John MacMurty (System Development Director)
Cc: Ann Hamilton (System Support Director)
Subject: Software Design Modifications for the Customer Support System

John, we're behind a few days in finalizing all of the design details, but I think we can make up the time as we move into implementation. It took longer than expected to develop dialogs and other user-interface specifications. Because those results didn't match up well with our initial expectations, we need to go back to the software design and make some modifications, which I expect to be completed by the end of the week.

Also, during a walkthrough with Ann, we discovered some inconsistencies and potential network bottlenecks in our original plans for distributing software modules across layers and computer systems. We've made appropriate modifications to the application architecture and identified some changes that will be needed in the network, particularly in our firewall and router configurations. Ann has scheduled those updates for the middle of June, which is well in advance of related software testing activities.

Let me know if you have any questions.

BH

The user interface consists of a set of input forms, output forms, and reports. Interactive user interfaces are usually based on a dialog between the user and the system and include a series of input and output forms. Every form must be displayed and the data retrieved somewhere in a module in the structure chart. As these forms are developed, the structure chart needs to be evaluated from three aspects:

- Are additional user-interface modules needed?
- Does the pseudocode in the interface modules need to be modified?
- Are additional data couples needed to pass data?

In earlier chapters, you learned that the entity-relationship diagram (ERD) must be consistent with the data stores found on the data flow diagrams. There is not necessarily a one-to-one correspondence between data stores and database tables, but the information on every data store must be somewhere in the database. In addition, every database table and field must be represented by a data store somewhere. During design, the project team performs this same type of analysis and makes appropriate changes to the structure chart.

The same three aspects—modules, pseudocode, and data couples—need to be evaluated for the database. If a database management system is being used, a common interface is usually provided. The designer can make the database accessible either by calling a database interface module or by embedding SQL (Structured Query Language) statements within the pseudocode.

Finally, the structure charts and system flowcharts must be checked for correspondence to the existing or planned network architecture. Because architectural design normally precedes or runs concurrently with detailed software design, system flowcharts and structure charts are normally developed with the proper assumptions about network architecture. However, changes can be made and new issues might be uncovered as additional details are added to the design. Thus, an important final step in detailed design is to reevaluate its correspondence to the network architecture, particularly with respect to required protocols, capacity, and security.

The linear nature of a textbook makes it necessary to present details of design activities in a fixed sequence. However, in real development projects, the order in which design activities are performed varies greatly. Some projects assign detailed design tasks such as software, user interface, system interface, and database design to multiple teams operating in parallel. If an iterative approach to the SDLC is used, detailed design tasks are completed for each iteration. Other projects may follow a more linear sequence due to lack of personnel or specific project characteristics. In projects that follow a relatively linear order, early detailed design decisions of all types must be reevaluated after later design tasks are completed.

THREE-LAYER DESIGN

Chapter 9 described three-layer design and its division of application software into the view, business logic, and data access layers. Structure charts and system flowcharts predate the development of three-layer architecture by at least a decade. Still, they can be used to describe design decisions and software structuring based on three-layer architecture.

Figure 10-19 shows a system flowchart for the RMO Customer order program. The flowchart divides processing according to a three-layer architecture of view layer, business logic layer, and data layer, as described in Chapter 9. Each layer communicates using well-defined protocols as noted on the flowchart, which enables the layers to be located on different machines, if desired. As described in Chapter 9, the choice of protocols, such as HTTPS and SQL, and the choice of deployment environments, such as Microsoft Internet Explorer, Java components, and an Oracle database management system (DBMS), are architectural design decisions made early in the design activities. Annotating a system flowchart with specifics of the deployment environment is one way of documenting the decision and communicating important constraints to other project participants.

However, a system flowchart doesn't necessarily describe where software layers execute. In Figure 10-19, the view layer is described as executing on a client workstation, but no locations are given for the business logic and data layers. They could also execute on the client workstation (an unusual arrangement), both could execute on a single server, or each could execute on a separate server or cluster of servers. Unless specific location information is included, a system flowchart should be assumed to describe the distribution of processing functions across programs or groups of modules, not across computer systems.

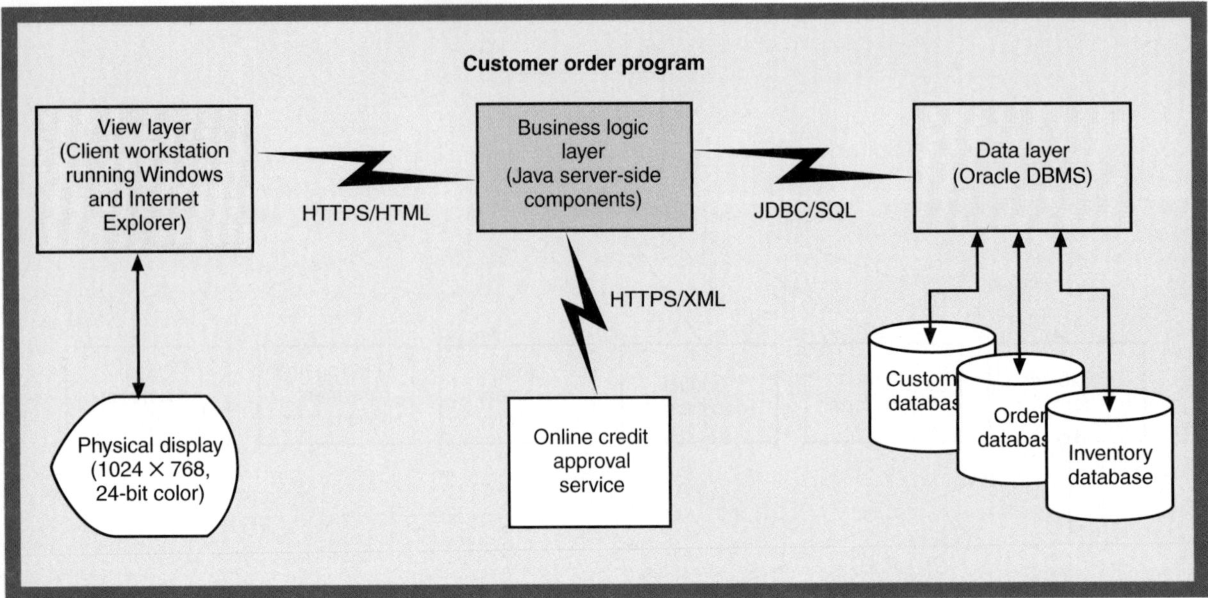

Figure 10-19
A system flowchart showing three-layer architecture for the Customer order program

Describing software structure with structure charts under a three-layer architecture can be quite different from the examples shown earlier in the chapter, depending on the deployment environment and development tools. As a point of comparison, consider the structure charts in Figures 10-15 and 10-16. Structure charts were developed in an era when application programs were functionally all encompassing. That is, they included modules to handle all operational aspects of the application task, including input/output (view layer), business logic, and interaction with data stored in files or databases (data layer). Figure 10-16 describes just such a program, as might be implemented in a traditional programming language such as C or COBOL and deployed in a centralized hardware architecture.

Figure 10-20 shows an alternate structure chart for a three-layer architecture for creating an order, based on a form-oriented dialog that will be discussed in Chapter 14. View layer modules are shown in yellow, business logic layer modules in red, and data layer modules in green. All view layer modules are shown, but only the business logic modules called by the Customer form are shown. Other view layer modules would require similar business logic layer modules. The data layer is composed of a DBMS, and business logic layer modules include code to generate appropriate database access commands and process responses.

Figure 10-21 shows the RMO Customer form used to find, add, or update customer data, as depicted on the structure chart. Figure 10-22 shows how some of the code attached to the Customer form might be implemented in the Visual Basic programming language to represent the view layer. The event procedure btnSearch_Click() executes when the user clicks the Search button. The CustomerID number typed into the form is passed to one of the functions in a code module representing the business logic layer. The view layer does not include the details of the customer search.

Figure 10-23 shows a template for the business logic layer function RetrieveExistingCustomer() that is called by the btnSearch_Click() event procedure. It indicates the code insertion point for database retrieval statements. Note that this function handles retrieval of existing customer data but not data entry for new customer data or database updates of new or existing customer data. Additional functions or procedures shown on the structure chart below the Customer form would also be included in the business layer to handle this functionality. The example demonstrates a clean division of program code between the view and business logic layers, even though they might execute within the same program on a single machine.

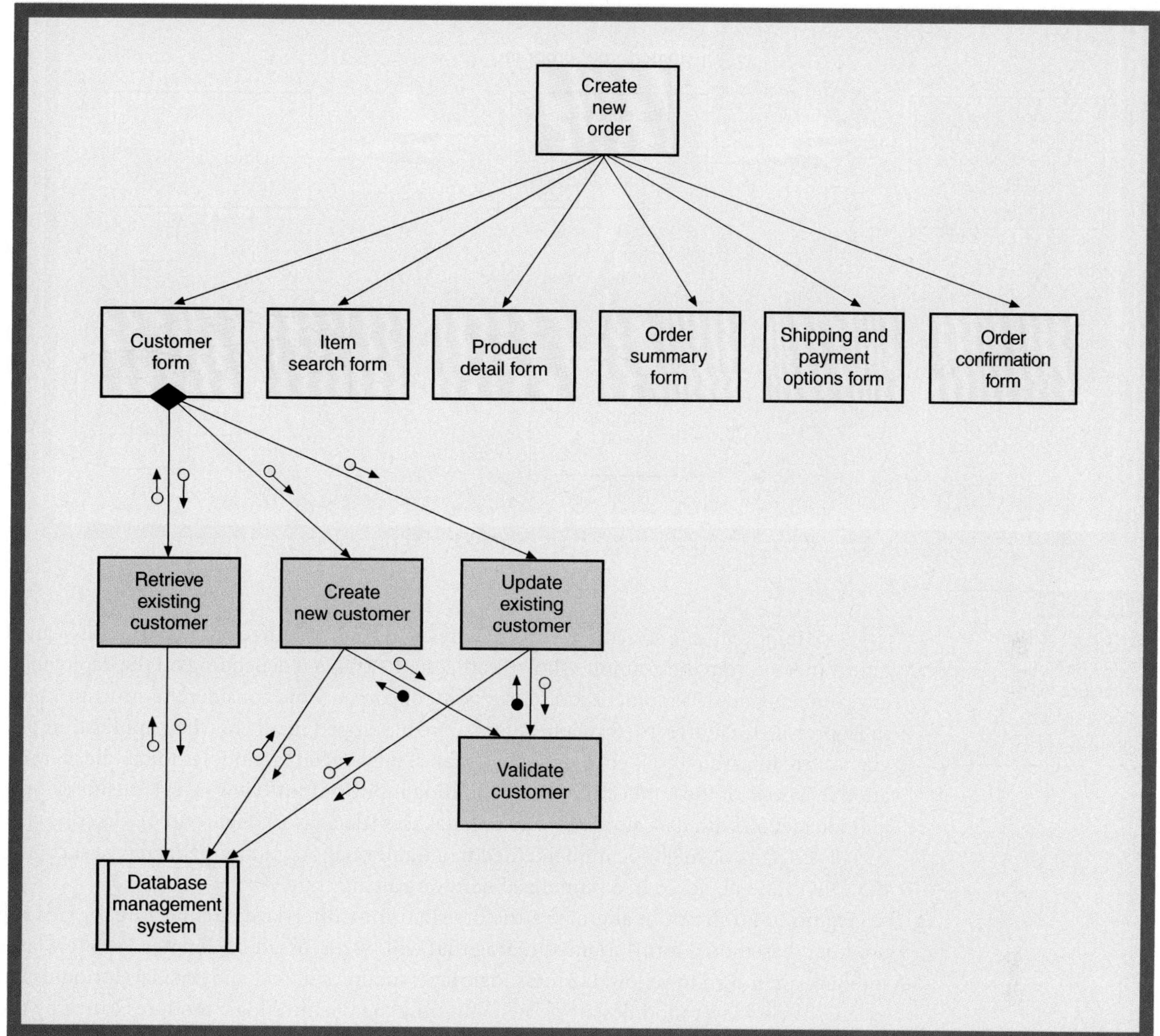

Figure 10-20

A structure chart showing three-layer architecture for the activity *Create new order*

When layers are distributed across multiple computer systems, programs are more specialized and numerous than in centralized architectures. Rather than combining user interface, business logic, and data access modules within a single program and structure chart, distributed three-layer architecture employs multiple programs. Some layers, such as the view layer, may not even be programs in the traditional sense, such as when a Web-based HTTP user interface is used. The layers that are written in traditional or OO programming languages are usually separate programs. For example, in a Web-based system, the top two layers of the structure chart in Figure 10-20—the yellow modules—would be implemented as a set of Web pages. The third-level modules *Retrieve existing customer*, *Create new customer*, and *Update existing customer*—the red modules—would probably be separate programs, stored on an application or Web server and executed via calls from a Web page. Thus, the single structure chart in Figure 10-20 would be decomposed into several smaller structure charts.

Note that independent programs executing on different computer systems can't communicate using function or procedure parameters represented as data couples on a system flowchart. Instead, modules in different layers communicate over real-time links using well-defined protocols. That form of communication is represented on the system flowchart, as shown in Figure 10-19. The exact format and content of messages passed among layers must be specified within module pseudocode or elsewhere.

Figure 10-21

A simple RMO form to find, add, or update Customer data

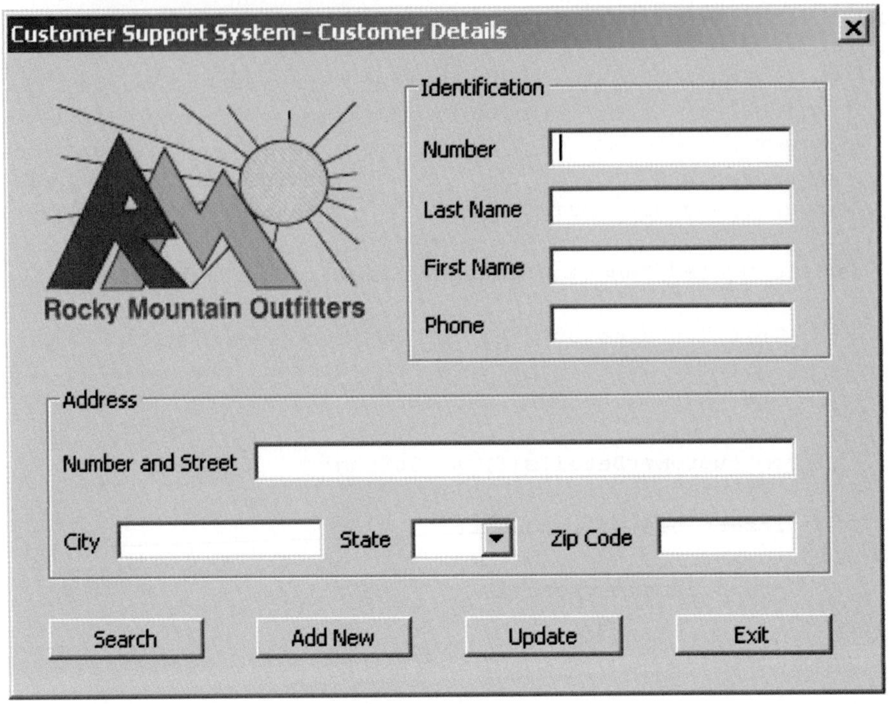

Customer Support System - Customer Details

Rocky Mountain Outfitters

Identification

Number

Last Name

First Name

Phone

Address

Number and Street

City State Zip Code

Search Add New Update Exit

Figure 10-22

Visual Basic code for the form shown in Figure 10-21

```
Public Class CustomerForm
    Inherits System.Windows.Forms.Form

Windows Form Designer generated code

    Private Sub btnSearch_Click(ByVal sender As System.Object, _
    ByVal e As System.EventArgs) Handles btnSearch.Click

        'when btnSearch is clicked, call the code module
        'in the Business Logic Layer using customerID
        'and get back array of customer details

        Dim custID As String
        Dim customerDetails(6) As String
        custID = txtCustomerID.Text
        Try
            customerDetails = OrderEntryModule.RetrieveExistingCustomer(custID)
            txtLastName.Text = customerDetails(0)
            txtFirstName.Text = customerDetails(1)
            txtStreetAddress.Text = customerDetails(2)
            txtCity.Text = customerDetails(3)
            txtState.Text = customerDetails(4)
            txtZipCode.Text = customerDetails(5)
            txtPhone.Text = customerDetails(6)
        Catch
            MessageBox.Show("Customer not found")
        End Try

    End Sub
```

```
Public Class OrderEntryModule

    'This module contains functions and procedures that are
    'called from event procedures on forms

    'It represents part of the Business Logic Layer for shared
    'order entry processing code.

    Public Shared Function RetrieveExistingCustomer(ByVal anID As String) As String()

        'This function queries the database based on the
        'customer ID to search for an existing customer
        'and then it returns the customer details

        Dim customerDetails(6) As String

        'Insert needed code here

        Return customerDetails

    End Function

    'Continue with other order entry functions and procedures
    '. . .

End Class
```

Figure 10-23

A Visual Basic code template to search for an existing customer

We've now covered architectural and detailed software design using traditional models and methods. In the next chapter, we'll cover those same design tasks using object-oriented tools and techniques. After that, we'll turn our attention to the remaining design activities—database, user interface, and system interface design.

SUMMARY

For the traditional structured approach to systems design, the primary input is the data flow diagram. The data flow diagram is first enhanced by the addition of a system boundary. The designer sketches the system boundary to show the overall system. He or she also sketches the boundary on the DFD fragments to show program boundaries at a lower level.

The designer describes processes within each DFD boundary using one or more structure charts. The designer develops structure charts using transaction analysis, transform analysis, or both. Transaction analysis is appropriate for the upper structure chart levels of a system that processes multiple input or transaction types. Transform analysis is appropriate for designing programs that transform a single transaction from its input form to an output. Structure charts may also be based on three-layer architecture, in which case modules will be clearly identified by layer and the structure chart may be decomposed into smaller structure charts if layers will execute on multiple computer systems.

A structured design may also include system flowcharts and module pseudocode. System flowcharts show the movement of data among programs, files, and manual processing steps, thus providing an overall view of an entire system. System flowcharts can also describe the interaction between layers of a multilayered system. Module pseudocode describes the internal logic of a structure chart module.

KEY TERMS

afferent data flow, p. 367
central transform, p. 367
computer program, p. 354
data couples, p. 362
efferent data flow, p. 367
module, p. 354
module cohesion, p. 369
module coupling, p. 369
program call, p. 361
pseudocode, p. 354
structure chart, p. 360
system flowchart, p. 354
transaction analysis, p. 364
transform analysis, p. 365

REVIEW QUESTIONS

1. Explain the relationship and differences between a module and a program.
2. What is the purpose of the automation system boundary? How do you develop one?
3. What is a system flowchart used for?
4. What symbols are used on a system flowchart?
5. What is the purpose of a structure chart?
6. What are the symbols used on a structure chart?
7. Explain *transaction analysis*.
8. Explain *transform analysis*. What is meant by the term *central transform*?
9. What is the difference between afferent and efferent data flow?
10. Explain module coupling and module cohesion. Why are these concepts important?
11. Describe how structure charts for three-layer architecture are different from those for all-encompassing programs that execute on a single computer system.

THINKING CRITICALLY

1. Given the data flow diagram shown in Figure 10-24, do the following: (a) draw a system boundary; (b) divide the DFD into program components such as real-time, monthly, daily, periodic, and so forth; and (c) draw a system flow-chart based on the division into program components.

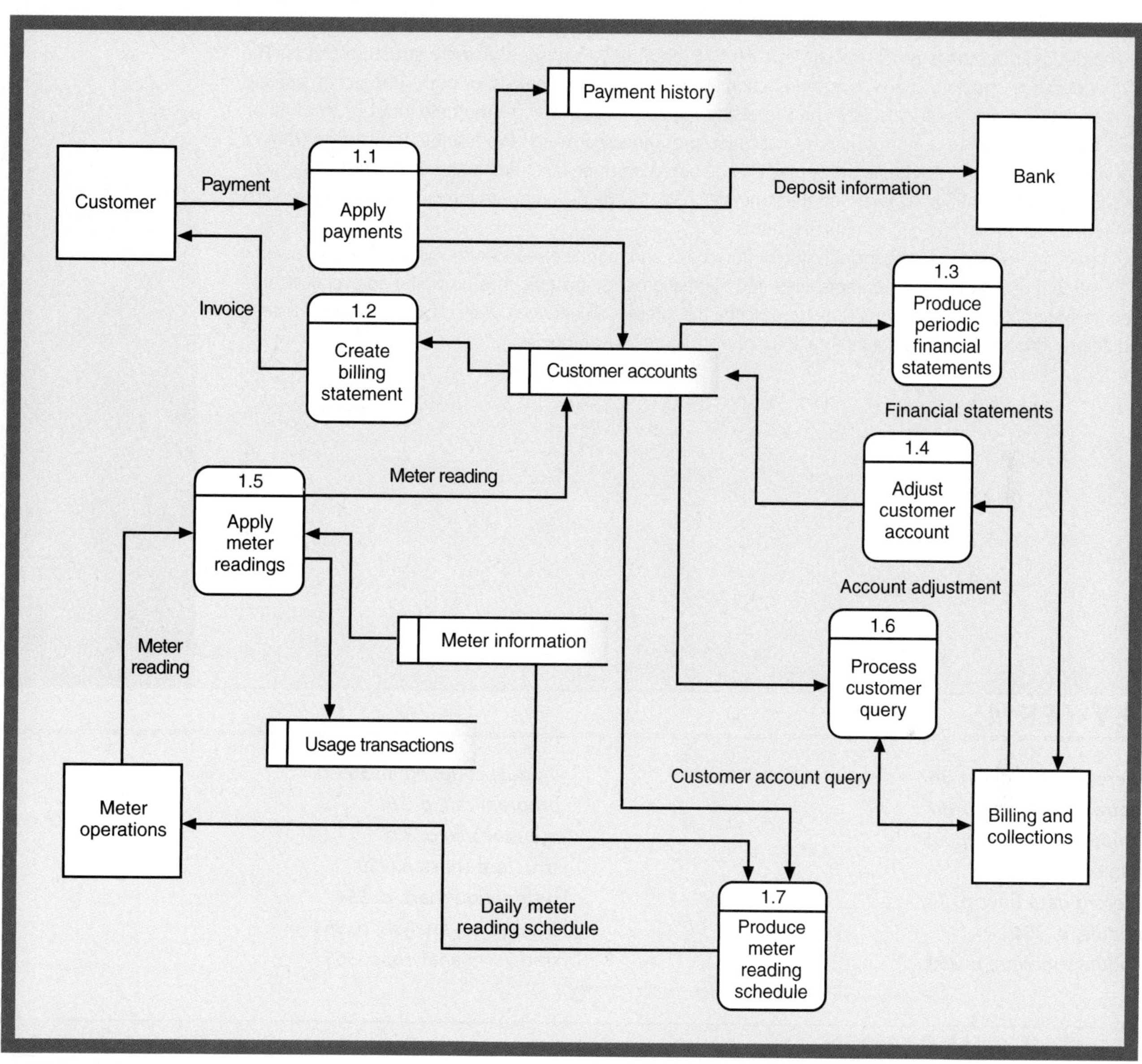

Figure 10-24

Electric company customer billing

2. Given the data flow diagram shown in Figure 10-25, and using transaction analysis, develop a structure chart.

Figure 10-25

Student registration program

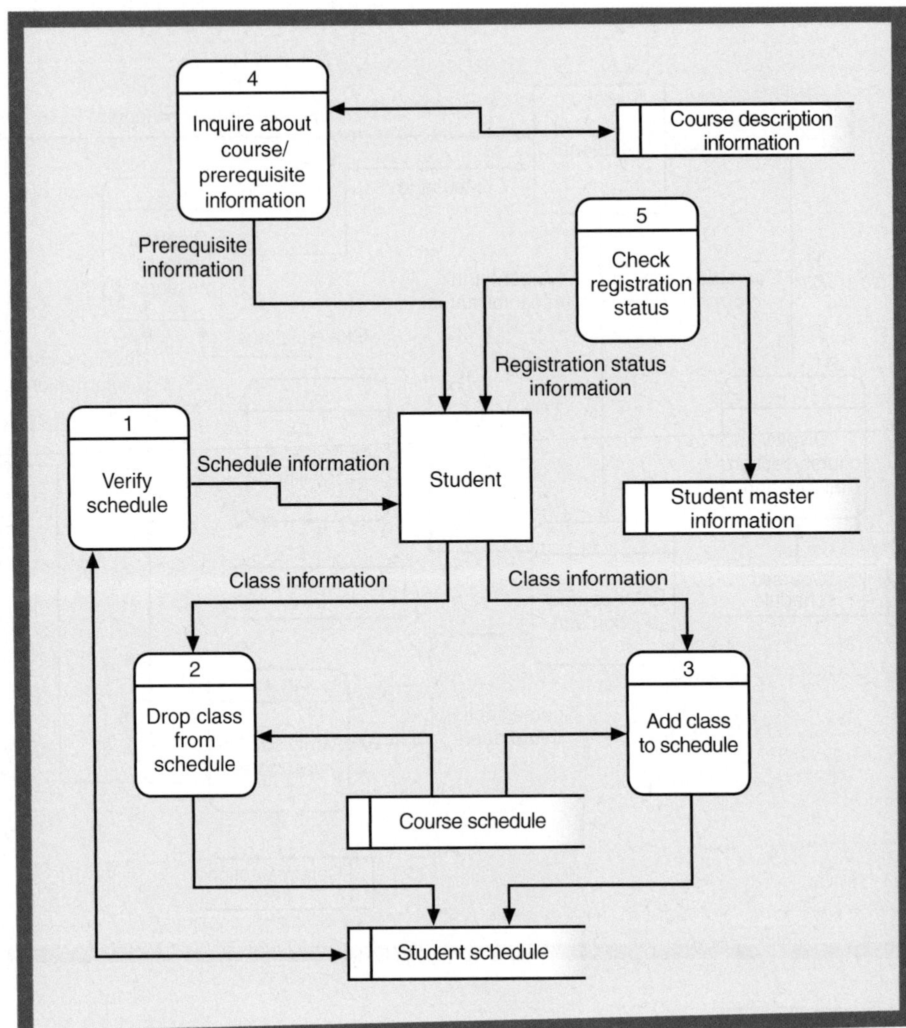

3. Given the data flow diagram shown in Figure 10-26, and using transform analysis, develop a structure chart.

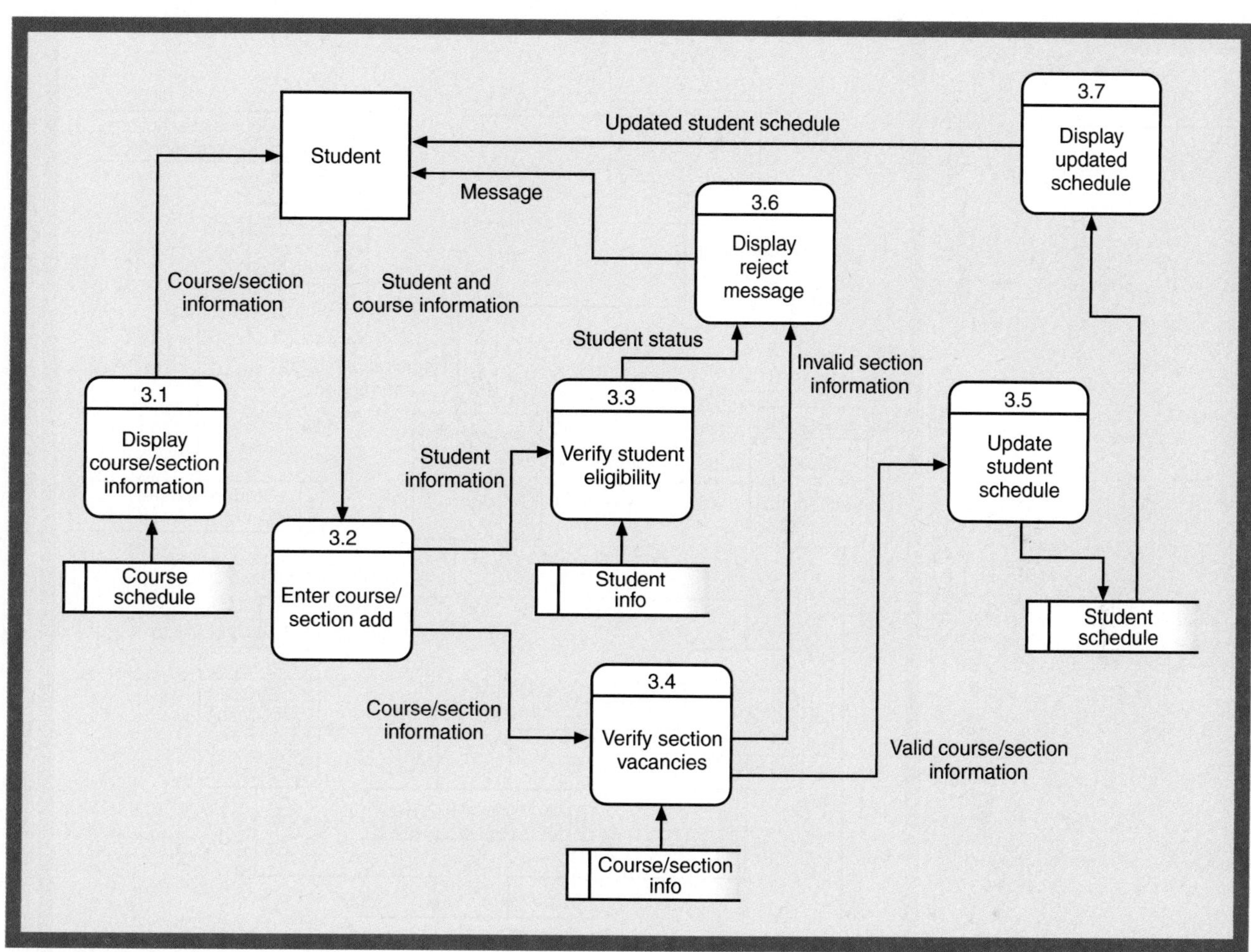

Figure 10-26

Explosion of *Add class to schedule*

4. Integrate the structure charts from problems 2 and 3 into a single structure chart.

5. Given the data flow diagram shown in Figure 10-27, and using transform analysis, develop a structure chart.

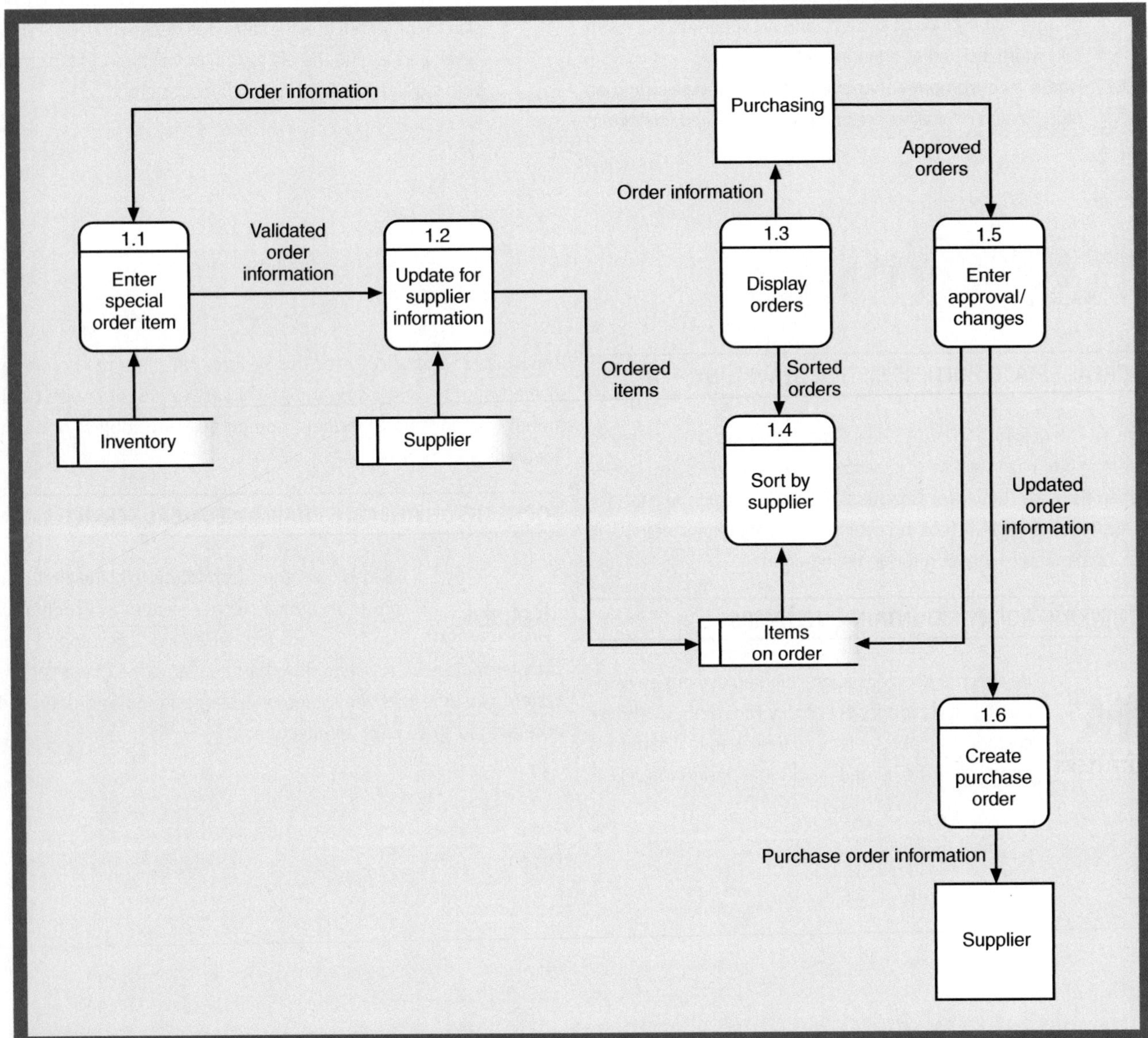

Figure 10-27

Special-order purchasing

6. Finish developing business logic layer modules for the view layer modules in Figure 10-20.

EXPERIENTIAL EXERCISES

1. Discuss the hierarchical nature of traditional structured design. What kinds of systems and architectures are naturally more inclined to a hierarchy?
2. Find a local company that is updating or redeveloping an older system originally developed with structured techniques including structured design. How does the development team plan to use (or not use) the structured design models? Will the developers extend the old system or will they (re)implement both old and new functions with modern tools?

CASE STUDIES

THE REAL ESTATE MULTIPLE LISTING SERVICE SYSTEM

Refer to the description of the Real Estate Multiple Listing Service system in the case studies of Chapter 5 and the DFDs you developed in the case studies for Chapter 6. Develop a structure chart for the system. Follow the steps indicated in this chapter, including any additional modules required for accessing data.

RETHINKING ROCKY MOUNTAIN OUTFITTERS

Review the decisions about the deployment environment and design for the Rocky Mountain Outfitters customer support system, as described in Chapters 8 and 9, and the related traditional design models in this chapter. Specifically for this system, what are the comparative advantages and disadvantages of software design with traditional methods and models compared with object-oriented methods and models?

FOCUSING ON RELIABLE PHARMACEUTICAL SERVICE

Based on the description of the Reliable Pharmaceutical Service system in Chapters 5 and 6 and the DFDs you developed for Chapter 6, develop a system flowchart and structure charts for the system. Assume that the system will be designed and deployed according to three-layer architecture.

FURTHER RESOURCES

Tom DeMarco, *Structured Analysis and System Specification.* Yourdon Press, 1979.

Meilir Page-Jones, *The Practical Guide to Structured Systems Design*, (2nd ed.). Yourdon Press, 1988.

Edward Yourdon, *Modern Structured Analysis.* Yourdon Press, 1989.

Edward Yourdon and Larry L. Constantine, *Structured Design: Fundamentals of a Discipline of Computer Program and Systems Design*. Prentice Hall, 1979.

CHAPTER 11

OBJECT-ORIENTED DESIGN: PRINCIPLES

LEARNING OBJECTIVES

After reading this chapter, you should be able to:

- Explain the purpose and objectives of object-oriented design
- Develop package diagrams and component diagrams
- Develop design class diagrams
- Use CRC cards to define class responsibilities and collaborations
- Explain the fundamental principles of object-oriented design

CHAPTER OUTLINE

Object-Oriented Design: Bridging from Analysis to Implementation

Object-Oriented Architectural Design

Fundamental Principles of Object-Oriented Detailed Design

Design Classes and the Design Class Diagram

Detailed Design with CRC Cards

Fundamental Detailed Design Principles

NEW CAPITAL BANK: PART 1

Despite some hiccups at the beginning of the project, things seemed to be under control now. Bill Santora, the project leader responsible for developing an integrated customer account system at New Capital Bank, had just finished a technical review of the new system's first-cut design with the review committee. This first-cut design focused on six core use cases, which had been chosen as the most fundamental to the business and would be implemented in the first development iteration.

New Capital Bank had been using object-oriented languages for quite a while, but it had been slower to adopt object-oriented analysis and design techniques. Bill had been involved in some early pilot projects that had used the Unified Process (UP) and the Unified Modeling Language (UML) to develop systems using object-oriented techniques. However, this development project was his first large-scale project that would be entirely object oriented.

As Bill collected his presentation materials, his supervisor, Mary Garcia, spoke. "Your technical review went very well, Bill. The committee found only a few minor items that need to be fixed. Even though I am not completely current on the new object-oriented techniques, it was easy for me to understand what you presented and how these core functions will work. I still find it hard to believe that you will have these six pieces implemented in the next few weeks."

"Wait a minute," Bill said, laughing. "It won't be ready for the users then. Getting these six core functions coded and running doesn't mean that we are almost done. This project is still going to take a year to complete."

"Yes, I know. But it is nice that we will have something to show after only two months. Not only do I feel more confident in this project, but the users love to see things developing."

"I know. Remember how much grief I got when I originally laid out this plan based on an iteration approach? It was difficult to detail the project schedule for the later iterations, so I had a hard time convincing everybody that the project schedule was not too risky. The upside is that because each iteration is only six weeks long, we have something to show right at the beginning. You don't know how relieved I am that the design passed the review! The team has done a lot of work to make sure the design was solid, and we all felt confident. It is good to get confirmation, though. And we really will have some basic pieces of the new system working in two or three more weeks."

"Well, building it incrementally makes a lot of sense and certainly seems to be working. I especially liked the diagrams you showed. It was terrific how you showed that the three-layer architectural design supported each use case. Even though I do not consider myself an advanced object-oriented technician, I could understand how the object-oriented design fit into the architecture. I think you wowed everybody when you demonstrated how you could use the same basic design to support both our internal bank tellers and a Web portal for our customers. Congratulations."

Bill's response reflected Mary's enthusiasm. "How about the design class diagrams? Don't they give a nice overview of the classes and the methods? We use them extensively as a focus for discussion on the team. They really help the programmers write good, solid code."

"By the way, have you scheduled a review with the users?" Mary asked.

"No, not yet. The architectural design is mostly technical stuff, and we are not quite ready to meet with the users. We next need to move into the detailed design. We will have some design meetings with the users when we start doing detailed design. However, even then most of the work is very technical. The users will help us by verifying our understanding of the information availability, but much of what we do is too technical for them to follow.

"We will design and then code six core use cases in this first development iteration," Bill continued. "We have scheduled six weeks for this first iteration. I hope it is enough time. We will be crunched to get it all done by then. But, of course, it will be nice to show the users some working pieces at that time. Then we will have another round of meetings with users to let them verify our work and to begin work on the next iteration."

"I am excited to see the first pieces run. It just makes so much sense to be able to test these core functions during the rest of the project. Let me congratulate you again," Mary said as she and Bill headed off to lunch together.

OVERVIEW

In Chapter 7 you learned how to do object-oriented analysis by developing functional requirements models. You learned that analysis consists of two parts—discovery and understanding. Understanding is taking the information gleaned from user interviews and constructing a set of interrelated and comprehensive models. Model building is an essential part of understanding the user needs and how they influence the proposed system. However, remember that the objective of analysis models is not to describe the new system, but only to understand, in precise terms, the requirements.

Chapter 9 then introduced you to the concepts of logical design. Figure 9-6 identified the system components that require detailed design specifications, and Figure 9-8 showed the detail design models used for object-oriented design. The focus of this chapter and the next is primarily how to develop these object-oriented design models, which are then used by the programmers to code the system.

This chapter focuses on two levels of design—architectural design, often referred to as high-level design, and detailed design, where the design of each use case is specified. The chapter starts by teaching the models and processes required to develop an overall architectural structure for the new system. Two types of model diagrams are used: component diagrams and deployment diagrams. This chapter takes the basic ideas you learned in Chapter 9 about the deployment environment, and extends them by teaching you how to do architectural design using the appropriate models.

In the latter part of the chapter, you will begin learning the process of detailed design. The discussion first explains design class diagrams, which are an extension of the problem domain class diagram with design information added. Next we explain Class-Responsibility-Collaboration (CRC) cards to begin teaching the details of use case centered, object-oriented design.

The chapter ends with an important discussion of design principles for good object-oriented design. Throughout this chapter and the next, we are concerned not only with teaching the basics of object-oriented design, but with teaching foundation principles so that the systems you build are well structured and maintainable. The design principles will provide you with a solid foundation for designing systems correctly.

OBJECT-ORIENTED DESIGN: BRIDGING FROM ANALYSIS TO IMPLEMENTATION

So what is object-oriented design? It is a process by which a set of detailed object-oriented design models are built and then used by the programmers to write and test the new system. System design is the bridge between user requirements and programming the new system. One strength of the object-oriented approach is that the design models are often just extensions of the requirements models. Obviously, it is much easier to extend an existing model than to create entirely different models for design. However, we emphasize that it is a good practice to create design models and not just jump into coding. Just as a builder does not build something larger than a doghouse or a shed without a set of blueprints, a system developer would never try to develop a large system without a set of design models. Students who are building personal Web pages or small systems for course assignments sometimes think that design models are unnecessary. Remember, however, that blueprints may not be necessary for a doghouse, but they certainly are for a home.

One tenet of the new adaptive approaches to development is to create models only if they have meaning and are necessary. Sometimes new developers misinterpret this guideline to mean they do not need to develop design models at all. The design models may not be formalized into a comprehensive set of documents and diagrams, but the models they develop are certainly necessary. Developing a system without doing design is comparable to writing a research paper without an outline. You could just sit down and start writing; however, if you want a paper that

is cohesive, complete, and comprehensive, you should write an outline first. You could write a complex paper without an outline, but in all probability it would be disjointed, hard to follow, and missing important points—and it would earn a low grade! The outline can be jotted down on paper, but the process of thinking it through and writing it down allows the writer to ensure that it is cohesive. Systems design provides the same type of framework.

One important point about adaptive approaches is that requirements and design are done incrementally within iterations. So, a complete set of design documents is not developed at one time. The requirements for a particular use case or several use cases may be developed, and then the design documents are developed for that use case. Immediately following the design of the solution, the programming can be done. Some people call this "just in time" system design.

OVERVIEW OF OBJECT-ORIENTED PROGRAMS

Before going further, let's quickly review how an object-oriented program works. Then we will discuss the design models and how they must be structured to support object-oriented programming.

An object-oriented program consists of a set of program objects that cooperate to accomplish a result. Each program object has program logic and any necessary attributes encapsulated into a single unit. These objects work together by sending each other messages and working in concert to support the functions of the main program.

Figure 11-1 depicts how an object-oriented program works. The program includes a window object that displays a form in which to enter student ID and other information. After the student ID is entered, the window object sends a message (number 2) to the Student class to tell it to create a new student object (instance) in the program, and to go to the database, get the student information, and put it in the object (message 3). Next, the new Student object sends the information back to the window object to display it on the screen. The student then enters the updates to her personal information (message 4), and another sequence of messages is sent to update the Student object in the program and the student information in the database.

Figure 11-1

Object-oriented event-driven program flow

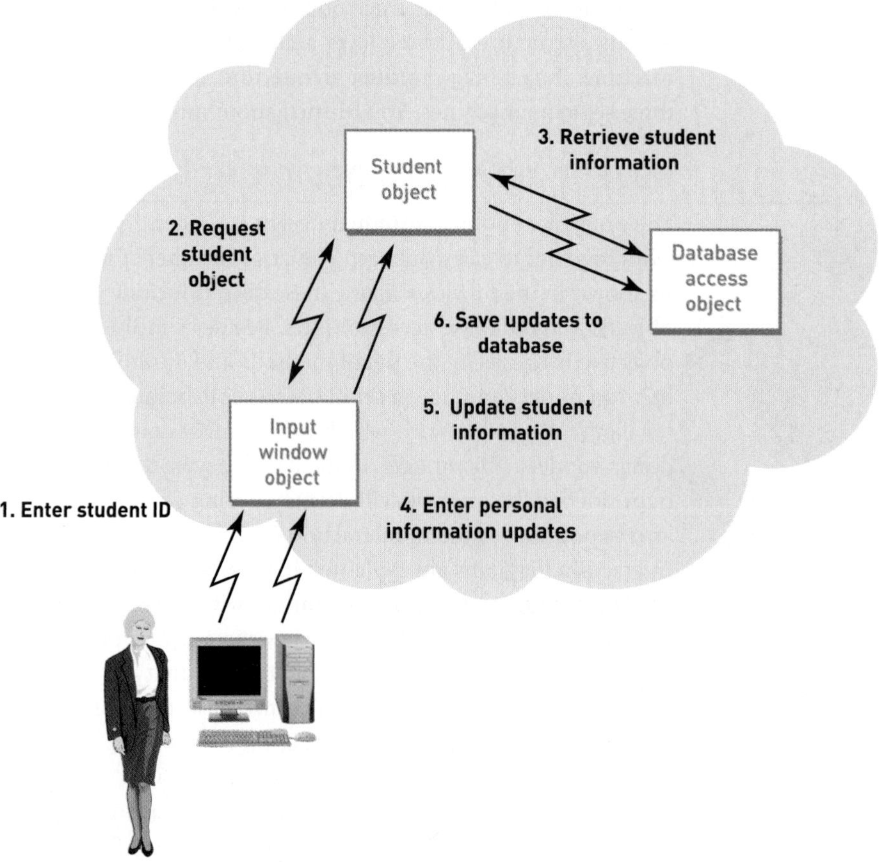

One common question about object-oriented programs is, "Who is in charge?" In a structured program, it is obvious who the boss module is and who controls the computing. In an object-oriented program, it is not as obvious. In fact, no one may be in charge. Yes, one program gets things started, but once the program is executing, no particular module or object has to be in charge.

Let's compare traditional structured programs to object-oriented programs through an analogy—a computer analogy, but one with which you may be familiar. A mainframe computer has a massive amount of computing capability. It may be connected to thousands of work terminals, and it controls them all. No individual terminal does work unless the main computer directly instructs it to do so. The mainframe also does all the database access and execution. This system is much like a traditional structured program.

In contrast, a network of personal computers consists of many individual computers connected by network cables. Each computer has its own capabilities, but if you are working at one, you can communicate with another on the network through a message and ask for assistance. For example, some individual computers on the network have large disk drives and special databases that you can access. These computers are called file servers. Other resources such as printers are also on the network. Typically, in most PC networks you have encountered, there is no single, coherent purpose for the individual computers. But sometimes it is desirable to have many computers working together for a specific purpose. Again, they can work together by sending messages to each other to fulfill the overall computing objective. This system is how an object-oriented system or program is designed.

An object-oriented system consists of sets of computing objects. Each object has data and program logic encapsulated within itself. Analysts define the structure of the program logic and data fields by defining a class. The class definition describes the structure or a template of what an executing object looks like. The object itself does not come into existence until the program begins to execute. This is called an **instantiation** of the class, or making an instance (an object) based on the template provided by the class definition.

instantiation

creation of an object based on the template provided by the class definition

Figure 11-1 illustrates three objects in this simple program execution. Each object also represents a structure of three-layer architecture. The three objects need not exist on the same machine. In fact, in a multitier architecture, the three classes of objects will generally exist on three separate machines. You learned about multitier architectures in Chapter 9.

OBJECT-ORIENTED DESIGN MODELS AND PROCESSES

The objective of object-oriented design is to identify and specify all of the objects that must work together to carry out each use case and where they reside in different computing nodes. As shown in the previous figure, these objects include user interface objects, problem domain objects, and database access objects. Besides simply identifying the classes, another design objective is to specify the detail methods and attributes within the classes so that a programmer can understand how a set of objects collaborate to execute a use case.

Figure 11-2 illustrates which requirements models are directly used to develop which design models. The models on the left side were developed during analysis, and those on the right side are the ones we will develop during design. As you might infer from the number of arrows pointing to them, interaction diagrams are the core diagrams used for detailed design. Interaction diagrams are explained in detail in Chapter 12.

At this point you should be familiar with the requirements models on the left side of the diagram. Let's take a minute to review the purpose of each model. The domain model class diagram identifies all the classes, or "things," that are important in the problem domain; for example, to the system users. The use case diagram identifies the elementary business processes that the system needs to support—in other words, all the ways users want to use the system to carry out procedures or processing goals. The set of activity diagrams are used to document the internal workflow of each use case. An activity diagram shows the individual steps necessary to carry out a particular use case. The system sequence diagram is closely

Figure 11-2

Design models with their respective input requirements models

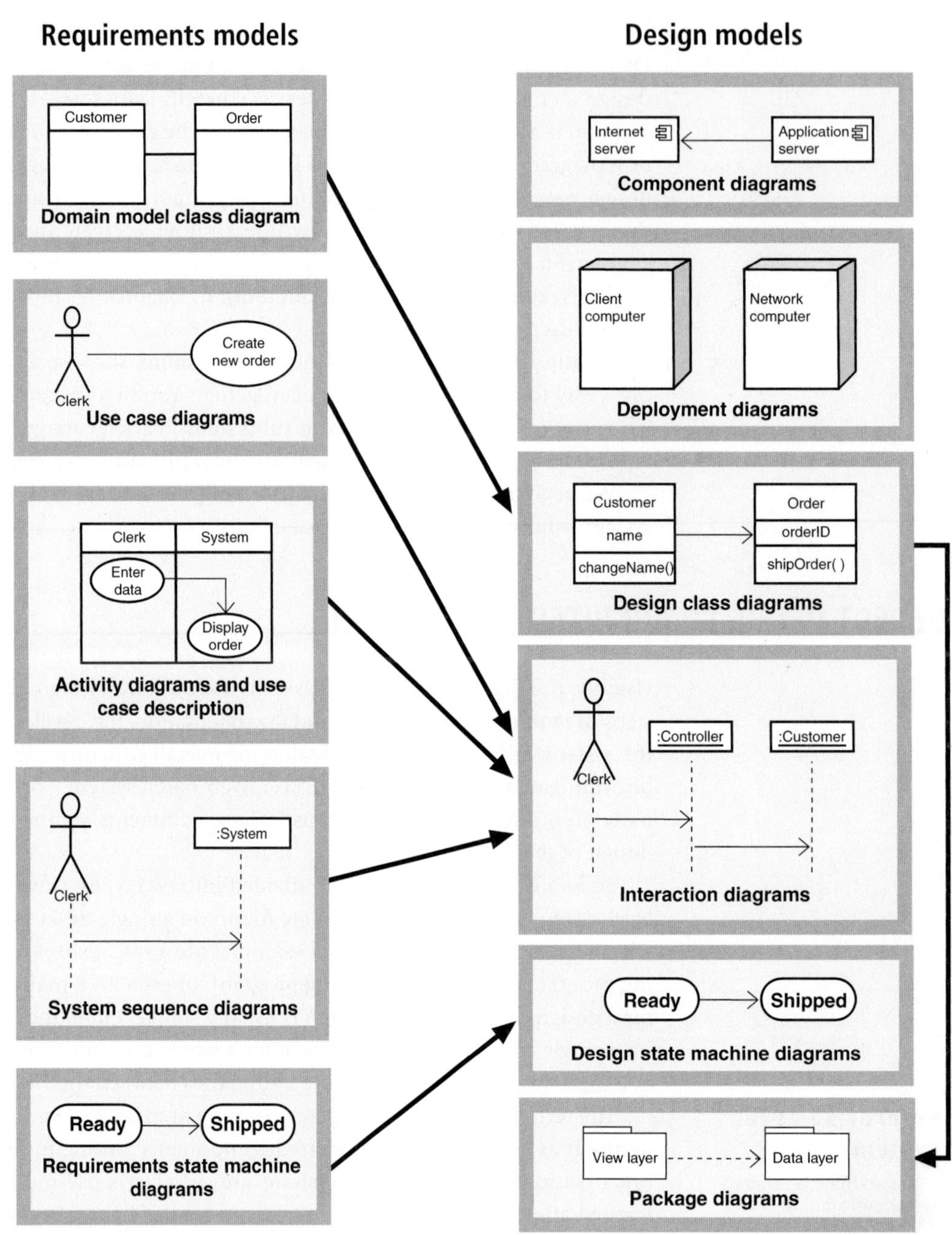

related to an activity diagram, except that it shows the messages or data that is sent back and forth between the user and the system during the steps of the use case. Finally, the state machine diagram keeps track of all status conditions for one particular class. It also shows the business rules that control the changing of one state (status condition) to another.

The right side of the figure shows the design models. Architectural design is one of the first steps in system design, inasmuch as it provides the big picture and overall structure of the new system. At the top of the right column are the component diagram and the deployment diagram. You learned a bit about the deployment environment in Chapter 9, but without using official UML notation. This chapter explains how to draw component and deployment diagrams. You also learned a little about multilayer software design in Chapter 9. This chapter explains how to use these two diagrams to document the architectural design of the software system.

Moving down the right column of Figure 11-2, we next see a design class diagram, or DCD, which is an expansion of the domain model class diagram. It is used to document the design elements of software classes, as you will learn later in this chapter. The next type of diagram is an interaction diagram, which can be either a sequence diagram or a communication diagram. You learned how to build system sequence diagrams (SSDs) in Chapter 7. The design version of sequence diagrams is much more detailed, and is used to carry out much of the detailed design activity. You will learn how to create interaction diagrams in the next chapter. State machine diagrams are also used by programmers to develop the detailed class methods. You already learned about those in Chapter 7. The design version is similar to the version used during analysis.

Finally, the square at the bottom of the column shows package diagrams, which are simply a way to group design elements together. You first saw a package symbol in Figure 7-5, when use cases were grouped into subsystems. Package diagrams can be used to group any type of design elements, but we will use them primarily to group design classes.

We begin system design by first thinking about the overall structure of the new solution system—the architecture of the system.

OBJECT-ORIENTED ARCHITECTURAL DESIGN

Usually, the first step in system design is architectural design. In most cases during the early steps of requirements gathering and documentation, the developers begin to think about how the system will be deployed and what the overall structure will look like. It is normal at the beginning of a project to say, "This is a Web-based system," or "This will only be used internally on our network and desktops." Those comments are the beginning of the architecture design of the solution system.

Software systems are generally divided into two types: single-user systems and enterprise-level systems. Single-user systems are found on a single desktop, or execute from a server but without sharing resources. Typical examples are a spreadsheet program, an engineering drawing program, a simple accounting program, or even an e-mail client program. The architectural design of a single-user system is usually simple. Often there is only one layer, and it runs on a single computer. However, even for a single-use system, it is wise to develop the system as a multilayer program so that the boundaries between the various levels are well defined.

enterprise-level system
a system that has shared resources among multiple people or groups in an organization

The term **enterprise-level system** can mean many different things. For our purposes, we define it as a system that has shared components among multiple people or groups in an organization. Enterprise-level systems almost always use multiple tiers of computers. You learned about n-layer or n-tiered architectures in Chapter 9. A typical example of this architecture is an internal networked client/server environment in which the client computers contain the view and domain layer programs, and the data access layer is on a central server. Characteristic of enterprise-level systems, the database and data access are on a central server because it is a shared resource throughout the organization. This configuration is a three-layer, two-tiered system. Because the central database is shared across the enterprise, it is placed on a central server that all users of the application program can share. Because local client computers are often powerful Macintosh or personal computers, both the view layer and domain logic can be executed locally.

Our definition of an enterprise-level system is a broad one. Two major categories of systems fit this definition in relation to systems design: (1) client/server network-based systems and (2) Internet-based systems. You may find that many people only think of the second category when they talk about enterprise-level systems, because so much new development is being done for the Web. Remember, however, that the broader definition is equally valid.

These two methods of implementing enterprise-level systems have many similar properties. Both require a network, both have central servers, and both have the view layer on the

client machines. However, some fundamental differences also exist in the design and implementation of these two approaches. The primary difference is in how the view layer interacts with the domain and data access layers. As developers, we must be able to distinguish between these two types of systems, because we must consider important design issues.

Figure 11-3 identifies three fundamental differences that affect the architectural design of the system—state, client configuration, and server configuration. The concept of state relates to the permanence of the connection between the client view layer and the server domain layer. If the connection is permanent, as in a client/server system, values in variables can be passed back and forth and are remembered by each component in the system. The view layer has direct access to the data fields in the domain layer. For example, data in an order, such as all of the line items and their prices, is displayed in the forms.

Figure 11-3

Differences between client/server and Internet systems

Design Issue	Client/Server Network System	Internet System
State	"Stateful" or state-based system, e.g., client/server connection is long term	Stateless system, e.g., client/server connection is not long term and has no inherent memory.
Client Configuration	Screens and forms that are programmed are displayed directly. Domain layer is often on the client or split between client and server machines.	Screens and forms are displayed only through a browser. They must conform to browser technology.
Server Configuration	Application or data server directly connects to client tier.	Client tier connects indirectly to the application server through a Web server.

In a stateless system, such as the Internet, the client view layer does not have a permanent connection to the server domain layer. The Internet was designed so that when a client requests a screen via a URL address typed in the browser, the server sends the appropriate document, and then the two disconnect. In other words, the client does not know the state of the server, and the server does not remember the state of the client. This transient connection makes it difficult to implement such things as an order in a shopping cart. To add more permanence to the stateless environment, Web designers have developed other techniques, such as cookies, session variables, and XML data transmission. As a systems designer, you must consider these additional components when designing an Internet enterprise-level system.

Concerning client configuration, the client side of a network-based system contains the view layer classes and often the domain layer classes. Formatting, displaying, and event processing within the screens are all directly controlled by the view layer and domain layer program logic. There is great flexibility in the design and programming of these electronic screens. The view layer classes and domain layer classes can communicate directly with each other. Even if the domain layer is split across tiers, a permanent communication link can be established—all under the program's control.

In an Internet-based system, all electronic screens are displayed by a browser. The formatting, displaying, and event processing all must conform to the capabilities of the browser being used. Special techniques and tools, such as scripting languages, applets, and style sheets, have been developed to simulate the network-based capability. However, as a designer, you will need to design for the environment.

The server configuration in a network-based system consists of data access layer classes and sometimes domain layer classes. These classes collaborate through direct communication and access to each other's public methods. In an Internet-based system, all communications from the client tier must go through the HTTP server. Communication is not direct, and methods and program logic are invoked indirectly through passed parameters. This indirect technique of accessing domain layer logic is more complex and requires additional care in designing the system.

The next two sections more precisely define both component diagrams and deployment diagrams. Component diagrams focus on logical components, while deployment diagrams focus on physical components. You will learn the UML notation used for both diagrams and how they are used to carry out the process of architectural design.

COMPONENT DIAGRAMS AND ARCHITECTURAL DESIGN

component diagram
a type of implementation diagram that shows the overall system architecture and the logical components within it

application program interface or API
the set of public methods that is available to the outside world

We have started to address the physical components of the system—how the system is partitioned into executable components. Earlier design discussions focused on identifying logical components. The **component diagram** identifies the logical, reusable, and transportable components of the system. The essential element of a component diagram is the component element with its interfaces.

A component is an executable module or program, and it consists of all the classes that are compiled into a single entity. It has well-defined interfaces, or public methods, that can be accessed by other programs or external devices. The set of all of these public methods that are available to the outside world is called the **application program interface**, or **API**. Figure 11-4 illustrates the UML notation for a component and its interfaces. It is not necessary to list all of the interfaces on a single component. Only those that are pertinent to the context of the diagram are listed. There are two ways to represent a component: either as a general class or as a specific instance. The same rules apply in this situation as with class and object notation—a general class uses the name of the component class, and an instance name is underlined. The name of the component is written inside.

Figure 11-4
Object-oriented component notation

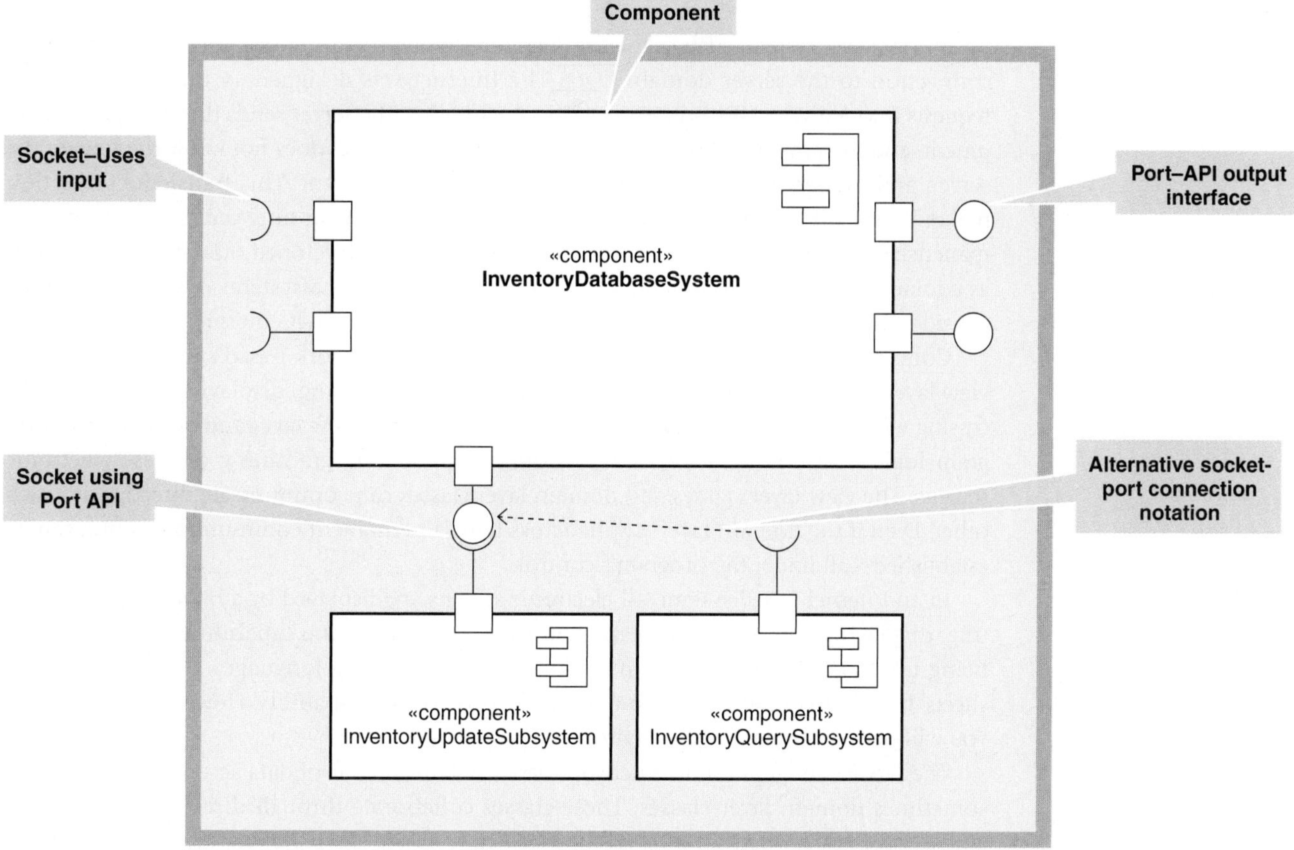

The top rectangle in the figure illustrates the notation for a component along with its interfaces. The component rectangle has an icon in the top right corner. The icon is a small rectangle with two plugs extending from the left side, signifying that it is a moveable, executable component, and is possibly reusable and pluggable.

The figure shows two types of interfaces: an output port and an input socket. The output port is similar to a programming interface—it defines the method names (for example, a portion of the API) that can be used by other components to access the component's functions. The input socket represents the services that the component needs from other components. Notice the ball and socket notation. They go together so that the input of one component precisely fits the output of another component.

The bottom portion of the figure shows how the port interfaces and sockets can be used in a component diagram. The InventoryDatabaseSystem presents an interface to the world, as denoted by the interface ball along its bottom edge. The InventoryUpdateSubsystem uses that interface to access the methods of the InventoryDatabaseSystem. In this figure, we show that the InventoryQuerySubsystem also accesses the same interface by connecting it to the interface via a dashed arrow.

In our design examples, we would like to show how to do multilayer design for a Web-based system, and we would like to illustrate the locations of various Web pages. In other words, we want to have some notation to show where Web pages reside and are deployed. Because the UML notation that we use does not have standard notation for a window, we extend the notation to include it. UML does have rules both for stereotyping a symbol and for extending the language. Figure 11-5 shows the notation we have invented.

Figure 11-5

Extension notation for window and Web page

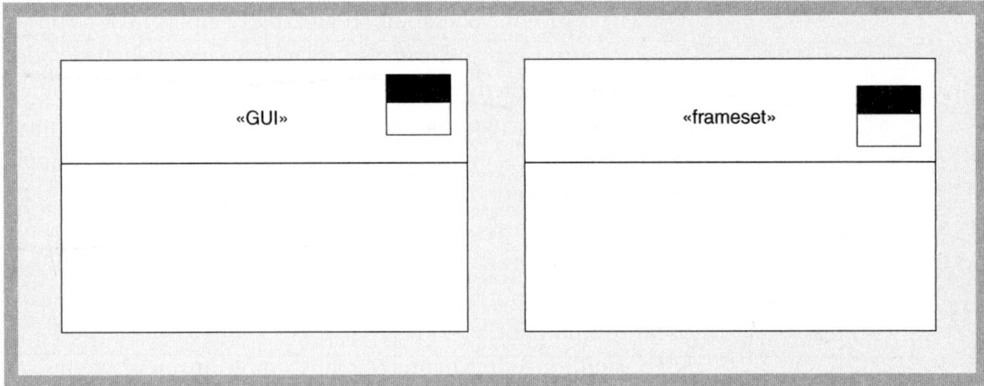

The figure simply displays a class notation with a stereotype notation such as «GUI» and «frameset», along with a small window icon in the top right corner. This notation will serve for either a desktop system window or a Web system frameset. A frameset is a high-level object that can hold items to be displayed by a browser. We will use a frameset notation and stereotype to indicate a Web page. You can think of a frameset as the window in a browser that can display frames or a set of frames.

One other object-oriented notation used for high-level design is the package notation. In Chapter 7 you learned that a package is a group of similar items. The notation is a box with a tab on top, much like a file folder. In Chapter 7, Figure 7-5, we used a notation to group use cases together into subsystems. However, it can also be used to group any other type of elements, such as classes.

Next let's use the component and window notation to do an architectural design of some straightforward Internet systems.

Two-Layer Architectural Design of Internet Systems

Many colleges have courses in Web development. Most of those courses fall into two categories: Web site design or Web programming languages. Both classes are important and beneficial for your education as a system developer.

Web programming courses teach you various programming languages and the ways to insert program logic into Web pages. You learn JavaScript, VBScript, PHP, and ASP (Active Server Pages). You may learn how to use advanced database tools, such as Cold Fusion, to access databases from your pages. You also learn how the browser and server work together to serve up pages that have sufficient programming logic to support the business application. Advanced versions of this course even teach you the Java or .NET environments so that you can configure an entire application.

We do not intend for this short section to replace that course. Instead, we introduce the architecture of these Web-based systems and provide a few principles of good design that you can apply as you develop skills in other courses. In Chapter 9, we explained three-layer design as one effective approach to developing robust, easily maintainable systems. But how can designers implement a three-layer design in a Web-based architecture? This question is particularly important if an organization wants to use the same problem domain logic for both types of enterprise systems: a client/server system and an Internet-based system.

Figure 11-6 illustrates a simple, generic Internet architecture. Remember, we are doing logical design at this point and are not yet concerned with the physical computer configuration. We will discuss the computer configuration when we learn more about deployment diagrams in the next section. Of course, because there is an Internet cloud between two components, we can naturally assume they are in different physical locations.

Figure 11-6 includes four recognizable components. The browser is an executable component whose purpose is to format, display, and execute active code such as JavaScript or ActiveX Controls. The Internet Server is another executable component whose purpose is to retrieve pages and invoke other components. This diagram shows two examples of components—executable programs in the Common Gateway Interface (CGI) and the Application Server—which also may invoke other components that are not shown in this diagram.

In the interest of brevity, the ports and sockets have been omitted from this diagram. The interfaces between these components are industry standard, and the unique port/socket combination does not need to be emphasized here. Every two-headed arrow, however, does represent two port/socket pairs.

As indicated in Chapter 9, many simple business systems can be designed as two-layer systems. These systems primarily capture information from the user and update a database. No complex domain layer logic is required. In those instances, the domain layer and data access layer are usually combined. The business logic in the domain layer frequently relates only to data formatting and to deciding which database table to update. Many business applications fall in this category. For example, a simple address book system could be easily designed as a two-layer system.

The CGI was the original way to process input data. The CGI directory contains compiled programs that are available to receive input data from the server. The programs in the CGI directory can be written in any compiled language, such as C++. This technique is effective and usually has quick response and processing times. The only downside is that these programs can be quite complex and difficult to write. They process the input data, access any required database, and format a response page in HTML, as indicated by the ResponsePage in the diagram.

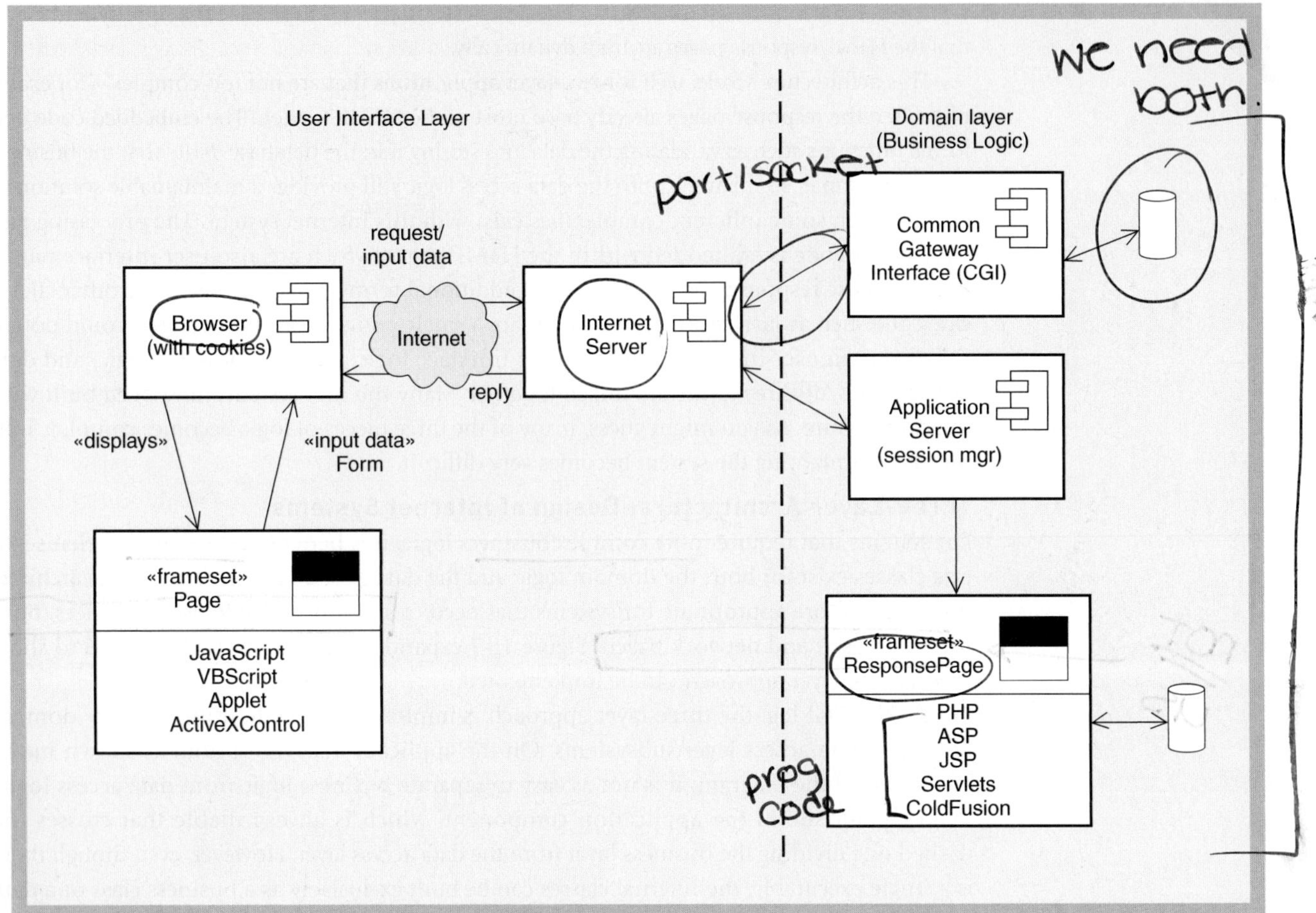

Figure 11-6

Two-layer Internet architecture

The other potential direction for input data is directly to a URL for a Web page with embedded program code. The extension shown on the ResponsePage indicates the type of program code embedded in the page—ASP for Active Server Pages, PHP for PHP Hypertext Preprocessor, JSP for Java Server Pages, CFM for Cold Fusion Pages, and so forth. Depending on the type of extension, an application server—which is the language processor—is invoked to process the embedded code. The embedded code, via the application server, can process the code, including reading and writing to a database. The application server, working with cookies on the browser, can manage sessions with the user. Session variables are set up to maintain information about the user across multiple page requests and forms. The application server also formats the response page based on the HTML statements and the code and forwards it back to the Internet server.

Even though we have referred to this design as two-layer architecture, the user interface classes often contain the business logic and data access. Due to the structure of Web servers, the program (defined as object-oriented classes) that processes the input forms also outputs the HTML code that is sent back to the client browser. For example, in Java-based systems, Java servlets receive the data from input on a Web form, process the data, and format the output HTML page. To process the data, the servlets usually include any required business logic and data access logic. The .NET environment is similar. For every Web form, there is a code-behind class written in Visual Basic, C#, J#, or some similar language. The code-behind object receives the data from the Web form, processes the data, and formats the output HTML page. So, it is often not clear whether the architecture is one-layer or

two-layer. However, we refer to this architecture as two-layer to emphasize that it is dynamic and that the HTML response pages are built dynamically.

This architecture works well for two-layer applications that are not too complex—for example, when the response pages already have most of the HTML written. The embedded code performs functions such as validating the data and storing it in the database. Note that the business logic is minimal, so mixing it with the data access logic still provides a maintainable solution.

However, some inherent complexities exist with this Internet system. The processing and data access code is embedded within the HTML pages, which are also user-interface pages. Because these response pages may contain additional forms, they may also have other client-side code such as JavaScript and VBScript. So, a single entity—the HTML page—could potentially contain user-interface controls, user-interface logic, problem domain logic, and data access logic. All three layers are mixed together. Many Internet systems have been built with this architecture. As you might guess, if any of the three pieces of logic becomes complex, testing and maintaining the system becomes very difficult.

Three-Layer Architectural Design of Internet Systems

For systems that require more complex business logic, it is better to add a layer so that separate classes exist for both the domain logic and the data access process. Three-layer architecture is also more appropriate for systems that need to support multiple user interfaces, both Internet-based and network-based. Figure 11-7 expands the diagram in Figure 11-6 to show how a three-layer approach can be implemented.

On the CGI leg, the three-layer approach is implemented by defining separate domain layer and data access layer subsystems. On the application server option, as shown in the lower part of the diagram, it is not as easy to separate business logic from data access logic. The diagram shows the application component, which is an executable that crosses the dashed line dividing the business layer from the data access layer. However, even though there is a single executable, the internal classes can be built exclusively as a business class or a data access class. This is illustrated by defining two packages of classes. Let's address two approaches for how this is done—the Java approach and the .NET approach.

For Java server pages, which have a .jsp extension, the application server invokes a Java servlet when the input form is received. A Java servlet is a Java program that executes like any other program. The Java servlet identified for the input form can be a special class that serves as a use case controller, which can then distribute the input message to other domain classes to process the request. After the request is processed, including any database access, the servlet takes control and formats the output response page. The difference, of course, is that the output is in HTML statements and must flow through the server and browser before it is displayed to the user.

For the .NET environment, the process is similar. The input data form is sent to the ASP.NET application server, which invokes the code-behind object (written in Visual Basic or C#) for the particular Web form. The code-behind object will then call appropriate methods in other objects. In the new .NET environment, the program modules are compiled into a common language called Common Language Runtime, a mid-level language that allows programs to be partially compiled and managed for faster, more efficient execution. As with Java, the program structure can use a three-layer design. One of the major differences in the .NET environment is that the .NET framework has an extensive library of standard components that does everything from automatically authenticating users to accessing data and populating forms. But this feature can be both an advantage and disadvantage. The advantage is that the high-level components provided by .NET make it easy to program and build forms. However, the disadvantage is that you can easily mix the layers, scattering business logic with user-interface processes and data access logic.

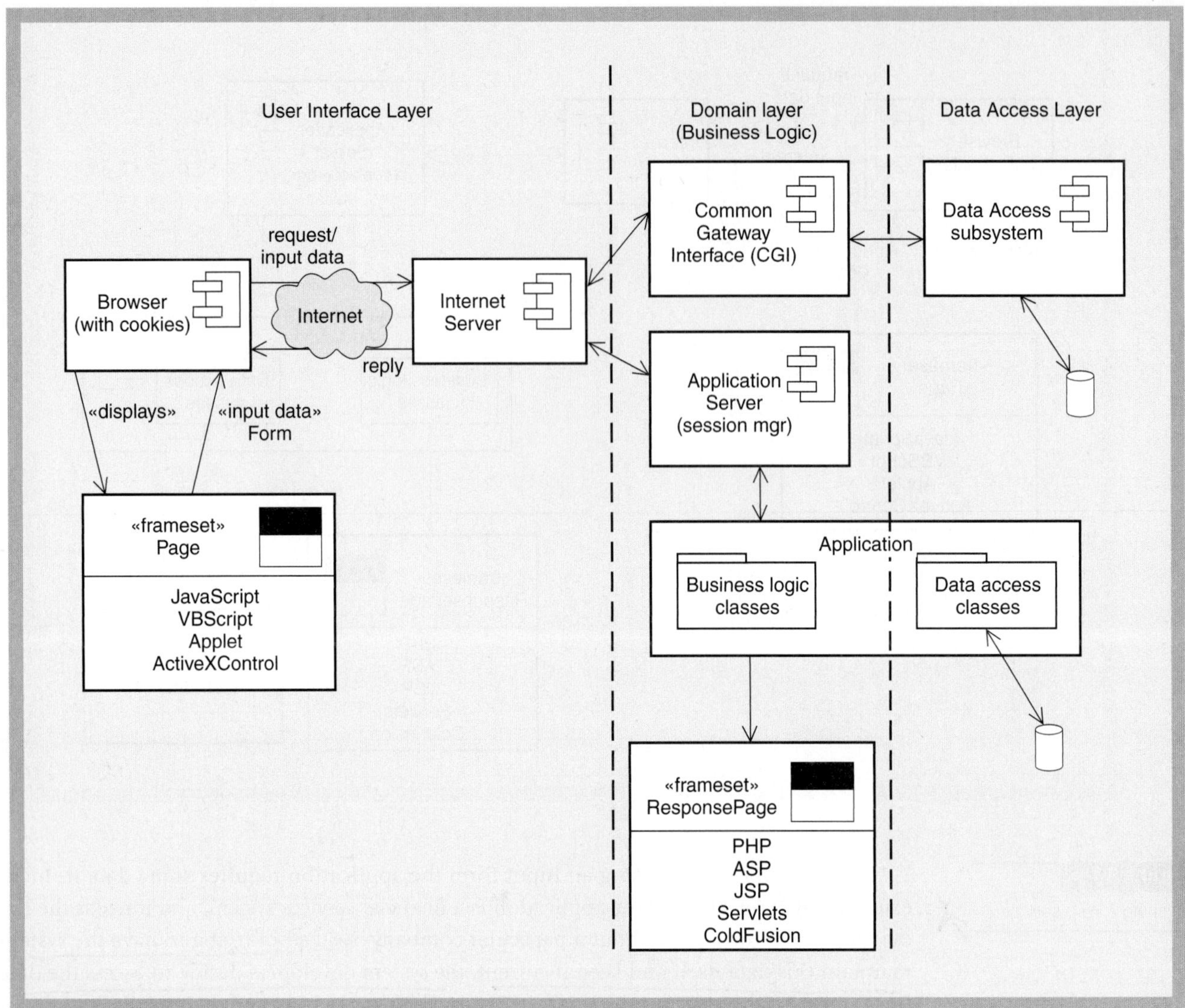

Figure 11-7

Three-layer Internet architecture

BEST PRACTICE

Be sure the problem domain classes are well defined first, before defining the Web forms and user interface pages.

Web Services

Finally, let's look at one last variation of an Internet architecture—a Web service. One new technique being used to develop Internet-based systems is through the use of Web services. We expect to hear and see more about Web services in the near future. So what are Web services? They are simply computer programs that provide services to other systems via the Internet and are posted in a directory so that other systems can find and use them. Figure 11-8 shows how a Web service might be used.

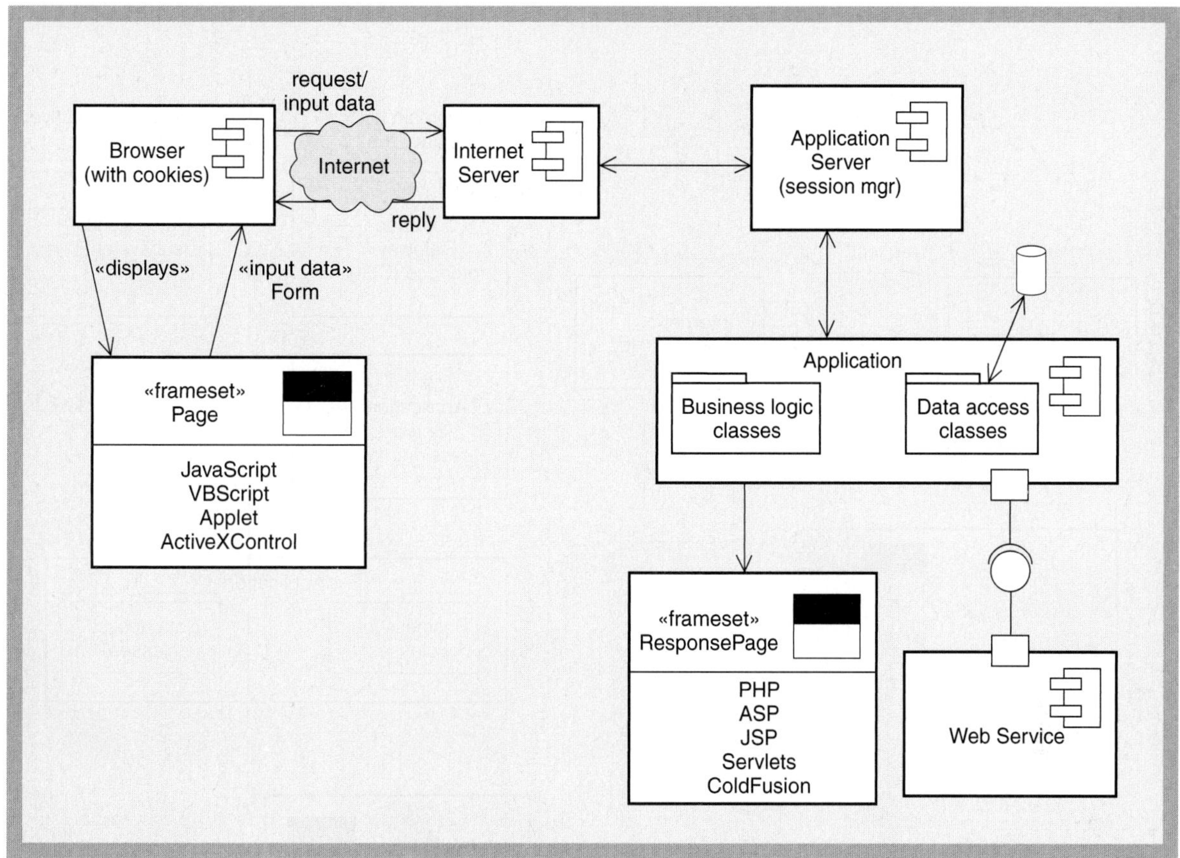

Figure 11-8

Invoking a Web service

Suppose that in processing an input form the application requires some data from an external database. Perhaps the application is a financial services system, and it needs the latest financial information about a particular company. Instead of trying to have the system maintain this data itself and keep it current, the system developers decide to access the data from some other service on the Internet. So, as the program is executing, it determines that it needs current financial data. It does not care where it gets the data, just that the data is current. First, the program sends a request for information. This request will go to a services directory called Universal Discovery, Description, and Integration (UDDI), which is an indexing service to help locate Web services. The request will be based on keywords that describe exactly what is desired. The UDDI provides an Internet address of a program that provides the service. The application then requests the desired information over the Internet.

This process sounds easy, but it was never possible before, for various reasons. One major obstacle was that the requestor and the provider had to use exactly the same format of data exchange. This fact is emphasized in the figure by making the port/socket interface visible between the two components. Defining the same format is easy if only two programs are involved. But to have a general-purpose format that any program can use has been a major problem. In Web services, however, all communication is based on XML (eXtensible Markup

Language), a text-based language much like HTML. The difference is that HTML has standard tags, whereas XML can include self-defining tags to describe any data you want. In other words, the sequence and format of the data are defined within the transmitted file itself. Thus, the recipient of the data can process it no matter what the sequence. Examples of XML are shown in Chapter 15 in the discussion of interface design.

The next section extends the concepts of architectural design by mapping the logical components to physical computers and locations. You first learned about physical design in Chapter 9 in the discussion of multitiered design. In the next section, you will learn about the notation of deployment diagrams and how to describe the multiple tiers of today's systems.

DEPLOYMENT DIAGRAMS

deployment diagram

a type of implementation diagram that shows the physical components across different locations

A deployment diagram shows the placement of various physical nodes (components) across different locations. A node can be thought of as a computer, or a bank of computers, representing a single computing resource. A node is a physical entity at a specific location. Figure 11-9 illustrates the symbol that is used for a node—a shaded rectangle. The shading is added to represent a real object that can cast a shadow. The name of the node is listed inside, either as a classification of node or as a single instance with the name underlined.

Figure 11-9

Notation for a node in a deployment diagram

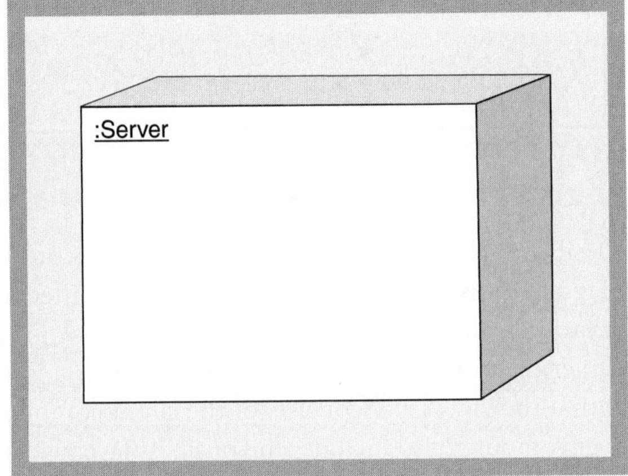

artifact

a class invented by a system designer to handle a needed system function

One other symbol used in deployment diagrams is the artifact symbol. The term artifact simply means something that is man-made. In this situation, an artifact means something tangible that is created and exists on a particular node. The symbol for an artifact is a rectangle with a document icon in the top right corner. The document icon is a piece of paper with the corner folded down. If a component in a component diagram defines a logical piece of a system, then an artifact can be thought of as the physical item that holds or specifies that component. Figure 11-10 illustrates both the symbol for an artifact and its relationship with a component. As shown, an artifact has a dependency relationship with a component. The standard UML notation for a dependency relationship is a dashed arrow. In other words, an artifact depends on the component, and if the component changes, the artifact also must change.

Figure 11-10

An artifact and its relationship to a component

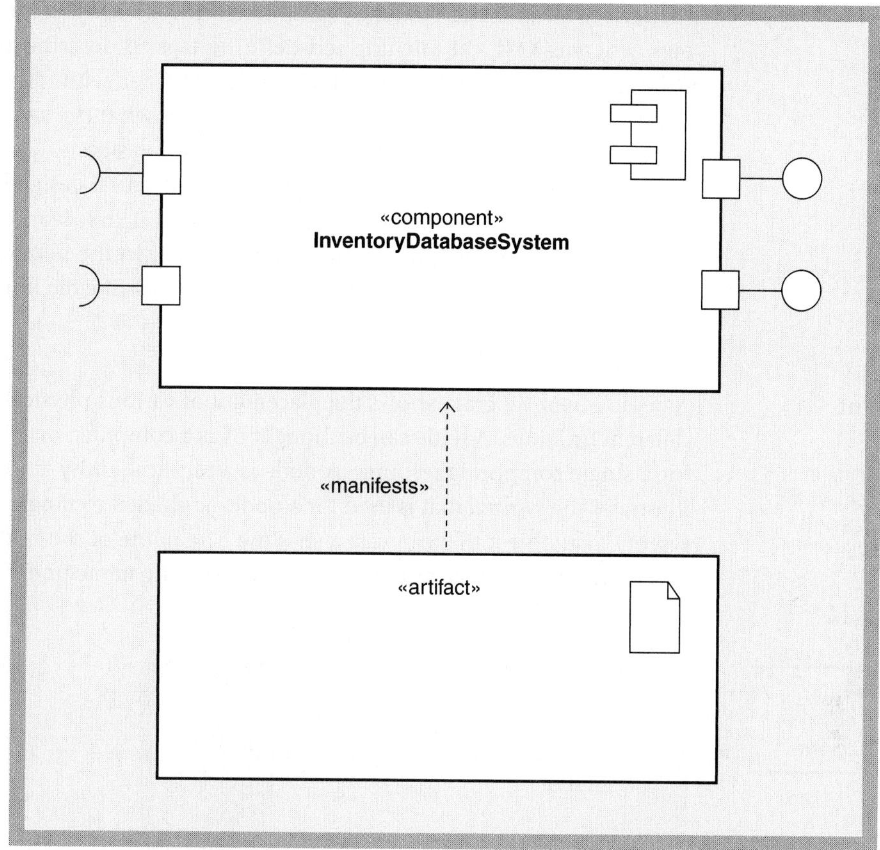

Figure 11-11 illustrates the use of nodes, artifacts, and components in a deployment diagram. This diagram illustrates how the three-layer architecture of Figure 11-7 might be deployed on various machines. The browser artifact is normally deployed on a client machine. A server computer is used to deploy both an Internet server and an application server. Notice that the view layer consists of the client computer as well as the server computer. The server computer also includes the business layer as a separate artifact. The far-right symbol is a logical component representing the database system. To interpret this part of the figure, you must understand that the physical implementation of the database system is either still open for discussion or is not important for this figure. It could be implemented on the same node or on a different node. This figure simply indicates that a database system is needed in the logical solution, but tells us nothing about where or how it will be done. It is not unusual to have logical components on a deployment diagram.

Figure 11-11

Example of a deployment diagram for an Internet-based system

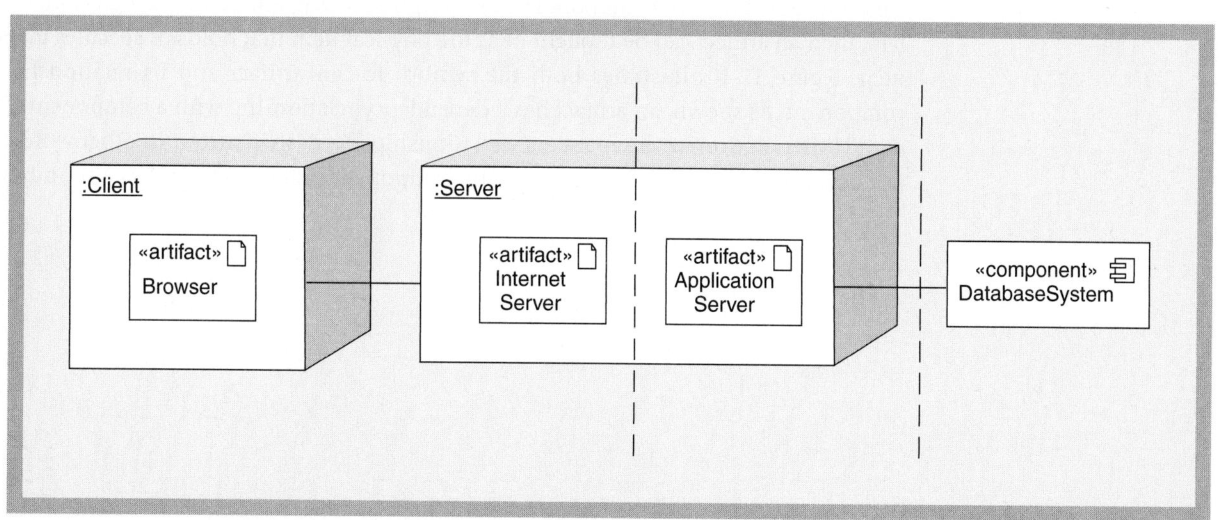

So far you have learned foundation principles for multilevel design of object-oriented systems. The examples in the text are fairly high-level and generic solutions. You might think that all solutions would look exactly alike, but they do not. As you focus on specific applications and a slightly more detailed level, you might see different configurations of logical components and computer nodes. For example, some systems run in conjunction with other systems, like Facebook and the separate Facebook applications. The applications run on a separate platform and may have their own application database servers. Some Facebook applications are also gateways into separate commercial applications on separate sites. By the time an entire application is configured, there may be several different nodes, each with separate artifacts representing the various logical components. (See problems 3 and 4 in the "Thinking Critically" section at the end of this chapter.)

Once the architectural design is determined, it is time to drill down to a lower level of abstraction. In other words, you stop treating the logical components as black boxes and start to look inside. Each component is an executable program and is made up of classes. So the next step in the application design is to begin defining the design classes. Designing at this level is usually called detailed design.

As noted in Chapters 1 and 2, RMO's new customer support system needed to support both a Web user interface and an internal desktop interface. It is critically important for the same back end—business logic and database access—to link with either user interface. Consequently, the design team must specify the architectural design in enough detail to ensure that the programmers implement a system that can support both user interfaces. Barbara Halifax has updated John MacMurty on her team's progress in completing the architectural design and creating the system's Web interface (see the memo).

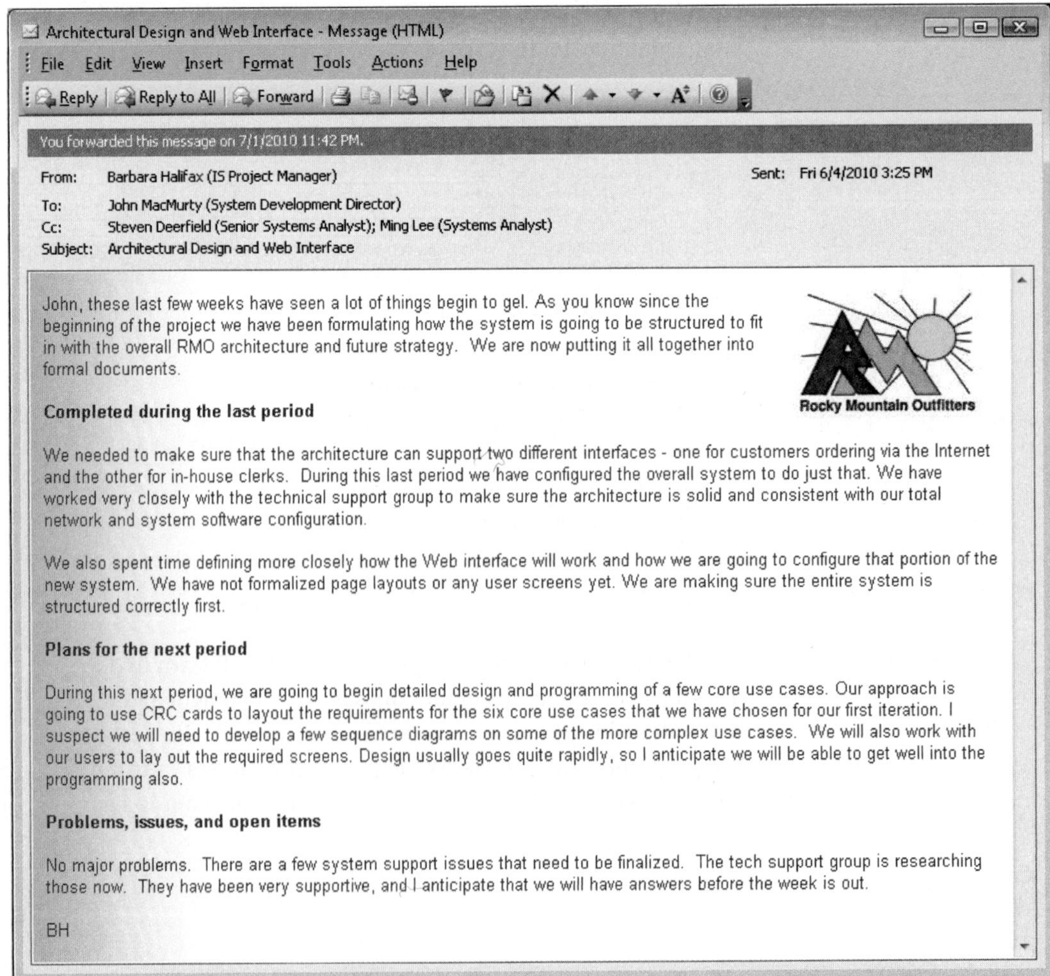
Architectural Design and Web Interface - Message (HTML)

File Edit View Insert Format Tools Actions Help

Reply | Reply to All | Forward

You forwarded this message on 7/1/2010 11:42 PM.

From: Barbara Halifax (IS Project Manager) Sent: Fri 6/4/2010 3:25 PM
To: John MacMurty (System Development Director)
Cc: Steven Deerfield (Senior Systems Analyst); Ming Lee (Systems Analyst)
Subject: Architectural Design and Web Interface

John, these last few weeks have seen a lot of things begin to gel. As you know since the beginning of the project we have been formulating how the system is going to be structured to fit in with the overall RMO architecture and future strategy. We are now putting it all together into formal documents.

Rocky Mountain Outfitters

Completed during the last period

We needed to make sure that the architecture can support two different interfaces - one for customers ordering via the Internet and the other for in-house clerks. During this last period we have configured the overall system to do just that. We have worked very closely with the technical support group to make sure the architecture is solid and consistent with our total network and system software configuration.

We also spent time defining more closely how the Web interface will work and how we are going to configure that portion of the new system. We have not formalized page layouts or any user screens yet. We are making sure the entire system is structured correctly first.

Plans for the next period

During this next period, we are going to begin detailed design and programming of a few core use cases. Our approach is going to use CRC cards to layout the requirements for the six core use cases that we have chosen for our first iteration. I suspect we will need to develop a few sequence diagrams on some of the more complex use cases. We will also work with our users to lay out the required screens. Design usually goes quite rapidly, so I anticipate we will be able to get well into the programming also.

Problems, issues, and open items

No major problems. There are a few system support issues that need to be finalized. The tech support group is researching those now. They have been very supportive, and I anticipate that we will have answers before the week is out.

BH

Now that we have learned about architectural design, we can turn to detailed design. If you refer back to Figure 11-2 and proceed down the right column, the next two diagrams to discuss are the design class diagram and the interaction diagrams (sequence diagrams and communication diagrams). These two diagrams are the most important for detailed design.

The objective of object-oriented detailed design is to identify and specify all of the objects that must work together to carry out each use case. As shown in Figure 11-1, there are user-interface objects, problem domain objects, and database access objects. Additional objects to perform specific services, such as logon authentication, may also be required. As you may suppose, a major responsibility of detailed design is to identify and describe each set of objects within each layer and to identify and describe the interactions or messages that are sent between these objects.

The most important model in object-oriented design is a sequence diagram—or its first cousin, a communication diagram. In Chapter 7, you learned to develop system sequence diagrams (SSDs) to model input and output requirements for a use case. The full sequence diagram is used for design and is a type of interaction diagram. A communication diagram is also a type of interaction diagram. During design, developers extend the SSD by modifying the single :System object to include all of the interacting user-interface, problem domain, and database access objects. In other words, they look inside the :System object to see what is happening inside the system. We will spend a good deal of time in the next chapter learning how to develop these detailed sequence diagrams. Figure 11-12 shows a simple sequence diagram based on Figure 11-1, which updates student information. A sequence diagram uses the same notation as an SSD, which you learned to develop in Chapter 7. In fact, a sequence diagram is simply an extension of a systems sequence diagram. We explain the details of sequence diagrams and how to develop them in the next chapter.

Figure 11-12

Sequence diagram for updating student name

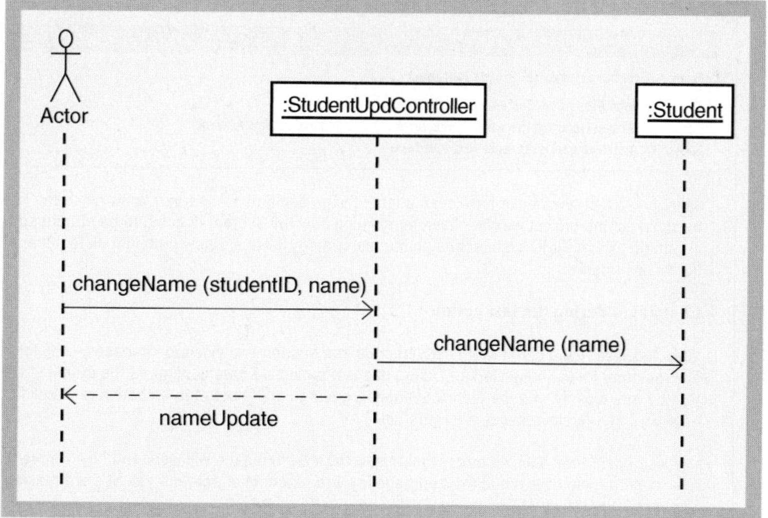

The other major design model, which you will learn to develop later in this chapter, is the design class diagram. Its main purpose is to document and describe the programming classes that will be built for the new system. They describe the set of object-oriented classes needed for programming, navigation between the classes, attribute names and properties, and method names and properties. A design class diagram is a summary of the final design that

was developed using the detailed sequence diagrams, and it is used directly when developing the programming code. Figure 11-13 shows the original domain model that was developed during analysis and the design class diagram version of that class. The design class version has a new compartment at the bottom that specifies the method signatures for the class. The attributes have also been enhanced. We explain the details of this notation in the next section. Detailed design is the process that takes the domain model to the design class model.

Figure 11-13

Student class examples for the domain class and the design class diagrams

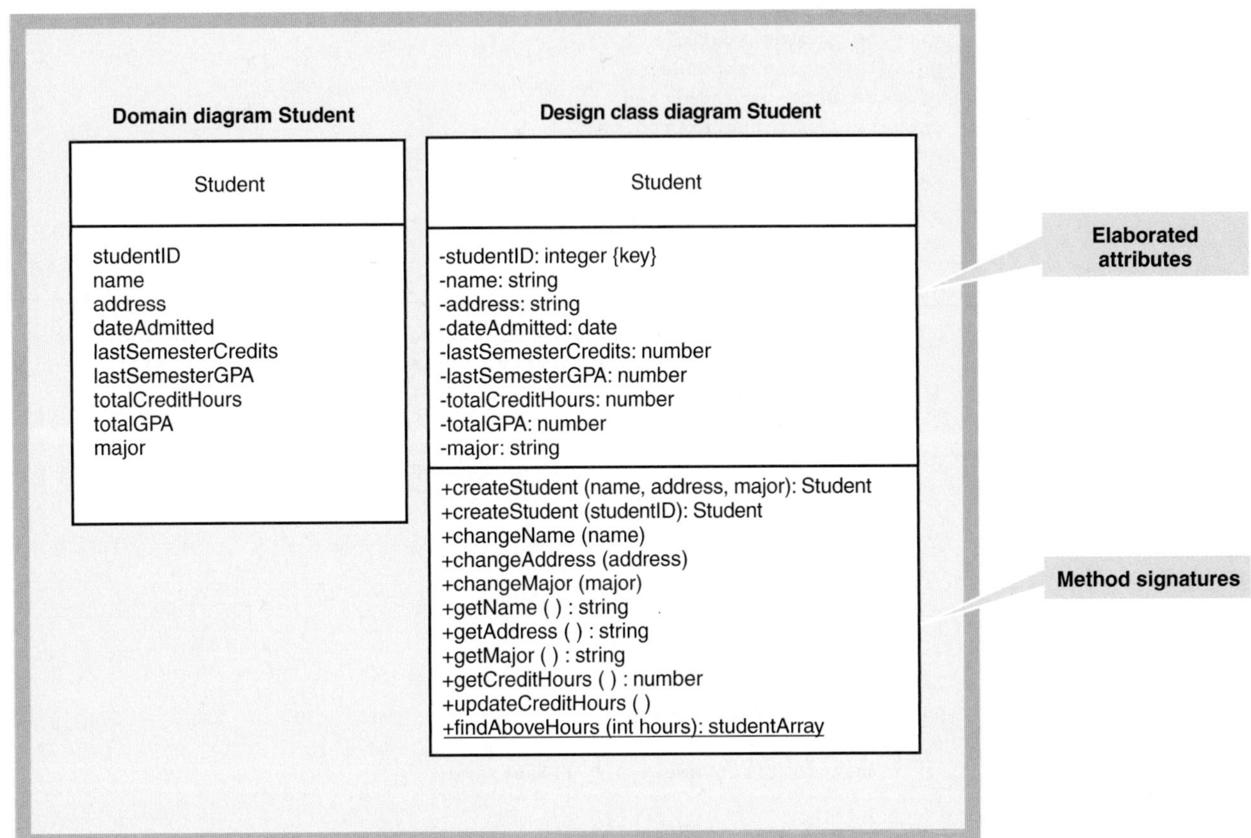

As an object-oriented system designer, you must provide enough detail so that a programmer can write the initial class definitions, including the method code. As you will see in the following sections, the primary components of the OO design are design class diagrams, interaction diagrams, and, for some classes, state machine diagrams. For example, a design class specification helps define an object's attributes and methods. Figure 11-14(a) illustrates some sample code, written in Java, for the Student class. Figure 11-14(b) shows the same example code written in Visual Basic .NET. Referring back to Figure 11-13, you should be able to see how the design class provides the input to write the code for Figure 11-14. Notice that the class name, the attributes, and the method names are derived from the design class notation. Of course, in the design class we took some liberties by abbreviating the first name and last name to *name* and by combining all the components of an address into one field called *address*. If there is any question whether programmers will know that they should break these shortened names out into the detailed fields, the designer should not take these shortcuts. Other code that needs to be added to the class definition can be derived from the other design models, including the interaction diagrams and the state machine diagrams.

```
public class Student
{
        //attributes
        private int studentID;
        private String firstName;
        private String lastName;
        private String street;
        private String city;
        private String state;
        private String zipCode;
        private Date dateAdmitted;
        private float numberCredits;
        private String lastActiveSemester;
        private float lastActiveSemesterGPA;
        private float gradePointAverage;
        private String major;

        //constructors
        public Student (String inFirstName, String inLastName, String inStreet,
              String inCity, String inState, String inZip, Date inDate)
        {
              firstName = inFirstName;
              lastName = inLastName;
              ...
        }
        public Student (int inStudentID)
        {
              //read database to get values
        }

        //get and set methods
        public String getFullName ( )
        {
              return firstName + " " + lastName;
        }
        public void setFirstName (String inFirstName)
        {
              firstName = inFirstName;
        }
        public float getGPA ( )
        {
              return gradePointAverage;
        }
        //and so on

        //processing methods
        public void updateGPA ( )
        {
              //access course records and update lastActiveSemester and
              //to-date credits and GPA
        }
}
```

Figure 11-14(a)

Example class definition in Java for Student class

```
Public Class Student

   'attributes
   Private studentID As Integer
   Private firstName As String
   Private lastName As String
   Private street As String
   Private city As String
   Private state As String
   Private zipCode As String
   Private dateAdmitted As Date
   Private numberCredits As Single
   Private lastActiveSemester As String
   Private lastActiveSemesterGPA As Single
   Private gradePointAverage As Single
   Private major As String

   'constructor methods
   Public Sub New(ByVal inFirstName As String, ByVal inLastName As String,
         ByVal inStreet As String, ByVal inCity As String, ByVal inState As String,
         ByVal inZip As String, ByVal inDate As Date)
      firstName = inFirstName
      lastName = inLastName
       ...
   End Sub

   Public Sub New(ByVal inStudentID)
      'read database to get values
   End Sub

   'get and set accessor methods
   Public Function GetFullName() As String
      Dim info As String
      info = firstName & " " & lastName
      Return info
   End Function

   Public Property firstName()
      Get
         Return firstName
      End Get
      Set (ByVal Value)
      firstName = Value
      End Set
   End Property

   Public ReadOnly Property GPA()
      Get
        Return gradePointAverage
      End Get
   End Property

   'Processing Methods
   Public Function UpdateGPA()
      'read the database and update last semester
      'and to date credits and GPA
   End Function

End Class
```

Figure 11-14(b)

Example class definition in VB .NET for Student class

OBJECT-ORIENTED DESIGN PROCESS

Object-oriented design is model driven and use case driven. As you saw in Figure 11-2, the design process takes the requirements models as input and produces the design models as output. Obviously, we need a method for organizing this activity, and it is focused around use cases. In other words, we develop the design models use case by use case. For example, a design interaction diagram is developed for each use case. After a group of them have been designed, the design class diagram is completed for that entire group of use cases. We can divide the process of design into five major steps, as summarized in Figure 11-15.

Figure 11-15

Object-oriented detailed design steps

Object-Oriented Detailed Design Steps	
1. Develop the first-cut design class diagram showing navigation visibility.	Chapter 11
2. Determine class responsibilities and class collaborations for each use case using Class-Responsibility-Collaboration (CRC) cards.	Chapter 11
3. Develop detailed sequence diagrams for each use case. (a) Develop the first-cut sequence diagrams. (b) Develop multilayer sequence diagrams.	Chapter 12
4. Update the design class diagram by adding method signatures and navigation information using CRC cards and/or sequence diagrams.	Chapter 12
5. Partition the solution into packages as appropriate.	Chapter 12

First, a preliminary version, or first-cut model, of the design class diagrams is created. Some basic information, such as attribute names, must be listed in the first-cut model to develop the interaction diagrams. This step provides a foundation for the second and third steps.

The second step often used by developers is to take each use case and develop a set of CRC cards. The development of CRC cards helps provide an overall understanding of the internal steps required for the system to support the use case. CRC cards provide a simple method to identify all of the objects involved in a particular use case and their responsibilities. The results of a CRC activity will be sets of cards that can be used to help develop a sequence diagram; if a use case is simple enough, the cards can be used to program the use case. In other words, for simple use cases, CRC cards may be sufficient to write the code. CRC cards are explained in detail later in the chapter.

The third step in detailed design is to develop interaction diagrams, resulting in one for each use case or scenario. Developing an interaction diagram is a multistep process of determining which objects work together and how they work together. The first part is to develop a sequence diagram that includes only the domain classes. Next, a multitier solution is developed that includes data access classes and view layer classes. Development of the interaction diagrams is the heart of object-oriented systems design. As shown in Figure 11-2, input models for interaction diagrams are use case diagrams, activity diagrams, and system sequence diagrams. Design class diagrams are also used in the process. The result of the development of these design models is called **realization of use cases**. Here, the term *realization* is the specification of the detailed processing that the system must perform to carry out the use case—in other words, to make a set of software blueprints. Just as object-oriented analysis was driven by use cases, so is object-oriented design.

realization of use cases

specification of all detailed system processing for each use case

The fourth step in OO detailed design is to return to the design class diagram and develop method names based on information developed during the design of the interaction diagrams. The navigation visibility and attribute information is also updated in this iteration of the design class diagram.

The final design step is to partition the design class diagram into related functions using package diagrams. A system might be partitioned in several ways, such as by subsystem or by layers. In Chapter 9, you learned about multilayer architectures and multiple tiers. This chapter explains how to partition design class diagrams into packages to represent the multiple layers in a multitier system. We focus on a basic multilayer design that consists of the view layer

(user-interface classes), the domain layer (problem domain classes from the domain model class diagram), and the data layer (database access classes). Note that several synonymous terms are used to denote the domain layer, including *problem domain layer* and *business logic layer*. Package diagrams provide the detailed class organization information that can be added to the architectural view of the final system.

DESIGN CLASSES AND THE DESIGN CLASS DIAGRAM

As shown in Figure 11-2, the design class diagrams and the detailed interaction diagrams work together. A first iteration of the design class diagram is created based on the domain model and on engineering design principles. The preliminary design class diagram is then used to help develop interaction diagrams. As design decisions are made during development of the interaction diagrams, the results are used to refine the design class diagram.

The domain model class diagram shows a set of problem domain classes and their associations. During analysis, because it is a discovery process, analysts generally do not worry much about the details of the attributes or the methods. However, in object-oriented programming, the attributes of a class must be declared as public or private, and each attribute must also be defined by its type, such as character or numeric. During detailed design, it is important to elaborate on these details, as well as to define parameters that are passed to the methods and return values from methods. Sometimes, developers also define the internal logic of each method at this point. We complete the design class diagram by integrating information from interaction diagrams and other models.

As developers build the design class diagrams, they add many more classes than were originally defined in the domain model. To build a complete object-oriented system, many other design classes must be identified and specified. Referring to Figure 11-1, the Input window objects and Database access objects are examples of additional classes that must be defined. As the classes are defined, designers usually document them on various class diagrams. The classes in a system can be partitioned into distinct categories, such as user-interface classes. At times, designers may also develop distinct class diagrams by subsystem. Whatever process they use, designers document their decisions with class diagrams, so class diagrams are used in different ways. We now turn to design class diagram notation and discuss the design principles used in developing the first iteration of the design class diagram.

DESIGN CLASS SYMBOLS

UML does not specifically distinguish between design class notation and domain model notation. However, practical differences occur simply because the objective of design modeling is distinct from that of domain modeling. Domain modeling shows things in the users' work environment and the naturally occurring associations among them. The classes at that point are not specifically software classes. After we start a design class diagram, though, we are specifically defining software classes. Because many different types of design classes are identified during the design process, UML has a special notation, called a *stereotype*, which allows designers to designate a special type of class. A stereotype is simply a way to categorize a model element as a certain type. A stereotype extends the basic definition of a model element by indicating that it has some special characteristic we want to highlight. The notation for a stereotype is the name of the type placed within printer's guillemets, like this: «Control». You were first exposed to a stereotype when you were developing a use case diagram. You learned that connecting lines between actors and use cases indicated a relationship and that a certain type of relationship existed between use cases, called the «includes» relationship.

stereotype

a way of categorizing a model element by its characteristics, indicated by guillemets (« »)

Four types of design classes are considered standard: an entity class, a control class, a boundary class or view class, and a data access class. Figure 11-16 shows the notation used to identify these four stereotypes. Two types of notation can be used for design classes. The class

entity class
a design identifier for a problem domain class

persistent class
an entity class that exists after a system is shut down

boundary class or view class
a class that exists on a system's automation boundary, such as an input window

rectangles on the left show the full symbols. Notice that the stereotypes are placed above the name in the name compartment. The circular symbols on the right are shorthand notation for these stereotypes, and are called *icons*. We will use the stereotype icons from time to time, but in most cases we prefer the full notation.

An **entity class** is the design identifier for a problem domain class. In other words, it comes from the domain model. These objects are normally passive, in that they wait for business events to occur before they do anything. They are also usually persistent classes. A **persistent class** is one that exists after the program quits. In other words, the data must persist after the system is shut down. Obviously, the way to make data persistent is to write it to a file or database.

A **boundary class**, or **view class**, is specifically designed to live on the system's automation boundary. In a desktop system, these classes would be the windows classes and all the other classes associated with the user interface.

Figure 11-16
Standard stereotypes found in design models

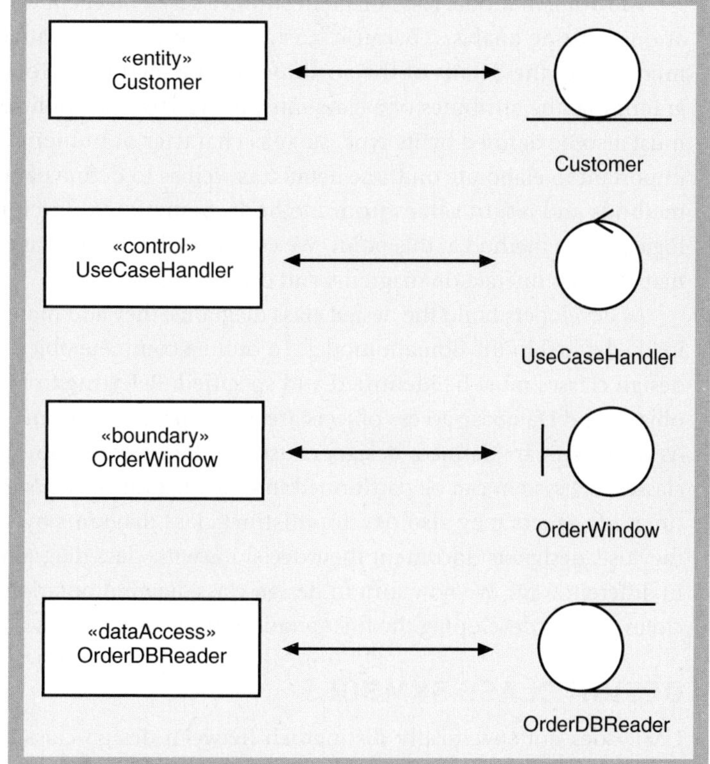

control class
a class that mediates between boundary classes and entity classes, acting as a switchboard between the view layer and domain layer

data access class
a class that is used to retrieve data from a database

A **control class** mediates between the boundary classes and the entity classes. In other words, its responsibility is to catch the messages from the boundary class objects and send them to the correct entity class objects. It acts as a kind of switchboard, or controller, between the view layer and the domain layer.

A **data access class** is used to retrieve data from and send data to a database. Rather than insert database access logic, including SQL statements, into the entity class methods, a separate layer of classes to access the database is often included in the design.

DESIGN CLASS NOTATION

Figure 11-17 shows the details within a design class symbol, as you first saw in the design class in Figure 11-13. The name compartment includes the class name and the stereotype information. The lower two compartments contain more details about the attributes and the methods.

Figure 11-17

Notation used to define a design class

«Stereotype Name» Class Name::Parent Class
Attribute list visibility name:type-expression = initial-value {property}
Method list visibility name (parameter list): type-expression

visibility

a notation of whether an attribute can be directly accessed by another object; indicated by plus or minus signs

method signature

a notation that shows all of the information needed to invoke, or call, the method

overloaded method

a method with one name but two or more parameter lists

The format that analysts use to define each attribute includes the following:

- Attribute visibility. **Visibility** denotes whether other objects can directly access the attribute. (A + sign indicates that an attribute is visible, or public, and a – sign means that it is not visible, or private.)
- Attribute name
- Type-expression (such as character, string, integer, number, currency, or date)
- Initial-value
- Property (within curly braces), such as {key}

The third compartment contains the method signature information. A **method signature** shows all of the information needed to invoke (or call) the method. It shows the format of the message that must be sent, which consists of the following:

- Method visibility
- Method name
- Method parameter list (incoming arguments)
- Type-expression (the type of the return parameter from the method)

In object-oriented programming, analysts use the entire signature to identify a method. Some OO languages allow multiple methods to have the same name as long as they have different parameter lists or return types. In those languages, both the method name and the parameter list are used to invoke the correct method. For example, suppose that we want to be able to find a customer record either by the customer ID number or by the customer name. We could identify two methods, each with the same name, such as getCustomer (customerID) and getCustomer (customerName). When a method such as getCustomer has the same name but different parameter lists, we say the method is an **overloaded method**. To know which method to invoke, the run-time environment must also note what parameters are included, and whether a number (customerID) or a text field (customerName) was entered.

The domain model attribute list contains all attributes discovered during analysis activities. The design class diagram includes more information on attribute types, initial values, and properties. It can also include a stereotype for clarification. As shown in Figure 11-13 in the Student design class diagram, the third compartment contains the method signatures for the class. Remember that UML is meant to be a general object-oriented notation technique and not specific to any one language. So, the notation will not be the same as programming method notation.

For example, for those of you with programming experience, the constructor notation we use is *createStudent (name, address, major): Student*. Remember that the constructor is the method that makes a new object for the class. In many programming languages, the constructor is given the same name as the class. However, in this situation we use a create statement to

follow more closely the message names used in interaction diagrams. Figure 11-13 also illustrates another constructor. In the second method line, only the student ID is passed in. This implies that a student with an ID exists, and the constructor itself must fill in the information about the student. This usually requires access to a database to get values for the fields.

The method called findAboveHours (int hours): studentArray, as denoted with an underline in Figure 11-13, is a special kind of method. Remember in the object-oriented approach that a class is a template to create individual objects or instances. Most of the methods apply to one instance of the class. However, frequently analysts need to look through all of the instances at once. Such a method is called a **class-level method**, which is denoted by an underline. In VB .NET it is called a *shared* method, and in Java it is a *static* method. This type of method is executed by the class instead of a specific object of the class. Because these methods are used at the class level, they do not depend on the existence of a particular object. If necessary, they can access data across all objects. In this example, the findAboveHours method looks through all the instances of the class and returns the ones that have more total hours than the input parameter.

class-level method

a method that is associated with a class instead of with objects of the class

Figure 11-18 is an example of design classes with attributes and methods; it shows how inheritance works for design classes. In Chapter 5 you learned about generalization/specialization. Generalization/specialization in the problem domain model becomes inheritance in the design model and in a programming language. The notation is the same as in the problem domain. In this case, we have used three separate arrows instead of one single arrow, but the meaning is identical. Each of the three subclasses inherits all of the attributes and methods of the parent Order class. Hence, each subclass has an orderID, an orderDate, and so forth. In this example, each subclass also has additional attributes that are unique to its own specific class. Each of the subclasses also has a unique attribute that is underlined, such as noOfPhoneOrders. Underlined attributes are **class-level attributes** and have the same characteristics as class-level methods. A class-level attribute is a static variable and it contains the same value in all instantiated objects of the same type.

class-level attribute

an attribute that contains the same value for all objects in the system

Each subclass also has an addItem() method and a makePayment() method, as well as all the other methods in the parent class. When those methods are executed for a particular subclass, such as PhoneOrder, the code in the parent class is executed. The InternetOrder class also has an additional method to send an e-mail and confirm the order. The MailOrder class overrides one of the methods in the parent Order class. The cancelOrder() method is an **overridden method**, in that the version in the subclass replaces the method in the parent class.

overridden method

a method in a subclass (with inheritance) that overrides the method in the parent class

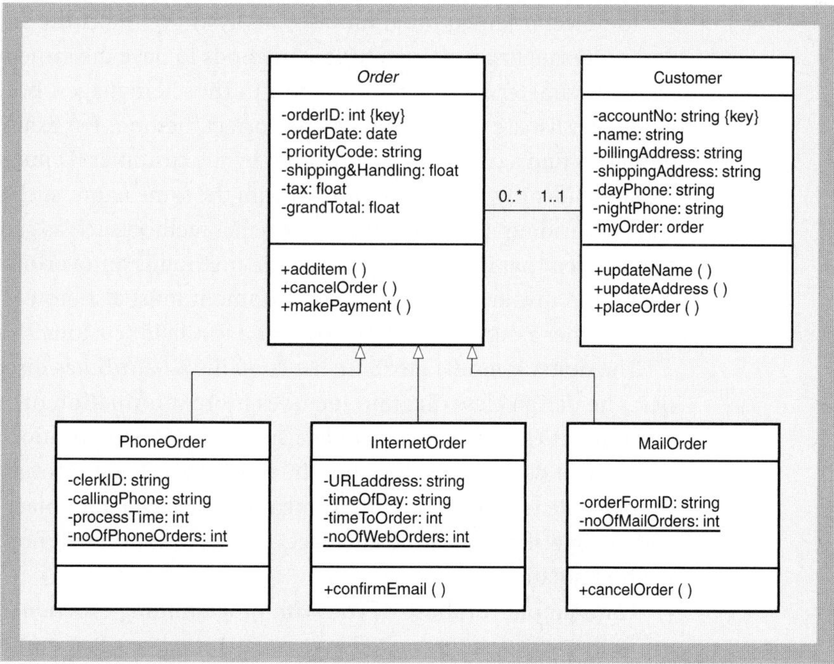

Figure 11-18

Order class with three subclasses showing inheritance

Not only are methods and attributes inherited by the subclasses, associations are also inherited. In the figure, the order object must be associated with exactly one customer. Each

subclass inherits the same association and must be associated with exactly one customer. Finally, if you look carefully, you will notice that the title of the Order class is italicized. An italicized class name indicates that it is an **abstract class**, a class that can never be instantiated. In other words, there are never any Order class objects. All orders in the system must be instantiated as one of the three subclasses. Every order in the system will be either a PhoneOrder, an InternetOrder, or a MailOrder. Each of the three subclasses is considered a **concrete class** because it can be instantiated (in other words, objects can be created). The purpose of an abstract class is illustrated by the figure. It provides a central holding place for all the attributes and methods that each of the three subclasses will need. This example demonstrates one way that object-oriented programming implements reuse. The methods and attributes in the abstract class only need to be written once in order to be reused by each of the subclasses.

abstract class

a class that can never be instantiated (no objects can be created of this type)

concrete class

a normal class that can be instantiated (objects can be created)

DEVELOPING THE FIRST-CUT DESIGN CLASS DIAGRAM

To start the design process, we develop a first-cut design class diagram based on the domain model. Figure 11-19 is a partial RMO domain model, as developed in Chapter 5 (see Figure 5-38).

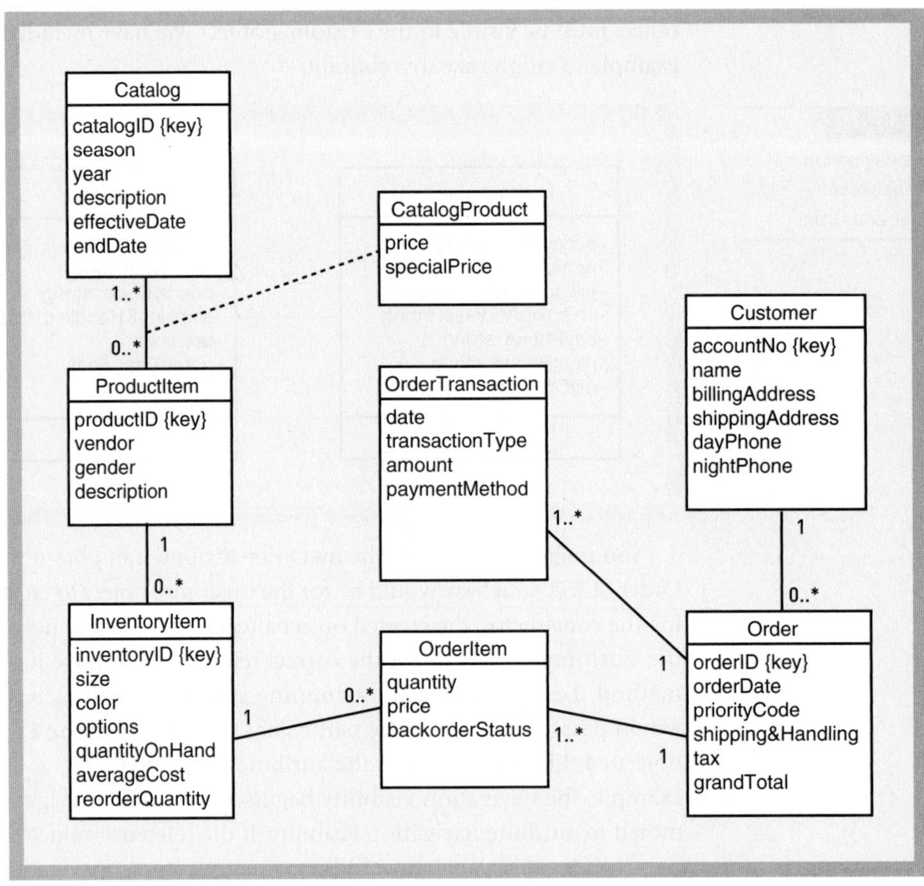

Figure 11-19

Partial RMO domain model class diagram

The first-cut design class diagram is developed by extending the domain model class diagram. It requires two steps: (1) elaborating on the attributes with type and initial value information and (2) adding navigation visibility arrows. As indicated earlier, object-oriented design is use case driven. So, let's choose a use case to start with and focus only on classes involved in that use case. It is always a good idea to begin with simple use cases, so we start with *Look up item availability*.

Elaboration of Attributes

The elaboration of the attributes is fairly straightforward. The type information is determined by the designer based on his or her expertise. In most instances, all attributes are kept invisible or private and are indicated with minus signs before them. We also need to add a new compartment to each class for the addition of method signatures.

Navigation Visibility

navigation visibility

a design principle in which one object has a reference to another object and thus can interact with it

As stated earlier, an object-oriented system is a set of interacting objects. The interaction diagrams developed during design document what interactions occur between which objects. However, for one object to interact with another by sending a message, the first object must be visible to the second object. Navigation visibility, in this context, refers to the ability of one object to be able to view and interact with another object. We use two types of navigation visibility during design: *attribute navigation visibility* and *parameter navigation visibility*. Attribute navigation visibility occurs when a class has an attribute that references another object. Visibility is obtained through the attribute reference. Parameter navigation visibility occurs when a class is passed a parameter that references another object. A parameter is usually passed through a method call. Sometimes developers refer to navigation visibility as just *navigation* or *visibility*. However, we prefer the term *navigation visibility* to distinguish the concept from public and private visibility on attributes and methods.

Figure 11-20 shows one-way attribute navigation visibility between the Customer class and the Order class. Notice the variable called myOrder in the Customer class. This variable holds a value to refer to an order instance. The navigation arrow indicates that an Order object must be visible to the Customer object. We have included the myOrder attribute in the example to emphasize this concept.

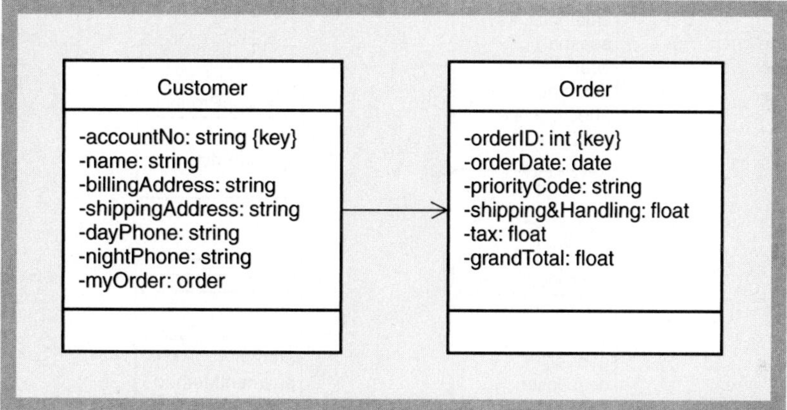

Figure 11-20

Attribute navigation visibility between Customer and Order

You might wonder how the myOrder attribute can obtain the correct value to refer to the Order object. One way would be for the Customer object to create the Order object. By invoking the constructor, the created object already contains the needed reference. Another way for the Customer object to get the correct reference is to have it passed in as a parameter on a method. For example, a programming statement such as *myCustomer.addOrder (anOrder)* would pass a reference to the particular Order object to the Customer object. The Customer object could then place it in the attribute called myOrder, as shown in Figure 11-20. In this example, the navigation visibility begins as parameter navigation visibility, but is then promoted to attribute navigation visibility. If the reference value for an Order is stored in a temporary or local variable, it would remain at parameter navigation visibility and would not be promoted to attribute navigation visibility.

Because interactions between objects can only be accomplished with navigation visibility, you must always be aware of it as a designer. In programming jargon, invoking a method on an object frequently requires dot notation to invoke the correct method on the correct object. Remember from your programming experience that dot notation qualifies the name of a method or attribute to correctly identify the object to which it applies, such as myOrder.changePriority(). One responsibility of a design is to specify which classes have navigation visibility to other classes. Attribute navigation visibility can be either one way or two way. For example, a Customer object may be able to view an Order object, which means the Customer object knows which orders a customer has placed. In programming terms, the Customer class has a variable, or an array of variables, that point to the Order object(s) for that customer. If navigation is two-way, then each Order object will also have a variable that refers to the Customer object. If the navigation is not two-way, then Order objects will not have a variable to point to the Customer object. In a design class diagram, attribute navigation visibility is identified by an arrow between the classes, where the arrow points to the visible class. Parameter navigation visibility is sometimes shown by a dashed arrow between the classes.

Now let's think about adding navigation visibility to the RMO class diagram. Remember that we are designing just the first-cut class diagram, so we might need to modify the navigation arrows as the design progresses. We ask the following basic question when building navigation visibility: Which classes need to have references to, or be able to access, which other classes? Here are a few general guidelines.

- One-to-many relationships that indicate a superior/subordinate relationship are usually navigated from the superior to the subordinate; for example, from Order to OrderItem. Sometimes these relationships form hierarchies of navigation chains; for example, from Catalog to ProductItem to InventoryItem.
- Mandatory relationships, in which objects in one class cannot exist without objects of another class, are usually navigated from the more independent class to the dependent class; for example, from Customer to Order.
- When an object needs information from another object, a navigation arrow might be required, pointing either to the object itself or to its parent in a hierarchy.
- Navigation arrows may also be bidirectional.

Figure 11-21 is a first-cut design class diagram for the use case *Process new order* based on the two steps described earlier in this section. The first step is to elaborate on the attributes with type information and visibility. The second step is to identify which classes may be involved and which classes require navigation visibility to other classes. We identify the classes that appear to be necessary to carry out the use case. We determine what other classes are necessary based on what information is needed. For example, price information is in the CatalogProduct class and description information is in the ProductItem class. One thing to remember about visibility is that the classes are programming classes, not database classes. So we are not thinking about foreign keys in a relational database. We are thinking about object references in a programming language.

Figure 11-21

First-cut RMO design class diagram for the *Process new order* use case

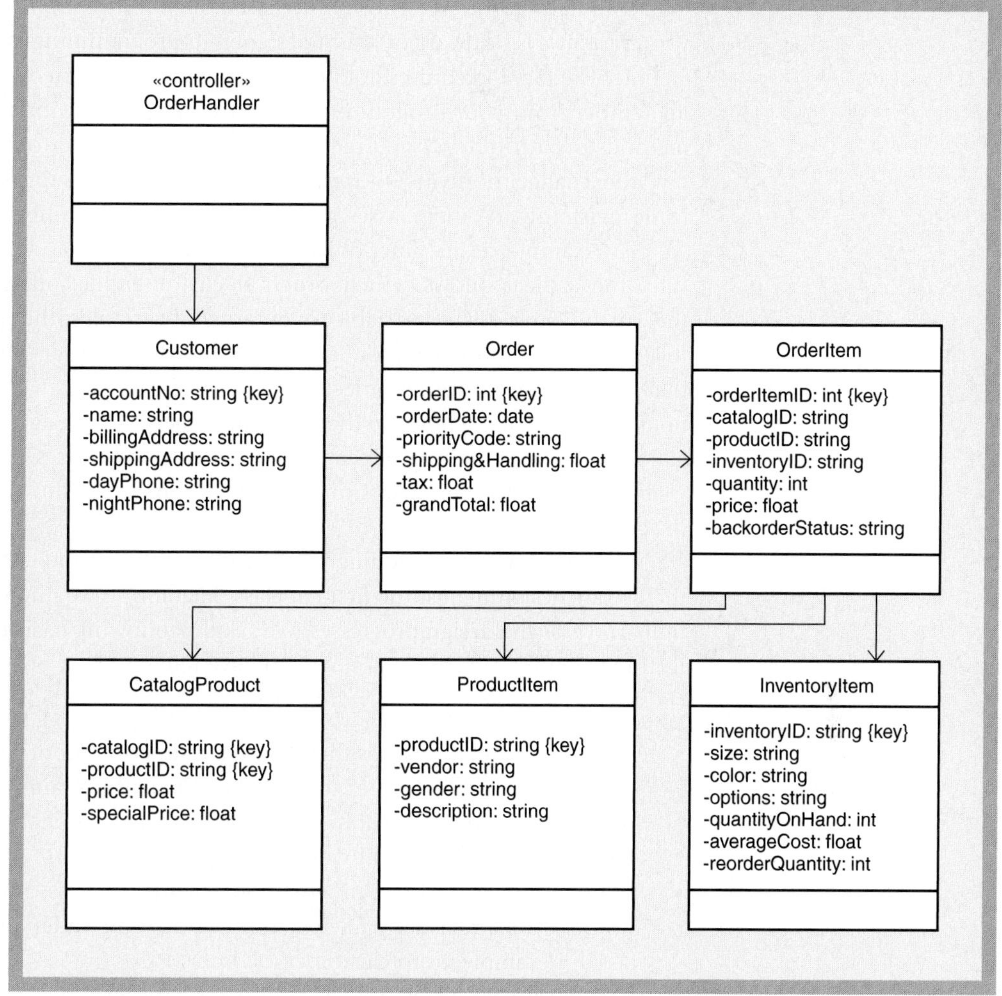

Figure 11-21 has one additional design class in the diagram for this use case, OrderHandler, which is stereotyped as a controller class. As mentioned previously, a controller class, or use case controller, is a utility class that helps in the processing of a use case. Notice that it has visibility at the top of the visibility hierarchy.

Three points are important to note. First, as detailed design proceeds use case by use case, we need to ensure that the interaction diagrams support and implement the navigation that was initially defined. Second, the navigation arrows need to be updated as design progresses to be consistent with the design details. Finally, method signatures will be added to each class based on the design decisions made when creating the interaction diagrams for the use cases.

As a preliminary step before developing interaction diagrams, many developers like to use CRC cards in brainstorming sessions to help identify the sets of classes involved in each use case. The next section explains how CRC cards can help with detailed object-oriented design.

DETAILED DESIGN WITH CRC CARDS

CRC stands for Class-Responsibility-Collaboration. It defines a brainstorming technique that is quite popular among object-oriented developers. Developers use it during analysis activities to help identify classes and the scope of each class. It is also used extensively during design to help identify responsibilities of the class and the sets of classes that collaborate for a particular use case.

A CRC card is simply a 3x5 or 4x6 index card with lines that partition it into three areas—class name, responsibility, and collaboration classes. Figure 11-22 illustrates the two sides of a

CRC card from the RMO customer support system. The card is partially filled out. Along the top of the card is the name of the class. The left partition lists the responsibilities for objects in this class. Responsibilities include information that the class maintains and actions that the class carries out in support of some use case. The right partition lists other classes with which this class collaborates in support of a particular use case. The information within parentheses is return information from the collaborating class to the main class. On the back of the card, you have the option of listing important attributes that are required for a particular use case.

Figure 11-22
Example CRC card

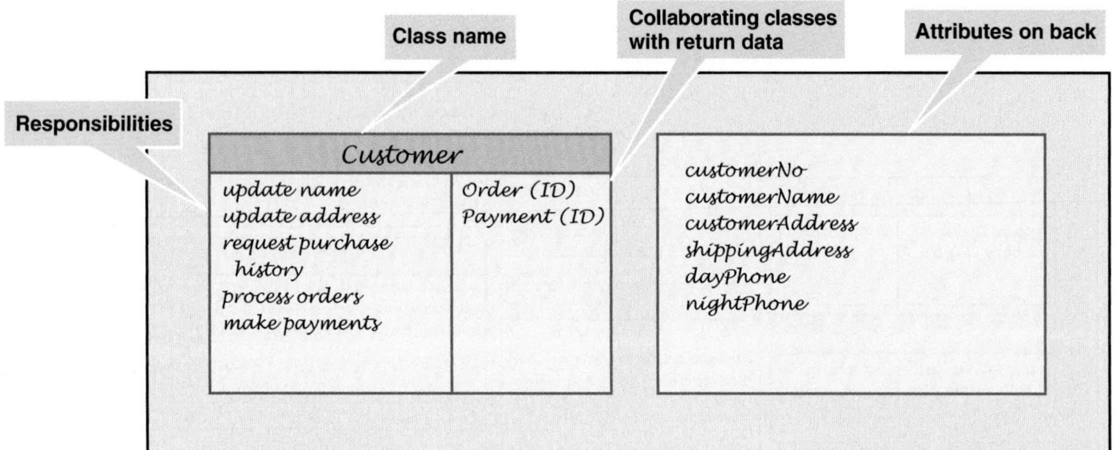

The process of developing a CRC model is usually done in a brainstorming session. A principal benefit of using CRC cards is that it requires a group effort, so the design is being reviewed and evaluated while it is being developed. For detailed design, this can be done with a couple of developers. A user may be invited to participate—particularly a technically oriented user—although it is not essential. In any event, a member of the group should be a domain expert, either as a user or a team member who has become proficient in the problem domain.

A design session using CRC cards already has substantial information from which to begin. Before starting the design session, each team member should have a copy of the domain model class diagram. Of course, the use case diagram or list of use cases also needs to be available. Other detailed information such as activity diagrams, system sequence diagrams, and use case descriptions should be provided, along with a stack of empty, CRC-formatted index cards. For each use case you need to design, the following process is done iteratively.

- **Select a use case.** Because the process is to design, or realize, a single use case, start with a set of unused CRC cards. Because we are doing multilayer design, make up one card as the use case controller card.
- **Identify the problem domain class that has responsibility for this use case.** This object will receive the first message from the use case controller. Using the domain model that was developed during analysis, select one class to take responsibility. Focus only on the problem domain classes. On the left side of the card, write the object's responsibility. For example, a customer object may take responsibility to make a new order, so one responsibility may be *Process a new order*.
- **Identify other classes that must collaborate with the primary object class to complete the use case.** Other classes will have required information. For our example of creating an order, we will need an order class card and order item class card, pricing information will probably come from the product class, and the inventory class will have to be checked for stock on hand. List these classes on the primary problem domain card. As you identify the other classes, write their responsibilities on their cards. Also, on the backs of all cards, write the pertinent information or attribute of each object class.

At the end of this process, you will have a small set of CRC cards that collaborate to support the use case. This process can be enhanced with several other activities. First, the CRC cards can be arranged on the table in the order they are executed or called. In other words, the calling order can be determined at this time. For example, the customer object creates an order object, the order object creates order item objects, and order item objects access product and inventory objects to get required information. Figure 11-23 illustrates a solution set of CRC cards for the use case *Process new order*.

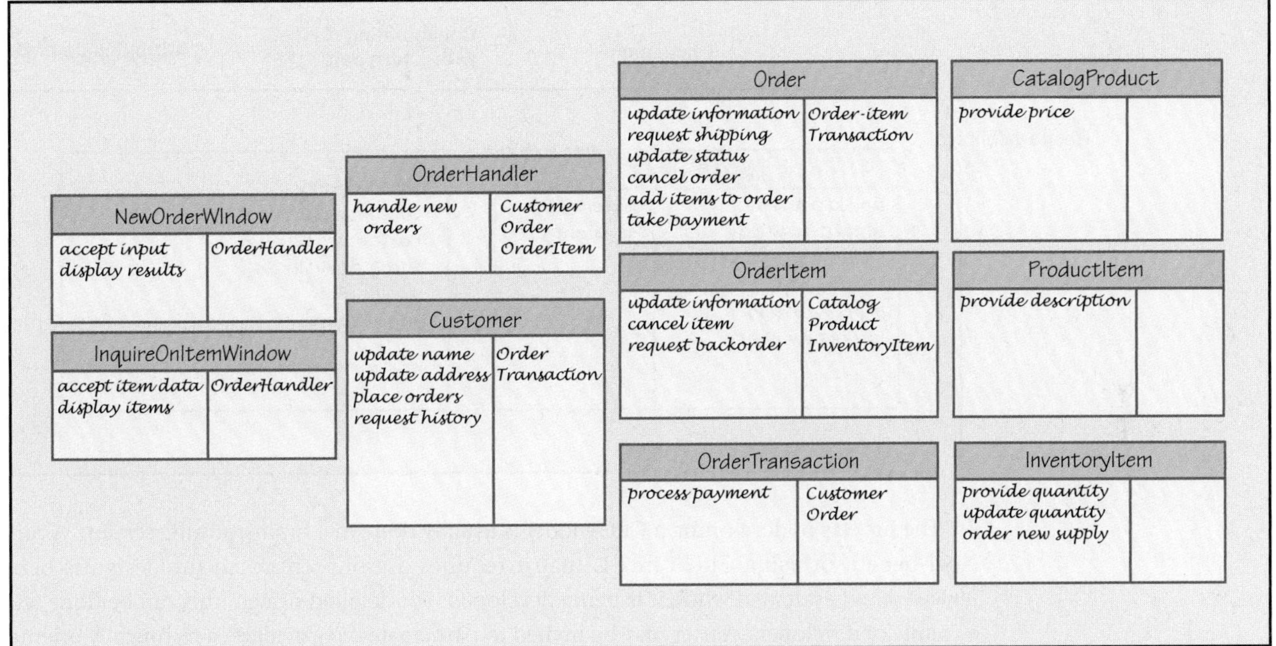

Figure 11-23

CRC cards model for *Process new order* use case

Another helpful step is to include the GUI classes. Chapter 12 explains the details of user interface design. It focuses heavily on the techniques and rules to develop effective computer screens for the system. If a user is part of the team, and if some preliminary work has been done on the user interface requirements, it could be effective to add CRC cards for all user interface window classes that are required for the use case. By including GUI classes, all of the input and output forms can be included in the design. Obviously, this is a much more complete design.

Any other required utility classes can also be added to the solution. For example, for a three-layer design, data access objects will be part of the solution. Each persistent class will have a data access class to read and write to the database. Other utility classes will be needed to be consistent with some design patterns, which will be discussed later. CRC cards for those classes can also be added to the solution.

At the end of the design for one use case, two other important tasks remain. Because the CRC cards only have data for a single use case, the information can be transferred to the design class diagram. The design class diagram then becomes a central repository for all information about every class in the new system.

A second task is to put an elastic band around the set of CRC cards for the next step. If the use case is a simple one, the CRC cards can be taken by a programmer and the use case can be implemented. If the use case is more complex, the set of cards can be given to a developer to be expanded into a sequence diagram for a more complete solution. Use case realization with sequence diagrams is covered in the next section.

To finish the example, let's go back to the DCD and update it based on the design information created during the brainstorming session. Figure 11-24 shows an updated DCD with several methods added and updates in the visibility. We first note that a new class has been

added. Evidently we overlooked the OrderTransaction class in the first-cut DCD. We also note that the OrderHandler needs visibility to the Order class to process a payment. Compare the responsibilities identified on the CRC cards and the method names described in each class. Note the close correlation.

Often when developers begin using CRC cards, they list many different responsibilities for a given class. For example, developers might say the OrderItem class should get the price. In reality, the CatalogProduct class provides the price and the OrderItem only uses it. In other words, it usually helps to think of responsibilities as being similar to methods—requests to do something, rather than random actions that need to occur.

Figure 11-24

Final DCD for the *Process new order* use case

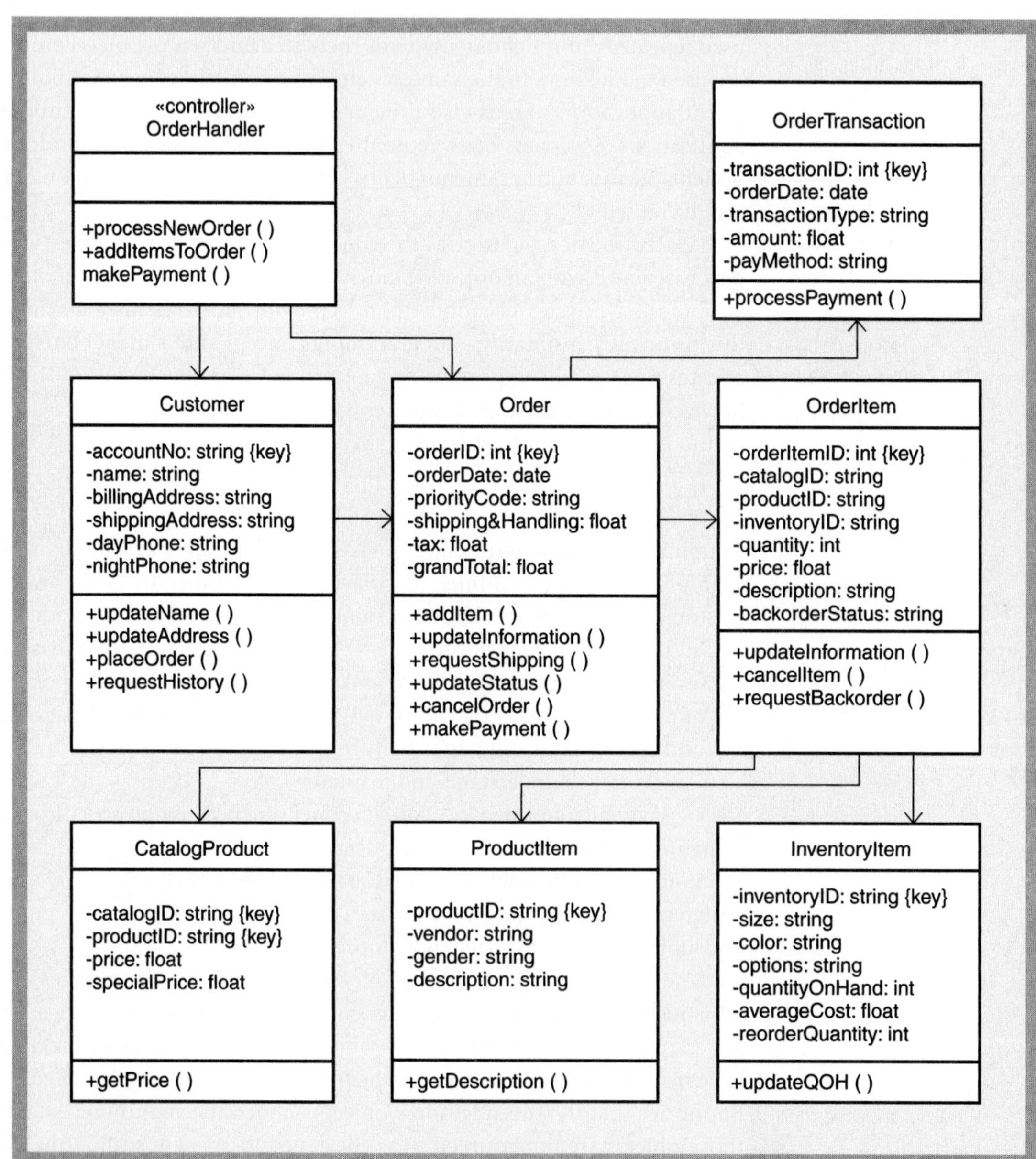

FUNDAMENTAL DETAILED DESIGN PRINCIPLES

Now that you understand how an object-oriented program works and you know the notation for a design class, let's review several basic principles that will guide design decisions. We used these principles throughout the chapter as we discussed the steps of object-oriented design because they are important to all parts of the process.

ENCAPSULATION AND INFORMATION HIDING

encapsulation

a design principle of objects in which both data and program logic are included within a single, self-contained unit

Encapsulation is the design principle that each object is a self-contained unit that includes both data and program logic. Each object internally carries its own data and provides a set of methods that access the data. Each object also provides a set of services that are invoked by calling the object's methods. One of the benefits of this approach is that a software developer can design the system in a building-block fashion. Nearly all engineering disciplines have standard units that serve as building blocks and that can be combined into a final design. Encapsulated objects are the software equivalent of building blocks.

object reuse

a design principle in which a set of standard objects can be used repeatedly within a system

Programmers also depend heavily on the benefits of encapsulation to support the idea of object reuse. Every object-oriented language comes with a set of standard objects that are used repeatedly throughout a system. These standard sets of objects provide basic services that are used many times in the same system—and sometimes even in multiple systems. One frequent application of reuse is in the design of the user interface, either for desktop or Web applications. Designers often reuse the same classes to develop windows and window components such as buttons, menus, icons, and so forth. Sometimes problem domain classes can also be reused.

information hiding

a design principle in which data associated with an object is not visible to the outside world, but methods are provided to access or change the data

Related to encapsulation is the concept of information hiding, which dictates that the data associated with an object is not visible to the outside world. In other words, the object's attributes are private. A set of methods is provided to access the data and modify it. Although this principle is primarily a programming concept and is most beneficial for programming and testing, several important design principles are based on it. The linkage or coupling between objects in a system is better if access to data attributes uses a standard interface of method names, as explained next.

COUPLING

Coupling is a general term that is derived from attribute navigation visibility. In the previous example in which Customer had navigation visibility to Order, we could also say that Customer and Order are coupled, or linked. Now, extend this same idea of visibility throughout all the classes in the entire system. Coupling is a qualitative measure of how closely the classes in a design class diagram are linked. A simple way to think about coupling is the number of navigation arrows on the design class diagram. Low coupling for the system is usually better than high coupling. In other words, fewer navigation visibility arrows indicate that a system is easier to understand and maintain.

coupling

a qualitative measure of how closely the classes in a design class diagram are linked

We say that coupling is a qualitative measure because no specific number measures coupling in a system. A designer must develop a feel for coupling—to recognize when there is too much or to know what is a reasonable amount of coupling. Coupling is evaluated as a design progresses use case by use case. Generally, if each use case design has a reasonable level of coupling, the entire system will, too.

Refer back to Figure 11-1 and observe the flow of messages between the objects. Obviously, objects that send messages to each other must have navigation visibility and thus are coupled. For the Input window object to send a message to the Student object, it must have navigation visibility to it. So, the Input window object is coupled to the Student object. But notice that the Input window object is not connected to the Database access object, so those objects are not coupled. If we designed the system so that the Input window object accessed the Database access object, the overall coupling for this use case would increase—that is, there would be more connections. Is that good or bad? In this simple example, it might not be a problem. But for a system with 10 or more classes, too many connections with navigation visibility can cause high levels of coupling, making the system more complex.

So why is high coupling bad? The main reason is that it adds unnecessary complexity to a system, making it hard to maintain. A change in one class ripples throughout the entire system. So, experienced analysts make every effort to simplify coupling and reduce ripple effects in the design of a new system.

COHESION

cohesion

a qualitative measure of the consistency of functions within a single class

Cohesion refers to the consistency of the functions within a single class. Cohesion is a qualitative measure of the focus or unity of purpose within a single class. For example, in Figure 11-1, you would expect the Student class to have methods—that is, functions—to enter student information such as identification number or name. That would represent a unity of purpose and a highly cohesive class. But what if that same object also had methods to make classroom assignments or assign professors to courses? The cohesiveness of the class would be reduced.

Classes with low cohesion have several negative effects. First, they are hard to maintain. Because they perform many different functions, they tend to be overly sensitive to changes within the system, suffering from ripple effects. Second, it is hard to reuse such classes. Because they have many different—and often unrelated—functions, it usually does not make sense to reuse them in other contexts. For example, a button class that processes button clicks can easily be reused. However, a button class that processes button clicks and user logons has limited reusability. A final drawback is that classes with low cohesion are usually difficult to understand. Frequently, their functions are intertwined and their logic is complex.

Although there is no firm metric to measure cohesiveness, we can think about classes as having very low, low, medium, or high cohesion. Remember, high cohesion is the most desirable. An example of very low cohesion is a class that has responsibility for services in different functional areas, such as a class that accesses both the Internet and a database. These two types of activities are different and accomplish different purposes. To put them together in one class causes very low cohesion.

An example of low cohesion is a class that has different responsibilities but in related functional areas; an example might be a class that does all database access for every table in the database. It would be better to have different classes to access customer information, order information, and inventory information. Although the functions are the same—that is, they access the database—the types of data passed and retrieved are very different. So, a class that is connected to the entire database is not as reusable as one that is only connected to the customer table.

An example of medium cohesion is a class that has closely related responsibilities, such as a single class that maintains customer information and customer account information. Two highly cohesive classes could be defined, one for customer information such as names and addresses. Another class or set of classes could be defined for customer accounts, such as balances, payments, credit information, and all financial activity. If the customer information and the account information are limited, they could be combined into a single class with medium cohesiveness. Either medium or highly cohesive classes can be acceptable in system design.

BEST PRACTICE

Good, experienced developers always think about how to keep coupling low and cohesion high. Always keep these concepts in mind when designing.

PROTECTION FROM VARIATIONS

protection from variations

a design principle in which parts of a system that are unlikely to change are segregated from those that will

One of the underlying principles of good design is protection from variations: the principle that parts of a system that are unlikely to change should be segregated (or protected) from those that will change. As you design systems, you should try to isolate the parts that will change from those that are more stable.

Protection from variations is a principle that drives the multilayer design pattern. Designers could mix all of the user-interface logic and business logic together in the same classes. In fact, in early user-oriented, event-driven systems such as those built with early versions of Visual Basic and PowerBuilder, the business logic was included in the view layer classes, often in the windows input forms. The problem with this design was that when an interface needed to be updated, all of the business logic had to be rewritten. A better

approach is to decouple the user-interface logic from the business logic. Then, the user interface can be rewritten without affecting the business logic. In other words, the business logic, being more stable, is protected from variations in the user interface.

Also, what if updates to the business logic require the addition of new classes and new methods? If the user-interface classes are tightly coupled to the business classes, there could be a ripple effect of changes throughout the user-interface classes. However, because the user interface can simply send all of its input messages to the use case controller class, changes to the methods or classes in the business logic and domain layer are isolated to the controller class. You will find that protection from variations affects almost every design decision, so you should watch for and recognize the application of this principle in all design activities.

INDIRECTION

indirection
a design principle in which an intermediate class is placed between two classes to decouple them but still link them

Indirection is a popular object-oriented design principle that can protect stable components from variations and reduce coupling. **Indirection** is the principle of decoupling two classes or other system components by placing an intermediate class between them to serve as a link. In other words, instructions don't go directly from A to B; they are sent through C first. Or, in message terminology, don't send a message from A to B. Let A send the message to C, and then let C forward it to B.

Although there are many ways to implement protection from variations, indirection is frequently used. Inserting an intermediate object allows any variations in one system to be isolated in that intermediate object. Indirection is also useful for many corporate security systems. For example, many companies have firewalls and proxy servers that receive and send messages between an internal network and the Internet. A proxy server appears as a real server, ready to receive messages such as e-mail and HTML page requests. However, it is a fake server, which catches all of the messages and redistributes them to the recipients. This step of indirection allows security controls to be put in place and protect the system.

OBJECT RESPONSIBILITY

object responsibility
a design principle in which objects are responsible for carrying out system processing

One of the most fundamental principles of object-oriented development is the idea of **object responsibility**—objects are responsible for carrying out the system processing. These responsibilities are categorized in two major areas: knowing and doing. In other words, what is an object expected to know, and what is an object expected to do or to initiate?

Knowing includes an object's responsibilities for knowing about its own data and knowing about other classes with which it must collaborate to carry out use cases. Obviously, a class should know about its own data, what attributes exist, and how to maintain the information in those attributes. It should also know where to go to get information when required. For example, during the initiation of an object, data that is not passed as parameters may be required. An object should know about, or have navigation visibility to, other objects that can provide the required information. In Figure 11-13, the first constructor method for the Student class does not receive a studentID value as a parameter. Instead, the Student class takes responsibility for creating a new studentID value based on some rules it knows.

Doing includes all the activities of an object to assist in executing a use case. Some of those activities include receiving and processing messages. Another responsibility is to instantiate, or create, new objects that may be required for completion of a use case. Classes must collaborate to carry out a use case, and some classes are responsible for coordinating the collaboration. For example, for the use case *Process new order*, the Order class has responsibility to create OrderItem objects. Another class, such as InventoryItem, is only responsible for providing information about itself.

SUMMARY

The primary creative activity of system developers is to write computer software that solves the business problem. So far, this textbook has focused on two major activities: to first understand the user's requirements (the business problem), and then to figure out and visualize a solution system. This chapter focused on how to configure and develop the solution system—in other words, design the system. System design is the bridge that puts business requirements in terms that the programmers can use to write the software that becomes the solution system.

Architectural design is the first step in configuring the new system. Its purpose is to determine the structure and configuration of the new system's various components. Two types of diagrams are used to document the architectural design: component diagrams and deployment diagrams. Component diagrams show the various executable components of the new system and how they relate to each other. Deployment diagrams show how and where the components are executed on various computing platforms. Together they define the system's configuration.

Many new systems are enterprise-level systems, in that they are used at locations throughout the entire organization. They also share resources, such as a common database of information.

Once the architectural design is known, then detailed design can begin. The objective of detailed design is to determine the objects and methods within individual classes to support the use cases. The process of detailed design is use case driven, in that it is done for each use case separately.

The process of detailed design can be divided into two major areas: developing a design class diagram (DCD), and developing the set of interacting classes and their methods for each use case via a sequence diagram. The DCD is usually developed in two steps. A first-cut DCD is created, but then it is refined and corrected as the sequence diagrams are developed. Sequence diagrams can also be developed in two steps. A preliminary idea of how the objects collaborate can be created using Class-Responsibility-Collaboration (CRC) cards. For simple use cases, a set of CRC cards may be sufficient to write code. For more complex use cases, the CRC cards serve as the beginning point for developing sequence diagrams.

One reason that we suggest a more formal system of design, rather than just starting to write code, is that the final system is much more robust and maintainable. Doing design as a rigorous activity builds better systems. Some fundamental principles should be considered as a system is developed; two critical ideas are coupling and cohesion. A good system has low coupling between the classes, and each of the classes has high cohesion. Another important principle is "protection from variations," meaning that some parts of the system should be protected from, and not tightly coupled to, other parts of the system that are less stable and subject to change. Being a good developer entails learning and following the principles of good design.

KEY TERMS

abstract class, p. 413
application program interface or API, p. 394
artifact, p. 401
boundary class or view class, p. 410
class-level attribute, p. 412
class-level method, p. 412
cohesion, p. 421
component diagram, p. 394
concrete class, p. 413
control class, p. 410
coupling, p. 420
data access class, p. 410
deployment diagram, p. 401
encapsulation, p. 420
enterprise-level system, p. 392
entity class, p. 410
indirection, p. 422
information hiding, p. 420
instantiation, p. 390
method signature, p. 411
navigation visibility, p. 414
object responsibility, p. 422
object reuse, p. 420
overloaded method, p. 411
overridden method, p. 412
persistent class, p. 410
protection from variations, p. 421
realization of use cases, p. 408
stereotype, p. 409
visibility, p. 411

REVIEW QUESTIONS

1. List the models that are used for systems design.
2. Which two models are used to do architectural design? What is the difference between the two?
3. What is an enterprise-level system? Why is it an important consideration in design?
4. What are some of the differences between a desktop system and a Web-based system?
5. What is an API? Why is it important?
6. What notation is used to identify the interface of a component?
7. What is meant by a three-layer design, and normally what are the three layers?
8. What is meant by Web services?
9. What is an artifact and how does it relate to a component?
10. What is the difference between the notation for problem domain classes and design classes?
11. In your own words, list the steps for doing detailed design.
12. What do we mean by use-case driven design, and what is use case realization?
13. What is a persistent class?
14. What is a class-level method?
15. Describe navigation visibility. Why is it important in detailed design?
16. List some typical conditions that dictate in which direction navigation visibility occurs.
17. What information is maintained on CRC cards?
18. What is the objective of a CRC card design session?
19. Compare and contrast the ideas of coupling and cohesion.
20. What is protection from variations and why is it important in detailed design?
21. What is meant by object responsibility and why is it important in detailed design?

THINKING CRITICALLY

1. Given the following system description, develop a component diagram for a desktop-operated internal network system (in other words, Internet access is not required).
2. Develop a deployment diagram for the network mentioned in the previous problem.

   The new Benefits for Employees, Spouses, and Dependents (BESD) system will be used primarily by the human resource department and will contain confidential information. Consequently, it will be built as a totally in-house system without any Internet elements. The database for the system is the human resource employee database (HRED), which is shared by several other systems within the company.

   There are two types of screens, from a systems design point of view: simple inquiry screens and complex inquiry/update screens. The simple inquiry screens just access the database, with no business logic required. The complex screens usually do fairly complicated calculations based on sophisticated business rules. These programs often have to access other data tables from other databases in the company.

   The database will always remain on a central database server. The application program itself will be installed on each person's desktop that is allowed access. However, authentication is a centralized process, and it will control which screens and program functions can be accessed by which users.
3. Develop a component diagram for the following description of a Facebook application.
4. Develop a deployment diagram for the following description of a Facebook application.

   The Facebook platform is available for entrepreneurs to develop applications for use among all Facebook users. A new application is being written that allows Facebook users to send gifts and greeting cards to their friends. (These are real gifts and greeting cards, not just electronic images.) The application running within Facebook is on its own server and has its own database of information, which includes a list of gifts and cards that have been sent or received. The actual retail store of gifts and cards to send must be located on a different server because it is part of a regular Internet sales storefront. This storefront maintains the database of inventory items to sell, and collects credit card payment information.
5. In the chapter we developed a first-cut DCD, a set of CRC cards, and final DCD for the *Process new order* use case for RMO. Create the same three drawings for the *Look up item availability* use case.
6. In Chapter 7, Problem 1, you developed a problem domain class diagram for a library system. Convert that class diagram to a first-cut DCD. Update the attributes and add navigation visibility as much as possible, using the rules explained in the chapter.
7. For the previous problem, use the DCD and identify a set of CRC cards that would be used for the *Check out book* use case. You might want to get a set of index cards and actually do a design session. Either scan your cards or type the information to give to your instructor. Once you have identified the responsibilities by class, update the DCD by adding method names to the classes in the DCD.

8. In Chapter 7, Problem 3, you developed a problem domain class diagram for a dentist office system. Convert your class diagram to a first-cut DCD. Update the attributes and add navigation visibility as much as possible, using the rules explained in this chapter.
9. For the previous problem, use the DCD and identify a set of CRC cards that would be used for the *Patient visits dentist* use case. Assume that the visit is a simple checkup and that no other procedures will be performed. In other words, entering specific procedures is done with a different use case. You might want to get a set of index cards and actually do a design session. Either scan your cards or type the information to give to your instructor. Once you have identified the responsibilities by class, update the DCD by adding method names to the classes in the DCD.
10. Problems 7 and 8 in Chapter 7 describe a car insurance system and identify various problem domain classes. Using the information in those problems, develop a first-cut DCD. Update the attributes and add navigation visibility as much as possible, using the rules explained in the chapter.
11. For the previous problem, use the DCD and identify a set of CRC cards that would support the *Add new vehicle to existing policy* use case. You might want to get a set of index cards and actually do a design session. Either scan your cards or type the information to give to your instructor. Once you have identified the responsibilities by class, update the DCD by adding method names to the classes in the DCD.

EXPERIENTIAL EXERCISES

1. Find a company that does object-oriented design using CRC cards. The information systems unit at your university often uses OO techniques. Sit in on a CRC design brainstorming session. Interview some of the developers about their feelings regarding the effectiveness of doing CRC design. Find out what documentation remains after the sessions and how it is used.
2. Find a company that has an internal system. (If you are working for a company, see what systems they use.) Analyze the system and develop a component diagram and a deployment diagram.
3. Find a system that was developed using Java. If possible, find one that has both an Internet user interface and a network-based user interface. Is it multilayer—three layer or two layer? Can you identify the view layer classes, the domain layer classes, and the data access layer classes?
4. Find a system that was developed using Visual Studio .NET (or Visual Basic). If possible, find one that has both an Internet user interface and a network-based user interface. Is it multilayer? Where is the business logic? Can you identify the view layer classes, the domain layer classes, and the data access layer classes?
5. Pick an object-oriented programming language with which you are familiar. Find a programming integrated development environment (IDE) tool that supports that language. Test its reverse-engineering capabilities to generate UML class diagrams from existing code. Evaluate how well it does and how easy the models are to use. Does it have any capability to input UML diagrams and generate skeletal class definitions? Write a report on how it works and what UML models it can generate.

CASE STUDIES

THE REAL ESTATE MULTIPLE LISTING SERVICE SYSTEM

In Chapter 7, you developed a use case diagram, a class diagram, and a system sequence diagram for the real estate company's use cases. First develop a three-layer architectural solution, which allows agents and other clients to browse the database via the Internet. Internal updates should be done by screens within an internal network system. Also, based on those solutions or others provided by your teacher, convert the domain class diagram to a first-cut design class diagram by type casting the attributes and adding navigation visibility. Next, using the use cases indicated by your teacher, develop a set of CRC cards for each one. Based on the class responsibilities identified on the CRC cards, make up method names for the appropriate classes and add those methods to the DCD classes.

THE STATE PATROL TICKET PROCESSING SYSTEM

In Chapter 7, you developed a use case diagram, a class diagram, and a system sequence diagram for the use cases *Recording a traffic ticket* and *Scheduling a court date*. Based on those solutions or others provided by your teacher, develop a first-cut DCD by type casting the attributes and adding navigation visibility. Then, for each use case, develop a set of CRC cards. Add method names to the classes in the DCD based on the responsibilities identified on the CRC cards.

THE DOWNTOWN VIDEOS RENTAL SYSTEM

In Chapter 7, you developed a use case diagram, a class diagram, and a system sequence diagram for the use cases *Rent movies* and *Return movies*. Based on those solutions or others provided by your teacher, develop a first-cut DCD by type casting the attributes and adding navigation visibility. Then, for each use case, develop a set of CRC cards. Add method names to the classes in the DCD based on the responsibilities identified on the CRC cards.

THEEYESHAVEIT.COM BOOK EXCHANGE

First, develop a three-layer architectural design of the book exchange system. In Chapter 7, you developed a use case diagram, a class diagram, and a system sequence diagram for the use cases *Add a seller* and *Record a book order*. Based on those solutions or others provided by your teacher, develop a first-cut DCD by type casting the attributes and adding navigation visibility. Then, for each use case, develop a set of CRC cards. Add method names to the classes in the DCD based on the responsibilities identified on the CRC cards.

RETHINKING ROCKY MOUNTAIN OUTFITTERS

This chapter presented the solutions for two use cases for RMO—*Look up item availability* and *Process new order*. Design three-layer solutions for two more use cases, *Create new order* and *Record order fulfillment*. Update the design class diagram for the problem domain classes with method signatures from these use case designs. Often, the sequence diagram to produce a report can be quite interesting. Do a three-layer design for the use case *Produce order fulfillment report*. Because you do not have detailed user requirements for this use case, you must first lay out a sample fulfillment report.

FOCUSING ON RELIABLE PHARMACEUTICAL SERVICE

In Chapter 7, you developed a use case diagram, a domain model class diagram, and detailed documentation for three use cases. In your detailed documentation, you generated a fully developed specification and a system sequence diagram. Based on that information and the guidelines in this chapter, design a three-layer architecture for the new system. Update the design class diagram with information on attribute types and navigation visibility. For the same three use cases, develop a set of CRC cards. Then update the DCD with method names derived from the class responsibilities on the CRC cards.

FURTHER RESOURCES

Grady Booch, James Rumbaugh, and Ivar Jacobson, *The Unified Modeling Language User Guide*. Addison-Wesley, 1999.

Grady Booch, et al., *Object-Oriented Analysis and Design with Applications, Third Edition*. Addison-Wesley, 2007.

E. Reed Doke, J. W. Satzinger, and S. R. Williams, *Object-Oriented Application Development Using Java*. Course Technology, 2002.

E. Reed Doke, J. W. Satzinger, and S. R. Williams, *Object-Oriented Application Development Using Microsoft Visual Basic .NET*. Course Technology, 2003.

Hans-Erik Eriksson, Magnus Penker, Brian Lyons, and David Fado, *UML 2 Toolkit*. John Wiley and Sons, 2004.

Martin Fowler, *UML Distilled Third Edition: A Brief Guide to the Standard Object Modeling Language*. Addison-Wesley, 2004.

Ivar Jacobson, Grady Booch, and James Rumbaugh, *The Unified Software Development Process*. Addison-Wesley, 1999.

Philippe Kruchten, *The Rational Unified Process, An Introduction*. Addison-Wesley, 2000.

Craig Larman, *Applying UML and Patterns: An Introduction to Object-Oriented Analysis and Design and the Unified Process* (3rd ed.). Prentice-Hall, 2004.

Jeffrey Putz, *Maximizing ASP.NET Real World, Object-Oriented Development.* Addison-Wesley, 2005.

James Rumbaugh, Ivar Jacobson, and Grady Booch, *The Unified Modeling Language Reference Manual*. Addison-Wesley, 1999.

CHAPTER

12

OBJECT-ORIENTED DESIGN: USE CASE REALIZATIONS

LEARNING OBJECTIVES

After reading this chapter, you should be able to:

- Explain the different types of objects and layers in a design
- Develop sequence diagrams for use case realization
- Develop communication diagrams for detailed design
- Develop updated design class diagrams
- Develop multilayer subsystem packages
- Explain design patterns and recognize various specific patterns

CHAPTER OUTLINE

Detailed Design of Multilayer Systems

Use Case Realization with Sequence Diagrams

Designing with Communication Diagrams

Updating and Packaging the Design Classes

Design Patterns

NEW CAPITAL BANK: PART 2

The integrated customer account system project for New Capital Bank was now two months old. The first development iteration had gone well, although there were a few snags because the team was still learning the ins and outs of object-oriented design. Most team members were beginning to feel more comfortable with detailed design techniques. Bill Santora, the project leader, was discussing some of the system's technical details with one of his team leaders, Charlie Hensen.

"How is the team feeling about doing detailed design? I know some of the programmers wanted to just start coding from the use case descriptions that were developed with the users. They were not very happy about taking the time to design. Is that still a problem?" Bill asked.

Charlie was one of the early critics of doing more formal design, so Bill was interested to hear his opinion. "It really has worked out quite well. As you know, I was skeptical at first and thought it would waste a lot of time. But instead it has enabled us all to work together better because we know what the other team members are doing. I also think the system is much more solid. We are all using the same approach and we also have discovered that there are quite a few classes we share. In other words, reusability really does work for our utility classes. Of course, we don't waste a lot of time making fancy drawings. We do document our designs with some quick drawings, but that is about as far as we take it."

"What would you say were the strengths and weaknesses of our approach? Or really, are there ways you think we could do it better in this next iteration?" As the project leader, Bill was always trying to improve the effectiveness of his team.

"I really like the approach to first do a rough design using CRC cards. It's nice to have a couple of users there with us to verify that our collaborations are correct. For the simple use cases, we then work with the users to lay out the user interface. Between the CRC cards and the user interface specifications, we often have enough to program from, especially now that we have the basic structure set up. Then for the more complex use cases, we go ahead and do a detailed design with sequence diagrams. Even then we need quite a bit of user involvement to get the GUI windows designed. The nice thing about the sequence diagrams is that they are detailed enough that we can give those designs to some of the junior programmers. It makes them much more effective in their team contributions." Charlie stopped a minute to think.

"So would you change our approach, or do you think it is working the right way?" Bill was still trying to look for ways to improve the process.

"Well, it really is working pretty well right now. One thing that I really like about it is that we have a common DCD that everyone can access and review. That really helps when you are ready to insert some code into a class to check and see what is already there. The central repository for all our code and for those diagrams that we do formalize is a great tool. I wonder if there is a way to get more use out of that tool. Other than that, I would say let's leave it for another iteration and then see if it needs to be changed." Charlie seems to have been converted to the benefits of creating design diagrams before actually generating code.

OVERVIEW

Chapter 11 explained the design concepts and models used for multilayer systems and their architectural design. The latter portion of the chapter introduced the concepts associated with object-oriented detailed design. You also learned how to begin the detailed design process by using CRC cards and design class diagrams to identify which classes collaborate to carry out use cases. Simple use cases can frequently be programmed from the design information developed with these two steps. In fact, most systems analysis and design textbooks go no further in teaching detailed design.

This chapter pursues object-oriented detailed design in much more depth and formality. Detailed design is a subject that can be addressed at multiple levels. For the beginner, a fairly straightforward, yet complete, process can be defined. It can also be pursued in more depth. This chapter focuses on the foundation principles, which are based on the concepts of use case realization using sequence diagrams and design patterns. Once you master these two subjects, you can consider yourself an accomplished object-oriented designer. Again, many good books are available on design patterns and design methods. If you are considering a more technical career as a systems developer, you will do further reading and studying. The resources at the end of the chapter provide a good starting point.

use case realization
the process of elaborating the detailed design with interaction diagrams of a particular use case

The method we use to extend the process of detailed design is called use case realization. In use case realization, you take each use case individually and determine all the classes that collaborate on it. As part of that process, we also determine what other utility or support classes may be required. These support classes may be user interface classes, data access classes, and other utility classes. We are careful during this process to define the classes so that the integrity of the multilayer architectural design is maintained. As we design the details of the classes, use case by use case, we update the design class diagram as necessary.

The last section of the chapter is a brief introduction to design patterns. As with any engineering discipline, certain standard procedures are tried and proven solutions. Even though object-oriented development is a relatively young discipline, it offers standard ways to design use cases that provide solid, well-constructed solutions. You will learn a few of those standard designs or patterns.

DETAILED DESIGN OF MULTILAYER SYSTEMS

The discussion of CRC cards in the previous chapter introduced you to the idea of collaborating objects to execute various use cases. However, the focus of design sessions using CRC cards is on the problem domain classes, with very little about multilayer issues as they affect the detailed design. This chapter goes into more depth to describe the detailed design of all layers of a multilayer system.

Referring back to Figure 11-1 in Chapter 11, we note that the three objects represent the three layers of a system. Each object also has certain responsibilities. The input window object has primary responsibility to format and present student information on the screen. Another responsibility is to accept input data, either student ID or changed information, and forward it into the system. The object probably does some editing of the input data as well. Where does this object come from? What are the attributes and methods of this object? Identifying and defining the window objects is part of the application design and the user interface design. This chapter discusses the more technical issues, while Chapter 14 discusses the human factors and human-computer interaction issues.

The student object represents the middle layer, or business logic layer, for the use case. A CRC design session will help you design the structure of the objects in this layer. However, you probably noted that CRC cards are quite informal, especially when trying to ascertain class methods from object responsibilities. CRC cards provide little direction in defining method signatures with appropriate input and output parameters. This chapter will formalize the process of precisely identifying methods and defining method signatures.

The database access object corresponds to the third layer in the multilayer design. Its responsibility is to connect to the database, read the student information, and send it back to the student object. It also has the responsibility of writing the student information back to the database when it needs to be updated. This object does not come from a problem domain class; it is a utility object created by the designer.

Several questions should come to mind as you review detailed systems design. The first question is: "How do all these objects get created in memory?" For example, how and when does the student object get created? How about the database access object? Other questions might be: Will other objects be necessary? What object does authentication? What is the life span of each of these objects? Maybe the student object should go away after the update, but how about the database access object?

DESIGN PATTERNS AND THE USE CASE CONTROLLER

Patterns, also called *templates*, are used repeatedly in everyday life. A chef uses a recipe, which is just another word for a pattern, to combine ingredients into a flavorful dish. A tailor uses a pattern to cut fabric for a great-fitting suit. Engineers take standard components and combine them into established configurations, or set patterns, to build buildings, sound systems, and thousands of other products. Patterns are created to solve problems. Over time, and with many attempts, people who work on a particular problem develop a set solution to the problem. The solution is general enough that it can be applied over and over again. As time passes, the solution is documented and published, and eventually it becomes accepted as the standard.

design patterns

standard design techniques and templates that are widely recognized as good practice

Standard design templates have become popular among software developers because they can speed OO design work. The formal name for these templates is **design patterns**. Design patterns became a widely accepted object-oriented design technique in 1996, with the publication of *Elements of Reusable Object-Oriented Software* by Eric Gamma, Richard Helm, Ralph Johnson, and John Vlissides. The four authors are now referred to as the Gang of Four (GoF). As you learn more about design patterns, you will often see references to a particular design pattern as a GoF pattern. In their book, the authors identified 23 basic design patterns. Today, scores of patterns have been defined, from low-level programming patterns to mid-level architectural patterns to high-level enterprise patterns. The two primary enterprise platforms, Java and .NET, both have sets of enterprise patterns, which are described in various books and publications.

In Chapter 11 you were introduced to the idea of a use case controller. In this chapter, we formalize the concept and explain its importance as a design pattern. For any particular use case, messages come from the external actor to a windows class (that is, an electronic input form) and then to a problem domain class. One issue in systems design is the question of which problem domain class should receive input messages to reduce coupling, maintain highly cohesive domain classes, and maintain independence between the user interface and the domain layer. Designers often define an intermediary class that acts as a buffer between the user interface and the domain classes. We call these classes use case controllers. For example, the use case *Create new order* might have a controller class named OrderHandler.

Figure 12-1 provides a more formal specification for the use case controller pattern. Note that this specification has five main elements:

- The pattern name
- The problem that requires a solution
- The solution, or explanation, of the pattern
- Example of the pattern
- The benefits and consequences of the pattern

You should read this specification to understand the important principles of the controller pattern, what problem it solves, how it works, and its benefits. This same template will be used later in the chapter with the other design patterns.

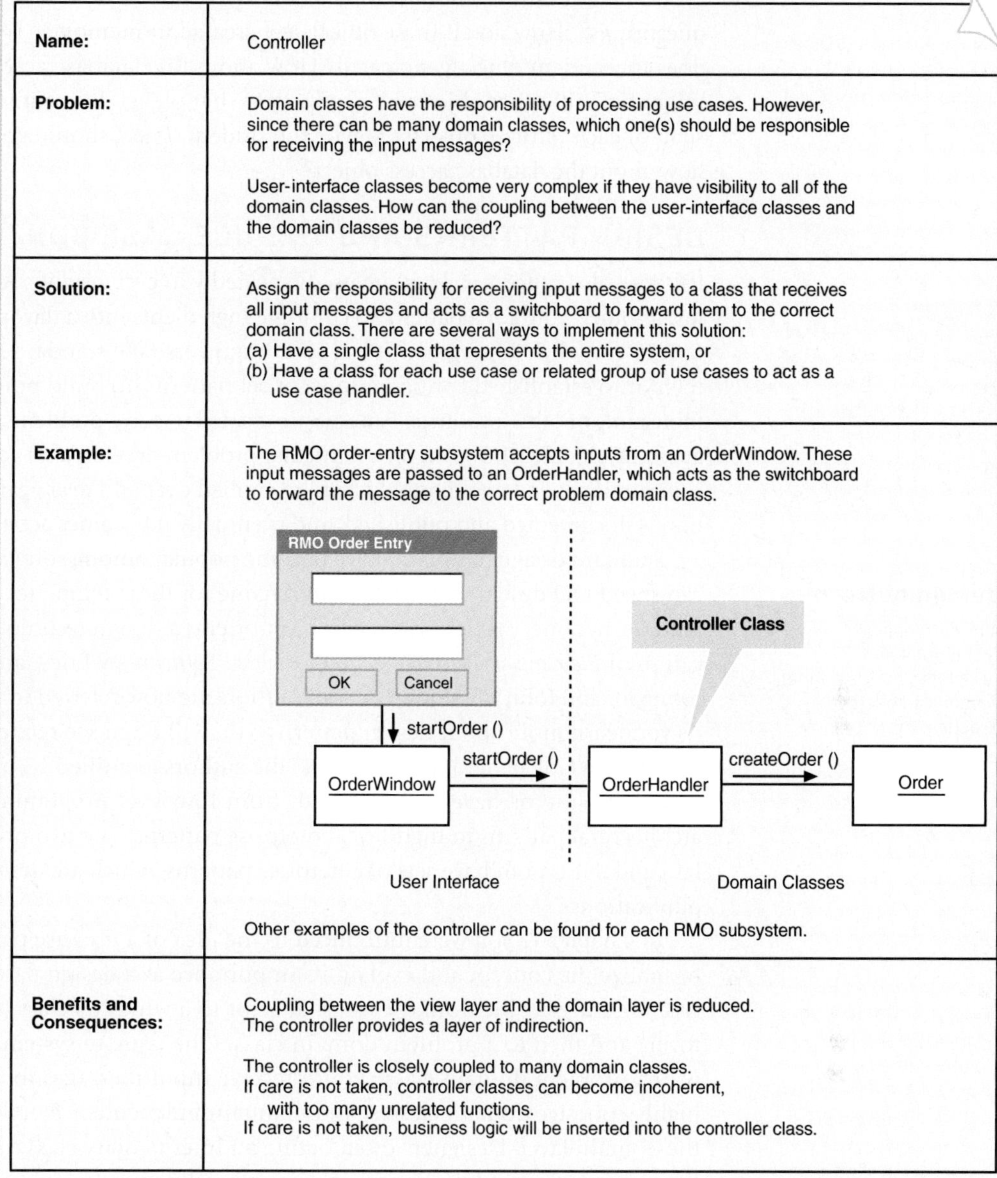

Name:	Controller
Problem:	Domain classes have the responsibility of processing use cases. However, since there can be many domain classes, which one(s) should be responsible for receiving the input messages? User-interface classes become very complex if they have visibility to all of the domain classes. How can the coupling between the user-interface classes and the domain classes be reduced?
Solution:	Assign the responsibility for receiving input messages to a class that receives all input messages and acts as a switchboard to forward them to the correct domain class. There are several ways to implement this solution: (a) Have a single class that represents the entire system, or (b) Have a class for each use case or related group of use cases to act as a use case handler.
Example:	The RMO order-entry subsystem accepts inputs from an OrderWindow. These input messages are passed to an OrderHandler, which acts as the switchboard to forward the message to the correct problem domain class. RMO Order Entry OK Cancel startOrder () Controller Class OrderWindow startOrder () OrderHandler createOrder () Order User Interface Domain Classes Other examples of the controller can be found for each RMO subsystem.
Benefits and Consequences:	Coupling between the view layer and the domain layer is reduced. The controller provides a layer of indirection. The controller is closely coupled to many domain classes. If care is not taken, controller classes can become incoherent, with too many unrelated functions. If care is not taken, business logic will be inserted into the controller class.

Figure 12-1
Pattern specification for the controller pattern

A use case controller acts as a switchboard, taking input messages and routing them to the correct domain class. In effect, the use case controller acts as an intermediary between the outside world and the internal system. What if a particular window object needs to send messages to several problem domain objects? To do so, it would need references to all of these objects. The coupling between the Input window object and the internal system would be very high—there would be many connections. The coupling between the user-interface objects and the problem domain objects could be reduced by making a single use case controller object to handle all of the input messages. A use case controller also contains logic that controls the flow of execution for the use case. In this way, domain layer design classes can remain more cohesive by focusing only on the precise functions that truly belong to that domain object.

In the examples that follow in this chapter, we define a controller class for each use case. This is a common practice, and many development environments (such as Java Struts) automatically define a controller class for each use case. Of course, this creates many artifact objects in a system. If there are 100 use cases, there would be 100 use case controller artifact objects. To reduce the number of controllers, developers sometimes will combine the control of several closely related use cases into a single use case controller. Either approach, if done judiciously, provides a good solution.

A use case controller is a completely artificial class created by the person doing the system design. Sometimes such classes are called artifacts or artifact objects. As we get deeper into the explanation of design, you will see the need to create many kinds of service classes as artifacts—classes that are needed to execute the use case but are not based on any domain model classes.

The process illustrated in the preceding paragraphs and in Chapter 11—to balance the design principles of coupling, cohesion, class responsibility, indirection, and protection from variations—is precisely the process of systems design. As you read this chapter, you will see the importance of the design principles discussed in Chapter 11. The next section explains exactly how use cases are realized—in other words, how the system is designed, using sequence diagrams.

BEST PRACTICE

Always identify appropriate controller classes as entry points into the domain layer.

USE CASE REALIZATION WITH SEQUENCE DIAGRAMS

Developing interaction diagrams is at the heart of object-oriented detailed design. The realization of a use case—determining what objects collaborate and the messages they send to each other to carry out the use case—is done through the development of an interaction diagram. Two types of interaction diagrams can be used during design: sequence diagrams or communication diagrams. This section shows you how to design with sequence diagrams; then, in the next section, we explain how communication diagrams also can be used for systems design.

Interaction diagrams are used both as the mechanism for the design activity and as the tool to document the results. Designers develop design class diagrams and interaction diagrams while doing software design. The diagrams communicate structural and behavioral details to programmers and other developers. But the diagrams are not an end in themselves. Instead, they represent the results of design decisions and facilitate the inclusion of well-established design principles such as coupling, cohesion, and separation of responsibilities. In other words, if a developer jumps right into coding without thinking about design principles, the created system often is poorly structured. Typically, the most effective designers develop rough drafts of diagrams and then evaluate their quality by assessing how well they reflect principles of good design. The diagrams may be refined many times as designers improve their quality and correct errors. The diagrams are both a scratchpad for the designers' thinking and a means to communicate the final result of that thinking to programmers.

As noted in the previous section, a set of CRC cards may be sufficient for simple use cases. For complex use cases, however, it is usually beneficial to develop a fully detailed sequence diagram. Input for the development of a sequence diagram can come from a set of CRC cards if the cards were done first. However, they are not necessary. Many developers do not use CRC cards, but come directly to designing with sequence diagrams.

The following sections explain in detail the steps and techniques required for use case realization. In the first section, we provide a complete three-layer sequence diagram for a fairly simple use case, *Create new customer*. This example illustrates the final form of a complete design. After this first example, the next sections use two separate use cases to explain how to design the domain layer (also called the business logic layer). The domain layer focuses exclusively on the problem domain classes. The final examples in this tutorial explain how to add the data access layer classes and the view layer classes. Each layer is illustrated with two examples using the same two use cases. The examples include sections that explain the design steps and guidelines.

UNDERSTANDING THREE-LAYER SEQUENCE DIAGRAMS

You first learned about sequence diagrams in Chapter 7 when you learned how to develop a system sequence diagram (SSD). By now you should feel comfortable reading, interpreting, and developing an SSD. Remember that an SSD is used to document the inputs to and outputs from the system for a single use case or scenario. An SSD captures the interactions between the system and the external world as represented by the actors. The system itself is treated as a single object named :System. The inputs to the system are messages from the actor, and the outputs are usually return variables showing the data being returned. Figure 12-2 is an SSD for the *Create new customer* use case. As shown in the figure, each message has a source and a destination. In an SSD, because only two objects have lifelines, the source and destination are constrained. When we get to detailed sequence diagrams, some of the most critical decisions involve the source and destination objects for the messages. Remember that the syntax of an input message, as discussed in Chapter 7, is:

* *[true/false condition] return-value := message-name (parameter-list)*

Figure 12-2

SSD for the *Create new customer* use case

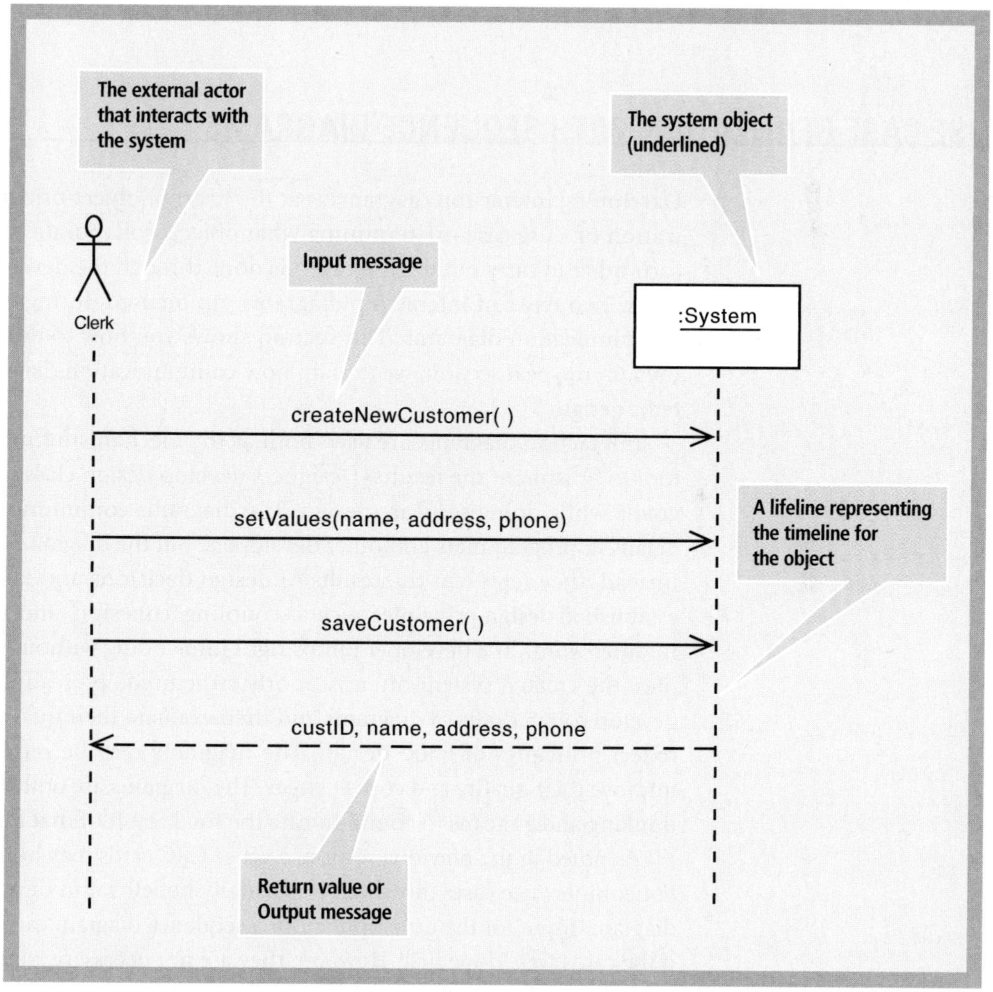

The starting point for the detailed design of a use case is always its SSD. Remember that the SSD only has two lifelines—one for the actor and one for the system. The most important information on an SSD is the sequence of messages between the actor and the system. Frequently there are both input and output messages. There may be a single input message or many. The input messages may have data parameters or not. There also may be Loop frames, Alt frames, and Opt frames and repeating inputs and outputs. A Loop frame denotes a set of messages within a loop. An Alt frame is similar to an if-then-else statement or switch statement, which allows the firing of different sets of messages. An Opt frame is an optional invoking of a set of messages.

The three input messages and one output message in Figure 12-2 characterize the sequence of actions required to add a customer. The first message might be to click a Create New Customer button on a main form. After the new customer screen appears, the second message is required to enter all the customer information. The third message might be a button click to save the data. Finally, the output message, which is also known as a return value and denoted by a dashed arrow, is simply a reformat of the input data. A new customer ID is added, along with a confirmation message that the customer was created and the database was updated.

Figure 12-3 illustrates a three-level detailed design for this use case. Callouts show the classes associated with each of the three layers. This use case has two view layer objects—the :MainMenu and :CustForm objects. Notice that the input messages from the external actor always go to the view layer objects. The purpose of the design process is to take each input message and determine what the system must do to respond to the message. For the first message, createNewCustomer(), the system simply opens the :CustForm screen. For the second message, setValues (...), the system accepts the data and perhaps edits it. At this point in the design, we will not worry about the required editing. The objective of the sequence diagram is primarily to identify which classes collaborate and what messages they must send to each other. The requirements and code for editing are usually deferred until programming begins.

Figure 12-3

Sequence diagram for *Create new customer* use case

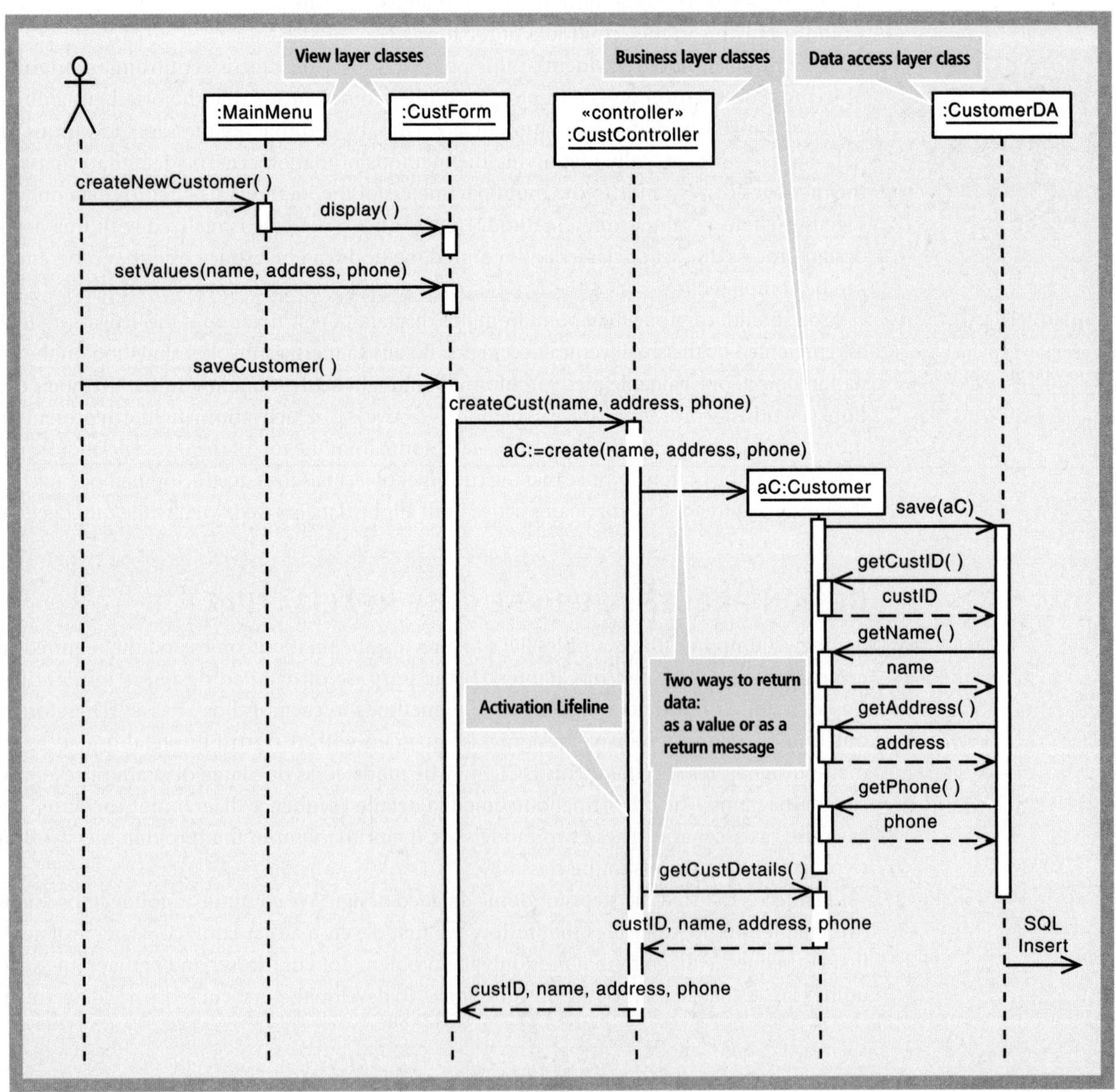

The third message, saveCustomer(), triggers the internal messages. After the :CustForm receives the saveCustomer() message, it sends a message to the :CustController, which in turn sends a message to the customer class, asking it to instantiate a new customer object. This message is sent directly to the object's rectangle, which is an optional but preferred notation to indicate a message that invokes a constructor. The label in the rectangle, aC:Customer, indicates that the box represents a customer object with a reference variable name of aC. Notice on the create message that the object reference is returned to the controller, which gives it visibility to the customer object. It will need this reference later to send the getCustDetails() message.

After the customer object is created, it has responsibility to save itself to the database. It sends a message to the :CustomerDA object, which is the data access object for all the customer objects, requesting that the data be saved to the database. Notice that the message also sends the customer reference, aC, so that :CustomerDA has visibility to be able to request the attribute values from the customer. Once the data access object has all the data, it invokes an SQL insert instruction to the database.

Once the customer object is created and saved to the database, the controller requests all of the newly generated data, including the system-generated custID. The data is returned to the controller, which then returns it to the :CustForm screen, which is visible to the external actor. Optionally, we could have shown another dashed line going to the external actor, which would simply mean the actor looks at the form.

One thing should be evident at this point. When a message is sent from an originating object to a destination object, in programming terms it means that the originating object is invoking a method on the destination object. Thus, by defining the messages to various internal objects, we are actually identifying the methods of that object. The data that is passed by the messages corresponds to the input parameters of the methods. The return data on a message is the return value from a method. Hence, once a use case is realized with this detailed design process, the set of classes and required methods can be extracted so that programming can be completed.

activation lifeline

a representation of the period in which a method of an object is alive and executing

You should learn one final point from this diagram. A new notation is the activation lifeline, as represented by the small vertical rectangles. Because a message invokes a method on the destination object, one valuable piece of information might be the duration of that method's execution (in other words, the time a method is active). The activation lifeline represents that information. That is why the input message is normally at the top of the rectangle and the return message is at the bottom. Notice that the customer object has the constructor method attached to the bottom of the object. It remains active until all the data is saved, even while other get methods are invoked.

DESIGN PROCESS FOR USE CASE REALIZATION

Before we jump into the examples, let's first review the final outcomes and the required steps to get there. As indicated in Chapter 11, the purpose of detailed design is to identify the classes required for the new system and the methods in each of those classes. Therefore, one outcome is a comprehensive design class diagram with the attributes elaborated and the method signatures specified. This DCD may be modeled as one large diagram or several subsystem diagrams. The other final outcome is a detailed sequence diagram for each use case or each use case scenario. These two models are the primary input that programmers will need to program the methods in the classes.

Figure 11-15 listed the steps for doing detailed design. We continue to follow those steps in this chapter. In the examples that follow, we first develop a first-cut DCD. The next step is a first-cut sequence diagram that uses only the problem domain classes and perhaps any specific utility classes that are part of the business logic. To develop the first-cut sequence diagram, each

input message is reviewed, one at a time, to determine what other internal messages and classes are required to fully process the input request. Once the processing with the problem domain classes is known, the data access layer and the view layer classes and messages are added to the diagram. Finally, the DCD is updated with method signatures from the details generated during use case realization.

In the next two sections, we realize two use cases, *Cancel an order* and *Create new phone order*. We address *Cancel an order* first because it is the simpler use case. The *Create new phone order* use case is a specific scenario of *Process new order*, which was done in Chapter 11 with CRC cards.

FIRST-CUT SEQUENCE DIAGRAM—*CANCEL AN ORDER* USE CASE

A detailed sequence diagram uses all of the elements that an SSD uses. The difference is that the :System object is replaced by all of the internal objects and messages within the system. We will identify the internal objects that collaborate and the messages they send to each other to carry out the use case or the use case scenario. Figure 12-4 is the SSD for this use case. The SSD has only one input message, cancelOrder (orderID), which passes the identifier for the order to be canceled. In this situation, we assume that another use case provided the processing to search for and identify the order to cancel.

Figure 12-4

SSD for *Cancel an order* use case

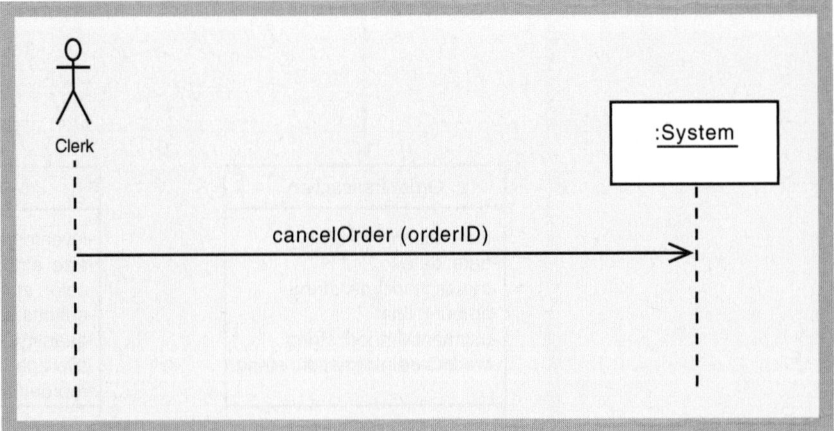

The next step is to look at the problem domain classes and determine which classes are required for this use case. Obviously, the Order class is needed. Assuming that inventory was reserved when the order was made, the order items and any individual inventory items need to be included. Let's assume a business rule that orders must be cancelled before they are shipped, and that a credit card payment is not initiated until the items are shipped. Therefore, no credit card refund will be required, but an order transaction is needed to specify the charge. The transaction and charge must both be canceled. With these classes, the next step is to develop the first-cut DCD of the problem domain classes. Figure 12-5 shows the first-cut DCD with visibility identified.

Based on Figures 12-4 and 12-5, we proceed with the detailed design of the *Cancel an order* use case. The first step in expanding an SSD is to place the problem domain objects in the diagram, along with the input messages from the SSD. Figure 12-6 shows this first step in the detailed design.

The next step is to determine the internal messages that must be sent between the objects, including which object should be the source and destination of each message. Decisions about what messages are required and which objects are involved are based on the design principles described earlier—coupling, cohesion, object responsibility, and controllers.

Figure 12-5

First-cut DCD for *Cancel an order* use case

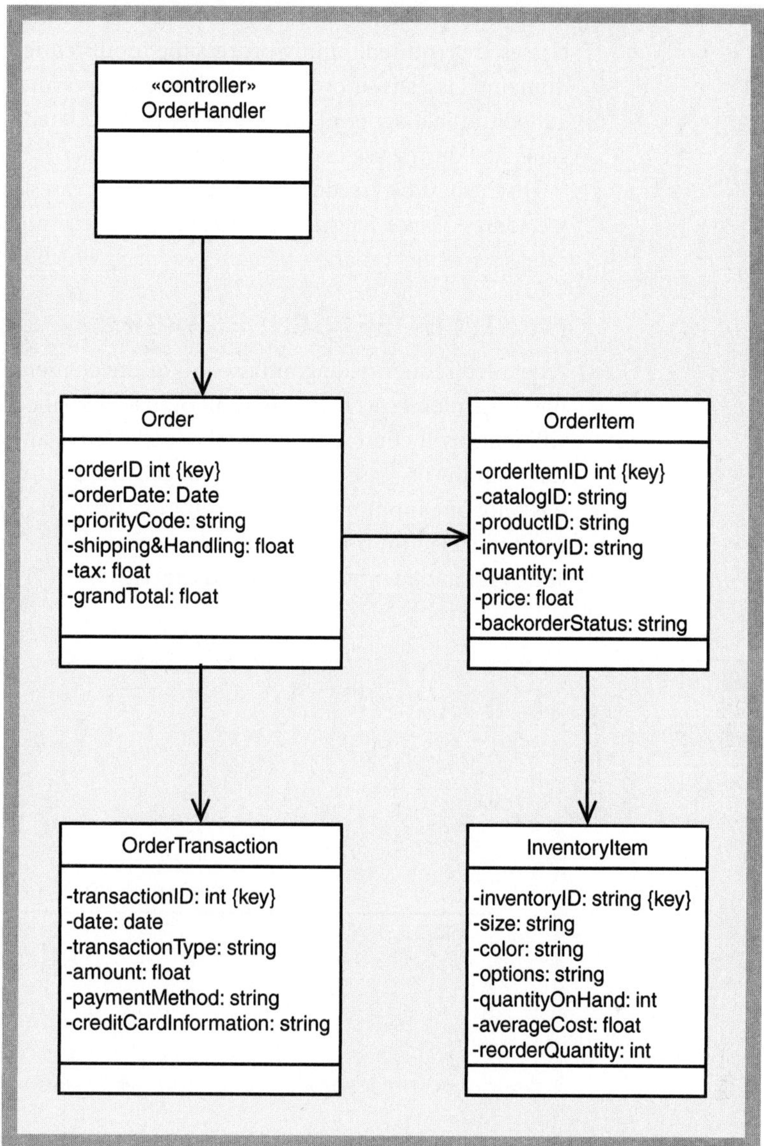

Figure 12-6

Objects included in *Cancel an order*

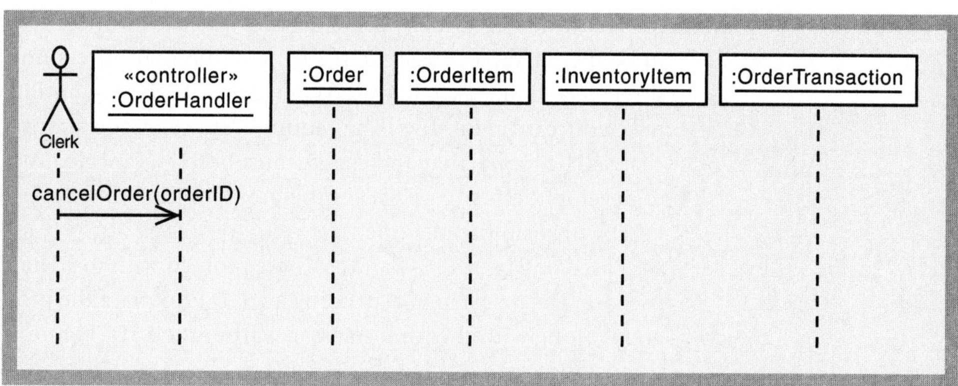

Figure 12-7 is the completed first-cut sequence diagram for the design of the *Cancel an order* use case. The :OrderHandler controller receives the input message, searches for the correct order object, and then forwards the cancelOrder message to the correct :Order object. The :Order object takes responsibility to do all the cleanup for the cancellation. It loops through

all of the order items associated with the order and sends each one a cancelItem() message. Each :OrderItem object in turn sends a quantity update to the appropriate :InventoryItem object. In other words, the quantity reserved for the order in the :OrderItem object is added back to the inventory item when the order is canceled. The rectangle loop frame indicates that these two messages are part of multiple occurring sets. Finally, the :Order object also sends the cancelPayment() message to the :OrderTransaction to stop the charges to the customer.

Notice the use of the activation lifeline. It is important on the :Order object because it indicates that both outgoing messages are part of the same execution. Activation lifelines help you to understand the time period when an object is executing some code. Some developers put them on all lifelines whenever there is a message arrowhead. Other developers only include them when there may be some ambiguity. Figure 12-7 shows both examples. If there is any chance of misreading the diagram, such as on the first three messages, the activation lifelines are included. The final two messages do not have activation lifelines because there is little chance of misunderstanding.

Figure 12-7

First-cut sequence diagram for *Cancel an order* use case

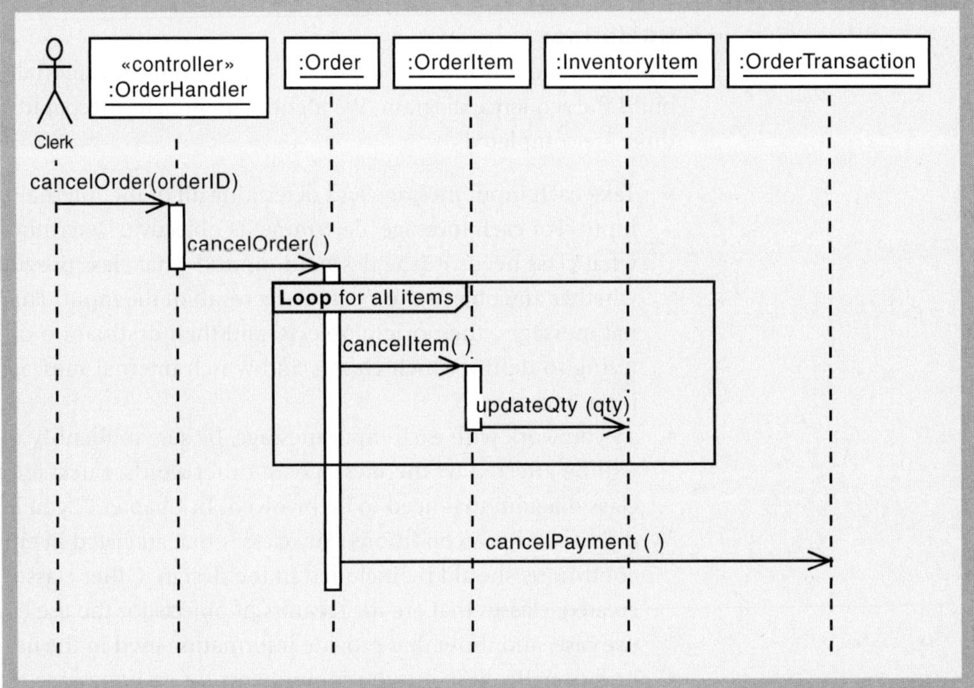

Two questions require careful consideration when identifying and creating messages: Which objects are involved and what should we name the message? First, we must determine the origin and destination objects for the message. The origin object is obviously the one that needs information or help in carrying out a responsibility. The destination object is the one that has the information to help in the solution. Second, what should we name the message? Because a message is requesting a service from the destination object, the message name should reflect the requested service. For example, when a quantity needs to be updated in the destination object, the message name indicates the requested process to update the quantity. Notice also that the input parameters provide the information that the destination object needs to be able to provide the service.

Before moving on, let's analyze this solution based on some principles of good design that we discussed previously—coupling, cohesion, object responsibility, and use case controllers.

The use case controller provides the link between the internal objects and the external environment. This intermediary limits the coupling to the external environment to the controller object. The responsibilities assigned to :OrderHandler are to catch incoming messages, distribute them to the correct internal domain objects, and return the required information

to the external environment. By using a use case controller as the switchboard, overall coupling is limited between the domain objects and the environment.

The responsibility assigned to :Order is to be in charge of canceling itself and to control all of the other required updates. The :OrderItem object cancels itself and has responsibility to notify the appropriate inventory item. Coupling is straightforward, being basically vertical on the hierarchy. Thus, the assignment of responsibilities and corresponding messages seems to conform to good design principles. Many other issues will need to be addressed as the design expands to include all three layers. We develop the domain layer with the data access layer later in the chapter.

GUIDELINES AND ASSUMPTIONS FOR PRELIMINARY SEQUENCE DIAGRAM DEVELOPMENT

Even though the example in Figure 12-7 was fairly simple, we can distill several tasks to help you learn to develop a design for a use case or scenario using sequence diagrams. Several assumptions are also implicit in this process.

Guidelines

Note that the following design tasks are not done sequentially but only when necessary to build the sequence diagram. We identify them here as separate tasks simply to ensure that all three are completed.

- Take each input message and determine all of the internal messages that result from that input. For each message, determine its objective. Determine what information is needed, what class needs it (the destination) and what class provides it (the source). Determine whether any objects are created as a result of the input. This will help you to define internal messages, their origin objects, and their destination objects. In other words, you are trying to define which classes and which internal messages are needed to support the input message.
- As you work with each input message, be sure to identify the complete set of classes that will be affected by the message. In other words, select all the objects from the domain class diagram that need to be involved. In Chapter 7, you learned about use case preconditions and postconditions. Any classes that are listed in either the preconditions or postconditions should be included in the design. Other classes to include are those that are created, classes that are the creators of objects for the use case, classes updated during the use case, and those that provide information used in the use case.
- Additionally, flesh out the components for each message. Add iteration, true/false conditions, return values, and passed parameters. The passed parameters should be based on the attributes found in the domain class diagram. Return values and passed parameters can be attributes, but they may also be objects from classes.

These three steps will produce the preliminary design. Refinements and modifications may be necessary; again, we focused only on the problem domain classes involved in the use case.

Assumptions

The development of the first-cut sequence diagram is based on several simplifying assumptions, including the following three:

- **Perfect technology assumption.** We first encountered this assumption in Chapter 5 when we identified business events. We continue that assumption here. We do not include steps such as the user having to log on or testing the availability of the network.
- **Perfect memory assumption**. You might have noticed our assumption that the necessary objects were in memory and available for the use case. We did not ask whether those objects were created in memory. We will change this assumption when we get to multilayer design. In multiple-layer design, we do include the steps necessary to create objects in memory.

Another way to think about this assumption is that memory is infinite, or that the computer's hard drives are just an extension of memory and we do not have to worry about how things get into memory. You may think this is not a realistic assumption; however, object-oriented languages that include embedded object-oriented databases do implement this assumption. Those languages include an underlying database; if you need an object, the system automatically moves it from hard drive storage to memory and back, without any requirement by the programmer.

- **Perfect solution assumption.** The first-cut sequence diagram assumes that there are no exception conditions. No logic is included to handle a situation in which the requested catalog or product is not found. More serious exception conditions, such as the failure of a credit check, might also be encountered. Many developers design the basic processing steps first, and later add the other messages and processes to handle the exception conditions. We do the same here.

FIRST-CUT SEQUENCE DIAGRAM—*CREATE NEW PHONE ORDER* USE CASE

Before moving ahead to multilayer design, let's work through a slightly more complex example of a first-cut diagram. Figure 12-8 is a simple version of an SSD for the *Create new phone order* scenario. This SSD is for the telephone scenario of the *Create new order* use case. As before, we will create a design for each input message in the SSD. The design components for all the messages are combined to provide a comprehensive sequence diagram for the entire use case.

Figure 12-8

SSD for the *Create new phone order* scenario

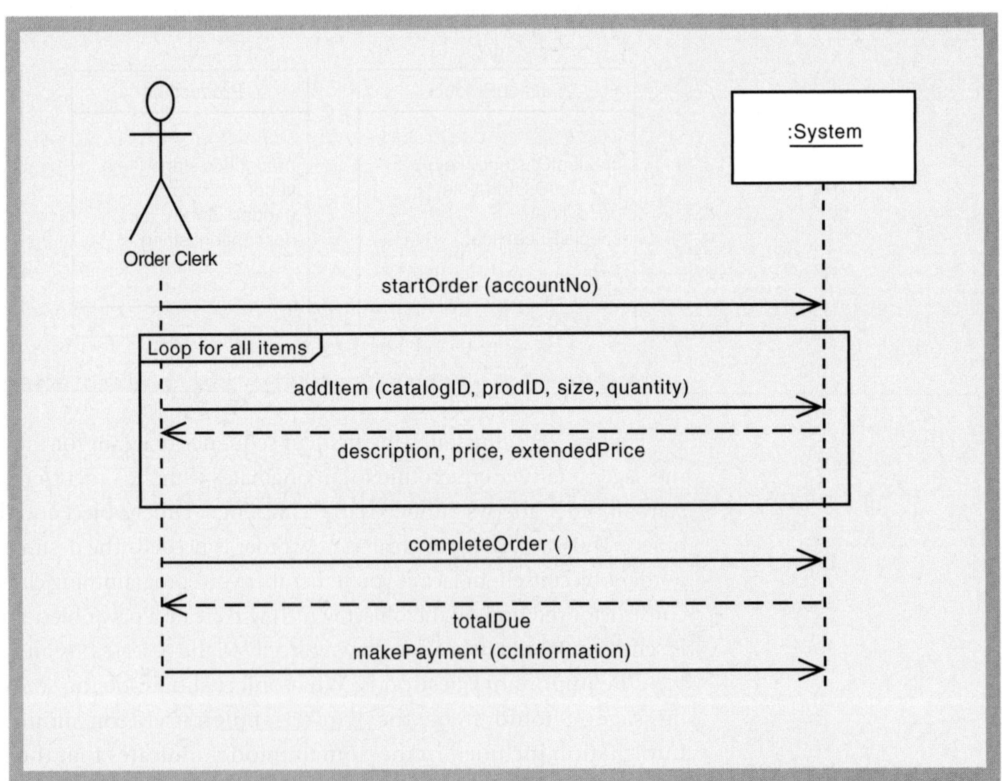

First we review the class diagram and create the first-cut design class diagram. We created a similar example with the CRC design in Chapter 11. Figure 12-9 illustrates the navigation visibility that corresponds to the rules and example from Chapter 11. Information from the SSD and the first-cut design class diagram will again be used to develop the sequence diagram. We will use the same controller object that we used for the *Cancel an order* use case. We anticipate that this controller may also serve for creating new orders and maintaining existing orders. We will decide whether this is the best design after designing other use cases and reviewing the design for good design principles.

Figure 12-9

First-cut DCD for *Create new phone order* scenario

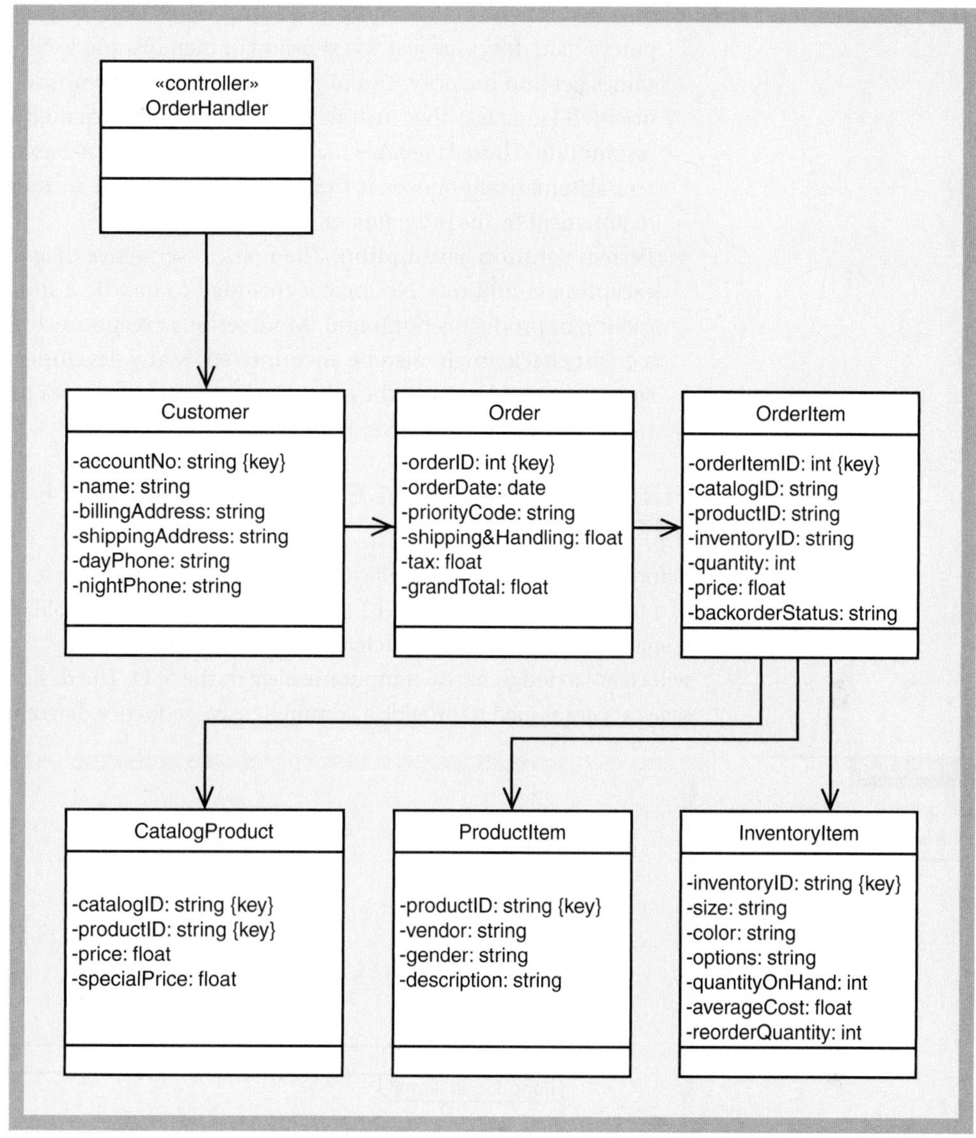

Figure 12-10 illustrates the first-cut sequence diagram for the first input message only. This message is startOrder(accountNo); it originates with the order clerk and is input to the :System. To start an order, the system needs to create a new Order object and then connect it to a Customer object. Thus, a message to create a new order is needed. The destination of the message will be the :Order object itself. In fact, if you remember your programming class, the create message invokes a constructor method on the object, which will create a new object. In UML, when a create message is sent to an object, it is often drawn directly to the object's box and not to the lifeline.

One important question is: What object should be the source object for the createOrder message? Should it be the :OrderHandler itself, or should it be some other object? Information included in the domain model indicates that the :Order object has a relationship or link with the :Customer object. This link could be built in several ways. One option would be to have the :OrderHandler object send the create message directly to the :Order object, then send another message to the :Customer object with a reference to the order. Another option is simply to let the :Customer object create the :Order object. Because order objects are not allowed unless a customer object exists, this option is one way to ensure that the customer existence precondition is met. Figure 12-10 shows the results using the second approach. Note that a specific identifier is given to the new Order object—anOrd. That reference is passed back to the Customer object, which passes it back to the :OrderHandler. We will see the need for this approach in later steps.

Figure 12-10

Sequence diagram for the first input message for the phone order scenario

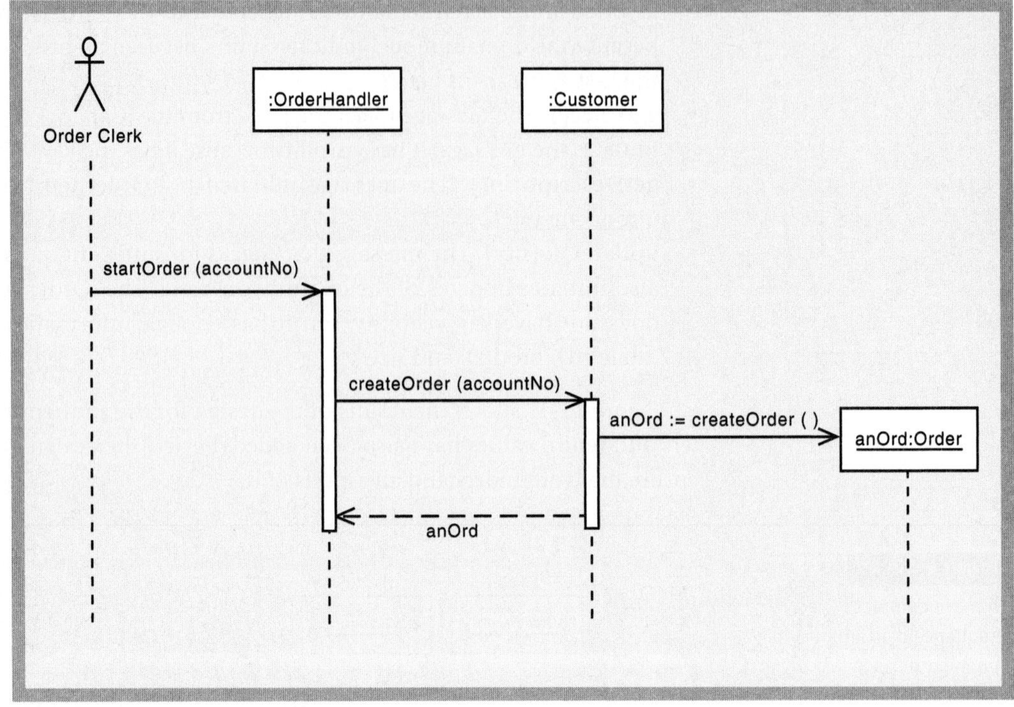

The next input message is addItem (catalogID, prodID, size, quantity), a repeating message to add line items to the order. This message is shown in Figure 12-11. By referencing the OrderItem class in the domain model, we find that the necessary attributes for an OrderItem are quantity, price, description, and back-order status. The price can be obtained from the CatalogProduct class. The description comes from the ProductItem class. The quantity is input by the clerk, although the system must check the InventoryItem class to see whether items are in stock. (The detailed description of the use case also indicates that inventory should be checked by the system.) So, the sequence diagram will also need objects for :OrderItem, :CatalogProduct, :ProductItem, and :InventoryItem.

As we identify the specific messages, along with source and destination and the passed parameters, we need to consider some critical issues. As before, an important question is: Which object is the source or initiator of a message? If the message is a query message, the source is the object that needs information. If the message is an update or create message, the source is the object that controls the other object or that has the information necessary for its creation.

Another important consideration is navigation visibility—to send a message to the correct destination object, the source object must have visibility to the destination object. Remember that the purpose of doing design is to prepare for programming. As a designer, you must think about how the program will work and consider programming issues. Given these two considerations and the source considerations discussed in the previous paragraph, we have determined that the following internal messages will be required. For each message, a source object and a destination object have been identified.

- **addItem().** Original message, from Order Clerk to :OrderHandler
- **addItem().** A forwarded version of the input message from :OrderHandler to :Order. Because :OrderItem objects are dependent on an order, :Order is the logical object to create :OrderItem objects. System has visibility to :Order from the previous return message, when anOrd was returned to the system.
- **createOrdItem().** The internal message from :Order to :OrderItem. Because the OrderItem will be responsible for obtaining the data for its attributes, it needs visibility to :CatalogProduct, :ProductItem, and :InventoryItem. As a result, those keys are sent as parameters. An alternate approach is to let :Order collect the required information, such

as price and backOrderStatus, and send it to the :OrderItem as parameters. However, because the domain model indicates a link between an order item and an inventory item, the first approach is better.

- **getPrice().** The message to get the price from the :CatalogProduct object. The :OrderItem initiates the message. It has visibility because it has the key values.
- **getDescription().** The message initiated by :OrderItem to get the description from :ProductItem.
- **updateQty(qty).** The message that checks for sufficient quantity on hand. This message also initiates updates of the quantity on hand. The :OrderItem initiates the message. It does not have key visibility, but it has enough information to search on an index of catalogID, prodID, and size.

Figure 12-11 shows the results of the design for the addItem message. The input parameters and return values have also been added. Review the design, including the parameters, to ensure that you understand all aspects of it.

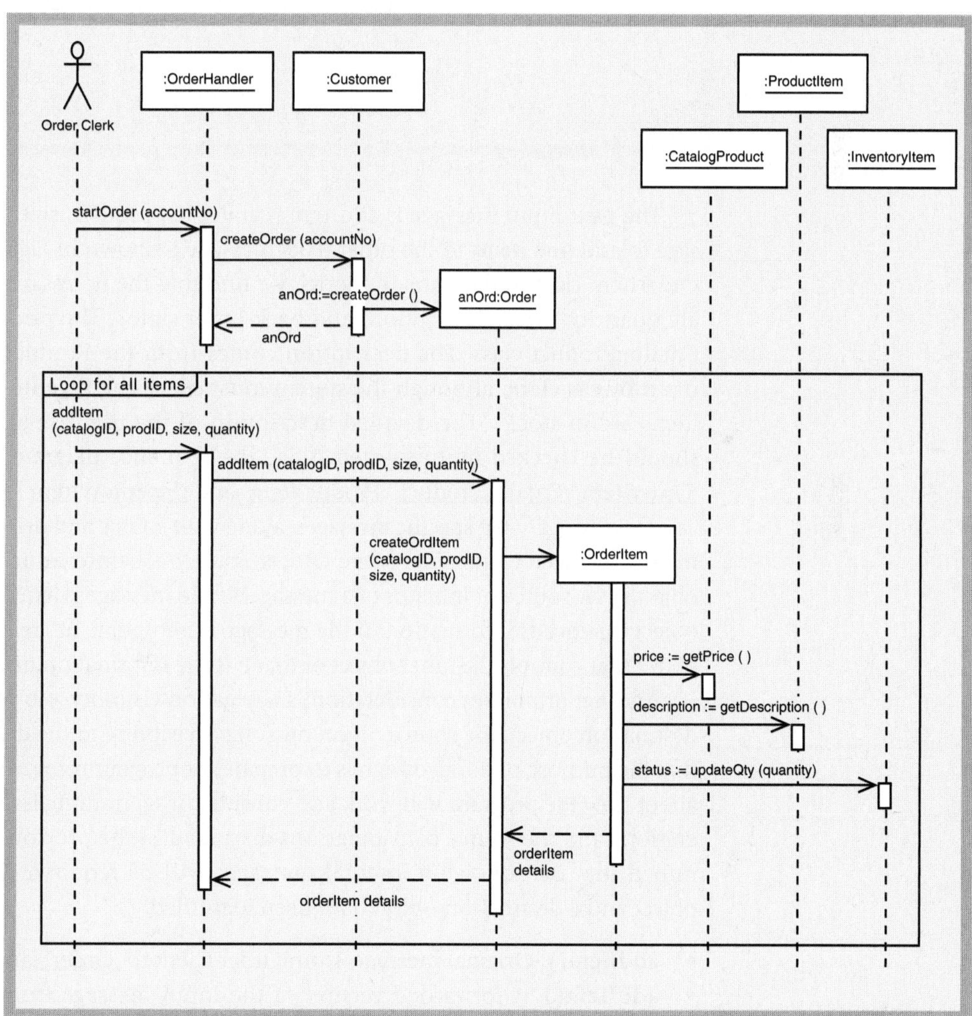

Figure 12-11

A more complete sequence diagram for the phone order scenario

After each item is added to the order, control is returned to the order clerk. The clerk will add another item or, at the end of the order, will send a completeOrder() message. This message has no parameters. Its purpose is simply to tell the order to calculate the total amount due. If we assume, as designers, that the :Order object keeps a running total of the individual line items that were added, then it simply calculates the appropriate tax and shipping and sends back a total amount. This is a valid and solid design. Another alternative is that the

:Order object does not keep a running total but must query each of the line items and accumulate a total. The second design requires additional detailed messages to be sent to the :OrderItem objects.

The final message on the SSD is makePayment (ccInformation). For simplicity's sake, we assume that payments are always via credit card. In reviewing the domain model, we see that a new object must be created—the OrderTransaction object. Because transactions are connected to orders in the model, an Order object should create a transaction. Thus, the system forwards the completion message to the :Order, which in turn creates a payment for :OrderTransaction. These new messages are shown in Figure 12-12.

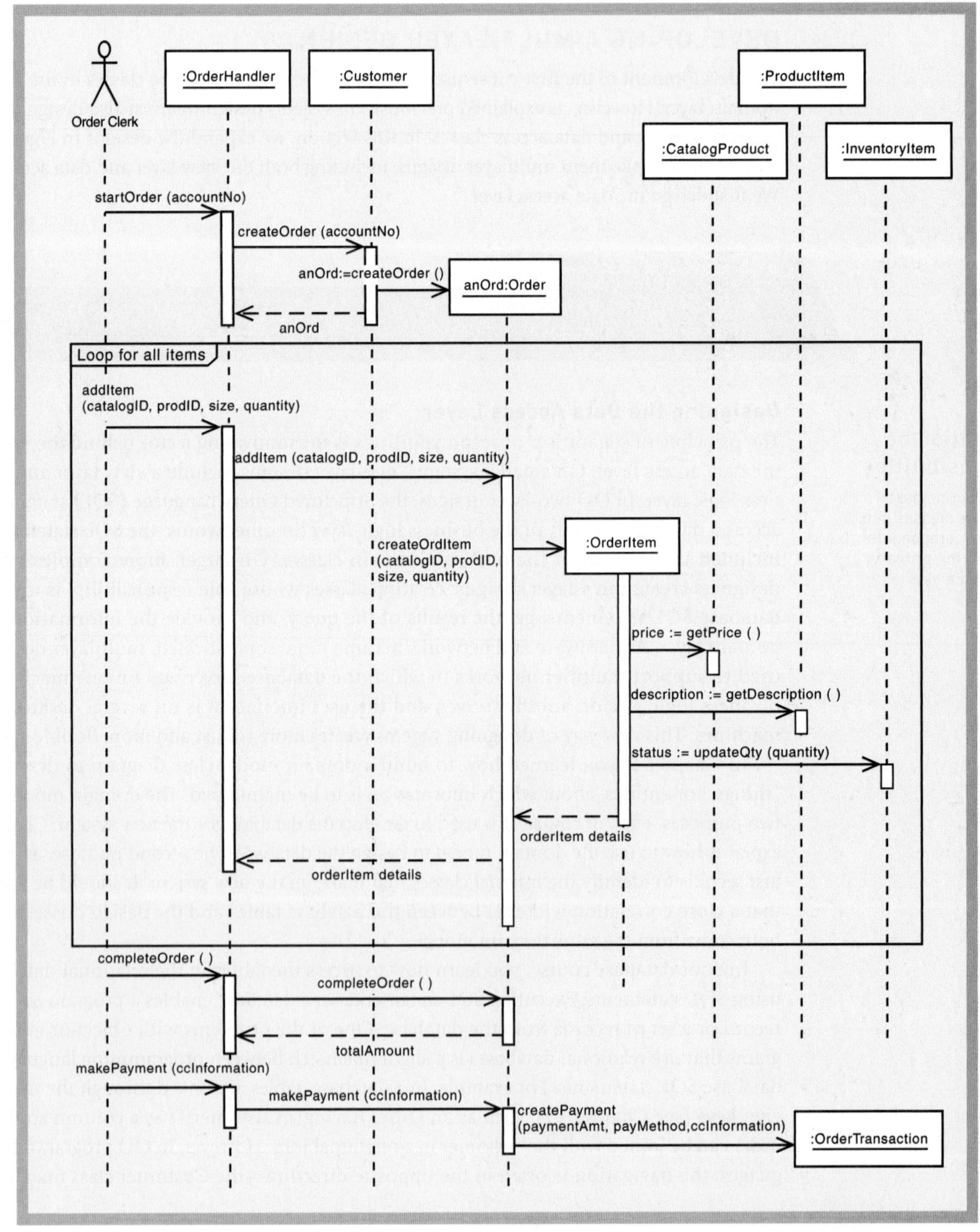

Figure 12-12

Sequence diagram for the *Create new phone order* scenario

Figure 12-12 contains the final design of the domain model classes and all the internal messages that are required to execute the use case. This section focused only on the classes from the domain model, plus one additional object called :OrderHandler. By focusing only on the domain classes, we could design the core processing for the use case without having to worry about the user interface or the database. Figure 12-12 is rather complex, even though it only contains domain objects. However, this design provides a solid base for programming. Working with design models enables the designer to think through all the requirements to process a use case without having to worry about code. More importantly, it enables the designer to modify and correct a design without having to throw away code and write new code. In the next section, we will add the view layer and data access layer objects to the telephone order scenario.

DEVELOPING A MULTILAYER DESIGN

The development of the first-cut sequence diagram focuses only on the classes in the problem domain layer. However, as explained previously, in systems design we must also design the user-interface classes and data access classes. In this section, we expand the designs in Figures 12-7 and 12-12 to make them multilayer designs, including both the view layer and data access layer. We first design the data access layer.

BEST PRACTICE

Be sure the use case design for the domain layer classes is solid before adding data access or view layer classes.

Designing the Data Access Layer

separation of responsibilities

a design principle to segregate classes into separate components based on the primary focus of the classes

The principle of separation of responsibilities is the motivating factor behind the design of the data access layer. On smaller systems, two-layer designs include a view layer and a business logic layer. In OO two-layer designs, the Structured Query Language (SQL) statements to access a database are part of the business logic layer. In other words, the SQL statements are included in methods of the problem domain classes. On larger, more complex systems, designers create three-layer designs, creating classes whose sole responsibility is to execute database SQL statements, get the results of the query, and provide the information to the domain layer. As hardware and networks became more sophisticated, multilayer design was used to support multitier networks in which the database server was on one machine, the business logic was on another server, and the user interface was on several desktop client machines. This new way of designing systems creates more robust and more flexible systems.

In Chapter 5, you learned how to build a domain model class diagram to describe the "things," or entities, about which information is to be maintained. The domain model serves two purposes. First, of course, it is used to develop the database for the new system. Chapter 13 explains how to use the domain model to design the database. The second purpose, as we have just seen, is to identify the internal classes that make up the new system. It should be apparent that a close correlation will exist between the database tables and the design classes because both come from the same domain model.

In your database course, you learn how to access the tables in the relational database by using SQL statements. Executing SQL statements on a database enables a program to access a record or a set of records from the database. One of the problems with object-oriented programs that use relational databases is a slight mismatch between programming languages and database SQL statements. For example, in a database, tables are linked through the use of foreign keys (see Chapter 13), such as an Order having a CustomerID as a column so that the order can be joined with the customer in a relational join. However, in OO programming languages, the navigation is often in the opposite direction—the Customer class may have an

array of references that point to the Order objects, which are in computer memory and are being processed by the system. In other words, design classes do not have foreign keys.

These differences between programming languages and database languages have partially driven the trend to a multilayer design. The design, programming, and maintenance of a system are easier if separate classes are defined to access the database and format the data so that it is conducive to computer processing. Rather than mix the business logic with the data access logic, it is better to define separate classes and let each focus on its primary responsibility. This idea is an application of the good design principles of highly cohesive classes with separate responsibilities.

In this chapter, we take a somewhat simplified design approach to teach the basic ideas without getting embroiled in the complexities of database access. Let us assume that every domain object will have a table in a relational database. There are several techniques, which provide different designs, to linking the domain layer to the data access layer. Within the constructor of each problem domain object, the data access object would be invoked to get the necessary information to complete the instantiation of the new object. Another way is to send a message to the data access layer object and have it read the database and then instantiate a new problem domain object. Either way works, and both are good solutions.

Figure 12-13 illustrates these two methods for instantiating a customer object in memory from the database. In method (a), the controller invokes the constructor for a :Customer and passes it the custID. The customer constructor invokes the :CustomerDA object, passing a reference to itself, aC. The :CustomerDA first gets the custID from the customer object. Using the custID, it then reads the database and populates all of the fields in aC. The reference to the new customer object, referred to as aC, is then passed to the controller, and it proceeds with other messages that need to be passed to the customer object. This method is used in the example shown in Figure 12-3.

In method (b), the controller sends a message to the :CustomerDA object, passing a key or some other field value. The :CustomerDA object uses the key to access the database, and then it invokes a constructor for :Customer with data for the fields sent as parameters. We use the shortcut "parameter list" to indicate that values for all the attributes are being passed in as parameters on the constructor. The controller then continues with the messages that need to be sent to the customer object.

The logical question is "Which method is best?" From a theoretical point of view, method (a) assigns responsibility for creating a customer object to the customer class. The new :Customer object then has the responsibility to populate itself by calling the :CustomerDA. However, method (b) works better when you need to create a list of objects from a database. In the example of reading an order and all of its line items, the order normally does not know how many objects need to be created. Therefore, it is better to send a message to the order item data access object and tell it to create as many as needed. For this reason, method (b) has become popular among programmers because it is efficient and easy. In fact, it has become a popular standard pattern, and it is also a good solution.

The Data Access Layer for *Cancel an order*

To design the data access layer, we no longer assume that the objects are automatically in memory when we need them—that is, we disregard the perfect memory assumption. The design of the use case now requires additional messages to get data from the database to instantiate classes. In this example, we use the second approach, as shown in method (b) of Figure 12-13, which works best for retrieving a list of items from the database.

Figure 12-14 illustrates the use case design with all of the data access classes included. Because the original solution included four problem domain objects, this solution has four additional classes for accessing the database. All the classes and messages may seem daunting at first, but accessing the database is really just the same pattern repeated several times.

Figure 12-13

Two methods for accessing the database and instantiating objects

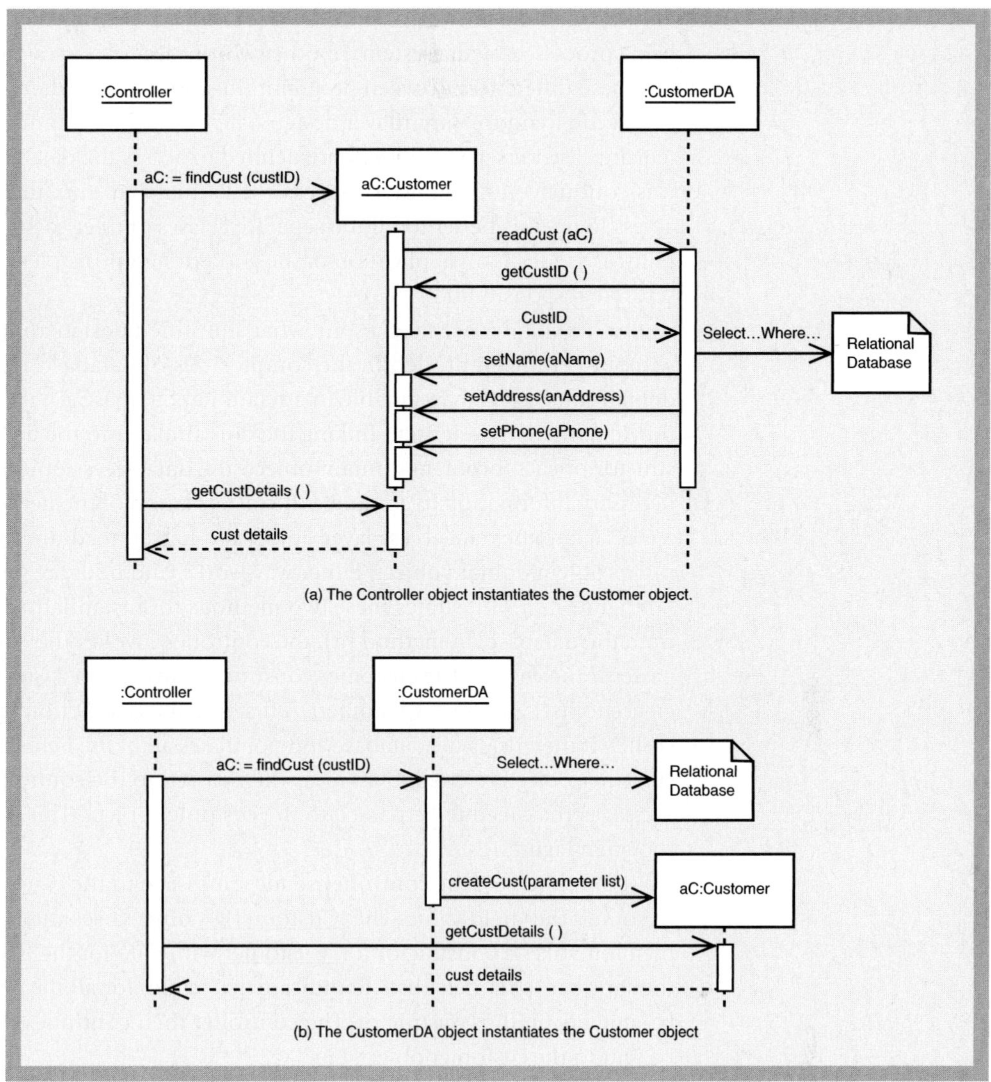

As you examine the figure, you will notice all of the original messages such as cancelOrder(), cancelItem(), and cancelPayment(), but each one is preceded by a pair of messages involving the data access class. The first part of the pair is a request to find the object, which means to read it from the database. The second part of the pair creates the object with the information from the database. The first part of the pair has a return value that is the reference to the created object. In the creation of :OrderItem objects, a list is returned because an order normally will have more than one item. The other addition in the figure is that each object is saved back to the database through the data access object after it has been updated.

One important point during this process is ensuring that source objects have navigation visibility to destination objects so that messages can be sent. We assume, but do not show, that the data access objects have global visibility. (In your programming class, you will learn that factory or singleton classes often are designed with global methods.) After the appropriate problem domain object is created, a reference to it is returned to the object that needs visibility. As you look closely at Figure 12-14, you will note that every object that sends a message to another object first has navigation visibility to that object. Remember this important design point as you develop your design solutions.

Let's next look at the problem domain layer and data access layer for the *Create new phone order* use case in Figure 12-12. Remember that this figure is already quite complex. To add the

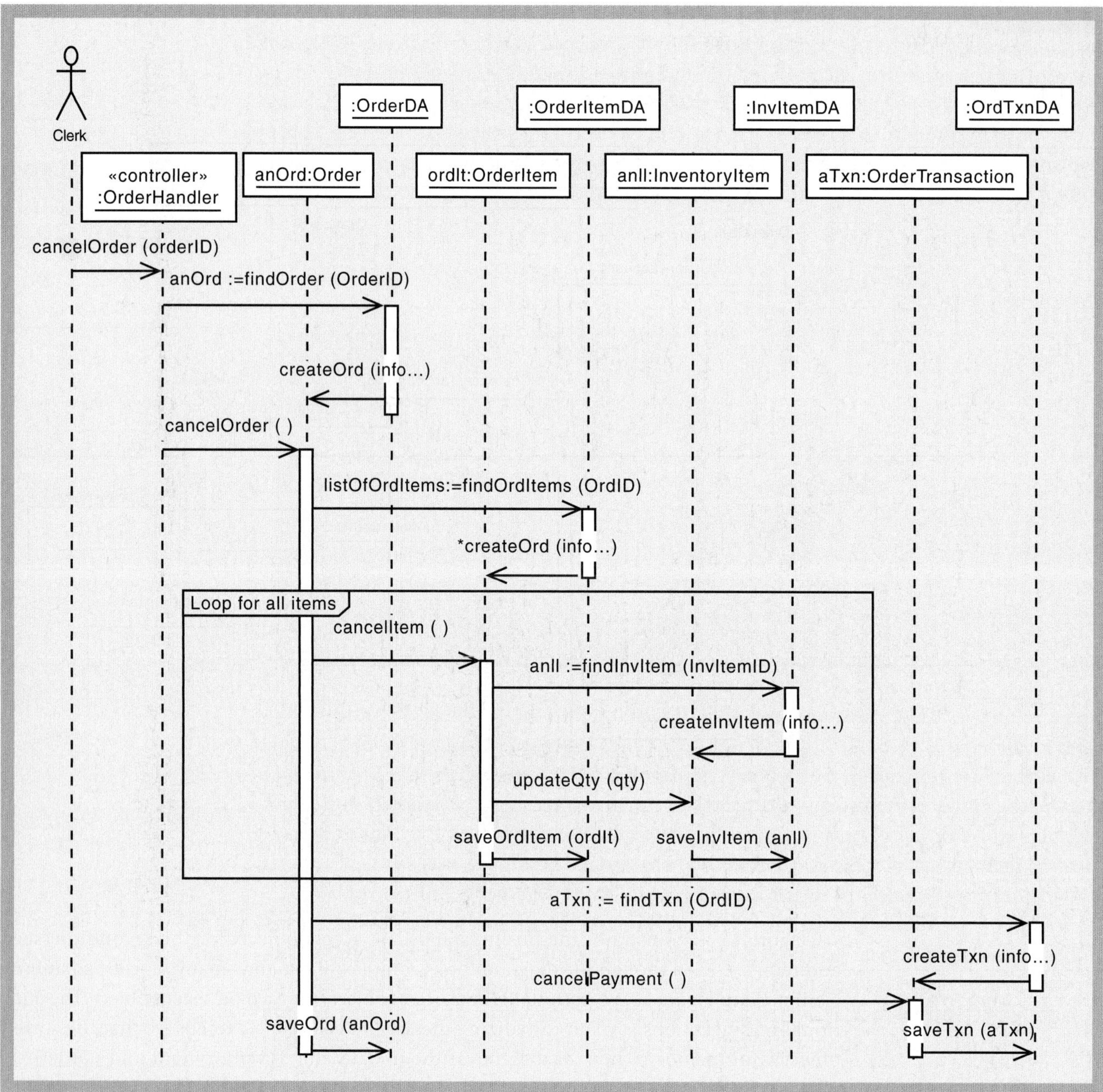

Figure 12-14

Problem domain and data access layers for *Cancel an order*

data access layer, the best approach is to focus on each input message separately, just as we did with the initial design. In these figures we use method (a) of the data access technique; that is, the initializing message goes first to the class that then instantiates a new object. After the object is instantiated, it sends a message to the data access object to obtain any required data. However, due to the complexity of these drawings, we do not show all of the get and set messages. You should review Figure 12-13(a) to ensure that you understand how method (a) works.

Figure 12-15 illustrates the sequence diagram with data access objects for the startOrder(...) message. The message name we chose to represent the creation of the required customer object in memory is findCustomer(). This creates an empty :Customer object, which then accesses the :CustomerDA object to read the database and populate the data fields with set messages. Inasmuch as the purpose of this use case is to create a new order, after a new :Order object is created, it calls the :OrderDA object to write it to the database.

Figure 12-15
Problem domain and data access layers for first input message for *Create new phone order*

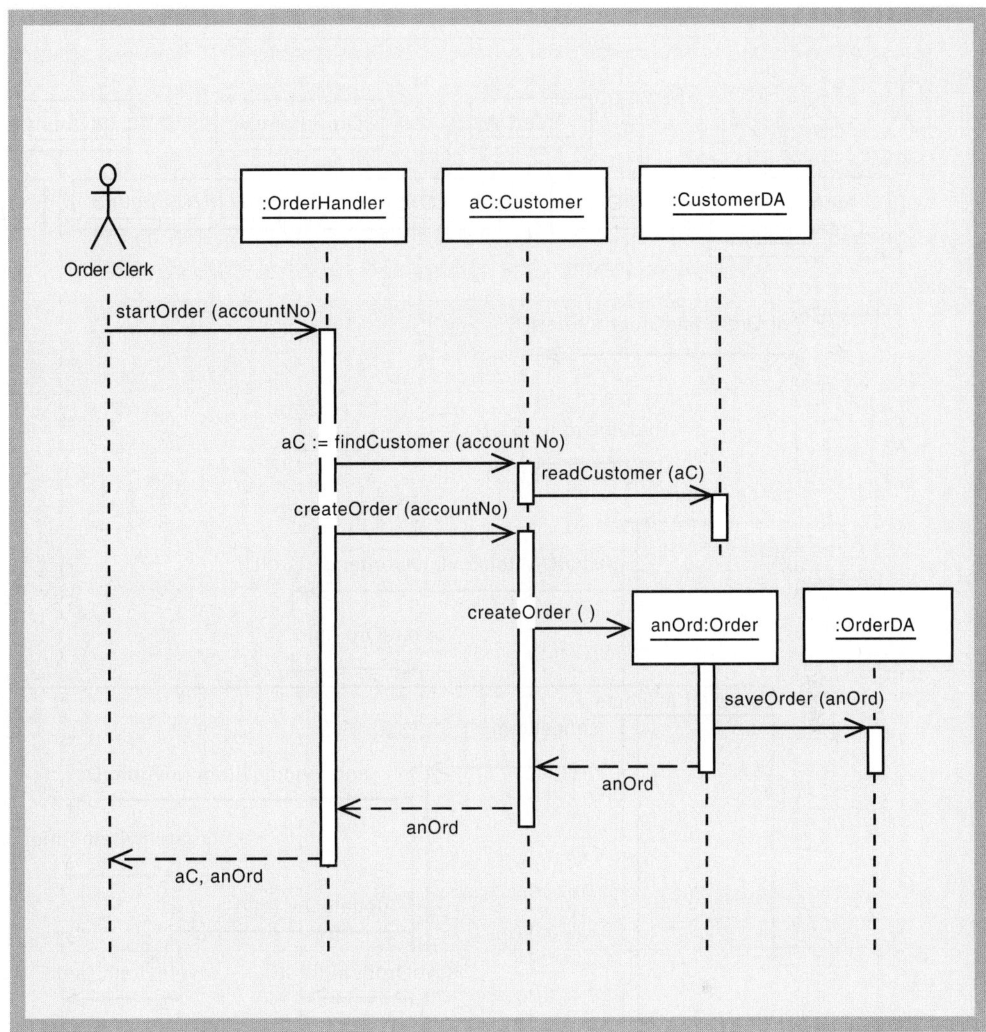

Figure 12-16 illustrates the business and data access objects for the second message, additem(...). Even though the diagram looks complex, it contains many repeating patterns. Compared to Figure 12-12, it contains data access objects for each persistent object. The additional messages consist of an initializing message to each object to ensure that the correct object is in memory. Then, at the end of the update, the :InventoryItem and :OrderItem objects are both updated with a save message to the appropriate data access object. It is important to pay special attention to the parameters passed with the message. To read the database, the data access object needs to have a reference to the object to be read, as shown in method (a) of Figure 12-13. Again, note that the get and set messages are omitted.

Figure 12-17 contains the problem domain and data access layers for the two final input messages. At this point in the use case processing, all the objects are in memory and none need to be initialized. Data access consists solely of writing out the :OrderTransaction object and the :Order object.

Designing the View Layer

The final step in multilayer design of a particular use case is to add the view layer. For many use cases, the view layer consists of a single user-interface window. Some use cases are more complex, and require multiple windows to enter and view the data associated with a use case or business transaction. At this point in the design, we do not go into great detail about the contents of the window class itself. Simply identifying the user-interface window is sufficient. Obviously, detailed design of the user interface is much more complex. The detailed design of the windows classes, including the window controls and form layout, is described in Chapters 14 and 15.

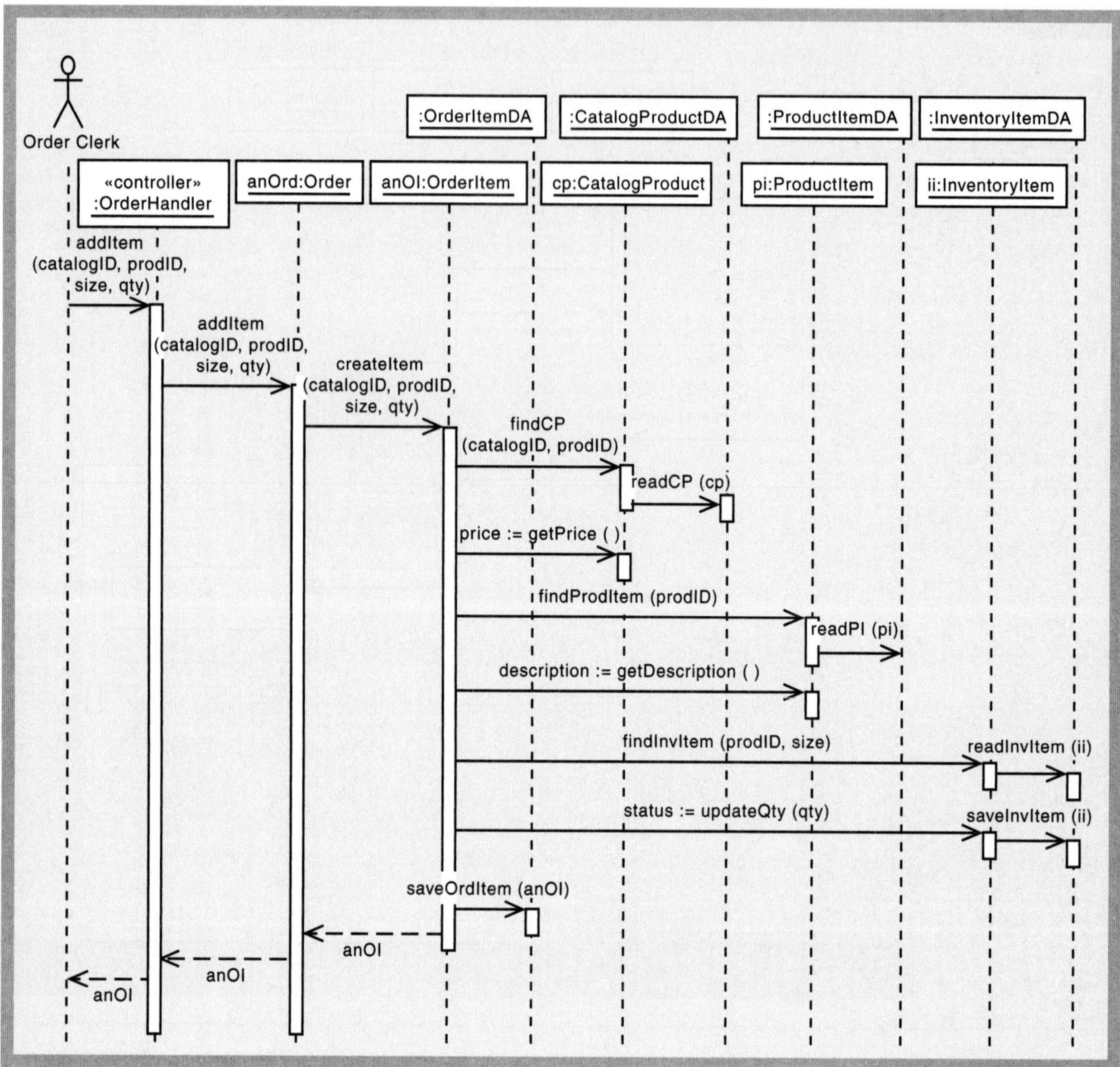

Figure 12-16

Business and data access layers for second input message for *Create new phone order*

In the early days of interactive systems and graphical user interfaces (GUIs), developers invented languages and tools that made it easy to develop systems with GUI elements such as windows and buttons. Early versions of languages such as Visual Basic, Delphi, and PowerBuilder were designed to make it easy to build interactive, event-driven, graphical systems. However, in these languages, the program logic was attached to the windows and other graphical components. So, to move these systems to other environments, such as browser-based systems, designers had to completely rewrite the system. In fact, systems developed in this way became good illustrations of the problems that follow when programmers violate design principles such as highly cohesive classes and separating responsibilities. When a class mixes user-interface functions and business logic, upgrading and maintaining the system become more difficult.

Figure 12-17

Business and data access layers for final input messages for *Create new phone order*

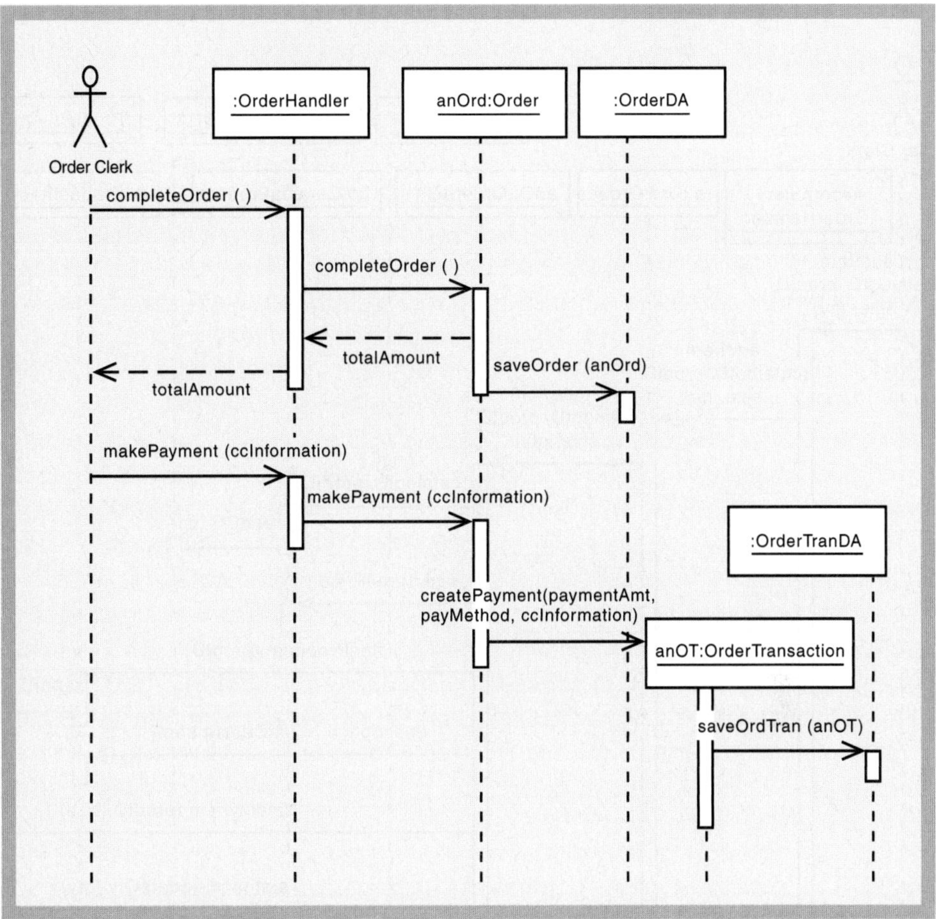

BEST PRACTICE

Don't be tempted to put business logic in the view layer classes.

As object-oriented programming became more prevalent and tools integrated both object-oriented programming and graphical interfaces, it became easier to build systems that could be partitioned and in which responsibilities could be separated. User-interface classes did not need to have business logic, other than edits on the input data. Designers could build multilayer systems that were more robust and easier to maintain, and they could apply the principles of good design. Tools such as Java Swing and Visual Studio .NET provide the capability to easily build GUIs as well as sophisticated problem domain classes.

As the dialog design for a use case progresses, windows classes are added to the sequence diagram—as the view layer. Typically, designers include one input form for all messages entering the system for the use case. If the messages are unique, however, then each may require its own input form. Each message from an external actor must be entered into the system in some way, and output messages must be displayed. One logical way to make this happen is through a window form class that can accept input data and possibly display output data. Returning to the *Cancel an order* use case, Figure 12-18 illustrates the result of defining a window class named :OrderWindowForm, which the clerk uses to start the use case. Remember that when the use case was first described, we noted that some other search forms and use case probably existed to help locate the correct order to cancel. In this figure, we maintain this simple approach and only show the single form to cancel an order. The input form merely allows the user to invoke the use case. It passes the message on to the controller. From there,

processing occurs as shown previously. Figure 12-18 only shows the :Order business object and the :OrderDA data access object. However, all the other objects are part of the final solution.

Figure 12-18

Partial sequence diagram for *Cancel an order* showing the view layer objects

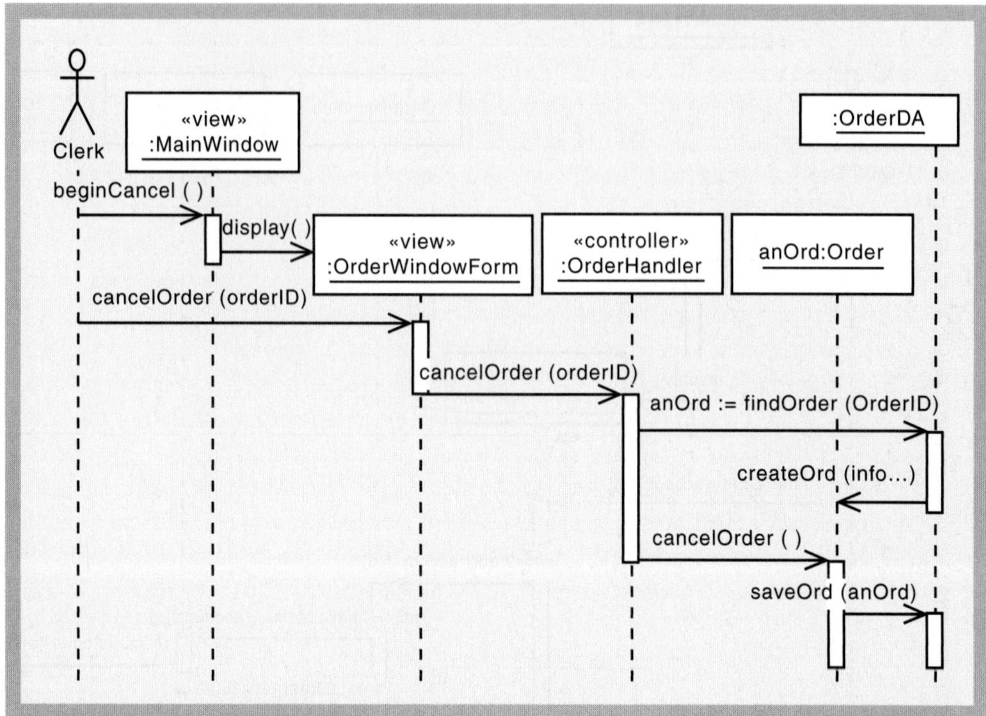

Although adding the user interface class may sound simple, it must be done in conjunction with the detailed design of the user-interface forms, as described in Chapter 14. For example, for the *Cancel an order* use case, are there two forms—one for input and another to display the results of the cancellation—or is a single form used both for input and display? In this design, we have assumed only one form. However, until the user sees one or two alternatives, the design must remain tentative. Often, the preliminary use case design deals with two layers—the business logic layer and the data access layer—and the user-interface layer design is done concurrently with the detailed user-interface design.

We next turn to the view layer for the *Create new phone order* use case. As we just indicated, identifying the user-interface classes is part of the user-interface design. In this case, we will show the user interface classes to start a new order. The first message is to a main window that will link to open other windows, including the new order window. The second input message will be to the new :OrderWindowForm to begin processing a new order. Figure 12-19 is an update of Figure 12-15, with the two view layer classes added and the data access classes included. The additional objects have been added to the use case design based on the user-interface decisions.

In the *Cancel an order* use cases, we first designed the business layer, then the data access layer, and finally the view layer. For *Create new phone order*, we only began the design of the view layer. We leave the rest of the view layer design and the data layer design as an exercise for the student. Developers always start with the business layer. However, the order of design for the data access layer and view layer varies. Some developers prefer doing the data access layer first, while others do the view layer. The order often depends on the schedule and the availability of users to help with the user interface.

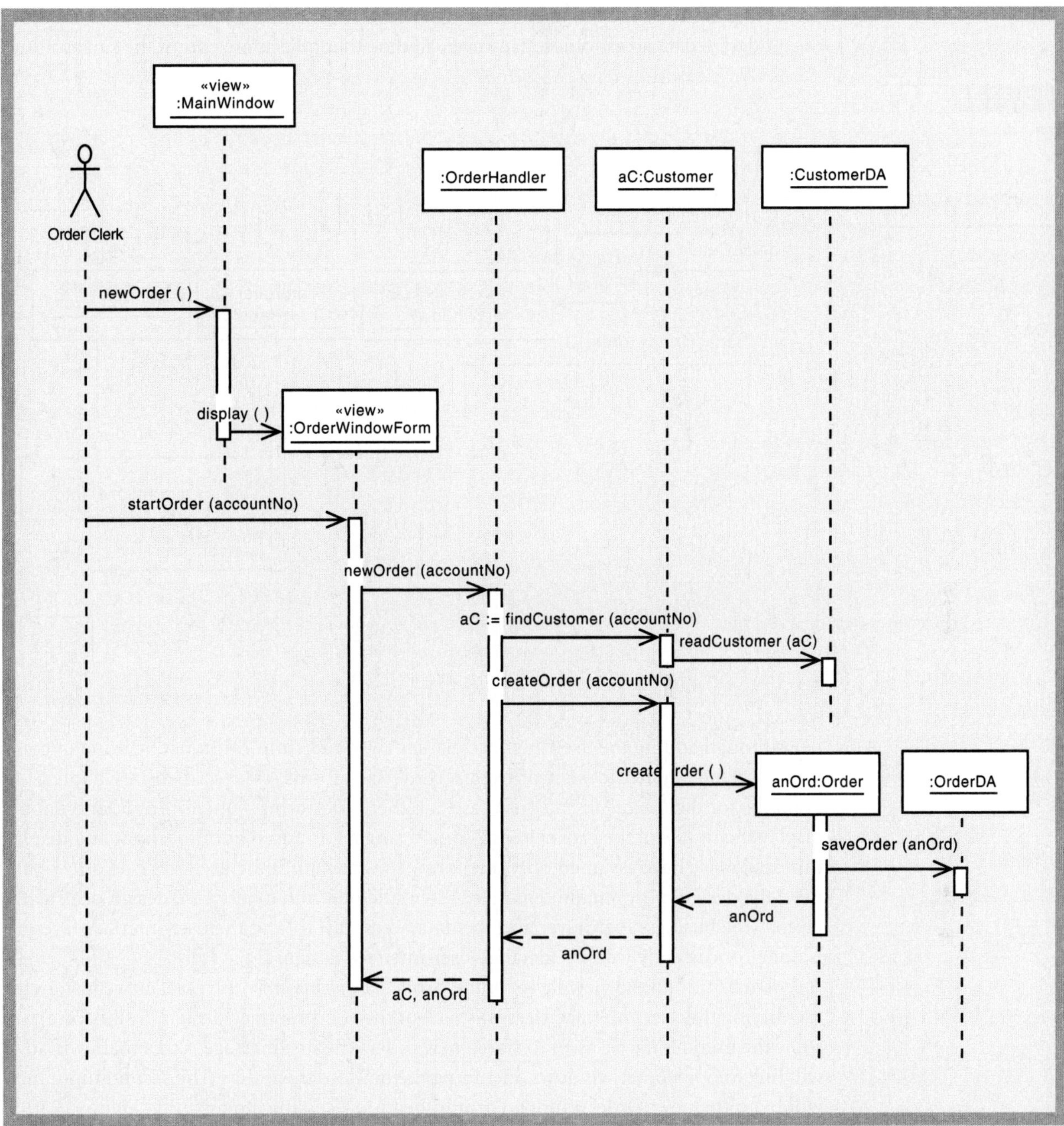

Figure 12-19

Partial *Create new phone order* with view layer

DESIGNING WITH COMMUNICATION DIAGRAMS

Communication diagrams and sequence diagrams are both interaction diagrams, and they capture the same information. The process of designing is the same whether you use communication diagrams or sequence diagrams. Which model you use for design is primarily your own personal preference. Many designers prefer to use sequence diagrams to develop the design because use case descriptions and dialog designs follow a sequence of steps. Communication diagrams are useful for showing a different view of the use case—one that

emphasizes coupling. Communication diagrams are also easier to use to sketch design ideas in a meeting, as they are easier to change and rearrange on the fly. We provide a brief introduction to communication diagrams in this section.

A communication diagram uses the same symbols for actors, objects, and messages as a sequence diagram. The lifeline and activation lifeline symbols are not used. However, a different symbol, the link symbol, is used. Figure 12-20 illustrates the four symbols used in most communication diagrams.

Figure 12-20

The symbols of a communication diagram

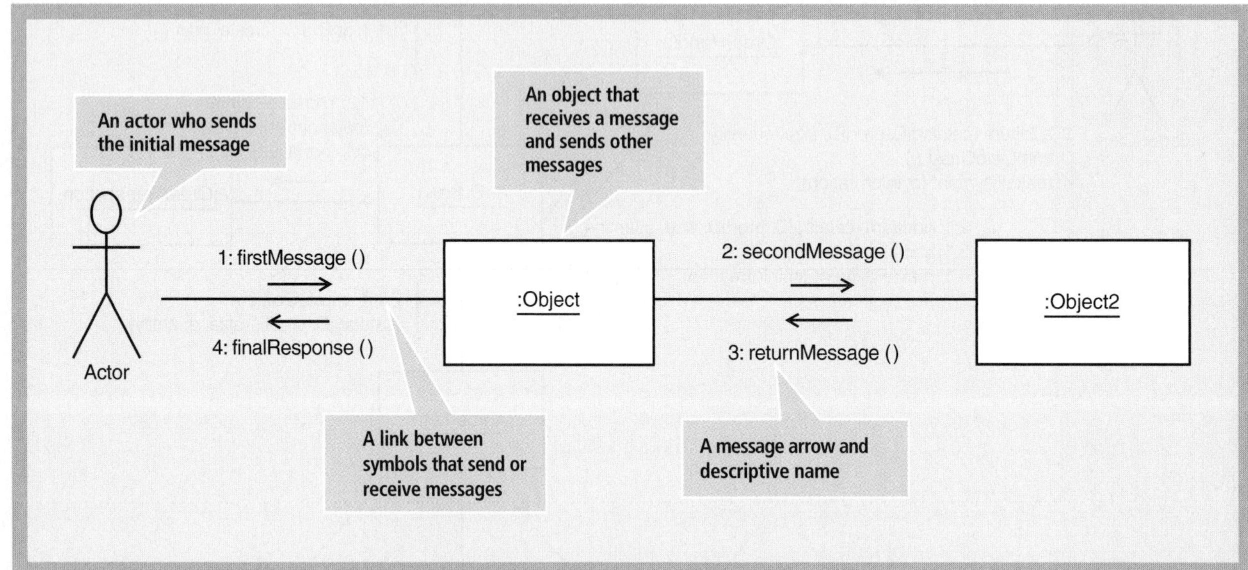

The format of the message descriptor for a communication diagram differs slightly from that for a sequence diagram. Because no lifeline shows the passage of time during a scenario, each message is numbered sequentially to indicate the order of the messages. The syntax of the message descriptor in a communication diagram is the following:

[true/false condition] sequence-number: return-value := message-name (parameter-list)

As you can see in Figure 12-20, a colon always directly follows the sequence number.

link

in a communication diagram, the connection between classes that indicates messages can be passed

The connecting lines between the objects or between actors and objects represent **links**. In a communication diagram, a link shows that two items share a message—that one sends a message and the other receives it. The connecting lines are essentially used only to carry the messages, so you can think of them as the wires used to transmit the messages.

Figure 12-21 presents a communication diagram for the RMO use case shown earlier with the sequence diagram in Figure 12-12, *Create new phone order*. This communication diagram contains only domain model objects and not the view layer or data access layer. However, multilayer design can be done just as effectively with communication diagrams as with sequence diagrams.

The numbers on the messages indicate the sequence in which the messages are sent. The hierarchical dot numbering scheme is used when messages depend on other messages. In this instance, the primary message, 1: startOrder (accountNo), is sent to :OrderHandler, which then forwards a similar message, 1.1: createOrder (accountNo), to :Customer. The second message is a direct result of the first, so it is numbered 1.1, as a subordinate to the primary message. Sometimes new designers struggle with determining when to number messages as subordinate and when to number them at the same level. For example, you could argue that the entire sequence of messages depends on the first one being sent and that the entire set should be subordinated to the initial message. One good way to determine how to number messages is to have the first message start with an external actor. Other messages triggered by the first message are dependent on it, and can be numbered with lower-level numbers. When control returns to the sender, a new hierarchy begins. The hierarchy of messages can go as

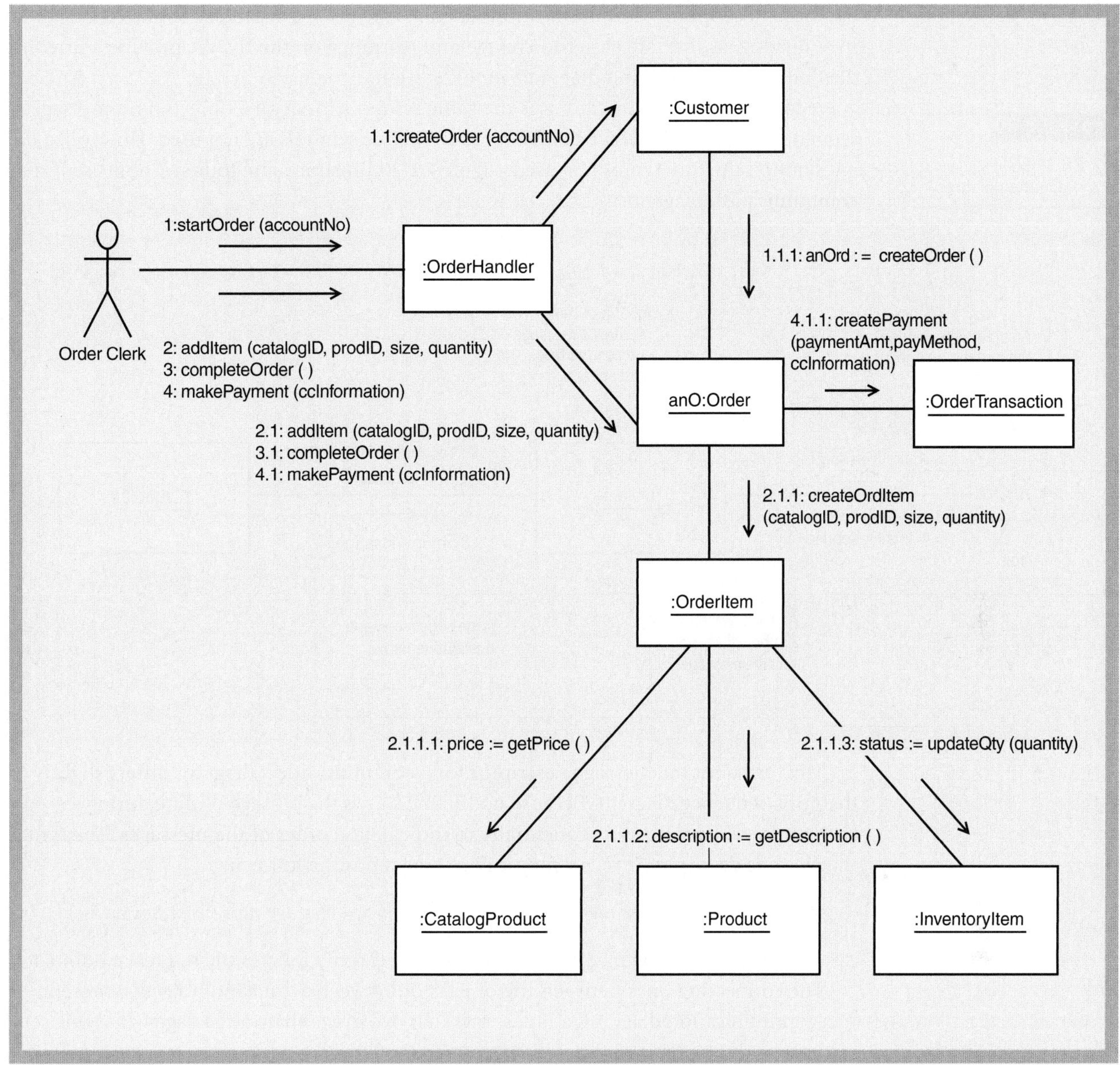

Figure 12-21

A communication diagram for *Create new phone order*

deep as required to indicate dependency. In Figure 12-21, the numbering sequence goes down several levels. Because multiple input messages initiate a series of other messages, this diagram is explicit in using the hierarchical numbering scheme.

When you compare the communication diagrams with the sequence diagrams, it should be evident that the focus of a communication diagram is on the objects themselves. Drawing a communication diagram is an effective way to get a quick overview of the objects that work together. However, as you look at the diagrams, you should see that it is more difficult to visualize the sequence of the messages. You have to hunt to find the numbers to see the sequence of the messages. On the other hand, a communication diagram is an effective way to get a quick overview of the collaborating objects.

Many designers use communication diagrams to sketch out a solution. If the use case is small and not too complex, a simple communication diagram may suffice. However, for more complex situations, a sequence diagram may be required to allow you to visualize the flow and sequence of the messages. It is not unusual to find a mix within the same set of specifications: some use cases are described by communication diagrams and others are shown with sequence diagrams. As a system developer, you should be comfortable using both types of diagrams.

UPDATING AND PACKAGING THE DESIGN CLASSES

Design class diagrams can now be developed for each layer. In the view layer and the data access layer, several new classes must be specified. The domain layer also has some new classes added for the use case controllers.

In Figure 12-5, we developed a partial first-cut design class diagram for the domain layer based on the *Cancel an order* use case. At that point in the development, no method signatures had been developed. Now that several sequence diagrams have been created, method information can be added to the classes. We also mentioned that the navigation arrows may need updating from the decisions that were made during sequence diagram development. In Chapter 11, we briefly introduced the idea of creating method names in the classes based on responsibilities identified on the CRC cards. However, at that point, we did not have enough information to rigorously define method signatures with names, return types, and parameter lists. Use case realization with sequence diagrams generates enough information to be rigorous in defining methods.

First, we add method signatures before finalizing visibility. Three types of methods are found in most classes: (1) constructor methods, (2) data get and set methods, and (3) use case specific methods. Remember that constructor methods create new instances of objects. Get and set methods retrieve and update attribute values. Because every class must have a constructor, and most usually have get and set methods, it is optional to include those method signatures in the design class diagram. In fact, to avoid information overload, most developers do not include those methods in the DCD. The third type of method—use case specific methods—must be included in the design class diagram.

As in sequence diagrams, every message has a source object and a destination object. When a message is sent to an object, it must be prepared to accept the message and initiate some activity. This process is nothing more than invoking or calling a method on an object. In other words, every message that appears in a sequence diagram requires a method in the destination object. In fact, the syntax for a message looks very much like the syntax for a method. Thus, the process of adding method signatures to a design class is to go through every sequence diagram and find the messages sent to that class. Each message indicates a method.

Let's work through one example based on the InventoryItem class. In Figure 12-14, two messages are sent to InventoryItem. The first is a constructor and the other is an update message, updateQty(qty). The update message return can be void, or it can return a success status code as an int. We need to add a method signature that corresponds to this message. Adding this method to the InventoryItem class is shown in Figure 12-22.

Figure 12-22

Design class diagram for the InventoryItem class showing a method signature

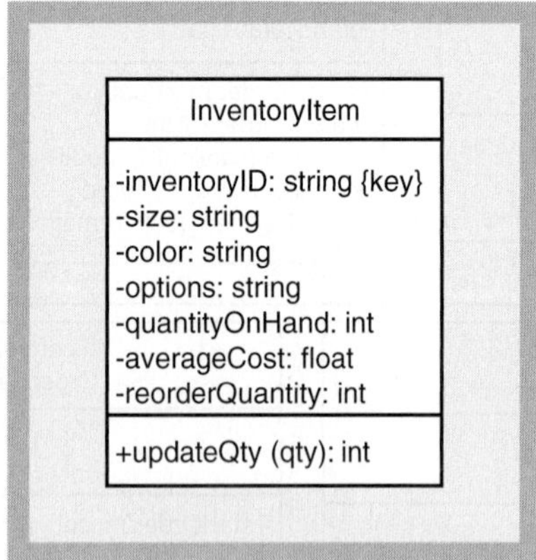

This process is continued for every class in the domain layer, including the added use case controller classes. Figure 12-23 contains the completed design class diagram for the domain layer classes. As you can see, this diagram provides excellent, thorough documentation of the design classes and serves as the blueprint for programming the system.

Figure 12-23

Updated design class diagram for the domain layer

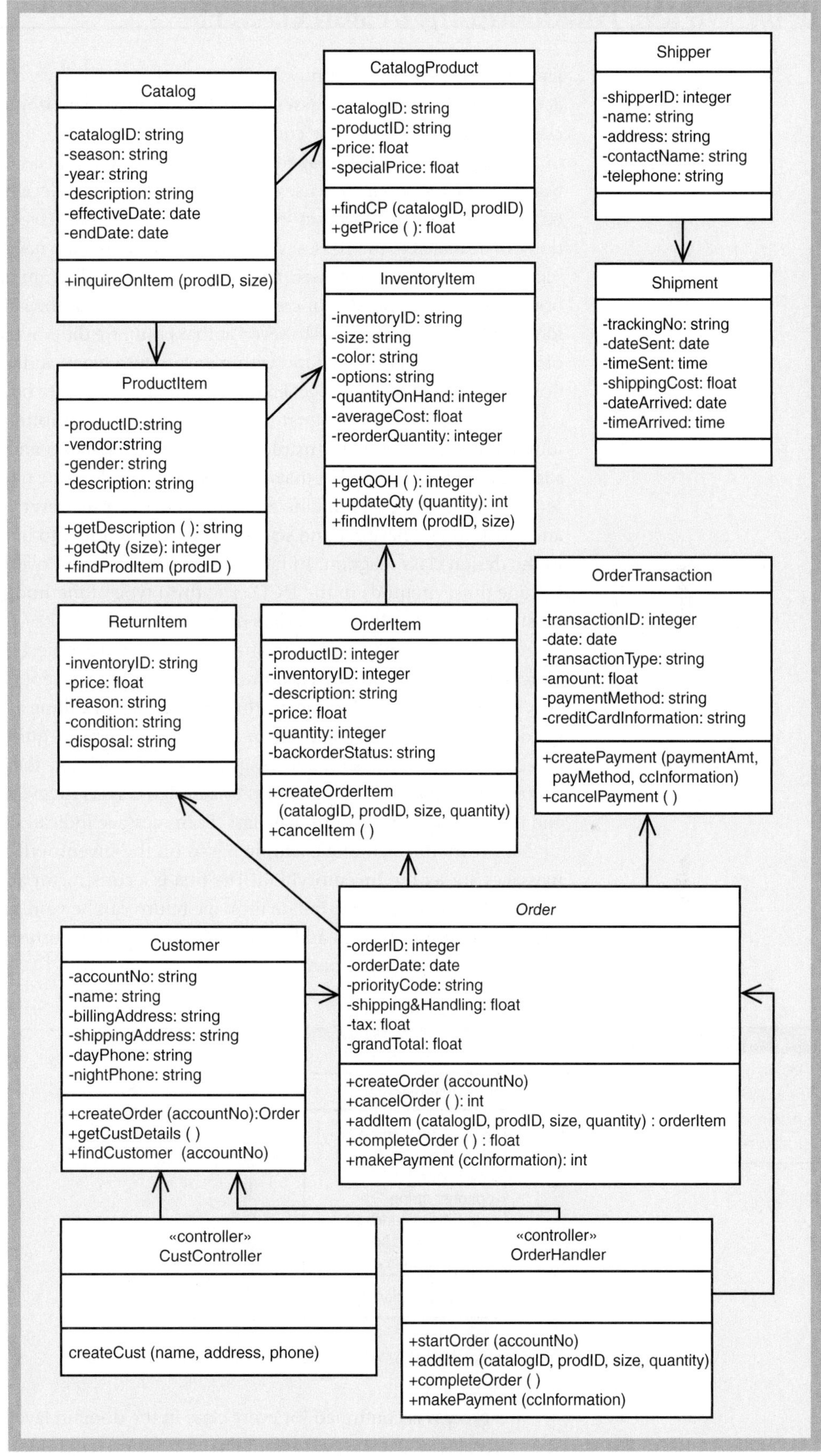

The two major additions to the domain layer classes are the two use case handlers. Additional navigation arrows have also been added to document which classes are visible from the controller classes. The other navigation arrows, which were defined during the first cut of the class diagram, have proved to be adequate for these two use cases. Additional use case development will enable us to add more navigation arrows, such as those to the Order, Shipment, and ReturnItem classes, as well as additional controller classes.

STRUCTURING THE MAJOR COMPONENTS WITH PACKAGE DIAGRAMS

As you learned previously, a package diagram in UML is simply a high-level diagram that allows designers to associate classes of related groups. The preceding sections illustrated three-layer design, which includes the view layer, the domain layer, and the data access layer. In the interaction diagrams, the objects from each layer were shown together in the same diagram. However, designers sometimes need to document differences or similarities in the objects' relationships in these different layers—perhaps separating or grouping them based on a distributed processing environment. This information can be captured by showing each layer as a separate package. Figure 12-24 illustrates how these layers might be documented.

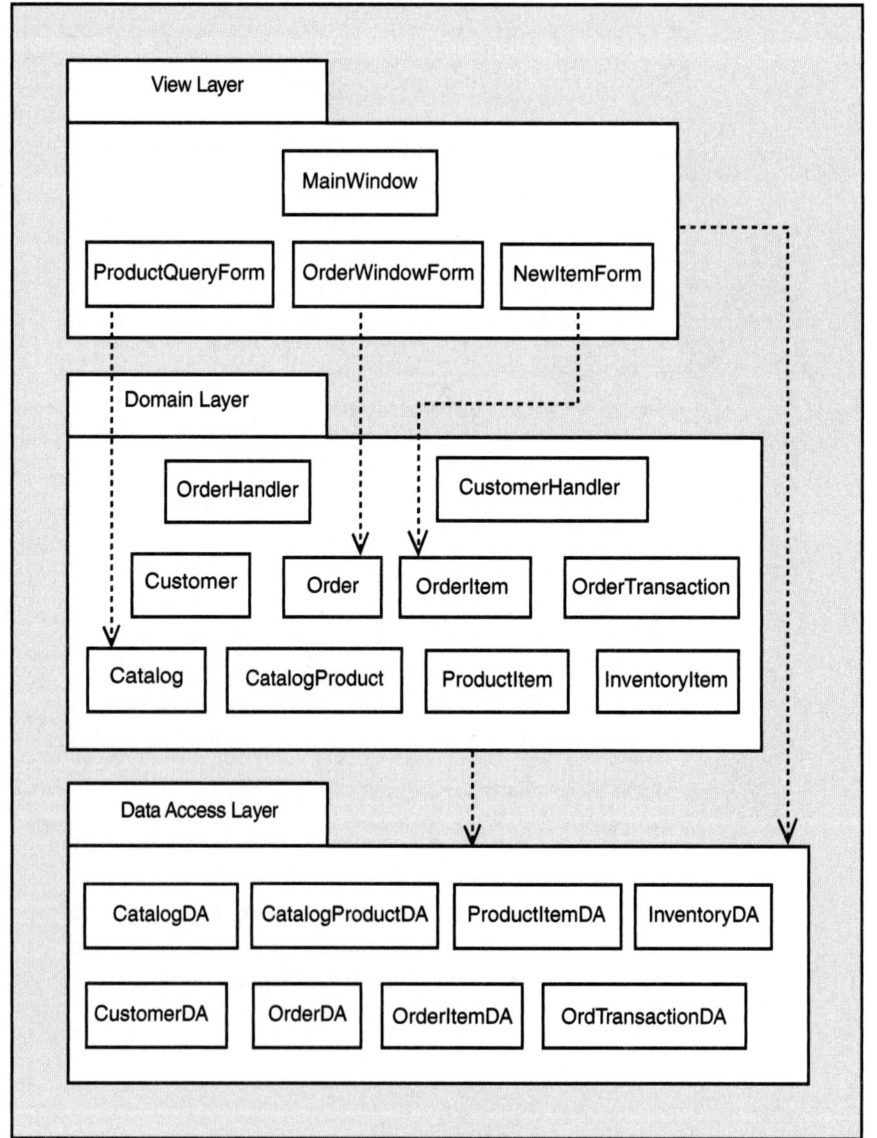

Figure 12-24
Partial design of three-layer package diagram for RMO

The package name is usually shown on the tab, although it also can be placed inside the package rectangle for high-level views if no details are shown inside the package. In this instance, the classes that belong to the package are placed inside the package rectangle.

The classes are placed inside the appropriate package based on the layer to which they belong. Classes are associated with different layers as they are developed in the interaction diagrams. To develop this package diagram, we simply extracted the information from design class diagrams and interaction diagrams for each use case. Figure 12-24 is only a partial package diagram, because the packages contain only the classes from the use case interaction diagrams that were developed in this chapter.

dependency relationship

a relationship between packages, classes, or use cases that indicates a change in the independent item will require a corresponding change in the dependent item

The other symbol used on a package diagram is a dashed arrow, which represents a **dependency relationship**. The arrow's tail is connected to the package that is dependent, and the arrowhead is connected to the independent package. Dependency relationships are used in package diagrams, class diagrams, and even interaction diagrams. A good way to think about a dependency relationship is that if one element changes (the independent element), the other, dependent element might also have to be changed. Dependency relationships can be between packages or between classes within packages. Figure 12-24 indicates that three classes in the view layer are dependent on classes in the domain layer. So, for example, if a change is made in the Order class, the OrderWindowForm class should be evaluated to capture that change. However, the reverse is not necessarily true. Changes to the view layer usually do not carry through to the domain layer.

Two examples of dependency relationships are given in Figure 12-24. The first, we have seen, is between classes. Another example is less detailed and indicates a dependency between packages. Figure 12-24 indicates that both the view layer and the domain layer depend on the data access layer. For some simple queries against the database, the view layer may directly access the data layer without requiring any involvement of the domain layer. These dependencies indicate that changes to the data structures, as reflected in the data access layer, usually require changes at the domain layer and the view layer.

Package diagrams can also be nested to show different levels of packages. Figure 12-25 indicates that the packages, and some of the classes contained within them, are all part of the order-entry subsystem. The RMO system can be divided into subsystems, and one way to document them is with package diagrams. A major benefit of this documentation is that different packages can be assigned to different teams of programmers to program the classes. The dependency arrows will help them recognize where communication among teams is needed to ensure an integrated system.

As shown in Figure 12-25, dependency is indicated from order fulfillment to order entry. Also, order entry depends on the customer maintenance and catalog maintenance subsystems. The order fulfillment subsystem may use the order classes defined in the order-entry subsystem, or information may be sent from the order-entry subsystem. In any event, if anything changes in the order-entry subsystem, the order fulfillment subsystem may also require modification. As the classes are added to individual packages, the use of objects in a class will determine the dependencies. For example, dependency arrows can also be determined after the classes are assigned to packages. The order-entry subsystem obviously needs to access the Customer class; thus, it depends on the customer maintenance package.

In summary, package diagrams show related components and dependencies. Generally, we use package diagrams to relate classes or other system components such as network nodes. The preceding figures show two uses of package diagrams—to divide a system into subsystems and to show nesting within packages.

Figure 12-25

RMO subsystem packages

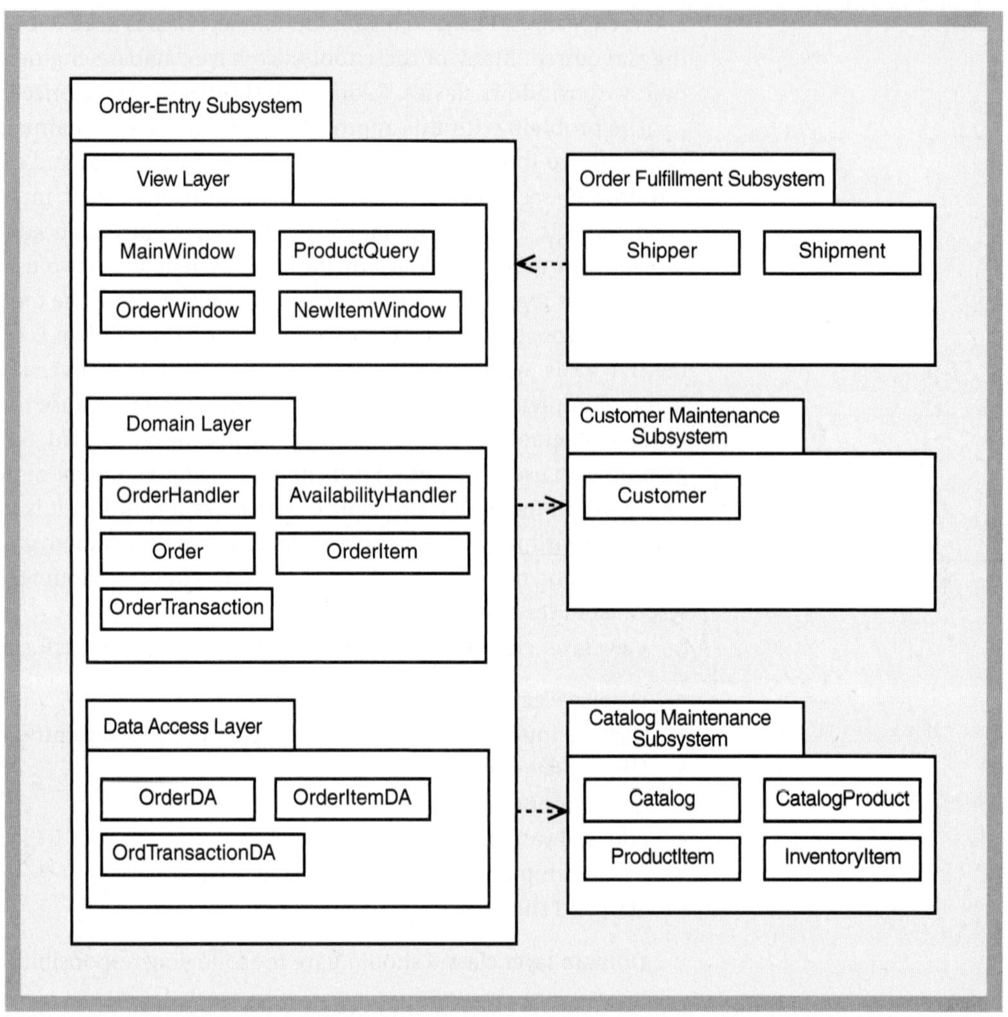

IMPLEMENTATION ISSUES FOR THREE-LAYER DESIGN

Using design class diagrams, interaction diagrams, and package diagrams, programmers can begin to build the components of a system. So, implementation in this sense means constructing the system with a programming language such as Java or VB. NET. Over the last few years, powerful integrated development environment (IDE) tools have been developed to help programmers construct systems. Such tools as Jbuilder and Eclipse for Java, Visual Studio for Visual Basic and C#, and C++Builder for C++ provide a high level of programming support, especially in building the view layer classes—the windows and window components of a system.

Unfortunately, these same tools have propagated some bad programming habits in some developers. The ease with which programmers can build GUI windows and automatically insert code has allowed them to put all of the code in the windows. Each window component has several associated events where code can be inserted. So, some programmers find it easy to build a window with an IDE tool, let the tool automatically generate the class definition,

and merely insert business logic code. No new classes need to be defined, and little other coding is required. Many of these tools also have database engines, so the entire system can be built with windows classes. Taking such shortcuts exacts a price later, however.

The problem with this approach is the difficulty of maintaining the system. Code snippets scattered throughout the GUI classes are hard to find and maintain. Plus, when the user-interface classes need to be upgraded, the programmer must also find and update the business logic. If a network-based system needs to be enhanced to include a Web front end, a programmer must rebuild nearly the entire system. Or, if two user interfaces are desired, all of the business logic is programmed twice. Finally, without the tool that generates the code, it is almost impossible to keep the system current. This problem is exacerbated by new releases of the IDE tools, which may not be compatible with earlier versions. Many programmers have had to completely rewrite the front end of a system because the new release of an IDE tool did not generate code the same way the previous release did. So, we advise analysts and programmers to use good design principles in the development of new systems.

Based on the design principle of object responsibility, it is possible to define which program responsibilities belong to each layer. If you follow these guidelines when writing code, a system will be much easier to maintain throughout its lifetime. Let's summarize the primary responsibilities of each layer.

View layer classes should have programming logic to perform the following:

- Display electronic forms and reports.
- Capture input events such as clicks, rollovers, and key entry.
- Display data fields.
- Accept input data.
- Edit and validate input data.
- Forward input data to the domain layer classes.
- Start and shut down the system.

Domain layer classes should have the following responsibilities:

- Create problem domain (persistent) classes.
- Process all business rules with appropriate logic.
- Prepare persistent classes for storage to the database.

Data access layer classes should include the following:

- Establish and maintain connections to the database.
- Contain all Structured Query Language (SQL) statements.
- Process result sets (the results of SQL executions) into appropriate domain objects.
- Disconnect gracefully from the database.

DETAILED OO DESIGN AT RMO

Barbara Halifax and the RMO project team are in the middle of the first iteration. They have been designing use cases based on the techniques explained earlier in this chapter. Each use case is assigned to a particular developer. The team usually meets in a brainstorming session to do a quick design for each use case using CRC cards. Then, if the developer in charge has any concerns about the complexity of the solution, he or she will create a sequence diagram for the use case. Before programming starts on a use case, the team will do one more walkthrough of the sequence diagrams. This may sound like a slow process, but the team works together so closely that a walkthrough of a given use case usually takes only 15 to 30 minutes. The team considers a walkthrough a small price to pay compared with the hours of programming time than can be lost from a serious mistake. The other benefits of walkthroughs are that all team members know what is going on, all programmers are using the same techniques and patterns, and new team members get good mentoring. The discussions often center not only on solutions, but on the patterns used to make the solutions better.

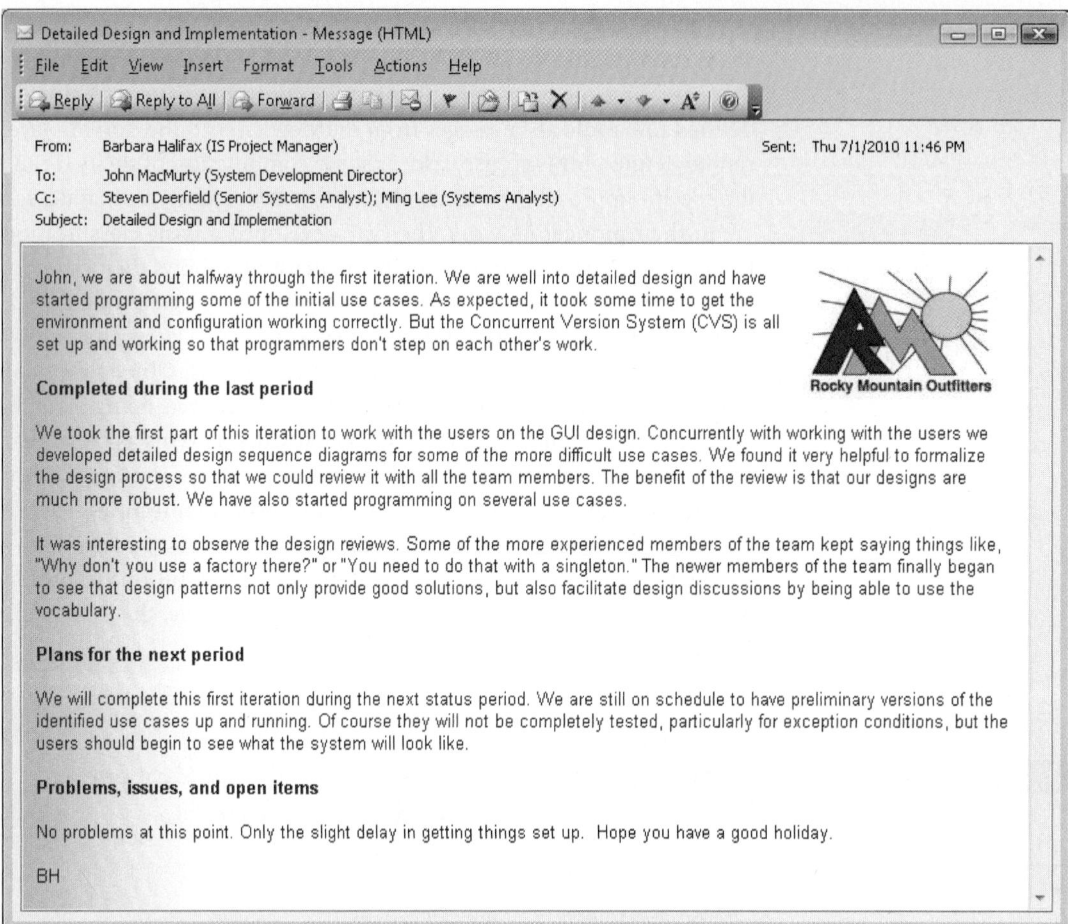
Detailed Design and Implementation - Message (HTML)

File Edit View Insert Format Tools Actions Help

Reply | Reply to All | Forward

From: Barbara Halifax (IS Project Manager) Sent: Thu 7/1/2010 11:46 PM

To: John MacMurty (System Development Director)

Cc: Steven Deerfield (Senior Systems Analyst); Ming Lee (Systems Analyst)

Subject: Detailed Design and Implementation

John, we are about halfway through the first iteration. We are well into detailed design and have started programming some of the initial use cases. As expected, it took some time to get the environment and configuration working correctly. But the Concurrent Version System (CVS) is all set up and working so that programmers don't step on each other's work.

Completed during the last period

We took the first part of this iteration to work with the users on the GUI design. Concurrently with working with the users we developed detailed design sequence diagrams for some of the more difficult use cases. We found it very helpful to formalize the design process so that we could review it with all the team members. The benefit of the review is that our designs are much more robust. We have also started programming on several use cases.

It was interesting to observe the design reviews. Some of the more experienced members of the team kept saying things like, "Why don't you use a factory there?" or "You need to do that with a singleton." The newer members of the team finally began to see that design patterns not only provide good solutions, but also facilitate design discussions by being able to use the vocabulary.

Plans for the next period

We will complete this first iteration during the next status period. We are still on schedule to have preliminary versions of the identified use cases up and running. Of course they will not be completely tested, particularly for exception conditions, but the users should begin to see what the system will look like.

Problems, issues, and open items

No problems at this point. Only the slight delay in getting things set up. Hope you have a good holiday.

BH

DESIGN PATTERNS

We have stated many times that developing software systems is difficult and complex. Historically, developers have not routinely developed systems that work correctly and solve the right business problem. In addition, every system requires constant modification and upgrading. Perhaps functions that were not added in the first version need to be added later, or errors need to be fixed. Or maybe the business requirements have changed and the system needs to be upgraded. The cost of maintaining a system over its lifetime can be many times the cost of the original development. However, systems that are based on good design principles are not only easier to develop and put into operation the first time, they are much easier to maintain. Such concepts as object responsibility, coupling, cohesion, protection from variations, and indirection were introduced in Chapter 11, and have been applied throughout the discussion in Chapter 12.

BEST PRACTICE

Continually think about protection from variations as you design—during user-interface design, program logic design, and database design.

You are also familiar with the concepts of design patterns and with two specific patterns—three-layer design and use case controllers. Patterns exist at various levels of abstraction. At a concrete level, a pattern may be a class definition that is written in code to be used by any developer. At the most abstract level, a pattern might only be an approach to solving a problem. For example, the multilayer design pattern tends to be more abstract, and states that it is better to separate system functions into three layers of classes: the GUI logic is placed in a set of view-layer classes that are separate and distinct from the domain layer and data access layer. So, multilayer design is an approach to building a system rather than a specific solution.

The use case controller pattern is more concrete. It defines a specific class or classes that act as the switchboard for all incoming messages from the environment. As with all patterns, there are multiple ways to implement the controller pattern. A single controller class can be defined to handle all messages from the view layer to the domain layer. Alternatively, a class can be defined for each use case, or some combination of the two can be used. Regardless of the specific approach, the controller pattern does require a separate, specified class.

In their pioneering work, the GoF developed a basic classification scheme for patterns. Figure 12-26 lists most of the original patterns and shows their classifications. The rows of the table identify the scope of the pattern—whether the design is a class-level or object-level pattern. Class-level patterns define solutions such as abstract classes that apply to static methods—in other words, that do not actually instantiate objects. Object-level patterns apply when the implementation of the pattern results in specific objects being instantiated from classes. The columns of the table in Figure 12-26 classify the patterns as creational, structural, or behavioral. Creational patterns help assign responsibilities to classes to instantiate new objects. Structural patterns provide solutions to meet the architectural needs of the system—that is, the set of classes and the ways they are related. Structural patterns help solve problems associated with indirection. Behavioral patterns provide solutions to problems that are related to the way internal system processes execute. For example, the iterator pattern solves the problem of how to process arrays and lists effectively.

Figure 12-26

Classification of GoF design patterns

Scope of pattern	Type of pattern		
	Creational	**Structural**	**Behavioral**
Class-level patterns	Factory method	Adapter	Interpreter Template Method
Object-level patterns	Abstract Factory Builder Prototype Singleton	Adapter Bridge Composite Decorator Façade Proxy	Chain of Responsibility Command Iterator Mediator Memento Flyweight Observer State Strategy Visitor

Even though GoF patterns are some of the most fundamental and important, many other patterns are also frequently used. We will limit our discussion to a few GoF patterns. To help you begin learning about design patterns, we present at least one pattern from each category. The following sections explain the adapter, factory, singleton, and observer patterns. As you continue your career, you can learn about the other patterns and add them to your knowledge and vocabulary.

ADAPTER

We start with the adapter pattern because the concept is straightforward. The adapter pattern is also a good example of the design principles of protection from variations and indirection. An adapter pattern is roughly akin to an electrical adapter used for international travel. For example, if you are traveling to England, you might decide to take your hair dryer with you. It has a switch for either 110 volts or 220 volts, so you think you can run it on either voltage. However, the plug on the end of the power cord has two flat prongs. Unfortunately, wall sockets in England have slots for three large prongs set at angles. You need something that can adapt the power cord's two prongs to the wall's three angled slots. Figure 12-27 shows a typical electrical adapter you might use.

Figure 12-27

Electrical adapter

The adapter design pattern works just like the electrical adapter; it plugs an external class into an existing system. The method signatures on the external class are different from the method names being called from within the system, so the adapter class is inserted to convert the method calls from within the system to the method names in the external class.

This pattern is a standard solution for protection from variations. The external class could be a variable class—that is, it could be replaced at any time by an upgrade or an entirely different class. This situation often occurs when commercial software libraries are purchased to provide special services. For example, in an organization's internal payroll system, designers might purchase a set of classes to calculate all of the complex income tax deductions. Because tax deductions are a specialized area, and most companies do not have tax specialists on their IT staff, an easy solution is simply to purchase those classes. However, knowing that the classes could be replaced later if the tax law changes, a designer would be wise to create an adapter class between the system and the tax calculation classes.

Figure 12-28 describes the details of the adapter design pattern. Be sure to read it carefully. The sample diagram has four UML classes. The one labeled System represents the entire system. The classes within the system use method names such as getSTax() and getUTax() to access the tax routines. The tax calculator class has method names of findTax1() and findTax2(). The two UML classes in the middle represent the adapter. The top class symbol represents an interface class; an interface is useful to specify the method names. Although not absolutely necessary, it is a simple way to specify and enforce the use of the correct method names. The adapter class then inherits those method names and provides the method logic for those methods. The body of each method simply extends a call to the final method name of findTax1() or findTax2(). In other words, it "adapts," or translates, the method names from one to the other.

Name:	Adapter
Problem:	A class must be replaced, or is subject to being replaced, by another standard or purchased class. The replacing class already has a predefined set of method signatures that are different from the method signatures of the original class. How do you link in the new class with a minimum of impact so that you don't have to change the names throughout the system to the method names in the new class?
Solution:	Write a new class, the adapter class, which serves as a link between the original system and the class to be replaced. This class has method signatures that are the same as those of the original class (and the same as those expected by the system). Each method then calls the correct desired method in the replacement class with the method signature. In essence, it "adapts" the replacement class so that it looks like the original class.
Example:	There are several places in the RMO system where class libraries were purchased to provide special processing. These purchased libraries provide specialized services such as tax calculations and shipping and postage rates. From time to time, these service libraries are updated with new versions. Sometimes a service library is even replaced with one from an entirely different vendor. The RMO systems staff applies protection from variations and indirection design principles by placing an adapter in front of each replaceable class. System → «interface» TaxCalculatorIF (getSTax (), getUTax ()) TaxCalcAdapter (getSTax (), getUTax ()) → ABCTaxCalculator (findTax1 (), findTax2 ())
Benefits and Consequences:	The adaptee class can be replaced as desired. Changes are confined to the adapter class and do not ripple through the system. Two classes are defined, an interface class and the adapter class. Passed parameters may add more complexity, and it is difficult to limit changes to the adapter class.

Figure 12-28
Adapter pattern template

As you become familiar with this design pattern, you will find that it has a multitude of uses. It is a powerful and elegant solution to making a system more maintainable. Experienced developers use this pattern frequently, both for foreign classes and for internally written classes that may need frequent upgrades. It is an excellent way to insulate the system from frequently changing classes.

FACTORY

In the discussions of detailed design, we have often expressed the need to have utility classes, which include the data access objects or controller classes. An adapter in an adapter pattern situation is also a utility class. What class should create these utility objects? In most situations it does not make sense for domain classes to create them, because it is not a listed responsibility

of domain classes. A popular solution in object-oriented programming is to have some classes that are factories. In other words, these classes instantiate objects from utility classes.

For example, an executing customer object may need to write some data. If the factory class is designed with static methods, which means they have global visibility, then the customer object can just say to the factory, "Get me a reference to a data access object for the customer table." The factory will create a new data access object and return the reference. If a customer data access object already exists in memory, it simply returns the reference. The customer object does not have to be concerned about creating objects to access the database. It just uses whatever is passed to it. This reduces coupling, enhances cohesion, and assigns responsibilities to the right classes. Figure 12-29 is an example of a factory class.

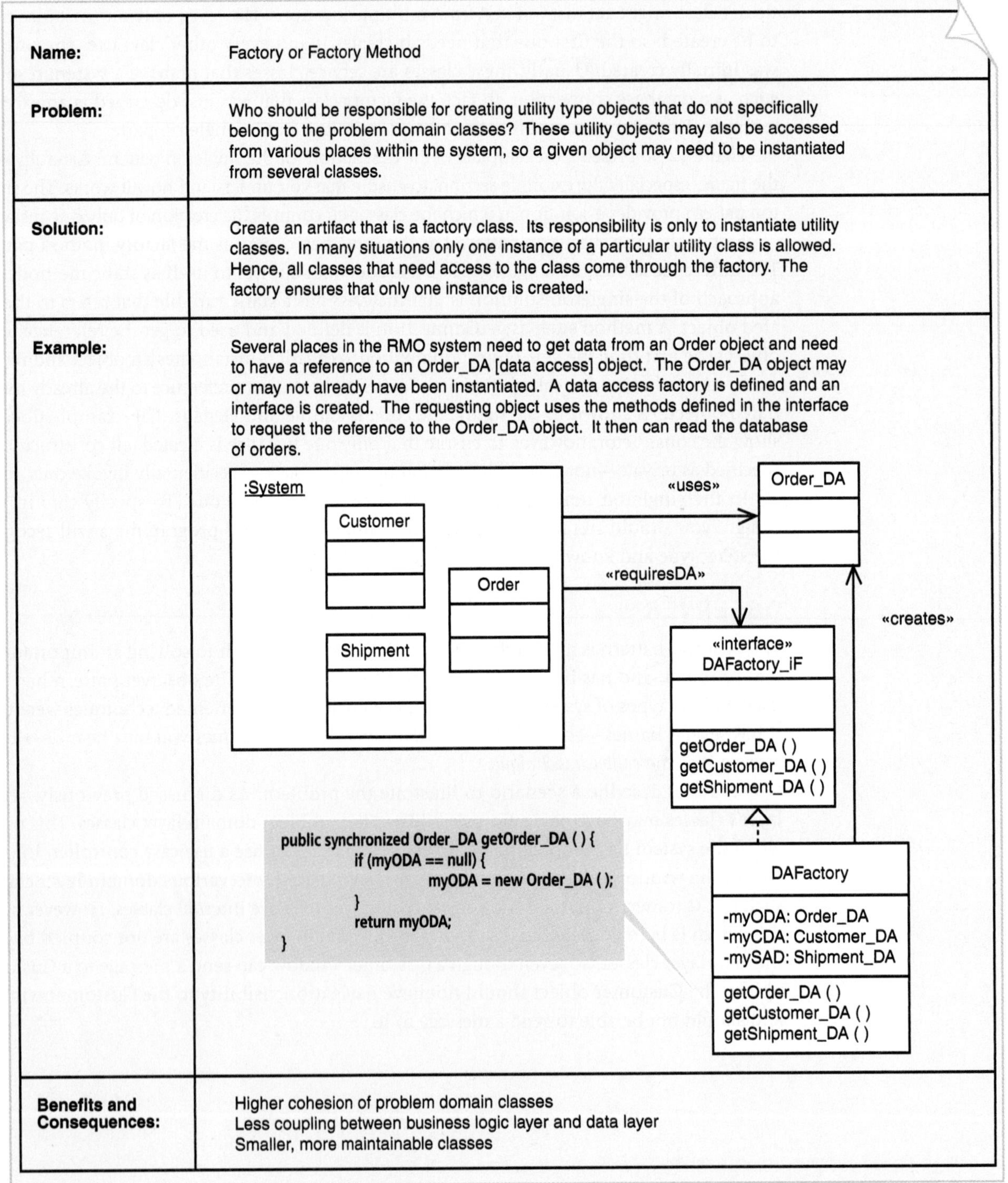

Name:	Factory or Factory Method
Problem:	Who should be responsible for creating utility type objects that do not specifically belong to the problem domain classes? These utility objects may also be accessed from various places within the system, so a given object may need to be instantiated from several classes.
Solution:	Create an artifact that is a factory class. Its responsibility is only to instantiate utility classes. In many situations only one instance of a particular utility class is allowed. Hence, all classes that need access to the class come through the factory. The factory ensures that only one instance is created.
Example:	Several places in the RMO system need to get data from an Order object and need to have a reference to an Order_DA [data access] object. The Order_DA object may or may not already have been instantiated. A data access factory is defined and an interface is created. The requesting object uses the methods defined in the interface to request the reference to the Order_DA object. It then can read the database of orders.
Benefits and Consequences:	Higher cohesion of problem domain classes Less coupling between business logic layer and data layer Smaller, more maintainable classes

Figure 12-29 Factory method pattern template

The factory class has private attributes to hold the references to the data objects that are created. When a request is made to get the reference to a data object, the method simply checks to see if the attribute is null. If so, it creates a new object, places its reference in the attribute, and returns the value. Otherwise, it just returns the parameter with the reference already in it.

SINGLETON

Some classes must have exactly one instance—for example, a factory class or the main window class. These classes have only one instance, but because they are instantiated from only one place, it is easy to limit the logic to create only one object.

Other classes must have exactly one instance, but cannot be easily controlled by having only one place to invoke the constructor. Depending on the flow of logic of the system, a particular class might get instantiated from multiple locations. However, only one instance needs to be created, so the first one that needs it creates it and every other class uses the one that was initially created. Usually, these classes are service classes that manage a system resource, such as a database connection. In fact, the factory class that was just described is an excellent example. This common problem has a standard solution: the singleton pattern.

Figure 12-30 presents the template of the description for the singleton pattern. Carefully study the figure, especially the example section, to ensure that you understand how it works. The singleton pattern provides a solution in which the class itself controls the creation of only one instance.

Notice that the singleton pattern has the same basic logic as the factory method pattern. The difference is that the singleton class has code that applies to itself as static methods. The approach of the singleton solution is that the class has a static variable that refers to the created object. A method such as getConnection is defined and used to get the reference to the object. The first time the getConnection method is called, it instantiates an object and returns a reference to it. On later calls to the method, it simply returns a reference to the already instantiated object. As shown in the figure, the code is simple and elegant. The example does not show the constructor; however, to ensure that only one instance is created, all constructors are specified as private—not accessible—so that no other class can accidentally invoke one.

In the singleton template, the pattern is represented by code. To specify this in your design, you should stereotype the class as a «singleton». Good programmers will recognize the stereotype and know exactly how to code the class.

OBSERVER

The observer pattern is a powerful and comprehensive approach to solving an important system problem, and has been used in practice for a long time. The observer pattern has been used with all types of system development, even before object-oriented techniques were used. It has several names—*observer, listener, publish/subscribe.* Sometimes you may hear developers refer to it as the *callback technique.*

Let's first describe a scenario to illustrate the problem. As discussed previously, a view layer's classes may have navigation visibility to the problem domain layer classes. This occurs when the system has simple updates to classes that do not use a use case controller. In other words, the windows classes know about and have visibility to various domain layer classes, such as Customer and Order, and can send messages to those internal classes. However, when the design is being developed, it is better if the domain layer classes are not coupled back to the view layer classes. So, even though a Customer window can send a message to a Customer object, the Customer object should not have navigation visibility to the Customer window and should not be able to send a message to it.

Name:	Singleton
Problem:	Only one instantiation of a class is allowed. The instantiation (new) can be called from several places in the system. The first reference should make a new instance, and later attempts should return a reference to the already instantiated object. How do you define a class so that only one instanceis ever created?
Solution:	A singleton class has a static variable that refers to the one instance of itself. All constructors to the class are private and are accessed through a method or methods, such as getInstance(). The getInstance() method checks the variable; if it is null, the constructor is called. If it is not null, then only the reference to the object is returned.

Example:	In RMO's system, the connection to the database is made through a class called Connection. However, for efficiency, we want each desktop system to open and connect to the database only once, and to do so as late as possible. Only one instance of Connection, that is, only one connection to the database, is desired. The Connection class is coded as a singleton. The following coding example is similar to C# and Java.
	
	`Class Connection`
	`{`
	`private static Connection conn = null;`
	`public synchronized static getConnection ( )`
	`    {`
	`      if (conn == null) {`
	`    conn = new Connection ( ) ;}`
	`      return conn;`
	`    }`
	`}`
	
	Another example of a singleton pattern is a utilities class that provides services for the system, such as a factory pattern. Since the services are for the entire system, it causes confusion if multiple classes provide the same services.
	
	An additional example might be a class that plays audio clips. Since only one audio clip should be played at one time, the audio clip manager will control that. However, for this to work, there must be only one instance of the audio clip manager.

Benefits and Consequences:	There are other times when only one instance of an object is needed, but if it is instantiated from only one place, then a singleton may not be required. The singleton object controls itself and ensures that only one instance is created—no matter how many times it is called and wherever the call occurs in the system. The code to implement the singleton is very simple, which is one of the desirable characteristics of a good design pattern.

Figure 12-30

Singleton pattern template

In this hypothetical example, let's use three classes: a Customer window, an Order window, and an Order class. The first two are windows classes and part of the view layer. The Order class is a domain layer class. The use case is *Create new order*. When the order was created, navigation visibility was provided from both the Customer window and the Order window to the Order class. Figure 12-31 illustrates the three classes, with two windows and the Order class. The navigation arrows show how messages can be sent.

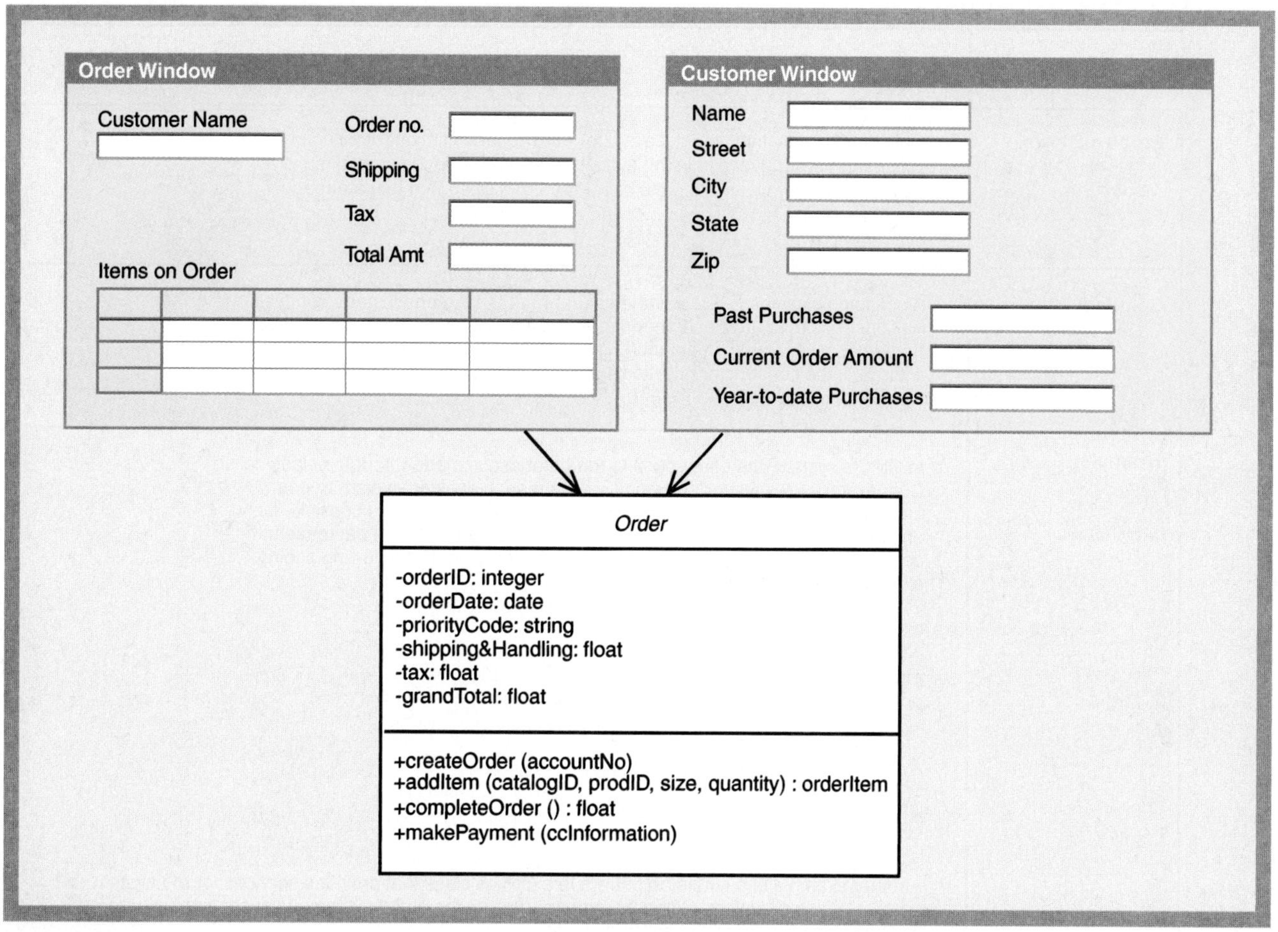

Order Window

Customer Name

Order no.

Shipping

Tax

Total Amt

Items on Order

Customer Window

Name

Street

City

State

Zip

Past Purchases

Current Order Amount

Year-to-date Purchases

Order

-orderID: integer
-orderDate: date
-priorityCode: string
-shipping&Handling: float
-tax: float
-grandTotal: float

+createOrder (accountNo)
+addItem (catalogID, prodID, size, quantity) : orderItem
+completeOrder () : float
+makePayment (ccInformation)

Figure 12-31

Three classes in the *Create new order* use case

Notice that the Customer window includes fields called Past Purchases, Current Order Amount, and Year-to-date Purchases. Because Rocky Mountain Outfitters (RMO) provides special discounts to customers when their year-to-date purchases exceed a certain amount, the telephone clerk watches these amounts as the order is being taken, and if necessary suggests additional purchases to take advantage of the discounts. As items are added to the order, the Current Order Amount and Year-to-date Purchases fields need to be updated. The data to update these fields is contained in the Order class. But, because Order does not have navigation visibility to the Customer window, how can it send information to that window? A novice designer would probably say, "Let's just put a reference to the Customer window in the Order object." But that violates the loose coupling principle. Instead, the observer pattern has been developed to handle this problem. Let's see how that solution works.

The concept of the observer, or listener, pattern is to have the Customer window "listen" for any changes to the Order object. When it "hears" of a change in an order, it updates the appropriate fields. For the listening to work, the Order class must contain mechanisms that (1) allow other classes to "subscribe" as listeners, and (2) "publish" the changes to the listeners when they occur.

The Order class must have the following components. First, it must have an array of object references that holds the list of all objects that have subscribed as listeners. The type of the array is an object array. Second, the Order class must have a method, usually something like addOrderListener, which can be called by potential listeners to subscribe. When a class wants to subscribe as a listener, it just calls the addOrderListener method and sends a reference to itself as a parameter. The logic of addOrderListener is simply to add the passed reference parameter to the array. Third, the Order class has a method named notify(), which iterates through the array and sends a message to each object referenced in the array. Normally, along with the Order class, a designer develops an interface class, which we call OrderListener, to

provide the method signature that will be called by the Order object—often this interface class is called the *publisher*. The listener class—that is, the CustomerWindow—can then inherit from the OrderListener interface to define the method signatures.

Figure 12-32 illustrates the changes required to implement the observer pattern. The changes have been added in bold. The Order class has a new array variable called orderListeners to hold the subscriber objects. Order has three new methods—one each to add and remove listeners, and one to notify OrderListener of any changes. The third method is private or invoked internally. The interface class OrderListener is used to specify the name of the method that will be called when the listener is called back. (This callback approach is why the pattern is sometimes called the *callback pattern*.)

Figure 12-32

Implementation of the observer pattern

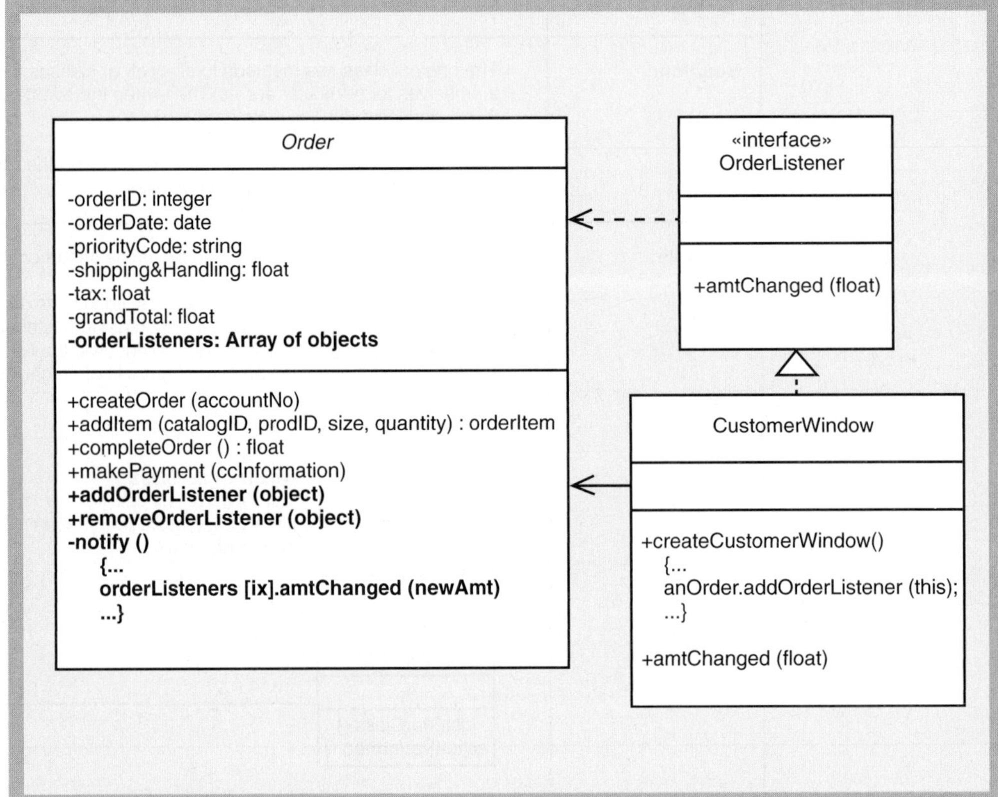

The CustomerWindow class has navigation visibility to the Order object—it is needed to be able to subscribe to the Order object. The CustomerWindow also inherits from the OrderListener interface so that it knows which method to implement. The OrderListener interface is dependent on the Order class. So, if the Order class changes, the OrderListener might need to change, too. The figure also includes some snippets of code to illustrate how the methods are invoked between the classes. You should carefully read the code to ensure that you understand how this works.

Because the link from the Order to the CustomerWindow is dynamic and temporary, it has no negative side effects on the code. In other words, if the CustomerWindow were no longer there, it would not affect the Order class. Or, if multiple windows wanted to listen to the order amount changing, no changes would be required in the Order class. This type of dynamic linking is a powerful and effective technique to avoid permanent coupling in places where it could cause problems.

Figure 12-33 is the pattern template for the observer pattern. The listener design pattern is used extensively as the technique to handle event processing by window objects. The class windows libraries for Java and .NET are all implemented using the listener pattern as the fundamental event-handling technique. The explanation in Figure 12-33 illustrates the names

used in the window's events. For example, the method on a window button to add a listener is addListener(). The method invoked on the subscriber is actionPerformed(). By using these standard names, the window's GUI classes become standard classes that can be reused in any application.

Name:	Observer
Problem:	One class has attributes that change or events that other classes need to know about. However, the original class does not know which other classes are interested in its internal activity. How do you allow classes to observe this behavior without coupling them?
Solution:	The original class has methods to allow other classes to dynamically register themselves for particular events. Then when the event occurs, it notifies all those classes that have registered. This pattern is also called the *listener pattern*, and the observer class is sometimes referred to as the *listener class.*
Example:	The entire windows event-handling system is based on this pattern. Windows components, such as text boxes and buttons, should not be directly coupled to classes and methods that process events, such as text entry or button clicks. Each component will have a method such as addListener () to allow other objects to register themselves. In windows, the listener method that notifies the listeners when an event occurs is often called *actionPerformed 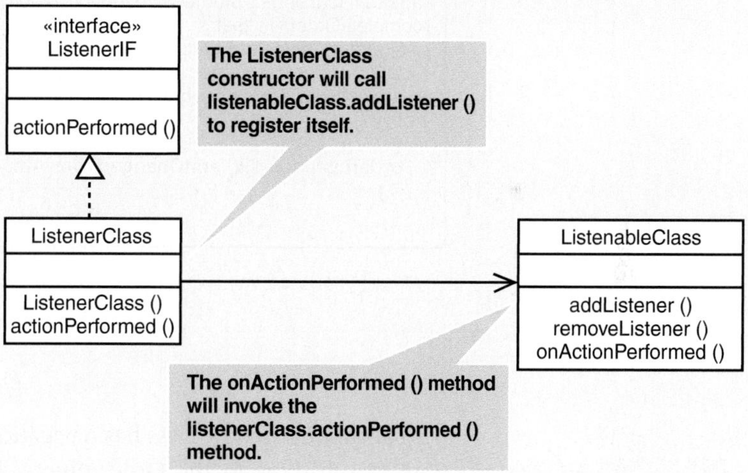 ().*
Benefits and Consequences:	This pattern is a variation of the publish/subscribe pattern. This pattern keeps coupling low because dynamic coupling does not cause maintenance problems. If a class has many observers, there can be a delay in notifying all of them.

Figure 12-33
Observer pattern template

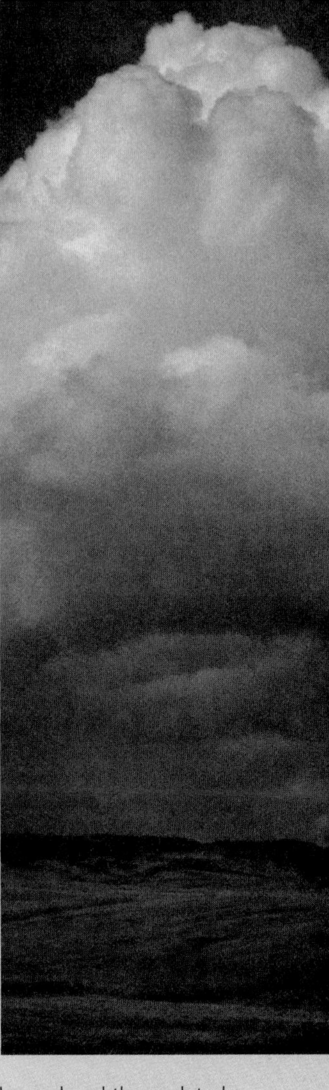

SUMMARY

Multilayer design of new systems is not limited to architectural design. Detailed object-oriented design also identifies the various levels in a system. The identification of classes and their responsibilities follows the three-layer pattern explained in Chapters 11 and 12. The three layers are the view layer, the business or logic layer, and the data access layer.

Three-layer design is part of the overall movement in systems design based on design patterns. A design pattern is a standard solution or template that has proven to be effective to a particular requirement in systems design. The other pattern, which was introduced in Chapter 11, is that of a use case controller. The use case controller pattern addresses the need to isolate the view layer from the business layer in a simple way that limits coupling between the two layers.

Detailed design is use case driven in that each use case is designed separately. This type of design is called use case realization. The two primary models used for detailed design are the design class diagram and the sequence diagram. Design class diagrams were introduced in Chapter 11.

Detailed design of use cases entails identifying problem domain classes that collaborate to carry out a use case. Using a sequence diagram or a communication diagram, each input message from an external actor is expanded to identify and define every internal message that is required to complete the event triggered by the input message. In the first cut, only the problem domain classes and their internal messages are identified. Next the solution is completed by adding the classes and messages for the view layer and the data access layer.

The final step is to convert each message, along with the passed parameters and return values, into method signatures located in the correct classes. This information is used to update the design class diagram. Changes are also made to the design class diagram to show required visibility between the classes in order to send messages in the sequence diagrams.

As classes are identified during the design process, they are added to the DCD. The DCD can also be partitioned into several layers, or into subsystems. Package diagrams are used to partition the DCD into appropriate packages. Dependency between the classes and the packages is also added to the package diagram.

Latter sections of this chapter expanded the idea of design patterns by introducing several popular ones: the adapter pattern, factory pattern, singleton pattern, and observer pattern. The adapter pattern implements the design principle of protection from variations by allowing a changing piece of the system simply to plug into a more stable part of the system. When the pluggable piece of the system needs to change, it can just be unplugged and the updated component can be plugged in.

The factory and singleton patterns have much in common. Both return a reference to a specific object. Both allow only one instance of that object to exist in the system. The difference is that the factory pattern enforces a single occurrence for utility classes and the singleton only enforces a single occurrence for itself.

The observer pattern is a standard technique upon which GUI event-driven windows are built. The basic idea is that when an event happens to an object, it notifies all of its observers that the event occurred. The observers can then take any appropriate action.

KEY TERMS

activation lifeline, p. 436
dependency relationship, p. 460
design patterns, p. 431
link, p. 455
separation of responsibilities, p. 446
use case realization, p. 430

REVIEW QUESTIONS

1. What is meant by the term *use case realization*?
2. What are the benefits of knowing and using design patterns?
3. What is the contribution to systems development from the Gang of Four?
4. What are the five components of a standard design pattern definition?
5. List five elements included in a sequence diagram.
6. How does a sequence diagram differ from a system sequence diagram?
7. What is the difference between doing design with CRC cards and designing with sequence diagrams?
8. Explain the syntax of a message on a sequence diagram.
9. What is the purpose of the first-cut sequence diagram? What kinds of classes are included?
10. What is the purpose of the use case controller?
11. What is meant by an activation lifeline? How is it used on a sequence diagram?
12. Describe the three major steps in developing the set of messages for the first-cut sequence diagram.
13. What assumptions do developers usually make while doing the initial use case realization?
14. When doing multilayer design, what is the order in which layers should be designed? Why?
15. What is the principle of separation of responsibilities?
16. Explain the two methods of accessing the database to create new objects in memory.
17. What symbols are used in a communication diagram, and what do they mean?
18. Explain the components of message syntax in a communication diagram. How does this syntax differ from that of a sequence diagram message?
19. Explain the method signature syntax on design classes.
20. What is meant by a dependency relationship? How is it indicated on a drawing?
21. List the major implementation responsibilities of each layer of a three-layer design.
22. What is the purpose of the adapter pattern?
23. What common element is found in the singleton pattern and the factory pattern? What is the basic difference between the two?
24. What common programming construct is based on the observer pattern?

THINKING CRITICALLY

Exercises 1, 2, 3, and 4 are based on the solutions you developed in Chapter 7 for "Thinking Critically" exercises 1 and 2, which are based on a university library system. Alternatively, your teacher may provide you with a use case diagram and a class diagram.

1. Figure 12-34 is a system sequence diagram for the use case *Check out books* in the university library system. Do the following:
   a. Develop a first-cut sequence diagram, which only includes the actor and problem domain classes.
   b. Add the view layer classes and the data access classes to your diagram from part a.
   c. Develop a design class diagram based on the domain class diagram and the results of parts a and b.
   d. Develop a package diagram showing a three-layer solution with view layer, domain layer, and data access layer packages.

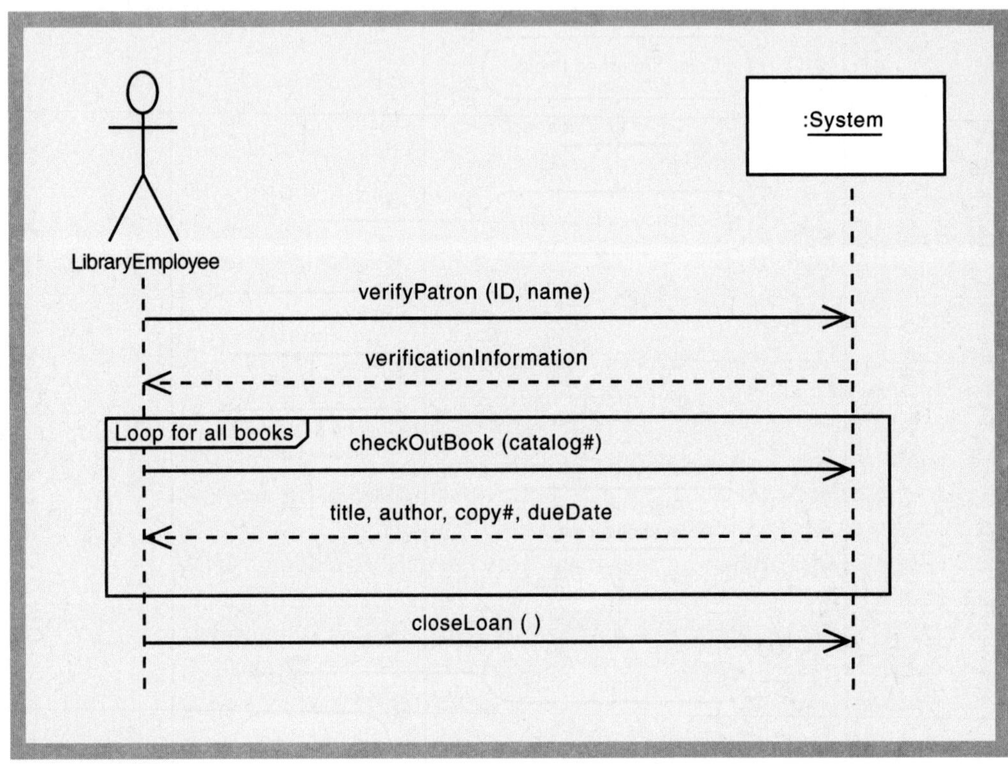

Figure 12-34

System sequence diagram for *Check out books*

2. Figure 12-35 is an activity diagram for the use case *Return books* in the university library system. Do the following:
   a. Develop a first-cut sequence diagram, which only includes the actor and problem domain classes.
   b. Add the view layer classes and the data access classes to your diagram from part a.
   c. Develop a design class diagram based on the domain class diagram and the results of parts a and b.
   d. Develop a package diagram showing a three-layer solution with view layer, domain layer, and data access layer packages.

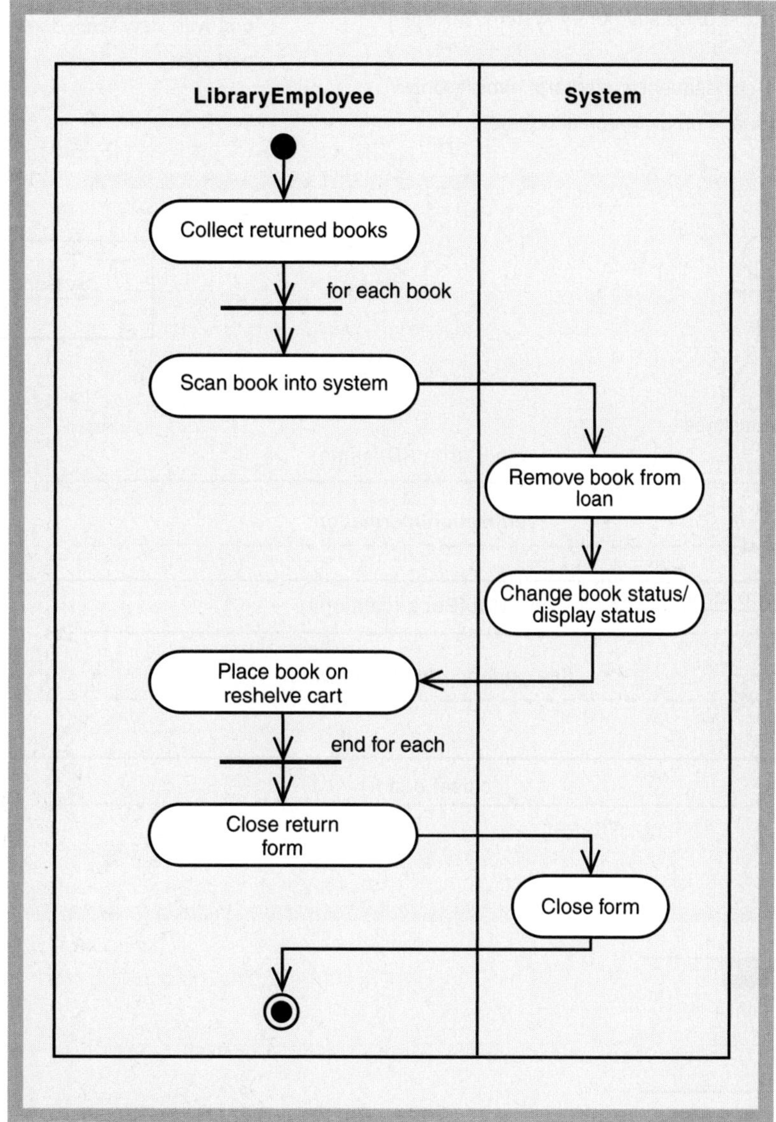

Figure 12-35

Activity diagram for *Return books*

3. Figure 12-36 is a fully developed use case description for the use case *Receive new book* in the university library system. Do the following:
   a. Develop a first-cut sequence diagram, which only includes the actor and problem domain classes.
   b. Add the view layer classes and the data access classes to your diagram from part a.
   c. Develop a design class diagram based on the domain class diagram and the results of parts a and b.
   d. Develop a package diagram showing a three-layer solution with view layer, domain layer, and data access layer packages.
4. Integrate the design class diagram solutions that you developed for exercises 1, 2, and 3 into a single design class diagram.

Use Case Name:	Receive new book	
Scenario:	Receive new book	
Triggering Event:	Newly purchased book arrives	
Brief Description:	The librarian decides on purchases of new books and places order (prior to this use case). Shipments of new books arrive. Each new book is assigned a library catalog number. Some books are simply additional copies of existing titles. Some books are new editions of existing titles. Some books are new titles and new physical books. The new book information is added to the system.	
Actors:	Library Employee	
Stakeholders:	Library Employee, Librarian	
Preconditions:	None	
Postconditions:	Book Title exists, Physical Book exists	
Flow of Activities:	Actor	System
	1. Collect new books from receipt of shipment. 2. For each book, research book category and catalog numbers. Assign tentative number. 3a. If new copy of existing title, enter book information and catalog number into system. 3b. If new edition of existing title, enter book information, edition information, and catalog number. 3c. If new title, assign general catalog number. Assign book copy number. 4. Mark book with number. 5. Place book on shelving cart. 6. Repeat for each book (back to step 2).	3a.1 Update catalog with new number.Verify that not duplicate. 3b.1 Update catalog with new number. Verify that not duplicate. 3c.1 Verify that catalog number not duplicate.
Exception Conditions:	Duplicate numbers require further research and reassignment of catalog numbers.	

Figure 12-36

Fully developed use case description for *Receive new book*

Exercises 5, 6, 7, and 8 are based on the solutions you developed for "Thinking Critically" exercises 3 and 4 in Chapter 7, which describe a dental clinic system. Alternatively, your teacher may provide you with a use case diagram and class diagram.

5. Figure 12-37 is a system sequence diagram for the use case *Record dental procedure* in the dental clinic system. Do the following:
   a. Develop a first-cut sequence diagram, which only includes the actor and problem domain classes.
   b. Add the view layer classes and the data access classes to your diagram from part a.
   c. Develop a design class diagram based on the domain class diagram and the results of parts a and b.
   d. Develop a package diagram showing a three-layer solution with view layer, domain layer, and data access layer packages.

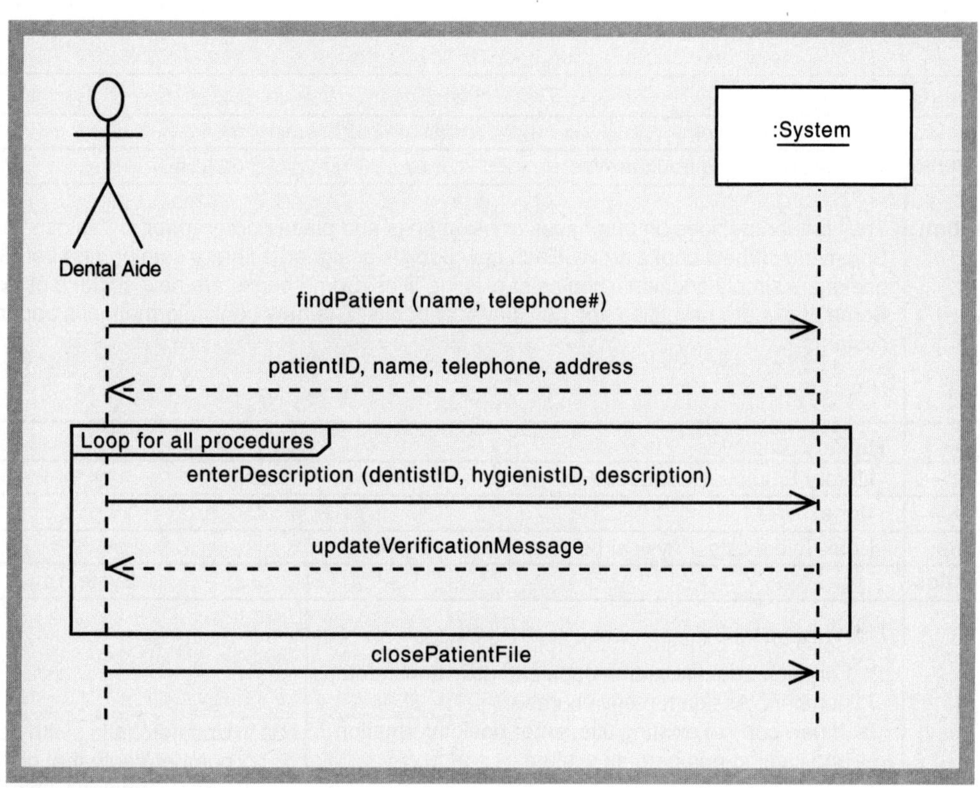

Figure 12-37

System sequence diagram for *Record dental procedure*

6. Figure 12-38 is an activity diagram for the use case *Enter new patient information* in the dental clinic system. Do the following:
   a. Develop a first-cut sequence diagram, which only includes the actor and problem domain classes.
   b. Add the view layer classes and the data access classes to your diagram from part a.
   c. Develop a design class diagram based on the domain class diagram and the results of parts a and b.
   d. Develop a package diagram showing a three-layer solution with view layer, domain layer, and data access layer packages.

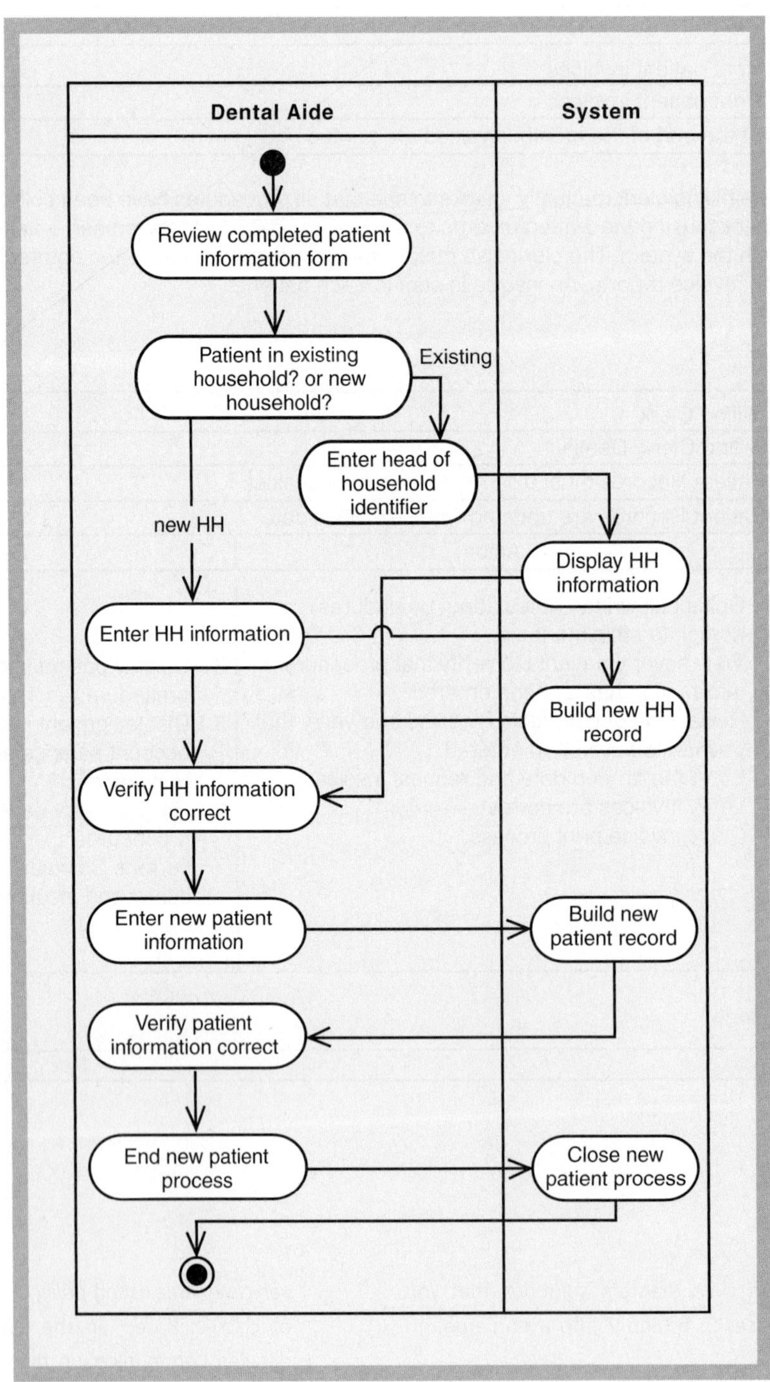

Figure 12-38

Activity diagram for *Enter new patient information*

7. Figure 12-39 is a fully developed use case description for the use case *Print patient invoices* in the dental clinic system. Do the following:
   a. Develop a first-cut sequence diagram, which only includes the actor and problem domain classes.
   b. Add the view layer classes and the data access classes to your diagram from part a.
   c. Develop a design class diagram based on the domain class diagram and the results of parts a and b.
   d. Develop a package diagram showing a three-layer solution with view layer, domain layer, and data access layer packages.

Use Case Name:	Print patient invoices	
Scenario:	Print patient invoices	
Triggering Event:	At the end of the month, invoices are printed	
Brief Description:	The billing clerk manually checks to see that all procedures have been collected. The clerk spot-checks, using the written records to make sure procedures have been entered by viewing them with the system. The clerk also makes sure all payments have been entered. Finally, he/she prints the invoice reports. An invoice is sent to each patient.	
Actors:	Billing Clerk	
Stakeholders:	Billing Clerk, Dentist	
Preconditions:	Patient Records must exist, Procedures must exist	
Postconditions:	Patient Records are updated with last billing date	
Flow of Activities:	Actor	System
	1. Collect all written notes about procedures completed this month. 2. View several patients to verify that procedure information has all been entered. 3. Review log of payments received and verify that payments have been entered. 4. Enter month-end date and request invoices. 5. Verify invoices are correct. 6. Close invoice print process.	2.1 Display patient information, including procedure records. 3.1 Display patient information, including account balance and last payment transactions. 4.1 Review every patient record. Find unpaid procedures. List on report as aged or current. Calculate and break down by copay and insurance pay.
Exception Conditions:	None	

Figure 12-39

Fully developed use case description for *Print patient invoices*

8. Integrate the design class diagram solutions that you developed for exercises 5, 6, and 7 into a single design class diagram.
9. In Chapter 7, "Thinking Critically" exercises 7, 8, and 10, you developed a system sequence diagram for *Add a new vehicle to an existing policy*. You were also provided a list of classes. Based on the SSD you created, develop a detailed communication diagram. Include only problem domain classes.

10. In Figure 12-40, the package on the left contains the classes in a payroll system. The package on the right is a payroll tax subsystem. What technique would you use to integrate the tax subsystem into the payroll system? Show how you would solve the problem by modifying the existing classes (in either figure). What new classes would you add? Use UML notation.

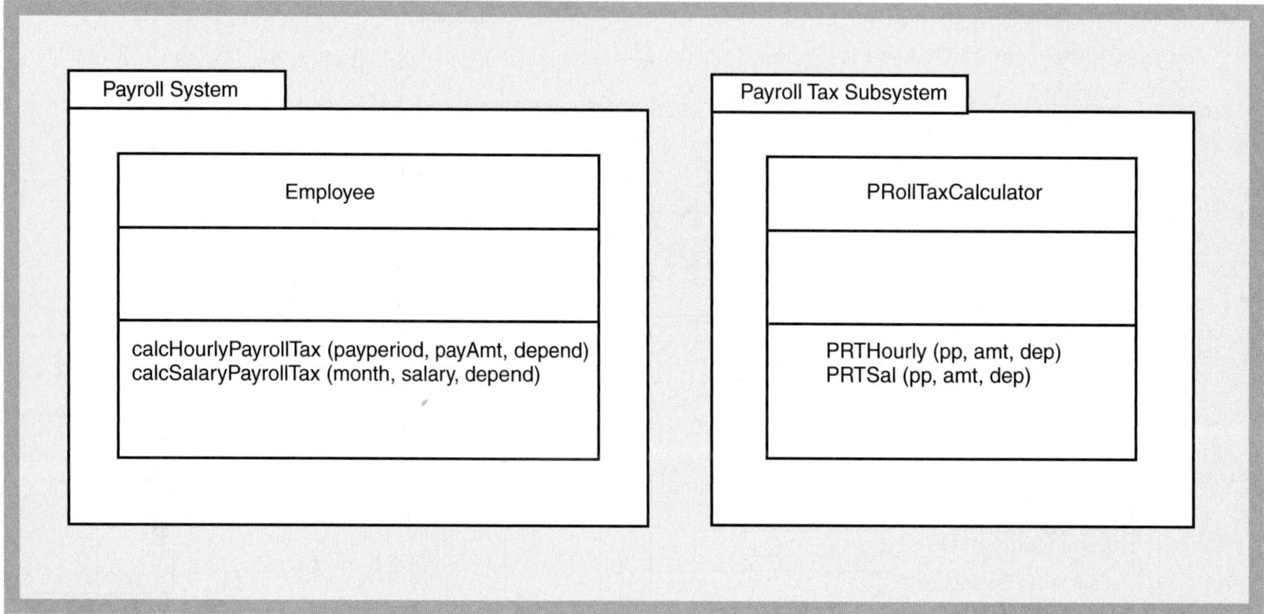

Figure 12-40
Payroll system packages and classes

11. The right side of Figure 12-41 shows a system that simulates the manufacture of computer chips. The equations in the simulation system are based on statistical probabilities of failures in the manufacturing. The package on the left illustrates a window (and its associated class definition) that will display these results dynamically. The values in the top five fields are obtained when the window is opened. However, the bottom three fields should be updated after every iteration, which takes about one second. From a design standpoint, the simulation system on the right should not be coupled to the user-interface system on the left. Show how you would solve this problem, including any class methods for existing classes, new classes, and new definitions that you would use. Use UML notation.

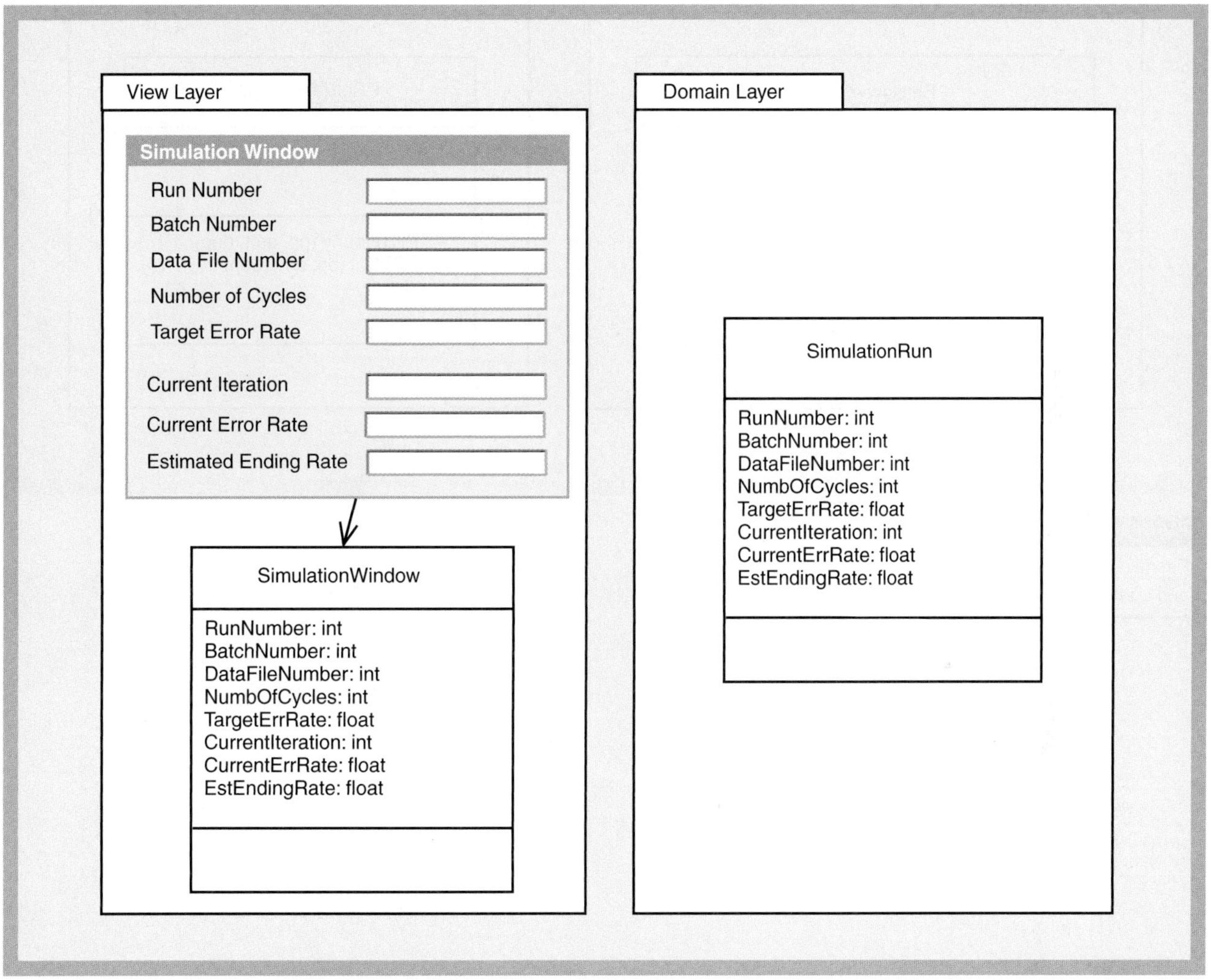

View Layer

Simulation Window

Run Number
Batch Number
Data File Number
Number of Cycles
Target Error Rate
Current Iteration
Current Error Rate
Estimated Ending Rate

SimulationWindow

RunNumber: int
BatchNumber: int
DataFileNumber: int
NumbOfCycles: int
TargetErrRate: float
CurrentIteration: int
CurrentErrRate: float
EstEndingRate: float

Domain Layer

SimulationRun

RunNumber: int
BatchNumber: int
DataFileNumber: int
NumbOfCycles: int
TargetErrRate: float
CurrentIteration: int
CurrentErrRate: float
EstEndingRate: float

Figure 12-41

Manufacture simulation system classes

12. Review the observer pattern description in Figure 12-42. Find the errors in the diagram. After you have identified the errors, state how you would fix them.

Figure 12-42

Sample observer pattern with errors

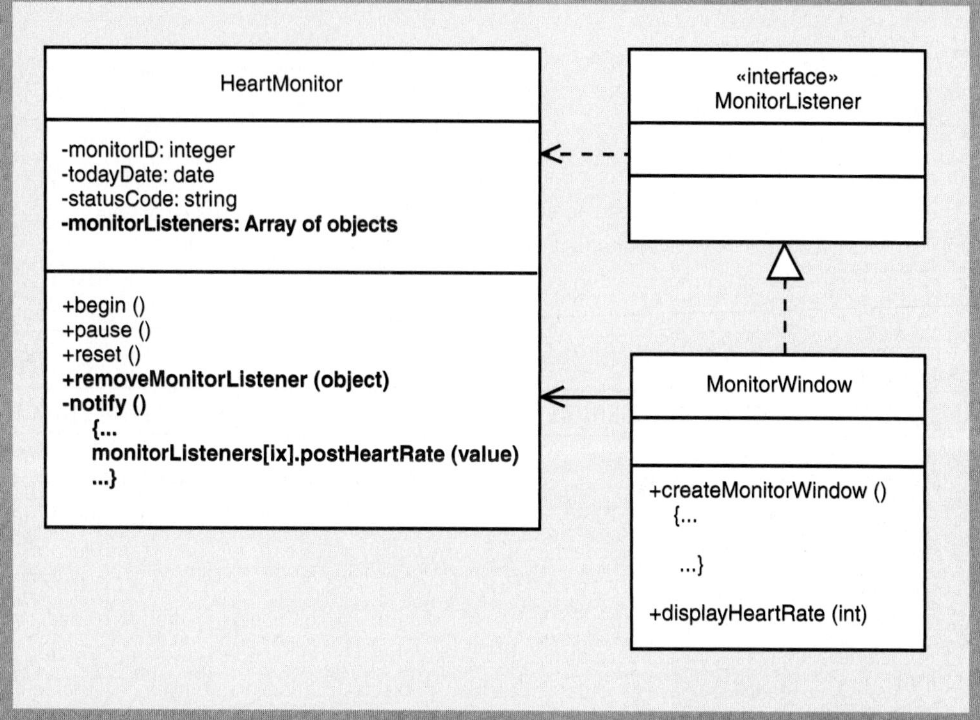

EXPERIENTIAL EXERCISES

1. Design patterns are a young but rich field of research and study. Locate the original GoF book on design patterns and briefly summarize two or three patterns discussed in that text. This exercise will begin your lifelong learning process of reading and understanding technical design material.
2. Find another book on design patterns and report on two or three of the patterns listed in that book. (See the "Further Resources" list later in this chapter for suggestions.) Some patterns are for network designs. Books on enterprise-level designs frequently discuss Internet design.
3. Do more research into the basics of Web services. Chapter 14 will also add to your understanding. Find some articles on the Internet that explain how Web services are implemented. Then find an example of a company that provides Web services and document what it provides and how it does so.
4. Find a system that was developed using Visual Studio .NET (or Visual Basic). Find one that has both an Internet user interface and a network-based user interface. Is it multilayer? Where is the business logic? Can you identify the view layer (user interface) classes, the domain layer classes, and the data access layer classes?

CASE STUDIES

THE REAL ESTATE MULTIPLE LISTING SERVICE SYSTEM

In Chapter 7, you developed a use case diagram, a fully developed use case description or activity diagram, and a system sequence diagram for the real estate company's use cases. In Chapter 11, you developed a three-layer design and a set of CRC cards for a particular use case. Based on those solutions or others provided by your teacher, develop a first-cut sequence diagram for the problem domain classes. Next, add view layer and data access layer objects to the sequence diagram. Convert the domain class diagram to a design class diagram by typing the attributes and adding method signatures.

THE STATE PATROL TICKET PROCESSING SYSTEM

In Chapter 7, you developed a use case diagram, a fully developed use case description, a system sequence diagram, and a statechart for the use cases *Recording a traffic ticket* and *Scheduling a court date*. In Chapter 11 you developed a first-cut DCD and a set of CRC cards for each use case. Based on those solutions or others provided by your teacher, develop a first-cut sequence diagram for the problem domain classes. Next, add view layer and data access layer objects to the sequence diagram. Convert the domain class diagram you developed in Chapter 5 to a design class diagram by typing the attributes and adding method signatures.

THE DOWNTOWN VIDEOS RENTAL SYSTEM

In Chapter 7, you developed a use case diagram, an activity diagram, and a system sequence diagram for the use cases *Rent movies* and *Return movies*. In Chapter 11 you developed a first-cut DCD and sets of CRC cards. Based on those solutions or others provided by your teacher, and the problem domain class diagram in Figure 12-43, develop a first-cut communication diagram for the problem domain classes. Next, add view layer and data access layer objects to the communication diagram. Convert the domain class diagram to a design class diagram by typing the attributes and adding method signatures.

THEEYESHAVEIT.COM BOOK EXCHANGE

In Chapter 7, you developed a use case diagram, a fully developed use case description, and a system sequence diagram for the use cases *Add a seller* and *Record a book order*. In Chapter 11 you developed a first-cut DCD and a set of CRC cards. Based on those solutions or others provided by your teacher, and the problem domain class diagram in Figure 12-44, develop a first-cut communication diagram for the problem domain classes. Next, add view layer and data access layer objects to the communication diagram. Convert the domain class diagram to a design class diagram by typing the attributes and adding method signatures.

Figure 12-43

Domain class diagram for the DownTown Videos Rental System

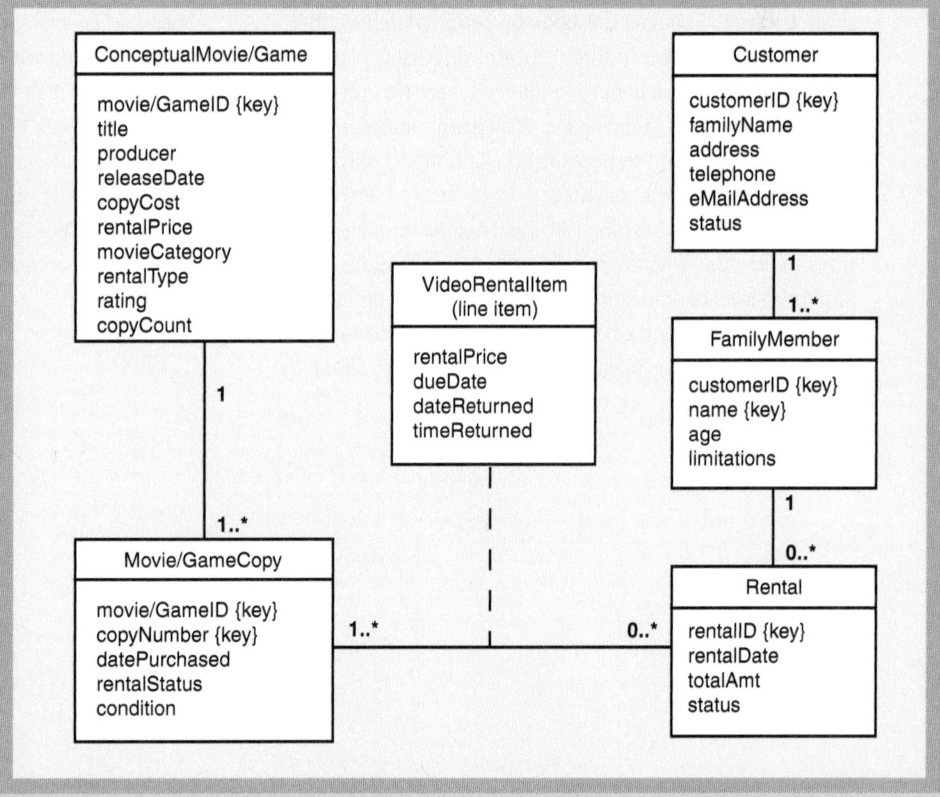

Figure 12-44
Domain class diagram for TheEyesHaveIt.com Book Exchange

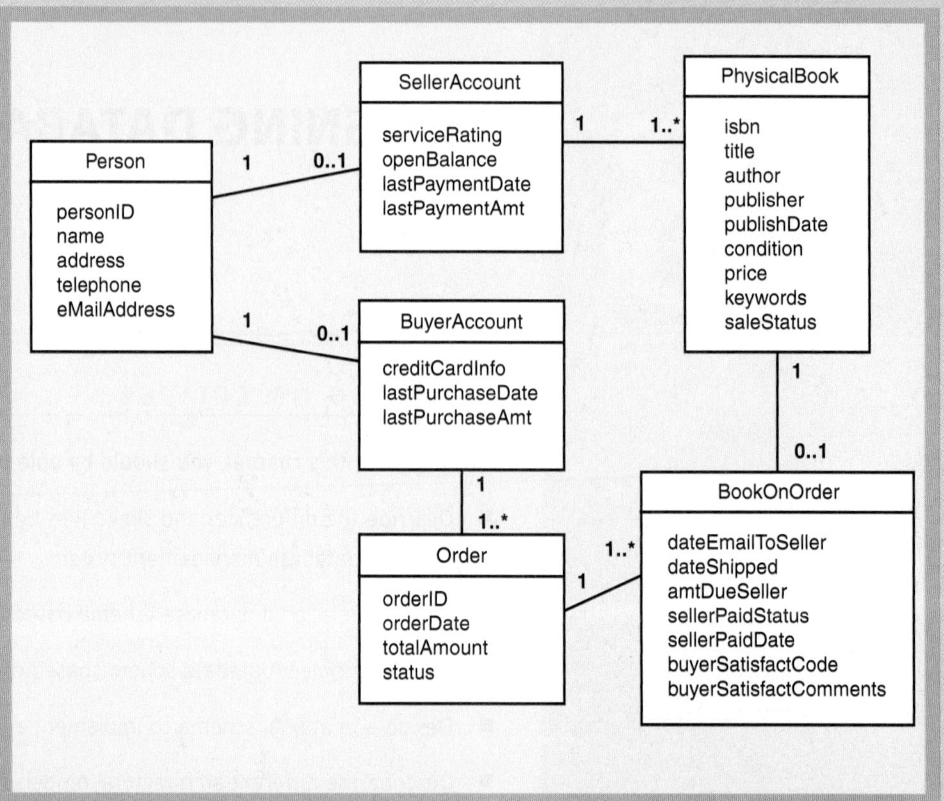

RETHINKING ROCKY MOUNTAIN OUTFITTERS

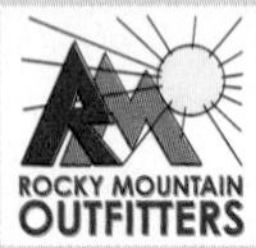

This chapter presented the solutions for the *Cancel an order* use cases. It also presented a partial solution for *Create new phone order*. Complete the solution for the *Create new phone order* use case by adding view and data layer classes to the sequence diagrams. In Chapter 11 you developed architectural designs and CRC cards for two more use cases, *Create order return* and *Record order fulfillment*. Develop three-layer sequence diagrams for these two use cases. Update the design class diagram for the problem domain classes with method signatures from these use case designs. Compare this solution to what you developed in Chapter 11.

FOCUSING ON RELIABLE PHARMACEUTICAL SERVICE

In Chapter 7, you developed a use case diagram, a domain model class diagram, and detailed documentation for three use cases. In Chapter 11 you developed a three-layer architectural solution and CRC cards. Develop three-layer sequence diagrams for those same use cases. Update the design class diagram with attribute information and method signatures derived from the sequence diagrams. Compare this to your DCD solution from Chapter 11.

FURTHER RESOURCES

Grady Booch, James Rumbaugh, and Ivar Jacobson, *The Unified Modeling Language User Guide*. Addison-Wesley, 1999.

Grady Booch, et al., *Object-Oriented Analysis and Design with Applications*, Third Edition. Addison-Wesley, 2007.

Frank Buschmann, R. Meunier, H. Rohnert, P. Sommerlad, and M. Stal, *Pattern-Oriented Software Architecture: A System of Patterns*. John Wiley and Sons, 1996.

Hans-Erik Eriksson, Magnus Penker, Brian Lyons, and David Fado, *UML 2 Toolkit*. John Wiley and Sons, 2004.

Alur Deepak, J. Crupi, and D. Malks, *Core J2EE Patterns: Best Practices and Design Strategies*. Sun Microsystems Press, 2001.

Erich Gamma, R. Helm, R. Johnson, and J. Vlissides, *Elements of Reusable Object-Oriented Software*. Addison-Wesley, 1995.

Mark Grand, *Patterns in Java*, Volumes I and II. John Wiley and Sons, 1999.

Craig Larman, *Applying UML and Patterns: An Introduction to Object-Oriented Analysis and Design and the Unified Process* (3rd ed.). Prentice-Hall, 2004.

David S. Linthicum, *Next Generation Application Integration: From Simple Information to Web Services*. Addison-Wesley, 2004.

James Rumbaugh, Ivar Jacobson, and Grady Booch, *The Unified Modeling Language Reference Manual*. Addison-Wesley, 1999.

CHAPTER

13

DESIGNING DATABASES

LEARNING OBJECTIVES

After reading this chapter, you should be able to:

- Describe the differences and similarities between relational and object-oriented database management systems
- Design a relational database schema based on an entity-relationship diagram
- Design an object database schema based on a class diagram
- Design a relational schema to implement a hybrid object-relational database
- Describe the different architectural models for distributed databases

CHAPTER OUTLINE

Databases and Database Management Systems

Relational Databases

Object-Oriented Databases

Hybrid Object-Relational Database Design

Data Types

Distributed Databases

NATIONWIDE BOOKS: DESIGNING A NEW DATABASE

The project leaders for Nationwide Books' (NB) new Web-based ordering system were meeting with NB's database administrator to discuss how they were going to tackle the database design. Present at the meeting were Sharon Thomas (who had led the project since its inception), Vince Pirelli (a contractor who had completed most of the analysis activities), and Bill Anderson (NB's database administrator, who hadn't directly participated in earlier phases of the project). Sharon started the meeting by saying, "When the project began, we planned to use the existing DB2 database. Maria Peña (the chief information officer) also wanted us to use this project to try out newer development methods and tools that we knew we'd need in the coming years. So we hired Vince to do the systems analysis using object-oriented (OO) methods. We also purchased a Java development tool and sent two of our programmers to a three-week training course."

Vince added, "I developed a traditional entity-relationship diagram as a basis for designing the interface between the new programs and the existing database. I checked the documentation for the current database schema. Most of the information needed by the new system is already there, although we need to add some new fields and we might have to change some table definitions. But now that we're looking at design and implementation, I'm concerned that we may be handicapping our new system with an outdated database management system."

Sharon added, "We decided to use this project as a pilot for OO development and implementation to speed up development. But I'm not sure that we'll actually achieve that if we use the existing DB2 database."

Bill said, "I understand your concerns about interfacing OO programs with a relational database. It sounds like trying to mix oil and water, but it's really not that difficult. We just hired Anna Jorgensen, a database developer who has experience writing Java programs that interface with relational databases. I had her look over the class diagrams and use cases for the new system, and she assures me that the interfaces to DB2 will be straightforward and simple. I can assign her to your project for a few months if you need the help."

Vince responded, "There's no question that an interface can be written, but it may not be as easy as Anna thinks. There are also problems with basic elements of an OO program such as inheritance and class methods. Those things simply can't be represented in a relational database. That'll force us to make some ugly compromises when designing and implementing our OO code. I'm afraid that it'll lengthen our design and implementation phases and make future system upgrades much more difficult."

"I've done some research, and there are quite a few commercial OO data packages available," said Sharon. "None of them have the track record of DB2, but then the technology is fairly new. An OO database management system would be the best fit with Java, and it would open the door to other new technologies."

Vince added, "It would also shorten the time we need to complete design, because an OO database is directly based on the class diagram."

Sharon continued, "I think that this project presents a good opportunity for us to make the leap to the next generation of database software."

Bill was taken aback by the suggestion but quickly replied, "Are you asking me to support two redundant databases based on two completely different database technologies? Management is already breathing down my neck about my budget. I've managed to save some money by switching to cheaper server hardware, but it's only a marginal improvement. Supporting another database management system will be a major effort. And we'd need new hardware to isolate the new DBMS from our existing database. I can't risk having a buggy new piece of software crash our operational databases. I'd also need to train my people to bring them up to speed on the new software. And how will we get data back and forth between the two databases?"

Sharon let the air clear for a bit before replying. "There's no question that it'll be a major undertaking. I'm not trying to downplay that fact or stretch your people and equipment to the breaking point. But there comes a time when we need to move on to newer technologies, and I think that time has arrived. I've already discussed the idea with Maria, and she thinks it deserves serious consideration."

Vince added, "I can do the database design either way. But we won't be building a base for the future if we use a relational database. I could design it to interface with indexed files on an old IBM mainframe if I had to. But why would we want to go backward instead of forward?"

OVERVIEW

Databases and database management systems are important components of a modern information system. Databases provide a common repository for data so that it can be shared by the entire organization. Database management systems provide designers, programmers, and end users with sophisticated capabilities to store, retrieve, and manage data. Sharing and managing the vast amounts of data needed by a modern organization simply would not be possible without a database management system.

In Chapter 5, you learned to construct conceptual data models. You also learned to develop entity-relationship diagrams (ERDs) for traditional analysis and domain model class diagrams for object-oriented (OO) analysis. To implement an information system, developers must transform a conceptual data model into a more detailed database model and implement that model within a database management system.

The process of developing a database model depends on the type of conceptual model and the type of data management software that will be used to implement the system. This chapter describes the design of relational and OO data models and their implementation using database management systems. We will use examples from Rocky Mountain Outfitters to show how information collected during analysis is used in database design.

DATABASES AND DATABASE MANAGEMENT SYSTEMS

database (DB)
an integrated collection of stored data that is centrally managed and controlled

database management system (DBMS)
system software that manages and controls access to a database

physical data store
the storage area used by a database management system to store the raw bits and bytes of a database

schema
a description of the structure, content, and access controls of a physical data store or database

A **database (DB)** is an integrated collection of stored data that is centrally managed and controlled. A database typically stores information about dozens or hundreds of entity types or classes. The information stored includes entity or class attributes (for example, names, prices, and account balances) as well as relationships among the entities or classes (for example, which orders belong to which customers). A database also stores descriptive information about data, such as field names, restrictions on allowed values, and access controls to sensitive data items.

A database is managed and controlled by a **database management system (DBMS)**. A DBMS is a system software component that is generally purchased and installed separately from other system software components (for example, operating systems). Examples of modern database management systems include Microsoft Access, Oracle, DB2, ObjectStore, and Gemstone.

DBMS COMPONENTS

Figure 13-1 illustrates the components of a typical database and its interaction with a DBMS, application programs, users, and administrators. The database consists of two related information stores: the physical data store and the schema. The **physical data store** contains the raw bits and bytes of data that are created and used by the information system. The **schema** contains descriptive information about the data stored in the physical data store, including the following:

- Access and content controls, including allowable values for specific data elements, value dependencies among multiple data elements, and lists of users allowed to read or update data element contents

Figure 13-1

The components of a database and database management system and their interaction with application programs, users, and database administrators

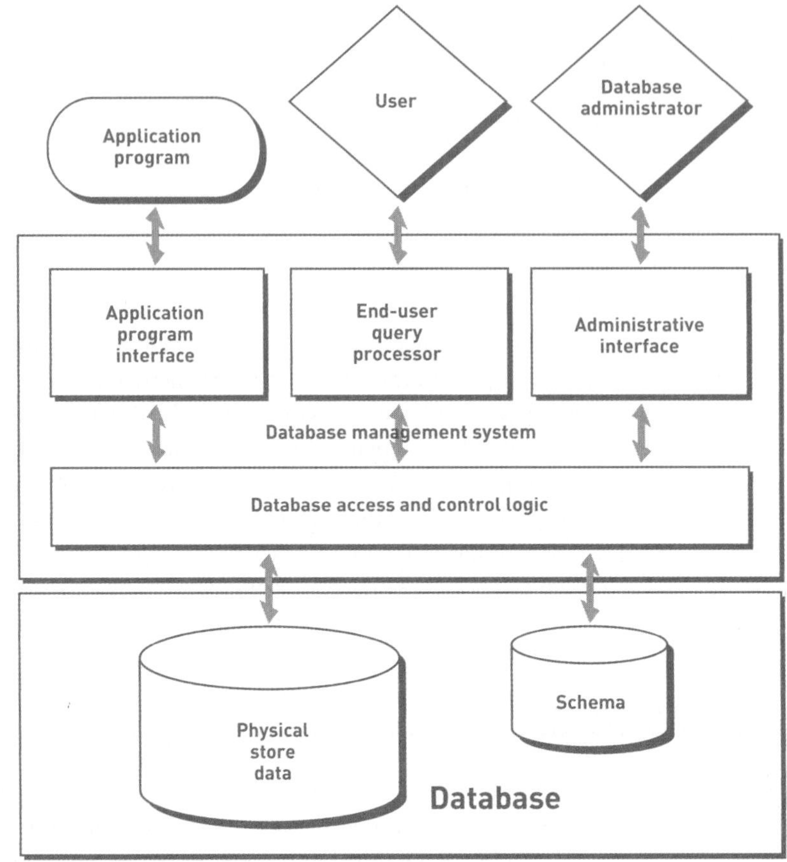

- Relationships among data elements and groups of data elements (for example, a pointer from data describing a customer to orders made by that customer)
- Details of physical data store organization, including type and length of data elements, the locations of data elements, indexing of key data elements, and sorting of related groups of data elements

A DBMS has four key components: an application program interface (API), a query interface, an administrative interface, and an underlying set of data access programs and subroutines. Application programs, users, and administrators never access the physical data store directly. Instead, they tell an appropriate DBMS interface what data they need to read or write, using names defined in the schema. The DBMS accesses the schema to verify that the requested data exists and that the requesting user has appropriate access privileges. If the request is valid, the DBMS extracts information about the physical organization of the requested data from the schema and uses that information to access the physical data store on behalf of the requesting program or user.

Databases and database management systems provide several important data access and management capabilities, including the following:

- Simultaneous access by many users and application programs
- Access to data without writing application programs (that is, via a query language)
- Application of uniform and consistent access and content controls

For these and other reasons, databases and DBMSs are widely used in modern information systems.

DATABASE MODELS

DBMSs have evolved through a number of technology stages since their introduction in the 1960s. The most significant change has been the type of model used to represent and access the content of the physical data store. Four such model types have been widely used:

- Hierarchical
- Network
- Relational
- Object-oriented

The hierarchical model was developed in the 1960s. It represented data using sets of records organized into a hierarchy. The network model also grouped data elements into sets of records but allowed those records to be organized into more flexible network structures. Relational DBMSs replaced hierarchical and network DBMSs by the end of the twentieth century, though a few of the older concepts live on in object-oriented DBMSs.

The remainder of this chapter describes design issues for the relational and object-oriented database models—the most widely used models for both existing and newly developed systems. Design issues for the hierarchical and network models are not described, because few students of information systems are likely to encounter DBMSs based on these models.

RELATIONAL DATABASES

relational database management system (RDBMS)
a database management system that stores data in tables

table
a two-dimensional data structure containing rows and columns; also called a *relation*

row
the portion of a table containing data that describes one entity, relationship, or object; also called *tuple* or *record*

field
a column of a relational database table; also called an *attribute*

field value
the data value stored in a single cell of a relational database table; also called an *attribute value* or *data element*

key
a field that contains a value that is unique within each row of a relational database table

The relational database model was first developed in the early 1970s. Relational databases were slow to be adopted because of the difficulties inherent in converting systems implemented with hierarchical and network DBMSs and because of the amount of computing resources required to implement them successfully. As with other theoretical advances in computer science, it took many years for the cost-performance characteristics of data storage and processing hardware to catch up to the new theory. Relational DBMSs now account for the vast majority of DBMSs currently in use.

A **relational database management system (RDBMS)** is a DBMS that organizes stored data into structures called **tables**, or *relations*. Relational database tables are similar to conventional tables—that is, they are two-dimensional data structures of columns and rows. However, relational database terminology is somewhat different from conventional table and file terminology. A single row of a table is called a **row**, *tuple*, or *record*, and a column of a table is called a **field**, or *attribute*. A single cell in a table is called a **field value**, *attribute value*, or *data element*.

Figure 13-2 shows the content of a table as displayed by the Microsoft Access relational DBMS. Note that the first row of the table contains a list of field names (column headings) and that the remaining rows contain a collection of field values that each describe a specific product. Each row contains the same fields in the same order.

BEST PRACTICE

Relational databases and SQL are two of the most important knowledge areas you will need to master as a system developer. This is equally true whether you are emphasizing the traditional or the object-oriented approach.

Each table in a relational database must have a unique **key**. A key is a field or set of fields, the values of which occur only once in all the rows of the table. If only one field (or set of fields) is

Figure 13-2
A partial display of a relational database table

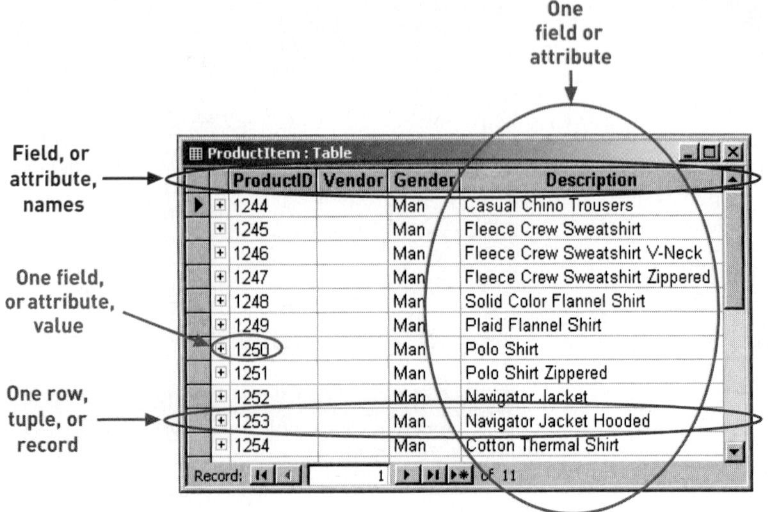

primary key
a key used to uniquely identify a row of a relational database table

unique, then that key is also called the table's **primary key**. If there are multiple unique fields (or sets of fields), the database designer must choose one of the possible keys as the primary key.

Key fields can be natural or invented. An example of a natural key field in chemistry is the atomic weight of an element in a table containing descriptive data about elements. Unfortunately, in business, few natural key fields are useful for information processing, so most key fields in a relational database are invented. Your wallet or purse probably contains many examples of invented keys, including your Social Security number, driver's license number, credit card numbers, and ATM card number. Some invented keys are externally assigned (for example, a Federal Express tracking number) and some are internally assigned (for example, ProductID in Figure 13-2). Invented keys are guaranteed to be unique because unique values are assigned by a user, application program, or the DBMS as new rows are added to the table.

Keys are a critical element of relational database design because they are the basis for representing relationships among tables. Keys are the "glue" that binds rows of one table to rows of another table—in other words, keys relate tables to each other. For example, consider the ERD fragment from the Rocky Mountain Outfitters example shown in Figure 13-3 and the tables shown in Figure 13-4. The ERD fragment shows an optional one-to-many relationship between the entities Product Item and Inventory Item. The upper table in Figure 13-4 contains data representing the entity type ProductItem. The lower table contains data representing the entity type InventoryItem.

Figure 13-3
A portion of the RMO entity-relationship diagram

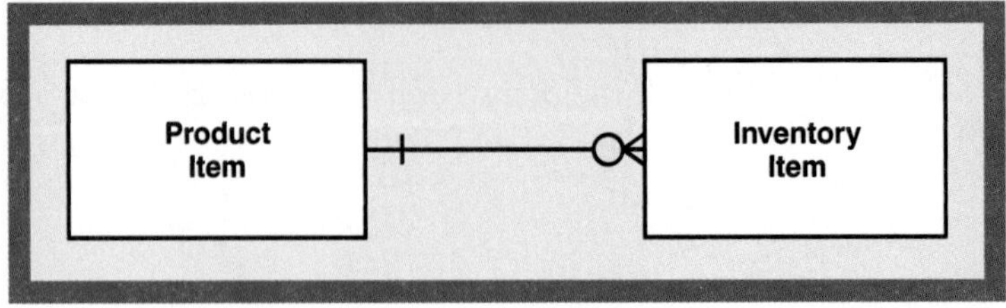

foreign key
a field value stored in one relational database table that also exists as a primary key value in another relational database table

The relationship between the entity types Product Item and Inventory Item is represented by a common field value within their respective tables. The field ProductID (the primary key of the ProductItem table) is also stored within the InventoryItem table, where it is called a foreign key. A **foreign key** is a field that duplicates the primary key of a different (or foreign) table. In Figure 13-4, the existence of the value 1244 as a foreign key within the InventoryItem table indicates that the values of Vendor, Gender, and Description in the first row of the ProductItem table also describe inventory items 86779 through 86790.

Figure 13-4

A relationship between data in two tables; the foreign key ProductID in the InventoryItem table refers to the primary key ProductID in the ProductItem table

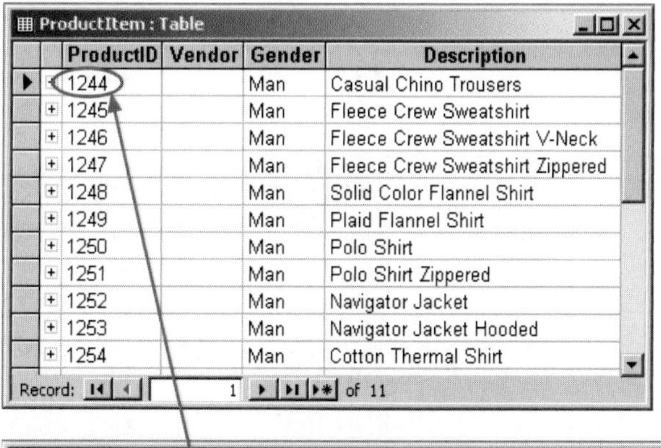

ProductItem : Table

ProductID	Vendor	Gender	Description
1244		Man	Casual Chino Trousers
1245		Man	Fleece Crew Sweatshirt
1246		Man	Fleece Crew Sweatshirt V-Neck
1247		Man	Fleece Crew Sweatshirt Zippered
1248		Man	Solid Color Flannel Shirt
1249		Man	Plaid Flannel Shirt
1250		Man	Polo Shirt
1251		Man	Polo Shirt Zippered
1252		Man	Navigator Jacket
1253		Man	Navigator Jacket Hooded
1254		Man	Cotton Thermal Shirt

Record: 1 of 11

InventoryItem : Table

InventoryID	ProductID	Size	Color	Options	QuantityOnHand	Average Cost	RecorderQuantity
86779	1244	30/30	Khaki		45	$12.75	100
86780	1244	30/30	Slate		10	$12.75	100
86781	1244	30/30	LightTan		17	$12.75	100
86782	1244	30/31	Khaki		22	$12.75	100
86783	1244	30/31	Slate		6	$12.75	100
86784	1244	30/31	LightTan		31	$12.75	100
86785	1244	30/32	Khaki		120	$12.75	100
86786	1244	30/32	Slate		28	$12.75	100
86787	1244	30/32	LightTan		21	$12.75	100
86788	1244	30/33	Khaki		7	$12.75	100
86789	1244	30/33	Slate		41	$12.75	100
86790	1244	30/34	LightTan		35	$12.75	50

Record: 1 of 12

DESIGNING RELATIONAL DATABASES

Relational database design begins with either an ERD or a class diagram. This section explains how to create a schema based on an ERD. Schema creation based on a class diagram is discussed later in this chapter.

To create a relational database schema from an ERD, follow these steps:

1. Create a table for each entity type.
2. Choose a primary key for each table (invent one, if necessary).
3. Add foreign keys to represent one-to-many relationships.
4. Create new tables to represent many-to-many relationships.
5. Define referential integrity constraints.
6. Evaluate schema quality and make necessary improvements.
7. Choose appropriate data types and value restrictions (if necessary) for each field.

The following subsections discuss each of these steps in detail.

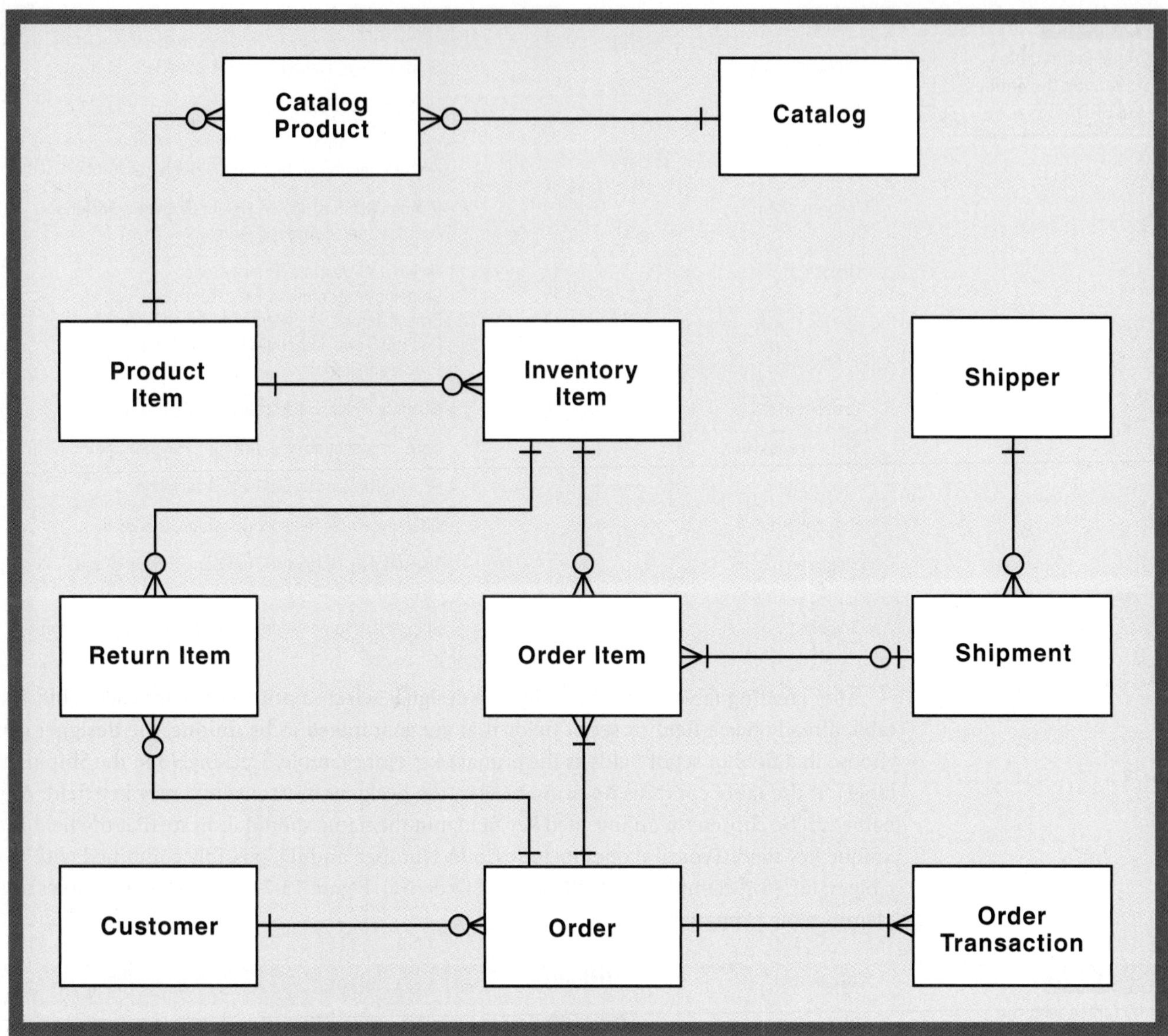

Figure 13-5

The RMO entity-relationship diagram

REPRESENTING ENTITIES

The first step to creating a relational DB schema is to create a table for each entity on the ERD. Figure 13-5 shows the ERD for the RMO customer support system. Eleven entities are represented, and a table is created for each entity. The data fields of each table will be the same as those defined for the corresponding entity on the ERD. To avoid confusion, table and field names should match the names used on the ERD and in the project data dictionary. Initial table definitions for the RMO case are shown in Figure 13-6.

Figure 13-6

An initial set of tables representing the entities on the ERD

Table	Attributes
Catalog	Season, Year, Description, EffectiveDate, EndDate
CatalogProduct	Price, SpecialPrice
Customer	AccountNo, Name, BillingAddress, ShippingAddress, DayTelephoneNumber, NightTelephoneNumber
InventoryItem	InventoryID, Size, Color, Options, QuantityOnHand, AverageCost, ReorderQuantity
Order	OrderID, OrderDate, PriorityCode, ShippingAndHandling, Tax, GrandTotal, EmailAddress, ReplyMethod, PhoneClerk, CallStartTime, LengthOfCall, DateReceived, ProcessorClerk
OrderItem	Quantity, Price, BackorderStatus
OrderTransaction	Date, TransactionType, Amount, PaymentMethod
ProductItem	ProductID, Vendor, Gender, Description
ReturnItem	Quantity, Price, Reason, Condition, Disposal
Shipment	TrackingNo, DateSent, TimeSent, ShippingCost, DateArrived, TimeArrived
Shipper	ShipperID, Name, Address, ContactName, Telephone

After creating tables for each entity, the designer selects a primary key for each table. If a table already has a field or set of fields that are guaranteed to be unique, the designer can choose that field or set of fields as the primary key (for example, TrackingNo in the Shipment table). If the table contains no primary keys, the designer must invent a new key field. Any name can be chosen for an invented key field, but the name should indicate that the field is a unique key field. Typical names include Code, Number, and ID, possibly combined with the table name (for example, ProductCode and OrderID). Figure 13-7 shows the entity tables and identifies the primary key of each.

Figure 13-7

Entity tables with the primary keys identified in bold

Table	Attributes
Catalog	**CatalogID**, Season, Year, Description, EffectiveDate, EndDate
CatalogProduct	**CatalogProductID**, Price, SpecialPrice
Customer	**AccountNo**, Name, BillingAddress, ShippingAddress, DayTelephoneNumber, NightTelephoneNumber
InventoryItem	**InventoryID**, Size, Color, Options, QuantityOnHand, AverageCost, ReorderQuantity
Order	**OrderID**, OrderDate, PriorityCode, ShippingAndHandling, Tax, GrandTotal, EmailAddress, ReplyMethod, PhoneClerk, CallStartTime, LengthOfCall, DateReceived, ProcessorClerk
OrderItem	**OrderItemID**, Quantity, Price, BackorderStatus
OrderTransaction	**OrderTransactionID**, Date, TransactionType, Amount, PaymentMethod
ProductItem	**ProductID**, Vendor, Gender, Description
ReturnItem	**ReturnItemID**, Quantity, Price, Reason, Condition, Disposal
Shipment	**TrackingNo**, DateSent, TimeSent, ShippingCost, DateArrived, TimeArrived
Shipper	**ShipperID**, Name, Address, ContactName, Telephone

REPRESENTING RELATIONSHIPS

Relationships are represented within a relational database by foreign keys. Which foreign keys should be placed in which tables depends on the type of relationship being represented. The RMO ERD in Figure 13-5 contains nine one-to-many relationships. There is one many-to-many relationship between Catalog and Product Item, which is represented by two one-to-many relationships and the associative entity Catalog Product. The rules for representing each relationship type are as follows:

- **One-to-many relationship**. Add the primary key field(s) of the "one" entity type to the table that represents the "many" entity type.
- **Many-to-many relationship**. If no associative entity exists for the relationship, create a new table to represent the relationship. If an associative entity does exist, use its table to represent the relationship. Use the primary key field(s) of the related entity types as the primary key of the table that represents the relationship.

Figure 13-8 shows the results of representing the nine one-to-many relationships within the tables from Figure 13-7. Each foreign key represents a single relationship between the table containing the foreign key and the table that uses that same key as its primary key. For example, the field AccountNo was added to the Order table as a foreign key representing the one-to-many relationship between the entities Customer and Order. The foreign key ShipperID was added to the Shipment table to represent the one-to-many relationship between Shipper and Shipment. When representing one-to-many relationships, foreign keys do not become part of the primary key of the table to which they are added.

Figure 13-9 expands the table definitions in Figure 13-8 by updating the CatalogProduct table to represent the many-to-many relationship between Catalog and ProductItem. The primary key of the CatalogProduct becomes the combination of CatalogID and ProductID. The old

Figure 13-8

Represent one-to-many relationships by adding foreign key attributes (shown in italic)

Table	Attributes
Catalog	**CatalogID**, Season, Year, Description, EffectiveDate, EndDate
CatalogProduct	**CatalogProductID,** Price, SpecialPrice
Customer	**AccountNo**, Name, BillingAddress, ShippingAddress, DayTelephoneNumber, NightTelephoneNumber
InventoryItem	**InventoryID**, *ProductID*, Size, Color, Options, QuantityOnHand, AverageCost, ReorderQuantity
Order	**OrderID**, *AccountNo*, OrderDate, PriorityCode, ShippingAndHandling, Tax, GrandTotal, EmailAddress, ReplyMethod, PhoneClerk, CallStartTime, LengthOfCall, DateReceived, ProcessorClerk
OrderItem	**OrderItemID**, *OrderID*, *InventoryID*, *TrackingNo*, Quantity, Price, BackorderStatus
OrderTransaction	**OrderTransactionID**, *OrderID*, Date, TransactionType, Amount, PaymentMethod
ProductItem	**ProductID**, Vendor, Gender, Description
ReturnItem	**ReturnItemID**, *OrderID*, *InventoryID*, Quantity, Price, Reason, Condition, Disposal
Shipment	**TrackingNo**, *ShipperID*, DateSent, TimeSent, ShippingCost, DateArrived, TimeArrived
Shipper	**ShipperID**, Name, Address, ContactName, Telephone

Figure 13-9

The table CatalogProduct is modified to represent the many-to-many relationship between Catalog and ProductItem

Table	Attributes
Catalog	**CatalogID**, Season, Year, Description, EffectiveDate, EndDate
CatalogProduct	***CatalogID***, ***ProductID***, Price, SpecialPrice
Customer	**AccountNo**, Name, BillingAddress, ShippingAddress, DayTelephoneNumber, NightTelephoneNumber
InventoryItem	**InventoryID**, *ProductID*, Size, Color, Options, QuantityOnHand, AverageCost, ReorderQuantity
Order	**OrderID**, *AccountNo*, OrderDate, PriorityCode, ShippingAndHandling, Tax, GrandTotal, EmailAddress, ReplyMethod, PhoneClerk, CallStartTime, LengthOfCall, DateReceived, ProcessorClerk
OrderItem	**OrderItemID**, *OrderID*, *InventoryID*, *TrackingNo*, Quantity, Price, BackorderStatus
OrderTransaction	**OrderTransactionID**, *OrderID*, Date, TransactionType, Amount, PaymentMethod
ProductItem	**ProductID**, Vendor, Gender, Description
ReturnItem	**ReturnItemID**, *OrderID*, *InventoryID*, Quantity, Price, Reason, Condition, Disposal
Shipment	**TrackingNo**, *ShipperID*, DateSent, TimeSent, ShippingCost, DateArrived, TimeArrived
Shipper	**ShipperID**, Name, Address, ContactName, Telephone

primary key CatalogProductID is discarded. The two fields that make up the primary key are also foreign keys. CatalogID is a foreign key from the Catalog table, and ProductID is a foreign key from the ProductItem table.

ENFORCING REFERENTIAL INTEGRITY

referential integrity
a consistent relational database state in which every foreign key value also exists as a primary key value

Now that we've described how foreign keys are used to represent relationships, we need to describe how to enforce restrictions on the values of those foreign key fields. The term **referential integrity** describes a consistent state among foreign key and primary key values. Each foreign key is a reference to the primary key of another table. In most cases, a database designer wants to ensure that these references are consistent. That is, foreign key values that appear in one table must also appear as the primary key value of the related table. A referential integrity constraint is a constraint on database content—for example, "an order must be from a customer" and "an order item must be something that we normally stock in inventory."

The DBMS usually enforces referential integrity automatically after the schema designer identifies primary and foreign keys. Automatic enforcement is implemented as follows:

- When a row containing a foreign key value is created, the DBMS ensures that the value also exists as a primary key value in the related table.
- When a row is deleted, the DBMS ensures that no foreign keys in related tables have the same value as the primary key of the deleted row.
- When a primary key value is changed, the DBMS ensures that no foreign key values in related tables contain the same value.

In the first case, the DBMS will simply reject any new row containing an unknown foreign key value. In the latter two cases, a database designer usually has some control over how referential integrity is enforced. When a row containing a primary key is deleted, the DBMS can be instructed to delete all rows in other tables with corresponding keys. Or, the designer can instruct the DBMS to set all corresponding foreign keys to NULL. A similar choice is available when a primary key value is changed. The DBMS can be instructed to change all corresponding foreign key values to the same value or to set foreign key values to NULL.

EVALUATING SCHEMA QUALITY

After creating a complete set of tables, the designer should check the entire schema for quality. Ironing out any schema problems at this point ensures that none of the later design effort will be wasted. A high-quality data model has the following features:

- Uniqueness of table rows and primary keys
- Lack of redundant data
- Ease of implementing future data model changes

Unfortunately, there are few objective or quantitative measures of database schema quality. Database design is the final step in a modeling process, and as such, it depends on the analyst's experience and judgment. Various formal and informal techniques for schema evaluation are described in the following sections. No one technique is sufficient by itself, but a combination of techniques can ensure a high-quality database design.

Row and Key Uniqueness

A fundamental requirement of all relational data models is that primary keys and table rows be unique. Because each table must have a primary key, uniqueness of rows within a table is guaranteed if the primary key is unique. Data access logic within programs usually assumes that keys are unique. For example, a programmer writing a program to view customer records will generally assume that a database query for a specific customer number will return one and only one row (or none if the customer doesn't exist in the database). The program will be designed around this assumption and will probably fail if the DBMS returns two records.

A designer evaluates primary key uniqueness by examining assumptions about key content, the set of possible key values, and the methods by which key values are assigned. Internally invented keys are relatively simple to evaluate in this regard because the system itself creates them. That is, an information system that uses invented keys can guarantee uniqueness by implementing appropriate procedures to assign key values to newly created rows.

It is common for several different programs in an information system to be capable of creating new database rows. Each program needs to be able to assign keys to newly created database rows. However, the importance of key uniqueness requires that key-creation procedures be consistently applied throughout the information system.

Because key creation and management are such pervasive problems in information systems, many relational DBMSs automate key creation. Such systems typically automate a special data type for invented keys (for example, the AutoNumber type in Microsoft Access). The DBMS automatically assigns a key value to newly created rows and communicates that value to the application program for use in subsequent database operations. Embedding this capability in the DBMS frees the IS developer from designing and implementing customized key-creation software modules.

Invented keys that aren't assigned by the information system must be given careful scrutiny to ascertain their uniqueness and usefulness over time. For example, employee databases in the United States commonly use Social Security numbers as keys. Because the U.S. Government has a strong interest in guaranteeing the uniqueness of Social Security numbers, the assumption that they will always be unique seems safe. But other assumptions concerning their use deserve closer examination. For example, will all employees who are stored in the database have a Social Security number? What if the company opens a manufacturing facility in Europe or South America?

Invented keys assigned by nongovernmental agencies deserve even more careful scrutiny. For example, Federal Express, UPS, and most shipping companies assign a tracking number to each shipment they process. Tracking numbers are guaranteed to be unique at any given point in time, but are they guaranteed to be unique forever (that is, are they ever reused)? Could reuse of a tracking number cause a primary key duplication in the RMO database? And what would happen if two different shippers assigned the same tracking number to two different shipments?

Uncertainties such as these make internally invented keys the safest long-term strategy in most cases. Although internally invented keys may initially entail additional design and development, they prevent one possible source of upheaval after the database is installed. Few changes have the pervasive and disruptive impact of a database key change in a large information system with terabytes of stored data and thousands of application programs and stored queries.

Data Model Flexibility

Database flexibility and maintainability were primary goals in the original specification of the relational database model. A database model is considered flexible and maintainable if changes to the database schema can be made with minimal disruption to existing data content and structure. For example, adding a new entity to the schema should not require redefining existing tables. Adding a new one-to-many relationship should only require adding a new foreign key to an existing table. Adding a new many-to-many relationship should only require adding a single new table to the schema.

Data redundancy plays a key role in determining the flexibility and maintainability of any database or data model. A truism of database processing is that "redundant storage requires redundant maintenance." That is, if data is stored in multiple places, each of those places must be found and manipulated when data is added, changed, or deleted. Of course, performing any of those actions on multiple data storage locations is more complex (and less efficient) than performing them on a single location. Failure to update, modify, or delete multiple copies of the same information creates a condition called *inconsistency*. By definition, inconsistency cannot occur if information is stored only once.

The relational data model deliberately stores key values multiple times (that is, redundantly) and non-key fields only once. Key values are stored once as a primary key and again each time they are used as a foreign key. The model requires such redundancy because correspondence between the primary and foreign key is the only way to represent relationships among tables, but using redundant key values adds complexity to processes that manipulate key fields.

normalization
a technique that ensures relational database schema quality by minimizing data redundancy

first normal form (1NF)
a relational database table structure that has no repeating fields or groups of fields

functional dependency
a one-to-one correspondence between two field values

second normal form (2NF)
a relational database table structure in which every non-key field is functionally dependent on the primary key

third normal form (3NF)
a relational database table structure in which no non-key field is functionally dependent on any other non-key field(s)

Relational DBMSs ensure consistency among primary and foreign keys by enforcing referential integrity constraints, but there are no automatic mechanisms for ensuring consistency among other redundant data items. Thus, the best way to avoid inconsistency in a relational database is to avoid redundancy in non-key fields. Database designers can avoid such redundancy by never introducing it into a schema—but it is all too easy to let redundancy slip in. If data redundancy is somehow introduced into the schema, it must be identified and removed. The most commonly used process to detect and eliminate redundancy is database normalization.

Database Normalization

Normalization is a formal technique used to evaluate the quality of a relational database schema. It determines whether a database schema contains any of the "wrong" kinds of redundancy and defines specific methods to eliminate them. Normalization is based on a concept called *functional dependency* and on a series of normal forms:

- **First normal form (1NF).** A table is in **first normal form** if it contains no repeating fields or groups of fields.
- **Functional dependency.** A **functional dependency** is a one-to-one relationship between the values of two fields. The relationship is formally stated as follows: *Field A is functionally dependent on field B if for each value of B there is only one corresponding value of A.*
- **Second normal form (2NF).** A table is in **second normal form** if it is in first normal form and if each non-key element is functionally dependent on the entire primary key.
- **Third normal form (3NF).** A table is in **third normal form** if it is in second normal form and if no non-key element is functionally dependent on any other non-key element.

SSN	Name	Department	Salary	Dependent1	Dependent2	Dependent3 ...	DependentN
111-22-3333	Mary Smith	Accounting	40,000	John	Alice	Dave	
222-33-4444	Jose Pena	Marketing	50,000				
333-44-5555	Frank Collins	Production	35,000	Jane	Julia		

Figure 13-10

An employee table with a repeating field (not in normal form)

Let's explain these concepts further.

First Normal Form First normal form defines a structural constraint on table rows. Repeating fields such as Dependent in Figure 13-10 are not allowed within any table in a relational database. Repeating groups of fields are also prohibited. In practice, this constraint is not difficult to enforce because relational DBMSs do not allow a designer to define a table containing repeating fields.

Functional Dependency Functional dependency is a difficult concept to describe and apply. The most precise way to determine whether functional dependency exists is to pick two fields in a table and insert their names in the italicized portion of the definition shown previously. For example, consider the fields ProductID and Description in the ProductItem table (see Figure 13-4). ProductID is an internally invented primary key that is guaranteed to be unique within the table. To determine whether Description is functionally dependent on ProductID, substitute Description for field A and ProductID for field B in the italicized portion of the functional dependency definition:

> Description is functionally dependent on ProductID if for each value of ProductID there is only one corresponding value of Description.

Now ask whether the statement is true for all rows that could possibly exist in the ProductItem table. If the statement is true, Description is functionally dependent on ProductID. As long as the invented key ProductID is guaranteed to be unique within the ProductItem table, the preceding statement is true. Therefore, Description is functionally dependent on ProductID.

A less formal way to analyze functional dependency of Description on ProductID is to remember that the ProductItem table represents a specific product sold by RMO. If that product can have only a single description in the database, Description is functionally dependent on the key of the table that represents products (ProductID). If it is possible for any product to have multiple descriptions, the field Description is not functionally dependent on ProductID.

Second Normal Form To evaluate whether the ProductItem table is in second normal form, we must first determine whether it is in first normal form. Because it contains no repeating fields, it is in first normal form. Then we must determine whether every non-key field is functionally dependent on ProductID (that is, consider each field in turn by substituting it for A in the functional dependency definition). If all the non-key fields are functionally dependent on ProductID, the ProductItem table is in 2NF. If one or more non-key fields are not functionally dependent on ProductID, the table is not in 2NF.

Verifying that a table is in 2NF is more complicated when the primary key consists of two or more fields. For example, consider the RMO table CatalogProduct shown in Figure 13-11. Recall that this table represents a many-to-many relationship between Catalog and ProductItem. Thus, the table representing this relationship has a primary key consisting of the primary keys of Catalog (CatalogID) and ProductItem (ProductID). The table also contains two non-key fields called Price and SpecialPrice.

If this table is in 2NF, then each non-key field must be functionally dependent on the *combination* of CatalogID and ProductID. We can verify the first functional dependency by substituting terms in the functional dependency definition and determining the truth or falsity of the resulting statement:

> Price is functionally dependent on the combination of CatalogID and ProductID if for each combination of values for CatalogID and ProductID there is only one corresponding value of Price.

Figure 13-11

A simplified RMO CatalogProduct table

CatalogProduct : Table

CatalogID	ProductID	Price	SpecialPrice
23	1244	$15.00	$12.00
23	1245	$15.00	$12.00
23	1246	$15.00	$13.00
23	1247	$15.00	$13.00
23	1248	$14.00	$11.20
23	1249	$14.00	$11.20
23	1252	$21.00	$16.80
23	1253	$21.00	$16.40
23	1254	$24.00	$19.20
23	1257	$19.00	$15.20

Record: 11 of 11

Analyzing the truth of the preceding statement is tricky, because you must consider all the possible combinations of key values that might occur in the CatalogProduct file. A simpler way to approach the question is to think about the underlying entities represented in the table. A product can appear in many different catalogs. If Price can be different in different current catalogs, the preceding statement is true. If a product's Price is always the same, regardless of the catalog in which it appears, the preceding statement is false and the table is not in 2NF. The correct answer doesn't depend on any universal sense of truth. Instead, it depends on RMO's normal conventions for setting product prices in different catalogs.

If a non-key field is functionally dependent on only part of the primary key, you must remove the non-key field from its present table and place it in another table. For example, consider a modified version of the CatalogProduct table, as shown in the upper half of Figure 13-12. The non-key field CatalogIssueDate is functionally dependent only on CatalogID, not on the combination of CatalogID and ProductID. Thus, the table is not in 2NF.

To correct the problem, you must remove CatalogIssueDate from the CatalogProduct table and place it in a table that uses CatalogID alone as the primary key. Because the Catalog

Figure 13-12

Decomposition of a first normal form table into two second normal form tables

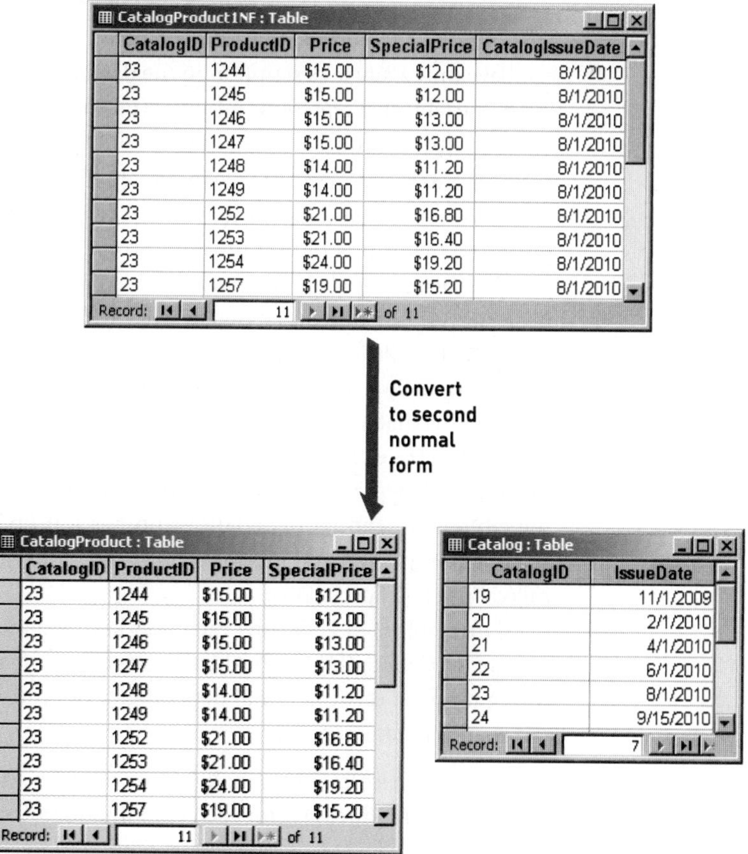

CatalogProduct1NF : Table

CatalogID	ProductID	Price	SpecialPrice	CatalogIssueDate
23	1244	$15.00	$12.00	8/1/2010
23	1245	$15.00	$12.00	8/1/2010
23	1246	$15.00	$13.00	8/1/2010
23	1247	$15.00	$13.00	8/1/2010
23	1248	$14.00	$11.20	8/1/2010
23	1249	$14.00	$11.20	8/1/2010
23	1252	$21.00	$16.80	8/1/2010
23	1253	$21.00	$16.40	8/1/2010
23	1254	$24.00	$19.20	8/1/2010
23	1257	$19.00	$15.20	8/1/2010

Record: 11 of 11

CatalogProduct : Table

CatalogID	ProductID	Price	SpecialPrice
23	1244	$15.00	$12.00
23	1245	$15.00	$12.00
23	1246	$15.00	$13.00
23	1247	$15.00	$13.00
23	1248	$14.00	$11.20
23	1249	$14.00	$11.20
23	1252	$21.00	$16.80
23	1253	$21.00	$16.40
23	1254	$24.00	$19.20
23	1257	$19.00	$15.20

Record: 11 of 11

Catalog : Table

CatalogID	IssueDate
19	11/1/2009
20	2/1/2010
21	4/1/2010
22	6/1/2010
23	8/1/2010
24	9/15/2010

Record: 7

table in Figure 13-9 uses CatalogID alone as its primary key, you should add CatalogIssueDate to that table. If a Catalog table did not already exist, you would need to create a new table to hold CatalogIssueDate, as shown in Figure 13-12.

Third Normal Form To verify that a table is in 3NF, we must check the functional dependency of each non-key element on every other non-key element. This can be cumbersome for a large table because the number of pairs that must be checked grows quickly as the number of non-key fields grows. The number of functional dependencies to be checked is $N \times (N - 1)$, where N is the number of non-key fields. Note that functional dependency must be checked in both directions (that is, A dependent on B, and B dependent on A).

In practice, you can simplify finding 3NF violations by concentrating on two common types of problems:

- Tables that store attributes describing two or more entities
- Computable fields

Consider the simple table shown in the upper half of Figure 13-13. Assume that AccountNo is the primary key and that all customers live in the United States. Because there are three non-key fields, you must check six functional dependencies:

- Is State functionally dependent on StreetAddress?
- Is StreetAddress functionally dependent on State?
- Is ZipCode functionally dependent on StreetAddress?
- Is StreetAddress functionally dependent on ZipCode?
- Is ZipCode functionally dependent on State?
- Is State functionally dependent on ZipCode?

Figure 13-13

Converting a second normal form table into two third normal form tables

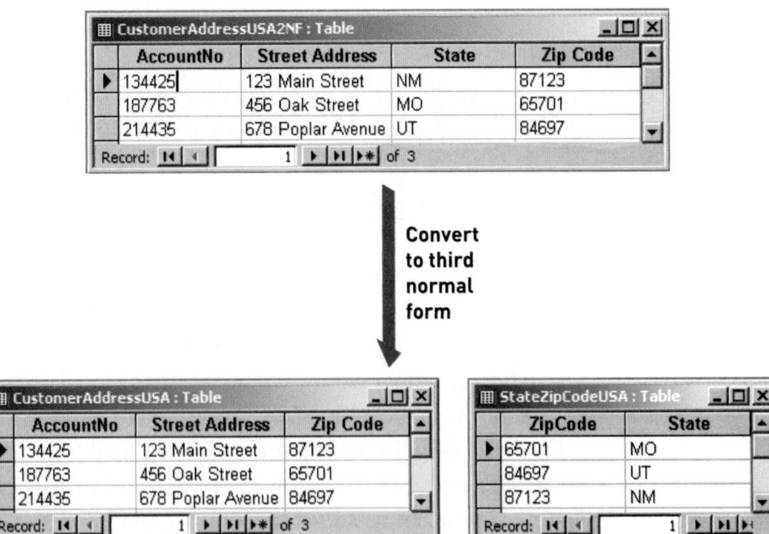
CustomerAddressUSA2NF : Table

AccountNo	Street Address	State	Zip Code
134425	123 Main Street	NM	87123
187763	456 Oak Street	MO	65701
214435	678 Poplar Avenue	UT	84697

Record: 1 of 3

CustomerAddressUSA : Table

AccountNo	Street Address	Zip Code
134425	123 Main Street	87123
187763	456 Oak Street	65701
214435	678 Poplar Avenue	84697

Record: 1 of 3

StateZipCodeUSA : Table

ZipCode	State
65701	MO
84697	UT
87123	NM

Record: 1

The answer to the first five statements is no, but the answer to the last is yes. In the United States, all zip codes are wholly contained within a single state. Thus, for each value of Zip Code there is only one corresponding value of State (for example, 87123 is always in New Mexico). Including both fields in this table is a form of redundancy. For example, if the addresses of 100 customers who live in the 87123 zip code are stored in the table, the fact that 87123 is located in New Mexico is (redundantly) stored 100 times.

In essence, the table combines information about two entities—Customer and Postal Delivery Area. Each entity has its own primary key (AccountNo for Customer and ZipCode for Postal Delivery Area) and its own non-key attributes (StreetAddress and Zip Code for Customer, and State for Postal Delivery Area).

Because State is functionally dependent on Zip Code, the table is not in 3NF. To correct this problem, you must remove State from the table. The relationship between states and zip codes must be stored somewhere in the database, or the information system can't generate complete mailing labels. The solution is to create a new table containing only ZipCode and State (see Figure 13-13) and remove State from the Customer table. ZipCode is the primary key of the new table, and State is its only non-key field. Programs or methods that print or display a complete mailing address must now use the value of Zip Code in the CustomerAddress table to look up the corresponding value of State in the newly created table.

A computable field stores a value that can be computed by a formula or algorithm that uses other database fields as inputs. Common examples of computable fields include subtotals, totals, and taxes. For example, consider the field GrandTotal in the Order table in Figure 13-9 and the formula:

$$GrandTotal = (\sum Quantity \times Price) + Shipping + Tax$$

Note that all of the inputs to the formula are not stored in the same table (see Figure 13-14). Unlike 3NF violations involving multiple entities stored in the same table, problems with computable fields can involve multiple tables. Shipping and Tax are stored in the Order table, and Quantity and Price are stored in related rows of the OrderItem table. An algorithm that computes GrandTotal for a particular invoice needs to extract all matching rows in the OrderItem table using the OrderID foreign key.

Figure 13-14

GrandTotal is computed from fields in two tables

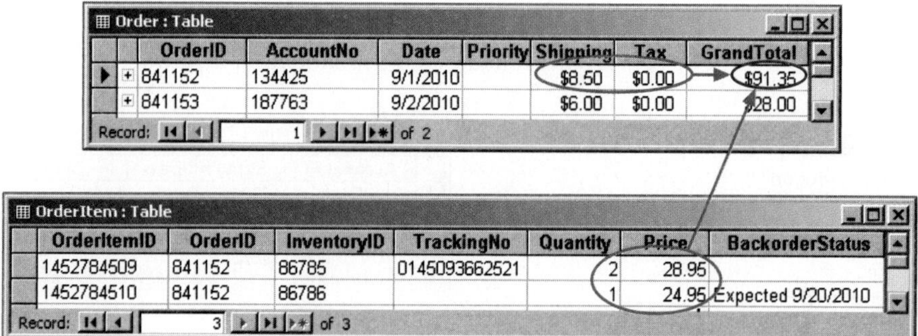
Order : Table

OrderID	AccountNo	Date	Priority	Shipping	Tax	GrandTotal
841152	134425	9/1/2010		$8.50	$0.00	$91.35
841153	187763	9/2/2010		$6.00	$0.00	$28.00

Record: 1 of 2

OrderItem : Table

OrderItemID	OrderID	InventoryID	TrackingNo	Quantity	Price	BackorderStatus
1452784509	841152	86785	0145093662521	2	28.95	
1452784510	841152	86786		1	24.95	Expected 9/20/2010

Record: 3 of 3

GrandTotal is functionally dependent on the combination of the other four fields. Computational dependencies are a form of redundancy because a change to the value of any input variable in the computation (for example, Shipping) also changes the result of the computation (in other words, GrandTotal).

The way to correct this type of 3NF violation is simple: Remove the computed field from the database. Eliminating the computed field from the database doesn't mean that its value is lost. For example, any program or method that needs GrandTotal can query the OrderItem table for matching values of Quantity and Price, sum the result of multiplying each Quantity and Price, and add Shipping and Tax.

Entity-Relationship Modeling and Normalization

Entity-relationship modeling and normalization are complementary techniques for relational database design. Note that the tables generated from the RMO ERD (see Figure 13-9) do not contain any violations of first, second, or third normal form. This is not a chance occurrence. Attributes of an entity are functionally dependent on any unique identifier (primary key) of that entity. Attributes of a many-to-many relationship are functionally dependent on unique identifiers of both participating entities. Thus, while creating an ERD, an analyst must directly or indirectly consider issues of functional dependency when deciding which attributes belong with which entities or relationships.

BEST PRACTICE

You can use an automated visual modeling tool to reverse engineer an existing relational database to an entity-relationship diagram (ERD) when you need to maintain or enhance an existing system.

We now turn our attention to the second type of database commonly in use today—the object-oriented database.

OBJECT-ORIENTED DATABASES

object database management system (ODBMS)
a database management system that stores data as objects or class instances

Object database management systems (ODBMSs) are a direct extension of the OO design and programming paradigm. ODBMSs are designed specifically to store objects and to interface with object-oriented programming languages. It is possible to store objects in files or relational databases. But there are many advantages to using an ODBMS, including direct support for method storage, inheritance, nested objects, object linking, and programmer-defined data types.

ODBMSs first appeared as research prototypes in the 1980s and later as fledgling commercial products in the early 1990s. Current commercial ODBMSs include GemStone, ObjectStore, and Objectivity. ODBMSs are the database technology of choice for newly designed systems implemented with OO tools, especially for scientific and engineering applications. ODBMSs are expected to supplant RDBMSs in more traditional business applications gradually over the next decade.

Object Definition Language (ODL)
a standard object database description language promulgated by the Object Database Management Group

Because ODBMSs are relatively new, there are few widely accepted standards for specifying an object database schema. In the late 1990s and early 2000s, the Object Database Management Group developed and refined a standard called the **Object Definition Language (ODL)**, a language for describing the structure and content of an object database. ODMG standards were the foundation of the Java Data Objects standard, which was adopted in 2003. ODMG standards are also the basis of some interfaces between ODBMSs and the C++ and SmallTalk programming languages. The schema examples in the sections that follow use ODL.

DESIGNING OBJECT DATABASES

To create an object database schema from a class diagram, follow these steps:

1. Determine which classes require persistent storage.
2. Define persistent classes.
3. Represent relationships among persistent classes.
4. Choose appropriate data types and value restrictions (if necessary) for each field.

Each of these steps is discussed in detail in the following sections.

Representing Classes

transient class
a class that doesn't need to store any attribute values between instantiations or method invocations

There are two broad types of classes for purposes of data management. Objects of a **transient class** exist only during the lifetime of a program or process. In a design that follows three-layer architecture, view layer objects such as windows and forms are usually transient. Transient objects are created each time a program or process is executed and then destroyed when a program or process terminates.

persistent class
a class that must store one or more attribute values between instantiations or method invocations

An object of a **persistent class** is not destroyed when the program or process that creates it ceases execution. Instead, the object continues to exist independently of any program or process. In a design that follows three-layer architecture, problem domain (or business) classes are usually persistent. Storing the object state to persistent memory (such as a magnetic or optical disk) ensures that the object exists between process executions. Objects can be persistently stored within a file or database management system.

An object database schema includes a definition for each class that requires persistent storage. ODL class definitions derive from the corresponding UML class diagram. Thus, classes already defined in UML are simply reused for the database schema definition.

For example, an ODL description of the RMO Customer class is:

```
class Customer {
 attribute string accountNo
 attribute string name
 attribute string billingAddress
 attribute string shippingAddress
 attribute string dayTelephoneNumber
 attribute string nightTelephoneNumber
}
```

This ODL class definition corresponds to the Customer table in Figure 13-9. A similar ODL class definition would be created for each RMO class shown in Figure 13-15. After defining each class, the analyst must define relationships among classes.

Representing Relationships

object identifier
a physical storage address or a reference that can be converted to a physical storage address at run time

Each object stored within an ODBMS is automatically assigned a unique object identifier. An **object identifier** may be a physical storage address or a reference that can be converted to a physical storage address at run time. In either case, each object has a unique identifier that can be stored within another object to represent a relationship.

An ODBMS represents relationships by storing the identifier of one object within related objects. Object identifiers provide navigation visibility among objects, as first described in Chapter 11. For example, consider a one-to-one relationship between the classes Employee and Workstation, as shown in Figure 13-16. Each Employee object has an attribute called *computer* that contains the object identifier of the Workstation object assigned to that employee. Each Workstation object has a matching attribute called *user* that contains the object identifier of the Employee who uses that workstation.

navigation
the process of accessing an object by extracting its object identifier from another (related) object

The ODBMS uses attributes containing object identifiers to find objects that are related to other objects. The process of extracting an object identifier from one object and using it to access another object is sometimes called **navigation**. For example, consider the following query posed by a user:

> List the manufacturer of the workstation assigned to employee Joe Smith.

An ODBMS query processor can find the requested employee object by searching all employee objects for the name attribute Joe Smith. The query processor can find Joe Smith's workstation object by using the object identifier stored in computer. The query processor can also answer the opposite query (list the employee name assigned to a specific workstation) by using the object identifier stored in user. A matched pair of attributes enables navigation in both directions.

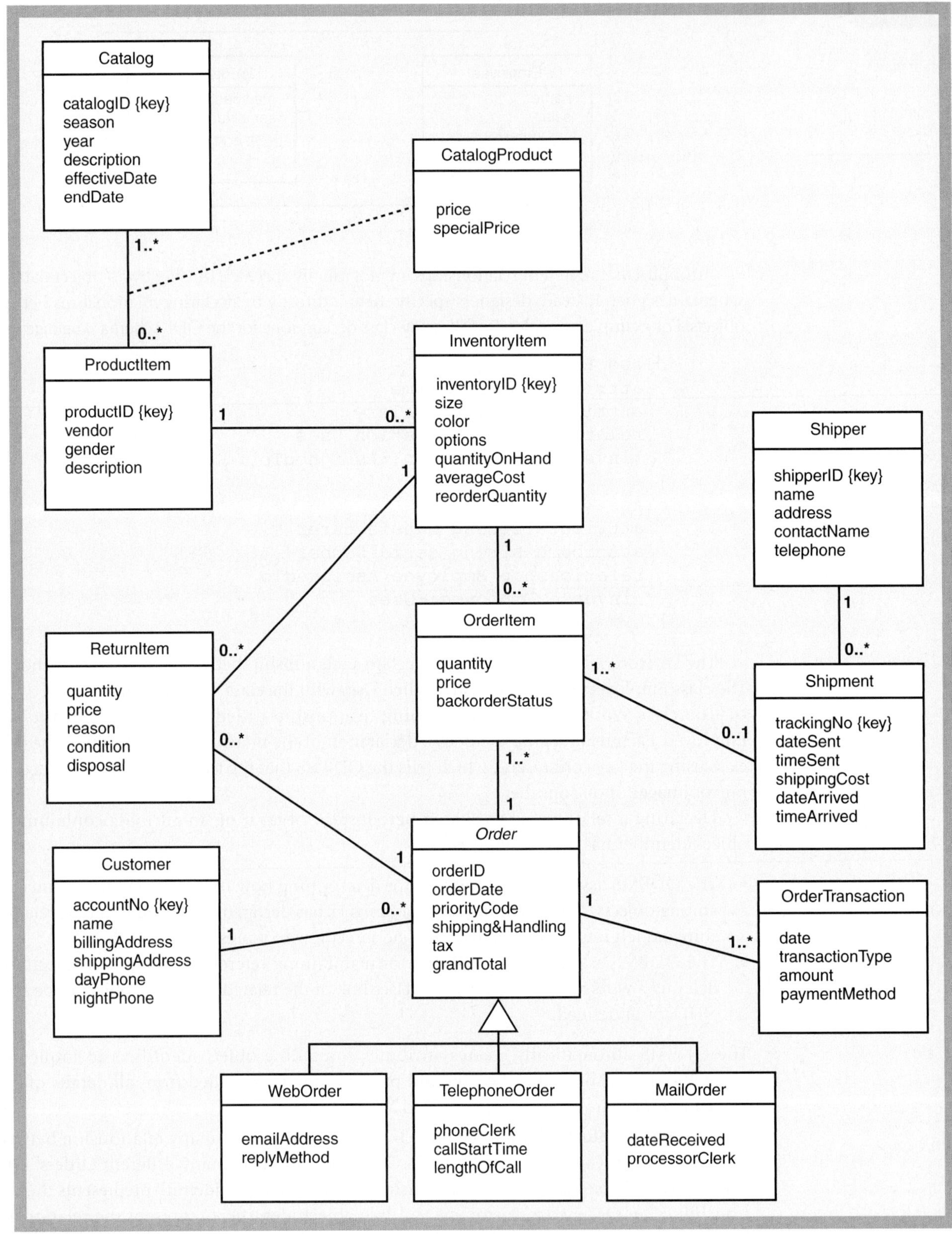

Figure 13-15

The RMO class diagram

Figure 13-16

A one-to-one relationship represented with attributes (shown in color) containing object identifiers

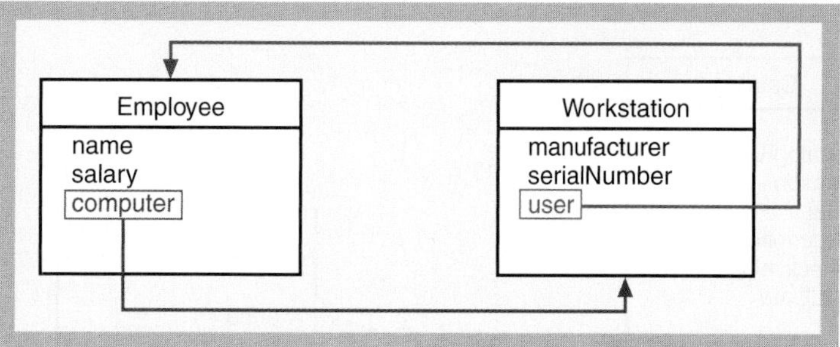

Attributes that represent relationships are not usually specified directly by an object database schema designer. Instead, designers specify them indirectly by declaring relationships between objects. For example, consider the following class declarations for the ODL schema language:

```
class Employee {
  attribute string name
  attribute integer salary
  relationship Workstation Uses
   inverse Workstation::AssignedTo
}
class Workstation {
  attribute string manufacturer
  attribute string serialNumber
  relationship Employee AssignedTo
   inverse Employee::Uses
}
```

The keyword *relationship* is used to declare a relationship between one class and another. The class Employee has a relationship called Uses with the class Workstation.

The class Workstation has a matching relationship called AssignedTo with the class Employee. Each relationship includes a declaration of the matching relationship in the other class using the keyword *inverse*, which tells the ODBMS that the two relationships are actually mirror images of one another.

Declaring a relationship as shown here instead of creating an attribute containing an object identifier has two advantages:

- The ODBMS assumes responsibility for determining how to implement the connection among objects. In essence, the schema designer has declared an attribute of type relationship and left it up to the ODBMS to determine how to represent that attribute.
- The ODBMS assumes responsibility for maintaining referential integrity. For example, deleting a workstation will cause the Uses link of the related Employee object to be set to NULL or undefined.

The ODBMS automatically creates attributes containing object identifiers to implement declared relationships. But the user and programmer are shielded from all details of how those identifiers are actually implemented and manipulated.

One-to-Many Relationships Figure 13-17 shows the one-to-many relationship between the RMO classes Customer and Order. A Customer can make many different Orders, but a single Order can be made by only one Customer. A single object identifier represents the relationship of an Order to a Customer. Multiple object identifiers represent the relationship between one Customer and many different Orders, as shown in Figure 13-18.

Partial ODL class declarations for the classes Customer and Order are as follows:

```
class Customer {
  attribute string accountNo
  attribute string name
  attribute string billingAddress
  attribute string shippingAddress
  attribute string dayPhone
```

Figure 13-17

The one-to-many relationship between the Customer and Order classes

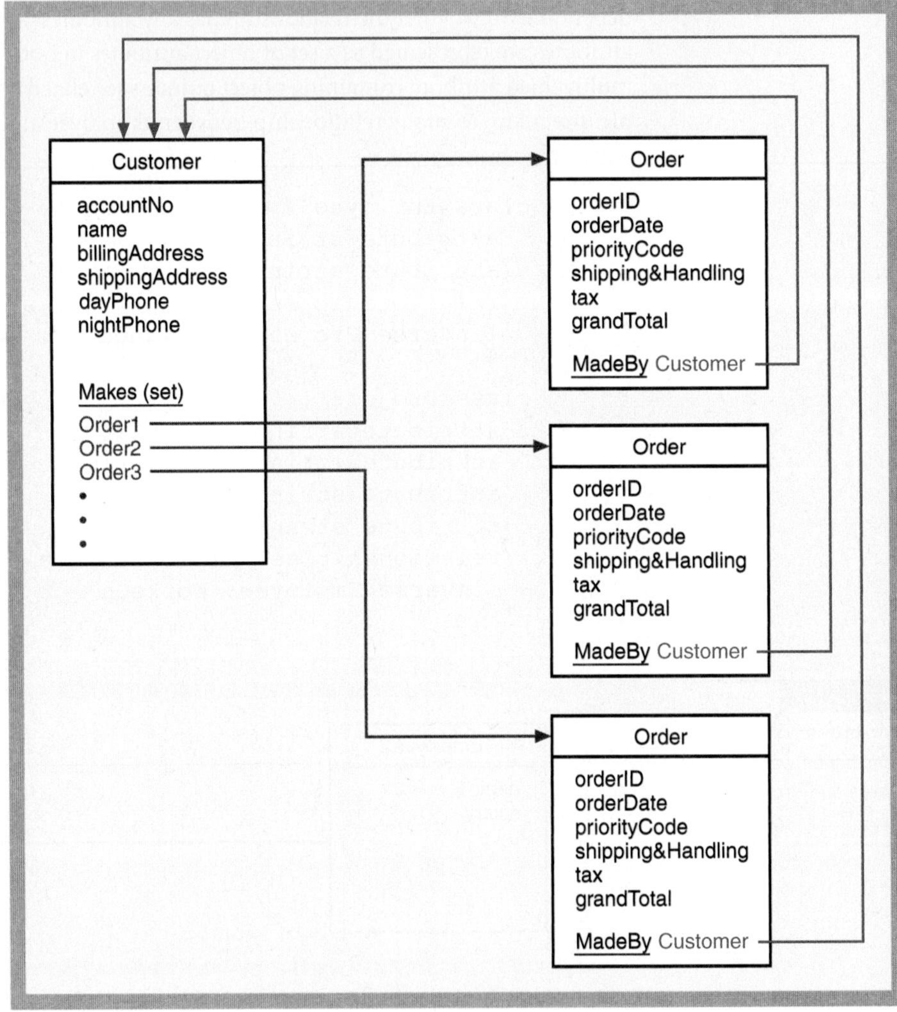

Figure 13-18

A one-to-many relationship represented with attributes containing object identifiers

```
  attribute string nightPhone
  relationship set<Order> Makes
   inverse Order::MadeBy
}
class Order {
  attribute string orderID
  attribute string orderDate
  attribute string priorityCode
  attribute real shipping&Handling
  attribute real tax
  attribute real grandTotal
  relationship Order MadeBy
   inverse Customer::Makes
}
```

The relationship Makes is declared between a single Customer object and a set of Order objects. By declaring the relationship as a set, you instruct the ODBMS to allocate as many Order object identifier attributes to each Customer object as are needed to represent relationship instances. The ODBMS dynamically adds or deletes object identifier attributes to the set as instances of the relationship are created or deleted.

The set of object identifier attributes can also be called a multivalued attribute. A **multivalued attribute**, also called a repeating group, is an attribute that contains zero or more instances of the same data type. Multivalued attributes are commonly supported in ODBMSs but are not supported in RDBMSs because they violate first normal form.

multivalued attribute
an attribute that contains zero or more instances of the same data type

Many-to-Many Relationships A many-to-many relationship is represented differently depending on whether the relationship has any attributes. Many-to-many relationships without attributes are represented as a set of object attributes in both related classes. Both classes have a multivalued attribute containing object pointers to related objects of the other class. For example, the many-to-many relationship between Employee and Project shown in Figure 13-19 is represented as follows:

```
class Employee {
  attribute string name
  attribute string salary
  relationship set<Project> WorksOn
   inverse Project::Assigned
}
class Project {
  attribute string projectID
  attribute string description
  attribute string startDate
  attribute string endDate
  relationship set<Employee> Assigned
   inverse Employee::WorksOn
}
```

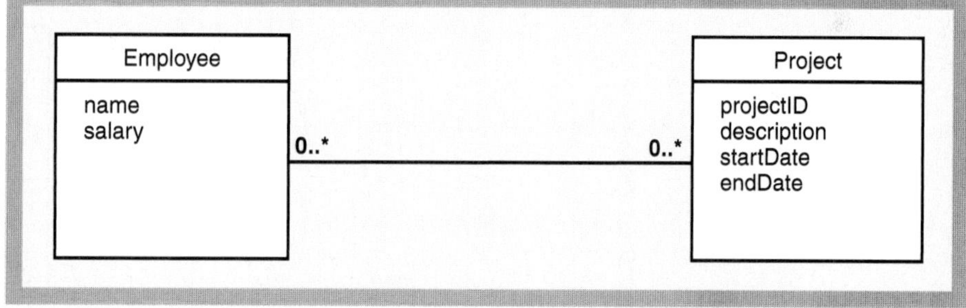

Figure 13-19
A many-to-many relationship between the Employee and Project classes

Representing a many-to-many relationship with attributes requires a more complex approach. The RMO class diagram has a many-to-many relationship between Catalog and ProductItem with an association class named CatalogProduct (see Figure 13-15). Recall from Chapter 5 that an *association class* is a class that stores the attributes of a many-to-many relationship.

To represent a many-to-many relationship with an association class, we must reorganize the relationship as shown in Figure 13-20. The many-to-many relationship between Catalog and ProductItem has been decomposed into a pair of one-to-many relationships between the original classes and the association class. The ODL schema descriptions are as follows:

```
class Catalog {
  attribute string season
  attribute integer year
  attribute string description
  attribute string effectiveDate
```

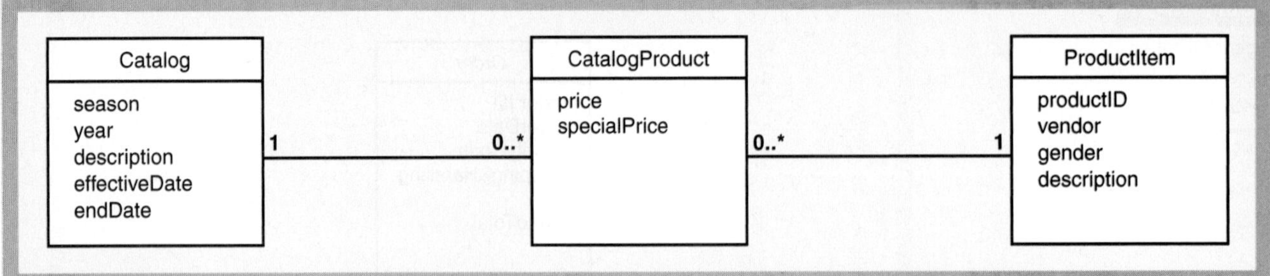

Figure 13-20

A many-to-many relationship represented with two one-to-many relationships

```
    attribute string endDate
    relationship set<CatalogProduct> Contains1
     inverse CatalogProduct::AppearsIn1
}
class ProductItem {
    attribute string productID
    attribute string vendor
    attribute string gender
    attribute string description
    relationship set<CatalogProduct> AppearsIn2
     inverse CatalogProduct::Contains2
}
class CatalogProduct {
    attribute real price
    attribute real specialPrice
    relationship Catalog AppearsIn1
     inverse Catalog::Contains1
    relationship ProductItem AppearsIn2
     inverse ProductItem::Contains2
}
```

Generalization Relationships Figure 13-21 shows the order generalization hierarchy from the RMO class diagram. WebOrder, TelephoneOrder, and MailOrder are each more specific versions of the class Order. The ODL class definitions that represent these classes and their interrelationships are as follows:

```
class Order {
    attribute string orderID
    attribute string orderDate
    attribute string priorityCode
    attribute real shipping&Handling
    attribute real tax
    attribute real grandTotal
}
class WebOrder extends Order {
    attribute string emailAddress
    attribute string replyMethod
}
class TelephoneOrder extends Order {
    attribute string phoneClerk
    attribute string callStartTime
    attribute integer lengthOfCall
}
class MailOrder extends Order {
    attribute string dateReceived
    attribute string processorClerk
}
```

Figure 13-21
A generalization hierarchy within the RMO class diagram

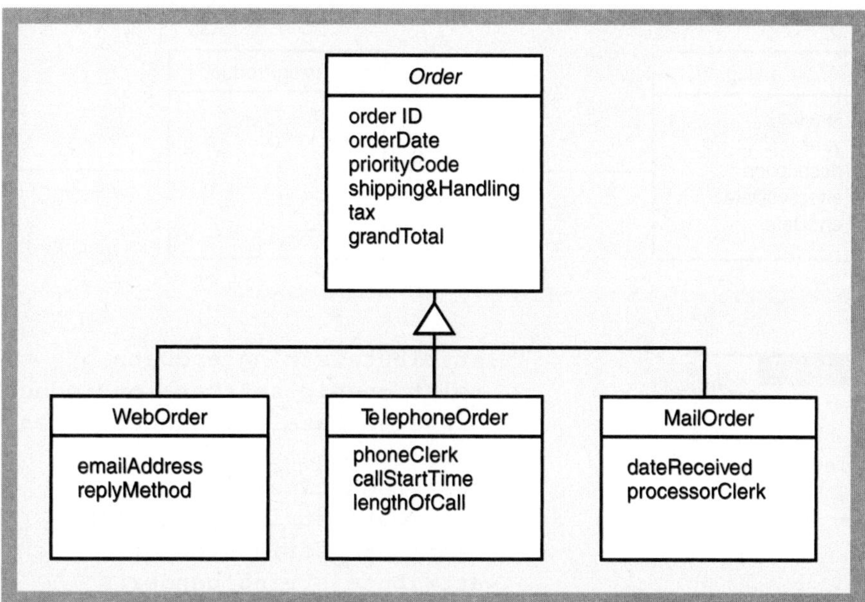

The keyword *extends* indicates that WebOrder, TelephoneOrder, and MailOrder derive from Order. When stored in an object database, objects of the three derived classes will inherit all of the attributes, methods, and relationships defined for the Order class.

Key Attributes

Key attributes are not required in an object database because referential integrity is implemented with object identifiers. However, key attributes are useful in object databases for a number of purposes, including guaranteeing unique object content and providing a means of querying database contents. The ODBMS automatically enforces uniqueness of key attributes in an object database. Thus, declaring an attribute to be a key guarantees that no more than one object in a class can have the same key value.

In addition to relational and object-oriented databases, a third type exists that mixes elements of both the relational and OO approaches. We discuss this hybrid next.

HYBRID OBJECT-RELATIONAL DATABASE DESIGN

OO development tools were first widely employed in the mid- to late-1980s. During this same time period, RDBMSs were widely used and had reached a mature stage of development. Many OO tool developers exploited the existing base of RDBMS tools and knowledge by using an RDBMS to store persistent object states. This made sense both as an economy measure (there was one less OO tool to develop) and because many newer OO systems needed to manipulate data stored in existing relational databases. There is no widely accepted name to describe object storage using an RDBMS, so we will invent one to use for the remainder of the text: hybrid object-relational DBMS (or simply hybrid DBMS).

hybrid object-relational DBMS
a relational database management system used to store object attributes and relationships; also called *hybrid DBMS*

The hybrid DBMS approach is currently the most widely employed approach for persistent object storage. Designing a hybrid database is essentially two design problems in one. That is, the designer must develop a complete relational database schema and an equivalent set of classes to represent the relational database contents within the OO programs. This second task is complex because the designer must bridge the differences between the object-oriented and relational views of stored data.

Following are the most important mismatches between the relational and OO views of stored data:

- Class methods cannot be directly stored or automatically executed within an RDBMS.
- ODBMSs can represent a wider range of relationship types than RDBMSs, including classification hierarchies and whole-part aggregations. Relationships in an RDBMS can only be represented using referential integrity.
- ODBMSs can represent a wider range of data types than RDBMSs. New classes can be defined to store application-specific data.

Because RDBMSs were developed prior to the OO paradigm, they have no features that can represent methods or inheritance. Programs that access the database must implement methods internally. Inheritance cannot be directly represented in an RDBMS because a classification hierarchy cannot be directly represented.

Although the relational and OO views of stored data have significant differences, they also have significant overlaps. Recall from Chapter 5 that "things" within a system can be conceptually modeled using an ERD (the basis for a relational database schema), a class diagram (the basis for an OO database schema), or both. There is considerable overlap among the two representations, including the following:

- Grouping of data items into entities or classes
- Defining one-to-one, one-to-many, and many-to-many relationships among entities or classes

This overlap provides a basis for representing classes and objects within a relational database.

CLASSES AND ATTRIBUTES

Designers can store classes and object attributes in an RDBMS by defining appropriate tables in which to store them. For a completely new system, a relational schema can be designed based on a class diagram—essentially the same process as for an ERD. Figure 13-22 describes the correspondence among OO, ER, and relational database concepts. A table is created to represent each class, and the fields of each table are the same as the attributes of the corresponding class. Each row holds the attribute values of a single object.

Figure 13-22
Correspondence among concepts in the object-oriented, entity-relationship, and relational database views of stored data

Object-oriented	Entity-relationship	Relational database
Class	Entity type	Table
Object	Entity instance	Row
Attribute	Attribute	Column

A key field (or group of fields) must be chosen for each table. As described earlier, a designer can choose a natural or invented key field from the existing attributes or add an invented key field. Primary key fields are needed to guarantee uniqueness within tables and to represent relationships using foreign keys.

Figure 13-23 shows a set of relational database tables that represent classes from the RMO class diagram in Figure 13-15. Note that the table definitions are identical to those in Figure 13-7, except for the addition of tables to represent the specialized classes MailOrder, TelephoneOrder, and WebOrder and corresponding changes to the Order table.

Figure 13-23

Class tables, with primary keys identified in bold

Table	Attributes
Catalog	**CatalogID**, Season, Year, Description, EffectiveDate, EndDate
CatalogProduct	**CatalogProductID**, Price, SpecialPrice
Customer	**AccountNo**, Name, BillingAddress, ShippingAddress, DayTelephoneNumber, NightTelephoneNumber
InventoryItem	**InventoryID**, Size, Color, Options, QuantityOnHand, AverageCost, ReorderQuantity
MailOrder	**MailOrderID**, DateReceived, ProcessorClerk
Order	**OrderID**, OrderDate, PriorityCode, ShippingAndHandling, Tax, GrandTotal
OrderItem	**OrderItemID**, Quantity, Price, BackorderStatus
OrderTransaction	**OrderTransactionID**, Date, TransactionType, Amount, PaymentMethod
ProductItem	**ProductID**, Vendor, Gender, Description
ReturnItem	**ReturnItemID**, Quantity, Price, Reason, Condition, Disposal
Shipment	**TrackingNo**, DateSent, TimeSent, ShippingCost, DateArrived, TimeArrived
Shipper	**ShipperID**, Name, Address, ContactName, Telephone
TelephoneOrder	**TelephoneOrderID**, PhoneClerk, CallStartTime, LengthOfCall
WebOrder	**WebOrderID**, EmailAddress, ReplyMethod

RELATIONSHIPS

ODBMSs use object identifiers to represent relationships among objects. But RDBMSs do not create object identifiers, so relationships among objects stored in a relational database must be represented using foreign keys. Foreign key values serve the same purpose as object identifiers in an ODBMS. That is, they provide a means for one "object" to refer to another.

To represent one-to-many relationships, designers add the primary key field of the class on the "one" side of the relationship to the table representing the class on the "many" side of the relationship. To represent many-to-many relationships, designers create a new table that contains the primary key fields of the related class tables and any attributes of the relationship itself. Note that these methods of representing relationships among objects are the same as previously described for representing relationships among entities.

Figure 13-24 extends the table definitions in Figure 13-23 by adding foreign keys representing the relationships shown in Figure 13-15. For example, the one-to-many relationship between the Customer and Order classes is represented by the foreign key AccountNo stored in the Order table. The many-to-many relationship between the Catalog and ProductItem classes is represented by the table CatalogProduct that contains the foreign keys CatalogID and ProductID.

Note that the tables in Figure 13-24 are identical to those in Figure 13-9 except for the content of the tables Order, MailOrder, TelephoneOrder, and WebOrder (these tables will be discussed shortly). The similarity is no accident; it follows from the similarity between the RMO entity-relationship and class diagrams. The diagrams are similar because they represent the same underlying reality. Thus, it should be no surprise that the relational database schemas derived from a class diagram and an ERD representing that underlying reality are similar. In fact, it would be surprising (and probably indicate an error) if they weren't similar.

Classification relationships such as the relationship among Order, MailOrder, TelephoneOrder, and WebOrder are a special case in relational database design. Just as a child class inherits the data and methods of a parent class, a table representing a child class inherits

Figure 13-24

Relationship information added to the class tables by adding foreign key attributes (shown in italic) to represent relationships

Table	Attributes
Catalog	**CatalogID**, Season, Year, Description, EffectiveDate, EndDate
CatalogProduct	***CatalogID, ProductID***, Price, SpecialPrice
Customer	**AccountNo**, Name, BillingAddress, ShippingAddress, DayTelephoneNumber, NightTelephoneNumber
InventoryItem	**InventoryID**, *ProductID*, Size, Color, Options, QuantityOnHand, AverageCost, ReorderQuantity
MailOrder	***OrderID***, DateReceived, ProcessorClerk
Order	**OrderID**, *AccountNo*, OrderDate, PriorityCode, ShippingAndHandling, Tax, GrandTotal
OrderItem	**OrderItemID**, *OrderID*, *InventoryID*, *TrackingNo*, Quantity, Price, BackorderStatus
OrderTransaction	**OrderTransactionID**, *OrderID*, Date, TransactionType, Amount, PaymentMethod
ProductItem	**ProductID**, Vendor, Gender, Description
ReturnItem	**ReturnItemID**, *OrderID*, *InventoryID*, Quantity, Price, Reason, Condition, Disposal
Shipment	**TrackingNo**, *ShipperID*, DateSent, TimeSent, ShippingCost, DateArrived, TimeArrived
Shipper	**ShipperID**, Name, Address, ContactName, Telephone
TelephoneOrder	***OrderID***, PhoneClerk, CallStartTime, LengthOfCall
WebOrder	***OrderID***, EmailAddress, ReplyMethod

some or all of its data from the table representing its parent class. This inheritance can be represented in two ways:

- Combine all the tables into a single table containing the superset of all class attributes but excluding any invented key fields of the child classes.
- Use separate tables to represent the child classes and substitute the primary key of the parent class table for the invented keys of the child class tables.

Either method is an acceptable approach to representing a classification relationship.

Figure 13-9 shows the definition of the Order table under the first method. All of the non-key fields from MailOrder, TelephoneOrder, and WebOrder have been added to the Order table. For any particular order, some of the field values in each row will be NULL. For example, a row representing a telephone order would have no values for the fields EmailAddress, ReplyMethod, DateReceived, and ProcessorClerk.

Figure 13-24 shows the table definitions for the RMO case using the second method for representing inheritance. The relationship among the three child order types and the parent Order table is represented by the foreign key OrderID in all three child class tables. The invented key of each table has been removed. Thus, in each case, the foreign key representing the inheritance relationship also serves as the primary key of the table representing the child class.

DATA ACCESS CLASSES

In Chapter 12, you learned how to develop an OO design based on three-layer architecture. Under that architecture, data access classes implement the bridge between data stored in program objects and in a relational database.

Figure 13-25 illustrates the interaction among the RMO problem domain class ProductItem, the data access class ProductItemDA, and the relational database. The data access class has methods that add, update, find, and delete fields and rows in the table or tables that represent the class. Data access class methods encapsulate the logic needed to copy values from the problem domain class to the database and vice versa. Typically, that logic is a combination of program code in a language such as C++ or Java and embedded relational database commands in Structured Query Language (SQL).

Figure 13-25

Interaction among a problem domain class, a data access class, and the DBMS

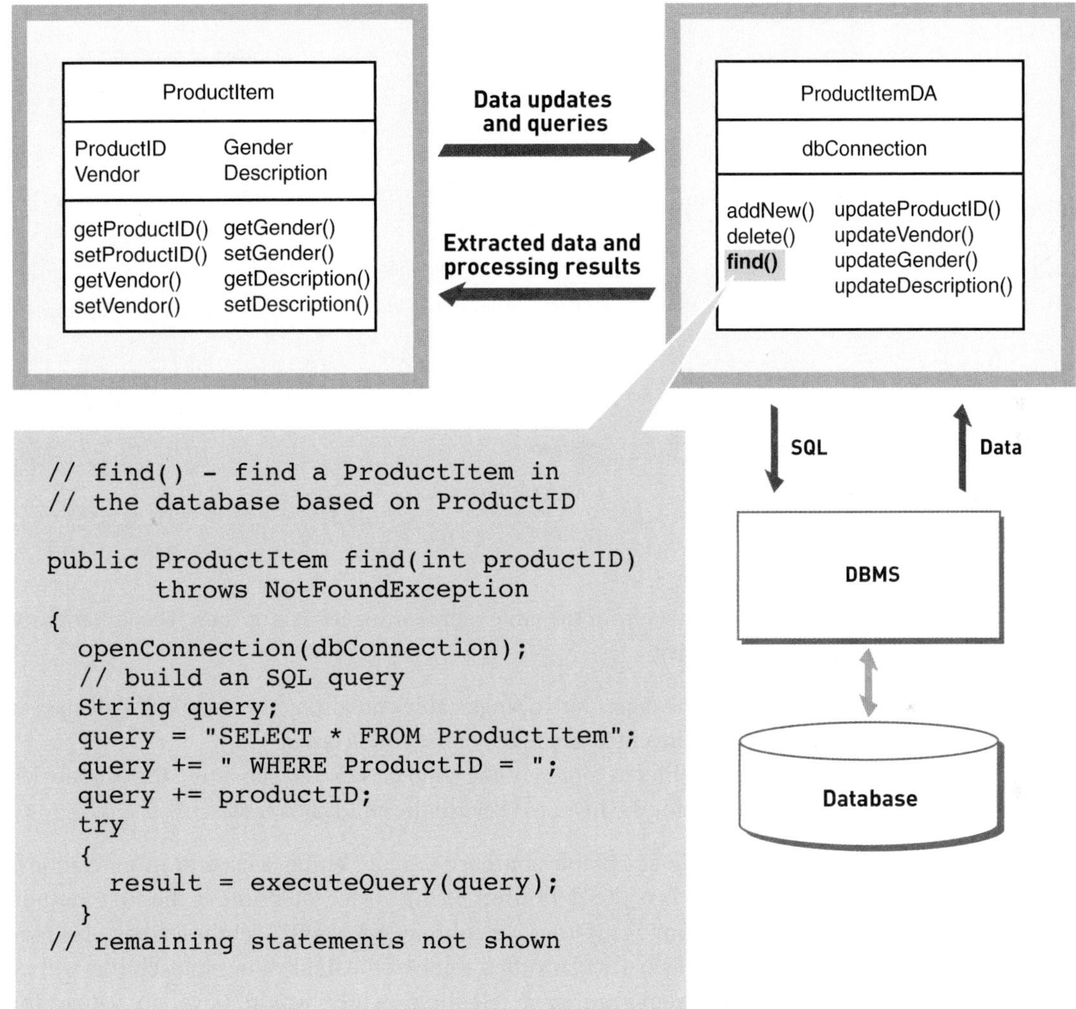

The lower-left part of Figure 13-25 shows a fragment of Java code with an embedded SQL statement that implements the find() method of ProductItemDA. Similar code is needed for all other methods in the data access class.

Now that we've covered the different approaches to database schema design, we consider the data types that are stored within that schema.

DATA TYPES

data type
the storage format and allowable content of a program variable or database field

A **data type** defines the storage format and allowable content of a program variable, object state variable, or database field or attribute. **Primitive data types** are data types that are supported directly by computer hardware and programming languages. Examples include memory address (a pointer), Boolean, integer, unsigned integer, short integer (one byte), long integer (multiple bytes), single characters, real numbers (floating-point numbers), double-precision

primitive data type
a storage format directly implemented by computer hardware or a programming language

complex data type
a data type not directly supported by computer hardware or a programming language; also called *user-defined data type*

(double-length) integers, and real numbers. In some procedural programming languages (such as C) and most OO languages, programmers can define additional data types using the primitive data types as building blocks.

As information systems have become more complex, the number of data types used to implement them has increased. Examples of modern data types include dates, times, currency (money), audio streams, still images, motion video streams, and uniform resource locators (URL or Web links). Such data types are sometimes called **complex data types** because they are usually defined as complex combinations of primitive data types. They may also be called *user-defined data types* because they may be defined by users during analysis and design or by programmers during design and implementation.

RELATIONAL DBMS DATA TYPES

The designer must choose an appropriate data type for each field in a relational database schema. For many fields, the choice of a data type is relatively straightforward. For example, designers can represent customer names and addresses using a set of fixed- or variable-length character arrays. Inventory quantities and item prices can be represented as integers and real numbers, respectively. A color can be represented by a character array containing the name of the color or by a set of three integers representing the intensity of the video-display colors red, blue, and green.

Modern RDBMSs have added an increasing number of new data types to represent the data required by modern information systems. Figure 13-26 contains a partial listing of some of the data types available in the Oracle RDBMS. Complex data types available in Oracle include DATE, LONG, and LONGRAW. LONG is typically used to store large quantities of formatted or unformatted text (such as a word-processing document). LONGRAW can be used to store large binary data values, including encoded pictures, sound, and motion video.

Figure 13-26
A subset of the data types available in the Oracle relational DBMS

Type	Description
CHAR	Fixed-length character array
VARCHAR	Variable-length character array
NUMBER	Real number
DATE	Date and time with appropriate checks of validity
LONG	Variable-length character data up to 2 gigabytes
LONGRAW	Binary large object (BLOB) with no assumption about format or content
ROWID	Unique six-byte physical storage address

Modern RDBMSs can also perform many validity and format checks on data as it is stored in the database. For example, a schema designer can specify that a quantity on hand cannot be negative, that a U.S. zip code must be five or nine digits long, and that a string containing a URL must begin with *http://*. All application programs that use the database then automatically share the validity and format constraints. Each program is simpler, and the possibility for errors from mismatches among data validation logic is eliminated. Application programs still have to provide program logic to recover from attempts to add "bad" data, but they are freed from actually performing validity checks.

OBJECT DBMS DATA TYPES

ODBMSs typically provide a set of primitive and complex data types comparable to those of an RDBMS. ODBMSs also allow a schema designer to define format and value constraints. But ODBMSs provide an even more powerful way to define useful data types and constraints. A schema designer can define a new data type and its associated constraints as a new class.

A class is a complex user-defined data type that combines the traditional concept of data with processes (methods) that manipulate that data. In most OO programming languages,

programmers are free to design new data types (classes) that extend those already defined by the programming language. Incompatibility between system requirements and available data types is not an issue, because the designer can design classes specifically to meet the requirements. To the ODBMS, instances of the new data type are simply objects to be stored in the database.

Class methods can perform many of the type- and error-checking functions previously performed by application program code and/or by the DBMS itself. In essence, the programmer constructs a "custom-designed" data type and all of the programming logic required to use it correctly. The DBMS is freed from direct responsibility for managing complex data types and the values stored therein. It indirectly performs validity checking and format conversion by extracting and executing programmer-defined methods stored in the database.

The flexibility to define new data types is one reason that OO tools are so widely employed in non–business information systems. In fields such as engineering, biology, and physics, stored data is considerably more complex than simple strings, numbers, and dates. OO tools enable database designers and programmers to design custom data types that are specific to a problem domain.

Another issue that must be considered during database design is the locations where data is stored and accessed. In today's networked information systems, organizations often use distributed databases.

DISTRIBUTED DATABASES

Rarely does an organization store all of its data in a single database. Instead, organizations typically store data in many different databases, often under the control of many different DBMSs. Reasons for employing a variety of databases and DBMSs include the following:

- Information systems may have been developed at different times using different DBMSs.
- Parts of an organization's data may be owned and managed by different organizational units.
- System performance improves when data is physically close to the applications that use it.

DISTRIBUTED DATABASE ARCHITECTURES

Chapter 9 described various approaches to organizing and computing information processing resources in a networked environment. Several architectures for distributing database services are possible, including the following:

- Single database server
- Replicated database servers
- Partitioned database servers
- Federated database servers

Combinations of these architectures are also possible.

Single Database Server

Figure 13-27 shows a typical single database server architecture. Clients on one or more LANs share a single database located on a single computer system. The database server may be connected to one of the LANs or directly to the WAN backbone (as shown in the figure). Connection directly to the WAN ensures that no one LAN is overloaded by all of the network traffic to and from the database server.

The primary advantage of single database server architecture is its simplicity. There is only one server to manage, and all clients are programmed to direct requests to that server. Disadvantages of the single database server architecture include susceptibility to server failure and possible overload of the network or server. A single server provides no backup capabilities in the event of server failure. All application programs that depend on the server are disabled whenever the server is unavailable (such as during a crash or during hardware maintenance). Thus, single database server architecture is poorly suited to applications that must be available on a seven-day, 24-hour basis.

Figure 13-27

A single database server architecture

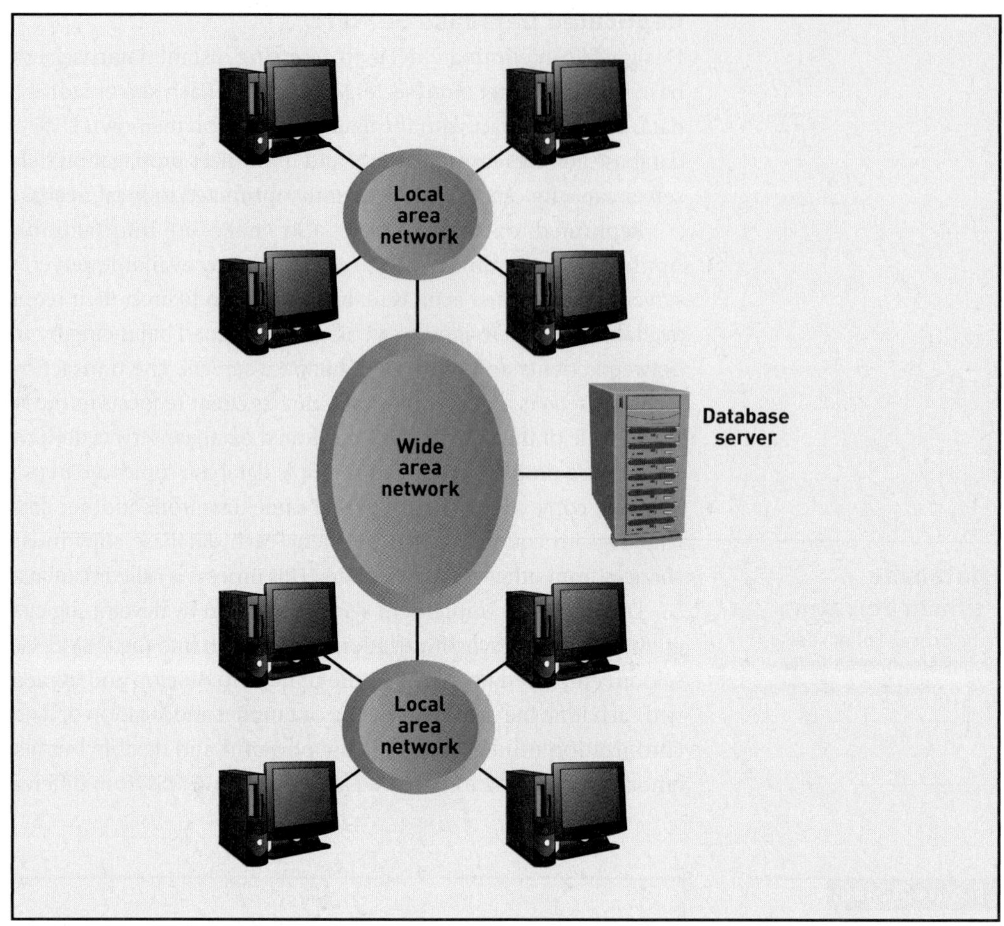

Performance bottlenecks can occur within a single database server or in the network segment to which the server is attached. As transaction volumes grow, the capabilities of a single database server may become insufficient to respond quickly to all of the service requests it receives. In an attempt to improve performance, a designer may employ a more powerful computer system as the database server. But in an era of multiterabyte databases, it is not unusual for the size and transaction volume of larger databases to exceed the capabilities of any single computer system. Employing the largest mainframes may also be impractical because of cost, system management, or network performance considerations.

Requests to and responses from a database server may traverse great distances across local and wide area networks. Database transactions must also compete with other types of network traffic (such as voice, video, and Web site access) for available transmission capacity. Thus, delays in accessing a remote database server may result from network congestion or propagation delay from client to server.

One way to reduce network congestion is to increase capacity of the entire network. But this approach is expensive and often impractical. Another approach, specifically geared to improving database access speed, is to locate database servers physically close to their clients (for example, on the same LAN segment). This approach minimizes the distance-related delay for requests and responses and removes a large amount of traffic from the WAN.

Moving a database server closer to its clients is a relatively simple matter when all of the clients are located close to one another. But what happens when clients are widely dispersed, as in a multinational corporation? In this case, no single location for the database server can possibly improve database access performance for all clients at the same time. Thus, the "distant" clients must pay a greater performance penalty for database access.

Replicated Database Servers

Designers can eliminate delay in accessing distant database servers by using a replicated database server architecture (see Figure 13-28). Each server stores a separate copy of the needed data. Clients interact with the database server on their own LAN. Such an architecture eliminates database accesses from the WAN and minimizes propagation delay. Local network and database server capacity can be independently optimized to local needs.

Replicated database servers also make an information system more fault tolerant. Applications can direct access requests to any available server, with preference to the nearest server. When a server is unavailable, clients can redirect their requests across the WAN to another available server. Designers can also achieve load balancing by interposing a transaction server between clients and replicated database servers. The transaction server monitors loads on all database servers and automatically directs client requests to the server with the lowest load.

In spite of their advantages, replicated database servers do have some drawbacks. Data inconsistency is a problem whenever multiple database copies are in use. When data is updated on one database copy, clients accessing that same data from another database copy receive an outdated response. To counteract this problem, each database copy must periodically be updated with changes from other database servers. This process is called database synchronization.

database synchronization
the process of ensuring consistency among two or more database copies

Designers can implement synchronization by developing customized synchronization programs or by using synchronization utilities built into the DBMS. Custom application programs are seldom employed because they are difficult to develop and because they would need to be modified each time the database schema or number and location of database copies change. DBMS synchronization utilities are generally powerful and flexible but also expensive. Incompatibilities among synchronization methods make using DBMSs from different vendors impractical.

Figure 13-28
A replicated database server architecture

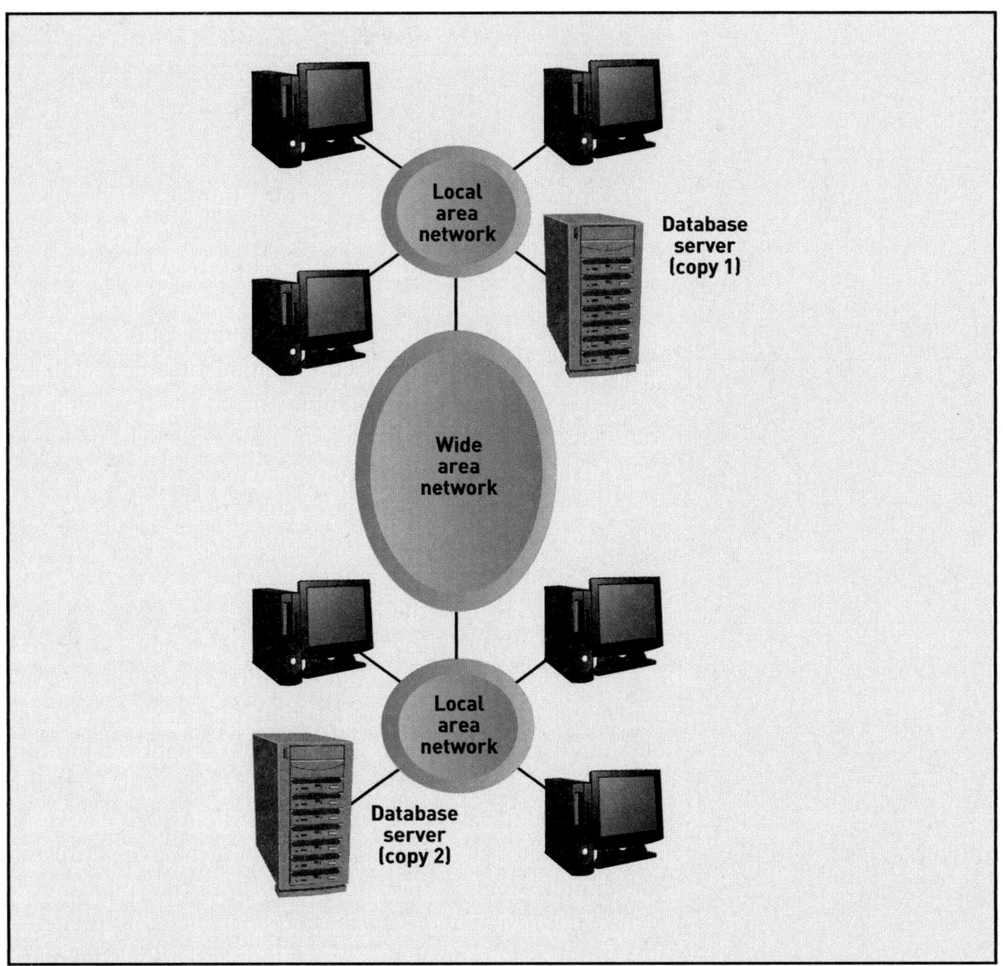

The time delay between an update to a database copy and the propagation of that update to other database copies is an important database design decision. During the time between the original update and the update of database copies, application programs that access outdated copies aren't receiving responses that reflect current reality. Designers can address this problem by reducing the synchronization delay. But shorter delays imply more frequent (or possibly continuous) database synchronization. Synchronization then consumes a substantial amount of database server capacity, and a large amount of network capacity among the related database servers must be provided. The proper synchronization strategy is a complex trade-off among cost, hardware and network capacity, and the need of application programs and users for current data.

Partitioned Database Servers

Designers can minimize the need for database synchronization by partitioning database contents among multiple database servers. Figure 13-29 shows the division of a hypothetical database schema into two partitions. A different group of clients accesses each partition. Figure 13-30 shows a partitioned database server architecture that maintains each partition on a separate database server. Traffic among clients and the database server in each group is restricted to a local area network.

Figure 13-29

Partitioning a database schema into client access subsets

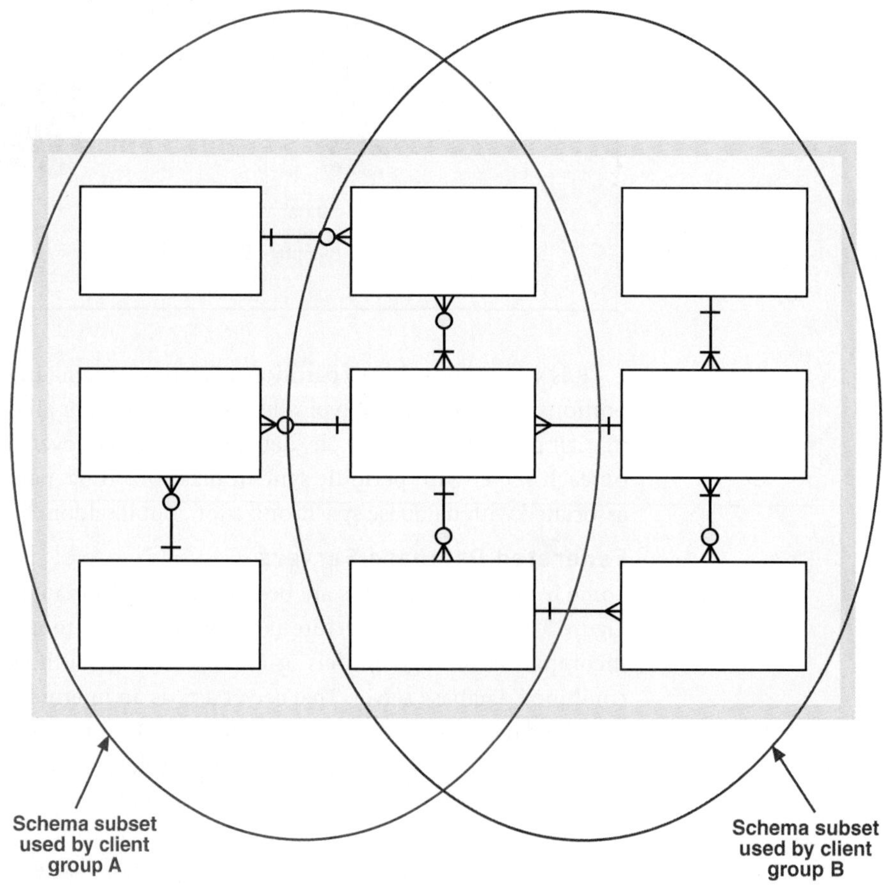

Partitioned database server architecture is feasible only when a schema can be cleanly partitioned among client access groups. Client groups must require access to well-defined subsets of a database (for example, marketing data rather than production data). In addition, members of a client access group must be located in small geographic regions. When a single access group is spread among multiple geographic sites (for example, order processing at three regional centers), a combination of replicated and partitioned database server architecture is usually required.

Figure 13-30
A partitioned database server architecture

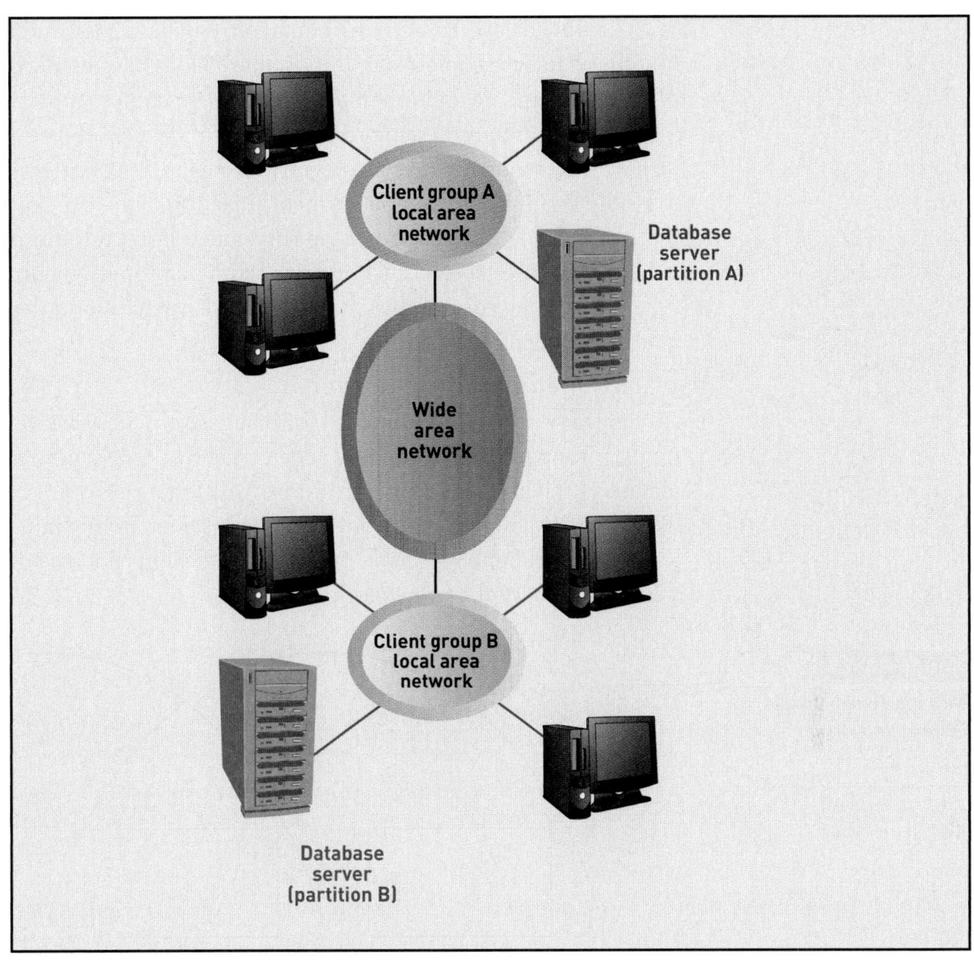

It is seldom possible to partition a database schema into mutually exclusive subsets. Some portions of a database are typically needed by most or all users, and those portions must exist in each partition. For example, data in the region of overlap in Figure 13-29 should be stored on each server with periodic synchronization. Thus, partitioning can reduce the problems associated with database synchronization, but it seldom eliminates them entirely.

Federated Database Servers

Some information systems are best served by a federated database architecture, as shown in Figure 13-31. This architecture is commonly used to access data stored in databases with incompatible storage models or DBMSs. A single unified database schema is created on a combined database server. That server acts as an intermediary between application programs and the databases residing on other servers. Database requests are first sent to the combined database server, which in turn makes appropriate requests of the underlying database servers. Results from multiple servers are combined and reformatted to fit the unified schema before the system returns a response to the client.

Federated database server architecture can be extremely complex. A number of DBMS products are available to implement such systems, but they are typically expensive and difficult to implement and maintain. Federated database architectures also tend to demand considerable computer hardware and network capacity, but their expense and management complexity are generally less than would be required to implement and maintain application programs that interact directly with all of the underlying databases.

data warehouse
a collection of data used to support structured and unstructured managerial decisions

A common use of a federated database server architecture is to implement a data warehouse. A **data warehouse** is a collection of data used to support structured and unstructured managerial decisions. Data warehouses typically draw their content from operational databases within an organization and multiple external databases (for example, economic and trade data from

Figure 13-31
A federated database server architecture

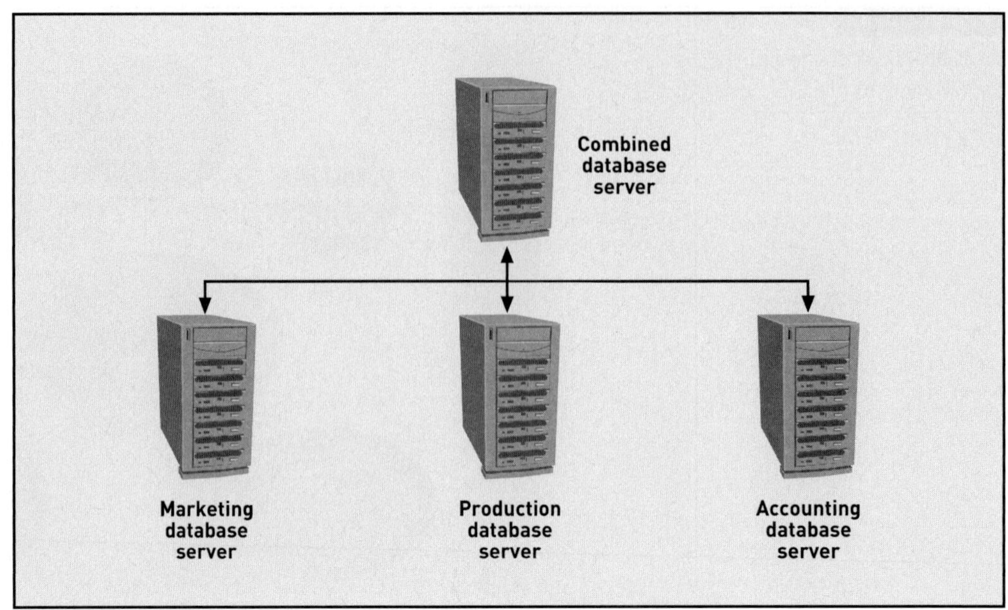

databases maintained by governments, trade industry associations, and private research organizations). Because data originates from a large number of incompatible databases, a federated architecture is typically the only feasible approach for implementing a data warehouse.

Now that we've discussed the basic issues underlying distributed database design, we will show how they come into play when making decisions for Rocky Mountain Outfitters' new customer support system.

RMO DISTRIBUTED DATABASE ARCHITECTURE

The starting point for designing a distributed database architecture is information about the data needs of geographically dispersed users. Some of this information for RMO was gathered as an analysis activity (see Figures 6-32, 6-33, and 6-34) and is summarized here:

- Warehouse staff (Portland, Salt Lake City, and Albuquerque) need to check inventory levels, query orders, record back orders and order fulfillment, and record order returns.
- The phone-order staff (Salt Lake City) need to check inventory levels; create, query, update, and delete orders; query customer account information; and query catalogs.
- The mail-order staff (Provo) need to check inventory levels, query orders, query catalog information, and update customer accounts.
- Customers (location not yet determined) need the same access capabilities as phone-order staff.
- Headquarters staff (Park City) need to query and adjust orders, query and adjust customer accounts, and create and query catalogs and promotions.

RMO has already decided to manage its database using the existing server cluster in the Park City data center. Thus, a WAN will be required to connect the server to LANs in the warehouses, phone-order center, mail-order center, headquarters, and data center. A connection will eventually be required for the Web servers used for direct customer ordering, although they probably will be located at an existing site (such as the data center).

A single-server architecture for RMO is shown in Figure 13-32. This architecture requires sufficient WAN capacity to carry database (and other) traffic from all locations. The primary advantage of this architecture is its simplicity. There are no partitions or database copies to manage, and only a single server must be maintained. The primary disadvantages are relatively high WAN capacity requirements and the susceptibility of the entire system to failure of the single server.

Figure 13-32

A single-server database architecture for RMO

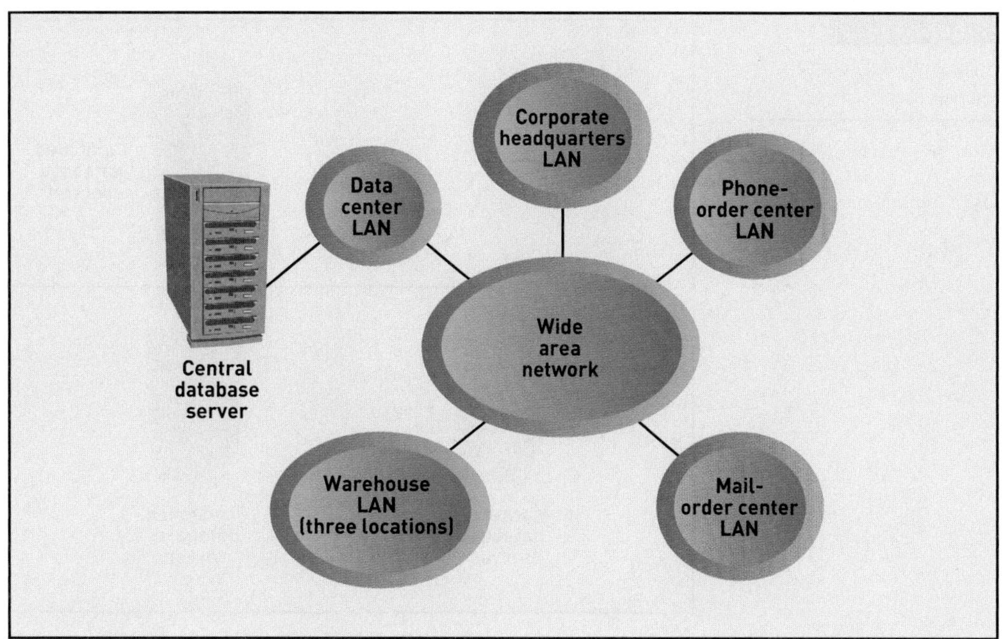

A more complex alternative is shown in Figure 13-33. Each remote location employs a combination of database partitioning and replication. A server at each warehouse stores a local copy of the order and inventory portions of the database. Servers in the phone- and mail-order centers store local copies of a larger subset of the database. Corporate headquarters relies on the central database server in the data center.

The primary advantages of this architecture are fault tolerance and reduced WAN capacity requirements. Each location could continue to operate if the central database server failed. However, as the remote locations continued to operate, their database contents would gradually drift out of synchronization. A synchronization strategy must be implemented to address both regular database updates and recovery from server failure. The strategy could vary by location.

Figure 13-33

A replicated and partitioned database server architecture for RMO

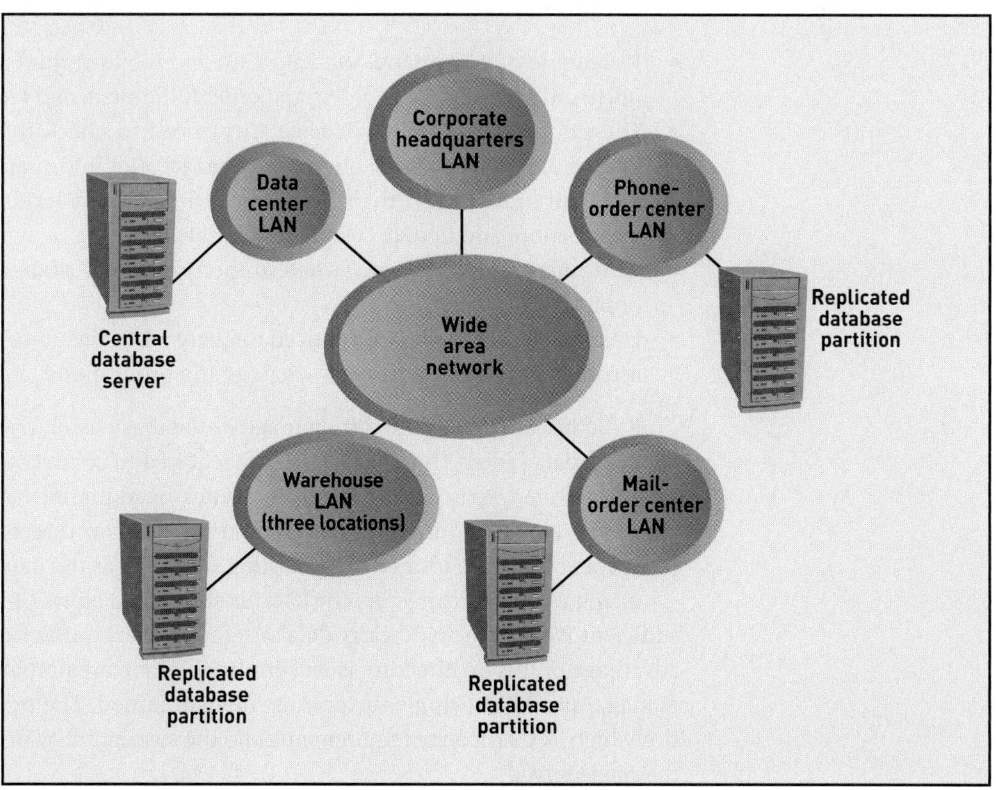

The primary disadvantages to the distributed architecture are cost and complexity. The architecture saves WAN costs through reduced capacity requirements but adds costs for additional database servers. The cost of acquiring, operating, and maintaining the additional servers would probably be much higher than the cost of adding greater WAN capacity.

So which alternative makes the most sense for RMO? The answer depends on some data that hasn't yet been gathered and on answers to some questions about desired system performance. RMO management must also determine its goals for system performance and reliability. The distributed architecture would provide higher performance and reliability but at substantially increased cost. Management must determine whether the extra cost is worth the expected benefits.

Additional data about network traffic is needed to precisely determine LAN and WAN communication requirements between clients and database servers. Estimates of transaction and query volume, including normal and peak demand, are required for each location. Such estimates may be gathered during analysis or design. The estimates are required to determine an optimal configuration of LAN, WAN, and database server capacity. The analysis of the estimates and the actual design of the networks and database architecture are complex endeavors that require highly specialized knowledge and experience. If possible, estimates should be tested and validated under realistic operating conditions. Barbara Halifax describes such a test in the accompanying RMO memo.

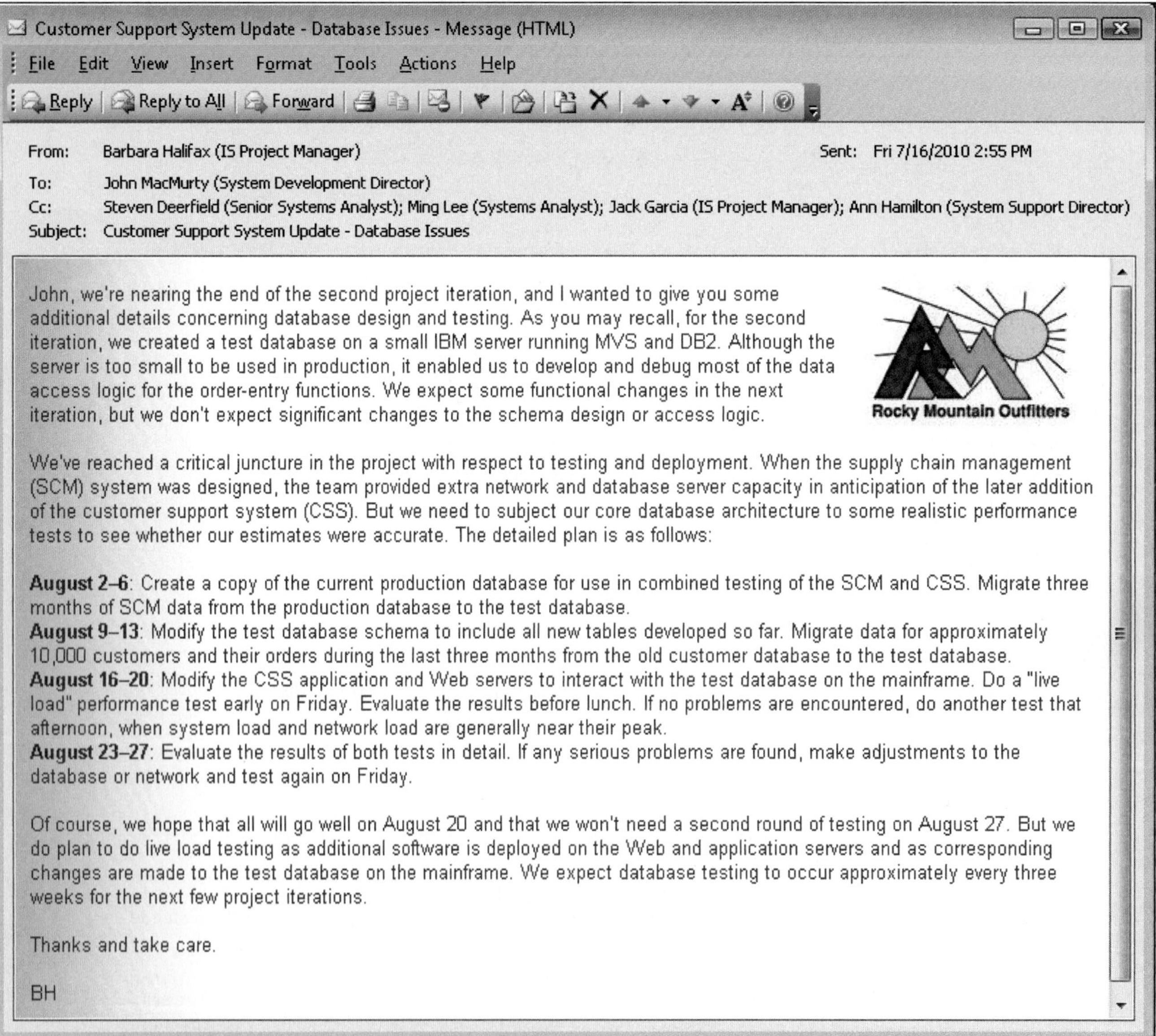
Customer Support System Update - Database Issues - Message (HTML)

File Edit View Insert Format Tools Actions Help

Reply | Reply to All | Forward

From: Barbara Halifax (IS Project Manager) Sent: Fri 7/16/2010 2:55 PM
To: John MacMurty (System Development Director)
Cc: Steven Deerfield (Senior Systems Analyst); Ming Lee (Systems Analyst); Jack Garcia (IS Project Manager); Ann Hamilton (System Support Director)
Subject: Customer Support System Update - Database Issues

John, we're nearing the end of the second project iteration, and I wanted to give you some additional details concerning database design and testing. As you may recall, for the second iteration, we created a test database on a small IBM server running MVS and DB2. Although the server is too small to be used in production, it enabled us to develop and debug most of the data access logic for the order-entry functions. We expect some functional changes in the next iteration, but we don't expect significant changes to the schema design or access logic.

We've reached a critical juncture in the project with respect to testing and deployment. When the supply chain management (SCM) system was designed, the team provided extra network and database server capacity in anticipation of the later addition of the customer support system (CSS). But we need to subject our core database architecture to some realistic performance tests to see whether our estimates were accurate. The detailed plan is as follows:

August 2–6: Create a copy of the current production database for use in combined testing of the SCM and CSS. Migrate three months of SCM data from the production database to the test database.
August 9–13: Modify the test database schema to include all new tables developed so far. Migrate data for approximately 10,000 customers and their orders during the last three months from the old customer database to the test database.
August 16–20: Modify the CSS application and Web servers to interact with the test database on the mainframe. Do a "live load" performance test early on Friday. Evaluate the results before lunch. If no problems are encountered, do another test that afternoon, when system load and network load are generally near their peak.
August 23–27: Evaluate the results of both tests in detail. If any serious problems are found, make adjustments to the database or network and test again on Friday.

Of course, we hope that all will go well on August 20 and that we won't need a second round of testing on August 27. But we do plan to do live load testing as additional software is deployed on the Web and application servers and as corresponding changes are made to the test database on the mainframe. We expect database testing to occur approximately every three weeks for the next few project iterations.

Thanks and take care.

BH

SUMMARY

Most modern information systems store data in a database and access and manage the data using a DBMS. Relational databases and DBMSs are most commonly used today, but object databases and DBMSs are increasing in popularity. One of the key activities of systems design is developing a relational or object database schema.

A relational database is a collection of data stored in tables. A relational database schema is normally developed from an entity-relationship diagram. Each entity is represented as a separate table. One-to-many relationships are represented by embedding foreign keys in entity tables. Many-to-many relationships are represented by creating additional tables containing foreign keys of the related entities.

An object database stores data as a collection of related objects. The design class diagram is the starting point for developing an object database schema. The database schema defines each class, and the ODBMS stores each object as an instance of a particular class. Each object is assigned a unique object identifier. Relationships among objects are represented by storing the object identifier of an object within related objects.

Objects can also be stored within a relational database. Object attributes and relationships among objects—including one-to-many, many-to-many, and generalization hierarchies—can be represented. However, an RDBMS cannot store methods and cannot directly represent inheritance.

Medium- and large-scale information systems typically use multiple databases or database servers in various geographic locations. Replicated database architecture employs multiple database copies on different servers, usually in different geographic locations. Partitioned database architecture employs partial database copies stored on different servers in proximity to a distinct user subset. Federated database architecture employs multiple databases (possibly of different types) and a special-purpose DBMS that provides a unified view of the databases and a single point of access.

KEY TERMS

complex data type, p. 515
database (DB), p. 488
database management system (DBMS), p. 488
database synchronization, p. 518
data type, p. 514
data warehouse, p. 520
field, p. 490
field value, p. 490
first normal form (1NF), p. 498
foreign key, p. 491
functional dependency, p. 498
hybrid object-relational DBMS, p. 510
key, p. 490
multivalued attribute, p. 508
navigation, p. 504
normalization, p. 498
object database management system (ODBMS), p. 503
Object Definition Language (ODL), p. 503
object identifier, p. 504
persistent class, p. 504
physical data store, p. 488
primary key, p. 491
primitive data type, p. 515
referential integrity, p. 496
relational database management system (RDBMS), p. 490
row, p. 490
schema, p. 488
second normal form (2NF), p. 498
table, p. 490
third normal form (3NF), p. 498
transient class, p. 503

REVIEW QUESTIONS

1. List the components of a DBMS and describe the function of each.
2. What is a database schema? What information does it contain?
3. Why have databases become the preferred method of storing data used by an information system?
4. List four different types of database models and DBMSs. Which are in common use today?

5. With respect to relational databases, briefly define the terms *row* and *field*.
6. What is a primary key? Are duplicate primary keys allowed? Why or why not?
7. What is the difference between a natural key and an invented key? Which type is most commonly used in business information processing?
8. What is a foreign key? Why are foreign keys used or required in a relational database? Are duplicate foreign key values allowed? Why or why not?
9. Describe the steps used to transform an ERD into a relational database schema.
10. How is an entity on an ERD represented in a relational database?
11. How is a one-to-many relationship on an ERD represented in a relational database?
12. How is a many-to-many relationship on an ERD represented in a relational database?
13. What is referential integrity? Describe how it is enforced when a new foreign key value is created, when a row containing a primary key is deleted, and when a primary key value is changed.
14. What types of data (or fields) should never be stored more than once in a relational database? What types of data (or fields) usually must be stored more than once in a relational database?
15. What is relational database normalization? Why is a database schema in third normal form considered to be of higher quality than an unnormalized database schema?
16. Describe the process of relational database normalization. Which normal forms rely on the definition of functional dependency?
17. Describe the steps used to transform a class diagram into an object database schema.
18. What is the difference between a persistent class and a transient class? Provide at least one example of each class type.
19. What is an object identifier? Why are object identifiers required in an object database?
20. How is a class on a class diagram represented in an object database?
21. How is a one-to-many relationship on a class diagram represented in an object database?
22. How is a many-to-many relationship without attributes represented in an object database?
23. What is an association class? How are association classes used to represent many-to-many relationships in an object database?
24. Describe the two ways in which a generalization relationship can be represented in an object database.
25. Does an object database require key fields or attributes? Why or why not?
26. Describe the similarities and differences between an ERD and a class diagram that models the same underlying reality.
27. How are classes and relationships on a class diagram represented in a relational database?
28. What is the difference between a primitive data type and a complex data type?
29. What are the advantages of having an RDBMS provide complex data types?
30. Does an ODBMS need to provide predefined complex data types? Why or why not?
31. Why might all or part of a database need to be replicated in multiple locations?
32. Briefly describe the following distributed database architectures: replicated database servers, partitioned database servers, and federated database servers. What are the comparative advantages of each?
33. What additional database management complexities are introduced when database contents are replicated in multiple locations?

THINKING CRITICALLY

1. The Universal Product Code (UPC) is a bar-coded number that uniquely identifies many products sold in the United States. For example, all copies of this textbook sold in the United States have the same UPC bar code on the back cover. Now consider how the design of the RMO database might change if all items sold by RMO were required by law to carry a permanently attached UPC (for example, on a label sewn into garments). How might the RMO relational database schema change under this requirement?
2. Assume that RMO plans to change its pricing policy. If two or more catalogs are in circulation at the same time, then all item prices in the catalogs must be the same. Prices can still rise or fall over time, and those changes will be recorded in the database and printed in newly issued catalogs. Any customer who makes an order will always be given the lower of the current price or the price in the current catalog. What changes to the tables shown in Figure 13-9 will be required to ensure that the RMO database is in 3NF after the pricing policy change?

3. Assume that RMO will begin asking a random sample of customers who order by telephone about purchases made from competitors. RMO will give customers a 15 percent discount on their current order in exchange for answering a few questions. To store and use this information, RMO will expand the ERD and class diagram with two new entities (classes) and three new relationships. The new entities (classes) are Competitor and ProductCategory. Competitor has a one-to-many relationship with ProductCategory, and the existing Customer entity (class) also has a one-to-many relationship with ProductCategory. Competitor has a single field (attribute) called Name. ProductCategory has four fields (attributes): Description, DollarAmountPurchased, MonthPurchased, and YearPurchased. Revise the relational database schema shown in Figure 13-9 to include the new entities and relationships. All tables must be in 3NF.
4. Assume that RMO is developing its database using object-oriented methods. Assume further that the database designers want to make some changes to the class diagram in Figure 13-15. Specifically, they want to make ProductItem an abstract parent class from which more specific product classes are specialized. Three specialized classes will be added: ClothingItem, EquipmentItem, and OtherItem. ClothingItem will add the attribute *color*, and that same attribute will be removed from the InventoryItem class. EquipmentItem will also add an attribute called *color* but will not have an attribute called *gender*. OtherItem will have both the *color* and *gender* attributes. Revise the relational database schema in Figure 13-24 to store the new ProductItem generalization hierarchy. Use a separate table for each of the specialized classes.
5. Assume that RMO will use a relational database, as shown in Figure 13-9. Assume further that a new catalog group located in Milan, Italy, will now create and maintain the catalog. To minimize networking costs, the catalog group will have a dedicated database server attached to its LAN. Develop a plan to partition the RMO database. Which tables should be replicated on the catalog group's local database server? Update Figure 13-33 to show the new distributed database architecture.
6. Revisit the issues raised in the Nationwide Books (NB) case at the beginning of the chapter. Should NB adopt an ODBMS for the new Web-based ordering system? Why or why not?

EXPERIENTIAL EXERCISES

1. This chapter did not discuss network databases in detail, but some database textbooks discuss them. Investigate the network database model and its use of pointers to represent relationships among record types. In what ways is the use of pointers in a network database similar to the use of object identifiers in an object database? Does the similarity imply that object databases are little more than a renamed version of an older DBMS technology?
2. Access the Object Database Management Group Web site (*www.odmg.org*) and gather information on the current status of the ODMG standard.
3. Investigate the student records management system at your school to determine what database management system is used. What database model is used by the DBMS? If the DBMS isn't object oriented, find out what plans, if any, are in place to migrate to an ODBMS. Why is the migration being planned (or not being planned)?
4. Visit the Web site of an online catalog vendor similar to RMO (such as *www.llbean.com*) or an online vendor of computers and related merchandise (such as *www.cdw.com*). Browse the online catalog and note the various types of information contained therein. Construct a list of complex data types that would be needed to store all of the online catalog information.

CASE STUDIES

REAL ESTATE MULTIPLE LISTING SERVICE SYSTEM

Refer to the description of the Real Estate Multiple Listing Service system in the Chapter 5 case studies. Using the ERD and domain model class diagram for that system as a starting point:

1. Develop a relational database schema in 3NF.
2. Develop an ODL database schema.

STATE PATROL TICKET PROCESSING SYSTEM

Refer to the description of the State Patrol ticket processing system in the Chapter 5 case studies. Using the ERD and domain model class diagram for that system as a starting point:

1. Develop a relational database schema in 3NF.
2. Develop an ODL database schema.

COMPUTER PUBLISHING, INCORPORATED

In only a decade, Computer Publishing, Incorporated (CPI) had grown from a small textbook publishing house into a large international company with significant market share in traditional textbooks, electronic books, and distance education courseware. CPI's processes for developing books and courseware were similar to those used by most other publishers, but those processes had proven cumbersome and slow in an era of rapid product cycles and multiple product formats.

Text and art were developed in a wide variety of electronic formats, and conversions among those formats were difficult and error-prone. Many editing steps were performed with traditional paper-and-pencil methods. Consistency errors within books and among books and related products were common. Developing or revising a book and all its related products typically took a year or more.

CPI's president initiated a strategic project to reengineer the way that CPI developed books and related products. CPI formed a strategic partnership with Davis Systems (DS) to develop software that would support the reengineered processes. DS had significant experience developing software to support product development in the chemical and pharmaceutical industries using the latest development tools and techniques, including object-oriented software and databases. CPI expected the new processes and software to reduce development time and cost. Both companies expected to license the software to other publishers within a few years.

A joint analysis team specified the workflows and high-level requirements for the software. The team developed plans for a large database that would hold all book and courseware content through all stages of production. Authors, editors, and other production staff would interact with the database in a variety of ways, including traditional word-processing programs and Web-based interfaces. When required, format conversions would be handled seamlessly and without error. All content creation and modification would be electronic—no text or art would ever be created or edited on paper, except as a printed book ready for sale.

Software would track and manage content through every stage of production. Content common to multiple products would be stored in the database only once. Dependencies within and across products would be tracked in the database. Software would ensure that any content addition or change would be reflected in all dependent content and products, regardless of the final product form. For example, a sentence in Chapter 2 that refers to a figure in Chapter 1 would be updated automatically if the figure were renumbered. If a new figure were added to a book, it would be added automatically to the related courseware presentation slides. Related courseware and study material on the Web site would automatically reflect changes to the answer to an end-of-chapter question.

1. Consider the contents of this textbook as a template for CPI's database content. Draw an ERD and class diagram that represents the book and its key content elements. Which diagram is a more accurate representation of book content? Expand your diagrams to include related product content such as a set of PowerPoint slides, an electronic book formatted as a Web site, and a Web-based test bank.
2. Develop a list of data types required to store the content of the book, slides, and Web sites. Are the relational DBMS data types listed in Figure 13-26 sufficient?
3. What features of an ODBMS, beyond or different from RDBMS features, might be useful when implementing CPI's database? Give examples of how they might be used.

RETHINKING ROCKY MOUNTAIN OUTFITTERS

The "Rethinking RMO" case in Chapter 5 asked you to consider additional things and relationships that would need to be modeled if RMO were to implement its own customer charge accounts. If you have not already done so, complete that exercise and update the ERD and class diagram accordingly, then complete the following tasks:

1. Update the RMO relational database design in Figure 13-9 based on the changes that you made to the ERD. Be sure that all your database tables are in 3NF.
2. Write ODL schema specifications for all new classes and relationships that you added to the class diagram.
3. Verify that the new classes and relationships are accurately represented in the updated relational database design that you developed for question 1.

FOCUSING ON RELIABLE PHARMACEUTICAL SERVICE

Use the ERD that you developed in Chapter 5 and the domain model class diagram that you developed in Chapter 7 to complete the following tasks:

1. Develop a relational database schema in 3NF.
2. Develop an ODL database schema.
3. Discuss the pros and cons of distributed database architecture. Which architectural approach (or combination of approaches) should Reliable employ in their new system after it is fully implemented?

FURTHER RESOURCES

The Object Database Management Group Web site, *www.odmg.org*.

Robert Orfali, Dan Harkey, and Jeri Edwards, *The Essential Client/Server Survival Guide* (3rd ed.). John Wiley & Sons, 1999.

Peter Rob and Carlos Coronel, *Database Systems: Design Implementation and Management* (7th ed.). Course Technology, 2007.

CHAPTER

14

DESIGNING THE USER INTERFACE

LEARNING OBJECTIVES

After reading this chapter, you should be able to:

- Describe the difference between user interfaces and system interfaces
- Explain why the user interface *is* the system to the users
- Discuss the importance of the three principles of user-centered design
- Describe the historical development of the field of human-computer interaction (HCI)
- Describe the three metaphors of human-computer interaction
- Discuss how visibility and affordance affect usability
- Apply the eight golden rules of dialog design when designing the user interface
- Define the overall system structure as a menu hierarchy
- Write user-computer interaction scenarios as dialogs
- Create storyboards to show the sequence of forms used in a dialog
- Design windows forms and browser forms that are used to implement a dialog
- List the key principles used in Web design

CHAPTER OUTLINE

Identifying and Classifying Inputs and Outputs

Understanding the User Interface

Guidelines for Designing User Interfaces

Documenting Dialog Designs

Guidelines for Designing Windows and Browser Forms

Guidelines for Designing Web Sites

Designing Dialogs for Rocky Mountain Outfitters

INTERFACE DESIGN AT AVIATION ELECTRONICS

Bob Crain was admiring the user interface for the manufacturing support system recently installed at Aviation Electronics (AE). Bob is the plant manager for AE's Midwest manufacturing facility, which is responsible for producing aviation devices used in commercial aircraft. These aviation devices provide guidance and control functions for flight crews, and they provide the latest safety and security features that pilots need when flying commercial aircraft.

The manufacturing support system is used for all facets of the manufacturing process, including product planning, purchasing, parts inventory, quality control, finished goods inventory, and distribution. Bob was involved extensively in the development of the system over a period of several years, including initial planning and development. The system reflected almost everything he knew about manufacturing. The information systems team that developed the system relied extensively on Bob's expertise. That was the easy part for Bob.

What particularly pleased Bob was the final user interface. Bob had insisted that the development team think "outside the box." He did not want just another cookie-cutter transaction processing system. He wanted a system that acted as a partner in the manufacturing process—with a look and feel that really fit the work the users were doing. After all, the facility produced devices whose major design goal was usability. Shouldn't the manufacturing support system be designed that way, too?

The first manager assigned to the project didn't want to discuss usability at all. "We'll add the user interface later, after we work out the accounting controls" was a typical comment. When Bob insisted that the project manager be replaced, the information systems department sent Sara Robinson to lead the project.

Sara had a completely different attitude; she started out asking about events that affect the manufacturing process and about cases in which users need support from the system. Although she had a team of analysts working on the accounting transaction details right from the beginning, she always focused on how the user would interact with the system. Bob and Sara conducted meetings to involve users in discussions about how they might use the system, even asking users to act out the roles of the user and the system in carrying on a conversation. That approach was outside the box.

At other meetings, Sara presented sketches of screens and asked users to draw on them, to indicate the information they wanted to see and options they wanted to be able to select. These sessions produced many ideas. For example, many users did not sit at their desks all day—they needed larger and more graphic displays they could see from across the room. Many users needed to refer to several displays, and they needed to be able to read them simultaneously. Several functions were best performed using graphical simulations of the manufacturing process. Users made sketches showing how the manufacturing process actually worked, and the team used these sketches later to define much of the interface. Sara and her team kept coming back every month or so with more examples to show, asking for more suggestions.

When the system was finally completed and installed, most users already knew how to use it because they had been so involved in its design. Bob knew everything the system could do, but he had his own uses for it. He sat at his desk and clicked the *review ongoing processes* button on the screen, and the manufacturing support system gave him his morning briefing.

OVERVIEW

Information systems capture inputs and produce outputs, and inputs and outputs occur where there are *interfaces* between the system and its environments. System interfaces handle inputs and outputs that require minimal human intervention. User interfaces handle inputs and outputs that involve a system user directly. This chapter differentiates between both types

of interfaces and then focuses on the design of user interfaces. Then Chapter 15 focuses on system interfaces, system outputs, and system controls.

One of the key systems design activities is to design the user interface for a system. Designing the user interface means designing the inputs and outputs involved when the user interacts with the computer to carry out a task. This chapter emphasizes the interaction between the user and the computer—called *human-computer interaction*, or HCI. For every input, a developer must consider the interaction between user and computer and design an interface to process the input. Similarly, for every output produced at the request of a user (an online report, for example), the developer must design the interaction. Because the interaction is much like a dialog between the user and the computer, user-interface design is often referred to as *dialog design*.

This chapter begins with discussion of the user interface by providing background on user-centered design, the development of the field of human-computer interaction, and several metaphors used to describe the user interface. Many guidelines are available to help ensure usability of the system, and some of the most important guidelines are discussed, including guidelines for Web-based development. Next, approaches to documenting dialog designs are presented. Examples are given throughout the chapter, including some dialog design examples for Rocky Mountain Outfitters that show Windows forms and Web pages. Remember, user-interface design is often completed using an iterative approach—addressing only a few use cases at a time. But it is important to establish an overall user-interface design concept early in the project so that the design of each dialog can be coordinated.

IDENTIFYING AND CLASSIFYING INPUTS AND OUTPUTS

Inputs and outputs of the system are an early concern of any system development project. The project plan lists key inputs and outputs that the analyst identified when defining the scope of the system. During the analysis phase, analysts also discussed inputs and outputs early and often with system stakeholders to identify external agents and actors that affect the system and that depend on information it produces. Requirements models produced during analysis also emphasize inputs and outputs. For example, the event table includes a trigger for each external event, and the triggers represent inputs. Outputs are shown as responses to external, state, and temporal events.

TRADITIONAL AND OO APPROACHES TO INPUTS AND OUTPUTS

In the traditional approach, inputs and outputs are shown as data flows on the context diagram, the data flow diagram (DFD) fragments, and the detailed DFDs. A data flow definition that lists all data elements describes each input and output in detail. During design, analysts add more detail about the data flows based on the choices they made when deciding on a design alternative. The question of whether an input is captured automatically or entered by a system user, for example, determines details about the design of the system. As discussed in Chapter 10, these details must be coordinated with the design of the application software.

In the object-oriented approach, inputs and outputs are defined by messages entering or leaving the system. Inputs and outputs are included in the event table as triggers and responses. Actors provide inputs for many use cases, and many use cases provide outputs to actors. The messages exchanged during a scenario define these inputs and outputs in more detail, and as the design of each scenario becomes more detailed, so does the specification of messages. They are reflected in interaction diagrams, in design class diagrams as methods, and in state machine diagrams. The system sequence diagram introduced in Chapter 7 first showed these inputs and outputs.

USER VERSUS SYSTEM INTERFACES

system interfaces
the parts of an information system involving inputs and outputs that require minimal human intervention

user interfaces
the parts of an information system requiring user interaction to create inputs and outputs

In both the traditional and object-oriented approaches, a key step in systems design is to classify the inputs and outputs for each event as either a system interface or a user interface. System interfaces involve inputs and outputs that require minimal human intervention. They might be inputs captured automatically by special input devices such as scanners, electronic messages from another system, or batch processing transactions compiled by another system. Many outputs are considered system interfaces if they primarily send messages or information to other systems or if they produce reports, statements, or documents for external agents or actors without much human intervention.

User interfaces involve inputs and outputs that more directly involve a system user. A user interface enables a user to interact with the computer to record a transaction, such as when a customer service representative records a phone order for an RMO customer. Sometimes outputs are produced after user interaction, such as the information displayed after a user query about the status of an order. In Web-based systems, a customer can interact directly with a system to request information, place an order, or look up the status of an order. At Rocky Mountain Outfitters, Barbara Halifax's regular status memo updates John MacMurty on some of the activities of user-interface design for the customer support system.

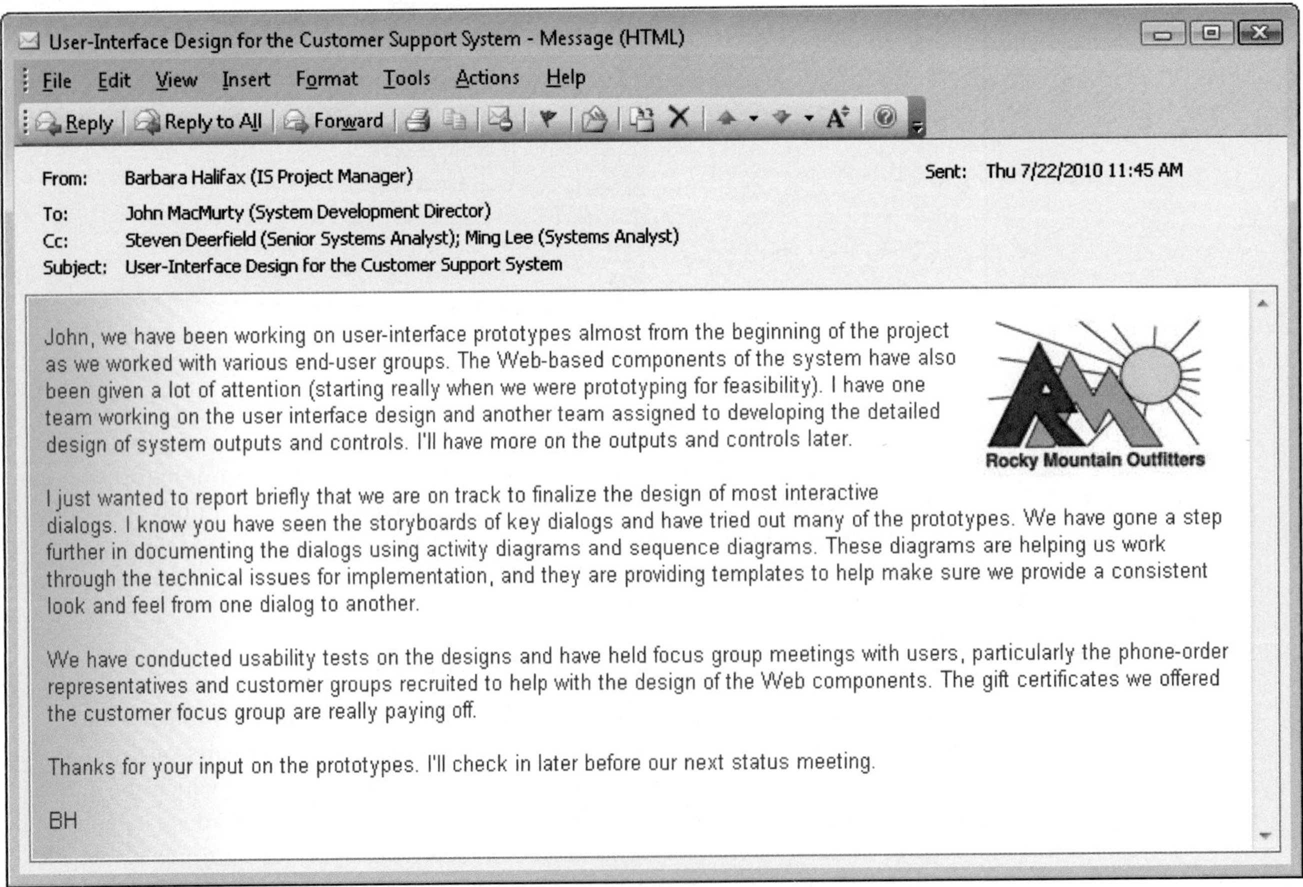

User-Interface Design for the Customer Support System - Message (HTML)
File Edit View Insert Format Tools Actions Help
Reply | Reply to All | Forward
From: Barbara Halifax (IS Project Manager)
Sent: Thu 7/22/2010 11:45 AM
To: John MacMurty (System Development Director)
Cc: Steven Deerfield (Senior Systems Analyst); Ming Lee (Systems Analyst)
Subject: User-Interface Design for the Customer Support System
John, we have been working on user-interface prototypes almost from the beginning of the project as we worked with various end-user groups. The Web-based components of the system have also been given a lot of attention (starting really when we were prototyping for feasibility). I have one team working on the user interface design and another team assigned to developing the detailed design of system outputs and controls. I'll have more on the outputs and controls later.
Rocky Mountain Outfitters
I just wanted to report briefly that we are on track to finalize the design of most interactive dialogs. I know you have seen the storyboards of key dialogs and have tried out many of the prototypes. We have gone a step further in documenting the dialogs using activity diagrams and sequence diagrams. These diagrams are helping us work through the technical issues for implementation, and they are providing templates to help make sure we provide a consistent look and feel from one dialog to another.
We have conducted usability tests on the designs and have held focus group meetings with users, particularly the phone-order representatives and customer groups recruited to help with the design of the Web components. The gift certificates we offered the customer focus group are really paying off.
Thanks for your input on the prototypes. I'll check in later before our next status meeting.
BH

In most system development projects, analysts separate design of system interfaces from design of user interfaces because they require different expertise and technology. But as with the design of any system component, considerable coordination is required. This chapter discusses user interfaces. The next chapter deals with system interfaces and system controls.

UNDERSTANDING THE USER INTERFACE

Many people think the user interface is developed and added to the system near the end of the development process. But the user interface of an interactive system is much more than that. The user interface is everything the end user comes in contact with while using the system—physically, perceptually, and conceptually (see Figure 14-1). To the end user of a system, the user interface *is* the system itself.

Figure 14-1

Physical, perceptual, and conceptual aspects of the user interface

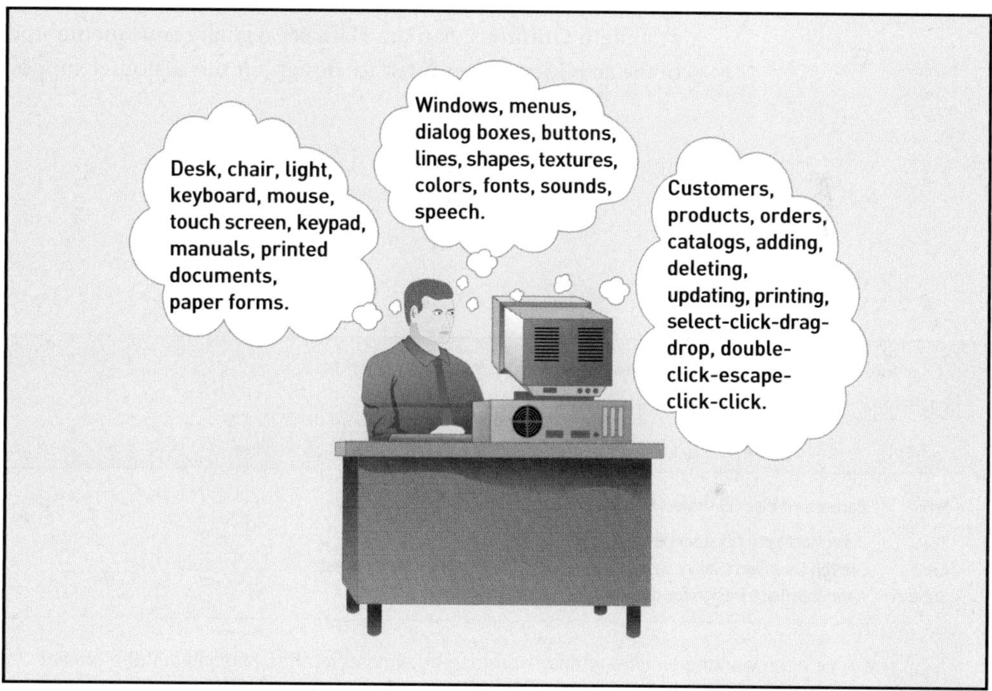

Many system developers, particularly those who work on highly interactive systems, echo this point of view in claiming that to design the user interface is to design the system. Therefore, consideration of the user interface should come very early in the development process. The term human-computer interaction (HCI) is generally used to refer to the study of end users and their interaction with computers.

human-computer interaction (HCI)
the study of end users and their interactions with computers

PHYSICAL ASPECTS OF THE USER INTERFACE

Physical aspects of the user interface include the devices the user actually touches, including the keyboard, mouse, touch screen, or keypad. But other physical parts of the interface include reference manuals, printed documents, and data-entry forms, which the end user works with while completing tasks at the computer. For example, a mail-order data-entry clerk at Rocky Mountain Outfitters works at a computer terminal but uses printed catalogs and handwritten order forms when entering orders into the system. The desk space, the documents, the available light, and the computer terminal hardware all make up the physical interface for this end user.

PERCEPTUAL ASPECTS OF THE USER INTERFACE

Perceptual aspects of the user interface include everything the end user sees, hears, or touches (beyond the physical devices). What the user sees includes all data and instructions displayed on the screen, including shapes, lines, numbers, and words. The user might rely on the sounds made by the system, even a simple beep or click that tells the user that the system recognizes a keystroke or selection. More recently, computer-generated speech makes it seem that the system is actually talking to the user, and with speech recognition software, the user can talk to the computer. The user "touches" objects such as menus, dialog boxes, and buttons on the screen using a mouse, but the user also touches objects such as documents, drawings, or records of transactions with a mouse when completing tasks.

CONCEPTUAL ASPECTS OF THE USER INTERFACE

Conceptual aspects of the user interface include everything the user knows about using the system, including all of the problem domain "things" in the system the user is manipulating, the operations that can be performed, and the procedures followed to carry out the operations. To use the system, the end user must know all about these details—not how the system is implemented internally, but what the system does and how to use it to complete tasks. This knowledge is referred to as the user's model of the system. Much of the user's model is a logical model of the system, as you learned in Chapters 5, 6, and 7. A logical model of the system requirements can be quite detailed, so the user must know quite a few details to operate the system. Recall also that a systems analyst relies on the end users to help define the requirements that the analyst captures in various models. The user's knowledge of the requirements for the system becomes the fundamental determinant of what the system is, and if the user's knowledge of the system is part of the interface, then the user interface must be much more than a component added near the end of the project.

user's model
what the user knows about using the system, including the problem domain "things" the user is manipulating, the operations that can be performed, and the procedures followed when carrying out tasks

BEST PRACTICE

Remember that to the user the user interface *is* the system itself.

USER-CENTERED DESIGN

Many researchers focus their attention on creating analysis and design techniques that place the user interface at the center of the development process because they recognize the importance of the user interface to system developers and system users. These techniques are often referred to collectively as user-centered design. User-centered design techniques emphasize three important principles:

user-centered design
a collection of techniques that place the user at the center of the development process

- Focus early on the users and their work.
- Evaluate designs to ensure usability.
- Use iterative development.

The early focus on users and their work is consistent with the approach to systems analysis in this text: Analysts must understand and identify the system users and their requirements for the system. The traditional approach to development focuses more on the requirements from the business point of view—what needs to be accomplished and what are the sources and destinations for data? The object-oriented approach, probably because most object-oriented systems are interactive, focuses more on users and their work by identifying actors, use cases, and scenarios followed when using the system. As discussed in Chapter 7, the automation boundary between the user and computer is defined very early during requirements modeling.

User-centered design goes much further in attempting to understand the users, however. What do they know? How do they learn? How do they prefer to work? What motivates them?

The amount of focus on users and their work varies with the type of system being developed. If the system is a shrink-wrapped desktop application marketed directly to end users, the focus on users and their preferences is intense.

usability
the degree to which a system is easy to learn and use

The second principle of user-centered design is to evaluate designs to ensure usability. **Usability** refers to the degree to which a system is easy to learn and use. Ensuring usability is not easy; there are many different types of users with differing preferences and skills to accommodate. Features that are easy to use for one person might be difficult for another. If the system has a variety of end users, how can the designer be sure that the interface will work well for all of them? If it is too flexible, for example, some end users might feel lost. On the other hand, if the interface is too rigid, some users will be frustrated.

But there is more to consider for ease of learning and ease of use. These concepts often conflict because an interface that is easy to learn is not always easy to use. For example, menu-based applications with multiple forms, many dialog boxes, and extensive prompts and instructions are easy to learn—indeed, they are self-explanatory. Easy-to-learn interfaces are appropriate for systems that end users use infrequently. But if office workers use the system all day, it is important to make the interface fast and flexible, with shortcuts, hot keys, and information-intensive screens. This second interface might be harder to learn, but it will be easier to use after it is learned. Office workers (with the support of their management) are willing to invest more time learning the system to become efficient users.

Developers employ many techniques to evaluate interface designs to ensure usability. User-centered design requires testing all aspects of the user interface. Some usability testing techniques collect objective data that can be statistically analyzed to compare designs. Some techniques collect subjective data about user perceptions and attitudes. To assess user attitudes, developers conduct formal surveys, focus group meetings, design walkthroughs, paper-and-pencil evaluations, expert evaluations, formal laboratory experiments, and informal observation.

The third principle of user-centered design is to use iterative development—doing some analysis, then some design, then some implementation, and then repeating the processes. After each iteration, the project team evaluates the work on the system to date. Iterative development keeps the focus on the user by continually returning to the user requirements during each iteration and by evaluating the system after each iteration. Iterative development is discussed throughout this text as applicable to both traditional and object-oriented approaches to development.

HUMAN-COMPUTER INTERACTION AS A FIELD OF STUDY

human factors engineering (ergonomics)
the study of human interaction with machines in general

User-interface design techniques and HCI as a field of study evolved from studies of human interaction with machines in general, referred to as **human factors engineering** or **ergonomics**. The formal study of human factors began during World War II, when aerospace engineers studied the effects on airplane pilots of rearranging controls in the cockpit. Pilots are responsible for controlling many devices as they fly, and the effectiveness of the interaction between the pilot and the devices is critical. If the pilot makes a mistake (that is, if he or she can't correctly use a device), the plane might crash. What the pilot does is the "human factor" that engineers realized was often beyond their control.

One story about the importance of the human factor involved a minor change to the design of the plane cockpit. The designers switched the locations of the throttle and the release handle for the ejection seat. The result was a dramatic increase in the number of unexplained pilot ejections. When under pressure, the pilots grabbed what they thought was the throttle and ejected themselves from the plane. Initially, designers dismissed the problem as the need for better training. But even with training, pilots under pressure continued to grab the wrong handle. It became apparent that the key to the "human factor" was to change the machine to accommodate the human rather than trying to change the human to accommodate the machine.

The field of human factors was first associated with engineering because engineers designed machines. But engineers, who are generally used to precise specifications and predictable behavior, often find the human factor frustrating. Gradually, specialists emerged who began to draw on many disciplines to understand people and their behavior. These disciplines include cognitive psychology, social psychology, linguistics, sociology, anthropology, and others, as shown in Figure 14-2. Information systems specialists with an interest in human-computer interaction study computers plus all of these disciplines.

Figure 14-2

The fields contributing to the study of HCI

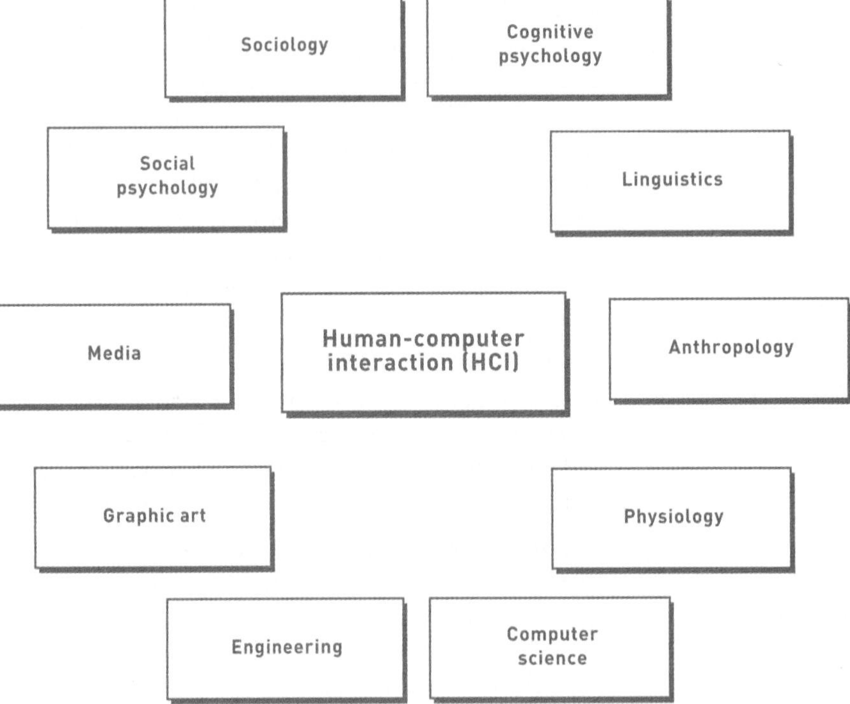

An important contribution to the development of the field of human-computer interaction began with the Xerox Corporation in the 1970s. Xerox produced high-speed photocopying machines that provided an increasing number of special options and capabilities that the human operator could specify. The designers of the photocopying machines recognized the importance of making the complex machines easy for the operators to learn and use. Xerox customers wanted minimal training time for their operators, and operator errors could be costly. For example, if a clerk began a large photocopying job but made a mistake in specifying details, it would be wasteful and delay the distribution of important documents. Therefore, Xerox emphasized the usability of its machines.

Xerox established a research and development laboratory, called the Xerox Palo Alto Research Center (Xerox PARC), to study issues that affect how humans operate machines. As a result of this investment, Xerox eventually offered photocopying machines with touch screen, menu-driven interfaces that displayed icons representing objects such as documents, stacks of paper, staples, and sorting bins.

Research and development at Xerox PARC also involved work on computers and object-oriented programming. The first pure object-oriented programming language, called Smalltalk, was created at Xerox PARC by Alan Kay and associates to facilitate the development of interactive user interfaces. In the early 1970s, Kay envisioned an advanced, portable personal computing platform (similar to today's ultralight notebook computers) called the Dynabook. Many researchers thought such a machine could not be built for three or four decades because the hardware required for the Dynabook was not available. Kay decided to work on the software that would run the machine in anticipation of the hardware, which led to Smalltalk.

Smalltalk includes classes that make up the key parts of windowing interfaces today—windows, menus, buttons, labels, text fields, and so forth. The design and programming philosophy used to describe and build these interfaces was developed along with the language—all 100 percent object oriented.

Because of the work at PARC, Xerox eventually developed and marketed one of the first general-purpose personal computers with a graphical user interface—the Xerox Star—in the late 1970s. Although it was ahead of its time and far too expensive, it is considered a landmark development in computing. Its key features were exploited in the early 1980s by a small company near Xerox PARC named Apple Computers. Apple first exploited the Xerox Star's features as the Apple Lisa and then as the Apple Macintosh. The work at Xerox PARC had a substantial impact on object-oriented programming, personal computers, and user-interface design.

Now that the object-oriented approach to system development is becoming more influential, user-interface design concepts and development techniques pioneered at labs such as Xerox PARC are becoming better integrated into system development methodologies used for business systems. The field of HCI has grown and now sponsors many academic journals, conferences, and book series devoted to research and practice. Undergraduate and graduate degree programs are also available to train HCI specialists.

METAPHORS FOR HUMAN-COMPUTER INTERACTION

There are many ways to think about human-computer interaction, including *metaphors* or *analogies*. Three alternatives are the direct manipulation metaphor, the document metaphor, and the dialog metaphor. Because each metaphor provides an analogy to a different concept, each has implications for the design of the user interface.

The Direct Manipulation Metaphor

direct manipulation
a metaphor of HCI in which the user interacts directly with objects on the display screen

Direct manipulation assumes that the user interacts with objects on the screen instead of typing commands on a command line. Objects that the user can interact with are made visible on the screen so the user can point at them and manipulate them with the mouse or arrow keys. The earliest direct manipulation interfaces were word processors that allowed users to type in words directly where desired in a document. By the early 1980s, electronic spreadsheet applications (first VisiCalc, then Lotus 1-2-3) became available for IBM DOS PCs. These applications used a direct manipulation approach—the user typed numbers, formulas, or text directly into cells on a spreadsheet. The spreadsheet on the screen was conceptually similar to a paper spreadsheet that was familiar to people working in accounting and finance. The familiarity and direct manipulation features made these applications easy to understand and natural to use, and end users could speed their work by including formulas to automatically do the calculations on the spreadsheets. These early direct manipulation DOS applications were an important reason for the success of the personal computer. Even though they did not have graphical user interfaces, they were very popular because they made interacting with a computer straightforward, natural, and useful.

The Smalltalk language developed at Xerox PARC extended direct manipulation to all objects on the screen. Some of these objects are interface objects such as buttons, check boxes, scroll bars, and slider controls, but other problem domain objects such as documents, schedules, file folders, and business records were also displayed as objects that the user could directly manipulate. For example, an interface might include a trash can object; to delete a document file, the user clicks on the document with the mouse and drags the document to the trash can. By directly manipulating the objects in this way, the user tells the computer to delete the document file.

desktop metaphor
a direct manipulation approach in which the display screen includes an arrangement of common objects found on a desk

Direct manipulation coupled with object-oriented programming eventually evolved into the **desktop metaphor**, in which the display screen includes an arrangement of common desktop objects—a notepad, a calendar, a calculator, and folders containing documents. Many desktops now also include a telephone, an answering machine, a CD player, and even a video

monitor. Interacting with any of these objects is similar to interacting with the real-world objects they represent. End users now expect all applications, including business information systems, to be as natural to work with as objects on the desktop. New larger displays and multiple display arrangements allow for many desktop applications to be arranged in front of the user, as shown in Figure 14-3.

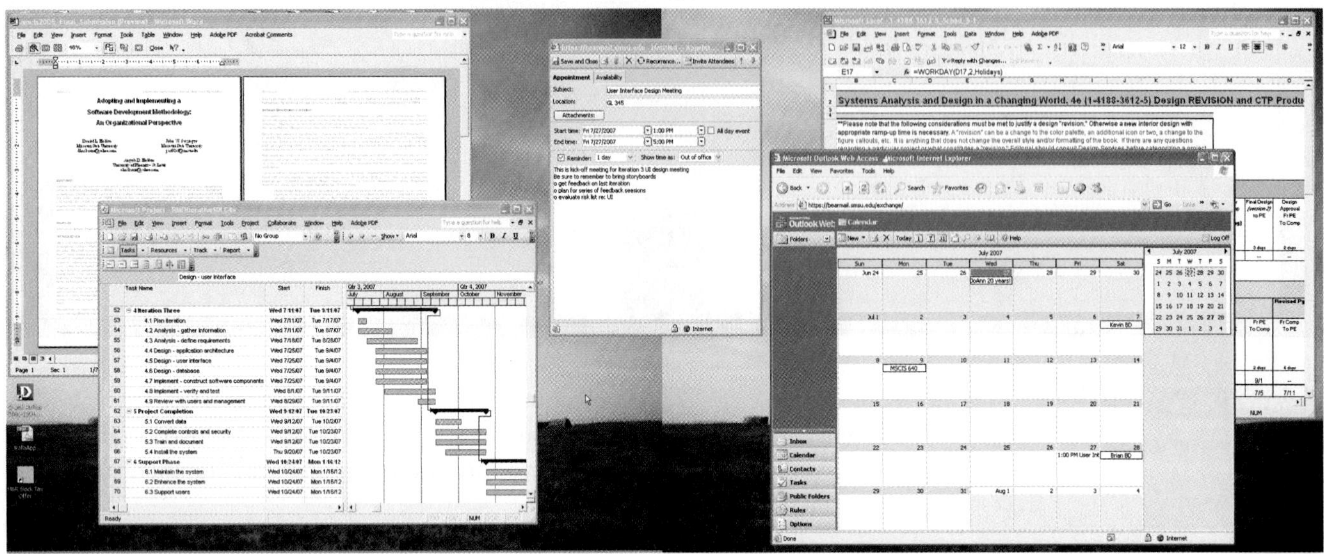

Figure 14-3

The desktop metaphor based on direct manipulation, shown on a double-wide display screen

document metaphor
a metaphor of HCI in which interaction with the computer involves browsing and entering data in electronic documents

hypertext
documents that allow the user to click on a link and jump to a different part of the document or to another document

hypermedia
technology that extends the hypertext concepts to include multimedia content such as graphics, video, and audio

dialog metaphor
a metaphor of HCI in which interacting with the computer is much like carrying on a conversation or dialog

The Document Metaphor

Another view of the interface is the **document metaphor**, in which interaction with the computer involves browsing and entering data in electronic documents. These documents are much like printed documents, but because the documents are electronic, they are more interactive. Electronic versions of documents can be organized differently from paper versions because the reader can jump around from place to place. **Hypertext** documents allow the user to click on a link and jump to a different part of the document or to another document entirely.

Most common desktop applications create and edit electronic documents, which are not limited to text and usually include word processing, spreadsheets, presentations, and graphics. All of these applications produce documents, but any one document can contain words, numbers, and graphics produced by any of these applications, making documents collections of all sorts of interrelated media. **Hypermedia** extends the hypertext concept to include multimedia content such as graphics, video, and audio that can be linked for navigation by the user in a document.

The World Wide Web is based on the document metaphor, because everything at a Web site is organized as pages that are linked as hypermedia (note that HTML means *Hypertext* Markup Language). A Web site processes transactions by selecting information in a Web page document. The document metaphor and the browser interface function as useful ways of describing and designing interactive systems, and they will continue to affect user interface designs. Figure 14-4 shows a wide display with two browsers, each containing a hypermedia document.

The Dialog Metaphor

The direct manipulation and document metaphors emphasize objects in the computer with which the user interacts. Another view of the interface is the **dialog metaphor**, in which interacting with the computer is much like carrying on a conversation or dialog. In fact, user-interface design is often referred to as *dialog design*. Carrying on a dialog, or conversation, with someone requires each person to listen to and respond to questions and comments from the other person, exchanging information in a sequence. The dialog metaphor is another way of thinking about human-computer interaction because the computer "listens to" and responds to user questions or comments, and the user "listens to" and responds to the computer's

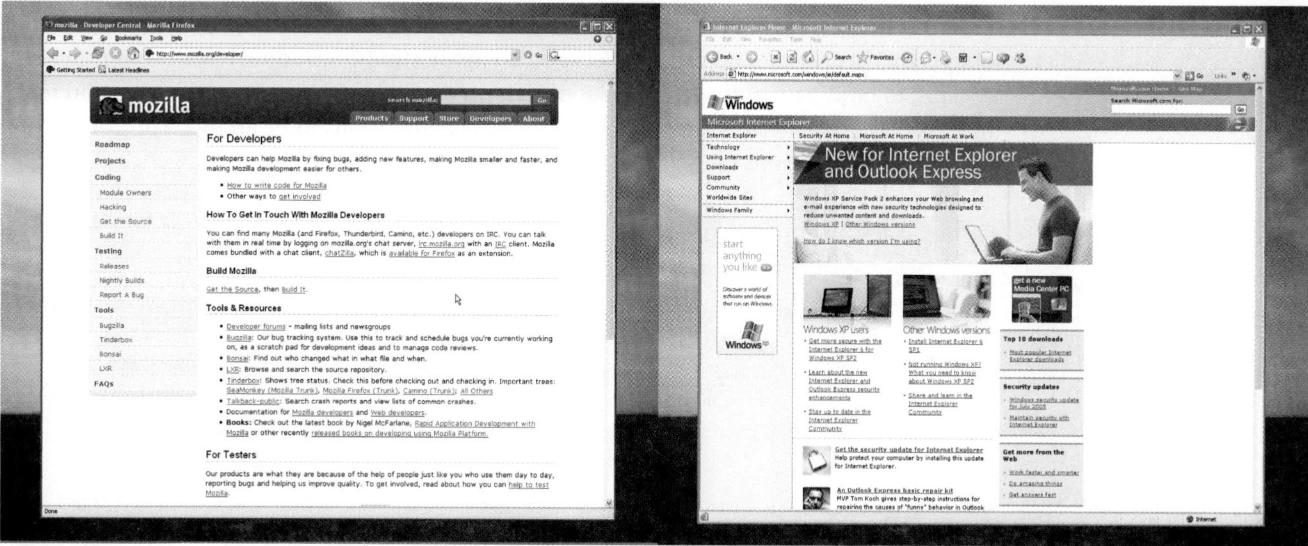

Figure 14-4

The document metaphor shown as hypermedia in Web browsers

questions and comments. Like the direct manipulation metaphor, the dialog metaphor is based on an object-oriented view of the system because communication involves messages from one "object" to another. Figure 14-5 shows how the user and the computer communicate by sending messages to each other.

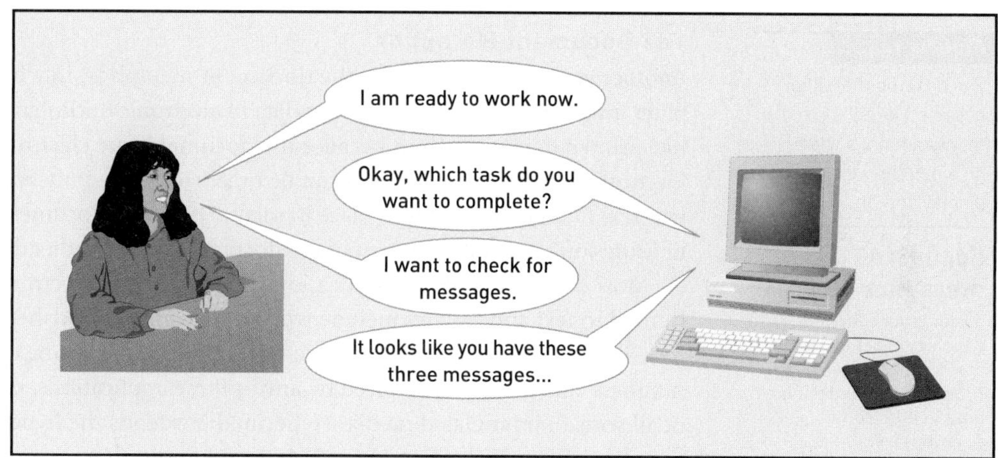

Figure 14-5

The dialog metaphor expresses the concept that the user and computer interact by sending messages

Consider the following dialog between a manager and an assistant:

Manager:	Did I get any messages while I was out?
Assistant:	Yes, you have three messages—from Bob, Mary, and Lim.
Manager:	What did Lim have to say?
Assistant:	Lim left a message at 8:15 last night regarding the meeting next Monday about the inventory management system. The message is, "Can we change the time for the meeting to 10:30? I'll be delayed by the testing session."
Manager:	I better respond. Say that the change is not a problem.
Assistant:	Okay, I'll leave him that message. Do you want the next message?
Manager:	What did Mary have to say?
Assistant:	She left a message at 8:15 this morning regarding lunch. She said, "Lunch is still on, but Joe will not be able to join us."
Manager:	Okay, no response. And that's all for now. Thanks.
Assistant:	Okay, you still have one message from Bob. I'll remind you later.

This dialog involves the manager and the assistant carrying on a conversation about messages. The questions asked by the manager and the responses and follow-up questions asked by the assistant seem clear and natural. Would the basic dialog be any different if this were an automated phone-answering service that responded to voice commands and replied in a computer-generated voice? Probably not. Would it be any different if this were a computer application simulating an intelligent "assistant"? Probably not. The basic dialog followed would be the same: a question, a response, another question, a response that might include a request for clarification, a response to the request for clarification, and a final response.

The basic dialog is also the same for a typical e-mail application, even though the user and computer send messages in different ways. The user selects a menu item for *read new mail*. The computer lists the new mail messages for the user to choose, the user chooses one message, then the computer displays the message. It might seem odd to think of interaction with an e-mail application as being similar to the dialog just presented, but the basic information exchanged and the sequence of actions are the same.

The user and the computer both send messages, but each is forced to use a different language because of limitations of both the user and the computer. The user cannot understand cryptic binary codes or plug in directly to the computer to interpret the electrical impulses the computer uses to represent the binary codes. The natural language of the computer just won't work for people. The computer has to adapt to the user and provide its messages in a form that is natural for the user—text and graphics that the user can see and read.

Similarly, the computer cannot understand complex voice messages, facial expressions, and body language that are the natural communication cues of the user, so the user has to adapt to the computer and provide messages by clicking the mouse, dragging objects, and typing words on the keyboard. Advances in computer technology are making it possible for the user to communicate in more natural ways, but the typical user interfaces today still rely on the mouse and keyboard. One reason is the need for silence and privacy in the office, so it is not clear whether voice commands will become common in computer applications.

The challenge of user-interface design is to construct a natural dialog sequence that allows the user and computer to exchange the messages required to carry out a task. Then the designer needs to develop the details of the language required for the user to send the messages to the computer (the user's language), plus the language needed for the computer to send messages to the user (the computer's language).

Figure 14-6 shows the earlier dialog between manager and assistant translated into the languages used by the user and the computer. Interface designers use a variety of informal diagrams and written narratives to model human-computer interaction. This is just one way the dialog design details can be modeled; you'll learn about additional techniques later in this chapter.

Figure 14-6

The user's language and the computer's language used to implement an e-mail application based on the natural dialog between manager and assistant

	Message	User's language	Computer's language
Manager	Did I get any messages while I was out?	Click the *read messages* menu item on the main menu.	
Assistant	Yes, you have three messages—from Bob, Mary, and Lim.		Look up new messages for the user and display a new message form with message headers listed in a list box.
Manager	What did Lim have to say?	Double-click the message from Lim in the list box.	
Assistant	Lim left a message at 8:15 last night regarding the meeting next Monday about the inventory management system. The message is, "Can we change the time for the meeting to 10:30? I'll be delayed by the testing session."		Look up the message body for the selected message and display it in message detail form.

(continued)

	Message	User's language	Computer's language
Manager	I better respond. Say that the change is not a problem.	Click the Reply button on the message detail form.	
			Display the new message form addressed to the sender.
		Type in the message, "Okay, that is not a problem."	
		Click the Send Button.	
Assistant	Okay, I'll leave him that message. Do you want the next message?		Display the Message Sent dialog box and redisplay the new messages form with message headers listed in the list box.
Manager	What did Mary have to say?	Double-click the message from Mary in the list box.	
Assistant	She left a message at 8:15 this morning regarding lunch. She said, "Lunch is still on, but Joe will not be able to join us."		Look up the message body for the selected message and display it in message detail form.
Manager	Okay, no response. And that's all for now. Thanks.	Click the *close message* button.	
			Redisplay a new message form with message headers listed in the list box.
		Click the *close new message form* button.	
Assistant	Okay, you still have one message from Bob. I'll remind you later.		Display the Closing Read New Mail dialog box, showing one unread message remaining.

Figure 14-6 cont.

GUIDELINES FOR DESIGNING USER INTERFACES

There are many published interface design guidelines to guide system developers. User-interface design guidelines range from general principles to very specific rules. This section describes some well-known guidelines for designing the user interface. Later, this chapter presents some of the guidelines and rules for designing windows forms and browser forms used with Web-based development. Some system development organizations adopt interface design standards—general principles and rules that an organization must follow when developing any system. Design standards help ensure that all user interfaces function well and that all systems developed by the organization have a similar look and feel.

interface design standards
general principles and rules that must be followed for the interface of any system developed by the organization

VISIBILITY AND AFFORDANCE

Donald Norman is a leading researcher in HCI who proposes two key principles to ensure good interaction between a person and a machine: visibility and affordance. These two principles apply to human-computer interaction just as they do for any other device.

visibility
a key principle of HCI that states all controls should be visible and provide feedback to indicate that the control is responding to the user's action

affordance
a key principle of HCI that states the appearance of any control should suggest its functionality

Visibility means that a control should be visible so users know it is available, and that the control should provide immediate feedback to indicate it is responding. For example, a steering wheel is visible to a driver, and when the driver turns it to the left, it is obvious that the wheel is responding to the driver's action. Similarly, a button that can be clicked by a user is visible, and when it is clicked, it changes to look as though it has been pressed to indicate it is responding. Some buttons make a clicking sound to provide feedback.

Affordance means that the appearance of any control should suggest its functionality—that is, the purpose for which the control is used. For example, a control that looks like a steering wheel suggests that the control is used for turning. On the computer, a button affords clicking, a scroll bar affords scrolling, and an item in a list affords selecting. Norman's principles apply to any objects on the desktop, such as those shown previously in the examples in Figures 14-3 and 14-4.

If user-interface designers make sure that all controls are visible and clear in what they do, the interface will be usable. Most users are familiar with the Windows interface and the common Windows controls. However, designers should be careful to apply these principles of visibility and affordance when designing Web pages. Many new types of controls are now possible at Web sites, but these controls are not always as visible and their effects are not always as obvious as they are in a standard Windows interface. More objects are clickable, but it is not always clear what is clickable, when a control has recognized the click, and what the click will accomplish. For example, sometimes you click on an image and a new page opens in the browser. Other times you click on an image and nothing happens.

EIGHT GOLDEN RULES

Ben Shneiderman, another leading researcher in HCI, proposes eight underlying principles that are applicable in most interactive systems (see "Further Resources" at the end of the chapter for Shneiderman's text). Although they are general guidelines rather than specific rules, he names them "golden rules" to indicate that they are the key to usability (see Figure 14-7).

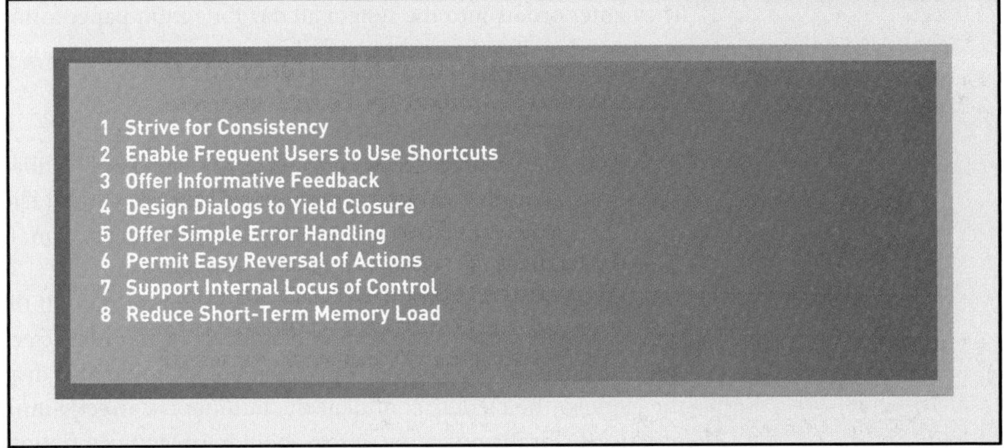

1 Strive for Consistency
2 Enable Frequent Users to Use Shortcuts
3 Offer Informative Feedback
4 Design Dialogs to Yield Closure
5 Offer Simple Error Handling
6 Permit Easy Reversal of Actions
7 Support Internal Locus of Control
8 Reduce Short-Term Memory Load

Figure 14-7
The eight golden rules for designing interactive interfaces

Strive for Consistency

Designing a consistent-appearing and -functioning interface is one of the most important design goals. The way that information is arranged on forms, the names and arrangement of menu items, the size and shape of icons, and the sequence followed to carry out tasks should be consistent throughout the system. Why? People are creatures of habit. After we learn one way of doing things, it is difficult to change. When we operate a computer application, many of our actions become automatic—we do not think about what we are doing. People who can touch-type do not have to think about each key press—their fingers just respond automatically. Consider what would happen to touch-typists if rows two and three on the keyboard were reversed. They would not be able to use the keyboard (and certainly wouldn't like it). If a new application comes along that has a different way of functioning, productivity suffers and users will not be happy.

The Apple Macintosh first emphasized the benefits of consistency in the 1980s. Apple provided applications for the Macintosh that set the standard for developers to follow when creating new applications. If new applications were consistent with these applications, Apple claimed that learning them would be easy. Apple also published a standards document to explain how to be consistent with the Macintosh interface. Similar examples and standards documents followed for the Microsoft Windows interface.

Business information systems are different from desktop applications originally produced for the Macintosh. Sometimes an application needs to be inconsistent with the original guidelines. For example, the original standards specified that every application include menus on the menu bar for File, Edit, and Format. All document-oriented applications—such as word

processors, spreadsheets, and graphics—need those menus. But many business systems do not have File, Edit, and Format functions. Most other guidelines and standards do apply, however.

Research has also shown that inconsistent interfaces sometimes are beneficial. If the user is interacting with multiple applications in separate windows, a different visual appearance may help the user differentiate them. In addition, when the user is learning several applications in one session, some differences in the interfaces may help the user remember which application is which. Inconsistencies introduced for these reasons should be subtle and superficial. The basic operation of the applications should be the same.

Enable Frequent Users to Use Shortcuts

Users who work with one application all day long are willing to invest the time to learn shortcuts. They rapidly lose patience with long menu sequences and multiple dialog boxes when they know exactly what they want to do. Therefore, shortcut keys reduce the number of interactions for a given task. Also, designers should provide macro facilities for users to create their own shortcuts.

Sometimes the entire interface should be designed for frequent users who do not need much flexibility. Consider the mail-order data-entry clerks for Rocky Mountain Outfitters. They enter orders into the system all day long from paper forms mailed by customers. These users need an interface that is simple, fast, and accurate. Long dialogs, multiple menus, and multiple forms would slow these users down.

Offer Informative Feedback

Every action a user takes should result in some type of feedback from the computer so the user knows that the action was recognized. Even keyboard clicks help the user, so an electronic "click" is included deliberately by the operating system. If the user clicks a button, the button should visually change and perhaps make a sound.

Feedback of information to the user is also important. If the Rocky Mountain Outfitters mail-order clerk enters a customer ID number in an order screen, the system should look up the customer to validate the ID number, but it should also display the name and address to the clerk so the clerk is confident the number is correct. Similarly, when the clerk enters a product ID for the order, the system should display a description of the product. As the clerk's attention shifts back and forth from the mail-order form to the computer screen, he or she compares the name and product description from the system with the information on the form to confirm that everything is correct. This sense of confirmation and the resulting confidence in the system are very important to users, particularly when they work with a system all day. But the system should not slow the user down by displaying too many dialog boxes to which the user must respond.

Sometimes feedback is provided to help the user in other ways. The phone-order representative at Rocky Mountain Outfitters needs information from the system just as the mail-order clerk does, but he or she also needs additional information. Phone customers may ask questions, so the information provided as feedback for the phone-order representative is more detailed and flexible. We discuss some designs for the phone-order representative at RMO later in this chapter.

Design Dialogs to Yield Closure

Each dialog with the system should be organized with a clear sequence—a beginning, middle, and end. Any well-defined task has a beginning, middle, and end, so users' tasks on the computer should also feel this way. If the user is thinking, "I want to check my messages," as in the earlier manager and assistant dialog example, the dialog begins with a request, exchanges information, and then ends. The user can get lost if it is not clear when a task starts and ends. In addition, the user often focuses intently on a task, so when it is confirmed that the task is complete, the user can clear his or her mind and get ready to focus on the next task.

If the system requirements are defined initially as events to which the system responds, each event leads to processing of one specific, well-defined activity. In the traditional, structured approach, each activity is defined by data flow diagrams and structured English. With the

object-oriented approach, each use case might be further defined as multiple scenarios, each with a flow of steps. Each scenario is a well-defined interaction; therefore, event decomposition sets the stage for dialogs with closure in both the traditional approach and the object-oriented approach.

Offer Simple Error Handling

User errors are costly, both in the time needed to correct them and in the resulting mistakes. If the wrong items are sent to a customer at Rocky Mountain Outfitters, it is a costly error. So, the systems designer must prevent the user from making errors whenever possible. A chief way to do this is to limit available options and allow the user to choose from valid options at any point in the dialog. Adequate feedback, as discussed previously, also helps reduce errors.

If an error does occur, the system needs mechanisms for handling it. The validation techniques discussed in Chapter 15 are useful for catching errors, but the system must also help the user correct the error. When the system does find an error, the error message should state specifically what is wrong and explain how to correct it. Error messages should not be judgmental. It is not appropriate to blame the user or make the user feel inadequate.

The system also should make it easy to correct the error. For example, if the user typed in an invalid customer ID, the system should tell the user that and then place the insertion point in the customer ID text box with the previously typed number displayed and ready to edit. This way, the user can see the mistake and edit it rather than having to retype the entire ID. Consider the following error message that occurs after a user has typed in a full screen of information about a new customer:

```
The customer information entered is not valid. Try again.
```

This message does not explain what is wrong or what to do next. Further, after this message appears, what if the system cleared the data-entry form and redisplayed it? The user would have to reenter everything previously typed, yet still have no idea what is wrong. The error message did not explain it, and now that the typed data has been cleared, the user cannot tell what might have been wrong. A better error message would say:

```
The date of birth entered is not valid. Check to be sure only numeric
characters in appropriate ranges are entered in the date of birth
field...
```

The input form should be redisplayed with all fields still filled in, and the insertion point should be placed at the field with invalid data, ready for the user to edit the field.

Permit Easy Reversal of Actions

Users need to feel that they can explore options and take actions that can be canceled or reversed without difficulty. This is one way that users learn about the system—by experimenting. It is also a way to prevent errors; as users recognize they have made a mistake, they cancel the action. In the game of checkers, a move is not final until the player takes his or her fingers off the game piece; it should be the same when a user drags an object on the screen. In addition, designers should be sure to include cancel buttons on all dialog boxes and allow users to go back one step at any time. Finally, when the user deletes something substantial—a file, a record, or a transaction—the system should ask the user to confirm the action.

Support Internal Locus of Control

Experienced users want to feel that they are in charge of the system and that the system responds to their commands. They should not be forced to do anything or made to feel as if the system is controlling them. Systems should make users feel that they are deciding what to do. Designers can provide much of this comfort and control through the wording of prompts and messages. Writing out a dialog like the manager and assistant message dialog given previously will lead to a design that conveys the feeling of control.

Reduce Short-Term Memory Load

People have many limitations, and short-term memory is one of the biggest. As discussed earlier in this book, people can remember only about seven chunks of information at a time. The interface designer cannot assume that the user will remember anything from form to form, or dialog box to dialog box, during an interaction with the system. If the user has to stop and ask, "Now what was the filename? The customer ID? The product description?", then the design places too much of a burden on the user's memory.

With these eight golden rules in mind, an interface designer can help ensure that user interactions are efficient and effective. We now turn to some basic techniques for documenting the design of the dialog.

DOCUMENTING DIALOG DESIGNS

Many techniques are available to help the designer think through and document dialog designs. The dialogs that must be designed are based on the inputs and outputs requiring user interaction, as discussed earlier. They are used to define a menu hierarchy that allows the user to navigate to each dialog. Storyboards, prototypes, and UML diagrams can be used to complete the designs.

USE CASES, SUBSYSTEMS, AND THE MENU HIERARCHY

Inputs and outputs are obtained from data flow diagrams (in the traditional approach) or use cases and scenarios (in the object-oriented approach). Generally, each input *obtained interactively from a user* requires a dialog design. In addition, each output produced *at the request of a user* requires a dialog design. So, each dialog is based on a use case documented early during the analysis process that is classified as requiring a user interface rather than a system interface.

Dialog design must be done simultaneously with other design activities. As shown in Chapter 10, the structure charts for subsystems (transaction analysis) include details about menu structure of the interactive parts of the system. In addition, the structure chart for each activity or use case (transform analysis of each DFD fragment) also includes details about the dialog with the user. The object-oriented approach also integrates dialog design very early, even during analysis tasks. Use case descriptions, activity diagrams, and system sequence diagrams (SSDs) include details about the dialog. Remember that menu design and dialog design are not done in isolation.

The available menus reflect the overall system structure from the standpoint of the user. Each menu contains a hierarchy of options, and they are often arranged by subsystem or by actions on objects. Rocky Mountain Outfitters' customer support system includes the order-entry subsystem, order fulfillment subsystem, customer maintenance subsystem, and catalog maintenance subsystem, as well as a reporting subsystem added during design. Menus might also be arranged based on objects—customers, orders, inventory, and shipments. Each menu might include duplicate functions, such as *Look up past orders*, under customers and under inventory.

Sometimes several versions of the menus are needed based on the type of user. For example, mail-order clerks at RMO do not need many of the options available—they process new orders only. The phone-order sales representatives need many more options, but they still do not need all system functions. And some options should be available only to managers, such as management reports and price adjustments.

Menus should also include options that are *not* activities or use cases from the event list—most important are options related to the system controls, which are discussed in Chapter 15. These include backup and recovery of databases in some cases, plus user account maintenance. In addition, user preferences are usually provided to allow the user to tailor the interface. Finally, menus should always include help facilities.

All use cases that lead to dialogs in the RMO customer support system are listed and grouped by subsystem in Figure 14-8. These groupings form one set of menu hierarchies. In addition, there are menu hierarchies for utilities, preferences, and help. The list in the figure is only one of many possible menu hierarchy designs—a starting place.

Figure 14-8

One overall menu hierarchy design for the RMO customer support system (not all users will have all of these options available)

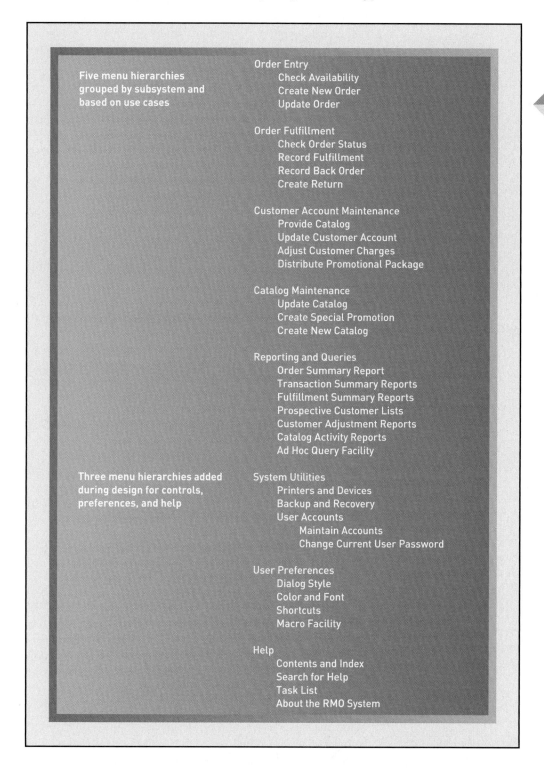

Five menu hierarchies grouped by subsystem and based on use cases

- Order Entry
  - Check Availability
  - Create New Order
  - Update Order
- Order Fulfillment
  - Check Order Status
  - Record Fulfillment
  - Record Back Order
  - Create Return
- Customer Account Maintenance
  - Provide Catalog
  - Update Customer Account
  - Adjust Customer Charges
  - Distribute Promotional Package
- Catalog Maintenance
  - Update Catalog
  - Create Special Promotion
  - Create New Catalog
- Reporting and Queries
  - Order Summary Report
  - Transaction Summary Reports
  - Fulfillment Summary Reports
  - Prospective Customer Lists
  - Customer Adjustment Reports
  - Catalog Activity Reports
  - Ad Hoc Query Facility

Three menu hierarchies added during design for controls, preferences, and help

- System Utilities
  - Printers and Devices
  - Backup and Recovery
  - User Accounts
    - Maintain Accounts
    - Change Current User Password
- User Preferences
  - Dialog Style
  - Color and Font
  - Shortcuts
  - Macro Facility
- Help
  - Contents and Index
  - Search for Help
  - Task List
  - About the RMO System

A dialog design is created for each of these menu options. After completing the dialog designs for all options, the designer can then redefine the structure of the menus for different types of users. Menu hierarchies can be rearranged easily as the design evolves.

DIALOGS AND STORYBOARDS

After identifying all required dialogs, the designers must document the dialogs. Many options exist; there are no *de facto* standards. One approach is to list the key steps followed for the dialog with a written description of what the user and computer do at each step. The format for writing these steps can follow the activity diagram described in Chapter 4 or the use case description format in Chapter 7, or they can be more free-form.

Designers can also document dialog designs by writing out a dialog like the one presented earlier in this chapter (for the manager and assistant), then documenting how the user and computer express themselves, as shown previously in Figure 14-6. Sometimes the designer can go right from the dialog to sketching a series of screens that make the design more visible. Consider the DownTown Videos rental system described in a case study at the end of Chapter 7. One use case is *Rent videos to customer*, and the dialog that goes on between the clerk and the system would follow this pattern:

System:	What DownTown Videos option do you want?
User:	I need to rent some videos.
System:	Okay, what is the customer's ID number for this rental?
User:	It is 201564.
System:	Fine, your customer is Nancy Wells at 1122 Silicon Avenue. Note that she has two kids, Bob and Sally, who can only rent PG-13 videos. Correct?
User:	Yes.
System:	What's the ID number of the first video she wants to rent?
User:	It is 5312.
System:	Fine, that is *Titanic* (the 1953 version with Barbara Stanwyck) with a rental charge of $1.00 for three days, rated PG. What's the ID number of the next video, or is that all?
User:	It is 8645.
System:	Fine, that is *Titanic* (the 2007 reissue version) with a rental charge of $3.00 for three days, rated PG-13. What's the ID number of the next video, or is that all?
User:	That's all.
System:	Okay, collect $4.00 plus $.24 tax, for a total of $4.24, and let me know when you are done.
User:	Yes, I'm done.
System:	Remind Nancy Wells that these videos are due back Wednesday, the seventh, before 11 p.m. closing.
User:	Okay. Thanks.

The designer thinks through the sequence of screens the user would need to carry on this dialog with the computer. The user needs to enter a customer ID and several video IDs. The computer would have to look up and display information on the customer and each video rented. Then the computer would calculate the total due and remind the user of the due date so the user can pass that information on to the customer.

storyboarding
a technique used to document dialog designs by showing a sequence of sketches of the display screen

One technique used to show the screens is called **storyboarding**—showing a sequence of sketches of the display screen during a dialog. The sketches do not have to be very detailed to show the basic design concept. Designers can implement a storyboard with a visual programming tool such as Visual Basic, but using simple sketches drawn with a graphics package can help keep the focus on the fundamental design ideas.

BEST PRACTICE

Use storyboarding to define user-interface requirements for each use case early in the project. Showing users storyboards gets feedback on important aspects of their work.

Figure 14-9 shows the storyboard for the *rent videos* dialog. The system has a menu hierarchy based on the event list plus needed controls, preferences, and help. The dialog uses one form and a few dialog boxes and adds more information to the form as the dialog progresses. Note that the prompt area at the bottom of the form displays the questions the computer asks, matching almost identically the phrases used in the written dialog. The user has a choice of either scanning or typing the few IDs that must be entered. Information provided to the user is shown in labels on the form. The information provided allows the user to confirm the identity of the customer, to see any restrictions that might apply, and to pass on to the customer any information about cost and return dates. In other words, the system helps the user do a better job of interacting with the customer by confirming information, providing feedback, and providing closure.

These approaches to dialog design provide only a framework to work from, and the resulting design remains fairly general. As working prototypes are produced, many details still have to be worked out. As the design progresses, reviewing the golden rules and other guidelines will help you keep the focus on usability.

DIALOG DOCUMENTATION WITH UML DIAGRAMS

The object-oriented approach provides UML diagrams that are useful for modeling user-computer dialogs. Use case descriptions, shown in Chapter 5, include a list of steps followed as the user and system interact. Activity diagrams, shown in Chapters 4 and 7, also document the dialog between user and computer for a use case. Both can be used to provide models of the user-computer interaction required in each dialog. In the object-oriented approach, objects send messages back and forth, listening to and responding to each other in sequence. People also send messages to objects and receive messages back from them. The system sequence diagram (SSD) described in Chapter 7 includes an actor (a user) sending messages to the system and the system returning information in the form of messages, shown in sequence. It basically shows a dialog between the user and the system. The SSD is based on the sequence of steps included in the use case description, so the dialog design for the use case begins very early and is refined continually.

The object-oriented approach involves adding more types of objects to class diagrams and interaction diagrams as the project moves from analysis to design, as discussed in Chapter 12. The additional classes of objects are packaged into three layers that contain user-interface classes, problem domain classes, and data access classes. Designers add user-interface classes and objects to these diagrams to show more detail about the design of the dialog between the user and computer. This design process was demonstrated in Chapter 12. The first step is to determine what window or Web forms are required for the dialog based on informal dialog design techniques described previously. Next, the sequence diagram for the scenario is expanded to show the user (an actor) interacting with the forms. The designer can then use a class diagram to model the user-interface classes that make up the forms. Finally, the sequence diagram can be further expanded to show the user interacting with specific objects that make up the form.

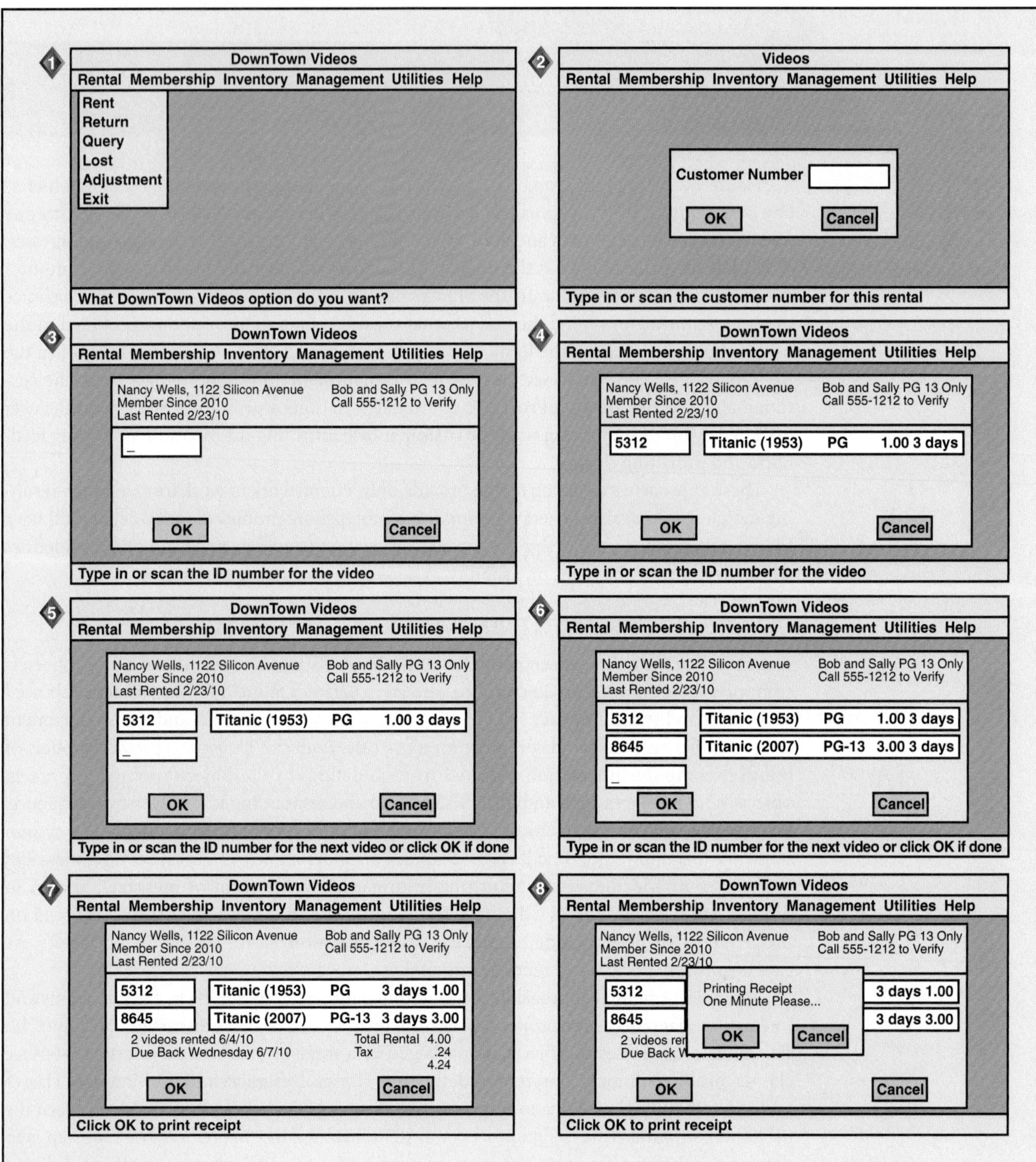

Figure 14-9

Storyboard for the DownTown Videos *rent videos* dialog

GUIDELINES FOR DESIGNING WINDOWS AND BROWSER FORMS

As with the previous activities of user-interface design, analysts must take care in designing the forms that users see on the screen. Each dialog might require several windows forms, and each form must be designed for usability. Almost all of the new business systems today are developed for an interactive Microsoft Windows, X-Windows (UNIX), or Macintosh environment. The underlying principles are the same for forms in any of these environments. In this section, when we refer to forms or windows, we mean any of these three environments. Within each windows environment, however, we need to consider two types of forms: windows and browsers.

browser forms

forms programmed using HTML and script languages that follow Internet conventions

Windows forms are programmed in a full-featured programming language, such as Visual Basic, C++, or Java. Because of this, windows forms have the advantage of being extremely flexible and capable of accessing data directly on a workstation. Browser forms, on the other hand, are programmed using HTML and script languages such as VBScript or JavaScript. Browser forms can be displayed using any Internet browser, which makes them accessible on a variety of platforms. Browser forms produced by Visual Studio .NET are now called *Web forms*, and they now provide the same design flexibility as windows forms. In addition, server-side processing using Active Server Pages (ASP) or Java servlets can add functionality. The advantage of browser forms is that the same forms can be used for both internal staff on company intranets and customers and suppliers on the Internet. As a result, many firms are designing user interfaces for their new systems as browser forms.

After identifying the objective of a form and its associated data fields, the system developer can construct the form using one of the many prototyping tools available. Earlier, developers spent tremendous amounts of time laying out a form on paper diagrams before beginning programming. Today, however, the process is streamlined with prototyping tools. Not only can developers design the content of the form, but they can design the look and feel of the form at the same time. Prototyping tools also permit users to be heavily involved in the development, and such involvement ensures that the user interface is in fact the *user's* interface, providing a strong sense of ownership and acceptance.

Categories of forms include input forms, input/output forms, and output forms. Input forms are used to record a transaction or enter data, although some portions of the form may display information from the system. The form used in the DownTown Videos storyboard in Figure 14-9 is an example of an input form. Input/output forms are generally used to update existing data. These forms display information about a single entity, such as a customer, and enable users to type over existing information to update it. Output forms are primarily for displaying information. The design of output forms is based on the same principles as report design, discussed later in Chapter 15. Input and input/output forms are closely related and are designed using similar principles. Before developing a form, the designer should carefully analyze the integrity controls required for data input, which are discussed in Chapter 15.

There are four major issues to consider in the design of these forms:

- Form layout and formatting
- Data keying and entry
- Navigation and support controls
- Help support

FORM LAYOUT AND FORMATTING

Form layout and formatting are concerned with the general look and feel of the form. You might have encountered systems with hard-to-use input forms—the font was too small, the labels were hard to understand, the colors were abrasive, the navigation buttons were not obvious, and so forth. These deficiencies can occur on both windows forms and browser forms. In contrast, forms that are easy to use are well laid out, with the fields easily identified and understood. One of the best methods to ensure that forms are well laid out is to prototype various alternatives and let users test them. Users will let you know which characteristics are helpful and which are distracting. As you design your input forms, you should think about the following:

- Consistency
- Headings, labels, and logos
- Distribution and order of data-entry fields and buttons
- Font sizes, highlighting, and colors

Consistency belongs at the top of the list because of its importance for ease of learning and use, as discussed previously. Some large systems require multiple input forms and several teams of programmers and analysts to develop them. Sometimes those teams don't coordinate their efforts, resulting in inconsistencies, and a system that is not consistent across all forms can be error-prone and difficult to use. To avoid these problems, all of the forms within a system need to have the same look and feel. A consistent use of function keys, shortcuts, control buttons, and even color and layout makes a system much more useful and professional looking. Cascading style sheets help designers control the consistency of Web forms. Design templates help designers control the consistency of windows forms. For example, with Microsoft Visual Studio .NET, a template form can be designed that is used as the superclass of all forms in the project.

The headings, labels, and logos on the form help to convey the purpose and use of the form. A clear, descriptive title at the top of the form helps to minimize confusion about a form's use. Labels should also be easy to identify and read.

The designer also should carefully place the data-entry fields around the form. Related fields are usually placed next to each other and can even be isolated with a fine-lined box. The designer also should carefully consider the tab order. If input is coming from a paper form, the tab order should follow the order of the paper form—left to right, top to bottom. Blank space should be used throughout the form so that the fields do not appear crammed together and are easy to distinguish and read. Normally navigation buttons are at the bottom of the window. *De facto* traditions for the placement of buttons are developing based on standards of the large development firms such as Apple, Microsoft, Sun, Oracle, and others. It is a good idea to be sensitive to these traditions as they change with technology upgrades.

The purpose of font size, highlighting, and color is to make the form easy to read. A careful mix of large and small fonts, bold and normal type, and different-colored fonts or background can help a user find important or critical information on the form. Too much variation makes the form cluttered and difficult to use. However, judicious use of these techniques makes the form more easily understood. Column headings and totals can be made slightly larger or boldfaced. The form can highlight negative or credit balances by changing font color. However, font color and background color should be used in concert to ensure that the field is readable. For example, neither red type on green or black backgrounds nor a dark color on a dark background is a good choice—some people with colorblindness cannot distinguish red from green or black.

Figure 14-10 is an example of a windows form designed for the Rocky Mountain Outfitters customer support system. This form is used to look up information about a product and to add it to an order. Notice how the title and labels make the form easy to read. The natural flow of the form is top to bottom, with related fields placed together. Navigation and close buttons are easily found but are not in the way of data-entry activity.

DATA KEYING AND DATA ENTRY

The heart of any input form is the entry of the new data. Even here, however, a primary objective is to require as little data entry as possible. Any information already in the computer, or that can be generated by the computer, should not be reentered. A generous use of selection lists, check boxes, automatic retrieval of descriptive fields, and so forth will speed up data entry and reduce errors. The Product Detail form for RMO (see Figure 14-10) shows many examples that reduce the need to enter data.

text box
an input control that accepts keyboard data entry

list box
an input control that contains a list of acceptable entries the user can select

spin box
a variation of the list box that presents multiple entries in a text box from which the user can select

combo box
another variation of the list box that permits the user to enter a new value or select from the entries

Several types of data-entry controls are widely used in windows systems today. A **text box** is the most common element used for data entry. A text box consists of a rectangular box that accepts keyboard data. In most cases, it is a good idea to add a descriptive label to identify what should be typed in the text box. Text boxes can be designed to limit the entry to a specified length on a single line or to permit scrolling with multiple lines of data.

Variations of a text box consist of a list box, a spin box, and a combo box. A **list box** contains a list of the acceptable entries for the box. The list usually consists of a predefined list of data values, and the user selects one from the list. The list can be presented either within a rectangular box or as a drop-down list. A variation of a list box is a spin box. A **spin box** presents the possible values within the text box itself. Two spinner arrows let the user scroll through all the values. A **combo box** also contains a predefined list of acceptable entries but permits the user to enter a new value when the list does not contain the desired value. Both a list box and combo box facilitate data entry by minimizing keystrokes and the corresponding possibility of errors.

Figure 14-10
The RMO Product Detail form used to look up information about a product, select size and color, and then add the product to an order

radio buttons (option buttons)
input controls that enable the user to select one option from a group

check boxes
input controls that enable the user to select more than one option from a group

Two types of input controls are used in groups: radio buttons (sometimes called *option buttons*) and check boxes. **Radio buttons** are associated as a group, and the user selects one and only one of the group. The system automatically turns off all other buttons in the group when one is selected. Because all of the possible values appear on the form, this control is used only when the list of alternatives is small and the values never change. **Check boxes** also work together as a group. However, check boxes permit the user to select as many values as desired within the group.

Browser forms contain similar controls. A major difference between a standard windows input form and a browser input form is that the windows form can easily perform edits field by field as the data is entered. In a basic browser input form, the edits are not performed until the entire form is transmitted to the server computer. However, as browser programs have become increasingly more sophisticated, more and more capabilities are being provided for data entry. Windows input forms and browser input forms now have very similar capabilities.

NAVIGATION AND SUPPORT CONTROLS

Standard window interfaces provide several controls for navigation and window manipulation. For Microsoft applications, these controls consist of the Minimize, Maximize, and Close buttons in the upper-right corner, horizontal and vertical scroll bars, record selection bar on the left panel, record navigation arrows at the bottom of the window, and so forth. To maintain consistency across systems, it is generally a good idea to utilize these navigation controls when possible. A well-designed user interface, however, should also include other controls or buttons. You can place buttons on the form to enable users to move to other relevant screens, to search and find data, and to close the open window. Browser forms also provide navigation and support controls. Each page might also include its own navigation buttons.

HELP SUPPORT

A primary objective in the design of each input form is that it should be intuitive so that users will not need help. However, even well-designed forms will be misunderstood, and access to online help is always recommended. Three types of help are common in today's systems: a tutorial that walks you through the use of the form, an indexed list of help topics, and the context-sensitive help.

Most systems provide tutorial help to assist in training new users. Tutorials can be organized by task, in which case the tutorial generally includes one dialog with a set of related forms. Every new system also should have an indexed list of help topics. This list can be invoked either through a keyword search or, as with many Microsoft systems, with a help wizard. The help wizard is simply a program that does an automatic keyword search based on words found in a question or sentence. The wizard returns several alternative help topics based on the results of the keyword searches.

Context-sensitive help can be based on the indexed list of help, but it is invoked differently. Context-sensitive help automatically displays the appropriate help topic based on the location of the insertion point. In other words, if the insertion point is within a certain text field on a form, and the user invokes context-sensitive help, the system displays the help for that text field.

GUIDELINES FOR DESIGNING WEB SITES

Web site design draws from the guidelines and rules for designing the windows forms and browser forms just presented. Many business systems today, including the RMO customer support system, make use of both technologies. Yet a business system such as the order-processing function for RMO is just part of the RMO Web site. Web sites also are used for corporate communication, customer information and service, online sales and distribution, and marketing. Because they are available 24 hours a day, 7 days a week, they need to interact seamlessly with customers. This section introduces some guidelines and lessons for Web design. A complete discussion of Web site design principles is beyond the scope of this book. Many excellent books are available, and we list some of them in the "Further Resources" at the end of this chapter.

TEN GOOD DEEDS IN WEB DESIGN

Jacob Nielsen is an HCI researcher who now focuses specifically on Web design. Like many useful guidelines, Nielsen's guidelines focus on general issues, including these "Ten Good Deeds in Web Design."

1. Place the organization's *name and logo* on every page and make the logo a link to the home page.
2. Provide a *search* function if the site has more than 100 pages.
3. Write straightforward and simple *headlines and page titles* that clearly explain what the page is about and that will make sense when read out of context in a search engine listing.
4. Structure the page to *facilitate reader scanning* and help users ignore large chunks of the page in a single glance. For example, use grouping and subheadings to break a long list into several smaller units.
5. Instead of cramming everything about a product or topic into a single, huge page, use *hypertext to structure the content space* into a starting page that provides an overview and several secondary pages that each focus on a specific topic.
6. Use *product photos*, but avoid cluttered and bloated product family pages with lots of photos. The primary product page must load quickly and function fast, so it should be limited to a thumbnail product shot.
7. Use *relevance-enhanced image reduction* when preparing small photos and images. Instead of simply reducing the original image to a tiny and unreadable thumbnail, zoom in on its most relevant detail by cropping and resizing the image.
8. Use *link titles* to provide users with a preview of where each link will take them, *before* they have clicked on it.
9. Ensure that all important pages are *accessible for users with disabilities*, especially visually impaired users.
10. *Do the same as everybody else*, because if most big Web sites do something in a certain way, users will expect other sites to work similarly.

WEB SITE DESIGN PRINCIPLES

Because Web sites include so many facets, many designers take a broader view of Web site design principles. A Web design book by Joel Sklar suggests that the designer focus on three broad aspects of Web design: (1) designing for the computer medium, (2) designing the whole site, and (3) designing for the user.

Designing for the Computer Medium

It is important to remember that the Web site will be displayed on a computer screen and not on paper. Designers can select from a wide array of video display fonts, colors, and layouts, but the look of the site should flow from its function and the organization's goals. Hypermedia allows the user to navigate through the site in nonlinear ways, so the designer should take advantage of new ways to organize information. Five guidelines to consider include:

- Craft the look and feel of the pages to take advantage of the medium.
- Make the design portable because it will be accessed with a wide range of technology.
- Design for low bandwidth because users will not want to wait for a page to load.
- Plan for clear presentation and easy access to information to ease users' navigation through the site.
- Reformat information for online presentation when it comes from other sources.

Designing the Whole Site

The entire site must have unifying themes and a structure, and the theme should reflect the impression the organization wants to convey. For example, a site for adult, business-oriented users should use subdued colors, familiar business-oriented fonts, and structured linear columns. A site for children should combine bright colors, an open and friendly dynamic structure, and simple, appealing graphics. Four guidelines to consider include the following:

- Craft the look and feel of the pages to match the impression desired by the organization.
- Create smooth transitions between Web pages so users are clear about where they have been and where they are going.
- Lay out each page using a grid pattern to provide visual structure for related groups of information.
- Leave a reasonable amount of white space on each page between groups of information.

Designing for the User

We discussed user-centered design previously in this chapter, and it is important to focus Web design efforts on the users and their needs. If a feature will annoy or distract users, do not include it. It is sometimes difficult to know who the Web users will be, but if the purpose and objectives of the whole site are defined carefully, the designer can make better judgments. Some guidelines to consider include the following:

- Design for interaction because Web users expect sites to be interactive and dynamic.
- Guide the user's eye to information on the page that is the most important.
- Keep a flat hierarchy so that the user does not have to drill down too deeply to find detailed information.
- Use the power of hypertext linking to help users move around and through the site.
- Decide how much content per page is enough based on the characteristics of the typical user; don't clutter the pages.
- Design for accessibility for a diverse group of users, including those with disabilities.

DESIGNING DIALOGS FOR ROCKY MOUNTAIN OUTFITTERS

Now that dialog design concepts and techniques have been discussed, we can demonstrate the process of designing one specific dialog for Rocky Mountain Outfitters, as well as part of the RMO Web site.

DIALOG DESIGN FOR THE RMO PHONE-ORDER REPRESENTATIVES

The Rocky Mountain Outfitters customer support system includes support for the phone-order sales representatives who process orders for customers. The dialog corresponds to the use case *Create new order* and more specifically to the scenario *Phone-order representative creates new order*. The target environment for this part of the system is the phone-order representative's desktop PC on a Windows platform.

The designer starts by referring to the models produced during analysis: either the data flow diagram fragment and corresponding detailed DFD for the activity or the UML sequence diagram for this scenario, depending on the approach used for analysis and design. The four basic steps followed in these models are as follows:

1. Record customer information.
2. Create a new order.
3. Record transaction details.
4. Produce order confirmation.

With the traditional approach to development, the designer would produce a structure chart, as shown in Chapter 10, to correspond to these steps. With the object-oriented approach, the system sequence diagram for this scenario would be expanded to include forms the user will need for interaction, as shown in Chapter 12. With either approach, the dialog design activity coordinates the user-interface design with the processing design for the activity or use case.

Based on the sequence of processing required, a basic dialog can be written to convey in more detail how the dialog will flow from the user's perspective. The details of the interface objects needed on each form should wait until this dialog is refined. The process of a new customer placing an order with a phone sales representative (the user) might resemble the following dialog:

Computer:	What customer support system option do you want?
User:	I need to create a new order.
Computer:	Okay, is it a new or existing customer?
User:	It's a new customer.
Computer:	Fine, give me the customer's name, address, phone, and so on.
User:	The customer is Ginny Dekker, 11980 Visual Blvd. . . .
Computer:	Okay, what is the first item she wants to order?
User:	Boots, Women's, Hiking Supreme Line in the Spring Fling catalog.
Computer:	They come in these sizes and colors, and we have all in stock except size 9.
User:	Okay, one pair of size 8 in tan.
Computer:	That comes to $65.50 plus tax, shipping, and handling. Anything else?
User:	Yes, Raincoat, Woman's, On the Run Line in the Spring Fling catalog.
Computer:	They come in these sizes and colors, and we have all in stock except medium and large blue.
User:	Okay, one coat size small in brown.
Computer:	That comes to $87.95 plus tax, shipping, and handling. Anything else?
User:	That's it.
Computer:	How does the customer want the order shipped?
User:	UPS two-day air.
Computer:	Does the customer want to use a credit card, or is there another payment approach for this order?
User:	Use the MasterCard number xx674-22-xxxx expiring January 2011.
Computer:	Okay, the order is recorded. To summarize, for customer Ginny Dekker, ship by UPS two-day mail Boots, Women's, Hiking Supreme Line, size 8, tan, at $65.50, and Raincoat, Woman's, On the Run Line, size small, brown, at $87.95. Total cost, $153.45 plus $9.20 tax and $13.40 shipping and handling—$176.05 charged to MasterCard xx674-22-xxxx.
User:	Thanks.

While working on this dialog, the designer can begin to refine the forms that will be required for the user and the computer. A list of required forms might include the following:

- Main menu form
- Customer form
- Item search form
- Product detail form
- Order summary form
- Shipping and payment options form
- Order confirmation form

The designer can use the list of forms to define a design concept for the flow of interaction from form to form. One approach is to show potential forms in sequence, as shown in Figure 14-11. After the main menu form, the customer form appears first; the user fills it in or updates it, and then the item search form displays to let the user search for an item. Product details are shown for the item, and then the item search form is shown again. When all items are selected, the order summary form is shown, and so on. The designer should concentrate on highlighting parts of the dialog that occur through each form rather than worrying about the physical design of each form. After considering what information is needed on each form, the designer can create a more detailed storyboard or implement prototype forms using a tool such as Visual Basic.

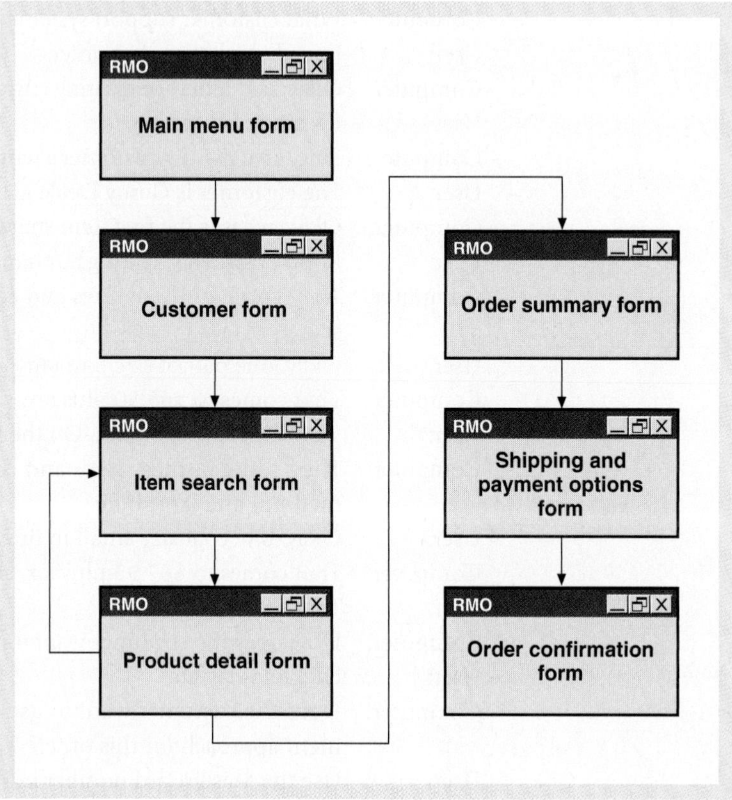

Figure 14-11 A design concept for the sequential approach to the *Create new order* dialog

This initial design is very sequential but reasonable for a first iteration. Phone sales representatives at RMO evaluated the storyboard and the prototype, and they suggested that the sequence was too rigid; it assumes the dialog always follows the same sequence. However, because phone-order representatives are on the phone with customers, they have to follow the customers' lead. Sometimes customers do not want to give information about themselves until the order is processed and confirmed, for example. Sometimes customers want to know the totals for the order or to review details about something already included in the order. But the sequential design assumes that the customer information always comes first. Although the sequential approach might work well for mail-order clerks, more flexibility is required for the phone-order representatives.

BEST PRACTICE

When storyboarding the user interface for each use case, you will find that you can reuse forms from one use case to another. Begin a list of shared forms early in the project. Reusing forms saves design and implementation effort, and it provides the additional benefit of consistency.

To address the flexibility and information needs of the users, the project team developed a second design concept. This design concept makes the order form the center of the dialog, with options to switch to other forms at will. After each action, the order form is redisplayed to the phone-order representative with the current order details. The order-centered design concept is shown in Figure 14-12. It allows the same sequence to be followed as the basic dialog, but it also allows flexibility when needed. It also shows the user more information about the order throughout the dialog in case the user needs the information.

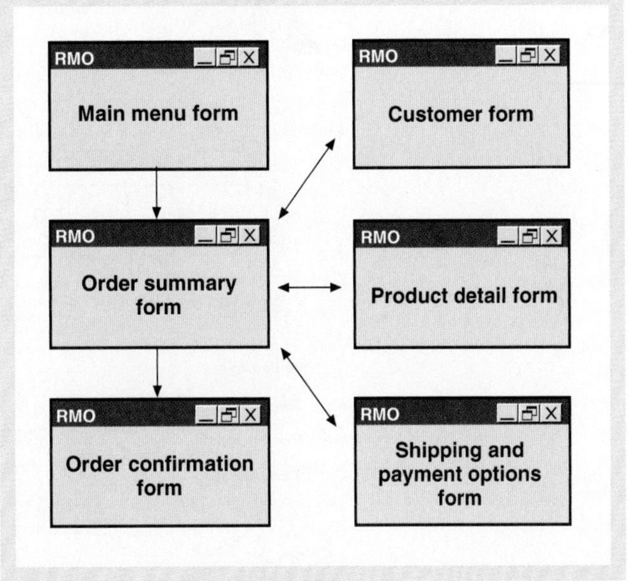

Figure 14-12
A design concept for an order-centered approach to the *Create new order* dialog

After adopting the order-centered concept, the project team designed the detailed forms. Some of the forms are shown in Figure 14-13. After the user selects *Place an order* from the main menu, the system displays the Order Summary form with a new order number assigned. The user can add the customer information immediately (either by searching for a previous customer based on customer number or name or by adding new customer information). The user can then search for a requested item and look up more details about the item on the Product Detail form. If the customer wants to order the item, the user adds it to the order, and the Order Summary form redisplays. The user can add another item to the order, make changes to the first item ordered, or select shipping and payment options. This flexibility and information display are what the phone-order representatives wanted. It took quite a few iterations and user evaluations to begin to achieve the best design.

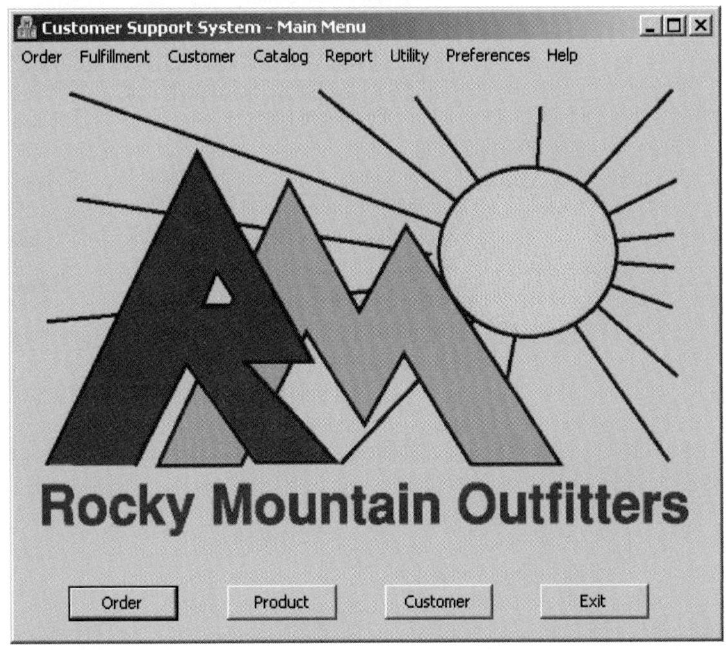

Figure 14-13
Prototype forms for an order-centered approach to the dialog

(a) The Main Menu form for an order-centered approach to the dialog

Figure 14-13 cont.

(b) The Order Summary form for an order-centered approach to the dialog beginning a new order

Customer Support System - Order Summary

Rocky Mountain Outfitters

Customer
Name
Number
Search
View Details

Order
Number 633856
Date 10/15/2010
New Order

Products
ID | Description | Size/Color | Price | Quantity | Extension
1
2
3
4
5 <-- Click here to display more order lines
Product Subtotal

Order Completion
Shipping Method
Payment Method
Account Number
Ship Edit
Payment Edit
Shipping
Tax
Credits
Total
Abandon
Confirm Order

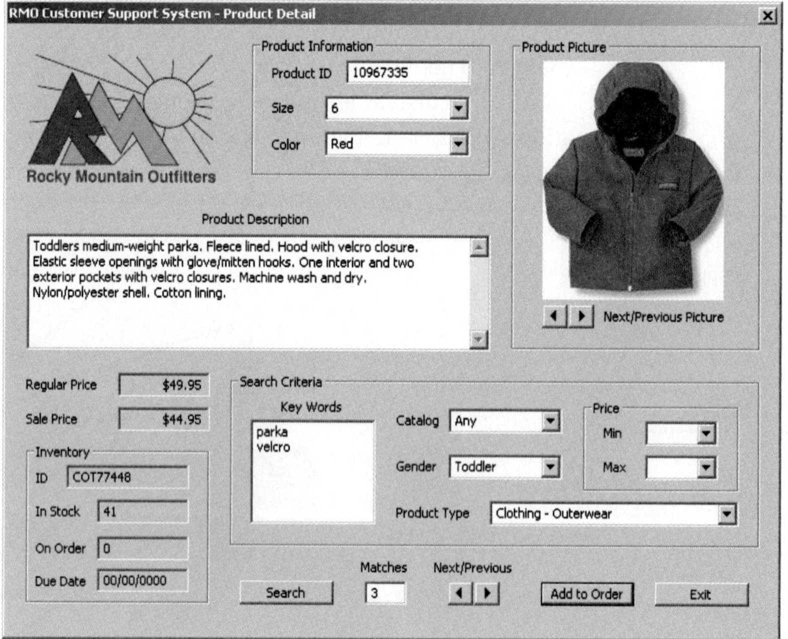

(c) The Product Detail form after the user has searched for a product

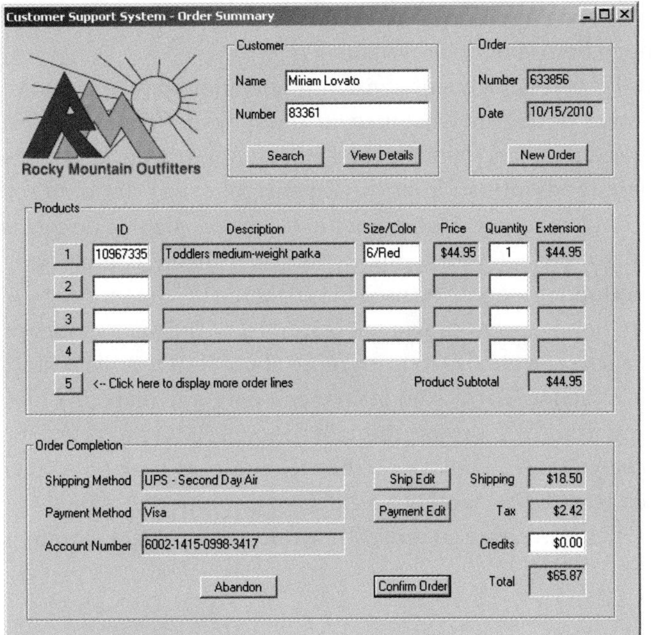

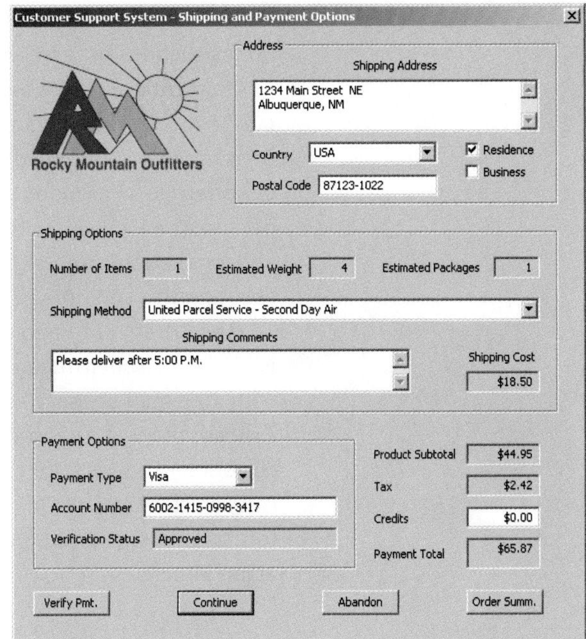

Figure 14-13 cont.

(d) The Order Summary form after the user adds the product

(e) The Shipping and Payment Option form for the completed order

DIALOG DESIGN FOR THE RMO WEB SITE

The RMO customer support system use case *Create new order* requires several different dialog designs (one for each scenario), including a scenario for the phone-order representatives just discussed and a scenario for the mail-order clerks. But another key system objective is to allow customers to interact with the system directly to place orders via the World Wide Web. The design of a complete company Web site is beyond the scope of this text, but some general rules and guidelines apply to direct customer orders.

The basic dialog between user and computer will be about the same as for the phone-order representative, but the Web site will have to provide even more information for the user, be even more flexible, and be even easier to use. First, customers might want to browse through all possible information about a product. More pictures will be needed, including pictures showing different colors and patterns of items. Although the phone-order representative needs detailed information on the screen to be able to answer questions, the customer will probably want even more. The information will need to be displayed differently, too. For example, the phone-order clerk is accustomed to a dense display of information and knows where to look for the details, but customers will need organized information to make it easy to locate details the first time they use the system.

The system will need to be very flexible because customers will have different preferences for interacting with the system. As discussed for the phone-order scenario, the sequence should be flexible—some customers will want to browse first and even select items to order before entering any information about themselves. Such options as reviewing past orders and reviewing shipping and payment approaches will also be required. If a customer wants to do something that the system does not allow, the customer will become frustrated. Unlike phone-order and mail-order employees, who will work through their frustration, the customer can simply log off and shop elsewhere. Finally, the system must be so easy to learn and use that the customer does not even have to think about it. The initial dialog options need to be very clear, and once a sequence starts, all options should be self-explanatory. Customers cannot be expected to sit through training just to use a Web site, nor should they have to look up instructions or help (even though both should be available).

As mentioned previously regarding the guidelines for visibility and affordance, it is important that controls used on a Web page be clear in what they do and how they are used. Most users are more concerned about speed than fancy graphics and animations, but novice designers often go overboard on graphics and animation at the expense of speed. In addition, because the Web site reflects the company image, it is important to involve graphic designers and marketing professionals in the design. A well-thought-out visual theme is important. Focus groups and other feedback techniques should be used in generating the design.

The Rocky Mountain Outfitters' home page is shown in Figure 14-14. The main emphasis is direct customer interaction. The user can choose to learn more about RMO, contact RMO with an e-mail message, or request a catalog. There is no main menu option to place an order. Instead, the Web site offers the opportunity to browse through pages of RMO products. The customer can search for products based on keywords or product ID numbers, or the customer can select a category of a product from a list. Weekly specials are also offered. When customers find something they want, they can create an order. They can also change their minds at any time.

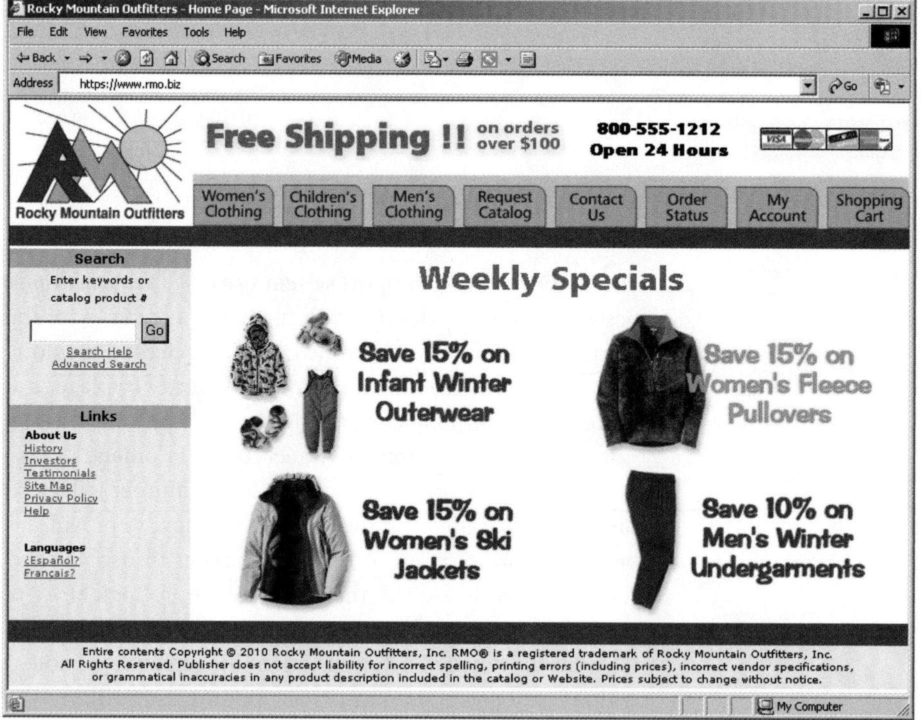

Figure 14-14

Rocky Mountain Outfitters' home page

The RMO Web site uses the shopping cart analogy for orders. After customers find a product they want, they select the quantity, size, and other options, and then they add the item to their shopping cart. They can view the shopping cart at any time and then continue browsing and adding items. When they are done, they check out, and the system confirms the order. Figure 14-15 shows a Product Detail page reached after a user navigates through the women's clothing option. Figure 14-16 shows the shopping cart with a summary of an order.

Figure 14-15

The Product Detail page from the Rocky Mountain Outfitters' Web site

Figure 14-16

The shopping cart page from Rocky Mountain Outfitters' Web site

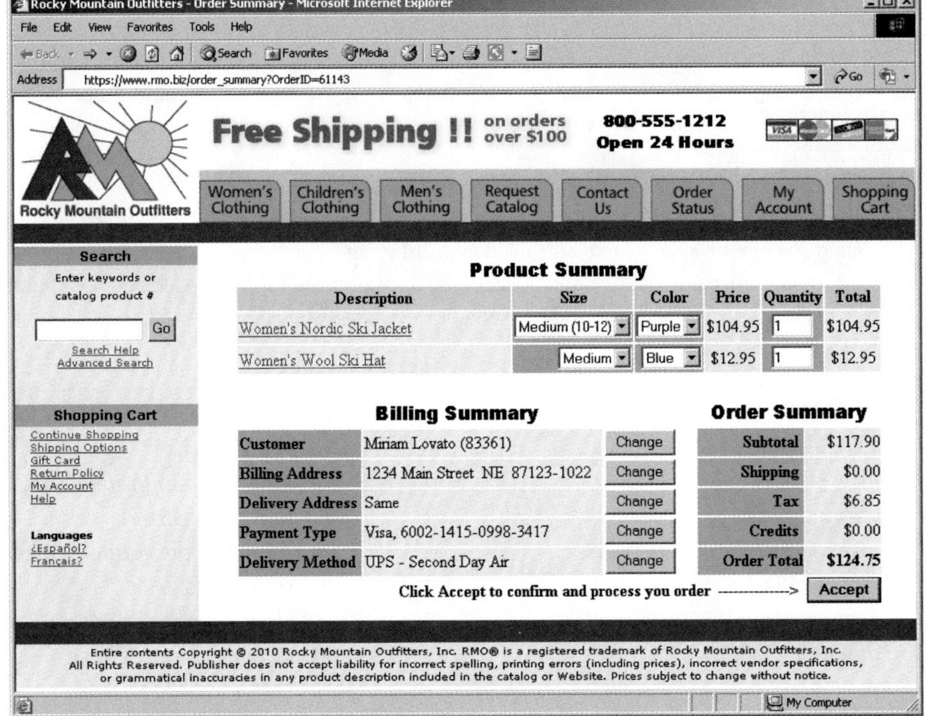

SUMMARY

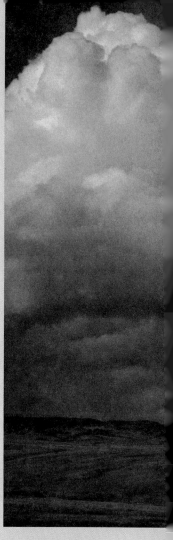

Inputs and outputs can be classified as system interfaces or user interfaces. This chapter focuses on user interfaces and describes concepts and techniques for designing the interaction between the user and the computer—human-computer interaction (HCI). Chapter 15 describes system interfaces, including system outputs and system controls.

The user interface is everything the user comes into contact with while using the system—physically, perceptually, and conceptually. To the user, the user interface *is* the system. The knowledge the user must have to use the system (the user's model) includes information about objects and functions available in the system—the kind of information that defines the requirements model the analyst works hard to uncover during systems analysis.

User-centered design refers collectively to techniques that focus early on users and their work, evaluate designs to ensure usability, and apply iterative development. Usability refers to the degree to which a system is easy to learn and use. Ensuring usability is a complex task because design choices that promote ease of learning versus those that promote ease of use often conflict. In addition, there are many different types of users to consider for any system. Human-computer interaction as a field of research grew out of human factors engineering (ergonomics) research that studies human interaction with machines in general.

There are many different ways to describe the user interface, including the desktop metaphor, the document metaphor, and the dialog metaphor. The dialog metaphor emphasizes the interaction between user and computer, and interface design is often called *dialog design* for that reason. Interface design guidelines and interface design standards are available from many sources. Norman's visibility and affordance guidelines state that controls should be visible, provide feedback indicating that they are working, and be obvious in their function. Shneiderman's eight golden rules are to strive for consistency, provide shortcuts, offer informative feedback, design dialogs to yield closure, offer simple error handling, permit easy reversal of actions, support internal locus of control, and reduce short-term memory load.

Dialog design starts with identifying dialogs based on activities or use cases. Additional dialogs are needed for integrity controls added during design, for user preferences, and for help. Menu hierarchies can be designed for different types of users, and after the dialogs are designed, menu hierarchies can be rearranged easily. Writing a dialog sequence much like a script can help a developer work out the key information that needs to be exchanged during the dialog. A storyboard showing sketches of screens in sequence can be drawn to convey the design for review with users, or prototypes can be created using a tool such as Visual Basic. The object-oriented approach provides UML models that can document dialog designs, including sequence diagrams, activity diagrams, and class diagrams.

Each form used in a dialog needs to be designed, and there are guidelines for the layout, selection of input controls, navigation, and help. These guidelines apply to windows forms and to browser forms used in Web-based systems. Designing a dialog for a Web site is similar to creating any other dialog, except users need more information and more flexibility. Additional Web design guidelines apply to designing for the computer medium, designing the whole site, and designing for the user. In addition, because a Web site reflects the company's image to customers, graphic designers and marketing professionals should be involved.

KEY TERMS

affordance, p. 540
browser forms, p. 549
check boxes, p. 551
combo box, p. 551
desktop metaphor, p. 536
dialog metaphor, p. 537
direct manipulation, p. 536
document metaphor, p. 537
human-computer interaction (HCI), p. 532
human factors engineering (ergonomics), p. 534
hypermedia, p. 537
hypertext, p. 537
interface design standards, p. 540
list box, p. 551
radio buttons (option buttons), p. 551
spin box, p. 551
storyboarding, p. 546
system interfaces, p. 531
text box, p. 551
usability, p. 534
user-centered design, p. 533
user interfaces, p. 531
user's model, p. 533
visibility, p. 540

REVIEW QUESTIONS

1. Why is interface design often referred to as *dialog design*?
2. What are the three aspects of the system that make up the user interface for a user?
3. What term is generally used to describe the study of end users and their interaction with computers?
4. What are some examples of physical aspects of the user interface?
5. What are some examples of perceptual aspects of the user interface?
6. What are some examples of conceptual aspects of the user interface?
7. What collection of techniques places the user interface at the center of the development process?
8. What are the three important principles emphasized by user-centered design?
9. What term refers to the degree to which a system is easy to learn and use?
10. What is it about the "human factor" that engineers find difficult? What is the solution to human factor problems?
11. What are some of the fields that contribute to the field of human-computer interaction?
12. What research center significantly influenced the nature of the computers we use today?
13. What are the three metaphors used to describe human-computer interaction?
14. A desktop on the screen is an example of which of the three metaphors used to describe human-computer interaction?
15. What type of document allows the user to click on a link and jump to another part of the document?
16. What type of document allows the user to click on links to text, graphics, video, and audio?
17. What is the name for general principles and specific rules that developers must always follow when designing the interface of a system?
18. What two key principles are proposed by Norman to ensure good interaction between a person and a computer?
19. List the eight golden rules proposed by Shneiderman.
20. What is the technique that shows a sequence of sketches of the display screen during a dialog?
21. What UML diagram can be used to show how the interface objects are plugged in between the actor and the problem domain classes during a dialog?
22. What are the three basic types of windows and browser forms used in business systems?
23. What are some of the input controls that can be used to select an item from a list?
24. What two types of input controls are included in groups?
25. What are three requirements for usability of a direct customer access Web site beyond those of a windows interface used by employees?
26. What popular analogy is used for direct customer access with a Web site when customers shop online?
27. What are three principles of Web design that guide designers?
28. What are four of the 10 good deeds of Web design?

THINKING CRITICALLY

1. Think of all of the software you have used. What are some examples of ease of learning conflicting with ease of use?
2. Visit some Web sites and identify all of the controls used for navigation and input. Are they all obvious? Discuss some differences in visibility and affordance of the controls.
3. Consider the human factor solution that states it is better to change the machine than to try to change the human to accommodate the machine. Are there machines (or systems) that you use in your daily life that still have room for improvement? Are the current generations of Windows PC and Apple Mac as usable as they might be? If not, what improvements can you suggest? Is the World Wide Web as usable as it might be? If not, what improvements would help? Are we just beginning to see some breakthroughs in usability, or have most of the big improvements already been made?
4. Review the dialog between the user and computer shown for DownTown Videos. Create a table like Figure 14-6 that shows how the dialog can be converted to the user's language and the computer's language. Discuss how moving from the dialog to the table starts with a logical model and then creates a physical model.
5. Refer again to the table shown in Figure 14-6. Create a storyboard for the e-mail system based on the information in the table.

6. Read through the following dialog, which shows a user trying to place an order with the system. Critique the dialog in terms of ease of learning and ease of use.

   User: I want to order a product.
   System: Okay. Enter your name and address.
   User: My name is Timothy Mudd, 5139 North Center Street, Los Angeles, CA 98210.
   System: Oh, we have all of that information on file, but thank you for entering it again.
   User: I want to order the Acme Drill Press with adjustable belt drive.
   System: Sure, continue with your request.
   User: I want the blue color and rubber feet but not the foot brake.
   System: Sure, anything else?
   User: I want it shipped priority with the special discount you offer.
   System: We hear you.
   User: Okay, that's all.
   System: We are sorry, but your transaction violated transaction code 312 and we must terminate the session.

7. Review all of the controls that come with Visual Basic that are visible when added to a form. Discuss how well each satisfies the visibility and affordance requirements.

EXPERIENTIAL EXERCISES

1. Review the sequential design concept and the order-centered design concept for the Rocky Mountain Outfitters *Create new order* dialog. Consider what information and navigation options are not needed by the mail-order clerks. Then design the forms that would be appropriate for clerks to use when entering mail orders for RMO. Either sketch the forms or implement the forms using a tool such as Visual Basic. Use input controls that minimize data entry.
2. Evaluate the course registration system at your university. List the basic steps followed through a dialog with the system. What are some of the problems with the system from the perspective of ease of learning and ease of use? In what ways is the system inflexible? In what ways is needed information not available? Is too much information provided that distracts from the task at hand?
3. Evaluate the online catalog system at your university library (see some of the questions and issues in exercise 2). Write a dialog that shows the interaction between the user and the system. Rewrite the dialog to improve it. Create a storyboard to show how your design would look and feel.
4. Locate a system that users work with all day. Ask some of the users of the system what they like and dislike about it. Ask how they learned the system. Ask them what typical errors they make when using the system and how these errors are prevented or corrected. Observe the users working with the system. Does it appear that any of the interaction is "automatic"?
5. Find a Web site with direct customer ordering. Browse through some product descriptions and note the design of the dialog and the Web pages. What do you like and dislike about the design? Evaluate the Web site based on visibility and affordance and on the eight golden rules. Evaluate the design based on the 10 good deeds and three principles of Web design.

CASE STUDIES

THE 21ST CENTURY UNIVERSITY SYSTEM

Consider the following dialog for a user interacting with the course registration system at 21st Century University. The user is a student registering for classes at the university.

System: What course registration function do you want to perform?
User: *Preregistration.*
System: Okay, what is your account ID and password?
User: *My account ID is SATZ791, and my password is xxxxxxxx.*
System: Okay, welcome back, Kevin. You last adjusted your schedule on 7/7.
What preregistration function do you want to perform?
User: *To add courses.*
System: These are the courses you are preregistered for now:
(The system displays a list of course information for three courses.)

Because you are a CIS major, these are all of the CIS courses you might add that have no time conflicts with your present schedule (note, though, that prerequisites have not been checked for these courses).

(The system displays a list of CIS courses available.)

Your options are as follows:

- Switch to a list of courses in another department.
- List these courses again but ignore time conflicts.
- List these courses again after checking prerequisites.
- Provide details of the indicated course.
- Add the indicated course to your schedule.
- Cancel.

User: *I want to switch to courses in another department.*

System: Fine. To what department do you want to switch?

User: *Art.*

System: These are the courses for which you are registered now:

(The system displays a list of course information for three courses.)

As requested, these are the art courses you might add that have no time conflicts with your present schedule (note, though, that prerequisites have not been checked for these courses).

(The system displays a list of art courses available.)

Your options are as follows:

- Switch to a list of courses in another department.
- List these courses again, but ignore time conflicts.
- List these courses again after checking prerequisites.
- Provide details of the indicated course.
- Add the indicated course to your schedule.
- Cancel.

User: *Add the second course in the list.*

System: (The system continues the dialog.)

1. Is this dialog a logical or a physical model? Could it be a phone registration system? Could a clerk at a desk handle the dialog manually? Discuss.
2. Discuss how well this dialog follows the eight golden rules.
3. Complete the dialog by adding likely exchanges between the student and the system as the student adds one more course. The student should request more details about a course and then decide to add the course. Include a request from the student to see final schedule details.
4. Design and implement either a storyboard or prototype for this dialog, using a tool such as Visual Basic and being as faithful to the dialog as possible. Make up some sample data to show in your design as needed.

THE DOWNTOWN VIDEOS RENTAL SYSTEM

This chapter includes an example of a storyboard for DownTown Videos, a case study first introduced in Chapter 7. The storyboard showed the *Rent videos to customer* dialog. Revisit the DownTown Videos case and complete the following:

1. Implement the storyboard in this chapter as a prototype using a tool such as Visual Basic.
2. Write a dialog, and then create a storyboard for the use case *Return rented videos*. Consider that one or more videos might be returned and that one or more of them might be late, requiring a late charge.
3. Using a tool such as Visual Basic, implement the storyboard as a prototype, and then ask several people to evaluate it. Discuss the suggestions made.

THE WAITERS ON CALL SYSTEM

Review the Chapter 5 opening case study that describes Waiters on Call, the restaurant meal-delivery service. The analyst found at least 14 use cases for the system, which are listed in the case.

1. Create a set of menu hierarchies for the system based on the use cases listed. Then add more menu hierarchies for system utilities based on controls, user preferences, and help.
2. The most important event listed in the case is *Customer calls in to place an order*, and the use case might be named *Process food order.* Write out a dialog between the user and the computer with a natural sequence and appropriate exchange of information.
3. Sketch a storyboard of the forms needed to implement the dialog.
4. Ask several people to evaluate the design and discuss any suggested changes.
5. Implement a prototype of the final dialog design using a tool such as Visual Basic.

THE STATE PATROL TICKET PROCESSING SYSTEM

Review the State Patrol ticket processing system introduced as a case study at the end of Chapter 5.

1. Create a set of menu hierarchies for the system based on the use cases in the case. Then add more menu hierarchies for system utilities based on controls, user preferences, and help.
2. Write out a dialog between the user and the computer for the use case *Record new ticket* with a natural sequence and appropriate exchange of information.
3. Sketch a storyboard of the forms needed to implement the dialog. Be sure to use input controls that minimize data entry, such as list boxes, radio buttons, and check boxes.
4. Ask several people to evaluate the design and discuss any suggested changes.

5. Implement a prototype of the final dialog design using a tool such as Visual Basic.

RETHINKING ROCKY MOUNTAIN OUTFITTERS

A few of the phone-order representatives at Rocky Mountain Outfitters were not completely satisfied with the order-centered dialog design presented in this chapter. They suggested that the design could be streamlined if only one form were displayed throughout the dialog—the order summary form. They thought the order summary form could expand to show additional information instead of switching to separate forms for customer information, product information, or shipping information. When the additional information is not needed, the form could contract. They thought this approach would be easier on the eyes, requiring less effort to focus and refocus on multiple forms that pop up. They requested that a prototype of the one-form design concept be created for their review. The interface might include both options, allowing users to choose the one they prefer.

1. Draw a storyboard to show how the one-form design concept might look as customer information, product information, and shipping information are added to the expanded form.
2. Implement a prototype of the storyboard using a tool such as Visual Basic.
3. Ask several people to evaluate the design, specifically comparing it with the order-centered design in the text, and discuss the results.
4. Can you describe or implement yet another alternative?

FOCUSING ON RELIABLE PHARMACEUTICAL SERVICE

The Reliable Pharmaceutical Service system has users who process orders in the Reliable offices and users who place orders and monitor order information in the nursing homes. Consider the events and use cases that apply to Reliable employees versus nursing home employees.

1. Design two separate menu hierarchies, one for Reliable employees and one for nursing home employees.
2. Write out the steps of the dialog between the user and the system for the use case *Place new order* for nursing home employees.
3. Create a storyboard for the dialog for *Place new order* by making a sketch of the sequence of interaction with Web pages. Implement a prototype for the Web pages using a Web development tool such as Dreamweaver or FrontPage.

FURTHER RESOURCES

Merlyn Holmes, *Web Usability and Navigation*. McGraw-Hill Osborn, 2002.

Patrick J. Lynch and Sarah Horton, *Web Style Guide: Basic Design Principles for Creating Web Sites*. Yale University Press, 1999.

Deborah J. Mayhew, *Principles and Guidelines in Software User Interface Design*. Prentice Hall, 1992.

Jakob Nielsen, *Designing Web Usability: The Practice of Simplicity*. New Riders Publishing, 2000.

Donald Norman, *The Design of Everyday Things*. Doubleday, 1990.

Jenny Preece, Yvonne Rogers, David Benyon, Simon Holland, and Tom Carey, *Human Computer Interaction*. Addison Wesley, 1994.

Ben Shneiderman, *Designing the User Interface* (3rd ed). Addison Wesley, 1998.

Joel Sklar, *Principles of Web Design* (3rd ed). Course Technology, 2006.

CHAPTER 15

DESIGNING SYSTEM INTERFACES, CONTROLS, AND SECURITY

LEARNING OBJECTIVES

After reading this chapter, you should be able to:

- Discuss examples of system interfaces found in information systems
- Define system inputs and outputs based on the requirements of the application program
- Design printed and on-screen reports appropriate for recipients
- Explain the importance of integrity controls
- Identify required integrity controls for inputs, outputs, data, and processing
- Discuss issues related to security that affect the design and operation of information systems

CHAPTER OUTLINE

Identifying System Interfaces

Designing System Inputs

Designing System Outputs

Designing Integrity Controls

Designing Security Controls

DOWNSLOPE SKI COMPANY: DESIGNING A SECURE SUPPLIER SYSTEM INTERFACE

Downslope Ski Company is a medium-sized manufacturer of skis and snowboards. The company started with the manufacture of downhill snow skis, hence its name. But a few years ago, it expanded into the production of snowboards and then into water skis. In the company's early years, ski manufacturing was fairly straightforward. However, with the introduction of advanced materials such as carbon-laced resins and other sophisticated compounds, manufacturing has become quite complicated, requiring exacting controls for the precise mixture of ingredients and temperature tolerances in baking furnaces. To maintain consistency in snowboard production, Downslope has been very demanding in the quality of the raw materials that it buys. Several times over the last few years, it has had to change suppliers to ensure an adequate supply of high-quality raw materials.

In addition to controlling the quality of materials for manufacturing, Downslope instituted a modified just-in-time (JIT) manufacturing process, which means that it does not stockpile a large quantity of raw materials. Generally, it keeps about a five-day supply on hand and depends on its suppliers to restock materials at least weekly. To facilitate quick ordering and delivery of its many raw materials, senior management at Downslope decided to permit its suppliers to access its inventory database. Depending on which types of boards were being produced, the various raw materials would be depleted at different rates. So, Downslope began developing and installing a complex inventory management system that was integral to the manufacturing process.

Nathan Lopez, Downslope's system development project manager, was fast recognizing that providing a system interface for suppliers to access the database was more complex than he originally thought. He reported to Downslope management the results of a two-month study to determine the feasibility of various alternatives for this electronic approach to supply chain management.

"I have met with each of our suppliers and determined what information they need and the formats they would like it in. As expected, there was little consistency in the desired formats. I have been able to consolidate some of their needs to narrow them down to three basic formats. Originally, we thought we could convince our suppliers to accept our output design—in other words, to make them conform to our output. However, that does not seem to be such a good idea anymore. If we do not build flexibility into our system, it will be more difficult for us to add or change suppliers. Instead, if we build the interface correctly, with several versions, it should be much easier for suppliers to gain access to the data that we allow them to see.

"Another critical issue that has surfaced is the integrity and security of our data and our systems. For example, our production process is unique and one of our competitive advantages. If the wrong company got access to our data, it could potentially analyze our usage patterns and not only discover what materials we used but perhaps even reverse-engineer our processes. To protect our data, we need to ensure that our computers allow only secure access. We also need to ensure that data is secure while being transmitted to our suppliers and is protected after arrival there. These security issues relate to both our systems and our suppliers."

The meeting lasted a long time, with considerable discussion about the opportunities and dangers of opening company systems to an outside system interface. The oversight committee finally decided that Nathan should study the situation for a couple more weeks and develop a list of every possible breach of security with potential solutions for each one. Only after these issues were addressed would the project move forward. Nathan knew that the development of this new type of system interface would be a very sticky problem. He hoped he would be able to find solutions for all of the issues that had been raised.

OVERVIEW

Most modern information systems involve extensive input and output (I/O), and many people and organizations require access to the data stored by a system. In Chapter 14, you read about human-computer interaction (HCI) and user interfaces, where I/O is the result of user interaction with the computer. But many system inputs and outputs do not involve much human interaction. Many of these system interfaces are not as obvious to end users. But systems analysts need a deep understanding of existing systems, databases, and network technologies where I/O occurs to design an information system that incorporates all I/O needs. We therefore discuss system interfaces separately from user interfaces in this chapter.

Many system interfaces are electronic transmissions or paper outputs to external agents, including reports, statements, and bills. They need to be identified and designed to suit their intended purpose. Frequently, the quality of the system outputs is a mark of the quality of the entire system and of the company that uses it. This chapter discusses the design of these outputs.

Because of the many and varied inputs and outputs, system developers need to design and implement integrity controls and security controls to protect the system and its data. Today more than ever, designing system controls is crucial because computer systems increasingly exist in an open environment. They are part of networks that provide broad access to many different people both within and outside an organization. So, one of the major considerations in systems design is how to ensure that errors or fraudulent transactions are not entered into a system. Integrity controls validate data when it is input and processed. Internal checks and cross-checks help ensure that data integrity is maintained. This chapter discusses techniques to provide the integrity controls to reduce errors, fraud, and misuse of system components.

Finally, outside threats to systems, business organizations, and individuals continue to be a major concern of companies that are connected to the Internet. Many companies use the Internet as a marketing and sales channel, so they need to let customers and prospective customers into their systems, yet keep intruders and malicious hackers out. In addition, e-commerce entails transmission of private information, such as credit-card numbers and financial transactions. The last section of the chapter discusses security controls and explains the basic concepts of data protection, digital certificates, and secure transactions.

IDENTIFYING SYSTEM INTERFACES

The user interface, as described in the previous chapter, includes inputs and outputs that directly involve system users. But there are many other system interfaces. We define *system interfaces* broadly as any inputs or outputs with minimal or no human intervention. Included in this definition are standard outputs, such as billing notices, reports, printed forms, and electronic outputs to other automated systems. Inputs that are automated or come from nonuser interface devices are also included. For example, inputs from automated scanners, bar-code readers, optical character recognition devices, and other computer systems are included as part of a system interface.

It often seems that user interfaces are the most common—and thus most important—interfaces to consider when analyzing and designing an information system. In fact, considerable progress has been made in understanding human-computer interaction and applying user-centered design principles to user-interface design. However, today's highly integrated and interconnected information systems increasingly go beyond user needs, requiring system interfaces to handle inputs and outputs faster, more efficiently, more accurately, and at any hour of the day or night.

System interfaces can process inputs, interact with other systems in real time, and distribute outputs with minimal human intervention. When researching and modeling the requirements for a system, analysts must look carefully for system interfaces that might not appear obvious at first. When designing the system, they should consider alternatives to HCI to automate the capture of inputs and the distribution of outputs. The full range of inputs and outputs in an information system is shown in Figure 15-1.

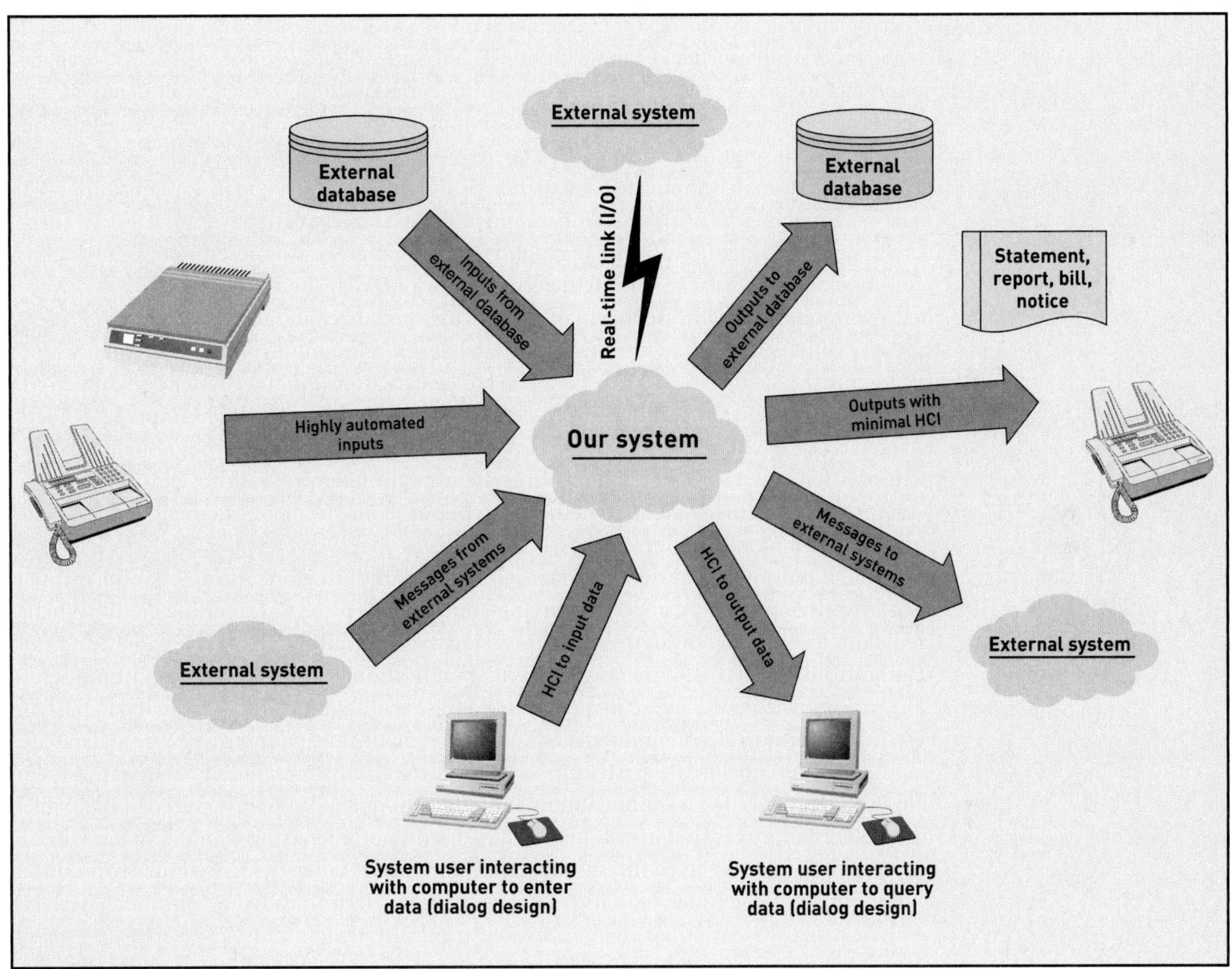

Figure 15-1

The full range of inputs and outputs in an information system

The following list provides some categories of system interfaces to aid in identifying I/O requirements and design possibilities:

- Inputs from other systems
- Highly automated inputs
- Inputs that are from data in external databases
- Outputs that are to external databases
- Outputs with minimal HCI
- Outputs to other systems
- Real-time connections (both input and output)

BEST PRACTICE

System interfaces are becoming more and more important as information systems are being linked and as they function in real time. Be sure to identify opportunities to automate system interfaces whenever possible because they have a large impact on system performance and return on investment.

Inputs can arrive directly from other information systems as network messages. Electronic data interchange (EDI) and many Web-based systems are integrated with other systems through direct messaging. The message received triggers system processing much the way that user interaction does. For example, in RMO's integrated supply chain management and customer support systems, the arrival of inventory items from a supplier might trigger the shipment of a back-ordered item to a customer. No human intervention is required, and as a result, the transaction can be processed immediately and with little chance of error. In Web-based systems, a separate shopping cart order-entry application might send a message to the order fulfillment system to process a new order. Analysts decide what is an input and what is an internal message by determining the scope of the system.

Highly automated input devices such as scanners can capture many system inputs. An item in a warehouse that is picked for shipment, for example, might have a machine-readable label that an attendant can scan. This highly automated process represents an input to a system. In some cases, a scanner might record the input as an item moves by on a conveyor belt—with no human interaction at all. We discuss specialized input devices later.

Many inputs may come from external databases. For example, one system might record transactions in a database, perhaps as a batch of transactions. Another system might periodically search those transactions and process one or more of them. For example, consider a charge-account system in which charges are stored in a database throughout the billing period. A separate billing and collection system might later process those transactions. As with inputs from other systems, whether these inputs come from an external database or are processed within one billing system is a question of scope.

Some inputs from an external database might occur during processing of another input, such as verifying credit history or verifying employment prior to extending credit. When a credit application is received, the user processing the application might have to rely on information from an external database before the application can be approved. Within the use case *Process credit application* is a requirement for a system input from an external database.

The output side of system interfaces mirrors the input side. Outputs to external databases might be required when the system produces large amounts of detailed data. Many system outputs are produced with minimal human intervention. Reports are produced and e-mailed to recipients or printed and distributed, but the user is not interacting with the system directly to obtain the output. Bills, notices, statements, form letters, and so on are similarly produced with minimal human intervention. They might be sent electronically or be printed. Messages sent to external systems, triggering processing, are also system outputs.

Sometimes system inputs and outputs must be real-time connections. RMO's real-time credit-card authorization is an example. Rather than accessing an external database, RMO's system establishes a real-time connection with another system that accepts inputs and provides outputs. A real-time connection is, therefore, both a system output and a system input much like a system-to-system dialog. In this aspect, real-time connections parallel user-interface functions, which use a dialog to enter data and look up data in the system (as shown in Figure 15-1).

Another mechanism for correct system input is to have a direct interface with another system. EDI reduces the need for user input. Purchase orders, invoices, inventory updates, and payment all are generated by one system sending transactions to another. With EDI, these transactions normally occur between systems in separate organizations. However, the same principle can be applied to systems within an organization.

One of the main challenges of EDI is in defining the format of the transactions. It is easy enough to design the format for a single type of transaction, but it becomes more complex when many different types of transactions are going to many different systems. The complexity and difficulty multiplies when many different companies are trying to work together. For example, General Motors, which was one of the early users of EDI, has literally thousands of suppliers with thousands of different transactions, each in a different format. To complicate the situation further, each of these suppliers might also be linked via EDI with tens or hundreds of customers, many of whom might also use EDI. So, a single type of transaction might have a dozen or more defined formats. It is easy to see why it is so costly to set up and maintain EDI systems. Even so, EDI is much more efficient and effective than paper transactions, which must be printed and reentered.

In recent years, there has been a move to develop a standard communication method between systems based on Hypertext Markup Language (HTML) concepts. As you know, HTML embeds beginning and ending markup codes within a text-based document to define the characteristics, such as formatting, of text or figures. HTML embeds formatting information within the document itself. Thus, a program that can read HTML reads a text document and then, using the embedded markup codes, can format the document correctly. Although this system is not extremely efficient from a computational point of view, it has many advantages due to its simplicity and readability by human beings.

A relatively new system-to-system interface that is gaining popularity is called *eXtensible Markup Language (XML)*. XML is an extension of HTML that embeds self-defining data structures within textual messages. So, a transaction that contains data fields can be sent with XML codes to define the meaning of the data fields. Many newer systems are using XML to provide a common system-to-system interface. Figure 15-2 illustrates a simple XML transaction that can be used to transfer customer information between systems.

Figure 15-2

A system-to-system interface based on XML

```
<customer record>
        <accountNumber>RM010989</accountNumber>
        <name>William Jones</name>
        <billingAddress>
                <street>120 Roundabout Road</street>
                <city>Los Angeles</city>
                <state>CA</state>
                <zip>98115</zip></billingAddress>
        <shippingAddress>
                <street>120 Roundabout Road</street>
                <city>Los Angeles</city>
                <state>CA</state>
                <zip>98115</zip></shippingAddress>
        <dayPhone>215.767.2334</dayPhone>
        <nightPhone>215.899.8763</nightPhone>
</customer record>
```

Just like HTML, XML is simple and readable by human beings. For it to work, both systems must recognize the markup codes, but after a complete set of codes is established, transactions can include many different formats and still be recognized and processed. The receiving system merely has to parse the incoming data stream and extract the values from the markup codes. XML is extremely scalable—it does not matter how many different companies or different transaction types there are. Every transaction can have its own format as long as it uses the standard markup codes.

Markup codes for XML are defined in a separate file called a *document type definition* (*DTD*) file or its successor *XML schema*. Many industries and specialty groups have now formed standards committees that are defining sets of markup codes. Standard codes already exist for general business, retailing, railroad, news media, and medical transactions. The list of groups forming and defining standard codes is extensive.

Chapter 9 introduced the idea of Web services. Web services are based on XML, so these business transactions can be sent over the Internet. In fact, XML was designed to take advantage of the Internet.

DESIGNING SYSTEM INPUTS

When designing inputs for a system, the system developer must focus on three areas:

- Identifying the devices and mechanisms that will be used to enter input
- Identifying all system inputs and developing a list with the data content of each
- Determining what kinds of controls are necessary for each system input

The first task, identifying devices and mechanisms, is a high-level review of the most up-to-date method of entering information into a system. In nearly all business systems, end users perform some input through electronic forms. However, in today's high-technology world, there are numerous ways to enter information into a system—among them are scanning, reading, and transmitting devices that are faster, more efficient, and less error-prone than user input.

The second task, developing the list of required inputs, provides the link between the design of the application software and the design of the user and system interfaces. As described earlier, the design of the system inputs and interfaces must be integrated with the design of the application. This activity accomplishes that purpose.

The third task is to identify the control points and the level of security required for the system being developed. The project team should develop a statement of policy and control requirements before beginning the detailed design of the electronic forms that make up the system interface. These concepts are discussed in the last two sections of the chapter.

INPUT DEVICES AND MECHANISMS

Often when analysts begin developing a system, they assume that all input will be entered via electronic, graphical forms because they are now so common on personal computers and workstations. However, as the design of the user inputs commences, one of the first tasks is to evaluate and assess the various alternatives for entering information into the system.

The primary objective of any form of data input is to enter or update error-free data into the system. The key term here is *error-free*. Several good practices can help reduce input errors:

- Capture the data as close to the originating source as possible.
- Use electronic devices and automatic entry whenever possible.
- Avoid human involvement as much as possible.
- If the information is available in electronic form anywhere, use it instead of reentering the information.
- Validate and correct information at the time and location it is entered.

Today, many systems enable the data to be captured electronically at the point that it is generated. For example, the old way of selling a life insurance policy is to have the applicant or the insurance agent fill out a paper policy application. Then the agent sends the application to a central office to be entered into the system. With this method, numerous errors can occur from indecipherable writing, key-entry errors, missing fields, and so forth. Currently, agents often carry laptop computers with easy-to-use electronic forms, so applicants can fill in the data themselves. Or the agent can enter the data while the applicant looks over the agent's shoulder and verifies that the information is accurate and complete. A portable printer can even be attached to the laptop to print the completed form for the applicant to review immediately. This new approach dramatically reduces the error rate and speeds the business process of new policy data entry.

The second and third practices, automating data entry and avoiding human involvement, are very closely related and often are essentially different sides of the same coin, although using electronic devices does not automatically avoid human involvement. When system developers think carefully about minimizing human input and using electronic input media, they can design a system with fewer electronic input forms and avoid some common data-entry problems. One of the most pervasive sources of erroneous data is typing mistakes by users. A few of the more prevalent devices used to avoid human keystroking are as follows:

- Magnetic card strip readers
- Bar-code readers
- Optical character recognition readers and scanners
- Radio-frequency identification tags
- Touch screens and devices
- Electronic pens and writing surfaces
- Digitizers, such as digital cameras and digital audio devices

We have all seen new electronic input devices. At the grocery store, electronic scanners identify and price each item from the printed UPC codes. Machines weigh and price the produce automatically at the checkout. Cash registers now read your check, including the amount and customer and bank information. New self-service checkout stations depend almost entirely on automated data-entry devices.

Historically, paper contracts and ink signatures were necessary for legally binding contracts. Today, new laws and regulations allow paper documents and signatures to be digitized. Credit-card purchases now also record digitized signatures to eliminate the need for paper vouchers. This technique conforms to the good practices stated previously—the information is captured at its source in electronic form, which eliminates many of the sources of errors.

The next principle of error reduction is to reuse the information already in the computer whenever possible. Some outdated systems require reentry of the same information multiple times. This practice not only generates errors but also creates multiple copies of the same information, which require more checks and balances—as well as computer resources—to synchronize the various copies. And when an error is discovered, it is very difficult to know which copy is correct. Also, when a change is required, it must be made to all copies of the data. One current high-tech example of using existing information is found in car rental systems. When a customer rents a car, the rental agency's system captures the customer's name and credit-card, car mileage, and fuel information. Then when the customer returns the car, an agent in the parking lot simply scans the contract ID and enters the return mileage and fuel-gauge level. The system calculates the charges and prints a credit-card receipt for the customer right in the parking lot. This solution was designed primarily to provide a higher level of customer service, but it also eliminates many problems and errors with data entry.

Eliminating input errors through various techniques, one of which is using electronic devices for data input, is critical, but another potential source of problems is the input of fraudulent information. Two problems need to be addressed with fraudulent data: access control and input control. First, access to the system must be controlled so that only authorized people or systems can gain access. Today, devices such as fingerprint readers, body temperature sensors, and iris scanners are used more and more often to provide additional security to standard password access. Second, input controls must be built into the system so that fraudulent data cannot easily be entered. Although it is impossible to completely eliminate the potential for fraud, the careful design of input controls will help minimize the risk. More discussion on security devices and input controls is provided later in the chapter.

DEFINING THE DETAILS OF SYSTEM INPUTS

The objective of this task is to ensure that the designer has identified all of the required inputs to the system and specified them correctly. In the previous chapter, you learned various methods of defining user inputs based on the analysis of activities or use cases. Those techniques defined the user interface through user-centered design. In this chapter, we focus on defining the system inputs, including user inputs, by analyzing models that were built during the analysis activities. As with other aspects of analysis and design, this task also provides a mechanism to cross-check the quality of both the user-centered design and the detailed information developed in the analysis models.

The fundamental approach that analysts use to identify user and system inputs is to identify all information flows that cross the system boundary for each activity or use case. The idea is the same for both traditional structured models and object-oriented models; however, the detailed techniques are unique to each model. We begin with the structured approach.

Using Traditional Structured Models

During systems design with structured techniques, one of the first tasks is to define the automation boundary. Figure 15-3, duplicated from Figure 10-2, is an example of an automation boundary on a data flow diagram. Several of the inputs to the system based on this data flow diagram are as follows:

- Time card information
- Updates to tax rate tables
- Updates to employee files

The updates to employee files are probably done with a graphical user interface screen. However, both the time card information and the updates to the tax tables are more likely done through a system interface that is not graphical. Time card information frequently

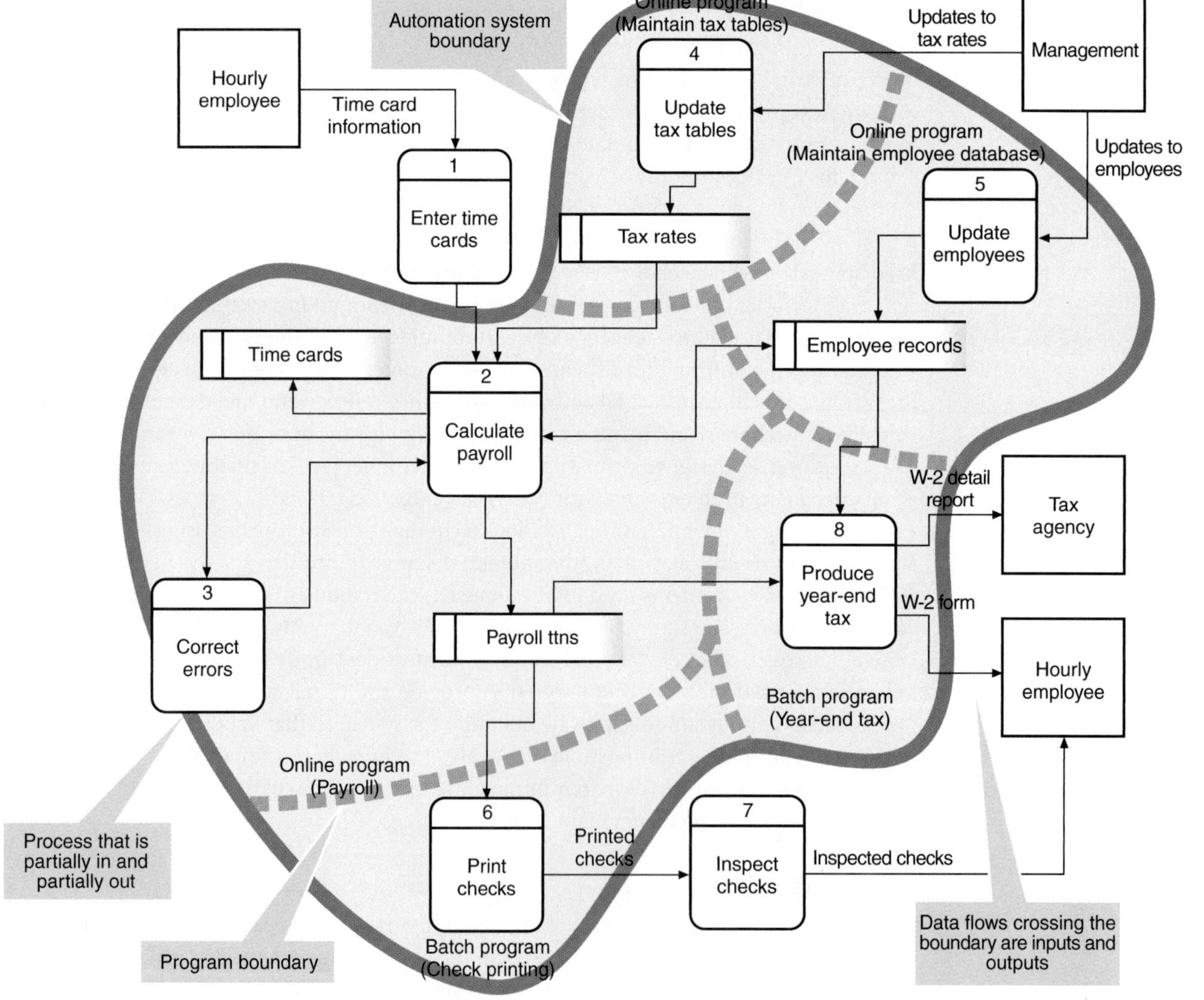

Figure 15-3 An automation boundary on a system-level DFD

comes through an automated employee check-in system. Tax updates can come directly from a government service bureau and can be provided electronically.

Even though it is possible to build an automation boundary on a high-level DFD and identify the inputs on this diagram, it usually is better to work from the DFD fragments or even more detailed DFDs. The high-level diagram frequently does not provide enough detail to discern many data flows and, hence, inputs to the system. For example, one of the processes on the diagram, *Correct errors*, is intersected by the system boundary. Thus, a view from a lower-level DFD, which provides more detailed process bubbles for the *Correct errors* process, would be necessary to discern what processes to include within the automated system and what data flows cross the boundary.

Chapter 6 explained how to build DFD fragments based on the events in the event table. For more complex models, you can define system inputs by looking at each DFD fragment and creating the system boundary on each fragment. The high-level DFD with an automation boundary gives a good overview, but the DFD fragments, or even the detailed DFD for each fragment, are easier to work with.

Figure 15-4 shows the *Create new order* detailed data flow diagram from RMO with the automation boundary superimposed. The input data flows crossing the boundary are clearly defined, so required inputs will be the new order information data flow and the real-time link from the credit bureau. The input for the user interface will be the new order information, and the real-time link to the credit bureau will be an electronic system interface.

Figure 15-4

The *Create new order* DFD with an automation boundary

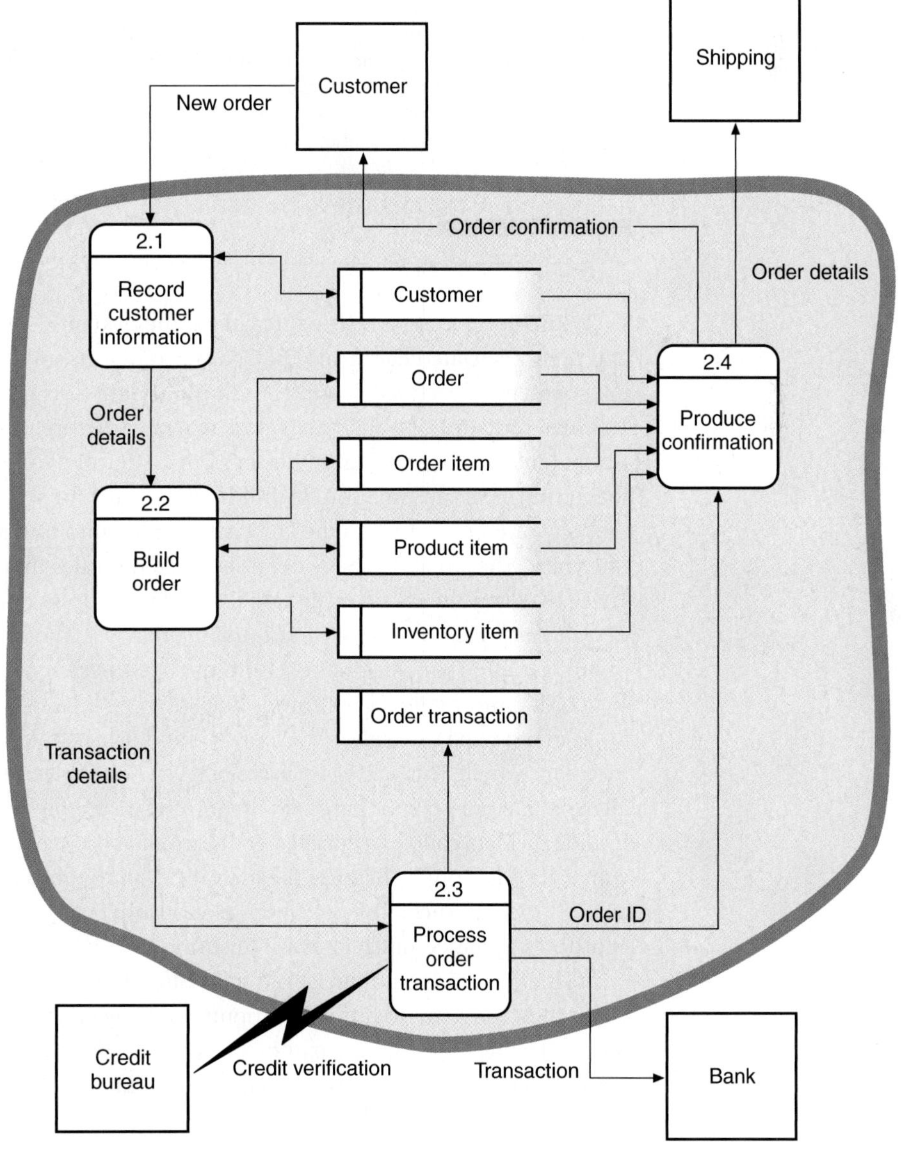

The designer analyzes each DFD fragment to determine the required inputs. The data flows that cross the boundary on the DFDs as inputs correspond to triggers for external events in the event table. The result of this task is a list of high-level inputs for the new system. Figure 15-5 is a list of inputs for RMO's customer support system as developed from the DFD fragments. To develop this list, the designer analyzes every DFD fragment in Figure 6-12, as well as all other DFD fragments for RMO. The purpose of this preliminary list is to provide a master control list of all system inputs and user inputs that need to be designed. It does not, however, provide quite enough detailed information to design the inputs themselves. The additional information that is needed is obtained from the data flow definitions, structure charts, and the user-centered design activities (for the user interface) as explained in Chapter 14.

Figure 15-5

List of inputs (system and user) for the customer support system

Item inquiry
New order information
Change order information
Order status inquiry
Order fulfillment notice
Back-order notice
Order return notice
Catalog request
Customer account update
Promotion package information
Customer change adjustment
Catalog update information
Special promotion information
New catalog information
Credit-card authorization

While developing the structure charts, the designer defines individual program modules and their associated data couples. Chapter 10 discussed the process of defining the detailed data content of each data couple. Each input data flow on a data flow diagram might translate into one or more physical inputs on the structure chart. In Figure 15-6, which derives from Figure 10-15, input modules have been defined for getting customer information and for getting order information, including the details for several order line items. In this figure, the New order data flow on the DFD is expanded into four separate data couples on the structure chart. The structure chart identifies three modules that get data from outside the system. These three modules and their associated data couples are named *Get customer information, Get order information,* and *Get credit-card information.* In other words, it requires three modules to provide all of the information from outside the system on the Customer information and New order information data flows.

The next step is to analyze each module and data couple and list the individual data fields for each data couple. This analysis consists of reviewing the elements in the data stores to ensure that all elements can be built based on the input data couples. Figure 15-7 expands Figure 15-5 to include the data couples associated with each data flow as well as the data fields to be associated with each data couple. Because the identification of the detailed data elements is based on the analysis models, including the entity-relationship data model, it provides a cross-check for the approach explained in Chapter 14. Ensuring that the two approaches yield the same data elements is a powerful technique to verify the quality of the design.

Each of the items identified as inputs made by end users and listed in the data couple column of Figure 15-7 becomes part of an electronic input data form or an input/output form. An input/output form permits users to enter key blank fields and then query the database to display

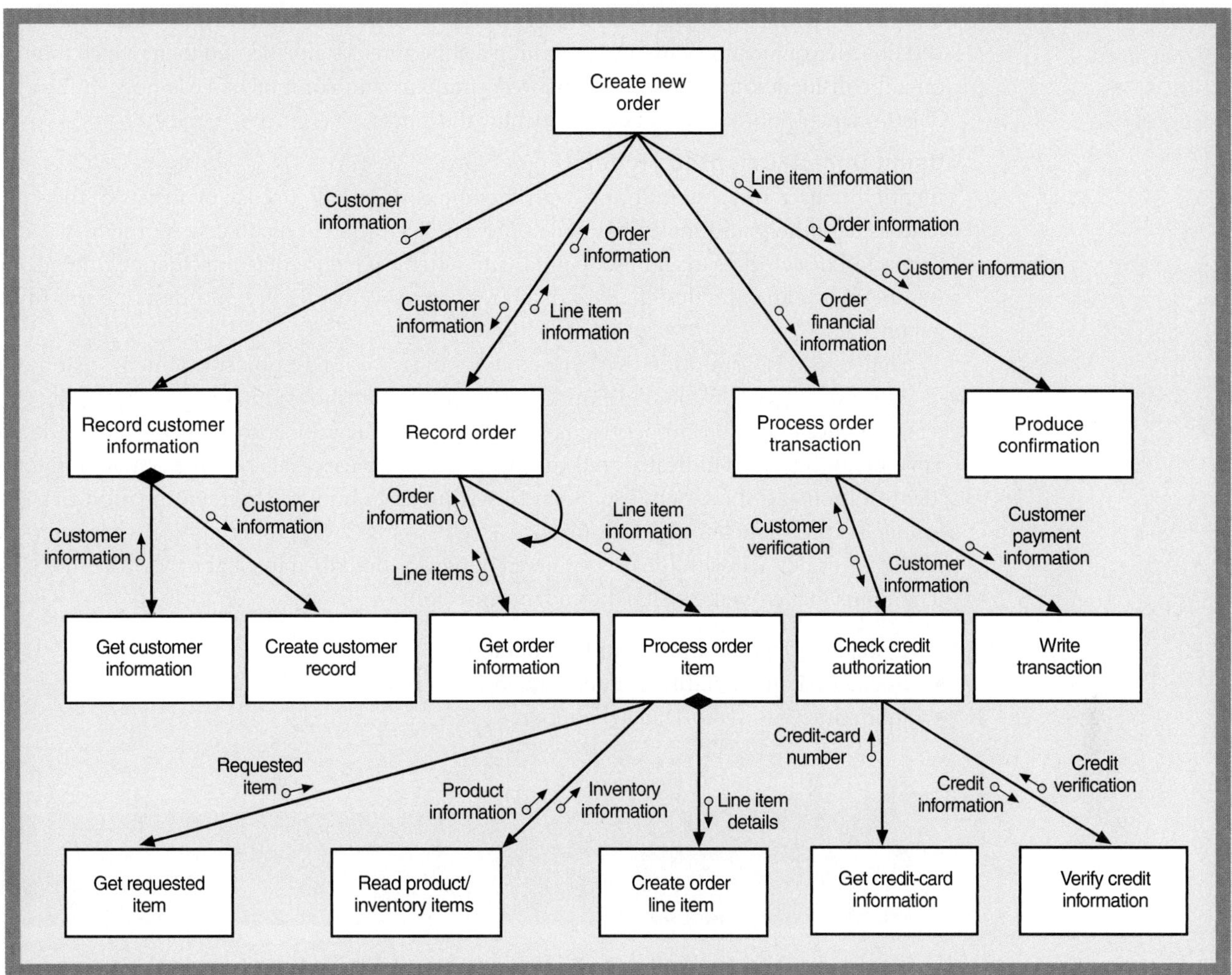

Figure 15-6

Structure chart for *Create new order*

Figure 15-7

The data flows, data couples, and data elements making up inputs

Data flow	Data couples	Data elements
New order	Customer information	Account number, Name, Billing address, Shipping address, Day phone, Night phone
	Order information	Order date, Priority code
	Line item information	Product ID, Color, Size, Quantity
	Credit-card number	Credit-card name, Credit-card number, Expiration date
Credit authorization (real time)	Credit information (output)	Credit-card number, Expiration date, Customer name, Amount
	Credit verification (input)	Accept/Reject code, Date/time, Authorization number, Amount

information such as product descriptions. These tasks associated with design of input forms are best done in conjunction with the application architecture design tasks and then refined through careful consideration of interaction between humans and computers. Let's now shift to the object-oriented approach to see how to identify the inputs.

Using Object-Oriented Models

Identifying user and system inputs with the object-oriented approach consists of the same tasks as with the traditional structured approach. The difference is that system sequence diagrams and design class diagrams are used. The system sequence diagrams identify the incoming messages, and the design class diagrams are used to identify and describe the input parameters.

Figure 15-8 is a simplified system sequence diagram for an object-oriented version of a payroll system (such as the traditional structured version shown in Figure 15-3). In this system sequence diagram, snippets from various use cases have been combined to illustrate the same major inputs as indicated in Figure 15-3. The messages that cross the system boundary identify inputs, both system inputs and user interface inputs. The identification of these inputs on the sequence diagram provides a cross-check with the graphical user interface (GUI) forms defined with the user-centered design, as described in Chapter 14. The three system inputs that cross the system boundary are as follows:

- updateEmployee (empID, empInformation)
- updateTaxRate (taxTableID, rateID, rateInformation)
- inputTimeCard (empID, date, hours)

BEST PRACTICE

Consider and plan for the system interfaces for inputs and outputs very early—when defining requirements with use case descriptions and system sequence diagrams.

The first input is part of the GUI and is detailed during the design of the user interface. The other two inputs, however, are from external systems and do not require user involvement. The information from the tax bureau can be sent as a set of real-time messages or in the form of an input file on a CD or some other electronic device. The time card information could come into the system in various formats. Perhaps physical time cards are entered via an electronic card reader. Or an input from a subsystem, such as an electronic employee ID card reader, might send time card information at the end of every workday. These last two input messages need to be precisely defined, including transmission method, content, and format.

Figure 15-9 is a variation of the system sequence diagram for the telephone order scenario of the *Create new order* use case, as originally shown in Figure 7-13. In the figure, the Order Clerk actor and the BankSystem package are external to the system. The messages that go from the Order Clerk to the system are part of the user interface. The messages that go between the external BankSystem package and the system are system inputs. In the object-oriented models, the boundary between actors and external packages with the internal objects is more explicit than in structured models. In Figure 15-9, four messages go from the Order Clerk actor to the system, and one input message enters the system boundary from the BankSystem package. The input messages, along with the actual message signatures as shown in the figure, are as follows:

- startOrder(accountNo)
- addItem(catalogID, prodID, size, quantity)
- completeOrder ( )
- makePayment(paymentAmt, ccInformation)
- returnVerification (creditCard#, verificationCode, amount)

Figure 15-8

Partial system sequence diagram for the payroll system use cases

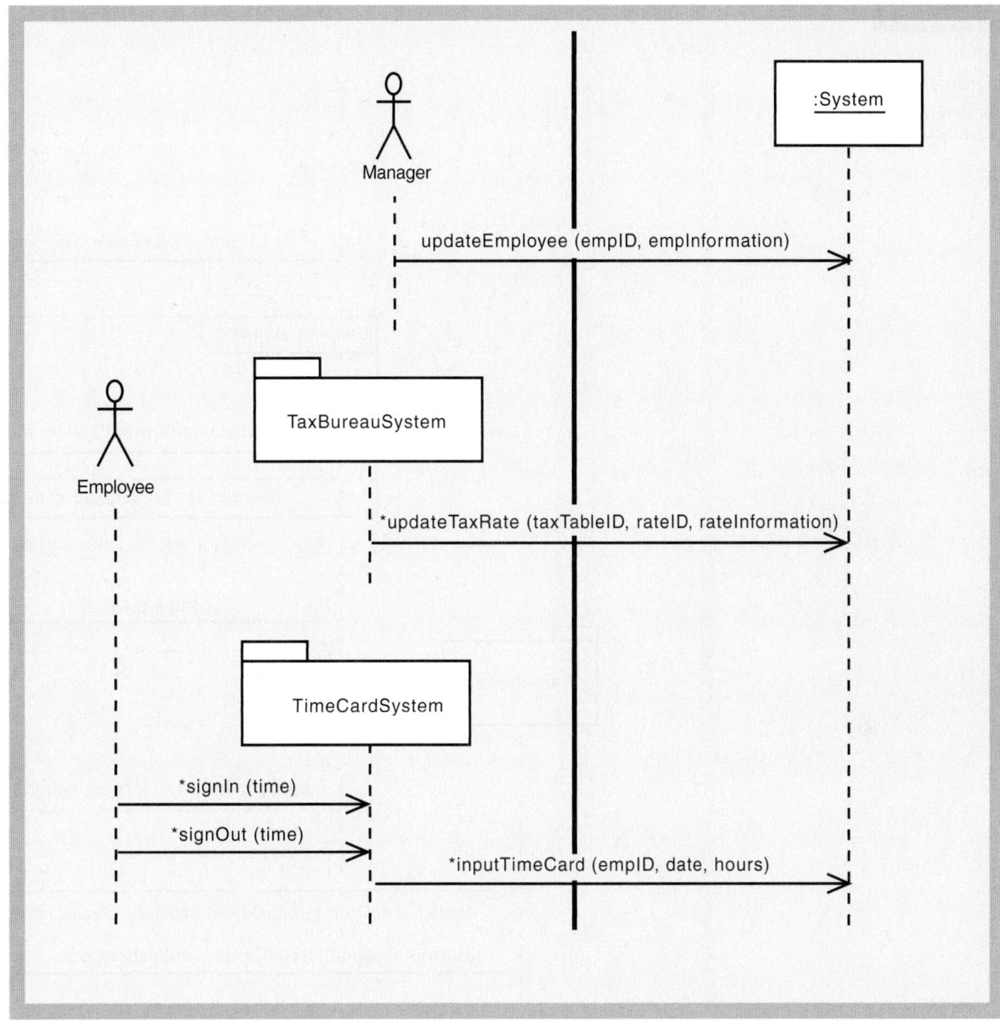

The system sequence diagram developed during the analysis activities identifies the series of steps that occur in the overall process to create a new order from a telephone call. Along with a sequence diagram, a detailed dialog is developed to highlight the communications between the user and computer, as explained in Chapter 14. The point to note here is that a sequence diagram provides a detailed perspective of the user and system inputs to support the use case and the corresponding business event.

Additional analysis of the messages themselves also supplies information about the data fields on the message. To obtain a more thorough analysis of the messages, the developer might need to consult the design class diagram. The actual parameters that are passed in on the messages need to be consistent with the attributes that are found in the design classes. Because the design class attributes are typed, the input parameters can also be typed to be consistent with the design class attributes.

Figure 15-10 lists each input message and the data fields that must be passed with the message. In the example, we show the data associated with every input message. However, the messages associated with the user interface might have already been precisely specified during the design of the user interface. In that case, only the system inputs need to be placed in the table. Not only is this analysis necessary to develop the details of system inputs, but it also provides a good check on the analysis. Notice that the table is more detailed to define the input parameters more precisely. Often this detail will be transferred back to the sequence diagram to make it more complete as well.

Figure 15-9

System sequence diagram for *Create new order*

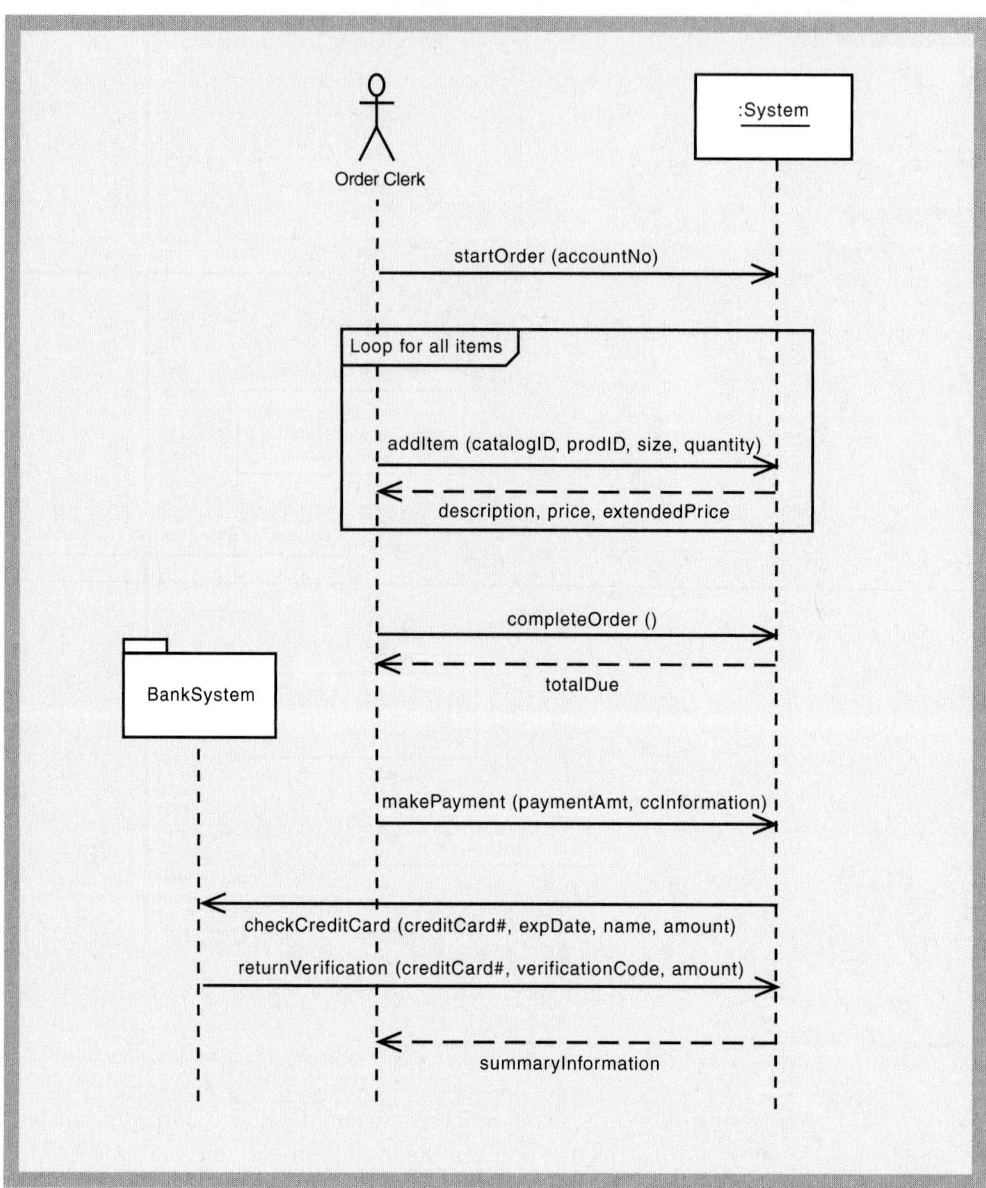

Figure 15-10

Input messages and data parameters from an RMO system sequence diagram

Message	Data parameters
startOrder	accountNo
addItem	catalogID, prodID, size, quantity
completeOrder	—
makePayment	paymentAmt, creditCard#, expDate, name
returnVerification	creditCard#, verificationCode, amount

DESIGNING SYSTEM OUTPUTS

The primary objective of system outputs is to present information in the right place at the right time to the right people. Historically, the most common form of output information has been printed textual reports. Although straight text and tables are still used extensively, new formats such as charts and diagrams provide many more options to present, emphasize, and summarize information.

As with input design, the tasks in this activity accomplish four objectives:

- Determine the type of each system output.
- Make a list of specific system outputs required based on application design.
- Specify any necessary controls to protect the information provided in the output.
- Design and prototype the output layout.

The purpose of the first two tasks is to evaluate the various alternatives and to design the most appropriate approach for each needed output. The list of required output reports is normally specified during the analysis activities as part of modeling system requirements. During design, the task is to coordinate the production of those outputs with the modules (structured techniques) and methods (object-oriented techniques) that are identified during the application architecture design.

The third task ensures that the designer evaluates the value of the information to the organization and protects it. Frequently, organizations implement controls on the inputs and system access but forget that output reports often have sensitive information. The upcoming section "Designing Integrity Controls" identifies several important controls that all outputs should include.

ad hoc reports
reports that are not predefined by a programmer but designed as needed by a user

Today users can develop their own reports using tools and preformatted templates. These reports, called **ad hoc reports**, have not been designed by programmer/analysts. Instead, an ad hoc report is the result of a new user query to a database in response to a specific question or need. In Chapter 13, you learned about relational databases and Structured Query Language (SQL). Many systems provide a simplified graphical tool to permit users to formulate queries in SQL to produce ad hoc reports. Obviously, analysts do not design those reports during system development. However, the tools and capability to support user requests do need to be built into the system. The report-design activity is a good time to ask whether the system requires an ad hoc reporting capability and to add it if necessary.

DEFINING THE DETAILS OF SYSTEM OUTPUTS

The objective of this task is to ensure that the designer has identified and specified all of the outputs for the new system. The technique is the same as that used for the definition of the system inputs. This model-based approach utilizes the information in the event table and other models to identify and define the detailed specifications of the outputs. Although many system inputs might be defined as part of the user-interface design, many outputs do not require dynamic human-computer interaction—for example, printed reports, turnaround documents, or simple screen displays. So, analysts must look to the analysis models they developed earlier for many more system outputs. For analysts using traditional structured techniques, the data flows from an internal process to an external agent or external process identify the outputs. For object-oriented techniques, messages that originate on internal classes and whose destination is an actor or another external system are the outputs.

Using Traditional Structured Models

To identify the outputs in the traditional approach, analysts look at data flows coming out of the system across the system boundary in the data flow diagrams and fragments. The payroll example shown in the DFD of Figure 15-3 contains several system outputs that are not normally part of the graphical user interface. Processes 6, *Print checks*, and 8, *Produce year-end tax*, have three outputs that are generated in a batch execution. The program to print checks prints all payroll checks at once for a specific payroll period. Similarly, the year-end tax program will print W-2 forms and a detail report. Batch-oriented reports are always classified as system outputs.

Figure 15-4, the *Create new order* DFD, shows three outputs: a confirmation to the customer, a notice to shipping, and a payment transaction that goes to the bank. As with system inputs, analysts should build a table of the DFD outputs, defining exactly what reports are needed and determining the data fields. Data flow definitions for each of these outputs should have been created and documented on a data flow diagram as part of the analysis of requirements. One additional system output shown in Figure 15-4, which is not a report, is the electronic credit verification going to the credit bureau.

As analysts build the table of DFD outputs, report definitions, and data fields, they add two more columns to the three identified in the input table. These two additional columns list (1) the files or data stores that are required to produce the report and (2) the number of records from which the report is generated—a single record or a set of records. Figure 15-11 is an example of the table of system outputs. The single/multiple record column generally indicates whether the report is printed immediately after each process or in batch at some other time. For example, from Figure 15-3, the W-2 reports will process multiple records to print the report.

Figure 15-11

A table of system outputs based on the traditional structured approach

DFD data flow	Structure chart data couple	Data content	Database files required	Single record/ multiple records
Order confirmation	Produce confirmation	Account-Number, Name, Billing-Address, Shipping-Address, Order-Number, Date, Priority-Code, Shipping-And-Handling, Tax, Total *(Product-Item-Number, Description, Size, Color, Options, Quantity, Price)	Customer, Order, Order Item	Single
Shipping order details	Produce confirmation	Account-Number, Name, Shipping Address, Order-Number, Date, Priority-Code	Customer, Order	Single
Transaction details	Write transaction	Account-Number, Shipping-And Handling, Tax, Total	Customer, Transaction	Single
Credit verification (real time)	Credit information	Credit-Card-Number, Expiration-Date, Customer-Name, Amount	Customer, Order	Single

To verify that the structure chart modules are consistent with the structure of the output report, analysts again look at the data couples and the report data requirements. An analysis of the data couple being sent to the module and the data fields on the output report will verify that the application has been designed correctly to generate the report.

Using Object-Oriented Models

In the object-oriented approach, outputs are indicated by messages in sequence diagrams that originate from an internal system object and are sent to an external actor. In Figure 15-9, the message with the parameters of description, price, extendedPrice is an example of an output message. This message is generated as a result of the input message addItem(catalogID, prodID, size, quantity) to the internal order object. A review of all the output messages generated across all sequence diagrams provides the consistency check against the required outputs identified during analysis activities.

Output messages that are based on an individual object (or record) are usually part of the methods of that object class. To report on all objects within a class, a class-level method is used. A class-level method is a method that works on the entire class of objects, not a single object. For example, a customer confirmation of an order is an output message that contains information for a single order object. However, to produce a summary report of all orders for the week, a class-level method looks at all the orders in the Order class and sends output information for each one with an order date within the week's time period.

The system sequence diagram in Figure 15-9 shows four output messages—three from the system to the Order Clerk actor, and one to the BankSystem package. Each of these messages must have a list of parameters that are transmitted with it, even though the completeOrder ( ) message in the diagram does not show them. Output design is a good time for analysts to elaborate on the messages to include all of the required parameters. Similar to the traditional approach, Figure 15-12 shows a table listing the output messages, the database files or tables required, and the number of objects (a single object or a set of objects) to be included in the report. Comparing this figure with Figure 15-11, you might notice that some outputs are not identified. Because Figure 15-12 derives from a single use case, the *Create order* use case, it does not include other messages such as shipping messages that are on other use cases.

Figure 15-12

A table of system outputs based on object-oriented messages

Output message	Data parameters	Classes or database tables	Single record/ multiple records
Response to addItem ()	description, price, extendedPrice	CatalogProduct, ProductItem, InventoryItem, OrderItem	Single
Response to completeOrder ( )	totalDue	Order, OrderItem	Single
checkCreditCard ( )	creditCard#, expDate, name, amount	Customer, OrderTransaction	Single
summary Information—response to makePayment ( )	customerName, billingAddress, shippingAddress, orderNumber, date, s&h, tax, totalAmount	Order, OrderItem, Customer	Single

DESIGNING REPORTS, STATEMENTS, AND TURNAROUND DOCUMENTS

With the advent of office automation and other business systems, businesspeople originally thought that paper reports would no longer be needed. In fact, just the opposite has happened. Business systems have made information much more widely available, with a proliferation of all types of reports, both paper and electronic. In fact, one of the major challenges to organizations and the designers of their information systems today is to organize the overwhelming amount of information so that it is meaningful. One of the most difficult aspects of output design is to decide what information to provide and how to present it to avoid a confusing mass of complex data.

Type of Output

Before looking at the different formats that analysts use in designing reports, let's discuss four types of output reports that users require: detailed reports, summary reports, exception reports, and executive reports.

detailed report
a report containing detailed transactions or records

Detailed reports are used to carry out the day-to-day processing of the business. They contain specific information on business transactions. Sometimes a report might be for a single transaction, such as an order confirmation sheet with details of a particular customer order. Other detailed reports list a set of transactions—for example, a list of all overdue accounts, with each line of the report presenting information about a particular account. A clerk could use this report to research overdue accounts and determine actions to collect past-due amounts. The purpose of detailed reports is to provide working documents for people in the company.

summary report
a report that recaps or summarizes detailed information over a period of time or some category

exception report
a report that contains only information about nonstandard, or exception, conditions

Summary reports are often used to recap periodic activity. An example of this report is a daily or weekly summary of all sales transactions, with a total dollar amount of sales. Middle managers often use this type of report to track departmental or division performance. Exception reports are also used to monitor performance. An **exception report** is produced only when a normal range of values is exceeded. When business is progressing normally, no report is needed. But when something exceeds an expected range, a report is produced to alert staff. An example is a report from a production line that lists rejected parts. If the reject rate is above a set threshold, a report is generated. Sometimes exception reports are produced regularly. The rejected parts report might be produced every day if the production line usually has some rejected parts. An aged accounts receivable report might be produced each month showing the accounts that are past due. Unfortunately, the organization may always have some accounts to list in such reports, so they are produced regularly.

executive report
a summary report from various information sources that is normally used for strategic decisions

Top management uses **executive reports** to assess overall organizational health and performance. These reports thus contain summary information from activities within the company. They might also contain comparative performance with industry-wide averages. Using these reports, executives can assess competitive strengths or weaknesses of their company.

Chapter 1 discussed the various types of information systems that systems analysts develop. You will note that some types of information systems focus on producing a particular type of report. Although there is no strict requirement for a system to produce only one type of report, we often categorize a system based on the type of report it produces. The next section looks at some examples of printed reports.

internal output
a printed report or document produced for use inside an organization

external output
printed documents—such as statements, notices, form letters, and legal documents—produced for use outside an organization

turnaround document
an external output that includes a portion that is returned to the system as an input

Internal Versus External Outputs

Printed outputs are classified as either internal outputs or external outputs. **Internal outputs** are produced for use within the organization. The types of reports just discussed fall under this category. **External outputs** include statements, notices, and other documents that are produced for people outside the organization. Because they are official business documents for an outside audience, they need to be produced with the highest-quality graphics and color. Some examples include monthly bank statements, late notices, order confirmation and packing slips (such as those provided to Rocky Mountain Outfitters' customers), and legal documents such as insurance policies. Some external outputs are referred to as **turnaround documents** because they are sent to a customer but include a tear-off portion that is returned for input later, such as a bill that contains a payment stub to be returned with a check. All of these printed outputs must be designed with care, but organizations have many more options for printed output. Today's high-speed color laser printers enable all types of reports and other outputs to be produced.

An example of a detailed report for an external output is shown in Figure 15-13. This report is produced from a Web order similar to that shown in Figure 14-16. Good user-interface design specifies that when a customer places an order over the Web, the system will be able to print the order information as a confirmation. Of course, a user can always print the Web screen display

Figure 15-13
RMO shopping cart order report

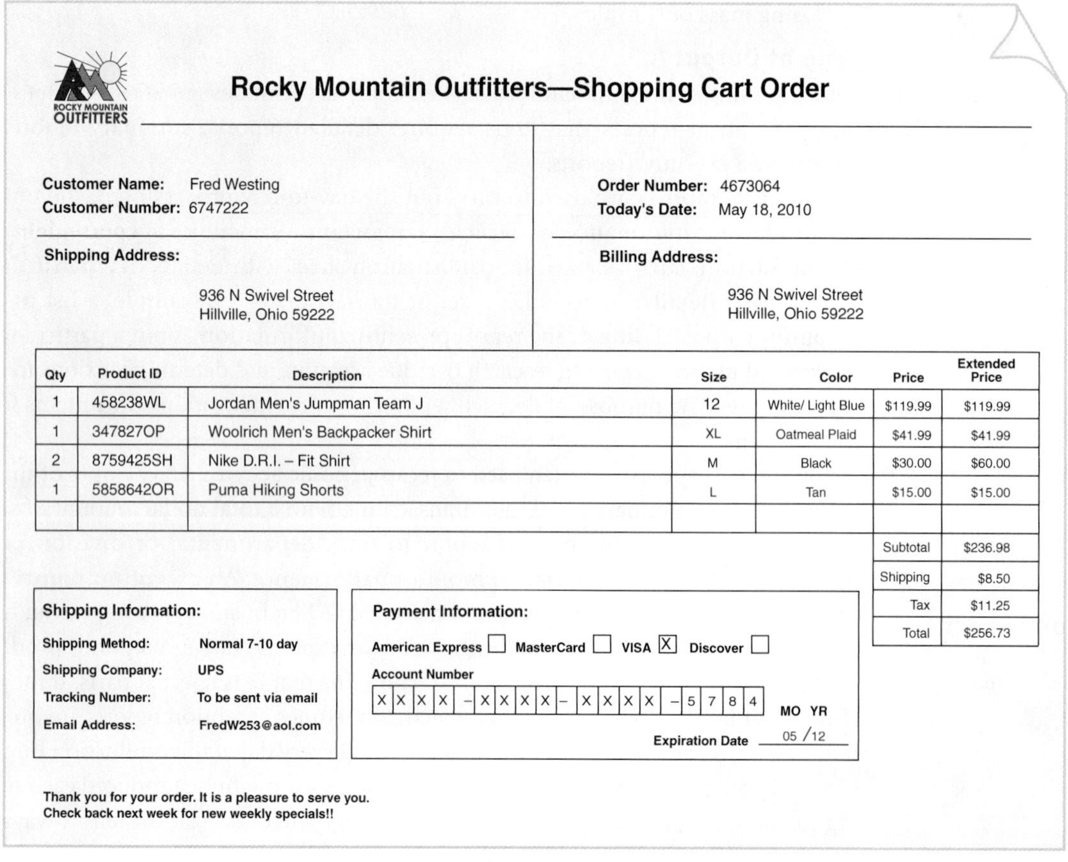
ROCKY MOUNTAIN OUTFITTERS

Rocky Mountain Outfitters—Shopping Cart Order

Customer Name: Fred Westing
Customer Number: 6747222

Order Number: 4673064
Today's Date: May 18, 2010

Shipping Address:
936 N Swivel Street
Hillville, Ohio 59222

Billing Address:
936 N Swivel Street
Hillville, Ohio 59222

Qty	Product ID	Description	Size	Color	Price	Extended Price
1	458238WL	Jordan Men's Jumpman Team J	12	White/ Light Blue	$119.99	$119.99
1	347827OP	Woolrich Men's Backpacker Shirt	XL	Oatmeal Plaid	$41.99	$41.99
2	8759425SH	Nike D.R.I. – Fit Shirt	M	Black	$30.00	$60.00
1	5858642OR	Puma Hiking Shorts	L	Tan	$15.00	$15.00
					Subtotal	$236.98
					Shipping	$8.50
					Tax	$11.25
					Total	$256.73

Shipping Information:
Shipping Method: Normal 7-10 day
Shipping Company: UPS
Tracking Number: To be sent via email
Email Address: FredW253@aol.com

Payment Information:
American Express ☐ MasterCard ☐ VISA ☒ Discover ☐
Account Number
X X X X – X X X X – X X X X – 5 7 8 4
MO YR
Expiration Date 05 /12

Thank you for your order. It is a pleasure to serve you.
Check back next week for new weekly specials!!

using the browser's print capability, but doing so is time consuming because it includes all of the graphics and index links on the page. It is much more user-friendly to provide shoppers a formatted order confirmation only. Figure 15-13 illustrates such a report. This type of report is based on the information of a single order. The data required to print this order are a customer record, an order record, and all of the line-item records for ordered items. Notice that it is nicely formatted for easy reading. Different pieces of information are grouped together and placed within boundaries. The report is comprehensive; it contains complete and current information about the order, including today's date, items on the order, payment details, and shipping details.

In contrast to the Web shopping cart order report, Figure 15-14 is an example of an internal output. This report is different in several ways. First, it is based on an entire set of records from the inventory database, whereas the shopping cart order is for a single order. The report includes

Figure 15-14
RMO inventory report

Rocky Mountain Outfitters — Products and Items

ID	***Season***	***Category***	***Supplier***	***Unit Price***	***Special Price***	***Discontinued***
RMO12587	*Spr/Fall*	*Mens C*	*8201*	*$39.00*	*$34.95*	*No*

Description *Outdoor Nylon Jacket with Lining*

Size	***Color***	***Style***	***Units in Stock***	***Reorder Level***	***Units on Order***
Small	Blue		691	150	
	Green		723	150	
	Red		569	150	
	Yellow		827	150	
Medium	Blue		722	150	
	Green		756	150	
	Red		698	150	
	Yellow		590	150	
Large	Blue		1289	150	
	Green		1455	150	
	Red		1329	150	
	Yellow		1370	150	
Xlarge	Blue		1498	150	
	Green		1248	150	
	Red		1266	150	
	Yellow		1322	150	

ID	***Season***	***Category***	***Supplier***	***Unit Price***	***Special Price***	***Discontinued***
RMO28497	*All*	*Footwe*	*7993*	*$49.95*	*$44.89*	*No*

Description *Hiking Walkers with Patterned Tread Durable Uppers*

Size	***Color***	***Style***	***Units in Stock***	***Reorder Level***	***Units on Order***
7	Brown		389	100	
	Tan		422	100	
8	Brown		597	100	
	Tan		521	100	
9	Brown		633	100	
	Tan		654	100	
10	Brown		836	100	
	Tan		954	100	
11	Brown		862	100	
	Tan		792	100	
12	Brown		754	100	
	Tan		788	100	
13	Brown		830	100	
	Tan		921	100	

control break report
a report that includes detailed and summary information

both a detailed and summary section, sometimes called a **control break report**. A control break is the data item that divides the report into groups. In this example, the control break is on the product item number—called *ID* on the report. Whenever a new value of the ID is encountered on the input records, the report begins a new control break section. The detailed section lists the transactions of records from the database, and the summary section provides totals and recaps of the information. The report is sorted and presented by product. However, within each product is a list of each inventory item showing the quantity currently on hand.

External outputs can consist of complex, multiple-page documents. A well-known example is the set of reports and statements that you receive with your car insurance statement. This statement is usually a multipage document consisting of detailed automobile insurance information and rates, summary pages, turnaround premium payment cards, and insurance cards for each automobile. Another example is a report of employment benefits with multiple pages of information customized to the individual employee. Sometimes the documents are printed in color with special highlighting or logos. Figure 15-15 is one page of an example report for survivor protection from an employee benefit booklet. The text is standard wording, and the numbers are customized to the individual employee.

Figure 15-15
A sample employee benefit report

Survivor Protection

In the event of your death while working for a participating employer, your designated beneficiaries could receive:

Lump Sum Benefits

$50,000	Basic Life Insurance
$230,000	Supplemental Life Insurance
$148,677	Thrift Plan
$31,686	Tax Sheltered Annuity (TSA) Plan
$255	Social Security for your eligible dependents
$460,618	Total*

You have not elected Universal Life Insurance. If you would like more information on this plan, please call 1-800-555-7772.

*Refer to page 7 for additional information on the amount of coverage needed to provide ongoing replacement income.

Accidental Death Benefits
If your death is due to an accident, your designated beneficiaries will receive the above benefits plus:

$100,000	24-Hour Accidental Death and Dismemberment Insurance
$100,000	Occupational Accidental Death and Dismemberment Insurance, if the accident is work related

Monthly Death Benefits
If you die before receiving the Master Retirement Plan benefits and you are vested and have a surviving spouse, your spouse may be eligible for a Qualified Pre-Retirement Survivor Annuity.

In addition, your family may be eligible for the following estimated monthly benefits from Social Security, not to exceed a maximum of $2,591 based on:

$1,110	for each child under age 18
$1,110	for a spouse with children under age 16; or
$1,058	for a spouse age 60 or older

Electronic Reports

Organizations use various types of electronic reports, each serving a different purpose, and each with its respective strengths and weaknesses. Electronic reports provide great flexibility in the organization and presentation of information. In some instances, screen output is formatted like a printed report but displayed electronically. However, electronic reports can also present information in many other formats: Some have detailed and summary sections, some show data and graphics together, others contain boldface type and highlighting, others can dynamically change their organization and summaries, and still others contain hotlinks to related information. An important benefit of electronic reporting is that it is dynamic—it can change to meet the specific needs of a user in a particular situation. In fact, many systems provide powerful ad hoc reporting capabilities so that users can design their own reports on the fly. For example, an electronic report can provide links to further information. One technique, called **drill down**, allows the user to activate a "hot spot hyperlink" on the report, which tells the system to display a lower-level report, providing more detailed information. For example, Figure 15-16 contains a summary valuation report of inventory on hand. The report provides a summary valuation for each product item. However, if the user clicks on the hotlink for any product, a detailed report pops up with the list of inventory items, the quantities on hand, and the valuation for each inventory item.

drill down
to link a summary field to its supporting detail and enable users to view the detail dynamically

Monthly Sales Summary

Year **2010** *Month* **January**

Category	*Season Code*	*Web Sales*	*Telephone Sales*	*Mail Sales*	*Total Sales*
Footwear	All	$ 289,323	$ 1,347,878	$ 540,883	$ 2,178,084
Men's Clothing	Spring	$ 1,768,454	$ 2,879,243	$ 437,874	$ 4,691,484
	Summer	213,938	387,121	123,590	724,649
	Fall	142,823	129,873	112,234	384,930
	Winter	2,980,489	6,453,896	675,290	10,109,675
	All	[illegible]	4,897,235	349,234	7,086,198
Totals			[illegible]	$ 1,698,222	$ 23,391,023
Women's Clothing	Spring				[illegible]
	Summer				
	Fall				
	Winter				
	All				
Totals					

Monthly Sales Detail

Year **2010** *Month* **January** *Category* **Men's Clothing** *Season* **Winter**

Product ID	*Product Description*	*Web Sales*	*Telephone Sales*	*Mail Sales*	*Total Sales*
RMO12987	Winter Parka	$ 1,490,245	$ 3,226,948	$ 337,640	$ 5,054,833
RMO13788	Fur-Lined Gloves	149,022	322,695	33,765	505,482
RMO23788	Wool Sweater	596,097	1,290,775	135,058	2,021,930
RMO12980	Long Underwear	298,050	645,339	68,556	1,003,005
RMO32998	Fleece-Lined Jacket	447,075	1,258,079	100,271	1,805,425
Total		$ 2,980,489	$ 6,743,836	$ 675,290	$ 10,394,615

Figure 15-16
An RMO summary report with drill down to the detailed report

Another variation of this hotlink capability lets the user correlate information from one report to related information in another report. Most people are familiar with hotlinks from using their Internet browsers. In an electronic report, hotlinks can refer to other information that correlates or extends the primary information. This same capability can be very useful in a business report, which, for example, links the annual statements of key companies in a certain industry.

Another dynamic aspect of electronic reports is the capability to view the data from different perspectives. For example, it might be beneficial to view sales commission data by region, by sales manager, by product line, or by time period or to compare the data with last season's data. Instead of printing all these reports, you can use an electronic format to generate the different views only as needed. Sometimes long or complex reports include a table of contents with hotlinks to the various sections of the report. Some report-generating programs provide electronic reporting capability that includes all of the functionality found on pages on the Internet, including frames, hotlinks, graphics, and even animation.

Graphical and Multimedia Presentation

The graphical presentation of data is one of the greatest benefits of the information age. Tools that permit data to be presented in charts and graphs have made information reporting much more user-friendly for printed and electronic formats. Information is being used more and more for strategic decision making as businesspeople examine their data for trends and changes. In addition, today's systems frequently maintain massive amounts of data—much more than people can review. The only effective way to use much of this data is by summarizing and presenting it in graphical form. Figure 15-17 illustrates a bar chart and a pie chart, which are two common ways to present summary data.

Figure 15-17

Sample bar chart and pie chart reports

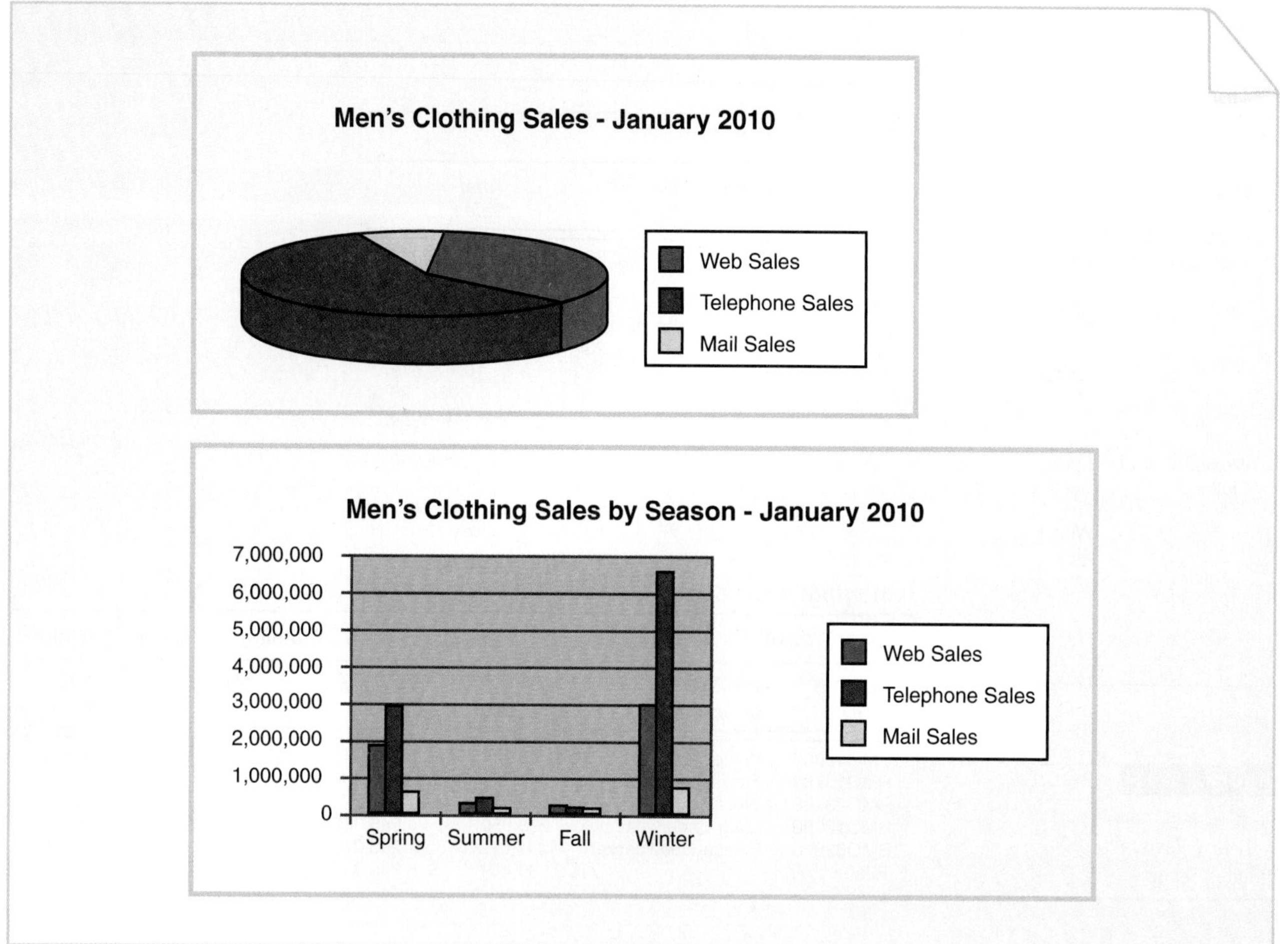

Multimedia outputs have become available recently as multimedia tool capabilities have increased. Today it is possible to see a graphical, and possibly animated, presentation of the information on a screen and have an audio description of the salient points. Combining visual and audio output is a powerful way to present information. (Of course, video games are pushing the frontier of virtual reality to include visual, audio, tactile, and olfactory outputs.)

As the design of the system outputs progresses, it is beneficial to evaluate the various presentation alternatives. Reporting packages can be designed into the system to provide a full range of reporting alternatives. Developers should carefully analyze each output report to determine the objective of the output and to select the form of the output that is most appropriate for the information and its use.

The previous examples illustrated some reports for RMO's customer support system. RMO has many options to present data to satisfy the needs of the different users. In Chapter 4, the RMO organization chart showed that many upper and middle managers had an interest in the new system. Each of the departments represented needs information from the system, and in most cases, each will want the information to be presented in a format unique to its needs. Barbara Halifax emphasized to her team the importance of this flexibility at various times during the development of the outputs (see the accompanying memo).

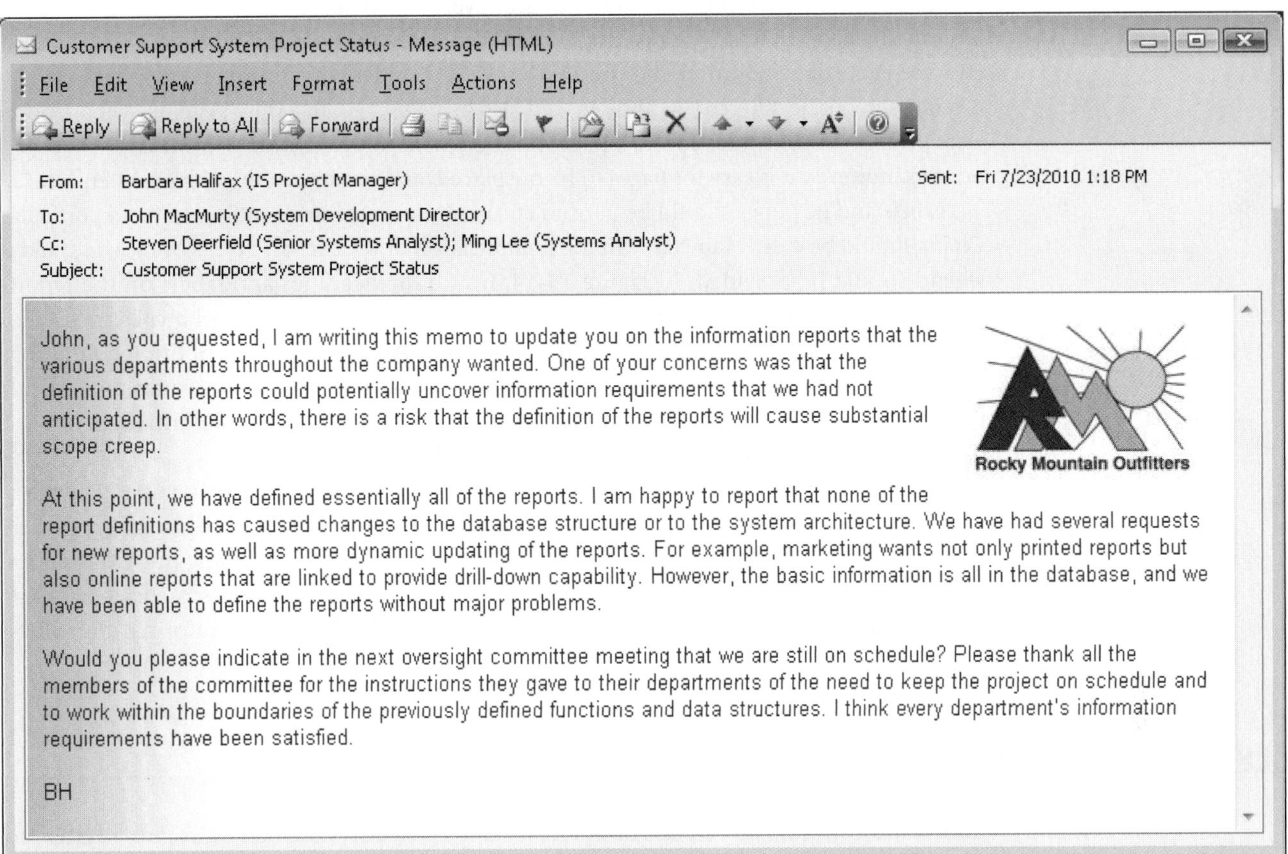
Customer Support System Project Status - Message (HTML)

File Edit View Insert Format Tools Actions Help

Reply | Reply to All | Forward

From: Barbara Halifax (IS Project Manager) Sent: Fri 7/23/2010 1:18 PM
To: John MacMurty (System Development Director)
Cc: Steven Deerfield (Senior Systems Analyst); Ming Lee (Systems Analyst)
Subject: Customer Support System Project Status

John, as you requested, I am writing this memo to update you on the information reports that the various departments throughout the company wanted. One of your concerns was that the definition of the reports could potentially uncover information requirements that we had not anticipated. In other words, there is a risk that the definition of the reports will cause substantial scope creep.

At this point, we have defined essentially all of the reports. I am happy to report that none of the report definitions has caused changes to the database structure or to the system architecture. We have had several requests for new reports, as well as more dynamic updating of the reports. For example, marketing wants not only printed reports but also online reports that are linked to provide drill-down capability. However, the basic information is all in the database, and we have been able to define the reports without major problems.

Would you please indicate in the next oversight committee meeting that we are still on schedule? Please thank all the members of the committee for the instructions they gave to their departments of the need to keep the project on schedule and to work within the boundaries of the previously defined functions and data structures. I think every department's information requirements have been satisfied.

BH

FORMATTING REPORTS

With all of the choices available today for output format, system designers have more flexibility in what they can offer users. But sifting through those options to provide workable reports can be challenging. Analysts must keep three principles in mind during the design of output reports:

- What is the objective of the report?
- Who is the intended audience?
- What is the medium for presentation?

The importance of these principles cannot be overemphasized when designing reports. In some instances users only need the reports to monitor progress. In others, however, the report might be a critical element in a strategic decision. As a system designer, you should be sure you understand who is going to use the report and how he or she is going to use it. Both the content and the format of the report should be decided based on the audience and the use of the report.

Without considering these factors, you could easily omit critical information or present it in an unwieldy format.

Often designers must decide on the level of detail for the format of the report. It can be tempting to produce reports that mirror the structure and format of the data in the database. Newer systems, however, maintain a tremendous amount of detail in the database. Without careful consideration, designers can easily produce reports that suffer from information overload. Information overload occurs when so much data is provided that it becomes difficult for the user to find and focus on the information that is important. Many people have this problem when searching the Internet—the search engine often returns an overwhelming number of results for the search. Careful design and presentation are required to prevent the same problem with output reports.

The format of the report is also important. Every report should have a meaningful title to indicate the data content. The heading should also list both the date the report was produced and a separate date indicating the effective date of the underlying information; sometimes the two dates might be different. Reports should also be paginated. In earlier systems, when reports were printed on continuous forms, page numbers were not as critical. With today's sheet-fed printers, however, it is easy for pages to be misplaced, and results can then be misinterpreted.

Labels and headings should be used to ensure the correct interpretation of the report data. Charts should be clearly labeled with the identification of the axes and units of measure, and a legend should be provided. In Figure 15-14, notice the headings and labels on the report, which help to ensure that the reader does not misinterpret the data. Control breaks are used to divide the data into meaningful pieces that can be easily referenced. Use of lines, boldfacing, and different-sized fonts makes the report easy to read. Generally, report design is not difficult if you remember that the objective of any report is to provide meaningful information, not just data, and to provide it in a format that is easy to read.

Designers often assume that reports will be printed on standard stock paper. However, that assumption might not be correct. As we just saw, electronic reports are also a very powerful method of producing output information, and the forms of electronic presentation are becoming more and more diverse, ranging from standard computer screens to wireless portable devices. Designers need to carefully consider whether output information will be accessed from nonstandard devices and transmitted via limited-bandwidth channels.

DESIGNING INTEGRITY CONTROLS

Information system controls are mechanisms and procedures that are built into a system to safeguard both the system and the information within it. Let's describe a few scenarios to illustrate the need for controls.

- A furniture store sells merchandise on credit with internal financing. An error was made to a customer balance. How do we ensure that only a manager, someone with authority to make adjustments to credit balances, can make the correction?
- A person in accounts payable uses the system to write checks to suppliers. How does the system ensure that the check is correct and that it is made out to a valid supplier? How does the system ensure that no one can commit fraud by writing checks to a bogus supplier? How does the system know that a given payment has been authorized?
- Many companies now have internal LAN networks or intranets. How does a company protect its sensitive data from being accessed by outsiders or even from disgruntled employees?
- Electronic commerce is expanding exponentially, and many companies are now providing e-commerce sites. How does a company ensure that the financial transactions of its customers are protected and secure? How does a company make sure that its systems and databases are protected from hackers who use the Internet access paths to break in?

- Many companies are now connected to the Internet to provide online access to external employees, such as salespeople, or to customers and suppliers. How does a company safeguard its systems from viruses, worms, and other malicious attacks?

All of these situations involve common business and system activities. Because a company's information is one of its most valuable assets, developers of a new system must consider how to protect and maintain information integrity. As illustrated in Figure 15-18, various locations must be protected with security measures and controls. Some of the controls must be integrated into the application programs that are being developed and the database that supports them. Other controls are part of the operating system and the network. Generally, controls that are integrated into the application and database are called **integrity controls**. The controls in the operating system and network are often referred to as *security controls*. This section explains integrity controls. Later sections discuss security controls.

integrity control
mechanisms and procedures that are built into an application system to safeguard information contained within it

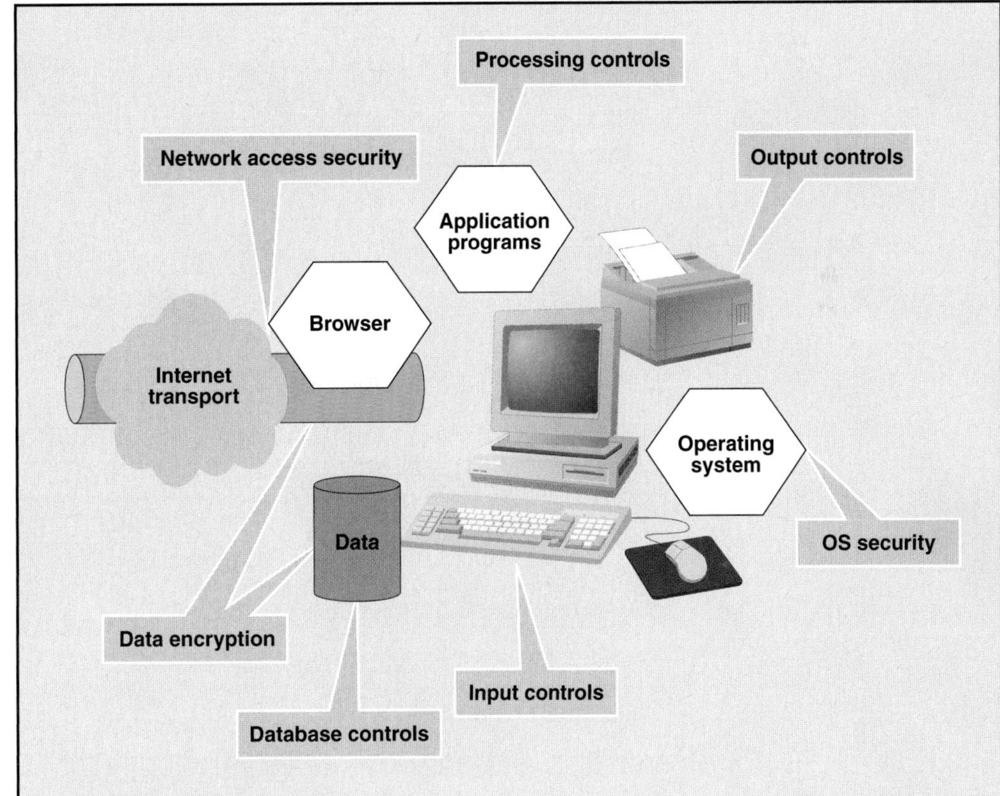

Figure 15-18
Points of security and integrity controls

Usually when considering integrity controls, system developers focus on avoiding problems with the application systems and the employees who rightly have access to those systems. Thus, the primary focus is internal—inside the organization. The primary objectives of integrity controls are to:

- Ensure that only appropriate and correct business transactions occur
- Ensure that the transactions are recorded and processed correctly
- Protect and safeguard the assets of the organization (including hardware, software, and information)

The first objective, to ensure that only appropriate and correct business transactions occur, focuses on the identification and capture of input transactions. Integrity controls must make sure that all important business transactions are included—that is, that none are lost or missing and that no fraudulent or erroneous transactions are entered.

The second objective, to ensure that the transactions are recorded and processed correctly, also relates to errors and fraudulent activities. Controls need to detect and alert users to data-entry errors and system bugs that cause problems in processing and recording data. An example of a fraudulent activity is a user who changes the dollar amount on an otherwise valid transaction.

The third objective, to protect and safeguard the assets of the organization, addresses loss of information from computer crashes or catastrophes. It also includes protection of important information on computer files that could be destroyed by a disgruntled employee or possibly even a hacker.

Frequently, system developers are so focused on designing the system software itself that they forget to develop the necessary controls. Because computer systems are so pervasive and companies depend on information systems so heavily, a development project that does not specifically include integrity controls is inviting disaster. The system will be subject to errors, fraud, and deceptive practices, making it unusable. One of the primary control points for ensuring correct data is at the point of data input.

INPUT INTEGRITY CONTROLS

Input integrity controls are used with all input mechanisms, from electronic devices to standard keyboard inputs. Input controls are an additional level of verification that helps reduce errors on input data. For example, a system might need a certain amount of information for a valid entry, but an input device cannot ensure that all the necessary fields have been entered. An additional level of verification, a control, is necessary to check for completeness.

The old computer systems adage of "garbage in, garbage out" relates to input controls, in which the objective is to reduce bad data within the system by limiting erroneous input. Historically, the most common control method to ensure correct input was to enter data twice. This technique, called *keypunch and verify*, was first developed for batch entry of large amounts of data. One person would enter the data, and a second person would reenter it on equipment that would then verify that the two inputs were the same. Today, that technique is not used as much because many high-volume transactions are scanned for data. Online systems also validate input as it is being entered. Here are the more common control techniques in use today:

field combination control
an integrity control that verifies the data in one field based on data in another field or fields

value limit control
an integrity control that identifies when a value in a field is too large or too small

completeness control
an integrity control to ensure that all necessary fields on an input form have been entered

data validation control
an integrity control to validate the input data for correctness and appropriateness

- **Field combination controls** review various combinations of fields to ensure that the correct data is entered. For example, on an insurance policy, the application date must be prior to the date the policy is placed in force.
- **Value limit controls** check numeric fields to make sure that the amount entered is reasonable. For example, the amount of a sale or the amount of a commission usually falls within a certain range of values.
- **Completeness controls** ensure that all the necessary fields are completed. This check can be executed as input occurs so that, depending on which fields are entered, additional required fields must also be entered. For example, if a dependent is entered on an insurance form, that person's birthday must also be entered.
- **Data validation controls** ensure that numeric fields that contain codes are correct. For example, bank account numbers might be created with a seven-digit field and a trailing check digit to make an eight-digit account number. The check digit is calculated based on the previous seven digits, and the system recalculates it as the data-entry person enters the account number with the check digit. If results do not match, an input error has occurred. Other data validation can be done online against internal tables or files. For example, a customer number can be validated against the customer file at the time a new order is entered. The systems designer can reduce the need for this type of control by designing a system to obtain the data for a particular field from other information already in the system.

DATABASE INTEGRITY CONTROLS

Most database management systems include integrity controls and security features that provide an additional layer of control. Five major areas of security and control can be implemented at the database level:

- Access control
- Data encryption
- Transaction control
- Update control
- Backup and recovery protection

Access Controls

access control
an integrity control that determines who has access to a system and its data

Access controls refer to the ability of a user to get access to the data. An operating system typically applies security and access controls on a file-by-file basis. A DBMS can apply these controls at a much finer level of detail. Controls can be defined on schema subsets such as groups of related tables or objects, single tables or objects, or single fields or attributes. For example, different controls might be applied to the name, Social Security number, and salary fields of an employee table. Also, controls on a single field might differ for read and write access.

A DBMS stores security access information within the schema and applies controls each time data is read or written. When the DBMS enforces security controls, it automatically enforces them for application programs that access the database. Some DBMSs rely on the operating system to identify the user who is attempting to access data, which relieves the user from having to identify himself or herself multiple times. Other DBMSs implement security controls independently of the operating system.

Encryption

Encryption is used both for data within a database and the transmission of data, especially over public carriers. Data within a database is normally encrypted with a single-key encryption method. More details on the various types of encryption are explained in the section on security controls.

Transaction Controls

transaction logging
a technique whereby all updates to a database are recorded with the information of who, when, and how the update was performed

Transaction logging is a technique by which any update to the database is logged with audit information such as user ID, date, time, input data, and type of update. The fundamental idea is to create an audit trail of all updates to the database that can trace any errors or problems that occur. The more advanced database systems—such as those that run on servers, workstations, and mainframes—include transaction logging as part of the DBMS software. However, several smaller DBMSs, particularly those that run on personal computers, do not include this capability, so design teams must add it directly to those applications.

Transaction logging achieves two objectives. First, it helps discourage fraudulent transactions. If a person knows that every transaction is logged, that person is less apt to attempt a fraudulent transaction. For example, if a person knows that her ID will be associated with every check request, that person is not likely to request a bogus payment.

The second objective of a logging system is to provide a recovery mechanism for erroneous transactions. A mid-level logging system maintains the set of all updates. The system can then recover from errors by "unapplying" the erroneous transactions. More sophisticated logging systems can provide a "before" and "after" image of the fields that are changed by the transaction, as well as the audit trail of all transactions. These sophisticated systems are typically used only for highly sensitive or critical data files, but they do represent an important control mechanism that is available when necessary.

Update Controls

Database management systems are designed to support many application programs simultaneously. Thus, several programs might want to access and update a record or field at the same time. Update controls within a DBMS provide record locking to protect against multiple updates that might conflict with or overwrite each other.

In addition, some transactions that are applied to the database have multiple parts, such as a financial transaction that must credit one account and debit a different account. Delaying commitment of the update until all updates have been verified is a technique used to protect the data from partial updates of these complex transactions.

Backup and Recovery

Backup and recovery procedures are designed to protect the database from all other types of catastrophes. Many database management systems provide various levels of backup and recovery. Partial or incremental backups are used to capture changes to the database during the time periods between total backups. A total backup is used only periodically to archive a complete copy of all the data. Frequently, this archive is placed in a secure off-site location to protect it against catastrophic threats, such as fire, earthquake, or terrorist attacks.

Another popular security measure used for systems that rely on up-to-the-minute data is a mirror database or mirror site. This technique completely duplicates the database and all transactions as they occur. Obviously, this approach can be expensive, but it is becoming more important as information becomes more and more critical to the daily operations of organizations.

OUTPUT INTEGRITY CONTROLS

As already discussed, output from a system comes in various forms, such as output that is used by other systems, printed reports, and data output on computer screens. The purpose of output controls is to ensure that output arrives at the proper destination and is accurate, current, and complete. It is especially important that reports with sensitive information arrive at the proper destination and that they not be accessed by unauthorized parties.

Destination Controls

In the past, when most output was in printed form, a distribution control desk collected all the printed reports from the nightly processing and distributed them to the correct departments and people. This control desk was important because some of the reports had sensitive, confidential information, and it was vital to keep those reports secure. Systems with good controls printed destination and routing information on a report cover page along with the report. Today, businesses accomplish the same function of a control desk by placing printers in each of the locations that need printed reports. It is still a good idea to print a cover sheet with destination and report heading information. Destination codes and routing capabilities are included during the design process to handle the distribution of reports to separate printing facilities. Controlling access to these reports then becomes an issue of physical access. These types of controls are called **destination controls**.

destination controls
integrity controls to ensure that output information is channeled to the correct people

Electronic output to other systems is usually provided in two forms: either an online transaction-by-transaction output or a single data file with a batch of output transactions. Each form has its own type of controls. If the system produces online transactions, it must ensure that each transaction includes the routing codes identifying the correct destination. Both systems need to work together to ensure that each transaction is sent and received correctly. The output transaction will have verification codes and bits to permit the receiving system to verify the accuracy of the transaction. The receiving system also responds with an acknowledgment of a successful receipt of the transaction. Many of these controls are now built into the network transmission protocols. However, during the design activities, the systems designers need to be aware of the network and operating system capability and supplement it where necessary to ensure that the data is received successfully.

Controls for output data files carefully identify the contents, version, date, and time of the file. Normally, a system produces a data file, either on magnetic tape or disk, and another system must find that data file and use it. The major control issue is how to ensure that the second system uses the correct data file. For example, to avoid serious problems, we want to make sure that Friday's transactions aren't run twice. Or, if by some processing quirk, two data files are produced for the same day—one for the first half of the day and another for the second—the system must use both data files. Or, if the second system had processing errors and needs to be rerun, it must be able to find the correct file to use on its rerun. Controls for this situation generally have special beginning and ending records that contain date, time, version, record counts, dollar control totals, processing period, and so forth. During systems design, provision must be made to accumulate the appropriate totals and to produce the necessary control records.

Destination controls for computer screen output are not as widely used as those for printed reports. Normally, the previously discussed user access controls manage the availability of information on computer screens. In some instances, however, destination controls limit what information can be displayed on which terminal. This extra safeguard is used primarily for military or other systems that house computer terminals in secure areas and provide access to the system's information to anyone who has access to the area. The design of these systems requires close coordination between the application program and the network security control system.

Completeness, Accuracy, and Correctness Controls

The completeness, accuracy, and correctness of output information are a function primarily of the internal processing of the system rather than any set of controls. System developers ensure completeness and accuracy by printing control fields on the output report. For example, every report should have a date and time stamp, both for the time the report was printed and for the date of the underlying data. Frequently, they are the same, but not always, especially when a report is reprinted because of a previous error. The following items are controls that should be printed on reports:

- Date and time of report printing
- Date and time of data in the report
- Time period covered by the report
- Beginning header with report identification and description
- Destination or routing information
- Pagination in the form "page __ of __"
- Control totals and cross footings
- An "End of Report" trailer
- The report version number and version date (such as those for special printed forms)

INTEGRITY CONTROLS TO PREVENT FRAUD

The preceding sections have identified several types of integrity controls that support the three control objectives. Many of those techniques are focused on preventing errors and protecting the system from foreign intrusion. However, an equally serious problem is the use of the system by authorized people to commit fraud against an organization.

Fraud is a problem that is reaching epidemic proportions in the United States and around the world. Almost every week we see newspaper articles describing fraud and other white-collar crime. The economic losses caused by fraudulent activity around the world are staggering. These losses reach into the billions of dollars and far exceed those from violent and personal crimes. In the last few years, several major corporations have been forced into bankruptcy or closure due to the fraudulent behavior of key executives. Obviously, software and system controls will not completely eliminate fraud. However, system developers should be aware of the fundamental elements that make fraud possible and incorporate system controls to combat it. The controls that we discussed previously—input controls, database controls, and output controls—are critical components in the battle against fraud, but several additional techniques should be considered in systems design to further increase protection.

Research into the perpetration of fraud indicates that three conditions are present in almost all fraud cases:

- Personal pressure, such as the desire to maintain an extravagant lifestyle
- Rationalization, such as a person's thoughts that "I will repay this money"
- Opportunity, such as unverified cash receipts

The objective of integrity controls is to reduce or eliminate the opportunity for fraud by having adequate manual controls and automated records of money and assets. Control of fraud requires both manual procedures and computer integrity controls. Neither component is sufficient by itself to reduce the opportunities for fraud. System developers need to work closely with business users who are knowledgeable about accounting principles to prevent fraud.

Sometimes system developers might think that integrity controls are not necessary because the system in development is not a financial or accounting system. However, an opportunity for fraud exists in almost every business system. Because most business systems track an organization's assets, someone could manipulate those assets, writing checks for incorrect amounts or to fictitious parties. Hence, almost every system requires some type of integrity controls.

Figure 15-19 contains several of the more important factors that increase the risk of fraud. This list is not comprehensive, but it does provide a foundation from which developers can design a computer system that reduces the opportunity for fraud. As a system developer, you should include discussions both with your users and within the project teams to ensure that adequate controls have been included to reduce fraud.

Figure 15-19

Fraud risks and prevention techniques
Source: Information in the table was provided by Dr. Marshall Romney of the School of Accountancy and Information Systems at Brigham Young University

Factors affecting fraud risk	Techniques to reduce risk
Separation of duties	Design separate electronic forms, with separate access controls, for request, approval, and generation of expenditures.
Inadequate audit trails	Include transaction logging. Avoid, or very tightly control, manual override capability that circumvents logs.
Inadequate records	Implement a comprehensive database with sufficient detail and logs.
Inadequate monitoring	Include manual procedures and automated routines to monitor patterns and out-of-bound conditions. Include exception reports. Implement third-person audit capability.
Easily removable assets	Include an easy-to-use capability to cross-check physical counts with automated records.
Inadequate security system	Supplement operating system security features with additional program and data-level security. Include automatic shutdown and lockup features. Include routines to analyze access patterns.

Now that we have an overview of input and output integrity controls, we turn our attention to security controls.

DESIGNING SECURITY CONTROLS

security control
mechanisms usually provided by the operating system or environment to protect the data and processing systems from malicious attack

Although the objective of security controls is to protect the assets of an organization from all threats, as indicated earlier, the primary focus is generally on external threats. In addition to the objectives enumerated earlier for integrity controls, security controls also have the following two objectives:

- Maintain a stable, functioning operating environment for users and application systems (usually 24 hours a day, seven days a week).
- Protect information and transactions during transmission outside the organization (public carriers).

The first objective, to maintain a stable operating environment, focuses on security measures to protect the organization's systems from external attacks, such as from hackers, viruses, worms, and message overloads. Most organizations today have gateways between their internal systems and the Internet. Every time someone in an organization sends a communication to or receives one from the Internet, there is the potential for a security violation and for undesirable access that could disrupt the internal systems. So, eliminating and controlling any undesirable access help avoid disruption of the system.

BEST PRACTICE

Be prepared to answer this question every time your system project is discussed with management: "Are you sure the company will be protected from all threats when this system is operating?"

The second objective, to protect transactions during transmission, focuses on the information that is sent or received via the Internet. More and more organizations utilize the Internet as a portal to their customers and to their suppliers. After a transaction is sent outside the organization, it could be intercepted, destroyed, or modified. So, security controls use techniques to protect data while it is in transit from the source to the destination.

Security controls can be implemented within different types of software, including the network and computer operating system, the database management system, or the application programs. The most common security control points are network and computer operating systems because they exercise direct control over assets such as files, application programs, and disk drives. All modern operating systems contain extensive security features that can identify users, restrict access to files and programs, and secure data transmission among distributed software components. Operating system security is the foundation of security for most information systems.

On some occasions, developers might implement security controls directly within application software. Developers can define their own security controls over individual data items or records when data is stored in files instead of a database. Developers can also implement security controls to prevent unauthorized users from performing certain functions such as deleting existing data or creating backup copies on removable storage media.

Most developers avoid implementing security controls within application software because of the complexity and importance of security functions. Most operating system and DBMS developers have a large programming staff dedicated exclusively to developing and maintaining security software. It is difficult for application developers to dedicate sufficient resources to implement system security controls correctly and fully. Thus, security-related implementation tasks in a typical information system development project are usually limited to configuring security software in the underlying operating system or DBMS.

SECURITY FOR ACCESS TO SYSTEMS

Modern operating systems, networking software, and Internet access all need implemented control mechanisms. These mechanisms can be used to control access to any resource managed by the operating system or network—including hardware, application programs, and data files.

System access controls are mechanisms that are established to restrict what portions of the computer system a person can use. This category includes controls to limit access to certain applications or functions within an application, restrict access to the computer system itself, and limit access to certain pieces of data.

With proper design and implementation, an information system can use access control functions embedded in system software. The advantage to this approach is that a consistent set of access controls is then applied to every resource on a hardware platform or network. Thus, the systems designer can implement a single access control scheme and apply it to every resource or information system.

The systems designer can also add controls over and above those already provided by system software. However, designing and implementing effective application-based access controls require technical expertise. Operating system and network software developers expend considerable energy and resources to develop reliable and efficient access controls, and it is difficult and expensive for a typical organization to duplicate these efforts. For these reasons, most information systems build on the access control already within system software.

Types of Users

System developers must consider different types of users when designing access controls. Figure 15-20 illustrates various types of users and the access that is appropriate for each. The following paragraphs explain the types of system access available to users.

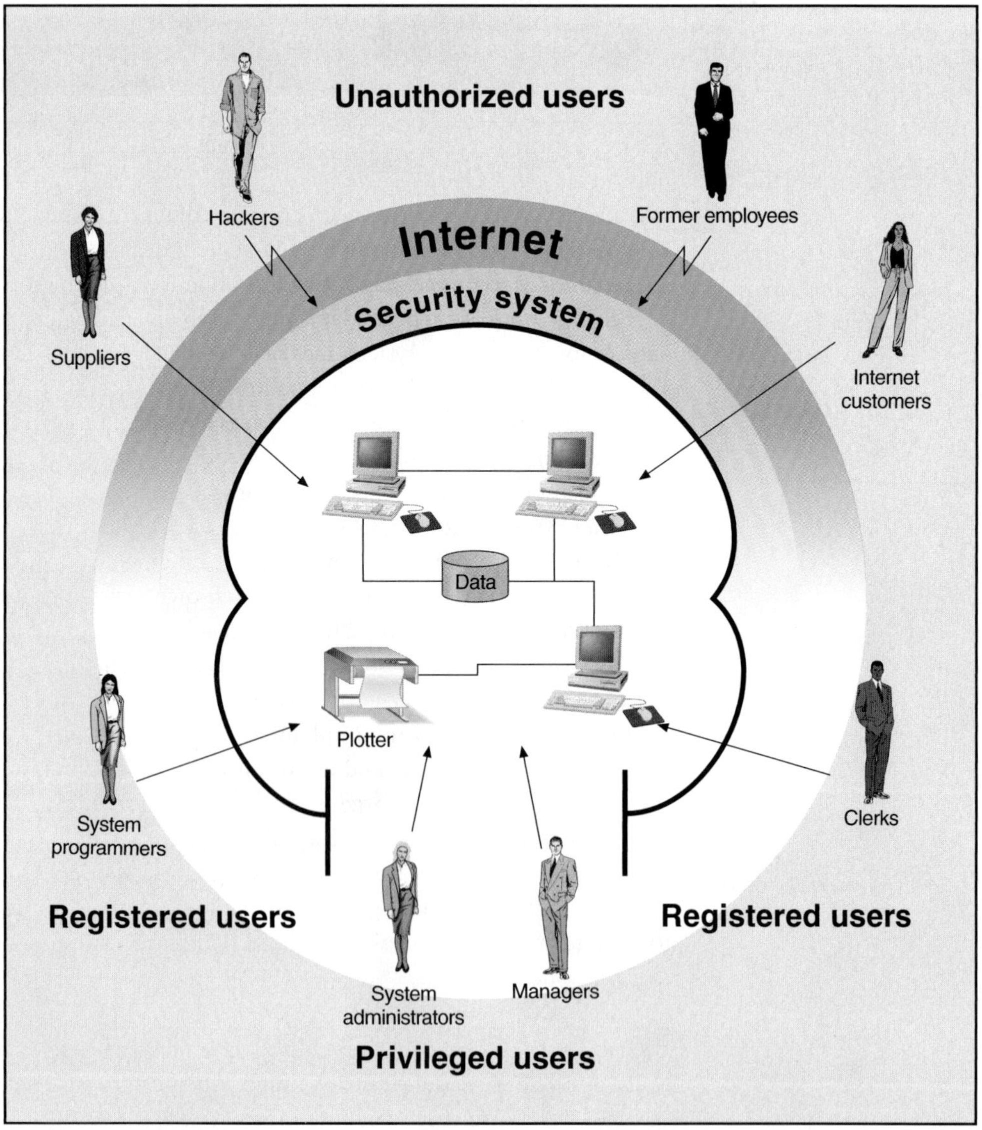

Figure 15-20

Users and access roles to computer systems

To begin development of access controls, designers first must identify and consider all three of these user categories: unauthorized users, registered users, and privileged users. **Unauthorized users** are people who are not allowed access to any part or functions of the system. Such users include employees who are prohibited from accessing the system, former employees who no longer are permitted to access the system, and outsiders such as hackers and intruders. Controls must be able to identify and exclude access from these people.

unauthorized user
a person who does not have authorized access to a system

registered user
a user who is registered or known to the system and is authorized to access some part of it

Registered users are those who are authorized to access the system. Normally, various levels of registered users are set up depending on what they are authorized to view and update. The different levels of access are defined during the design of the new system. For example, some users might be allowed to view data but not update it, and other users can update only certain data fields. Some screens and functions of the new system might be hidden from other levels of registered users. The important point for systems designers to recognize is that there might be multiple levels of registered users. **Authorization** is the process of determining whether a user is permitted to access a specific resource for a particular purpose. In other words, it is the process of deciding whether a user should be a registered user. The security system stores an access control list for each protected resource. An **access control list** is a list of users or user groups that can access a resource and the permitted access type(s).

authorization
the process of determining whether a user is permitted to have access to the system and data

access control list
the list of users who have rights to access the system and data

privileged user
a user who has special security access privileges to a system

Privileged users include people who have access to the source code, executable program, and database structure of the system. These people include system programmers, application programmers, operators, and system administrators, and they might also have differing levels of security access. Usually, system programmers have full access to all components of the systems and data. Application programmers have access to the applications themselves but often not to the secure libraries and data files used for the systems in production. System administrators have access to all functions of the system and can control and establish the various levels of registration and register users. A system administrator also usually has software programs to help control access and to monitor access attempts.

Passwords and Smart Cards

authentication
the process of identifying a user to verify that he can have access to the system

Authentication is the process of identifying users (that is, the authorized or registered users) who request access to sensitive resources. Authentication is the basis of all security because security controls are useless unless the user is correctly identified. In many operating systems, authentication requires the user to enter a user name and password. The user is authenticated if the password he or she enters matches the password stored in the security database.

Two techniques are used to define passwords. The computer can randomly generate and assign passwords, or each user can define his or her own password. There are advantages to both techniques. The first creates passwords that are usually longer and more random, but they tend to be hard for users to remember. Most users would have a hard time trying to remember a password such as a3x7869bts21. User-developed passwords are easier to remember, but they are usually not as complex and, therefore, not quite as secure. Some restrictions can be placed on the syntax of the password to ensure at least a minimum level of security.

Of course, one of the problems with passwords is remembering what they are. It is not uncommon for heavy computer users to have 5 or even 10 different passwords for different systems that they access. One alternative is to use the same password for all systems, but if someone determines the password, all the systems are compromised. Most often, the security system should be organized so that all resources can be accessed with the same unique identifier and password combination. In other words, only one user ID/password combination should be required for access to the different systems throughout the organization. When users have to remember different IDs and passwords to access different systems, they often write them down and post them near the computer. Obviously, this practice defeats the purpose of user verification security.

smart card
a computer-readable plastic card with security information embedded within it

A **smart card** is a computer-readable plastic card with a small amount of stored security data that can be read by a card scanner in much the same way a credit or debit card can be read at a supermarket checkout counter. The smart card stores an encrypted version of the user's password, fingerprint, retinal scan, or voice characteristics. To authenticate himself or herself, the user scans the card and then enters the password or submits to a fingerprint, retinal, or voice scan. Such a system enhances security because the user must possess both the card and the appropriate identifying information to be authenticated. Only the security subsystem knows the key, which prevents potential intruders from using cards with altered data.

A final security step is to make sure the system keeps a record of attempted logons, especially unsuccessful ones. An unsuccessful logon might simply indicate that the user mistyped or forgot a password, but it might also indicate an attempted breach of security, which should be investigated.

Biometric Devices

Authentication can also be based on other forms of personal identification, including keystroke patterns, fingerprints, retinal scans, and voice characteristics. When a user enters a password or other keystroke sequence, the timing and force of each keystroke are unique. Some security systems use both the password and the keystroke pattern to authenticate the user, which prevents someone with a stolen password from accessing system resources.

Many companies are now experimenting with a new form of security based on biometric devices. The principle behind use of a biometric device is that the person himself becomes the password or gateway into a secure system. These more sophisticated security systems can scan fingerprints, retinal blood vessels, or voices, which are unique for every person. With the advent of very small computer chips with very high memory densities and logic circuitry, biometric devices can be built into almost any of the normal hardware components of a computer. In addition, the complex logic necessary to do sophisticated pattern matching of fingerprints, hand vein patterns, retinas, iris patterns, or complete facial patterns can be located right in the micro-sized biometric device itself.

Biometric fingerprint devices are now being embedded in such components as a computer mouse, computer keyboard, and small touch pads. Other biometric scanners, such as very small cameras, can be embedded in the computer monitor. Such a device might do an iris or facial scan of the person looking at the monitor. Figure 15-21 illustrates a computer mouse with an embedded touch pad to test fingerprints. Other types of mouse devices have the sensor on the side so that the thumb must be placed on it and authorization can be performed before every mouse action.

Figure 15-21
Biometric mouse

Security based on biometric devices can also be multilevel. Security verification can be done when the user first tries to log on. Higher levels of security can later be activated within a given program to obtain additional authorization to access specific forms or database records. Obviously, each individual must be authorized and appropriate information stored for the level of security allowed.

DATA SECURITY

In addition to the need for controlling access to an organization's systems and internal network, it is frequently important to make the data itself secure. For example, user IDs and passwords are important information that must be secret. Frequently, the password information is even kept secret from the system administrators. They can assign a new password to a user, but they cannot read or access the current password. So, if a user forgets her password, the administrator assigns a new one.

Many other types of files are also kept confidential. Some examples include files that contain the following:

- Financial information
- Credit-card numbers, bank account numbers, payroll information, and other personal data
- Strategies and plans for products and other mission-critical data
- Government and sensitive military information

Some operating systems, especially UNIX and its derivations, have built-in security for each file in the system. Each UNIX file has security corresponding to three types of users: the owner of the file, other members of the owner's workgroup, and all other users. The security for each user is also further divided into three levels: read access, update (create, update, and delete) access, and execute access. Execute access determines whether the file is executable (such as an .exe file in Windows), and the security level determines who is allowed to execute the file.

Data that resides on an internal system needs to be protected, but data that is being transmitted outside the organization is especially subject to snooping and even modification. With the increasing acceptance of electronic communications, more and more organizations are transmitting and receiving transactions via the Internet. On the sales and distribution side of business, customers are viewing catalogs, ordering products, making payments, and tracking shipments via the Internet. On the supply side, organizations are ordering inventory, monitoring receivables, sending purchase orders, and making financial transactions through the Internet. Because this information is being transmitted via the public Internet, the raw data is available to anybody who has tools to listen and intercept information packets.

encryption
the process of altering data so that it is unreadable by unauthorized users

decryption
the process of converting encrypted data back into a readable format

encryption algorithm
a complex mathematical formula and process that encrypts or decrypts data

encryption key
a binary field that the encryption algorithm uses to transform the data

symmetric key encryption
an encryption process that uses the same key to encrypt and to decrypt the data

The primary method of maintaining the security of data, both on internal systems and transmitted data, is by encrypting the data. **Encryption** is the process of altering data so that unauthorized users cannot view it. **Decryption** is the process of converting encrypted data back to its original state. Data stored in files or a database on hard drives or other storage devices can be encrypted to protect it against theft. Data sent across a network can be encrypted to prevent eavesdropping or theft during transmission. A thief or eavesdropper who steals or intercepts encrypted data receives a meaningless group of bits that are difficult or impossible to convert back into the original data.

An **encryption algorithm** is a complex mathematical transformation that encrypts or decrypts binary data. An **encryption key** is a binary input to the encryption algorithm—typically a long string of bits. The encryption algorithm varies the data transformation based on the encryption key so that data can be decrypted only with the same key or a compatible decryption key. Many encryption algorithms are available, and a few, including Data Encryption Standard (DES) and several algorithms developed by RSA Security, are widely deployed governmental or Internet standards. An encryption algorithm must generate encrypted data that is difficult or impossible to decrypt without the encryption key. Decryption without the key becomes more difficult as key length is increased. Both sender and receiver must use the same or compatible algorithms.

Figure 15-22 is an example of **symmetric key encryption**, where the same key encrypts and decrypts the data. A significant problem with symmetric key encryption is that both sender and receiver use the same key, which must be created and shared in a secure manner. Security is compromised if the key is transmitted over the same channel as messages encrypted with the key. Also, sharing a key among many users increases the possibility of key theft.

Figure 15-22
Symmetric key encryption

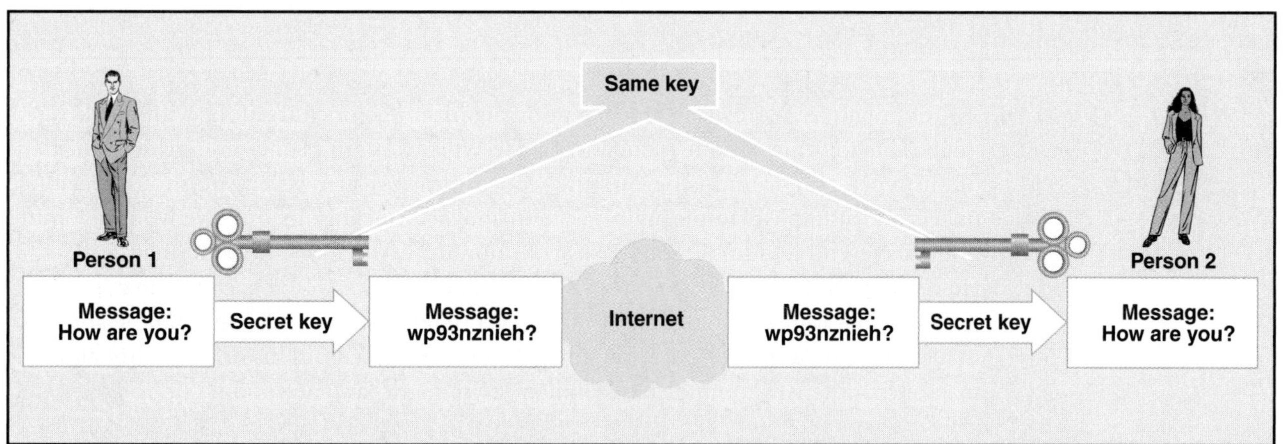

asymmetric key encryption
an encryption process that uses one key to encrypt and a different key to decrypt the data

public key encryption
an asymmetric key method in which one key is publicized and the other key is kept private

Asymmetric key encryption uses different but compatible keys to encrypt and decrypt data. **Public key encryption** is a form of asymmetric key encryption that uses a public key for encryption and a private key for decryption. The two keys are like a matched pair. After information is encrypted with the public key, it can be decrypted only with the private key. It cannot be decrypted with the same public key that encrypted it. Organizations that use this technique broadcast their public key so that it is freely available to anybody who wants it. Then when some entity—for example, someone who wants to order something from the vendor—wants to transmit a secure message to a vendor, that customer reads the vendor's public key from a public source such as a Web site. The customer encrypts the message with the public key and sends the message to the vendor. The vendor decrypts the message with the private key. Because no one else has the private key, no one else can decrypt the message.

Some asymmetric encryption methods can encrypt and decrypt messages in both directions. That is, in addition to encrypting a message with the public key that can be decrypted with the private key, an organization can also encrypt a message with the private key and decrypt it with the public key. Notice that both keys must still work as a pair, but the message can go forward or backward through the encryption/decryption pair. This second technique is the basis for digital signatures and certificates, which are explained in the next section. Figure 15-23 illustrates an asymmetric key encryption transmittal.

Figure 15-23
Asymmetric key encryption

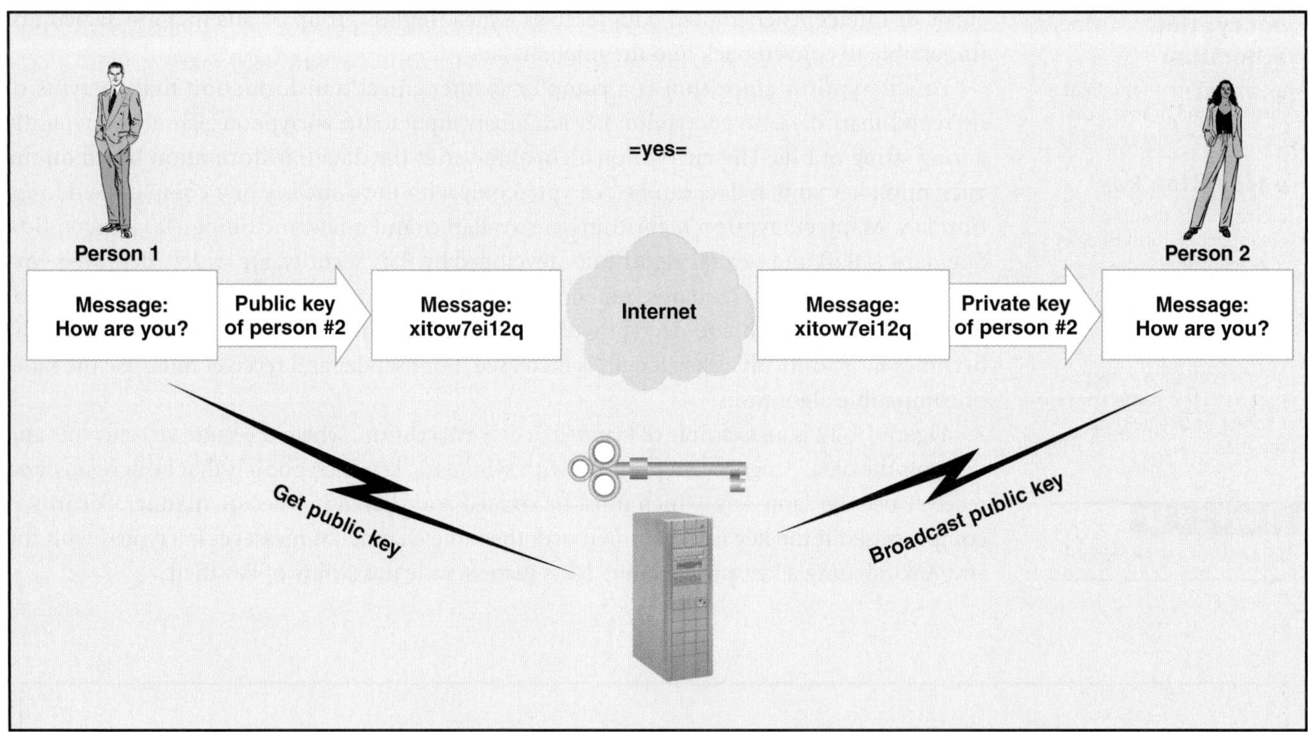

You might ask, "How can an encryption algorithm go one direction (with one key) and not be able to come back the same way (be decrypted with the same key)?" The mathematics of this type of algorithm is beyond the scope of this text. However, you should be able to understand a simple example: multiplication and factoring. If someone gives you two or three numbers, even big numbers, and asks you to multiply them, you can do that fairly easily. However, if someone gives you one very big number and asks you to factor it (that is, find the numbers that were originally multiplied to get that number), you would not be able to do that easily. It would take you a long time. Algorithms based on this one-directional mathematical characteristic form the basis of many asymmetric key encryption routines.

DIGITAL SIGNATURES AND CERTIFICATES

digital signature
a technique in which a document is encrypted using a private key to verify who wrote the document

certificate, or digital certificate
a text message that is encrypted by a verifying authority and used to broadcast an organization's name and public key

certifying authority
a well-known third party that sells digital certificates to organizations

The encryption of messages is an effective technique to enable a secure exchange of information between two entities who have appropriate keys. However, how do you know that the entity on the other end of the communication is really who you think it is? A **digital signature** is a technique in which a document is encrypted using a private key to verify who wrote the document. If you have the public key of an entity, and that entity sends you a message with its private key, you can decode it with the public key. You know that the party is the one you want to communicate with because that entity is the only one who can encode a message with that private key. The encoding of a message with a private key is called *digital signing*.

Taking the example one step further, you can ask the question, "How do I know that the public key I have is the correct public key and not some counterfeit key?" In other words, maybe someone is impersonating another entity and is passing out false public keys to be able to intercept encoded messages (such as financial transactions) and steal information. In essence, the problem is ensuring that the key that is purported to be the public key of some institution is, in fact, that institution's public key. The solution to that problem is a certificate.

A **certificate**, or **digital certificate**, is an institution's name and public key (plus other information such as address, Web site URL, and validity date of the certificate) that is encrypted and certified by a third party. Many third parties are very well known and widely accepted **certifying authorities**, such as VeriSign or Equifax. In fact, they are so well known that their public keys are built right into Netscape and Microsoft Internet Explorer. As shown in Figure 15-24, you can know that the entities with whom you are communicating are, in fact, who they say they are and that you do have their correct public key.

Figure 15-24
Using a digital certificate

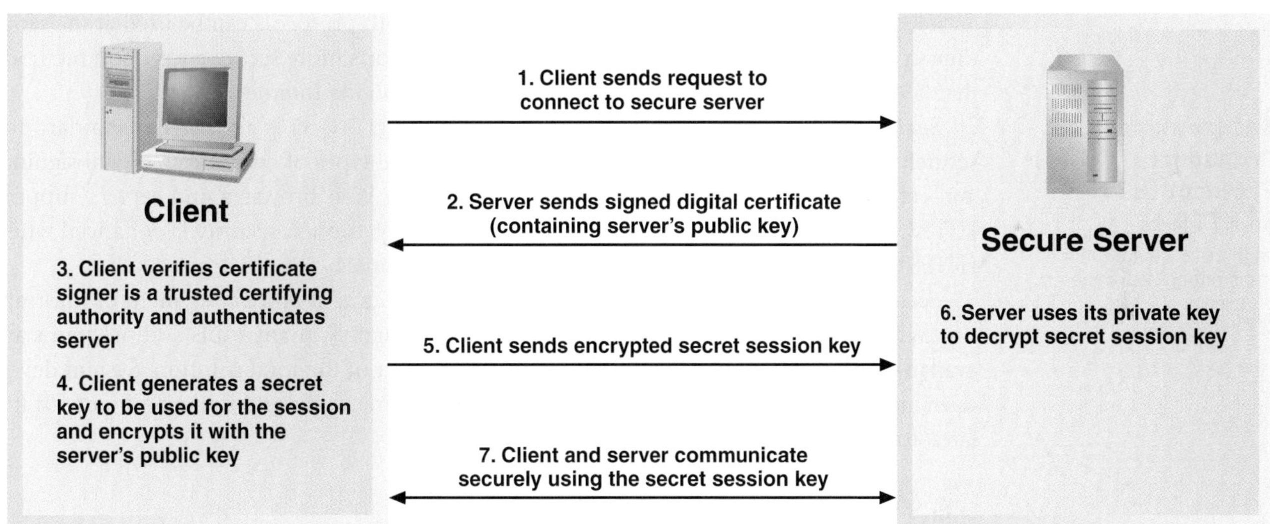

An entity who wants a certificate with its name and public key goes to a certifying authority and buys a certificate. The certifying authority encrypts the data with its own private key (signs the data) and gives the data back to the original entity. Now when someone, such as a customer, asks the entity for its public key, it sends the certificate. The customer receives the certificate and opens it with the certifying authority's public key. Again, the certifying authority is so well known that its public key is built into everyone's browser and is essentially impossible to counterfeit. Now the customer can be sure that he or she is communicating with the original entity and can do so with encrypted messages using the entity's public key.

A variation of this scenario occurs when the buyer and seller transmit their certificates to one another. Each participant can decrypt the certificate using the certifying authority's public key to extract information such as name and address. However, to ensure that the public key contained within the certificate is valid, the certificates are transmitted to the certifying authority for verification. The authority stores certificate data, including public keys, within its database and verifies transmitted certificates by matching their content against the database.

SECURE TRANSACTIONS

Secure Sockets Layer (SSL)
a standard protocol to connect and transmit encrypted data

Transport Layer Security (TLS)
an updated version of SSL

Secure electronic transactions require a standard set of methods and protocols that address authentication, authorization, privacy, and integrity. Netscape originally developed the **Secure Sockets Layer (SSL)** to support secure transactions. SSL was later adopted as an Internet standard and renamed **Transport Layer Security (TLS)**, though the original name, SSL, is still widely used.

TLS is a protocol for a secure channel to send messages over the Internet. The sender and receiver first establish a connection using ordinary Internet protocols and then ask each other to create a TLS connection. The sender and receiver then verify each other's identity by exchanging and verifying identity certificates as explained previously. At this point, either or both have exchanged public keys, so they can send secure messages. Because asymmetric encryption is quite slow and difficult, the two entities agree on a protocol and encryption method, usually a single-key encryption method. Of course, all of the messages to establish a secure connection are sent using the public key/private key combination. After the encryption technique has been decided and the secret, single key has been transmitted, all subsequent transmission is done using the secret, single key.

IP Security (IPSec) is a newer Internet standard for secure message transmission. IPSec is implemented at a lower layer of the network protocol stack, which enables it to operate with greater speed. IPSec can replace or complement SSL. Both protocols can be used at the same time to provide an extra measure of security. IPSec supports more secure encryption methods than SSL, but these methods are not yet fully deployed on the Internet.

Secure Hypertext Transport Protocol (HTTPS or HTTP-S)
an Internet standard for transmitting Web pages securely

Secure Hypertext Transport Protocol (HTTPS or HTTP-S) is an Internet standard for securely transmitting Web pages. HTTPS supports several types of encryption, digital signing, and certificate exchange and verification. All modern Web browsers and servers support HTTPS. It is a complete approach to Web-based security, though security is enhanced when HTTPS documents are sent over secure TLS or IPSec channels.

Security is an important consideration in the development and deployment of information systems in today's networked environment. Fortunately, many tools and programs are available and can be integrated into new systems as part of the total solution. System developers need to be aware of the need to include security measures and to be familiar with the latest tools and techniques.

SUMMARY

The chapter began with a discussion of identifying and then designing system interfaces. System interfaces include all inputs and outputs except those that are part of the interactive user interface.

Designing the inputs to the system is a three-step process:

- Identify the devices and mechanisms that will be used to enter input.
- Identify all system inputs and develop a list with the data content of each.
- Determine what kinds of controls are necessary for each system input.

To develop a list of the inputs to the system, designers use diagrams that were developed during the analysis and application design activity. For the traditional structured approach, DFDs, data flow definitions, and structure charts are used. For the object-oriented approach, sequence diagrams are the primary source of information, but the design class diagram is used to ensure that designers provide the correct data fields and the correct methods that produce the outputs.

The process to design the outputs from the system consists of the same steps as for input design. For output design, the DFDs and sequence diagrams are used to identify data flows and messages that exit the system. New technology provides numerous ways to present output with charts, graphs, and multimedia. Before deciding on output media, the designer should carefully consider the intended audience and the purpose of the output.

This chapter next discussed the concepts of integrity controls in systems. The objectives of integrity controls are to:

- Ensure that only appropriate and correct business transactions occur
- Ensure that the transactions are recorded and processed correctly
- Protect and safeguard the assets (including information) of the organization

Integrity controls are concerned with defining who has access to the various components of the system and the database. Access controls identify various classifications of users—such as unauthorized users, registered users, and privileged users—to ensure that systems are safeguarded. Additional integrity controls are concerned with reducing errors, preventing fraud, and maintaining the correctness of the data in the system.

The last section of the chapter introduced the basic concepts of security for systems that have access to public networks (primarily the Internet). Security is becoming more and more important, and various techniques should be considered when developing new information systems. The underlying technology in many of the security approaches is based on public key systems that have public and private key components. Encryption and public key systems are the basis for digital signatures, digital certificates, secure connection, and secure transaction implementations.

KEY TERMS

access control, p. 595
access control list, p. 601
ad hoc reports, p. 583
asymmetric key encryption, p. 604
authentication, p. 601
authorization, p. 601
certificate, or digital certificate, p. 605
certifying authority, p. 605
completeness control, p. 594
control break report, p. 588
data validation control, p. 594
decryption, p. 603
destination controls, p. 596
detailed report, p. 585
digital signature, p. 605
drill down, p. 589
encryption, p. 603
encryption algorithm, p. 603
encryption key, p. 603
exception report, p. 585
executive report, p. 586
external output, p. 586
field combination control, p. 594
integrity control, p. 593
internal output, p. 586
privileged user, p. 601
public key encryption, p. 604
registered user, p. 601
Secure Hypertext Transport Protocol (HTTPS or HTTP-S), p. 606
Secure Sockets Layer (SSL), p. 606
security control, p. 599
smart card, p. 601

summary report, p. 585
symmetric key encryption, p. 603
transaction logging, p. 595
Transport Layer Security (TLS), p. 606
turnaround document, p. 586
unauthorized user, p. 601
value limit control, p. 594

REVIEW QUESTIONS

1. What does XML stand for? Explain how XML is similar to HTML. Also discuss the differences between XML and HTML.
2. Compare the strengths and weaknesses of using a DFD to define inputs with using a system sequence diagram to define inputs. Which do you like the best? Why?
3. Explain the system boundary. Why was one used on a DFD but not on a system sequence diagram?
4. What additional information does the structure chart provide that is not obtained from a DFD in the development of input forms?
5. How are the data fields identified using the traditional structured approach?
6. How are the data fields identified using UML and the object-oriented approach?
7. Explain four types of integrity controls for input forms. Which have you seen most frequently? Why are they important?
8. What protection does transaction logging provide? Should it be included in every system?
9. What are the different considerations for output screen design and output report design?
10. What is meant by *drill down*? Give an example of how you might use it in a report design.
11. What is the danger from information overload? What solutions can you think of to avoid it?
12. Describe what kinds of integrity controls you would recommend to place on all output reports. Why?
13. What are the objectives of integrity controls in information systems? In your own words, explain what each of the three objectives means. Give an example of each.
14. What are the four types of input controls used to reduce input errors? Describe how each works.
15. Explain what is meant by update controls for a database management system.
16. What is the basic purpose of transaction logging? Microsoft Access does not have automatic transaction logging. Is this a deficiency, or is it not really an important consideration in database integrity?
17. On a printed output report, what is the difference between the date the report was printed and the date of the data?
18. What are the two primary objectives of security controls?
19. Explain the three categories of user access privileges. Is three the right number, or should there be more or fewer than three? Why or why not?
20. How does single-key (symmetric) encryption work? What are its strengths? What are its weaknesses?
21. How does public key (asymmetric) encryption work? What are its strengths? What are its weaknesses?
22. What is a digital certificate? What role do certifying authorities play in security systems?
23. What is a digital signature? What does it tell a user?

THINKING CRITICALLY

1. The chapter described various situations that emphasized the need for controls. In the first scenario presented, a furniture store sells merchandise on credit. Based on the descriptions of controls given in the chapter, identify the various controls that should be implemented in the system to ensure that corrections to customer balances are made only by someone with the correct authorization.
2. In the second scenario illustrating the need for controls, an accounts payable clerk uses the system to write checks to suppliers. Based on the information in the chapter, what kinds of controls would you implement to ensure that checks are written only to valid suppliers, that checks are written for the correct amount, and that all payouts have the required authorization? How would you design the controls if different payment amounts required different levels of authorization?
3. The executives of a company have asked for a special decision support system report on corporate financials. They want this report to be based on actual financial data for the past several years. The report is to have several input parameters so that the executives can do "what-if" analysis of future sales based on past performance. They want the report to be viewable online and in printed form. What kinds of controls would you implement to ensure that (1) only authorized executives can request the report, (2) the executives understand the basis (past and projected data) for a given report, and (3) executives are aware of the sensitive nature of the information and treat it as confidential?
4. A payroll system has a data-entry subsystem that is used to enter time card information for hourly employees. What kinds of controls would you implement to ensure that the data is correct and error-free? What other controls would you include to ensure that a data-entry clerk (who might be a friend of an employee) does not inflate the hours on the time card (after it was approved by a supervisor)?
5. Based on the DFD (Figure 10-26) given in Chapter 10, "Thinking Critically" problem 3, *Add class to schedule*, and the structure chart you developed there, identify the set of input and output screens for the system. Include the data fields that will be required.
6. Based on the DFD (Figure 10-27) given in Chapter 10, "Thinking Critically" problem 5, *Special-order purchasing*, and the structure chart you developed there, identify the set of inputs and outputs required. Develop the list of data fields for each screen and report.
7. A university library system is depicted in Figure 15-25, with partial system sequence diagrams for two use cases, *Check out a book* and *Return a book*. Based on the figure, construct four tables showing inputs and outputs, as shown in Figures 15-10 and 15-12: (1) Inputs for the Library System, (2) Outputs for the Library System, (3) Inputs for the Student Record System, and (4) Outputs for the Student Record System.
8. You work for a grocery chain that always has many customers in the stores. To facilitate and speed checkout, the company wants to develop self-service checkout stands. Customers can check their own groceries and pay by credit card or cash. How would you design the checkout register and equipment? What kinds of equipment would you use to make it easy and intuitive for the customers, make sure that prices are entered correctly, and ensure that cash or credit-card payments are done correctly? In other words, what equipment would you have at the checkout station? In your solution, you can use existing state-of-the-art solutions or invent new devices.

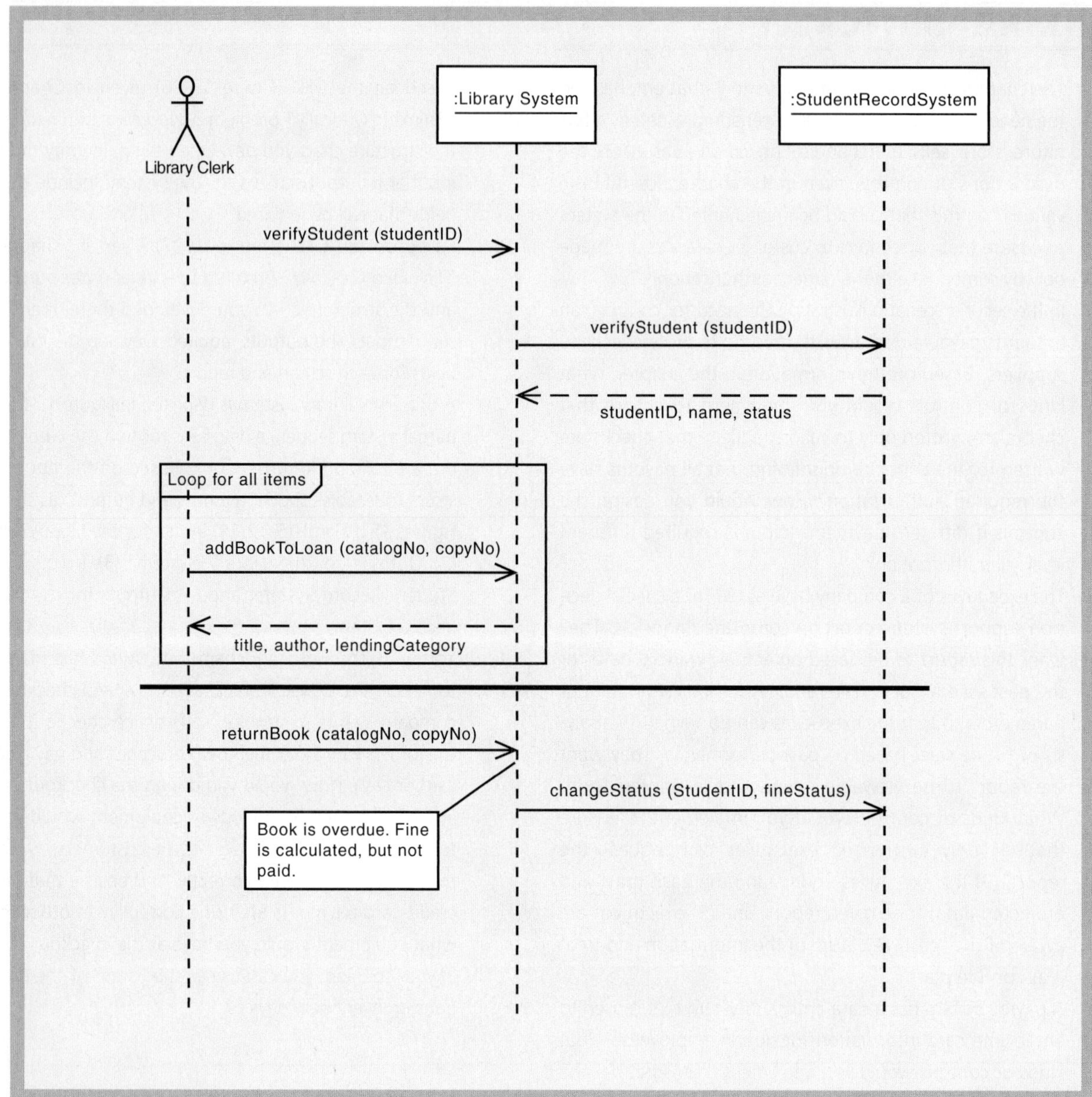

Figure 15-25

Partial system sequence diagram for the university library system

EXPERIENTIAL EXERCISES

1. Look on the Web for an e-commerce site (for example, Amazon.com or eBay). Evaluate the effectiveness of the screens. What kind of security and controls are integrated into the system? Do you see potential problems with the integrity controls? Evaluate the design of the individual screens. How easy are they to read and use? What suggestions would make them easier to use? How effective are they in minimizing data-entry errors?
2. Examine the information system of a local business (a fast-food restaurant, doctor's office, video store, grocery store, and so forth). Evaluate the screens (and reports, if possible) for ease of use and effectiveness. What kind of integrity controls are in place? How easy are the screens to use? What kinds of improvements would you make?
3. Find and research a system that is being constructed or has recently been constructed. You may work for a company that has a development project in progress or have a friend who works for such a company. Another source of development projects is your university or college itself. Interview one of the developers. Ask about integrity controls, methodology for screen design, and guidelines to ensure consistency across the user interface. Ask about the number and scope of the input and output design tasks (for example, how many screens or hours required) and the method used to lay out the screens and reports (such as prototyping, visual modeling tools, and so forth).
4. If your university uses Java, find out about the JSwing class library. Write a one-page description of the JSwing library, its purpose, and ways to use it. Your objective is to demonstrate that you understand the concept of JSwing and the way it is used to build windows and input screens in a Windows environment.
5. If your university uses Studio .NET from Microsoft, find out about using the .NET class library to build user interfaces. Write a one-page description of the .NET library, its purpose, and ways to use it. Your objective is to demonstrate that you understand the concept of .NET forms design and the way it is used to build windows and input screens in a Windows environment.
6. Go to the Internet and find out what you can about Pretty Good Privacy. What is it? How does it work? Research what you can about a passphrase. What does it mean? Here are two sites that you can use to start your research: *www.pgpi.org/* and *web.mit.edu/network/pgp.html*.

CASE STUDIES

ALL-SHOP SUPERSTORES

All-Shop Superstores is a regional chain of superstores in the Boston, New York, and Washington, D.C., corridor. These stores compete with other giants, such as Wal-Mart, Kmart, Target, and other budget retailers. The stores contain large grocery stores as well as domestics, clothing, automobile, and home improvement products. Overall, the margins in this portion of the retail industry are very small. Grocery profits have always been small, in the range of 5 to 10 percent. The margin for domestics, clothing, and other goods is a little higher, but to compete with Wal-Mart, All-Shop must keep all margins low.

To reduce operating costs as much as possible, All-Shop has decided to move very heavily into electronic data interchange (EDI) with its suppliers. All-Shop is aware that several of its more advanced competitors allow their suppliers to manage inventory levels in the stores themselves. For example, paper hygiene products such as disposable diapers and toilet paper are high-volume products that require very close monitoring of inventory levels. All-Shop has already installed sophisticated sales and inventory systems that track activity of each individual item (using the UPC code) daily. These systems not only capture daily activity but also maintain histories in a data warehouse to support online data analysis.

The first step for All-Shop was to enable its major suppliers to have access to its daily sales and inventory database. That way, the suppliers could monitor sales activities and check inventory to ensure that deliveries are made on time to maintain optimal inventory levels. The system should also permit each supplier to access and check the status of its individual accounts and a history of past payment activity. Obviously, All-Shop must control all of this information so that suppliers cannot observe each other's information.

1. Based on what you have learned in this and previous chapters, develop a use case diagram identifying the use cases that apply to the supplier as an actor. Even though this is really a system-to-system interface, the supplier system can be considered an actor. Identify two lists of controls that you consider necessary for this interface. On the first list, identify overall controls for the entire EDI interface. Then, for the second list, for each identified use case, develop a specific set of controls that will be necessary. Base your analysis on the types of controls discussed in the chapter as well as the three primary objectives of integrity controls. In other words, your assignment is to develop a statement of required controls that the system developers can use to ensure that the system adequately protects the assets and information of All-Shop.

2. All-Shop is considering a plan to provide supplier access to its data warehouse to enable executives to analyze past trends and help design promotions to increase overall sales and those of individual products. In other words, All-Shop is building partnerships with its suppliers to maximize its presence in the retail marketplace. One major concern of All-Shop executives is how to ensure that the suppliers treat this information with maximum security and not damage All-Shop. How can they ensure that the suppliers do not use this information to benefit All-Shop's competitors inadvertently, as suppliers also work with these competitors?
3. Do you think this second step is a wise move for All-Shop? If not, why not? If so, what kinds of controls and contractual arrangements should be made to protect All-Shop? You can see how a narrow focus on integrity controls might be inadequate to protect proprietary information. A broader view and understanding of controls and their objectives are required in this instance.

REAL ESTATE MULTIPLE LISTING SERVICE SYSTEM

Based on the DFD fragments you developed in Chapter 6 and the structure charts from Chapter 10, develop a table of inputs along with the associated data couples and data fields for each input. Also, develop a table of outputs with the required data fields.

THEEYESHAVEIT.COM BOOK EXCHANGE SYSTEM

Based on the system sequence diagrams you developed in Chapter 7, develop a list of inputs and outputs required for this system. Also, identify any specific controls that may be necessary to ensure that information is entered accurately.

DOWNTOWN VIDEOS RENTAL SYSTEM

Using the system sequence diagrams you developed in Chapter 7, develop a list of inputs and outputs, along with the necessary data fields, for the system.

RETHINKING ROCKY MOUNTAIN OUTFITTERS

The RMO event table lists six system reports that are part of the new system:

- Order summary
- Transaction summary
- Fulfillment summary
- Prospective customer activity
- Customer adjustments
- Catalog activity

For each of these six reports, answer the following questions:

1. Identify the data fields that each report should include.
2. What questions will users want each report to answer?
3. What type of report is it: detailed, summary, or exception?
4. How might graphics be used? What about drill-down capabilities?
5. How would you prepare a mock-up of each report, assuming a printed output and also an online output?
6. What output controls should be associated with each report?

FOCUSING ON RELIABLE PHARMACEUTICAL SERVICE

One of the challenges of a pharmaceutical company is keeping current with new drugs and changes to existing drugs. New drugs are continually being developed and approved. In addition, generic drugs are often available to compete with brand-name drugs. One of the services that Reliable provides is to try to find the least expensive alternative to fulfill a prescription. This cost-saving service is one of the marketing advantages that the nursing homes can use to promote their services. Obviously, this service builds tremendous loyalty between Reliable and its customers.

To keep current with these changes, Reliable subscribes to an online drug-update service. The service provides updates in several formats, one of which is an XML file.

1. Based on the content of your design class diagrams that you developed in Chapter 11, illustrate a sample XML input file that could be used to update drug information in the Reliable database.
2. In earlier chapters, the case description indicated that a case manifest was produced for each patient whenever prescriptions needed to be filled and delivered. Based on the data found in your class diagrams, design a case manifest. Consider that a patient might have multiple prescriptions that are being filled on the same delivery.
3. Each month, Reliable produces a statement for each nursing home. The statement lists each patient who received prescriptions during the month. All the filled prescriptions are listed. For each prescription, the following information is listed: the price, the amount billed to the patient's insurance provider, the amount paid by the insurance provider, and the co-pay amounts due from the patients. Design this monthly statement. Also, identify and highlight output controls that you believe are appropriate for this type of report.
4. In the preceding chapter, you defined an input form to be used to collect orders from the nursing homes. Go back and analyze that input form and identify all of the input controls that you think are necessary to ensure that the prescriptions are correct. What other procedures or controls would you recommend to make sure that there are no mistakes on the prescriptions?

FURTHER RESOURCES

David Benyon, Diana Bental, and Thomas Green, *Conceptual Modeling for User Interface Development*. Springer-Verlag, 1999.

Elfriede Dustin, Jeff Rashka, Douglas McDiarmid, and Jakob Nielson, *Quality Web Systems: Performance, Security, and Usability.* Addison-Wesley, 2001.

Simson Garfinkel, Gene Spafford, and Debby Russell, *Web Security, Privacy, & Commerce*. O'Reilly Publishing, 2001.

Anup K. Ghosh, *E-Commerce Security: Weak Links, Best Defenses.* John Wiley & Sons, 1997.

IS Audit and Control Association, *IS Audit and Control Journal*, Volume I. 1995.

Brenda Laurel, *The Art of Human-Computer Interface Design*. Addison-Wesley, 1990.

Ben Shneiderman, *Designing the User Interface: Strategies for Effective Human-Computer Interaction*. Addison-Wesley-Longman, 1998.

Donald Warren Jr. and J. Donald Warren, *The Handbook of IT Auditing*. Warren Gorham & Lamont, 1998.

Donald A. Wayne and Peter B. B. Turney, *Auditing EDP Systems*. Prentice Hall, 1990.

PART 4

IMPLEMENTATION AND SUPPORT

CHAPTER 16
Making the System Operational

CHAPTER 17
Current Trends in System Development

CHAPTER

16

MAKING THE SYSTEM OPERATIONAL

LEARNING OBJECTIVES

After reading this chapter, you should be able to:

- Describe implementation and support activities
- Choose an appropriate approach to program development
- Describe various types of software tests and explain how and why each is used
- List various approaches to data conversion and system installation and describe the advantages and disadvantages of each
- Describe different types of documentation and the processes by which they are developed and maintained
- Describe training and user support requirements for new and operational systems

CHAPTER OUTLINE

Program Development

Quality Assurance

Data Conversion

Installation

Documentation

Training and User Support

Maintenance and System Enhancement

TRI-STATE HEATING OIL: JUGGLING PRIORITIES TO BEGIN OPERATION

It was 8:30 on Monday morning, and Maria Grasso, Kim Song, Dave Williams, and Rajiv Gupta were just about to begin the weekly project status review meeting. Tri-State Heating Oil had started developing a new customer order and service call scheduling system five months earlier. The target completion date was 10 weeks away, but the project was behind schedule. Analysis and design had taken eight weeks longer than anticipated because key users had disagreed on what new system requirements to include and the system scope was larger than expected.

Maria began the meeting by saying, "We've gained a day or two since our last meeting due to better-than-expected unit testing results. All of the methods developed last week sailed through unit testing, so we won't need any time this week to fix errors in that code."

Kim spoke, "I wouldn't get too cocky just yet. All of the nasty surprises in my last project came during integration testing. We're completing the user-interface classes this week, so we should be able to start integration testing with the business classes sometime next week."

Dave nodded enthusiastically and said, "That's good! We have to finish testing those user-interface classes as quickly as possible because we're scheduled to start user training in three weeks. I need that time to develop the training materials and work out the final training schedule with the users."

Rajiv replied, "I'm not sure that we should be trying to meet our original training schedule with so much of the system still under development. What if integration testing shows major bugs that require more time to fix? And what about the unfinished business and database classes? Can we realistically start training with a system that's little more than a user interface with half a system behind it?"

Dave replied, "But we have to start training in three weeks. We contracted for a dozen temporary workers so that we could train our staff on the new system. Half of them are scheduled to start in two weeks and the rest two weeks after that. It's too late to renegotiate their start dates. We can extend the time they'll be here, but delaying their starting date means we'll be paying for people we aren't using."

Maria said, "I think that Rajiv's concerns are valid. It's not realistic to start training in three weeks with so little of the system completed and tested. We're at least five weeks behind schedule, and there's no way we'll recapture more than four or five days of that during the next few weeks. I've already looked into rearranging some of the remaining coding to give priority to the work most critical to user training. There are a few batch processes that can be back-burnered for awhile. Kim, can you rearrange your test plans to handle all of the interactive applications first?"

Kim replied, "I'll have to go back to my office and take another look at the dependencies among those programs. Offhand, I'd say yes, but I need a few hours to make sure."

Maria replied, "Okay, let's proceed under the assumption that we can rearrange coding and testing to complete a usable system for training in five weeks. I'll confirm that by e-mail later today as soon as Kim gets back to me. I'll also schedule a meeting with the CIO to deliver the bad news about temporary staffing costs."

After a few moments of silence, Rajiv asked, "So what else do we need to be thinking about?"

Maria replied, "Well, let's see. . . . There's user documentation, hardware delivery and setup, operating system and DBMS installation, importing data from the old database, the network upgrade, and stress testing for the distributed database accesses."

Rajiv smiled and said to Maria, "You must have been a juggler in your youth, and it was good practice for keeping all of these project tasks up in the air. Does management pay you by the ball?"

Maria chuckled and replied, "I do think of myself as a juggler sometimes. And if management paid me by the ball, I could retire as soon as this project is finished!"

OVERVIEW

This chapter focuses on the implementation and support activities of the systems development life cycle (see Figure 16-1). Activities that occur before the system is turned over to its users are collectively called implementation. Activities that occur after the system becomes operational are collectively called support.

Figure 16-1
Implementation and support activities

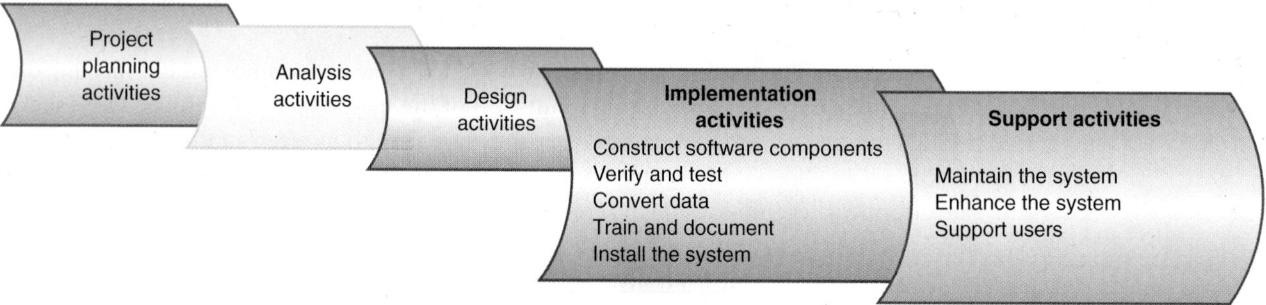

Implementation and support activities are often considered straightforward and dull—they don't attract the same attention or enthusiasm as analysis and design activities. The situation is analogous to the difference between architecture and construction. An architect gets most of the credit for creating a new building, even though his or her job essentially ends with the

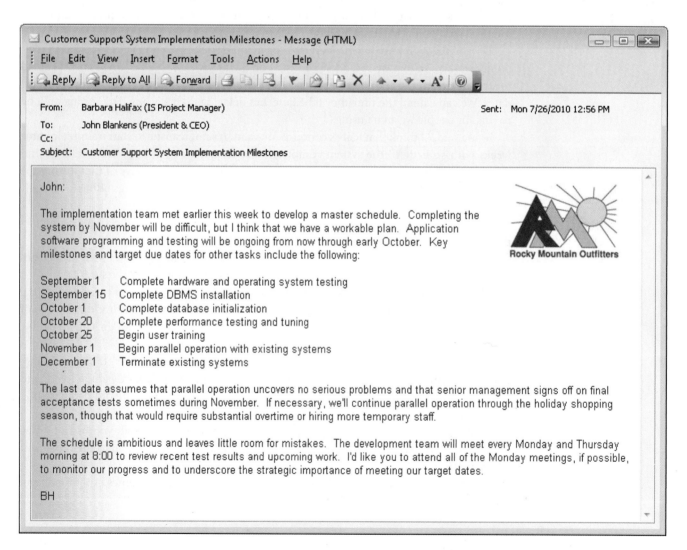

Customer Support System Implementation Milestones - Message (HTML)

File Edit View Insert Format Tools Actions Help

Reply | Reply to All | Forward

From: Barbara Halifax (IS Project Manager) Sent: Mon 7/26/2010 12:56 PM
To: John Blankens (President & CEO)
Cc:
Subject: Customer Support System Implementation Milestones

Rocky Mountain Outfitters

John:

The implementation team met earlier this week to develop a master schedule. Completing the system by November will be difficult, but I think that we have a workable plan. Application software programming and testing will be ongoing from now through early October. Key milestones and target due dates for other tasks include the following:

September 1	Complete hardware and operating system testing
September 15	Complete DBMS installation
October 1	Complete database initialization
October 20	Complete performance testing and tuning
October 25	Begin user training
November 1	Begin parallel operation with existing systems
December 1	Terminate existing systems

The last date assumes that parallel operation uncovers no serious problems and that senior management signs off on final acceptance tests sometimes during November. If necessary, we'll continue parallel operation through the holiday shopping season, though that would require substantial overtime or hiring more temporary staff.

The schedule is ambitious and leaves little room for mistakes. The development team will meet every Monday and Thursday morning at 8:00 to review recent test results and upcoming work. I'd like you to attend all of the Monday meetings, if possible, to monitor our progress and to underscore the strategic importance of meeting our target dates.

BH

blueprints. Yet, the vast majority of the effort that goes into making the building a reality occurs after the blueprints are finished. The same is true of information system development.

Implementation consumes substantially more time and resources than other system development life cycle (SDLC) activities. A large number of people are required to perform implementation activities—particularly software construction and testing. In addition, implementation activities are highly interdependent. Project management complexity is at its greatest during the implementation activities because so many people and tasks must be coordinated. The RMO project progress memo on the preceding page illustrates the complexity of project management and the many tasks that must be completed.

Information systems are the lifeblood of a modern organization. Thus, supporting those systems is one of the most important jobs in an organization. Support activities ensure that the system and its users function efficiently and effectively for years after installation. Most organizations spend much more money maintaining and supporting existing systems than they do building new ones.

PROGRAM DEVELOPMENT

Developing a complex system is an inherently difficult process. Consider the complexity of manufacturing automobiles. Tens of thousands of parts must be fabricated or purchased. Parts are assembled into small subcomponents (such as dashboard instruments, wiring harnesses, and brake assemblies), which are, in turn, assembled into larger subcomponents (such as instrument clusters, engines, and transmissions) and eventually into a complete automobile. Parts and subcomponents must be constructed, tested, and passed on to subsequent assembly steps. There are tens or hundreds of thousands of individual production steps. The effort, timeliness, cost, and output quality of each step depend on all of the preceding steps.

Program development is similar in many ways to automobile manufacturing. Requirements and design specifications have already been determined. What remains is a complex production and assembly process that must ensure efficient resource use, minimal construction time, and maximum product quality. But unlike automobile manufacturing, the process is not designed once and then used to build thousands of similar units. Instead, a software manufacturing process must be redeveloped for each new project to match that project's unique characteristics.

When most people think of system development, they primarily think of programming. Programming isn't the only development activity, but it is clearly one of the most important. Its importance arises from several factors, including the following:

- Required resources
- Managerial complexity
- System quality

Program development consumes more resources than any other system development activity. Program development (including unit testing) typically accounts for at least one-third of all development labor. Program development also accounts for between one-third and one-half of the project development schedule. The magnitude of resources and time consumed during program development clearly warrants careful planning and management attention.

ORDER OF IMPLEMENTATION

One of the most basic decisions to be made about program development is the order in which program components will be developed. Several orders are possible, including the following:

- Input, process, output
- Top-down
- Bottom-up

Each project must adapt one or a combination of these approaches to specific project requirements and constraints.

Input, Process, Output Development Order

input, process, output (IPO) development
a development order that implements input modules first, process modules next, and output modules last

The **input, process, output (IPO) development** order is based on data flow through a system or program. Programs or modules that obtain external input are developed first. Programs or modules that process the input (that is, transform it into output) are developed next. Programs or modules that produce output are developed last.

For structured designs and programs, an analyst can determine IPO ordering by examining the system flowchart and structure charts. For example, consider the payroll system flowchart in Figure 16-2. The programs Maintain tax tables and Maintain employee database obtain and modify data inputs for other programs, so they would be the first programs implemented. The Payroll program combines input and processing, so it would be implemented next. The Check printing and Year-end tax programs produce system outputs, so they would be implemented last.

Figure 16-3 shows a structure chart for the Payroll program. An analyst can apply IPO implementation order to modules within a program by classifying them as input, process, and output modules. If the analyst developed the structure chart using transform analysis,

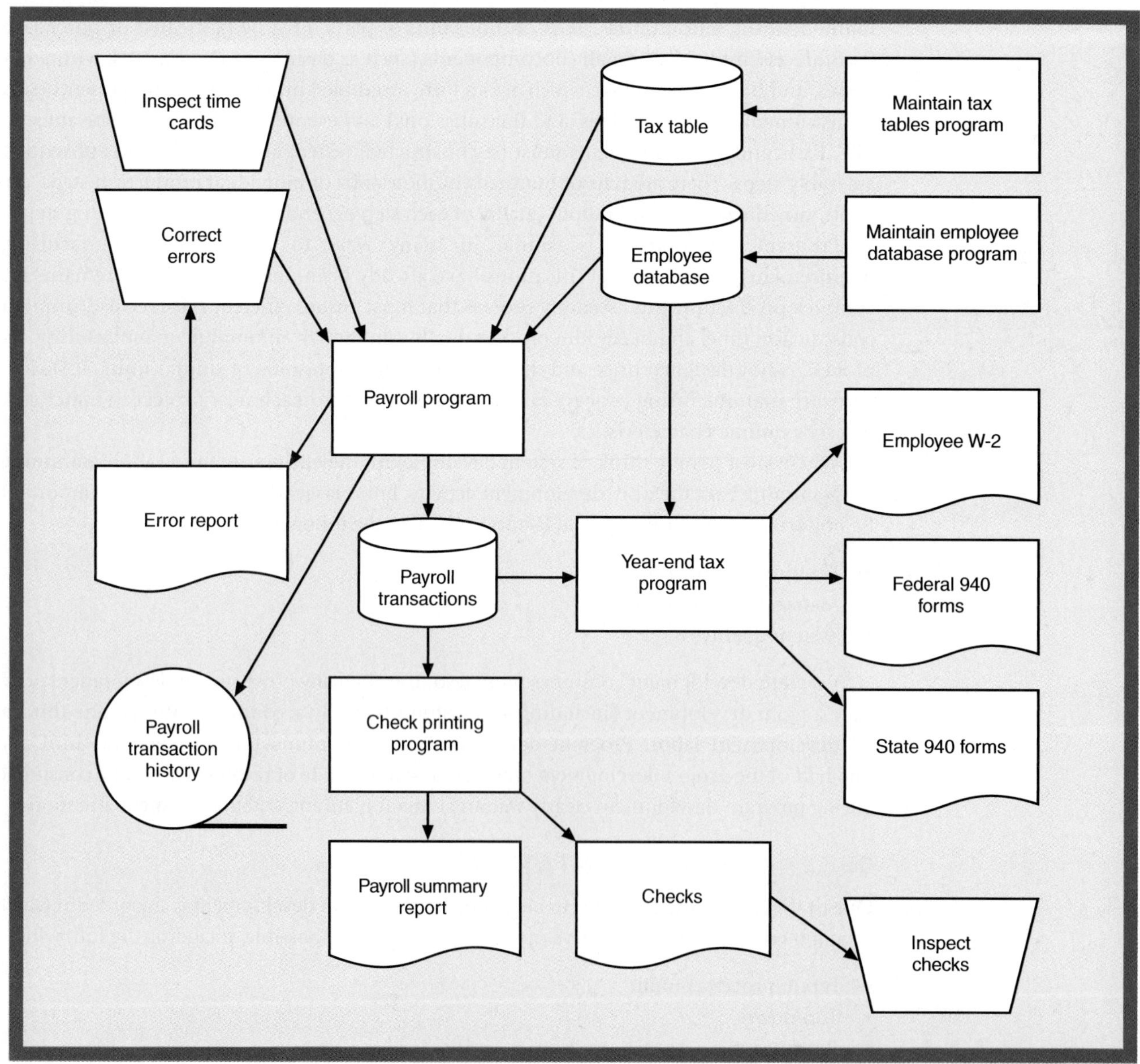

Figure 16-2
A system flowchart for a payroll system

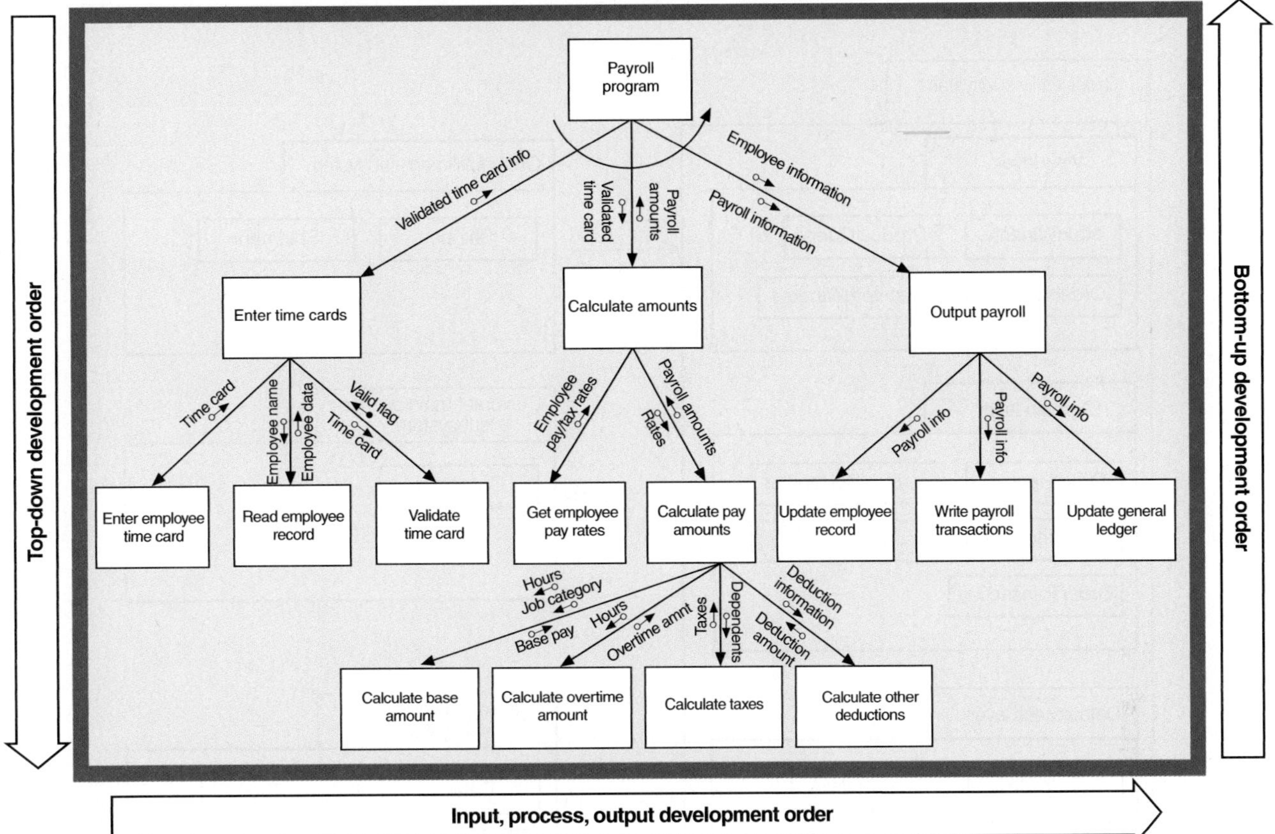

Figure 16-3

A structure chart for the Payroll program in Figure 16-2

modules will be clearly organized into afferent (input), central transform (process), and efferent (output) "legs." (See the section "Developing a Structure Chart" in Chapter 10 for a review of structure chart organization.) For this program, the modules in the afferent leg of the structure chart (those below and including *Enter time cards*) would be implemented first, followed by the modules in the transform leg (those below and including *Calculate amounts*), and finally the modules in the efferent leg (those below and including *Output payroll*).

IPO development order can also be applied to object-oriented (OO) designs and programs. The key issue to analyze is dependency—that is, which classes and methods capture or generate data that is needed by other classes or methods. Dependency information is documented in package, sequence, and class diagrams.

For example, the package diagram in Figure 16-4 shows that the customer and catalog maintenance subsystems are not dependent on each other or on either of the other two subsystems. The order-entry subsystem is dependent on both the customer and catalog maintenance subsystems, and the order fulfillment subsystem is dependent on the order-entry subsystem.

Data dependency among the packages (subsystems) implies data dependency among their embedded classes. Thus, the classes Customer, Catalog, and Package have no data dependency on the remaining RMO classes. Under IPO development order, those three classes are implemented first.

The chief advantage of the IPO development order is that it simplifies testing. Because input programs and modules are developed first, they can be used to enter test data for process and output programs and modules. The need to write special-purpose programs to generate or create test data is reduced, thus speeding the development process.

IPO development order is also advantageous because important user interfaces (for example, data-entry routines) are developed early. User interfaces are more likely to require change during development than other portions of the system, so early development allows for early testing and user evaluation. If changes are needed, there is still plenty of time to make them. Early development of user interfaces also provides a head start for related activities such as training users and writing documentation.

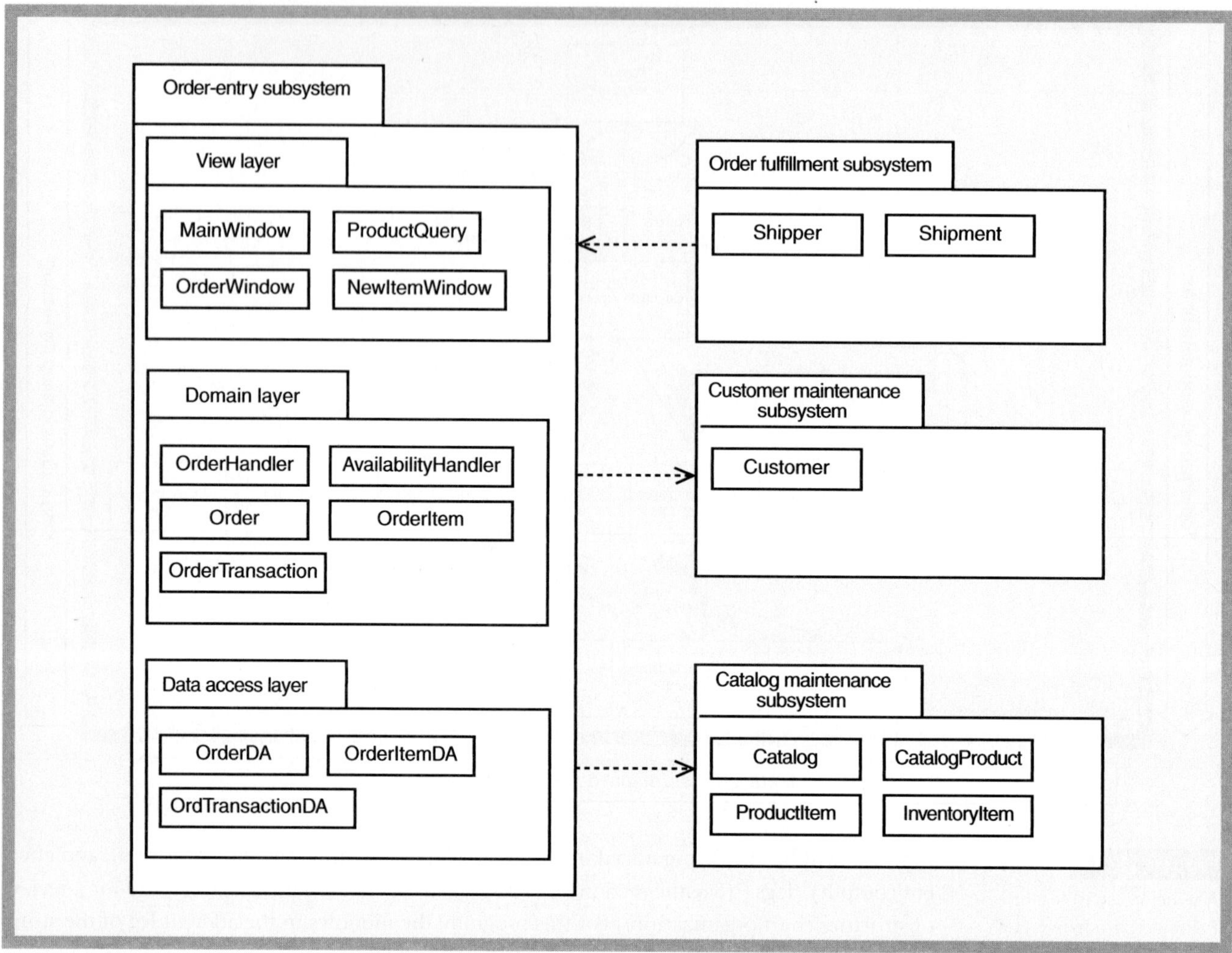

Figure 16-4

A package diagram for the four RMO subsystems

A disadvantage of IPO development order is the late implementation of outputs. Output programs are useful for testing process-oriented modules and programs; analysts can find errors in processing by manually examining printed reports or displayed outputs. IPO development defers such testing until late in development. However, analysts can usually generate alternate test outputs by using the query processing or report writing capabilities of a database management system (DBMS). If such outputs can be quickly and easily defined, the disadvantage of late implementation of output routines is substantially mitigated.

Top-Down and Bottom-Up Development Order

top-down development
a development order that implements modules at the top of a structure chart first

bottom-up development
a development order that implements modules at the bottom of a structure chart first

The terms *top-down* and *bottom-up* have their roots in traditional structured design and structured programming. Both terms describe the order of implementation with respect to a module's location within a structure chart. For example, consider the structure chart in Figure 16-3. Top-down development begins with the module at the top of the structure chart (Payroll program). Bottom-up development begins with the set of modules at the lowest level of the structure chart.

Top-down and bottom-up program development can also be applied to OO designs and programs, although a visual analogy is not as obvious with OO diagrams as with structure charts. The key issue is method dependency—that is, which methods call which other methods. Within an OO subsystem or class, method dependency can be examined in terms of navigation visibility, as discussed in Chapter 11.

For example, consider the three-layer design of part of the RMO order-entry subsystem shown in Figure 16-5. The arrows among packages and classes show navigation visibility requirements. Methods in the view layer call methods in the domain layer, which, in turn, call methods in the data access layer. Top-down implementation would implement the view

Figure 16-5

A package diagram for a three-layer OO design

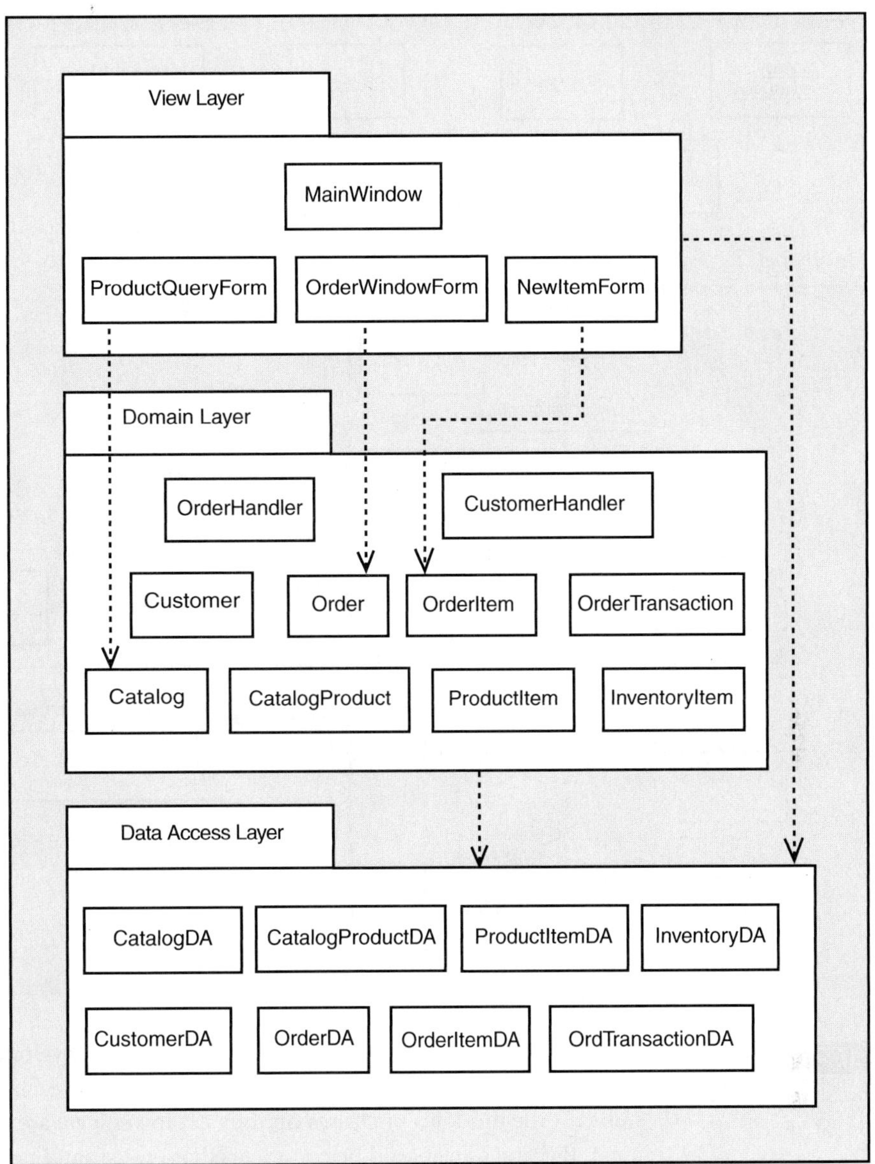

layer classes and methods first, the domain layer classes and methods next, and the data access layer classes and methods last. Bottom-up implementation would reverse the top-down implementation order.

Method dependency can also be documented in a sequence diagram. For example, in Figure 16-6, method dependency is documented in the left-to-right flow of messages among objects. Rotating the figure 90 degrees clockwise creates a top-down and bottom-up visual analogy similar to a structure chart. Top-down development would start with ProductQuery, AvailabilityHandler, CatalogDA, and Catalog. Bottom-up development would start with InventoryItem, InventoryDA, CatalogProduct, and CatalogProductDA.

The primary advantage of top-down development is that there is always a working version of a program. For example, top-down development of the program in Figure 16-3 would begin with a complete version of the topmost module and dummy (or stub) versions of its three subordinate modules (stub modules are discussed later in the "Unit Testing" section). This set of modules is a complete program that can be compiled, linked, and executed, although at this point it wouldn't do very much when executed. Top-down development of the three-layer design in Figure 16-5 would begin with a complete version of the view layer classes and dummy (or stub) versions of the domain layer classes.

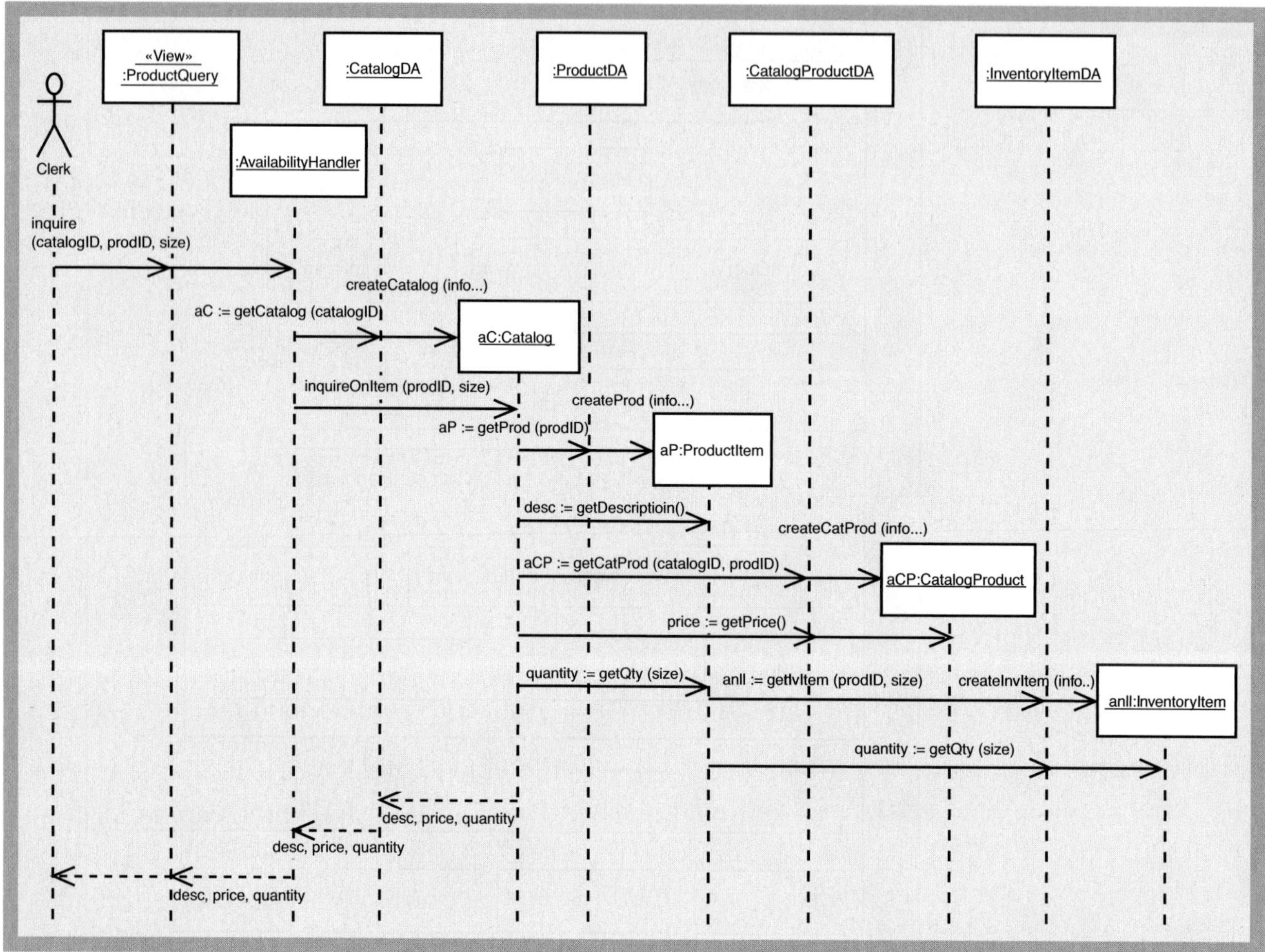

Figure 16-6

A sequence diagram for the event *Look up item availability*

After the topmost module or layer is completed, development proceeds downward to the next level of the structure chart or package diagram. As each module or class is implemented, stubs for the modules or classes on the next lower level are added. At every stage of development, the program should be complete (that is, it should be able to be compiled, linked, and executed). Its behavior becomes more complex and realistic as development proceeds.

The primary disadvantage of top-down development order is that it doesn't use programming personnel very efficiently at the beginning of software development. Development has to proceed through two or three levels before a significant number of modules or methods can be developed simultaneously. However, if the first few levels of the program can be completed quickly, the disadvantage is minimal.

The primary advantage of bottom-up development is that many programmers can be put to work immediately. In addition, lower-level modules are often the most complex and difficult to write, so early development of those modules allows more time for development and testing. Unfortunately, bottom-up development also requires writing a large number of driver programs to test bottom-level modules, which adds additional complexity to the development and testing process (this issue is discussed further in the "Testing" section). Also, the entire system isn't assembled until the topmost modules are written. Thus, testing of the system as a whole is delayed.

Other Development Order Considerations

IPO, top-down, and bottom-up development are only a starting point for creating a software development plan. Other factors that must be considered include user feedback, training, documentation, use case driven development, and testing. User feedback, training, and documentation all depend heavily on the user interfaces of the system. Early implementation of user interfaces enables early user training and early development of user documentation. It also gathers early feedback on the quality and usability of the interface. Note the important role that this issue played in the opening case of this chapter.

In projects that follow the Unified Process, use cases are one of the primary bases for dividing a development project into iterations. In most UP projects, developers choose a set of related use cases for a single iteration and complete all requirements, design, implementation, and testing activities for those use cases. The choice of use cases might be based on many factors, including minimizing project risk, efficiently using nontechnical staff, or installing some parts of the system earlier than others. For example, use cases with uncertain requirements or high technical risks are typically addressed in early iterations.

As individual software components (such as modules or methods) are constructed, they must be tested. Programmers must find and correct errors as soon as possible because they become much harder to find and more expensive to fix as the construction process proceeds. It's important both to identify portions of the software that are susceptible to errors and to identify portions of the software where errors can pose serious problems that affect the system as a whole. These portions of the software must be built and tested early regardless of where they fit within the basic approaches of IPO, top-down, or bottom-up development.

Testing and construction are highly interdependent. For this reason, a formal plan covering both testing and construction is normally created before either activity begins. The construction and test plan covers many specifics, including the following:

- Development order
- Testing order
- Data used to test modules, module groups, methods, classes, programs, and subsystems
- Acceptance criteria
- Personnel assignments (construction and testing)

Testing is discussed in detail later in this chapter. But for now, keep in mind that construction and testing go hand in hand. Their interdependence and complexity necessitate formal planning and regular comparisons between the plan and actual performance.

FRAMEWORK DEVELOPMENT

When implementing a large OO system, it is not unusual to build an object framework (or set of foundation classes) that covers most or all of the domain and data access layer classes. For example, when implementing an OO account maintenance system for a bank, developers might build a set of classes to represent and store customers and various types of bank accounts (for example, savings accounts, checking accounts, and certificates of deposit).

Foundation classes are typically reused in many parts of the system and across many different applications. Because of this reuse, they are a critical system component. Errors in a foundation class can affect every program in the system. In addition, later changes to foundation classes might require significant changes throughout the system.

Foundation classes are typically implemented first to minimize the impact of errors and changes. They are typically assigned to the best programmers and are tested more thoroughly than other classes. Early and thorough testing guarantees that bugs or other problems will be discovered before other code that depends on the foundation classes has been written.

TEAM-BASED PROGRAM DEVELOPMENT

A team of programmers normally works on program development. Using multiple programmers compresses the development schedule by allowing many portions of the system to be developed simultaneously. However, team-based program development introduces its own set of management issues, including the following:

- Organization of programming teams
- Task assignment to specific teams or members
- Member and team communication and coordination

You can organize an implementation team in many different ways. Some commonly used organizational models include the following:

- Cooperating peer
- Chief developer
- Collaborative specialist

Figure 16-7

A comparison and summary of development team types

Figure 16-7 summarizes the characteristics of each team type and the types of projects and tasks best suited to each.

Team type	Team characteristics	Task and project types
Cooperating peers	Equal skill levels Overlapping specialities Consensus-based decision making	Experimentation Creative problem solving
Chief developer	Organized as a military platoon or squad One leader makes all important decisions	Well-defined objectives Well-defined path to completion
Collaborative specialists	Wide variation in skill and experience Minimal overlap in technical specialities Leader is primarily an administrator Consensus-based decision making	Diagnosis or experimentation Creative and integrative problem solving Wide range of technology

cooperating peer team
a team with members of roughly equal skill and experience and with overlapping areas of specialization

chief developer team
a team with a single leader who makes all important decisions

collaborative specialist team
a team with members who have wide variation in and minimal overlap of skills and experience

A **cooperating peer team** includes members of roughly equal skill and experience with overlapping areas of specialization. Members are considered equals, although they might be assigned tasks of varying importance or complexity. Decisions are primarily made by consensus, and the team frequently meets to exchange information and build consensus.

A **chief developer team** is similar to a small military unit. An assigned leader performs a number of functions, including technical consulting, team coordination, and task assignment. In this type of team, there is much less communication than with a cooperating peer team. The chief developer makes most of the important decisions, although he or she might seek input from members individually or collectively.

A **collaborative specialist team** is similar to a cooperating peer team, but its members have wide variation in and minimal overlap of skills and experience. Such teams are often composed of members from different organizational subunits. The team might have an appointed leader, but his or her leadership covers only administrative functions, such as scheduling, coordinating, and interacting with external constituencies. Technical decisions are generally made by consensus, although member opinions usually carry extra weight within the member's own area of expertise. In large projects, a collaborative specialist team might be formed to "float" among other teams to deal with complex problems as they arise.

Some common principles of team organization underlie all development projects and organizational structures. One is that team size should be kept relatively small (no more than 10 members). Larger teams tend to be inefficient because of the inherent complexity of communication and coordination in large groups. When more than 10 developers are assigned to a project, it is best to break them up into small teams (approximately five members each). Each team should be assigned a relatively independent portion of the project. One member of each team should be designated to handle coordination and communication with other teams. Having a single point of contact simplifies communication and provides for some specialization of functions within each team.

Another common principle of team organization is that team structure should be matched to the task and project characteristics. Teams with a well-defined implementation task that does not push the limits of member knowledge or technical feasibility are usually best organized as chief programmer teams.

Teams assigned to tasks that require experimentation or a high level of creativity are better served by a cooperating peer or collaborative specialist model. Because of their more open communication, cooperating peer teams are especially well suited to tackling tasks that require generating and evaluating a large number of ideas. Overlap in skill, specialty, and experience allows thorough evaluation of each idea.

Collaborative specialist teams are well suited to projects that span a wide range of cutting-edge technology. They are also well suited to tackling projects that require integrated problem solving (for example, diagnosing and fixing bugs in an existing complex system). However, success depends on a true collaborative process, which is sometimes difficult to achieve among members with wide variation in skill and experience.

Member skills must be appropriately matched to the tasks at hand. Skill matching is a fairly obvious requirement with respect to technical skills such as database management, user interfaces, and numeric algorithms. Skill matching is less obvious but no less important for nontechnical skills. Teams need a mix of nontechnical skills and traits, including the ability to generate new ideas, build consensus, manage details, and communicate with external constituencies. The project manager should perform a skills inventory early in the project so that gaps can be filled to avoid project delays and inappropriate personnel assignments.

SOURCE CODE CONTROL

source code control system (SCCS)

an automated tool for tracking source code files and controlling changes to those files

Development teams need tools to help coordinate their programming tasks. A source code control system (SCCS) is an automated tool for tracking source code files and controlling changes to those files. An SCCS stores project source code files in a repository. The SCCS acts the way a librarian would—it implements check-in and checkout procedures, tracks which programmer has which files, and ensures that only authorized users have access to the repository.

Programmers can manipulate files in the repository as follows:

- Check out a file in read-only mode
- Check out a file in read/write mode
- Check in a modified file

A programmer checks out a file in read-only mode when he or she wants to examine the code without making changes (for example, to examine a module's interfaces to other modules). When a programmer needs to make changes to a file, he or she checks out the file in read/write mode. The SCCS allows only one programmer to check out a file in read/write mode. The file must be checked back in before another programmer can check it out in read/write mode.

Figure 16-8 shows the main display of Microsoft Visual SourceSafe. Various source code files from the RMO customer support system are shown in the display. Some files are currently checked out by programmers. For each file checked out in read/write mode, the program lists the programmer who checked it out, the date and time of checkout, and the current location of the file. The icon for each checked-out file is displayed with a red border and check mark.

An SCCS prevents multiple programmers from updating the same file at the same time, thus preventing inconsistent changes to the source code. Source code control is an absolute necessity when programs are developed by multiple programmers. It prevents inconsistent changes and automates coordination among programmers and teams. The repository also serves as a common facility for backup and recovery operations.

VERSIONING

Medium- and large-scale systems are complex and constantly changing. Changes occur rapidly during implementation and more slowly afterward. System complexity and rapid change create a host of management problems—particularly for testing and support. Testing

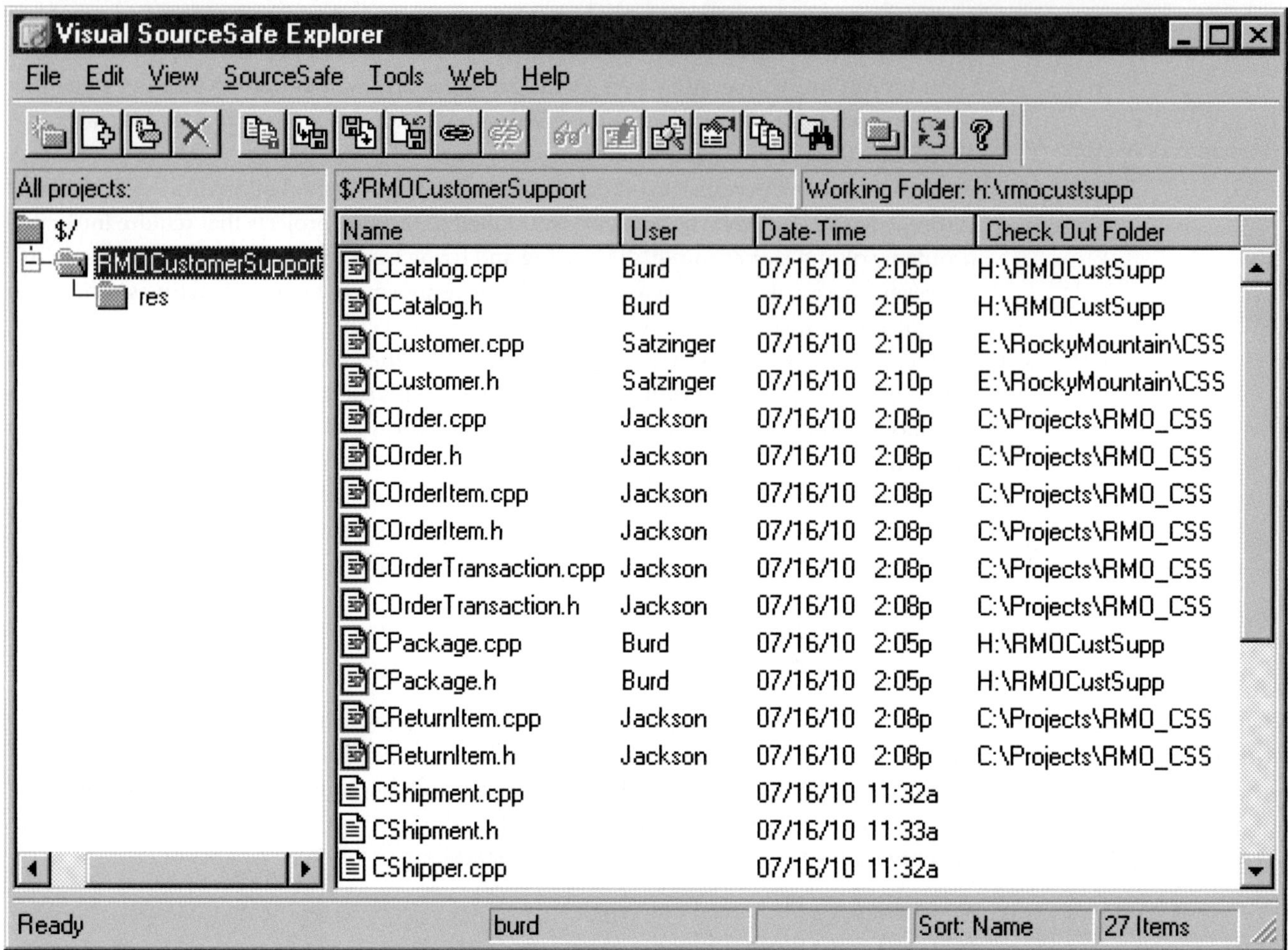

Figure 16-8

Project files managed by a source code control system

is problematic in such an environment because the system is a moving target. By the time an error is discovered, the code that caused the error might already have been moved, altered, or deleted. Support is complex for similar reasons. Support personnel need to know the state of the system as it is installed on a user's computer system to respond properly to a bug report or request for help.

Complex systems are developed, installed, and maintained in a series of versions to simplify testing and support. It is not unusual to have multiple versions of a system deployed to end users and more versions in different stages of development. A system version created during development is called a *test version*. A test version contains a well-defined set of features and represents a concrete step toward final completion of the system. Test versions provide a static system snapshot and a checkpoint to evaluate the project's progress.

An **alpha version** is a test version that is incomplete but ready for some level of rigorous testing. Multiple alpha versions might be built depending on the size and complexity of the system. The lifetime of an alpha version is typically short—days or weeks.

A **beta version** is a test version that is stable enough to be tested by end users. A beta version is produced after one or more alpha versions have been tested and known problems have been corrected. End users test beta versions by using them to do real work. Thus, beta versions must be more complete and less prone to disastrous failures than alpha versions. Beta versions are typically tested over a period of weeks or months.

A system version created for long-term release to users is called a **production version**, **release version**, or **production release**. A production version is considered a final product, although software systems are rarely "finished" in the usual sense of that term. Minor production releases (sometimes called **maintenance releases**) provide bug fixes and minor changes to existing features. Major production releases add significant new functionality and might be the result of rewriting an older release from the ground up.

alpha version
a system that is incomplete but ready for some level of rigorous testing

beta version
a system that is stable enough to be tested by end users

production version, release version, or production release
a system that is formally distributed to users or made operational

maintenance release
a system update that provides bug fixes and minor changes to existing features

Figure 16-9 shows a series of possible test and production versions for the RMO customer support system. Each version is described in Figure 16-10. The system is delivered in two major production releases—versions 1.0 and 2.0. Each initial production release is preceded by one or more alpha and beta test versions. Each version adds or updates functionality and includes bug fixes for the previous version. Version 1.1 is a maintenance, or minor production, release of version 1.0. Note that the time line for developing version 2.0 overlaps maintenance changes to version 1.0. Overlapping older production versions with test versions of future production releases is typical.

Figure 16-9

A time line of test and production versions for the RMO customer support system

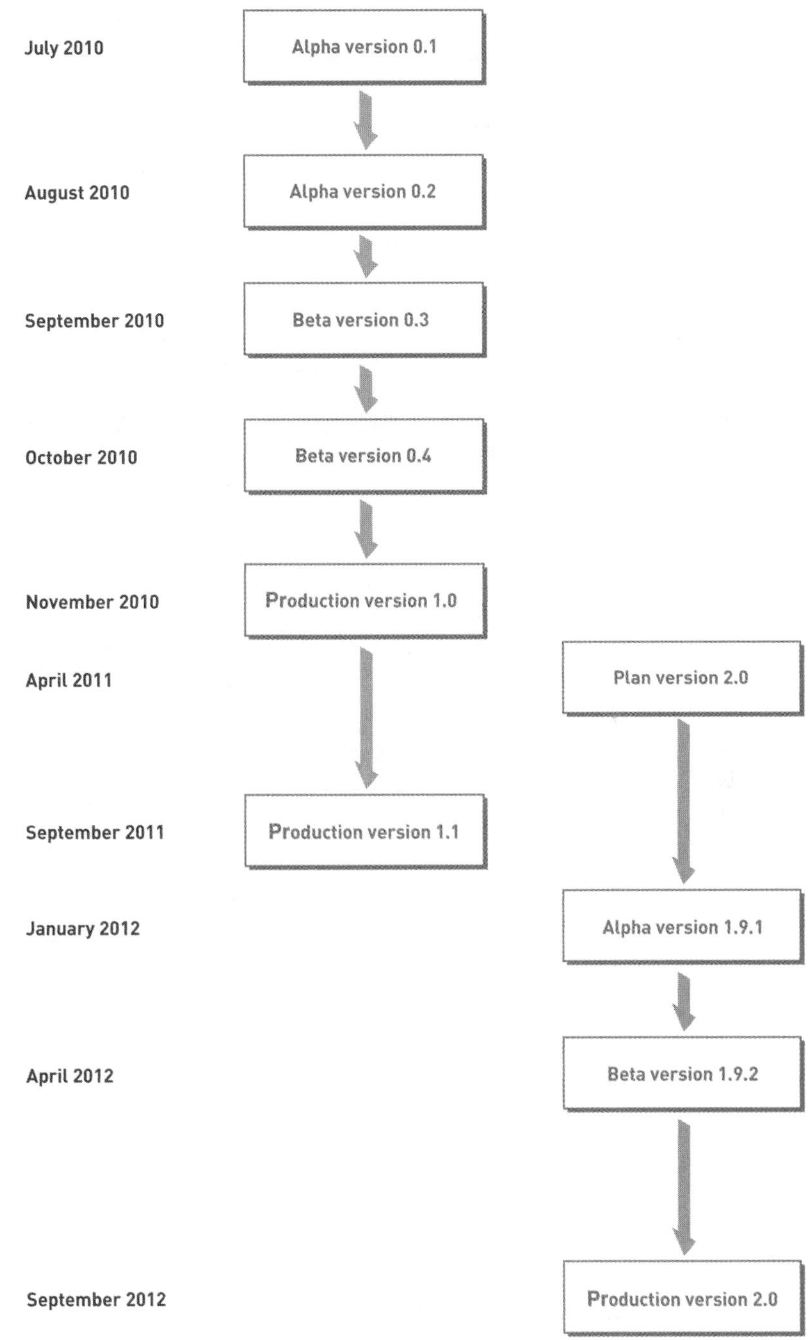

Keeping track of versions is complex. Each version needs to be uniquely identified for users and testers. In applications designed to run under Windows, users typically view the version information by choosing the About item of the standard Help menu, as shown in Figure 16-11. Users seeking support or reporting errors use this feature to report the system version to testers or support personnel.

Figure 16-10

Description of versions in Figure 16-9 for the RMO customer support system

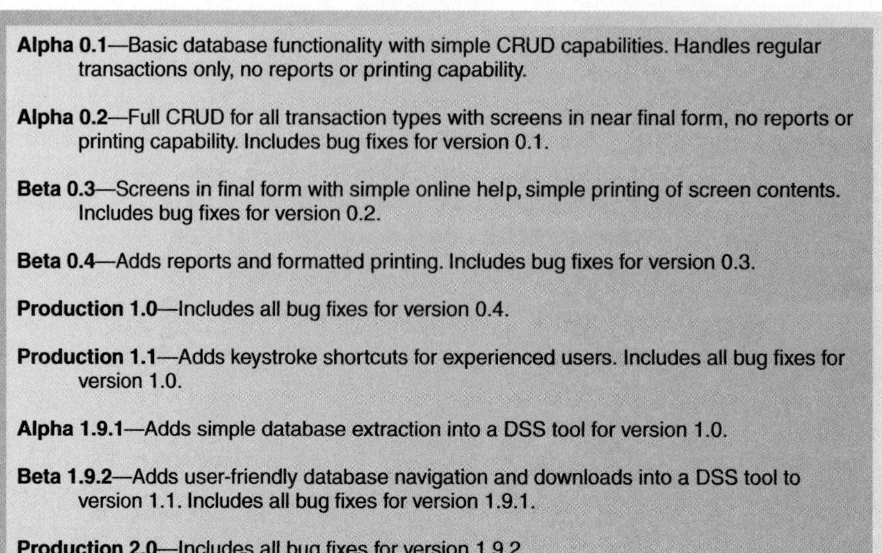

Alpha 0.1—Basic database functionality with simple CRUD capabilities. Handles regular transactions only, no reports or printing capability.

Alpha 0.2—Full CRUD for all transaction types with screens in near final form, no reports or printing capability. Includes bug fixes for version 0.1.

Beta 0.3—Screens in final form with simple online help, simple printing of screen contents. Includes bug fixes for version 0.2.

Beta 0.4—Adds reports and formatted printing. Includes bug fixes for version 0.3.

Production 1.0—Includes all bug fixes for version 0.4.

Production 1.1—Adds keystroke shortcuts for experienced users. Includes all bug fixes for version 1.0.

Alpha 1.9.1—Adds simple database extraction into a DSS tool for version 1.0.

Beta 1.9.2—Adds user-friendly database navigation and downloads into a DSS tool to version 1.1. Includes all bug fixes for version 1.9.1.

Production 2.0—Includes all bug fixes for version 1.9.2.

Figure 16-11

The About box of a typical Windows application

About This Software

roxio

By Sonic Solutions

Roxio Creator DE
Version: 9.0.116
Build: 901B16J, DL2
Serial Number: DZ-2M9X6-F2CNG-0S08G
Copyright: Sonic Solutions 2006

Module Name: PX Engine
Build: 3.6.36a, 5
Copyright: Sonic Solutions 2006

Module Name: Roxio Core
Version: 3.3.0
Build: 331B17F, DLD
Copyright: Sonic Solutions 2006

Module Name: Roxio Creator Audio

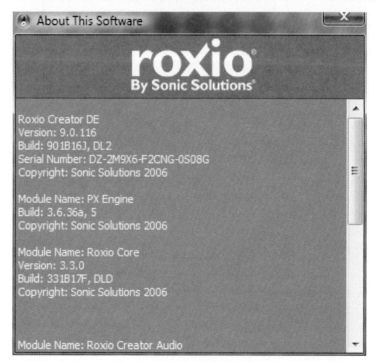

Controlling multiple versions of the same system requires sophisticated version control software. Version control capabilities are normally built into an SCCS. Programmers and support personnel can extract the current version or any previous version for execution, testing, or modification. Modifications are saved under a new version number to protect the accuracy of the historical snapshot.

Beta and production versions must be stored as long as they are installed on any user machines. Stored versions are used to evaluate future bug reports. For example, when a user reports a bug in version 1.0, support personnel extract that release from the archive, install it, and attempt to replicate the user's error. Feedback provided to the user will be specific to version 1.0 even if the most recent production release is a higher-numbered version.

BEST PRACTICE

To help programmers and support personnel, assign version numbers and enable users to display them.

QUALITY ASSURANCE

quality assurance (QA)
the process of ensuring that an information system meets minimal quality standards

As with any business procedure or system, quality is a major concern with information systems. Quality assurance (QA) is the process of ensuring that an information system meets minimal quality standards as determined by users, implementation staff, and management. QA is sometimes equated with finding bugs in program code, but this view is narrow and incomplete. QA is a set of activities that are performed throughout the SDLC to build systems correctly from the start and to detect and fix errors as soon as possible. Integrating quality assurance into early project activities allows many programming errors to be completely avoided. It also ensures that the system that is actually developed meets the needs of the users and the organization.

QA activities during analysis concentrate on identifying gaps or inconsistencies in system requirements. QA activities during design concentrate on satisfying stated requirements and on making design decisions that will lead to easily implemented, bug-free programs. QA activities during implementation consist primarily of testing. However, design and implementation overlap in many projects. Thus, quality assurance activities for design are typically integrated with testing activities.

QA activities are often shortchanged during design and especially during implementation. This lapse occurs for several reasons, including the following:

- Schedule pressures can build as the project progresses. QA and testing activities might be bypassed in an ill-fated attempt to speed up the project.
- QA activities require development personnel to open their work to thorough examination and criticism by others. Many people are reluctant to do this.
- Many people view testing and test personnel as the bearers of bad news. They mistakenly believe that no news is good news and that bypassing testing is a way of avoiding bad news.

There is a simple way to prevent human nature and schedule pressure from derailing QA activities: Formally integrate QA into the project and schedule from the beginning and never abandon it. Quality standards should be clearly stated, be measurable, and require a product that doesn't meet those standards to be fixed no matter what the effect on the schedule or budget. This approach to QA requires an organizational commitment from top management to the lowest levels. Unfortunately, top management is often the source of pressure to shortchange QA activities to speed up a project.

Another key factor in firmly establishing QA in the development process is to build an environment of openness, collegiality, and mutual respect among project participants. Personnel must be receptive to suggestions and constructive criticism and be willing to provide suggestions and criticisms to others. QA cannot be effective if it is allowed to devolve into an exercise in destructive criticism, finger-pointing, and blame assignment.

The cost of fixing an error grows as development proceeds. Errors are best detected during analysis or design and, thus, never committed to program code. Errors in programs are much easier to fix in the early stages of implementation than during later acceptance testing or, in the worst case, after the system is operational. This economic reality makes QA efforts throughout the SDLC well worth their cost.

TECHNICAL REVIEWS

Most programmers have had many experiences in which they were unable to correct an error because they "couldn't see it." But when the source code containing the error is shown to another programmer, the other programmer spots it immediately. Common examples include misspelled keywords, malformed if statement conditions, and illegal or spurious characters in source code. Such errors happen to programmers at all skill and experience levels.

technical review
a formal or informal review of design or construction details by a group of developers

A **technical review** opens the design and implementation process to input from other people. Technical reviews provide an opportunity for other people to find problems and offer constructive criticism. Technical reviews vary widely from one organization to another and sometimes among projects within an organization. Some organizations use informal processes, while others adopt formal procedures.

A *walkthrough* is a review by two or more people of the accuracy and completeness of a model or program. Walkthroughs are most often used during analysis and design, although they can be used during implementation. During design and implementation, a walkthrough is a technical review in which two or more developers review work to assess and improve its quality. Typically, one of the participants has already created a model or module before the walkthrough. The developer describes its underlying assumptions and operation, and the other participants provide comments and suggestions.

inspection
a formal review of design or construction details by a group of developers, where each person plays a specific role

An **inspection** is a more formal version of a walkthrough. Participants review and analyze materials before they meet as a group. Review materials include the model or code to be inspected, related models (for example, a structure chart or class diagram), and notes on specific types of errors that could occur. Group meetings usually follow a standard format.

When the group meets, participants play specific roles, including presenter, critic, and secretary. The presenter (usually the developer of the model or code) summarizes the material being inspected. The critics describe errors or concerns they found before the meeting, and the errors are discussed by all members of the group. Additional errors or problems might be uncovered during the discussion. The participants discuss possible solution strategies and agree on a specific approach. The secretary records all of the errors and the agreed-upon solution strategies.

Walkthroughs and inspections are important QA processes because they can detect errors *before* code has been written. Studies have shown that technical reviews accomplish the following:

- Reduce the number of errors that reach testing by a factor of 5 to 10
- Reduce testing costs by approximately 50 percent

Technical reviews reduce development costs and shorten the development schedule because a large number of errors never are passed along to be coded, tested, diagnosed, or fixed.

Testing and technical reviews each find between 50 and 75 percent of errors. But some errors are more easily detected by one method or the other. Some errors are rarely found by one technique but are easily found by the other. Thus, the two techniques are more effective jointly than individually.

BEST PRACTICE

Combine technical reviews and testing to maximize their effectiveness.

TESTING

Testing is the process of examining a product to determine what defects it contains. To conduct a test, programmers must have already constructed the software and have well-defined standards for what constitutes a defect. The developers can test products by reviewing their construction and composition or by exercising their function and examining the results. This section concentrates on the latter type of testing. This process is shown in Figure 16-12.

Software Testing

An information system is an integrated collection of software components. Components can be tested individually or in groups, or the entire system can be tested as a whole. Testing components individually is called *unit testing*. Testing components in groups is called *integration testing*. Testing entire systems is called *system testing*. Each type of testing is described in detail later in this section.

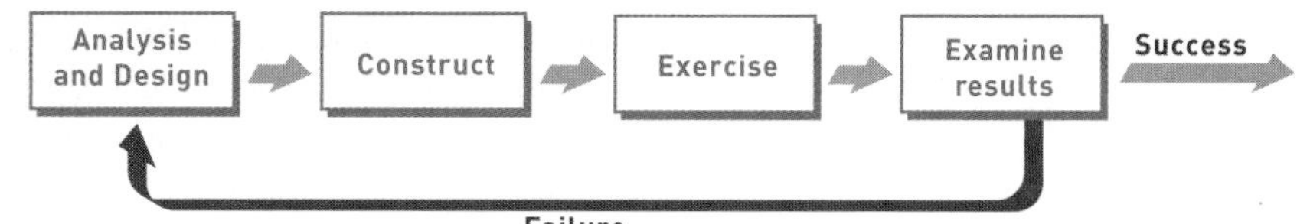

Figure 16-12
A generic model of software testing

The three testing types are each correlated to specific SDLC activities, as shown in Figure 16-13. A system test examines the behavior of an entire system with respect to technical and user requirements. These requirements are determined during analysis activities. During high-level design, the division of the system into high-level components and the structural design of those components are determined. Integration testing tests the behavior of related groups of software components. Low-level design is concerned with the internal construction of individual components. Unit testing tests each individual software component in isolation.

Figure 16-13
The correspondence between SDLC activities and various types of testing

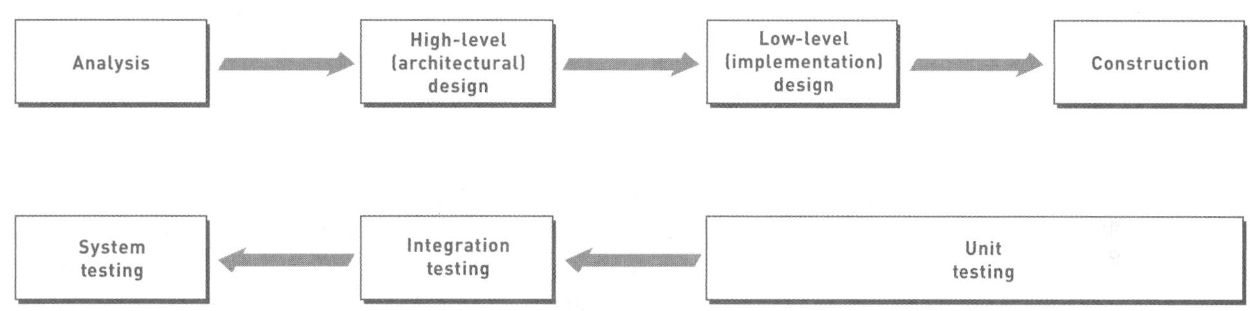

Because each testing level is related to a specific SDLC activity, testing can be spread throughout the life cycle, as illustrated in Figure 16-14. Planning for each type of testing can occur during its related SDLC activity, and development of specific tests can occur after the planning is complete. Tests cannot be conducted until relevant portions of the system have been constructed, however.

Figure 16-14
SDLC activities and related testing activities

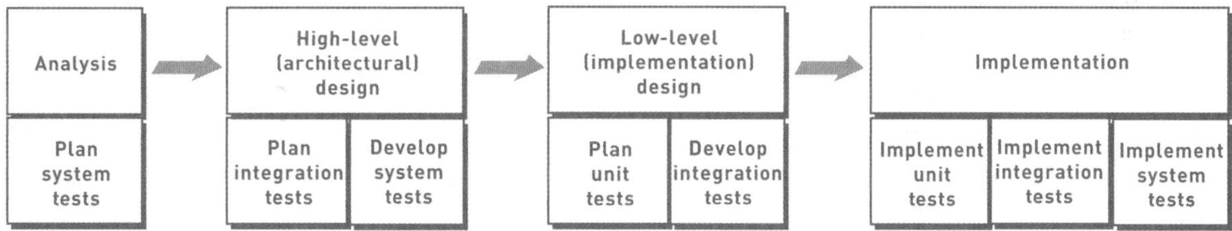

test case
a formal description of a starting state, one or more events to which the software must respond, and the expected response or ending state

test data
a set of starting states and events used to test a module, a group of modules, or an entire system

An important part of developing tests is specifying test cases and data. A test case is a formal description of the following:

- A starting state
- One or more events to which the software must respond
- The expected response or ending state

Both starting state and events are represented by a set of test data.

For example, the starting state of a system might represent a particular set of database contents (such as the existence of a particular customer and an order placed by that customer). The event might be represented by a set of input data items (such as a customer account number and order number used to query order status). The expected response might be a described behavior (such as the display of certain information) or a specific state of stored data (such as a canceled order).

Preparing test cases and data is a tedious and time-consuming process. At the program or module level, every instruction must be executed at least once. Ensuring that all instructions are executed during testing is a complex problem. Fortunately, automated tools based on proven mathematical techniques are available to generate a complete set of test cases. See the Watson and McCabe article in the "Further Resources" section for a thorough discussion of this topic.

Analysis documentation is useful when preparing test cases. If the system was analyzed and designed with OO techniques, developers prepare test cases for each use case and scenario. Many test cases representing both normal and exceptional processing situations should be prepared for each scenario.

The correspondence between traditional analysis models and test cases is less clear-cut. Developers can use the data flow diagrams and event table as the primary guide to prepare test cases. Developers should prepare multiple test cases for each event, and every process on every detailed data flow diagram should be exercised by at least one test case.

Unit Testing

unit testing, or module testing
testing of individual code modules or methods before they are integrated with other modules

Unit testing is the process of testing individual code modules before they are integrated with other modules. Unit testing is sometimes called **module testing**, although that term implies that software units are structured programming modules. In fact, unit testing can be applied to structured or OO software, and the unit being tested can be a function, subroutine, procedure, or method (for the remainder of this section, we will use the term *module* to refer to any of these programming constructs). Units can also be relatively small groups of interrelated modules that are always executed as a group. The goal of unit testing is to identify and fix as many errors as possible before modules are combined into larger software units (such as programs, classes, and subsystems). Errors become much more difficult and expensive to locate and fix when many modules are combined.

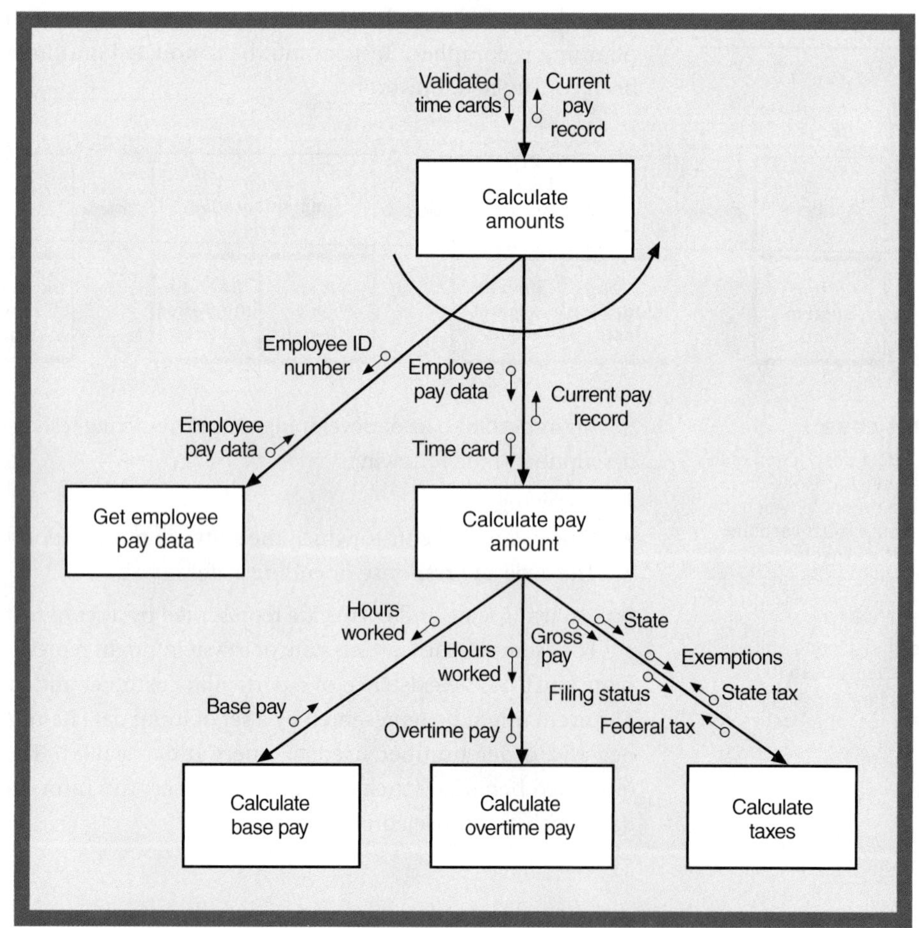

Figure 16-15
A portion of a structure chart for a program to calculate payroll

Few modules are designed to operate in isolation. Instead, groups of modules are designed to execute as an integrated whole. Modules can call other software units to perform tasks or can be called by other modules. This relationship is easily seen in a structure chart (see Figure 16-15), although it also exists among methods in OO software. For example, the module *Calculate pay amount* is called by the module *Calculate amounts,* which, in turn, calls the three modules below it in the structure chart.

driver
a module, developed for unit testing, that simulates the calling behavior of a module that hasn't yet been developed

If *Calculate pay amount* is being tested in isolation, two types of testing modules are required. The first module type is called a *driver*. A **driver** simulates the calling behavior of a module. A driver module implements the following functions:

- Sets the value of input parameters
- Calls the tested module, passing it the input parameters
- Accepts return parameters from the tested module and prints or displays them

Figure 16-16 shows a simple driver module for testing *Calculate pay amount*. A more complex driver module might use test data consisting of hundreds or thousands of module inputs and

Figure 16-16
A driver module for testing *Calculate pay amount*

```
module main()

// Driver Module to Test CalculatePayAmount()
{
   //  Declare Module Parameters

   record EmployeePayData   {
      integer EmployeeIDNumber;
      boolean SalariedEmployee;
      real PayRate;
      char[2] State;
      integer FilingStatus;
      integer Exemptions;
   }
   record TimeCard   {
      integer EmployeeIDNumber;
      date StartDate;
      array[7] of real HoursWorked;
   }
   record CurrentPayRecord   {
      real BasePay;
      real OvertimePay;
      real FederalTax;
      real StateTax;
   }

   // Set Input Parameter Values

   EmployeeData.EmployeeNumber=123456789;
   EmployeeData.SalariedEmployee=false;
   EmployeeData.PayRate=32.50;
   EmployeeData.State="AZ";
   EmployeeData.FillingStatus=1;
   EmployeeData.Exemptions=5;
   TimeCard.EmployeeIDNumber=123456789;
   TimeCard.StartDate=05/21/2005;
   TimeCard.HoursWorked[0]=0.0;
   TimeCard.HoursWorked[1]=0.0;
   TimeCard.HoursWorked[2]=8.5;
   TimeCard.HoursWorked[3]=7.5;
   TimeCard.HoursWorked[4]=8.0;
   TimeCard.HoursWorked[5]=8.0;
   TimeCard.HoursWorked[6]=9.0;

   // Call Tested Module

   call CalculatePayAmount (EmployeeData, TimeCard, CurrentPayRecord);

   // Print Results

   print(EmployeeData, TimeCard, CurrentPayRecord);
}
```

correct outputs stored in a file or database. The driver would loop through the test data and repeatedly call *Calculate pay amount*, check the return parameter against the expected value, and print or display warnings of any discrepancy.

Using a driver enables testing of a subordinate module before modules that call it have been written. Drivers are used extensively in bottom-up development because child modules (or methods) are developed and unit-tested before their parents are developed.

stub
a module, developed for testing, that simulates the execution or behavior of a module that hasn't yet been developed

The second type of testing module used to perform unit tests is called a *stub*. A **stub** simulates the behavior of a called module that hasn't yet been written. A unit test of *Calculate pay amount* would require three stub modules, one for each of the modules that appear below it in Figure 16-15. Stubs are relatively simple modules that usually have only one or two lines of executable code. Each of the stubs used to test *Calculate pay amount* can be implemented as a statement that simply returns a constant regardless of the parameters passed as input. Figure 16-17 shows sample code for each of the three stub modules.

```
Module CalculateBasePay(HoursWorked,BasePay)

// Stub Module

array[7] of real HoursWorked;
real BasePay;
{
   BasePay=1000.00;
   return;
}

Module CalculateOvertimePay(HoursWorked,OvertimePay)

// Stub Module

array[7] of real HoursWorked;
real OvertimePay;
{
   OvertimePay=125.00;
   return;
}

Module CalculateTaxes(GrossPay,FilingStatus,State,Exemptions,FederalTax,StateTax)

// Stub Module

real GrossPay;
integer FilingStatus;
char[2] State;
integer Exemptions;
real FederalTax;
real StateTax;
{
   FederalTax=275.00;
   StateTax=75.00;
   return;
}
```

Figure 16-17
Stub modules used for testing *Calculate pay amount*

Stubs are needed for top-down development. In fact, top-down development often begins by writing a stub for every module or method in a program or class. Individual stub modules and methods are then replaced with fully implemented code as it is developed.

integration test
a test of the behavior of a group of modules or methods

Integration Testing

An **integration test** tests the behavior of a group of modules or methods. The purpose of an integration test is to identify errors that were not or could not be detected by unit-testing individual modules or methods. Such errors might result from a number of problems, including the following:

- **Interface incompatibility.** For example, a caller module passes a variable of the wrong data type to a subordinate module.

- **Parameter values.** A module is passed or returns a value that was unexpected (such as a negative number for a price).
- **Run-time exceptions.** A module generates an error such as "out of memory" or "file already in use" due to conflicting resource needs.
- **Unexpected state interactions.** The states of two or more modules interact to cause complex failures (such as an order class method that operates correctly for all possible customer object states except one).

These are some of the most common integration testing errors, but there are many other possible errors and causes.

After an integration error has been detected, the responsibility for incorrect behavior must be traced to a specific module or modules. The person responsible for performing the integration test is generally also responsible for identifying the cause of the error. After the error has been traced to a particular module, the programmer who wrote the module is asked to rewrite it to correct the error.

Integration testing of structured software is straightforward but not necessarily easy to implement. Most structured modules are called by only a single parent module. In addition, most structured modules do not store a permanent state within themselves. Internal variables are always reinitialized to the same values each time the module is called. The combination of these characteristics allows test personnel (often with the assistance of automated testing tools) to generate test cases and data that exercise all possible control paths through the software being tested. Confidence that testing has revealed important errors increases with the number of control paths that are tested.

In contrast, integration testing of OO software is much more complex and not as well understood. There is no clear hierarchical structure to an OO program. An OO program consists of a set of interacting objects that can be created or destroyed during execution. Object interactions and control flow are dynamic and complex.

Additional factors that complicate OO integration testing include the following:

- Methods can be (and usually are) called by many other methods, and the calling methods can be distributed across many classes.
- Classes can inherit methods and state variables from other classes.
- The specific method to be called is dynamically determined at run time based on the number and type of message parameters.
- Objects can retain internal variable values (that is, the object state) between calls. The response to two identical calls can be different due to state changes that result from the first call or occur between calls.

The combination of these factors makes it difficult to determine an optimal testing order. The factors also make it difficult to predict the behavior of a group of interacting methods and objects. Thus, developing and executing an integration test plan for OO software are much more complex than for structured software. Specific methods and techniques for dealing with that complexity are well beyond the scope of this book. See the "Further Resources" for OO software testing references.

system test
a test of the behavior of an entire system or independent subsystem

A system test is an integration test of an entire system or independent subsystem. System testing is normally first performed by developers or test personnel to ensure that the system does not malfunction in obvious ways and that the system fulfills the developers' understanding of user requirements. Later testing by users confirms whether the system does indeed fulfill their requirements. If a system is developed in many iterations, system testing is usually performed at the end of each iteration to identify significant issues such as performance problems that must be addressed in the next iteration.

build and smoke test
a system test that is performed daily

A build and smoke test is a system test that is typically performed daily. The system is completely compiled and linked (built), and a battery of tests is performed to see whether anything malfunctions in an obvious way ("smokes"). Build and smoke tests are commonly associated with iterative or rapid development. However, build and smoke tests can also be used in more traditional projects if top-down development is employed.

Build and smoke tests are valuable because they provide rapid feedback regarding significant problems. Any problem that occurs during a build and smoke test must be the result of code modified or added since the previous test. Daily testing ensures that errors are found quickly and that they can be easily tracked to their source. Less frequent testing provides rapidly diminishing benefits because more code has changed and errors are more difficult to track to their source.

Usability Testing

usability test
a test to determine whether a module, method, class, subsystem, or system meets user requirements

A usability test is a test to determine whether a module, method, class, subsystem, or system meets user requirements. Because there are many types of requirements, both functional and nonfunctional, many types of usability tests are performed at many different times.

The most common type of usability test evaluates functional requirements and the quality of a user interface. Users interact with a portion of the system to determine whether it functions as expected and whether the user interface is easy to use. Such tests are conducted frequently as user interfaces are developed to provide rapid feedback to developers for improving the interface and correcting any errors in the underlying software components.

performance test
a system test that determines whether a system can meet time-based performance criteria

response time
the desired or maximum allowable time limit for software response to a query or update

throughput
the desired or minimum number of queries and transactions that must be processed per minute or hour

A performance test is a system test that determines whether a system can meet time-based performance criteria such as response time or throughput. Response time requirements specify desired or maximum allowable time limits for software responses to queries and updates. Throughput requirements specify the desired or minimum number of queries and transactions that must be processed per minute or hour.

Performance testing can be conducted with unit or integration testing, but it is more commonly integrated with system testing. Performance tests are complex because they can involve multiple programs, subsystems, computer systems, and network infrastructure. They require a large suite of test data to simulate system operation under normal or maximum load. Diagnosing and correcting performance test failures are also complex. Bottlenecks and underperforming components must first be identified. Corrective actions can include application software tuning or reimplementation, hardware or system software reconfiguration, and upgrade or replacement of underperforming components.

acceptance test
a system test that determines whether the system fulfills user requirements

An acceptance test is a system test that determines whether the system fulfills user requirements. Acceptance testing is typically the last round of testing before a system is handed over to its users. Acceptance testing is a very formal activity in most development projects. Details of acceptance tests are sometimes included in the request for proposal (RFP) and procurement contract when a new system is built by or purchased from an external party.

Who Tests Software?

There are many participants in the testing process. Their exact number and role depend on the size of the project and other project characteristics. Specific participants include these people:

- Programmers
- Users
- Quality assurance personnel

testing buddy
a programmer assigned to test code written by another programmer

Programmers are generally responsible for unit-testing their own code prior to integrating it with modules written by other programmers. In some organizations, programmers are assigned a testing buddy to help them test their own code. The name derives from programmers who are assigned the specific responsibility for testing their buddy's code prior to integration testing. Having a different programmer test the code usually results in more errors being found.

Users are primarily responsible for beta testing and acceptance testing. When beta versions are developed, they are distributed to a group of users for testing over a period of days, weeks, or months. Volunteers are frequently used, although they are not always desirable because they tend to be more computer literate and have a higher tolerance for malfunctions than ordinary users. These characteristics might result in higher-quality feedback for some problems but a lack of feedback for other problems.

Acceptance testing is normally conducted by users with assistance from IS development or operations personnel. The rigor and importance of acceptance tests require participation by a large number of users across a wide range of user levels (for example, data-entry clerks and the managers who will "own" the system). Although IS personnel can perform system setup and troubleshooting functions, it is ultimately up to the users to accept or reject the system.

In a large system development project, a separate quality assurance group or organization is usually formed. The QA group is responsible for all aspects of testing except unit testing and acceptance testing. The QA group's responsibilities and activities typically include the following:

- Developing a testing plan
- Performing integration and system testing
- Gathering and organizing user feedback on alpha and beta software versions and identifying needed changes to the system design or implementation

To maintain objectivity and independence, the QA group normally reports directly to the project manager or to a permanent IS manager.

DATA CONVERSION

An operational system requires a fully populated database to support ongoing processing. For example, the RMO order-entry subsystem relies on stored information about catalogs, products, customers, and previous orders. Implementation staff must ensure that such information is present in the database at the moment the subsystem becomes operational.

Data needed at system startup can be obtained from the following sources:

- Files or databases of a system being replaced
- Manual records
- Files or databases of other systems in the organization
- User feedback during normal system operation

REUSING EXISTING DATABASES

Most new information systems replace or augment an existing manual or automated system. In the simplest form of data conversion, the old system database is used directly by the new system with little or no change to the database structure. Reusing an existing database is fairly common because of the difficulty and expense of creating new databases from scratch, especially when a single database often supports multiple information systems as in today's enterprise application systems.

Although old databases are commonly reused in new or upgraded systems, some changes to database content are usually required. Typical changes include adding new classes or entities, adding new attributes or relationships, and modifying existing attributes or relationships. Modern database management systems (DBMSs) usually allow database administrators to modify the structure of a fully populated database. Simple changes such as adding new attributes or changing attribute types can be performed entirely by the DBMS.

RELOADING DATABASE CONTENTS

More complex changes to database structure might require reloading data after the change. In that case, implementation staff must develop programs to alter data after the database has been modified. Figure 16-18 shows two possible approaches to reloading data. The first approach initializes a new database and copies the contents of the old database to it. The conversion program translates data stored within the former database structure into the newly modified database structure.

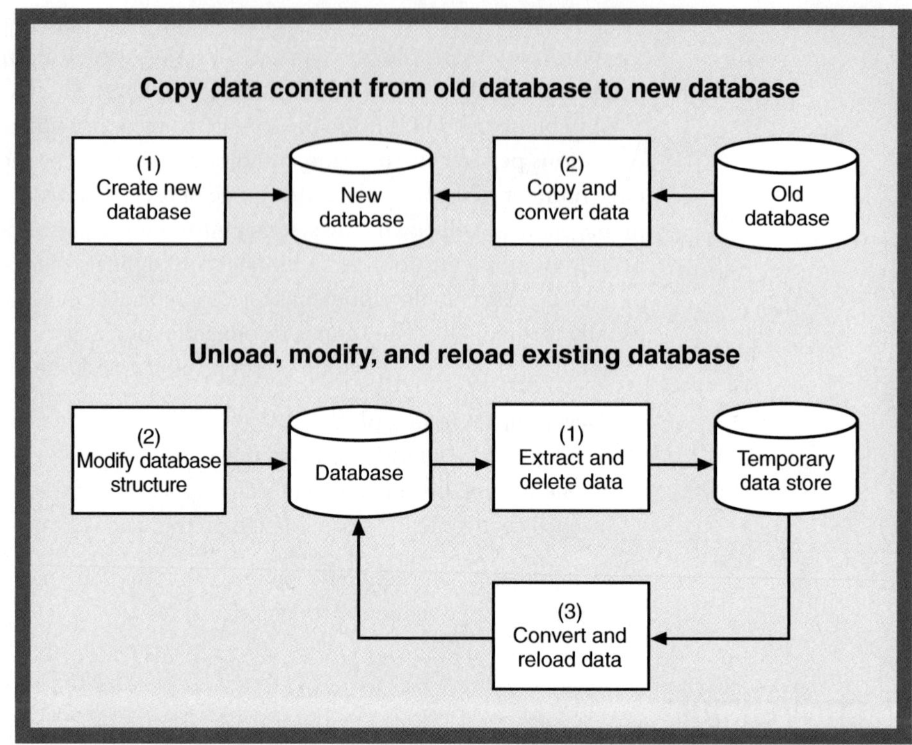

Figure 16-18

Two approaches to reloading database content after a structural modification

The second approach uses a program or DBMS utility to extract and delete data from an existing database and store it in a temporary data store. The database structure is then modified, and a second DBMS utility or program is used to reload the modified database. The first approach is simpler than the second, but it requires sufficient data storage to hold both databases temporarily. The second approach is required if there is insufficient data storage to hold two complete sets of data.

Many DBMSs provide a rich set of import utilities to extract and load data from existing databases, files, or scanned documents. DBMS developers provide such utilities because system developers are more likely to adopt a DBMS that eases the process of importing data from other sources. If DBMS import and export utilities are inadequate for data conversion, developers must construct conversion programs that will be used only once. Although conversion programs are not part of the operational system, they must be constructed and tested in the same manner as operational software.

CREATING NEW DATABASES

If the system being developed is entirely new or if it replaces a manual system, initial data must be obtained from manual records or from other automated systems in the organization. Data from manual records can be entered using the same programs being developed for the operational system. In that case, data-entry programs are usually developed and tested as early as possible. Initial data entry can be structured as a user training exercise. In addition, data from manual records can also be scanned into an optical character recognition program and then entered into the database using custom-developed conversion programs or a DBMS import utility.

Some data might already be stored in other automated systems within the organization. For example, when implementing a new order-entry system, some product data might already be present in a manufacturing planning and control system, and some customer data might already be present in an existing billing system. Copying such data to a new database is similar to reloading a modified database from an old database or backup data store.

Figure 16-19 shows a complex data conversion process that draws input from a variety of sources. Data is input using a mix of manual data entry, optical character recognition, conversion programs, and DBMS import and export utilities. Data conversion processes of this complexity are common in large system development projects.

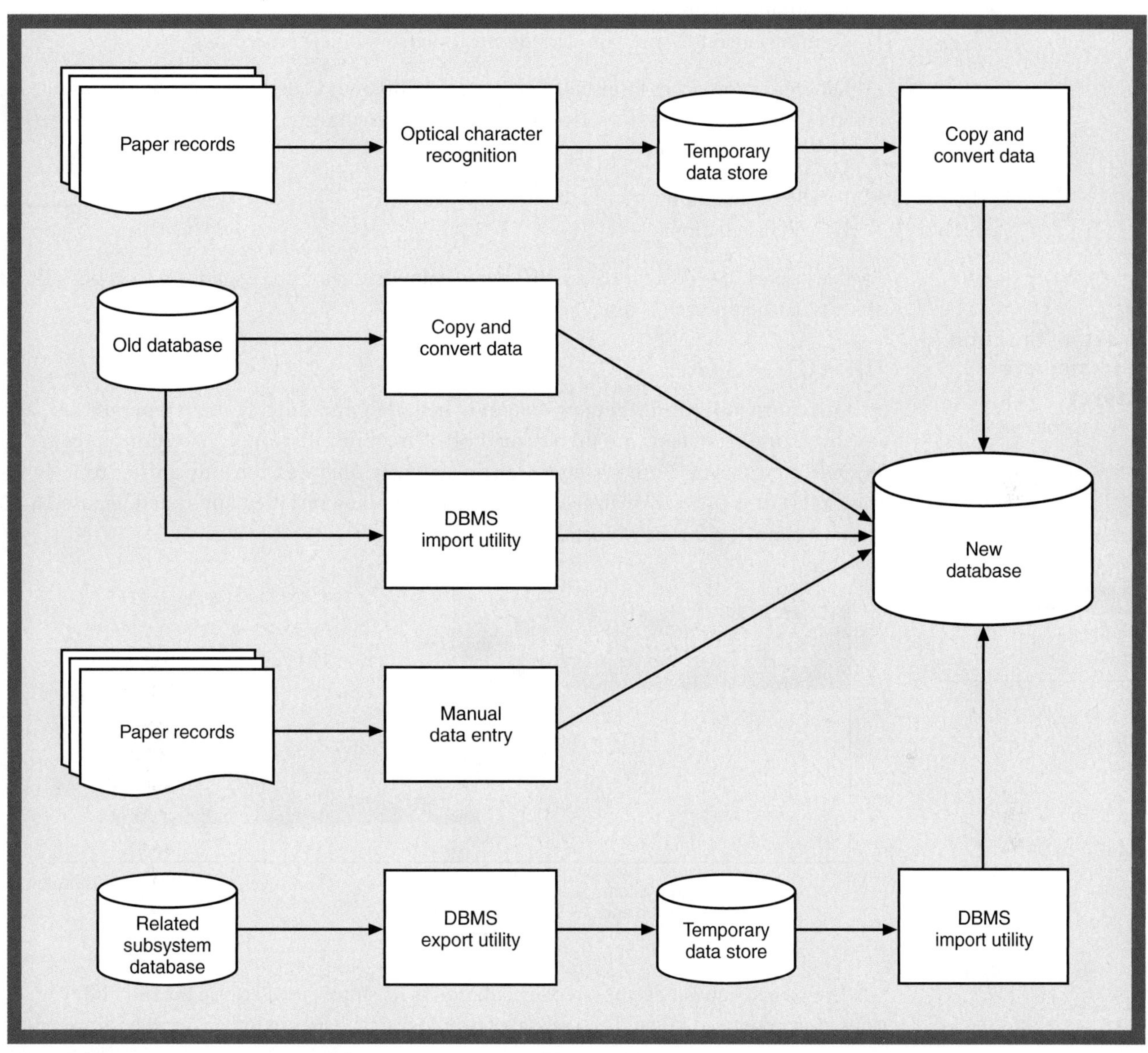

Figure 16-19
A complex data conversion example

In some cases, it might be possible to begin system operation with a partially or completely empty database. For example, a customer order-entry system need not have existing customer information loaded into the database. Customer information could be added the first time a customer places an order, based on a dialog between a telephone order-entry clerk and the customer. Adding data as it is encountered reduces the complexity of data conversion but at the expense of slower processing of initial transactions.

INSTALLATION

After a new system has been developed and tested, it must be installed and placed into operation. Installing a system and making it operational are complex because there are many conflicting constraints, including cost, customer relations, employee relations, logistical complexity, and overall exposure to risk. Some of the more important issues to consider when planning installation include the following:

- Incurring costs of operating both systems in parallel
- Detecting and correcting errors in the new system
- Potentially disrupting the company and its IS operations
- Training personnel and familiarizing customers with new procedures

Different approaches to installation represent different trade-offs among cost, complexity, and risk. The most commonly used installation approaches are as follows:

- Direct installation
- Parallel installation
- Phased installation

Each approach has different strengths and weaknesses, and no one approach is best for all systems. Each approach is discussed in detail here.

direct installation, or immediate cutover
an installation method that installs a new system, quickly makes it operational, and immediately turns off any overlapping systems

DIRECT INSTALLATION

In a **direct installation**, the new system is installed and quickly made operational, and any overlapping systems are then turned off. Direct installation is also sometimes called **immediate cutover**. Both systems are concurrently operated for only a brief time (typically a few days or weeks) while the new system is being installed and tested. Figure 16-20 shows a time line for direct installation.

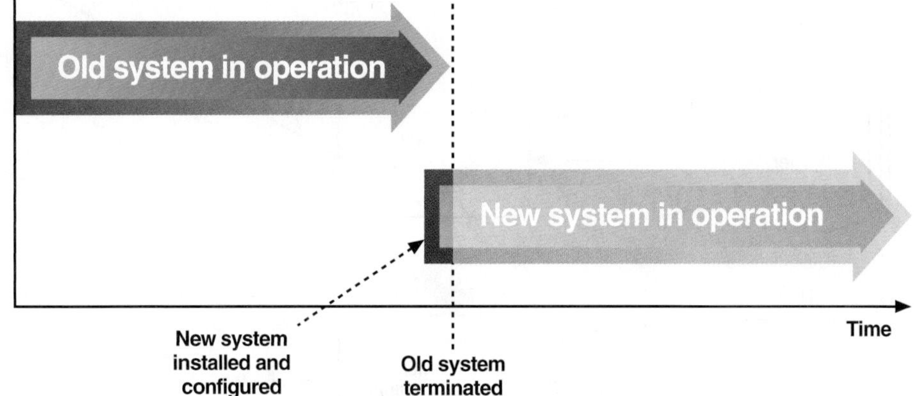

Figure 16-20
Direct installation and cutover

The primary advantage of direct installation is its simplicity. Because the old and new systems aren't operated in parallel, there are fewer logistical issues to manage and fewer resources required. The primary disadvantage of direct installation is its risk. Because older systems are not operated in parallel, there is no backup in the event that the new system fails. The magnitude of the risk depends on the nature of the system, the cost of workarounds in the event of a system failure, and the cost of system unavailability or less-than-optimal system function.

Direct installation is typically used under one or both of the following conditions:

- The new system is not replacing an older system (automated or manual).
- Downtime of days or weeks can be tolerated.

If neither condition applies, parallel or phased installation is usually preferable to minimize the risk of system unavailability.

parallel installation
an installation method that operates both the old and new systems for an extended time period

PARALLEL INSTALLATION

In a **parallel installation**, the old and new systems are both operated for an extended period of time (typically weeks or months). Figure 16-21 illustrates the time line for parallel installation. Ideally, the old system continues to operate until the new system has been thoroughly tested and determined to be error-free and ready to operate independently. As a practical matter, the time allocated for parallel operation is often determined in advance and limited to minimize the cost of dual operation.

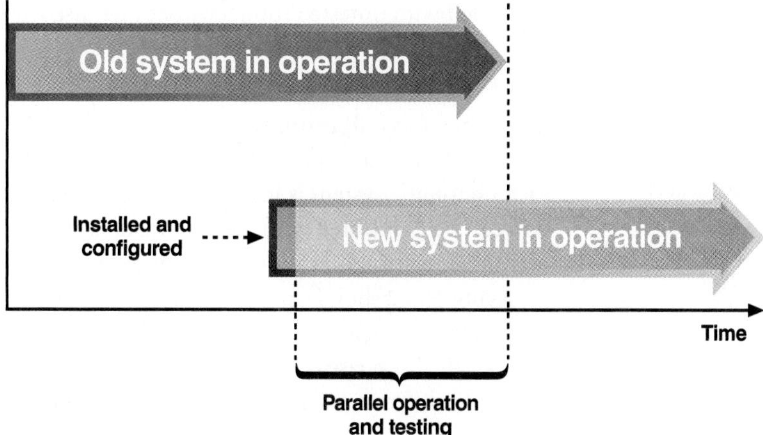

Figure 16-21

Parallel installation and operation

The primary advantage of parallel installation is a relatively low risk of system failure and the negative consequences that might result from that failure. If both systems are operated completely (that is, using all data and exercising all functions), the old system functions as a backup for the new system. Any failure in the new system can be mitigated by relying on the old system.

The primary disadvantage of parallel installation is cost. During the period of parallel operation, the organization pays to operate both systems. Extra costs associated with operating two systems in parallel include the following:

- Hiring temporary personnel or temporarily reassigning existing personnel
- Acquiring extra space for computer equipment and personnel
- Increasing managerial and logistical complexity

Unless the operational costs of the new system are substantially less than that of the old system, the combined operating cost is typically 2.5 to 3 times the cost of operating the old system alone.

Parallel operation is generally best when the consequences of a system failure are severe. Parallel operation substantially reduces the risk of a system failure through redundant operation. The risk reduction is especially important for "mission-critical" applications such as customer service, production control, basic accounting functions, and most forms of online transaction processing. Few organizations can afford any significant downtime in such important systems.

Full parallel operation might be impractical for any number of reasons, including the following:

- Inputs to one system might be unusable by the other, and it might not be possible to use both types of inputs.
- The new system might use the same equipment as the old system (for example, computers, I/O devices, and networks), and there might not be sufficient capacity to operate both systems.
- Staffing levels might be insufficient to operate or manage both systems at the same time.

When full parallel operation is not possible or feasible, a partial parallel operation might be employed instead. Possible modes of partial parallel operation include the following:

- Processing only a subset of input data in one of the two systems. The subset could be determined by transaction type, geography, or sampling (for example, every 10th input transaction).

- Performing only a subset of processing functions (such as updating account history but not printing monthly bills).
- Performing a combination of data and processing function subsets.

Partial parallel operation always entails the risk that significant errors or problems will go undetected. For example, parallel operation with partial input increases the risk that errors associated with untested inputs will not be discovered.

PHASED INSTALLATION

phased installation
an installation method that installs a new system and makes it operational in a series of steps or phases

In a **phased installation**, the system is installed and brought into operation in a series of steps or phases. Each phase adds components or functions to the operational system. During each phase, the system is tested to ensure that it is ready for the next phase. Phased installation can be combined with parallel installation, particularly when the new system will take over the operation of multiple existing systems.

Figure 16-22 shows a phased installation with both direct and parallel installation of individual phases. The new system replaces two existing systems. The installation is divided into three phases. The first phase is a direct replacement of one of the existing systems. The second and third phases are different parts of a parallel installation that replace the other existing system.

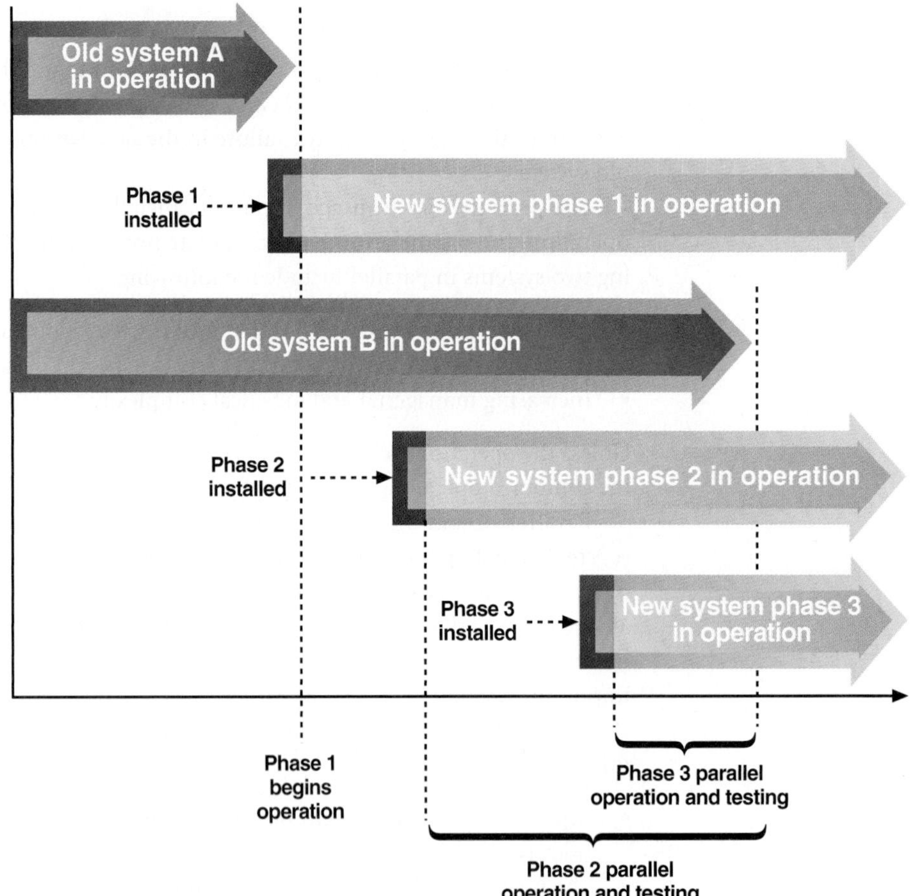

Figure 16-22
Phased installation with direct cutover and parallel operation

There is no single method for performing phased installation. Installation details such as the composition of specific phases and their order of installation vary widely from one system to another. These specifics determine the number of installation phases, the order of installation, and the parts of the new system that are operated in parallel with existing systems.

BEST PRACTICE

To reduce risk, combine phased and parallel installation whenever possible.

The primary advantage of phased installation is reduced risk. Risk is reduced because failure of a single phase is less problematic than failure of an entire system. The primary disadvantage of phased installation is increased complexity. Dividing the installation into phases creates more activities and milestones, thus making the entire process more complex. However, each individual phase contains a smaller and more manageable set of activities. If the entire system is simply too big or complex to install at one time, the reduced risks of phased installation outweigh the increased complexity inherent in managing and coordinating multiple phases.

Phased installation is most useful when a system is large, complex, and composed of relatively independent subsystems. If the subsystems are not substantially independent, it is difficult or impossible to define separate installation phases. System size and complexity might also be too great for an "all at once" installation to be feasible. In that case, there is really no choice but to use phased installation.

PERSONNEL ISSUES

Installing a new system places significant demands on personnel throughout an organization. Installation typically involves demanding schedules, rapid learning and adaptation, and high stress. Planning should anticipate these problems and take appropriate measures to mitigate their effects.

New system installation usually stretches IS personnel to their limits. Many tasks must be performed in little time. The problem is most acute in parallel installation, in which personnel must operate both the old and new systems. Often, development and customer support personnel must be temporarily reassigned to provide sufficient manpower to operate both systems. Reassignment may reduce progress on other ongoing projects and reduce support and maintenance activities for other systems.

Temporary and contract personnel can be hired to increase available manpower during an installation. Two types are particularly useful:

- Personnel with experience in hardware and software installation and configuration
- Personnel with experience (or who can be trained) to operate the old system

Installation may require technical skills that are in short supply within the organization. In that case, hiring contractors to assist in hardware and software installation is a necessity. Hardware and software specialists can be contracted directly from vendors or from IS consulting firms.

Hiring temporary employees to operate the old system during a parallel installation has several benefits. First, it provides the extra manpower needed to operate both systems. Second, it frees permanent employees for training and new system operations. Temporary personnel are often hired several months in advance, trained to operate the old system, and employed for the duration of parallel operation. If problems occur with the new system, employment contracts can be extended until the old system can be safely phased out.

Another personnel issue that must be considered is employee productivity. All new systems have a learning curve for users and system operators. Both require training before the installation begins, but no matter how good the training, users and operators require some time (typically a few months) to reach their peak efficiency with a new system. Manpower requirements are higher during that time, as is the general level of employee stress.

DOCUMENTATION

Preparing documentation is an important but frequently overlooked activity during implementation. Documentation provides information to users on how to operate and maintain a system. Documentation also provides information needed for future modifications or reimplementation.

Rapid technology changes over the last two decades have altered the nature of documentation. Prior to the 1980s, most documentation was printed on paper and organized into bound or looseleaf books. Automated documentation is now the norm, and formats include the following:

- Electronic documents, such as documents stored in Microsoft Word or Adobe Acrobat files
- Hyperlinked documents, such as documents formatted for Web browser viewing with embedded links among documentation components
- Online documentation stored on a vendor Web site that can be viewed with a Web browser or downloaded and installed on a local computer system
- Embedded documentation, such as manuals, tutorials, and multimedia presentations included on a CD or DVD and installed as an integral part of an application
- Electronic system models, such as text and graphics formatted and stored in graphics file formats such as GIF, JPEG, and Visio
- Tool-specific system models, such as those developed with integrated programming environments and DBMS tools

Electronic documents can be distributed in a number of standard formats, including Adobe Acrobat, Windows help files, and standard Web pages. Choosing a standard format ensures that users have or can easily obtain the software to view the documentation. Hyperlinked documents enable users to navigate rapidly among related topics. Storing documentation on Web sites allows vendors to make updates easily and allows users to share a single copy. Embedded documentation enables users to access information through the application and provides features such as context-sensitive help.

Electronic and tool-specific system models are primarily intended for software developers' use. Generic model formats (such as ordinary text and GIF images) can be formatted as any type of electronic format. Tool-specific models must generally be accessed via specific software tools (for example, viewing a model generated by a DBMS usually requires a viewer supplied with the DBMS). However, most development tools allow models to be exported to other formats (such as Acrobat or Microsoft Word).

Documentation can be loosely classified into two types:

system documentation
descriptions of system functions, architecture, and construction details, as used by maintenance personnel and future developers

user documentation
descriptions of how to interact with and maintain the system, as used by end users and system operators

- System documentation—descriptions of system functions, architecture, and construction details
- User documentation—descriptions of how to interact with and maintain the system

System documentation is generated throughout the SDLC as outputs of each life cycle activity. User documentation is created during implementation. The development team cannot create user documentation earlier because many details of the user interface and system operation either haven't yet been determined or may change during development.

SYSTEM DOCUMENTATION

System documentation serves one primary purpose: providing information to designers and developers who will maintain or reimplement the system. Most or all of the documentation needed for this purpose is generated as a by-product of analysis, design, and implementation activities. Figure 16-23 shows three SDLC activities and the system documentation produced or modified in each. The documentation produced by each activity is useful for future maintenance or upgrades.

Life cycle activity	System documentation	
	Traditional approach	**Object-oriented approach**
Analysis	Entity-relationship diagram Data flow diagram Process description Data flow and element definition	Domain class diagram Use case Activity diagram System sequence diagram
	Event list	
Design	System flowchart Structure chart	Design class diagram Interaction diagram Communication diagram Package diagram State machine diagram
	Module or method pseudocode Database schema diagram	
Implementation	Program source code Database schema source code Test data	

Figure 16-23

Life cycle activities and related system documentation

Source code is the most frequently used documentation because it is the most direct link to the system's executable software. Direct changes to binary code are complex and expensive, so changing and recompiling source code is the only realistic method of altering a system's behavior. After a system is changed, test data is used to check the system. Rerunning old tests with old test data helps determine whether a change in one part of the system has accidentally "broken" some other part of the system.

Source code can be difficult (and, thus, inefficient) for human beings to use as documentation because it is entirely textual and often poorly commented. Important types of system information—such as how programs interact and what user needs a program satisfies—are usually not documented within source code. Yet such information is needed when evaluating significant changes in systems design or function and when tracing errors that flow from one program to another via shared data. Information needed to perform these tasks is readily available in analysis and design models.

Design models tend to be used more frequently than analysis models because design parameters change more often than system requirements. Examples of maintenance changes that require design models but not analysis models include redeploying existing programs or databases to new hardware, fixing bugs in individual programs, and optimizing the performance of an existing distributed system. Such changes alter the corresponding design models (for example, the system flowchart or package diagram) but do not change analysis models.

Analysis models do change when user requirements are altered. For example, adding a new transaction type or an entirely new processing subsystem changes the data flow diagram or domain class diagram. It also changes other analysis models such as the entity-relationship diagram, event list, and use cases.

System documentation must be actively managed to remain effective. It must be stored in an accessible location and form, retrieved when necessary for maintenance changes, and updated after changes have been implemented. In large organizations with many information systems, managing the documentation is a very formal process. Large organizations typically have one person responsible for archiving and retrieving documentation and for enforcing documentation standards.

Failure to adequately maintain system documentation compromises the value of a system. Systems with inadequate documentation are difficult or impossible to maintain, thus increasing the likelihood that a system will be prematurely scrapped or reimplemented. Maintaining documentation extends the useful life of a productive asset.

System documentation mirrors the system itself. That is, any information contained within system documentation can also be obtained by directly examining the system. For example, programmers can determine the entities and relationships of a relational database by examining the SQL statements that describe the database schema. Programmers can also determine the modular structure of a traditional program or the classes within an OO program by directly examining the program source code. If the source code is unavailable, the program structure can also be determined from executable code, although the process is much more difficult.

As changes are made to the system, its documentation must also be updated, however. If documentation is not updated, it is inconsistent with the system and useless to future designers and maintenance programmers. Making documentation an integral part of the installed system minimizes or eliminates inconsistency because updates to the system automatically update the documentation. Some tools—in particular, visual modeling tools and reverse-engineering tools—can simplify documentation and help ensure its accuracy.

With a visual modeling tool, the system is built automatically from design models and stored by the visual modeling tool. Design models, in turn, are built automatically (or nearly so) based on analysis models. To implement a system change, a programmer modifies an analysis or design model and then regenerates the installed system. The visual modeling tool automatically maintains consistency among the installed system and the models. As long as only the models are changed (instead of the source or executable code), the models and system will always be consistent.

A reverse-engineering tool can generate system models by examining source code. For example, such a tool can generate a class diagram by examining OO programs, and it can generate a structure chart by examining a program written in a procedural programming language. If a reverse-engineering tool is powerful and reliable enough to generate all types of system documentation, there is no need to maintain a separate store of documentation. The source code itself is the documentation, and the reverse-engineering tool generates other forms of documentation on demand.

Both visual modeling and reverse-engineering tools are highly specialized to specific operating environments (such as programming languages, database management systems, and operating systems). They also tend to be expensive and to have steep learning curves. As a result, they aren't used as often as you might think. Thus, for many systems, system documentation must still be maintained separately and manually.

USER DOCUMENTATION

User documentation provides ongoing support for end users of the system. It primarily describes routine operation of the system, including functions such as data entry, output generation, and periodic maintenance. Topics typically covered include the following:

- Software startup and shutdown
- Keystroke, mouse, or command sequences required to perform specific functions
- Program functions required to implement specific business procedures (for example, the steps followed to enter a new customer order)
- Common errors and ways to correct them

For ease of use, user documentation includes a table of contents, a general description of the purpose and function of the program or system, a glossary, and an index.

User documentation for modern systems is almost always electronic and is usually an integral part of the application. Most modern operating systems provide standard facilities to support embedded documentation. Figure 16-24 shows electronic user documentation of

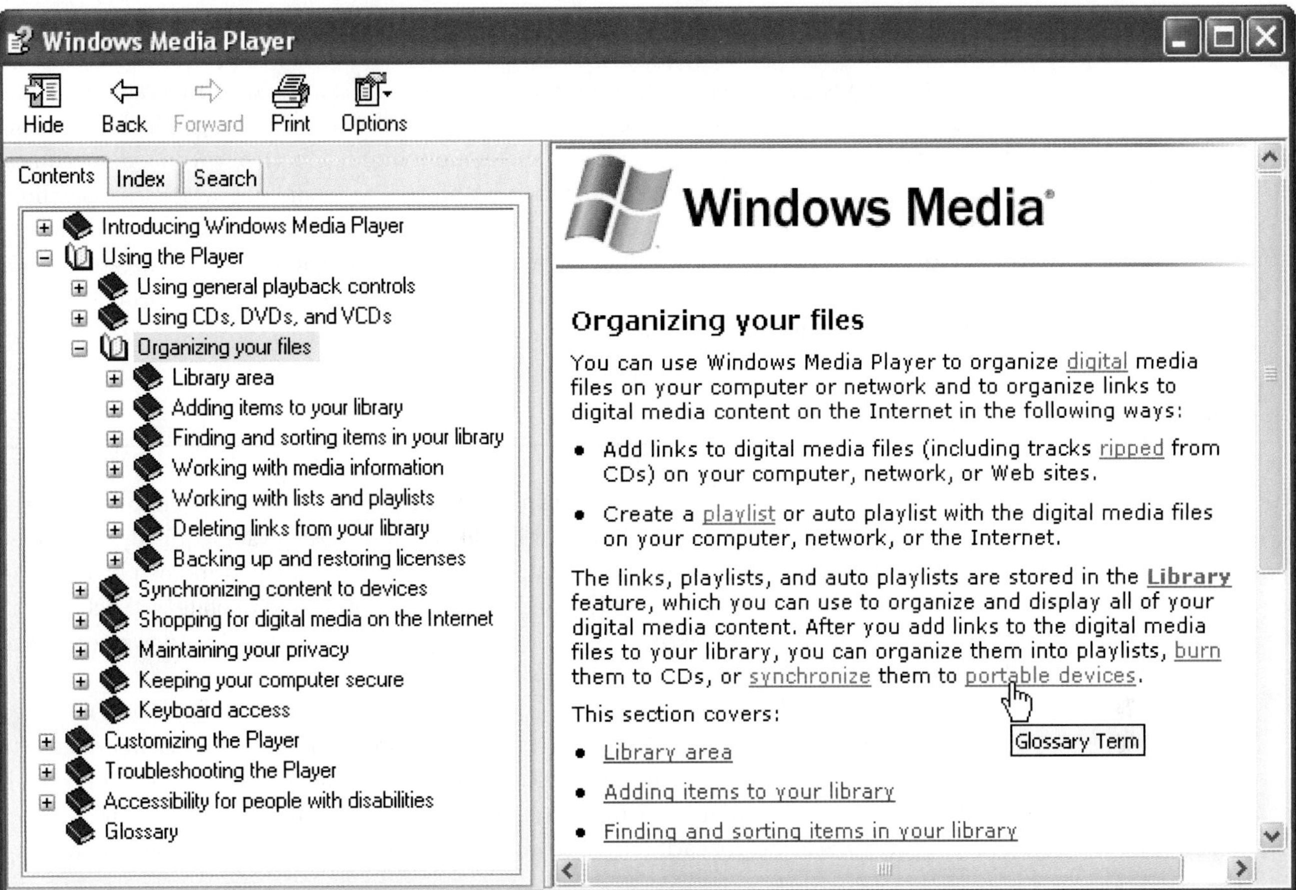

Figure 16-24

Sample Windows Help display

a typical Windows application. The left pane displays the table of contents, and the user can access an index or search engine by clicking the appropriate tab at the top. The right pane displays individual pages of user documentation. The sample page shows an embedded glossary definition (in green) and two hyperlinks (in blue).

User documentation is an important organizational asset. Unfortunately, many organizations fail to prepare comprehensive high-quality user documentation for internally developed systems. Some of the reasons for this problem include the following:

- The assumption that trained programmers can examine source code, figure out how the system works, and train users as needed
- The assumption that the users trained during system implementation will informally pass on their knowledge to future users
- The lack of resources and special skills required to develop documentation and keep it up to date

As discussed in the previous section, source code is a poor form of system documentation. But it is an even worse form of indirect user documentation. Although source code provides a detailed instruction-by-instruction view of how pieces of a system work, it provides little or no information about how those pieces interact and how the entire system functions within a specific context. Supplementing source code with other forms of system documentation does provide other critical information. But even then, figuring out how a system works based only on system documentation is slow and error-prone.

Knowledge of how to use a system is as important an asset as the system itself. After initial training is completed, that practical knowledge is stored in the minds of end users. But experience such as that is difficult to maintain or effectively transfer to other users. Employee turnover, reassignment, and other factors make direct person-to-person transfer of operational knowledge difficult and uncertain. In contrast, written or electronic documentation is easier to access and far more permanent.

Developing good user documentation requires special skills and considerable time and resources. Writing clearly and concisely, developing effective presentation graphics, organizing information for easy learning and access, and communicating effectively with a nontechnical audience are skills for which there is high demand and limited supply. Development takes time, and high-quality results are achieved only with thorough review and testing. Unfortunately, preparing user documentation is often left to technicians lacking in one or more necessary skills. Also, preparation time, review, and testing are often shortchanged because of schedule overruns and the last-minute rush to tie up all the loose ends of implementation.

TRAINING AND USER SUPPORT

Good documentation can reduce training needs as well as the frequency of support requests. But some training before and support after installation is almost always required. Remember, users are part of the system, too! Without training, users would slowly work their own way up the learning curve, error rates would be high, and the system would operate well below peak efficiency. Training allows users to be productive as soon as the system becomes operational. Support activities ensure continuing user productivity long after installation.

There are two classes of users—end users and system operators—who must be considered for documentation, training, and support. End users are people who use the system from day to day to achieve the system's business purpose. System operators are people who perform administrative functions and routine maintenance to keep the system operating. Figure 16-25 shows representative activities for each role. In smaller systems, a single person might fill both roles.

Figure 16-25

Typical activities of end users and system operators

End user activities	System operator activities
Creating records or transactions	Starting or stopping the system
Modifying database contents	Querying system status
Generating reports	Backing up data to archive
Querying database	Recovering data from archive
Importing or exporting data	Installing or upgrading software

Training and support activities vary with the target audience. Audience characteristics that affect training include the following:

- Frequency and duration of system use
- Need to understand the system's business context
- Existing computer skills and general proficiency
- Number of users

In general, end users use the system frequently and for extended periods of time, and system operators interact with the system infrequently and usually for short periods. End users solve a particular business problem with the system or implement specific business procedures. System operators are usually computer professionals with limited knowledge of the business processes that the system supports. End-user computer skill levels vary widely, whereas system operators typically have higher and more uniform skill levels. Also, the number of end users is generally much larger than the number of system operators.

Training for end users must emphasize hands-on use and application of the system for a specific business process or function, such as order entry, inventory control, or accounting. If the users are not already familiar with those procedures, training must also include them. Widely varying skill and experience levels call for at least some hands-on training, including practice exercises, questions and answers, and one-on-one tutorials. Self-paced training materials can fill some of this need, but complex systems usually require some face-to-face training also. The relatively large number of end users makes group training sessions feasible, and a subset of well-qualified end users can be trained and then pass on their knowledge to other users.

System operator training can be much less formal when the operators are not end users. Experienced computer operators and administrators can learn most or all they need to know by self-study. Thus, formal training sessions might not be required. Also, the relatively small number of system operators makes one-on-one training feasible, if it is necessary.

Determining the best time to begin formal training can be difficult. On one hand, early training provides plenty of time for learning and can ensure that users "hit the ground running." On the other hand, starting early can be frustrating to both users and trainers because the system might not be stable or complete. End users can quickly become upset when trying to learn on a buggy, crash-prone system with features and interfaces that are constantly changing.

In an ideal world, training doesn't begin until the interfaces are finalized and a test version has been installed and fully debugged. But the typical end-of-project crunch makes that approach a luxury that is often sacrificed. Instead, training materials are normally developed as soon as the interfaces are reasonably stable, and end-user training begins as soon as possible thereafter. It is much easier to provide training if system interfaces are developed early and if the top-down modular development approach is employed.

ONGOING TRAINING AND USER SUPPORT

The term *user support* covers training and user assistance that occur after the system is up and running. Some of the activities are the same as preinstallation training activities. For example, new users must be trained periodically due to employee turnover. Other activities such as refresher training and help desk operation are unique to support.

User support can be provided by a number of methods, including the following:

- Online documentation and troubleshooting
- Resident experts
- A help desk
- Technical support

Online documentation and troubleshooting have surged as a support method in recent years. Much of this support is built into the application, although Web sites are also commonly employed. The goal of online support is to minimize the need for human support by putting useful information into the hands of end users when they need it. Achieving that goal, however, requires well-designed support materials that are comprehensive and easy to use.

Resident experts are the most common form of user support, and their help is usually provided informally. A resident expert can be an on-site IS staff member or (more frequently) a business area staff member or user who assists other users. The position of resident expert is often informal. A person frequently grows into that position simply by displaying exceptional computer literacy or knowledge of software. Over time, all other users begin to approach that person first with questions or problems.

A help desk is a permanent IS department that provides end-user support for a wide range of systems and software. Help desks are staffed by personnel trained to install, operate, and troubleshoot application software, including off-the-shelf products (such as word processors). A help desk serves as a central contact point for users. Help desk staff are trained to handle the majority of user problems and questions. Those who require further assistance are forwarded to technical support.

Technical support is typically a specific function or department within IS maintenance because of the close relationship between user support, change requests, and system error reporting. If help desk personnel can't solve a user's problem, there's a good chance that an error has been discovered or that there is a gap between system capability and user needs. If the problem is a system error, maintenance needs to be notified quickly to investigate the cause and correct it if it is critical. Noncritical errors and unmet user needs must also be brought to the attention of maintenance, but timeliness is less critical. In either case, technical support is the bridge between users and maintenance activities.

MAINTENANCE AND SYSTEM ENHANCEMENT

software maintenance
modification of a software product after delivery to correct faults, improve performance or other attributes, or adapt the product to a changed environment

The Institute of Electrical and Electronics Engineers (IEEE) and the American National Standards Institute (ANSI) have defined software maintenance as modification of a software product after delivery to accomplish at least one of the following objectives:

- Correct faults
- Improve performance or other attributes
- Adapt the product to a changed environment

The term *maintenance* covers virtually everything that happens to a system after delivery except total replacement or abandonment.

In most organizations, the cost of maintaining existing systems is at least as great as the cost of developing new ones. Existing systems are an organizational asset and must be actively managed to preserve their value and utility. In that sense, maintaining software is similar to maintaining other types of capital assets such as buildings and equipment.

Maintenance involves change—to adapt to a new environment, to adapt to changing user requirements, and to fix problems as they occur or are discovered. But change is risky. Making changes to an operational system is much more difficult than making changes to a system under development. When a change causes a developmental system to crash, there are no frantic calls to the support desk and no immediate financial impact. But changes to operational systems have an immediate impact on users, customers, and the organization as a whole.

Failure of an operational system can be disastrous. Thus, software maintenance differs greatly from new system development. New system development generally occurs in a relatively open environment where change is expected, new ideas are tried out, and risk taking is tolerated, if not encouraged. In contrast, maintenance is very conservative—change is tolerated as a necessary evil, and risk taking is strongly discouraged.

Maintenance activities include the following:

- Tracking modification requests and error reports
- Implementing changes
- Monitoring system performance and improving performance or increasing capacity
- Upgrading hardware and system software
- Updating documentation to reflect maintenance changes

Maintenance and new system development do have many activities in common, including analysis, design, construction, testing, and documentation. However, implementation of those activities differs in many ways, including scope and detail, triggering events, and implementation constraints. Each maintenance activity is described in detail in the following sections.

SUBMITTING CHANGE REQUESTS AND ERROR REPORTS

To manage the risks associated with change, most organizations adopt formal control procedures for all operational systems. Formal controls are designed to ensure that potential changes are adequately described, considered, and planned before being implemented. Typical change control procedures include the following:

- Standard change request forms
- Review of requests by a change control committee
- Extensive planning for design and implementation

Figure 16-26 shows a sample change request form that has been completed by a user or system owner and submitted to the change control committee for consideration. The change control committee reviews the change request to assess the impact on existing computer hardware and software, system performance and availability, security, and operating budget. The recommendation of the change control committee is formally recorded in a format such as the sample shown in Figure 16-27. Approved changes are added to the list of pending changes for budgeting, scheduling, planning, and implementation.

Figure 16-26

A sample change request form

Change Request				
Request Date	2/1/2010	**Change Type**	☐	**Error Correction**
Requested By	Wen-Hsu Chang, Comptroller		☒	**Modification**
Target System	Customer Accounts - Refunds		☐	**New Function**

Change (or Error) Description

U.S. check formats will soon change due to a recently enacted federal law. The new format reserves an area to the right of the current routing number to be used for a security bar code checksum.

The law requires the new checksum to be printed on checks dated on or after January 1, 2011.

We currently use a portion of the area in question to print a multicolored security symbol. The security symbol will need to be moved or eliminated, and the security bar code checksum will have to be added.

Change Request ID	2010-11		
Date Received	2/2/2010	**Review Date**	2/7/2010, 0930-1100
Review Participants	W. Chang (Comptroller), R. Brooks (IS Operations), J. Hernandez (IS Security), G. Weeks (IS Change Coordinator)		

Bugs can be reported using a standard change request form, but many computing organizations use a different form and procedure because they need to fix such bugs immediately. Bug reports can come from many sources, including end users, computer operators, or IS support staff. Bug reports are typically routed to a single person or organization for logging and follow-up.

IMPLEMENTING A CHANGE

Change implementation follows a miniature version of the system development life cycle. Most of the same activities are performed, although they might be reduced in scope or sometimes completely eliminated. In essence, a maintenance change is an incremental development

Change Review				
Change Request ID	2010-11	Date Reviewed	2/7/2010	
Priority	☐ **Critical**	☒ **Necessary**	☐ **Optional**	
Hardware Implications need to verify ability of current printers to write a security bar code in mandated area				
Software Implications database will need to be modified to store the security bar code with other check information check writing program must be modified to generate and print the security bar code				
Performance Implications none				
Operating Budget Implications none				
Other Implications none				
Disposition	☒ **Approved**	☐ **Rejected**	☐ **Suspended**	
Reason				
Latest Implementation Date	12/31/2010			
Reevaluation Date	n/a	**Signature**		

Figure 16-27

A sample change review form

project in which the user and technical requirements are fully known in advance. Thus, analysis activities are typically skimmed or skipped.

Planning for a change includes the following activities:

- Identify what parts of the system must be changed.
- Secure resources (such as personnel) to implement the change.
- Schedule design and implementation activities.
- Develop test criteria and a testing plan for the changed system.

System documentation is reviewed by design, development, and operations staff to determine the scope of the change. Test criteria and plans for the existing system are the starting point for testing the new system. The testing plan is simply updated to account for changed or added functions, then the modified plan and test data is archived for use in future change projects.

Design can be combined with planning if the change is relatively simple. For more complex changes, separate design activities are used. The existing system design is evaluated and modified as necessary to implement the proposed changes. As with test plans and data, the revised design is archived for use in future change projects.

production system
the version of the system used from day to day

test system
a copy of the production system that is modified to test changes

Implementation activities are normally performed on a copy of the operational system. The production system is the version of the system used day to day. The test system is a copy of the production system that is modified to test changes. The test system can be developed and tested on separate hardware or on a redundant system. The test system becomes the operational system only after complete and successful testing.

BEST PRACTICE

Test all changes on a test system before deploying them to a production system.

UPGRADING COMPUTING INFRASTRUCTURE

Computer hardware, system software, and networks must be periodically upgraded for many reasons, including the following:

- Software maintenance releases
- Software version upgrades
- Declining system performance

Like application software, system software such as operating and database management systems must periodically be changed to correct errors and add new functions. System software developers typically distribute maintenance releases several times per year. The frequency of maintenance update distribution has increased in recent years in part because of the convenience of Internet-based software distribution. In some cases (such as virus checkers and operating system security subsystems), updates might be released weekly or even more frequently.

As with internally generated changes, system software updates are risky. Application software that worked well with an older software version might fail when that software is updated. For this reason, system software updates are extensively tested before they are applied to operational systems. In many cases, maintenance and version updates are simply ignored to reduce risk. Unless errors related to system software have already been encountered, there is little immediate benefit to an upgrade. Operational system maintenance usually follows the old engineering maxim, "If it isn't broken, don't fix it!"

BEST PRACTICE

If an operational system isn't broken, don't fix it!

Increases in transaction volume or support of new systems on existing hardware and networks sometimes reduce performance to unacceptable levels. So, an infrastructure upgrade might be required to add capacity or address a performance-related problem. Infrastructure upgrades are implemented like any other change. The primary difference is how a performance upgrade is initiated.

Input from users or IS staff might indicate the need for a performance upgrade. But a final determination of whether an upgrade is needed and what exactly should be upgraded requires thorough investigation and research. Computer and network performance is complex and highly technical, so what appears to be a performance problem might have little to do with hardware or network capacity. If the problem is ultimately traced to hardware or networks, the specific cause must be identified and a suitable upgrade chosen.

Performance problems require careful diagnosis to determine the best approach to address the problem. Staff with solid technical backgrounds who can understand all of the relevant trade-offs should diagnose the problem. Larger IS organizations may have permanent staff with such skills, but many organizations must rely on contract personnel or consultants to diagnose performance problems, recommend corrective measures, and install or configure those measures. The skills come at a high price, but they can prevent an organization from wasting larger amounts of money buying hardware or network capacity that isn't really needed.

SUMMARY

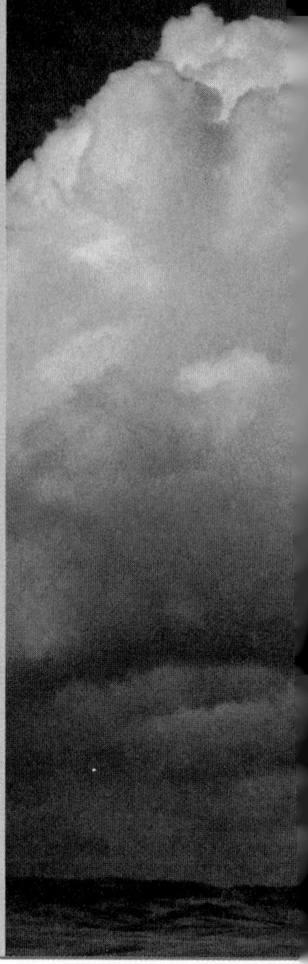

System development activities that occur after design and before system delivery are collectively called *implementation*. Implementation is complex because it consists of many interdependent activities, including programming, quality assurance, hardware and software installation, documentation, and training. Implementation is difficult to manage because activities must be properly sequenced and progress must be continually monitored. Implementation is risky because it requires significant time and resources and because it often affects systems vital to the daily operation of an organization.

Programming and testing are two of the most interdependent implementation activities. Software components must be constructed in an order that minimizes the use of development resources and maximizes the ability to test the system and correct errors. Unfortunately, those two goals often conflict. Thus, a program development plan is a trade-off among available resources, available time, and the desire to detect and correct errors prior to system installation.

Data conversion, installation, documentation, and training are activities that normally follow program development. They are highly interdependent because an installed and documented system is a prerequisite for complete training, and a fully populated database is needed to begin operation. Manpower utilization and the number of directly affected personnel generally peak during these activities.

Support activities occur after a system is made operational. User support activities ensure that an organization realizes the full benefit of the system. Maintenance and system enhancement activities ensure that the system functions at peak efficiency and that needed changes are implemented with minimal disruption to the organization. For most systems, the resources required for support are greater than the resources required to develop the system. Because of high resource requirements and greater operational risk, support activities are normally implemented in a formal and carefully managed fashion.

KEY TERMS

acceptance test, p. 638
alpha version, p. 628
beta version, p. 628
bottom-up development, p. 622
build and smoke test, p. 637
chief developer team, p. 626
collaborative specialist team, p. 626
cooperating peer team, p. 626
direct installation, or immediate cutover, p. 642
driver, p. 635
input, process, output (IPO) development, p. 620
inspection, p. 632
integration test, p. 636
maintenance release, p. 628
parallel installation, p. 642
performance test, p. 638
phased installation, p. 644
production system, p. 654
production version, release version, or production release, p. 628
quality assurance (QA), p. 631
response time, p. 638
software maintenance, p. 652
source code control system (SCCS), p. 627
stub, p. 636
system documentation, p. 646
system test, p. 637
technical review, p. 632
test case, p. 633
test data, p. 633
test system, p. 654
testing buddy, p. 638
throughput, p. 638
top-down development, p. 622
unit testing, or module testing, p. 634
usability test, p. 638
user documentation, p. 646

REVIEW QUESTIONS

1. List and briefly describe the three basic approaches to program development order. What are the advantages and disadvantages of each?
2. How can the concepts of top-down and bottom-up development order be applied to object-oriented software?

3. Describe three approaches to organizing programming teams. For what types of projects or development activities is each approach best suited?
4. What is a source code control system? Why is such a system necessary when multiple programmers build a program or system?
5. Define the terms *alpha version*, *beta version*, and *production version*. Are there well-defined criteria for deciding when an alpha version becomes a beta version or a beta version becomes a production version?
6. List and briefly describe QA activities during implementation other than software testing. What is the effect of not performing such QA activities?
7. What are the characteristics of good test cases?
8. Define the terms *acceptance test*, *integration test*, *system test*, and *unit test*. In what order are these tests normally performed? Who performs (or evaluates the results of) each type of test?
9. What is a driver? What is a stub? With what type of test is each most closely associated? With what development order is each most likely to be used?
10. What factors make testing object-oriented programs more complex than testing structured programs?
11. List possible sources of data used to initialize a new system database. Briefly describe the tools and methods used to load initial data into the database.
12. Briefly describe direct, parallel, and phased installation. What are the advantages and disadvantages of each installation approach?
13. Why are additional personnel generally required during the later stages of system implementation?
14. What are the differences between documentation for end users and system operators?
15. How or why is system documentation redundant with the system itself? What are the practical implications of this redundancy?
16. List the types of documentation needed to support maintenance activities. Which documentation types are needed most frequently? Which are needed least frequently?
17. How do training activities differ between end users and system operators?
18. How does implementing a maintenance change differ from developing a new system? How are they similar?
19. Why might system software upgrades not be installed? What are the costs of not installing them?

THINKING CRITICALLY

1. Examine the system flowchart for Rocky Mountain Outfitters in Figure 10-5. Develop a preliminary development plan based on IPO development order. Which programs are difficult to classify as input, process, or output? Is a straightforward application of IPO development order appropriate for this system? If not, what changes should be made to the preliminary development plan?
2. This chapter discussed top-down and bottom-up development order for transform-oriented structure charts. Can these development orders also be applied to transaction-oriented structure charts such as the one shown in Figure 10-10? If so, how?
3. Describe the process of testing software developed using both top-down and bottom-up development order. Which method results in the fewest resources required for testing? What types of errors are likely to be discovered earliest under each development order? Which development order is best as measured by the combination of required testing resources and the ability to capture important errors early in the testing process?
4. Assume that the Rocky Mountain Outfitters' customer support system will be developed as described in your answer to question 1. Assume that 14 people are available for programming and testing. What sizes and types of teams are best suited to the project?
5. Consider the issue of documenting a system using only electronic models developed with a full life-cycle development tool. The advantages are obvious (for example, the analyst modifies the models to reflect new requirements and automatically generates an updated system). Are there any disadvantages? Hint: The system might be maintained for a decade or more.
6. Some types of system documentation (such as models developed during analysis activities) are seldom looked at after the system is made operational. What are the advantages and disadvantages of not keeping such documentation types?

EXPERIENTIAL EXERCISES

1. Assume that you and five of your classmates are charged with developing the Customer order program shown in Figure 10-16. Create a development and testing plan to write and test the required modules. Assume that you have three weeks to complete all tasks.
2. Implement a formal QA process in one of your programming or system development classes (be sure to obtain permission from your instructor first). Form a group of students to implement an inspection process. Have one or all members prepare presentation materials for code that they've written and distribute them prior to a group meeting. If possible, create a buddy system for testing each other's code. Evaluate the results in terms of time required for code development and quality of the final product.
3. Examine the end-user documentation supplied with a typical personal or office productivity package such as Microsoft Office. Compare the documentation to the categories described in this chapter. Which categories of documentation are supplied? How might documentation for a business application differ in content or format?
4. Talk with a computer center or IS manager about the testing process used with a recently installed system or subsystem. What types of tests were performed? How were test cases and test data generated? What types of teams developed and implemented the tests?
5. Talk with an end user at your school or work about the documentation and training provided with a recently installed or distributed business application. What types of training and documentation were provided? Did the user consider the training to be sufficient? Does the user consider the documentation to be useful and complete?

CASE STUDIES

HUDSONBANC BILLING SYSTEM UPGRADE

Two regional banks with similar geographic territories merged to form HudsonBanc. Both banks had credit-card operations and operated billing systems that had been internally developed and upgraded over three decades. The systems performed similar functions, and both operated primarily in batch mode on IBM mainframes. Merging the two billing systems was identified as a high-priority, cost-saving measure.

HudsonBanc initiated a project to investigate how to merge the two billing systems. Upgrading either system was quickly ruled out because the existing technology was considered old, and the costs of upgrading the system were estimated to be too high. HudsonBanc decided that a new system should be built or purchased. Management preferred the purchase option because it was assumed that a purchased system could be brought online more quickly and cheaply. An RFP was prepared, many responses were received, and after months of analysis and investigation, a vendor was chosen.

Hardware for the new system was installed in early January. Software was installed the following week, and a random sample of 10 percent of the customer accounts was copied to the new system. The new system was operated in parallel with the old system for two months. To save costs involved with complete duplication, the new system computed but did not actually print billing statements. Payments were entered into both systems and used to update parallel customer account databases. Duplicate account records were checked manually to ensure that they were the same.

After the second test billing cycle, the new system was declared ready for operation. All customer accounts were migrated to the new system in mid-April. The old system was turned off on May 1, and the new system took over operation. Problems occurred almost immediately. The system was unable to handle the greatly increased volume of transactions. Data entry slowed to a crawl, and payments were soon backed up by several weeks. The system was not handling certain types of transactions correctly (for example, charge corrections and credits for overpayment). Manual inspection of the recently migrated account records showed errors in approximately 50,000 accounts.

It took almost six weeks to manually adjust the incorrect accounts and to update functions to correctly handle all transaction types. On June 20, the company attempted to print billing statements for the 50,000 corrected customer accounts. The system refused to print any information for transactions more than 30 days old. A panicked consultation with the vendor concluded that fixing the 30-day restriction would require more than a month of work and testing. It was also concluded that manual entry of account adjustments followed by billing within 30 days was the fastest and least risky way to solve the immediate problem.

Clearing the backlog took two months. During that time, many incorrect bills were mailed. Customer support telephone lines were continually overloaded. Twenty-five people were reassigned from other operational areas, and additional phone lines were added to provide sufficient customer support capacity. System development personnel were reassigned to IS operations for up to three months to assist in clearing the billing backlog. Federal and state regulatory authorities stepped in to investigate the problems. HudsonBanc agreed to allow customers to spread payments for late bills over three months without interest charges. Setting up the payment arrangements further aggravated the backlog and staffing problems.

1. What type of installation did HudsonBanc use for its new system? Was it an appropriate choice?
2. How could the operational problems have been avoided?

THE DOWNTOWN VIDEOS RENTAL SYSTEM

Using the design class diagram you developed in Chapter 12 for the DownTown Videos rental system, develop an implementation and testing plan. Specify the order in which classes and their methods will be implemented and the groups of methods and classes that will be tested during integration testing.

RETHINKING ROCKY MOUNTAIN OUTFITTERS

Assume that it is late April 2010 and that customer support system (CSS) analysis activities are nearly completed. Design is scheduled to finish by June 15 and implementation is scheduled to finish by November 1. RMO wants to use the new CSS during the holiday sales peak—roughly between Thanksgiving and Christmas—during which 40 percent of annual sales normally occur. RMO wants to announce the new Web ordering system in an upcoming catalog. New catalog mailings are scheduled for June 15, September 1, October 31, and December 10.

1. Describe the risks associated with planning the new CSS implementation and announcing the new Web ordering system to customers. Remember that the new CSS will replace the current telephone, mail-order, and Web order systems. How conservative should RMO be with respect to testing, installation, and customer announcements? What is the cost of being too conservative?
2. What fallback strategies should be developed, if any? What should the "drop dead" date be for deciding whether to use the new CSS to process holiday orders?
3. Develop an installation plan and schedule. Justify your approach(es) and your timetable based on your previous risk analysis.
4. Analyze the training requirements and develop a training plan and schedule. How can training, data conversion, and testing activities be overlapped or combined? What about training and support for customers using the Web ordering system?

FOCUSING ON RELIABLE PHARMACEUTICAL SERVICE

Using the structure chart you developed in Chapter 10 for Reliable Pharmaceutical Service, develop an implementation and testing plan. Specify the order in which modules will be implemented and the groups of modules that will be tested during integration testing.

FURTHER RESOURCES

Robert V. Binder, *Testing Object-Oriented Systems: Models, Patterns, and Tools*. Addison-Wesley, 2000.

Barry Boehm, *Software Engineering Economics*. Prentice Hall, 1981.

Robert G. Ebenau and Susan H. Strauss, *Software Inspection Process*. McGraw-Hill, 1994.

William Horton, *E-Learning Tools and Technologies: A Consumer's Guide for Trainers, Teachers, Educators, and Instructional Designers*. John Wiley & Sons, 2003.

William Horton, *Designing Web-Based Training: How to Teach Anyone Anything Anywhere Anytime*. John Wiley & Sons, 2000.

William Horton, *Illustrating Computer Documentation: The Art of Presenting Information Graphically on Paper and Online*. John Wiley & Sons, 1991.

International Association of Information Technology Trainers (ITrain) Web site—*http://itrain.org/*.

Edward Kit, *Software Testing in the Real World: Improving the Process*. ACM Press, 1996.

David Kung, Jerry Gao, Pei Hsia, Yasufumi Toyoshima, Chris Chen, Young-Si Kim, and Young-Kee Song, "Developing an Object-Oriented Software Testing and Maintenance Environment," *Communications of the ACM*, volume 38:10, October, 1995, pp. 75–87.

Steve McConnell, *Code Complete*. Microsoft Press, 1995.

Arthur H. Watson and Thomas J. McCabe, "Structured Testing: A Testing Methodology Using the Cyclomatic Complexity Metric," NIST Special Publication 500-235, National Institute of Standards and Technology, September, 1996, *http://hissa.ncsl.nist.gov/HHRFdata/Artifacts/ITLdoc/235/mccabe.html*.

CHAPTER

17

CURRENT TRENDS IN SYSTEM DEVELOPMENT

LEARNING OBJECTIVES

After reading this chapter, you should be able to:

- Explain the foundations for the adaptive development methodologies
- List and describe the features of the Unified Process system development methodology
- List and describe the features of Agile Modeling
- Compare and contrast the features of Extreme Programming and Scrum development
- Explain the importance of Model-Driven Architecture on enterprise-level development
- Describe frameworks and components, the process by which they are developed, and their impact on system development

CHAPTER OUTLINE

Software Principles and Practices

Adaptive Methodologies to Development

Model-Driven Architecture—Generalizing Solutions

Frameworks, Components, and Services

VALLEY REGIONAL HOSPITAL: MEASURING A PROJECT'S PROGRESS

Claire Haskell, the vice president of technology at Valley Regional Hospital (VRH), listened quietly to Henry Williams's progress report on the new patient records system. Henry was the project leader for a team that was developing the patient records system for VRH. Also in the meeting were the project's sponsor, Charlie Montgomery, who was the director of patient information and records, and Jason Smith, the director of software development. Months before, Jason and Henry had approached Claire and asked her to try a new development approach called *Agile Development* for this recently approved project. They had already spoken with Charlie, and he had agreed to try Agile Development. Claire had approved the project and their request to try the new approach, even though she knew very little about it.

During his presentation, Henry kept talking about how wonderfully the team worked together and how much fun they were having. Although she was glad that the team was functioning well, Claire wanted more specifics. She wanted to know whether the new system was on schedule and within budget. After about 20 minutes of patient listening, she couldn't wait any longer; she asked Henry directly to show her the schedule and to report on the team's progress. He flipped up a schedule, but that did little to help—the schedule had no familiar milestones such as analysis, design, and programming. Instead, she saw other terms: *iteration*, *user stories*, and *refactoring*.

At this point, Claire really became worried. So she turned to Charlie and said pointedly, "Exactly how is the project progressing from your viewpoint?" His answer surprised her.

Charlie said, "The records administrators and I are extremely pleased with the demos we are seeing. We are also satisfied with the quality of the system we saw during our acceptance testing. From what we have seen so far, the system seems to be exactly what we need. But as far as the schedule is concerned, I'm not certain whether the entire system will be delivered on time. I think so, but I'm not involved in the day-to-day development."

Claire felt a little better. At least the system was doing what it needed to do so far. But she still wanted reassurance from the project leader. "Henry, are we going to hit the completion date? The system needs to be ready on time."

Henry responded, "We are progressing on schedule so far and everything looks fine. But no, I can't show you a traditional schedule—one with major milestones. But here is a short-term schedule for the next two months of work."

Claire wasn't satisfied. She asked Henry to stay and talk with her privately after the meeting ended. She became agitated and said, "Henry, we need more accountability for this project. The only solution I can see is to meet with you frequently to monitor its progress. I want a rough schedule for the rest of the project on my desk on Monday morning. That gives you three days to develop one. Then I want you to meet with me every Monday from here out so that we can be sure we are on track and hit the delivery date."

Although he was not pleased with Claire's suggestion, Henry reluctantly agreed.

OVERVIEW

In previous chapters, you have been introduced to many concepts and skills that are necessary to develop robust information systems to solve real business needs. You have learned so-called "soft" skills associated with managing projects, interacting in teams, gathering information, and making presentations. You have also learned "hard" skills—those associated with problem solving, building requirements models, and designing new systems. You might have covered the traditional approach or the object-oriented approach, or both. And you have learned many important concepts about projects, iterative development, and implementation alternatives. In short, you have developed a solid working knowledge of system development and obtained a bag of tools to get you started developing information systems for businesses and other organizations.

As you have learned through your exposure to information systems, the IS discipline is dynamic and ever changing. Many of the tools and techniques that were commonly used just a few years ago have disappeared, replaced by newer approaches. In addition, the reach of today's systems has broadened, encompassing systems that are enterprisewide, distributed, interactive, and interconnected; that support both desktop and Internet-based computing; and that run on all types of computers and mobile devices. These more complex system requirements have necessitated a whole new set of programming languages and tools. As part of the continual reinvention of the discipline, IS professionals are also creating and developing new techniques for building systems; that is, new methodologies to support system development.

In Chapter 2, you were introduced to predictive and adaptive development methodologies. Historically, predictive approaches have dominated the field. Today, however, many developers are inventing different, more adaptive approaches to system development. In this chapter, we introduce you to some of the newest approaches to organizing and running a system development project, both predictive and adaptive:

- We explain three methodologies introduced in Chapter 2 that are adaptive in nature, including the Unified Process, Extreme Programming, and Scrum. These three methodologies share many ideas but also have some distinct features.
- We introduce Model-Driven Architecture (MDA), which consists of ideas for enterprise-level integration of systems. MDA is applicable to predictive approaches and to the Unified Process, Extreme Programming, and Scrum.
- We present basic ideas associated with object frameworks and components. These two technologies provide additional support to increase developer productivity, speed the rate of development, and improve the quality of the final system.

The material in this chapter is only introductory. Your interest might be piqued, and you might want to investigate one or more of these approaches in more detail. If so, you will be contributing both to your own professional development and to the advancement of the field of system development.

SOFTWARE PRINCIPLES AND PRACTICES

ubiquitous computing
the current trend of using computer technology in every aspect of our lives

The last 50 years have seen tremendous advances in all areas of computing. Moore's law, which states that hardware computing power doubles approximately every 18 months, has yet to be disproved. Computing capability still continues to advance at an astounding rate. In addition, all sorts of new uses and devices for computing have been created, including cell phones with digital cameras, handheld PCs, Internet-enabled telephones, appliances with embedded computer chips, and radio-frequency ID chips on products in retail stores. Large-scale systems, although not as obvious to consumers, drive many of the basic activities on which our society is based. Such activities as money transactions between banks and other financial institutions support our banking and credit-card industries. Behind the scenes, supply chain management systems trigger instant production of new products based on sales and inventory levels halfway around the world. Transportation scheduling and information sharing between carriers and shippers enable people and goods to travel worldwide with minimal disruptions. Today we speak of ubiquitous computing, meaning that computer technology is everywhere and affects almost every aspect of our lives. Just as one set of computing and system problems is solved, an entirely new need or desire is uncovered. The quest for improved business and consumer computing solutions continues unabated. Because of these trends, the long-term outlook for information systems specialists is incredibly positive. The downside for the industry is that the effort to keep current is extremely demanding.

So how are we able to address the multifaceted computing needs of industry and society? Obviously, no single solution or technology can satisfy all needs. Solutions come about because thousands of people are working on and solving individual problems. However, some principles and practices can advance the industry. Many come from the field of computer science, which focuses on the theories and principles of computing. Others develop from the discipline of information systems, which applies the principles of computing to everyday business problems. Of course, there is substantial overlap between these two disciplines, which can be compared with the relationship between scientists and engineers—for example, the chemists who do research and the chemical engineers who apply that research. Collaboration between people in all of the disciplines of computer science, information systems, decision sciences, and mathematics provides the fuel to propel the technology of our society forward.

Before we discuss the current trends, let's look back at this textbook for a moment. If you were asked to summarize the contents of this textbook in one or two sentences, what would you say? What are the main ideas you learned from the previous chapters in this book? We hope you would include these two points:

1. You learned how to build models—models to capture and explain needs, requirements, and solutions. In a broad definition of a model, we would also include writing the code, which is a model of a real-world process.
2. You learned the processes or steps necessary to build a solution—processes to both manage and conduct a system development project.

In this chapter, we also focus on those two primary areas, by describing current trends in modeling and in development processes. In addition, we discuss some current tools and techniques that support these trends.

To begin, we should review five important software principles and practices before discussing the details of current trends:

- Abstraction
- Models and modeling
- Patterns
- Reuse
- Methodologies

ABSTRACTION

Abstraction is the process by which we extract and distill core principles from a set of facts or statements. You learned about this principle when you learned to identify an abstract class—one that has no instances. An abstract class serves as a repository of generalized attributes and methods from which other classes inherit. You also learned to think abstractly when you built models to define user requirements. Thinking abstractly is a difficult skill to learn. Most people learn new concepts by seeing examples—that is, concrete instances. However, as you become more sophisticated in your thinking ability, you learn to think in abstract terms.

Abstraction is important in the field of computing. Many advances in computing have been developed because computer scientists were able to think abstractly and in fact to raise the level of abstraction. For example, in the early days of computing, developers wrote systems using assembly language, which is essentially machine language. Then they thought more abstractly and invented programming languages, such as Fortran, COBOL, C, Java, and Visual Basic. To do so, they had to think in the abstract about the characteristics of computer languages and language compilers. Then, thinking about user requirements led to the invention of models and diagrams to represent those requirements. Again, the process required thinking in the abstract about the characteristics of a good model and inventing class diagrams and sequence diagrams and their properties. Today, we think more abstractly by defining

metamodel
a model that describes the characteristics of another model

metamodels. A **metamodel** is a model that describes another model. For example, could you define a class diagram that would generalize and describe the components of any class diagram? As you will see when we discuss Model-Driven Architecture, abstraction is an important idea in advancing the body of knowledge for computing and system development.

MODELS AND MODELING

The second important software principle is models and modeling. You learned what a model is early in the book, and you practiced building models throughout the course. A model is an abstraction of something in the real world, representing a particular set of properties. There are two primary reasons developers build models. First, we must understand a process or thing by identifying and explaining its key characteristics. Modeling helps us crystallize our thinking so that we are more precise. In many instances, we cannot truly understand something until we try to model it. Second, we use models to document ideas that we need to remember and to communicate those ideas to other people. You will see later in this chapter that some of the current trends focus on how to better use models in software development—by doing either more or less modeling.

PATTERNS

The third software principle—patterns—is closely associated with abstraction and modeling. A pattern is a standard solution to a given problem or a template that can be applied to a problem. As we begin to think more abstractly and build models, we recognize that problems or issues that at first seemed very different in fact have similar characteristics when viewed at a generalized level. Patterns begin to form, both in the problems and in the solutions. In Chapter 12, we presented some ideas about design patterns. We noted that different industries have standard patterns to solve recurring problems. For example, banking systems serve the fundamental purpose of recording and processing financial transactions. So, design patterns exist not only for system structure, but also for entire systems. Standard design patterns are becoming widely accepted as they are identified and refined to solve problems. We will not revisit the discussion of patterns in this chapter, but you should recognize that they are a driving force in improving the quality of systems and speeding system development.

REUSE

The fourth software principle—reuse—is an outgrowth of the previous principles. Because developers have discovered patterns, they are building standard solutions and components that can be used over and over again. Through the principle of reuse, today's developers have become more productive. For example, to develop a graphical user interface for the Windows platform, nearly all developers use standard class libraries of forms, buttons, menus, drop-down boxes, text boxes, and radio buttons. If we instead had to write the code to display a button every time we put one on a screen, it would take a very long time to create a graphical system. Windows components are examples of reuse at the code level. Developers also like to think more abstractly and invent higher-level components that can be reused. Industry experts now indicate that many new systems are built primarily by integrating operating systems, communication systems, and applications into a single system. So, system developers today often work to integrate components into a complete solution. Reuse is a driving force in technologies such as Web services, .NET, and enterprise resource planning (ERP) systems. We will discuss the idea of reuse in more detail when we discuss component libraries and frameworks.

METHODOLOGIES AND PROCESSES

The fifth and final principle, methodologies, was also introduced earlier in the book. By methodology, we mean a set of processes—including the rules, guidelines, and techniques—that defines how systems are built and development projects are managed. By now you should be familiar with the differences between predictive and adaptive projects. You should also be able to recognize how the level of formality for a project will also dictate how a project is managed.

In most cases, the discussions throughout the textbook were designed to teach foundation principles that could be applied to either predictive or adaptive development methodologies as well as be used for formal projects or informally managed projects. In this chapter, as we discuss current trends and future directions, we discover that there are two seemingly opposed trends for project development. The agile philosophy of system development tends to have less formality, less documentation, and even less use of models. The model-driven architecture philosophy, of course, requires more formality, more documentation, and more use of models. We will investigate both in the following sections.

In the next section, we discuss the various adaptive approaches to systems development. First the Unified Process (UP) is explained in detail. Even though the UP is often considered an adaptive approach, in reality it should be considered both an adaptive and formal approach. As you will see, it contains specific processes and disciplines to ensure that a project is executed successfully. Following that discussion, we provide more detail about agile concepts before explaining two other approaches—Extreme Programming and Scrum—which are considered to be completely within the camp of less formal agile methods.

In her routine report concerning Rocky Mountain Outfitters' customer support system, Barbara Halifax recapped her team's basic approach to system development—one based on the adaptive Unified Process—and its progress (see Barbara's memo).

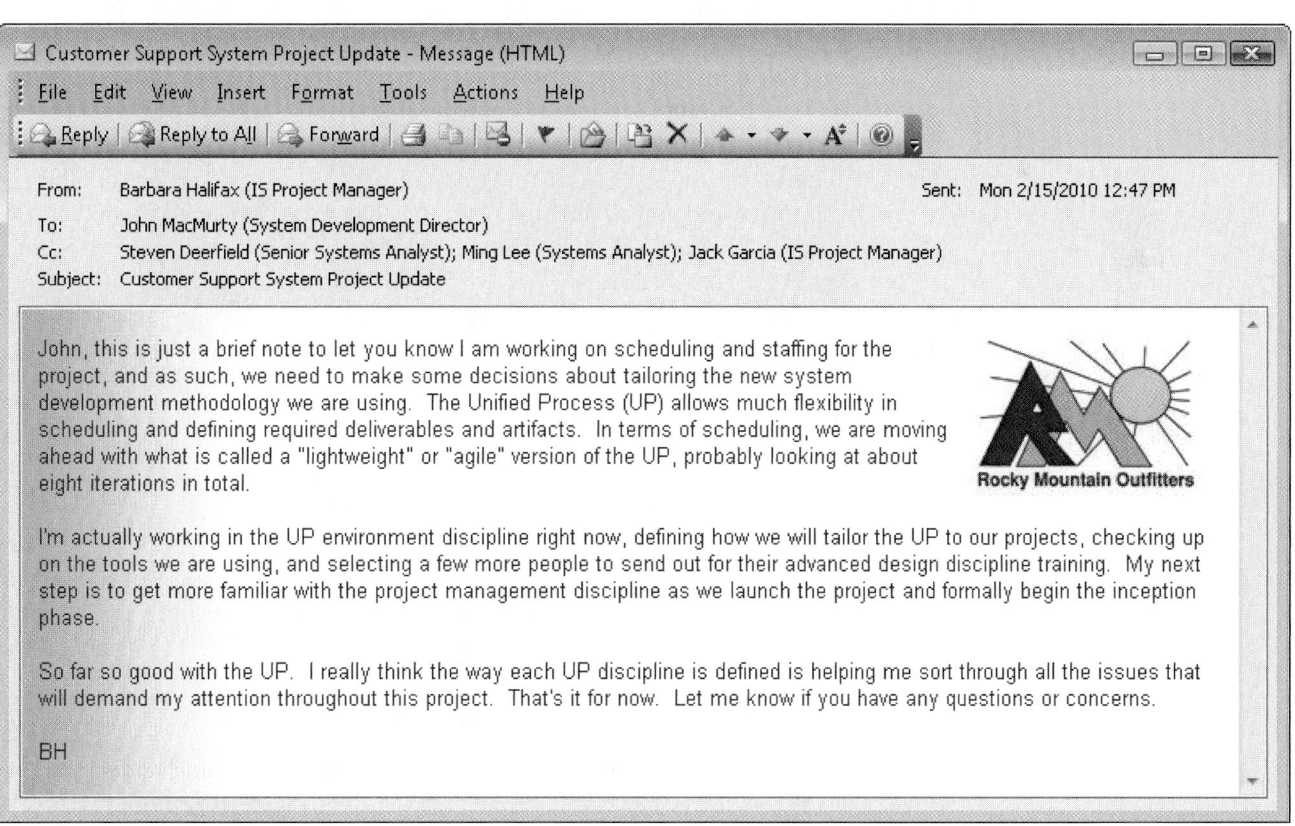

Customer Support System Project Update - Message (HTML)

File Edit View Insert Format Tools Actions Help

Reply | Reply to All | Forward

From: Barbara Halifax (IS Project Manager) Sent: Mon 2/15/2010 12:47 PM
To: John MacMurty (System Development Director)
Cc: Steven Deerfield (Senior Systems Analyst); Ming Lee (Systems Analyst); Jack Garcia (IS Project Manager)
Subject: Customer Support System Project Update

John, this is just a brief note to let you know I am working on scheduling and staffing for the project, and as such, we need to make some decisions about tailoring the new system development methodology we are using. The Unified Process (UP) allows much flexibility in scheduling and defining required deliverables and artifacts. In terms of scheduling, we are moving ahead with what is called a "lightweight" or "agile" version of the UP, probably looking at about eight iterations in total.

Rocky Mountain Outfitters

I'm actually working in the UP environment discipline right now, defining how we will tailor the UP to our projects, checking up on the tools we are using, and selecting a few more people to send out for their advanced design discipline training. My next step is to get more familiar with the project management discipline as we launch the project and formally begin the inception phase.

So far so good with the UP. I really think the way each UP discipline is defined is helping me sort through all the issues that will demand my attention throughout this project. That's it for now. Let me know if you have any questions or concerns.

BH

ADAPTIVE METHODOLOGIES TO DEVELOPMENT

Adaptive methodologies to system development allow for uncertainty. In development projects that focus on entirely new applications, many times the user's requirements are not well understood and cannot be described in detail. Because the scope of a new system might not be well defined, it is difficult, if not impossible, for analysts to create a detailed project plan. The best way to carry out a system development project in this situation is to identify core objectives early and to develop detailed work plans as the project progresses.

Two forces drive the increased interest in adaptive development. First, as you learned in Chapter 3, the low success rate of system development projects has always been troubling. Developing software is difficult, and success has always been elusive. In Chapter 3 we discussed the importance of good project management skills to improve processes and increase the chance of success. In addition, industry experts and developers have invented various adaptive development approaches to improve success rates.

The other force behind adaptive approaches is the volatility of today's business climate. Yesterday's more stable environment relied mainly on controlling costs and on tight management of internal procedures. In contrast, business success today depends on flexibility and rapid response to changes in the marketplace. As a result, a rigid system development process that defines system requirements 12, 24, or more months in advance is not necessarily flexible enough to meet the accelerated pace of change. But the newer, adaptive approaches allow for critical changes in business needs.

From a theoretical viewpoint, the development of any item, either a physical item or a software item, is completed by following a process. Every process needs controls to ensure that it stays on track. Process controls can be categorized into two types: predictive and empirical. Predictive controls define the steps to monitor a process in great detail. If a process gets off track, more detailed steps—for example, a work breakdown structure—and descriptions can be used to control it. Predictive control works well when more planning can provide more detail. However, for processes that are unpredictable, adding more detail and more controls only exacerbates the problem; trying to control the uncontrollable results in further loss of money and time.

Empirical controls, in contrast, describe processes that are variable and unpredictable. These processes are best controlled by handling each variation as it occurs and determining the best way to correct the deviation. In other words, empirical controls monitor progress and then make corrections on the fly, based on the specific situation. Because many software development projects contain a high amount of uncertainty, an empirical process might be a better choice for them.

All adaptive methodologies use empirical controls and have their own sets of rules and guidelines. However, they do share a few characteristics:

- Less emphasis on up-front analysis, design, and documentation
- More focus on incremental development
- More user involvement in project teams
- Reduced detailed planning, which is used for near-term work phases only; downstream phases might have high-level plans
- Tightly controlling schedules by fitting work into discrete time boxes
- More use of small work teams that are self-organizing

First, the adaptive team does not spend a lot of time analyzing, designing, and documenting because these activities are a means to an end; they are simply tools for writing executable code. Second, the only way to develop code quickly is to do it in small chunks and to develop the system incrementally. Third, developing code quickly requires the user to be completely involved with the project team—to become part of the project team and work side by side with the developers in creating the solution so that it fits the business needs.

Reducing detailed planning and scheduling the work to fit in time boxes help control the team's progress. Each iteration defines a specific function that will be added to the system, and the function is small enough so that a meaningful detailed schedule can be developed. Then the team completes the work in the time allotted. As developers become more experienced, they become proficient at knowing what can be done in a small time box of three or four weeks.

Consistent with time boxing, under adaptive methodologies, each team develops its own schedule for each iteration and organizes itself and its work productively. The only external schedules that are imposed on the work are those that require coordination with other projects or deliverables.

In the following sections, we look at each of the three adaptive methodologies: the Unified Process, Extreme Programming, and Scrum. We begin with one that is considered by many to be the foundation for adaptive software development—the Unified Process (UP). Next we elaborate on the concept of Agile Development to set the stage for the more highly adaptive and lightweight methodologies (for example, Extreme Programming and Scrum) that are increasingly popular with developers.

THE UNIFIED PROCESS

The Unified Process (UP) is an object-oriented system development methodology originally offered by Rational Software, which is now part of IBM. Developed by Grady Booch, James Rumbaugh, and Ivar Jacobson—the three pioneers who are also behind the success of the Unified Modeling Language (UML)—the UP is their attempt to define a complete methodology that uses UML for system models and describes a new, adaptive system development life cycle. In the UP, the term development *process* is synonymous with development *methodology*.

The UP is now widely recognized as a standard system development methodology for object-oriented development, and many variations are in use. The original version of UP defined an elaborate set of activities and deliverables for every step of the development process. More recent versions are streamlined, with fewer activities and deliverables, simplifying the methodology.

As discussed previously, adaptive methodologies—including the UP—are all based on an iterative approach to development. You learned in Chapter 2 (see Figure 2-7) that each iteration is like a miniproject, in which requirements are defined based on analysis tasks, system components are designed, and those components are then implemented, at least partially, through programming and testing. One of the big questions in adaptive development, however, is what the focus of each iteration should be. In other words, do iterations early in the project have the same objectives and focus as those done later? The UP answers this question by dividing a project into four major phases. The Unified Process life cycle, with phases and disciplines, was first shown in Figure 2-20 in Chapter 2.

UP Phases

A phase in the UP can be thought of as a goal, or major emphasis for a particular portion of the project. The four phases of the UP life cycle are named *inception, elaboration, construction,* and *transition,* as shown in Figure 17-1.

Each phase of the UP life cycle describes the *emphasis* or *objectives* of the project team members and their activities at that point in time. So, the four phases provide a general framework for planning and tracking the project over time. Within each phase, several iterations are planned to allow the team flexibility to adjust to problems or changing conditions. The emphases or objectives of the project team in each of the four phases are described briefly in Figure 17-2.

Figure 17-1
The Unified Process system development life cycle

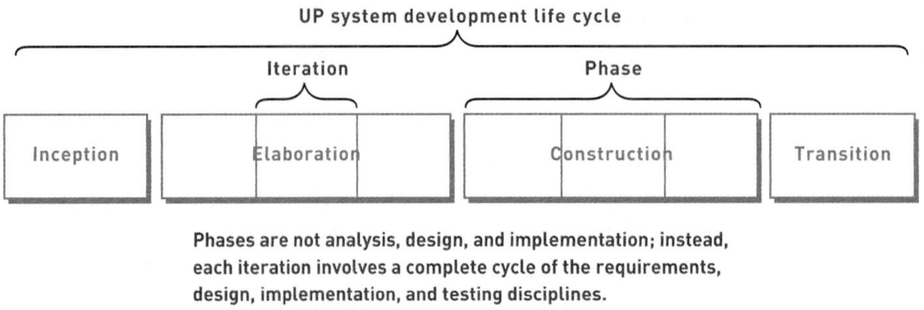

Figure 17-2
UP phases and objectives

UP phase	Objective
Inception	Develop an approximate vision of the system, make the business case, define the scope, and produce rough estimates for cost and schedule.
Elaboration	Define the vision, identify and describe all requirements, finalize the scope, design and implement the core architecture and functions, resolve high risks, and produce realistic estimates for cost and schedule.
Construction	Iteratively implement the remaining lower-risk, predictable, and easier elements and prepare for deployment.
Transition	Complete the beta test and deployment so users have a working system and are ready to benefit as expected.

Inception Phase As in any project planning phase, in the inception phase the project manager develops and refines a vision for the new system to show how it will improve operations and solve existing problems. The project manager makes the business case for the new system, meaning that the benefits of the new system must outweigh the cost of development. The scope of the system must also be defined so that it is clear what the project will accomplish. Defining the scope includes identifying many of the key requirements for the system.

The inception phase is usually completed in one iteration, and as with any iteration, parts of the actual system might be designed, implemented, and tested. As software is developed, team members must confirm that the system vision still matches user expectations or that the technology will work as planned. Sometimes prototypes are discarded after proving that point.

Elaboration Phase The elaboration phase usually involves several iterations, and early iterations typically complete the identification and definition of all of the system requirements. Because the UP is an adaptive approach to development, the requirements are expected to evolve and change after work starts on the project.

Elaboration phase iterations also complete the analysis, design, and implementation of the core architecture of the system. Usually the aspects of the system that pose the greatest risk are identified and implemented first. Until developers know exactly how the highest-risk aspects of the project will work out, they cannot determine the amount of effort required to complete the project. By the end of the elaboration phase, the project manager should have more realistic estimates for a project's cost and schedule, and the business case for the project can be confirmed. Remember that the design, implementation, and testing of key parts of the system are completed during the elaboration phase. The elaboration phase is not at all the same as the traditional SDLC's analysis phase.

BEST PRACTICE

Be sure *not* to confuse the UP phases with the waterfall approach to the SDLC. Elaboration is not at all like the traditional analysis phase.

Construction Phase The construction phase involves several iterations that continue the design and implementation of the system. The core architecture and highest-risk aspects of the system are already complete. Now the focus of the work turns to the routine and predictable parts of the system; for example, detailing the system controls such as data validation, fine-tuning the user interface's design, finishing routine data maintenance functions, and completing the help and user preference functions. The team also begins to plan for deployment of the system.

Transition Phase During the transition phase, one or more final iterations involve the final user-acceptance and beta tests, and the system is made ready for operation. After the system is in operation, it will need to be supported and maintained.

UP Disciplines

As we mentioned earlier, the four UP phases define the project sequentially by indicating the emphasis of the project team at any point in time. To make iterative development manageable, the UP defines disciplines to use within each iteration. A **discipline** is a set of functionally related activities that together contribute to one aspect of the development project. UP disciplines include business modeling, requirements, design, implementation, testing, deployment, configuration and change management, project management, and environment. Each iteration usually involves activities from all disciplines.

discipline
a set of functionally related activities that together contribute to the development process of a UP project

Figure 17-3 shows how the UP disciplines are involved in each iteration, which is typically planned to span four weeks. The size of the shaded area under the curve for each discipline indicates the relative amount of work included in each discipline during the iteration. The amount and nature of the work differs from iteration to iteration. For example, in iteration 2 much of the effort focuses on business modeling and defining requirements, with much less effort focused on implementation and deployment. In iteration 5, very little effort is focused

Figure 17-3
UP disciplines used in varying amounts in each iteration

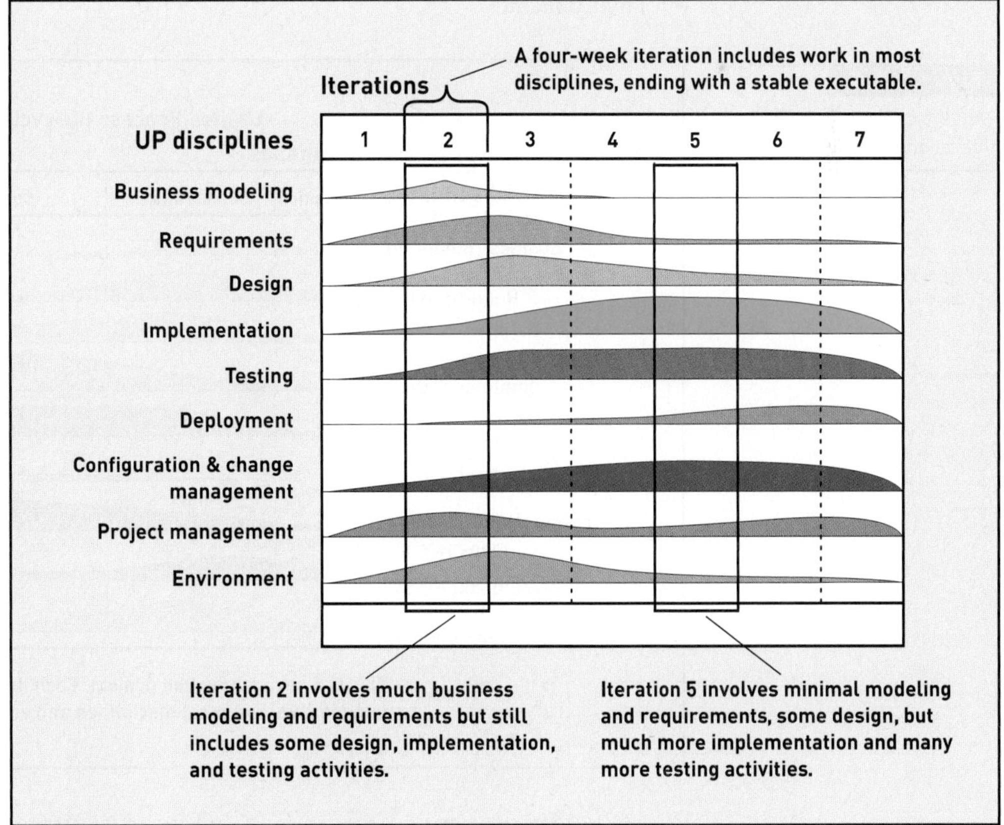

A four-week iteration includes work in most disciplines, ending with a stable executable.
Iterations
UP disciplines
1
2
3
4
5
6
7
Business modeling
Requirements
Design
Implementation
Testing
Deployment
Configuration & change management
Project management
Environment
Iteration 2 involves much business modeling and requirements but still includes some design, implementation, and testing activities.
Iteration 5 involves minimal modeling and requirements, some design, but much more implementation and many more testing activities.

on modeling and requirements, with much more effort focused on implementation, testing, and deployment. But most iterations involve some work in all disciplines.

Figure 17-4 shows the entire UP life cycle—phases, iterations, and disciplines. This figure includes all of the key UP life cycle features and is useful for understanding how a typical information system development project is managed.

The previous figures illustrate how the phases include activities from each discipline. But what about the detailed activities that occur within each discipline? The disciplines can be divided into two main categories: system development activities and project management activities. The six main UP development disciplines are as follows:

- Business modeling
- Requirements
- Design
- Implementation
- Testing
- Deployment

Recall that each iteration is similar to a miniproject, which completes a portion of the system. For each iteration, the project team must understand the business environment (business modeling), define the requirements that the portion of the system must satisfy (requirements), design a solution for that portion of the system that satisfies the requirements (design), write and integrate the computer code that makes the portion of the system actually work (implementation), thoroughly test the portion of the system (testing), and then in some cases put the portion of the system that is completed and tested into operation for users (deployment).

Three additional support disciplines are necessary for planning and controlling the project:

- Configuration and change management
- Project management
- Environment

Figure 17-4

UP life cycle with phases, iterations, and disciplines

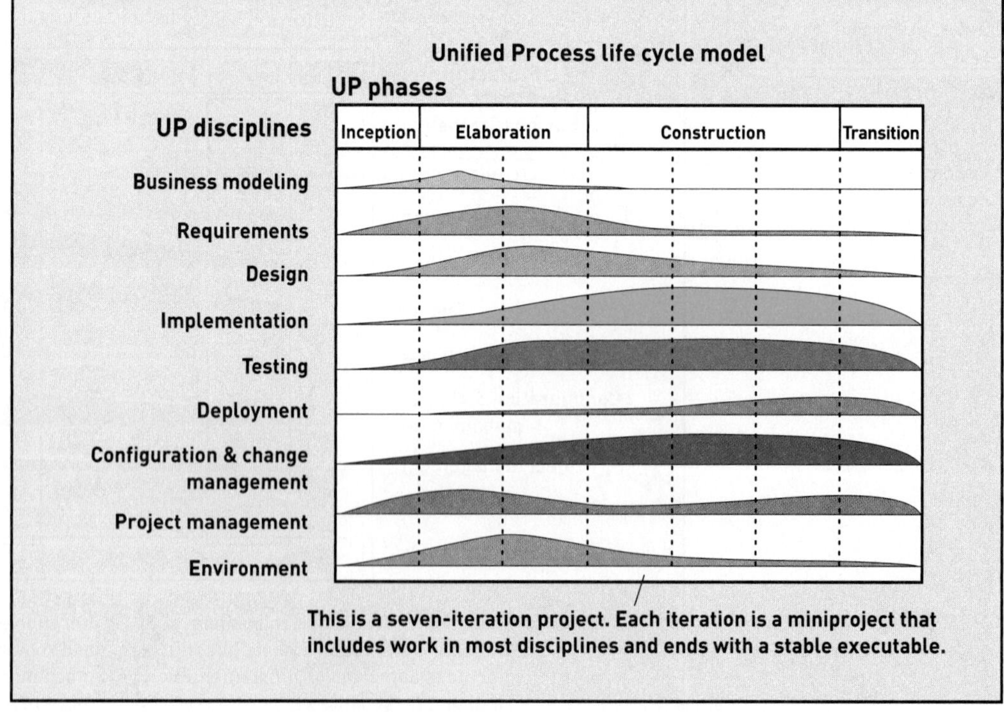

All nine UP disciplines are employed throughout the lifetime of a project but to different degrees. For example, in the inception phase there is one iteration. During the inception phase iteration, the project manager might complete a model showing some aspect of the system environment (the business modeling discipline). The scope of the system is delineated by defining many of the key requirements of the system and listing use cases (the requirements discipline). To prove technological feasibility, some technical aspect of the system might be designed (the design discipline), programmed (the implementation discipline), and tested to make sure it will work as planned (the testing discipline). In addition, the project manager is making plans for handling changes to the project (the configuration and change management discipline), working on a schedule and cost/benefit analysis (the project management discipline), and tailoring the UP phases, iterations, deliverables, and tools to match the needs of the project (the environment discipline).

The elaboration phase includes several iterations. In the first iteration, the team works on the details of the domain classes and use cases addressed in the iteration (the business modeling and requirements disciplines). At the same time, they might complete the description of all use cases to finalize the scope (the requirements discipline). The use cases addressed in the iteration are designed by creating design class diagrams and interaction diagrams (the design discipline), programmed using Java or Visual Basic .NET (the implementation discipline), and fully tested (the testing discipline). The project manager works on the plan for the next iteration and continues to refine the schedule and feasibility assessments (the project management discipline), and all team members continue to receive training on the UP activities they are completing and the system development tools they are using (the environment discipline).

By the time the project progresses to the construction phase, most of the use cases have been designed and implemented in their initial form. The focus of the project turns to satisfying other technical, performance, and reliability requirements for each use case, finalizing the design, and implementation. These requirements are usually routine and lower risk, but they are key to the success of the system. The effort focuses on designing system controls and security and on implementing and testing these aspects.

The Unified Process as a system development methodology must be tailored to the development team and specific project. Choices must be made about which deliverables to produce and the level of formality to be used. Sometimes a project requires formal reporting and controls. Other times, it can be less formal. The UP should always be tailored to the project.

BEST PRACTICE

Be sure to tailor the UP disciplines to fit the project.

THE AGILE DEVELOPMENT PHILOSOPHY AND AGILE MODELING

The highly volatile marketplace has forced businesses to respond rapidly to new opportunities. Sometimes new opportunities appear in the middle of implementing another business initiative. To survive, businesses must be agile. Agility—being able to change directions rapidly, even in the middle of a project—is the keystone of Agile Development. Agile Development is a philosophy and set of guidelines for developing software in an unknown, rapidly changing environment. It provides an overarching philosophy to be used with a specific development methodology such as the Unified Process. For example, we identified the UP as being somewhat adaptive, but being adaptive is not the same as being agile. Some UP projects may adopt many agile philosophies, and others may use fewer.

Related to Agile Development, Agile Modeling is a philosophy about how to build models, some of which are formal and detailed and others sketchy and minimal. Figure 17-5 illustrates the relationships among an Agile Development philosophy, specific adaptive methodologies, and use of Agile Modeling.

Figure 17-5

Adaptive methodologies using Agile Development and Agile Modeling

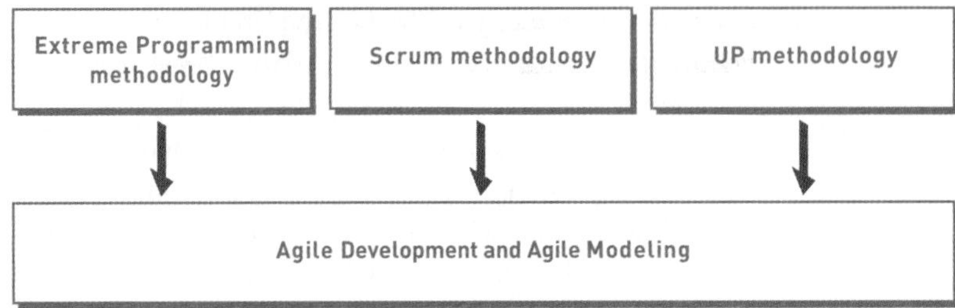

Agile Development Philosophy and Values

The "Manifesto for Agile Software Development" (see the "Further Resources" section) identifies four basic values, which represent the core philosophy of the agile movement. The four values emphasize:

- Responding to change over following a plan
- Individuals and interactions over processes and tools
- Working software over comprehensive documentation
- Customer collaboration over contract negotiation

Note that each of the phrases in the list prioritizes the value on the left over the value on the right. The people involved in system development, whether as team members, users, or other stakeholders, all need to accept these priorities for a project to be truly agile. Adopting an agile approach is not always easy.

chaordic
a term used to describe adaptive projects, which are both chaotic and ordered

Some industry leaders in the agile movement coined the term **chaordic** to describe an agile project. *Chaordic* comes from two words, *chaos* and *order*. The first two values in our list—responding to change over following a plan and individuals and interactions over processes and tools—do seem to be a recipe for chaos. But they recognize that software projects inherently have many unknowns and unpredictable elements, and hence a certain amount of chaos. Developers need to accept the chaos but also need to use the specific methodologies discussed later to impose order on this chaos to move the project ahead.

Managers and executive stakeholders frequently struggle to accept this less rigid point of view, often wanting to impose more controls on development teams and to enforce detailed plans and schedules. However, the agile philosophy takes the opposite approach, providing more flexibility in project schedules and letting the project teams plan and execute their work as the project progresses.

Another important value of Agile Development is that customers must continually be involved with the project team. They do not sit down with the project team for a few sessions to develop the specifications and then go their separate ways. Instead, customers collaborate with and become part of the technical team. Because working software is being developed throughout the project, customers are continually involved in defining requirements and testing components.

Contracts also take on an entirely different flavor. Fixed prices and fixed deliverables do not make sense. Contracts take more of a collaborative tack but include options for the customer to cancel if the project is not progressing, as measured by the incremental deliverables. Incremental deliverables in agile projects are working pieces of the new system, not documents or specifications.

Models and modeling are critical to Agile Development, so we look next at Agile Modeling. Many of the core values are illustrated in the principles and practices of building models.

Agile Modeling Principles

Your first impression might be that an agile approach means less modeling, or maybe even no modeling. Agile Modeling (AM) is not about doing less modeling but about doing the right kind of modeling at the right level of detail for the right purposes. Early in this chapter, we identified two primary reasons to build models: (1) to understand what you are building and (2) to communicate important aspects of the solution system. AM consists of a set of principles and practices that reinforce these two reasons for modeling. AM does not dictate which models to build or how formal to make those models but instead helps developers to stay on track with their models—by using them as a means to an end instead of building models as end deliverables. AM's basic principles express the attitude that developers should have as they develop software. Figure 17-6 summarizes Agile Modeling principles. We discuss those principles next.

Figure 17-6
Agile Modeling principles

Agile Modeling principles

- Develop software as your primary goal.
- Enable the next effort as your secondary goal.
- Minimize your modeling activity—few and simple.
- Embrace change, and change incrementally.
- Model with a purpose.
- Build multiple models.
- Build high-quality models and get feedback rapidly.
- Focus on content rather than representation.
- Learn from each other with open communication.
- Know your models and how to use them.
- Adapt to specific project needs.

BEST PRACTICE

Be sure your models are a means to an end and not an end themselves. All models should have a purpose.

Develop Software as Your Primary Goal The primary goal of a software development project is always to produce high-quality software. The primary measurement of progress is working software, not intermediate models of system requirements or specifications. Modeling is always a means to an end, not the end itself. Any activity that does not directly contribute to the end goal of producing software should be questioned and avoided if it cannot be justified.

Enable the Next Effort as Your Secondary Goal Focusing only on working software can also be self-defeating, so developers must consider two important objectives. First, requirements models might be necessary to develop design models. So, do not think that if the model cannot be used to write code, it is unnecessary. Sometimes several intermediate steps are needed before the final code can be written. Second, although high-quality software is the primary goal, long-term use of that code is also important. So, some models might be necessary to support maintenance and enhancement of the system. Yes, the code is the best documentation, but some architectural design decisions might not be easily identified from the code. Look carefully at what other artifacts might be necessary to produce high-quality systems in the long term.

Minimize Your Modeling Activity—Few and Simple Create only the models that are necessary. Do just enough to get by. This principle is not a justification for sloppy work or inadequate analysis. The models you create should be clear, correct, and complete. But do not create unnecessary models. Also, keep each model as simple as possible. Normally, the simplest solution is the best solution. Elaborate solutions tend to be difficult to understand and maintain. However, we emphasize again that simplicity is not justification for being incomplete.

Embrace Change and Change Incrementally Because the underlying philosophy of Agile Modeling is that developers must be flexible and respond quickly to change, a good agile developer willingly accepts—and even embraces—change. Change is seen as the norm, not the exception. Watch for change and have procedures ready to integrate changes into the models. The best way to accept change is to develop incrementally. Take small steps and address problems in small bites. Change your model incrementally, and then validate it to make sure it is correct. Do not try to accomplish everything in one big release.

Model with a Purpose We indicated earlier that the two reasons to build models are to understand what you are building and to communicate important aspects of the solution system. Make sure that your modeling efforts support those reasons. Sometimes developers try to justify building certain models by claiming that (1) the development methodology mandates the development of the model, (2) someone wants a model, even though the person does not know why it is important, or (3) a model can replace a face-to-face discussion of issues. Always identify a reason and an audience for each model you develop. Then develop the model in sufficient detail to satisfy the reason and the audience. Incidentally, the audience might be you.

Build Multiple Models UML, along with other modeling methodologies, has several models to represent different aspects of the problem at hand. To be successful—in understanding or communication—you will need to model various aspects of the required solution. Don't develop all of them; be sure to minimize your modeling, but develop enough models to make sure you have addressed all the issues.

Build High-Quality Models and Get Feedback Rapidly Nobody likes sloppy work. It is based on faulty thinking and introduces errors. One way to avoid error in models is to get feedback rapidly, while the work is still fresh. Feedback comes from users, as well as technical team members. Others will have helpful insights and different ways to view a problem and identify a solution.

Focus on Content Rather than Representation Sometimes a project team has access to a sophisticated visual modeling tool. Visual modeling tools can be helpful, but at times they are distracting because developers spend time making the diagrams pretty. Be judicious in the use of tools. Some models need to be well drawn for communication or contracts or even to handle expected changes and updates. In other cases, a hand-drawn diagram might suffice.

Learn from Each Other with Open Communication All of the adaptive approaches emphasize working in teams. Do not be defensive about your models. Other team members have good suggestions. You can never truly master every aspect of a problem or its models.

Know Your Models and How to Use Them Being an agile modeler does not mean that you are not skilled. If anything, you must be *more* skilled to know the strengths and weaknesses of the models, including how and when to use them. An expert modeler applies the previous principles of simplicity, quality, and development of multiple models.

Adapt to Specific Project Needs Every project is different because it exists in a unique environment; involves different users, stakeholders, and team members; and requires a different development environment and deployment platform. Adapt your models and modeling techniques to fit the needs of the business and the project. Sometimes models can be informal and simple. For other projects, more formal, complicated models might be required. An agile modeler is able to adapt to each project.

Agile Modeling Practices

The following practices support the AM principles just expressed. The heart of AM is in its practices, which give the practitioner specific modeling techniques. Figure 17-7 summarizes the Agile Modeling practices. We discuss each of the practices next.

Figure 17-7
Agile Modeling practices

Agile Modeling practices

- Iterative and incremental modeling
  - Use the right models.
  - Create several models in parallel.
  - Iterate frequently.
  - Model in small increments.
- Teamwork
  - Model with others.
  - Involve users and other stakeholders.
  - Share ownership of the models.
  - Display the models publicly.
- Simplicity
  - Create simple content.
  - Depict the models simply.
  - Use simple tools.
- Validation
  - Prove it with code.
- Documentation
  - Discard temporary models.
  - Formalize contract models.
  - Update only when it hurts.
- Motivation
  - Model to communicate.
  - Model to understand.

Iterative and Incremental Modeling Remember that modeling is a support activity, not the end result of software development. As a developer, you should create small models frequently to help you understand or solve a problem. New developers sometimes have difficulty deciding which models to select. You should continue to learn about models and expand your repertoire. UML has a large set of models that cover a lot of analysis and design territory. However, they are not the only models you might find useful. Many developers still use data flow diagrams and decomposition diagrams from the traditional structured approach. The point is that models are a tool, and as a professional, you should have a large set of tools.

Teamwork As shown in Figure 17-5, AM supports various development methodologies. One of the tenets in all of these methodologies is that developers work together in small teams of two to four members. In addition, users should be integrally involved in modeling exercises. For example, suppose the task at hand is to understand how a purchase order is created and processed. Good AM practice says to get the right players together, including team members and users, and develop a detailed model of the process, possibly on a whiteboard. Other teams could then take a digital photograph of the whiteboard and post it in a repository on the project's network server. The model then becomes public; no one owns it, and all can access it. If it later needs to be corrected, it can be annotated with software and reposted. An alternative method, especially if the model will become a permanent document, is to develop the model using a drawing tool such as Visio with a laptop and a projector. This process is not quite as flexible as a whiteboard, but it yields a more permanent model. In any case, the model is again posted for all to use, review, and update.

Simplicity The previous purchase order example illustrated an approach that is simple and easy to support. Also, developers should create a set of models to help them understand or solve a narrow problem. In the purchase order example, the model focused on one business process or use case. In the first iteration, developers should only focus on the typical process, one without all of the possible variations. Then later iterations can add exception conditions, security and control requirements, and other details.

Validation Between modeling sessions, the team can begin to write code for the solutions already conceived so that they can validate the models. Simplicity supports frequent validation. Do not create too many or complex models until the simple ones have been validated with code.

Documentation Many models are temporary working documents that are developed to solve a particular problem. These models quickly become obsolete as the code evolves and improves. Do not try to keep them up to date. Discard them. If they were posted to a repository, date them so that everyone knows they show a history of decisions and progress but are not now in sync with the code. Updating only when it hurts is a guideline that tells us not to waste time trying to keep temporary models synchronized. During the first iteration, when many models are developed concurrently, they should be consistent. However, as development progresses, some models will become working documents that no longer relate well to other models. Remember that the objective of the project is to develop software, not to have a set of pretty models. Only update when it hurts—that is, when the project team can't work effectively without the information.

Motivation Remember the basic objectives of modeling. Only build a model if it helps you understand a process or solve a problem or if you need to record and communicate something. For example, the team members in a design session might make some design decisions. To communicate these decisions, the team posts a simple model to make it public. The model can be a very effective tool to document the decisions and ensure that all have a common understanding and reference point. Again, a model is simply used as a tool for communication, not as an end in itself.

Now that we have explored the basic philosophy, principles, and practices underlying Agile Development, we turn to two methodologies that employ agile concepts: Extreme Programming and Scrum.

EXTREME PROGRAMMING

Extreme Programming (XP) is an adaptive, agile development methodology that was created in the mid-1990s. The word *extreme* sometimes makes people think that it is completely new and that developers who embrace XP are radicals. However, XP is really an attempt to take the best practices of software development and extend them "to the extreme." Extreme programming has the following characteristics:

- Takes proven industry best practices and focuses on them intensely
- Combines those best practices (in their intense form) in a new way to produce a result that is greater than the sum of the parts

Figure 17-8 lists the core values and practices of XP. In the following sections we first present the four core values of XP. Then we explain its 12 primary practices. Finally, we describe the basic structure of an XP project and the way XP is used to develop software.

XP Core Values

The four core values of XP—communication, simplicity, feedback, and courage—drive its practices and project activities. You will recognize the first three as best practices for any development project. With a little thought, you should also see that the fourth is a desired value for any project, even though it might not be stated explicitly.

Communication One of the major causes of project failure has been a lack of open communication with the right players at the right time and at the right level. Effective communication involves not only documentation but also open verbal discussion. The practices and methods of XP are designed to ensure that open, frequent communication occurs.

Figure 17-8

XP core values and practices

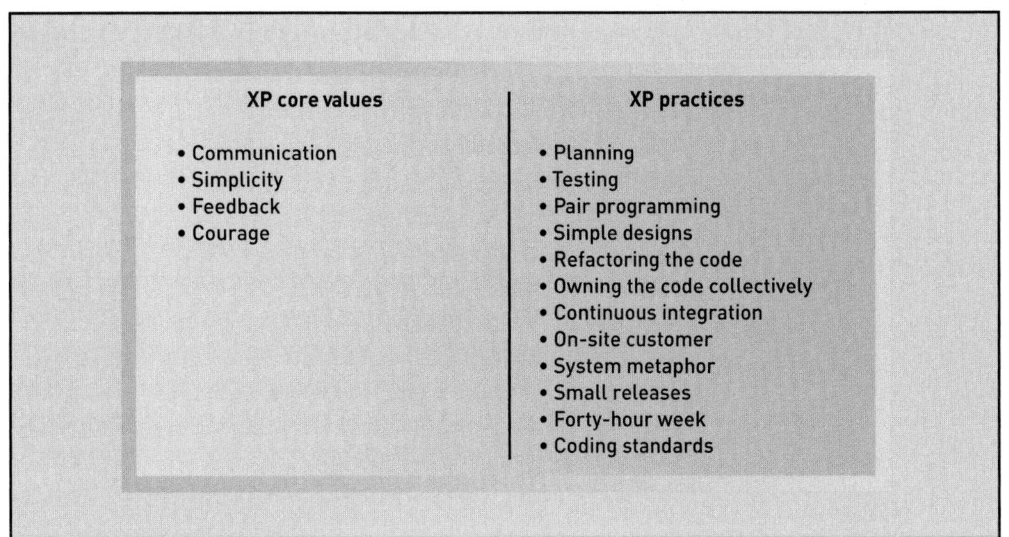

XP core values
• Communication
• Simplicity
• Feedback
• Courage
XP practices
• Planning
• Testing
• Pair programming
• Simple designs
• Refactoring the code
• Owning the code collectively
• Continuous integration
• On-site customer
• System metaphor
• Small releases
• Forty-hour week
• Coding standards

Simplicity Even though developers have always advocated keeping solutions simple, they do not always follow their own advice. XP includes techniques to reinforce this principle and make it a standard way of developing systems.

Feedback As with simplicity, getting frequent, meaningful feedback is recognized as a best practice of software development. Feedback on functionality and requirements should come from the users, feedback on designs and code should come from other developers, and feedback on satisfying a business need should come from the client. XP integrates feedback into every aspect of development.

Courage Developers always need courage to face the harsh choice of doing things right or throwing away bad code and starting over. But all too frequently they have not had the courage to stand up to a too-tight schedule, resulting in bad mistakes. XP practices are designed to make it easier to give developers the courage to "do it right."

XP Practices

XP's 12 practices embody the basic values just presented. These practices are consistent with the agile principles explained earlier in the chapter.

Planning Some people describe XP as glorified hacking or as the old "code and fix" methodology that was used in the 1960s. That is not true; XP does include planning. However, XP is an adaptive technique that recognizes that you cannot know everything at the start. As indicated earlier, XP embraces change. XP planning focuses on making a rough plan quickly and then refining it as things become clearer. This reflects the Agile Development philosophy that change is more important than detailed plans. It is also consistent with the idea that individuals—and their abilities—are more important than an elaborate process.

The basis of an XP plan is a set of stories that users develop. A story simply describes what the system needs to do. XP does not use the term *use case*, but a user story and a use case express a similar idea. Planning involves two aspects: business issues and technical issues. In XP the business issues are decided by the users and clients, while technical issues are decided by the development team. The plan, especially in the early stages of the project, consists of the list of stories—from the users—and the estimates of effort, risk, and work dependencies for each story—from the development team. As in Agile Development, the idea is to involve the users heavily in the project, rather than requiring them simply to sign off on specifications.

Testing Every new piece of software requires testing, and every methodology includes testing. XP intensifies testing by requiring that the tests for each story be written first—before the solution is programmed. There are two major types of tests: unit tests, which test the correctness of a small piece of code, and acceptance tests, which test the business function. The developers write the unit tests, and the users write the acceptance tests. Before any code can be integrated into the library of the growing system, it must pass the tests. By having the tests written first, XP automates their use and executes them frequently. Over time, a library of required tests is created, so when requirements change and the code needs to be updated, the tests can be rerun quickly and automatically.

pair programming
XP practice in which two programmers work together on designing, coding, and testing

Pair Programming This practice, more than any other, is one for which XP is famous. Instead of simply requiring one programmer to watch another's work, **pair programming** divides up the coding work. First, one programmer might focus more on design and double-checking the algorithms while the other writes the code. Then they switch roles so that both think about design, coding, and testing. XP relies on comprehensive and continual code reviews. Interestingly, research has shown that pair programming is actually more efficient than programming alone. It takes longer to write the initial code, but the long-term quality is higher. Errors are caught quickly and early, two people become familiar with every part of the system, all design decisions are developed by two brains, and fewer "quick-and-dirty" shortcuts are taken. The quality of the code is always higher in a pair-programming environment.

BEST PRACTICE

XP's use of pair programming develops high-quality code more efficiently than programming alone.

Simple Designs Opponents say that XP neglects design, but that is not true. XP conforms to the principles of Agile Modeling expressed earlier by avoiding the "Big Design Up Front" approach. Instead, it views design as so important that it should be done continually, but in small chunks. As with everything else, the design must be verified immediately by reviewing it along with coding and testing.

So what is a simple design? It is one that accomplishes the desired result with as few classes and methods as possible and that does not duplicate code. Accomplishing all that is often a major challenge.

refactoring
revising, reorganizing, and rebuilding part of a system so that it is of higher quality

Refactoring the Code **Refactoring** is the technique of improving the code without changing what it does. XP programmers continually refactor their code. Before and after adding any new functions, XP programmers review their code to see whether there is a simpler design or a simpler method of achieving the same result. Refactoring produces high-quality, robust code.

Owning the Code Collectively This practice requires all team members to have a new mindset. In XP, everyone is responsible for the code. No one person can say, "This is my code." Someone can say, "I wrote it," but everyone owns it. Collective ownership allows anyone to modify any piece of code. However, because unit tests are run before and after every change, if programmers see something that needs fixing, they can run the unit tests to make sure that the change did not break something. This practice embodies the team concept that developers are building a system together.

Continuous Integration This practice embodies XP's idea of "growing" the software. Small pieces of code—which have passed the unit tests—are integrated into the system daily or even more often. Continuous integration highlights errors rapidly and keeps the project moving ahead. The traditional approach of integrating large chunks of code late in the project often resulted in tremendous amounts of rework and time lost while developers tried to determine just what went wrong. XP's practice of continuous integration prevents that.

On-Site Customer As with all adaptive approaches, XP projects require continual involvement of users who can make business decisions about functionality and scope. Based on the core value of communication, this practice keeps the project moving ahead rapidly. If the customer is not ready to commit resources to the project, the project will not be very successful.

System Metaphor This practice is XP's unique and interesting approach to defining an architectural vision. It answers the questions, "How does the system work? What are its major components?" To answer those questions, developers identify a metaphor for the system. For example, Big Three automaker Chrysler's payroll system was built as a production-line metaphor, with its system components using production-line terms. Everyone at Chrysler understands a production line, so a payroll transaction was treated the same way—developers started with a basic transaction and then applied various processes to complete it. Of course, the metaphor should be easily understood or well known to the members of the development team. A system metaphor can guide members toward a vision and help them understand the system.

Small Releases A release is a point at which the new system can be turned over to users for acceptance testing, and sometimes even for productive use. Consistent with the entire philosophy of growing the software, small and frequent releases provide upgraded solutions to the users and keep them involved in the project. They also facilitate other practices, such as immediate feedback and continual integration.

Forty-Hour Week and Coding Standards These final two practices set the tone for how the developers should work. The exact number of hours a developer works is not the issue. The issue is that the project should not be a death march that burns out every member of the team. Neither is the project a haphazard coding exercise. Developers should follow standards for coding and documentation. XP uses just the engineering principles that are appropriate for an adaptive process based on empirical controls.

XP Project Activities

Figure 17-9 shows an overview of the XP system development approach. The XP development approach is divided into three levels—system (the outer ring), release (the middle ring), and iteration (the inner ring). System-level activities occur once during each development project. A system is delivered to users in multiple stages called *releases*. Each release is a fully functional system that performs a subset of the full system requirements. A release is developed and tested within a period of no more than a few weeks or months. The activities in the middle ring cycle multiple times—once for each release. Releases are divided into multiple iterations. During each iteration, developers code and test a specific functional subset of a release. Iterations are coded and tested in a few days or weeks. There are multiple iterations within each release, so the iteration ring (inner) cycles multiple times.

The first XP development activity is creating user stories, which are similar to use cases in OO analysis. A team of developers and users quickly documents all of the user stories that the system will support. Developers then create a class diagram to represent objects of interest within the user stories.

Developers and users then create a set of acceptance tests for each user story. Releases that pass the acceptance tests are considered finished. The final system-level activity is to create a development plan for a series of releases. The first release supports a subset of the user stories, and subsequent releases add support for additional stories. Each release is delivered to users and performs real work, thus providing an additional level of testing and feedback.

The first release-level activity is planning a series of iterations. Each iteration focuses on a small (possibly just one) system function or user story. The iterations' small size allows developers to code and test them within a few days. A typical release is developed using a few to a few dozen iterations.

After the iteration plan is complete, work begins on the first iteration-level activity. Code units are divided among multiple programming teams, and each team develops and tests its own code. XP recommends a test-first approach to coding. Test code is written before system code. As code modules pass unit testing, they are combined into larger units for integration testing. When an iteration passes integration testing, work begins on the next iteration.

When all iterations of a release have been completed, the release undergoes acceptance testing. If a release fails acceptance testing, the team returns it to the iteration level for repair. Releases that pass acceptance testing are delivered to end users, and work begins on the next release. When acceptance testing of the final release is completed, the development project is finished.

Figure 17-9

The XP development approach

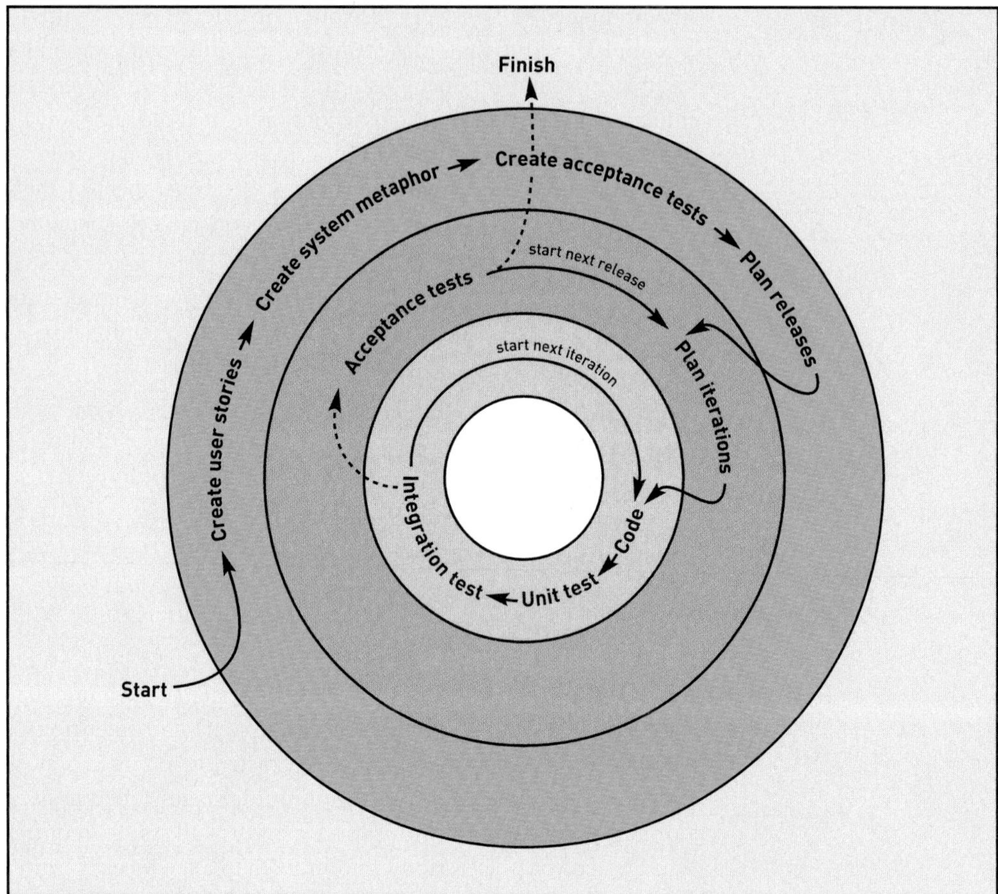

SCRUM

Scrum is another adaptive development methodology. The term refers to rugby's system for getting an out-of-play ball back into play. Rugby players get together in a big mass, the referee drops the ball, and then the scrum participants pass the ball backward through their legs to a waiting runner. The name stuck due to many similarities between the sport and the system development approach: Both are quick, adaptive, and self-organizing. The basic idea behind Scrum is to respond to a current situation as rapidly and positively as possible. Scrum can be described as a truly empirical process control approach to developing software. The Scrum software development process is shown in Figure 17-10. There are three important concepts that describe Scrum: (1) its philosophy, (2) its organization, and (3) its practices.

Scrum Philosophy

The Scrum philosophy is based on the Agile Development principles described earlier. Scrum is responsive to a highly changing, dynamic environment in which users might not know exactly what is needed and might also change priorities frequently. In this type of environment, changes are so numerous that projects can bog down and never reach completion. Scrum excels in this type of situation.

BEST PRACTICE

Consider using Scrum for a system in a highly dynamic environment in which users may change priorities frequently.

Scrum focuses primarily on the team level. It is a type of social engineering that emphasizes individuals more than processes and describes how teams of developers can work

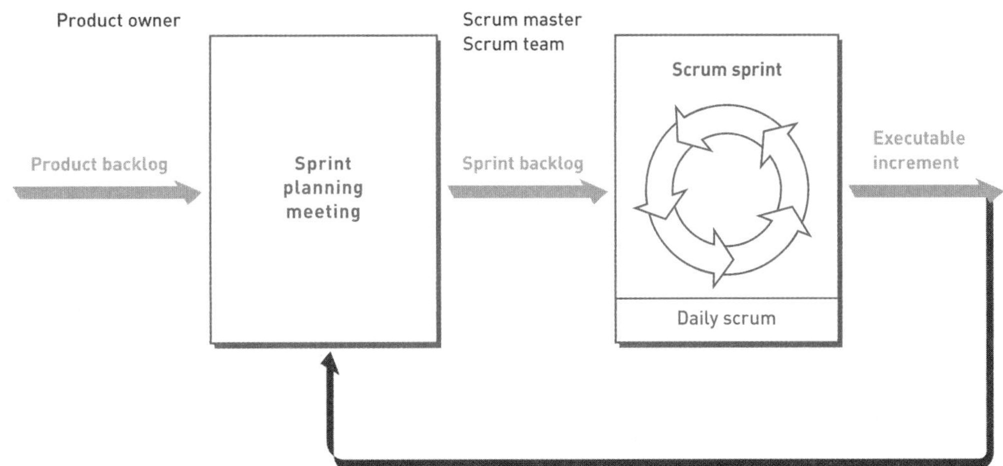

Figure 17-10
Scrum software development process

together to build software in a series of short miniprojects. Key to this philosophy is the complete control a team exerts over its own organization and its work processes. Software is developed incrementally, and controls are imposed empirically—by focusing on the things that can be accomplished.

The basic control mechanism for a Scrum project is a list of all the things the system should include and address. This list, called the **product backlog**, includes user functions (such as use cases), features (such as security), and technology (such as platforms). The product backlog list is continually being prioritized, and only a few of the high-priority items are worked on at a time, according to the current needs of the project and its sponsor.

product backlog
a prioritized list of user requirements used to choose work to be done during a Scrum project

Scrum Organization

The three main organizational elements that affect a Scrum project are the product owner, the Scrum master, and the Scrum team or teams.

The **product owner** is the client—that is, the person who is buying the result—but the product owner has additional responsibilities. Remember that in Agile Development the user and client are closely involved in the project. In Scrum, the product owner maintains the product backlog list. For any function to be included in the final system, it must first be placed on the product backlog. Because the product owner maintains that list, any request must first be approved and agreed to by the product owner. In traditional development projects, the proj-ect team initiates the interviews and other activities to identify and define requirements. In a Scrum project, the primary client controls the requirements. This forces the client and user to be intimately involved in the project. Nothing can be accomplished until the product owner creates the backlog.

product owner
the client stakeholder for whom a system is being built

The **Scrum master** enforces Scrum practices and helps the team complete its work. A Scrum master is comparable to a project manager in other approaches. However, because the team is self-organizing and there is no overall project schedule, the Scrum master's duties are slightly different. He or she is the focal point for communication and progress reporting, just as in a traditional project. But the Scrum master does not set the schedule or assign tasks. The team does. One of the primary duties of the Scrum master is to remove impediments so that the team can do its work. In other words, the Scrum master is a facilitator.

Scrum master
the person in charge of a Scrum project, similar to a project manager

The **Scrum team** is a small group of developers, typically five to nine people, who work together to produce the software. For projects that are very large, the work should be partitioned and delegated to smaller teams. If necessary, the Scrum masters from all the teams can coordinate multiple team activities.

Scrum team
the team members working on a Scrum project

The Scrum team sets its own goal for what it can accomplish in a specific period of time. It then organizes itself and parcels out the work to members. In a small team it is much easier to sit around a table, decide what needs to be done, and have members of the team volunteer or accept pieces of work.

Scrum Practices

The Scrum practices are the mechanics of how a project progresses. Of course, the practices are based on the Scrum philosophy and organization. The basic work process is called a *sprint*, and all other practices are focused on supporting a sprint.

sprint a time-controlled miniproject that implements a specific portion of a system

A Scrum **sprint** is a firm 30-day time box, with a specific goal or deliverable. At the beginning of a sprint, the team gathers for a one-day planning session. In this session, the team decides on the major goal for the sprint. The goal draws from several items on the prioritized product backlog list. The team decides how many of the highest-priority items it can accomplish within the 30-day sprint. Sometimes lower-priority items can be included for very little additional effort and can be added to the deliverables for the sprint.

After the team has agreed on a goal and has selected items from the backlog list, it begins work. The scope of that sprint is then frozen, and no one can change it—neither the product owner nor any other users. If users do find new functions they want to add, they put them on the product backlog list for the next sprint. If team members determine that they cannot accomplish everything in their goal, they can reduce the scope for that sprint. However, the 30-day period is kept constant.

Every day during the sprint, the Scrum master holds a daily Scrum, which is a meeting of all members of the team. The objective is to report progress. The meeting is limited to 15 minutes or some other short time period. Members of the team answer only three questions:

- What have you done since the last daily Scrum (last 24 hours)?
- What will you do by the next daily Scrum?
- What got in your way, or is in your way, preventing you from completing your work?

The purpose of this meeting is simply to report issues, not to solve them. Individual team members collaborate and resolve problems after the meeting as part of the normal workday. One of the major responsibilities of the Scrum master is to note the impediments and see that they are removed. A good Scrum master clears impediments rapidly. The Scrum master also protects the team from any intrusions. The team members are then free to accomplish their work. Team members do talk with users to obtain requirements, and users are involved in the sprint's work. However, users cannot change the items being worked on from the backlog list or change the intended scope of any item without putting it on the backlog list.

At the end of each sprint, the agreed-on deliverable is produced. A final half-day review meeting is scheduled to recap progress and identify changes that need to be made for following sprints. By time boxing these activities—the planning, the sprint, the daily Scrum, and the Scrum review—the process becomes a well-defined template to which the team easily conforms, which contributes to the success of Scrum projects.

PROJECT MANAGEMENT AND ADAPTIVE METHODOLOGIES

As indicated in the Valley Regional Hospital case at the beginning of the chapter, the adaptive methodologies to system development can sometimes frustrate executives who are accustomed to having a complete schedule at the start of a project and tracking the progress against this schedule. But with adaptive methodologies, developers create the schedule as the project progresses. This approach can seem chaotic and uncontrolled to those who are not used to it. In fact, some might think that advocates of adaptive development are throwing project management out the window. Project management is an integral part of adaptive approaches, however. As shown in Chapter 3, project management activities are an integral part of all adaptive approaches. Each iteration can be treated as a miniproject with its own set of project management tasks.

BEST PRACTICE

Good project management techniques are important no matter what methodology you use. They are an integral part of adaptive software methodologies.

In Chapter 3 we also identified several criteria of successful projects. Included on the list were the following:

- Clear system requirement definitions
- Substantial user involvement
- Support from upper management
- Thorough and detailed project plans
- Realistic work schedules and milestones

Let's review the eight primary areas in the project management body of knowledge to see how project management changes for adaptive projects.

Project time management is radically changed in adaptive methodologies. Because projects operate under uncertain conditions, the project team does not attempt to make a complete, detailed project schedule. However, as we saw, time and the schedule are managed differently. First, in some approaches such as Scrum, each cycle must conform to a firm time box. In other approaches such as XP or the UP, the length of iterations is more flexible, but each has its own schedule. In fact, the schedule can be very detailed and is usually more accurate because of the smaller scope and focused nature of each iteration. So, time management is still an important skill for a project manager using adaptive approaches. One key success element, that of realistic work schedules, is much more evident in adaptive projects than in purely predictive approaches.

Project scope management is also radically altered. With predictive development, one of the project manager's primary responsibilities is to control the project scope. The most difficult tasks of the project manager are ensuring that the requirements are correct, seeing that the users are involved, and preventing the scope from growing uncontrollably. In contrast to this control-from-the-top viewpoint, adaptive development makes the users or clients part of the team and gives them responsibility for the scope. The backlog list in Scrum is the responsibility of the client. The scope of an iteration, or sprint, is not allowed to change. The scope of the project can only change through a very controlled mechanism, which includes the approval of the client. One potential problem with this approach, however, is that project iterations could just go on forever. So, scope control for adaptive projects consists of controlling the iterations. For a UP project, the elaboration iterations would be stopped, and the team would move on to implementation, testing, and deployment. Scope control is still necessary; it just takes a different form.

Time and scope are always interdependent. Changes in scope influence the schedule and time required to complete project deliverables. Monitoring and control are still critical to a successful project. Within an XP iteration or a Scrum sprint, monitoring and control techniques are still needed to keep the project on schedule. If the team is self-organizing, as in the Scrum approach, the members need to establish project management tasks. Deliverables and time frames are still required. So, the project can still include metrics to measure progress and predict completion dates.

Project cost management is still important in adaptive approaches. Total project costs might be harder to predict because the complete project schedule is unknown. Executives will feel uneasy about the lack of both an overall project schedule and a total budget, so the team will need to reassure them by emphasizing iteration control and scope control.

Project communication management is critical in adaptive approaches because users are heavily involved in all aspects of the project. Because requirements are identified, defined, implemented, and tested in short iterations, the success factors associated with defining requirements and eliciting user support are emphasized. Open verbal communication and collaborative work are the primary tools for defining the business need. Communication throughout the entire team is a must for adaptive projects.

Project quality management is a continual focus of adaptive projects, and in fact, more tools are provided to the project manager and the team to ensure a high-quality system. Two major techniques are available. First, testing is conducted throughout the project—from writing test plans first to continual checks and integration. Second, time is allocated to refactor the system as it is built, so the resulting code is simple and solid. The project manager needs to ensure that adequate time is scheduled in each iteration for these important activities.

Project risk management is also enhanced in adaptive approaches. Early iterations should address the high-risk aspects of the system. As a result, the project team and client find out early in the project whether there are insurmountable obstacles that could seriously undermine the project's success.

Project human resource management is as challenging in adaptive approaches as in any project. Good project management in both predictive and adaptive approaches emphasizes small teams that are self-managed. The major difference is that the adaptive techniques have a built-in mechanism for teams to organize themselves for each iteration or sprint. So project managers are less tempted to take control in adaptive approaches.

Project procurement management must address the same issues for all projects. Issues such as integrating purchased elements into the overall project, verifying the quality of the purchased components, and satisfying contractual commitments must be completed in both approaches.

We next turn to new ideas for integrating all system development work across a large organization or enterprise. The adaptive and agile concepts that were just discussed focus on individual projects within an organization. Rather than focus on individual efforts, Model-Driven Architecture focuses on enterprise-level activities. It describes a use of models and modeling that is different from that of Agile Modeling.

MODEL-DRIVEN ARCHITECTURE—GENERALIZING SOLUTIONS

One of the primary problems medium-sized and large organizations face is how to build enterprise-level systems that work together seamlessly or that are at least able to communicate. Organizations frequently have legacy mainframe systems, coupled with UNIX-based systems, coupled with Windows platform systems. Each of these systems has its own operating systems, interaction standards, middleware systems, and specific business application systems. The question facing many organizations today is how to make all of these systems work together.

Middleware includes such services as messaging and e-mail, HTML servers, directory and name servers, database and data servers, component registries, and transaction and event processing handlers. Middleware is typically based on standards that define message format, message content, security, and the way applications and servers discover one another. It is very difficult for a large organization to establish a single middleware platform because different groups and divisions often have unique needs, which led them to adopt a variety of middleware solutions based on different standards. Some of the most widely used middleware environments today are based on standards such as CORBA, Java 2 Web Services, XML, SOAP, and .NET. These environments are explained later in the chapter.

The question system developers must answer is how to capture and extract the information from all of these different enterprise-level middleware environments and use it independently of the middleware systems themselves. In other words, how can a company define its architectural needs independently of any vendor or platform?

One solution proposed by the Object Management Group (OMG) is Model-Driven Architecture. The OMG, which is a consortium of more than 800 companies and organizations, sets standards for interoperability that are independent of language, platform, and vendor. Some of these standards are developed and established by the OMG itself. One such example is Common Object Request Broker Architecture (CORBA), which describes a standard communication architecture for enterprise-level interfaces. Other standards are developed in conjunction with outside groups. UML, which we have been discussing throughout this text, was first proposed as a standard by Grady Booch, Ivar Jacobson, and James Rumbaugh. It has since been adopted as a standard and incorporated into the OMG procedures for updating and approval.

Model-Driven Architecture (MDA) is an OMG initiative that is built on the principles of abstraction, modeling, reuse, and patterns to provide companies with an additional tool to help them understand and extend their enterprise-level systems. MDA provides a framework to identify and classify all the system development work being done in an enterprise. The official motto of the MDA is "the architecture of choice for a changing world." As we look at the details of the MDA, you will recognize many of its concepts. MDA helps describe the concepts that you have been learning throughout this course.

Figure 17-11 is a diagram that depicts the typical process for software development. This model applies to both predictive processes and adaptive, iterative processes. In predictive processes, the development team gathers all the requirements, does a comprehensive design, and then codes the entire system. In the adaptive approaches, the flow is repeated many times in several iterations. But note the rectangles on the right side of the diagram. The top rectangle, showing descriptive text, represents the documents that describe the users' needs. These documents are usually notes, rough sketches, outlines, and other unorganized descriptions of the business processes and user activities.

platform-independent model (PIM)
a model describing system characteristics that are not specific to any deployment platform

platform-specific model (PSM)
a model describing system characteristics that include deployment platform requirements

The second rectangle in the diagram, called the **platform-independent model (PIM)**, models information about the business that is independent of how it will be built. One example of PIM that you will recognize is a UML class diagram. A UML class diagram describes the information requirements of a system, regardless of whether it is implemented using a relational database system such as SQL Server or a hierarchical database system such as IBM's DB2.

The third rectangle, called the **platform-specific model (PSM)**, provides detailed information that includes computer platform and implementation specifics. So, for example, a PSM would contain a relational data model showing the data tables, the keys, the foreign keys, and the type information of individual fields for a SQL Server database. It provides the details about how to implement the PIM.

The other rectangles are the specific implementations of the models in programming code and routines. At this point, you might be asking, "What is so great about the MDA? So far, it just seems to be putting fancy names on activities that we already know how to do." What the MDA does is provide a mechanism by which organizations can extract critical features and information about each of their current systems and combine them into a PIM. As we indicated earlier, most organizations have multiple systems running in many different platforms and programming languages. To describe those systems, individual PSMs are used, all of which are different. So, extracting that information into a PIM allows an organization to analyze the combined PIM to determine where duplication, inconsistencies, and conflicts in technologies exist. In addition, new systems can be designed to conform to existing systems.

For an organization to use an MDA strategy, it must first have a common modeling system and language to describe the PIMs. A key component of the PIM modeling language is the UML, which you have learned in this textbook.

The other required component is a set of standard transformations to move from the code to a PSM and from a PSM to a PIM. In fact, it would be ideal if the transformations could be automated. Then a company could use the automated tools to read existing code, generate

Figure 17-11
Software development and MDA

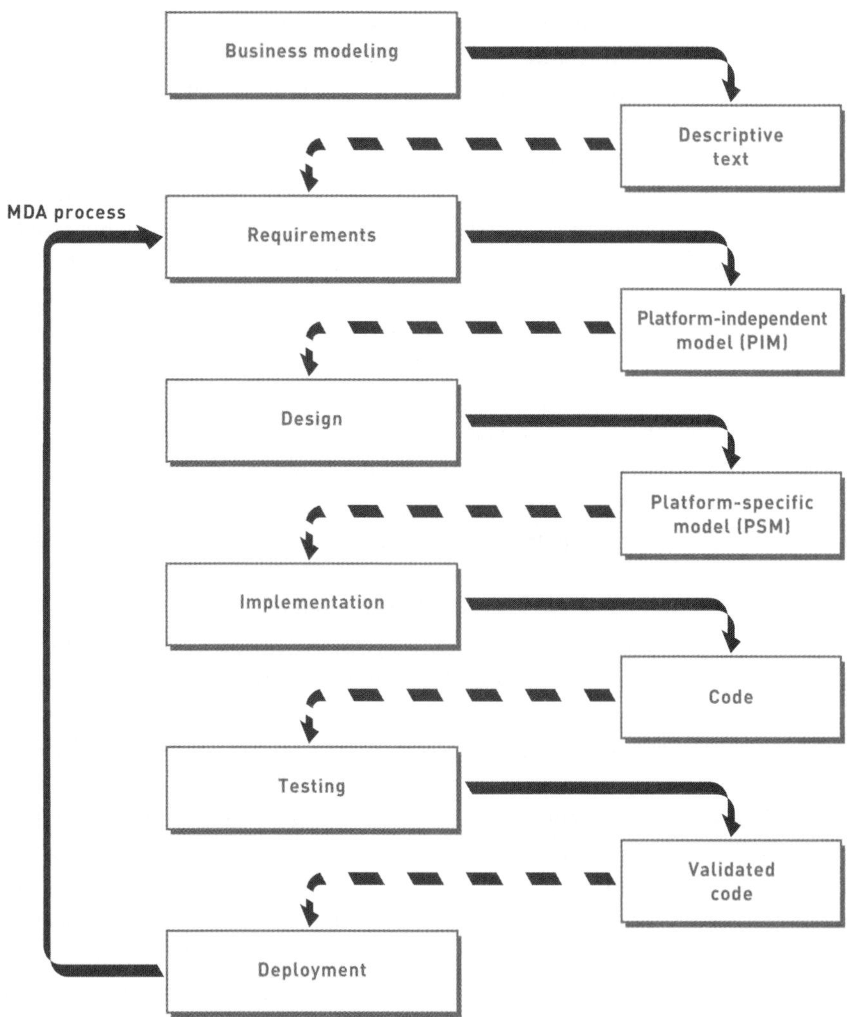

PSMs, analyze the PSMs, and generate PIMs. After those tools were in place, they could also assist in the transformation in the other direction—from PIM to PSM to code.

The role of the OMG is to help define these standard transformations. The creation and sale of specific tools to support this activity is usually done by commercial companies.

Figure 17-12 illustrates how this process occurs. This figure shows different types of metamodels. Recall from earlier in the chapter that a metamodel is a model that describes the characteristics of another model. To build tools that automatically convert one model into another model, developers need to describe each of those models in precise mathematical terms. That mathematical description is a *metamodel*.

Figure 17-12
Metamodels and transitions between PIM, PSM, and code

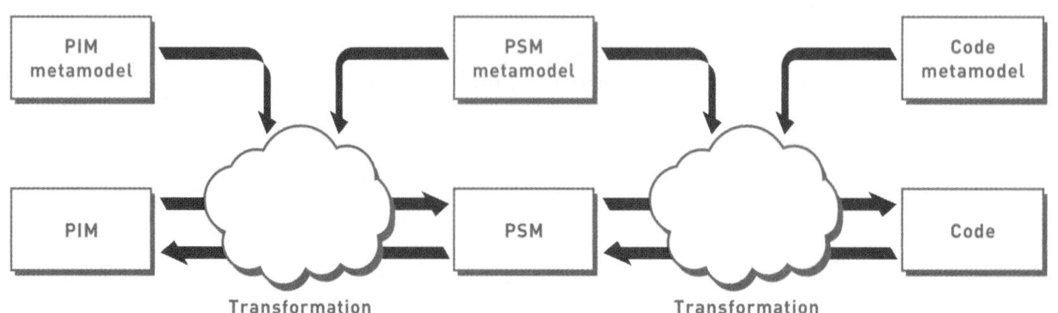

For example, if we were to describe a UML design class diagram, we would say that it is made up of boxes and lines. A box represents a class. The boxes have three compartments, with the top being a class name, the middle a list of attributes, and the bottom a list of methods. The lines represent relationships. Each line is connected to a box. Each connection has a numeric text field that describes the multiplicity of the connection. Figure 17-13 shows a partial class diagram metamodel that describes a UML design class diagram.

Figure 17-13

Partial metamodel of a UML class diagram

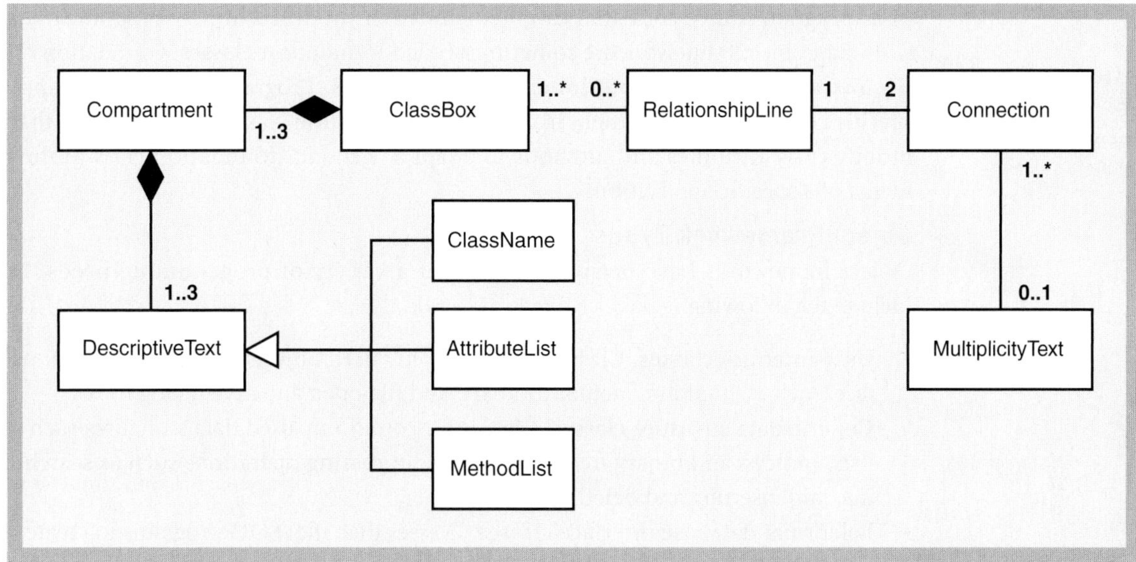

With a metamodel for each model and a defined set of transformations, organizations can automatically transform the models. The OMG is currently finalizing the first release of the MDA standards to define these models and the transformations. This initiative is ongoing, with many organizations contributing to the standard and with commercial companies beginning to build automated tools.

The MDA was originally defined to fit into a more predictive development approach. The adaptive approaches minimize documentation and model building. So, are the two ideas incompatible or contradictory?

The primary benefit of MDA is to give a big-picture view of the architecture of the entire enterprise. Any new system should fit into the existing data and platform configuration. It makes sense during system development, even with adaptive approaches, to maintain consistency with the existing infrastructure. So, adaptive development approaches can benefit from utilizing principles of MDA. The important point is to use each technique in its appropriate situation to maximize productivity and organizational benefits.

FRAMEWORKS, COMPONENTS, AND SERVICES

Similar functions are embedded in many different types of systems. For example, the graphical user interface (GUI) is nearly ubiquitous in modern software. Many features of a GUI—such as drop-down menus, help screens, and drag-and-drop manipulation of on-screen objects—are used in many or most GUI applications. Other functions—such as searching, sorting, and simple text editing—are also common to many applications. Higher-level functions such as credit verification and shipment creation are reused in many different applications.

Reusing software to implement such common functions is a decades-old development practice. But such reuse was awkward and cumbersome with older programming languages and before ubiquitous networks. Object orientation includes two powerful techniques, frameworks and components, that support software reuse. Modern networks and standards provide another method of software reuse: network-based services.

OBJECT FRAMEWORKS

object framework
a set of classes that are designed to be reused in a variety of programs

foundation classes
the classes within an object framework

An object framework is a set of classes that are specifically designed to be reused in a wide variety of programs. The object framework is supplied to a developer as a precompiled library or as program source code that can be included or modified in new programs. The classes within an object framework are sometimes called foundation classes. Foundation classes are organized into one or more inheritance hierarchies. Programmers develop application-specific classes by deriving them from existing foundation classes. Programmers then add or modify class attributes and methods to adapt a "generic" foundation class to the requirements of a specific application.

Object Framework Types

Object frameworks have been developed for a variety of programming needs. Examples include the following:

- **User-interface classes.** Classes for commonly used objects within a graphical user interface, such as windows, menus, toolbars, and file open and save dialog boxes.
- **Generic data structure classes.** Classes for commonly used data structures such as linked lists, indices, and binary trees and related processing operations such as searching, sorting, and inserting and deleting elements.
- **Relational database interface classes.** Classes that allow OO programs to create database tables, add data to a table, delete data from a table, or query the data content of one or more tables.
- **Classes specific to an application area.** Classes specifically designed for use in application areas such as banking, payroll, inventory control, and shipping.

General-purpose object frameworks typically contain classes from the first three categories. Classes in these categories can be reused in a wide variety of application areas. Application-specific object frameworks provide a set of classes for use in a specific industry or type of application. Third parties usually design application-specific frameworks as extensions to a general-purpose object framework. An application- or company-specific framework requires a significant development effort typically lasting several years. The effort is repaid over time through continuing reuse of the framework in newly developed systems.

The Impact of Object Frameworks on Design and Implementation Tasks

Developers need to consider several issues when determining whether to use object frameworks. Object frameworks affect the process of systems design and development in several different ways:

- Frameworks must be chosen early in the project—within the first development iteration.
- Systems design must conform to specific assumptions about application program structure and operation that the framework imposes.
- Design and development personnel must be trained to use a framework effectively.
- Multiple frameworks might be required, necessitating early compatibility and integration testing.

The process of developing a system using one or more object frameworks is essentially one of adaptation. The frameworks supply a template for program construction and a set of classes that provide generic capabilities. Systems designers adapt the generic classes to the specific requirements of the new system. Frameworks must be chosen early so that designers know the application structure imposed by the frameworks, the extent to which needed classes can be adapted from generic foundation classes, and the classes that cannot be adapted from foundation classes and thus must be built from scratch.

Of the three object layers typically used in OO system development (view, business logic, and data access), the view and data layers most commonly derive from foundation classes. User interfaces and database access tend to be the areas of greatest strength in object frameworks, and they are typically the most tedious classes to develop from scratch. It is not unusual for 80 percent of a system's code to be devoted to view and data classes. Thus, constructing view and data classes from foundation classes provides significant and easily obtainable code reuse benefits. Adapting view classes from foundation classes has the additional benefit of ensuring a similar look and feel of the user interface across systems and across application programs within systems.

Successful use of an object framework requires a great deal of up-front knowledge about its class hierarchies and program structure. That is, designers and programmers must be familiar with a framework before they can successfully use it. Thus, a framework should be selected as early as possible in the project, and developers must be trained in use of the framework before they begin to implement the new system.

COMPONENTS

component
a standardized and interchangeable software module that is fully assembled and ready to use and that has well-defined interfaces to connect it to clients or other components

A **component** is a software module that is fully assembled and tested, is ready to use, and has well-defined interfaces to connect it to clients or other components. Components can be single executable objects or groups of interacting objects. A component can also be a non-OO program or system "wrapped" in an OO interface. Components implemented with non-OO technologies must still implement objectlike behavior. In other words, they must implement a public interface, respond to messages, and hide their implementation details.

Components are standardized and interchangeable software parts. They differ from objects or classes because they are binary (executable) programs, not symbolic (source code) programs. This distinction is important because it makes components much easier to reuse and reimplement than source code programs.

For example, consider the grammar-checking function in most word processing programs. A grammar-checking function can be developed as an object or as a subroutine. Other parts of the word processing program can call the subroutine or object methods via appropriate source code constructs (for example, a C++ method invocation or a BASIC subroutine call). The grammar-checking function's source code is integrated with the rest of the word processor's source code during program compilation and linking. The executable program is then delivered to users.

Now consider two possible changes to the original grammar-checking function:

- The developers of another word processing program want to incorporate the existing grammar-checking function into their product.
- The developers of the grammar-checking function discover new ways to implement the function that result in greater accuracy and faster execution.

To integrate the existing function into a new word processor, the word processor developers must be provided with the source code of the grammar-checking function. They then code appropriate calls to the grammar checker into their word processor source code. The combined program is then compiled, linked, and distributed to users. When the developers of the grammar checker revise their source code to implement the faster and more accurate function, they deliver the source code to the developers of the word processors. Both development teams integrate the new grammar-checking source code into their word processors, recompile and relink the programs, and deliver a revised word processor to their users.

So what's wrong with this scenario? Nothing in theory, but a great deal in practice. The grammar-checker developers can provide their function to other developers only as source code, which opens up a host of potential problems concerning intellectual property rights and software piracy. Of greater importance, the word processor developers must recompile and relink their entire word processing programs to update the embedded grammar checker.

The revised binary program must then be delivered to users and installed on their computers. This is an expensive and time-consuming process. Delivering the grammar-checking program in binary form would eliminate or minimize most of these problems.

A component-based approach to software design and construction solves both of these problems. Component developers, such as the developers of the grammar checker, can deliver their product as a ready-to-use binary component. Users, such as the developers of the word processing programs, can then simply plug in the component. Updating a single component doesn't require recompiling, relinking, and redistributing the entire application. Perhaps applications already installed on user machines could query an update site via the Internet each time they started and automatically download and install updated components.

At this point, you might be thinking that component-based development is just another form of code reuse. But systems design, object frameworks, and client/server architecture all address code reuse in different ways. The following points are what make component-based design and construction different:

- Components are reusable packages of executable code. Systems design and object frameworks are methods of reusing source code.
- Components are executable objects that advertise a public interface (that is, a set of methods and messages) and hide (encapsulate) the implementation of their methods from other components. Client/server architecture is not necessarily based on OO principles. Component-based design and construction are an evolution of client/server architecture into a purely OO form.

Components provide an inherently flexible approach to systems design and construction. Developers can design and construct many parts of a new system simply by acquiring and plugging in an appropriate set of components. They can also make newly developed functions, programs, and systems more flexible by designing and implementing them as collections of components. Component-based design and construction have been the norm in the manufacturing of physical goods (such as cars, televisions, and computer hardware) for decades. However, it has only recently become a viable approach to designing and implementing information systems.

COMPONENT STANDARDS AND INFRASTRUCTURE

Interoperability of components requires standards to be developed and readily available. For example, consider the video display of a typical IBM-compatible personal computer. The plug on the end of the video signal cable follows an interface standard. The plug has a specific form, and each connector in the plug carries a well-defined electrical signal. Years ago, a group of computer and video display manufacturers defined a standard that describes the physical form of the plug and the type of signals carried through each connector. Adherence to this standard guarantees that any video display unit will work with any compatible personal computer and vice versa.

Components might also require standard support infrastructure. For example, video display units are not internally powered. Thus, they require not only a standard power plug but also an infrastructure to supply power to the plug. A component might also require specific services from an infrastructure. For example, a cellular telephone requires the service provider to assign a transmission frequency with the nearest cellular radio tower, to transfer the connection from one tower to another as the user moves among telephone cells, to establish a connection to another person's telephone, and to relay all voice data to and from the other person's telephone via the public telephone grid. All cellular telephones require these services.

Software components have a similar need for standards. Components could be hardwired together, but this reduces their flexibility. Flexibility is enhanced when components can rely on standard infrastructure services to find other components and establish connections with them.

In the simplest systems, all components execute on a single computer under the control of a single operating system. Connection is more complex when components are located on different machines running different operating systems and when components can be moved from one location to another. In this case, a network protocol independent of the hardware platform and operating system is required. In fact, a network protocol is desirable even when all components execute on the same machine because such a protocol guarantees that systems can be used in different environments—from a single machine to a network of computers.

Modern networking standards have largely addressed the issue of common hardware and communication software to connect distributed software components. Internet protocols are a nearly universal standard and thus provide a ready means of transmitting messages among components. Internet standards can also be used to exchange information between two processes executing on the same machine. However, Internet standards alone do not fully supply a component connection standard. The missing pieces are the following:

- A definition of the format and content of valid messages and responses
- A means of uniquely identifying each component on the Internet and routing messages to and from that component

To address these issues, some organizations have developed and continue to modify standards for component development and reuse.

CORBA

Common Object Request Broker Architecture (CORBA)
a standard for software component connection and interaction developed by the Object Management Group (OMG)

object request broker (ORB)
a CORBA service that provides component directory and communication services

Internet Inter-ORB Protocol (IIOP)
a CORBA protocol for communication among objects and object request brokers

The Common Object Request Broker Architecture (CORBA) was developed by the OMG, a consortium of computer software and hardware vendors. CORBA was designed as a platform- and language-independent standard. The core elements of the CORBA standard are the object request broker (ORB) service and the Internet Inter-ORB Protocol (IIOP) for component communication. A component user contacts an ORB server to locate a component and determine its capabilities and interface requirements. Messages that are sent between a component and its user are routed through the ORB, which performs any necessary translation services.

COM+

Component Object Model Plus (COM+)
a standard for software component connection and interaction developed by Microsoft

The Component Object Model Plus (COM+) is a Microsoft-developed standard for component interoperability. It is widely implemented in Windows-based application software, and it is often used in older three-tier distributed applications based on Microsoft Internet Information Server and Transaction Server. Most Windows office suites, such as Microsoft Office, are constructed as a cooperating set of COM+ components.

COM+ components are registered by individual computer systems within the Windows registry, which limits COM+ components to computer systems running Windows operating systems. After components locate one another through the registry, they communicate directly using a network protocol or Windows interprocess communication facilities.

Enterprise JavaBeans

Java is an OO programming language developed by Sun Microsystems. Most people have heard of Java in connection with applets that execute on Web pages. Java differs from other OO programming languages in several important ways, including the following:

- Java programs are compiled into object code files that can execute on many hardware platforms under many operating systems.
- The Java language standard includes an extensive object framework, called the *Java Development Kit (JDK)*, which includes classes for GUIs, database manipulation, and internetworking.

The JDK defines a number of classes and naming conventions that support component development. One class enables a Java object to convert its internal state into a sequence of bytes that can be stored or transmitted across a network. Other classes allow components to enumerate a Java object's internal variables. Naming conventions allow components to deduce the names of methods that manipulate those variables. An object of a class that implements all of the required component methods and follows the required naming conventions is called a **JavaBean**.

An **Enterprise JavaBean (EJB)** is a JavaBean that can execute on a server and communicate with clients and other components using CORBA. EJBs provide additional capabilities beyond JavaBeans, including the following:

JavaBean
an object that implements the required component methods and follows the required naming conventions of the JavaBean standard

Enterprise JavaBean (EJB)
a JavaBean that can execute on a server and communicate with clients and other components using CORBA

- Multicomponent transaction management
- Packaging of multiple components into larger run-time units
- Sophisticated object storage and retrieval in relational or object DBMSs
- Component and object access controls

COMPONENTS AND THE DEVELOPMENT LIFE CYCLE

Component purchase and reuse is a viable approach to speeding completion of a system. Two development scenarios involve components:

- Purchased components can form all or part of a newly developed or reimplemented system.
- Components can be designed in-house and deployed in a newly developed or reimplemented system.

Each scenario has different implications for system development, as explored in the following sections.

Purchased Components

Components change the project inception activities because they affect the way the system will be implemented. Purchasing and using components is generally cheaper and takes much less time than building equivalent software. Purchased components can also solve technical problems that developers could not easily or inexpensively solve themselves.

The search for suitable components must begin during the first iteration of the development cycle, but it cannot begin until user requirements are understood well enough to evaluate their match to component capabilities. When developers purchase entire software packages, the match between component capabilities and user requirements is seldom exact. Thus, developers might need to refine user requirements based on the capabilities of available components, particularly if the development project has a short schedule.

Components operate within an extensive infrastructure based on standards such as CORBA or EJBs. Many system software packages implement key parts of each standard. Thus, choosing a component isn't simply a matter of choosing an application software module. Developers must also choose compatible hardware and system software to support components.

The reliance of purchased components on a particular infrastructure has several implications for development activities, including the following:

- The standards and support software required by purchased components must become part of technical requirements definition.
- A component's technical support requirements restrict the options considered during software architectural design.
- Hardware and system software that provide component services must be acquired, installed, and configured before testing begins.
- The components and their support infrastructure must be maintained after system deployment.

Many development projects, particularly large ones, might use components from many different vendors, which raises compatibility issues. The component search and selection process must carefully consider compatibility—often eliminating some choices and altering the desirability of others. Preliminary testing might have to be conducted early to verify component performance and compatibility before the architectural design is structured around those components and their support infrastructure. Maintenance is also more complicated because significant portions of the system are not under the direct control of the system owner or the in-house IS staff.

System Performance

Component-based software is usually deployed in a distributed environment. Components typically are scattered among client and server machines and among local area network (LAN) and wide area network (WAN) locations. Distributing components across machines and networks raises the issue of performance. System performance depends on the location of the components (that is, component topology), the hardware capacity of the computers on which they reside, and the communications capacity of the networks that connect the computers. Performance also depends on the demands on network and server capacity made by other applications and communication traffic, such as telephone, video, and interactions among traditional clients and servers.

The details of analyzing and fine-tuning computer and network performance are well beyond the scope of this text. But anyone planning to deploy a distributed component-based system should be aware of the performance issues. These issues must be carefully considered during systems design, implementation, and deployment.

Steps developers should take to ensure adequate performance include the following:

- Examine component-based designs to estimate network traffic patterns and demands on computer hardware.
- Examine existing server capacity and network infrastructure to determine their ability to accommodate communication among components.
- Upgrade network and server capacity prior to development and testing.
- Test system performance during development and make any necessary adjustments.
- Continuously monitor system performance after deployment to detect emerging problems.
- Redeploy components, upgrade server capacity, and upgrade network capacity to reflect changing conditions.

Implementing these steps requires a thorough understanding of computer and network technology, as well as detailed knowledge of existing applications, communications needs, and infrastructure capability and configuration. Applying this knowledge to real-world problems is a complex task typically performed by highly trained specialists.

SERVICES

The era of the Internet and high-speed networks has enabled a new method of software reuse described by various names, including Web services and service-oriented architecture (SOA). Unlike object frameworks that are inserted into an application when it is compiled or components that are dynamically or statically linked to an application before execution, an application interacts with a service via the Internet or a private network during execution. Like object frameworks and components, services rely on a suite of standards that has significant implications for software design, development, and performance.

Service Standards

Service standards have evolved from distributed object standards such as CORBA and EJBs to include standards such as SOAP, .NET, and J2WS. The primary difference between service standards and earlier distributed object standards is a decrease in the amount of information that must be compiled or linked into an executable application and an increased reliance on Web-based data interchange standards such as XML.

Simple Object Access Protocol (SOAP)
a standard for component communication over the Internet using HTTP and XML

Simple Object Access Protocol (SOAP) is a service standard based on existing Internet protocols, including Hypertext Transport Protocol (HTTP) and eXtensible Markup Language (XML). Messages between objects are encoded in XML and transmitted using HTTP, which enables the objects to be located anywhere on the Internet. Because SOAP components communicate using

XML, they can be easily incorporated into applications that use a Web-browser interface. Complex applications can be constructed using multiple SOAP components that communicate via the Internet.

Figure 17-14 shows an application and service communicating with SOAP messages. The SOAP encoder/decoder and HTTP connection manager are standard components of a SOAP programmer's toolkit. Applications can also be embedded scripts that use a Web server to provide SOAP message-passing services.

Figure 17-14
Communication using SOAP

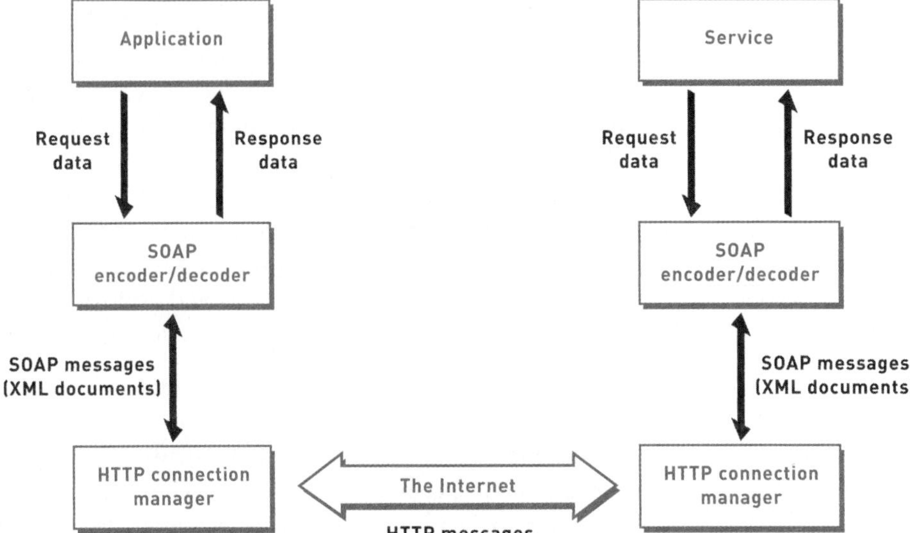

Microsoft .NET
a Microsoft service standard based on SOAP

Java 2 Web Services (J2WS)
a service standard for implementing applications and services in Java

Microsoft .NET is a service standard based on SOAP. The .NET applications and services communicate using SOAP protocols and XML messages, and these applications and services are installed on Microsoft's Web/application server and rely on Microsoft's Active Directory for various naming, location, and security capabilities. The .NET applications and services can be developed in several programming languages, including Visual BASIC and C#.

Java 2 Web Services (J2WS) is a service standard for implementing applications and services in Java. J2WS extends SOAP and several other standards to define a Java-specific method of implementing communication among applications and servers. Although Java is the only programming language, supporting infrastructure components such as Web servers and security software are nonproprietary.

Services and the Development Life Cycle

The impact of services on the development life cycle parallels the impact of object frameworks and components:

- External services must be identified early in the project—their implementation details constrain later design and development tasks.
- Service standards and infrastructure must be chosen early in the project.
- Application and service design must conform to specific assumptions about program structure and operation that the service standards impose.
- Design and development personnel must be trained to use the service standards effectively.
- Developers must carefully consider both network design and application/service component deployment to ensure adequate performance and security.

SUMMARY

The focus of this chapter is on new techniques that are advancing the way systems are being developed. The chapter begins by reviewing five important principles in software development: abstraction, models and modeling, patterns, reuse, and methodologies. Together, these principles form the foundation on which object-oriented development is based. System developers are using these principles to devise new, unique approaches to developing systems.

One of the most active trends in software development is adaptive development methodologies. The Unified Process (UP) is an example of an adaptive methodology that is also influencing many more radical approaches. The idea behind adaptive methodologies is that software projects need to be agile, or flexible, because the business world is so unpredictable and changes so rapidly. The "Manifesto for Agile Software Development" describes four philosophical principles for software projects. These principles emphasize:

- Responding to change over following a plan
- Individuals and interactions over processes and tools
- Working software over comprehensive documentation
- Customer collaboration over contract negotiation

Agile Modeling, another core concept in Agile Development, proposes several guidelines on how models and modeling should be carried out within a development project. The essence of this approach is to remember that models are a means to an end and not the end. Hence, the Agile Modeling philosophy views models as a tool—for example, for understanding a user requirement or for designing a specific function—rather than as elaborate, formal diagrams that are important by themselves.

Extreme Programming and Scrum are unique methodologies that embody agile principles. Core elements of XP are that the system tests are written first and that programmers work in pairs to design, code, and test the software. So, when a function is completed, it has not only been designed and coded, but it has also been reviewed and tested.

The Scrum approach defines a specific goal that can be completed within four weeks. During the four-week sprint, the project team is protected from all outside distractions so that they can complete the defined goal. A product backlog of all outstanding requests is maintained by the client, and changes to the work the team is doing are only allowed between sprints.

Model-Driven Architecture is an initiative of the OMG to provide techniques for large organizations to integrate all software and all software development across the entire enterprise. At this point, the Model-Driven Architecture is primarily a set of principles and ideas. For the MDA initiative to be used, specific tools need to be developed by tool vendors. The MDA defines models at various levels, including a platform-independent model (PIM) and a platform-specific model (PSM), which can provide a comprehensive view of all enterprise-level systems. The MDA is a framework in which all new development can be done so that the organization is able to maintain an integrated, consistent operating environment.

Software reuse is a fundamental approach to rapid development. It has a long history, although it has been applied with greater success since the advent of object-oriented programming, object frameworks, and component-based design and development. Object frameworks provide a means of reusing existing software through inheritance. They provide a library of reusable source code, and inheritance provides a means of quickly adapting that code to new application requirements and operating environments.

Components are units of reusable executable code that behave as distributed objects. They are plugged into existing applications or combined to make new applications. Like the concept of software reuse, component-based design and implementation are not new, but the standards and infrastructure required to support component-based applications have only recently emerged. Thus, components are only now entering the mainstream of software development techniques.

KEY TERMS

chaordic, p. 672
Common Object Request Broker Architecture (CORBA), p. 691
component, p. 689
Component Object Model Plus (COM+), p. 691
discipline, p. 669
Enterprise JavaBean, p. 692
foundation classes, p. 688
Internet Inter-ORB Protocol (IIOP), p. 691
Java 2 Web Services (J2WS), p. 694
JavaBean, p. 692
metamodel, p. 664
Microsoft .NET, p. 694

object framework, p. 688
object request broker (ORB), p. 691
pair programming, p. 678
platform-independent model (PIM), p. 685
platform-specific model (PSM), p. 685
product backlog, p. 681
product owner, p. 681
refactoring, p. 678
Scrum master, p. 681
Scrum team, p. 681
Simple Object Access Protocol (SOAP), p. 693
sprint, p. 682
ubiquitous computing, p. 662

REVIEW QUESTIONS

1. Identify the five important principles and practices that are driving many of the current trends in software development. Briefly explain each.
2. What are the driving forces that are moving many companies to adopt more adaptive approaches to system development?
3. Explain the difference between a predictive control process and an empirical control process.
4. List the six fundamental characteristics of adaptive projects.
5. What are the elements of the "Manifesto for Agile Software Development"? Explain what each means.
6. What does *chaordic* mean? What implications does it have for development projects?
7. What are the four UP phases and what is the objective of each?
8. What are the six UP development disciplines?
9. What are the three UP support disciplines?
10. List the basic principles of Agile Modeling.
11. Why is the word *extreme* included as part of Extreme Programming?
12. List the core values of XP.
13. List the XP practices.
14. What is the product backlog used for in a Scrum project?
15. Explain how a Scrum sprint works.
16. Explain the difference in project time management and project scope management for projects using agile methods.
17. What is a PIM? What is a PSM? How are they related?
18. What are the potential benefits of Model-Driven Architecture?
19. What is a metamodel? How is a metamodel used?
20. What is an object framework? How is it different from a library of components?
21. For which layers of an OO program are off-the-shelf components most likely to be available?
22. What is a software component?
23. Why have software components only recently come into widespread use?
24. In what ways do components make software development faster?
25. What is a service? How does a service differ from a component? How are services similar to components?
26. On what standards is service-oriented architecture based?

THINKING CRITICALLY

1. Consider the capabilities of the programming language and development tools used in your most recent programming or software development class. Are they powerful enough to implement developmental prototypes for single-user software on a personal computer? Are they sufficiently powerful to implement developmental prototypes in a multiuser, distributed, database-oriented, and high-security operating environment? If they were used with a tool-based development approach, what types of user requirements might be sacrificed because they didn't fit language or tool capabilities?
2. The Unified Process (UP) was first developed by a company called Rational, which is now owned by IBM. On the IBM Web site, find any information about UP tools available through IBM/Rational. Briefly describe the suite of tools available. Also look on the IBM Web site and other Web sites (such as the Agile Modeling Web site) for opinions on the relationships and commonality between the UP and Agile Modeling. Report your findings.
3. Consider XP's team-based programming approach in general and its principle of allowing any programmer to modify any code at any time in particular. No other development

approach or programming management technique follows this particular principle. Why not? In other words, what are the possible negative implications of this principle? How does XP minimize these negative implications?

4. Visit the Web sites of the Agile Alliance (*www.agilealliance.com/home*) and Agile Modeling (*www.agilemodeling.com/*). Find some articles on project management in an agile environment. Summarize key points that you think make project management more difficult in this environment than in a traditional, predictive project. Do the same for key points that make project management easier for an agile project.
5. The chapter discussed the benefits of using Agile Development techniques. List and explain the conditions under which it would be unwise to use an Agile Development methodology such as XP or Scrum.
6. Visit the Web site of the World Wide Web Consortium (*www.w3.org*) and review recent developments related to the SOAP standard. What new capabilities have been added, and what is the effect of those capabilities on the standard's complexity and infrastructure requirements?
7. Compare and contrast object frameworks, components, and service-oriented architecture in terms of ease of modification before system deployment, ease of modification after system deployment, and overall cost savings from code reuse. Which approach is likely to yield greater benefits for a unique application system, such as a distribution management system that is highly specialized to a particular company? Which approach is likely to yield greater benefits for general-purpose application software, such as a spreadsheet or virus-protection program?
8. Consider the similarities and differences between component-based design and construction of computer hardware (such as personal computers) and design and construction of computer software. Can the "plug-compatible" nature of computer hardware ever be achieved with computer software? Does your answer depend on the type of software (for example, system or application software)? Do differences in the expected lifetime of computer hardware and software affect the applicability or desirability of component-based techniques?

EXPERIENTIAL EXERCISES

1. Talk with someone at your school or place of employment about a recent development project that was canceled because of slow development. What development approach was employed for the project? Would a different development approach have resulted in faster development?
2. Find a company in your community that uses the UP or some other adaptive method as its development methodology (variations of the UP are okay). Learn how it has applied the UP and how it applies UP principles and practices. Also research what development tools it uses and how well UP is supported.
3. Find someone in your community who is working on a software development project that is using agile principles. How was the team trained to use Agile Development? How was this approach adopted in the organization? What is the general feeling about its success? What aspects does this developer like? What aspects does he/she find frustrating or difficult to use?
4. Consider a project to replace the student advisement system at your school with one that employs modern features (for example, Web-based interfaces, instant reports of degree program progress, and automatic course registration based on a long-term degree plan). Now consider how such a project would be implemented using tool-based development. Investigate alternative tools, such as Visual Studio, PowerBuilder, and Oracle Forms, and determine (for each tool) what requirements would need to be compromised for the sake of development speed if the tool were chosen.
5. Examine the capabilities of a modern programming environment such as Microsoft Visual Studio .NET, IBM WebSphere Studio, or Borland Enterprise Studio. Is an object framework or component library provided? Does successful use of the programming environment require a specific development approach? Does successful use require a specific development methodology?
6. Examine the technical description of a complex end-user software package such as Microsoft Office. In what ways was component-based software development used to build the software?
7. Examine the architecture of a typical consumer-oriented e-commerce Web site such as Amazon.com. How is service-oriented architecture employed within the site?

CASE STUDIES

MIDWESTERN POWER SERVICES

Midwestern Power Services (MPS) provides natural gas and electricity to customers in four Midwestern states. Like most power utilities, over the last several years, MPS has seen significant changes in federal and state regulations. Several years ago, federal deregulation opened the floodgates of change but provided little guidance or restriction on the future shape of the industry. State legislatures also changed their laws and regulations significantly. The industry went through tremendous upheaval, with significant problems created by power shortages at several California power companies and the Enron debacle. Now regulations such as the Sarbanes-Oxley Act are changing the scenario again. These new regulations seriously affect all areas of business, including accounting, record keeping, power purchases, distribution agreements, and customer consumption and billing.

New and proposed regulations seek to increase controls and expand competition for electricity and natural gas. The final form these regulations will take is unknown, and the exact details will probably vary from state to state.

MPS needed to prepare its systems rapidly for these new regulations. Three systems are most directly affected—one for purchasing wholesale natural gas, one for purchasing wholesale electricity, and one for billing customers for combined gas and electric services. The billing system is not currently structured to separate supply and distribution charges, and it has no direct ties to the natural gas and electricity purchasing systems. MPS's general ledger accounting system is also affected because it is used to account for MPS's own electricity-generating operations.

MPS plans to restructure its accounting, purchasing, and billing systems to match the proposed regulation framework:

- Customer billing statements will clearly distinguish between charges for supply and distribution of both gas and electricity. The wholesale suppliers of each power commodity will determine prices for supply. Revenues will be allocated to appropriate companies (for example, distribution charges to MPS and supply charges to wholesale providers).
- MPS will create a new payment system for wholesale suppliers to capture per-customer revenues and to generate payments from MPS to wholesale suppliers. Daily payments will be made electronically based on actual payments by customers.
- MPS will restructure its own electricity-generating operations into a separate profit center, similar to other wholesale power providers. Revenues from customers who choose MPS as their electricity supplier will be matched to generation costs.

MPS's current systems were all developed internally. The general ledger accounting and natural gas purchasing systems are mainframe based. They were developed in the mid-1990s, and incremental changes have been made ever since. All programs are written in COBOL, and DB2 (a relational DBMS) is used for data storage and management. There are approximately 50,000 lines of COBOL code.

The billing system was also rewritten from the ground up in the mid-1990s and has been slightly modified since that time. The system runs on a cluster of servers using the UNIX operating system. The latest version of Oracle (a relational DBMS) is used for data storage and management. Most of the programs are written in C++, although some are written in C and others use Oracle Forms. There are approximately 80,000 lines of C and C++ code.

MPS has a network that is used primarily to support terminal-to-host communications, Internet access, and printer and file sharing for personal computers. The billing system relies on the network for communication among servers in the cluster. The mainframe that supports the accounting and purchasing systems is connected to the network, although that connection is primarily used to back up data files and software to a remote location. The company has experimented with Web-browser interfaces for telephone customer support and online statements. However, no functioning Web-based systems have been completed or installed.

MPS is currently in the early stages of planning the system upgrades. It has not yet committed to specific technologies or development approaches. MPS has also not yet decided whether to upgrade individual systems or replace them entirely. The target date for completing all system modifications is three years from now, but the company is actively seeking ways to shorten that schedule.

1. Describe the pros and cons of the UP approach versus XP and Scrum development approaches to upgrading the existing systems or developing new ones. Do the pros and cons change if the systems are replaced instead of upgraded? Do the pros and cons vary by system? If so, should different development approaches be used for each system?
2. Is component-based development a viable development approach for any of the systems? If so, identify the system(s) and suggest tools that might be appropriate. For each tool suggested, identify the types of requirements likely to be sacrificed because of a poor match to tool capabilities.
3. Assume that all systems will be replaced with custom-developed software. Will an object framework be valuable for implementing the replacements? Is an application-specific framework likely to be available from a third party? Why or why not?
4. Assume that all systems will be replaced with custom-developed software. Should MPS actively pursue component-based design and development? Why or why not? Does MPS have sufficient skills and infrastructure to implement a component-based system? If not, what skills and infrastructure are lacking?

RETHINKING ROCKY MOUNTAIN OUTFITTERS

Now that you have studied the material in this textbook, you'll be able to make more informed and in-depth choices regarding development approach and techniques for the RMO customer support system (CSS). Review the CSS system capabilities in Chapter 3, and the "Rethinking RMO" cases at the end of Chapters 2 and 3. You may also need to look at other RMO material from Chapters 2 and 3 to answer the following questions:

1. Consider the criteria discussed in this chapter for choosing among the adaptive approaches to system development. Which CSS project characteristics favor a predictive approach? Which favor the UP? What characteristics might indicate use of a more agile approach? Which approach is best suited to the CSS development project?
2. Should RMO consider using purchased components within the new CSS? If so, when should it begin looking for components? How will a decision to use components affect the requirements, design, and implementation phases? If purchased components are used, should the portions of the system developed in-house also be structured as components? Will a decision to pursue component-based design and development make it necessary to adopt OO analysis and design methods?

FOCUSING ON RELIABLE PHARMACEUTICAL SERVICE

Reread the Reliable Pharmaceutical Service cases in Chapters 2 and 3. Armed with the new knowledge that you've gained from reading this chapter, answer the following questions:

1. Which of the development approaches described in this chapter seem best suited to the project? Why? Plan the first six weeks of the project under your chosen development approach.
2. What role will components play in the system being developed for Reliable? Does it matter on which component-related standards they're based? Why or why not?

FURTHER RESOURCES

Agile Alliance Web site, *www.agilealliance.com/home.*

Scott W. Ambler, *Agile Modeling: Effective Practices for Extreme Programming and the Unified Process.* John Wiley and Sons Publishing, 2002.

Ken Auer and Roy Miller, *Extreme Programming Applied: Playing to Win.* Addison-Wesley Publishing Company, 2002.

Kent Beck, *Extreme Programming Explained: Embrace Change.* Addison-Wesley Publishing Company, 1999.

Ivar Jacobson, Grady Booch, and James Rumbaugh, *The Rational Unified Process.* Addison-Wesley, 1999.

Anneke Kleppe, Jos Warmer, and Wim Bast, *MDA Explained: The Model Driven Architecture: Practice and Promise.* Addison-Wesley Publishing Company, 2003.

Craig Larman, *Agile and Iterative Development: A Manager's Guide.* Addison-Wesley Publishing Company, 2004.

Scott M. Lewandowski, "Frameworks for Component-Based Client/Server Computing." *ACM Computing Surveys*, volume 30:1 (March 1998), pp. 3–27.

"Manifesto for Agile Software Development," the Agile Alliance, *http://agilemanifesto.org.*

Pete McBreen, *Questioning Extreme Programming.* Addison-Wesley Publishing Company, 2003.

Stephen Mellor, Kendall Scott, Axel Uhl, and Dirk Weise, *MDA Distilled: Principles of Model-Driven Architecture.* Addison-Wesley Publishing Company, 2004.

Ken Schwaber and Mike Beedle, *Agile Software Development with Scrum.* Prentice-Hall, 2002.

Steve Sparks, Kevin Benner, and Chris Faris, "Managing Object-Oriented Framework Reuse." *Computer*, volume 29:9 (September 1996), pp. 53–61.

INDEX

1NF (first normal form), 498–499
2NF (second normal form), 498–501
3NF (third normal form), 498, 501–502

A

abstract classes, 413
abstract design patterns, 463
abstraction, 663–664
acceptance tests, 638, 639
access control list, 601
access controls, 595
accounting, career opportunities, 15
accounting and financial management (AFM) systems, 10
accounting/finance systems (sample scenario), 24
accuracy controls, 597
action-expression, 262
activation lifeline, 436
activities in schedules, 90
activity-data matrix, 231, 233
activity diagrams
  definition, 141
  describing use cases, 249–252
  documenting workflow, 141–144
  examples, 142, 143, 144
  swimlanes, 141
  symbols, 142
  synchronization bars, 141
activity-location matrix, 230, 232
actors, 171, 242–243
  See also users
adapter pattern, 465–466
adaptive methodologies
  See also Agile Development; Agile Modeling; predictive methodologies; Scrum; UP; XP
  at Bestway Fuel Systems (sample scenario), 73
  definition, 39
  description, 42–45
  empirical controls, 666
  MPS (Midwestern Power Sources) (sample scenario), 698
  overview, 666–667
  predictive controls, 666
  process controls, 666
  project management, 682–684
  Reliable Pharmaceutical Service (sample scenario), 699
  Rocky Mountain Outfitters (sample scenario), 699
ad hoc reports, 583
afferent data flow, 367
affordance, 540–541
AFM (accounting and financial management) systems, 10
aggregation, 190
Agile Development
  at Bestway Fuel Systems (sample scenario), 73
  chaordic, 672
  characteristics of, 82
  definition, 81
  philosophy and values, 672
  project management, 81–82
Agile Modeling, 672–676
  See also models
AITP (Association for Information Technology Professionals), 33
Ajax Corporation (sample scenario), 37
All-Shop Superstores (sample scenario), 611–612
alpha versions, 628–630
alternatives, generating and evaluating
  conformance to strategic plan, 288
  cultural feasibility, 288
  description, 121
  economic feasibility, 288
  key criteria, 287–288
  key question, 121
  operational feasibility, 288
  resource feasibility, 288
  Rocky Mountain Outfitters (sample scenario), 289–291
  schedule feasibility, 288
  technological feasibility, 288
analysis models, as documentation, 647
analysis phase, 40
analysis phase activities
  alternatives, generating and evaluating, 121
  feasibility assessment, 121
  information gathering, 119–120
  key questions, 122
  overview, 45–46
  prototyping, 121
  recommendations, management review, 121, 307–308
  requirements definition, 120
  requirements prioritization, 120
API (application program interface), 394
application architecture
  design activities, 331–332
  key questions, 332
  plan, 17
  traditional techniques. *See* traditional design
application deployment environment. *See* deployment environment
applications, 47
application software, 331–332
application-specific classes, 688
architectural design
  See also systems design
  applications. *See* application architecture
  centralized architecture, 341
  clustered architecture, 340
  definition, 326
  deployment environment. *See* deployment environment architecture
  Internet applications, 345–347
  object-oriented. *See* OOD
  technology architecture plan, 17
  traditional. *See* traditional design
  Web-based applications, 345–347
artifacts, 401
Association for Information Technology Professionals (AITP), 33
associative entities, 184
assumptions, use case realization, 440–441
asymmetric key encryption, 604
attribute navigation visibility, 414–415
attributes
  class-level, 412
  elaboration of, 413

hybrid object-relational DBMS, 511–512
things in the problem domain, 181
visibility, 411
authentication, 601
authorization, 601
automation boundary
definition, 8
system interface design, 576–580
traditional design, 355–357
use case diagrams, 244–245
Aviation Electronics (sample scenario), 529

B

backup and recovery controls, 596
balancing DFDs, 219–220
Beck, Kent, 63
behavioral design patterns, 463
benchmarking, 306–307
Bestway Fuel Systems (sample scenario), 73
beta versions, 628–630
binary relationships, 181
biometric devices, 602
black holes, 220–221
Booch, Grady, 61, 667
boss module, 390
bottom-up development, 622–624
boundary classes, 409–410
brainstorming, 416–419
breakeven point, 105
browser forms, 549–552
build and smoke tests, 637–638
business benefits, project planning, 87
business expertise, required of analysts, 12–13
business intelligence systems, 10
business logic layer, 344–345
business processes. *See* information gathering, business processes
business process reengineering, 16
business users, stakeholders, 130

C

callback technique, 468–472
calling structure, 363
cardinality, 180
careers in systems analysis, 14–16
centralized architecture, 341–342
central transform, 367
certificates, 605–606
certifying authority, 605–606
change requests, 652–653
chaordic, 672
charts. *See* diagrams and charts
check boxes, 551–552
chief developer team, 626
class diagrams, 60
See also DCDs
classes
abstract, 413
application-specific, 688
boundary, 409–410
concrete, 413
control, 409–410
data access, 409–410, 513–514
defining. *See* DCDs
definition, 60
entity, 409–410
external, adapting, 465–466
foundation, 688–689
generic data structure, 688
hybrid object-relational DBMS, 511–514
identifying, 416–419
limited to one instance, 468
modeling. *See* domain model class diagrams
ODBMS, 503–504
persistent, 504
relational database interface, 688
scoping, 416–419
transient, 503
user interface, 688
utility, 466–468
view, 409–410
class-level attributes, 412
class-level design patterns, 463
class-level methods, 412
Class-Responsibility-Collaboration (CRC) cards, 416–419, 426
class types, 409–410
clients (computer), 342–344
clients (human), 76, 130
client/server architecture
business logic layer, 344–345
clients, 342–344
client/server interactions, 343
data layer, 344–345
definition, 342
flexibility, 344
issues, 342
maintainability, 344
network-based systems, 392–394
pros and cons, 344
scalability, 344
servers, 342–344
three-layer, 344–345
view layer, 344–345
closed-ended questions, 145
closure, 542–543
clustered architecture, 340–341
cohesion, 421
collaboration, computer support for, 320
collaboration software, JAD (joint application design), 148
collaboration support system (CSS), 10
collaborative specialist teams, 626–627
COM+ (Component Object Model Plus), 691
combo boxes, 551
Common Object Request Broker Architecture (CORBA), 691
communication diagrams, 454–456, 484
communication management
Agile Development, 674
computer support for collaborative work, 320
information repositories, 320–321
project manager skills, 82
steps in, 319–321
XP (Extreme Programming), 676
communication protocols, 338
completeness controls, 594, 597
complex data types, 515
component-based development. *See* software, components
component diagrams
API (application program interface), 394
definition, 394
symbols, 394–395
three-layer Internet system, 398–399, 426
two-layer Internet system, 396–398
UML notation, 394–395
Web services, 399–401
composite state, 262–263
composition, 190
compound attributes, 181

computer networks
  application deployment environment, 291–292
  designing. *See* design activities, computer networks
  key question, 331
Computer Publishing (sample scenario), 527
computer support for collaborative work, 320
concrete classes, 413
concrete design patterns, 463
concurrency, 262–263
concurrent state, 262–263
consistency, user interface design, 541–542
Consolidated Concepts (sample scenario), 37
Consolidated Refineries (sample scenario), 3
construction phase, 61–62, 669
content over representation, 674
context diagram, 87–89, 208–209, 212–213
contracts with vendors, 307
control break reports, 588
control classes, 409–410
control flags, 362
control points
  inputs, 594
  output, 596–597
  security control, 599
control, security. *See* system controls
  user interface, 551–552
cooperating peer teams, 626
CORBA (Common Object Request Broker Architecture), 691
correctness controls, 597
cost/benefit analysis, 102
cost management, 82
coupling, 420
courage, XP (Extreme Programming), 677
CRC (Class-Responsibility-Collaboration) cards, 416–419, 426
creational design patterns, 463
critical path, 95
CRM (customer relationship management) system, 9
CRUD technique
  definition, 161
  Rocky Mountain Outfitters (sample scenario), 231, 233
  use case diagrams, 249
CSS (collaboration support system), 10
CSS (customer support system) (sample scenario), 26–27
cultural feasibility, 288
custom-built software, 299–300
Custom Load Trucking, 113–114

D

data
  redundancy, 498
  security, 602–604
  storage requirements, modeling, 182–187
  validation controls, 594
data, conversion
  creating databases, 640–641
  reloading databases, 639–640
  reusing databases, 639
data access classes, 409–410, 513–514
data access layer, 446–450
data access layer classes, 462
database management systems (DBMS). *See* DBMS
databases (DBs). *See* DBs
data couples, 362
data dictionary, 227–228
data element definitions, 227–228, 229
data entities, 182
data entry, user interface design, 551–552
data flow, 206
data flow consistency, 219–220
data flow definitions, 226–227, 229
data flow diagrams (DFDs). *See* DFDs
data layer, 344–345
data model flexibility, 498
data store, 207, 227
data types, 514–516
data warehouses, 520–521
DBMS (database management systems)
  components, 488–489
  definition, 488
  hierarchical model, 490
  models, 490. *See also* hybrid object-relational DMBS; ODBMS; RDBMS
  network model, 490
  object databases. *See* ODBMS
  physical data stores, 488
  relational databases. *See* RDBMS
  schema, 488–489
DBs (databases)
  Computer Publishing (sample scenario), 527
  creating, 640–641
  definition, 488
  design activities, 333
  integrity controls, 595–596
  key questions, 333
  managing. *See* DBMS
  Nationwide Books (sample scenario), 487–488
  Real Estate Multiple Listing Service (sample scenario), 526
  Reliable Pharmaceutical Service (sample scenario), 527
  reloading, 639–640
  reusing, 639
  Rocky Mountain Outfitters (sample scenario), 527
  role in information systems, 10–11
  State Patrol ticketing process (sample scenario), 526
  synchronizing, 518
DBs (databases), distributed
  database synchronization, 518
  data warehouses, 520–521
  federated database servers, 520–521
  partitioned database servers, 519–520
  replicated database servers, 518–519
  Rocky Mountain Outfitters (sample scenario), 521–523
  single database server, 516–517
DCDs (design class diagrams)
  abstract classes, 413
  attribute navigation visibility, 414–415
  attribute visibility, 411
  boundary classes, 409–410
  class-level attributes, 412
  class-level methods, 412
  class types, 409–410
  concrete classes, 413
  control classes, 409–410
  data access classes, 409–410
  description, 409
  Downtown Videos (sample scenario), 426
  elaboration of attributes, 413
  entity classes, 409–410
  first-cut, 413–416
  inheritance, 412
  method signatures, 411

navigation visibility, 414–416
notation, 409–413
overloaded methods, 411
overridden methods, 412
parameter navigation visibility, 414–415
persistent classes, 410
Real Estate Multiple Listing Service (sample scenario), 425
Rocky Mountain Outfitters (sample scenario), 415–416
shared methods, 412
State Patrol ticketing process (sample scenario), 425
static methods, 412
stereotypes, 409
symbols, 409–410
updating and packaging, 457–463
view classes, 409–410
decision tables and trees, 223–226
decryption, 603
dependencies, in schedules, 92–93
dependency relationships, 460
deployment diagrams, 401–403
deployment environment
architectural plan conformance, 292
characteristics, 292–293
costs, 292
definition, 291
development tools, 292–293
external system interfaces, 292
hardware, 291–292
networks, 291–292
Rocky Mountain Outfitters (sample scenario), 293–296, 311, 384
schedules, 292
strategic plan conformance, 292
system requirements compatibility, 292
system software, 291–292
deployment environment architecture
centralized, 341–342
clustered, 340–341
distributed, 341–342
Internet applications, 345–347
middleware, 348
multicomputer, 340–341
multitier, 340–341
Rocky Mountain Outfitters (sample scenario), 384
single-computer, 340–341
Web-based applications, 345–347
Web services, 347–348
deployment environment architecture, client/server
business logic layer, 344–345
clients, 342–344
client/server interactions, 343
data layer, 344–345
definition, 342
flexibility, 344
issues, 342
maintainability, 344
pros and cons, 344
scalability, 344
servers, 342–344
three-layer, 344–345
view layer, 344–345
descriptive models, 126
design activities
See also architectural design; systems design; systems development
application architecture, 331–332
application software, 331–332
databases, 333
definition, 40
key questions, 331
overview, 46–47
prototyping, 333–334
system controls, 334
system interfaces. *See* system interface design
user interfaces. *See* user interface design
design activities, computer networks
capacity, 338–339
communication protocols, 338
definition, 335
describing, 337
extranets, 336
integration, 337
Internet, 336
intranets, 336
key question, 331
LAN (local area network), 335
network diagrams, 337, 351
overview, 331, 335–336
Real Estate Multiple Listing Service (sample scenario), 351
Rocky Mountain Outfitters (sample scenario), 338–339, 351
routers, 335
virtual organizations, 336–337
VPN (virtual private network), 336–337
WAN (wide area network), 335
WWW (World Wide Web), 336
design class diagrams (DCDs). *See* DCDs
design models, as documentation, 647
design patterns
abstract level, 463
adapter, 465–466
behavioral, 463
callback technique, 468–472
classifying, 463
class-level, 463
concrete-level, 463
creational, 463
definition, 431
factory, 466–468
levels of abstraction, 463
listener, 468–472
object-level, 463
observer, 468–472
principles, 664
publish/subscribe, 468–472
singleton, 468, 469
structural, 463
use case controller, 431–433
desktop metaphor, 536–537
destination, events, 169
destination controls, 596–597
destination state, 261
detail design, 326
See also OOD (object-oriented design), detailed
detailed reports, 585
developing systems. *See* architectural design; design activities; systems development
development costs, 102–103
development environment, 292
development tools, 292–293
DFD fragments, 210, 214
DFDs (data flow diagrams)
balancing, 219–220
black holes, 220–221
complexity, minimizing, 218–219
context diagrams, 208–209, 212–213
data dictionary, 227–228
data element definitions, 227–228, 229
data flow, 206
data flow consistency, 219–220
data flow definitions, 226–227, 229
data store, 207
data store definitions, 227
decision tables and trees, 223–226
definition, 56, 206
diagram 0, 210–211, 215

event-partitioned system model, 210–211, 215
external agents, 206
information overload, 218–219
levels of abstraction, 208–211
logical, 216–218
Miller's Number, 219
minimization of interfaces, 219
miracles, 220
object-oriented, 205
physical, 216–218
process descriptions, 222–226
processes, 206
quality, evaluating, 218–221
Real Estate Multiple Listing Service (sample scenario), 235
Reliable Pharmaceutical Service (sample scenario), 236
Rocky Mountain Outfitters (sample scenario), 211–216, 235–236
rule of 7 ± 2, 219
San Diego Periodicals (sample scenario), 203
State Patrol ticketing process (sample scenario), 235
into structure charts, 367–369
structured English, 222–223
traditional design, 205, 229
diagram 0, 210–211, 215
diagrams and charts
business processes. *See* use case diagrams
class definition. *See* DCDs; domain model class diagrams
class state. *See* state machine diagrams
component relationships. *See* package diagrams
data flow. *See* system flowcharts
data storage requirements. *See* ERDs
dependency relationships. *See* package diagrams
Gantt charts, 93
hierarchical program organization. *See* structure charts
interaction, 433. *See also* communication diagrams; SSDs
internal workflows. *See* activity diagrams; communication diagrams; SSDs
location of physical components. *See* deployment diagrams
PERT/CPM charts, 93
physical system components. *See* component diagrams; deployment diagrams
program control flow. *See* system flowcharts
subsystems. *See* system flowcharts
system, external interactions. *See* SSDs
things (entities) in the problem domain. *See* domain model class diagrams; ERDs
top-down design. *See* structure charts
use case realization. *See* communication diagrams; SSDs
use cases, input/output requirements, modeling. *See* communication diagrams; SSDs
diagrams and charts, I/O
object-oriented user interface design. *See* DCDs; interaction diagrams; SSDs; state machine diagrams
system interface. *See* communication diagrams; SSDs
traditional user interface design. *See* context diagrams; DFD fragments; DFDs
dialog metaphor, 537–540
See also user interface design, dialog documentation
digital certificates, 605–606
digital signatures, 605–606
direct installation, 642
direct manipulation metaphor, 536
disciplines, UP (Unified Process), 669–671
distributed architecture, 341–342
distributed databases
database synchronization, 518
data warehouses, 520–521
federated database servers, 520–521
partitioned database servers, 519–520
replicated database servers, 518–519
Rocky Mountain Outfitters (sample scenario), 521–523
single database server, 516–517
documentation
analysis models as, 647
design models as, 647
implementation, 647–650
menu hierarchy, 544–545
online, 651
source code as, 647
storyboarding, 546–547
subsystems, 544–545
system, 646–648
system scope, 87–88
UML diagrams, 547–548
use cases, 544–545
user, 646, 648–650
user interface dialogs, 544–548
workflow, 141–144
document metaphor, 537
document type definition (DTD), 573
domain layer classes, use case realization, 462
domain model, 242
domain model class diagrams
aggregation, 190
composition, 190
definition, 187
generalization/specialization hierarchies, 189–190
inheritance, 190
multiplicity, 187–189
notation, 187–189
object-oriented system requirements, 269–270
Rocky Mountain Outfitters (sample scenario), 192–193
subclasses, 189
superclasses, 189
whole-part hierarchies, 190–192
Downslope Ski Company (sample scenario), 569–570
Downtown Video Rentals (sample scenario)
communication diagrams, 484
CRC (Class-Responsibility-Collaboration) cards, 426
DCDs (design class diagrams), 426
object-oriented system requirements, 276–277
QA (quality assurance), 659
system interface design, 612
drill down, 589
drivers, 635–636
DTD (document type definition), 573

E

EBP (elementary business process), 161
economic feasibility. *See* feasibility assessment, economic
efferent data flow, 367

eight golden rules, user interface design, 541–544
EJB (Enterprise JavaBean), 691–692
elaboration of attributes, 413
elaboration phase, 61–62, 668
electronic reports, 588–590
Electronics Unlimited (sample scenario), 239
Elements of Reusable Object-Oriented Software, 431
empirical controls, 666
encapsulation, 420
encryption, 603
encryption algorithm, 603
encryption controls, 595
encryption key, 603
enterprise-level systems, OOD (object-oriented design), 392–394
entities, RDBMS, 493–494
entities in the problem domain. *See* problem domain, things in
entity classes, 409–410
entity relationship modeling, 502–503
ERDs (entity-relationship diagrams)
  associative entities, 184
  data entities, 182
  definition, 57
  modeling data storage requirements, 182–187
  notation, 182–185
  Rocky Mountain Outfitters (sample scenario), 185–187
ergonomics (human factors engineering), 534–536
ERP (enterprise resource planning), 10–11, 298–299
error handling, user interface design, 543
error reduction, 574–575
error reports, 652–653
ethics, 13–14
event decomposition, 162
event-driven program flow, 389–390
event-partitioned system model, 210–211, 215
events
  definition, 162
  destination, 169
  external, 163–164
  identifying, 165–167
  perfect technology assumption, 167
  versus prior conditions and responses, 165–166
  response, 169
  resulting use cases, 168–171
  Rocky Mountain Outfitters (sample scenario), 167–168
  sequence of, 166
  source, 169
  state, 165
  system controls, 166–167
  technology-dependent, 166–167
  temporal, 164
  transaction life cycle, 166
  triggers, 163–165, 169
  types of, 163–165
event tables, 168–171, 247–248
exception reports, 585
executive reports, 586
executive users, stakeholders, 130
eXtensible Markup Language (XML) interfaces, 573
external agents, 206
external events, 163–164
external outputs, 586–588
external system interfaces, 292
external users, stakeholders, 130
extranets, design activities, 336
Extreme Programming (XP). *See* XP

F

facilities management, 298
factory design patterns, 466–468
failure factors, 75, 631
feasibility assessment
  cultural, 288
  operational, 288
  organizational, 100–101
  resources, 82, 288
  schedules, 101–102, 288
  system requirements, 121
  technological, 101, 288
feasibility assessment, economic
  automation scope, 288
  breakeven point, 105
  cost/benefit analysis, 102
  development costs, 102–103
  financial calculations, 105–106
  intangible benefits, 106
  NPV (net present value), 105
  ongoing operational costs, 103
  payback period, 105
  ROI (return on investment), 105
  sources of benefits, 104–105
  sources of funds, 107
  tangible benefits, 106
federated database servers, 520–521
feedback
  Agile modeling, 674
  user interface design, 542
  XP (Extreme Programming), 677
field combination controls, 594
fields, RDBMS, 490–491
field values, RDBMS, 490–491
financial calculations, 105–106
finish-finish relationships, 93
finish-start relationships, 93
first-cut DCDs, 413–416
first-cut SSDs, 437–440, 441–446
first normal form (1NF), 498–499
flags, 371
float, 96
flowcharts. *See* structure charts; system flowcharts
foreign keys, RDBMS, 491–492
formality
  project management, 80–82
  systems design, 327
formatting
  forms, 550–551
  reports, 591–592
forms, designing
  browser forms, 549–552
  formatting, 550–551
  layout, 550–551
  windows forms, 549–552
forty-hour week, 679
foundation classes, 625, 688–689
fragments of DFDs, 210, 214
framework development, 625
fraud prevention controls, 597–598
functional decomposition, 7
functional dependency, 498–499
functional requirements, 122–123, 301–302

G

Gamma, Eric, 431
Gang of Four, 431, 464
Gantt charts, 93
generalization relationships, 509–510
generalization/specialization hierarchies, 189–190
general requirements, implementation alternatives, 301

generic data structure classes, 688
graphical models, 126–127
grouping input controls, 551–552
GSS (group support system), 148
guard-condition, 261

H

hard skills, 13
hardware, application deployment environment, 291–292
hardware maintenance, 655
HCI (human-computer interaction)
  See also usability; user interface design
  definition, 532
  desktop metaphor, 536–537
  dialog metaphor, 537–540
  direct manipulation metaphor, 536
  document metaphor, 537
  field of study, 534–536
  hypermedia, 537
  hypertext, 537
  usability tests, 638
  Xerox PARC, 535–536
Helm, Richard, 431
help desk, 49, 651
Help support, user interface design, 552
hierarchical DBMS, 490
highly cohesive modules, 55
HRM (human resource management) system, 10
HTTPS or HTTP-S (Secure Hypertext Transport Protocol), 606
HudsonBanc (sample scenario), 658–659
human factors engineering (ergonomics), 534–536
human factors engineers, 332
human resource management, 82
human resources (sample scenario), 24
hybrid object-relational DBMS, 510–514
hypermedia, 537
hypertext, 537

I

IBM, Rational Software, 61, 65–66
IDE (integrated development environment), 51
identifiers (keys), things in the problem domain, 181
IE (information engineering), 58–59
IIOP (Internet Inter-ORB Protocol), 691
immediate cutover, 642
implementation
  data conversion, 639–641
  documentation, 647–650
  installation, 642–645
  Reliable Pharmaceutical Service (sample scenario), 659
  Rocky Mountain Outfitters (sample scenario), 659
  training, 650–652
implementation, alternatives
  custom-built software, 299–300
  ERP (enterprise resource planning), 298–299
  facilities management, 298
  functional requirements, 301–302
  general requirements, 301
  identifying criteria, 300–301
  overview, 297–298
  packaged software, 298–299
  selecting, 300–304
  service providers, 298
  technical requirements, 302–304
  turnkey systems, 298–299
implementation, program development
  alpha versions, 628–630
  beta versions, 628–630
  bottom-up, 622–624
  chief developer teams, 626
  collaborative specialist teams, 626–627
  cooperating peer teams, 626
  foundation classes, 625
  framework, 625
  IPO (input, process, output) order, 620–622
  maintenance releases, 628–630
  order of, 619–625
  overview, 47–48
  production releases, 628–630
  production versions, 628–630
  release versions, 628–630
  SCCS (source code control system), 627
  source code control, 626–627
  team-based, 625–627
  test versions, 628–630
  top-down, 622–624
  Tri-State Heating Oil (sample scenario), 617
  versioning, 627–630
implementation, QA (quality assurance)
  acceptance tests, 638, 639
  build and smoke tests, 637–638
  cost of error correction, 631
  definition, 631
  Downtown Video Rentals (sample scenario), 659
  drivers, 635–636
  failure factors, 631
  inspections, 632
  integration tests, 636–638
  interface incompatibility, 636
  module tests, 634–636
  parameter values, 637
  performance tests, 638
  Reliable Pharmaceutical Service (sample scenario), 659
  response time, 638
  run-time exceptions, 637
  software testing, 632–633
  stubs, 636
  system tests, 637
  technical reviews, 631–632
  test cases, 633–634
  test data, 633–634
  testers, 638–639
  testing, 632–639
  testing buddy, 638
  throughput, 638
  unexpected state interactions, 637
  unit tests, 634–636
  usability tests, 638
implementation phase, 40, 47–48
inception phase, 61, 668
<<includes>> relationships, 245–247
incremental change, Agile Development, 674
incremental development, 44–45
incremental modeling, 675
indirection, 422
  See also adapter pattern
information engineering (IE), 58–59
information gathering
  See also system requirements
  business processes, 134–135, 140–144
  from existing documents, 135–136
  JAD (joint application design) sessions, 147–148

overview, 119–120
prototypes, 144–145
questionnaires, 145–147
question themes, 134–135
relation to modeling, 134
sources of information, 135–136
user interviews, 137–140
vendor solutions, 149
information hiding, 420
information overload, 218–219
information repositories, 320–321
information systems
AFM (accounting and financial management), 10
business intelligence, 10
CRM (customer relationship management), 9
CSS (collaboration support system), 10
databases, 10–11
definition, 6–7
ERP (enterprise resource planning), 10–11
HRM (human resource management), 10
KMS (knowledge management system), 10
manufacturing management, 10
at Reliable Pharmaceutical Service, 33–34
Rocky Mountain Outfitters (sample scenario), 21, 23–26
SCM (supply chain management), 9
strategic planning, 17–18, 24–26, 33
types of, 9–10
information users, stakeholders, 130
infrastructure, software components, 690–692
inheritance, 190, 412
initiating projects, 83–86
input, process, output (IPO) order, 620–622
input controls, 551–552, 594
input/output classification, 530
inputs, designing. *See* system interface design, inputs
inspections, 632
installation, 642–645
instances of use cases, 171–172
instantiation, 390
intangible benefits, 106
integrated development environment (IDE), 51
integration, continuous, 678
integration management, 83
integration tests, 636–638
integrity, 13–14
integrity controls
See also validation
access, 595
accuracy, 597
backup and recovery, 596
completeness, 594, 597
correctness, 597
databases, 595–596
data validation, 594
definition, 593
destination, 596–597
encryption, 595
field combination, 594
fraud prevention, 597–598
input integrity, 594
need for, 592–593
objectives, 593–594
output integrity, 596–597
transaction logging, 595
update, 596
value limit, 594
interaction diagrams, 433
See also communication diagrams; SSDs
interactions, 252
interface designers, 332
interface design standards, user interface, 540
interface incompatibility, 636
internal events. *See* state events
internal locus of control, 543
internal outputs, 586–588
Internet, design activities, 336
Internet applications, 345–347
Internet-based systems, OOD (object-oriented design), 392–394
Internet Inter-ORB Protocol (IIOP), 691
interpersonal skills, required of analysts, 13
interviewing users. *See* users, interviewing
intranets, design activities, 336
invented keys, 491, 494, 497–498
I/O. *See* input; output; SSDs
IPO (input, process, output) order, 620–622
iteration, 43–44
iterative modeling, 675

J

J2WS (Java 2 Web Services), 694
Jacobson, Ivar, 61, 667
JAD (joint application design) sessions, 147–148
JavaBeans, 691–692
John and Jacob, Inc (sample scenario), 156
Johnson, Ralph, 431

K

Kay, Alan, 535
keys
attributes, 510
foreign, 491–492
invented, 491, 494, 497–498
natural, 491
primary, 491–492, 494
RDBMS, 490–492, 494, 497–498
schema quality, 497–498
uniqueness, 497
KMS (knowledge management system), 10

L

LAN (local area network), 335
launching a project, 107–108
level of abstraction, 208
life cycle
systems development. *See* SDLC
UP (Unified Process), 61–62
lifeline, 253
links, use case realization, 455
list boxes, 551
listener design patterns, 468–472
location diagrams, 230
location information, gathering, 230–233
logical DFDs, 216–218
logical models, 120, 127–128
loosely coupled modules, 55

M

mail order system (sample scenario), 24
maintenance. *See* software, maintenance
maintenance releases, 628–630
management activities *versus* development activities, 79–80
management tasks *versus* development tasks, 79
management users, stakeholders, 130
manufacturing management systems, 10
many-to-many relationships, 495–496, 508–510
Martin, James, 58
mathematical models, 125–126
MDA (model-driven architecture), 684–687
menu hierarchy, 544–545
message event, 261
messages, 242, 252
metamodel, 664
methodologies, systems development, 49
methods, 205, 411–412
method signatures, 411
Microsoft NET, 694
middleware, 348, 684
milestones, 96
Miller's Number, 219
minimization of interfaces, 219
miracles, 220
mock-ups, 145
modeling
  See also use cases
  classes. *See* domain model class diagrams
  purposeful, 674
  relation to information gathering, 134
  visual, tools for, 51, 64–66
modeling, system requirements
  data storage requirements, 182–187
  ERDs (entity-relationship diagrams), 182–187
  object-oriented, 194
  Real Estate Multiple Listing Service (sample scenario), 198–199
  Reliable Pharmaceutical Service (sample scenario), 200
  Rocky Mountain Outfitters (sample scenario), 199–200
  Spring Breaks 'R' Us (sample scenario), 198
  State Patrol ticketing process (sample scenario), 199
  traditional design, 194
  Waiters on Call (sample scenario), 159–160
models
  See also Agile Modeling
  in analysis and design, 127–128
  DBMS (database management systems), 490
  definition, 50
  descriptive, 126
  domain, 242. *See also* domain model class diagrams
  graphical, 126–127. *See also* diagrams and charts
  integration, 269–270
  logical, 120, 127–128
  mathematical, 125–126
  multiple, 674
  physical, 120
  PIM (platform-independent model), 685–687
  project planning, 50
  PSM (platform-specific model), 685–687
  purpose of, 124–125
  quality, 674
  schematic. *See* diagrams and charts
  software, 664
  spiral model, 42–43
  system components, 50
  system interface design, 576–580, 583–584
  types of, 125–127
  visual modeling tools, 51, 64–66
  waterfall model, 40–41
models, object-oriented
  design principles, 390–392
  integrating, 269–270
  system interface design, 580–582, 584–585
modular programming, 55
module algorithms, traditional design, 371–372
module cohesion, 369–370
module coupling, 369
modules, traditional design, 354
module tests, 634–636
motivation, Agile Development, 676
Mountain States Motor Sports (sample scenario), 117
MPS (Midwestern Power Sources) (sample scenario), 698
MS Project, 93–97
multicomputer architecture, 340–341
multilayer design, 446–454
multilayer systems, use case realization, 430–433
multiplicity, 180, 187–189
multitier architecture, 340–341
multivalued attribute, 508

N

n-ary relationships, 181
Nationwide Books (sample scenario), 487–488
natural keys, RDBMS, 491
navigation, ODBMS (object database management system), 504
navigation controls, 552
navigation visibility, 414–416
network DBMS, 490
network diagrams
  definition, 337
  Rocky Mountain Outfitters (sample scenario), 338, 351
networks. *See* computer networks
New Capital Bank (sample scenario), 429
Nielsen, Jacob, 553
nonfunctional requirements, 123
normalization, 498–503
Norman, Donald, 540
notation
  component diagrams, 394–395
  DCDs (design class diagrams), 409–413
  domain model class diagrams, 187–189
  ERDs (entity-relationship diagrams), 182–185
  SSDs (system sequence diagrams), 252–255
  structure charts, 361
  system flowcharts, 358
NPV (net present value), 105

O

object database management system (ODBMS). *See* ODBMS
Object Definition Language (ODL), 503
object frameworks, 688–689
object identifiers, 504
object-level design patterns, 463
object lifeline, SSDs, 253
Object Management Group (OMG), 240
object-oriented analysis (OOA), 60
object-oriented databases. *See* hybrid object-relational DBMS; ODBMS
object-oriented design (OOD). *See* OOD
object-oriented models
  design principles, 390–392
  integrating, 269–270
  system interface design, 580–582, 584–585
object-oriented programming (OOP), 60, 535
object-oriented system requirements
  domain model. *See* domain model class diagrams
  Downtown Videos (sample scenario), 276–277
  Electronics Unlimited (sample scenario), 239
  inputs and outputs. *See* SSDs
  messages, 242
  object behavior. *See* state machine diagrams
  object-oriented model integration, 269–270
  overview, 240–242
  Real Estate Multiple Listing Service (sample scenario), 276
  Reliable Pharmaceutical Service (sample scenario), 277–278
  Rocky Mountain Outfitters (sample scenario), 277
  State Patrol ticketing process (sample scenario), 276
  system activities. *See* use case diagrams
  TheEyesHaveItcom (sample scenario), 277
  versus traditional, 241
object-oriented systems development
  See also OOD
  benefits of, 60
  class diagrams, 60
  classes, 60
  definition, 59
  history of, 59–60
  objects, definition, 59
  OOA (object-oriented analysis), 60
  OOP (object-oriented programming), 60, 535
object-oriented user interface design, 530
object request broker (ORB), 691
object responsibility, 422, 462
object reuse, 420
objects, 59
observer design patterns, 468–472
ODBMS (object database management system)
  classes representation, 503–504
  data types, 515–516
  definition, 503
  generalization relationships, 509–510
  key attributes, 510
  many-to-many relationships, 508–510
  multivalued attribute, 508
  navigation, 504
  object databases, designing, 503–510
  object identifiers, 504
  ODL (Object Definition Language), 503
  one-to-many relationships, 506–508
  persistent classes, 504
  relational hybrid. *See* hybrid object-relational
  relationships, 504–510
  transient classes, 503
ODL (Object Definition Language), 503
office systems (sample scenario), 24
OMG (Object Management Group), 240
one-to-many relationships, 495–496, 506–508
one-to-one relationships, RDBMS, 494
ongoing operational costs, 103
online documentation, 651
on-site customers, 678
OOA (object-oriented analysis), 60
OOD (object-oriented design)
  class diagrams, 60
  combining with traditional, 328–329
  definition, 60
  DFDs, 205
  history of, 59–60
  model integration, 269–270
  models, 390–392
  processes, 390–392
  programs, 389–390
OOD (object-oriented design), architectural
  artifacts, 401
  client/server network-based systems, 392–394
  definition, 60
  deployment diagrams, 401–403
  description, 388–389
  enterprise-level systems, 392–394
  Internet-based systems, 392–394
  single-user systems, 392
OOD (object-oriented design), architectural component diagrams
  API (application program interface), 394
  definition, 394
  symbols, 394–395
  three-layer Internet system, 398–399, 426
  two-layer Internet system, 396–398
  UML notation, 394–395
  Web services, 399–401
OOD (object-oriented design), detailed
  See also use case realization
  brainstorming, 416–419
  class definitions. *See* DCDs
  classes, identifying, 416–419
  classes, scoping, 416–419
  cohesion, 421
  components of, 405
  coupling, 420
  CRC (Class-Responsibility-Collaboration) cards, 416–419
  encapsulation, 420
  indirection, 422
  information hiding, 420
  multilayer systems, 430–433
  objective, 404
  object responsibility, 422
  object reuse, 420
  principles, 404–407, 419–422
  process steps, 408–409
  protection from variations, 421–422
  SSDs (system sequence diagrams), 404

OOP (object-oriented programming), 60, 535
open-ended questions, 146
operational feasibility, 288
option (radio) buttons, 551–552
ORB (object request broker), 691
organizational feasibility, 100–101
origin state, 261
output classification, 530
output controls, 596–597
outputs, designing. *See* system interface design, outputs
overlapping phases, 41–42
overloaded methods, 411
overridden methods, 412
oversight committee, 77
owning the code, 678

P

package diagrams, 459–461
packaged software, 298–299
packages, 245
pair programming, 678
parallel installation, 642–644
parameter navigation visibility, 414–415
parameter values, QA (quality assurance), 637
partitioned database servers, 519–520
passwords, 601
paths, state machine diagrams, 262
patterns. *See* design patterns
payback period, 105
perceptual aspects of user interface design, 533
perfect memory assumption, 440–441
perfect solution assumption, 441
perfect technology assumption, 167, 440
performance
  requirements, 123
  software components, 693
  testing, 638
persistent classes, 410, 504
PERT/CPM charts, 93
phased installation, 644–645
phases
  project planning, 40
  UP (Unified Process), 61–62, 667–669
phases, SDLC (systems development life cycle)
  analysis, 40, 45–46
  definition, 40
  design, 40, 46–47
  implementation, 40, 47–48
  overlapping, 41–42
  project planning, 40, 45
  support, 40, 48–49
phone order system (sample scenario), 24
physical aspects of user interface design, 532
physical data stores, 488
physical DFDs, 216–218
physical models, 120
PIM (platform-independent model), 685–687
Pinnacle Manufacturing (sample scenario), 37
planning
  SDLC (systems development life cycle). *See* project management; project planning
  XP (Extreme Programming), 677
PMBOK (Project Management Body of Knowledge), 82–83
PMI (Project Management Institute), 82
postconditions, 174–175
preconditions, 174
predecessor tasks, 92
predictive controls, 666
predictive methodologies
  See also adaptive methodologies
  analysis activities, 40
  definition, 39
  design activities, 40
  MDA (model-driven architecture), 684–687
  phases, 39–42
  project planning activities, 39–42
  Reliable Pharmaceutical Service (sample scenario), 699
  support phase, 40
  waterfall model, 40–42
presentations, 590–591
primary keys, RDBMS, 491–492, 494
primitive data types, 515
prioritizing projects, 83
prioritizing requirements
  automation level, determining, 284–287
  description, 120
  project scope, 284
  Rocky Mountain Outfitters (sample scenario), 285–287
privileged users, 601
problem definition, project planning, 87–90
problem domain, 46
problem domain, things in
  attributes, 181
  binary relationships, 181
  cardinality, 180
  compound attributes, 181
  developing a list of, 179
  identifiers (keys), 181
  multiplicity, 180
  n-ary relationships, 181
  overview, 176
  relationships among, 178, 180–181
  ternary relationships, 181
  types of, 177
  unary (recursive) relationships, 181
problem solving, 4–6
process controls, 666
process descriptions, 222–226
processes, 206
procurement management, 83
product backlog, 681
production releases, 628–630
production systems, 654
production versions, 628–630
product owner, Scrum, 681
program calls, 361–362
program development. *See* architectural design; design activities; systems development
Project, 93–97
project management
  See also project planning
  adaptive methodologies, 682–684
  Agile Development, 81–82
  body of knowledge, 82–83
  communications, 82, 319–321
  cost management, 82
  definition, 75
  human resource management, 82
  integration management, 83
  issue management, 323
  launching the project, 107–108
  level of formality, 80–82
  management activities *versus* development activities, 79–80
  management tasks *versus* development tasks, 79

organizational feasibility, 100–101
procurement management, 83
progress, measuring (sample scenario), 661
quality management, 82
Reliable Pharmaceutical Service, 114
resource feasibility, 82
risk management, 83, 99–100, 323
schedule feasibility, 101–102
scope management, 82
in the SDLC, 78–80
staffing, 107–108
teams, 317–318, 324
technological feasibility, 101
time management, 82

project management, economic feasibility
breakeven point, 105
cost/benefit analysis, 102
development costs, 102–103
financial calculations, 105–106
intangible benefits, 106
NPV (net present value), 105
ongoing operational costs, 103
payback period, 105
ROI (return on investment), 105
sources of benefits, 104–105
sources of funds, 107
tangible benefits, 106

Project Management Body of Knowledge (PMBOK), 82–83

Project Management Institute (PMI), 82

project managers
Custom Load Trucking, 113–114
external responsibilities, 76
internal responsibilities, 76
oversight committee, 77
role of, 75–78

project planning
See also project management
business benefits, 87
context diagram, 87–89
definition, 39–40, 45
models, 50
monitoring the plan, 321–323
problem definition, 87–90
proof of concept prototype, 87
in the SDLC, 45
system scope document, 87–88

project planning, schedules
activities, 90
critical path, 95
dependencies, 92–93
for an entire SDLC, 97–99
feasibility assessment, 101–102
finish-finish relationships, 93
finish-start relationships, 93
float, 96
Gantt charts, 93
launching the project, 107–108
milestones, 96
in MS Project, 93–97
PERT/CPM charts, 93
predecessor tasks, 92
resource requirements, 96–97
slack time, 96
staffing plan, 96–97
start-start relationships, 93
successor tasks, 92
tasks, 90
WBS (work breakdown structure), 90–92, 93–96

project planning phase, 40

projects
definition, 38
failure factors, 75
initiation, 83–86
prioritizing, 83
success factors, 75
weighted scoring, 83

proof of concept prototype, 87

protection from variations, 421–422
See also adapter pattern

prototypes, 42, 144–145

prototyping
design activities, 333–334
for feasibility, 121
key question, 334
proof of concept, 87

pseudocode, traditional design, 354–355, 371–372

pseudostate, 260

PSM (platform-specific model), 685–687

public key encryption, 604

publish/subscribe design patterns, 468–472

purchased components, 692–693

purposeful modeling, 674

Q

QA (quality assurance)
acceptance tests, 638, 639
build and smoke tests, 637–638
cost of error correction, 631
definition, 631
Downtown Video Rentals (sample scenario), 659
drivers, 635–636
failure factors, 631
inspections, 632
integration tests, 636–638
interface incompatibility, 636
module tests, 634–636
parameter values, 637
performance tests, 638
Reliable Pharmaceutical Service (sample scenario), 659
response time, 638
run-time exceptions, 637
software testing, 632–633
stubs, 636
system tests, 637
technical reviews, 631–632
test cases, 633–634
test data, 633–634
testers, 638–639
testing, 632–639
testing buddy, 638
throughput, 638
unexpected state interactions, 637
unit tests, 634–636
usability tests, 638

quality management, 82

questionnaires, 145–147

question themes, information gathering, 134–135

R

radio (option) buttons, 551–552

Rational Software, 61, 65–66

RDBMS (relational database management system)
data types, 515
definition, 490
designing, 492–493
entity representation, 493–494
fields, 490–491

field values, 490–491
foreign keys, 491–492
history of, 490
interface classes, 688
invented keys, 491, 494
keys, 490–492, 494
managing. *See* RDBMS
many-to-many relationships, 495–496
natural keys, 491
object-oriented hybrid, 512–513
one-to-many relationships, 495–496
organization, 490–491
primary keys, 491–492, 494
referential integrity, 496
relationships, 495–496
rows, 490–491
schema quality, 497–503
tables, 490–491

Real Estate Multiple Listing Service (sample scenario)
databases, 526
DCDs (design class diagrams), 425
design activities, 351
DFDs (data flow diagrams), 235
object-oriented system requirements, 276
requirements modeling, 198–199
structure charts, 384
system interface design, 612
use case realization, 484

realization of use cases. *See* use case realization
recommendations, management review, 121, 307–308
recursive (unary) relationships, 181
refactoring code, 678
referential integrity, RDBMS, 496
registered users, 601
relational DBMSs. *See* hybrid object-relational DBMS; RDBMS
relationships
binary, 181
entity, modeling, 502–503
generalization, 509–510
hybrid object-relational DBMS, 512–513
many-to-many, 495–496, 508–510
n-ary, 181
ODBMS, 504–510
one-to-many, 495–496, 506–508
one-to-one, 494
in the problem domain, 178, 180–181
RDBMS, 495–496, 502–503
ternary, 181
unary (recursive), 181

release versions, 628–630
reliability requirements, 123
Reliable Pharmaceutical Service (sample scenario)
adaptive methodologies, 699
CRC (Class-Responsibility-Collaboration) cards, 426
databases, 527
DFDs (data flow diagrams), 236
implementation, 659
information systems, 33–34
modeling, 200
object-oriented system requirements, 277–278
project management, 114
QA (quality assurance), 659
structure charts, 384
system flowcharts, 384
system interface design, 612
system requirements, 157
systems design, 351
systems development approach, 70
three-layer architecture, 426
user interface design, 566

replicated database servers, 518–519
reports, designing, 585–592
repository, 64
requirements, functional, 122–123, 301–302
See also system requirements
resident experts, 651
resource feasibility, 82, 288
resource requirements, scheduling, 96–97
response, events, 169
response time, 638
retail store systems (sample scenario), 24
return on investment (ROI), 105
reuse, 664
reversal of actions, 543
RFP (request for proposal)
definition, 305
Real Estate Multiple Listing Service (sample scenario), 311
Reliable Pharmaceutical Service (sample scenario), 312
Tropic Fish Tales (sample scenario), 281, 311
with vendors, 305

risk management, 83, 99–100, 323
Rocky Mountain Outfitters (sample scenario)
accounting/finance systems, 24
activity-data matrix, 233
activity-location matrix, 232
adaptive methodologies, 699
application architecture plan, 25
application deployment environment, 293–296, 311
automation level, determining, 284–287
component-based development, 699
CSS (customer support system), 26–27
databases, 527
DCDs (design class diagrams), 415–416
deployment environment, 384
description, 17–18
design activities, 338–339, 351
DFDs (data flow diagrams), 211–216, 235–236
distributed databases, 521–523
domain model class diagram, 192–193
ERDs (entity-relationship diagrams), 185–187
events, 167–168
existing systems, 24
human resources, 24
implementation, 659
information systems department, 21, 23
information systems strategic plan, 24–26, 33
locations, 21, 23
mail order system, 24
modeling, 199–200
network capacity, 339
network diagrams, 338
object-oriented system requirements, 277
office systems, 24
organizational structure, 21, 22
phone order system, 24
problem definition, 89–90
project feasibility assessment, 114
project initiation, 84–86

project planning, 109–110
project teams, 324
reports, 586–587, 612
retail store systems, 24
SCM (supply chain management), 24, 84–86
stakeholders, 131–133
state machine diagrams, 265–269
strategic issues, 19–21
strategic planning, 24–26, 33
support, 659
system flowcharts, 359–360, 375–377
system interface design, 612
system requirements, 156–157
systems development approach, 69
technology architecture plan, 25
three-layer architecture, 375–377, 426
use case realization, 462–463
user interface design, 554–559, 559–561, 566
Web-based catalog and order, 24
ROI (return on investment), 105
roles. *See* actors
routers, 335
rows, RDBMS, 490–491
row uniqueness, 497
rule of 7 ± 2, 219
Rumbaugh, James, 61, 667
run-time exceptions, 637

S

SADT (structural analysis and design technique), 53
San Diego Periodicals (sample scenario), 203
SCCS (source code control system), 627
scenarios, 171–172
schedule feasibility, 101–102, 288
schedules
application deployment environment, 292
project planning. *See* project planning, schedules
schemas
DBMS, 488–489
quality, 497–503
RDBMS, 497–503
schematic models. *See* diagrams and charts
SCM (supply chain management), 9, 24, 84–86
scope management, 82
Scrum, 63, 680–682
Scrum master, 681
Scrum team, 681
SDLC (systems development life cycle)
See also systems development
adaptive approach, 39, 42–45
component-based development. *See* software, components
definition, 38
incremental development, 44–45
iteration, 43–44
predictive approach, 39–42
project management, 78–80
projects, 38
prototypes, 42
spiral model, 42–43
versus UP (Unified Process), 62
waterfall model, 40–41
SDLC (systems development life cycle), phases
See also specific phases
analysis, 40, 45–46
definition, 40
design, 40, 46–47
implementation, 40, 47–48
overlapping, 41–42
project planning, 39–40, 45
support, 40, 48–49
second normal form (2NF), 498–501
Secure Hypertext Transport Protocol (HTTPS or HTTP-S), 606
Secure Sockets Layer (SSL), 606
secure transactions, 606
security
Internet applications, 346
requirements, 123
Web-based applications, 346
security controls
See also system controls
access control list, 601
asymmetric key encryption, 604
authentication, 601
authorization, 601
biometric devices, 602
certifying authority, 605–606
control points, 599
data, 602–604
decryption, 603
definition, 599
digital certificates, 605–606
digital signatures, 605–606
encryption, 603
encryption algorithm, 603
encryption key, 603
HTTPS or HTTP-S (Secure Hypertext Transport Protocol), 606
objectives, 599
passwords, 601
privileged users, 601
public key encryption, 604
registered users, 601
secure transactions, 606
smart cards, 601–602
SSL (Secure Sockets Layer), 606
symmetric key encryption, 603
system access, 599–602
TLS (Transport Layer Security), 606
unauthorized users, 601
user types, 600–601
separation of responsibilities, 446
servers, 342–344
service provider solutions, 298
services, 693–694
service standards, 693–694
shared methods, 412
Shneiderman, Ben, 541
shortcuts, user interface design, 542
short-term memory load, 544
single-computer architecture, 340–341
single database server, 516–517
singleton design patterns, 468, 469
single-user systems, OOD (object-oriented design), 392
Sklar, Joel, 553
slack time, 96
Smalltalk language, 59, 535
smart cards, 601–602
SOAP (Simple Object Access Protocol), 693–694
soft skills, 13
software
abstraction, 663–664
application-specific classes, 688
development. *See* architectural design; design activities; systems development
foundation classes, 688–689
generic data structure classes, 688
metamodel, 664

methodologies. *See* adaptive methodologies; predictive methodologies
models and modeling, 664
object frameworks, 688–689
patterns, 664
relational database interface classes, 688
reuse, 664
testing, 632–633
ubiquitous computing, 662
user-interface classes, 688
software, components
COM+ (Component Object Model Plus), 691
CORBA (Common Object Request Broker Architecture), 691
definition, 689
description, 689–690
EJB (Enterprise JavaBean), 691–692
IIOP (Internet Inter-ORB Protocol), 691
infrastructure, 690–692
J2WS (Java 2 Web Services), 694
JavaBeans, 691–692
Microsoft NET, 694
ORB (object request broker), 691
purchased, 692–693
Rocky Mountain Outfitters (sample scenario), 699
SDLC (systems development life cycle), 692–693
services, 693–694
service standards, 693–694
SOAP (Simple Object Access Protocol), 693–694
standards, 690–692
system performance, 693
software, maintenance
change requests, 652–654
definition, 652
error reports, 652–653
hardware, 655
HudsonBanc (sample scenario), 658–659
implementing changes, 653–654
maintenance releases, 628–630
production systems, 654
test systems, 654
source, events, 169
source code, as documentation, 647
source code control, 626–627
source code control system (SCCS), 627
sources of benefits, 104–105
sources of funds, 107
spin boxes, 551
spiral model, 42–43
spiral SDLC model, 42–43
Spring Breaks 'R' Us (sample scenario), 198
sprints, 682
SSDs (system sequence diagrams)
See also communication diagrams; use case realization, SSDs
definition, 242
developing, 255–259
interaction diagrams, 252
interactions, 252
lifeline, 253
messages, 252
notation, 252–255
object lifeline, 253
OOD (object-oriented design), 404
true/false condition, 255
SSL (Secure Sockets Layer), 606
staffing, 107–108
staffing plan, scheduling, 96–97
stakeholders
business users, 130
clients, 130
definition, 128
executive users, 130
external users, 130
information users, 130
management users, 130
Rocky Mountain Outfitters (sample scenario), 131–133
technical, 131
users, 129–130
start-start relationships, 93
state
definition, 260
unexpected interactions, 637
state events, 165
state machine diagrams
action-expression, 262
composite state, 262–263
concurrency, 262–263
concurrent state, 262–263
destination state, 261
developing, rules for, 263–265
guard-condition, 261
message event, 261
naming conventions, 260
origin state, 261
paths, 262
pseudostate, 260
Rocky Mountain Outfitters (sample scenario), 265–269
state, 260
transition, 260
statements, designing, 585–592
State Patrol ticketing process (sample scenario)
databases, 526
DCDs (design class diagrams), 425
DFDs (data flow diagrams), 235
object-oriented system requirements, 276
requirements modeling, 199
use case realization, 484
static methods, 412
stereotypes, DCDs (design class diagrams), 409
Stimula language, 59
storyboarding, 546–547
strategic planning
application architecture plan, 17
definition, 17
information systems, 17–18
Rocky Mountain Outfitters (sample scenario), 19–21, 24–26, 33
systems analyst's role in, 16–18
technology architecture plan, 17
structural analysis and design technique (SADT), 53
structural design patterns, 463
structure charts
See also system flowcharts
afferent data flow, 367
calling structure, 363
central transform, 367
control flags, 362
data couples, 362
definition, 55, 360
developing, 364–369
from DFDs (data flow diagrams), 367–369
efferent data flow, 367
examples, 363, 364, 365, 366, 367–368
flags, 371
integrating user interface, database, and networks, 373–374
module cohesion, 369–370
module coupling, 369
notation, 361

program calls, 361–362
quality, evaluating, 369–371
Real Estate Multiple Listing Service (sample scenario), 384
Reliable Pharmaceutical Service (sample scenario), 384
symbols, 361
three-layer architecture, 374–378
transaction analysis, 364–365
transform analysis, 365, 367–369

structured
analysis, 56–57
application architecture. *See* traditional design
approach, 53. *See also* traditional design
design, 55. *See also* traditional design
English, DFDs (data flow diagrams), 222–223
programming, 53–55, 390
walkthroughs, 150–152

stubs, 636
subclasses, 189
subsystems, 7, 544–545
success factors, 75
successor tasks, 92
summary reports, 585
superclasses, 189
supply chain management (SCM), 9, 24, 84–86

support
hardware maintenance, 655
help desks, 651
online documentation, 651
resident experts, 651
Rocky Mountain Outfitters (sample scenario), 659
software maintenance, 652–654
technical support, 651–652
training users, 650–652
troubleshooting, 651

support controls, 552
support phase, 40, 48–49
swimlanes, 141

symbols
activity diagrams, 142
component diagrams, 394–395
DCDs (design class diagrams), 409–410
structure charts, 361
system flowcharts, 358

symmetric key encryption, 603
synchronization bars, 141
system access control, 599–602
system boundary, definition, 8
system controls, 166–167, 334
See also integrity controls; security controls

system documentation, 646–648
system enhancement. *See* software, maintenance

system flowcharts
See also structure charts
definition, 354
description, 357
examples, 358–359
notation, 358
Reliable Pharmaceutical Service (sample scenario), 384
Rocky Mountain Outfitters (sample scenario), 359–360, 375–377
symbols, 358
three-layer architecture, 374–378

system interface design
See also user interface design
All-Shop Superstores (sample scenario), 611–612
design activities, 333
Downslope Ski Company (sample scenario), 569–570
DTD (document type definition), 573
identifying the interfaces, 570–574
key questions, 333
Reliable Pharmaceutical Service (sample scenario), 612
XML (eXtensible Markup Language) interfaces, 573

system interface design, inputs
automation boundary, 576–580
control point, 594
devices and mechanisms, 574–575
Downtown Video Rentals (sample scenario), 612
error reduction, 574–575
objectives, 574
object-oriented models, 580–582
Real Estate Multiple Listing Service (sample scenario), 612
required, 576–582
TheEyesHaveItcom (sample scenario), 612
traditional structured models, 576–580

system interface design, outputs
ad hoc reports, 583
control break reports, 588
control point, 596–597
detailed reports, 585
Downtown Video Rentals (sample scenario), 612
drill down, 589
electronic reports, 588–590
exception reports, 585
executive reports, 586
external, 586–588
formatting reports, 591–592
internal, 586–588
objectives, 583
object-oriented models, 584–585
presentations, 590–591
reports, 585–592
required, 583–585
Rocky Mountain Outfitters (sample scenario), 612
statements, 585–592
summary reports, 585
TheEyesHaveItcom (sample scenario), 612
traditional structured models, 583–584
turnaround documents, 586
types of, 585–586

system interfaces
categories of, 571
definition, 570
versus user interfaces, 531–532

system metaphor, 678–679

system requirements
See also analysis phase activities; models
alternatives. *See* alternatives
definition, 122
feasibility assessment, 121
functional, 122–123
information sources. *See* information gathering; stakeholders
I/O. *See* system interface design, inputs; system interface design, outputs
John and Jacob, Inc (sample scenario), 156
modeling. *See* modeling, system requirements; use cases
Mountain States Motor Sports (sample scenario), 117

nonfunctional, 123
object-oriented. *See* object-oriented system requirements
performance, 123
prioritizing. *See* prioritizing requirements
prototyping, 121
recommendations, management review, 121, 307–308
reliability, 123
Reliable Pharmaceutical Service (sample scenario), 157
Rocky Mountain Outfitters (sample scenario), 156–157
security, 123
structured walkthroughs, 150–152
technical, 123
traditional. *See* DFDs
transactions, 130
unit of work, 130
usability, 123
validating, 150–152
system requirements, modeling
data storage requirements, 182–187
ERDs (entity-relationship diagrams), 182–187
object-oriented, 194
Real Estate Multiple Listing Service (sample scenario), 198–199
Reliable Pharmaceutical Service (sample scenario), 200
Rocky Mountain Outfitters (sample scenario), 199–200
Spring Breaks 'R' Us (sample scenario), 198
State Patrol ticketing process (sample scenario), 199
traditional design, 194
Waiters on Call (sample scenario), 159–160
systems, 6–7
systems analysis, 4, 14–16
systems analysts
career opportunities, 14–16
at Consolidated Refineries (sample scenario), 3
definition, 4
problem solving, 4–6
role in strategic planning, 16–18
skills requirements, 11–14
system scope document, 87–88
systems design
See also architectural design; design activities; systems development
components, 325–326
definition, 4
detail, 326
formality, 327
inputs, 326–329
levels of, 325–326
project structure, 327
Reliable Pharmaceutical Service (sample scenario), 351
systems development
See also architectural design; design activities; systems design
alpha versions, 628–630
beta versions, 628–630
bottom-up, 622–624
chief developer teams, 626
collaborative specialist teams, 626–627
component-based. *See* software, components
cooperating peer teams, 626
current trends, 61–63
foundation classes, 625
framework development, 625
IDE (integrated development environment), 51
IPO (input, process, output) order, 620–622
maintenance releases, 628–630
methodologies, 49, 665. *See also* adaptive methodologies; predictive methodologies
models, 50
object-oriented. *See* object-oriented systems development
phases of. *See* SDLC
production releases, 628–630
production versions, 628–630
release versions, 628–630
SCCS (source code control system), 627
source code control, 626–627
team-based, 625–627
techniques, 51–52
test versions, 628–630
tools, 51, 63–66
top-down, 622–624
traditional. *See* traditional design
Tri-State Heating Oil (sample scenario), 617
versioning, 627–630
visual modeling tools, 51, 64–66
systems development life cycle (SDLC). *See* SDLC
system sequence diagrams (SSDs). *See* communication diagrams; SSDs; use case realization, SSDs
system software, application deployment environment, 291–292
system tests, 637

T

tables, RDBMS, 490–491
tangible benefits, 106
tasks, 90, 92
teams
program development, 625–627
project, 317–318, 324
teamwork, Agile Development, 675
technical expertise, required of analysts, 11–12
technical requirements, 123, 302–304
See also system requirements
technical reviews, 631–632
technical stakeholders, 131
technical support, 651–652
techniques, 12, 51–52
technological feasibility, 101, 288
technology architecture plan, 17, 25
technology-dependent events, 166–167
templates. *See* design patterns
temporal events, 164
ternary relationships, 181
test cases, 633–634
test data, 633–634
testers, 638–639
testing, 632–639, 677–678
testing buddy, 638
test systems, 654
test versions, 628–630
text boxes, 551
TheEyesHaveItcom (sample scenario)
communication diagrams, 484
CRC (Class-Responsibility-Collaboration) cards, 426
DCDs (design class diagrams), 426
object-oriented system requirements, 277
system interface design, 612
things in the problem domain. *See* problem domain, things in
third normal form (3NF), 498, 501–502
three-layer architecture
client/server, 344–345
Rocky Mountain Outfitters (sample scenario), 375–377

  - structure charts, 374–378
  - system flowcharts, 374–378
- three-layer implementation issues, 461–462
- three-layer Internet system, component diagrams, 398–399, 426
- three-layer SSDs (system sequence diagrams), 434–436
- throughput, 638
- time management, 82
- TLS (Transport Layer Security), 606
- tools
  - Rational Software Development, 65–66
  - required expertise, 12
  - system development, 51, 63–66
  - Visible Analyst, 64–65
  - Visio, 64–65
  - visual modeling, 51, 64–66
- top-down development, 622–624
- top-down programming, 54–55
- traditional design
  - *See also* structure charts; system flowcharts
  - automation system boundary, 355–357
  - combining with object-oriented, 328–329
  - computer programs, 354
  - definition, 53
  - DFDs (data flow diagrams), 56, 205, 229
  - ERDs (entity-relationship diagrams), 57
  - highly cohesive modules, 55
  - IE (information engineering), 58–59
  - loosely coupled modules, 55
  - modular programming, 55
  - module algorithms, 371–372
  - modules, 354
  - pseudocode, 354–355, 371–372
  - SADT (structural analysis and design technique), 53
  - structure charts, 55
  - structured analysis, 56–57
  - structured approach, 354–355
  - structured design, 55
  - structured programming, 53–55
  - system interface design, 576–580, 583–584
  - top-down programming, 54–55
  - user interface, 530
  - weaknesses, 58
- training users, 650–652
- transaction analysis, 364–365
- transaction life cycle, 166
- transaction logging, 595
- transactions, 130
- transform analysis, 365, 367–369
- transient classes, 503
- transition, 260
- transition phase, 61–62
- transitive phase, 669
- Transport Layer Security (TLS), 606
- triggers, events, 163–165, 169
- Tri-State Heating Oil (sample scenario), 617
- troubleshooting, 651
- true/false condition, SSDs (system sequence diagrams), 255
- turnaround documents, 586
- turnkey systems, 298–299
- two-layer Internet system, component diagrams, 396–398

U

- ubiquitous computing, 662
- UML diagrams, user interface design, 547–548
- UML notation, component diagrams, 394–395
- unary (recursive) relationships, 181
- unauthorized users, 601
- undoing actions, 543
- units of work, 130
- unit tests, 634–636
- UP (Unified Process)
  - best practices, 61
  - construction phase, 61–62, 669
  - disciplines, 669–671
  - elaboration phase, 61–62, 668
  - history of, 61
  - inception phase, 61, 668
  - life cycle phases, 61–62
  - overview, 667
  - phases, 61–62, 667–669
  - *versus* traditional SDLC, 62
  - transition phase, 61–62
  - transitive phase, 669
  - use cases, 61–62
- update controls, 596
- usability
  - *See also* HCI
  - consultants, 332
  - definition, 534
  - requirements, 123
  - tests, 638
- use case controller design patterns, 431–433
- use case description, 171
- use case diagrams
  - activity diagrams, 249–252
  - automation boundary, 244–245
  - CRUD technique, 249
  - definition, 242
  - developing, 248–249
  - *versus* event tables, 247–248
  - example, 243
  - <<includes>> relationships, 245–247
  - organization, 244–245
  - packages, 245
- use case instances, 171
- use case models, 242–243
- use case realization
  - communication diagrams, 454–456, 484
  - data access layer classes, 462
  - definition, 430
  - domain layer classes, 462
  - interaction diagrams, 433. *See also* communication diagrams; SSDs
  - links, 455
  - multilayer systems, 430–433
  - New Capital Bank (sample scenario), 429
  - object responsibility, 462
  - Rocky Mountain Outfitters (sample scenario), 462–463
  - templates. *See* design patterns
  - three-layer implementation issues, 461–462
  - view layer classes, 462
- use case realization, SSDs (system sequence diagrams)
  - *See also* SSDs
  - activation lifeline, 436
  - assumptions, 440–441
  - data access layer, 446–450
  - DCDs (design class diagrams), updating and packaging, 457–463
  - dependency relationships, 460
  - design process, 436–437

first-cut, 437–440, 441–446
guidelines, 440
multilayer design, 446–454
overview, 433
package diagrams, 459–461
perfect memory assumption, 440–441
perfect solution assumption, 441
perfect technology assumption, 440
Real Estate Multiple Listing Service (sample scenario), 484
separation of responsibilities, 446
State Patrol ticketing process (sample scenario), 484
three-layer, 434–436
view layer, 450–454

use cases
See also modeling
actors, 171
CRUD technique, 161
definition, 61–62, 160, 242–243
EBP (elementary business process), 161
event decomposition, 162
event tables, 168–171
examples, 172–173, 174–175
instances, 171–172
level of detail, 172–176
postconditions, 174–175
preconditions, 174
scenarios, 171–172
user goal technique, 160–161
user interface design, 544–545

use cases, events
definition, 162
destination, 169
external, 163–164
identifying, 165–167
perfect technology assumption, 167
versus prior conditions and responses, 165–166
response, 169
resulting use cases, 168–171
Rocky Mountain Outfitters (sample scenario), 167–168
sequence of, 166
source, 169
state, 165
system controls, 166–167
technology-dependent, 166–167
temporal, 164
transaction life cycle, 166
triggers, 163–165, 169
types of, 163–165

use cases, things in the problem domain
attributes, 181
binary relationships, 181
cardinality, 180
compound attributes, 181
developing a list of, 179
identifiers (keys), 181
multiplicity, 180
n-ary relationships, 181
overview, 176
relationships among, 178, 180–181
ternary relationships, 181
types of, 177
unary (recursive) relationships, 181

user-centered design, 533–534
user documentation, 646, 648–650
user goal technique, 160–161
user-interface classes, 688

user interface design
See also HCI; system interface design; system interfaces
Aviation Electronics (sample scenario), 529
conceptual aspects, 533
consultants, 332
definition, 531
human factors engineering (ergonomics), 534–536
human factors engineers, 332
input/output classification, 530
interface designers, 332
key questions, 332
object-oriented approach, 530
perceptual aspects, 533
physical aspects, 532
Reliable Pharmaceutical Service (sample scenario), 566
Rocky Mountain Outfitters (sample scenario), 554–559, 559–561, 566
versus system interfaces, 531–532
traditional approach, 530
usability, 534
usability consultants, 332
user-centered design, 533–534
user's model, 533

user interface design, dialog documentation
menu hierarchy, 544–545
storyboarding, 546–547
subsystems, 544–545
UML diagrams, 547–548
use cases, 544–545

user interface design, guidelines
affordance, 540–541
browser forms, 549–552
check boxes, 551–552
closure, 542–543
combo boxes, 551
consistency, 541–542
data entry, 551–552
eight golden rules, 541–544
error handling, 543
feedback, 542
formatting forms, 550–551
form layout, 550–551
grouping input controls, 551–552
Help support, 552
input controls, 551–552
interface design standards, 540
internal locus of control, 543
list boxes, 551
navigation controls, 552
radio (option) buttons, 551–552
reversal of actions, 543
shortcuts, 542
short-term memory load, 544
spin boxes, 551
support controls, 552
text boxes, 551
visibility, 540–541
Web sites, 552–554
Web sites (sample scenario), 559–561
windows forms, 549–552

users
See also actors
definition, 77
interviewing, 138–140
stakeholders, 129–130

user's model, 533
<<uses>> relationships. *See* <<includes>> relationships

V

validation
See also integrity controls
Agile Development, 676
system requirements, 150–152

Valley Regional Hospital (sample scenario), 661
value limit controls, 594
vendors, 305–307
vendor solutions, 149
  See also software, components
versioning, 627–630
view classes, 409–410
view layer, 344–345, 450–454
view layer classes, 462
virtual organizations, 336–337
visibility
  attribute, 411
  attribute navigation, 414–415
  definition, 411
  navigation, 414–416
  parameter navigation, 414–415
  user interface design, 540–541
visual modeling tools, 51, 64–66
Vlissides, John, 431
VPN (virtual private network), 336–337

W

Waiters on Call (sample scenario), 159–160
WAN (wide area network), 335
waterfall model, 40–41
WBS (work breakdown structure), 90–92, 93–96
Web-based applications, 24, 345–347
Web services, 347–348, 399–401
Web sites, user interface design, 552–554, 559–561
Web technology, systems analysis, 15
weighted scoring, 83
whole-part hierarchies, 190–192
window forms, designing, 549–552
workflow, documenting, 141–144
WWW (World Wide Web), 336

X

Xerox PARC, 535–536
XML (eXtensible Markup Language) interfaces, 573
XP (Extreme Programming)
  coding standards, 679
  communication, 676
  continuous integration, 678
  core values, 676–677
  courage, 677
  feedback, 677
  forty-hour week, 679
  on-site customers, 678
  overview, 63
  owning the code, 678
  pair programming, 678
  planning, 677
  practices, 677–679
  project activities, 679–680
  refactoring code, 678
  simplicity, 677, 678
  small releases, 679
  system metaphor, 678–679
  testing, 677–678